1974

**W. B. SAUNDERS COMPANY**

PHILADELPHIA
LONDON
TORONTO

# CHEMICAL SEPARATIONS AND MEASUREMENTS

## Theory and Practice of Analytical Chemistry

**Dennis G. Peters, John M. Hayes, and Gary M. Hieftje**

Department of Chemistry,
Indiana University,
Bloomington, Indiana

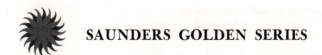

**SAUNDERS GOLDEN SERIES**

Based on the third edition of *Quantitative Chemical Analysis* by Fischer and Peters.

W. B. Saunders Company: West Washington Square
Philadelphia, Pa. 19105

12 Dyott Street
London, WC1A 1DB

833 Oxford Street
Toronto, Ontario M8Z 5T9, Canada

About the Cover:

The superimposed images of a high-resolution gas chromatogram and the glass capillary column tubing essential in its production. The chromatogram happens to be that of marijuana smoke. The full recording, not shown here, reveals well over 200 components. Only 2.5 ml of unenriched smoke was taken as the sample. The column temperature was programmed from $-70°$ to $+130°$ C. The average width at half height of the peaks is 9 seconds. In the chromatogram as shown, time is increasing linearly from right to left, with retention times varying from about 20 to 60 min. Chromatogram and column provided by M. V. Novotny and M. L. Lee, Indiana University. (See M. V. Novotny and M. L. Lee: Detection of Marijuana Smoke in the Atmosphere of a Room. Experientia 29:1038, 1973.)

Chemical Separations and Measurements:
The Theory and Practice of Analytical Chemistry

ISBN 0-7216-7203-5

Last digit is the print number:     9   8   7   6   5   4   3   2   1

# PREFACE

Virtually all of "modern chemistry" has been developed through the application of new analytical technology. Taken together, these analytical techniques form a fascinating subject for study and investigation. There is a very strong synergism in the unified study of analytical chemistry—it is immensely more fruitful to focus on the objective of accurate chemical measurement than to attempt to separate techniques according to their present applications. It is, for example, well known that the methods useful to organic chemists today might be taken up by inorganic chemists next year (or *vice versa*). In addition, workers in related fields such as biology, geology, and engineering often have use for analytical chemical techniques. Workers in these "client" disciplines have frequently made important contributions to the science of chemical analysis, and the subject has become an important and stimulating interface between a wide variety of research fields. Our goal has been to capture this diversified vitality in a way which will stimulate students.

It would be nice, we think, to have a textbook that introduced students to genuine modern practice without first leading them through too much history. The composition of such a textbook is made very difficult by the breadth of topics to be considered, and the task is compounded by our philosophy that any topic chosen for discussion must be treated in enough depth to make it truly understandable. Our solution has been to edit very carefully the content of a normal analytical chemistry text and to supplement it with material generally new to the first analytical chemistry course. The latter material has been chosen with care, and we have made every effort to avoid merely listing techniques. In the areas of chemical separations and spectrochemistry, for example, we have provided highly generalized introductions and then have moved quickly to the detailed consideration of a few examples. In all cases, we have tried to avoid a preoccupation with metal ions in aqueous solutions and to introduce more organic, biochemical, solid-state, and gas-phase technology.

In its organization, the present text falls naturally into five units:
- I. Introduction and statistics
- II. Chemical equilibria and their analytical applications
  - A. General aspects
  - B. Acids and bases
  - C. Complexes
  - D. Precipitates
- III. Redox equilibria, techniques, and electrochemistry
- IV. Separations, particularly chromatography
- V. Spectrochemical analyses

These units are not strongly interdependent, and instructors should find it easy to adapt the text to their individual needs. Section I might become the introductory unit of the laboratory course or be covered in lectures. The chapter on statistics is designed to introduce the student to principles, not formulas. The generally available tables of the normal distribution and of the statistic $t$ are used rather than any special tables available only in this text. Sections II and III have been completely rewritten from the earlier text (Fischer and Peters, *Quantitative Chemical Analysis,* third edition), with, for example, the Brønsted-Lowry definitions of acids and bases being used throughout. An extensive treatment of potentiometric measurements with ion-selective electrodes is provided. The four new chapters on separation techniques emphasize chromatography, certainly one of the most widely applied modern analytical tools. In order to allow treatment in some depth, only three examples are discussed, but these represent a partition technique, an adsorption technique, chromatography with a liquid mobile phase and with a gaseous mobile phase. Like the section on separations, the new material on spectrochemical analysis begins with a broadly general introductory chapter. A generalized spectrochemical instrument based on transducer definitions is described, and various techniques are classified according to whether they involve absorption, luminescence, or emission, and whether the species involved are atomic or molecular. The remaining chapters deal with specific examples within this unified context.

Each chapter is abundantly provided with illustrative problems, a number of which are quite challenging. Answers are compiled in an appendix. A program of experiments appropriate to an analytical chemistry laboratory can be obtained from the companion text, *Chemical Separations and Measurements: Background and Procedures for Modern Analysis,* by W. E. Harris and B. Kratochvil (W. B. Saunders Co., 1974).

We would like to acknowledge the assistance and invaluable cooperation of a number of our friends, colleagues, and students. First of all, this book would probably not even exist were it not for the preceding texts principally authored by Dr. Robert B. Fischer, now at California State College, Dominguez Hills. His cooperation in the creation of this new text is greatly appreciated. *Quantitative Chemical Analysis,* third edition, by Fischer and Peters, was very carefully reviewed by Professors Byron Kratochvil, University of Alberta; Richard Ramette, Carleton College; and George Wilson, University of Arizona. Their extensive comments were very helpful in the preparation of the present text, which was in turn reviewed by Professors Stanley Crouch, Michigan State University; A. James Diefenderfer, Lehigh University; Dennis Evans, University of Wisconsin; W. E. Harris and Byron Kratochvil, University of Alberta; Milos Novotny, Indiana University; and Harry Pardue, Purdue University. Kaye Fichman and Robert Yount, two excellent students who were subjected to manuscript versions of text, provided enormously helpful reviews. The tasks of production were eased by the assistance of Norman Clampitt, Larry Games, Janice Hayes, Terry Hunter, Gary McNamee, and Robert Sydor, all of whom helped in the preparation of figures and tables; and by the cooperation of Pam Hieftje, who figured out how to keep the third author at the table. At W. B. Saunders, John Vondeling moved things along with the finesse of a Roman galleymaster and the tactics of Niccolo Machiavelli (he succeeded, however), and Joan Garbutt mediated with cheerful patience, for which we are most grateful.

The entire manuscript was typed twice by Shirley Williams and Sue Hughes and relentlessly reproduced by Dan Logan. Ann Forsee prepared the index. Their extraordinary speed, skills, and cooperation are very sincerely appreciated. Draft versions of the text have been used for two years at Indiana University, and we sincerely appreciate the tolerance and very helpful cooperation of our students.

We hope the text will be worthy of revision, and that the present version is lively enough to stimulate some comments. We invite letters of comment and correction from all readers, especially students.

DENNIS G. PETERS

JOHN M. HAYES

GARY M. HIEFTJE

*Bloomington, Indiana*

# CONTENTS

Chapter 3

## WATER, SOLUTES, AND CHEMICAL EQUILIBRIUM

Chapter 4

## AQUEOUS ACID-BASE REACTIONS

Chapter 5

## ACID-BASE REACTIONS IN NONAQUEOUS SOLVENTS

Chapter 6

## COMPLEXOMETRIC TITRATIONS

Chapter 7

## FORMATION AND DISSOLUTION OF PRECIPITATES

Chapter 8

## ANALYTICAL APPLICATIONS OF PRECIPITATION REACTIONS

## Chapter 9

## PRINCIPLES AND THEORY OF OXIDATION-REDUCTION METHODS

## Chapter 10

## ANALYTICAL APPLICATIONS OF OXIDATION-REDUCTION REACTIONS

Chapter 11

**DIRECT POTENTIOMETRY AND POTENTIOMETRIC TITRATIONS**

## Chapter 12

## ELECTROGRAVIMETRIC AND COULOMETRIC METHODS OF ANALYSIS

Chapter 13

## POLAROGRAPHY AND AMPEROMETRIC TITRATIONS

Chapter 18

# INTRODUCTION TO SPECTROCHEMICAL METHODS OF ANALYSIS

Chapter 19

# SPECTROSCOPY IN THE ULTRAVIOLET AND VISIBLE REGIONS—INSTRUMENTATION AND MOLECULAR ANALYSIS

Chapter 20

## SPECTROSCOPY IN THE ULTRAVIOLET AND VISIBLE REGIONS—ATOMIC ELEMENTAL ANALYSIS

Chapter 21

**INFRARED AND RAMAN VIBRATIONAL SPECTROMETRY**

# NATURE OF
# ANALYTICAL CHEMISTRY

1

In this book we hope to provide an introductory coverage of the field of analytical chemistry. Yet even an introduction will require us to deal in some depth with a broad range of topics. This is because analytical chemistry, like most scientific disciplines, has become so intertwined with other sciences that it is difficult to define the field.

## WHAT IS ANALYTICAL CHEMISTRY?

Analytical chemists working in industrial control laboratories are often responsible for maintaining, by means of appropriate chemical measurements, the quality of an outgoing product or an incoming raw material. In addition, there are analytical chemists employed in clinical laboratories, where important tests are performed to establish the condition of health of a patient or to serve as an indispensable aid to medical diagnosis. Most of these chemists concentrate on applying the knowledge and findings of chemistry to the analysis of real samples. Generally, the tools and techniques utilized are well known and established, with little room for innovation or error.

Another kind of activity is the development of new analytical techniques and the improvement of existing ones. New discoveries and instrumentation in chemistry, physics, or engineering may frequently lead to new analytical techniques. For example, the mass spectrometer, originally developed as a tool of nuclear physics, has been adapted to a wide variety of chemical uses ranging from structure determinations in organic chemistry to quantitative analysis of isotopes in tracer experiments. Similarly, both x-ray spectroscopy and electron spectroscopy are presently being utilized primarily in chemical applications.

Development of a new analytical method often occurs in response to a specific need. In many instances, the new technique takes the form of an extension of some existing principle of measurement. Thus, for the detection, determination, and monitoring of environmental pollutants, chromatographic and electrochemical techniques have been reshaped and made more sensitive as dictated by the particular analytical problem. Improvements in existing methods can be especially worthwhile if they improve the accuracy or lower the unit cost of an analysis. Very frequently, both of the preceding objectives can be met through automation of an already established technique. Control of an experiment as well as the processing of data can be taken over by a small computer which costs less and performs more reliably than any alternative system.

There is more to technique development than just engineering. To be sure, once the concept and design underlying a method have been formulated, engineering can play a major role in the successful execution of experiments. However, the design of a system for a chemical measurement requires a familiarity both with the basic principles and capabilities of the technique being developed and with the goals and practical aspects of the chemical experiments. Thus, an analytical chemist involved in technique development provides, for example, an interface between physicists who have perfected a mass spectrometer and organic chemists interested in determining the structure of organic compounds. Such a role is stimulating and important, for the development of new investigative techniques is often a crucial step in the discovery of new chemical knowledge.

Today, analytical chemists are confronted with such a bewildering array of established, modified, and new techniques that it is practically impossible to keep up with the entire field. For this reason, most analytical chemists at the advanced level choose one subdiscipline such as analytical separations, spectrochemical analysis, or electrochemical analysis to which they devote their major attention, while still maintaining a working knowledge of other subdisciplines. Within each of the numerous but still broad subdisciplines, active areas of research are usually even more narrowly defined. For example, an electroanalytical chemist may opt to study the mechanisms of organic electrode reactions, to apply ion-selective electrodes to pollution monitoring or biochemical analysis, or to perfect new coulometric techniques for nonaqueous titrations of organic compounds. Someone working in the field of analytical separations might concentrate his attention on designing novel detectors for use in liquid chromatography, on the development of ion-exchange resins to separate optically active compounds, or on the chromatographic examination of body fluids containing drug metabolites. Spectrochemical analysis might center on fundamental investigations of how processes occurring in flames affect the precision and accuracy of atomic absorption spectrometry, on spectroscopic methods for the rapid and sensitive measurement of substances of biological and pharmaceutical interest, or on the construction and evaluation of new instrumentation for the spectroscopic determination of molecular structure.

## ORGANIZATION OF THE TEXTBOOK

A glance at the table of contents of this textbook shows that we have placed emphasis on practical chemical measurements and on separation techniques. This selection of topics has broad utility for chemists who go on to concentrate in any of the non-theoretical areas of chemistry, and forms an introductory coverage of the field for chemists who plan to specialize in chemical instrumentation or other aspects of problem-solving methodology.

Because the treatment and interpretation of data are so vital to all kinds of chemical experiments, Chapter 2 describes in detail how to express the accuracy and precision of an analytical result and how to assess errors in measurements through the application of rigorous statistical and mathematical concepts; in addition, this material provides a solid base for the discussion of chromatographic separations in later chapters. In Chapter 3, the behavior of solutes in aqueous media is examined, and some of the principles of chemical equilibrium upon which material in subsequent chapters will depend are discussed. Chapters 4 and 5 cover the topics of acid-base reactions in aqueous and nonaqueous systems; this treatment is essential to the quantitative descriptions of the solubility of precipitates in various solvents and of the chemical interactions involved in analytical methods based on complexation and solvent-extraction phenomena. In Chapter 6, the theory and analytical uses of complex-formation reactions are considered, and a foundation is laid for applying these concepts to such analytical techniques as direct potentiometry, coulometric titrations, polarography, and chromatography. Analytical methods based on the formation of precipitates are discussed in Chapters 7 and 8.

A thorough description of oxidation-reduction processes is presented in Chapters 9 through 13; starting with a discussion of galvanic cells, the material progresses to titrimetric redox methods involving both inorganic and organic species and then to surveys of electrometric procedures, including potentiometry, coulometry, and polarography. Chapters 14 through 17 offer an excellent introduction to all aspects of analytical separations; there is information about the theory and practice of distillation, the nature of phase equilibria and extraction, and the applications of various chromatographic techniques to the resolution of mixtures of inorganic, organic, and biological species. In the final portion of the textbook, Chapters 18 through 21 present some of the most prominent techniques employed in modern spectrochemical analysis—interactions of ultraviolet and visible radiation with atomic and molecular species leading to absorption, emission, and fluorescence; and applications of infrared and Raman spectroscopy for the determination of molecular structure.

Some of the topics in this textbook seem far less glamorous than many of the activities of analytical chemists listed earlier. However, in order to engage in productive and meaningful work, an analytical chemist must be well versed in the fundamental principles upon which so many modern techniques rely. This then is the rationale for the existence of the present textbook.

## LITERATURE OF ANALYTICAL CHEMISTRY

If there is any doubt that analytical chemistry is a diverse and active area of endeavor, one need only examine the large number of general and specialized scientific journals that publish both review articles and original research papers in the field to realize the tremendous importance and impact of modern analytical chemistry. All the specialized areas of analytical chemistry—including clinical analysis, spectrochemical analysis, electrochemical analysis, chromatography, instrumentation, computer applications, kinetics, mass spectrometry, and the collection, storage, retrieval, and processing of experimental data—are represented by at least several specialized journals that publish theoretical and practical aspects of a subdiscipline. In addition, there are a number of journals that bring together papers covering innovations in all these specialized areas as well as new developments in gravimetric and volumetric methods of analysis. It is undoubtedly conservative to assert that approximately 15,000 articles pertaining to analytical chemistry are published annually.

A professional analytical chemist intending to keep abreast of the rapid advances in his field must develop the habit of regularly reading the literature. We can only urge the student beginning a first serious excursion into analytical chemistry to adopt a similar attitude as well as to recognize the virtue of consulting several books and articles on any one subject in order to reach a balanced point of view.

Among the prominent journals that cater to all branches of analytical chemistry are *The Analyst*, *Analytica Chimica Acta*, *Analytical Chemistry*, *Chimie Analytique* (French), *Zhurnal Analiticheskoi Khimii* (Russian, but available in English translation), *Talanta*, and *Zeitschrift für analytische Chemie* (German). Biennial reviews published in the April issue of *Analytical Chemistry* in each even-numbered year offer comprehensive summaries and bibliographies of fundamental research written by experts in almost every subdiscipline of analytical chemistry; an April issue in each odd-numbered year is devoted to a series of reviews of analytical applications in such fields as air pollution, clinical chemistry, food, petroleum, pharmaceuticals and drugs, and water analysis. *Analytical Letters* is an international journal for the rapid dissemination of new results in all branches of analytical chemistry, and *Analytical Abstracts* is a monthly publication that presents short summaries of papers appearing in journals around the world in all subdisciplines of analytical chemistry.

It is almost impossible to cite all the specialty journals in the field of analytical chemistry, but some idea of the scope of subdisciplinary research—and of the important way that analytical chemistry impinges upon and contributes to other sciences—is provided by the following list of publications, the titles of which are self-descriptive: *Acta Crystallographica*, *Analytical Biochemistry*, *Applied Spectroscopy*, *Chemical Instrumentation*, *Chromatographic Reviews*, *Clinica Chimica Acta*, *Clinical Chemistry*, *Journal of Chromatographic Science*, *Journal of Electroanalytical Chemistry and Interfacial Electrochemistry*, *Journal of the Electrochemical Society*, *Organic Mass Spectrometry*, and *Spectrochimica Acta*.

Several monthly magazines regularly present articles as well as informative advertisements on the latest developments and trends in spectroscopy, electroanalytical chemistry, instrumentation, and computers. Of particular interest are *Industrial Research* (published by Industrial Research, Inc., Chicago, Illinois), *American Laboratory* (published by American Laboratory, Inc., Greens Farms, Connecticut), and *Research/Development* (published by Technical Publishing Company, Barrington, Illinois).

Subjects discussed in the monthly feature sections as well as in the regularly contributed articles of the *Journal of Chemical Education*, published by the Division of Chemical Education of the American Chemical Society, frequently include topics which are of interest to analytical chemists.

Many textbooks and reference works deal with one or more specific areas of analytical chemistry. Some of these are listed as suggestions for additional reading at the ends of the chapters in this book. Among the individual books and series of volumes of general interest throughout many areas of analytical chemistry are the following:

1. *Standard Methods of Chemical Analysis*, sixth edition, D. Van Nostrand Company, Princeton, New Jersey. Volume I (The Elements) was edited by N. H. Furman and

appeared in 1962. Volume II (Industrial and Natural Products and Noninstrumental Methods) and Volume III (Instrumental Methods) were both edited by F. J. Welcher and were published in 1963 and 1966, respectively. These books serve as a reliable source of analytical information for practical use in the laboratory.

2. *Handbook of Analytical Chemistry*, edited by L. Meites, McGraw-Hill Book Company, New York, 1963. This one-volume compendium provides summaries of fundamental data and practical laboratory procedures.

3. *Treatise on Analytical Chemistry*, edited by I. M. Kolthoff and P. J. Elving, Wiley-Interscience Publishers, New York (Volume 1 of Part I in 1959 and others subsequently). This three-part, multivolume series offers a thorough coverage of virtually every phase of analytical chemistry, including theory and practice of analytical techniques, analytical chemistry of the elements, and analysis of industrial products.

4. *Comprehensive Analytical Chemistry*, edited by C. L. Wilson and D. W. Wilson, Elsevier Publishing Company, Amsterdam (Volume I-A in 1959 and others subsequently). This is another multivolume series dealing with most aspects of the theory and practice of analytical chemistry.

5. *Advances in Analytical Chemistry and Instrumentation*, edited by C. N. Reilley and others, Wiley-Interscience Publishers, New York (Volume 1 in 1960 and others subsequently). Early volumes in this series contain a selection of reviews by outstanding authorities in various fields of analytical chemistry. Since 1968 each volume has been devoted to a different single field of analytical chemistry, with several experts contributing chapters in their own fields of expertise.

# TREATMENT OF ANALYTICAL DATA

# 2

Imagine that you are determining by analysis some variable quantity $x$. In two sets of quadruplicate measurements on two different materials, $A$ and $B$, you obtain the data shown in Figure 2–1. The question is, does material $A$ really differ from material $B$? An accurate, but useless, answer is "Maybe." *It is possible to assess quantitatively the confidence with which you might instead answer "Yes" or "No."* It is pathetic to guess at the answer to such questions—incredible wastes of time and money can result. For example, the materials $A$ and $B$ might be two batches of a manufactured product. If they truly differ, one must be discarded at great expense. Alternatively, the materials $A$ and $B$ might be products of a chemical reaction carried out under two slightly different sets of conditions. If the difference in conditions has actually affected the composition of the product, a new theory is proven (or required). Anyone who undertakes chemical measurements can either understand the statistical methods which allow useful answers to practical questions, or can accept the role of an analytical instrument, a mindless performer of laboratory procedures.

Consider a second example: the quantity $x$ has been determined by measurement of $a$, $b$, $c$, and $d$.

$$x = ab + \frac{c}{d}$$

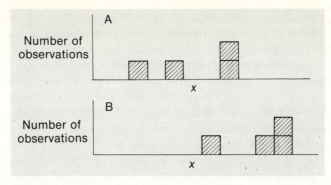

*Figure 2–1.* Histograms showing results for the determination of the variable x in two materials A and B.

How do uncertainties in *a*, *b*, *c*, and *d* affect the result? How can the uncertainty in *x* be computed from the known uncertainties in the measured quantities?

The questions posed above are very practical. The techniques which provide useful answers are easily mastered, and this chapter offers a brief treatment of some basic statistical techniques. A knowledge of calculus is helpful but by no means required for the understanding of this material, and we beg the reader unfamiliar with calculus not to faint at the first differential, but instead to get a "translation" from the instructor.

## NATURE OF QUANTITATIVE MEASUREMENTS

### Significant Figures

In the examples below, the number of significant figures in each quantity is given in parentheses.

| | | | | | | |
|---|---|---|---|---|---|---|
| **1** | 12.270 g | (5) | | **6** | 43,062 ml | (5) |
| **2** | 12.3 g | (3) | | **7** | 43,100 ml | (3) |
| **3** | 10 g | (1) | | **8** | 40,000 ml | (1) |
| **4** | 0.00524 $M$ | (3) | | **9** | 100.00 | (5) |
| **5** | 0.005 $M$ | (1) | | **10** | 1.00 | (3) |
| | | | | **11** | 0.010 | (2) |

The number of significant figures is always equal to or greater than the number of nonzero digits. When zeroes are used only to locate the decimal point, they do not count as significant figures. Thus, in **1**, **9**, **10**, and **11**, the trailing zeroes are not required to locate the decimal point but instead indicate that the measurement has been made with the indicated accuracy. Similarly, in **6** the interior zero is significant.

In **3**, **7**, and **8**, the trailing zeroes may be significant, or may be used only to locate the decimal point. Scientific notation furnishes the most convenient way of overcoming this ambiguity, although an alternative used by some workers is the marking of significant trailing zeroes; for example, $40\overline{0}$ indicates that all three figures are significant. In scientific notation, the same number would be written $4.00 \times 10^2$. If there were only one significant figure, the number would be written $4 \times 10^2$.

To overcome uncertainties associated with various round-off conventions, some workers include the first insignificant digit in numerical data. The insignificant figure is frequently written as a subscript. For example, in the number $4.05_2$, the 2 is insignificant. (Caution: number bases are sometimes noted in the same way—thus, $624_8$ might represent an octal rather than a decimal number.)

*Absolute uncertainty and relative uncertainty.* Uncertainty in measured values may be considered from two distinct viewpoints. *Absolute uncertainty* is expressed directly in units of the measurement. A weight expressed as 10.2 g is presumably valid within a tenth of a gram, so the absolute uncertainty is one tenth of a gram. Similarly, a volume measurement written as 46.26 ml indicates an absolute uncertainty of one hundredth of a milliliter. Absolute uncertainties are expressed in the same units as the quantity being measured—grams, liters, and so on.

*Relative uncertainty* is expressed in terms of the magnitude of the quantity being measured. The weight 10.2 g is valid to within one tenth of a gram and the entire quantity represents 102 tenths of a gram, so the relative uncertainty is about one part in 100 parts. The volume written as 46.26 ml is correct to within one hundredth of a milliliter in 4626 hundredths of a milliliter, so the relative uncertainty is one part in 4626 parts, or about 0.2 part in a thousand. It is customary, but by no means necessary, to express relative uncertainties as parts per hundred (per cent), as parts per thousand, or as parts per million. Relative uncertainties do not have dimensions because a relative uncertainty is simply a ratio between two numbers having the same dimensional units.

To distinguish further between absolute and relative uncertainty, consider the weighing of two different objects on an analytical balance, one object weighing 0.0021 g and the other 0.5432 g. As written, the absolute uncertainty of each number is one ten-thousandth of a gram; yet, the relative uncertainties differ widely—one part in 20 for the first weight and one part in approximately 5000 for the other value.

*Significant figures in mathematical operations.* *In addition and subtraction the absolute uncertainty in the result must be equal to the largest absolute uncertainty among the components.* Consider three examples:

| | | |
|---:|---:|---:|
| 10.0051 | | 0.5362 |
| 1.9724 | 42598 | 0.0014 |
| +0.0003 | −42595 | +0.25 |
| 11.9778 | 3 | 0.79 |

In the first two cases, all the component numbers have the same absolute uncertainty, and determination of the correct number of significant figures in the result is a simple matter, although it has the somewhat surprising consequence that in the first case a component with only one significant figure contributes to a result with six, and, in the second case, two components with five significant figures yield a result with only one. In the third example, the absolute uncertainties are not equal, and the number of significant figures in the result is determined by the absolute uncertainty in the third number added.

*In multiplication and division, the relative uncertainty in the result must be equal to the largest relative uncertainty among the components.* For example, in the operation $0.12 \times 9.678234$, the correct product is 1.2. Expressing the result as 1.1614 would be unjustified because the relative uncertainty in the first factor is one part in twelve.

## Precision and Accuracy

The terms, *precision* and *accuracy*, are not synonymous, and are defined here by contrasting examples. A balance which on repeated trials gives the weight of an object as 1.307, 1.308, 1.305, and 1.307 g, is more *precise* than one which gives the weights 1.302, 1.316, 1.305, and 1.310 g. *Precision* relates to the degree of scatter in a set of data, the scatter in these examples clearly being greater in the second set. A balance which gives the weight of a known 10.000-g standard as 10.001 g is more *accurate* than one which gives the weight of the same standard as 10.008 g. *Accuracy* relates to the difference between a measured quantity and its true value.

### Errors

*Systematic errors.* Systematic errors are frequently related to improper design or adjustment of experimental apparatus; such errors reduce accuracy by systematically skewing or offsetting the observed data. A careful study of measurements of some standard can reveal the nature of the error, and accuracy can be regained by adjusting the apparatus or applying some correction. For example, if a balance showed the weight of a ten-gram standard to be 10.080 g and of a one-gram standard to be 1.008 g, we could conclude that the balance was misadjusted so that it weighed 0.80 per cent high. It would be possible to apply a correction to all the data, but it would be far better to repair the balance. Alternatively, a balance might show the weight of a one-gram standard as 1.050 g and the weight of a ten-gram standard as 10.050 g. In this case, it could be concluded that the error was not proportional to the weight, but was, instead, a constant 0.050-g offset. Again, a correction could be applied or, preferably, the balance could be repaired. (These examples also illustrate the value of systematic observations during instrument troubleshooting, because the nature of the errors observed would naturally lead to certain types of adjustments. The constant 0.050-g offset, for example, must be due to a misplaced zero point on the mass scale; the 0.80 per cent relative error is probably due to improper positioning of the balance arm.)

*Random errors.* Random errors result from insufficiently controlled variations in measurement conditions. Many different effects acting together lead to small variations in the observed value. Repeated observations will scatter randomly around the true value, and it is thus clear that the size and frequency of these random errors will determine the precision of a given measurement. For example, in the case of the analytical balance, sources of random error include fluctuations in room temperature and humidity, small variations in the placement of weights, variations in the positioning of the knife edge on its bearing, and the subjectivity of the operator who reads the final weight from some uniformly graduated scale.

Careful consideration of the conditions of measurement can substantially reduce random errors, but, unlike systematic errors, they cannot be eliminated, nor can some formula be derived to correct an observation for their influence. As we shall learn in the following sections, random errors are amenable to statistical treatment, and repeated measurements of the same variable can have the effect of reducing their importance.

## BASIC STATISTICAL CONCEPTS

### Frequency Distributions

*Random errors and the normal distribution.* Consider the quantitative determination of, for example, iron in ferric oxide. Figure 2–2 shows a sequence of three graphs illustrating the effect of random errors. In the first case, only four quite

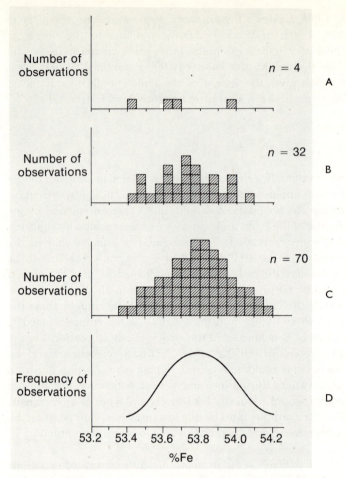

*Figure 2–2.* A, B, and C are histograms showing the distribution of results for a series of iron analyses. D shows the normal distribution which would be obtained if an infinite number of observations could be made.

widely scattered observations were obtained. More observations were obtained in the second and third cases, and it is clear that the distribution of results is tending toward the smooth curve drawn at the bottom of the figure. The smooth curve represents the **universe** of all possible determinations of iron in iron oxide, and each of the smaller sets of observations is a **sample** drawn from that universe. The height of the universe distribution at any given value of per cent iron is a measure of the *frequency* with which observations of that value will be obtained during sampling. The relationship is known generally as a *frequency distribution*. The shape of the frequency distribution is given by the normal law of error, and the curve is known as the **normal distribution**. Many other names are also applied, but indicate exactly the same curve. Some of the most commonly used synonyms are "normal curve of error," "Gaussian distribution," and "probability distribution."

*Other frequency distributions.* In Figure 2–3, two other fundamentally different frequency distributions are shown in order to further emphasize the concept. If dice are thrown or cards are cut, and the dice are not loaded and the deck is not stacked, it is equally likely that any one of the six sides of a die will come up or that any one of the thirteen possible playing-card denominations will be cut. The frequency distribution representing these situations is flat. If pictures are taken at random intervals of a swinging pendulum, the observed pendulum positions will be distributed as shown (where $\theta$ is the angle between the pendulum and its center of swings). This frequency distribution arises because the pendulum is most likely to be observed where its velocity is low (at the ends of its swing) and is least likely to be caught by the camera at its center of swing, where its velocity is at a maximum. These distributions apply to specialized situations. The normal distribution applies in nearly all cases where random errors occur.

### Average and Measures of Dispersion

For any set of data, our aim is to summarize it quickly and usefully. Everyone is familiar with summarizing a set of measurements by computing an average, but this is strictly a measurement of *central tendency*, and tells nothing about the *distribution* of the measurements. In order to summarize a set of data, we require in addition some measurement of spread or dispersion of the individual data points.

**Central tendency.** The most common measurement of central tendency is the **mean** or average. Usually, we denote the mean value of some variable by placing a bar over its symbol; thus,

$$\bar{x} = \frac{\sum x}{n}$$

where $\bar{x}$ is the mean of $n$ observations of $x$. We must distinguish between $\bar{x}$, the mean

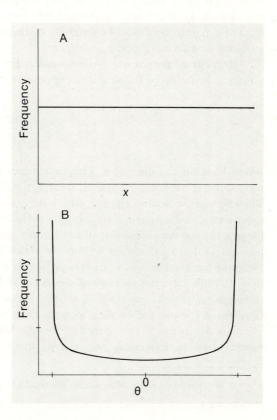

*Figure 2–3.* Frequency distributions for A, the cutting of cards or rolling of a die, and B, observations of a pendulum at random time intervals. (See text for discussion.)

of some finite sample, and $\mu$, the true value or universe mean. Notice that

$$\lim_{n \to \infty} \bar{x} = \mu$$

Other useful measures of central tendency are the **median**, or middle value of some sample, and the **mode**, or most frequently observed value of some sample. In particular, the appearance of more than one mode in a distribution requires careful study. The existence of a bimodal distribution might indicate that a sample is, in fact, drawn from two different universes. For example, a bimodal distribution in examination grades is frequently observed in classes which mix undergraduate and graduate students.

*Measures of dispersion.* The simplest measure of dispersion is the **range** of values found in some sample. It is, however, a poor indicator of the shape of the distribution because it depends only on the highest and lowest values. By far the best and most useful measure of dispersion is given by the **variance**, which is the mean square deviation of all observations, where the "deviation" is the difference between an observation and the mean:*

$$\sigma_x^{\ 2} = \frac{\sum (x - \mu)^2}{n}. \tag{2-1}$$

The variance has the dimensions of $x^2$, and its value does not convey any feeling for the amount of scatter in $x$. Thus, the **standard deviation**, or root-mean-square deviation, is often used. It is assigned the symbol $\sigma_x$ and is also defined by equation (2-1). Notice that the formula can be used only when the universe mean, $\mu$, is known. Thus, equation (2-1) relates only to the universe variance and standard deviation, $\sigma_x^{\ 2}$ and $\sigma_x$. These parameters must be carefully distinguished from the sample variance and standard deviation, $s_x^{\ 2}$ and $s_x$, which are discussed below.

*Degrees of freedom.* An expression for the calculation of the variance and standard deviation of the *sample* (as opposed to those of the *universe*) is

$$V_x = s_x^{\ 2} = \frac{\sum (x - \bar{x})^2}{n - 1} \tag{2-2a}$$

where $V_x$ is the variance of $x$. This expression differs from that given for $\sigma_x^{\ 2}$ in two ways. First, $\bar{x}$ has replaced $\mu$ and, second, the denominator is $(n - 1)$ rather than $n$. The effect of the second change, which decreases the denominator, is to increase the value for $s_x$. This is appropriate, because we have "stacked the deck" by substituting $\bar{x}$ for $\mu$, a change which has the effect of minimizing the numerator. This comes about because we have calculated $\bar{x}$ from the individual $x$ values; thus, $\bar{x}$ represents the center of our sample, but not necessarily that of the universe. More fundamentally, $(n - 1)$ represents the number of **degrees of freedom**, or independent deviation calculations, which are possible within the sample after $\bar{x}$ has been calculated. Consider a sample with $n$ data points. First, we calculate $\bar{x}$ and then we begin to calculate deviations from the mean, $(x - \bar{x})$. When we reach the $n$th data point, the comparison is no longer an independent one, because the $x$ value for that $n$th point

---

* A better representation of the numerator would be given by $\sum_{i=1}^{n} (x_i - \mu)^2$, but it is conventional in unambiguous cases to omit the subscripts and interval notations for $\Sigma$.

could be calculated before we even saw it, given $\bar{x}$ and the $(n - 1)$ preceding data points. In general,

$$\varphi = \text{degrees of freedom} = [n - (\text{number of constants calculated from data})]$$

$$s^2 = \frac{\text{sum of (deviations)}^2}{\varphi} \tag{2-3}$$

The sample variance is an *estimate* of the universe variance. The relationship between $s_x^2$ and $\sigma_x^2$ is similar to that between $\bar{x}$ and $\mu$:

$$\lim_{n \to \infty} s_x^2 = \sigma_x^2$$

$$\lim_{n \to \infty} \frac{\sum (x - \bar{x})^2}{n - 1} = \frac{\sum (x - \mu)^2}{n}$$

**Alternative expressions for the standard deviation.** The expressions given above are inconvenient for the rapid computation of the sample variance. The following relations are entirely equivalent and permit much more rapid calculation:

$$s_x^2 = \frac{(\sum x^2) - n\bar{x}^2}{n - 1} = \frac{\sum x^2 - \frac{(\sum x)^2}{n}}{n - 1} \tag{2-2b, 2-2c}$$

*Example 2-1.* Compute the sample standard deviation for the following results obtained for the analysis of carbon in lunar soil (Apollo 11, fine-grained material): 130, 162, 160, 122 ppm. It is convenient to arrange the data in a small table, as shown below:

| $x$ | $x^2$ |
|---|---|
| 130 | 16,900 |
| 162 | 26,244 |
| 160 | 25,600 |
| 122 | 14,884 |
| $\sum x = 574$ | $\sum x^2 = 83,628 \qquad n = 4$ |

Substituting these results in equation (2-2c), we obtain

$$s_x^2 = \frac{\sum x^2 - \frac{(\sum x)^2}{n}}{n - 1} = \frac{83,628 - \frac{(574)^2}{4}}{3}$$

$$= \frac{83,628 - 82,369}{3} = 419.7 \text{ (ppm)}^2$$

$$s_x = \sqrt{419.7} = 20.5 \text{ ppm}$$

Notice that the numerator is the difference between two large numbers. Thus, many significant figures must be carried in the calculation. Slide-rule accuracy is usually inadequate.

***Standard deviation calculated from multiple samples.*** The standard deviation is generally a function of the method of analysis, and is not dependent on the specific material analyzed. Consider the use of a particular technique to analyze ten different materials, each in triplicate. Thirty analyses have been performed, and much experience about the precision of the method has been obtained. If a standard deviation is calculated for each individual sample, ten different standard deviations are obtained; probably some are high, some are low, and each is based on only two degrees of freedom. In this situation, anyone would be tempted, whether he knew much about statistics or not, to calculate an "average standard deviation" in order to obtain a better estimate of $\sigma_x$. It happens that for the case mentioned, with equal numbers of observations in each sample, a simple averaging procedure would give a perfectly correct answer. In general, when means $\bar{x}_1, \bar{x}_2, \bar{x}_3, \ldots$ have been calculated from $n_1, n_2, n_3, \ldots$ observations,

$$s_x^2 = \frac{\sum (x_1 - \bar{x}_1)^2 + \sum (x_2 - \bar{x}_2)^2 + \sum (x_3 - \bar{x}_3)^2 + \cdots}{(n_1 - 1) + (n_2 - 1) + (n_3 - 1) + \cdots} \tag{2-4a}$$

$$s_x^2 = \frac{\left[\sum x_1^2 - \frac{(\sum x_1)^2}{n_1}\right] + \left[\sum x_2^2 - \frac{(\sum x_2)^2}{n_2}\right] + \left[\sum x_3^2 - \frac{(\sum x_3)^2}{n_3}\right] + \cdots}{(n_1 - 1) + (n_2 - 1) + (n_3 - 1) + \cdots} \tag{2-4b}$$

$$s^2 = \frac{\text{sum of (deviations)}^2}{\varphi} \tag{2-3}$$

The last expression has been given before, and is repeated here to stress its general applicability. The reader who takes time to understand the equivalence of all three expressions above will be well rewarded. In addition to allowing a better understanding of the standard deviation, these expressions demonstrate that many statistical expressions which *appear* complex when written out in abstract form are, in essence, very simple. An illustration of the application of this technique coupled with data coding (see below) is provided by Example 2–2.

***Coding.*** The labor required in the calculation of a standard deviation can be considerably reduced by subtracting some constant from each observation in order to obtain smaller numbers. In other calculations, we often do this mentally. For example, anyone would find the average of 61, 63, 65, and 67 by adding $1 + 3 + 5 + 7$, dividing by four, and adding the result to 60. This type of coding can also be applied to computation of the standard deviation without affecting the result in any way. Also, it is sometimes convenient to multiply or divide by some constant factor. This, too, is permissible, but the reverse operation must be performed after the calculations are complete in order to obtain a correct $s_x$. An example is given below.

*Example 2–2.* The same procedure for the determination of chloride is used to analyze four different samples. Estimate the standard deviation of the analytical procedure. Data: Sample 1, 11.25, 11.30, 11.31, 11.31% Cl; Sample 2, 14.26, 14.27, 14.30, 14.32% Cl; Sample 3, 18.72, 18.65, 18.60% Cl; Sample 4, 16.50, 16.45, 16.42% Cl.

The coded data can be conveniently tabulated as follows:

$$\text{Sample 1: } x_1 = (\% \text{ Cl} - 11.25)100$$
$$\text{Sample 2: } x_2 = (\% \text{ Cl} - 14.25)100$$
$$\text{Sample 3: } x_3 = (\% \text{ Cl} - 18.60)100$$
$$\text{Sample 4: } x_4 = (\% \text{ Cl} - 16.40)100$$

Given this coding, we have

$$s_x = 100 \, s_{Cl}$$

| Sample 1 | | Sample 2 | | Sample 3 | | Sample 4 | |
|---|---|---|---|---|---|---|---|
| $x_1$ | $x_1{}^2$ | $x_2$ | $x_2{}^2$ | $x_3$ | $x_3{}^2$ | $x_4$ | $x_4{}^2$ |
| 0 | 0 | 1 | 1 | 12 | 144 | 10 | 100 |
| 5 | 25 | 2 | 4 | 5 | 25 | 5 | 25 |
| 6 | 36 | 5 | 25 | 0 | 0 | 2 | 4 |
| 6 | 36 | 7 | 49 | 17 | 169 | 17 | 129 |
| 17 | 97 | 15 | 79 | | | | |

Applying equation (2–4b), we have

$$s_x{}^2 = \frac{\left[97 - \dfrac{(17)^2}{4}\right] + \left[79 - \dfrac{(15)^2}{4}\right] + \left[169 - \dfrac{(17)^2}{3}\right] + \left[129 - \dfrac{(17)^2}{3}\right]}{(4-1) + (4-1) + (3-1) + (3-1)}$$

$$= \frac{(97-72) + (79-56) + (169-96) + (129-96)}{3 + 3 + 2 + 2}.$$

$$= \frac{25 + 23 + 73 + 33}{10} = 15.4$$

$$s_x = \sqrt{15.4} = 3.9$$

$$s_{Cl} = \frac{s_x}{100} = 0.039 \text{ per cent}$$

For a similar problem, see the first part of Example 2–13.

## Probabilities Derived from the Normal Distribution

Putting it crudely, the normal distribution must be good for something. Imagine that we had some analytical procedure which had been applied hundreds of times, and that we could now regard $s_x$ as essentially equivalent to $\sigma_x$. Could we not then state the probability that an individual measurement was within some given distance of $\mu$? This is easily possible, but the method requires some introduction.

*Mathematical expression for the normal distribution.* The form of the normal distribution curve is given by

$$\frac{dN}{N} = \left\{\frac{1}{\sqrt{2\pi} \, \sigma_x} \exp\left[-\frac{(x-\mu)^2}{2\sigma_x{}^2}\right]\right\} dx \tag{2–5}$$

where* $dN/N$, the fraction of the universe between $x$ and $(x + dx)$, is the probability that $x$ will take a value between $x$ and $(x + dx)$. An expression cast in terms of $x$, $\mu$, and $\sigma_x$ is not as conveniently evaluated as one with a single variable. Furthermore,

---

* The notation "exp" indicates that the argument is the power to which $e$ must be raised. Thus, $\exp(u) = e^u$. When fractional exponents occur, this notation eliminates much confusion.

the quantities will vary with each situation to which the normal distribution is applied. However, the *shape* does not change, whatever the values of $x$, $\mu$, and $\sigma_x$, and this constancy is seen in the fact that the expression can be cast in terms of a single reduced variable, $u$.

Let

$$u = \frac{(x - \mu)}{\sigma_x} \tag{2-6}$$

Differentiating equation (2–6) with respect to $x$, we obtain

$$dx = \sigma_x \, du \tag{2-7}$$

Substituting equations (2–6) and (2–7) in equation (2–5), we obtain

$$\frac{dN}{N} = \left[ \frac{1}{\sqrt{2\pi}\,\sigma_x} \exp\left( -\frac{u^2}{2} \right) \right] \sigma_x \, du = \frac{1}{\sqrt{2\pi}} \exp\left( -\frac{u^2}{2} \right) du \tag{2-8}$$

The use of $u$, in effect, expresses all deviations in terms of the standard deviation.

*Example 2–3.* In the series of iron analyses noted above (Figure 2–2), $\mu = 53.78\%$ Fe and $\sigma_x = 0.20\%$. Calculate $u$ for $x = 53.58\%$ Fe.

Substitution in equation (2–6) yields

$$u = \frac{(x - \mu)}{\sigma_x} = \frac{(53.58 - 53.78)}{0.20} = \frac{-0.20}{0.20} = -1.0$$

The value $x = 53.58\%$ Fe would be termed "one standard deviation out," that is, one standard deviation, one $u$ unit, away from the mean.

*Example 2–4.* For the final examination in a certain chemistry course, $\mu = 75$ points and $\sigma_x = 10$ points. Calculate $u$ for $x = 100$ points.

Substituting the appropriate values in equation (2–6), we obtain

$$u = \frac{(x - \mu)}{\sigma_x} = \frac{(100 - 75)}{10} = \frac{25}{10} = 2.5$$

On this examination a perfect score is "2.5 standard deviations out." We shall soon see what the probability of this occurrence is.

**Relationship between probability and area.** Every observation in any universe lies somewhere under the normal distribution curve for that universe. Stating this fact mathematically, we can write the integral directly below:

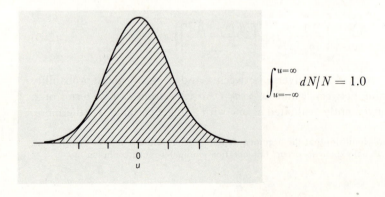

$$\int_{u=-\infty}^{u=\infty} dN/N = 1.0$$

The small sketch at the left illustrates the area given by the integral. Similarly, we could observe that *half* of the observations will lie on each side of the mean:

$$\int_{u=-\infty}^{u=0} dN/N = 0.5 = \int_{u=0}^{u=\infty} dN/N$$

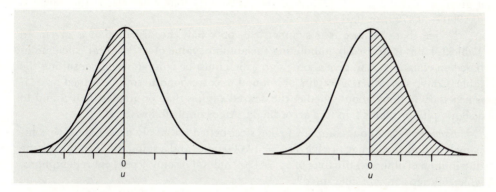

The small sketches illustrate the areas given by the two integrals. Because the distribution is symmetrical, the probability is 0.5 that $x$ will lie somewhere on the positive side of $\mu$ and 0.5 that $x$ will lie somewhere on the negative side of $\mu$. What is the probability that $x$ will take some value such that $u \geq 2.0$? That is, what is the chance that $x$ will be two or more standard deviations out on the positive side? This probability is represented by the area under the normal distribution curve in the range $2.0 \leq u \leq \infty$, and can be determined by means of either procedure A or procedure B below:

A:  probability that $x \geq (\mu + 2\sigma_x) = \int_{u=2.0}^{u=\infty} dN/N$

B:  probability that $x \geq (\mu + 2\sigma_x) = 0.5 - \int_{u=0}^{u=2.0} dN/N$

The areas representing the integrals in A and B are shown in Figure 2–4. Either procedure is correct, depending on which form of tabulation of the normal distribution is available. Although most mathematical handbooks list only one or the other, we present both forms in Table 2–1. For a tabulation of form A, the probability that $x$ will lie two or more standard deviations above the mean is given directly by the integral representing the area under the curve from $u = 2.0$ to $u = \infty$. This area, or

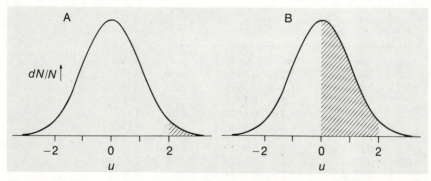

*Figure 2–4.* Normal error curves showing the areas which are given in the two alternative tabulation forms for the normal distribution. (See text for discussion.)

value of the integral, is listed at $u = 2.0$ in tables of form A, and is given in Table 2–1 as 0.0227, providing directly the information that there is a 2.27 per cent chance that an observation will fall two or more standard deviations away from the mean. When a table of form B is consulted, the area (or value of the integral) tabulated at $u = 2.0$ represents the probability that $x$ will lie between $\mu$ and $(\mu + 2\sigma_x)$. This area is given in Table 2–1 as 0.4773, and the probability that $x$ lies *outside* this range is $(0.5000 - 0.4773)$ or 0.0227.

Notice that, because it is symmetrical, only half the distribution is given. In Table 2–1 this is stressed by tabulating the absolute value of $u$. Practical calculations based on equation (2–6) give $u$ some sign which must be ignored when consulting the table. Conversely, when a certain $|u|$ is found to correspond to some specified area, it is necessary to think about whether the desired $x$ is less than or greater than $\mu$ and to assign a plus or minus sign to $u$ accordingly. An example follows.

*Example 2–5.* Your company makes steel-belted radial ply tires. Extensive company tests find an average tire life of $\mu = 50,000$ miles and a standard deviation on the distribution of observed tire lives of $\sigma_x = 4300$ miles. For only 1 per cent redemptions, what mileage can you guarantee?

We require $X$ to be chosen so that the shaded area in the sketch on the next page is 0.01, or 1 per cent of the total. We begin by consulting a table of the normal distribution. If it is of form A, we look for area = 0.0100. If it is of form B, we look for

Table 2–1.  The Normal Distribution

$$\text{Form A: area} = \frac{1}{\sqrt{2\pi}} \int_u^\infty \exp\left(-\frac{u^2}{2}\right) du$$

| $|u|$ | area | $|u|$ | area | $|u|$ | area | $|u|$ | area |
|-----|------|-----|------|-----|------|-----|------|
| 0.0 | 0.5000 | 1.0 | 0.1587 | 2.0 | 0.0227 | 3.0 | $1.3 \times 10^{-3}$ |
| 0.1 | 0.4602 | 1.1 | 0.1357 | 2.1 | 0.0179 | 3.2 | $6.9 \times 10^{-4}$ |
| 0.2 | 0.4207 | 1.2 | 0.1151 | 2.2 | 0.0139 | 3.5 | $2.3 \times 10^{-4}$ |
| 0.3 | 0.3821 | 1.3 | 0.0968 | 2.3 | 0.0107 | 4.0 | $3.2 \times 10^{-5}$ |
| 0.4 | 0.3446 | 1.4 | 0.0808 | 2.4 | 0.0082 | 4.5 | $3.4 \times 10^{-6}$ |
| 0.5 | 0.3085 | 1.5 | 0.0668 | 2.5 | 0.0062 | 5.0 | $2.9 \times 10^{-7}$ |
| 0.6 | 0.2743 | 1.6 | 0.0548 | 2.6 | 0.0047 | 5.5 | $1.9 \times 10^{-8}$ |
| 0.7 | 0.2420 | 1.7 | 0.0446 | 2.7 | 0.0035 | 6.0 | $9.9 \times 10^{-10}$ |
| 0.8 | 0.2119 | 1.8 | 0.0359 | 2.8 | 0.0026 | 8.0 | $6.2 \times 10^{-16}$ |
| 0.9 | 0.1841 | 1.9 | 0.0287 | 2.9 | 0.0019 | 10.0 | $7.6 \times 10^{-24}$ |

$$\text{Form B: area} = \frac{1}{\sqrt{2\pi}} \int_0^u \exp\left(-\frac{u^2}{2}\right) du$$

| $|u|$ | area | $|u|$ | area | $|u|$ | area |
|-----|------|-----|------|-----|------|
| 0.0 | 0.0000 | 1.0 | 0.3413 | 2.0 | 0.4773 |
| 0.1 | 0.0398 | 1.1 | 0.3643 | 2.1 | 0.4821 |
| 0.2 | 0.0793 | 1.2 | 0.3849 | 2.2 | 0.4861 |
| 0.3 | 0.1179 | 1.3 | 0.4032 | 2.3 | 0.4893 |
| 0.4 | 0.1554 | 1.4 | 0.4192 | 2.4 | 0.4918 |
| 0.5 | 0.1915 | 1.5 | 0.4332 | 2.5 | 0.4938 |
| 0.6 | 0.2258 | 1.6 | 0.4452 | 2.6 | 0.4953 |
| 0.7 | 0.2580 | 1.7 | 0.4554 | 2.7 | 0.4965 |
| 0.8 | 0.2881 | 1.8 | 0.4641 | 2.8 | 0.4974 |
| 0.9 | 0.3159 | 1.9 | 0.4713 | 3.0 | 0.4987 |

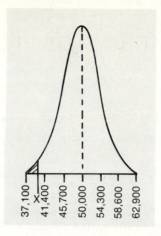

37,100 | X | 41,400 | 45,700 | 50,000 | 54,300 | 58,600 | 62,900

area = 0.4900. In either case, we find $|u| = 2.33 = (X - \mu)/\sigma_x$. Then, for $X < \mu$, $X = \mu - 2.33\sigma_x = 40,000$ miles.

*Example 2–6.* For the examination mentioned in Example 2–4, what fraction of the students can be expected to obtain a perfect score?

We have already found $u$ to be 2.5. Consulting a table of form A, we find the area to be 0.0062, indicating that 0.62 per cent of the students will get a perfect score. The area found in a table of form B is 0.4938.

*Example 2–7.* What is the probability that any single data point lies within two standard deviations of the mean?

This probability is given by the integral

$$\int_{u=-2.0}^{u=2.0} dN/N = 2\int_{u=0}^{u=2.0} dN/N = 2\left[0.5 - \int_{u=2.0}^{u=\infty} dN/N\right]$$

The second integral is listed in a table of form B, from which we find that the area is 0.4773. Thus, the probability that a single observation lies within two standard deviations of the mean is 2(0.4773), or 95.46 per cent. We could obtain the same result by using a table of form A, in which for $u = 2.0$ the area is 0.0227.

Notice that we can restate this result by saying "we are 95 per cent confident that $\mu$ is within $2\sigma_x$ of any individual $x$." Thus, if we have, for example, $x = 25.0$ and $\sigma_x = 0.1$, we can be 95 per cent confident that $(x - 2\sigma_x) \leq \mu \leq (x + 2\sigma_x)$ or $24.8 \leq \mu \leq 25.2$. These *95 per cent confidence limits* are often expressed for an experimental result in this way: $25.0 \pm 0.2$. Confidence limits are invaluable in the interpretation of experimental data, but we will not stress the concept at this point because the relations we have thus far derived require knowledge of the true $\sigma_x$. This is unlikely. In practical situations we usually deal with $s_x$; and a useful technique for the computation of confidence limits based on $s_x$ will be given in a later section of this chapter.

## PROPAGATION OF ERRORS

### The General Case

Consider some result $w$, obtained as a function of three experimentally determined independent variables, $x, y,$ and $z$. The variances of $x, y,$ and $z$ are known. How can they be combined to determine the variance of $w$? An approximate expression is given by

$$V_w = \left(\frac{\partial w}{\partial x}\right)^2 V_x + \left(\frac{\partial w}{\partial y}\right)^2 V_y + \left(\frac{\partial w}{\partial z}\right)^2 V_z \tag{2-9}$$

## Specific Cases

*Addition and subtraction.* Consider the case $w = f(x, y, z) = x + y + z$. Then

$$\frac{\partial w}{\partial x} = \frac{\partial w}{\partial y} = \frac{\partial w}{\partial z} = 1$$

and the general relationship given in equation (2–9) takes the form

$$V_w = (1)^2 V_x + (1)^2 V_y + (1)^2 V_z = V_x + V_y + V_z \tag{2-10a}$$

Rewriting equation (2–10a) in terms of standard deviations, we obtain

$$s_w^2 = s_x^2 + s_y^2 + s_z^2 \tag{2-10b}$$

$$s_w = \sqrt{s_x^2 + s_y^2 + s_z^2} \tag{2-10c}$$

Notice that $s_w$ is *not* $(s_x + s_y + s_z)$, but is instead considerably smaller. This occurs because it is unlikely that $x$, $y$, and $z$ will simultaneously take values far above their means. For example, at the same time $x$ happens to be two standard deviations high, $y$ and $z$ are likely to be much nearer their mean values, and possibly even negative with respect to their means. This opportunity for random errors in one variable to offset random errors in another variable is responsible for the form of the expression for $V_w$.

When coefficients are involved, the calculation is only slightly more complex:

$$w = ax + by + cz$$

$$\frac{\partial w}{\partial x} = a; \qquad \frac{\partial w}{\partial y} = b; \qquad \frac{\partial w}{\partial z} = c$$

$$V_w = a^2 V_x + b^2 V_y + c^2 V_z \tag{2-11}$$

Note that, if the signs of any of the coefficients happened to be negative (*i.e.*, if we were dealing with subtraction), there would be no difference in the result because the derivatives are always squared.

*Example 2–8.* Suppose that the weight of a sample $(w_s)$ is determined by noting first the weight of a sample bottle plus sample $(w_{s+b})$, and then the weight of the sample bottle alone $(w_b)$:

$$w_s = w_{s+b} - w_b$$

Assume that the standard deviation of single-weight observations on the balance used is known to be $s_w = 1$ mg. Calculate $s_{w_s}$.

In accord with equation (2–10a), we can write

$$V_{w_s} = V_{w_{s+b}} + V_{w_b}$$

Since $w_{s+b}$ and $w_b$ are both single-weight observations, $V_{w_{s+b}} = V_{w_b} = s_w{}^2 = 1$ mg$^2$. Thus, $V_{w_s} = 1 + 1 = 2$ mg$^2$, and $s_{w_s} = \sqrt{2} = 1.4$ mg.

*Example 2-9.* An analysis is made by means of a single measurement, $X$, from which $3B$ (where $B$ is a background correction factor) must be subtracted. We can write

$$P = X - 3B$$

where $P$ is the analytical result. Suppose that the variable $B$ has been quite precisely measured ($s_B = 0.01$), whereas the quantity $X$ is known more poorly ($s_X = 0.1$). Calculate $s_P$.

Using equation (2-11), we obtain

$$V_P = (1)^2 V_X + (3)^2 V_B$$
$$V_P = 10^{-2} + 9(10^{-4}) = 1.09 \times 10^{-2}$$
$$s_P \sim 1 \times 10^{-1}$$

**Standard deviation of the mean.** An interesting application of these expressions allows calculation of the standard deviation of some mean, $\bar{x}$, derived from $n$ observations:

$$\bar{x} = f(x_1, x_2, x_3, \ldots x_n) = \frac{1}{n}(x_1 + x_2 + x_3 + \cdots x_n)$$

$$\frac{\partial \bar{x}}{\partial x_1} = \frac{\partial \bar{x}}{\partial x_2} = \frac{\partial \bar{x}}{\partial x_3} = \frac{\partial \bar{x}}{\partial x_n} = \frac{1}{n}$$

$$V_{\bar{x}} = \left(\frac{1}{n}\right)^2 V_{x_1} + \left(\frac{1}{n}\right)^2 V_{x_2} + \left(\frac{1}{n}\right)^2 V_{x_3} + \cdots + \left(\frac{1}{n}\right)^2 V_{x_n} \qquad (2\text{-}12a)$$

Because all the $x$ observations are from the same universe,

$$V_{x_1} = V_{x_2} = V_{x_3} = V_{x_n} \equiv V_x \qquad (2\text{-}13)$$

Substituting equation (2-13) in equation (2-12a), we obtain

$$V_{\bar{x}} = n\left(\frac{1}{n}\right)^2 V_x = \frac{V_x}{n} \qquad (2\text{-}12b)$$

and

$$s_{\bar{x}} = \frac{s_x}{\sqrt{n}} \qquad (2\text{-}12c)$$

As the number of observations in a sample increases, the standard deviation of the mean decreases. In this sense, four measurements are twice as good as one, but 16 measurements are required if the precision is to be doubled again. The inverse square-root dependence of $s_{\bar{x}}$ on $n$ establishes this sequence of diminishing returns.

*Example 2-10.* In Example 2-1 above, the standard deviation ($s_x$) of a sample of lunar carbon measurements was calculated to be 20.5 ppm. Calculate the standard deviation of the mean, $\bar{x} = 574/4 = 144$ ppm. The number of observations, $n$, is four.

Applying equation (2-12c), we obtain

$$s_{\bar{x}} = \frac{s_x}{\sqrt{n}} = \frac{20.5}{2} = 10.2 \text{ ppm}$$

*Functions involving multiplication and division.* Consider as an example the expression

$$w = f(x, y, z) = \frac{(ax)(by^2)}{cz}$$

First, we can write

$$\frac{\partial w}{\partial x} = \frac{a(by^2)}{cz} = \frac{w}{x}; \qquad \frac{\partial w}{\partial y} = \frac{(ax)(2by)}{cz} = \frac{2w}{y}; \qquad \frac{\partial w}{\partial z} = -\frac{(ax)(by^2)}{cz^2} = -\frac{w}{z}$$

Following the general expression (2–9), we obtain

$$V_w = \left(\frac{w}{x}\right)^2 V_x + \left(\frac{2w}{y}\right)^2 V_y + \left(-\frac{w}{z}\right)^2 V_z$$

Notice that values for $x$, $y$, and $z$ must be inserted. The magnitude of $V_w$ is not independent of these values because they determine the extent to which each variable can affect $w$.

*Example 2–11.* Suppose that the overall formation constant, $\beta_2$, of the complex $MX_2$ is determined by making measurements of the ratio ($[MX_2]/[M] = r$) and of $[X]$:

$$\beta_2 = \frac{[MX_2]}{[M][X]^2} = \frac{r}{[X]^2}$$

For solutions in which $r$ is 10 with a standard deviation of 0.5 and $[X]$ is 0.1 $M$ with a standard deviation of 0.01 $M$, what is the standard deviation of the calculated overall formation constant, $\beta_2$?

Following the general expression (2–9), we obtain

$$V_{\beta_2} = \left(\frac{\partial \beta_2}{\partial r}\right)^2 V_r + \left(\frac{\partial \beta_2}{\partial [X]}\right)^2 V_{[X]}$$

Evaluating the partial derivatives, we find

$$\frac{\partial \beta_2}{\partial r} = \frac{1}{[X]^2} = \frac{\beta_2}{r} \qquad \frac{\partial \beta_2}{\partial [X]} = -2\frac{r}{[X]^3} = -2\frac{\beta_2}{[X]}$$

Substitution in the general expression yields

$$V_{\beta_2} = \left(\frac{\beta_2}{r}\right)^2 V_r + \left(-2\frac{\beta_2}{[X]}\right)^2 V_{[X]}$$

From information given above, $\beta_2 = r/[X]^2 = (10)/(0.1)^2 = 10^3$; insertion of this result, along with the values for other terms, gives

$$V_{\beta_2} = \left(\frac{10^3}{10}\right)^2 (0.5)^2 + \left(-\frac{2 \times 10^3}{0.1}\right)^2 (0.01)^2$$

$$V_{\beta_2} = s_{\beta_2}^2 = (25 \times 10^2) + (4 \times 10^4)$$

$$s_{\beta_2} = 2.1 \times 10^2$$

# THE *t*-DISTRIBUTION AND ITS APPLICATIONS

Although we have been careful to define the difference between $s_x$ and $\sigma_x$, we have performed a number of calculations in which it was assumed that $\sigma_x$ was known. If 1000 (or even 100) observations have been made, this is effectively true. The $s_x$ calculated from such a large sample is bound to be a very good estimate of $\sigma_x$. However, in the typical situation where a sample contains only two, three, or four observations, the calculated $s_x$ is a very uncertain estimate of $\sigma_x$. When trying to calculate confidence limits, for example, this uncertainty in $s_x$ must be taken into account, and the *t*-distribution provides the only way of doing so.

## Confidence Limits on the Means of Small Samples

*Confidence limits using $\sigma_x$.* In Example 2-7 above, we calculated 95 per cent confidence limits based on $\sigma_x$. In order to stress this important concept, let us repeat the procedure here in a slightly different and more formal way.

1. We begin by asking, "What limits must be placed on $u$ in order to include 95 per cent of the area under the normal distribution?"

2. See Figure 2–5. The shaded area totals $2\alpha$. The unshaded area is, therefore, $1.0 - 2\alpha$.

3. In order to include 95 per cent of the universe, we want to choose $u_\alpha$ such that $1.0 - 2\alpha = 0.95$; therefore $\alpha = 0.025$.

4. Searching a table of form A for area = 0.025, or one of form B for area = 0.4750, we would find that $u_\alpha = 1.96$; this limit is more accurate than the approximate one of $u = 2.0$ given in Example 2–7.

5. Thus, for any single observation drawn from the universe, we can be 95 per cent confident that $\mu$ is within $1.96\sigma_x$ of the observed $x$ and we can write the 95 per cent confidence limits on $x$ as

$$x \pm u_{0.025}\sigma_x = x \pm 1.96\sigma_x$$

There are only 5 chances in 100, or 1 chance in 20, that $\mu$ will be outside the 95 per cent confidence limits.

6. In general, we can write the $100(1 - 2\alpha)$ per cent confidence limits as

$$x \pm u_\alpha\sigma_x \tag{2-14}$$

7. An entirely equivalent expression holds for confidence limits on the mean. With reference to equation (2–12c), this relation can be formulated as

$$\bar{x} \pm u_\alpha\sigma_{\bar{x}} = \bar{x} \pm \frac{u_\alpha\sigma_x}{\sqrt{n}} \tag{2-15}$$

*Figure 2–5.* A view of the normal distribution used to clarify the concept of confidence limits. (See discussion in text.)

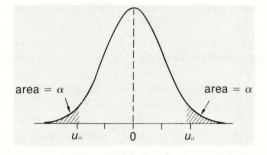

area = $\alpha$        area = $\alpha$

$u_\alpha$    0    $u_\alpha$

***Confidence limits using $s_x$.*** When $s_x$ replaces $\sigma_x$, a true value is replaced by an estimate, which may be in error. In order to take the uncertainty in $s_x$ into account, this substitution of $s_x$ for $\sigma_x$ requires an increase in the coefficient $u_\alpha$ in equations (2–14) and (2–15) in order to widen the confidence limits. If $n$ is large, the new coefficient can be about the same value as $u_\alpha$, because a large $n$ provides a relatively accurate estimate of $\sigma_x$. When $n$ is small, and $s_x$ is a relatively poor estimate of $\sigma_x$, the coefficient must be considerably larger than $u_\alpha$. The new coefficient is $t_{\alpha,\varphi}$, a function not only of $\alpha$, but also of $\varphi$, the number of degrees of freedom. For a sample from which one mean value has been calculated, $\varphi = n - 1$. The general expressions written above for the $100(1 - 2\alpha)$ per cent confidence limits can be rewritten as

$$x \pm t_{\alpha,\varphi}s_x \tag{2-16}$$

and

$$\bar{x} \pm t_{\alpha,\varphi}s_{\bar{x}} = \bar{x} \pm \frac{t_{\alpha,\varphi}s_x}{\sqrt{n}} \tag{2-17}$$

A tabulation of the $t$-distribution is given in Table 2–2. It is instructive to examine how $t_{0.025,\varphi}$ changes as $\varphi$ varies. Note first that, for $\varphi = \infty$, $t_{0.025,\infty} = 1.96$, the same value that is obtained from the normal distribution. This occurs, of course, because $s_x = \sigma_x$ when $\varphi = \infty$. In general, $t_{\alpha,\infty} = u_\alpha$. As $\varphi$ decreases, $t$ increases, although $t$ is still within 10 per cent of $u$ for $\varphi = 13$. Further decreases in $\varphi$ result in sharp increases in $t$. In the extreme case where $\varphi = 1$ (that is, where $s_x$ is determined from a pair of observations), the confidence limits correctly calculated using $(t_{0.025,1})(s_x)$ exceed those which are sometimes *incorrectly calculated* using $(u_{0.025})(s_x)$ by a factor of $(12.706/1.960)$ or 6.5.

*Example 2–12.* Calculate the 90, 95, and 99 per cent confidence limits on the mean carbon content in Apollo 11 lunar fines. The required data are in Examples 2–1 and 2–10 ($\bar{x} = 144$ ppm, $s_{\bar{x}} = 10.2$ ppm, $n = 4$).

An expression for the confidence limits is given by equation (2–17). In this case, $\varphi = n - 1 = 3$. For the 90 per cent confidence limits, $\alpha = 0.05$; for the 95 per cent confidence limits, $\alpha = 0.025$; and, for the 99 per cent confidence limits, $\alpha = 0.005$. The desired results are as follows:

| Confidence Level | $t_{\alpha,3}$ | Confidence Limits |
|---|---|---|
| 90% | 2.353 | $\pm(2.353)(10.2) = \pm24$ ppm |
| 95% | 3.182 | $\pm(3.182)(10.2) = \pm32$ ppm |
| 99% | 5.841 | $\pm(5.841)(10.2) = \pm60$ ppm |

Unless otherwise noted, the 95 per cent confidence limits are used in practical work.

## Significance Testing

***Confidence limits on the difference between two means.*** In Figure 2–1 at the beginning of this chapter, two small data samples are shown, and the question of whether they truly differ is posed. From each sample a mean can be calculated. Call these means $\bar{x}_1$ and $\bar{x}_2$, and note that each is an estimate of a universe mean, $\mu_1$ and $\mu_2$. The original question can be rephrased: does $\mu_1 = \mu_2$? It is entirely

**Table 2–2. The _t_-Distribution**

| φ | α = 0.05 | 0.025 | 0.005 | 0.0005 |
|---|---|---|---|---|
| 1 | 6.314 | 12.706 | 63.657 | 636.62 |
| 2 | 2.920 | 4.303 | 9.925 | 31.598 |
| 3 | 2.353 | 3.182 | 5.841 | 12.924 |
| 4 | 2.132 | 2.776 | 4.604 | 8.610 |
| 5 | 2.015 | 2.571 | 4.032 | 6.869 |
| 6 | 1.943 | 2.447 | 3.707 | 5.959 |
| 7 | 1.895 | 2.365 | 3.499 | 5.408 |
| 8 | 1.860 | 2.306 | 3.355 | 5.041 |
| 9 | 1.833 | 2.262 | 3.250 | 4.781 |
| 10 | 1.812 | 2.228 | 3.169 | 4.587 |
| 11 | 1.796 | 2.201 | 3.106 | 4.437 |
| 12 | 1.782 | 2.179 | 3.055 | 4.318 |
| 13 | 1.771 | 2.160 | 3.012 | 4.221 |
| 14 | 1.761 | 2.145 | 2.977 | 4.140 |
| 15 | 1.753 | 2.131 | 2.947 | 4.073 |
| 20 | 1.725 | 2.086 | 2.845 | 3.850 |
| 30 | 1.697 | 2.042 | 2.750 | 3.646 |
| 60 | 1.671 | 2.000 | 2.660 | 3.460 |
| ∞ | 1.645 | 1.960 | 2.576 | 3.291 |

possible that $\bar{x}_1$ and $\bar{x}_2$ could differ, that is, $\bar{x}_1 - \bar{x}_2 = \Delta\bar{x} \neq 0$, simply because of random scatter in the data. This difference might occur even if the two samples were drawn from the same universe, that is, $\mu_1 = \mu_2$. Note that $\Delta\bar{x}$ is an estimate of $\Delta\mu$, and rephrase the question again: does $\Delta\mu = 0$? We can answer this question by determining whether the confidence limits for $\Delta\bar{x}$ are larger than $\Delta\bar{x}$ itself. If they are, there is a good chance that $\Delta\bar{x}$ is zero and that $\Delta\mu = 0$, since $[\Delta\bar{x} - (\text{confidence limit})] \leq \Delta\mu \leq [\Delta\bar{x} + (\text{confidence limit})]$.

In general,

$$(\text{confidence limits for } \textit{anything}) = \pm t_{\alpha,\varphi} s_{anything} \qquad (2\text{–}18)$$

where _anything_ has $\varphi$ degrees of freedom. In this case, we are interested in the confidence limits for $\Delta\bar{x}$, which is defined by equation (2–19):

$$\Delta\bar{x} = \bar{x}_1 - \bar{x}_2 \qquad (2\text{–}19)$$

Our first task is to define $s_{\Delta\bar{x}}$, which here takes the role of $s_{anything}$. Considering equation (2–19) and equation (2–10a), we obtain

$$V_{\Delta\bar{x}} = V_{\bar{x}_1} + V_{\bar{x}_2} \qquad (2\text{–}20a)$$

In accord with equation (2–12b), the individual variances are given by

$$V_{\bar{x}_1} = \frac{V_{x_1}}{n_1} \quad \text{and} \quad V_{\bar{x}_2} = \frac{V_{x_2}}{n_2} \qquad (2\text{–}21a, \ 2\text{–}21b)$$

where $n_1$ and $n_2$ are the numbers of observations in each sample. When results from the analyses of two materials are being compared, it is quite generally true that $V_{x_1} = V_{x_2} = V_x$, because this variance is a function of the method of analysis only. Then, substituting equations (2–21a) and (2–21b) in (2–20a), we obtain

$$V_{\Delta\bar{x}} = \frac{V_x}{n_1} + \frac{V_x}{n_2} = V_x\left(\frac{1}{n_1} + \frac{1}{n_2}\right) \qquad (2\text{–}20\text{b})$$

Recasting equation (2–20b) in terms of standard deviations, we get

$$s_{\Delta\bar{x}} = s_x\left(\frac{1}{n_1} + \frac{1}{n_2}\right)^{1/2} \qquad (2\text{–}20\text{c})$$

Following equation (2–18), the confidence limits for $\Delta\bar{x}$ are given by

$$(\text{confidence limits for } \Delta\bar{x}) = \text{C.L.} = \pm t_{\alpha,\varphi} s_{\Delta\bar{x}} = \pm t_{\alpha,\varphi} s_x\left(\frac{1}{n_1} + \frac{1}{n_2}\right)^{1/2} \qquad (2\text{–}22)$$

where $\varphi = n_1 + n_2 - 2$, two constants ($\bar{x}_1$ and $\bar{x}_2$) having been calculated from the data in this case. In such cases, $\alpha$ is conventionally set equal to 0.025. Then, if the confidence limits are smaller than $\Delta\bar{x}$, the chances are 19 in 20 that $|\Delta\mu| > 0$, $\Delta\mu$ is said to be *significant*, and it is judged that there is a real difference between the samples ($\mu_1 \neq \mu_2$).

*Example 2–13.* Evidence for the identity of an organic compound can be obtained by measuring the time required for passage of the compound through a chromatographic column, and comparing this "retention time" to the retention time of a known standard. Equal retention times suggest, but do not prove, that the unknown and known compounds are identical. Suppose that an unknown compound is passed through a chromatographic column three times and that retention times of 10.20, 10.35, and 10.25 min are obtained. In addition, suppose that standard *n*-octane has been run eight times, and that the observed retention times are 10.24, 10.28, 10.31, 10.32, 10.34, 10.36, 10.36, and 10.37 min. Might the unknown be *n*-octane?

(1) *Determination of $s_x$.*
In each case, code the data such that $x = (t_R - 10.20)100$.

| | unknown | | | standard *n*-octane | | |
|---|---|---|---|---|---|---|
| | $t_{R_1}$ | $x_1$ | $x_1^2$ | $t_{R_2}$ | $x_2$ | $x_2^2$ |
| | 10.20 | 0 | 0 | 10.24 | 4 | 16 |
| | 10.35 | 15 | 225 | 10.28 | 8 | 64 |
| | 10.25 | 5 | 25 | 10.31 | 11 | 121 |
| $\bar{t}_{R_1} =$ | 10.27 | 20 | 250 | 10.32 | 12 | 144 |
| | $n_1 = 3$ | | | 10.34 | 14 | 196 |
| | | | | 10.36 | 16 | 256 |
| | | | | 10.36 | 16 | 256 |
| | | | | 10.37 | 17 | 289 |
| | | | | $\bar{t}_{R_2} = 10.32$ | 98 | 1342 |
| | | | | $n_2 = 8$ | | |

Using equation (2–4b), we obtain

$$s_x^2 = \frac{\left[250 - \frac{(20)^2}{3}\right] + \left[1342 - \frac{(98)^2}{8}\right]}{(3-1) + (8-1)} = \frac{258}{9}$$

$$s_x = 100 s_{t_R} = \sqrt{28.7} = 5.4$$

$$s_{t_R} = 0.054 \text{ min}$$

(2) *Determination of confidence limits for* $\Delta t_R$.
If we utilize equation (2–22), the result is as follows:

$$\text{C.L.} = \pm t_{\alpha,\varphi} s_{t_R} \left(\frac{1}{n_1} + \frac{1}{n_2}\right)^{1/2}$$

$$\varphi = n_1 + n_2 - 2 = 9; \qquad \text{for } 95\% \text{ C.L., } \alpha = 0.025$$

$$\text{C.L.} = \pm(2.262)(0.054)(\tfrac{1}{3} + \tfrac{1}{8})^{1/2} = \pm(2.262)(0.054)(0.677) = \pm 0.08 \text{ min}$$

(3) *Comparison of the calculated confidence limits with* $\Delta t_R$.

$$\Delta t_R = t_{R_1} - t_{R_2} = 10.27 - 10.32 = -0.05 \text{ min}$$

Since $|-0.05| < 0.08$, there is a good chance that $\Delta t_R = 0$. Accordingly, the un-known could be $n$-octane. Alternatively, we could note that the indicated confidence interval is $-0.13 \leq \Delta t_R \leq 0.03$ min, and that this range indicates the possibility of $\Delta t_R = 0$.

**The t-test.** In the preceding section, we have judged the significance of $\Delta \bar{x}$ by computing its confidence limits. The essence of this approach can be summarized as follows:

If the confidence limits for
$\Delta \bar{x}$ are greater than $\Delta \bar{x}$, . . . then the difference between
the true means might be zero    (2–23a)

If $t_{\alpha,\varphi} s_{\Delta \bar{x}} \geq |\Delta \bar{x}|$ . . . . . . . then possibly $\Delta \mu = 0$    (2–23b)

The arithmetic of significance testing is made simpler if this approach is slightly changed. A value for $t$ (no subscripts) is computed from the relation

$$t = \frac{\Delta \bar{x}}{s_{\Delta \bar{x}}} = \frac{\Delta \bar{x}}{s_x \left(\frac{1}{n_1} + \frac{1}{n_2}\right)^{1/2}} \tag{2–24}$$

Compare equations (2–23b) and (2–24). Note that, if $t < t_{\alpha,\varphi}$, the confidence limits will be greater than $\Delta \bar{x}$ and, therefore, $\Delta \mu = 0$. If $t \geq t_{\alpha,\varphi}$, $\mu_1 \neq \mu_2$. This method of calculating a test value for $t$ and comparing it to some critical value listed in a table is known as the *t-test*.

*Example 2–14.* Two limestone samples are analyzed for their magnesium content. The results obtained are Sample 1: 1.22, 1.25, 1.26% Mg; Sample 2: 1.31, 1.34, 1.35% Mg. Do the samples differ significantly?

(1) *Calculation of $s_{Mg}$, the standard deviation of the analysis*. Code the data.

$$\text{Sample 1: } x_1 = (\% \text{ Mg} - 1.20)100$$

$$\text{Sample 2: } x_2 = (\% \text{ Mg} - 1.30)100$$

| Sample 1 | | | | Sample 2 | | |
|---|---|---|---|---|---|---|
| % Mg | $x_1$ | $x_1^2$ | | % Mg | $x_2$ | $x_2^2$ |
| 1.22 | 2 | 4 | | 1.31 | 1 | 1 |
| 1.25 | 5 | 25 | | 1.34 | 4 | 16 |
| 1.26 | 6 | 36 | | 1.35 | 5 | 25 |
| $\overline{\text{Mg}} = \overline{1.24\%}$ | $\overline{13}$ | $\overline{65}$ | | $\overline{\text{Mg}} = \overline{1.33\%}$ | $\overline{10}$ | $\overline{42}$ |

Using equation (2–4b), we obtain

$$s_x^2 = (100 s_{Mg})^2 = \frac{\left[65 - \dfrac{(13)^2}{3}\right] + \left[42 - \dfrac{(10)^2}{3}\right]}{(3-1) + (3-1)}$$

$$s_{Mg} = 0.021 \text{ per cent}$$

(2) *Calculation of t according to equation (2–24)*.

$$t = \frac{\Delta \overline{\text{Mg}}}{s_{Mg}\left(\dfrac{1}{n_1} + \dfrac{1}{n_2}\right)^{1/2}} = \frac{1.33 - 1.24}{0.021\left(\dfrac{1}{3} + \dfrac{1}{3}\right)^{1/2}}$$

$$t = \frac{0.09}{0.017} = 5.29$$

(3) *Comparison of t to critical values.* In this case, two constants have been calculated from the data, and the number of degrees of freedom is given by $\varphi = n_1 + n_2 - 2 = 3 + 3 - 2 = 4$.

At the 95 per cent confidence level, $\alpha = 0.025$, and the critical value of $t_{\alpha,\varphi}$ is found in Table 2–2. In this case, $t_{0.025, 4} = 2.776$. Since 2.776 is less than 5.29 (the $t$ value calculated above), the difference between the samples is significant at the 95 per cent confidence level. That is, there is less than one chance in twenty that a difference this large could occur between two samples from the same universe.

Higher levels of possible significance can be tested by decreasing $\alpha$. For the 99 per cent confidence level, $\alpha = 0.005$. Table 2–2 shows $t_{0.005, 4} = 4.604$. Since $4.604 < 5.29$, the difference between samples is significant at the 99 per cent confidence level.

Table 2–2 shows $t_{0.0005, 4} = 8.610$. Since $8.610 > 5.29$, the observed difference is not significant at the 99.9 per cent confidence level. That is, differences this great will occur between two samples from the same universe more frequently than one time out of 1000.

**Detection limits.** Particularly in the analysis of trace constituents, the minimum detectable amount or concentration of some analyzed component is of great interest. It is asked, "What is the minimum quantity which produces a result significantly different from zero?" When an analytical technique is utilized near its detection limits, it is always observed that not only are the results on legitimate samples scattered by the effects of random error but also the results on "blank" samples are scattered about zero by the effects of random error. The determination of detection limits thus requires consideration of the minimum significant difference between two uncertain numbers: the sample result and the "blank" result.

If the average "blank" result is denoted as $\bar{x}_b$ and the average sample result as $\bar{x}_s$, the difference between them will be $\Delta\bar{x}_{s-b}$ and its confidence limits are given by

$$\text{C.L.} = \pm t_{\alpha,\varphi} s_{\Delta\bar{x}_{s-b}} = \pm t_{\alpha,\varphi} \sqrt{\frac{s_{x_s}^2}{n_s} + \frac{s_{x_b}^2}{n_b}} \qquad (2\text{--}25)$$

where $s_{x_s}$ and $s_{x_b}$ are the standard deviations of the sample and blank, respectively, and $n_s$ and $n_b$ are the number of observations on the sample and blank, respectively. Near the limit of detection, $s_{x_s}$ will be approximately equal to $s_{x_b}$, which can be conveniently measured by repeated observation of blank results. The preceding expression then takes the form

$$\text{C.L.} = \pm t_{\alpha,\varphi} s_{x_b} \sqrt{\frac{1}{n_s} + \frac{1}{n_b}} \qquad (2\text{--}26)$$

The minimum detectable $\Delta\bar{x}_{s-b}$ can be denoted by $\Delta\bar{x}_{\lim}$; to determine its value, we need only specify that it exceed the confidence limits set by equation (2–26),

$$\Delta\bar{x}_{\lim} > t_{\alpha,\varphi} s_{x_b} \sqrt{\frac{1}{n_s} + \frac{1}{n_b}} \qquad (2\text{--}27)$$

where $\varphi$ is the number of degrees of freedom used in the determination of $s_{x_b}$.

*Example 2–15.* Ten blank experiments using a new technique for the determination of traces of sulfur dioxide in air gave values of $-2, +3, +12, -1, +1, 0, +5, +10, +11,$ and $+8$ ppb (parts per billion) of sulfur dioxide in air. (a) Determine the minimum concentration of $SO_2$ detectable (at the 99 per cent confidence level) by means of this technique if a single sample measurement is compared with a single blank measurement. (b) What is the detection limit at the 99 per cent confidence level for the situation where triplicate observations are made of the blank and of the sample?

(1) *Determination of $s_{x_b}$.* Calculation of the standard deviation in the usual way (see Example 2–1) yields the result, $s_{x_b} = 5.2$ ppb.

(2) *Determination of detection limits.* In part (a), we have $n_s = n_b = 1$ and $\alpha = 0.005$. From the preceding calculation, we obtained $s_{x_b} = 5.2$ ppb, and $\varphi = 9$. Insertion of these values into equation (2–27) gives

$$\Delta\bar{x}_{\lim} > (3.25)(5.2)\sqrt{\tfrac{1}{1} + \tfrac{1}{1}} = 24 \text{ ppb}$$

Thus, any concentration of sulfur dioxide greater than 24 ppb will differ significantly (at the 99 per cent confidence level) from zero. In part (b), the only change is that $n_s = n_b = 3$; thus, the detection limit is computed to be

$$\Delta\bar{x}_{\lim} > (3.25)(5.2)\sqrt{\tfrac{1}{3} + \tfrac{1}{3}} = 14 \text{ ppb}$$

## REGRESSION

It often happens that an analytical instrument is calibrated with a series of standard samples. A graph of the instrument output versus known sample input is constructed by drawing a line through the calibration points. A typical example is shown in Figure 2–6. The simple graphic solution of such problems in instrument

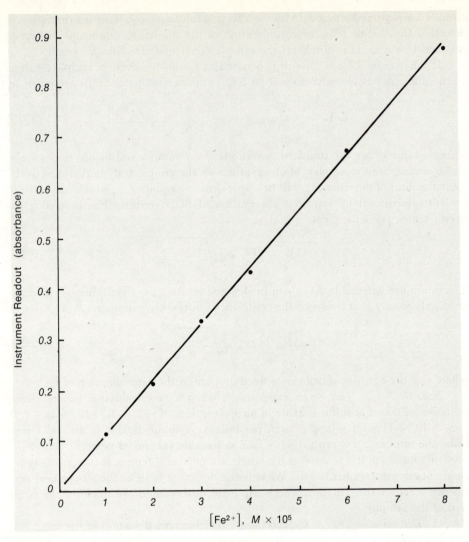

*Figure 2–6.* A typical instrument calibration graph; in this case, it is a plot of iron(II) concentration in an aqueous solution containing a complexing agent which gives an intense color with iron(II) versus the instrument readout (absorbance) at a certain wavelength. The statistical derivation of the calibration line is discussed in the text.

calibration is unsatisfactory in several respects. First, the graph must be very large if more than two significant figures are to be determined. Second, the placement of the line is an emotional process. The experimenter is very likely not to draw the line which best fits the points, and an improper fit can cause large errors. It is, in all cases, better to derive a mathematical relationship which expresses the instrument response function. Comparison of the derived coefficients on a day-to-day basis allows the experimenter to monitor instrument performance, and the unbiased mathematical procedure offers the only way to achieve line placement without prejudice.

The problem of deriving such relationships comes under the general heading of *regression analysis* in statistics. Here we shall be concerned with only the simplest case—the fitting of straight-line relationships by the **method of least squares**.

## The Distribution of Calibration Data

*The known chemical inputs.* The chemical input (for example, the iron(II) solutions of known concentration used in establishing the graph of Figure 2–6) is regarded as a variable which is free of error; $\sigma = 0$. Everything which follows concerning the least-squares determination of calibration lines assumes that the standard chemical inputs are known with perfect accuracy, or, at least, that errors on this axis are insignificant when compared with errors on the instrument output axis. If this is not true, a different approach to data analysis is required.

*The observed instrument response points.* Each instrument output reading is a single observation drawn from the universe of all possible instrument outputs for a given chemical input. This situation is depicted in Figure 2–7. For the input standard with value 0.3, for example, the small normal distribution curve which is sketched on the graph represents the universe of all possible instrument outputs for a chemical input of 0.300. Point A, observed during calibration, is near the center of this distribution, well within the $\pm 2\sigma$ limits marked by the shading. In fact, we were quite lucky to obtain a calibration point so close to the universe mean. Even so, *we do not want the calibration line to go through that point*, but instead through the exact *center of the distribution*, which is marked by the arrow. Similar comments apply to the distributions indicated for standard inputs of 0.600 and 0.900.

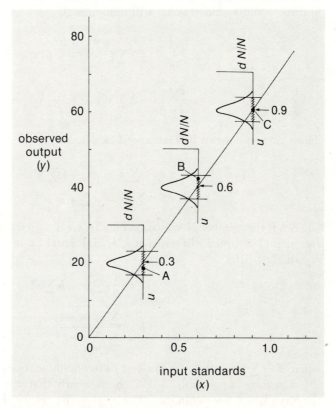

*Figure 2–7.* Another view of an instrument calibration graph. The small plots of $dN/N$ versus $u$ represent the normal distribution of outputs for each of three particular inputs. Therefore, they refer only to the $dN/N$ and $y$ axes, and do not have any significance on the $x$ axis. (See text for further discussion.)

## Principle of Operation

We might obtain estimates of the center of each distribution by recording several output observations for each input, but this is not necessary to obtain fullest possible accuracy from the method of least squares, which determines the locus of universe means by minimizing $\sum$ (deviations)$^2$ between the calibration data points and the line expressed by the function

$$y = a + bx \qquad (2\text{-}28)$$

The sum of squares of the deviations is minimized by adjusting $a$ and $b$; and, *if there is a linear functional relationship* between $x$ and $y$, this has the effect of putting the regression line through the best estimate of the true mean values. Thus, points B and C as well as point A aid in the determination of $\mu_{0.3}$, and so on.

Mathematically, we derive solutions for $a$ and $b$ by expressing $\sum$ (deviations)$^2$, which we denote by $Q$, as a function of $a$ and $b$:

$$Q = \sum \text{(deviations of experimental points from line)}^2 \qquad (2\text{-}29a)$$

$$Q = \sum [y - (a + bx)]^2 \qquad (2\text{-}29b)$$

Next, the partial derivatives of the latter relation are taken, first with respect to $a$ and then with respect to $b$. The resulting expressions are set equal to zero in order to find the $a$ and $b$ values which will minimize $Q$:

$$\frac{\partial Q}{\partial a} = -2 \sum [y - (a + bx)] = 0 \qquad (2\text{-}30a)$$

$$\frac{\partial Q}{\partial b} = -2 \sum x[y - (a + bx)] = 0 \qquad (2\text{-}31a)$$

These expressions can be rearranged as follows:

$$\sum y = na + b \sum x \qquad (2\text{-}30b)$$

$$\sum xy = a \sum x + b \sum x^2 \qquad (2\text{-}31b)$$

where $n$ is the number of $x, y$ pairs. Finally, we obtain solutions for $a$ and $b$ by solving equation (2–30b) for $a$ in terms of $y$, $b$, and $x$ and by substituting the result for $a$ in equation (2–31b),

$$a = \frac{1}{n} \left( \sum y - b \sum x \right) \qquad (2\text{-}30c)$$

$$b = \frac{\sum xy - n\bar{x}\bar{y}}{\sum x^2 - n\bar{x}^2} \qquad (2\text{-}32a)$$

where $\bar{x} = \sum x/n$ and $\bar{y} = \sum y/n$, just as in one-dimensional analysis.

The form of equation (2–29b) shows clearly that it is only deviations parallel to the $y$ axis which are minimized. The $x$ values are regarded as fixed. This is the origin of the requirement mentioned above, that the input standards be very carefully prepared. When it happens that significant errors can occur on either axis, the reader is referred to the excellent (and simple) explanation given by York in the paper listed at the end of this chapter.

## Method of Calculation

*General expressions for the sums of squares.* The data are presented as $n$ pairs of $x$ and $y$ values. For convenience in calculation, we define the following relationships:

$$\sum U^2 \equiv \sum (x - \bar{x})^2 = \sum x^2 - n\bar{x}^2 = \sum x^2 - \left(\sum x\right)^2/n \qquad (2\text{-}33)$$

$$\sum V^2 \equiv \sum (y - \bar{y})^2 = \sum y^2 - n\bar{y}^2 = \sum y^2 - \left(\sum y\right)^2/n \qquad (2\text{-}34)$$

$$\sum UV \equiv \sum (x - \bar{x})(y - \bar{y}) = \sum xy - n\bar{x}\bar{y} = \sum xy - \sum x \sum y/n \qquad (2\text{-}35)$$

*Solutions for a and b.* Substitution of equations (2–33) and (2–35) in equation (2–32a) gives a convenient expression for the calculation of $b$:

$$\text{slope} = b = \frac{\sum UV}{\sum U^2} \qquad (2\text{-}32\text{b})$$

After $b$ has been calculated, it is substituted in equation (2–30c) from which $a$ can be found.

## Uncertainties

*Confidence limits for b.* It often happens that the slope of a regression line is predicted by theoretical considerations to have some given value. Occasionally, a set of observations which should define a flat, straight line ($b = 0$), indicating that $y$ is independent of $x$, is examined to see if some finite value of $b$ is found. In either case, it is vital to have a method for the assignment of confidence limits for $b$.

The first quantity required is termed the **variance about the regression**, defined as follows:

$$s_{y \cdot x}^2 = \frac{\sum V^2 - b^2 \sum U^2}{n - 2} \qquad (2\text{-}36)$$

Then, the **variance of the regression coefficient**, or **standard deviation of b**, is given by

$$s_{b_{y \cdot x}}^2 = \frac{s_{y \cdot x}^2}{\sum U^2} \qquad (2\text{-}37)$$

Confidence limits for $b$ are assigned in the usual way

$$\text{C.L.} = \pm t_{\alpha, \varphi} s_{b_{y \cdot x}} \qquad (2\text{-}38)$$

where $\varphi = n - 2$.

*Confidence limits for a regression estimate.* In practice, the regression line is used to determine some estimate, $x_k$, of the chemical input which is responsible for an observed instrument output, $y_k$. The variance of an $x_k$ value which has been determined by the observation of $m$ outputs is given by the expression

$$s_{x_k}^2 = \frac{s_{y \cdot x}^2}{b^2} \left[ \left( \frac{1}{m} + \frac{1}{n} \right) + \frac{(\bar{y}_k - \bar{y})^2}{b^2 \sum U^2} \right] \qquad (2\text{-}39)$$

where, for clarity, the following points should be stressed:

(1) $x_k$ is the (unknown) chemical input corresponding to the observed instrument output. For example, in Figure 2–6, $x_k$ is the iron(II) concentration corresponding to a particular observed solution absorbance.

(2) $\bar{y}_k$ is $\sum y_k/m$, the average instrument output obtained from $m$ observations of the unknown chemical input. Frequently, only a single reading is taken, and therefore $m = 1$.

(3) The quantities $s_{y \cdot x}^2$, $b$, $n$, $\bar{y}$, and $\sum U^2$ all relate to the instrument calibration data and have the same values as in equations (2–29) through (2–36).

Notice that $s_{x_k}^2$ becomes large as $\bar{y}_k$ gets farther away from $\bar{y}$. This is nothing but a mathematical expression of the sensible fact that the uncertainty in any regression estimate is lowest near the center of the calibration data, and that extrapolations are a particularly chancy business. The confidence limits for the desired $x_k$ are given by the relation

$$\text{C.L.} = \pm t_{\alpha . \varphi} s_{x_k} \tag{2–40}$$

where $\varphi = (n - 2) \neq f(m)$.

*Example 2–16.* Use the method of least squares to construct a calibration line for the spectrophotometric determination of iron(II) described earlier.

(1) Actual data (which are plotted in Figure 2–6) appear in the table below:

| Standard Solution Concentration, $c$ moles/liter | Observed Absorbance, $A$ |
|---|---|
| $1.00 \times 10^{-5}$ | 0.114 |
| $2.00 \times 10^{-5}$ | 0.212 |
| $3.00 \times 10^{-5}$ | 0.335 |
| $4.00 \times 10^{-5}$ | 0.434 |
| $6.00 \times 10^{-5}$ | 0.670 |
| $8.00 \times 10^{-5}$ | 0.868 |

(2) The calibration line must fit an equation of the form $y = a + bx$, where $y$ is proportional to the absorbance and $x$ is proportional to the concentration. In order to simplify the arithmetic, we will take $y = 10A$ and $x = 10^5 c$. With $n = 6$, we obtain from equations (2–33), (2–34), and (2–35):

| | | | |
|---|---|---|---|
| $\sum x = 24.00$ | $\bar{x} = 4.00$ | $\sum x^2 = 130.00$ | $\sum U^2 = 34$ |
| $\sum y = 26.33$ | $\bar{y} = 4.3883$ | $\sum y^2 = 156.0845$ | $\sum V^2 = 40.5397$ |
| | | $\sum xy = 142.43$ | $\sum UV = 37.11$ |

In these calculations we have carried as many significant figures as the calculator will allow. This is done to minimize round-off errors and is a required procedure! It is not "synthetic precision" unless it is used in reporting some result.

(3) The coefficients $a$ and $b$ are calculated using equations (2–32b) and (2–30c):

$$b = \frac{\sum UV}{\sum U^2} = 1.0914_7$$

$$a = \frac{1}{n}\left(\sum y - b \sum x\right) = 0.0224_5$$

The equation for the regression line thus takes the form

$$y = 0.0224_5 + 1.09_1 x$$

Substituting the coded values of $x$ and $y$ in the latter relation, we obtain the desired equation for the regression line:

$$10A = 0.0224_5 + (1.09_1 \times 10^5 c)$$

$$c = (-2.05_8 \times 10^{-7}) + (9.16_6 \times 10^{-5} A)$$

(4) To determine the confidence limits, we first obtain the variance about the regression according to equation (2–36). Because the numerator of equation (2–36) is the difference between two large numbers, many significant figures must be carried in the computation:

$$s_{y \cdot x}^2 = \frac{\sum V^2 - b^2 \sum U^2}{n - 2} = 8.825 \times 10^{-3}$$

The standard deviation of $b$ can be calculated from equation (2–37):

$$s_{b_{y \cdot x}}^2 = \frac{s_{y \cdot x}^2}{\sum U^2} = 2.596 \times 10^{-4}; \qquad s_{b_{y \cdot x}} = 0.016$$

Next we determine confidence limits for $b$ by using equation (2–38) with $\alpha = 0.025$ and $\varphi = 4$:

$$\text{C.L.} = \pm t_{\alpha, \varphi} s_{b_{y \cdot x}} = \pm 0.044$$

Therefore, $b = 1.09_1 \pm 0.04_4$.

As an example of the calculation of confidence limits on a regression estimate, imagine

(a) That a single unknown sample gave an absorbance of 0.527.

(b) That five replicate samples gave an average absorbance of 0.527.
For either of the two situations:

$$c = (-2.05_8 \times 10^{-7}) + (9.16_6 \times 10^{-5} A) = 4.81_0 \times 10^{-5} M$$

The standard deviation of the regression estimate is obtained from equation (2–39), where, in case (a), $m = 1$; and in case (b), $m = 5$. In both situations, $\bar{y}_k = 10A = 5.27$. Inserting appropriate values in equation (2–39) for each of the two cases, we get

for (a): $\qquad s_{x_k}^2 = 8.804 \times 10^{-3}; \qquad s_{x_k} = 0.093_8$

and

for (b): $\qquad s_{x_k}^2 = 2.878 \times 10^{-3}; \qquad s_{x_k} = 0.053_6$

The confidence limits for the regression estimates are both computed from equation (2–40) with $\varphi = 4$:

(a) $\qquad \text{C.L.} = \pm t_{0.025, 4} s_{x_k} = \pm 0.26; \qquad c = (4.81 \pm 0.26) \times 10^{-5} M$

(b) $\qquad \text{C.L.} = \pm t_{0.025, 4} s_{x_k} = \pm 0.15; \qquad c = (4.81 \pm 0.15) \times 10^{-5} M$

Note that, because of the coding, the confidence limits for $c$ are $10^{-5}$ times those for the regression estimates.

If nothing else, these results, in which the correctly calculated uncertainties amount to about 4 per cent, demonstrate that the method of least squares is no substitute for quality in the calibration data themselves. Many chemists learn and use the method of least squares only up to the point of calculating $b$ and $a$, and completely omit step (4) above. In this way, they misuse the statistical method and frequently deceive themselves and others by reporting too many significant figures in their results.

## REJECTION OF AN OBSERVATION

Consider this set of analytical results: 15.25, 15.28, 15.30, 16.43 per cent. It is logical to suspect that something is wrong with the fourth observation, and the temptation to reject it before calculating the mean or the standard deviation is bound to be very strong. Such points are called **outliers**, and, though we have provided a flagrant example for the sake of illustration, they can be the source of considerable agony, particularly when the case is not so clear-cut and when dropping one or two data points would make the result coincide with some expected value. In such cases, any human being is tempted to look around for some statistical oil with which to anoint his whim, but this should be attempted only as a last resort, since the statistical tests for the exclusion of outliers are reasonably unsatisfactory.

The first thing to do is go back and thoroughly recheck all the calculations which led to the offending result. Second, check the data. Did the same sample bottle which weighed 15.145 g for samples 1, 2, and 3 weigh 14.145 g for sample 4? If so, the latter weight must be incorrectly recorded. Is there a note that the fourth sample included a peculiar green lump? These questions are aimed at uncovering some clear-cut *assignable cause* for the large deviation. If such a cause can be found, it can either be corrected, in the case of some arithmetic blunder, or used as a perfectly good reason for excluding the outlier.

Sometimes no clear-cut assignable cause is found, and the experimenter asks himself, "Don't I remember spilling a bit of that solution?" or, "I seem to remember that the analytical balance was acting up." This is only more evidence of our humanity, but in any real situation it has to be met with cold-hearted honesty. A test of the sincerity of such suggestions can be made if the experimenter will ask himself whether he is willing to resolve at that moment to discard any future result which might have been similarly affected, *regardless of its value*. Another useful policy to consider is the promise to discard any future sample which might have been similarly affected before the analysis is even completed.

When there are only three or four observations in a sample, it is nearly impossible to provide a useful statistical test for the exclusion of outliers. When there are five or more, it is sometimes helpful to practice the technique of **interior averaging**, in which *both* the highest and lowest data points in the sample are discarded.

## SUMMARY

In this chapter, we have attempted to provide a solid foundation for the understanding and use of statistical techniques in chemical experimentation. The discussions are not mathematically rigorous, but neither are they useless trivializations which force the reader into memorizing statistical formulas if anything is to be gained. We hope the reader will avoid memorization and instead work to understand a few cardinal points:

1. The differences between a sample and a universe, between $s$ and $\sigma$, between $\bar{x}$ and $\mu$

2. The generalization $s^2 = \dfrac{\text{sum of (deviations)}^2}{\varphi}$

3. The use of the normal distribution tables

4. For any

$$w = f(x, y, z),$$

$$V_w = \left(\frac{\partial w}{\partial x}\right)^2 V_x + \left(\frac{\partial w}{\partial y}\right)^2 V_y + \left(\frac{\partial w}{\partial z}\right)^2 V_z$$

5. Why the $t$-distribution is needed, and the generalization "confidence limits on *anything*" $= \pm t_{\alpha, \varphi} s_{anything}$

6. The principle of regression analysis, and the concept that poor data cannot be improved by its use.

7. The fact that outliers *must* have some assignable cause; the idea is to find it, not to throw away data.

## QUESTIONS AND PROBLEMS

1. Define and illustrate each of the following terms: significant figures, precision, accuracy, systematic error, random error, average, deviation, confidence limits, absolute uncertainty, relative uncertainty, $t$ test.

2. Assuming that the following quantities are all determined to within $\pm 1$ digit in the least significant figure, express the relative uncertainty in each in parts per thousand, or, where appropriate, in parts per million: (a) 0.104 g, (b) 1204.3 ml, (c) 1.007825 atomic mass units (the mass of a $^1$H atom), (d) 56.9354 atomic mass units (the mass of a $^{57}$Fe atom), (e) $4.56 \times 10^9$ yr (the age of the earth), (f) 5730 yr (the half life of $^{14}$C).

3. Are equations (2–1) and (2–3) entirely equivalent? If not, why not? What about equations (2–2a) and (2–3)?

4. Construct a simple algebraic proof of the equality of equations (2–2a), (2–2b), and (2–2c).

5. The following results were obtained for replicate determinations of the percentage of chloride in a solid chloride sample: 59.83, 60.04, 60.45, 59.88, 60.33, 60.24, 60.28, 59.77. Calculate (a) the arithmetic mean, (b) the standard deviation, and (c) the relative standard deviation (in per cent).

6. Assuming the sample of problem 5 was pure sodium chloride, calculate the absolute and relative errors of the arithmetic mean.

7. Working from equation (2–3), prove that, for $k$ sets of duplicates, $s^2 = \Sigma w^2 / 2k$, where $w$ is the difference between the two measurements in a duplicate pair.

8. Five students carry out the same analytical procedure on five different trace-iron samples. The results are given below. Assuming an absence of systematic errors, calculate the standard deviation (a) of the procedure and (b) of the mean of three measurements on a single sample. Iron concentrations (in ppm): student 1, sample C, 43, 48, 47; student 2, sample A, 38, 38, 42, 40, 41; student 3, sample D, 51, 54, 57, 58; student 4, sample E, 35, 36, 38, 41; student 5, sample B, 45, 48, 52, 54.

9. (Drill problems on the use of the normal distribution tables.) Given a normally distributed universe of quantitative observations, what is the probability that (a) some observation will fall more than $0.8\sigma$ above the mean, (b) some observation will fall more than $5\sigma$ below the mean, (c) some observed $x$ will be in the range $(\mu - 0.4\sigma) \leq x \leq (\mu + 1.3\sigma)$, (d) an observation will fall in the range $-1.60 \leq u \leq 0.24$?

10. A 10.0000-g object is weighed repeatedly using a procedure with a known $\sigma = 1.0$ mg. What fraction of the observations will be greater than or equal to 10.0016 g?

11. Given a population with a birth rate of $10^7$/yr, and an average height at full growth of 175 cm with a standard deviation of 15 cm, what will be the average time interval between births of individuals destined to grow to 225 cm?

12. The mass of a particular lunar rock sample is determined by adding the masses of seven fragments. The standard deviation of a single weighing is 3 mg. What is the standard deviation of the resulting sum?

13. A result, $w$, is calculated from the formula $w = (x/y) + z$. Given $s_x = 0.1$, $s_y = 0.2$, $s_z = 0.5$, $x = 8$, $y = 2$, and $z = 1$, calculate $s_w$.

14. In a certain semi-micro volumetric determination, the data obtained with their standard deviations are as follows: initial buret reading, 0.23 ml, $s = 0.02$ ml; final buret reading, 8.76 ml, $s = 0.03$ ml; sample weight, 50.0 mg, $s = 0.2$ mg. From the data, find the relative standard deviation of the final result obtained using the equation below (the atomic weights used to calculate the equivalent weight are known to within 1 part in $10^4$).

$$\%X = \frac{(\text{titrant volume})(\text{equivalent weight})(100)}{(\text{sample weight})}$$

15. For a single weighing by a particular procedure, $\sigma_x = 1.0$ mg. (a) Calculate $\sigma_{\bar{x}}$ for the mean of four weighings. (b) How many weighings are required for $\sigma_{\bar{x}} = 0.1$ mg? (c) What is the probability that the mean of five weighings is within 0.3 mg of the true weight?

16. A manufacturer supplying glass electrodes for an instrument which continuously monitors pH finds and advertises that his electrodes have an average operating life of 8000 hours with a standard deviation of 200 hours. (a) If you were monitoring the pH of an important production process and wished to prevent breakdowns, yet avoid the expense of replacing electrodes which were still perfectly good, how long could you use an electrode and expect no greater than (i) 5 per cent probability of failure, (ii) 1 per cent probability of failure? (b) If you get a shipment of four electrodes and find that their average life is 7700 hours, do you have cause for complaint? Why? (c) What is the probability that a given electrode will last at least 8400 hours?

17. Five observations of the chloride content of a potable water sample give an average of 29 ppm $Cl^-$ with a standard deviation of 3.4 ppm. What are the 95 per cent confidence limits of the mean?

18. Six measurements of the percentage of $TiO_2$ in a carload of titanium ore average 58.6% $TiO_2$ with an estimated standard deviation, $s$, of 0.7% $TiO_2$. (a) Find the 90 per cent confidence limits for the average result. (b) Compare with the 95 per cent confidence limits. (c) What would the 90 per cent confidence limits have been had the average been based on only four samples from the carload?

19. A continuous process is operated for the production of dichlorobutadiene from chlorobutadiene. A small amount of (undesirable) trichlorobutene is always formed. Long operating experience has shown that the product averages 1.60 per cent trichlorobutene. An experimental change in operating conditions is made and the analysis of six samples, taken at five-hour intervals, gives the following results: 1.46, 1.62, 1.37, 1.71, 1.52, and 1.40%. Does the change in operating conditions lead to any real change in the trichlorobutene content of the product?

20. The following determinations were made of the atomic weight of carbon: 12.0080, 12.0095, 12.0097, 12.0101, 12.0102, 12.0106, 12.0111, 12.0113, 12.0118, and 12.0120. Calculate (a) the arithmetic mean, (b) the standard deviation, (c) the standard deviation of the mean, (d) the 99 per cent confidence limits of the mean.

21. An iron determination by a gravimetric procedure yielded an average of 46.20% Fe for six trials; and four trials according to a volumetric procedure gave an average of 46.02% Fe, the standard deviation being 0.08% Fe in each case. Is there a significant difference in the results obtained from the two methods?

22. The mass of a crucible was determined to be 18.2463 g, with a standard deviation of 0.0003 g when weighed by 11 different students in one class, whereas eight students in another class obtained an average of 18.2466 g with the same standard deviation. Is there a significant difference in the results found by the two groups of students?

23. By the method of least squares, calculate the equation of the best-fit straight line representing the following spectrophotometric analysis:

| Concentration (ppm) | 0.20 | 0.40 | 0.60 | 0.80 | 1.00 |
|---|---|---|---|---|---|
| Absorbance | 0.077 | 0.126 | 0.176 | 0.230 | 0.280 |

24. (a) Determine the equation of the best-fit straight line for the flame emission photometric data given below:

| Na concentration (ppm) | 0.02 | 0.20 | 0.40 | 0.70 | 1.00 | 1.50 |
|---|---|---|---|---|---|---|
| output (relative units) | 1.20 | 2.60 | 4.40 | 7.60 | 10.80 | 15.60 |

(b) An unknown sample gives an instrument output of 5.35 units. What is the apparent concentration of Na in the sample? (c) Calculate (i) the standard deviation of the result calculated in (b) above, (ii) the 95 per cent confidence limits for the result calculated in (b) above.

25. The top of a peak in a mass spectrum (plot of mass $vs.$ ion abundance) in a particular mass spectrometer is supposed to be perfectly flat. If the top is skewed (*i.e.*, tilts toward one side or the other of the peak), the instrument requires painstaking readjustment. The instrument is scanned very slowly across the top of the peak at mass 44 and the following results are obtained :

| mass | 43.95 | 43.96 | 43.97 | 43.98 | 43.99 | 44.00 |
|---|---|---|---|---|---|---|
| peak height | 898 | 902 | 902 | 900 | 906 | 906 |

| mass | 44.01 | 44.02 | 44.03 | 44.04 | 44.05 |
|---|---|---|---|---|---|
| peak height | 905 | 910 | 907 | 909 | 912 |

Does the slope of the best-fit line differ significantly from zero? Calculate (a) the slope, (b) the standard deviation of the slope, (c) the 95 per cent confidence limits for the slope.

## SUGGESTIONS FOR ADDITIONAL READING

1. P. R. Bevington: *Data Reduction and Error Analysis for the Physical Sciences.* McGraw-Hill Book Company, New York, 1969.
2. O. L. Davies, ed.: *Statistical Methods in Research and Production.* Third edition, Hafner, New York, 1967.
3. W. J. Dixon and F. J. Massey, Jr.: *Introduction to Statistical Analysis.* Third edition, McGraw-Hill Book Company, New York, 1969.
4. R. A. Fisher: *Statistical Methods for Research Workers.* Fourteenth edition, Hafner, New York, 1969.
5. E. B. Wilson, Jr.: *An Introduction to Scientific Research.* McGraw-Hill Book Company, New York, 1952.

*Paper*

1. D. York: Least squares fitting of a straight line. *Canad. Jour. Phys.*, *44*:1079, 1966.

# WATER, SOLUTES, AND CHEMICAL EQUILIBRIUM

③

Since a large proportion of the systems encountered in analytical chemistry are aqueous solutions, it is appropriate to devote a little attention to the properties of water as a solvent and to the behavior of solutes in an aqueous medium. Accordingly, we shall begin this chapter with a brief discussion of the nature of water. Then, we will examine how dissolved solutes, especially ionic species, affect the solvent and how, in turn, the aqueous environment influences the behavior of the solutes.

In the final part of this chapter, we shall consider the nature of chemical equilibrium as well as the factors which influence the equilibrium state of a system.

## NATURE OF WATER

Water has unexpectedly high melting and boiling points and unusually large heats of fusion and vaporization. Furthermore, liquid water possesses a high dielectric constant, surface tension, and thermal conductivity, and is second in heat capacity only to liquid ammonia. Doubtless the most significant structural feature of water in the liquid and solid states is the existence of intermolecular hydrogen bonds. Hydrogen bonds hold molecules of water together, preventing them, for example, from melting or boiling as easily as might be anticipated. Let us consider the structures of solid and liquid water in more detail.

## Structures of Ice and Liquid Water

At atmospheric pressure, ice exists as a layered network of puckered hexagonal rings of oxygen atoms, as illustrated in Figure 3–1. Between every pair of oxygen atoms is a hydrogen atom, covalently bonded to one oxygen atom and hydrogen-bonded to the other. In addition, the individual layers of hexagonally arranged oxygen atoms are held together by hydrogen bonds, and each layer is the mirror image of the layers immediately above and below it.

An average of two hydrogen bonds, with a total energy of 9 kcal mole$^{-1}$, anchors each water molecule in the ice lattice. Since the heat of fusion of ice is 1.44 kcal mole$^{-1}$, it would appear that only 16 per cent of the hydrogen bonds originally present in ice are broken during the melting process. Thus, one picture of liquid water is that of partially broken-down ice consisting of a mixture of large, hydrogen-bonded, ice-like aggregates along with free or unbound water molecules. Presumably, the aggregates would be in equilibrium with individual water molecules, hydrogen bonds being ruptured at some sites while being simultaneously formed at other locations. Other investigators prefer a structure for liquid water that is a mixture of icelike clusters and other, more dense, possibly quartzlike aggregates of water molecules. Another theory suggests that when ice melts the hydrogen bonds between water molecules do not break, but merely become flexible, bent, or distorted, in which event liquid water may be envisioned as a homogeneous, fluid modification of ice.

Another model for the structure of liquid water envisages a mixture of so-called

*Figure 3–1.* Diagram showing the structure of a portion of an ice crystal.

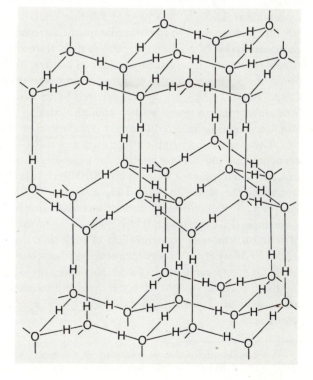

"flickering clusters" of hydrogen-bonded water molecules in the midst of free or unbound water. An important feature of this "flickering cluster" theory is that structural descriptions of the aggregates of water molecules are deliberately avoided, except for the statement that the large clusters continually appear and disappear as hydrogen bonds are formed or broken. Moreover, an unbound water molecule is free only in the sense that it is not associated with a hydrogen-bonded cluster because, in a liquid consisting of dipolar molecules such as water, there must always be strong intermolecular attractions.

Regardless of which of the preceding pictures is correct, or whether a yet undiscovered model eventually proves more acceptable, it is essential to recognize that intermolecular hydrogen bonding leads to significant structural order in liquid water. Much additional information about water can be found in the monograph by Horne listed at the end of this chapter.

## INTERACTIONS BETWEEN SOLUTES AND WATER

Dissolution of a substance in water is a remarkably complex process, resulting in the evolution or absorption of heat and leading to profound changes in the environment of the solute species and in the structure and properties of the solvent.

### Methods of Expressing Concentrations of Solutions

In analytical chemistry, it is convenient to specify the molar concentration or the formal concentration of a particular substance in solution. We may define the **molarity** of a solution, or the **molar concentration** of a solute, as the number of gram molecular weights (moles) of solute per liter of *solution*, the symbol $M$ being used to designate this concentration unit. For example, the molecular weight of sodium hydroxide is 40.00, so a 0.5000 $M$ solution contains 20.00 gm of sodium hydroxide per liter of solution. On the other hand, the **formality** of a solution, or the **formal concentration** of a solute, is expressed in terms of the number of gram formula weights of solute dissolved in one liter of *solution*, and is indicated by the symbol $F$. Since the formula weight of acetic acid, represented by the chemical formula $CH_3COOH$, is 60.05, a solution prepared by dissolution of 60.05 gm, or 1.000 gram formula weight, of acetic acid in enough water to provide a final volume of exactly one liter would be referred to as a 1.000 $F$ solution of acetic acid.

There is a subtly important distinction to be made between molar and formal concentration units. If we specify, for example, that the concentration of acetic acid in a particular aqueous solution is 0.001000 $M$, it is explicitly meant that the concentration of the *molecular species* $CH_3COOH$ is 0.001000 mole per liter of solution. However, acetic acid undergoes dissociation into hydrogen ions[*] and acetate ions. Therefore, if we start with 0.001000 mole or formula weight of acetic acid per liter of solution, the true concentration of molecular acetic acid at equilibrium is only 0.000876 $M$ whereas, as a consequence of dissociation, the concentrations of $H^+$ and $CH_3COO^-$ are each 0.000124 $M$. However, the solution still contains the original 0.001000 formula weight of acetic acid, although both molecular and dissociated acetic acid are present, and can correctly be specified as a 0.001000 $F$ solution of acetic acid.

---

[*] A complete description of the nature of the hydrated proton, or hydronium ion, in aqueous media is presented in the introductory section of Chapter 4.

Specifying a *formal* concentration provides definite information about how a solution is originally prepared, but it does not necessarily imply what happens to the particular solute after it dissolves. *Molarity* refers specifically to concentrations of actual molecules and ions present in solution at equilibrium, and might or might not be synonymous with formality. Because of the complex nature of aqueous solutions, we prefer to use *formal* rather than *molar* concentrations wherever possible, since it is frequently difficult, if not impossible, to identify the actual species present in solution. Consequently, solutions of acids, bases, and salts will be identified and discussed in terms of *formal* concentrations throughout this book. When it is necessary or desirable to refer to specific ions or molecules in equilibrium calculations, for example, we shall employ *molar* concentrations.

### Effects of Solutes on Water Structure

Every ionic species in solution is surrounded by a layer of water molecules which, under the influence of short-range ion-dipole forces, are highly oriented and tightly bound to form the **primary hydration shell**. Within the primary hydration layer, water molecules are drawn by the electrostatic field around the ion into a new, more densely packed structure than the hydrogen-bonded arrangement found in pure liquid water.

Apart from aquated transition metal cations, it is difficult to specify precisely how many water molecules comprise the primary hydration shell of a particular ionic species. However, we can define the **hydration number** of an ion as the number of solvent molecules—not always the same ones—aligned and held by electrostatic forces as the ion migrates through the solution. In general, ions with the highest charge density—that is, small ions having high charges—are most extensively hydrated. Among the common alkali metal cations, lithium ion exhibits the largest hydration number, between 4 and 6 water molecules, because it possesses a smaller ionic radius (0.60 Å) and a higher charge density than sodium and potassium ions, which have ionic radii of 0.95 and 1.33 Å and average hydration numbers of 4 and 3, respectively. Magnesium ion ($Mg^{2+}$) is similar in size (0.61 Å) to the lithium ion but, because it is dipositive, has a greater charge density and a hydration number between 6 and 12. Anions are characteristically less heavily hydrated than cations. For example, the average hydration numbers for fluoride, chloride, bromide, and iodide ions lie between 1 and 4, perhaps reflecting the fact that the ionic radii for these species have the relatively large values of 1.36, 1.81, 1.95, and 2.16 Å, respectively.

Beyond the primary solvent sheath is a secondary zone, in which the normal water structure has been disrupted by the still-effective electrostatic field of the central ion. However, as the distance from the central ion increases, the structure of the secondary region becomes less and less perturbed by the electrostatic field, and more and more like normal liquid water. To distinguish the primary and secondary solvent layers, it is customary to refer to the **structure-formed region**, where the electrostatic field of an ion disrupts the normal arrangement of water and creates a tightly organized shell of water molecules around itself, and the **structure-broken region**, in which the residual electrostatic field of the ion acts to break down the regular, hydrogen-bonded solvent structure without reorienting the disrupted water dipoles.

A dissolved ion is termed a **water-structure maker** or a **water-structure breaker**, respectively, according to whether it promotes structure-formed regions or structure-broken zones more effectively. Ions with the highest charge density—the same property that leads to a high hydration number—are the best water-structure makers, because they generate strong electrostatic fields capable of attracting, orienting, and stabilizing a large primary hydration shell. On the other hand, species with large ionic radii and correspondingly small charge densities cause structure-broken zones which are bigger than the structure-formed regions. Lithium ion is an excellent water-structure maker, sodium ion has a lesser tendency to favor the production of structure-formed zones in water, whereas the larger potassium, rubidium, and cesium ions are increasingly potent water-structure breakers. Magnesium, calcium, and strontium ions, all dipositive cations, are even better water-structure makers than lithium ion. Chloride, bromide, and iodide anions are water-structure breakers.

Large ionic species with organic substituents show unusual solute properties. Whereas the ammonium ion ($NH_4^+$) is a modest water-structure breaker comparable to the potassium ion, tetraalkylammonium cations such as $N(CH_3)_4^+$, $N(C_2H_5)_4^+$, and $N(C_3H_7)_4^+$ become more and more outstanding water-structure makers as the size of the alkyl group increases. This trend seems contrary to all expectations, since the charge density and water-structure making ability of an ion should decrease as its radius increases. Apparently, the observed behavior—namely, the presence of a large structure-formed region around a tetraalkylammonium ion—is due to the fact that a highly ordered cage of water molecules surrounds the cationic species in order to isolate the nonpolar, water-insoluble alkyl groups from the polar solvent.

## Effects of Solutes on Properties of Solutions

One of the properties of liquid water most affected by its structural organization is viscosity or fluidity. If we compare pure water and a dilute sodium chloride solution, the viscosity of the latter is found to be larger and to increase as the concentration of the solute becomes greater. An aqueous solution of lithium chloride is approximately 50 per cent more viscous or less fluid than a sodium chloride solution of the same concentration. However, a solution of potassium chloride has slightly more fluidity than pure water, and a solution of rubidium chloride or cesium chloride exhibits a smaller viscosity than a potassium chloride medium.

Thus, there is a perfect correlation between the effect of a solute on the viscosity of a solution and the tendency of that solute to act as a water-structure maker or a water-structure breaker. Water-structure makers increase the viscosity of an aqueous medium relative to pure water, whereas water-structure breakers have the opposite effect.

Because the conduction of electricity through a solution involves the movement of ions, the smallest ionic species would be expected to possess the greatest mobility and to cause the highest conductivity. When the conductivities of separate solutions containing identical concentrations of LiCl, NaCl, and KCl are measured, the potassium chloride medium exhibits the highest conductivity and the solution of the lithium salt shows the least conductivity. At first glance, these observations seem contrary to the order of increasing ionic radii for the alkali metal cations—lithium (0.60 Å), sodium (0.95 Å), and potassium (1.33 Å). However, these radii pertain to *unhydrated* ions, whereas the migrating species are *ions with more-or-less firmly attached primary hydration shells*. Reasonable values for the radii of the *hydrated* alkali metal ions are 3.40 Å for $Li^+$, 2.76 Å for $Na^+$, and 2.32 Å for $K^+$. Thus, lithium ion, with the highest charge density and the largest primary hydration layer of the three unhydrated cations, is actually bigger and slower moving in an aqueous medium than either sodium or potassium ion.

## Activities and Activity Coefficients

If a solution containing 0.1 $M$ hydrogen ion and 0.1 $M$ chloride ion behaved in a perfectly ideal manner—that is, if each hydrogen ion and chloride ion was totally independent of its environment—we would expect each of these species to act as if its concentration were 0.1 $M$. On the other hand, if there is mutual electrostatic attraction between a certain fraction of the hydrogen ions and chloride ions, the solution will exhibit the properties of one in which the effective or apparent concentrations of the ions are less than 0.1 $M$. As we shall discover, depending on the relative importance of factors such as ion-ion attractions and ion-solvent interactions, hydrogen ion or chloride ion in a 0.1 $F$ hydrochloric acid solution may behave as if its effective concentration is less than, equal to, or greater than 0.1 $M$.

To distinguish between the molar or formal concentration of a substance and the effective concentration of a substance, which accounts for its nonideal behavior, the latter is defined as the **activity** of the species. It is customary to relate the activity of a species to its concentration through the expression

$$a_i = f_i C_i$$

in which $a_i$ is the activity of substance $i$, $f_i$ is the **activity coefficient** of substance $i$, and $C_i$ is the molar or formal concentration of substance $i$. For example, we may write the activity of hydrogen ion as

$$a_{H^+} = f_{H^+}[H^+]$$

and that for chloride ion as

$$a_{Cl^-} = f_{Cl^-}[Cl^-]$$

where $[H^+]$ and $[Cl^-]$ represent the molar concentrations of hydrogen ion and chloride ion, respectively. Activity and concentration are always expressed in such a way that both terms have identical units, usually moles per liter, so the activity coefficient is a dimensionless parameter.

According to the third preceding equation, a substance dissolved in water may exhibit any of three kinds of behavior. First, if the activity $a_i$ of a solute is exactly equal to its concentration $C_i$, the activity coefficient $f_i$ is unity and the substance is said to behave in an *ideal* manner. Since the activity of a species reflects all possible physical and chemical interionic interactions which can occur in a solution and since these interactions become vanishingly small when the concentrations of ionic species approach zero, $f_i$ should be unity only in infinitely dilute solutions of electrolytes. Second, if the activity $a_i$ of a substance is smaller than its concentration $C_i$, the activity coefficient $f_i$ is less than unity and the substance is said to exhibit *negative* deviations from ideal behavior. Third, if the activity $a_i$ of a substance is larger than its concentration $C_i$, the activity coefficient $f_i$ is greater than unity and the substance is said to display *positive* deviations from ideal behavior. From a plot of the activity coefficient $f_i$ of a species as a function of its concentration, we should be able to tell at a glance for any given concentration whether that substance exhibits nearly ideal behavior, negative deviations from ideality, or positive deviations from ideality.

If any doubt exists that interionic attractions and repulsions or ion-solvent interactions can cause marked differences between activity and concentration, the data in Table 3–1 are especially impressive. This table lists various concentrations of hydrochloric acid and, for each concentration, the corresponding activity of hydrochloric acid in that solution. Obviously, the activities of hydrochloric acid solutions do differ from the concentrations of such solutions, especially at the highest concentration levels, and serious errors may arise if one attempts to make predictions based on

**Table 3–1. Concentrations and Activities of Hydrochloric Acid Solutions**

| Concentration, $F$ | Activity, $F$ |
|---|---|
| 0.00500 | 0.00465 |
| 0.0100 | 0.00906 |
| 0.0200 | 0.0176 |
| 0.0500 | 0.0417 |
| 0.100 | 0.0799 |
| 0.200 | 0.154 |
| 0.500 | 0.382 |
| 1.00 | 0.828 |
| 2.00 | 2.15 |
| 3.00 | 4.47 |
| 5.00 | 15.6 |
| 6.00 | 28.2 |
| 8.00 | 89.6 |
| 10.0 | 254. |
| 12.0 | 695. |

concentrations instead of activities. Figure 3–2 shows a plot of the quantity $f_{\pm}$ versus the concentration of hydrochloric acid, the data for the graph having been derived from Table 3–1. Notice that instead of $f_i$ we have plotted on the ordinate of this graph a new parameter, $f_{\pm}$, which is called the **mean activity coefficient**. Although the significance of the mean activity coefficient will become evident later, it suffices to mention here that $f_{\pm}$ is employed because the individual behavior of hydrogen ion and chloride ion in a hydrochloric acid medium cannot be sorted out, since each ion influences the activity of the other species. However, the use of $f_{\pm}$ does not alter our interpretation of the behavior of hydrochloric acid solutions.

Figure 3–2 confirms that the mean activity coefficient $f_{\pm}$ for hydrochloric acid is unity at infinite dilution, that is, at a hydrochloric acid concentration of virtually zero. However, as the hydrochloric acid concentration is increased, the activity coefficient decreases, reaching a minimum value of approximately 0.76 at a hydrochloric acid concentration near $0.5\,F$. If one continues to increase the concentration of hydrochloric acid above $0.5\,F$, the mean activity coefficient climbs steadily until it attains a value of unity at a concentration slightly less than $2\,F$. For concentrations greater than $2\,F$, the activity coefficient rises very abruptly, indicating that such solutions exhibit large positive deviations from ideality.

One feature of the behavior of hydrochloric acid is that the mean activity coefficient is unity for an approximately $2\,F$ hydrochloric acid solution. This observation could be interpreted to mean that the solution behaves ideally. Actually, the activity coefficient is unity only because two opposing phenomena offset each other perfectly. First, ion-ion attractions between hydrogen ion and chloride ion cause the activity to be *less* than the concentration of hydrochloric acid and the activity coefficient to be *less* than unity. Second, the solvation of hydrogen ions and chloride ions by water molecules decreases the available quantity of free or unbound solvent, thereby tending to make the activity *greater* than the concentration and the activity coefficient *greater* than unity. These effects just happen to counteract each other in $2\,F$ hydrochloric acid. For hydrochloric acid solutions less concentrated than $2\,F$, the first factor is of greater importance, whereas the second factor predominates in hydrochloric acid media more concentrated than $2\,F$. However, the increase of the activity coefficient in acid solutions more concentrated than $0.5\,F$ indicates that the second effect has begun to overcome the first factor.

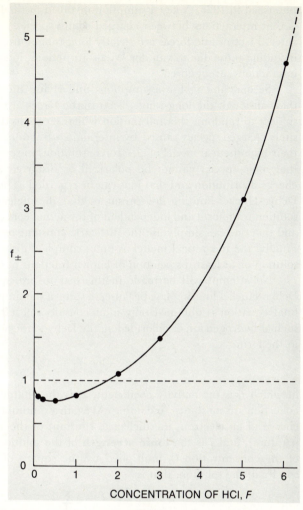

*Figure 3–2.* Mean activity coefficient $f_\pm$ for hydrochloric acid as a function of the acid concentration. (See text for discussion.)

Another interesting observation is that $12\,F$ hydrochloric acid, which is the commercially available concentrated reagent, behaves as if the hydrochloric acid concentration is really $695\,F$ (see Table 3–1). However, this is by no means extreme behavior. For example, a $20\,F$ solution of lithium bromide in water acts as if the lithium ion and bromide ion concentrations were each approximately $10{,}000\,M$, whereas the hydrogen ion activity in a $12\,F$ perchloric acid solution is $6000\,M$!

It should be emphasized that the behavior shown in Figure 3–2 for hydrochloric acid is typical of all ionic solutes. A plot of activity coefficient versus the concentration of an electrolyte usually exhibits a curve which is qualitatively similar to that for hydrochloric acid. Naturally, the absolute values of the activity coefficients, as well as the solute concentration at which the activity coefficient is minimal and at which the activity coefficient rises to unity again, depend on the nature of the solute.

## Debye-Huckel Limiting Law

One of the most significant contributions to our understanding of the behavior of electrolyte solutions is the Debye-Hückel limiting law, the derivation of which was published in 1923. A key assumption involved in the formulation of the limiting law is that interactions between charged solute species are purely electrostatic in character. Electrostatic forces are strictly long-range interactions. All short-range forces, including those due to van der Waals attractions, ion-pair formation, and ion-dipole interactions, are ignored.

Second among the assumptions underlying the Debye-Hückel limiting law is that, although the long-range electrostatic forces are operative, ions in solution are subject to random, thermal motion which serves to disrupt the orientation of oppositely charged species caused by interionic attractions. Debye and Hückel simplified their hypothetical model of electrolyte solutions in several other respects by proposing that ionic species cannot be polarized or distorted and, therefore, have spherical charge distribution and that ions can be regarded as point charges. Furthermore, the Debye-Hückel limiting law presumes that the dielectric constant of an electrolyte solution is uniform and independent of the actual concentration of the dissolved solute and that one can simply use the dielectric constant of pure water in all calculations. Finally, the theoretical model assumes complete dissociation of all electrolytes in the solution or at least dissociation of known fractions of these electrolytes.

No attempt will be made in this text to present the actual derivation of the Debye-Hückel limiting law because, in spite of the simplifications discussed above, the final equation is obtained only after considerable effort. For a single ionic species such as hydrogen ion or chloride ion, the Debye-Hückel limiting law is usually written in the form

$$-\log f_i = A z_i^2 \sqrt{I}$$

in which $f_i$ is the activity coefficient, $z_i$ is the charge of the ion of interest, $A$ is a collection of constants (including Avogadro's number, Boltzmann's constant, the charge of an electron, the dielectric constant of the solvent, and the absolute temperature), and $I$ is the **ionic strength** of the solution, defined as one-half the sum of the concentration $C_i$ multiplied by the square of the charge $z_i$ for each ionic species in the solution, that is

$$I = \tfrac{1}{2} \sum_i C_i z_i^2$$

In effect, the ionic strength, $I$, measures the total population of ions in a solution.

For water at 25°C, the value of $A$ is very close to 0.512, so the Debye-Hückel limiting law for aqueous solutions at this temperature becomes

$$-\log f_i = 0.512 z_i^2 \sqrt{I}$$

Although the latter relation may be employed to evaluate the activity coefficient for any single ionic species such as $Na^+$ or $Cl^-$, this expression cannot be tested experimentally because it is impossible to prepare a solution containing just one kind of ion. For example, to obtain a solution of $Na^+$ one might use sodium sulfate, and potassium chloride crystals could be dissolved in water for the preparation of a chloride solution. Obviously, the interaction of sulfate ion with $Na^+$ and of potassium ion with chloride will influence the behavior of $Na^+$ and $Cl^-$, respectively. Thus, any technique which one utilizes to measure activity coefficients for these systems will yield, in reality, the activity coefficient of $Na^+$ in the presence of one half as much sulfate ion or the activity coefficient of $Cl^-$ in the presence of an equal concentration of potassium ion. Fortunately, by defining a new term, called the **mean activity coefficient** $(f_\pm)$, we obtain a parameter which can be both experimentally and theoretically evaluated. For a *binary* salt, $M_m N_n$, whose cation M has a charge

$z_M$ and whose anion N has a charge $z_N$, the mean activity coefficient is related to the single-ion activity coefficients $f_M$ and $f_N$ by means of the equation

$$(f_\pm)^{m+n} = (f_M)^m (f_N)^n$$

If we take the negative logarithm of each side of this expression, the result is

$$-(m+n)\log f_\pm = -m\log f_M - n\log f_N$$

When the Debye-Hückel limiting law for single ions is substituted into each term on the right-hand side of this relation, one obtains

$$-(m+n)\log f_\pm = 0.512\, m z_M^2 \sqrt{I} + 0.512\, n z_N^2 \sqrt{I}$$

which upon rearrangement and simplification yields

$$-\log f_\pm = 0.512 \sqrt{I} \left[ \frac{m z_M^2 + n z_N^2}{m+n} \right]$$

Since the absolute values of $m z_M$ and $n z_N$ are equal

$$-\log f_\pm = 0.512 z_M z_N \sqrt{I}$$

It is important to note here that only the *absolute* magnitudes of $z_M$ and $z_N$ are used in this equation.

At this point it is instructive to consider some actual calculations involving the use of the Debye-Hückel limiting law. For example, let us determine the mean activity coefficient for $0.10\,F$ hydrochloric acid. First, we must evaluate the ionic strength of the solution. Hydrogen ion and chloride ion are the only species present, so it follows that

$$I = \tfrac{1}{2} \sum_i C_i z_i^2 = \tfrac{1}{2}[C_{H^+} z_{H^+}^2 + C_{Cl^-} z_{Cl^-}^2]$$
$$I = \tfrac{1}{2}[(0.10)(1)^2 + (0.10)(-1)^2] = 0.10$$

Strictly speaking, ionic strength should be expressed in concentration units, such as moles per liter; but, in most reference books and research publications, ionic strength is reported as a dimensionless number. Next, since the absolute values of the charges on hydrogen ion and chloride ion are unity, the mean activity coefficient may be obtained from the expression

$$-\log f_\pm = 0.512 z_M z_N \sqrt{I} = 0.512(1)(1)\sqrt{0.10}$$
$$-\log f_\pm = 0.512(1)(1)(0.316) = 0.162$$
$$\log f_\pm = -0.162 = 0.838 - 1.000$$
$$f_\pm = 0.689$$

For purposes of comparison, the mean activity coefficient for $0.10\,F$ hydrochloric acid has been experimentally measured and found to be 0.799.

As a second example, we shall calculate the mean activity coefficient for $0.10\,F$ aluminum chloride solution. Notice that the ionic strength

$$I = \tfrac{1}{2}[C_{Al_3^+} z_{Al_3^+}^2 + C_{Cl^-} z_{Cl^-}^2]$$
$$I = \tfrac{1}{2}[(0.10)(3)^2 + (0.30)(-1)^2] = 0.60$$

is six times greater than the concentration of the aluminum chloride solution. Substitution of the values for ionic strength and for the charges of the ions into the
Debye-Hückel limiting law yields

$$-\log f_{\pm} = 0.512(3)(1)\sqrt{0.60} = 1.189$$

$$\log f_{\pm} = -1.189 = 0.811 - 2.000$$

$$f_{\pm} = 0.0647$$

This predicted mean activity coefficient is more than five times smaller than the
experimentally determined value of 0.337.

Numerous similar calculations were performed for other concentrations of
hydrochloric acid and aluminum chloride, and the results are shown diagram-
matically in Figure 3–3. Included in the plot are the actual experimental values of
the mean activity coefficients. Perhaps the most important conclusion to be drawn
from Figure 3–3 is that the Debye-Hückel limiting law is quantitatively useful only

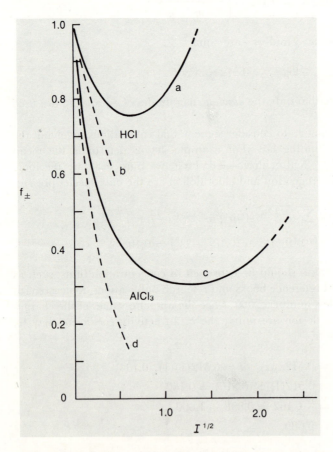

*Figure 3–3.* Comparison of the observed mean activity coefficients for hydrochloric acid (HCl) and aluminum chloride
(AlCl$_3$) with the theoretical mean activity coefficients calculated
from the Debye-Hückel limiting law, plotted as a function of the
square root of ionic strength. Curve *a*: observed mean activity
coefficients for HCl. Curve *b*: calculated mean activity coefficients for HCl. Curve *c*: observed mean activity coefficients for
AlCl$_3$. Curve *d*: calculated mean activity coefficients for AlCl$_3$.

in very dilute solutions. A second distinctive feature of the plot is that the mean activity coefficients calculated from the limiting law become significantly smaller than the true or observed values as ionic strength increases. Furthermore, the Debye-Hückel limiting law cannot account at all for the upward trend in activity coefficients at relatively high ionic strengths. Nevertheless, the theoretical and experimental curves do converge at low ionic strengths and extrapolate to a mean activity coefficient of unity at zero ionic strength, so the assumptions made by Debye and Hückel are valid for small electrolyte concentrations—particularly the ideas that ions act as point charges in a solution of uniform dielectric constant and that ions undergo simple electrostatic interactions with their neighbors.

In general, activity coefficients for salts consisting of singly charged species can be evaluated by means of the Debye-Hückel limiting law with an uncertainty not exceeding about 5 per cent for ionic strengths up to 0.05. For salts which consist of doubly charged ions, the Debye-Hückel limiting law gives results reliable to within several per cent up to ionic strengths of perhaps 0.01, whereas, for triply charged species like $Al^{3+}$, the limiting law fails above ionic strengths of approximately 0.005. However, these limits concerning the applicability of the Debye-Hückel limiting law must be regarded only as rough guidelines, because the specific nature of the ions in question has a great influence on the behavior of an electrolyte.

Why is the limiting law inadequate and how can the theoretical model be modified so that a more accurate equation may be formulated? Two of the assumptions used by Debye and Hückel were that ionic species are affected only by long-range electrostatic forces and that thermal motion causes the ions to behave essentially independently of each other. In a dilute solution, where one may envisage a given ion to be insulated from other ions by a sheath of water molecules, these assumptions appear to be valid. In more concentrated solutions, ions are close to each other, so a number of highly specific interactions such as ion-pair formation may occur. In the case of multiply charged species, the formation of ion pairs, triplets, and other aggregates is especially important, owing to the much stronger electrostatic forces which prevail. In addition, Debye and Hückel assumed that the dielectric constant of an electrolyte solution is uniform and equal to that of pure water. However, the dielectric constant of a solution is known to vary drastically in the vicinity of an ion. Hydrogen bonding between adjacent water molecules in pure water promotes the formation of aggregates in which the individual molecular dipole moments reinforce each other, causing the unusually high dielectric constant of water (78.5 at 25°C). In the presence of an ion, strong ion-dipole interactions occur and the hydrogen-bonded water structure is broken down. Accordingly, the dielectric constant of the water may be only 4 to 10 within a radius of a few angstroms around an ionic species.

It is difficult to incorporate the phenomena which we have been discussing into the general theory of electrolyte solutions. All of these effects tend to be highly specific; they vary considerably depending on the geometry, charge, and electronic configuration of an ion, as well as on the other species which may be present.

## Modifications of the Simple Debye-Hückel Theory

One of the assumptions originally used by Debye and Hückel was that ions are point charges. However, if we consider the fact that every ion has its own characteristic radius, it should be apparent that the electrostatic fields around an ion of finite size and a point charge will be different. Furthermore, two ions cannot approach each other any closer than the sum of their radii, so the coulombic force between them is smaller than if they were point charges. Accordingly, the **extended Debye-Hückel equation**

$$- \log f_{\pm} = 0.512 \, z_{\mathrm{M}} \, z_{\mathrm{N}} \left[ \frac{\sqrt{I}}{1 + Ba\sqrt{I}} \right]$$

has been derived to incorporate the effect of finite ion size on activity coefficients. This expression resembles the limiting law except for two new factors in the denominator of the term in brackets—$B$, a function $(50.3/\sqrt{DT})$ of the absolute temperature $(T)$ and the dielectric constant $(D)$ of the solution, and $a$, an *empirically adjustable* number, called the **ion-size parameter** and expressed in units of angstroms, which is said to correspond to the mean distance of closest approach between the solvated cation and anion of an electrolyte. If we take the absolute temperature to be 298°K and the dielectric constant to be the value for pure water (78.5), $B$ is 0.328.

For ionic strengths not exceeding 0.1, experimental studies and theoretical calculations have indicated that the ion-size parameters for many hydrated inorganic species range from 3 to 11 Å, as Table 3–2 shows. To compute the mean activity coefficient for a binary electrolyte by means of the extended Debye-Hückel equation, the ion sizes of the cation and anion which comprise the solute may be averaged. Thus, for chloride ion and hydrogen ion, the respective ion sizes are 3 and 9 Å, so that 6 Å can be employed as the ion-size parameter of hydrochloric acid. As long as the ionic strength is less than 0.1, use of the ion-size parameters compiled in Table 3–2 allows one to obtain good agreement between experimentally measured mean activity coefficients and those calculated from the extended Debye-Hückel equation. In the last five columns of Table 3–2 are listed single-ion activity coefficients for

**Table 3–2. Single-Ion Activity Coefficients Based on the Extended Debye-Hückel Equation***

| Ion | Ion Size $a$, Å | Ionic Strength | | | | |
|---|---|---|---|---|---|---|
| | | 0.001 | 0.005 | 0.01 | 0.05 | 0.10 |
| $H^+$ | 9 | 0.967 | 0.933 | 0.914 | 0.86 | 0.83 |
| $Li^+$ | 6 | 0.965 | 0.930 | 0.909 | 0.845 | 0.81 |
| $Na^+$, $IO_3^-$, $HCO_3^-$, $HSO_3^-$, $H_2PO_4^-$, $H_2AsO_4^-$ | 4 | 0.964 | 0.927 | 0.901 | 0.815 | 0.77 |
| $K^+$, $Rb^+$, $Cs^+$, $Tl^+$, $Ag^+$, $NH_4^+$, $OH^-$, $F^-$, $SCN^-$, $HS^-$, $ClO_3^-$, $ClO_4^-$, $BrO_3^-$, $IO_4^-$, $MnO_4^-$, $Cl^-$, $Br^-$, $I^-$, $CN^-$, $NO_3^-$ | 3 | 0.964 | 0.925 | 0.899 | 0.805 | 0.755 |
| $Mg^{2+}$, $Be^{2+}$ | 8 | 0.872 | 0.755 | 0.69 | 0.52 | 0.45 |
| $Ca^{2+}$, $Cu^{2+}$, $Zn^{2+}$, $Sn^{2+}$, $Mn^{2+}$, $Fe^{2+}$, $Ni^{2+}$, $Co^{2+}$ | 6 | 0.870 | 0.749 | 0.675 | 0.485 | 0.405 |
| $Sr^{2+}$, $Ba^{2+}$, $Ra^{2+}$, $Cd^{2+}$, $Pb^{2+}$, $Hg^{2+}$, $S^{2-}$, $CO_3^{2-}$, $SO_3^{2-}$ | 5 | 0.868 | 0.744 | 0.67 | 0.465 | 0.38 |
| $Hg_2^{2+}$, $SO_4^{2-}$, $S_2O_3^{2-}$, $CrO_4^{2-}$, $HPO_4^{2-}$ | 4 | 0.867 | 0.740 | 0.660 | 0.445 | 0.355 |
| $Al^{3+}$, $Fe^{3+}$, $Cr^{3+}$, $Ce^{3+}$, $La^{3+}$ | 9 | 0.738 | 0.54 | 0.445 | 0.245 | 0.18 |
| $PO_4^{3-}$, $Fe(CN)_6^{3-}$ | 4 | 0.725 | 0.505 | 0.395 | 0.16 | 0.095 |
| $Th^{4+}$, $Zr^{4+}$, $Ce^{4+}$, $Sn^{4+}$ | 11 | 0.588 | 0.35 | 0.255 | 0.10 | 0.065 |
| $Fe(CN)_6^{4-}$ | 5 | 0.57 | 0.31 | 0.20 | 0.048 | 0.021 |

* Taken from the paper by J. Kielland: J. Amer. Chem. Soc., 59: 1675, 1937.

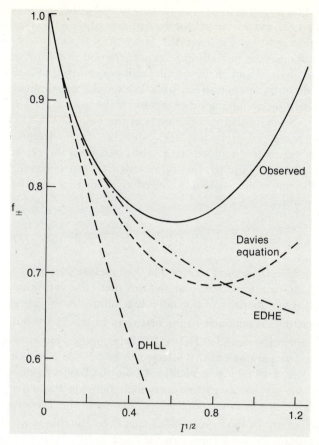

*Figure 3–4.* Comparison of mean activity coefficients for hydrochloric acid computed from the Debye-Hückel limiting law (DHLL), the extended Debye-Hückel equation (EDHE), and the Davies equation with observed mean activity coefficients. Results based on the extended Debye-Hückel equation correspond to $a = 6$ Å.

ionic strengths of 0.001, 0.005, 0.01, 0.05, and 0.1. For each of these ionic strengths, the tabulated single-ion activity coefficients can be used, along with the general relationship between $f_{\pm}$ and $f_i$ values presented earlier, to determine mean activity coefficients for binary salts.

To determine how well the predictions of the extended Debye-Hückel equation follow the true behavior of hydrochloric acid, we constructed the graph shown in Figure 3–4, which repeats some of the information given in Figure 3–3. As anticipated, the extended Debye-Hückel equation (EDHE) provides better correlation between theory and experiment than does the limiting law (DHLL).

However, there remain some disconcerting problems connected with the concept of an ion-size parameter. To obtain the best agreement between calculated and measured activity coefficients for a particular electrolyte, the ion-size parameter must be varied, often over a considerable range, as the solute concentration changes. For high concentrations of some electrolytes, experimental and theoretical activity coefficients coincide only if the ion-size parameter is chosen to have physically impossible (negative) values. For example, ion-size parameters of 13.8, −411.2, and

—14.8 Å, respectively, are needed to make the calculated mean activity coefficients for 1, 2, and 3 $F$ hydrochloric acid accord with observed $f_\pm$ values. Thus, although one should acknowledge the finite sizes of ions in perfecting the theory of ionic solutions, it is unrealistic to expect the ion-size parameter alone to reconcile all of the physical and chemical interactions occurring in a solution, many of which have little or no relation to ion sizes. We should seek improvements in the Debye-Hückel theory by examining other factors.

Several other modifications of the basic Debye-Hückel relationships have been proposed. In every instance, however, these changes take the form of empirical correction terms added to or subtracted from the extended Debye-Hückel equation in order to improve the agreement between observed and calculated activity coefficients. One well-known version of such a modified expression is the Davies equation,

$$-\log f_\pm = 0.512\, z_M\, z_N \left[ \frac{\sqrt{I}}{1 + \sqrt{I}} - 0.2I \right]$$

in which the second term inside the brackets seeks to correct for the combined effects of ion-pair formation, nonuniform dielectric constant, ion polarizability, and any other peculiarities of an individual solute on the mean activity coefficient. Notice that the denominator of the first term in brackets is not $(1 + Ba\sqrt{I})$, as usually appears in the extended Debye-Hückel equation, but $(1 + \sqrt{I})$, the latter arising if the ion-size parameter ($a$) is selected to be 3.05 Å. Mean activity coefficients for hydrochloric acid solutions calculated from the Davies equation are plotted in Figure 3–4. Obviously, although the agreement between the observed and calculated activity coefficients is not completely satisfactory, the Davies equation does simulate the minimum in the $f_\pm$-versus-$\sqrt{I}$ curve. By suitable arbitrary choice of the numerical constant for the second term in brackets, the Davies equation could be forced to fit the observed activity coefficients even better. However, this relation cannot adequately account for the tremendous rise in the activity coefficient of hydrochloric acid at very high solute concentrations.

We can understand most clearly why activity coefficients become much larger than unity in very concentrated electrolyte solutions by recognizing the interactions between dissolved ions and water molecules. For example, 12 $F$ hydrochloric acid behaves as if the acid concentration is 695 $F$, indicating that the mean activity coefficient is approximately 58 (Table 3–1). Let us consider the following crude picture. If we have one liter of 12 $F$ hydrochloric acid, the solution will contain 12 moles of the acid and, to a rough first approximation, 56 moles of water. If the hydrochloric acid is completely dissociated, 12 moles each of hydrogen ion and chloride ion will be present, and these ions will be solvated by water. Then, if we assume arbitrarily that each ion is solvated on the average by 2.3 water molecules, the total quantity of *bound* water will be $(24)(2.3)$ or 55.2 moles, leaving only 0.8 mole of water not bound to the ions. It must be emphasized that the water coordinated to these ions belongs to the solute species and is definitely not part of the solvent. Now, if 56 moles of water occupies one liter, 0.8 mole would occupy a volume of about 14 ml. Imagine that one dissolved 12 moles of hydrochloric acid in only 14 ml of water—the acid concentration would be 857 $F$, which is not so far from 695 $F$. Despite the fact that this model is terribly oversimplified, very high activity coefficients in concentrated solutions do arise because most of the solvent is actually bound to the solute species. It is almost impossible to specify the number of water molecules associated with a particular ion, so that the quantitative aspects of the hydration effect are not easily formulated. Furthermore, it would be unwise to attribute enormous activity coefficients only to ion-solvent interactions, just as it is invalid to expect the ion-size parameter to accommodate all of the solution phenomena occurring at moderate ionic strengths.

## Activity Coefficients of Neutral Molecules

If one considers the behavior of neutral molecules or nonelectrolytes in the light of the Debye-Hückel theory, such substances, being uncharged, should have activity coefficients of unity. In general, the activity of a neutral solute in pure water is within 1 per cent of its analytical concentration.

However, in the presence of electrolytes the activity coefficient of an uncharged solute varies in accordance with the equation

$$\log f = k I$$

where $f$ is the activity coefficient, $I$ is the ionic strength, and $k$, the "salting coefficient," may be either positive or negative. For small molecules, such as nitrogen, oxygen, and carbon dioxide, $k$ is usually positive with a value near 0.1, but for sugars and proteins $k$ may be larger. If $k$ is positive, $\log f$ is always greater than zero provided that an electrolyte is present, so the activity coefficient $f$ will be greater than unity and the activity will exceed the analytical concentration of the neutral solute. Solutes which have dielectric constants less than water behave in this manner.

Hydrogen cyanide is an example of a substance which exhibits reverse behavior, for the dielectric constant of hydrocyanic acid (111 at 25°C) is much greater than that of water. Thus, the value of $k$ is negative for this solute, and its activity is always less than its analytical concentration.

Finally, it may be noted that, if $k$ is positive, the solubility of a neutral solute in an aqueous medium decreases with an increase in the ionic strength. Conversely, solutes having negative $k$ values become more soluble at higher ionic strengths.

## CHEMICAL EQUILIBRIUM

Before we place chemical equilibrium on a firm mathematical foundation, it is worthwhile to examine qualitatively some of the characteristics of chemical systems in order to see what an equilibrium state is and how to recognize that a system has reached a state of equilibrium.

Let us consider the reversible reaction between iron(III) and iodide ion,

$$2 Fe^{3+} + 3 I^- \rightleftharpoons 2 Fe^{2+} + I_3^-$$

to form iron(II) and triiodide. Suppose that a reaction mixture containing 0.200 $M$ iron(III) and 0.300 $M$ iodide ion is prepared and that we wish to determine *when* and *if* chemical equilibrium is established. As the reaction proceeds in the forward direction, the concentrations of iron(III) and iodide will diminish and, simultaneously, the concentrations of iron(II) and triiodide will build up. If we follow the increase in the triiodide concentration as a function of time, by observing, for example, the intensity of the characteristic yellow-orange color of this species, the concentration of triiodide reaches a maximum value of 0.0866 $M$ and any further net formation of triiodide ceases within a fraction of a second (curve A of Figure 3–5). Therefore, on the basis of the observation that no detectable change in the triiodide concentration occurs after this time, one is tempted to conclude that equilibrium has been established.

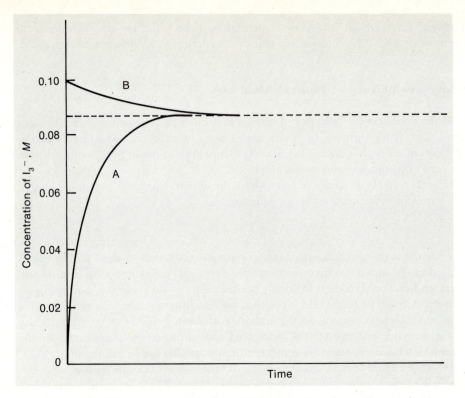

*Figure 3–5.* Hypothetical rate of change of triiodide concentration as the reaction

$$2 \, Fe^{3+} + 3 \, I^- \rightleftharpoons 2 \, Fe^{2+} + I_3^-$$

approaches a state of equilibrium. Curve A: approach to equilibrium in the *forward* direction. Curve B: approach to equilibrium in the *backward* direction. The horizontal, dashed line indicates the equilibrium triiodide concentration of 0.0866 *M*. (See text for data on initial concentrations of reagents and for additional discussion.)

However, such a conclusion is invalid unless the same final concentration of triiodide ion is approached when reaction occurs in the reverse direction, that is, when only iron(II) and triiodide are initially present. Accordingly, we should prepare another solution, one 0.200 *M* in iron(II) and 0.100 *M* in triiodide, being certain that the *total* amounts of iron and iodine are identical to those in the preceding experiment, and study the changes with time in the concentrations of the four species. If, as before, the intensity of the yellow-orange color of triiodide is monitored to indicate the progress of the reaction, the triiodide concentration is found to decrease rapidly, although exponentially, to a value of 0.0866 *M*, as shown in curve B of Figure 3–5. Since there is no decrease in the triiodide concentration below this value and since the final state of the system appears to be independent of the direction of reaction, we can now assert that chemical equilibrium has been definitely attained. To point out the fact that the rate of approach to equilibrium from the two opposite directions is usually different, curves A and B in Figure 3–5 have been purposely drawn so that equilibrium is reached somewhat more rapidly from the forward direction than from the backward direction. However, depending upon the initial concentrations of the various species, equilibrium can be attained more quickly from either direction.

It is important to stress that the absence of a detectable change in the concentrations of reactants and products does not constitute by itself a criterion for the state of chemical equilibrium. Some familiar reactions proceed toward equilibrium at such infinitesimally slow rates that no concentration variations occur even after long periods of time. For example, a two-to-one mixture of hydrogen and oxygen gases should combine to form water,

$$2 H_2 + O_2 \rightleftharpoons 2 H_2O$$

the position of equilibrium lying so far to the right that virtually no unreacted hydrogen or oxygen should remain. Nevertheless, in the absence of a suitable catalyst, such as spongy platinum, the reaction does not proceed at any measurable rate. Yet the mixture of hydrogen and oxygen must not be regarded as the true equilibrium state of the system. One should always verify the existence of true chemical equilibrium by showing that the same equilibrium state is reached from two directions.

Sometimes, when the forward and backward reactions are so slow that it is inconvenient to wait for true equilibrium to be established, one may attempt to extrapolate a set of experimental results to obtain information about the equilibrium state. Notice that curves A and B in Figure 3–5 both approach asymptotically a triiodide concentration of 0.0866 $M$. If one makes measurements of the triiodide concentration as a function of time for the forward and backward reactions and then plots the experimental data as in Figure 3–5, it may be feasible to extrapolate each curve to the equilibrium triiodide concentration, even though true equilibrium has not yet been reached. In order for the extrapolation procedure to be meaningful, however, one must obtain data reasonably close to the equilibrium state so that the asymptotic limits of the experimental curves can be accurately identified.

Experimental techniques employed to follow the progress of a chemical reaction toward equilibrium are extremely varied, but they do rely on the specific nature of the reactants and products as well as on the environment in which the reaction occurs. Above all, it is essential to select for observation or measurement a property of the chemical system which has a value at equilibrium distinctly different from that for any other state of the system. For the iron(III)-iodide equilibrium just discussed, the spectrophotometric measurement of the concentration of yellow-orange triiodide ion is especially suitable. In addition, iron(III) and iron(II) both exhibit characteristic absorption bands in the visible and ultraviolet regions of the electromagnetic spectrum which might serve as a means of following the iron(III)-iodide reaction.

We shall now turn our attention from an essentially qualitative picture to a more formal mathematical representation of chemical equilibrium. Two approaches may be employed to derive a fundamental relation pertaining to the state of equilibrium—the first of these methods involves a consideration of reaction kinetics, and the second approach is based upon the concepts of thermodynamics.

## Kinetic Approach to Equilibrium

Fundamentally, the rate of a chemical reaction depends on only two factors—the total number of collisions per unit time between the reacting species and the fraction of these collisions which is fruitful in accomplishing reaction. In turn, the fraction of collisions which is successful in promoting reaction is dependent upon the temperature of the system as well as the presence of catalysts. When such conditions are held constant, a reaction rate is simply a function of the total number of collisions per unit time between the reactants. If the number of collisions doubles, the rate of the reaction is likewise doubled, whereas, if the number of collisions is decreased by a factor of ten, the reaction rate is decreased ten-fold.

Consider the simple bimolecular reaction

$$A + B \rightarrow C + D$$

whose mechanism, we shall assume, involves the collision of a single particle of species A with a single particle of B to form one particle each of substances C and D. If the concentration of A is suddenly doubled, that is, if the number of A particles in a fixed volume is doubled, the number of collisions per unit time between A and B will momentarily double and the rate of reaction will be doubled accordingly. Similarly, if the concentration of A is maintained constant and the concentration of B is doubled, the reaction rate will be doubled. Now, if the concentrations of both A and B are doubled, the number of collisions per unit time between A and B particles is quadrupled, so that the rate of reaction is quadrupled. These considerations suggest that the rate or the velocity of the forward reaction, $v_f$, may be expressed by the equation

$$v_f = k_f[\text{A}][\text{B}]$$

where $k_f$ is the rate constant and the molar concentrations of A and B are represented by [A] and [B], respectively.

As mentioned earlier, demonstration that equilibrium has been attained requires approaching the equilibrium state from both directions. Hence, reactions which reach equilibrium must possess some degree of reversibility. Consequently, it is better to formulate the previous reaction as

$$\text{A} + \text{B} \rightleftharpoons \text{C} + \text{D}$$

to indicate that the backward reaction between one particle of C and one particle of D has definite significance. If the assumption is made that the mechanism of this process entails collision of one particle of C with one particle of D, the rate of the backward reaction, $v_b$, can be expressed by the equation

$$v_b = k_b[\text{C}][\text{D}]$$

When A and B are first mixed with no C or D present, the reaction will proceed toward the right at a finite rate. As this forward reaction continues, the concentrations of A and B diminish, and the rate of the forward reaction, $v_f$, decreases. Initially, the rate of the backward reaction, $v_b$, is zero since neither C nor D is present. However, as the forward reaction proceeds, the concentrations of C and D build up, so that the rate of the backward reaction begins to increase. Thus, $v_f$ starts at its maximum value and continues to diminish, whereas, during the same time, $v_b$ is initially zero but exhibits a steady increase. Sooner or later, the two rates become identical and the system is then said to be in a **state of equilibrium**. Thereafter, the individual concentrations of A, B, C, and D remain constant. Substances A and B are, from that time on, being used up at the same rate at which they are produced, while C and D are likewise consumed at the same rate at which they are being formed.

It must be emphasized that the forward and backward reactions do not stop when the equilibrium condition is achieved, but rather continue at equal rates. This means that the system is in a *dynamic* state of equilibrium, not a static one. To illustrate how the dynamic nature of equilibrium can be demonstrated, let us recall the iron(III)-iodide reaction.

Suppose that we know, in advance, just one of the infinite number of sets of iron(III), iodide, iron(II), and triiodide concentrations corresponding to the state of chemical equilibrium, and suppose that we mix solutions of the four species together in such a way that equilibrium prevails immediately at the start of the experiment. If we used solutions of radioactive iodide ion and nonradioactive triiodide ion for the preparation of the equilibrium mixture, the radioiodine would eventually distribute itself randomly between triiodide and iodide ions, indicating that the forward and backward reactions

$$2\,Fe^{3+} + 3\,I^- \rightleftharpoons 2\,Fe^{2+} + I_3^-$$

occur at equal rates, even though a state of chemical equilibrium was preserved throughout the experiment. On the other hand, if chemical equilibrium were a stationary state, all of the radioiodine would remain in the form of iodide ion.

Returning now to the reaction between A and B to form C and D, we may represent the equilibrium state mathematically by equating the rates of the forward and backward reactions

$$k_f[A][B] = k_b[C][D]$$

and by combining the individual rate constants $k_f$ and $k_b$ into a single term through rearrangement of the preceding relation

$$\frac{[C][D]}{[A][B]} = \frac{k_f}{k_b} = K$$

where $K$ is called the **equilibrium constant** for the reaction of interest.

We have assumed that in this process only one particle of each reactant combines in the forward reaction, and one particle of each product in the backward reaction. Let us now extend this treatment to the process

$$2\,A + B \rightleftharpoons C + 2\,D$$

which could conceivably occur according to the following two mechanistic steps:

$$A + B \rightleftharpoons Z + D$$
$$A + Z \rightleftharpoons C + D$$

For the first step,

$$v_f = k_f[A][B] \quad \text{and} \quad v_b = k_b[Z][D] \quad \text{and} \quad K_1 = \frac{[Z][D]}{[A][B]}$$

where $K_1$ is the equilibrium constant for this first step. For the second step,

$$v_f = k_f[A][Z] \quad \text{and} \quad v_b = k_b[C][D] \quad \text{and} \quad K_2 = \frac{[C][D]}{[A][Z]}$$

where $K_2$ is the equilibrium constant for the second step. Now, let us multiply $K_1$ by $K_2$:

$$K_1 \times K_2 = \frac{[Z][D]}{[A][B]} \times \frac{[C][D]}{[A][Z]}$$

$$K = \frac{[C][D]^2}{[A]^2[B]}$$

Note that the latter relation is exactly the same as what we would have derived for the overall process if we had assumed that the mechanism of the forward reaction consisted of simultaneous collision of two particles of A with one particle of B and that the backward reaction involved simultaneous collision of two particles of D with one particle of C. Thus, we are led to the **law of chemical equilibrium**, which states that for the general reaction

$$aA + bB \rightleftharpoons cC + dD$$

in which $a$, $b$, $c$, and $d$ signify the stoichiometric numbers of particles of A, B, C, and D, respectively, involved in the completely balanced reaction, the equilibrium constant may be expressed by the relation

$$K = \frac{[C]^c[D]^d}{[A]^a[B]^b}$$

regardless of the detailed mechanism of the reaction. Even though the true reaction mechanism may differ markedly from the stoichiometry of the overall process, the equilibrium properties of a system do not depend upon the pathway by which equilibrium is reached.

In this discussion we have used the concentrations of the various species both in rate equations and in equilibrium-constant expressions. For any given chemical system, the several concentrations in a rate expression are essentially independent of each other, and each one may assume any value. However, the concentrations of A, B, C, and D in the equilibrium-constant relation are the values which exist only at equilibrium. There is an infinite number of sets of concentrations which correspond to the equilibrium state, but no one concentration can vary independently of the concentrations of the other substances without destroying the equilibrium condition.

Attention should be directed to the convention—originally chosen quite arbitrarily, but now a matter of international agreement—that the numerical value of an equilibrium constant refers to the ratio in which the concentrations of the *products* of a reaction appear in the *numerator* and the concentrations of the *reactants* in the *denominator*. With this convention in mind, one can frequently tell by glancing at the numerical magnitude of the equilibrium constant whether a reaction has a tendency to proceed far toward the right, or only slightly so, before reaching equilibrium.

For a reaction having a large $K$, equilibrium is attained only after the reaction has proceeded far to the right. A two-to-one mixture of hydrogen and oxygen gases, as we have previously mentioned, has a strong tendency to form water vapor

$$2 \, H_2 + O_2 \rightleftharpoons 2 \, H_2O$$

although the reaction does not occur readily without a catalyst. At 25°C the equilibrium expression for the process is

$$\frac{[H_2O(g)]^2}{[H_2(g)]^2[O_2(g)]} = 1.7 \times 10^{80}$$

so that the numerator of the relation is exceedingly large compared to the denominator, indicating that very little hydrogen and oxygen remain unreacted at equilibrium.

Some familiar chemical processes have $K$ values much smaller than unity, so that equilibrium is reached before the reactions have gone very far toward the right. For example, the dissolution of silver chloride in water may be represented by the reaction

$$AgCl(s) \rightleftharpoons Ag^+ + Cl^-$$

for which the equilibrium expression* at 25°C is

$$[Ag^+][Cl^-] = 1.78 \times 10^{-10}$$

* The concentration (activity) of solid silver chloride does not appear in this equilibrium expression because, as described in Chapter 7, the concentration of a pure solid phase is *defined* as unity.

Such an equilibrium constant indicates that the concentrations of silver ion and chloride ion (equal to each other in a saturated aqueous solution of silver chloride) are quite small, which simply confirms our knowledge of the low solubility of silver chloride.

Some words of caution should be interjected at this point. Although the magnitude of an equilibrium constant is a criterion by which to judge whether a chemical reaction has a pronounced or only a slight tendency to proceed from left to right as written, the initial concentrations of the reactants and products provide a second and equally important criterion for deciding the direction in which a reaction will actually go in order to attain a state of equilibrium. Thus, a chemical reaction may have a *large* equilibrium constant, yet it may proceed from *right* to *left* if the initial concentrations of species on the right-hand side of the chemical equation happen to be much greater than the concentrations of substances on the left-hand side of the reaction. A system we have already discussed, namely the iron(III)-iodide reaction,

$$2 \, Fe^{3+} + 3 \, I^- \rightleftharpoons 2 \, Fe^{2+} + I_3^-$$

illustrates these considerations. Because the equilibrium constant for this reaction is moderately large ($5.55 \times 10^4$), the position of equilibrium lies toward the right, so the formation of iron(II) and triiodide ion is favored. Nevertheless, curve B of Figure 3–5 clearly confirms that, if the original concentrations of iron(II) and triiodide are very large compared to the initial concentrations of iron(III) and iodide, the reaction actually proceeds from *right* to *left*.

In conclusion of this section, it should be reëmphasized that the concept of chemical equilibrium tells us absolutely nothing about either the rate at which the equilibrium state has been reached or, once that state is reached, the direction from which it was approached. Furthermore, the characteristics of the state of equilibrium do not depend in any way on the detailed mechanism of the process whereby equilibrium is attained. No matter how complicated the mechanisms of the reactions leading to equilibrium are, the equilibrium-constant expression can always be written from inspection of the balanced chemical equation.

## Thermodynamic Concept of Equilibrium

One of the fundamental laws of nature is that a physical or chemical system always tends to undergo a spontaneous and irreversible change from some initial, nonequilibrium state to a final, equilibrium state. Once the system reaches a state of equilibrium, no additional change will occur unless some new stress is placed upon the system.

In describing equilibrium in thermodynamic terms, one must define a property of a chemical system which can be related to the concentrations of the various species involved in an equilibrium state—the thermodynamic parameter most useful in this respect is $G$, the **Gibbs free energy**. A chemical system which is initially in some nonequilibrium state has a tendency to change spontaneously until the free energy for the system becomes minimal and the system itself reaches equilibrium. It is strictly correct to speak only of a *tendency* for change to occur, since thermodynamics offers no information at all about the rate of a particular process. We can measure the driving force which causes a system to pass from an initial state to a certain final state by the free-energy change for that process. In turn, the free-energy change is defined as the difference between the free energies of the final and initial states, and is symbolized by $\Delta G$ when the system undergoes a finite change of state, that is,

$$\Delta G = G_{\text{final}} - G_{\text{initial}}$$

Our objectives are to evaluate the free-energy change for a particular chemical reaction and to relate the free-energy change to the state of equilibrium for that process.

*Standard states and free energies of formation.* In general, the free energy of a chemical species is dependent upon temperature and pressure, as well as upon the nature and quantity of the substance itself. Therefore, standard-state conditions have been established for the various kinds of chemical species, so that values for the absolute free energy of an element or the free energy of formation of a compound can be precisely specified and tabulated.

Our choice of **standard state** for a substance depends on whether it is a gaseous, liquid, solid, or solute species; however, all standard states pertain to a temperature of 298°K or 25°C. A gaseous substance, if ideal, exists in its standard state when present at a pressure of exactly one atmosphere. For a liquid or solid, the standard state is taken to be the *pure* liquid or *pure* solid, respectively, at a pressure of one atmosphere. Dissolved or solute species, including electrolytes, nonelectrolytes, and individual ions, are in their standard states if their activities are unity, again at a pressure of one atmosphere.

By convention, the most stable form of any pure element in its standard state is assigned a free energy of zero. For a chemical compound, one must measure or determine the free-energy *change* for the reaction by which that compound is formed from its elements, when all reactants and products are in their standard states. This free-energy change is termed the **standard free energy of formation**, $\Delta G_f^0$, for the substance of interest. In Table 3–3 are listed standard free energies of formation

**Table 3–3. Standard Free Energies of Formation ($\Delta G_f^0$) for Various Substances at 298°K (Kilocalories/Mole)**

| Gases | | Solids | |
|---|---|---|---|
| $H_2O$ | −54.64 | AgI | −15.81 |
| $H_2O_2$ | −24.7 | $BaSO_4$ | −323.4 |
| $O_3$ | 39.06 | CaO | −144.4 |
| HBr | −12.72 | $CaCO_3$ | −269.8 |
| HI | 0.31 | $Ca(OH)_2$ | −214.3 |
| $SO_2$ | −71.79 | | |
| $SO_3$ | −88.52 | **Aqueous Ions** | |
| NO | 20.72 | | |
| $NO_2$ | 12.39 | $H^+$ | 0.0 |
| $NH_3$ | −3.97 | $Ag^+$ | 18.43 |
| CO | −32.81 | $Ba^{2+}$ | −134.0 |
| $CO_2$ | −94.26 | $Ca^{2+}$ | −132.18 |
| $C_2H_4$ | 16.28 | $OH^-$ | −37.59 |
| $C_2H_6$ | −7.86 | $I^-$ | −12.35 |
| | | $SO_4^{2-}$ | −177.34 |

for some representative species. Standard free energies of formation may be determined experimentally from the direct measurement of either equilibrium constants or the electromotive force of galvanic cells, and are expressed in kilocalories per mole of the substance formed. It should be recognized that the free-energy change attending the formation of $n$ moles of a substance under standard-state conditions corresponds to a free-energy change of $n(\Delta G_f^0)$. For example, the standard free energy of formation of water vapor is listed in Table 3–3 as −54.64 kilocalories per mole,

$$H_2(g) + \tfrac{1}{2} O_2(g) \rightleftharpoons H_2O(g); \qquad \Delta G_f^0 = -54.64 \text{ kcal}$$

and, since this process is spontaneous, although kinetically slow, it can be stated that *if $\Delta G_f^0$ is negative the reactants will be spontaneously converted into products*, all species being in their standard states. On the other hand, the standard free energy of formation of nitric oxide has a value of $+20.72$ kilocalories per mole,

$$\tfrac{1}{2} N_2(g) + \tfrac{1}{2} O_2(g) \rightleftharpoons NO(g); \qquad \Delta G_f^0 = +20.72 \text{ kcal}$$

*the positive sign for $\Delta G_f^0$ revealing that the conversion of nitrogen and oxygen in their standard states to nitric oxide in its standard state is nonspontaneous.* Some nitric oxide will be formed, but the reaction between nitrogen and oxygen, each at one atmosphere pressure, will not proceed far enough to yield this substance in its standard state at one atmosphere pressure.

***Free-energy change for a reaction.*** From information such as that compiled in Table 3–3, it is possible to compute the **standard free-energy change**, $\Delta G^0$, for each of a large number of chemical reactions. First, we should recall the simple equation presented in the introduction to this section, namely,

$$\Delta G = G_{\text{final}} - G_{\text{initial}}$$

Now, if we allow reactants in their standard states (the initial state of the system) to go to products in their standard states (the final state of the system), the preceding general expression can be rewritten as

$$\Delta G^0 = \sum \Delta G_f^0 \, (\text{products}) - \sum \Delta G_f^0 \, (\text{reactants})$$

where the summation signs permit us to consider as many reactants and products as necessary. According to this defining equation, the standard free-energy change for a reaction is the sum of the standard free energies of formation of the products minus the sum of the standard free energies of formation of the reactants. Thus, for the general process

$$a A + b B \rightleftharpoons c C + d D$$

we may write

$$\Delta G^0 = c[\Delta G_f^0(C)] + d[\Delta G_f^0(D)] - a[\Delta G_f^0(A)] - b[\Delta G_f^0(B)] \qquad (3\text{–}1)$$

However, in most chemical systems, we encounter reactant and product species under conditions of concentration different from those specified for their standard states. For example, gaseous compounds may be present at pressures other than one atmosphere, whereas the activities of dissolved substances are very likely not equal to unity.

To formulate an equation for the free-energy change associated with the previously mentioned general reaction, we must employ $\Delta G$ and $\Delta G_f$ rather than $\Delta G^0$ and $\Delta G_f^0$, since the latter terms pertain only to processes occurring under standard-state conditions. Therefore,

$$\Delta G = c[\Delta G_f(C)] + d[\Delta G_f(D)] - a[\Delta G_f(A)] - b[\Delta G_f(B)] \qquad (3\text{–}2)$$

Subtraction of equation (3–1) from equation (3–2), followed by rearrangement of terms, yields

$$\Delta G - \Delta G^0 = c[\Delta G_f(C) - \Delta G_f^0(C)] + d[\Delta G_f(D) - \Delta G_f^0(D)]$$

$$- a[\Delta G_f(A) - \Delta G_f^0(A)] - b[\Delta G_f(B) - \Delta G_f^0(B)] \qquad (3\text{–}3)$$

From the second law of thermodynamics, we can obtain an expression to relate $\Delta G_f$, the free energy of formation of a substance at *any* activity, to the standard free energy of formation, $\Delta G_f^0$, which is

$$\Delta G_f = \Delta G_f^0 + RT \ln a$$

or

$$\Delta G_f - \Delta G_f^0 = RT \ln a$$

where $R$, the universal gas constant, is 0.001987 kilocalorie per mole-degree, $T$ is the absolute temperature ($298°K$), and $\ln a$ is the natural logarithm of the activity of the species of interest in moles per liter. Applying the latter expression to each of the four substances A, B, C, and D, and substituting the results into equation (3–3), one obtains

$$\Delta G - \Delta G^0 = cRT \ln a_C + dRT \ln a_D - aRT \ln a_A - bRT \ln a_B$$

Further rearrangement and combination of logarithmic terms may be performed as follows:

$$\Delta G - \Delta G^0 = RT \ln (a_C)^c + RT \ln (a_D)^d - RT \ln (a_A)^a - RT \ln (a_B)^b$$

$$\Delta G = \Delta G^0 + RT \ln \frac{(a_C)^c (a_D)^d}{(a_A)^a (a_B)^b} \qquad (3\text{–}4)$$

Thus, the free-energy change for a chemical reaction is dependent upon two factors—first, the standard free-energy change for the reaction which is related to the standard free energies of formation of the substances involved in the reaction and, second, the activities of the reactants and products.

**Free energy and the equilibrium state.** We have already noted that free-energy change is a measure of the driving force which causes a chemical reaction to tend to proceed spontaneously from a nonequilibrium state to an equilibrium state. When the system is at equilibrium, the free-energy change, $\Delta G$, is zero, and equation (3-4) becomes

$$0 = \Delta G^0 + RT \ln \frac{(a_C)^c (a_D)^d}{(a_A)^a (a_B)^b}$$

or

$$\Delta G^0 = -RT \ln \frac{(a_C)^c (a_D)^d}{(a_A)^a (a_B)^b} \qquad (3\text{–}5)$$

Recognizing that the ratio of factors in the logarithmic term of equation (3–5) contains the *equilibrium* activities of all reactants and products involved in the preceding general chemical reaction, we may identify this ratio as a **thermodynamic equilibrium constant**, $\mathscr{K}$. Thus,

$$\mathscr{K} = \frac{(a_C)^c (a_D)^d}{(a_A)^a (a_B)^b}$$

and

$$\Delta G^0 = -RT \ln \mathscr{K}$$

Since the activity of a chemical species is defined as its concentration multiplied by an activity coefficient ($a_i = f_i C_i$), the relation for the thermodynamic equilibrium constant can be written as

$$\mathscr{K} = \frac{(a_C)^c (a_D)^d}{(a_A)^a (a_B)^b} = \frac{(f_C[C])^c (f_D[D])^d}{(f_A[A])^a (f_B[B])^b} = \frac{f_C^c f_D^d}{f_A^a f_B^b} \cdot \frac{[C]^c [D]^d}{[A]^a [B]^b}$$

or

$$\mathscr{K} = \frac{f_C^c f_D^d}{f_A^a f_B^b} \cdot K$$

where $K$ is the equilibrium constant, based on concentration units, obtained from our previous consideration of kinetics. Because $\mathscr{K}$ is a thermodynamic constant, it does not depend on such factors as ion-ion attractions or ion-solvent interactions, so that the *product* of the terms on the right-hand side of the preceding equation is likewise constant. Of course, the individual activity coefficients for the several reactants and products do change as the composition and ionic strength of a chemical system vary, which means that $K$ is not truly constant.

When rigorously correct equilibrium calculations are to be performed, use of the thermodynamic equilibrium constant ($\mathscr{K}$) and the activities of the reactants and products is mandatory. Unfortunately, as we have discovered earlier, activity coefficients (and, consequently, activities) are difficult to evaluate accurately, particularly for aqueous solutions of analytical interest, which usually have relatively high ionic strengths and often contain more than one electrolyte, such that the solute-solute and solute-solvent interactions cannot be assessed quantitatively. Due to these uncertainties, we shall follow throughout most of this textbook the common practice of ignoring activity coefficients as well as the distinction between $\mathscr{K}$ and $K$. Instead, we will be satisfied to assume the constancy of $K$ and to utilize equilibrium-constant expressions involving concentrations rather than activities. Such an approach is quite adequate for many approximate calculations, including those made to predict the feasibility of analytical procedures.

It is appropriate to illustrate the concepts which have been presented in the last two subsections of this chapter.

*Example 3–1.* Calculate the equilibrium constant at 25°C for the reaction

$$Ca(OH)_2(s) \rightleftharpoons Ca^{2+} + 2\,OH^-$$

from information given in Table 3–3.

First, let us formulate the expression for the standard free-energy change of this process:

$$\Delta G^0 = \Delta G_f^0(Ca^{2+}) + 2\,\Delta G_f^0(OH^-) - \Delta G_f^0(Ca(OH)_2(s))$$

Since each of the required standard free energies of formation is available in Table 3–3, we can write

$$\Delta G^0 = -132.18 + 2(-37.59) - (-214.3) = +6.9 \text{ kcal}$$

Next, the equilibrium constant can be determined by substitution of the latter result into the (approximate) relation

$$\Delta G^0 = -RT \ln K$$

along with the values for $R$ and $T$ of 0.001987 kilocalorie per mole-degree and 298°K, respectively. Thus,

$$6.9 = -(0.001987)(298) \ln K$$

$$\ln K = -\frac{6.9}{(0.001987)(298)} = -11.7$$

If we convert natural logarithms to base-ten logarithms, taking note of the fact that $\ln K = 2.303 \log K$, the preceding equation becomes

$$\log K = -\frac{11.7}{2.303} = -5.1 = 0.9 - 6.0$$

and

$$K = 8 \times 10^{-6}$$

Further consideration of this result—particularly in the light of subsequent discussion in Chapter 7—reveals that we have just calculated the solubility product for calcium hydroxide from free-energy data.

## Factors Which Influence Chemical Equilibrium

In this section are discussed the effects of several factors, which may be experimentally adjusted and controlled, upon systems in a state of chemical equilibrium. Qualitative predictions of the influence of specific variables can be based on the **principle of Le Châtelier**, which states that, *when a stress is applied to a system at equilibrium, the position of equilibrium tends to shift in such a direction as to diminish or relieve that stress.* On the other hand, quantitative changes in the equilibrium properties of a system due, for example, to pressure or temperature variations may be evaluated from consideration of the equilibrium-constant expression or other thermodynamic relationships.

*Effect of concentration.* An increase or decrease in the concentration of a reactant or product in a system which is initially in an equilibrium state will cause that system to shift in one direction or the other to establish a new position of equilibrium. Such a change results in different concentrations of all species when equilibrium is restored, but the numerical value of the equilibrium constant remains unaltered. Le Châtelier's principle provides a qualitative guide to the direction in which an equilibrium is shifted by a change in the concentration of one of the chemical species. Since the stress in this case is a change in concentration, the position of equilibrium will shift so as to minimize this change. Thus, if the concentration of one of the components of the equilibrium mixture is diminished, the system will react to restore in part the concentration of the component which was withdrawn. Consider the system involving the overall dissociation of the silver-diammine complex in an aqueous medium:

$$Ag(NH_3)_2{}^+ \rightleftharpoons Ag^+ + 2\,NH_3$$

Removal from the solution of some of the silver ion by addition of chloride ion will cause further dissociation of the $Ag(NH_3)_2{}^+$ complex in order to reëstablish a condition of equilibrium. Similarly, when an additional amount of one of the components is introduced to a system at equilibrium, the system shifts toward a new position of equilibrium by consuming part of the added material. Thus, introduction of some silver ion, in the form of silver nitrate, would promote the formation of more silver-diammine complex, the reaction proceeding from right to left as written.

*Effect of catalysts.* By influencing the rate of a reaction, a catalyst hastens the approach of a chemical system to a state of equilibrium. However, catalysis always affects the rates of forward and backward reactions to the same extent, and never under any circumstance changes the value of the equilibrium constant for a specific reaction. In general, the catalyst enters into a chemical or physical process at some step in the reaction mechanism and is regenerated in a subsequent step. Although the physical properties of a catalyst may be modified, it undergoes no permanent chemical change. Thus, the catalyst does not appear as a reactant or product in the overall reaction.

Earlier, we mentioned that the rate of reaction between hydrogen and oxygen gases to form water, as well as the time required for equilibrium to be established, is very markedly influenced by the presence of a catalyst. Another example of catalysis is the isotope-exchange reaction occurring in the vapor phase between water and deuterium molecules,

$$H_2O + D_2 \rightleftharpoons HD + HDO$$

A mixture of $H_2O$ and $D_2$ gases in a clean vessel reacts only slowly, typically requiring hours to attain a state of equilibrium. In the presence of finely divided platinum oxide, however, the same initial mixture reaches equilibrium in a few minutes. Nevertheless, both equilibrium mixtures contain exactly the same concentrations of each of the four species, whether the catalyst is present or not.

*Effect of temperature.* Because chemical reactions may absorb or liberate heat as they proceed, increases or decreases in temperature—which, respectively, supply or withdraw thermal energy from a system—may produce substantial changes in the numerical values of equilibrium constants. Usually, the quantity of heat liberated or absorbed during occurrence of a chemical reaction is referred to as the **enthalpy change** and is given the symbol $\Delta H$. In a manner similar to the treatment of free energy, each element in its standard state is assigned an enthalpy of zero, and a **standard enthalpy of formation**, $\Delta H_f^0$, is associated with the production of one mole of a compound in its standard state from the appropriate elements in their standard states. Furthermore, the **standard enthalpy change**, $\Delta H^0$, for any process is the sum of the standard enthalpies of formation for the products minus the sum of the standard enthalpies of formation for the reactants. Processes for which $\Delta H$ (or $\Delta H^0$) is negative are heat-liberating or **exothermic**, and the products of such reactions tend to be stable relative to the reactants. Conversely, **endothermic** reactions, for which the products are apt to be less stable than the reactants, absorb thermal energy and have positive $\Delta H$ (or $\Delta H^0$) values.

Le Châtelier's principle predicts that, if the temperature of a system is increased, the position of equilibrium will shift in such a direction as to consume at least part of this extra thermal energy. Accordingly, when the temperature is raised, an endothermic reaction will be favored over an exothermic one. If the process of interest is exothermic when it occurs in the forward direction, the backward reaction will be endothermic and, thus, favored by a temperature rise; therefore, the equilibrium constant becomes smaller as the temperature increases, but larger as the temperature declines. On the other hand, if the forward reaction is endothermic, the backward process is exothermic, so that the equilibrium constant increases with a rise in temperature and decreases when the temperature drops.

Thermodynamics allows us to formulate an explicit equation to describe quantitatively how variations in temperature affect the magnitude of the equilibrium constant for a reaction, although it is sufficient here just to present the result of the derivation. If $K_1$ and $K_2$ are equilibrium constants at absolute temperatures $T_1$ and $T_2$, respectively, we discover that

$$\ln \frac{K_2}{K_1} = -\frac{\Delta H^0}{R}\left(\frac{1}{T_2} - \frac{1}{T_1}\right)$$

where $\Delta H^0$ is the standard enthalpy change for the reaction of interest and $R$ is the universal gas constant. If the equilibrium constant for a reaction at one temperature is known, along with the value of $\Delta H^0$, one can determine the equilibrium constant at another temperature. It is necessary to assume that $\Delta H^0$ remains constant within the temperature range of concern; however, for most purposes no serious error is incurred, because $\Delta H^0$ changes only slightly with temperature. Let us apply these considerations to a specific chemical system.

*Example 3–2.* Given that the standard enthalpy change ($\Delta H^0$) for the reaction

$$2\,HI(g) \rightleftharpoons H_2(g) + I_2(g)$$

is $+3.03$ kilocalories and that the equilibrium constant is 0.0162 at 667°K, what is the value of the equilibrium constant at 760°K?

If $T_1 = 667°K$, $T_2 = 760°K$, and $K_1 = 0.0162$, and if these values are substituted into the equation

$$\ln \frac{K_2}{K_1} = -\frac{\Delta H^0}{R}\left(\frac{1}{T_2} - \frac{1}{T_1}\right)$$

we have

$$\ln \frac{K_2}{(0.0162)} = -\frac{(3.03)}{(0.001987)}\left(\frac{1}{760} - \frac{1}{667}\right) = 0.276$$

If we convert to base-ten logarithms, $K_2$ (the equilibrium constant at 760°K) is obtained as follows:

$$\ln \frac{K_2}{(0.0162)} = 2.303 \log \frac{K_2}{(0.0162)} = 0.276$$

$$\log \frac{K_2}{(0.0162)} = \frac{0.276}{2.303} = 0.120$$

$$\frac{K_2}{(0.0162)} = 1.32$$

$$K_2 = 0.0214$$

Let us now examine the effect of temperature on the free-energy change for a chemical process. From the second law of thermodynamics, we can obtain the **Gibbs-Helmholtz equation**,

$$\Delta G = \Delta H - T\Delta S$$

which relates the free-energy change ($\Delta G$) for a reaction to the enthalpy change ($\Delta H$) and to the product of the absolute temperature ($T$) and $\Delta S$, the change in a quantity called the **entropy** of the system. A definite amount of entropy, like free energy and enthalpy, is associated with each chemical substance. More specifically, entropy is a measure of the *degree of disorder* in a substance. For example, the absolute entropies of helium, nitrogen, and *n*-pentane ($C_5H_{12}$) at 25°C are 30.1, 45.7, and 83.4 calories per mole-degree (entropy units), respectively. Helium, which has the lowest entropy of the three gases, is monatomic and is capable only of translational motion. However, as a diatomic molecule, nitrogen has a larger entropy because both translational and rotational movement can contribute to its degree of disorder. Finally, *n*-pentane, which consists of a chain of five carbon atoms, may exhibit large vibrational and rotational as well as translational disorder. Such factors as the weakness of the attractive forces between individual atoms or molecules and the high mobility of the molecules or atoms tend to increase the entropy of a substance. Entropy is greater for a given species in the gaseous state, where the particles are highly disordered with respect to each other, than for the same substance in the form of a crystalline solid, which consists of a well-organized structure.

According to the Gibbs-Helmholtz equation, the entropy change is a factor which contributes to the driving force of a chemical reaction. This fact may be interpreted as signifying that there is a natural tendency for a system to become disorganized. Such a tendency to attain maximal entropy is actually the major driving force for some chemical processes. As mentioned earlier, numerical values of $\Delta H$ for various reactions are essentially independent of temperature. Likewise, values of $\Delta S$ for numerous processes are largely insensitive to temperature. Therefore, the $T\Delta S$ term is directly proportional to temperature, and $\Delta G$ is very strongly temperature-dependent. At ordinary room temperatures, on the order of 25°C, values of $T\Delta S$ for most reactions are much smaller than the magnitudes of $\Delta H$. This means that the dominant component of $\Delta G$ is the $\Delta H$ term, which agrees with the fact that most exothermic processes ($\Delta H$ negative) are spontaneous ($\Delta G$ negative). However, at room temperature for a few reactions, and at higher temperatures for many reactions, the $T\Delta S$ term is sufficiently large compared to $\Delta H$, and is of the appropriate sign, that an exothermic process may be nonspontaneous and an endothermic process may be spontaneous.

At a pressure of one atmosphere and a temperature of 298°K, the reaction

$$4\,Ag(s) + O_2(g) \rightleftharpoons 2\,Ag_2O(s)$$

is exothermic, having a $\Delta H^0$ value of $-14.62$ kilocalories, whereas the standard entropy change ($\Delta S^0$) is $-0.0316$ kilocalorie per degree. It is of interest to determine the influence of temperature on the direction of this reaction, which we can accomplish by evaluating the free-energy change at various temperatures with the aid of

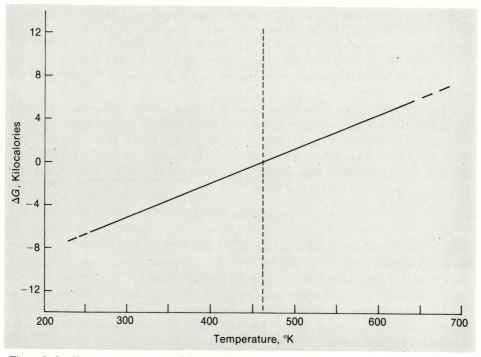

*Figure 3–6.* Free-energy change ($\Delta G$) for the reaction

$$4\,Ag(s) + O_2(g) \rightleftharpoons 2\,Ag_2O(s)$$

as a function of the absolute temperature. (See text for discussion.)

the Gibbs-Helmholtz equation. Figure 3–6 presents the results of a series of such calculations, which were based on the assumption that the values of $\Delta H^0$ and $\Delta S^0$ at 298°K are applicable over a relatively wide temperature range. At temperatures below 463°K, the reaction is spontaneous as written, because the free-energy change is negative. However, for temperatures above 463°K, the value of $T\Delta S$ exceeds the magnitude of $\Delta H$, so that $\Delta G$ becomes positive and the *reverse* process occurs spontaneously at one atmosphere pressure.

*Effect of pressure.* Changes in pressure may exert either a considerable influence upon the position of a chemical equilibrium, or almost none at all. As should be expected, the effects are most pronounced for gas-phase systems. For example, a rise in the total pressure of a system in which the gas-phase equilibrium

$$2\,SO_2(g) + O_2(g) \rightleftharpoons 2\,SO_3(g)$$

prevails would be anticipated to cause a marked change in the position of equilibrium. Such a result is explicable on the basis of Le Châtelier's principle, interpreted for the present situation to state that *any increase in pressure favors a shift in the equilibrium position in whichever direction tends to reduce the pressure.* From the properties of gases, we know that this will occur if the total number of gas molecules in the system is decreased. Notice that three molecules of reactants—two molecules of sulfur dioxide and one molecule of oxygen—combine to form two molecules of sulfur trioxide. Consequently, any rise in pressure will favor the formation of a higher percentage of sulfur trioxide, because the system can most readily relieve this stress through a decrease in the total number of molecules.

For a gas-phase equilibrium such as

$$2\,HI(g) \rightleftharpoons H_2(g) + I_2(g)$$

in which the *same* total number of molecules appears on both the reactant and product side of the reaction, no change in the number of molecules accompanies the decomposition of hydrogen iodide at a specified temperature and pressure. Therefore, the position of equilibrium is not influenced by variations in the total pressure of the system.

Equilibria occurring in condensed phases, such as aqueous solutions, are usually not greatly affected by variations in pressure because liquids are so much less compressible than gaseous systems. However, slight changes in the position of an equilibrium which do occur follow the principle of Le Châtelier—that is, a system always responds to an increase in pressure by undergoing a decrease in volume. Thus, the autoprotolysis or self-ionization of water*

$$H_2O \rightleftharpoons H^+ + OH^-$$

takes place with a very small decrease in volume because the orientation of water molecules around the ionic species, $H^+$ and $OH^-$, leads to a more compact structure than exists hypothetically before any autoprotolysis occurs. If the pressure upon a sample of water is raised from one atmosphere to one thousand atmospheres at 25°C, the extent of autoprotolysis of water only increases by a factor of 1.5.

We will now consider a sample problem which illustrates some of the principles introduced in this section.

*Example 3–3.* At a temperature of 1000°K, the gas-phase reactions listed below have standard enthalpy changes and equilibrium constants as indicated.

---

* A discussion of the autoprotolysis of water is presented in the opening section of Chapter 4.

| Reaction | $\Delta H^0$, kcal | $K$ |
|---|---|---|
| (i) $CO_2(g) + C(s) \rightleftharpoons 2\,CO(g)$ | $+40.8$ | $1.7$ |
| (ii) $H_2(g) + I_2(g) \rightleftharpoons 2\,HI(g)$ | $-3.2$ | $29.1$ |
| (iii) $2\,SO_3(g) \rightleftharpoons 2\,SO_2(g) + O_2(g)$ | $+46.4$ | $0.30$ |

(a) Which of these reactions is exothermic? Since an exothermic process is one which has a negative standard enthalpy change ($\Delta H^0$), only reaction (ii) falls into this category.

(b) Which of these reactions has a negative $\Delta G^0$ value? We must recall that the relationship between the standard free-energy change and the equilibrium constant for a reaction is essentially

$$\Delta G^0 = -RT \ln K$$

In order for $\Delta G^0$ to be negative, the value of $\ln K$ must be a positive number and $K$ itself must be greater than unity. Consequently, both reactions (i) and (ii) have negative standard free-energy changes.

(c) For which of the preceding reactions will the equilibrium constant increase with increasing temperature? According to our earlier discussion, a reaction which has a positive standard enthalpy change ($\Delta H^0$), and which is endothermic, will have its equilibrium constant increase with increasing temperature. Therefore, reactions (i) and (iii) satisfy this requirement.

(d) For which of the preceding reactions will the equilibrium constant increase with increasing pressure? If ideal gas behavior is assumed, changes in pressure do not influence the magnitude of an equilibrium constant.

(e) For which of the preceding reactions will the position of equilibrium shift toward the right if the volume of the system is increased at constant temperature? Since an *increase* in volume at constant temperature corresponds to a *decrease* in total pressure, there should be a shift in the equilibrium position which favors the formation of a greater total number of molecules. Reaction (i) involves conversion of one molecule of carbon dioxide to two molecules of carbon monoxide; *solid* carbon, the standard state of this element, has invariant concentration or activity and does not affect the position of equilibrium. Reaction (iii) shows decomposition of two molecules of sulfur trioxide into two sulfur dioxide molecules and one oxygen molecule. Both of these reactions will shift toward the right when the volume of the system is increased; however, there will be no net change in the position of equilibrium for reaction (ii), since it involves an identical number of molecules on both the reactant and product sides.

## Introduction to Equilibrium Calculations

Numerical calculations involving chemical equilibria vary widely in their degree of difficulty. Nevertheless, all problems can be solved by means of a four-step plan of attack.

STEP 1. Write a balanced chemical reaction representing the state of equilibrium involved in the problem and, if two or more chemical equilibria prevail simultaneously, write the chemical reaction for each.

STEP 2. Formulate the equilibrium-constant expression corresponding to each of the chemical reactions written in Step 1.

STEP 3. Write additional equations from available information and data until the number of mathematical relations, including those of Step 2, equals the number of unknowns contained therein. Each of these equations must be independent of the others, in the sense that no two relationships can merely be rearrangements of one another. One source of information consists of numerical data given in the statement of the problem. A second source is the compilation of data in various reference tables and books, particularly listings of equilibrium and other thermodynamic constants.

A third source of information for additional equations is the **mass-balance relationship**. For example, if a $0.100\,F$ solution of silver nitrate is mixed with an equal volume of an aqueous ammonia solution, the total concentration of all dissolved silver-bearing species must be $0.0500\,M$. However, silver ion exists partly in the form of $Ag^+$ and partly as the two silver-ammine complexes, $Ag(NH_3)^+$ and $Ag(NH_3)_2^+$, so

$$0.0500 = [Ag^+] + [Ag(NH_3)^+] + [Ag(NH_3)_2^+]$$

Another example is provided by the dissolution of 0.0100 mole of potassium chromate $(K_2CrO_4)$ in one liter of aqueous hydrochloric acid, in which some of the chromate remains as the $CrO_4^{2-}$ anion, a fraction of the chromate is protonated to yield $HCrO_4^-$,

$$CrO_4^{2-} + H^+ \rightleftharpoons HCrO_4^-$$

and part of the chromate undergoes dimerization to form dichromate ion:

$$2\,CrO_4^{2-} + 2\,H^+ \rightleftharpoons Cr_2O_7^{2-} + H_2O$$

Thus, the mass-balance equation is

$$0.0100 = [CrO_4^{2-}] + [HCrO_4^-] + 2[Cr_2O_7^{2-}]$$

the concentration of dichromate being multiplied by a factor of *two* because that species contains *two* chromium atoms.

A fourth source, used to treat ionic equilibrium, is the **charge-balance relationship** which states that *the total concentration of positive charge must equal the total concentration of negative charge*. For example, in an aqueous potassium chloride medium, the cationic species are $K^+$ and $H^+$, whereas the anions are $Cl^-$ and $OH^-$. Accordingly the charge-balance equation is

$$[K^+] + [H^+] = [Cl^-] + [OH^-]$$

Now, consider the hydrochloric acid solution of potassium chromate mentioned above, in which the positively charged species are $H^+$ and $K^+$ and the negatively charged species are $Cl^-$, $OH^-$, $CrO_4^{2-}$, $HCrO_4^-$, and $Cr_2O_7^{2-}$. In this situation, we may write the charge-balance relation as

$$[K^+] + [H^+] = [Cl^-] + [OH^-] + 2[CrO_4^{2-}] + [HCrO_4^-] + 2[Cr_2O_7^{2-}]$$

where the concentrations of chromate and dichromate ions are multiplied by *two* because these species bear doubly negative charges. Finally, if we have an aqueous solution of lanthanum chloride $(LaCl_3)$, the charge-balance expression is

$$3[La^{3+}] + [H^+] = [Cl^-] + [OH^-]$$

in which it is necessary to multiply the concentration of the triply charged lanthanum cation by *three*.

STEP 4. Solve these equations for the desired unknown quantity or quantities. As long as Steps 2 and 3 result in the same number of equations as unknowns, it is possible to manipulate these relations algebraically to obtain a numerical value for any or all of the unknown quantities.

Mathematical operations encountered in Step 4 frequently become complex and time-consuming. To simplify the work, it is often possible to make assumptions so as to minimize the time and effort involved in the solution of a problem. It must be stressed, however, that the desire to simplify a calculation is *never a sufficient* reason for making such an assumption. Each assumption must be clearly justified or justifiable, either in advance or by subsequent checking, to ensure that the resultant error is fully within the limit of uncertainty which is acceptable in light of the precision of the data and the desired use of the end result of the calculation. Several methods of doing this will be encountered as we progress through this textbook.

We will now consider some problems which illustrate the use of these principles.

*Example 3–4.* Calculate the equilibrium constant for the iron(III)-iodide system discussed earlier, in which the *initial* iron(III) concentration was 0.200 $M$, the *initial* iodide concentration was 0.300 $M$, and the *equilibrium* triiodide concentration was 0.0866 $M$; ignore activity coefficients.

We may write the *pertinent chemical reaction*

$$2\ Fe^{3+} + 3\ I^- \rightleftharpoons 2\ Fe^{2+} + I_3^-$$

and the corresponding *equilibrium expression*

$$K = \frac{[Fe^{2+}]^2[I_3^-]}{[Fe^{3+}]^2[I^-]^3}$$

Since no iron(II) was originally present, the stoichiometry of the reaction tells us that the equilibrium iron(II) concentration must be *twice* that of the triiodide ion, two $Fe^{2+}$ ions being produced for every triiodide ion. Therefore,

$$[I_3^-] = 0.0866\ M$$
$$[Fe^{2+}] = 0.173\ M$$

To obtain the equilibrium concentration of iron(III), it is necessary to use the *mass-balance relation* for the total iron concentration:

$$[Fe^{2+}] + [Fe^{3+}] = 0.200\ M$$
$$[Fe^{3+}] = 0.200 - [Fe^{2+}]$$
$$[Fe^{3+}] = 0.200 - 0.173 = 0.027\ M$$

A *mass-balance expression* for the total iodine concentration may be utilized to compute the concentration of iodide ion at equilibrium. Thus,

$$[I^-] + 3[I_3^-] = 0.300\ M$$
$$[I^-] = 0.300 - 3[I_3^-]$$
$$[I^-] = 0.300 - 3(0.0866) = 0.040\ M$$

If we now substitute the values for the concentrations of all four species into the equilibrium expression, the result is

$$K = \frac{(0.173)^2(0.0866)}{(0.027)^2(0.040)^3} = 5.55 \times 10^4$$

*Example 3–5.* For the dissociation of hydrogen fluoride (HF) gas into the elemental gases, $H_2$ and $F_2$, the equilibrium constant is $1.00 \times 10^{-13}$ at 1000°C. If 1.00 mole of hydrogen fluoride is allowed to reach dissociation equilibrium in a 2.00-liter container at this temperature, what will be the final molar concentration of fluorine gas?

As a starting point for the solution of this problem, we should write the *pertinent reaction*

$$2\,HF(g) \rightleftharpoons H_2(g) + F_2(g)$$

along with the *equilibrium expression*

$$K = \frac{[H_2(g)][F_2(g)]}{[HF(g)]^2} = 1.00 \times 10^{-13}$$

Based on the stoichiometry of the reaction which must occur before a state of equilibrium is reached, we can conclude that the final concentrations of hydrogen and fluorine gases are equal:

$$[H_2(g)] = [F_2(g)]$$

Next, we can formulate the *mass-balance relationship*

$$[HF(g)] + 2[F_2(g)] = \frac{1.00\ \text{mole}}{2.00\ \text{liters}} = 0.500\ M$$

which states that the *total* concentration of fluorine in both of its forms must be 0.500 $M$. From this equation it follows that the equilibrium concentration of hydrogen fluoride gas is

$$[HF(g)] = 0.500 - 2[F_2(g)]$$

If the expressions for the concentrations of hydrogen and hydrogen fluoride are inserted into the equilibrium-constant relation, we have

$$K = \frac{[F_2(g)]^2}{(0.500 - 2[F_2(g)])^2} = 1.00 \times 10^{-13}$$

This equation can be solved mathematically to obtain a numerical value for the concentration of $F_2$. However, we can simplify the expression by recognizing that, as a consequence of the very small equilibrium constant, only a tiny fraction of the HF dissociates. This means that the concentration of $F_2$ is far less than the concentration of HF, so the term $2[F_2(g)]$ may be neglected in the denominator, and the last preceding relation may be rewritten as

$$\frac{[F_2(g)]^2}{(0.500)^2} = 1.00 \times 10^{-13}$$

When solved for the fluorine concentration, this equation yields

$$[F_2(g)]^2 = (1.00 \times 10^{-13})(0.250) = 2.50 \times 10^{-14}$$
$$[F_2(g)] = 1.58 \times 10^{-7}\ M$$

It is necessary to verify the simplifying assumption which was made; twice $1.58 \times 10^{-7}\ M$ is indeed much smaller than 0.500 $M$.

*Example 3–6.* If the equilibrium constant for the reaction

$$PCl_5(g) \rightleftharpoons PCl_3(g) + Cl_2(g)$$

is 33.3 at 760°K, calculate the equilibrium concentration of each species in the system, if the initial concentrations of phosphorus pentachloride and phosphorus trichloride are 0.050 and 5.00 $M$, respectively.

First, we can state that the *equilibrium* concentration of phosphorus pentachloride is equal to the *initial* concentration minus the concentration of $PCl_5$ which decomposes. This relationship can be correctly expressed as

$$[PCl_5] = 0.050 - [Cl_2]$$

Notice that the equilibrium concentration of chlorine is a measure of how much phosphorus pentachloride decomposes because $PCl_5$ is the *only* source of $Cl_2$. Next, the following expression can be written for the equilibrium concentration of $PCl_3$:

$$[PCl_3] = 5.00 + [Cl_2]$$

This equation indicates that the final concentration of phosphorus trichloride is larger than its original value of 5.00 $M$ by a term that is equal to the concentration of phosphorus pentachloride which decomposes. However, we have already concluded that the chlorine concentration is equivalent to the concentration of decomposed $PCl_5$, so the use of $[Cl_2]$ in the last relationship is justified.

When we make appropriate substitutions into the equilibrium expression, the result is

$$K = \frac{[PCl_3][Cl_2]}{[PCl_5]} = \frac{(5.00 + [Cl_2])[Cl_2]}{(0.050 - [Cl_2])} = 33.3$$

If *all* the $PCl_5$ decomposed, the final concentrations of chlorine and phosphorus trichloride would be 0.050 and 5.05 $M$, respectively. Since the maximal concentration of $PCl_3$ is not much larger than the original value of 5.00 $M$, we will neglect $[Cl_2]$ which appears in the term $(5.00 + [Cl_2])$ in the preceding equation. This simplification, which corresponds to saying that the equilibrium $PCl_3$ concentration is essentially 5.00 $M$, leads to

$$\frac{(5.00)[Cl_2]}{(0.050 - [Cl_2])} = 33.3$$

which can be easily solved for the chlorine concentration:

$$5.00[Cl_2] = 1.67 - 33.3[Cl_2]$$

$$38.3[Cl_2] = 1.67$$

$$[Cl_2] = 0.0435 \ M$$

Now, when the concentrations of the other two species are calculated, we have

$$[PCl_5] = 0.050 - [Cl_2] = 0.050 - 0.0435 = 0.0065 \ M$$

$$[PCl_3] = 5.00 + [Cl_2] = 5.00 + 0.0435 = 5.04 \ M$$

Our earlier assumption that the equilibrium concentration of $PCl_3$ was very close to 5.00 $M$ is seen to be justified, so we can have confidence that the calculations are valid.

## QUESTIONS AND PROBLEMS

1. Calculate the formal concentration or formality of each of the following solutions:
   (a) 500.0 ml of solution containing 2.183 gm of silver nitrate ($AgNO_3$)
   (b) 250.0 ml of solution containing 4.708 gm of potassium permanganate ($KMnO_4$)
   (c) 20.00 ml of solution containing 8.879 mg of sodium chloride (NaCl)
   (d) 500.0 ml of solution containing 12.63 gm of hydrogen bromide (HBr)
   (e) 25.00 ml of solution containing 372.1 mg of potassium sulfate ($K_2SO_4$)

2. What weight of solute in grams is needed to prepare each of the following solutions?
   (a) 500.0 ml of a 0.005124 $F$ perchloric acid solution
   (b) 1.000 liter of a 0.1034 $F$ barium hydroxide solution
   (c) 50.00 ml of a 0.06663 $F$ potassium iodide solution
   (d) 250.0 ml of a 0.2767 $F$ copper(II) nitrate solution
   (e) 100.0 ml of a 0.04839 $F$ ammonium sulfate solution

3. Assume that you prepare a solution corresponding to the equilibrium state of the system described in Example 3–4 on page 73 by mixing solutions of iron(III), iron(II), iodide, and triiodide to give the appropriate equilibrium concentrations. If all the iodide ion is initially radioactive, what will be the final concentration of radioactive iodide ion after a true state of dynamic equilibrium is established?

4. A decision regarding the identity or formula of the gold(I) ion in aqueous solutions can be reached by equilibrium calculations. When *solid metallic gold* is shaken (equilibrated) with an acidic solution of gold(III) ion, which exists as the (aquated) species $Au^{3+}$, gold(I) ions are formed. Two possible reactions can occur:

$$2\,Au(s) + Au^{3+} \rightleftharpoons 3\,Au^+$$

if $Au^+$ is the species of gold(I) present, or

$$4\,Au(s) + 2\,Au^{3+} \rightleftharpoons 3\,Au_2^{2+}$$

if the dimeric species $Au_2^{2+}$ is present. When two such equilibration experiments (shaking solid gold metal with $Au^{3+}$ solutions) were performed and the solutions were finally analyzed for $Au^{3+}$ and for gold(I) species, the following results were obtained: Experiment I, $[Au^{3+}] = 1.50 \times 10^{-3}\,M$ and the *total* analytical concentration of gold(I) $= 1.02 \times 10^{-4}\,M$; Experiment II, $[Au^{3+}] = 4.50 \times 10^{-3}\,M$ and the *total* analytical concentration of gold(I) $= 1.47 \times 10^{-4}\,M$. Write the equilibrium-constant expressions for the two proposed reactions. Designate the equilibrium constants for the reactions as $K_1$ and $K_2$, respectively. Using the equilibrium data given, what conclusions do you draw regarding the formula for the gold(I) species?

5. Write the equilibrium-constant expression for each of the following reactions:
   (a) $H_3PO_4(aq) + PO_4^{3-}(aq) \rightleftharpoons H_2PO_4^-(aq) + HPO_4^{2-}(aq)$
   (b) $N_2(g) + 3\,H_2(g) \rightleftharpoons 2\,NH_3(g)$
   (c) $As_2S_3(s) + 6\,H_2O(l) \rightleftharpoons 2\,H_3AsO_3(aq) + 3\,H_2S(aq)$
   (d) $2\,CO_2(g) \rightleftharpoons 2\,CO(g) + O_2(g)$
   (e) $H^+(aq) + CO_3^{2-}(aq) \rightleftharpoons HCO_3^-(aq)$
   (f) $Ag^+(aq) + 2\,NH_3(aq) \rightleftharpoons Ag(NH_3)_2^+(aq)$
   (g) $I_2(aq) + 2\,S_2O_3^{2-}(aq) \rightleftharpoons 2\,I^-(aq) + S_4O_6^{2-}(aq)$
   (h) $MgNH_4AsO_4(s) \rightleftharpoons Mg^{2+}(aq) + NH_4^+(aq) + AsO_4^{3-}(aq)$
   Use terms involving molar concentrations and note that (aq), (g), (s), and (l) denote species present in an aqueous phase, in a gas phase, as a pure solid, and as a pure liquid, respectively.

6. For each of the following gas-phase reactions
   (a) $2\,HI(g) \rightleftharpoons H_2(g) + I_2(g)$
   (b) $PCl_5(g) \rightleftharpoons PCl_3(g) + Cl_2(g)$
   (c) $N_2(g) + 3\,H_2(g) \rightleftharpoons 2\,NH_3(g)$
   predict the direction in which the position of equilibrium will be shifted if (i) the volume of the system is decreased or (ii) the total pressure of the system is decreased.

7. If the system in which the reaction

$$Ag^+ + 2\ CN^- \rightleftharpoons Ag(CN)_2^-$$

occurs is at equilibrium, predict the effect of each of the following on the molar concentration of the $Ag(CN)_2^-$ complex: (a) solid silver nitrate is dissolved in the solution; (b) ammonia gas, $NH_3$, which can form a complex, $Ag(NH_3)_2^+$, with silver ion, is passed into the solution; (c) solid sodium iodide is dissolved in the solution, AgI being insoluble.

8. An equilibrium mixture at 377°C for the gas-phase system

$$2\ NH_3(g) \rightleftharpoons N_2(g) + 3\ H_2(g)$$

is found to contain, in one liter, 0.0100 mole of $NH_3$, 0.100 mole of $N_2$, and 0.162 mole of $H_2$. Evaluate the equilibrium constant, based on molar concentration units, for the reaction.

9. For the gas-phase reaction

$$2\ SO_2(g) + O_2(g) \rightleftharpoons 2\ SO_3(g)$$

the equilibrium constant, based on molar concentration units, is 5800 at 600°C. If 0.150 mole of sulfur dioxide and 6.00 moles of oxygen are admitted to a previously evacuated 6.00-liter chamber at 600°C, what will be the equilibrium concentration of $SO_3$?

10. If the equilibrium concentrations of sulfur trioxide and sulfur dioxide are equal to each other in the system of the preceding problem, what must be the concentration of $O_2$ at equilibrium?

11. If the reaction

$$N_2O_4(g) \rightleftharpoons 2\ NO_2(g)$$

has an equilibrium constant, based on molar concentration units, of 0.0510 at 320°K, how many grams of $NO_2$ are present in 5.00 liters of an equilibrium mixture which contains 10.0 grams of $N_2O_4$ gas?

12. If the equilibrium constant, based on molar concentration units, for the dissociation of hydrogen fluoride gas into elemental hydrogen and fluorine

$$2\ HF(g) \rightleftharpoons H_2(g) + F_2(g)$$

is $1.00 \times 10^{-13}$ at 1000°C, what will be the equilibrium molar concentration of fluorine if 1.00 mole of $H_2$ and 1.00 mole of $F_2$ are allowed to reach equilibrium in a 2.00-liter container at this temperature?

13. Standard enthalpy changes ($\Delta H^0$) for the decomposition of nitric oxide (NO) and for the formation of nitrous oxide ($N_2O$) at 25°C and a total pressure of 1.00 atmosphere are as follows:
  (i)  $2\ NO(g) \rightleftharpoons O_2(g) + N_2(g)$; $\Delta H^0 = -43.2$ kcal
  (ii) $N_2(g) + \frac{1}{2}\ O_2(g) \rightleftharpoons N_2O(g)$; $\Delta H^0 = +19.5$ kcal

Assume that each of these reactions is initially in a state of equilibrium at 25°C and a pressure of 1.00 atmosphere. For each of these two reactions, indicate whether the position of equilibrium will not shift at all, will shift toward the right, or will shift toward the left as each of the following occurs:
  (a) At 25°C additional oxygen is added to the system, and the total pressure is maintained at one atmosphere.
  (b) While the pressure is kept at one atmosphere, the temperature is increased to 250°C.
  (c) While the temperature is maintained at 25°C, the pressure is increased to 10 atmospheres.

14. For the decomposition reaction

$$2\ NOCl(g) \rightleftharpoons 2\ NO(g) + Cl_2(g)$$

the equilibrium constant, based on molar concentration units, is $4.66 \times 10^{-4}$ at $500°K$. If 0.200 mole of NOCl and 1.00 mole of $Cl_2$ are mixed in a 2.00-liter container at $500°K$, how many grams of the original NOCl will have decomposed when equilibrium is attained?

15. From the data listed in Table 3–3 on page 62, compute the standard free-energy change ($\Delta G^0$) for each of the following reactions:
    (a) $3 O_2(g) \rightleftharpoons 2 O_3(g)$
    (b) $CaCO_3(s) \rightleftharpoons CaO(s) + CO_2(g)$
    (c) $SO_2(g) + \frac{1}{2} O_2(g) \rightleftharpoons SO_3(g)$
    (d) $C_2H_4(g) + H_2(g) \rightleftharpoons C_2H_6(g)$
    (e) $H^+ + OH^- \rightleftharpoons H_2O(g)$
    (f) $2 H_2O(g) + O_2(g) \rightleftharpoons 2 H_2O_2(g)$
    (g) $BaSO_4(s) \rightleftharpoons Ba^{2+} + SO_4^{2-}$

16. Evaluate the equilibrium constant at $25°C$ for each of the following reactions using information given in Table 3–3 on page 62: ·
    (a) $6 NH_3(g) + 7 O_3(g) \rightleftharpoons 6 NO_2(g) + 9 H_2O(g)$
    (b) $3 H_2(g) + N_2(g) \rightleftharpoons 2 NH_3(g)$
    (c) $CO_2(g) + NO(g) \rightleftharpoons CO(g) + NO_2(g)$
    (d) $H_2(g) + Br_2(l) \rightleftharpoons 2 HBr(g)$
    (e) $AgI(s) \rightleftharpoons Ag^+ + I^-$
    (f) $C_2H_4(g) + 3 O_2(g) \rightleftharpoons 2 CO_2(g) + 2 H_2O(g)$
    (g) $2 NH_3(g) + 5 SO_3(g) \rightleftharpoons 2 NO(g) + 5 SO_2(g) + 3 H_2O(g)$

17. For the dissociation of $N_2O_4$ gas

$$N_2O_4(g) \rightleftharpoons 2 NO_2(g)$$

the equilibrium constant, based on molar concentration units, is 0.0125 at $35°C$ and is 0.171 at $75°C$. Is the dissociation of $N_2O_4$ an exothermic or an endothermic process? Explain.

18. At a temperature of $727°C$, the five gas-phase reactions listed below have enthalpy changes and equilibrium constants (based on molar concentration units) as indicated.

| Reaction | $\Delta H^0$, kcal | $K$ |
|---|---|---|
| (i) $CO_2(g) + CF_4(g) \rightleftharpoons 2 COF_2(g)$ | +10.4 | 0.472 |
| (ii) $CO(g) + H_2O(g) \rightleftharpoons CO_2(g) + H_2(g)$ | −8.3 | 1.44 |
| (iii) $2 NO(g) + Br_2(g) \rightleftharpoons 2 NOBr(g)$ | −11.1 | $1.31 \times 10^{-2}$ |
| (iv) $2 NH_3(g) \rightleftharpoons 3 H_2(g) + N_2(g)$ | +26.3 | $4.31 \times 10^2$ |
| (v) $2 H_2(g) + S_2(g) \rightleftharpoons 2 H_2S(g)$ | +43.1 | $4.10 \times 10^{-4}$ |

(a) Which, if any, of the reactions is (or are) exothermic?
(b) For which, if any, of the reactions is $\Delta G^0$ negative?
(c) For which, if any, of the reactions will the value of $K$ decrease with decreasing temperature?
(d) For which, if any, of the systems will reaction proceed toward the right when, with the system initially at equilibrium, the total pressure is decreased by an increase in the volume at constant temperature?
(e) For which, if any, of the reactions will the value of $K$ decrease with increasing pressure?

19. If the equilibrium constant for the reaction

$$PCl_5(g) \rightleftharpoons PCl_3(g) + Cl_2(g)$$

is $1.8 \times 10^{-7}$ at $25°C$ and 1.8 at $250°C$, determine the enthalpy change for the process.

20. At $1000°K$ the enthalpy changes for two different reactions are as follows:

$$H_2(g) + CO_2(g) \rightleftharpoons H_2O(g) + CO(g); \Delta H = +8.3 \text{ kcal}$$

$$2 H_2S(g) \rightleftharpoons 2 H_2(g) + S_2(g); \qquad \Delta H = -43.1 \text{ kcal}$$

Suppose that each of these systems is at equilibrium at a temperature of 1000°K and a total pressure of 0.500 atmosphere. For each of the following experimental changes, state (i) whether no net reaction occurs, (ii) whether net reaction occurs to the right, or (iii) whether net reaction occurs to the left:

(a) Additional hydrogen gas is introduced into the system at 1000°K, and the total pressure is maintained constant by an appropriate increase in the volume.

(b) While the total pressure is kept at 0.500 atmosphere, the temperature of the equilibrium mixture is lowered to 500°K.

(c) Without any change in the temperature, the total pressure is increased to 5.00 atmospheres.

21. At a temperature of 760°K, the reaction for the gas-phase decomposition of phosphorus pentachloride into phosphorus trichloride and chlorine

$$PCl_5(g) \rightleftharpoons PCl_3(g) + Cl_2(g)$$

has an equilibrium constant (based on molar concentration units) of 33.3. If 2.00 moles of phosphorus pentachloride is admitted to a previously evacuated 1.00-liter vessel, what will be the equilibrium concentrations of all three species?

22. (a) For the hypothetical all-gas reaction

$$aA + bB \rightleftharpoons cC + dD$$

in which all substances behave as ideal gases, show that the equilibrium constant ($K$) based on molar concentration units

$$K = \frac{[C]^c[D]^d}{[A]^a[B]^b}$$

can be related to the equilibrium constant ($K_p$) based on pressure (in atmospheres)

$$K_p = \frac{(p_C)^c(p_D)^d}{(p_A)^a(p_B)^b}$$

by means of the relation

$$K_p = K(RT)^{c+d-a-b}$$

where $R$ is the universal gas constant (0.0821 liter atm mole$^{-1}$ degree$^{-1}$) and $T$ is the absolute temperature.

(b) At equilibrium the pressures of $NO_2$, NO, and $O_2$ were found to be 0.200, 0.00026, and 0.600 atmosphere, respectively, in a 1.00-liter container at 158°C. For the reaction

$$2 NO_2(g) \rightleftharpoons 2 NO(g) + O_2(g)$$

calculate the values of $K_p$ and $K$.

23. Calculate the ionic strength of each of the following systems:

(a) a solution containing 0.00300 $F$ lanthanum chloride

(b) a solution containing 0.00500 $F$ sodium sulfate and 0.00300 $F$ lanthanum nitrate

(c) a solution containing 0.0200 $F$ sodium nitrate, 0.0500 $F$ potassium sulfate, and 0.0300 $F$ cadmium chloride

(d) a solution containing 0.0600 $F$ aluminum sulfate and 0.1000 $F$ sodium ferricyanide

24. Using single-ion activity coefficients based on the extended Debye-Hückel equation which are listed in Table 3–2 on page 52, evaluate the mean activity coefficient for the salt in each of the following aqueous solutions:
    (a) $0.05000\ F$ LiCl
    (b) $0.001667\ F$ Al(ClO$_4$)$_3$
    (c) $0.005000\ F$ Na$_4$Fe(CN)$_6$
    (d) $0.03333\ F$ K$_2$HPO$_4$
    (e) $6.667 \times 10^{-5}\ F$ La$_2$(SO$_4$)$_3$

25. Calculate the mean activity coefficient for a $0.200\ F$ calcium chloride solution using (a) the Debye-Hückel limiting law, (b) the extended Debye-Hückel equation, and (c) the Davies equation. Compare the results with the experimentally measured value of $0.472$.

26. (a) Using the Davies equation, show that the relationship between the common solubility product $(K_{sp})$ and the thermodynamic activity product $(K_{ap})$ for barium sulfate is

$$pK_{sp} = pK_{ap} - 4.10\left[\frac{\sqrt{I}}{1 + \sqrt{I}} - 0.2\,I\right]$$

where $K_{sp} = [\text{Ba}^{2+}][\text{SO}_4{}^{2-}]$, $K_{ap} = (\text{Ba}^{2+})(\text{SO}_4{}^{2-})$, $pK_{sp} = -\log K_{sp}$, and $pK_{ap} = -\log K_{ap}$.

   (b) Noting that the solubility of solid barium sulfate in an inert electrolyte solution is measured by the molar concentration of barium ion, $[\text{Ba}^{2+}]$, and that $pK_{ap}$ is 10.01 at 25°C, calculate the solubility of barium sulfate in a $0.0200\ F$ magnesium chloride (MgCl$_2$) medium at 25°C. Compare the result to the value for the solubility of barium sulfate in pure water at 25°C.

27. At 25°C the thermodynamic activity product $(K_w)$ for the autoprotolysis (self-ionization) of water

$$\text{H}_2\text{O} \rightleftharpoons \text{H}^+ + \text{OH}^-$$

may be formulated as

$$K_w = (\text{H}^+)(\text{OH}^-) = 1.008 \times 10^{-14}$$

where $(\text{H}^+)$ and $(\text{OH}^-)$ represent the activities of hydrogen ion and hydroxide ion, respectively, and where the activity of water is taken to be unity.

   (a) Using the Davies equation, show that

$$pH = 6.998 - 0.512\left[\frac{\sqrt{I}}{1 + \sqrt{I}} - 0.2\,I\right]$$

where $pH = -\log [\text{H}^+]$.

   (b) Using the result of the preceding derivation, construct a plot of $pH$ versus $\sqrt{I}$ for potassium chloride solutions of the following concentrations at 25°C: 0, 0.2, 0.4, 0.6, 0.8, 1.0, 1.2, 1.4, 1.6, 1.8, and $2.0\ F$. How does the pH of an aqueous $2.00\ F$ potassium chloride medium vary as the solution is extensively diluted with water? How could you best observe this variation in pH experimentally?

28. Use the extended Debye-Hückel equation as well as information presented in Table 3–2 to compute the mean activity coefficient for the specified binary salt in each of the following solutions:
    (a) $0.0250\ F$ calcium iodide (CaI$_2$) in water
    (b) $0.00100\ F$ barium chloride (BaCl$_2$) in an aqueous $0.0300\ F$ potassium nitrate solution
    (c) $0.00500\ F$ aluminum chloride (AlCl$_3$) in an aqueous $0.0100\ F$ sodium nitrate solution
    (d) $0.0150\ F$ sodium bicarbonate (NaHCO$_3$) in water
    (e) $0.0300\ F$ potassium chromate (K$_2$CrO$_4$) in an aqueous solution containing $0.100\ F$ sodium hydroxide and $0.200\ F$ lithium perchlorate

29. (a) Using the Davies equation, show that the relationship between the common solubility product ($K_{sp}$) and the thermodynamic activity product ($K_{ap}$) for silver chloride at 25°C is

$$pK_{sp} = pK_{ap} - 1.024\left[\frac{\sqrt{I}}{1+\sqrt{I}} - 0.2\,I\right]$$

where $K_{sp} = [Ag^+][Cl^-]$, $K_{ap} = (Ag^+)(Cl^-)$, $pK_{sp} = -\log K_{sp}$, and $pK_{ap} = -\log K_{ap}$.

(b) In a study of the solubility of silver chloride in pure water and in various potassium nitrate solutions at 25°C, Popoff and Neuman [J. Phys. Chem , *34:* 1853, 1930] obtained the following data:

| Concentration of $KNO_3$, $F$ | Solubility of AgCl, $F$ |
|---|---|
| 0 | $1.278 \times 10^{-5}$ |
| 0.0005090 | $1.311 \times 10^{-5}$ |
| 0.001005 | $1.325 \times 10^{-5}$ |
| 0.004972 | $1.385 \times 10^{-5}$ |
| 0.009931 | $1.427 \times 10^{-5}$ |

Assuming that the solubility of silver chloride is identical to the molar concentration of dissolved silver ion, $[Ag^+]$, calculate the ionic strength ($I$) and evaluate the common solubility product ($K_{sp}$) for each of the five solutions. Construct a plot of $pK_{sp}$ versus the term inside the square brackets of the equation derived in part (a), and determine the value of the thermodynamic activity product ($K_{ap}$) for silver chloride at 25°C.

30. For an aqueous solution containing a one-to-one electrolyte such as NaCl, it is often assumed that the single-ion activity coefficients for the cation and anion, *e.g.*, $f_{Na^+}$ and $f_{Cl^-}$, have the same value. In addition, the activity coefficient of any single ion in *two different solutions having identical ionic strengths* presumably must have the same value. For sodium chloride solutions, the following data have been obtained:

| Concentration of NaCl, $F$ | 0.050 | 0.100 | 0.200 | 0.300 | 0.400 | 0.500 |
|---|---|---|---|---|---|---|
| Mean Activity Coefficient, $f_{\pm}$ | 0.822 | 0.778 | 0.735 | 0.710 | 0.693 | 0.681 |

If the mean activity coefficient ($f_{\pm}$) of a 0.100 $F$ solution of disodium hydrogen phosphate ($Na_2HPO_4$) in water is 0.480, what are the single-ion activity coefficients for $Na^+$ and $HPO_4^{2-}$ in this medium?

## SUGGESTIONS FOR ADDITIONAL READING

1. A. J. Bard: *Chemical Equilibrium.* Harper & Row, New York, 1966.
2. T. R. Blackburn: *Equilibrium: A Chemistry of Solutions.* Holt, Rinehart and Winston, New York, 1969.
3. J. O'M. Bockris and A. K. N. Reddy: *Modern Electrochemistry.* Plenum Publishing Company, New York, 1970, Volume 1, pp. 45–286.
4. J. N. Butler: *Ionic Equilibrium.* Addison-Wesley Publishing Company, Reading, Massachusetts, 1964.
5. G. M. Fleck: *Equilibria in Solution.* Holt, Rinehart and Winston, New York, 1966.
6. H. Freiser and Q. Fernando: *Ionic Equilibria in Analytical Chemistry.* John Wiley and Sons, New York, 1963, pp. 9–34.

7. R. A. Horne: *Marine Chemistry*. Wiley-Interscience, New York, 1969.
8. H. A. Laitinen: *Chemical Analysis*. McGraw-Hill Book Company, New York, 1960, pp. 5–21, 107–122.
9. T. S. Lee: Chemical equilibrium and the thermodynamics of reactions. *In* I. M. Kolthoff and P. J. Elving, eds.: *Treatise on Analytical Chemistry*. Part I, Volume 1, Wiley-Interscience, New York, 1959, pp. 185–275.
10. O. Robbins, Jr.: *Ionic Reactions and Equilibria*. The Macmillan Company, New York, 1967.
11. L. G. Sillén: Graphic presentation of equilibrium data. *In* I. M. Kolthoff and P. J. Elving, eds.: *Treatise on Analytical Chemistry*. Part I, Volume 1, Wiley-Interscience, New York, 1959, pp. 277–317.
12. W. Stumm and J. J. Morgan: *Aquatic Chemistry*. Wiley-Interscience, New York, 1970.

# AQUEOUS ACID-BASE REACTIONS

Aqueous acid-base reactions are inherently more easily characterizable than any other kind of chemical interaction in water, because they essentially involve only the transfer of hydrogen ions from proton donors to proton acceptors. Thus, it is possible to acquire detailed information about the identities of the reactant and product species in a proton-transfer reaction and to know with some reliability the equilibria and equilibrium constants which govern the behavior of an acid-base system. In the present chapter, we shall consider the nature of acids and bases in water, the methods used to perform equilibrium calculations for acid-base systems, and the theory and techniques of acid-base titrimetry.

## FUNDAMENTAL CONCEPTS OF ACIDITY AND BASICITY

*Definitions of acids and bases.* From the point of view of chemical analysis, the most useful definition of acids and bases is the one suggested independently in 1923 by Brønsted and by Lowry. A **Brønsted-Lowry acid** is defined as a species having a tendency to lose or donate a proton; a **Brønsted-Lowry base** is a substance having a tendency to accept or gain a proton. Loss of a proton by a Brønsted-Lowry acid gives rise to the formation of a corresponding Brønsted-Lowry base, called the **conjugate base** of the parent acid. Addition of a proton to any Brønsted-Lowry base causes the formation of the **conjugate acid** of the original base. Accordingly, $H_2O$ is the conjugate base of the Brønsted-Lowry acid $H_3O^+$, but the conjugate acid of the Brønsted-Lowry base $OH^-$. Similarly, $HCO_3^-$ is the conjugate acid of $CO_3^{2-}$, whereas $NH_3$ is the conjugate base of $NH_4^+$. Two species such as $H_3O^+$ and $H_2O$ comprise a so-called conjugate acid-base pair.

A proton does not exist in solution by itself, that is, independently of its surrounding environment. In aqueous media the proton appears to be associated with four water molecules to give the **hydronium ion**, $H_9O_4^+$, which has the structure

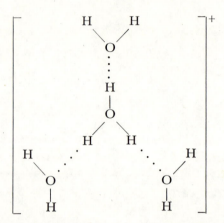

showing the proton covalently bonded to a central water molecule to form $H_3O^+$, each hydrogen of which is linked to another molecule of solvent by a hydrogen bond.* Usually, the hydronium ion is symbolized in the abbreviated form $H_3O^+$, which still conveys the idea that the proton is hydrated. In addition, the simpler designation $H^+$ is commonly used in chemical reactions and equilibrium expressions for aqueous systems where the hydronium ion is understood to be present; invariably when $H^+$ is employed, and often when $H_3O^+$ appears, the term **hydrogen ion** is introduced as a synonym for hydronium ion. In an aqueous medium, all other ions and molecules are hydrated (including hydroxide ion, $OH^-$, the conjugate base of the solvent)—many of them to an extent comparable to, if not greater than, the proton—even though the solvent is not ordinarily written as part of the chemical formulas of these dissolved species.

One of the important consequences of the nonexistence of bare protons is that acid-base reactions occur by the direct transfer of a proton from a Brønsted-Lowry acid (proton donor) to a Brønsted-Lowry base (proton acceptor) to form the conjugate base of the acid and the conjugate acid of the base.

*Autoprotolysis of water.* Pure water is slightly ionized, because one water molecule is capable of accepting a proton from a second water molecule:

$$H_2O + H_2O \rightleftharpoons H_3O^+ + OH^-$$

---

* Spatially, the shape of the hydronium ion appears to resemble that of ammonia; the three hydrogen atoms bonded to the central oxygen are positioned near three of the four corners of a regular tetrahedron, with a lone electron pair at the fourth corner. Moreover, this means that the three hydrogen-bonded water molecules lie at three of the four corners of a (larger) tetrahedron; the lone electron pair on the central oxygen may form a somewhat weaker (electrostatic) bond with a fourth solvent (water) molecule.

Such a proton-transfer reaction involving only solvent molecules is termed **auto-protolysis**, and the equilibrium expression, written in terms of activities, is

$$K = \frac{(H_3O^+)(OH^-)}{(H_2O)^2}$$

This equation can be considerably simplified, however, if the thermodynamic convention, that a pure liquid (such as water) is defined as having unit activity, is assumed to be valid for relatively dilute aqueous solutions. Furthermore, if values of unity are used for the activity coefficients of the hydronium and hydroxide ions in the relatively dilute solutions of analytical interest, we can replace the activities of $H_3O^+$ and $OH^-$ by their respective concentrations. Making these changes and writing $H^+$ to denote the hydronium ion, we arrive at the familiar ion-product expression for water:

$$K_w = [H^+][OH^-]$$

Although the ion-product constant, $K_w$, also known as the autoprotolysis constant, is temperature dependent (ranging from $1.14 \times 10^{-15}$ at $0°C$ through $1.01 \times 10^{-14}$ at $25°C$ and $5.47 \times 10^{-14}$ at $50°C$ to approximately $5.4 \times 10^{-13}$ at $100°C$), it will be assumed for all of the numerical calculations in this book that the temperature is $25°C$ and that the ion-product constant is precisely $1.00 \times 10^{-14}$.

**Definition of pH.**  As a convenient way to specify the degree of acidity of an aqueous solution, pH is usually employed. Throughout the present chapter, we will define the pH of an aqueous solution as the negative base-ten logarithm of the hydrogen ion concentration. Thus,

$$pH = -\log [H^+]$$

This definition is quite satisfactory for purposes such as the calculation and construction of acid-base titration curves, the selection of end-point indicators, and decisions concerning the feasibility of proposed titration methods. Actually, the pH of a solution is probably more accurately measured by the negative base-ten logarithm of the hydrogen ion *activity*, $a_{H^+}$. However, further discussion of the problems involved in the precise definition and measurement of pH will be deferred until we consider the subject of direct potentiometry in Chapter 11.

If we return to the ion-product expression for water,

$$K_w = [H^+][OH^-] = 1.00 \times 10^{-14}$$

and take the negative logarithm of each member of this equation, the result is

$$-\log K_w = -\log [H^+] - \log [OH^-] = 14.00$$

or

$$pK_w = pH + pOH = 14.00$$

where $pK_w$ is the negative logarithm of the ion-product constant and pOH is the negative logarithm of the hydroxide ion concentration. This equation provides a simple way to calculate the pOH of a solution if the pH is determined and, conversely, the pH can be obtained if the pOH is known.

In pure water, or in an aqueous medium free from any acidic or basic solutes, autoprotolysis leads to the formation of equal concentrations of hydrogen ion and hydroxide ion; that is, $[H^+] = [OH^-]$. Substitution of this condition into the ion-product expression gives

$$[H^+]^2 = [OH^-]^2 = K_w = 1.00 \times 10^{-14}$$

$$[H^+] = [OH^-] = 1.00 \times 10^{-7} \, M$$

$$pH = pOH = -\log(1.00 \times 10^{-7}) = 7.00$$

At 25°C an aqueous solution in which the concentrations of hydrogen ion and hydroxide ion are both $1.00 \times 10^{-7} \, M$, and for which the pH is 7.00, is stated to be *neutral*. However, if $[H^+]$ exceeds $[OH^-]$, the hydrogen ion concentration is larger than $1.00 \times 10^{-7} \, M$, the pH is below 7.00, and the solution is *acidic*. On the other hand, the solution is *basic* or *alkaline*, the hydrogen ion concentration is smaller than $1.00 \times 10^{-7} \, M$, and the pH is above 7.00, if $[H^+]$ is less than $[OH^-]$.

*Strengths of acids and bases in water.* In aqueous solutions, the strength of an acid is determined by the extent to which it loses protons to water, whereas the strength of a base is governed by how completely it captures protons from water—and a wide range of acid and base strength is observed. Quantitatively, the strength of an acid or base is measured by the magnitude of the equilibrium constant for its proton-transfer reaction with the solvent—the larger the equilibrium constant, the greater the strength of the acid or base. In writing these equilibrium constants, one customarily uses $K_a$ for acids and $K_b$ for bases.

In water, the hydronium ion (the conjugate acid of the solvent) is the strongest acidic species that can exist and remain stable. Any acid stronger than the hydronium ion reacts completely and instantaneously with the solvent (acting as a proton acceptor or a Brønsted-Lowry base) to yield essentially stoichiometric quantities of $H_3O^+$ and the conjugate base of the original acid. Perchloric ($HClO_4$), sulfuric ($H_2SO_4$),* hydriodic (HI), nitric ($HNO_3$), hydrobromic (HBr), and hydrochloric (HCl) acids, along with the less well known hexafluorophosphoric ($HPF_6$) and fluoboric ($HBF_4$) acids, exhibit this kind of behavior. Hydrogen chloride dissolved in water

$$HCl + H_2O \rightleftharpoons Cl^- + H_3O^+; \qquad K_a \gg 1$$

is representative of these strong acids. Donation of a proton by HCl to $H_2O$ is far more successful than the donation of a proton by $H_3O^+$ to $Cl^-$. Thus, the position of equilibrium for the above proton-transfer reaction lies well toward the right ($K_a \gg 1$), and we say that HCl is a very strong acid in water.

Most acidic species are much weaker in an aqueous medium than the strong acids mentioned in the last paragraph. For example, the proton-donating ability of acetic acid is much less than that of the hydronium ion, as reflected by the small equilibrium constant for the reaction

$$CH_3COOH + H_2O \rightleftharpoons CH_3COO^- + H_3O^+; \qquad K_a = 1.75 \times 10^{-5}$$

In other words, acetate ion (the conjugate base of the parent acid) is a better proton acceptor than water; therefore, the proton stays with the acetate ion, and formation of $H_3O^+$ occurs only to a slight extent. Ammonium ion is even weaker than acetic acid,

$$NH_4^+ + H_2O \rightleftharpoons NH_3 + H_3O^+; \qquad K_a = 5.55 \times 10^{-10}$$

---

* Although sulfuric acid is diprotic, only the loss of the first proton to water occurs completely to yield $HSO_4^-$ and $H_3O^+$; the second proton is transferred to water with much more difficulty, inasmuch as it is necessary to separate a positively charged proton ($H^+$) from sulfate anion ($SO_4^{2-}$), a stronger Brønsted-Lowry base than $HSO_4^-$.

the equilibrium constant ($K_a$) for its reaction with water being more than 30,000 times smaller than that for the acetic acid-water system. Clearly, $NH_3$ holds onto a proton much better than acetate ion or water. Finally, the hydrogen sulfide anion, $HS^-$ (which is the conjugate base of the weak acid $H_2S$), is an exceedingly feeble acid, showing almost no tendency at all to transfer a proton to water:

$$HS^- + H_2O \rightleftharpoons S^{2-} + H_3O^+; \qquad K_a = 1.1 \times 10^{-15}$$

Thus, sulfide ion is an excellent proton acceptor compared to water as well as ammonia and acetate ion.

Two conclusions emerge from the preceding discussion. First, the family of Brønsted-Lowry acids includes cationic ($NH_4^+$), anionic ($HS^-$), and uncharged ($CH_3COOH$) members. Second, the strength of a solute acid (HB) is fundamentally determined by competition for protons between the solvent and the conjugate base ($B^-$) of the solute acid, as shown by the generalized equilibrium

$$HB + H_2O \rightleftharpoons B^- + H_3O^+$$

If water is a much better proton acceptor than $B^-$, the proton of HB will be donated to $H_2O$, and HB will appear to be a strong or completely dissociated acid. As the proton-accepting ability of $B^-$ increases, and eventually surpasses that of $H_2O$, the acid strength (dissociation) of HB decreases because the proton is retained rather than donated to the weaker base $H_2O$.

Indeed, a general consequence of the Brønsted-Lowry concept of acid-base behavior is that, as the proton-donating ability of an acid *increases*, the proton-accepting tendency of its conjugate base *decreases*. Thus, the conjugate base ($Cl^-$) of a strong acid (HCl) is an exceedingly weak proton acceptor. Conversely, the conjugate base ($S^{2-}$) of an extremely weak acid ($HS^-$) is a very strong proton acceptor. It follows that, if the relative proton-donating abilities of a series of Brønsted-Lowry acids are known, we can immediately list the relative proton-accepting abilities of the corresponding conjugate bases. Since the relative strengths of the four acids discussed in the second-preceding paragraph are $HCl > CH_3COOH > NH_4^+ > HS^-$, the relative strengths of the conjugate bases must be $Cl^- < CH_3COO^- < NH_3 < S^{2-}$. This predicted order of relative base strength can be confirmed by inspection of the $K_b$ value for the proton-transfer reaction between water and each of the four bases:

$$Cl^- + H_2O \rightleftharpoons HCl + OH^-; \qquad K_b \text{ is immeasurably small}$$

$$CH_3COO^- + H_2O \rightleftharpoons CH_3COOH + OH^-; \qquad K_b = 5.71 \times 10^{-10}$$

$$NH_3 + H_2O \rightleftharpoons NH_4^+ + OH^-; \qquad K_b = 1.80 \times 10^{-5}$$

$$S^{2-} + H_2O \rightleftharpoons HS^- + OH^-; \qquad K_b = 9.1$$

Sulfide ion is an especially strong base; for, if one attempts to prepare a solution containing 1.0 $M$ sulfide ion by dissolving one formula weight of sodium sulfide ($Na_2S$) in a liter of water, most of the sulfide reacts with the solvent to produce $HS^-$ and $OH^-$, only 0.09 $M$ sulfide ion remaining unprotonated at equilibrium. Similar efforts to obtain a 0.01 $M$ sulfide ion solution result in almost complete transfer of protons from water to sulfide ions, since the equilibrium sulfide concentration is but $1.1 \times 10^{-5}$ $M$. Dissolution in water of sodium oxide ($Na_2O$), the salt of a much stronger base ($O^{2-}$) than sulfide or even hydroxide ion, is a violently exothermic process leading to the quantitative conversion of oxide ion to hydroxide ion:

$$O^{2-} + H_2O \rightleftharpoons OH^- + OH^-; \qquad K_b \gg 1$$

Thus, the strongest Brønsted-Lowry base capable of stable existence in an aqueous medium is the hydrated hydroxide ion—the conjugate base of the solvent.

*Relation between $K_a$ and $K_b$ for conjugate acid-base pairs.* A fundamental quantitative relationship exists between the equilibrium constant ($K_a$) for the proton-transfer reaction between water and a Brønsted-Lowry acid, and the equilibrium constant ($K_b$) for the proton-transfer reaction between water and the conjugate base of that acid. For example, if we consider acetic acid and its conjugate base, acetate ion, the appropriate equilibria are

$$CH_3COOH + H_2O \rightleftharpoons CH_3COO^- + H_3O^+; \qquad K_a = \frac{[CH_3COO^-][H^+]}{[CH_3COOH]}$$

(where [H$^+$] symbolizes the concentration of the hydronium ion) and

$$CH_3COO^- + H_2O \rightleftharpoons CH_3COOH + OH^-; \qquad K_b = \frac{[CH_3COOH][OH^-]}{[CH_3COO^-]}$$

If the equilibrium expressions for $K_a$ and $K_b$ are multiplied together, the result is

$$K_a \cdot K_b = \frac{[CH_3COO^-][H^+]}{[CH_3COOH]} \cdot \frac{[CH_3COOH][OH^-]}{[CH_3COO^-]} = [H^+][OH^-]$$

and, since the product of the hydrogen ion and hydroxide ion concentrations is the ion-product constant ($K_w$) for water, we reach the significant and useful conclusion that

$$K_a K_b = K_w$$

This expression is applicable to any Brønsted-Lowry conjugate acid-base pair; if $K_a$ for an acid is available, the value of $K_b$ for its conjugate base can be computed, and, if $K_b$ for a base is known, $K_a$ for its conjugate acid is readily obtainable.

*Polyprotic acids and their conjugate bases.* Thus far, we have considered acids which possess only one transferable proton—that is, **monoprotic acids**—and bases which can accept only one proton.* However, there is an important class of acids, called **polyprotic acids**, which have two or more ionizable protons. Included in the list of polyprotic acids are phosphoric acid ($H_3PO_4$), arsenic acid ($H_3AsO_4$), oxalic acid ($H_2C_2O_4$), carbonic acid ($H_2CO_3$), malonic acid (HOOC—$CH_2$—COOH), tartaric acid (HOOC—(CHOH)$_2$—COOH), and glycinium ion ($H_3N^+$—$CH_2COOH$)—the latter being the conjugate acid of glycine, one of the amino acids.

Phosphoric acid typifies the behavior of these substances, undergoing three stepwise dissociations as indicated by the following equilibria:

$$H_3PO_4 + H_2O \rightleftharpoons H_2PO_4^- + H_3O^+; \qquad K_{a1} = 7.5 \times 10^{-3}$$

$$H_2PO_4^- + H_2O \rightleftharpoons HPO_4^{2-} + H_3O^+; \qquad K_{a2} = 6.2 \times 10^{-8}$$

$$HPO_4^{2-} + H_2O \rightleftharpoons PO_4^{3-} + H_3O^+; \qquad K_{a3} = 4.8 \times 10^{-13}$$

From the magnitude of the first equilibrium constant ($K_{a1}$), we can see that phosphoric acid itself is a moderately strong acid, being much stronger than acetic acid

---

* Sulfide ion is the one exception, since two protons can be added to form $H_2S$.

but not so strong as hydrochloric acid. Dihydrogen phosphate ($H_2PO_4^-$) and hydrogen phosphate ($HPO_4^{2-}$) ions are increasingly weak acids, primarily because separation of a positively charged proton from a doubly charged anion ($HPO_4^{2-}$) and from a triply charged anion ($PO_4^{3-}$) in the second and third steps of dissociation, respectively, requires that a successively larger electrostatic force of attraction be overcome.

Conjugate bases of polyprotic acids (often in the form of solutions of alkali metal salts) are capable of accepting two or more protons. Thus, sulfide, carbonate, tartrate, malonate, hydrogen phosphate, hydrogen arsenate, and glycinate anions can each combine with two hydrogen ions, whereas phosphate and arsenate ions may react with up to three protons. Oxalate ion can interact with water (or some other Brønsted-Lowry acid) to form the hydrogen oxalate ion

$$C_2O_4^{2-} + H_2O \rightleftharpoons HC_2O_4^- + OH^-; \qquad K_{b1} = \frac{K_w}{K_{a2}} = 1.64 \times 10^{-10}$$

which can accept another proton to become oxalic acid:

$$HC_2O_4^- + H_2O \rightleftharpoons H_2C_2O_4 + OH^-; \qquad K_{b2} = \frac{K_w}{K_{a1}} = 1.54 \times 10^{-13}$$

Equilibrium constants for these two processes indicate that neither $C_2O_4^{2-}$ nor $HC_2O_4^-$ is a particularly strong base, since the conjugate acids of these anions—$HC_2O_4^-$ and $H_2C_2O_4$, respectively—are both stronger than acetic acid.

*Amphiprotic substances.* A species is said to be **amphiprotic** if it can accept a proton from a Brønsted-Lowry acid or donate a proton to a Brønsted-Lowry base. Water is the classic example of such a substance, because it can act as either a proton donor (acid) or a proton acceptor (base). Moreover, many solutes display amphiprotic character in water.

Species derived from the ionization of a polyprotic acid are included among the list of amphiprotic substances—hydrogen carbonate or bicarbonate ($HCO_3^-$), hydrogen phosphate ($HPO_4^{2-}$), dihydrogen arsenate ($H_2AsO_4^-$), hydrogen sulfide ($HS^-$), hydrogen oxalate ($HC_2O_4^-$), hydrogen sulfite ($HSO_3^-$), dihydrogen phosphite ($H_2PO_3^-$), hydrogen tartrate ($HOOC—(CHOH)_2—COO^-$), and hydrogen malonate ($HOOC—CH_2—COO^-$). Thus, bicarbonate can behave as an acid

$$HCO_3^- + H_2O \rightleftharpoons CO_3^{2-} + H_3O^+; \qquad K_{a2} = 4.68 \times 10^{-11}$$

or as a base

$$HCO_3^- + H_2O \rightleftharpoons H_2CO_3 + OH^-; \qquad K_{b2} = \frac{K_w}{K_{a1}} = 2.24 \times 10^{-8}$$

but an aqueous solution of sodium bicarbonate ($NaHCO_3$) will be distinctly alkaline —its pH will be greater than 7—because the equilibrium constant for the second reaction (which produces $OH^-$ ions) is larger than that for the first reaction (which yields $H_3O^+$ ions). A solution of sodium hydrogen oxalate ($NaHC_2O_4$) is acidic, on the other hand, because the reaction showing $HC_2O_4^-$ as a proton donor

$$HC_2O_4^- + H_2O \rightleftharpoons C_2O_4^{2-} + H_3O^+; \qquad K_{a2} = 6.1 \times 10^{-5}$$

has an equilibrium constant $(K_{a2})$ much greater than that $(K_{b2})$ for the equilibrium representing $HC_2O_4^-$ as a proton acceptor:

$$HC_2O_4^- + H_2O \rightleftharpoons H_2C_2O_4 + OH^-; \qquad K_{b2} = \frac{K_w}{K_{a1}} = 1.54 \times 10^{-13}$$

A substance such as ammonium acetate is amphiprotic in an aqueous medium, since ammonium ion is a Brønsted-Lowry acid,

$$NH_4^+ + H_2O \rightleftharpoons NH_3 + H_3O^+; \qquad K_a = 5.55 \times 10^{-10}$$

whereas acetate ion is a Brønsted-Lowry base:

$$CH_3COO^- + H_2O \rightleftharpoons CH_3COOH + OH^-; \qquad K_b = 5.71 \times 10^{-10}$$

Solutions of ammonium acetate in water are just barely alkaline, because $K_b$ for acetate ion is slightly greater than $K_a$ for ammonium ion.

Another important group of amphiprotic compounds is the family of amino acids, which are especially interesting because they exist predominantly as **zwitterions** or dipolar ions in aqueous media. Alanine is representative of these species:

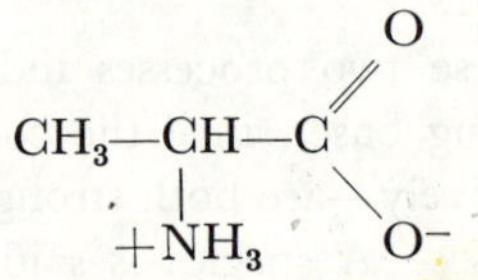

zwitterion form of alanine

Alanine, as well as other amino acids, is amphiprotic because the zwitterion may be protonated or deprotonated according to the following equilibria:

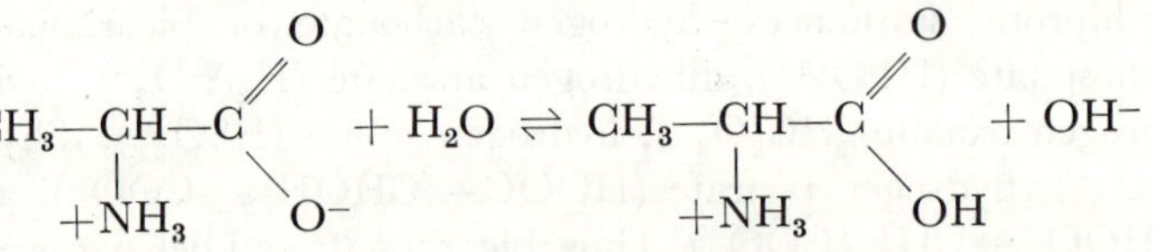

conjugate acid
of alanine

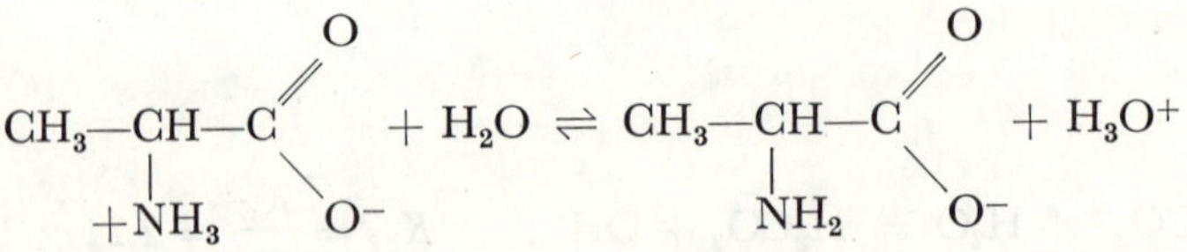

conjugate base
of alanine

*Metal cations as Brønsted-Lowry acids.* All metal cations are hydrated in aqueous media, so it is realistic to speak about species such as $Ce(H_2O)_6^{4+}$, $Fe(H_2O)_6^{3+}$, $Cr(H_2O)_6^{3+}$, $Cu(H_2O)_4^{2+}$, $Mg(H_2O)_4^{2+}$, and $Na(H_2O)_4^+$—although it is not always perfectly clear how many water molecules are coordinated to a particular metal ion. As a general rule, the acid strength of a hydrated metal ion *increases* with increasing charge and decreasing radius of the metal ion.

Singly positive metal cations such as $Ag(H_2O)_2^+$, $Na(H_2O)_4^+$, and $K(H_2O)_4^+$ are too weak as proton donors to display significant acidic properties in water, and so have no effect on the pH of an aqueous solution (aside from the secondary effect of ionic strength on the autoprotolysis of water). Cations of the alkaline earth elements —$Mg(H_2O)_4^{2+}$, $Ca(H_2O)_4^{2+}$, $Sr(H_2O)_4^{2+}$, and $Ba(H_2O)_4^{2+}$— are essentially neutral and do not influence the pH of aqueous solutions. However, dipositive cations of the transition metals such as $Fe(H_2O)_6^{2+}$, $Cu(H_2O)_4^{2+}$, $Cd(H_2O)_4^{2+}$, $Zn(H_2O)_4^{2+}$, and $Ni(H_2O)_4^{2+}$ are roughly comparable in acid strength to the ammonium ion, so that proton-transfer reactions such as

$$Cu(H_2O)_4^{2+} + H_2O \rightleftharpoons Cu(H_2O)_3OH^+ + H_3O^+$$

occur to some extent.

Tripositive cations, especially those of the transition-metal series, show considerable acid strength in water. For example, $Fe(H_2O)_6^{3+}$ is a much stronger proton donor than acetic acid

$$Fe(H_2O)_6^{3+} + H_2O \rightleftharpoons Fe(H_2O)_5OH^{2+} + H_3O^+; \qquad K_{a1} = 9.1 \times 10^{-4}$$

and the conjugate base, $Fe(H_2O)_5OH^{2+}$, is one of the species, along with $Fe(H_2O)_4$-$(OH)_2^+$, which impart the characteristic yellow-brown color to solutions of iron(III) in dilute acid media. Other cations which show analogous behavior are $Cr(H_2O)_6^{3+}$, $Al(H_2O)_6^{3+}$, and $Bi(H_2O)_6^{3+}$. Quadripositive cations such as $Ti(H_2O)_6^{4+}$ and $Ce(H_2O)_6^{4+}$ are so acidic that it is practically impossible for these species to exist in aqueous solutions other than very concentrated sulfuric, nitric, and perchloric acid media. Therefore, proton-transfer processes such as

$$Ce(H_2O)_6^{4+} + H_2O \rightleftharpoons Ce(H_2O)_5OH^{3+} + H_3O^+$$

and

$$Ce(H_2O)_5OH^{3+} + H_2O \rightleftharpoons Ce(H_2O)_4(OH)_2^{2+} + H_3O^+$$

occur extensively, and precipitation of the hydrous oxides of these elements, $TiO_2 \cdot 2\,H_2O$ and $CeO_2 \cdot 2\,H_2O$, can easily take place in an acidic medium, as exemplified by the following overall reaction for titanium(IV):

$$Ti(H_2O)_6^{4+} + 4\,H_2O \rightleftharpoons TiO_2 \cdot 2\,H_2O + 4\,H_3O^+ + 2\,H_2O$$

*Neutral species.* Dissolution of a number of inorganic salts in water does not influence the acidity or basicity of the resulting solution (other than an effect on the activity coefficients of hydrogen ion and hydroxide ion). Among these compounds are the chlorides, bromides, iodides, nitrates, sulfates, and perchlorates of lithium, sodium, and potassium—which consist of the anions of the very strong mineral acids and of the cations of very strong bases. An aqueous solution containing no solute other than one of these substances has a pH of essentially 7 at room temperature. Addition of such a salt to a solution already containing an acid or base does not alter the pH of that acid or base solution. Furthermore, although a cation or anion of one of these neutral substances often accompanies a weak acid ($NH_4Cl$) or a weak base ($NaCH_3COO$), the cation or anion has no effect on the pH of the solution.

## EQUILIBRIUM CALCULATIONS FOR SOLUTIONS OF ACIDS AND BASES IN WATER

This section of the chapter is intended to illustrate methods for calculating the pH of aqueous solutions containing monoprotic and polyprotic acids and their conjugate bases, amphiprotic substances, and buffers. As stated earlier, it is presumed that the temperature is 25°C and that the ion-product constant for water is exactly $1.00 \times 10^{-14}$. In addition, molar concentrations, rather than activities, will be employed, first, because the accurate calculation of activity coefficients is practically impossible for the complicated aqueous media of analytical interest and, second, because the results of approximate computations are usually quite satisfactory for judging the feasibility of an acid-base titration.

### Solutions of Strong Acids and Bases

In simple terms, a strong acid or base is one which is fully dissociated in an aqueous medium. Among the familiar strong acids are hydrochloric, hydrobromic, hydriodic, nitric, sulfuric, and perchloric acids, and the most common strong bases are the hydroxides of sodium, potassium, lithium, and barium.

*Strong acid solutions.* For a solution containing a strong monoprotic acid which is fully ionized, the hydrogen ion concentration is equal to the original molar concentration of the acid. For example, a solution containing 0.1 mole of hydrogen chloride per liter is 0.1 $M$ in hydrogen (hydronium) ions and 0.1 $M$ in chloride ions.

*Example 4-1.* Calculate the pH of each of the following solutions: (a) 0.00150 $F$ HCl and (b) 1.000 liter of solution containing 0.1000 gm of HCl.

(a) $[H^+] = 1.50 \times 10^{-3} M$

$$pH = -\log [H^+] = 2.82$$

(b) Number of moles of HCl $= 0.1000 \text{ gm} \times \dfrac{1 \text{ mole}}{36.46 \text{ gm}}$

$$= 0.00274 \text{ mole}$$

Concentration of HCl $= 0.00274$ mole/liter

$$[H^+] = 2.74 \times 10^{-3} M$$

$$pH = -\log [H^+] = 2.56$$

*Strong base solutions.* In a strong base solution, the hydroxide ion concentration is directly related to the original concentration of the base in a manner analogous to the situation for a strong acid solution, and the pH may be calculated from the hydroxide ion concentration and the ion-product constant for water.

*Example 4-2.* Calculate the pH of a 0.0023 $F$ barium hydroxide solution. Since two moles of hydroxide ion are formed from the ionization of each gram formula weight of barium hydroxide originally present,

$$[OH^-] = 4.6 \times 10^{-3} M$$

$$pOH = -\log [OH^-] = 2.34$$

$$pH = 14.00 - pOH = 11.66$$

### Solutions of Weak Acids and Bases

Weak acids and bases are incompletely dissociated in aqueous solution. As a result, the hydrogen ion concentration or the hydroxide ion concentration is, respectively, always less than the original concentration of the weak acid or weak base.

Calculation of the pH of a weak acid or weak base solution relies upon knowledge of the equilibrium constant ($K_a$ or $K_b$) for the proton-transfer reaction between the solute and water. Values of $K_a$ and $K_b$ for various species are tabulated in Appendix 2.

**Weak acid solutions.** Let us consider several procedures by which the pH of an aqueous solution of a monoprotic weak acid can be calculated. Acetic acid, formic acid, and the ammonium ion will serve as typical examples.

*Example 4–3.* Calculate the pH of a 0.100 $F$ acetic acid solution.

We must first identify the chemical reactions which can produce hydrogen ions in this system, namely the dissociation of acetic acid

$$CH_3COOH \rightleftharpoons H^+ + CH_3COO^-; \quad K_a = 1.75 \times 10^{-5}$$

and the autoprotolysis of water:

$$H_2O \rightleftharpoons H^+ + OH^-; \quad K_w = 1.00 \times 10^{-14}$$

Notice that each of these equilibria is an abbreviated form of the true acid-base reaction which occurs. Thus, for acetic acid we should show the transfer of a proton from acetic acid to water, whereas the autoprotolysis of water proceeds through the transfer of a proton from one water molecule to another. However, as long as we perform equilibrium calculations involving acid-base reactions in the *same* solvent— so that the same proton acceptor (water) is always involved—abbreviated versions of acid-base equilibria as well as the corresponding equilibrium expressions can be employed.

Although some hydrogen ions do come from the autoprotolysis of water, acetic acid is such a stronger acid that water may be regarded as an insignificant source of hydrogen ions (except for extremely dilute acetic acid solutions). Furthermore, the hydrogen ions from acetic acid repress the autoprotolysis of water through the common ion effect. Accordingly, we will assume that the dissociation of acetic acid is the only important equilibrium

$$\frac{[H^+][CH_3COO^-]}{[CH_3COOH]} = K_a = 1.75 \times 10^{-5}$$

but it will be essential to justify the assumption after the calculations are completed. Each molecule of acetic acid which undergoes dissociation produces one hydrogen ion and one acetate ion. Therefore,

$$[H^+] = [CH_3COO^-]$$

At equilibrium the concentration of acetic acid is equal to the initial acid concentration (0.100 $F$) minus the amount dissociated (which is expressible by either the concentration of hydrogen ion or the concentration of acetate ion):

$$[CH_3COOH] = 0.100 - [H^+]$$

Substitution of the latter two relations into the equilibrium expression for acetic acid yields

$$\frac{[H^+]^2}{0.100 - [H^+]} = 1.75 \times 10^{-5}$$

There are at least two ways to solve this equation for the hydrogen ion concentration. Since this relation is a quadratic equation, the formula for the solution of a quadratic equation* may be employed, from which the answer is found to be $[H^+] = 1.32 \times 10^{-3}\ M$ or pH = 2.88. There is, however, an alternate and much simpler procedure for solving the previous equation for $[H^+]$.

Let us assume that the equilibrium concentration of acetic acid is 0.100 $M$. In effect, we are saying that the value of $[H^+]$ is negligible in comparison to 0.100 $M$ or, in other words, that very nearly all of the acetic acid remains undissociated. Since a weak acid is, by definition, one that is only slightly dissociated, this assumption should be expected to be reasonable. Ultimately, the validity of this assumption in any particular case must be demonstrated as shown below. On the basis of the approximation that

$$[CH_3COOH] = 0.100\ M$$

the equilibrium expression for acetic acid takes the form

$$\frac{[H^+]^2}{0.100} = 1.75 \times 10^{-5}$$

which can be solved as follows:

$$[H^+]^2 = 1.75 \times 10^{-6}$$

$$[H^+] = 1.32 \times 10^{-3}\ M; \qquad pH = 2.88$$

Results obtained by means of the quadratic formula and by the second, approximate approach are identical to the third significant figure. Notice that the hydrogen ion concentration ($1.32 \times 10^{-3}\ M$) is only 1.3 per cent of the original concentration of acetic acid ($0.100\ F$). Therefore, the actual equilibrium acetic acid concentration would be

$$[CH_3COOH] = 0.100 - 0.00132 = 0.099\ M$$

which is not significantly different from the assumed approximate value of 0.100 $M$. In general, if the original concentration of weak, monoprotic acid is at least 10,000 times greater than $K_a$, the approximate method will be accurate to within about 1 per cent. Furthermore, if the original acid concentration is only 1000 times larger than $K_a$, the approximate calculation will be in error by less than 2 per cent, which is an acceptable uncertainty for most purposes, especially since activity coefficients have been ignored.

Another assumption made at the beginning of this problem was that the contribution of the autoprotolysis of water to the hydrogen ion concentration is negligible.

---

* An equation of the type $aX^2 + bX + c = 0$ may be solved by means of the quadratic formula, which is

$$X = \frac{-b \pm \sqrt{b^2 - 4ac}}{2a}$$

In the present example, the equilibrium equation may be rearranged to $[H^+]^2 + 1.75 \times 10^{-5}[H^+] - 1.75 \times 10^{-6} = 0$. Applying the quadratic formula, we get

$$[H^+] = \frac{-(1.75 \times 10^{-5}) \pm \sqrt{(1.75 \times 10^{-5})^2 - 4(1)(-1.75 \times 10^{-6})}}{2} = 1.32 \times 10^{-3}\ M$$

(The other root is negative, an impossible situation.)

If the hydrogen ion concentration is taken to be $1.32 \times 10^{-3}\ M$, the value of $[OH^-]$ can be obtained from the relation

$$[OH^-] = \frac{K_w}{[H^+]} = \frac{1.00 \times 10^{-14}}{1.32 \times 10^{-3}} = 7.5 \times 10^{-12}\ M$$

However, since water is the only source of hydroxide ion in this system and since one hydrogen ion and one hydroxide ion are formed for each water molecule ionized, the $[H^+]$ furnished by water is only $7.5 \times 10^{-12}\ M$.

One other technique used to solve equilibrium problems is called the **method of successive approximations**. We shall now use this procedure to consider the dissociation of formic acid.

*Example 4–4.* Calculate the pH of a $0.0250\ F$ formic acid solution.

As in the case of acetic acid, the only important source of hydrogen ions is the dissociation of the weak acid, for which the abbreviated equilibrium is

$$HCOOH \rightleftharpoons H^+ + HCOO^-; \qquad K_a = 1.76 \times 10^{-4}$$

Accordingly, we can set the concentrations of hydrogen ion and formate ion equal to each other:

$$[H^+] = [HCOO^-]$$

At equilibrium, the concentration of undissociated formic acid is

$$[HCOOH] = 0.0250 - [H^+]$$

and the complete equilibrium expression for the dissociation of formic acid takes the form

$$\frac{[H^+][HCOO^-]}{[HCOOH]} = \frac{[H^+]^2}{0.0250 - [H^+]} = 1.76 \times 10^{-4}$$

To solve this equation by the method of successive approximations, we shall first ignore tentatively the $[H^+]$ term in the denominator of the equation (in spite of the fact that the initial formic acid concentration is only about 440 times larger than $K_a$). Therefore, we obtain

$$\frac{[H^+]^2}{0.0250} = 1.76 \times 10^{-4}$$

$$[H^+]^2 = 4.40 \times 10^{-6}$$

$$[H^+] = 2.10 \times 10^{-3}\ M$$

Obviously, this hydrogen ion concentration is hardly negligible in comparison to the initial formic acid concentration, but, instead of abandoning this approach and resorting to the quadratic formula, we shall continue. A closer approximation to the formic acid concentration would be to subtract the $[H^+]$ from the original concentration of formic acid, *i.e.*,

$$[HCOOH] = 0.0250 - 0.0021 = 0.0229\ M$$

This new value for the concentration of formic acid is now substituted back into the equilibrium expression and the equation solved once again for $[H^+]$:

$$\frac{[H^+]^2}{0.0229} = 1.76 \times 10^{-4}$$

$$[H^+]^2 = 4.03 \times 10^{-6}$$

$$[H^+] = 2.01 \times 10^{-3} \, M$$

Through this method of successive approximations, the second approximation to the formic acid concentration (0.0229 $M$) has led to a self-consistent set of values for the concentrations of hydrogen ion and formic acid. In other words, if the formic acid concentration is assumed to be 0.0229 $M$ at equilibrium, the calculated $[H^+]$ is correct because, when the hydrogen ion concentration of 0.00201 $M$ is subtracted from the original formic acid concentration of 0.0250 $M$, the result (0.0230 $M$) agrees with the *assumed* value for [HCOOH]. Therefore, the value for $[H^+]$ of 2.01 $\times$ $10^{-3} \, M$ may be accepted as an accurate result, and the pH is 2.70.

*Example 4–5.* Calculate the pH of a 0.100 $F$ ammonium chloride solution.

Again, there are two sources of hydrogen ions. One is the autoprotolysis of water,

$$H_2O \rightleftharpoons H^+ + OH^-$$

whereas the second source is the dissociation of the ammonium ion,

$$NH_4^+ + H_2O \rightleftharpoons NH_3 + H_3O^+; \qquad K_a = 5.55 \times 10^{-10}$$

which may be represented in abbreviated form as follows:

$$NH_4^+ \rightleftharpoons NH_3 + H^+$$

However, the dissociation constant for ammonium ion ($K_a$) is more than 50,000 times greater than the ion-product constant for water ($K_w$), so the autoprotolysis of water should not contribute significantly to the hydrogen ion concentration. Dissociation of the ammonium ion produces stoichiometrically equal concentrations of $NH_3$ and $H^+$; therefore, neglect of the autoprotolysis of water leads to the relation

$$[NH_3] = [H^+]$$

In addition, ammonium ion is such a weak acid that we may safely approximate its equilibrium concentration as 0.100 $M$.

To calculate the concentration of hydrogen ion, and then the pH, it is only necessary to substitute the relations deduced from the previous arguments into the equilibrium expression:

$$\frac{[NH_3][H^+]}{[NH_4^+]} = \frac{[H^+]^2}{0.100} = 5.55 \times 10^-$$

Solution of this equation gives the results

$$[H^+]^2 = 5.55 \times 10^{-11}$$

$$[H^+] = 7.45 \times 10^{-6} \, M$$

$$pH = 5.13$$

Note that our neglect of the autoprotolysis of water was justified; for, if we accept the hydrogen ion concentration as $7.45 \times 10^{-6} \, M$, then

$$[OH^-] = \frac{K_w}{[H^+]} = \frac{1.00 \times 10^{-14}}{7.45 \times 10^{-6}} = 1.34 \times 10^{-9} \, M$$

which is equivalent to the quantity of hydrogen ion furnished by the autoprotolysis of water. Thus, the hydroxide ion concentration is indeed negligible compared to the concentration of hydrogen ion arising from dissociation of ammonium ions. Furthermore, the true equilibrium concentration of ammonium ion is not exactly $0.100 \, M$, but is smaller by an amount essentially equal to the hydrogen ion concentration ($7.45 \times 10^{-6} \, M$); however, the difference is negligible.

**Weak base solutions.** Equilibrium calculations involving solutions of weak bases differ little from those we have already considered. Pyridine and benzoate ion are representative of the behavior of Brønsted-Lowry bases in water.

*Example 4–6.* Calculate the pH of an aqueous solution of $0.0150 \, F$ pyridine.

In problems concerned with solutions of weak bases in water, we must identify the sources of hydroxide ion. In terms of structural formulas, the proton-transfer reaction between water and pyridine to yield the pyridinium ion and hydroxide ion is

but, for simplicity in the computations which follow, this equilibrium will be written as

$$C_5H_5N + H_2O \rightleftharpoons C_5H_5NH^+ + OH^-$$

A second source of hydroxide ions is the autoprotolysis of water:

$$H_2O \rightleftharpoons H^+ + OH^-; \qquad K_w = 1.00 \times 10^{-14}$$

Although pyridine may be classed as a very weak base, it is still sufficiently stronger as a base than water that we can consider pyridine the only important source of hydroxide ions. Neglecting the autoprotolysis of water, we may equate the concentrations of the pyridinium and hydroxide ions formed from the pyridine-water reaction:

$$[C_5H_5NH^+] = [OH^-]$$

At equilibrium the concentration of pyridine is given by

$$[C_5H_5N] = 0.0150 - [OH^-]$$

Substitution of the latter two relations into the equilibrium expression gives

$$\frac{[C_5H_5NH^+][OH^-]}{[C_5H_5N]} = \frac{[OH^-]^2}{0.0150 - [OH^-]} = 1.70 \times 10^{-9}$$

This equation can be readily solved if we provisionally let the equilibrium concentration of pyridine be $0.0150 \, M$. Such an approximation seems reasonable in view of the weakness of pyridine as a base and the relatively high concentration of pyridine. Therefore,

$$\frac{[OH^-]^2}{0.0150} = 1.70 \times 10^{-9}$$

$$[OH^-]^2 = 2.55 \times 10^{-11}; \qquad [OH^-] = 5.05 \times 10^{-6}\ M$$

$$pOH = -\log[OH^-] = 5.30$$

$$pH = 14.00 - pOH = 8.70$$

Let us check our approximations. Is the autoprotolysis of water a negligible source of hydroxide ions? *Yes;* the concentration of hydroxide ion from water is equal to the hydrogen ion concentration

$$[H^+] = \frac{K_w}{[OH^-]} = \frac{1.00 \times 10^{-14}}{5.05 \times 10^{-6}} = 1.98 \times 10^{-9}\ M$$

since water is the only proton source in this system. Is the equilibrium concentration of pyridine essentially 0.0150 M? *Yes;* the amount of pyridine which reacts is equivalent to the concentration of hydroxide ion, and $5.05 \times 10^{-6}\ M$ can be neglected in comparison to 0.0150 M.

We shall next consider the situation of a very weak base, present at a low concentration.

*Example 4–7.* Calculate the pH of a $2.00 \times 10^{-5}\ F$ sodium benzoate solution.

Although sodium ion is a neutral species, benzoate anion is a very weak base, and the pertinent proton-transfer process can be written as

or

$$C_6H_5COO^- + H_2O \rightleftharpoons C_6H_5COOH + OH^-$$

Once again, the autoprotolysis of water can provide hydroxide ions:

$$H_2O \rightleftharpoons H^+ + OH^-; \qquad K_w = 1.00 \times 10^{-14}$$

Let us attempt to calculate the pH of the sodium benzoate solution according to the method followed in the preceding problem (Example 4–6), by neglecting the autoprotolysis of water and by assuming that the equilibrium concentration of benzoate anion is $2.00 \times 10^{-5}\ M$. Thus,

$$\frac{[C_6H_5COOH][OH^-]}{[C_6H_5COO^-]} = \frac{[OH^-]^2}{2.00 \times 10^{-5}} = 1.59 \times 10^{-10}$$

$$[OH^-]^2 = 3.18 \times 10^{-15}; \quad [OH^-] = 5.64 \times 10^{-8}\ M; \quad pOH = 7.25; \quad pH = 6.75$$

These results make no sense at all. If benzoate anion were absent and we had pure water, we would expect $[OH^-] = 1.00 \times 10^{-7}\ M$ and $pOH = 7.00$. However, a small concentration of benzoate ion is present; so, although benzoate is a very weak base, the solution must be very slightly *alkaline*, not acidic as the approximate calculations above have suggested. Consequently, we cannot ignore the autoprotolysis of water.

In order to solve this problem correctly, we must note that one $H^+$ ion and one $OH^-$ ion are formed for each molecule of water which ionizes, and that each benzoate ion which reacts with water results in the production of one $C_6H_5COOH$ molecule and one $OH^-$ ion. Therefore, the true hydroxide ion concentration is the *sum* of the concentrations of benzoic acid and hydrogen ion:

$$[OH^-] = [C_6H_5COOH] + [H^+]$$

Upon rearrangement, this expression becomes

$$[C_6H_5COOH] = [OH^-] - [H^+]$$

Note that the benzoic acid concentration does *not* equal the hydroxide ion concentration as we had assumed previously when the autoprotolysis of water was neglected. Using the ion-product expression for water, we can rewrite the last equation as

$$[C_6H_5COOH] = [OH^-] - \frac{K_w}{[OH^-]}$$

If the latter relation for the benzoic acid concentration is substituted into the equilibrium expression for the benzoate-water reaction and if, because $K_b$ is so small, the approximation is retained that the equilibrium concentration of benzoate ion is $2.00 \times 10^{-5}\ M$, we have

$$\frac{[C_6H_5COOH][OH^-]}{[C_6H_5COO^-]} = \frac{\left([OH^-] - \dfrac{K_w}{[OH^-]}\right)[OH^-]}{2.00 \times 10^{-5}} = 1.59 \times 10^{-10}$$

$$[OH^-]^2 - K_w = 3.18 \times 10^{-15}$$

$$[OH^-]^2 = 1.32 \times 10^{-14}$$

$$[OH^-] = 1.15 \times 10^{-7}\ M; \qquad pOH = 6.94$$

$$[H^+] = \frac{K_w}{[OH^-]} = \frac{1.00 \times 10^{-14}}{1.15 \times 10^{-7}} = 8.70 \times 10^{-8}\ M; \qquad pH = 7.06$$

To check the validity of the approximation regarding the benzoate concentration at equilibrium, we can compute the concentration of benzoic acid and compare the result to $2.00 \times 10^{-5}\ M$. Earlier it was deduced that

$$[C_6H_5COOH] = [OH^-] - [H^+]$$

Substitution of the values for $[OH^-]$ and $[H^+]$ into this expression yields

$$[C_6H_5COOH] = (1.15 \times 10^{-7}) - (8.70 \times 10^{-8}) = 2.8 \times 10^{-8}\ M$$

Thus, only 0.14 per cent of the original quantity of benzoate ion reacted with water to form benzoic acid, so our calculations are reliable.

## Buffer Solutions

Examples of buffer solutions include mixtures of acetic acid and sodium acetate, ammonium nitrate and ammonia, pyridinium chloride and pyridine, and the sodium salts of dihydrogen phosphate and monohydrogen phosphate, the latter being one of the principal buffers in human blood.

In the discussions which follow, we will consider the calculation of pH values for two representative buffer solutions, and we will examine the more important questions concerning what changes occur in the pH of a buffer upon dilution or upon addition of other acids and bases.

*Example 4–8.* Calculate the pH of a solution initially $0.100\,F$ in acetic acid and $0.100\,F$ in sodium acetate.

For the proton transfer between acetic acid and water,

$$CH_3COOH + H_2O \rightleftharpoons CH_3COO^- + H_3O^+; \qquad K_a = 1.75 \times 10^{-5}$$

we may write abbreviated versions for both the reaction

$$CH_3COOH \rightleftharpoons CH_3COO^- + H^+$$

and the equilibrium expression:

$$\frac{[H^+][CH_3COO^-]}{[CH_3COOH]} = 1.75 \times 10^{-5}$$

To determine the hydrogen ion concentration from the latter equation, and thus the pH, we must obtain expressions for the acetate and acetic acid concentrations. Because acetic acid is a much stronger acid than water and is present in relatively large amount, we can ignore the autoprotolysis of water as a source of hydrogen ions. Therefore, the equilibrium acetic acid concentration is

$$[CH_3COOH] = 0.100 - [H^+]$$

At equilibrium, the concentration of acetate ion is the sum of the acetate added in the form of sodium acetate $(0.100\,M)$ and the acetate formed from the dissociation of acetic acid (given by the hydrogen ion concentration):

$$[CH_3COO^-] = 0.100 + [H^+]$$

Substitution of these relationships into the equilibrium expression gives the following equation:

$$\frac{[H^+](0.100 + [H^+])}{0.100 - [H^+]} = 1.75 \times 10^{-5}$$

It is reasonable to expect that the extent of dissociation of acetic acid (represented by $[H^+]$) is small compared to the original acid concentration because the concentration of acetic acid is much larger than $K_a$ and because the acetate ion furnished by the sodium acetate represses the dissociation of acetic acid. Accordingly, the above equation can be simplified to

$$\frac{[H^+](0.100)}{0.100} = 1.75 \times 10^{-5}$$

from which it is obvious that $[H^+] = 1.75 \times 10^{-5}\ M$. Our assumption that $[H^+]$ is small in comparison to 0.100 $M$ is well justified, and the pH of the solution is 4.76.

*Example 4–9.* Calculate the pH of 100 ml of a solution containing 0.0100 mole of ammonium nitrate and 0.0200 mole of ammonia.*

Ammonia, which can accept a proton from the solvent,

$$NH_3 + H_2O \rightleftharpoons NH_4^+ + OH^-$$

is a stronger base than water, so the autoprotolysis of water does not contribute appreciably to the hydroxide ion concentration. We can solve the equilibrium expression

$$\frac{[NH_4^+][OH^-]}{[NH_3]} = K_b = 1.80 \times 10^{-5}$$

for $[OH^-]$ after appropriate substitutions for $[NH_4^+]$ and $[NH_3]$ are made. From the information given in the statement of the problem, the initial concentrations of ammonium ion and ammonia are 0.100 $M$ and 0.200 $M$, respectively. After equilibrium is established, the ammonium ion concentration is greater than 0.100 $M$ by the amount of $NH_4^+$ formed from ammonia (which is measured in this case by $[OH^-]$)

$$[NH_4^+] = 0.100 + [OH^-]$$

and the concentration of ammonia is

$$[NH_3] = 0.200 - [OH^-]$$

However, as in the previous example of the acetic acid-sodium acetate buffer, these corrections to the ammonium ion and ammonia concentrations are insignificant, so the equilibrium expression may be written

$$\frac{(0.100)[OH^-]}{0.200} = 1.80 \times 10^{-5}$$

Therefore, the solution to this problem is as follows:

$$[OH^-] = 3.60 \times 10^{-5}\ M$$

$$pOH = 4.44$$

$$pH = 9.56$$

---

* In an aqueous medium, dissolved ammonia exists as a hydrated species, conceivably with four water molecules hydrogen bonded to it — the oxygen atoms of three $H_2O$ molecules bonded to the three hydrogens of $NH_3$, with a fourth $H_2O$ molecule sharing one of its hydrogen atoms with the lone electron pair on nitrogen.

*Effect of dilution.* Let us recall the problem discussed in Example 4–8, in which the pH of a solution 0.100 $F$ in both acetic acid and sodium acetate was calculated. We represented the acetic acid concentration at equilibrium as

$$[CH_3COOH] = 0.100 - [H^+]$$

and the acetate ion concentration as

$$[CH_3COO^-] = 0.100 + [H^+]$$

but we assumed and later justified that the term $[H^+]$ in each of these relations was negligibly small, being $1.75 \times 10^{-5}$ $M$. What would happen to the pH if the solution were diluted one-hundred-fold, that is, if the original concentrations of acetic acid and sodium acetate each became 0.00100 $F$?

In this new problem, the rigorous expressions for the acetic acid and acetate ion concentrations would be as follows:

$$[CH_3COOH] = 0.00100 - [H^+]$$
$$[CH_3COO^-] = 0.00100 + [H^+]$$

If the $[H^+]$ term in each relation were negligible compared to 0.00100 $M$, we could assert that no pH change would be observed. This statement must be tested. If the hydrogen ion concentration is taken to be $1.75 \times 10^{-5}$ $M$, we discover that the latter rigorous relations yield the results

$$[CH_3COOH] = 0.00100 - 0.00002 = 0.00098 \ M$$
$$[CH_3COO^-] = 0.00100 + 0.00002 = 0.00102 \ M$$

provided that we round off the hydrogen ion concentration to $2 \times 10^{-5}$ $M$ for these calculations. What we can now do is substitute these new values for $[CH_3COOH]$ and $[CH_3COO^-]$ into the equilibrium expression

$$\frac{[H^+][CH_3COO^-]}{[CH_3COOH]} = 1.75 \times 10^{-5}$$

and calculate a new value for the hydrogen ion concentration:

$$\frac{[H^+](0.00102)}{(0.00098)} = 1.75 \times 10^{-5}$$

$$[H^+] = 1.68 \times 10^{-5} \ M; \quad pH = 4.77$$

This result must now be carefully interpreted.

First, the calculation is valid because the values of $[CH_3COOH]$ and $[CH_3COO^-]$ are essentially the same (0.00098 and 0.00102 $M$, respectively) regardless of which hydrogen ion concentration is employed to correct the original concentrations (0.00100 $M$) of the reagents. However, the difference in the hydrogen ion concentration is significant in two ways. Our calculations have shown that the pH of a buffer solution composed of 0.1 $M$ reagents is 4.76, and that dilution by a factor of 100 causes the pH to increase only to 4.77. This result reveals how well a buffer solution does stabilize the pH even upon great dilution. Yet the small change in pH cautions against indiscriminate dilution. One can foresee considerable difficulty if the buffer is so extensively diluted that the concentrations of acetic acid and acetate ion approach the value of $K_a$.

**Effect of adding acid or base.** Let us assume that we have exactly 100 ml of a buffer solution which is 0.100 $F$ in both acetic acid and sodium acetate. In Example 4–8, we found that the hydrogen ion concentration was $1.75 \times 10^{-5}\ M$, corresponding to a pH of 4.76.

*Example 4–10.* Calculate the pH of the solution which results from the addition of 10.0 ml of 0.100 $F$ HCl to 100 ml of the buffer solution described in Example 4–8.

First, we note that the position of equilibrium of the reaction

$$CH_3COOH \rightleftharpoons H^+ + CH_3COO^-$$

will be shifted toward the left by the addition of the hydrochloric acid. In this case the large excess of acetate ions can consume the hydrochloric acid through the formation of the much weaker acetic acid.

We can calculate the hydrogen ion concentration from the expression

$$\frac{[H^+][CH_3COO^-]}{[CH_3COOH]} = 1.75 \times 10^{-5}$$

In 100 ml of the buffer solution, there are initially $(100)(0.100)$ or 10.0 millimoles of acetic acid and also 10.0 millimoles of acetate ion. Addition of 10.0 ml of 0.100 $F$ hydrochloric acid is equivalent to 1.00 millimole of strong acid, which we can assume reacts virtually completely with acetate ion to form acetic acid. Therefore, after this reaction, 11.0 millimoles of acetic acid are present and 9.0 millimoles of acetate remain. Since we are dealing with the *ratio* of two concentrations, $[CH_3COOH]$ and $[CH_3COO^-]$, it is unnecessary to convert the number of millimoles of each species into molar concentration. Therefore, substitution of these values into the equilibrium expression yields

$$\frac{[H^+](9.0)}{11.0} = 1.75 \times 10^{-5}$$

from which we calculate that $[H^+] = 2.14 \times 10^{-5}\ M$ and pH = 4.67.

*Example 4–11.* Calculate the pH of the solution which results from the addition of 10.0 ml of 0.100 $F$ NaOH to 100 ml of the buffer solution described in Example 4–8.

In this situation the principal equilibrium

$$CH_3COOH \rightleftharpoons H^+ + CH_3COO^-$$

will be shifted toward the right because the addition of strong base (NaOH) will consume or neutralize some of the acetic acid with the formation of an equivalent quantity of acetate ion.

Addition of 10.0 ml of a 0.100 $F$ sodium hydroxide solution to the 100-ml sample of the buffer introduces 1.00 millimole of strong base which reacts with a stoichiometric quantity of acetic acid. Consequently, at equilibrium only 9.0 millimoles of acetic acid remain, whereas a total of 11.0 millimoles of acetate ion is present. From the equilibrium equation

$$\frac{[H^+][CH_3COO^-]}{[CH_3COOH]} = \frac{[H^+](11.0)}{9.0} = 1.75 \times 10^{-5}$$

the hydrogen ion concentration may be found to be $1.43 \times 10^{-5}\ M$ and the pH is 4.84.

These calculations demonstrate how well a buffer solution can maintain constancy of pH—the pH of the solution, initially 4.76, changed only to 4.67 and to 4.84 by the addition, respectively, of 10 ml of $0.1\ F$ hydrochloric acid and 10 ml of a $0.1\ F$ sodium hydroxide solution.

There is a definite limit—called the **buffer capacity**—as to how much acid or base can be added to a given buffer solution before any appreciable change in pH results. Buffer capacity is set by the original amounts of the weak acid and its conjugate base, since one or the other is consumed by reaction with the added acid or base. In the preceding problem (Example 4–11), the original buffer solution contained 10 millimoles of acetic acid, the strong base added reacting with this species. Obviously, the addition of 10 millimoles of strong base would have neutralized all of the acetic acid, and no buffering action would have remained. Calculations similar to those of Examples 4–10 and 4–11 reveal that the pH changes at a rate which increases more and more rapidly as the limiting buffer capacity is approached.

Another closely related point concerning buffer action is that a specific buffer is useful only in a narrow pH range around $pK_a$ for the conjugate acid-base pair which comprises the buffer. For the dissociation of acetic acid, $pK_a$ is 4.76, and the buffering action of an acetic acid-sodium acetate mixture is most effective in a narrow range near pH 4.76, where the concentrations of acetic acid and acetate ion are equal. A formic acid-sodium formate buffer works best in a pH range centered upon 3.75, the numerical value of $pK_a$; since $pK_a$ for the ammonium ion is 9.26, an ammonium nitrate-ammonia buffer is most suitable for controlling the pH near this value.

## Amphiprotic Substances

Earlier in this chapter, we identified several classes of amphiprotic compounds—species capable of behaving as either acids (proton donors) or bases (proton acceptors). We must now establish methods by which pH values for aqueous solutions of these substances can be determined.

*Example 4–12.* Calculate the pH of a $0.100\ F$ sodium bicarbonate ($NaHCO_3$) solution.

This system can be described by means of three competing equilibria—proton donation to water by $HCO_3^-$,

$$HCO_3^- + H_2O \rightleftharpoons CO_3^{2-} + H_3O^+; \qquad K_{a2} = \frac{[H^+][CO_3^{2-}]}{[HCO_3^-]} = 4.68 \times 10^{-11}$$

proton donation to $HCO_3^-$ by water,

$$HCO_3^- + H_2O \rightleftharpoons H_2CO_3 + OH^-; \qquad K_{b2} = \frac{K_w}{K_{a1}} = \frac{[H_2CO_3][OH^-]}{[HCO_3^-]} = 2.24 \times 10^{-8}$$

and the autoprotolysis of water:

$$H_2O \rightleftharpoons H^+ + OH^-; \qquad K_w = [H^+][OH^-] = 1.00 \times 10^{-14}$$

In order to solve this problem in a rigorous fashion, we must utilize charge-balance and mass-balance relationships, as introduced in Chapter 3. Accordingly, because the concentration of positive charges in an aqueous solution of $NaHCO_3$ equals the concentration of negative charges, we can write

$$[Na^+] + [H^+] = [HCO_3^-] + [OH^-] + 2[CO_3^{2-}]$$

but, since the sodium ion concentration is 0.100 $M$,

$$0.100 + [H^+] = [HCO_3^-] + [OH^-] + 2[CO_3^{2-}]$$

In addition, the mass-balance expression requires that the *sum* of the concentrations of $H_2CO_3$, $HCO_3^-$, and $CO_3^{2-}$ be equal to the original concentration of $NaHCO_3$, which is the only source for these three species. Thus,

$$0.100 = [H_2CO_3] + [HCO_3^-] + [CO_3^{2-}]$$

Now let us subtract the mass-balance equation from the charge-balance relation:

$$[H^+] = [CO_3^{2-}] + [OH^-] - [H_2CO_3]$$

Next, we may re-express the preceding equation in terms of the hydrogen ion concentration, the bicarbonate concentration, and the various equilibrium constants:

Rewriting,
$$[H^+] = \frac{K_{a2}[HCO_3^-]}{[H^+]} + \frac{K_w}{[H^+]} - \frac{[H^+][HCO_3^-]}{K_{a1}}$$

$$[H^+]^2 = K_{a2}[HCO_3^-] + K_w - \frac{[H^+]^2[HCO_3^-]}{K_{a1}}$$

collecting terms,

$$[H^+]^2\left\{1 + \frac{[HCO_3^-]}{K_{a1}}\right\} = K_{a2}[HCO_3^-] + K_w$$

and solving for the hydrogen ion concentration, we obtain

$$[H^+] = \sqrt{\frac{K_{a2}[HCO_3^-] + K_w}{1 + \frac{[HCO_3^-]}{K_{a1}}}}$$

We must know the equilibrium concentration of $HCO_3^-$ to employ this relation for calculation of the hydrogen ion concentration. For the present situation, we can conclude that $[HCO_3^-]$ is essentially 0.100 $M$, because the equilibrium constants $K_{a2}$ and $K_{b2}$ for reactions leading to the consumption of bicarbonate are exceedingly small relative to the initial $HCO_3^-$ concentration.

Before we attempt to use this equation, let us examine two simplifications. First, if $[HCO_3^-]$ is significantly greater than $K_{a1}$ in the denominator, the "1" will be negligible in comparison to $[HCO_3^-]/K_{a1}$. Second, the numerator becomes simpler if $K_{a2}[HCO_3^-]$ is larger than $K_w$. If $[HCO_3^-] = 1.0$ $M$, then $K_{a2}[HCO_3^-]$ is 4680 times greater than $K_w$. However, if $[HCO_3^-]$ is only 0.010 $M$, $K_{a2}[HCO_3^-]$ is still 46.8 times greater than $K_w$. For $[HCO_3^-] = 0.010$ $M$, we can neglect the $K_w$ term in the numerator and incur an error of just 1 per cent, which is acceptable (especially when we intend to convert $[H^+]$ to pH). *In conclusion, under the specific condition that the bicarbonate ion concentration is no smaller than* $0.010\ M$, the complicated expression for $[H^+]$ becomes

$$[H^+] = \sqrt{K_{a1}K_{a2}}$$

and the pH of the solution is given by the expression

$$pH = \tfrac{1}{2}(pK_{a1} + pK_{a2}) = \tfrac{1}{2}(6.35 + 10.33) = 8.34$$

It is noteworthy that the pH is independent of the bicarbonate ion concentration, provided that this concentration is at least $0.010\ M$.

*Example 4–13.* Calculate the pH of a $0.0250\ F$ solution of glycine in water.

Glycine, an amino acid which exists predominantly as a zwitterion in aqueous media, displays the same kind of amphiprotic behavior as the hydrogen carbonate ion, either donating a proton to water

$$H_3N^+\!-\!CH_2\!-\!COO^- + H_2O \rightleftharpoons H_2N\!-\!CH_2\!-\!COO^- + H_3O^+;$$

$$K_{a2} = 2.5 \times 10^{-10}$$

or accepting a proton from the solvent:

$$H_3N^+\!-\!CH_2\!-\!COO^- + H_2O \rightleftharpoons H_3N^+\!-\!CH_2\!-\!COOH + OH^-;$$

$$K_{b2} = \frac{K_w}{K_{a1}} = 2.2 \times 10^{-12}$$

From the magnitudes of the respective equilibrium constants, two conclusions can be drawn. First, the equilibrium concentration of glycine will be very close to $0.0250\ M$, because $K_{a2}$ and $K_{b2}$ are so small. Second, an aqueous solution of glycine will have a pH below 7, because $K_{a2}$ for the reaction producing hydrogen ions is larger than $K_{b2}$ for the process yielding hydroxide ions.

Using the charge-balance relation as in the preceding example, we can derive the following equation for the hydrogen ion concentration in a solution of glycine:

$$[H^+] = \sqrt{\frac{K_{a2}[H_3N^+\!-\!CH_2\!-\!COO^-] + K_w}{1 + \dfrac{[H_3N^+\!-\!CH_2\!-\!COO^-]}{K_{a1}}}}$$

However, since $[H_3N^+\!-\!CH_2\!-\!COO^-] > K_{a1}$ and $K_{a2}[H_3N^+\!-\!CH_2\!-\!COO^-] > K_w$, we obtain the approximate, but still accurate relation

$$[H^+] = \sqrt{K_{a1}K_{a2}}$$

from which it follows that

$$pH = \tfrac{1}{2}(pK_{a1} + pK_{a2}) = \tfrac{1}{2}(2.35 + 9.60) = 5.97$$

At a pH of 5.97—called the **isoelectric point**—glycine has no *net* charge (although mainly in the form of a zwitterion) and, therefore, does not undergo migration when a solution containing it is placed in an electric field.

*Example 4–14.* Calculate the pH of a $0.0100\ F$ ammonium formate solution.

This amphiprotic substance differs somewhat from $HCO_3^-$ or glycine, in that the cation of the salt is a weak acid

$$NH_4^+ + H_2O \rightleftharpoons NH_3 + H_3O^+; \quad K_a = \frac{[NH_3][H^+]}{[NH_4^+]} = 5.55 \times 10^{-10}$$

and the anion is a weak base:

$$HCOO^- + H_2O \rightleftharpoons HCOOH + OH^-;$$

$$K_b = \frac{K_w}{K_a'} = \frac{[HCOOH][OH^-]}{[HCOO^-]} = 5.68 \times 10^{-11}$$

Note that $K_a'$ represents the acid dissociation constant of formic acid, the conjugate acid of formate anion. We can expect an aqueous solution of ammonium formate to have a pH less than 7, because the acid strength of ammonium ion exceeds the base strength of formate ion.

A single charge-balance relation

$$[NH_4^+] + [H^+] = [HCOO^-] + [OH^-]$$

and two mass-balance expressions

$$0.0100 = [NH_4^+] + [NH_3]$$

$$0.0100 = [HCOO^-] + [HCOOH]$$

are required for the present problem. If we equate the two mass-balance equations

$$[NH_4^+] + [NH_3] = [HCOO^-] + [HCOOH]$$

and subtract the result from the charge-balance relation, we obtain

$$[H^+] = [NH_3] + [OH^-] - [HCOOH]$$

We can rewrite the latter equation in terms of $[H^+]$, $[NH_4^+]$, and $[HCOO^-]$, as well as appropriate equilibrium constants, by employing the two equilibria mentioned above:

$$[H^+] = \frac{K_a[NH_4^+]}{[H^+]} + \frac{K_w}{[H^+]} - \frac{[HCOO^-][H^+]}{K_a'}$$

Solution of this expression for the hydrogen ion concentration leads to

$$[H^+] = \sqrt{\frac{K_a[NH_4^+] + K_w}{1 + \frac{[HCOO^-]}{K_a'}}}$$

Notice the close similarity between this result and the equations presented earlier for the concentration of hydrogen ion in $NaHCO_3$ and glycine solutions.

It is necessary to know the equilibrium concentrations of $NH_4^+$ and $HCOO^-$ in order to use this relation to calculate the hydrogen ion concentration. Inasmuch as

$K_a$ and $K_b$ are both exceedingly small, the reactions of $NH_4^+$ and $HCOO^-$ with water scarcely proceed at all. Does this not indicate that the equilibrium concentrations of $NH_4^+$ and $HCOO^-$ are equal to the original values of $0.0100\,M$? Before we can answer *yes*, we must consider the proton-transfer reaction

$$NH_4^+ + HCOO^- \rightleftharpoons NH_3 + HCOOH$$

for which the equilibrium expression is

$$\frac{[NH_3][HCOOH]}{[NH_4^+][HCOO^-]} = \frac{K_a K_b}{K_w} = \frac{(5.55 \times 10^{-10})(5.68 \times 10^{-11})}{(1.00 \times 10^{-14})} = 3.15 \times 10^{-6}$$

Obviously, the magnitude of the equilibrium constant shows this reaction to be much more prominent than the other proton-transfer processes. Using the latter equilibrium expression, let us determine the extent of reaction between $NH_4^+$ and $HCOO^-$. At equilibrium, $[NH_4^+] = [HCOO^-]$ and $[NH_3] = [HCOOH]$, so substitution yields

$$\frac{[NH_3][HCOOH]}{[NH_4^+][HCOO^-]} = \frac{[NH_3]^2}{[NH_4^+]^2} = 3.15 \times 10^{-6}$$

or

$$\frac{[NH_3]}{[NH_4^+]} = 1.78 \times 10^{-3}$$

This small ratio indicates that less than 0.2 per cent of the $NH_4^+$ and $HCOO^-$ will be converted, respectively, into $NH_3$ and $HCOOH$ (regardless of the initial concentration of ammonium formate). Thus, the equilibrium concentrations of $NH_4^+$ and $HCOO^-$ are essentially $0.0100\,M$.

Now we should seek to simplify the rigorous equation for $[H^+]$. First, if $[HCOO^-]$ is larger than $K_a'$, the "1" in the denominator can be neglected. Second, if $K_a[NH_4^+]$ is greater than $K_w$ in the numerator, the latter term may be deleted. When $[HCOO^-]$ $= [NH_4^+] = 0.0100\,M$, $[HCOO^-]/K_a' = (0.01)/(1.76 \times 10^{-4}) = 56.8$ (as compared to 1) and $K_a[NH_4^+] = (5.55 \times 10^{-10})(0.01) = 5.55 \times 10^{-12}$ (as compared to $1.00 \times 10^{-14}$). If the numerator and denominator are modified as suggested, we obtain

$$[H^+] = \sqrt{K_a K_a'}$$

where $K_a$ is the acid dissociation constant for the ammonium ion and $K_a'$ is the acid dissociation constant for formic acid. If this simple equation is compared to the rigorous expression for calculation of the hydrogen ion concentration in a $0.0100\,F$ ammonium formate solution, the results ($3.13 \times 10^{-7}$ and $3.10 \times 10^{-7}\,M$, respectively) show that the approximate relation is only 1 per cent in error.

Since we are interested in calculating the pH of the ammonium formate solution, the last equation can be rewritten as

$$pH = \tfrac{1}{2}(pK_a + pK_a')$$

Therefore,

$$pH = \tfrac{1}{2}(9.26 + 3.75) = 6.50$$

As long as the approximations used to obtain the simplified relationships are valid—which requires in the present situation that the concentrations of $NH_4^+$ and $HCOO^-$ be not much less than 0.01 $M$—the pH of an aqueous solution of ammonium formate (or any other similar compound) will be independent of concentration.

## Polyprotic Acids and Their Conjugate Bases

Solutions of polyprotic acids and their conjugate bases may contain several species, and the relative abundances of these species can vary tremendously as the pH changes. Therefore, it is useful to introduce a graphical method which can show at a glance how the composition of a polyprotic acid system is affected by pH.

As an example, let us consider the phosphoric acid system, which involves three proton-transfer equilibria:

$$H_3PO_4 + H_2O \rightleftharpoons H_2PO_4^- + H_3O^+; \quad K_{a1} = \frac{[H^+][H_2PO_4^-]}{[H_3PO_4]} = 7.5 \times 10^{-3}$$

$$H_2PO_4^- + H_2O \rightleftharpoons HPO_4^{2-} + H_3O^+; \quad K_{a2} = \frac{[H^+][HPO_4^{2-}]}{[H_2PO_4^-]} = 6.2 \times 10^{-8}$$

$$HPO_4^{2-} + H_2O \rightleftharpoons PO_4^{3-} + H_3O^+; \quad K_{a3} = \frac{[H^+][PO_4^{3-}]}{[HPO_4^{2-}]} = 4.8 \times 10^{-13}$$

For a given solution, we shall represent by $C$ the sum of the concentrations of all phosphate-containing species:

$$C = [H_3PO_4] + [H_2PO_4^-] + [HPO_4^{2-}] + [PO_4^{3-}]$$

Next, we will define the concentration of each individual phosphate species to be a certain fraction, $\alpha_n$, of the total concentration, $C$, the subscript $n$ designating the number of protons lost by the parent acid to form the species of interest. Thus,

$$[H_3PO_4] = \alpha_0 C, \quad [H_2PO_4^-] = \alpha_1 C, \quad [HPO_4^{2-}] = \alpha_2 C, \quad [PO_4^{3-}] = \alpha_3 C$$

where

$$\alpha_0 + \alpha_1 + \alpha_2 + \alpha_3 = 1$$

By proper combination of the equilibrium expressions for the stepwise dissociation of phosphoric acid, we can obtain relations for the concentrations of the three anionic phosphate species in terms of $[H^+]$, $[H_3PO_4]$, and the equilibrium constants:

$$[H_2PO_4^-] = \frac{K_{a1}[H_3PO_4]}{[H^+]}$$

$$[HPO_4^{2-}] = \frac{K_{a1}K_{a2}[H_3PO_4]}{[H^+]^2}$$

$$[PO_4^{3-}] = \frac{K_{a1}K_{a2}K_{a3}[H_3PO_4]}{[H^+]^3}$$

Substitution of these results term-by-term into the earlier mass-balance equation yields

$$C = [H_3PO_4] + \frac{K_{a1}[H_3PO_4]}{[H^+]} + \frac{K_{a1}K_{a2}[H_3PO_4]}{[H^+]^2} + \frac{K_{a1}K_{a2}K_{a3}[H_3PO_4]}{[H^+]^3}$$

which can be rearranged to give

$$[H_3PO_4] = \frac{C[H^+]^3}{[H^+]^3 + K_{a1}[H^+]^2 + K_{a1}K_{a2}[H^+] + K_{a1}K_{a2}K_{a3}}$$

Inserting the latter expression into the relation for the definition of $\alpha_0$, we get

$$\alpha_0 = \frac{[H^+]^3}{[H^+]^3 + K_{a1}[H^+]^2 + K_{a1}K_{a2}[H^+] + K_{a1}K_{a2}K_{a3}}$$

It can be similarly shown that

$$\alpha_1 = \frac{K_{a1}[H^+]^2}{[H^+]^3 + K_{a1}[H^+]^2 + K_{a1}K_{a2}[H^+] + K_{a1}K_{a2}K_{a3}}$$

$$\alpha_2 = \frac{K_{a1}K_{a2}[H^+]}{[H^+]^3 + K_{a1}[H^+]^2 + K_{a1}K_{a2}[H^+] + K_{a1}K_{a2}K_{a3}}$$

and

$$\alpha_3 = \frac{K_{a1}K_{a2}K_{a3}}{[H^+]^3 + K_{a1}[H^+]^2 + K_{a1}K_{a2}[H^+] + K_{a1}K_{a2}K_{a3}}$$

Notice that the denominator for each $\alpha_n$ value is the same function of the hydrogen ion concentration and the various dissociation constants; although the denominator will be different for every pH, it will be identical for each $\alpha_n$ value. Furthermore, the relations for the various $\alpha_n$ values emphasize the important fact that the fraction of each phosphate-containing species depends upon pH, but not on the total concentration, $C$.

We will now examine a typical situation which illustrates the use of these concepts.

*Example 4–15.* Calculate the individual concentrations of all phosphate species in 1.00 liter of solution containing a total of 0.100 mole of phosphate species at pH 8.00.

To begin, let us evaluate each of the four terms which appear in the various $\alpha_n$ expressions:

$$[H^+]^3 = (1.00 \times 10^{-8})^3 = 1.00 \times 10^{-24}$$

$$K_{a1}[H^+]^2 = (7.5 \times 10^{-3})(1.00 \times 10^{-8})^2 = 7.50 \times 10^{-19}$$

$$K_{a1}K_{a2}[H^+] = (7.5 \times 10^{-3})(6.2 \times 10^{-8})(1.00 \times 10^{-8}) = 4.65 \times 10^{-18}$$

$$K_{a1}K_{a2}K_{a3} = (7.5 \times 10^{-3})(6.2 \times 10^{-8})(4.8 \times 10^{-13}) = 2.23 \times 10^{-22}$$

Before proceeding, we should note that the first term is proportional to $[H_3PO_4]$, the second term to $[H_2PO_4^-]$, the third term to $[HPO_4^{2-}]$, and the fourth term to $[PO_4^{3-}]$. Since the magnitudes of these terms, from largest to smallest, are in the order, third, second, fourth, and first, we can immediately conclude that the relative concentrations of the four phosphate species are $[HPO_4^{2-}] > [H_2PO_4^-] > [PO_4^{3-}] > [H_3PO_4]$.

We obtain the denominator of each $\alpha_n$ value by adding the four terms together; the result is

$$[H^+]^3 + K_{a1}[H^+]^2 + K_{a1}K_{a2}[H^+] + K_{a1}K_{a2}K_{a3} = 5.40 \times 10^{-18}$$

Each $\alpha_n$ value is, in turn, computed as follows:

$$\alpha_0 = \frac{1.00 \times 10^{-24}}{5.40 \times 10^{-18}} = 1.85 \times 10^{-7}$$

$$\alpha_1 = \frac{7.50 \times 10^{-19}}{5.40 \times 10^{-18}} = 0.139$$

$$\alpha_2 = \frac{4.65 \times 10^{-18}}{5.40 \times 10^{-18}} = 0.861$$

$$\alpha_3 = \frac{2.23 \times 10^{-22}}{5.40 \times 10^{-18}} = 4.13 \times 10^{-5}$$

Since a total of 0.100 mole of phosphate species is present in 1.00 liter of solution, $C = 0.100\ M$. Finally, by utilizing the fundamental definitions for the $\alpha_n$ values, we can calculate the desired concentrations of the individual phosphate species:

$$[H_3PO_4] = \alpha_0 C = (1.85 \times 10^{-7})(0.100) = 1.85 \times 10^{-8}\ M$$

$$[H_2PO_4^-] = \alpha_1 C = (0.139)(0.100) = 1.39 \times 10^{-2}\ M$$

$$[HPO_4^{2-}] = \alpha_2 C = (0.861)(0.100) = 8.61 \times 10^{-2}\ M$$

$$[PO_4^{3-}] = \alpha_3 C = (4.13 \times 10^{-5})(0.100) = 4.13 \times 10^{-6}\ M$$

Calculations of $\alpha_n$ values for the phosphoric acid system have been performed, according to the procedure described, for the entire pH range from 0 to 14. Results of these computations are seen in Figure 4–1 as four curves, each showing how the

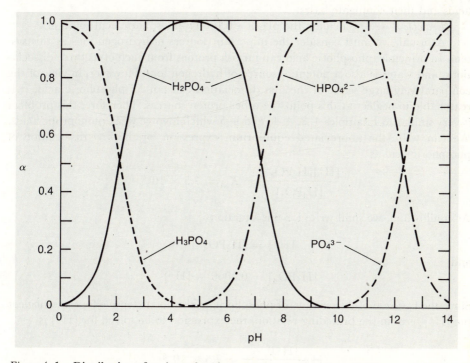

*Figure 4–1.* Distribution of various phosphate species as a function of pH.

fraction of one of the phosphate species changes with pH. Once such a distribution diagram is constructed for a polyprotic acid system, it is possible to tell by inspection which species are predominant at any pH as well as their relative concentrations. It is clear that, between pH 0 and 4.6, the only important species are $H_3PO_4$ and $H_2PO_4^-$. In the region from pH 4.6 to 9.8, only $H_2PO_4^-$ and $HPO_4^{2-}$ are present in significant amounts, whereas $HPO_4^{2-}$ and $PO_4^{3-}$ are the most abundant species between pH 9.8 and 14. In addition, notice that the curves for a conjugate acid-base pair intersect at $\alpha_n = 0.5$ and at a pH value numerically equal to $pK_a$ for the pertinent equilibrium. Thus, the curves for $H_2PO_4^-$ and $HPO_4^{2-}$ cross at pH 7.2, which is the numerical value of $pK_{a2}$ for the reaction

$$H_2PO_4^- + H_2O \rightleftharpoons HPO_4^{2-} + H_3O^+; \qquad K_{a2} = 6.2 \times 10^{-8}, \quad pK_{a2} = 7.2$$

As another illustration of the usefulness of Figure 4–1, let us consider the following problem.

*Example 4–16.* If 150 ml of an aqueous solution at pH 3.00 contains a total of 0.0250 mole of phosphate species, what are the concentrations of the *predominant* species?

Inspection of Figure 4–1 reveals that $H_3PO_4$ and $H_2PO_4^-$ are the only major phosphate species at pH 3.00, and that $\alpha_0$ and $\alpha_1$ are 0.12 and 0.88, respectively. In 150 ml of solution, there is 0.0250 mole of phosphate species, corresponding to a total concentration $(C)$ of 0.167 $M$. Therefore,

$$[H_3PO_4] = \alpha_0 C = (0.12)(0.167) = 0.020 \ M$$

and

$$[H_2PO_4^-] = \alpha_1 C = (0.88)(0.167) = 0.147 \ M$$

In addition to the preceding examples, many equilibrium problems concerning polyprotic acid systems involve concepts already considered for weak monoprotic acids and their conjugate bases.

*Example 4–17.* Calculate the pH of a 0.1000 $F$ phosphoric acid solution.

As usual, we must consider the important sources of hydrogen ions in this system. In principle, phosphoric acid can furnish protons from each of its three dissociations, and water is also a potential source of hydrogen ions. However, in view of the comparatively large value of the first dissociation constant of phosphoric acid, it is reasonable to neglect at this point the other proton sources. Therefore, this problem is very similar to Examples 4–3, 4–4, and 4–5 which involve weak monoprotic acids. We can write the abbreviated equilibrium expression for the first dissociation of phosphoric acid as

$$\frac{[H^+][H_2PO_4^-]}{[H_3PO_4]} = K_{a1} = 7.5 \times 10^{-3}$$

At equilibrium, we shall write for this situation

$$[H^+] = [H_2PO_4^-]$$

and

$$[H_3PO_4] = 0.1000 - [H^+]$$

Since the first step of dissociation of phosphoric acid is too extensive for us to neglect the $[H^+]$ term in the preceding relation, the expression to be solved for $[H^+]$ is

$$\frac{[H^+]^2}{0.1000 - [H^+]} = 7.5 \times 10^{-3}$$

This last equation may be solved by means of the quadratic formula or the method of successive approximations, and the result is [H⁺] = 2.39 × 10⁻² *M*; so pH = 1.62.

Dissociation of $H_2PO_4^-$ is the next most likely source of hydrogen ions. We can show that it and, hence, the other sources of protons can be neglected. If we provisionally take [H⁺] = 2.39 × 10⁻² *M* and $[H_2PO_4^-]$ = 2.39 × 10⁻² *M*, the concentration of $HPO_4^{2-}$ can be calculated from a rearranged form of the second dissociation-constant expression for phosphoric acid. Thus,

$$[HPO_4^{2-}] = K_{a2} \frac{[H_2PO_4^-]}{[H^+]} = 6.2 \times 10^{-8} \frac{(0.0239)}{(0.0239)} = 6.2 \times 10^{-8} \, M$$

Since the equilibrium concentration of $HPO_4^{2-}$ is so small, it is apparent that an insignificant quantity of $H_2PO_4^-$ has dissociated under these conditions. Consequently, the pH is 1.62 as calculated.

Examples of other problems commonly encountered with polyprotic acid systems include (a) calculation of the pH of a mixture of $H_3PO_4$ and $H_2PO_4^-$ or a mixture of $HPO_4^{2-}$ and $PO_4^{3-}$, which is analogous to a calculation of the pH of an acetic acid-sodium acetate buffer (Example 4–8), and (b) calculation of the pH of a solution of $Na_2CO_3$ or $Na_3PO_4$, which is essentially identical to a calculation of the pH of a pyridine or sodium benzoate solution (Examples 4–6 and 4–7).

## ACID-BASE TITRIMETRY

In a **titration**, the amount of a substance is determined by measurement of the quantity of a reagent—called the **titrant**—required to react stoichiometrically with that substance. Operationally, a titration involves the carefully measured and controlled addition of titrant to a solution of the substance to be determined until the reaction between the two species is judged to be complete. In classical volumetric analysis, the titrant is a solution of precisely known concentration—that is, a **standard solution**—and the amount of a substance is calculated from knowledge of the concentration and volume of the standard solution used in the titration. For the analysis of solutions containing acidic or basic constituents by means of acid-base titrimetry, standard solutions of acids and bases must be available.

That point during the course of a titration at which the amount of titrant added is stoichiometrically equivalent to the substance being determined is defined as the theoretical **equivalence point**. In practice, one looks for an abrupt change in some property of the solution being titrated to indicate that the equivalence point has been reached; the point at which this sharp change occurs is termed the experimental **end point**. Among the solution properties that may undergo change and thereby serve as a basis for end-point detection are color, refractive index, conductivity, and temperature; in addition, changes in the potential difference between indicator and reference electrodes immersed in a solution are often used to locate end points. Because the change in a particular solution property is not always adequately sharp, and because an abrupt change which does occur may not be detected properly, the end point seldom coincides perfectly with the theoretical equivalence point, although one strives to minimize the **titration error**—the difference between the end point and the equivalence point—as much as possible.

A chemical reaction must fulfill four requirements if it is to serve as the basis of a titrimetric method of analysis. First, the reaction must proceed rapidly, so that one can recognize without unnecessary waiting whether the end point has been reached or whether the titration should be continued. Second, the reaction must proceed nearly to completion—that is, it must have a large equilibrium constant—so that an adequately sharp end-point signal will be observed. Third, the reaction must proceed according to well-defined stoichiometry, so that the amount of substance being determined can be calculated from the titration data. Fourth, a convenient method of end-point detection must be available.

Acid-base systems meet all of these requirements exceptionally well. Proton transfer between an acid and a base is an exceedingly fast process which, by its very nature, is stoichiometrically exact. Moreover, equilibrium constants for many acid-base reactions are sufficiently large that sharp and reproducible end points are easily attainable. Finally, there are numerous acid-base indicators which undergo distinct color changes with a variation in pH and which can be employed for the accurate location of equivalence points.

## Preparation and Standardization of Solutions of Strong Acids

Titrant solutions of sulfuric acid, nitric acid, perchloric acid, and hydrochloric acid are prepared by dilution of the concentrated, commercially available acids and can be standardized by a variety of methods. Sulfuric acid is not suitable, however, for titrations of solutions containing a cation that forms an insoluble sulfate precipitate because the end point may be obscured. Dilute nitric acid solutions near the $0.1\,F$ concentration level and perchloric acid solutions as concentrated as $1.0\,F$ are stable for long periods of time and make excellent standard acid solutions. However, hydrochloric acid is the most preferred titrant, because of the purity of the concentrated, commercially available reagent and because of the long shelf-life of dilute solutions. In addition, there are numerous highly precise methods for the accurate standardization of hydrochloric acid titrants.

Any one of several primary standard bases may be used for the standardization of a solution of hydrochloric acid or one of the other strong acids. Reagent-grade sodium carbonate is commercially available in very high purity for standardization purposes. An accurately weighed sample of sodium carbonate is dissolved in water, and the resulting solution is titrated with hydrochloric acid to form carbonic acid,

$$CO_3{}^{2-} + 2\,H^+ \rightleftharpoons H_2CO_3$$

the end point being easily determined with the aid of an acid-base indicator.

Tris(hydroxymethyl)aminomethane, sometimes abbreviated THAM and also known as "tris," is available in analytical-reagent-grade purity, and is an excellent primary standard. It reacts with hydrochloric acid, much the same as does ammonia, according to the equation

$$H^+ + (CH_2OH)_3CNH_2 \rightleftharpoons (CH_2OH)_3CNH_3{}^+$$

Other primary standards which may be used for the standardization of hydrochloric acid solutions include calcium carbonate and borax $(Na_2B_4O_7 \cdot 10\,H_2O)$. Alternatively, a hydrochloric acid solution may be standardized gravimetrically by precipitating and weighing the chloride as silver chloride.

Another method to obtain a standard hydrochloric acid solution involves its preparation from constant-boiling hydrochloric acid. When concentrated $(12\,F)$ hydrochloric acid is boiled, hydrogen chloride and water molecules are distilled away until eventually the remaining solution approaches a definite, precisely known composition (close to $6.1\,F$) which is fixed by the prevailing barometric pressure.

Thereafter, this constant-boiling mixture continues to distill off, and it can be collected and stored for future use. Although the concentration of the constant-boiling acid is rather high, quantitative dilution with distilled water suffices to bring it to a suitable concentration.

## Preparation and Standardization of Solutions of Strong Bases

Standard base solutions are usually prepared from sodium hydroxide. Two less important, but occasionally used, reagents for the preparation of standard base solutions are potassium hydroxide and barium hydroxide. None of these compounds is available in a pure form; solid sodium hydroxide, for example, is invariably contaminated by moisture and by small amounts (up to 2 per cent) of sodium carbonate as well as sodium chloride. Consequently, a sodium hydroxide solution of approximately the required concentration must be prepared and standardized.

Solutions of barium hydroxide absorb carbon dioxide from the atmosphere with the resultant precipitation of insoluble barium carbonate and the simultaneous decrease in the concentration of hydroxide ion:

$$CO_2 + 2\,OH^- \rightleftharpoons CO_3^{2-} + H_2O$$

$$Ba^{2+} + CO_3^{2-} \rightleftharpoons BaCO_3$$

Sodium and potassium hydroxide solutions absorb carbon dioxide as well, but the carbonate which is formed remains soluble. In the case of a sodium or potassium hydroxide titrant, the absorbed carbon dioxide does not necessarily cause a change in the effective concentration of the strong base solution, because a carbonate ion can react with two protons just as the two hydroxide ions which are neutralized when the carbon dioxide molecule is originally absorbed. However, sometimes an end-point indicator is employed which changes color at a pH corresponding only to the titration of carbonate to bicarbonate ion, so that the effective concentration of the base solution is smaller than expected. This difficulty can be remedied, however, if one is careful to select an end-point indicator which does change color at a pH corresponding to the addition of *two* hydrogen ions to the carbonate ion.

It is highly desirable to remove carbonate ion from a sodium hydroxide solution prior to its standardization, and once standardized the solution should be stored away from contact with the atmosphere. An excellent method for preparation of a carbonate-free solution takes advantage of the insolubility of sodium carbonate in very concentrated sodium hydroxide solutions. If a 50 per cent sodium hydroxide solution is prepared, the sodium carbonate impurity is virtually insoluble and will settle to the bottom of the container within a day or so. Then, the clear, carbonate-free sodium hydroxide syrup may be carefully decanted and diluted appropriately with freshly boiled and cooled distilled water. Since glass containers are attacked by sodium hydroxide solutions, they should not be employed as storage vessels; however, screw-cap polyethylene bottles are especially serviceable for the short-term storage and use of these solutions.

There are several primary standard acids, any one of which may be used to standardize a solution of sodium, potassium, or barium hydroxide. Potassium acid phthalate ($KHC_8H_4O_4$) has the advantages of high equivalent weight and purity, stability on drying, and ready availability. On the other hand, it has the disadvantage of being a weak acid so that it is generally suitable for standardization of carbonate-free base solutions only. Sulfamic acid ($NH_2SO_3H$) is a very good primary standard, being readily available in a pure form, rather inexpensive, and a strong acid. Oxalic acid dihydrate ($H_2C_2O_4 \cdot 2\ H_2O$) and benzoic acid ($C_6H_5COOH$) are occasionally used as primary standard acids.

It is unnecessary to standardize both a hydrochloric acid and a sodium hydroxide solution against primary standards, since the concentration of either one may be established by titration against the other.

## Acid-Base Indicators for End-Point Detection

Acid-base indicators are in reality nothing more than highly colored organic dye molecules possessing acidic or basic functional groups. In a solution containing one of these substances, variations in pH cause the acid-base indicator molecule to lose or gain a proton and, because substantial structural rearrangement accompanies the proton-transfer process, a dramatic color change occurs.

As a matter of fact, the behavior of an acid-base indicator can be understood in terms of an equilibrium involving a weak monoprotic acid and its conjugate base

$$HIn + H_2O \rightleftharpoons In^- + H_3O^+$$

where HIn symbolizes the acid form of the indicator and $In^-$ denotes the conjugate-base form of the indicator. Needless to say, the colors of the two forms are different. Although we have represented the acid form as a neutral molecule, it often has a single negative charge, in which case the conjugate base is a doubly charged anion; the important point is that the conjugate-base form always bears one more negative charge than the acid form. An increase in the hydrogen ion concentration—or a decrease in pH—favors the formation of HIn and the appearance in solution of the color of the acid form. Conversely, an increase in pH shifts the proton-transfer equilibrium toward production of more $In^-$, which will impart its color to the solution.

Let us write the equilibrium expression for the proton-transfer reaction of the acid-base indicator

$$\frac{[In^-][H^+]}{[HIn]} = K_a$$

where $K_a$ is the acid dissociation constant for the indicator. Simple rearrangement of this relation yields

$$\frac{[In^-]}{[HIn]} = \frac{K_a}{[H^+]}$$

which shows that the ratio of the acid and base forms of an indicator varies with the hydrogen ion concentration. To the average human eye, a solution containing two colored species, HIn and $In^-$, will exhibit only the color of HIn if the concentration of HIn is at least 10 times greater than that of $In^-$. On the other hand, if the concentration of $In^-$ is 10 or more times larger than the HIn concentration, the solution will appear to have the color of $In^-$. Therefore, a solution of an acid-base indicator will show the color of the acid form if

$$\frac{[In^-]}{[HIn]} \leq 0.1$$

whereas the color of the base form will be observed if

$$\frac{[\text{In}^-]}{[\text{HIn}]} \geq 10$$

Only when the pH varies so that the ratio of $[\text{In}^-]$ to $[\text{HIn}]$ shifts from at least 0.1 to 10 will the human eye detect a color change. For values of the ratio between these two limits, the solution will appear to have an intermediate color—a mixture of the colors of HIn and In$^-$.

Substitution of the preceding concentration ratios into the equilibrium expression for the acid-base indicator allows us to determine what variation in the hydrogen ion concentration is necessary to produce a color change. When a solution of the indicator exhibits the color of the *acid* form

$$\frac{[\text{In}^-]}{[\text{HIn}]} = \frac{K_a}{[\text{H}^+]} = 0.1; \quad [\text{H}^+] = 10 K_a$$

but when the solution has the color of the *base* form of the indicator

$$\frac{[\text{In}^-]}{[\text{HIn}]} = \frac{K_a}{[\text{H}^+]} = 10; \quad [\text{H}^+] = 0.1 K_a$$

These results may be conveniently expressed in terms of $pK_a$ and pH as follows:

If $\text{pH} = pK_a - 1$, or less, the solution will have the color of the acid form (HIn).

If $\text{pH} = pK_a + 1$, or more, the solution will have the color of the base form (In$^-$).

An acid-base indicator which has, for example, a dissociation constant $(K_a)$ of $1.00 \times 10^{-8}$, or a $pK_a$ value of 8.00, will undergo a color change when the pH of a solution containing it varies from 7 to 9; the opposite color change occurs if the pH varies from 9 to 7.

A list of acid-base indicators is presented in Table 4–1, along with the colors of the acid and base forms of each indicator and the pH value at which one or the other form is sufficiently predominant to impart its color to the solution.

It is of interest to demonstrate the validity of the previous discussion by reference to a specific indicator. Methyl red, one of the azo-dye class of acid-base indicators, is an excellent example. According to Table 4–1, the red-colored acid form predominates at pH 4.2 and below, whereas the yellow conjugate base is the important form at and above pH 6.2. Structures of the two forms of methyl red, and the fundamental acid-base equilibrium, are given by

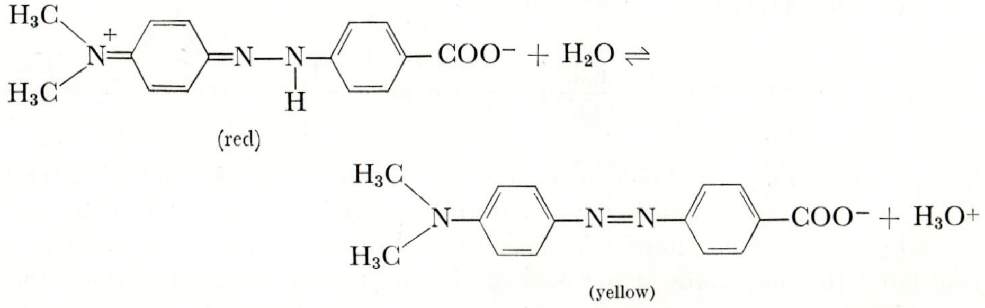

(red)

(yellow)

**Table 4–1. Selected Acid-Base Indicators**

| Indicator | Color Change | | pH Transition Interval | |
| --- | --- | --- | --- | --- |
| | Acid Form | Base Form | Acid Form Predominant at pH | Base Form Predominant at pH |
| Picric acid | colorless | yellow | 0.1 | 0.8 |
| Paramethyl red | red | yellow | 1.0 | 3.0 |
| 2,6-Dinitrophenol | colorless | yellow | 2.0 | 4.0 |
| Bromphenol blue | yellow | blue | 3.0 | 4.6 |
| Congo red | blue | red | 3.0 | 5.0 |
| Methyl orange | red | yellow | 3.1 | 4.4 |
| Ethyl orange | red | yellow | 3.4 | 4.5 |
| Alizarin red S | yellow | purple | 3.7 | 5.0 |
| Bromcresol green | yellow | blue | 3.8 | 5.4 |
| Methyl red | red | yellow | 4.2 | 6.2 |
| Propyl red | red | yellow | 4.6 | 6.6 |
| Methyl purple | purple | green | 4.8 | 5.4 |
| Chlorophenol red | yellow | red | 4.8 | 6.4 |
| Paranitrophenol | colorless | yellow | 5.0 | 7.0 |
| Bromcresol purple | yellow | purple | 5.2 | 6.8 |
| Bromthymol blue | yellow | blue | 6.0 | 7.6 |
| Brilliant yellow | yellow | orange | 6.6 | 8.0 |
| Neutral red | red | amber | 6.7 | 8.0 |
| Phenol red | yellow | red | 6.7 | 8.4 |
| Metanitrophenol | colorless | yellow | 6.7 | 8.6 |
| Phenolphthalein | colorless | pink | 8.0 | 9.6 |
| Thymolphthalein | colorless | blue | 9.3 | 10.6 |
| 2,4,6-Trinitrotoluene | colorless | orange | 12.0 | 14.0 |

and the equilibrium expression may be written as

$$\frac{[In^-][H^+]}{[HIn]} = K_a = 7.9 \times 10^{-6}; \quad pK_a = 5.1$$

where [HIn] is the concentration of the red acid form and [In$^-$] is the concentration of the yellow base form. We can calculate the ratio [HIn]/[In$^-$] at pH 4.2 and at pH 6.2 as follows:

At pH 4.2, [H$^+$] = 6.3 $\times$ 10$^{-5}$ $M$ and

$$\frac{[HIn]}{[In^-]} = \frac{6.3 \times 10^{-5}}{7.9 \times 10^{-6}} = 8.0$$

At pH 6.2, [H$^+$] = 6.3 $\times$ 10$^{-7}$ $M$ and

$$\frac{[HIn]}{[In^-]} = \frac{6.3 \times 10^{-7}}{7.9 \times 10^{-6}} = 0.080 = \frac{1}{12.5}$$

Thus, in the case of methyl red, we observe the red color of the indicator when there is 8.0 times more acid form than base form present. However, in order for us to see the indicator in its yellow form, there must be present 12.5 times more base form than acid form. In other words at pH 4.2 (and below) the human eye recognizes that methyl red is present as the red acid form and that it exists as the yellow base form if the pH becomes 6.2 (and above). This range of pH values over which the acid-base indicator is seen to change color is usually called the **pH transition interval**. It should be noted that the pH transition interval for the indicators listed in Table 4–1 is often about two pH units, although it may be smaller than this for some indicators because the human eye responds more readily to some colors than to others and because one form of an indicator may be more intensely colored than the other, even at identical concentrations.

*Effect of concentration of indicator.* Most of the indicators listed in Table 4–1 possess acid and base forms which are both colored. If the total concentration of a two-color indicator is increased, the individual concentrations of the acid and base forms will increase proportionally, and the pH transition interval should remain unchanged, even though the color intensities are increased. Several of the acid-base indicators included in Table 4–1 are so-called one-color indicators.

Phenolphthalein is the most familiar example of such an indicator. It has a colorless, acid form and a pink-colored base form in equilibrium with each other as follows:

(colorless)              (pink)

This indicator is widely employed for titrations of strong acids with strong bases, the end point being signaled by the first permanent appearance of a pink color throughout the solution. At the end point, the concentration of the pink base form, $[In^-]$, which is discernible to the eye is very small and quite reproducible regardless of the total concentration of phenolphthalein present or the concentrations of the acids and bases involved in the titration. Let us write the equilibrium expression for the dissociation of the colorless acid form of phenolphthalein as

$$\frac{[In^-][H^+]}{[HIn]} = K_a$$

or as

$$[H^+] = \frac{K_a}{[In^-]} \cdot [HIn]$$

where $K_a$ is the dissociation constant and $[In^-]$ is the "minimum detectable concentration" of the pink base form, which we may assume to be constant. We can see from the last equation that the hydrogen ion concentration at which the pink end-point color appears will depend on $[HIn]$, the concentration of phenolphthalein in its acid form. If more indicator is present, $[H^+]$ at the end point will be higher and the pH lower. Conversely, if less indicator is present, the end point will shift toward higher pH values.

There is one further point in regard to the concentration of the indicator—namely, that the conversion of an indicator from one form to another actually uses up some titrant. Clearly, the volume of titrant thus consumed must be negligible in comparison to the amount of acid or base involved in the main titration reaction; otherwise, a significant "indicator error" is introduced. Therefore, the indicator concentration should not be any larger than required to render the end-point color change visible. This "indicator error" is often of slight but measurable magnitude, and it is one reason why experimental end points and theoretical equivalence points may not coincide perfectly. A correction for the "indicator error" can be achieved by means of a **blank titration**. A solution of the same composition as the real sample, containing the indicator but not the acid or base being determined, is titrated to the same end point as in the real titration, and the small volume of titrant is subsequently subtracted from the total volume used in the real titration.

## Strong Acid-Strong Base Titrations

A titration curve for an acid-base reaction consists of a plot of pH versus the volume of titrant added. Such a curve may be constructed by a consideration of the pertinent acid-base equilibria according to the methods developed earlier in this chapter. Theoretical titration curves are of importance because they provide information concerning the feasibility of a titration, and they are useful for purposes of choosing what end-point indicator should be employed.

*Titration curve for a strong acid.* Let us consider the titration of 50.00 ml of 0.1000 $F$ hydrochloric acid with a 0.1000 $F$ sodium hydroxide solution. Prior to the equivalence point of the titration, there are two sources of hydrogen ions. These are the hydrochloric acid itself and the autoprotolysis of water. In general, the hydrogen ion concentration furnished by the hydrochloric acid is the number of millimoles of acid *remaining untitrated* divided by the solution volume in milliliters. Now, since the autoprotolysis of water yields equal concentrations of hydrogen ion and hydroxide ion, and since water is the *only* source of hydroxide ions, the hydroxide concentration is a direct measure of the hydrogen ion concentration from water. Therefore,

$$[H^+] = \frac{\text{millimoles of acid untitrated}}{\text{solution volume in milliliters}} + [OH^-]$$

or

$$[H^+] = \frac{\text{millimoles of acid untitrated}}{\text{solution volume in milliliters}} + \frac{K_w}{[H^+]}$$

Prior to the addition of any sodium hydroxide titrant, the hydrogen ion concentration contributed by hydrochloric acid is 0.1000 $M$. If we assume that the *total* hydrogen ion concentration is essentially 0.1000 $M$, the hydrogen ion contributed from water is exactly equal to the hydroxide ion concentration and is given by

$$[OH^-] = \frac{K_w}{[H^+]} = \frac{1.00 \times 10^{-14}}{0.1000} = 1.00 \times 10^{-13} \, M$$

Quite obviously, the concentration of hydrogen ions from the autoprotolysis of water is very unimportant. In fact, the autoprotolysis of water does not become significant until one has added *more* than 49.999 ml of the exactly 50 ml of titrant required, so we will neglect the contribution of water. Therefore, at the start of the titration, before any sodium hydroxide titrant has been added, the hydrogen ion concentration is 0.1000 $M$ and the pH is 1.00.

After the addition of 10.00 ml of 0.1000 $F$ sodium hydroxide (1.000 millimole of base), 4.000 millimoles of acid remain untitrated in 60.00 ml of solution, so

$$[H^+] = \frac{4.000 \text{ millimoles}}{60.00 \text{ milliliters}} = 6.67 \times 10^{-2}\ M; \qquad pH = 1.18$$

Similar calculations can be performed for the addition of various volumes of the sodium hydroxide titrant up to, and including, 49.999 ml. In Table 4–2 are presented the results of seven such calculations.

Table 4–2. Variation of pH During Titration of 50.00 ml of 0.1000 $F$ Hydrochloric Acid with 0.1000 $F$ Sodium Hydroxide

| Volume of NaOH, ml | pH |
|---|---|
| 0 | 1.00 |
| 10.00 | 1.18 |
| 20.00 | 1.37 |
| 30.00 | 1.60 |
| 40.00 | 1.95 |
| 49.00 | 3.00 |
| 49.90 | 4.00 |
| 49.99 | 5.00 |
| 49.999 | 6.00 |
| 50.000 | 7.00 |
| 50.001 | 8.00 |
| 50.01 | 9.00 |
| 50.10 | 10.00 |
| 51.00 | 11.00 |
| 55.00 | 11.68 |
| 60.00 | 11.96 |
| 65.00 | 12.12 |

After the addition of 50.00 ml of sodium hydroxide, which corresponds to the *equivalence point* of the titration, the hydrogen ion concentration is $1.00 \times 10^{-7}\ M$, water now being the only source of hydrogen ions, and the pH = 7.00.

Beyond the equivalence point of the titration, we are simply adding excess sodium hydroxide titrant to a solution of sodium chloride. There are two sources of hydroxide ions—the added sodium hydroxide solution and the autoprotolysis of water—and the *total* hydroxide ion concentration is given by the following equations:

$$[OH^-] = \frac{\text{millimoles of excess base}}{\text{solution volume in milliliters}} + [H^+]$$

$$[OH^-] = \frac{\text{millimoles of excess base}}{\text{solution volume in milliliters}} + \frac{K_w}{[OH^-]}$$

For volumes of sodium hydroxide titrant equal to or larger than 50.001 ml, the autoprotolysis of water is unimportant, so the second term on the right-hand side of each of the two preceding relations may be omitted. After the addition of 50.001 ml of

0.1000 $F$ sodium hydroxide, there is 0.0001000 millimole of excess base in essentially 100.0 ml of solution; thus,

$$[OH^-] = \frac{0.0001000 \text{ millimole}}{100.0 \text{ milliliters}} = 1.00 \times 10^{-6} \, M$$

$$pOH = 6.00; \qquad pH = 8.00$$

Six more calculations, for different volumes of titrant up to 65.00 ml, provided the remaining pH values listed in Table 4–2.

In Figure 4–2 is shown the complete titration curve constructed from the data in Table 4–2. In addition, a titration curve for the neutralization of 50.00 ml of 0.001000 $F$ hydrochloric acid with 0.001000 $F$ sodium hydroxide is depicted. Notice that the change in pH near the equivalence point for the latter curve is approximately four pH units smaller than observed with reagents which are one hundred times more concentrated.

*Selection of indicators.* In choosing end-point indicators, we should bear in mind several principles. First, in order for the indicator color change to be very sharply defined, the pH versus volume curve for the proposed titration must exhibit a steeply rising portion in the vicinity of the equivalence point. Second, this steep section of the titration curve must encompass an interval of pH values at least as large as the pH transition interval of an indicator. Third, the pH transition interval of the proposed end-point indicator must coincide with the steep portion of the titration curve. In addition, the direction in which a titration is performed may influence

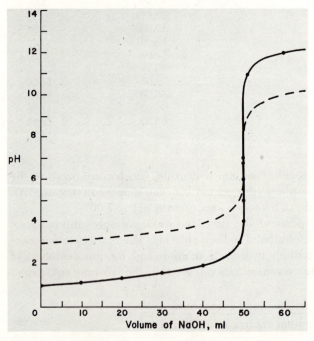

*Figure 4–2.* Titration curves for strong acid-strong base reactions. The solid line represents the titration of 50.00 ml of 0.1000 $F$ hydrochloric acid with a 0.1000 $F$ sodium hydroxide solution; titration data are listed in Table 4–2. The dashed line is the titration curve for the titration of 50.00 ml of 0.001000 $F$ hydrochloric acid with 0.001000 $F$ sodium hydroxide.

the choice of an end-point indicator. In the titration of a strong acid *with* a strong base, the indicator is initially present in its acid form, so the end point is signaled by the sudden appearance of the color of the *base* form of the indicator. In the reverse titration of a strong base *with* a strong acid, the indicator starts out in its base form, and the end point will be identified when the *acid* color of the indicator appears.

Which indicators listed in Table 4–1 would be suitable to detect the end point of the titration of 50.00 ml of 0.1000 *F* hydrochloric acid with 0.1000 *F* sodium hydroxide, if we do not want the titration error to exceed $\pm 0.1$ per cent? Accordingly, the permissible range of titrant volumes is from 49.95 to 50.05 ml. Using procedures discussed previously for the construction of titration curves, we can determine that the pH values after addition of 49.95 and 50.05 ml of the titrant are 4.30 and 9.70, respectively. Thus, to ensure the desired titration accuracy, the sudden appearance of the color of the *base* form of the indicator must occur between pH 4.30 and 9.70.

Although a number of indicators in Table 4–1 have *upper* extremes of their pH transition intervals between pH 4.30 and 9.70, not all indicators in this category are equally useful. To understand this latter statement, let us examine Figure 4–3, which shows magnified portions of the equivalence-point regions for the titration curves in Figure 4–2. Superimposed upon the curves are the pH transition intervals for five common acid-base indicators—methyl orange, methyl red, bromthymol blue, phenol red, and phenolphthalein. Bromthymol blue and phenol red are excellent indicators for this titration because their pH transition intervals lie almost in the middle of the

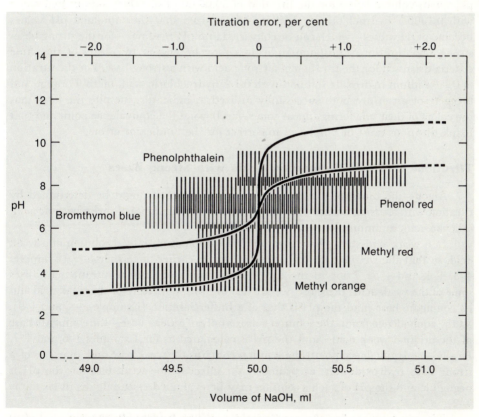

*Figure 4–3.* Magnified portions of equivalence-point regions for titration curves shown in Figure 4–2, along with pH transition intervals for methyl orange, methyl red, bromthymol blue, phenol red, and phenolphthalein.

steep vertical part of the titration curve. Methyl red and phenolphthalein are satisfactory indicators, since their pH transition intervals coincide reasonably well with the steeply rising portion of the titration curve. However, methyl orange begins to change color as early as one per cent before the equivalence point; although the *upper* limit of its pH transition interval is within the optimum pH range of 4.30 to 9.70, the color change from red to yellow would be too gradual to be acceptable. Choices of other acid-base indicators in Table 4–1 can be similarly evaluated.

For the titration of 50.00 ml of 0.001000 $F$ hydrochloric acid with 0.001000 $F$ sodium hydroxide solution, calculations reveal that, in order to restrict the titration error to $\pm 0.1$ per cent, an acid-base indicator must change to the color of its *base* form within the rather narrow range between pH 6.29 and 7.71. This information is conveyed graphically in Figure 4–3. Bromthymol blue, as well as several other indicators in Table 4–1, is a useful end-point indicator for this titration, whereas phenolphthalein, methyl orange, and methyl red are unsatisfactory. From Table 4–1 we see that phenol red would provide an end-point pH of 8.4. An inspection of Figure 4–3 reveals that a titration error of $+0.5$ per cent would result. However, if a blank titration were performed to correct for the "indicator error," phenol red could undoubtedly be employed as an end-point indicator.

***Titration curves and indicators for strong base-strong acid titrations.*** Calculation and construction of a titration curve for the titration of a strong base *with* a strong acid differ little from the preceding description for strong acid-strong base titrations, except that the roles of the strong acid and strong base are reversed. In fact, the data listed in Table 4–2 can immediately be employed to construct the pH-versus-volume curve for the titration of 50.00 ml of 0.1000 $F$ sodium hydroxide with 0.1000 $F$ hydrochloric acid, if it is recognized that the tabulated pH values become pOH values—which can be converted into pH readings—for the strong base-strong acid titration. End-point indicators may be chosen by means of the same criteria discussed for the titration of a strong acid with a strong base. For the titration of 0.1 $F$ sodium hydroxide solution with 0.1 $F$ hydrochloric acid, methyl orange and bromcresol green are both successfully utilized as indicators, despite the facts that the colors of their *acid* forms appear somewhat beyond the equivalence point and that blank titrations must be employed to correct for the "indicator errors."

## Titrations of Weak Monoprotic Acids with Strong Bases

Among the numerous weak monoprotic acids which might be determined by titration with a strong base are formic acid, acetic acid, propanoic acid, benzoic acid, salicylic acid, anilinium ion, and pyridinium ion.

In a weak acid-strong base titration, the sample is initially a solution of a weak acid, so that the pH may be calculated as shown earlier in this chapter (Examples 4–3, 4–4, and 4–5). Prior to the equivalence point, addition of titrant neutralizes some of the weak acid; therefore, we are dealing with a mixture of the weak acid and its conjugate base, and the pH is that of a buffer solution (Examples 4–8 and 4–9). At the equivalence point, the solution consists only of a weak base—the conjugate base of the original weak acid—and the pH is calculated as for Examples 4–6 and 4–7. Beyond the equivalence point, the solution contains the weak base and an excess of a strong base, hydroxide ion; accordingly, the effect of the weak base on the pH is negligible and the pH of such a solution may be evaluated essentially as in Example 4–2.

***Titration curve for the acetic acid–sodium hydroxide reaction.*** Let us construct the theoretical titration curve for the titration of 25.00 ml of 0.1000 $F$ acetic acid with 0.1000 $F$ sodium hydroxide titrant.

Initially, before any titrant is added, the pH of a 0.1000 $F$ acetic acid solution is 2.88.

As each volume increment of sodium hydroxide solution is added, it reacts virtually completely with acetic acid to produce a stoichiometrically equivalent amount of acetate

$$CH_3COOH + OH^- \rightleftharpoons CH_3COO^- + H_2O$$

and the concentration of hydrogen ion is determined by the ratio of the acetic acid and acetate ion concentrations from the expression

$$[H^+] = K_a \frac{[CH_3COOH]}{[CH_3COO^-]}$$

Since we have the *ratio* of two concentrations in this last equation, it suffices to substitute just the numbers of millimoles of acetic acid and acetate ion present at equilibrium after each portion of titrant is introduced. For example, after the addition of 5.00 ml of 0.1000 $F$ sodium hydroxide solution, 2.000 millimoles of acetic acid remain untitrated and 0.500 millimole of acetate ion is formed; hence,

$$[H^+] = 1.75 \times 10^{-5} \frac{(2.000)}{(0.500)} = 7.00 \times 10^{-5} M; \qquad pH = 4.15$$

In Table 4–3 are presented the results of other similar computations for titrant volumes up to 24.90 ml; in particular, note that the pH of the solution after addition of 12.50 ml of titrant is numerically equal to the $pK_a$ value of 4.76 for acetic acid, because at this point exactly one-half of the acetic acid has been titrated and the concentration ratio of acetic acid to acetate is precisely unity. In performing these calculations, we neglected the autoprotolysis of water and we avoided any points on the titration curve exceedingly close to the equivalence point, where the concentration of acetic acid becomes so small that we would have to take account of the decrease in $[CH_3COOH]$ and the increase in $[CH_3COO^-]$ due to the dissociation of acetic acid.

**Table 4–3. Variation of pH During Titration of 25.00 ml of 0.1000 $F$ Acetic Acid with 0.1000 $F$ Sodium Hydroxide**

| Volume of NaOH, ml | pH |
|---|---|
| 0 | 2.88 |
| 5.00 | 4.15 |
| 10.00 | 4.58 |
| 12.50 | 4.76 |
| 15.00 | 4.93 |
| 20.00 | 5.36 |
| 24.00 | 6.14 |
| 24.90 | 7.15 |
| 25.00 | 8.73 |
| 26.00 | 11.29 |
| 30.00 | 11.96 |
| 40.00 | 12.36 |

At the equivalence point, corresponding to the addition of 25.00 ml of sodium hydroxide, the solution is identical to one which could be prepared by dissolving 0.002500 mole of sodium acetate in 50.00 ml of water. Using the method discussed in Example 4–6, we calculate that the pH of a 0.05000 $F$ sodium acetate solution is 8.73.

After the addition of 26.00 ml of 0.1000 $F$ sodium hydroxide, there is an excess of 0.1000 millimole of strong base in a total volume of 51.00 ml. Therefore, $[OH^-] = 1.96 \times 10^{-3} M$, pOH = 2.71, and pH = 11.29. Results of this and two more such calculations are listed in Table 4–3.

In Figure 4–4 is plotted the complete titration curve. On examining it closely, we note several characteristics which distinguish it from the strong acid-strong base titration curve of Figure 4–2. Prior to the equivalence point, the titration curve for a weak acid is displaced toward higher pH values (lower hydrogen ion concentrations) because only a small fraction of the weak acid dissociates to furnish hydrogen ions, and the pH at the equivalence point is on the basic side of pH 7. In addition, the steep vertical portion of the curve encompasses a much smaller pH range than the analogous strong acid-strong base titration curve in Figure 4–2. There is one important similarity between the two titrations; beyond the equivalence point for a weak acid-strong base titration, where we are adding excess titrant, the system is essentially identical to a strong acid-strong base system.

***Selection of indicators.*** In view of the characteristics of the titration curve for the acetic acid-sodium hydroxide system, it is of interest to consider which end-point indicators can be employed without causing a titration error greater than one part per thousand. Hence, we must calculate the pH of the solution 0.1 per cent before and after the equivalence point. One-tenth per cent *before* the equivalence point, 99.9 per cent of the original acetic acid has been converted to acetate ion, and

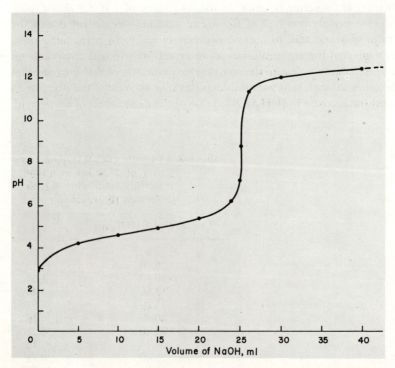

*Figure 4–4.* Titration curve for the titration of 25.00 ml of 0.1000 $F$ acetic acid with a 0.1000 $F$ sodium hydroxide solution. Titration data are listed in Table 4–3.

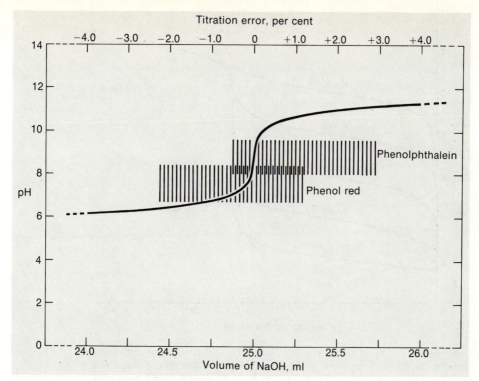

*Figure 4–5.* Magnified portion of equivalence-point region for titration of 25.00 ml of 0.1000 $F$ acetic acid with 0.1000 $F$ sodium hydroxide, showing pH transition intervals for phenolphthalein (the preferred end-point indicator) and phenol red.

0.1 per cent remains as acetic acid molecules. Since the ratio of acetic acid to acetate ion is 1/999, or approximately 1/1000, we can write

$$[H^+] = K_a \frac{[CH_3COOH]}{[CH_3COO^-]} = \frac{1.75 \times 10^{-5}}{1000} = 1.75 \times 10^{-8}\,M; \qquad pH = 7.76$$

One-tenth per cent *after* the equivalence point, there is an excess of 0.025 ml of 0.1000 $F$ sodium hydroxide in a solution whose volume is essentially 50.00 ml, which corresponds to a pH of 9.70. Thus, to be suitable, an indicator must change abruptly from its acid form to its base form within the pH range from 7.76 to 9.70.

Figure 4–5 shows the magnified equivalence-point region of the titration curve for the acetic acid–sodium hydroxide system, along with the pH transition intervals for phenol red and phenolphthalein. It is obvious why phenolphthalein is the almost universal choice of indicator for the titration. Although the color of the base form of phenol red appears at pH 8.4, the gradual color change makes this indicator less satisfactory than phenolphthalein. Three other indicators listed in Table 4–1— brilliant yellow, neutral red, and metanitrophenol—are comparable in behavior to phenol red.

**Feasibility of titrations.** As Figure 4–6 reveals, the magnitude of the dissociation constant of a weak acid has a profound influence on the size and sharpness of the pH change near the equivalence point of a titration. Furthermore, decreases in the concentrations of the weak acid and the titrant base diminish the pH change. Since

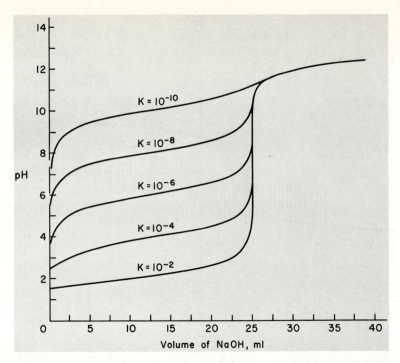

*Figure 4–6.* Family of titration curves for weak acid-strong base titrations, showing the effect of the value of the dissociation constant on the shape and position of the curve. For each titration curve, 25.00 ml of 0.1000 $F$ weak acid is titrated with 0.1000 $F$ sodium hydroxide.

most acid-base indicators change color over a two-pH-unit interval, the weakest acid which can be successfully titrated is one whose $K_a$ is about $10^{-7}$.

However, to obtain accurate analytical results, one must follow a special procedure. Suppose that it is desired to determine carbonic acid ($H_2CO_3$)—a very weak diprotic acid with a first dissociation constant ($K_{a1}$) of $4.47 \times 10^{-7}$—by titration with a standard sodium hydroxide solution, according to the reaction

$$H_2CO_3 + OH^- \rightleftharpoons HCO_3^- + H_2O$$

Phenolphthalein is the usual choice for end-point indicator. At the end point of the titration, the solution is indistinguishable from one prepared by dissolving sodium bicarbonate in water. From the results of the first titration, it is possible to calculate the concentration of sodium bicarbonate present at the end point. Now, in a separate titration flask, we reproduce artificially, but as accurately as possible, the conditions at the end point by adding the proper amounts of water, sodium bicarbonate, and phenolphthalein indicator. What we have, then, is a comparison solution in which the end-point pH is correct and in which a definite concentration of phenolphthalein is present in its pink base form. In subsequent titrations the titrant is carefully added until the end-point colors in the real and comparison titration flasks are matched as closely as possible.

## Titrations of Weak Bases with Strong Acids

By means of a titration with a standard solution of a strong acid, many weak bases can be determined; ammonia, piperidine, ethanolamine, and triethylamine, as well

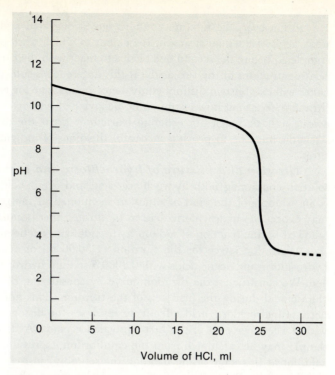

*Figure 4–7.* Titration curve for the titration of 25.00 ml of 0.01000 *F* sodium phenoxide solution with 0.01000 *F* hydrochloric acid.

as cyanide, borate, and phenoxide anions, are representative of this group of substances. In principle, the construction of theoretical titration curves for weak base-strong acid systems involves calculations similar to those already discussed for weak acid-strong base titrations. Figure 4–7 depicts the titration curve for the titration of 25.00 ml of 0.01000 *F* sodium phenoxide solution with 0.01000 *F* hydrochloric acid. Although detailed calculations and discussions of end-point indicators and titration errors are reserved for a problem at the end of this chapter, it should be apparent that the pH at the equivalence point of a weak base-strong acid titration is always less than 7—it is 6.13 for the particular titration curve illustrated—and that an end-point indicator must be selected which changes to the color of its *acid* form in the near vicinity of the equivalence point.

## Titrations of Polyprotic Acids and Mixtures of Monoprotic Acids

It is not unusual for two or more acidic or basic species to be present in the same solution. Consider a mixture of two acids which is to be titrated with a strong base. If both acids are strong, such as a mixture of hydrochloric acid and perchloric acid, the titration curve would have a shape similar to that in Figure 4–2, and it would be impossible to determine the concentrations of the individual acids. A second possibility is that we have a mixture of two weak acids with nearly identical dissociation constants. Here, too, the two weak acids would behave quite similarly as hydrogen ion sources. Therefore, the resulting titration curve would exhibit the general characteristics of the curve shown in Figure 4–4, and it would not be feasible to analyze the mixture for its individual components.

If, however, the mixture contains one strong acid and one weak acid, or two weak acids of unequal strength, the extent to which titration of the stronger acid is completed before the second one begins to react with the titrant is determined by the relative strengths of the two acids. If this difference is sufficiently great, the titration curve will exhibit two distinct equivalence points, one for each of the two successive titrations. We shall now examine the behavior of systems containing more than one acidic or basic species in order to learn how great the difference in acid or base strengths must be to provide successful titrations of individual components in mixtures.

*Titration of a mixture of hydrochloric and acetic acids.* Titration of a solution containing both hydrochloric acid and acetic acid with sodium hydroxide is an example of the kind of situation mentioned in the preceding paragraph. We may expect the hydrochloric acid to be titrated first because it is the stronger acid, followed by the reaction of sodium hydroxide with acetic acid.

A titration curve for the titration of 50.00 ml of a solution $0.1000 F$ in both hydrochloric and acetic acids with $0.2000 F$ sodium hydroxide is presented in Figure 4–8. We constructed the titration curve by considering the presence of only hydrochloric acid during the first step of the titration. This approximation is reasonable because hydrochloric acid effectively represses the dissociation of acetic acid up to within a few per cent of the first equivalence point. At the first equivalence point, the pH may be calculated from the equilibrium expression for dissociation of acetic acid, for at this stage of the titration the solution consists of acetic acid and sodium chloride only. Beyond the first equivalence point, the titration curve is simply that of a weak acid-strong base system and is analogous to Figure 4–4.

For our present purposes, the important characteristic of the titration curve is the absence of a sharply defined equivalence point at pH 3, due to the fact that acetic acid starts to react before all of the hydrochloric acid has been titrated. Although the quality of the titration curve would probably permit determination of the individual concentrations of hydrochloric acid and acetic acid with an error no worse than 1 per cent, the combination of $0.1 F$ hydrochloric acid and $0.1 F$ weak acid of $K_a = 10^{-5}$ represents a rough limit for a strong acid-weak acid mixture that can be successfully analyzed with reasonable accuracy. A titration curve for a mixture of $1.0 F$ strong acid and $1.0 F$ weak acid with a $K_a$ of $10^{-4}$ would be qualitatively similar to Figure 4–8 as would that for a solution containing $0.01 F$ strong acid and $0.01 F$ weak acid of $K_a = 10^{-6}$.

Experimentally, the location of the first equivalence point could be best accomplished graphically from a complete plot of the titration curve. Alternatively, an acid-base indicator which changes to the color of its *base* form at pH 3 might be employed, but it would be advisable to prepare a color comparison solution containing the appropriate concentrations of acetic acid and indicator in order to approximate the equivalence-point conditions as closely as possible. Phenolphthalein could be used to signal the second equivalence point.

*Titration of phosphoric acid with sodium hydroxide solution.* Earlier in this chapter, the stepwise dissociation of phosphoric acid was represented by the following set of proton-transfer equilibria:

$$H_3PO_4 + H_2O \rightleftharpoons H_2PO_4^- + H_3O^+; \quad K_{a1} = 7.5 \times 10^{-3}$$
$$H_2PO_4^- + H_2O \rightleftharpoons HPO_4^{2-} + H_3O^+; \quad K_{a2} = 6.2 \times 10^{-8}$$
$$HPO_4^{2-} + H_2O \rightleftharpoons PO_4^{3-} + H_3O^+; \quad K_{a3} = 4.8 \times 10^{-13}$$

A titration curve for the titration of 25.00 ml of $0.5000 F$ phosphoric acid with $0.5000 F$ sodium hydroxide solution is shown in Figure 4–9. We shall not pursue the

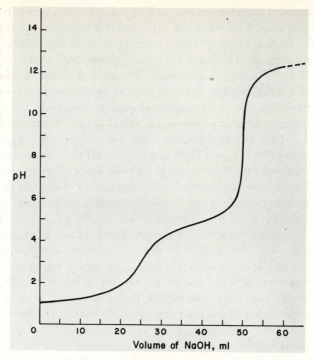

*Figure 4–8.* Titration curve for the titration of 50.00 ml of a mixture of 0.1000 $F$ hydrochloric acid and 0.1000 $F$ acetic acid with 0.2000 $F$ sodium hydroxide. The first step of the titration curve corresponds to the neutralization of the hydrochloric acid and the second step to the neutralization of acetic acid.

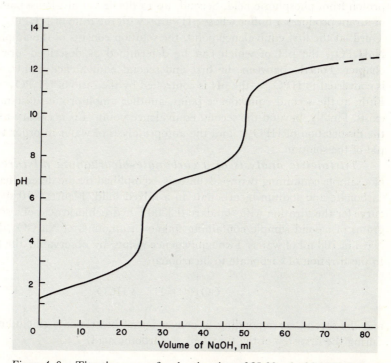

*Figure 4–9.* Titration curve for the titration of 25.00 ml of 0.5000 $F$ phosphoric acid with a 0.5000 $F$ sodium hydroxide solution. Although the stepwise titration of $H_3PO_4$, first to $H_2PO_4^-$ and then to $HPO_4^{2-}$, is evident, the titration of the third proton of phosphoric acid cannot be observed because $HPO_4^{2-}$ is too weak an acid in water.

detailed calculations required to construct this curve, but rather will emphasize the important features of the titration. It is important to realize that the present system is identical in every respect to an aqueous solution containing a mixture of three weak monoprotic acids, each at an initial concentration of $0.5\ F$, with dissociation constants of $K_{a1}$, $K_{a2}$, and $K_{a3}$, respectively, as listed above. A mixture of $0.5\ F$ concentrations of a weak diprotic acid and a weak monoprotic acid would likewise mimic the behavior of the phosphoric acid-sodium hydroxide titration if the dissociation constants had the same numerical values as those for phosphoric acid.

Only two equivalence points are observed, corresponding to the successive titrations of $H_3PO_4$ and $H_2PO_4^-$; because $HPO_4^{2-}$ is an exceedingly weak acid, the titration of the third proton of phosphoric acid does not proceed to any significant extent. For the case of phosphoric acid, the curve showing the titration of the first and second protons is resolved into two well-defined steps because $pK_{a1}$ and $pK_{a2}$ differ by slightly more than five $pK$ units ($pK_{a1} = 2.13$ and $pK_{a2} = 7.21$).

As a general rule, if the $pK_a$ values of two weak acids differ by four or more $pK$ units, one can obtain two distinct equivalence points for the stepwise titration of a mixture of these acids. Of course, as Figure 4–6 reveals, in order to observe a satisfactory titration curve for the weaker of the two acids, $pK_a$ for the latter must be *smaller* than approximately 8. If the two weak acids in a mixture have $pK_a$ values which differ by less than four units, the reaction of the weaker acid will be well underway before the titration of the stronger one is complete, and the first equivalence point will be poorly defined.

Because the acidities of $H_3PO_4$ and $H_2PO_4^-$ are substantially different, construction of the titration curve in Figure 4–9 is a relatively straightforward procedure. First, the initial pH of the solution is governed essentially by dissociation of the first proton from phosphoric acid. Second, up to the first equivalence point, the addition of titrant produces a buffer whose pH varies with the ratio of $[H_3PO_4]$ to $[H_2PO_4^-]$. Third, at the first equivalence point, the solution consists of the amphiprotic species $NaH_2PO_4$, the pH of which can be determined as described previously in this chapter. Fourth, between the first and second equivalence points, when $H_2PO_4^-$ is converted to $HPO_4^{2-}$, the pH is controlled by the ratio of $[H_2PO_4^-]$ to $[HPO_4^{2-}]$. Fifth, at the second equivalence point, another amphiprotic substance ($Na_2HPO_4$) exists. Finally, beyond the second equivalence point, it is necessary to consider both the dissociation of $HPO_4^{2-}$ and the autoprotolysis of water in order to calculate the pH of the solution.

***Titrimetric analysis of a carbonate-bicarbonate mixture.*** An analysis of a sample containing two weak bases is exemplified by the determination of sodium carbonate and sodium bicarbonate in a mixed solid. Figure 4–10 shows a titration curve for the titration with standard $0.3500\ F$ hydrochloric acid of a solution of $1.200$ grams of a solid sample containing unknown amounts of $Na_2CO_3$, $NaHCO_3$, and $NaCl$ in 100 ml of water. Two equivalence points are observable, the first pertaining to the titration of carbonate to bicarbonate

$$CO_3^{2-} + H^+ \rightleftharpoons HCO_3^-$$

and the second corresponding to conversion of bicarbonate (including that formed during the first step of the titration) to carbonic acid:

$$HCO_3^- + H^+ \rightleftharpoons H_2CO_3$$

Inspection of Figure 4–10 reveals that the pH change at the first equivalence point is inadequately sharp for ordinary use of an acid-base indicator. However,

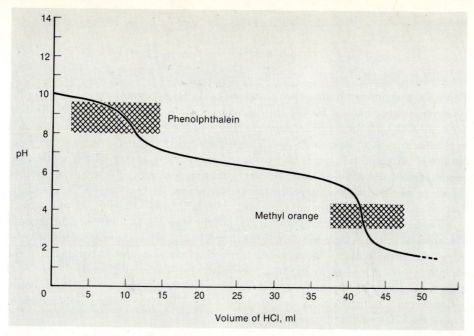

*Figure 4–10.* Titration curve for analysis of a carbonate-bicarbonate mixture with 0.3500 *F* hydrochloric acid. The first step of the titration corresponds to the formation of bicarbonate ion ($HCO_3^-$), and the second step pertains to the conversion of bicarbonate (including the $HCO_3^-$ produced during the first step) to carbonic acid. Phenolphthalein may be employed to indicate the first end point if a special procedure is utilized, whereas methyl orange is used to signal the second end point.

phenolphthalein may be employed if a color comparison solution is prepared in the manner described in our earlier discussion of the feasibility of weak acid-strong base titrations (page 128). Thus, a solution is prepared having a volume and containing concentrations of phenolphthalein and $NaHCO_3$ as close as possible to the sample solution at the first equivalence point. Then, in an actual titration of the sample, hydrochloric acid is introduced until the color of the solution titrated matches that of the comparison solution. One can obtain results accurate to within $\pm 0.5$ per cent by means of this technique. Methyl orange provides a good indication of the second equivalence point, the end point being taken as the first perceptible change from the pure yellow color of the base form to an intermediate orange color. Alternatively, it is possible to follow the entire course of the titration with the aid of a glass-electrode pH meter. A complete plot of pH versus volume of titrant can be constructed from the experimental data, and the titration curve can be analyzed graphically to locate the two equivalence points precisely. Furthermore, using a glass-electrode pH meter, one can titrate to the exact pH of the first equivalence point.

In analogy to the titration of a pair of weak acids, the difference between $pK_b$ values for two weak bases must be at least four $pK$ units in order to determine accurately the amounts of the individual components by acid-base titrimetry. For carbonate and bicarbonate, the equilibria which govern their behavior as proton acceptors are

$$CO_3^{2-} + H_2O \rightleftharpoons HCO_3^- + OH^-; \qquad K_{b1} = \frac{K_w}{K_{a2}} = 2.14 \times 10^{-4}$$

and

$$HCO_3^- + H_2O \rightleftharpoons H_2CO_3 + OH^-; \qquad K_{b2} = \frac{K_w}{K_{a1}} = 2.24 \times 10^{-8}$$

Thus, for the present system, $pK_{b1}$ and $pK_{b2}$ differ by almost exactly four units. However, the presence of $HCO_3^-$ in the original sample solution shifts the titration curve for $CO_3^{2-}$ to lower pH values, thereby causing the pH change near the first equivalence point to be smaller and less pronounced than predicted from the magnitudes of $pK_{b1}$ and $pK_{b2}$; this effect will not occur if the weaker base is *not* the conjugate acid of the stronger base. Between the first and second equivalence points, the shape of the titration curve for the conversion of $HCO_3^-$ to $H_2CO_3$ is exactly that expected for the titration of the conjugate base of any weak monoprotic acid.

To calculate the percentage of $Na_2CO_3$ in the mixed solid, note that the amount of carbonate in the sample solution is measured by the volume and concentration of the standard hydrochloric acid added up to the first equivalence point. Since the volume of titrant is close to 10.6 ml, the quantity of carbonate—and, therefore, of sodium carbonate in the original solid sample—is $(10.6)(0.350)$ or 3.71 millimoles; and, inasmuch as the formula weight of $Na_2CO_3$ is 106, the sodium carbonate content of the 1200-milligram sample is

$$\frac{3.71 \text{ millimoles} \cdot 106 \dfrac{\text{milligrams}}{\text{millimole}}}{1200 \text{ milligrams}} \times 100 = 32.8 \text{ per cent}$$

Although 31.2 ml of hydrochloric acid is added between the first and second equivalence points to convert $HCO_3^-$ to $H_2CO_3$, 10.6 ml of titrant reacts with the $HCO_3^-$ formed from the original $CO_3^{2-}$, so only 20.6 ml of hydrochloric acid is equivalent to the $NaHCO_3$ originally present. Thus, the number of millimoles of $NaHCO_3$ is $(20.6)(0.350)$ or 7.21 and, since the formula weight of $NaHCO_3$ is 84.0, the amount of $NaHCO_3$ in the solid sample is

$$\frac{7.21 \text{ millimoles} \cdot 84.0 \dfrac{\text{milligrams}}{\text{millimole}}}{1200 \text{ milligrams}} \times 100 = 50.5 \text{ per cent}$$

It follows that there must be 16.7 per cent of sodium chloride in the mixed solid.

## QUESTIONS AND PROBLEMS*

1. Convert the following pH values to $[H^+]$: 6.37; 4.83; 11.22; 8.54; 3.13.
2. Convert the following pOH values to $[H^+]$: 3.28; 9.86; 2.47; 12.19; 6.71.
3. Convert the following molar hydrogen ion concentrations to pH values: 0.0139; $2.07 \times 10^{-5}$; $7.34 \times 10^{-11}$; 2.41; $4.12 \times 10^{-7}$.

* A comprehensive table of dissociation constants for acids and bases is given in Appendix 2.

4. In a series of experiments, it was determined that the position of equilibrium for each of the proton-transfer reactions listed below lies well toward the right.

$$HCN + OH^- \rightleftharpoons H_2O + CN^-$$

$$HClO_4 + Cl^- \rightleftharpoons HCl + ClO_4^-$$

$$H_2S + CN^- \rightleftharpoons HCN + HS^-$$

$$(C_6H_5)_3C^- + C_4H_4NH \rightleftharpoons (C_6H_5)_3CH + C_4H_4N^-$$

$$CH_3COOH + HS^- \rightleftharpoons H_2S + CH_3COO^-$$

$$O^{2-} + (C_6H_5)_3CH \rightleftharpoons (C_6H_5)_3C^- + OH^-$$

$$C_4H_4N^- + H_2S \rightleftharpoons C_4H_4NH + HS^-$$

Using the above information as well as some semiquantitative knowledge about the relative strengths of common acids (but without consulting tables of acid dissociation constants), arrange the following acids in order of *decreasing* acidity: (a) $HCN$ (b) $HCl$ (c) $HClO_4$ (d) $CH_3COOH$ (e) $(C_6H_5)_3CH$ (f) $C_4H_4NH$ (pyrrole) (g) $H_2S$ (h) $OH^-$ (i) $H_2O$.

5. Calculate the pH of each of the following systems, neglecting activity effects:
   (a) 40 ml of solution containing 1.00 gm of nitric acid
   (b) 30 ml of solution containing 120 mg of barium hydroxide
   (c) 0.180 $F$ benzoic acid solution
   (d) 0.250 $F$ sodium dihydrogen phosphate solution
   (e) 0.700 $F$ arsenic acid solution
   (f) 0.200 $F$ ammonium cyanide solution
   (g) a solution prepared by dissolution of 0.267 mole of pyridine in 600 ml of water
   (h) a solution prepared by dissolution of 0.0235 mole of alanine (an amino acid) in 700 ml of water
   (i) 2.00 $\times$ 10$^{-5}$ $F$ potassium acetate solution
   (j) 0.120 $F$ chloroacetic acid solution
   (k) a solution prepared by dissolution of 0.500 gm of o-phthalic acid in 125 ml of water
   (l) a solution prepared by dissolution of 0.200 mole of potassium nitrite and 0.00030 mole of potassium hydroxide in 800 ml of water

6. Calculate the hydrogen ion concentration and the pH of a solution prepared by addition of 0.100 ml of 0.0100 $F$ perchloric acid ($HClO_4$) to 10.0 liters of pure water.

7. In what ratio must acetic acid and sodium acetate be mixed to provide a solution of pH 6.20?

8. In what ratio must ammonia and ammonium chloride be mixed to provide a solution of pH 8.40?

9. Consider 100 ml of a 0.100 $F$ acetic acid solution at 25°C.
   (a) Will the addition of 1 gm of pure acetic acid cause the extent (percentage) of dissociation of acetic acid to increase, decrease, or remain the same?
   (b) Will the addition of 1 gm of sodium acetate cause the pH of the solution to increase, decrease, or remain the same?
   (c) Will the addition of 1 gm of sodium acetate cause the hydroxide ion concentration to increase, decrease, or remain the same?

10. Consider 50 ml of a 0.100 $F$ aqueous ammonia solution at 25°C.
    (a) Will the solution at the end point of the titration of the ammonia solution with 0.100 $F$ hydrochloric acid have a pH less than 7, equal to 7, or greater than 7?
    (b) Will the addition of 1 gm of pure acetic acid cause the pH to increase, decrease, or remain the same?
    (c) Will the addition of 1 gm of ammonium chloride cause the hydroxide ion concentration to increase, decrease, or remain the same?

11. Suppose you wish to prepare 1 liter of a buffer solution, containing sodium carbonate and sodium bicarbonate, having an initial pH of 9.70, and of such

composition that, upon the formation of 60 millimoles of hydrogen ion during the course of a chemical reaction, the pH will not decrease below 9.30. What are the minimal original concentrations of sodium carbonate and sodium bicarbonate needed to prepare the desired buffer solution?

12. In what ratio must pyridine and pyridinium chloride be mixed to prepare a buffer solution with a pH of 5.00? If one liter of this buffer contains an initial pyridine concentration of 0.200 $M$, what will be the final pH of the solution after a chemical reaction occurring in the system consumes 50 millimoles of hydrogen ion?

13. Calculate the pH and the equilibrium concentration of $HPO_4^{2-}$ in a 0.500 $F$ solution of phosphoric acid ($H_3PO_4$) in water.

14. Calculate the equilibrium concentrations of $H^+$, $NH_4^+$, and $NH_3$ in a solution prepared by dissolution of 0.0400 mole of ammonia in 150 ml of water.

15. Calculate the equilibrium concentrations of $NH_4^+$ and $NH_3$ and the pH of the solution obtained from dissolution of 0.085 mole of ammonium nitrate and 0.000060 mole of nitric acid in 300 ml of water.

16. Calculate the equilibrium concentrations of $H^+$, $C_2H_5COO^-$, and $C_2H_5COOH$ in a solution prepared by dissolution of 0.0400 mole of propanoic acid in 150 ml of water.

17. Calculate the equilibrium concentrations of $C_2H_5COO^-$ and $C_2H_5COOH$ and the pH of the solution obtained by dissolution of 0.085 mole of sodium propanoate and 0.000050 mole of sodium hydroxide in 250 ml of water.

18. If 0.826 mole of pyridine and 0.492 mole of hydrochloric acid are mixed with water to give 500 ml of solution, what will be the pH of the resulting system?

19. Calculate the equilibrium concentrations of $H^+$, $NO_2^-$, and $HNO_2$ in a solution prepared by dissolution of 0.100 mole of nitrous acid in 200 ml of water.

20. Calculate the equilibrium concentrations of $F^-$ and $HF$ and the pH of the solution prepared by dissolution of 0.100 mole of sodium fluoride in 150 ml of water.

21. Calculate the equilibrium concentrations of all ionic and molecular species except water, $i.e.$, $NH_3$, $NH_4^+$, $H^+$, and $OH^-$, in a solution which initially contains 0.00400 mole of ammonia in 150 ml of water.

22. Oxalic acid undergoes the following stepwise dissociation reactions:

$$H_2C_2O_4 \rightleftharpoons H^+ + HC_2O_4^-; K_{a1} = 6.5 \times 10^{-2}$$

$$HC_2O_4^- \rightleftharpoons H^+ + C_2O_4^{2-}; K_{a2} = 6.1 \times 10^{-5}$$

(a) Construct a distribution diagram showing how the fraction, $\alpha_n$, of each oxalate-containing species varies as a function of pH.

(b) For a solution in which the total concentration of all oxalate-containing species is 0.0100 $F$, what would the concentration of $HC_2O_4^-$ be at pH 3.50? For the same solution, what would be the concentration of $H_2C_2O_4$ at pH 2.35, and what would be the concentration of $C_2O_4^{2-}$ at pH 5.68?

(c) Suppose that 250 mg of anhydrous sodium oxalate ($Na_2C_2O_4$) is dissolved in 500 ml of an aqueous solution which is well buffered at a pH of 4.00. Calculate the equilibrium concentration of each oxalate-containing species.

23. Citric acid, which behaves as a triprotic acid in water and which may be abbreviated as $H_3Cit$, undergoes the following stepwise dissociation reactions:

$$H_3Cit \rightleftharpoons H^+ + H_2Cit^-; K_{a1} = 7.4 \times 10^{-4}$$

$$H_2Cit^- \rightleftharpoons H^+ + HCit^{2-}; K_{a2} = 1.7 \times 10^{-5}$$

$$HCit^{2-} \rightleftharpoons H^+ + Cit^{3-}; \quad K_{a3} = 4.0 \times 10^{-7}$$

(a) Construct a distribution diagram to show how the fraction, $\alpha_n$, of each of the four citrate-containing species varies with pH.

(b) If 500 ml of solution containing a total of 0.200 mole of citrate species has a pH of 4.32, what are the individual concentrations of the four citrate species?

(c) If 0.100 mole of solid sodium hydroxide is dissolved in the solution described in (b) above, what will be the pH of the system and the concentrations of the various citrate-containing species?

24. Construct a distribution diagram, a plot of $\alpha_n$ versus pH, to show how the composition of an aqueous solution of alanine (one of the amino acids) varies with changes in hydrogen ion concentration. What is the pH at the isoelectric point for an aqueous solution of alanine? Calculate the individual concentrations of the various forms of alanine in a solution containing a total alanine concentration of $0.0100\,F$ at the isoelectric point.

25. Suppose that a solution of sodium hydroxide is permitted to absorb large amounts of carbon dioxide after originally being standardized. Would the original value for the concentration of the sodium hydroxide titrant be high, low, or correct? Explain.

26. If constant-boiling hydrochloric acid, at a pressure of 745 mm of mercury, contains 20.257 per cent HCl by weight, what weight of the constant-boiling solution must be distilled at this pressure for the preparation of exactly 1 liter of $1.000\,F$ hydrochloric acid?

27. If 48.37 ml of a sodium hydroxide solution was required to titrate a 3.972-gm sample of pure potassium acid phthalate, what was the concentration of the titrant?

28. Construct the titration curve for each of the following systems:
   (a) 50.00 ml of a $0.5000\,F$ sodium hydroxide solution titrated with $0.5000\,F$ hydrochloric acid
   (b) 50.00 ml of a $0.1000\,F$ sodium borate solution titrated with $0.1000\,F$ perchloric acid solution
   (c) 50.00 ml of a $0.1500\,F$ aqueous ethanolamine solution titrated with $0.1500\,F$ hydrochloric acid
   (d) 50.00 ml of $0.05000\,F$ acetic acid titrated with $0.1000\,F$ aqueous ammonia solution
   (e) 50.00 ml of a mixture of $0.1000\,F$ perchloric acid and $0.05000\,F$ propanoic acid titrated with $0.2000\,F$ potassium hydroxide solution
   (f) 50.00 ml of a $0.2000\,F$ disodium hydrogen phosphate solution titrated with $0.5000\,F$ perchloric acid

29. Calculate and plot the theoretical titration curve for the titration of 25.00 ml of $0.05000\,F$ formic acid with $0.1000\,F$ sodium hydroxide solution. Include in your plot the points that correspond to the following volumes in milliliters of titrant added: 0, 10.00, 12.45, 12.50, 13.00, and 20.00. What acid-base indicator would you select for this titration? Why? What would be the titration error in per cent with the proper sign if chlorophenol red were used as the indicator? What would be the titration error in per cent with the proper sign if phenolphthalein were used as the indicator?

30. Consider the titration of 25.00 ml of $0.2500\,F$ aqueous ammonia solution with $0.2500\,F$ hydrochloric acid. Calculate and plot the complete titration curve, including the points that correspond to the following volumes in milliliters of titrant added: 0, 5.00, 10.00, 15.00, 20.00, 25.00, 30.00, and 40.00. What acid-base indicator would you select for this titration? What would be the titration error in per cent with the proper sign if methyl orange were used as the indicator? What would be the titration error in per cent with the proper sign if bromthymol blue were used as the indicator?

31. Suppose that the following eight acid-base indicators listed in Table 4–1 are available: paramethyl red, methyl orange, bromcresol green, methyl red, chlorophenol red, phenol red, phenolphthalein, and thymolphthalein. Which of these indicators would indicate accurately (within perhaps two or three parts per thousand) the equivalence points of the following titrations? More than one indicator may be suitable, and all suitable ones should be cited.
   (a) $0.25\,F$ hydrochloric acid titrated with $0.25\,F$ sodium hydroxide
   (b) $0.5\,F$ phosphoric acid titrated to $NaH_2PO_4$ with $0.5\,F$ sodium hydroxide
   (c) $0.5\,F$ phosphoric acid titrated to $Na_2HPO_4$ with $0.5\,F$ sodium hydroxide
   (d) $0.2\,F$ sodium carbonate solution titrated to $NaHCO_3$ with $0.2\,F$ hydrochloric acid
   (e) $0.001\,F$ hydrochloric acid titrated with $0.1\,F$ sodium hydroxide

(f) The first equivalence point in the titration of a mixture of $0.05\,F$ hydrochloric acid and $0.05\,F$ propanoic acid with $0.10\,F$ sodium hydroxide

(g) The second equivalence point in the titration of part (f)

32. Suppose that one desires to determine the concentration $C$ of unknown hydrochloric acid solutions by titration with a standard sodium hydroxide solution of the same $C$ concentration. The titration could be followed with a glass-electrode pH meter. If the end point is taken when the pH of the solution exactly reaches the equivalence-point pH, if the uncertainty in the pH meter readings is $\pm 0.02$ pH unit, and if one should not want to incur a titration error greater than $\pm 0.10$ per cent, what would be the minimum concentration $C$ of hydrochloric acid which could be titrated by means of this technique?

33. A 25.00-ml aliquot of a benzoic acid solution was titrated with standard $0.1000\,F$ sodium hydroxide, methyl red being used erroneously as the end-point indicator. The observed end point occurred at pH 6.20, after the addition of 20.70 ml of the titrant.

(a) Calculate the titration error in per cent with the proper sign.

(b) Calculate the equivalence-point pH for this titration.

(c) Calculate the concentration of the original benzoic acid solution.

34. A 1.250-gm sample of a weak monoprotic acid (HZ) was dissolved in 50 ml of water. The volume of a $0.0900\,F$ sodium hydroxide solution required to reach the equivalence point was 41.20 ml. During the course of the titration, it was observed that, after the addition of 8.24 ml of titrant, the solution pH was 4.30.

(a) Calculate the formula weight of HZ.

(b) Calculate the acid dissociation constant of HZ.

(c) Calculate the theoretical equivalence-point pH for the titration.

35. Suggest an experimental procedure whereby the dissociation constant for an unknown weak monoprotic acid can be determined from the titration curve for that acid, assuming the availability of a pure sample of the unknown acid.

36. How could the acid dissociation constant for phenol red indicator (which has a pH transition range from 6.7, where the yellow acid form predominates, to 8.4, where the red base form is predominant) be determined experimentally?

37. Why is a solution of an acid salt, such as potassium hydrogen tartrate or sodium hydrogen carbonate, a convenient pH standard?

38. Consider the titration of 50.00 ml of $1.000 \times 10^{-4}\,F$ hydrochloric acid with $1.000 \times 10^{-3}\,F$ sodium hydroxide solution. Calculate the pH of the solution after the addition of the following volumes of titrant: (a) 0 ml, (b) 1.000 ml, (c) 2.500 ml, (d) 4.900 ml, (e) 4.999 ml, (f) 5.000 ml, (g) 5.555 ml, and (h) 6.000 ml. What means would you employ to locate the equivalence point of the titration?

39. What weight of vinegar should be taken for analysis so that the observed buret reading will be exactly 10 times the percentage of acetic acid in the vinegar if a $0.1000\,F$ sodium hydroxide solution is used as the titrant?

40. A 1.000-gm sample of pure oxalic acid dihydrate, $H_2C_2O_4 \cdot 2\,H_2O$, required 46.00 ml of a sodium hydroxide solution for its neutralization. Calculate the concentration of the sodium hydroxide solution.

41. The titration of 50.00 ml of a solution containing both sulfuric acid and phosphoric acid with a $1.000\,F$ potassium hydroxide solution was followed with a glass-electrode pH meter. Among other data, two of the points on the titration curve were:

| Volume of KOH added, ml | pH |
| --- | --- |
| 26.00 | 7.21 |
| 30.00 | 9.78 |

Calculate the concentrations of sulfuric and phosphoric acids in the original sample solution.

42. A 2.500-gm crystal of calcite, $CaCO_3$, was dissolved in excess hydrochloric acid, the carbon dioxide removed by boiling, and the excess acid titrated with

a standard base solution. The volume of hydrochloric acid used was 45.56 ml, and only 2.25 ml of the standard base was required to titrate the excess acid. In a separate titration, 43.33 ml of the base solution neutralized 46.46 ml of the hydrochloric acid solution. Calculate the concentrations of the acid and base solutions.

43. A 0.6300-gm sample of a pure organic diprotic acid was titrated with 38.00 ml of a 0.3030 $F$ sodium hydroxide solution, but 4.00 ml of a 0.2250 $F$ hydrochloric acid solution was needed for back titration, at which point the original organic acid was completely neutralized. Calculate the formula weight of the organic acid.

44. A sample of vinegar, weighing 11.40 gm, was titrated with a 0.5000 $F$ sodium hydroxide solution, 18.24 ml being required. Calculate the percentage of acetic acid in the vinegar.

45. A hydrochloric acid solution was standardized against sodium carbonate. Just 37.66 ml of the acid was required to titrate a 0.3663-gm sample of the sodium carbonate. Later, it was discovered that the sodium carbonate was really the hydrated compound, $Na_2CO_3 \cdot 10\ H_2O$, and not the anhydrous compound as expected. What were the reported and the corrected concentrations of the hydrochloric acid solution?

46. A 1.500-gm sample of impure calcium oxide (CaO) was dissolved in 40.00 ml of 0.5000 $F$ hydrochloric acid. Exactly 2.50 ml of a sodium hydroxide solution was required to titrate the excess, unreacted hydrochloric acid. If 1.00 ml of the hydrochloric acid corresponds to 1.25 ml of the sodium hydroxide solution, calculate the percentage of calcium oxide in the sample.

47. The sulfur in a 1.000-gm steel sample was converted to sulfur trioxide, which was, in turn, absorbed in 50.00 ml of a 0.01000 $F$ sodium hydroxide solution. The excess, unreacted sodium hydroxide was titrated with 0.01400 $F$ hydrochloric acid, 22.65 ml being required. Calculate the percentage of sulfur in the steel.

48. Malonic acid, $HOOC—CH_2—COOH$ (abbreviated $H_2M$), is a diprotic acid which undergoes the following dissociation reactions:

$$H_2M \rightleftharpoons HM^- + H^+; \ pK_{a1} = 2.86$$

$$HM^- \rightleftharpoons M^{2-} + H^+; \ pK_{a2} = 5.70$$

A 20.00-ml aliquot of a solution containing a mixture of disodium malonate ($Na_2M$) and sodium hydrogen malonate (NaHM) was titrated with 0.01000 $F$ hydrochloric acid. The progress of the titration was followed with a glass-electrode pH meter. Two specific points on the titration curve were as follows:

| Volume of HCl added, ml | pH |
|---|---|
| 1.00 | 5.70 |
| 10.00 | 4.28 |

Calculate the total volume in milliliters of hydrochloric acid required to reach the malonic acid ($H_2M$) equivalence point.

49. A solution was known to contain only disodium hydrogen phosphate, $Na_2HPO_4$, and sodium dihydrogen phosphate, $NaH_2PO_4$. Its initial pH was found to be 8.15. When a 25.00-ml aliquot of this solution was titrated with 0.1000 $F$ hydrochloric acid, the pH of the solution after the addition of 33.00 ml of the hydrochloric acid was 3.15. What were the molar concentrations of $HPO_4^{2-}$ and $H_2PO_4^-$ in the original solution?

50. A solution was known to contain only sodium carbonate and sodium bicarbonate, and its initial pH was found to be 10.63. When a 25.00-ml aliquot of this solution was titrated with 0.1000 $F$ hydrochloric acid, the pH of the solution after the addition of 21.00 ml of the acid was 6.35. What were the molar concentrations of sodium carbonate and sodium bicarbonate in the original solution?

# SUGGESTIONS FOR ADDITIONAL READING

1. C. Ayers: Acidimetry and alkalimetry. *In* C. L. Wilson and D. W. Wilson, eds.: *Comprehensive Analytical Chemistry*. Volume IB, Elsevier, New York, 1960, pp. 203–221.
2. R. G. Bates: Concept and determination of pH. *In* I. M. Kolthoff and P. J. Elving, eds.: *Treatise on Analytical Chemistry*. Part I, Volume 1, Wiley-Interscience, New York, 1959, pp. 361–404.
3. R. G. Bates: *Determination of pH*. John Wiley and Sons, New York, 1964.
4. R. P. Bell: *Acids and Bases*. John Wiley and Sons, New York, 1952.
5. S. Bruckenstein and I. M. Kolthoff: Acid-base strength and protolysis curves in water. *In* I. M. Kolthoff and P. J. Elving, eds.: *Treatise on Analytical Chemistry*. Part I, Volume 1, Wiley-Interscience, New York, 1959, pp. 421–474.
6. E. J. King: *Acid-Base Equilibria*. The Macmillan Company, New York, 1965.
7. I. M. Kolthoff: Concepts of acids and bases. *In* I. M. Kolthoff and P. J. Elving, eds.: *Treatise on Analytical Chemistry*. Part I, Volume 1, Wiley-Interscience, New York, 1959, pp. 405–420.
8. I. M. Kolthoff and V. A. Stenger: *Volumetric Analysis*. Second edition, Volume 2, Wiley-Interscience, New York, 1947, pp. 49–235.
9. C. A. VanderWerf: *Acids, Bases, and the Chemistry of the Covalent Bond*. Reinhold Publishing Corporation, New York, 1961.

# ACID-BASE REACTIONS IN NONAQUEOUS SOLVENTS

**5**

In the preceding chapter, we identified the hydronium ion, $H_9O_4^+$, as the form of the hydrated proton and as the strongest acid which can exist in water, although this species is more often written as $H_3O^+$. In nonaqueous solvents, such as ethanol or glacial acetic acid, the nature of the proton is less well characterized, but it is clear that the proton is solvated to a certain extent. By analogy to the hydronium ion in water, we shall represent the solvated proton in ethanol and in glacial acetic acid by the formulas $C_2H_5OH_2^+$ and $CH_3COOH_2^+$, respectively. It follows that $C_2H_5OH_2^+$ is the strongest acid to be encountered in ethanol and that $CH_3COOH_2^+$ is the strongest proton donor found in glacial acetic acid.

One of the virtues of the Brønsted-Lowry concept of acids and bases is that proton-transfer reactions in water and in nonaqueous solvents may be discussed in identical terms. If we consider the titration of ammonia ($NH_3$) with a strong acid in water and in glacial acetic acid, the two reactions may be represented by the following equilibria:

$$H_3O^+ + NH_3 \rightleftharpoons H_2O + NH_4^+ \tag{1}$$

$$CH_3COOH_2^+ + NH_3 \rightleftharpoons CH_3COOH + NH_4^+ \tag{2}$$

As each of these reactions proceeds toward equilibrium from left to right, $NH_3$ is accepting a proton, whereas a solvent molecule (water or acetic acid) accepts a proton as the reverse reaction occurs. For the present examples, the final positions of equilibrium are determined by the relative strengths as Brønsted-Lowry bases of $NH_3$ and $H_2O$ in reaction 1 and of $NH_3$ and $CH_3COOH$ in reaction 2. Since water is considerably more basic than acetic acid, the equilibrium position of reaction 2 lies much farther to the right than that of reaction 1. In other words, $NH_3$ appears to be a stronger base (or a better proton acceptor) in glacial acetic acid than it is in water because $CH_3COOH_2^+$ is a much better proton donor than $H_3O^+$.

This brief introductory discussion leads us to the important conclusion that many bases too weak to be titrated with a strong acid in water can be titrated successfully if glacial acetic acid is chosen as the solvent. In glacial acetic acid, such weak bases appear to be much stronger. Other solvents which, like glacial acetic acid, are less basic or more acidic than water could be similarly employed. Furthermore, weak acids which cannot be titrated in water with strong bases may be determined if the titrations are performed in solvents more basic than water. A solvent which is more basic than water will enhance the strength of a weak acid because such a solvent has a greater affinity for protons. Another key reason why nonaqueous solvents are of analytical interest is that many organic compounds having acidic or basic functional groups are insoluble in water and can be titrated only if they are dissolved in organic solvents or perhaps in mixtures of water and an organic solvent.

In the present chapter, we shall consider several factors which affect acid-base titrations in nonaqueous solvents. In addition, some titrations in three representative solvents—glacial acetic acid, anhydrous ethylenediamine, and methyl isobutyl ketone—will be discussed.

## SOME PROPERTIES OF SOLVENTS

In general, the acid-base behavior of any solute dissolved in a given solvent will depend on the interplay of at least three factors—the acid-base properties of the solvent relative to the solute, the autoprotolysis constant of the solvent, and the dielectric constant of the solvent.

*Classification of solvents.* An **amphiprotic solvent** is one capable of acting as either a Brønsted-Lowry acid or base. Perhaps the most familiar example is water, its amphiprotic character being shown by the reaction between one water molecule (acting as an acid) and a second molecule of water (acting as a base):

$$H_2O + H_2O \rightleftharpoons H_3O^+ + OH^-$$

This proton-transfer process is the so-called **autoprotolysis** of water. Most alcohols, such as methanol and ethanol, undergo analogous autoprotolysis reactions and have acid-base properties similar to water:

$$CH_3OH + CH_3OH \rightleftharpoons CH_3OH_2^+ + CH_3O^-$$

$$C_2H_5OH + C_2H_5OH \rightleftharpoons C_2H_5OH_2^+ + C_2H_5O^-$$

Glacial acetic acid represents a somewhat different kind of amphiprotic solvent. Although glacial acetic acid does undergo the usual type of autoprotolysis reaction,

$$CH_3COOH + CH_3COOH \rightleftharpoons CH_3COOH_2^+ + CH_3COO^-$$

it is distinctly more acidic than water because all bases appear to be stronger in acetic acid than in water. Other solvents such as liquid ammonia

$$NH_3 + NH_3 \rightleftharpoons NH_4^+ + NH_2^-$$

and ethylenediamine

$$H_2NCH_2CH_2NH_2 + H_2NCH_2CH_2NH_2$$
$$\rightleftharpoons H_2NCH_2CH_2NH_3^+ + H_2NCH_2CH_2NH^-$$

exhibit amphiprotic behavior but, compared to water, these are much more basic since they cause all acids to be stronger.

There is another class of solvents, called **inert** or **aprotic solvents**, which do not show any appreciable acid or base properties. Benzene, chloroform, and carbon tetrachloride are typical aprotic solvents, for they have no ionizable protons and they have little or no tendency to accept protons from other substances. In general, it is unrealistic to propose an autoprotolysis reaction for such a solvent.

Still another group of solvents includes those which possess definite base properties but which are without acid properties. Consequently, for these solvents an auto-protolysis reaction cannot be written. Such solvents are frequently included in the list of aprotic or inert solvents. For example, pyridine ($C_5H_5N$) can accept a proton from a Brønsted-Lowry acid such as water,

but the acid properties of pyridine are virtually nonexistent. Ethers and ketones are capable of accepting protons from strong Brønsted-Lowry acids such as sulfuric acid, but like pyridine they have no ionizable protons.

***Leveling effect and differentiating ability of a solvent.*** In water, the so-called mineral acids—perchloric acid, sulfuric acid, hydrochloric acid, hydrobromic acid, hydriodic acid, and nitric acid—all appear to be equally strong; it is impossible to decide whether any real differences in acid strength exist for these acids. Thus, if we consider the acid strengths of perchloric acid and hydrochloric acid in water, the following acid-base equilibria are involved:

$$HClO_4 + H_2O \rightleftharpoons ClO_4^- + H_3O^+$$
$$HCl + H_2O \rightleftharpoons Cl^- + H_3O^+$$

Since $HClO_4$ and $HCl$ are both much stronger acids than the hydronium ion, $H_3O^+$ (or because water is a stronger base than either $ClO_4^-$ or $Cl^-$), the position of equilibrium in each case lies so far to the right that it is impossible experimentally to distinguish any difference between the two equilibrium positions. Because the strengths of perchloric and hydrochloric acids appear to be identical in water, we speak of water as exerting a **leveling effect** on these two acids.

If the same acids, $HClO_4$ and $HCl$, are compared in glacial acetic acid as solvent, the pertinent acid-base reactions are

$$HClO_4 + CH_3COOH \rightleftharpoons ClO_4^- + CH_3COOH_2^+$$
and
$$HCl + CH_3COOH \rightleftharpoons Cl^- + CH_3COOH_2^+$$

Earlier, we stated that glacial acetic acid is a much weaker base than water and that the protonated acetic acid molecule ($CH_3COOH_2^+$) is a stronger acid than the hydronium ion ($H_3O^+$). Therefore, neither of these acid-base reactions proceeds so far toward the right in acetic acid as in water. However, the superior acid strength of perchloric acid shows up in the fact that its reaction with the glacial acetic acid solvent attains a greater degree of completion than the reaction of hydrochloric acid with the solvent. Thus, we find that glacial acetic acid possesses the capability to differentiate the acid strengths of $HClO_4$ and $HCl$, so we call it a **differentiating solvent**.

Clearly, the *leveling* or *differentiating* effect of a solvent depends to a large extent on the acid-base properties of the solvent relative to the dissolved solutes. Although water is a very poor choice of solvent to differentiate the acid strengths of $HClO_4$ and $HCl$, it is a good solvent to differentiate a mineral acid such as $HClO_4$ or $HCl$ from the much weaker acetic acid. However, a strongly basic solvent such as liquid ammonia would fail to differentiate a mineral acid from acetic acid because the reactions

$$HClO_4 + NH_3 \rightleftharpoons ClO_4^- + NH_4^+$$

and

$$CH_3COOH + NH_3 \rightleftharpoons CH_3COO^- + NH_4^+$$

would both proceed virtually to completion because of the great base strength of $NH_3$.

*Autoprotolysis constant.* Another factor which affects the differentiating ability of a given solvent is its autoprotolysis constant. If we denote an amphiprotic solvent as SH, the general form of the autoprotolysis reaction becomes

$$SH + SH \rightleftharpoons HSH^+ + S^-$$

and the corresponding equilibrium expression is

$$\frac{(HSH^+)(S^-)}{(SH)^2} = K$$

where the parentheses denote activities. By convention the activity of the pure solvent (SH) is taken to be unity. In addition, for the majority of solvents the extent of autoprotolysis is quite small, so that it is permissible to replace activities of ions by their analytical concentrations. Thus, the usual simplified form for the autoprotolysis equilibrium expression is

$$[HSH^+][S^-] = K_s$$

in which $K_s$ is the **autoprotolysis constant**. For water the ion-product constant $K_w$ and the autoprotolysis constant $K_s$ are synonymous:

$$[H_3O^+][OH^-] = [H^+][OH^-] = K_w = K_s = 1.00 \times 10^{-14}$$

Values of $pK_s$, the negative base-ten logarithm of the autoprotolysis constant, for several different solvents at 25°C are as follows: water, 14.0; ethanol, 19.1; methanol, 16.7; glacial acetic acid, 14.5; formic acid, 6.2; and ethylenediamine, 15.3.

How is the magnitude of the autoprotolysis constant related to the differentiating ability of a solvent? To establish a basis for discussion, let us consider the familiar case of water, which has an autoprotolysis constant of $1.00 \times 10^{-14}$ at 25°C. Upon inspection of Figures 4–6 and 4–9 in the previous chapter, we observe that an acid with $K_a$ less than $10^{-10}$ in water does not exhibit a titration curve, primarily because water competes so successfully with the weak acid as a proton donor at pH values above approximately 11. Furthermore, below pH 3, where we are interested in titrating acids with $K_a$ values greater than $10^{-2}$, it becomes difficult to distinguish a so-called weak acid from hydronium ion itself, as a comparison of Figures 4–2 and 4–6 shows. Below pH 3 and above pH 11, water exerts a definite leveling effect on acids and bases dissolved in it, but the range of pH values between 3 and 11 represents the analytically useful pH range of water. Note that these two pH extremes correspond, respectively, to the pH of a $10^{-3} F$ strong acid solution and to the pH of a $10^{-3} F$ strong base solution.

Compare these observations for water with the situation for ethanol. Ethanol has an autoprotolysis reaction of the form

$$C_2H_5OH + C_2H_5OH \rightleftharpoons C_2H_5OH_2{}^+ + C_2H_5O^-$$

and an autoprotolysis-constant expression given by the relation

$$[C_2H_5OH_2{}^+][C_2H_5O^-] = K_s = 7.9 \times 10^{-20}$$

In ethanol, the $C_2H_5OH_2{}^+$ ion is the counterpart of the hydronium ion in an aqueous medium, and the ethoxide ion ($C_2H_5O^-$) is the equivalent of the hydroxide ion. Let us assume that pH in ethanol represents the negative logarithm of the $C_2H_5OH_2{}^+$ concentration. However, obvious differences in the chemical and physical properties of water and ethanol make it almost impossible to relate the pH scales for the two solvents to each other in a quantitative way. If $10^{-3}$ mole of strong acid is dissolved in a liter of ethanol, the predominant proton-containing species will be the $C_2H_5OH_2{}^+$ ion and the pH will be 3.0. If we dissolve $10^{-3}$ mole of sodium ethoxide ($Na^+C_2H_5O^-$) in one liter of ethanol, the concentration of $C_2H_5O^-$ will be $10^{-3} M$, and the concentration of $C_2H_5OH_2{}^+$ and the pH may be calculated as follows:

$$[C_2H_5OH_2{}^+] = \frac{K_s}{[C_2H_5O^-]} = \frac{7.9 \times 10^{-20}}{1.0 \times 10^{-3}} = 7.9 \times 10^{-17} M$$

$$pC_2H_5OH_2 = pH = 16.1$$

Therefore, in ethanol the pH range encompassed by a $10^{-3} F$ strong acid solution and a $10^{-3} F$ strong base solution is $(16.1 - 3.0)$ or 13.1 pH units, significantly greater than the useful pH range of eight units in water.

We can conclude from these calculations that the useful pH range for a solvent increases as the autoprotolysis constant decreases. In other words, the smaller the autoprotolysis constant, the greater is the range of acid or base strengths which can exist in a solvent and the greater is the likelihood that it will be a differentiating solvent. Formic acid, with an autoprotolysis constant of $6.3 \times 10^{-7}$ ($pK_s = 6.2$), is a poor differentiating solvent because only a very narrow range of acid strengths can exist in it. Thus, the differentiating ability of a particular solvent must be carefully assessed in terms of both its autoprotolysis constant and its acid-base properties relative to the solutes dissolved in it. Water, with an autoprotolysis constant of $1.00 \times 10^{-14}$ at 25°C, cannot differentiate the strong mineral acids because it is a relatively basic solvent. On the other hand, glacial acetic acid has an autoprotolysis constant of similar magnitude ($3.2 \times 10^{-15}$), but the mineral acids are differentiated in it because of the acidic character of glacial acetic acid.

***Effect of the dielectric constant.*** Another factor to be reckoned with in nonaqueous acid-base titrimetry is that, compared to water, most other solvents have low dielectric constants.

We can describe the reaction of an *uncharged* solute acid BH with a solvent SH in terms of two discrete processes—*ionization* and *dissociation*:

$$BH + SH \underset{\text{ionization step}}{\rightleftharpoons} \{B^-HSH^+\} \underset{\text{dissociation step}}{\rightleftharpoons} B^- + HSH^+$$

In the ionization step, a proton is transferred from the solute acid BH to the solvent SH, giving rise to the ion pair $B^-HSH^+$. There is a definite separation of charges in such an ion pair, although the individual ions $B^-$ and $HSH^+$ are very close to each other. Dissociation occurs when additional solvent molecules attack the ion pair and separate it completely into the solvated $B^-$ and $HSH^+$ ions. According to Coulomb's law, the force of attraction between two oppositely charged ions is inversely proportional to the dielectric constant of the medium (solvent) in which the ions exist. For solvents of high dielectric constant such as water ($D = 78.5$ at $25°C$), the force of attraction between the ions of an ion pair is relatively small and the dissociation step will be virtually complete. However, for solvents having low dielectric constants (ethanol, $D = 24.3$; glacial acetic acid, $D = 6.1$; ethylenediamine, $D = 12.5$; methyl isobutyl ketone, $D = 13.1$; acetone, $D = 20.7$; and benzene, $D = 2.3$), there is considerable ion pairing as well as formation of larger ion aggregates.

We can understand the qualitative effect of dielectric constant on the strength of a solute acid by contrasting the behavior of acetic acid in water and in ethanol, two solvents of *similar* base strength. Since the ionization and dissociation of acetic acid molecules in a given solvent lead to the formation of protonated solvent molecules ($HSH^+$) and negatively charged, solvated acetate ions ($CH_3COO^-$), the acid strength of acetic acid would be enhanced in a solvent of high dielectric constant, such as water, which facilitates the separation of charged species. However, in ethanol which has a lower dielectric constant, the dissociation of the few ion pairs of the type $CH_3COO^{-+}H_2OC_2H_5$ that do exist would not be favored, so the strength of acetic acid should be less in ethanol than in water. On the other hand, the strength of acetic acid is much greater in ethylenediamine than in water, because ethylenediamine is a much more basic solvent than water even though the former has a smaller dielectric constant.

One would expect the acidity of a *positively charged* solute of the type $BH^+$ to be largely unaffected by variations in the dielectric constant of the solvent, because the transfer of a proton from $BH^+$ to SH does not produce an ion pair:

$$BH^+ + SH \underset{\text{ionization step}}{\rightleftharpoons} \{B\ HSH^+\} \underset{\text{dissociation step}}{\rightleftharpoons} B + HSH^+$$

In the absence of ion pairs, neither the ionization step nor the dissociation step involves the separation of charges. This line of reasoning might be expected to apply in the case of species such as ammonium ion ($NH_4^+$) or pyridinium ion ($C_5H_5NH^+$). However, the picture is not quite so simple, since the $NH_4^+$ species, for example, would be added to the solvent in the form of an ammonium salt, the anion of which would undoubtedly form ion pairs with $NH_4^+$ and thereby influence its acid strength.

A low dielectric constant for a given solvent (SH) has a pronounced effect on the autoprotolysis constant $K_s$ of that solvent. As shown by the following reaction scheme, the detailed process of autoprotolysis should include an ion-pair intermediate,

$$SH + SH \underset{\text{ionization step}}{\rightleftharpoons} \{S^-HSH^+\} \underset{\text{dissociation step}}{\rightleftharpoons} S^- + HSH^+$$

so that the autoprotolysis constant is the product of an *ionization constant* and a *dissociation constant*:

$$K_s = [HSH^+][S^-] = K_{ioniz}K_{diss}$$

If the magnitude of the dissociation constant, $K_{diss}$, is small because the dielectric constant of the solvent is low, the autoprotolysis constant $K_s$ is likely to be small. This effect seems to be operative for glacial acetic acid which has a surprisingly low dielectric constant of 6.1 and an autoprotolysis constant of $3.5 \times 10^{-15}$. In sharp contrast to acetic acid is anhydrous formic acid. Although formic acid is chemically very similar to acetic acid, its autoprotolysis constant is $6.3 \times 10^{-7}$. Such a relatively large autoprotolysis constant is partly explicable on the basis of the high dielectric constant of 58.5 for formic acid which promotes the dissociation of ion pairs of the type $HCOO^-HCOOH_2^+$. Water is a unique solvent due to its unusually high dielectric constant and its relatively small autoprotolysis constant.

## TITRATIONS IN BASIC SOLVENTS

In general, the titration of weakly acidic substances in aqueous media is limited to those species whose acid dissociation constants are not smaller than approximately $10^{-7}$. Acids which possess $K_a$ values below about $10^{-7}$ in water are too weak to provide adequately sharp titration curves for titrations performed in an aqueous medium with the common strong titrant bases such as sodium or potassium hydroxide. On the other hand, the use of solvents more basic than water can greatly improve the sharpness of titration curves and the accuracy of the titrations themselves by enhancing the strengths of the solute acids.

*Typical solvents and titrants.* Among the common basic solvents are butylamine, ethylenediamine, pyridine, and liquid ammonia (which can be used only at low temperatures). Methanol-benzene mixtures are sometimes used for nonaqueous acid-base titrations. In addition, dimethylformamide (though not as basic as amine solvents) is excellent in titrations of all except very weak organic acids.

Let us examine some acid-base reactions in ethylenediamine, a typical strongly basic solvent with a much greater viscosity than water and a pungent odor resembling that of ammonia. Solutions of both hydrochloric and benzoic acids in ethylenediamine are ionized virtually completely according to the reactions

$$HCl + H_2NCH_2CH_2NH_2 \rightleftharpoons H_2NCH_2CH_2NH_3^+Cl^-$$

and

$$C_6H_5COOH + H_2NCH_2CH_2NH_2 \rightleftharpoons H_2NCH_2CH_2NH_3^+{}^-OOCC_6H_5$$

However, because ethylenediamine (as well as the other basic solvents named) has a low dielectric constant ($D = 12.5$ at 25°C), these two acids exist almost entirely as undissociated ion pairs. Benzoic acid, which is rather weak in water ($K_a = 6.3 \times 10^{-5}$), is leveled to the strength of a strong acid in ethylenediamine because of the great affinity of this solvent for protons. Other acids, which are much weaker than benzoic acid and far too weak to be titrated successfully with a strong base in water, become sufficiently strong in ethylenediamine to make their titrations feasible. For example, phenol ($C_6H_5OH$) is such a weak acid in water ($K_a = 1.1 \times 10^{-10}$) that the reaction scarcely proceeds at all. However, when phenol is dissolved in ethylenediamine, the transfer of a proton from the solute acid to the solvent base occurs to a large extent,

$$\text{C}_6\text{H}_5\text{OH} + H_2NCH_2CH_2NH_2 \rightleftharpoons H_2NCH_2CH_2NH_3^{+-}O\text{—C}_6\text{H}_5$$

with the resultant formation of ion pairs which can be titrated with a strong base. Figures 5–1 and 5–2 compare and contrast curves for the titrations of benzoic acid and phenol with sodium hydroxide in water and for the titrations of these acids with sodium aminoethoxide ($Na^{+-}OCH_2CH_2NH_2$) as a titrant in ethylenediamine. An inspection of these titration curves confirms that benzoic acid behaves as a strong acid in ethylenediamine because its titration curve is extremely well defined, much like that of a strong acid-strong base titration in water. Similarly, the titration curve for phenol in ethylenediamine is as sharply defined as that for the titration of benzoic acid with sodium hydroxide in an aqueous medium.

Selection of a titrant base for a nonaqueous acid-base titration is as important as the choice of solvent itself. For best results, the base strength of the titrant must be comparable to the base strength of the conjugate base of the solvent. To use a weak titrant base in a strongly basic solvent is to defeat the whole purpose of the nonaqueous titration. Ideally, one should employ a salt of the conjugate base of the solvent as the titrant, since this is the strongest base which is capable of existence in a given solvent.

A wide variety of titrant bases has been employed for nonaqueous titrations. When simple alcohols such as methanol or ethanol are used as solvents for the titrimetric determinations of carboxylic acids, the usual titrants are sodium methoxide or sodium ethoxide in the corresponding alcoholic solvent. Such titrants are prepared by the reaction between pure sodium metal and the anhydrous alcohol. Occasionally, it is possible to use alcoholic solutions of sodium or potassium hydroxide as titrants. Ethylenediamine is one of the outstanding basic solvents for use in the titration of

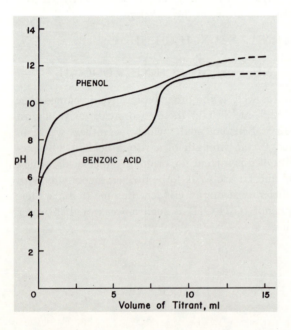

*Figure 5–1.* Titration curves for the titrations of benzoic acid and phenol in water with a 0.5 *F* sodium hydroxide solution. (Redrawn, with permission, from the paper by M. L. Moss, J. H. Elliott, and R. T. Hall: Anal. Chem., *20*: 784, 1948.)

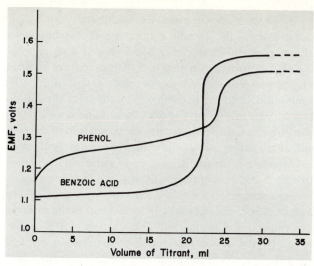

*Figure 5–2.* Titration curves for the titrations of benzoic acid and phenol in ethylenediamine (a strongly basic solvent). The titrant was 0.2 *F* sodium aminoethoxide dissolved in a mixture of ethanolamine and ethylenediamine. The use of a much more basic titrant-solvent system enhances the acidity of benzoic acid and phenol and makes the titration curves more sharply defined than in Figure 5–1. Although expressed in different units, the size of the vertical scale is the same as in Figure 5–1. (Redrawn, with permission, from the paper by M. L. Moss, J. H. Elliott, and R. T. Hall: Anal. Chem., *20*: 784, 1948.)

weak and very weak acids. However, the sodium salt of the conjugate base of ethylenediamine is not sufficiently soluble in the parent solvent to permit its use as the titrant. As an alternative, sodium aminoethoxide ($Na^+{}^-OCH_2CH_2NH_2$) in ethylenediamine-ethanolamine solvent mixtures has been recommended as a suitable titrant. Another class of titrant bases which have gained in popularity in recent years is the tetraalkylammonium hydroxides. For example, tetrabutylammonium hydroxide, $(C_4H_9)_4N^+OH^-$, may be prepared in a variety of solvents, including alcohols and benzene-alcohol mixtures, if one passes a solution of tetrabutylammonium iodide through an anion-exchange column in the hydroxide form.

A few words should be added concerning the handling and storage of strongly basic solvents and titrant solutions. Even more than aqueous sodium hydroxide solutions, they quickly absorb carbon dioxide and other acidic impurities, and so they should be protected from the atmosphere. Furthermore, the absorption of moisture can be equally detrimental to the proper behavior of nonaqueous titrants and solvents because the presence of even a trace of water in these nonaqueous systems can exert a pronounced leveling effect on the strengths of solute acids—and perhaps prevent them from being titrated at all.

**End-point detection.** End-point detection in nonaqueous titrations is by no means trivial. For one thing the pH ranges over which acid-base indicators change color in nonaqueous solvents are different from water. Frequently, the color changes are not the same as in water, probably because of ion-pairing phenomena and other structural (electronic) changes in the indicator molecules caused by the solvent environment. Nevertheless, several of the more familiar acid-base indicators, including methyl violet, methyl red, phenolphthalein, and thymolphthalein, have been found to function satisfactorily in certain nonaqueous titrations.

In practice, the progress of a nonaqueous acid-base titration may be best followed potentiometrically. Details concerning the techniques of potentiometric titrations are described in Chapter 11. Briefly stated, a potentiometric acid-base titration consists of measuring, as a function of the volume of titrant added, the potential developed between a reference electrode and a suitable indicator electrode, the latter being sensitive to the hydrogen ion activity of the solution. A titration curve such as those shown in Figure 5–2 may be obtained, and the equivalence point can be located by visual inspection or graphical analysis of the complete titration curve. Design and construction of suitable reference and indicator electrodes for nonaqueous potentiometric titrations can pose some difficult problems. When immersed in strongly basic, nonaqueous solvents, the familiar glass-membrane electrode used for pH measurements and titrations in aqueous media may not perform properly, because the presence of water molecules within the surface layers of the glass membrane is essential to the mechanism of operation of a glass electrode, and this water is removed by the desiccating action of nonaqueous solvents to which the electrode is exposed. Therefore, the antimony oxide electrode, the stainless steel electrode, and even several forms of the classic hydrogen gas electrode have been employed instead of the glass electrode; the roster of reference electrodes for nonaqueous titrations includes the aqueous saturated calomel electrode (SCE) as well as several nonaqueous versions of the calomel electrode.

***Examples of substances titrated.*** Two classes of weakly acidic organic compounds which can be titrated in basic solvents are carboxylic acids and phenols. Carboxylic acids are sufficiently strong, having $pK_a$ values in water in the neighborhood of 5 or 6, that only modestly basic solvents such as methanol or ethanol are needed to obtain nicely defined titration curves with sharp end points. On the other hand, the determination of phenols, which have $pK_a$ values of approximately 10 in water, requires a much more basic solvent than an alcohol. Anhydrous ethylenediamine is an excellent choice, provided that care is taken to remove and exclude traces of water from the system.

Several other types of weak acids that have not been previously mentioned may be titrated in basic solvents; they are sulfonamides, barbituric acids, amino acids, and certain salts of amines. Sulfonamides, known more familiarly as the sulfa drugs, possess the $-SO_2NH-$ group whose acidity is considerably enhanced in basic solvents such as butylamine or ethylenediamine. For example, sulfanilamide dissolved in butylamine

can be determined by titration with sodium methoxide:

Mixtures of sulfonamides can be analyzed by taking advantage of the different acid strengths of these compounds in various solvents.

A variety of amine salts can be determined by titration in basic solvents. One representative example of such a titration involves butylamine hydrochloride, $C_4H_9NH_3^+Cl^-$. Although the acid strength of $C_4H_9NH_3^+$ is indeed slight in a solvent such as water, the addition of ethylenediamine to an aqueous suspension of butylamine hydrochloride not only dissolves the amine salt but increases the apparent acid strength of $C_4H_9NH_3^+$ to the point where it can be titrated with sodium methoxide:

$$C_4H_9NH_3^+Cl^- + CH_3O^-Na^+ \rightleftharpoons C_4H_9NH_2 + CH_3OH + Na^+Cl^-$$

One requirement for the successful titration of amine salts is that the anion of the salt cannot be too strong a base. If it is, it may compete so well with the titrant base for the proton that a poorly defined end point is obtained. Among the other amine salts which have been successfully titrated are methylamine hydrochloride ($CH_3NH_3^+Cl^-$), pyridinium perchlorate ($C_5H_5NH^+ClO_4^-$), and quinine sulfate.

## TITRATIONS IN GLACIAL ACETIC ACID

In attempting to titrate a variety of bases of widely differing strength with a strong acid in an aqueous solution, one discovers that only those bases which have considerably greater affinity for protons than the solvent (water) can be satisfactorily determined. Fundamentally, the titration reaction involves the transfer of a proton from the conjugate acid of the solvent $HSH^+$ to the solute base B:

$$HSH^+ + B \rightleftharpoons SH + BH^+$$

If the affinity of the solvent SH for protons happens to be comparable to or greater than that of the base B, the titration reaction can never attain any substantial degree of completion and the titration will fail. Therefore, to achieve success in the titration of weak and very weak bases, one must select an acidic solvent with little affinity for protons to enhance as much as possible the meager basicity of the compounds of interest. Note that the solvent requirement is just the opposite of that for the titrimetric determination of weakly acidic substances.

Glacial acetic acid has received overwhelming attention as a titration medium for weak bases. Ion-pairing phenomena are extremely important in any detailed consideration of acid-base equilibria in this solvent because of its low dielectric constant ($D = 6.1$ at 25°C). Its autoprotolysis constant is $3.5 \times 10^{-15}$ which means that, as an acidic solvent, glacial acetic acid can differentiate a reasonably wide range of weak bases. On the other hand, glacial acetic acid exerts a leveling effect on all strong bases.

Perchloric acid dissolved in glacial acetic acid is almost universally used as a titrant for weak bases. When the exclusion or removal of water from the titrant or solvent is desirable in order to prevent leveling effects and improve the sharpness of titration curves, the common practice is to add approximately stoichiometric quantities of acetic anhydride, which consumes water to form more acetic acid by the reaction

$$(CH_3CO)_2O + H_2O \rightarrow 2 CH_3COOH$$

End-point detection can be accomplished on an empirical basis with the aid of acid-base indicators. However, a better approach is to employ the previously mentioned technique of potentiometric titration. In the case of glacial acetic acid, both the glass electrode and the aqueous saturated calomel reference electrode have been found to function very well, and so some of the problems encountered with potentiometric titrations in basic solvents do not appear in glacial acetic acid.

There are several classes of weak bases which may be determined titrimetrically in glacial acetic acid—amines, amino acids, alkaloids, antihistamines, and anions of weak acids. Primary, secondary, and tertiary amines can be titrated in anhydrous acetic acid with perchloric acid. As an example, the titration of aniline, a primary amine, proceeds according to the reaction

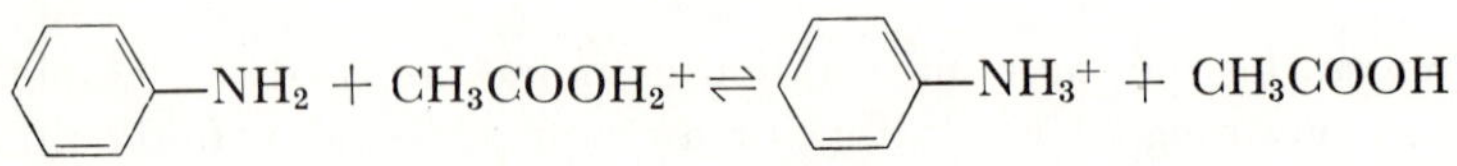

$$\langle\!\!\bigcirc\!\!\rangle\!-NH_2 + CH_3COOH_2^+ \rightleftharpoons \langle\!\!\bigcirc\!\!\rangle\!-NH_3^+ + CH_3COOH$$

where $CH_3COOH_2^+$ is the solvated proton in glacial acetic acid.

In the preceding chapter on aqueous acid-base chemistry, the amphiprotic character of alanine (a typical amino acid) was described. Alanine may accept or lose a proton as shown by the equilibria

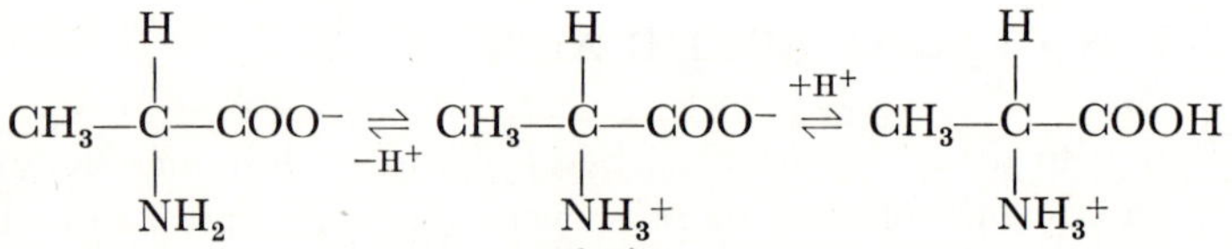

$$\underset{NH_2}{CH_3-\overset{\overset{\displaystyle H}{|}}{\underset{|}{C}}-COO^-} \underset{-H^+}{\overset{}{\rightleftharpoons}} \underset{\underset{\text{alanine}}{NH_3^+}}{CH_3-\overset{\overset{\displaystyle H}{|}}{\underset{|}{C}}-COO^-} \overset{+H^+}{\rightleftharpoons} \underset{NH_3^+}{CH_3-\overset{\overset{\displaystyle H}{|}}{\underset{|}{C}}-COOH}$$

However, in an acidic solvent such as glacial acetic acid, the carboxylic acid group (—COOH) is rendered unreactive, and so the titration of an acetic acid solution of alanine with perchloric acid may be represented as

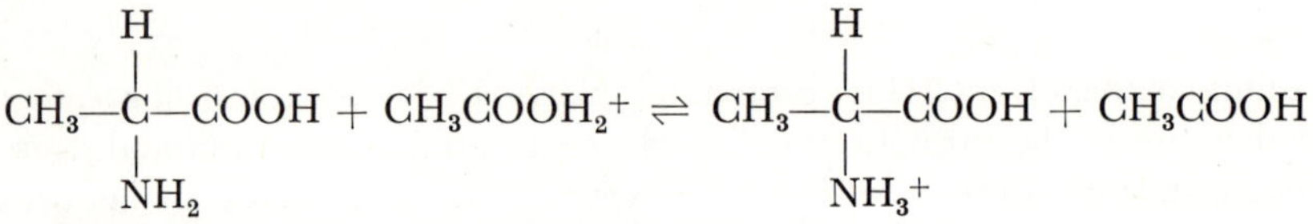

$$\underset{NH_2}{CH_3-\overset{\overset{\displaystyle H}{|}}{\underset{|}{C}}-COOH} + CH_3COOH_2^+ \rightleftharpoons \underset{NH_3^+}{CH_3-\overset{\overset{\displaystyle H}{|}}{\underset{|}{C}}-COOH} + CH_3COOH$$

Thus, the titration of amino acids in glacial acetic acid is analogous to the determination of amines. It is interesting to note that alanine exists as a dipolar ion or *zwitterion* in a solvent of high dielectric constant such as water. However, the low dielectric constant of glacial acetic acid does not favor the separation of charge, so the form of alanine shown in the above reaction is predominant.

Anions of weak acids are sufficiently basic in glacial acetic acid to be titrated. Just a few of the anions which can be determined by titration with standard perchloric acid are bicarbonate, carbonate, acetate, bisulfite, cyanide, sulfate, and barbiturate derivatives.

## TITRATIONS IN METHYL ISOBUTYL KETONE

Any of the solvents discussed up to this point has certain limitations as to its usefulness in the determination of a broad range of acidic or basic compounds. For example, ethylenediamine is best used in the determination of various weak acids, but it exerts a powerful leveling effect on acids stronger than acetic acid. It is of no use at all in the titration of basic substances. Similarly, although glacial acetic acid is an excellent solvent for the determination of weak bases, strong bases cannot be differentiated and acidic species cannot be determined. Several interrelated characteristics of these two solvents impose the limitations on their wide-range usefulness. First, the basic and acidic characters of ethylenediamine and acetic acid, respectively, influence acid and base strengths of solutes dissolved in them. Second, each of these solvents is amphiprotic, and thus undergoes an autoprotolysis reaction. Although the autoprotolysis constant for glacial acetic acid is relatively small ($3.5 \times 10^{-15}$), it is nevertheless large enough that, coupled with the acidity of the solvent, we can expect a definite limit on the total range of base strength which can be seen. A similar statement can be made concerning ethylenediamine.

In view of these limitations, it would be desirable to have available a versatile solvent whose properties would permit the existence in it of a much broader range of acids and bases than the former solvents. Ideally, such a solvent should not exhibit any acid or base properties of its own; that is, it should be aprotic or inert. An inert or aprotic solvent undergoes no autoprotolysis reaction, and allows dissolved substances to exhibit as nearly as possible their intrinsic acid or base properties. In fact, a truly inert solvent does not participate at all in the transfer of a proton when a dissolved acid reacts with titrant base or vice versa.

Methyl isobutyl ketone, $CH_3COCH_2CH(CH_3)_2$, is a solvent which seems to possess many of the desired qualities of a wide-range titration medium. However, it is not a true aprotic solvent because the oxygen atom can attract a proton by sharing a lone pair of electrons with $H^+$. On the other hand, the loss of a proton by methyl isobutyl ketone cannot occur readily, so this solvent does not undergo the usual type of autoprotolysis. As we shall see, a remarkably wide variety of acids is differentiable in this solvent. In addition, many types of bases can be titrated in methyl isobutyl ketone.

Although the commercially available solvent is suitable for use without additional purification, one can remove trace acidic impurities by passing the methyl isobutyl ketone through a small column packed with activated alumina ($Al_2O_3$). A solution of tetrabutylammonium hydroxide in isopropanol is the usual titrant for acids, and perchloric acid dissolved in dioxane may be employed to titrate bases. Titrations can be performed in open vessels without precautions to prevent the absorption of atmospheric moisture and carbon dioxide. However, it is advisable to use comparatively concentrated titrant solutions to avoid the leveling effects which would occur if large volumes of titrant were added. Acid-base titrations in methyl isobutyl ketone are invariably followed by means of the potentiometric technique mentioned earlier in this chapter. A glass indicator electrode and an aqueous calomel reference electrode both function well in methyl isobutyl ketone.

**Titrations of strong acids.** Figure 5–3 shows how well an inert solvent can resolve or differentiate the mineral acids. This figure presents plots of the potential of a glass electrode versus a saturated calomel electrode as a function of titrant volume for four different strong-acid mixtures. Curve *A* shows clearly the resolution of perchloric and sulfuric acids, and reveals the diprotic character of sulfuric acid. A mixture of perchloric and nitric acids can be resolved even better, as demonstrated by curve *B*. Curve *C* indicates that a mixture of sulfuric acid and hydrochloric acid can be successfully analyzed for each component; although the first equivalence point relates to the sum of the sulfuric acid and hydrochloric acid, the volume of titrant added between the first and second equivalence points corresponds only to the sulfuric acid, so the quantity of hydrochloric acid in the sample can be calculated by difference. Curve *D* establishes that a mixture of nitric and hydrochloric acids cannot be resolved in methyl isobutyl ketone.

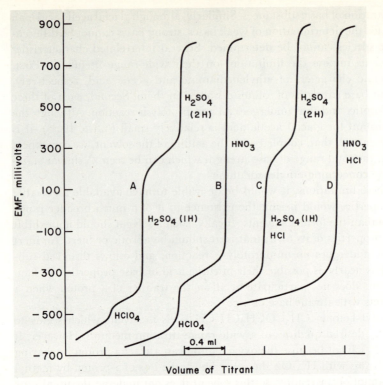

*Figure 5–3.* Titration curves for various mixtures of strong acids in methyl isobutyl ketone with 0.2 *F* tetrabutylammonium hydroxide (in isopropanol) as the titrant. A glass indicator electrode and an aqueous saturated calomel reference electrode were employed to measure the emf values. (Redrawn, with permission, from the paper by D. B. Bruss and G. E. A. Wyld: Anal. Chem., *29*: 232, 1957.)

Two interesting conclusions can be drawn from Figure 5–3. First, notice that the initial potential of the glass indicator electrode for the titration of perchloric acid is approximately −700 mv versus the saturated calomel reference electrode, whereas the potential in the presence of excess titrant base is around +800 mv versus the reference electrode. This total interval of at least 1500 mv is related to the useful range of acid strengths which can exist in methyl isobutyl ketone. Since 60 mv is equivalent to one pH unit, the overall operating range in this solvent is equivalent to at least 25 pH units, which may be contrasted with the useful range in water of approximately eight pH units.

A second point concerning Figure 5–3 is that the order of the strengths of the mineral acids as based on the positions of the different titration curves is $HClO_4 > H_2SO_4 > HCl > HNO_3$, although the difference in the strengths of hydrochloric and nitric acids is only very slight. This order is exactly the same as that established on the basis of conductance measurements. Thus, the relative strengths of the mineral acids can be determined from titration curves obtained in an inert solvent.

***Titration of complex acid mixtures.*** A more impressive demonstration of the resolving power of methyl isobutyl ketone is the titration curve shown in Figure 5–4 for a five-component acid mixture, containing species that range all the way from the strongest mineral acid, perchloric acid, to the very weak acid phenol (which has a dissociation constant in water of $1.1 \times 10^{-10}$ and which cannot be titrated in an aqueous medium). Because of the use of a different reference electrode, the potential

*Figure 5-4.* Titration curve showing the resolution of a mixture of strong, weak, and very weak acids in methyl isobutyl ketone. The titrant was 0.2 $F$ tetrabutylammonium hydroxide in isopropanol. A glass indicator electrode and a platinum wire (sealed into the buret tip) reference electrode were used. (Redrawn, with permission, from the paper by D. B. Bruss and G. E. A. Wyld: Anal. Chem., *29*: 232, 1957.)

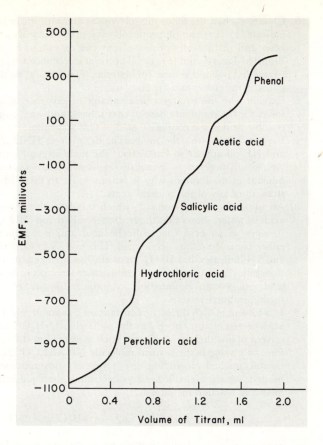

scale in Figure 5-4 varies slightly from Figure 5-3. To record the titration curve in Figure 5-4, a platinum wire sealed directly into the buret tip was used as a reference electrode, and the tip of the buret was kept immersed in the titration vessel (along with the glass indicator electrode) to complete the electrical circuit to the potentiometer.

## QUESTIONS AND PROBLEMS

1. Define or identify each of the following terms: amphiprotic solvent, aprotic solvent, leveling effect of a solvent, differentiating ability of a solvent, autoprotolysis constant, dielectric constant, ion pair.

2. There are at least three factors which can influence or affect the strength of an acid in a given solvent. These are the acid or base properties of the solvent, the autoprotolysis constant for the solvent, and the dielectric constant of the solvent. In three paragraphs, discuss how each of these factors relates to acid strength. Use specific examples to illustrate your conclusions and statements.

3. Which of the following species exhibits amphiprotic character? (a) $HCO_3^-$, (b) $Cl^-$, (c) $NH_4^+$, (d) $H_3O^+$, and (e) $CH_3COO^-$

4. Among the following solvents, which one would succeed best in making acetic acid, benzoic acid, hydrochloric acid, and perchloric acid all appear to have equal acid strengths? (a) pure water, (b) syrupy (85 per cent) phosphoric acid, (c) liquid ammonia, (d) concentrated sulfuric acid, and (e) methyl isobutyl ketone

5. Classify each of the following solvents as amphiprotic or aprotic. In addition, indicate (1) if each amphiprotic solvent is predominantly acidic or basic in character and (2) if each aprotic solvent can or cannot accept a proton from a Brønsted-Lowry acid. (a) glacial acetic acid, (b) dioxane, (c) ethylenediamine, (d) methyl isobutyl ketone, (e) benzene, (f) water, (g) diethyl ether, (h) isopropanol, (i) acetone, and (j) butylamine

6. Explain why the strongest base capable of existence in a given amphiprotic solvent is the conjugate base of that solvent and why, similarly, the strongest possible acid is the conjugate acid of that solvent.

7. Explain briefly why the mineral acids, such as $HNO_3$, $HCl$, $HClO_4$, $HBr$, and $HI$, all appear to have about the same strength in water. Explain how you would proceed to establish *experimentally* the *relative* strengths of these mineral acids. Explain why it is impossible to establish the *true* or *intrinsic* strength of any of the mineral acids.

8. An inspection of Appendix 2, which tabulates the dissociation constants for acids in water, shows that the hydrogen sulfate ion ($HSO_4^-$) is about 10 times stronger as an acid than salicylic acid and is significantly stronger than either benzoic acid or acetic acid. However, an examination of Figures 5–3 and 5–4 indicates that $HSO_4^-$ in methyl isobutyl ketone is a much weaker acid than salicylic acid and of only comparable strength to benzoic acid and acetic acid. Suggest an explanation to account for this reversal of behavior in methyl isobutyl ketone.

9. In a solvent of high dielectric constant such as water, glycine (one of the amino acids) exists almost entirely as the zwitterion $^+NH_3CH_2COO^-$, whereas in a solvent of low dielectric constant such as glacial acetic acid the predominant form of glycine is the neutral molecule $NH_2CH_2COOH$. Suggest an experimental method, including the types of measurements and calculations needed, which might be used to determine the equilibrium constant for the following reaction in water:

$$^+NH_3CH_2COO^- \rightleftharpoons NH_2CH_2COOH$$

10. Suppose that a solution containing hydrobromic acid (HBr) and a large excess of sodium bromide (NaBr) in ethylenediamine as solvent is titrated with an ethylenediamine solution of the sodium salt, $Na^+{}^-NHCH_2CH_2NH_2$. Although hydrobromic acid is a strong electrolyte in water, it exists predominantly as the ion pair $H_2NCH_2CH_2NH_3^+Br^-$ in ethylenediamine. To what factor can such behavior be attributed?

11. Suggest a solvent and a titrant which would be suitable for the nonaqueous acid-base titration of each of the following substances:

(a)

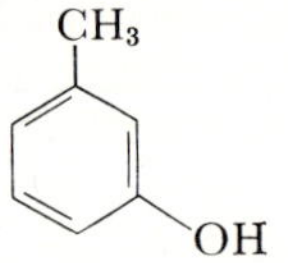

*m*-cresol (used in disinfectants, fumigants, photographic developers, and explosives)

(b)

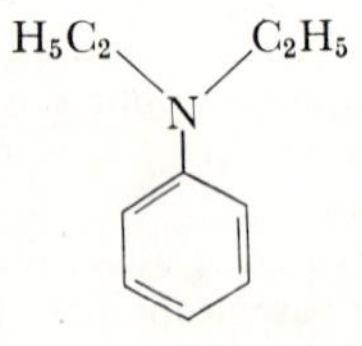

diethylaniline (an intermediate for the synthesis of dyestuffs)

(c)

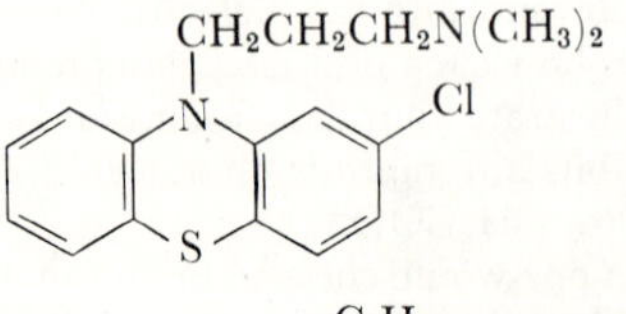

chlorpromazine (a sedative and antiemetic)

(d)

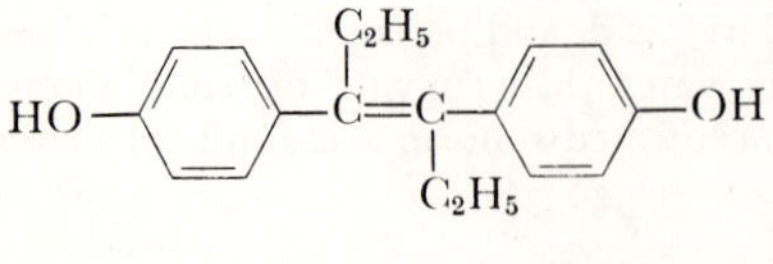

diethylstilbestrol (a growth hormone for cattle)

(e)

$$H_3C$$
$$CH-CH_2-CH-COOH \qquad \text{leucine (an amino acid)}$$
$$H_3C \qquad\qquad NH_2$$

(f) $H_2N-\text{⟨⟩}-SO_2NH-\text{[thiazole ring with S, N]}$    sulfathiazole (a sulfa drug effective against diseases due to streptococci)

12. A mixture of ethylamine, diethylamine, and triethylamine may be differentiated and analyzed by means of a set of nonaqueous acid-base reactions:

(a) Treatment of a mixture of ethylamine, diethylamine, and triethylamine with acetic anhydride in glacial acetic acid converts the primary and secondary amines into essentially neutral acetylation products:

$$C_2H_5NH_2 + (CH_3CO)_2O \rightarrow C_2H_5NHCOCH_3 + CH_3COOH$$

$$(C_2H_5)_2NH + (CH_3CO)_2O \rightarrow (C_2H_5)_2NCOCH_3 + CH_3COOH$$

However, the tertiary amine is unaffected. After the acetylation reaction is complete, the solution is titrated with a standard solution of perchloric acid in glacial acetic acid to determine the quantity of triethylamine.

(b) If a sample of the three amines is dissolved in a one-to-one mixture of ethylene glycol and isopropanol which contains an excess of salicylaldehyde, the latter reagent reacts with the ethylamine to form a Schiff base:

$$C_2H_5NH_2 + \text{[salicylaldehyde: benzene ring with OH and CHO]} \longrightarrow \text{[benzene ring with OH and C=NC}_2H_5\text{]} + H_2O$$

Since the Schiff base is too weak to react with hydrochloric acid, titration of the mixture with a standard solution of hydrochloric acid in the glycol-alcohol solvent yields the sum of the quantities of diethylamine and triethylamine.

In the analysis of a pure mixture of ethylamine, diethylamine, and triethylamine, separate 1.154-gm samples were treated according to each of the two procedures described above. For procedure (a), the final titration required 13.28 ml of 0.4781 $F$ perchloric acid titrant. For procedure (b), the titration required 10.69 ml of 0.9536 $F$ hydrochloric acid titrant. Calculate the weight percentages of each of the amines in the mixture.

13. A mixture of naphthalene, 2-naphthoic acid, and 1-hydroxy-2-naphthoic acid was analyzed by means of a nonaqueous acid-base titration. A 0.1402-gm sample of the solid mixture was dissolved in approximately 50 ml of methyl isobutyl ketone. Titration with a 0.179 $F$ solution of tetrabutylammonium hydroxide in anhydrous isopropanol yielded a potentiometric titration curve with two sharply defined equivalence points, the first occurring after addition of 3.58 ml of titrant and the second after a total of 5.19 ml of titrant had been introduced. Using the titration data along with knowledge that can be gained by inspection of Figure 5–4, calculate the weight percentages of 2-naphthoic acid and 1-hydroxy-2-naphthoic acid in the solid material.

14. Sulfonamides, known more familiarly as sulfa drugs, contain the —SO$_2$NH— functional group. Sulfanilamide

$$H_2N-\text{⟨⟩}-SO_2NH_2$$

is the simplest member of this class of compounds and is a relatively weak acid. Replacement of one of the hydrogen atoms of the sulfonamide group with any of a variety of heterocyclic groups results in the formation of many derivatives of sulfanilamide. These more complex sulfa drugs are usually more effective than sulfanilamide for the treatment of cocci and bacterial infections. In addition, the acidity of the remaining hydrogen atom of the sulfonamide group is considerably enhanced by the presence of the heterocyclic group. Thus, sulfathiazole

is a much stronger acid than sulfanilamide. This difference in acidity can be exploited to analyze mixtures of sulfanilamide and sulfathiazole. First, a sample is treated with dimethylformamide, and the resulting solution is titrated with sodium methoxide in benzene-methanol, thymol blue being used as indicator; in dimethylformamide, only the more acidic proton of sulfathiazole can be titrated. Second, another sample is treated with butylamine, and the solution is titrated with the same titrant as before to an azo violet endpoint; in butylamine, sulfanilamide behaves as a sufficiently strong acid that both it and sulfathiazole can be titrated. In the analysis of a sulfa ointment containing sulfathiazole and sulfanilamide, a 1.500-gm sample was extracted with dimethylformamide; the extracts were combined and titrated with 0.1033 $F$ sodium methoxide in benzene-methanol, 2.893 ml of titrant being needed to reach a thymol blue endpoint. Another 1.500-gm sample was treated with butylamine; the butylamine extract was titrated to an azo violet endpoint with 7.250 ml of the sodium methoxide reagent. Calculate the weight percentages of sulfanilamide and sulfathiazole in the ointment.

15. (a) Assuming that ion-pair formation can be neglected, calculate the useful pH range encompassed by a $10^{-3} F$ strong acid solution and a $10^{-3} F$ strong base solution for each of the following solvents: methanol ($pK_s = 16.7$), liquid ammonia ($pK_s = 37.7$ at $-50°C$), and formic acid ($pK_s = 6.2$). What do your computations reveal about the differentiating ability of these solvents?

(b) Assuming that ion-pair formation can be neglected, calculate the useful pH range encompassed by a $10^{-3} F$ strong acid solution and a $10^{-3} F$ strong base solution for glacial acetic acid, which has an autoprotolysis constant ($pK_s = 14.45$) comparable to that of water. Actually, due to the small dielectric constant for glacial acetic acid ($D = 6.1$), ion-pair formation cannot be ignored. How would ion-pair formation affect the useful pH range for glacial acetic acid?

16. Most nonaqueous acid-base titrations are performed in solvents having relatively low dielectric constants. In such solvents, all ionic substances (including salts, acids, and bases which are strong electrolytes in water) exist largely as undissociated ion-paired species in solution. As an example, for a univalent-univalent salt such as sodium bromide, relatively small quantities of dissociated sodium and bromide ions are in equilibrium with $Na^+Br^-$ ion pairs according to the reaction

$$Na^+ + Br^- \rightleftharpoons Na^+Br^-$$

We can write an ion-pair formation constant expression in the form

$$K_{NaBr} = \frac{(Na^+Br^-)}{(Na^+)(Br^-)}$$

where $K_{NaBr}$ is the formation constant for the ion pair and the terms in parentheses denote the activities of the various species.

(a) By assuming that the activity coefficient of the uncharged ion pair is unity and that (for reasonably large analytical concentrations of dissolved sodium bromide) the equilibrium concentration of $Na^+Br^-$ ion pairs is approximately equal to the total original sodium bromide concentration ($C_{NaBr}$), show that

$$(Na^+) = (Br^-) = \left(\frac{C_{NaBr}}{K_{NaBr}}\right)^{1/2}$$

(b) Calculate the activity of sodium ion ($Na^+$) in a 0.05 $F$ solution of sodium bromide in ethylenediamine, in which $K_{NaBr}$ is 3950. How good is the assumption that the concentration of $Na^+Br^-$ ion pairs is simply equal to the total analytical concentration of sodium bromide? How large must $C_{NaBr}$ be in order that 99 per cent of the $Na^+Br^-$ ion pairs remain associated?

(c) Derive an exact expression for the activity of sodium ion ($Na^+$) for the situation in which dissociation of ion pairs is *not* neglected. Include activity coefficients by letting $f_\pm$ be the mean activity coefficient of the ions $Na^+$ and $Br^-$.

17. When an acid such as hydrogen bromide is dissolved in ethylenediamine (a strongly basic solvent with a low dielectric constant), the following reaction occurs:

$$HBr + H_2NCH_2CH_2NH_2 \rightleftharpoons H_2NCH_2CH_2NH_3^+Br^-$$

Note that the species $H_2NCH_2CH_2NH_3^+$ is the counterpart of $H_3O^+$ in water. However, whereas ion-pair formation is relatively unimportant in water, the ion pair $H_2NCH_2CH_2NH_3^+Br^-$ is overwhelmingly predominant in ethylenediamine. Abbreviating the latter ion pair as $H^+Br^-$, derive the expression

$$(H^+) = \left(\frac{C_{HBr}}{K_{HBr}}\right)^{1/2}$$

where $C_{HBr}$ is the total analytical concentration of hydrogen bromide dissolved in ethylenediamine and $K_{HBr}$ is the formation constant for $H_2NCH_2CH_2NH_3^+Br^-$ or $H^+Br^-$ ion pairs. For what circumstances is the above relation valid?

18. It is known that ion-pair formation must be taken into account in order to explain the shapes of acid-base titration curves in solvents of low dielectric constant. Consider the titration of a simple acid HX with a base MB, where $B^-$ is the conjugate base of the very slightly ionized solvent BH:

$$HX + MB \rightleftharpoons MX + BH$$

Show in a logical order all the steps in the derivation of the following correct expression for ($H^+$) prior to the equivalence point of such a titration. (Assume that you are the first person to derive this equation and are intending to publish the complete derivation and the reasons for each step in a scientific journal.)

$$(H^+) = (1 - F)\left(\frac{C_{HX}^\circ}{K_{HX}}\right)^{1/2}\sqrt{\frac{K_{MX}}{K_{MX} + F(K_{HX} - K_{MX})}}$$

In this expression, dissociation of ion pairs is ignored; that is to say, the slight changes in the analytical concentrations of ion-paired species caused by dissociation is ignored. In addition, $F$ is the fraction of acid titrated, $K_{HX}$ is the ion-pair formation constant for the acid, $K_{MX}$ is the ion-pair formation constant for the salt, and $C_{HX}^\circ$ is the original concentration of the acid. Dilution may be neglected. (Hint: In this problem, the only salt MX present is that formed by the titration reaction, so you must relate the salt concentration to the original concentration of acid.)

# SUGGESTIONS FOR ADDITIONAL READING

1. R. G. Bates: *Determination of pH*. John Wiley and Sons, New York, 1964, pp. 172–229.
2. J. F. Coetzee and C. D. Ritchie: *Solute-Solvent Interactions*. Dekker, New York, 1969.
3. M. M. Davis: *Acid-Base Behavior in Aprotic Organic Solvents*. Nat. Bur. Stds., Monograph 105, Washington, 1968.
4. J. S. Fritz: *Acid-Base Titrations in Nonaqueous Solvents*. G. F. Smith Chemical Company, Columbus, Ohio, 1952.
5. I. Gyenes: *Titrations in Non-Aqueous Media*, translated by D. Cohen and I. T. Millar. Van Nostrand, Princeton, New Jersey, 1967.
6. W. Huber: *Titrations in Nonaqueous Solvents*. Academic Press, New York, 1967.
7. I. M. Kolthoff and S. Bruckenstein: Acid-base equilibria in nonaqueous solutions. *In* I. M. Kolthoff and P. J. Elving, eds.: *Treatise on Analytical Chemistry*. Part I, Volume 1, Wiley-Interscience, New York, 1959, pp. 475–542.
8. J. Kucharsky and L. Safarik: *Titrations in Non-Aqueous Solvents*. Elsevier, Amsterdam, 1965.
9. H. A. Laitinen: *Chemical Analysis*. McGraw-Hill Book Company, New York, 1960, pp. 57–83, 101–105.
10. H. H. Sisler: *Chemistry in Non-Aqueous Solvents*. Reinhold Publishing Corporation, New York, 1961.
11. T. C. Waddington, ed.: *Non-Aqueous Solvent Systems*. Academic Press, New York, 1965.
12. R. A. Zingaro: *Nonaqueous Solvents*. Heath, Lexington, Massachusetts, 1968.

# COMPLEXOMETRIC TITRATIONS

A **complex ion** or **coordination compound** is a species in which a metal atom or cation is covalently bonded to one or more electron-donating groups. Usually, the metal atom or cation is called the **central atom**, whereas the term **ligand** is used to designate the electron-donating group. In order for a complex ion to be formed from one or more ligands and a central atom, each ligand must possess at least one unshared pair of electrons, and the central atom must be able to accept an electron pair from each ligand. Thus, a ligand shares a pair of electrons with the central atom in the formation of a covalent bond.

To serve as the basis for a volumetric method of analysis, a complex-formation reaction must be rapid, must proceed according to exact stoichiometry, and must have an equilibrium constant large enough that the titration curve has a sharply defined equivalence point. These are the same requirements that must be met for all volumetric analytical procedures.

Formation of a complex ion by titration of a solution of a metal cation with a standard solution of a complexing agent or ligand has gained importance as a method of volumetric analysis in the last twenty-five years. This is because of the advent of a unique class of ligands—the aminopolycarboxylic acids—that have several electron-donating groups on the same molecule and that form unusually stable one-to-one complexes with many metal ions. To begin this chapter, we shall examine some of the characteristics of complex-formation reactions. Then, the theory and practice of complexometric titrations with ethylenediaminetetraacetic acid (EDTA), the most well known member of the aminopolycarboxylic acid family, will be discussed.

## METAL ION COMPLEXES

An example of the formation of a metal complex is the reaction between the hexaaquochromium(III) ion, $Cr(H_2O)_6^{3+}$, and cyanide to form the hexacyanochromate(III) anion:

$$\left[ \begin{array}{c} OH_2 \\ H_2O \quad OH_2 \\ Cr \\ H_2O \quad OH_2 \\ OH_2 \end{array} \right]^{3+} + 6\ CN^- \rightleftharpoons \left[ \begin{array}{c} CN \\ NC \quad CN \\ Cr \\ NC \quad CN \\ CN \end{array} \right]^{3-} + 6\ H_2O$$

Although the product species, canary-yellow $Cr(CN)_6^{3-}$, is a very stable complex ion, the rate of reaction between violet $Cr(H_2O)_6^{3+}$ and $CN^-$ to form this complex is very slow under conditions suitable for a titration—that is, at room temperature and for stoichiometric quantities of the reactants. This reaction is slow because cyanide ions cannot readily displace the water molecules bound to chromium(III). Therefore, although the formation of the $Cr(CN)_6^{3-}$ ion might be expected to be suitable as a titrimetric method on the basis of its stability, the slow rate of reaction prohibits the use of this reaction for analytical purposes. A complex ion characterized by a slow rate of ligand exchange is called an inert or **nonlabile** species. Among the metal ions whose complexes frequently display relatively slight reactivity are chromium(III), cobalt(III), and platinum(IV). There is another group of metals that characteristically form reactive or **labile** complexes, including cobalt(II), copper, lead, bismuth, silver, cadmium, nickel, zinc, mercury, and aluminum. Many complexes of iron(II) and iron(III) are labile, but the cyanide complexes $Fe(CN)_6^{4-}$ and $Fe(CN)_6^{3-}$ are familiar examples of nonlabile species.

Each of the chromium(III) complexes in the preceding reaction has perfect octahedral geometry. In other words, the ligands are symmetrically arranged around the central atom and the six chromium(III)-ligand bond lengths are identical. A ligand may be a neutral molecule, such as water, or it may be an ion, such as cyanide ion. A complex ion itself may be positively charged, as is $Cr(H_2O)_6^{3+}$, or it may be negatively charged, as is $Cr(CN)_6^{3-}$. In other instances, the complex may be a neutral molecule. In the present example, the central atom, chromium, is in the tripositive oxidation state in both complexes. It is possible for the central atom in a complex to be in the zero oxidation state, such as nickel in $Ni(CN)_4^{4-}$ or as iron in $Fe(CO)_5$.

All chromium(III) complexes of the general structure shown in the preceding examples have six ligands, each of which is linked to the central atom by a covalent bond. Accordingly, we can state that the **coordination number** of chromium(III) is 6. On the other hand, the coordination number of silver ion in the silver-diammine complex, $Ag(NH_3)_2^+$, is 2. In general, the coordination number is dependent upon the electronic structure of the metal atom as well as upon the structure, charge, and electronegativity of the ligands.

Metal cations have several available orbitals for bond formation with complexing agents; for example, the zinc ion has four coordination sites. However, numerous common ligands, including chloride, bromide, iodide, cyanide, thiocyanate, hydroxide, and ammonia, can occupy only one coordination position of a metal ion. In other words, each of these ligands donates one unshared pair of electrons to the central atom. Such a species is called a **monodentate ligand**, from the Latin word *dentatus*, meaning "toothed." Therefore, a single zinc ion can react with a maximum of four chlorides, four cyanides, four hydroxides, or four ammonia molecules to form $ZnCl_4^{2-}$, $Zn(CN)_4^{2-}$, $Zn(OH)_4^{2-}$, or $Zn(NH_3)_4^{2+}$. Complexes containing more than one kind of ligand, such as $Zn(NH_3)_2(H_2O)_2^{2+}$ and $Zn(H_2O)_2(CN)_2$, can be formed.

There are other ligands, known as **multidentate ligands**, which can contribute two or more electron pairs to a central atom in forming a complex. To identify those multidentate species that donate, for example, two or three electron pairs to the central atom, we speak of *bidentate* or *terdentate* ligands, respectively. In some instances, a ligand may be *quadridentate*, *quinquidentate*, or *sexadentate*. Ethylenediamine, $NH_2$-$CH_2CH_2NH_2$, is a bidentate ligand because each nitrogen atom possesses one unshared pair of electrons. Copper(II) can be complexed by two ethylenediamine ligands in much the same way as it is complexed by four ammonia molecules:

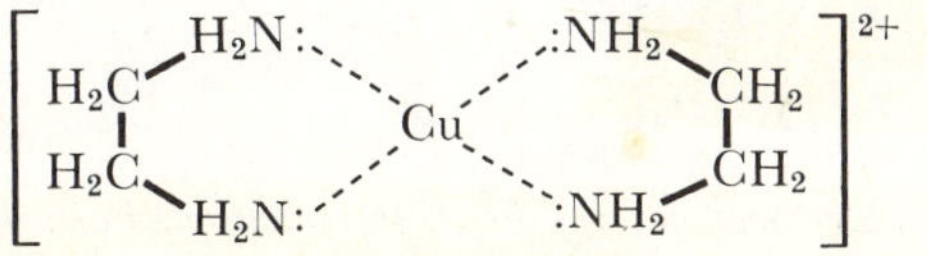

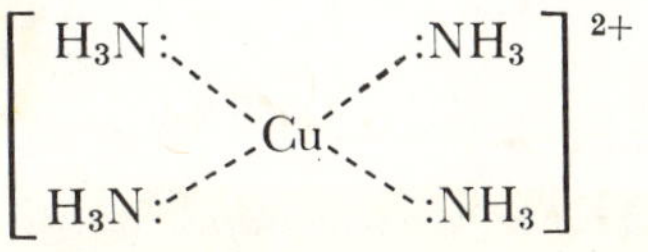

A complex composed of a central metal atom and one or more multidentate ligands is called a **chelate**, or chelate compound, after a Greek word meaning "claw." In a manner of speaking, the two or more electron-donating groups of the ligand act as a claw in bonding to the central atom.

A particularly interesting species is the complex of silver ion with ethylenediamine. Typically, silver ion has two coordination sites, as in $Ag(NH_3)_2^+$ and $Ag(CN)_2^-$; each of these ions has a linear structure. If both nitrogen atoms of a single ethylenediamine molecule were to be bonded to a silver ion, the resulting complex would be structurally strained and exceedingly unstable. However, *two* silver ions can be coordinated to *two* ethylenediamine molecules in such a geometrical arrangement that the complex is reasonably stable:

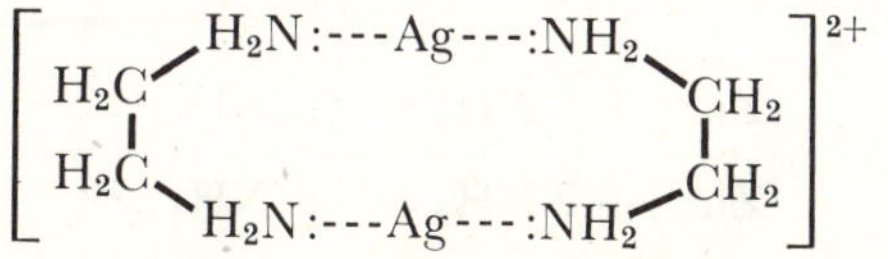

This is an example of a **binuclear** complex, one which contains two atoms or ions of the central element. All of the other complexes considered in this chapter are **mononuclear**.

## Stepwise and Overall Formation Constants

Let us consider the interaction between the hydrated zinc ion and ammonia in an aqueous medium. This reaction, as well as others involving monodentate ligands, proceeds in a stepwise fashion according to the following equilibria:

$$Zn(H_2O)_4{}^{2+} + NH_3 \rightleftharpoons Zn(NH_3)(H_2O)_3{}^{2+} + H_2O$$

$$Zn(NH_3)(H_2O)_3{}^{2+} + NH_3 \rightleftharpoons Zn(NH_3)_2(H_2O)_2{}^{2+} + H_2O$$

$$Zn(NH_3)_2(H_2O)_2{}^{2+} + NH_3 \rightleftharpoons Zn(NH_3)_3(H_2O)^{2+} + H_2O$$

$$Zn(NH_3)_3(H_2O)^{2+} + NH_3 \rightleftharpoons Zn(NH_3)_4{}^{2+} + H_2O$$

These reactions represent the stepwise formation of the monoammine, diammine, triammine, and tetraammine complexes of zinc(II), although as discussed later an aqueous solution of zinc(II) and ammonia will usually contain at least several of the zinc ammines in equilibrium with each other.

A **stepwise formation constant** is associated with each equilibrium and is designated by $K_n$, where, in the present example, the subscript $n$ takes an integral value to indicate the addition of the $n$th ammonia ligand to a complex containing $(n-1)$ ammonia molecules. Thus, the *first stepwise formation constant, $K_1$,* for the formation of the $Zn(NH_3)(H_2O)_3{}^{2+}$ complex pertains to the equilibrium expression

$$K_1 = \frac{[Zn(NH_3)^{2+}]}{[Zn^{2+}][NH_3]}$$

*Notice that the concentration (activity) of water does not appear in this relation. Furthermore, although water frequently functions as a ligand in aqueous solutions of metal ions, it is customary for simplicity in writing reactions and equilibrium expressions to omit water from the chemical formula of a hydrated cation or complex ion.* We will follow this practice throughout the remainder of this chapter. For the *fourth stepwise formation constant,* we can write

$$K_4 = \frac{[Zn(NH_3)_4{}^{2+}]}{[Zn(NH_3)_3{}^{2+}][NH_3]}$$

For the zinc(II)-ammonia system at 25°C, the numerical values for $K_1$, $K_2$, $K_3$, and $K_4$ are, respectively, 186, 219, 251, and 112.

Sometimes **overall formation constants** are used to characterize the equilibria in systems containing complex ions. Omitting the water molecules from the chemical formulas of all complexes, we can write the equilibria corresponding to the overall reactions for the zinc(II)-ammonia system as follows:

$$Zn^{2+} + \quad NH_3 \rightleftharpoons Zn(NH_3)^{2+}$$

$$Zn^{2+} + 2\,NH_3 \rightleftharpoons Zn(NH_3)_2{}^{2+}$$

$$Zn^{2+} + 3\,NH_3 \rightleftharpoons Zn(NH_3)_3{}^{2+}$$

$$Zn^{2+} + 4\,NH_3 \rightleftharpoons Zn(NH_3)_4{}^{2+}$$

Overall formation constants are designated by $\beta_n$, where the subscript $n$ gives the total number of ligands added to the original aquated cation. For example, the *third overall formation constant, $\beta_3$,* refers to the addition of three ammonia ligands to $Zn(H_2O)_4{}^{2+}$ to form $Zn(NH_3)_3(H_2O)^{2+}$, and the pertinent equilibrium expression is

$$\beta_3 = \frac{[Zn(NH_3)_3{}^{2+}]}{[Zn^{2+}][NH_3]^3}$$

There is a simple relationship between the stepwise formation constants and the overall formation constants for any particular system. This may be illustrated if we multiply the numerator and denominator of the equation defining $\beta_3$ by the quantity $[Zn(NH_3)^{2+}][Zn(NH_3)_2^{2+}]$

$$\beta_3 = \frac{[Zn(NH_3)_3^{2+}]}{[Zn^{2+}][NH_3]^3} \cdot \frac{[Zn(NH_3)^{2+}][Zn(NH_3)_2^{2+}]}{[Zn(NH_3)^{2+}][Zn(NH_3)_2^{2+}]}$$

and rearrange the expression as follows:

$$\beta_3 = \frac{[Zn(NH_3)^{2+}]}{[Zn^{2+}][NH_3]} \cdot \frac{[Zn(NH_3)_2^{2+}]}{[Zn(NH_3)^{2+}][NH_3]} \cdot \frac{[Zn(NH_3)_3^{2+}]}{[Zn(NH_3)_2^{2+}][NH_3]}$$

Note that the three terms on the right side of this equation correspond, in order, to $K_1$, $K_2$, and $K_3$. Consequently,

$$\beta_3 = K_1 K_2 K_3$$

In similar fashion it is possible to derive the relationship between each of the overall formation constants and the stepwise formation constants; the results are

$$K_1 = \beta_1$$
$$K_1 K_2 = \beta_2$$
$$K_1 K_2 K_3 = \beta_3$$
$$K_1 K_2 K_3 K_4 = \beta_4$$
$$K_1 K_2 K_3 K_4 \cdots K_n = \beta_n$$

For the zinc(II)-ammonia system, the overall formation constants have the following values: $\beta_1 = 186$, $\beta_2 = 4.08 \times 10^4$, $\beta_3 = 1.02 \times 10^7$, and $\beta_4 = 1.15 \times 10^9$.

## Distribution of Metal Among Several Complexes Involving Monodentate Ligands

For most metal ions, and the majority of monodentate ligands, the relative stabilities of the various complexes within a family are such that several species may coexist in the same solution for certain concentrations of the ligand. Let us confirm this fact graphically for the zinc(II)-ammonia system by showing how the fraction of each zinc-ammine complex varies with the concentration of free (uncomplexed) ammonia. We can employ the same approach used in Chapter 4 (pages 109 to 111) for polyprotic acid systems.

In an ammoniacal solution, the sum of the concentrations of all zinc-bearing species will have a fixed value:

$$C_{Zn} = [Zn^{2+}] + [Zn(NH_3)^{2+}] + [Zn(NH_3)_2^{2+}] + [Zn(NH_3)_3^{2+}] + [Zn(NH_3)_4^{2+}]$$

This equation can be rewritten if we utilize the equilibrium expressions involving overall formation constants:

$$C_{Zn} = [Zn^{2+}] + \beta_1[Zn^{2+}][NH_3] + \beta_2[Zn^{2+}][NH_3]^2$$
$$+ \beta_3[Zn^{2+}][NH_3]^3 + \beta_4[Zn^{2+}][NH_3]^4$$

Now, the concentration of each complex will be some fraction ($\alpha$) of the total concentration, $C_{Zn}$:

$$\alpha_{Zn^{2+}} = \frac{[Zn^{2+}]}{C_{Zn}}$$

$$\alpha_{Zn(NH_3)^{2+}} = \frac{[Zn(NH_3)^{2+}]}{C_{Zn}} = \frac{\beta_1[Zn^{2+}][NH_3]}{C_{Zn}}$$

$$\alpha_{Zn(NH_3)_2^{2+}} = \frac{[Zn(NH_3)_2^{2+}]}{C_{Zn}} = \frac{\beta_2[Zn^{2+}][NH_3]^2}{C_{Zn}}$$

$$\alpha_{Zn(NH_3)_3^{2+}} = \frac{[Zn(NH_3)_3^{2+}]}{C_{Zn}} = \frac{\beta_3[Zn^{2+}][NH_3]^3}{C_{Zn}}$$

$$\alpha_{Zn(NH_3)_4^{2+}} = \frac{[Zn(NH_3)_4^{2+}]}{C_{Zn}} = \frac{\beta_4[Zn^{2+}][NH_3]^4}{C_{Zn}}$$

Note that the sum of the five $\alpha$ values must be unity. Substitution of the relation for $C_{Zn}$ into each of the five preceding equations, followed by cancellation of $[Zn^{2+}]$ terms, yields

$$\alpha_{Zn^{2+}} = \frac{1}{1 + \beta_1[NH_3] + \beta_2[NH_3]^2 + \beta_3[NH_3]^3 + \beta_4[NH_3]^4}$$

$$\alpha_{Zn(NH_3)^{2+}} = \frac{\beta_1[NH_3]}{1 + \beta_1[NH_3] + \beta_2[NH_3]^2 + \beta_3[NH_3]^3 + \beta_4[NH_3]^4}$$

$$\alpha_{Zn(NH_3)_2^{2+}} = \frac{\beta_2[NH_3]^2}{1 + \beta_1[NH_3] + \beta_2[NH_3]^2 + \beta_3[NH_3]^3 + \beta_4[NH_3]^4}$$

$$\alpha_{Zn(NH_3)_3^{2+}} = \frac{\beta_3[NH_3]^3}{1 + \beta_1[NH_3] + \beta_2[NH_3]^2 + \beta_3[NH_3]^3 + \beta_4[NH_3]^4}$$

$$\alpha_{Zn(NH_3)_4^{2+}} = \frac{\beta_4[NH_3]^4}{1 + \beta_1[NH_3] + \beta_2\,NH_3]^2 + \beta_3[NH_3]^3 + \beta_4[NH_3]^4}$$

It is evident from these expressions that the fraction of each species is dependent upon the ammonia concentration but not the total concentration ($C_{Zn}$) of zinc-containing complexes. Using the five final equations for the $\alpha$ values, we computed the fraction of each zinc(II) complex as a function of the logarithm of the concentration of ammonia. Results of these calculations are plotted in Figure 6–1. For ammonia concentrations below $10^{-3}\,F$, the predominant species is the hydrated cation, $Zn(H_2O)_4^{2+}$, whereas the zinc-tetraammine complex, $Zn(NH_3)_4^{2+}$, is the major component of the system when the concentration of ammonia exceeds $10^{-2}\,F$. However, for ammonia concentrations between $10^{-3}$ and $10^{-2}\,F$, significant concentrations of all five zinc(II) species are present. This latter situation differs from that encountered in Chapter 4 for polyprotic acids. There we found, for example, that of the four phosphate-bearing substances, $H_3PO_4$, $H_2PO_4^-$, $HPO_4^{2-}$, and $PO_4^{3-}$, only one or two species are major contributors to the total phosphate concentration at any given pH (see Figure 4–1, page 111).

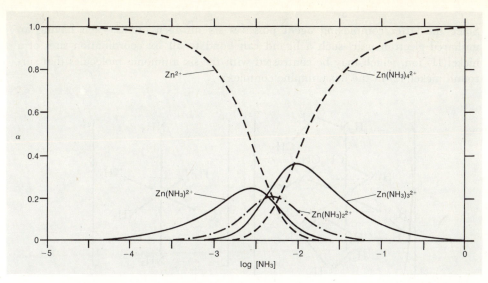

*Figure 6–1.* Distribution of $Zn(H_2O)_4^{2+}$ and the various zinc-ammine complexes as a function of the logarithm of the free (uncomplexed) ammonia concentration.

### Advantages of Multidentate Ligands for Complexometric Titrations

There are two interrelated reasons why complexes formed from monodentate ligands are generally unsuitable for complexometric titrations. To understand the difficulties, let us refer once again to the zinc(II)-ammonia system. What problems are encountered if we attempt to titrate a solution of $Zn(H_2O)_4^{2+}$ with a standard solution of ammonia in water? First, because no single zinc-ammine species is more stable than the others—*the stepwise formation constants are similar in magnitude*—the addition of ammonia produces a changing mixture of complexes, and no simple stoichiometric ratio between ammonia and zinc(II) ever exists. Second, because of the modest stabilities of the zinc ammines—*the stepwise formation constants are relatively small*—the quantitative formation of even the most extensively complexed $Zn(NH_3)_4^{2+}$ ion requires a large excess of free ammonia, which can only be achieved well beyond the theoretical equivalence point. Thus, during the addition of titrant, the concentrations of free ammonia and the various zinc-ammine complexes change so gradually that the resulting titration curve never exhibits an equivalence point.

When a complexometric titration is performed with a standard solution of a multidentate ligand, the difficulties associated with the use of monodentate ligands are avoided—other than the possibility of a slow rate of reaction. Most multidentate ligands employed in titrimetry coordinate at several or all positions around a central atom; therefore, only one-to-one metal-ligand complexes are formed and the stoichiometry of the titration reaction is simple. In addition, multidentate ligands usually form much more stable complexes than do chemically similar monodentate ligands, so that well defined titration curves are obtainable.

To illustrate the last point, let us compare the stabilities of the nickel(II)-hexaammine complex, $Ni(NH_3)_6^{2+}$, and the one-to-one complex formed between nickel(II) and the ligand N,N,N′,N′-tetra(2-aminoethyl)ethylenediamine (often called penten) whose structural formula is

$$
\begin{array}{ccc}
H_2\ddot{N}\text{—}CH_2CH_2 & \qquad & CH_2CH_2\text{—}\ddot{N}H_2 \\
& \ddot{N}\text{—}CH_2CH_2\text{—}\ddot{N} & \\
H_2\ddot{N}\text{—}CH_2CH_2 & \qquad & CH_2CH_2\text{—}\ddot{N}H_2
\end{array}
$$

Since the latter complexing agent possesses six nitrogen atoms, each having an unshared electron pair, such a ligand can bond to all six coordination sites of a nickel(II) ion, which may be contrasted with the six ammonia molecules that surround nickel(II) in the hexaammine complex:

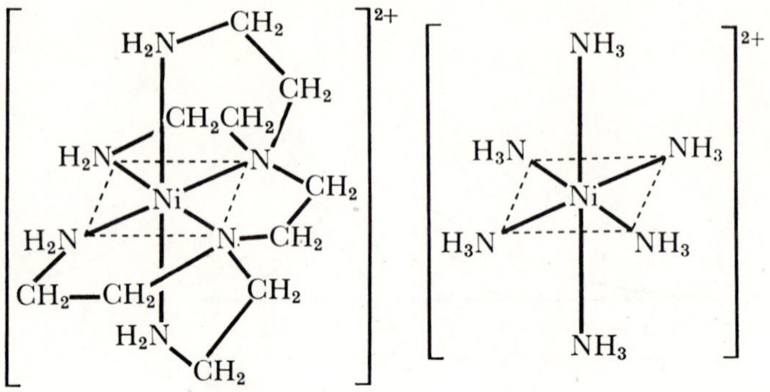

For the one-to-one nickel(II)-penten complex the formation constant is $2.0 \times 10^{19}$, whereas the overall formation constant ($\beta_6$) for the nickel(II)-hexaammine species is only $3.2 \times 10^8$. Since the strengths of the nickel(II)-nitrogen bonds in the two complexes are similar, this factor cannot account for the difference between the formation constants. However, the big difference in stabilities can be understood in terms of the so-called **chelate effect**. Replacement of the first water molecule of Ni-$(H_2O)_6^{2+}$ by one ammonia molecule or by one of the nitrogen atoms of penten is equally probable and energetically almost identical. On the other hand, replacement of the second, third, fourth, fifth, and sixth water molecules of the nickel complex by nitrogen atoms of the same penten ligand is *much more probable*—that is, the process has a more positive entropy change—than replacement of each water molecule by an ammonia ligand from the solution. This chelate effect is due to the fact that the penten is already attached to nickel(II) and the uncoordinated nitrogens of the penten molecule are in close proximity to the water molecules to be replaced. It is this enhanced probability of formation that manifests itself in the higher stability of Ni(penten)$^{2+}$ relative to Ni(NH$_3$)$_6^{2+}$.

## TITRATIONS WITH ETHYLENEDIAMINETETRAACETIC ACID (EDTA)

Today complexometric titrations are performed almost exclusively with a standard solution of one of the family of aminopolycarboxylic acids, ethylenediaminetetraacetic acid being by far the most popular choice. In their anionic forms, these reagents are usually either quadridentate or sexadentate, although some of the aminopolycarboxylate species are octadentate. These ligands combine in a one-to-one ratio with practically every metal cation in the periodic table to yield quite stable complexes possessing several five-membered chelate rings.

### Aminopolycarboxylic Acids

In Figure 6–2 are shown the structural formulas for five aminopolycarboxylic acids which have been used in complexometric titrimetry. These species are all

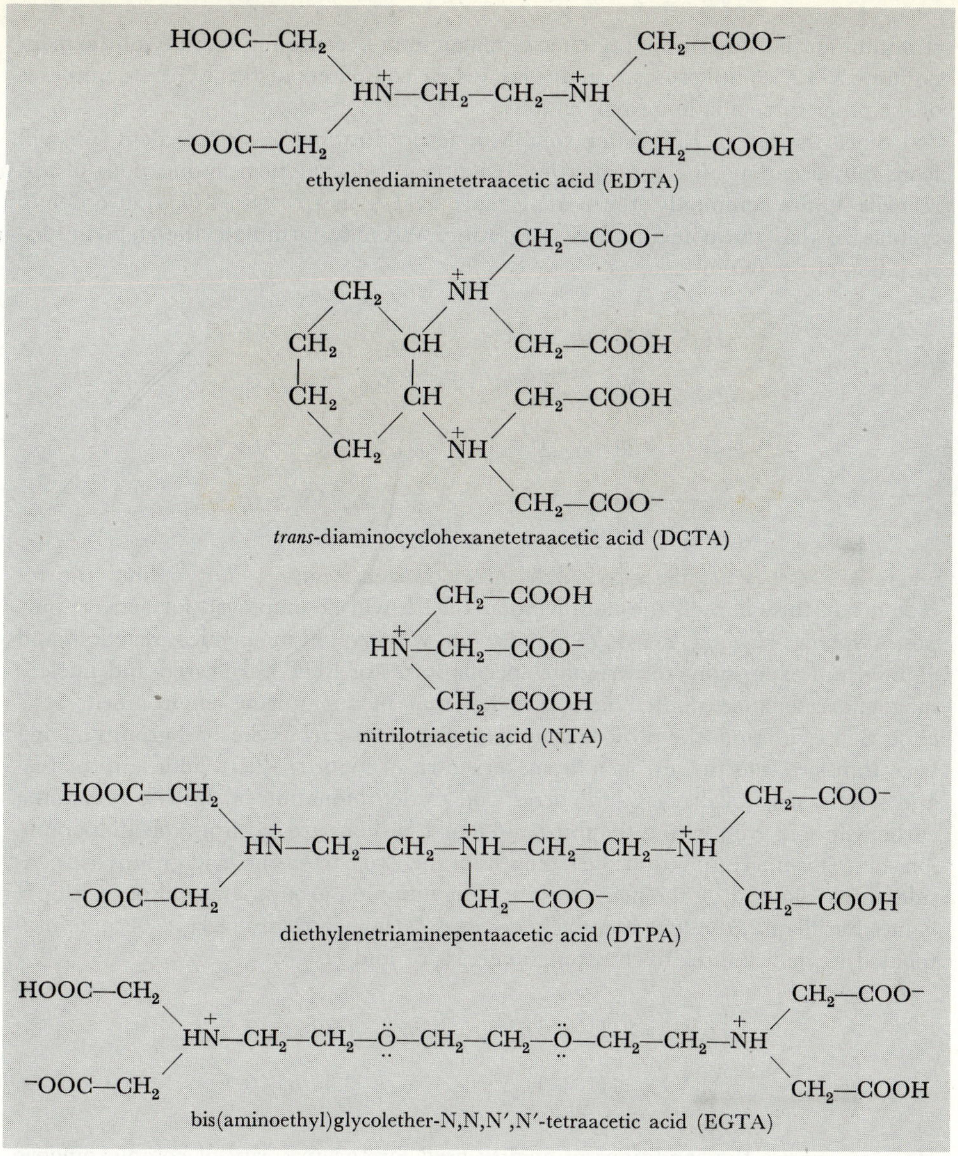


*Figure 6–2.* Structural formulas for five aminopolycarboxylic acids used in complexometric titrimetry.

zwitterions, a characteristic discussed below in greater detail. Because ethylenediaminetetraacetic acid (EDTA) is a suitable titrant for more than 95 per cent of all analytical applications, the other aminopolycarboxylic acids have not been employed to any significant extent. Although *trans*-diaminocyclohexanetetraacetic acid (DCTA) forms more stable metal ion complexes than does EDTA, the formation and dissociation of metal-DCTA complexes tend to be sluggish. On the other hand, nitrilotriacetic acid (NTA) forms much weaker complexes with metal ions than does EDTA and is seldom used for complexometric titrations. One of the virtues of diethylenetriaminepentaacetic acid (DTPA) is the high stability of its octadentate complexes with cations of the lanthanide and actinide elements. Finally, bis(aminoethyl)-glycolether-N,N,N′,N′-tetraacetic acid (EGTA) is useful for the titration of calcium, strontium, or barium in the presence of magnesium because the stability of the magnesium-EGTA complex is unusually low when compared to the EGTA complexes of the other three alkaline earth cations.

Since the use of EDTA for complexometric titrations is so prevalent, we will focus our attention on the properties, reactions, and analytical applications of this reagent. Quite commonly, the parent acid (EDTA) is written as $H_4Y$ in order to emphasize that the compound is tetraprotic. We may formulate the stepwise dissociation of the parent acid as

$$H_4Y \rightleftharpoons H^+ + H_3Y^-; \qquad K_{a1} = 1.00 \times 10^{-2}$$

$$H_3Y^- \rightleftharpoons H^+ + H_2Y^{2-}; \qquad K_{a2} = 2.16 \times 10^{-3}$$

$$H_2Y^{2-} \rightleftharpoons H^+ + HY^{3-}; \qquad K_{a3} = 6.92 \times 10^{-7}$$

$$HY^{3-} \rightleftharpoons H^+ + Y^{4-}; \qquad K_{a4} = 5.50 \times 10^{-11}$$

where $Y^{4-}$ represents the ethylenediaminetetraacetate anion. Throughout the remainder of this chapter, the abbreviation EDTA will be employed for general purposes, whereas $H_4Y$, $H_3Y^-$, $H_2Y^{2-}$, and so on, will be used in chemical reactions and equilibrium expressions to designate specific forms of EDTA. Infrared and nuclear magnetic resonance studies have revealed that in an aqueous environment $H_4Y$ exists as a zwitterion, the protons from two of the four carboxylic acid groups having been transferred to the nitrogen atoms as shown in Figure 6–2. In addition, the first and second steps of dissociation correspond to deprotonation of the two remaining carboxylic acid groups, and the third and fourth protons are lost from the ammonium ion sites. It is interesting that the strengths of the two carboxylic acid groups are considerably enhanced by the neighboring ammonium ion groups. In solutions with pH values less than 2, the two negatively charged carboxylate sites of $H_4Y$ become protonated to yield the relatively strong acids $H_5Y^+$ and $H_6Y^{2+}$:

$$H_6Y^{2+} \rightleftharpoons H^+ + H_5Y^+; \qquad K = 1.26 \times 10^{-1}$$

$$H_5Y^+ \rightleftharpoons H^+ + H_4Y; \qquad K' = 2.51 \times 10^{-2}$$

However, under the conditions of acidity used for the majority of complexometric titrations with EDTA, the $H_6Y^{2+}$ and $H_5Y^+$ species are not present in significant amounts.

Inasmuch as the parent acid ($H_4Y$) is only sparingly soluble in water, it is not ordinarily used for complexometric titrations. On the other hand, the disodium salt is relatively soluble and, being commercially available in the form $Na_2H_2Y \cdot 2 H_2O$, serves as the starting material for the preparation of the standard EDTA solutions employed in titrimetry. Since the predominant species in solution is the $H_2Y^{2-}$ ion, the pH of the titrant is very close to $\frac{1}{2}(pK_{a2} + pK_{a3})$ or 4.42.

## Composition of EDTA Solutions as a Function of pH

Changes in the hydrogen ion concentration cause enormous variations in the composition of EDTA solutions. Figure 6–3 shows how the fraction ($\alpha$) of EDTA present as each of the seven possible species, $H_6Y^{2+}$ through $Y^{4-}$, varies as a function

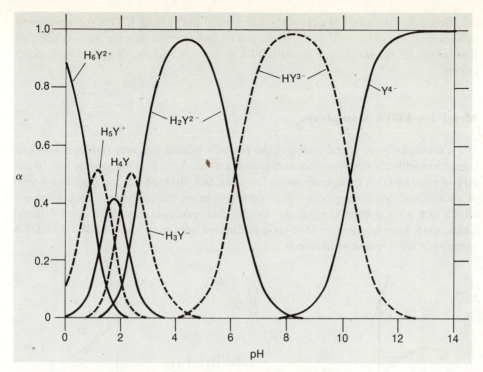

*Figure 6–3.* Distribution of EDTA species as a function of pH.

of pH. In constructing this diagram, we followed exactly the same procedure*
employed to obtain Figure 4–1 (page 111) for the phosphoric acid system. From an
examination of Figure 6–3, it is apparent that $H_6Y^{2+}$ is the predominant form of
EDTA below pH 1 and that $H_5Y^+$ is the major species between pH 1 and 1.6. Only
between pH 1.6 and 2 is $H_4Y$ the most important component of the solution, whereas
$H_3Y^-$ is the major constituent from pH 2 to 2.7. At pH values greater than 2.7, the
behavior of the EDTA system resembles that of phosphoric acid insofar as there are
wide pH ranges within which one species predominates—$H_2Y^{2-}$ between pH 2.7
and 6.2, $HY^{3-}$ between pH 6.2 and 10.2, and $Y^{4-}$ above pH 10.2.

---

* To obtain the data needed to plot Figure 6–3, one first defines the total concentration of EDTA
species:

$$C_{\text{EDTA}} = [H_6Y^{2+}] + [H_5Y^+] + [H_4Y] + [H_3Y^-] + [H_2Y^{2-}] + [HY^{3-}] + [Y^{4-}]$$

Next, the fraction of EDTA present as each species is written, *e.g.*,

$$\alpha_{Y^{4-}} = \frac{[Y^{4-}]}{C_{\text{EDTA}}}$$

Then, the concentration of each species other than $H_6Y^{2+}$ is expressed in terms of $[H^+]$, $[H_6Y^{2+}]$,
and the six equilibrium constants identified in the preceding section ($K$, $K'$, $K_{a1}$, $K_{a2}$, $K_{a3}$, and $K_{a4}$).
Proceeding as outlined on page 110, one obtains the $\alpha$ value for each species as a function only of $[H^+]$
and the equilibrium constants, *e.g.*,

$$\alpha_{Y^{4-}} = \frac{KK'K_{a1}K_{a2}K_{a3}K_{a4}}{[H^+]^6 + K[H^+]^5 + KK'[H^+]^4 + KK'K_{a1}[H^+]^3 + KK'K_{a1}K_{a2}[H^+]^2} \\ + KK'K_{a1}K_{a2}K_{a3}[H^+] + KK'K_{a1}K_{a2}K_{a3}K_{a4}$$

Finally, each $\alpha$ is computed for many different pH values and the results are plotted.

Because of the effect of pH on the fractions of the various EDTA species present in a solution, we shall see that the hydrogen ion concentration is an important factor that influences the stability of metal-EDTA complexes and the sharpness of titration curves.

## Metal Ion-EDTA Complexes

Practically every metal cation in the periodic table forms very stable, one-to-one complexes with the ethylenediaminetetraacetate ion, $Y^{4-}$. This one-to-one stoichiometry of metal-EDTA complexes arises from the fact that the $Y^{4-}$ ion possesses a total of six electron-donating groups—four carboxylate groups and two amine groups— which can occupy four, five, or six coordination positions around a central metal atom. One known example of a six-coordinated species is the cobalt(III)-EDTA complex, $CoY^-$, whose structure is

Metal-EDTA complexes gain particular stability from the five-membered chelate rings which are formed. In the cobalt(III)-EDTA complex, there are five such five-membered rings. A five-membered ring is an especially stable configuration because the bond angles allow all five atoms in the ring to lie in a plane. In some instances, only four or five of the six electron-donating groups of the $Y^{4-}$ anion are bound to a metal ion, and the remaining positions around the metal are occupied by monodentate ligands such as water, hydroxide, or ammonia.

We can write the general reaction for the formation of metal-EDTA complexes as

$$M^{n+} + Y^{4-} \rightleftharpoons MY^{(n-4)+}$$

where $M^{n+}$ denotes a hydrated metal cation. For the preceding process the equilibrium expression is

$$K_{MY} = \frac{[MY^{(n-4)+}]}{[M^{n+}][Y^{4-}]}$$

in which $K_{MY}$ is the formation constant for the $MY^{(n-4)+}$ complex. In Table 6–1 are listed formation constants for a number of metal-EDTA complexes.

## Table 6-1. Formation Constants of Metal-EDTA Complexes*

| Element | Cation | log $K_{MY}$ | Element | Cation | log $K_{MY}$ |
|---------|--------|--------------|---------|--------|--------------|
| Aluminum | $Al^{3+}$ | 16.1 | Nickel | $Ni^{2+}$ | 18.6 |
| Barium | $Ba^{2+}$ | 7.8 | Scandium | $Sc^{3+}$ | 23.1 |
| Bismuth | $Bi^{3+}$ | 27.9 | Silver | $Ag^+$ | 7.3 |
| Cadmium | $Cd^{2+}$ | 16.5 | Strontium | $Sr^{2+}$ | 8.6 |
| Calcium | $Ca^{2+}$ | 10.7 | Thallium | $Tl^{3+}$ | 21.5 |
| Cobalt | $Co^{2+}$ | 16.3 | Thorium | $Th^{4+}$ | 23.2 |
| Copper | $Cu^{2+}$ | 18.8 | Titanium | $Ti^{3+}$ | 21.3 |
| Gallium | $Ga^{3+}$ | 20.3 | | $TiO^{2+}$ | 17.3 |
| Indium | $In^{3+}$ | 24.9 | Uranium | $U^{4+}$ | 25.5 |
| Iron | $Fe^{2+}$ | 14.3 | Vanadium | $V^{2+}$ | 12.7 |
| | $Fe^{3+}$ | 25.1 | | $V^{3+}$ | 25.9 |
| Lead | $Pb^{2+}$ | 18.0 | | $VO^{2+}$ | 18.8 |
| Magnesium | $Mg^{2+}$ | 8.7 | | $VO_2^+$ | 15.6 |
| Manganese | $Mn^{2+}$ | 13.8 | Yttrium | $Y^{3+}$ | 18.1 |
| Mercury | $Hg^{2+}$ | 21.8 | Zinc | $Zn^{2+}$ | 16.5 |

* Constants are valid for a temperature of 20°C and an ionic strength of 0.1. Data are from G. Schwarzenbach and H. Flaschka: *Complexometric Titrations*. Translated by H. M. N. H. Irving, second English edition, Methuen, London, 1969, p. 10.

As the data presented in Table 6-1 show, metal-EDTA complexes exhibit a great range of stability. An obvious trend is that the formation constants become larger as the charge of the cation increases. However, the stability of a metal-EDTA species can be significantly *decreased* by changes in the composition of a solution. For example, if an acid is added to a solution containing the nickel-EDTA complex, the dissociation of the latter is enhanced because protons combine with the $Y^{4-}$ anion to yield $HY^{3-}$ or $H_2Y^{2-}$ according to the equilibria

$$NiY^{2-} + H^+ \rightleftharpoons Ni^{2+} + HY^{3-}$$

and

$$NiY^{2-} + 2\,H^+ \rightleftharpoons Ni^{2+} + H_2Y^{2-}$$

Alternatively, to the solution of the nickel-EDTA species, one can add a ligand that forms a different complex with nickel(II). Thus, in the presence of ammonia, the reaction

$$NiY^{2-} + 6\,NH_3 \rightleftharpoons Ni(NH_3)_6^{2+} + Y^{4-}$$

causes more dissociation of the nickel-EDTA complex than occurs in the absence of ammonia—although the formation constant for the nickel-EDTA species is much larger than that for the nickel-hexaammine complex. In each case the effective stability of the nickel-EDTA complex is diminished by a reagent that reacts with either nickel(II) or the $Y^{4-}$ anion.

## Conditional Formation Constants

Let us now extend the preceding considerations to the subject of complexometric titrations with EDTA. Most of these titrations are performed in neutral or in mildly acidic or alkaline media that are well buffered. Typically, the solution to be titrated is buffered with acetic acid and sodium acetate (pH 4 to 5) or with ammonium chloride and ammonia (pH 9 to 10); citrate and tartrate buffers are occasionally employed. Since the predominant form of EDTA under these conditions is $H_2Y^{2-}$ or $HY^{3-}$ (Figure 6–3), the titration process leading to the formation of a metal-EDTA complex might be formulated as

$$M^{n+} + H_2Y^{2-} \rightleftharpoons MY^{(n-4)+} + 2\,H^+$$

or

$$M^{n+} + HY^{3-} \rightleftharpoons MY^{(n-4)+} + H^+$$

However, we have seen that the production of free hydrogen ion can be deleterious; if the pH becomes too low, the formation of the metal-EDTA complex will be incomplete. One of the purposes of the buffer is to consume the hydrogen ion liberated during the titration. In a number of instances, this buffer plays a second role. For example, cadmium, copper, nickel, and zinc ions may be precipitated as insoluble hydroxides near pH 7 or 8. Fortunately, these metal cations form stable ammine complexes, so that the use of an ammonium chloride-ammonia buffer prevents the undesired precipitation of these hydroxides. A buffer that performs these special functions for metal-EDTA titrations is called an **auxiliary complexing agent**.

Although an auxiliary complexing agent helps to ensure the success of a complexometric titration, such a buffer system decreases the stability of the metal-EDTA complex. We can understand this effect qualitatively by examining two hypothetical experiments. In the first experiment, let us prepare a solution containing known initial concentrations of $M^{n+}$ and $Y^{4-}$. Using the formation constant, $K_{MY}$, we can calculate what concentration of the metal-EDTA complex, $MY^{(n-4)+}$, would exist at equilibrium. In the second experiment, let us prepare a solution with the same initial concentrations of $M^{n+}$ and $Y^{4-}$ as before but containing an auxiliary complexing agent. Now, if the auxiliary complexing agent interacts with the metal cation, the effective concentration of $M^{n+}$ will be lowered. Similarly, at pH values provided by most auxiliary complexing agents, the effective concentration of $Y^{4-}$ will be reduced because of the formation of $HY^{3-}$ and $H_2Y^{2-}$. Since less $M^{n+}$ and $Y^{4-}$ are available to form the metal-EDTA complex, the equilibrium concentration of the latter must be smaller than in the first experiment. Thus, in the presence of the auxiliary complexing agent, the *effective* formation constant for the metal-EDTA complex is smaller, even though the same total concentrations of metal ion and EDTA are used in both experiments. Our next task is to develop the quantitative aspects of this effect.

Consider the formation of the zinc-EDTA complex in a buffer consisting of ammonium chloride and ammonia. For such a system, the total concentration of zinc(II) species *not complexed by EDTA* is given by the expression presented on page 165; that is,

$$C_{Zn} = [Zn^{2+}] + [Zn(NH_3)^{2+}] + [Zn(NH_3)_2^{2+}] + [Zn(NH_3)_3^{2+}] + [Zn(NH_3)_4^{2+}]$$

In addition, the concentration of hydrated zinc(II) can be written as

$$[Zn^{2+}] = \alpha_{Zn^{2+}} C_{Zn}$$

where $\alpha_{Zn^{2+}}$ can be computed as described earlier in this chapter from the concentration of free ammonia and the formation constants for the zinc-ammine species. This latter relation stipulates that, whatever the total concentration of zinc(II) *not complexed by EDTA*, the concentration of $Zn^{2+}$ is *a fixed fraction of the total*. Next, we can formulate an equation* for the total concentration of EDTA *not complexed by zinc(II)*

$$C_{EDTA} = [H_6Y^{2+}] + [H_5Y^+] + [H_4Y] + [H_3Y^-] + [H_2Y^{2-}] + [HY^{3-}] + [Y^{4-}]$$

and we can state that, *in an ammonium chloride-ammonia buffer of known pH, a constant fraction of the uncomplexed EDTA exists as* $Y^{4-}$:

$$[Y^{4-}] = \alpha_{Y^{4-}}C_{EDTA}$$

Let us substitute the above relations for $[Zn^{2+}]$ and $[Y^{4-}]$ into the expression for the formation constant for the zinc-EDTA complex:

$$K_{ZnY^{2-}} = \frac{[ZnY^{2-}]}{[Zn^{2+}][Y^{4-}]} = \frac{[ZnY^{2-}]}{(\alpha_{Zn^{2+}}C_{Zn})(\alpha_{Y^{4-}}C_{EDTA})}$$

Inasmuch as $\alpha_{Zn^{2+}}$ and $\alpha_{Y^{4-}}$ are constants, we can include them on the left side of the equation:

$$K_{ZnY^{2-}}\alpha_{Zn^{2+}}\alpha_{Y^{4-}} = \frac{[ZnY^{2-}]}{C_{Zn}C_{EDTA}}$$

It is customary to call the product of the three terms on the left side of the last equation a **conditional formation constant**, $K'_{ZnY^{2-}}$:

$$K'_{ZnY^{2-}} = K_{ZnY^{2-}}\alpha_{Zn^{2+}}\alpha_{Y^{4-}}$$

This conditional formation constant is a measure of the effective stability of the zinc-EDTA complex for a particular set of conditions—that is, $K'_{ZnY^{2-}}$ has a specific value for every pH and free ammonia concentration. Obviously, $K'_{ZnY^{2-}}$ will change if a buffer having a different pH or a different ammonia concentration is employed or if a new auxiliary complexing agent is used. Since $\alpha_{Zn^{2+}}$ and $\alpha_{Y^{4-}}$ are both less than unity under most conditions, we see that $K'_{ZnY^{2-}}$ is almost always smaller than $K_{ZnY^{2-}}$, which confirms our previous conclusion that a metal-EDTA complex has a lower stability in the presence than in the absence of an auxiliary complexing agent.

Because the stabilities of metal-EDTA complexes are all affected by the pH of the solution, the conditional formation constant for every such complex will contain the $\alpha_{Y^{4-}}$ term. However, some cations do not interact significantly with the components of the auxiliary complexing agent. For example, magnesium and calcium ions form no ammine complexes in an ammoniacal buffer. Consequently, the conditional formation constants for the EDTA complexes of these two cations in an ammonium chloride-ammonia buffer have simpler forms, $K'_{MgY^{2-}} = K_{MgY^{2-}}\alpha_{Y^{4-}}$ and $K'_{CaY^{2-}} = K_{CaY^{2-}}\alpha_{Y^{4-}}$, respectively. In other words, $\alpha_{Mg^{2+}}$ and $\alpha_{Ca^{2+}}$ are both unity.

In the next section, we shall evaluate the conditional formation constant for the zinc-EDTA complex in an ammoniacal buffer and we will use the conditional formation constant to construct the titration curve for this system.

## Titration Curves for the Zinc(II)-EDTA Reaction in an Ammoniacal Solution

Let us consider the titration of a 50.00-ml sample of 0.001000 $F$ zinc nitrate with a 0.1000 $F$ $Na_2H_2Y$ solution. We will assume that the zinc(II) solution contains

---

* See footnote on page 171.

0.10 $F$ ammonia and 0.10 $F$ ammonium chloride. Note that the EDTA titrant is 100 times more concentrated than the zinc(II) solution. Therefore, dilution of the sample medium during the titration can be ignored. Because only 0.5000 ml of titrant is required for the titration, it would be performed with the aid of a micro-buret.

*Evaluation of the conditional formation constant.* To calculate the conditional formation constant, defined by the relation

$$K'_{ZnY^{2-}} = K_{ZnY^{2-}}\alpha_{Zn^{2+}}\alpha_{Y^{4-}}$$

we must first determine $\alpha_{Zn^{2+}}$ and $\alpha_{Y^{4-}}$. Since the original concentration of ammonia is 0.10 $F$ and the total zinc(II) concentration is only 0.001000 $M$, it is reasonable to assert that the free ammonia concentration remains essentially constant at 0.10 $F$ despite the formation of zinc-ammine complexes. Figure 6–1 shows that the predominant zinc(II) species in this solution is $Zn(NH_3)_4^{2+}$ and that very little $Zn^{2+}$ is present. Taking the concentration of ammonia to be 0.10 $F$ and using the relationship derived on page 166, we can show that

$$\alpha_{Zn^{2+}} = \frac{1}{1 + \beta_1[NH_3] + \beta_2[NH_3]^2 + \beta_3[NH_3]^3 + \beta_4[NH_3]^4}$$

$$\alpha_{Zn^{2+}} = \frac{1}{1 + 186(0.1) + 4.08 \times 10^4(0.1)^2 + 1.02 \times 10^7(0.1)^3 + 1.15 \times 10^9(0.1)^4}$$

$$\alpha_{Zn^{2+}} = \frac{1}{1.26 \times 10^5} = 7.94 \times 10^{-6}$$

If the procedure outlined on page 101 is employed to compute the pH and hydrogen ion concentration for a buffer consisting of 0.10 $F$ concentrations of ammonia and ammonium chloride, we find that pH = 9.26 and $[H^+] = 5.55 \times 10^{-10}$ $M$. Substitution of the latter result, along with values for appropriate equilibrium constants, into the expression* for $\alpha_{Y^{4-}}$ yields

$$\alpha_{Y^{4-}} = \frac{KK'K_{a1}K_{a2}K_{a3}K_{a4}}{[H^+]^6 + K[H^+]^5 + KK'[H^+]^4 + KK'K_{a1}[H^+]^3 + KK'K_{a1}K_{a2}[H^+]^2 + KK'K_{a1}K_{a2}K_{a3}[H^+] + KK'K_{a1}K_{a2}K_{a3}K_{a4}}$$

$$\alpha_{Y^{4-}} = \frac{2.57 \times 10^{-24}}{\begin{cases}(2.92 \times 10^{-56}) + (6.64 \times 10^{-48}) + (3.00 \times 10^{-40}) + (5.40 \times 10^{-33}) \\ + (2.08 \times 10^{-26}) + (2.59 \times 10^{-23}) + (2.57 \times 10^{-24})\end{cases}}$$

$$\alpha_{Y^{4-}} = \frac{2.57 \times 10^{-24}}{2.86 \times 10^{-23}} = 8.99 \times 10^{-2}$$

Combining the values of $\alpha_{Zn^{2+}}$ and $\alpha_{Y^{4-}}$ with $K_{ZnY^{2-}}$ found in Table 6–1, we can evaluate the conditional formation constant for the zinc-EDTA complex:

$$K'_{ZnY^{2-}} = K_{ZnY^{2-}}\alpha_{Zn^{2+}}\alpha_{Y^{4-}}$$

$$= (3.16 \times 10^{16})(7.94 \times 10^{-6})(8.99 \times 10^{-2})$$

$$= 2.25 \times 10^{10}$$

---

* See footnote on page 171.

Although the conditional formation constant is more than one million times smaller than the true formation constant, the former is still large enough to yield a well defined titration curve.

**Construction of the titration curve.** To assess the feasibility of the titration of zinc(II) with EDTA, it is desirable to construct the complete titration curve—that is, a plot of pZn (the negative logarithm of the concentration of $Zn^{2+}$) as a function of the volume of titrant. Such a curve enables us to predict the accuracy of the titration and to select an appropriate method for locating the equivalence point.

AT THE START OF THE TITRATION. Before any of the EDTA titrant has been added, the *total* concentration of zinc(II), $C_{Zn}$, is $1.00 \times 10^{-3} M$. It follows that

$$[Zn^{2+}] = \alpha_{Zn^{2+}} C_{Zn} = (7.94 \times 10^{-6})(1.00 \times 10^{-3}) = 7.94 \times 10^{-9} M$$

and that pZn is 8.10.

BEFORE THE EQUIVALENCE POINT. When the EDTA titrant is introduced into the zinc(II) solution, the principal reaction may be written as

$$Zn(NH_3)_4^{2+} + HY^{3-} \rightleftharpoons ZnY^{2-} + NH_4^+ + 3\,NH_3$$

Because the conditional formation constant for the zinc-EDTA complex is large ($K'_{ZnY^{2-}} = 2.25 \times 10^{10}$), each increment of titrant may be assumed to react completely with an equivalent amount of zinc(II). Furthermore, in calculating pZn, we can safely neglect the tiny amount of zinc(II) produced by dissociation of the zinc-EDTA complex. After the addition of 0.0500 ml of $0.1000 F$ $Na_2H_2Y$ titrant, or 0.00500 millimole, there will remain 0.04500 millimole of zinc(II) *uncomplexed by EDTA* in approximately 50 ml of solution. Therefore, the *total* concentration of zinc(II) uncomplexed by EDTA, $C_{Zn}$, is essentially $9.00 \times 10^{-4} M$. Proceeding as before, we can write

$$[Zn^{2+}] = \alpha_{Zn^{2+}} C_{Zn} = (7.94 \times 10^{-6})(9.00 \times 10^{-4}) = 7.15 \times 10^{-9} M$$

and

$$pZn = 8.15$$

Results for six more similar calculations are listed in Table 6–2.

Table 6–2. Variation of pZn During the Titration of 50.00 ml of 0.001000 $M$ Zinc(II) in a 0.1 $F$ NH$_3$-0.1 $F$ NH$_4$Cl Buffer with 0.1000 $F$ EDTA Solution

| Volume of EDTA, ml | $[Zn^{2+}]$, $M$ | pZn |
|---|---|---|
| 0 | $7.94 \times 10^{-9}$ | 8.10 |
| 0.050 | $7.15 \times 10^{-9}$ | 8.15 |
| 0.100 | $6.35 \times 10^{-9}$ | 8.20 |
| 0.200 | $4.76 \times 10^{-9}$ | 8.32 |
| 0.300 | $3.18 \times 10^{-9}$ | 8.50 |
| 0.400 | $1.59 \times 10^{-9}$ | 8.80 |
| 0.450 | $7.94 \times 10^{-10}$ | 9.10 |
| 0.490 | $1.59 \times 10^{-10}$ | 9.80 |
| 0.500 | $1.68 \times 10^{-12}$ | 11.78 |
| 0.510 | $1.76 \times 10^{-14}$ | 13.75 |
| 0.550 | $3.52 \times 10^{-15}$ | 14.45 |
| 0.600 | $1.76 \times 10^{-15}$ | 14.75 |
| 0.700 | $8.80 \times 10^{-16}$ | 15.06 |

AT THE EQUIVALENCE POINT. A calculation of pZn at the equivalence point requires use of the conditional formation constant for the zinc-EDTA complex:

$$K'_{ZnY^{2-}} = \frac{[ZnY^{2-}]}{C_{Zn}C_{EDTA}} = 2.25 \times 10^{10}$$

At the equivalence point, the concentration of the zinc-EDTA complex is $1.00 \times 10^{-3} M$ minus the total concentration of zinc(II) not complexed by EDTA, $C_{Zn}$:

$$[ZnY^{2-}] = (1.00 \times 10^{-3}) - C_{Zn}$$

Because the stoichiometry of the complexation process is one-to-one and because no reaction is ever 100 per cent complete, we can state that the total concentration of zinc(II) not complexed by EDTA at the equivalence point is, by definition, equal to the total concentration of EDTA not complexed by zinc(II):

$$C_{Zn} = C_{EDTA}$$

If the two preceding relations are substituted into the expression for $K'_{ZnY^{2-}}$, we have

$$K'_{ZnY^{2-}} = \frac{[ZnY^{2-}]}{C_{Zn}C_{EDTA}} = \frac{(1.00 \times 10^{-3}) - C_{Zn}}{(C_{Zn})^2} = 2.25 \times 10^{10}$$

In solving this equation, let us neglect provisionally the $C_{Zn}$ term in the numerator because the zinc-EDTA complex is so stable. Accordingly,

$$(C_{Zn})^2 = \frac{1.00 \times 10^{-3}}{2.25 \times 10^{10}} = 4.45 \times 10^{-14}; \qquad C_{Zn} = 2.11 \times 10^{-7} M$$

Note that $C_{Zn}$ is indeed negligible in comparison to $1.00 \times 10^{-3} M$. Finally, we obtain the result

$$[Zn^{2+}] = \alpha_{Zn^{2+}}C_{Zn} = (7.94 \times 10^{-6})(2.11 \times 10^{-7}) = 1.68 \times 10^{-12}; \qquad pZn = 11.78$$

AFTER THE EQUIVALENCE POINT. Consider the calculation of pZn after the addition of 0.510 ml of the EDTA titrant. This corresponds to an *excess* of (0.010)(0.1000) or $1.00 \times 10^{-3}$ millimole of EDTA in a volume that we shall take to be 50 ml. Therefore, the total concentration of EDTA not complexed by zinc(II), $C_{EDTA}$, is $2.00 \times 10^{-5} M$. Since dilution of the solution is ignored, the concentration of the zinc-EDTA complex, $[ZnY^{2-}]$, is essentially $1.00 \times 10^{-3} M$. When these two conditions are inserted into the relation for $K'_{ZnY^{2-}}$, the result is as follows:

$$K'_{ZnY^{2-}} = \frac{[ZnY^{2-}]}{C_{Zn}C_{EDTA}} = \frac{1.00 \times 10^{-3}}{C_{Zn}(2.00 \times 10^{-5})} = 2.25 \times 10^{10}$$

$$C_{Zn} = \frac{1.00 \times 10^{-3}}{(2.00 \times 10^{-5})(2.25 \times 10^{10})} = 2.22 \times 10^{-9} M$$

$$[Zn^{2+}] = \alpha_{Zn^{2+}}C_{Zn} = (7.94 \times 10^{-6})(2.22 \times 10^{-9}) = 1.76 \times 10^{-14} M$$

$$pZn = 13.75$$

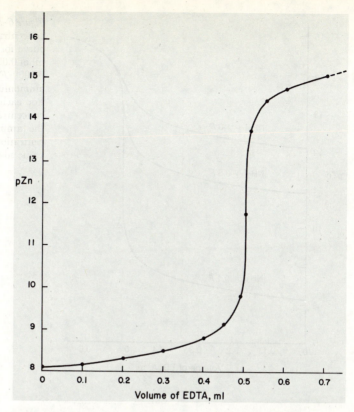

*Figure 6–4.* Titration curve for the titration of 50.00 ml of 0.001000 $M$ zinc(II) solution with 0.1000 $F$ EDTA. In addition, the zinc(II) solution is 0.1 $F$ in ammonia and 0.1 $F$ in ammonium chloride.

Table 6–2 includes data obtained from three more calculations of this type, and the complete titration curve is portrayed in Figure 6–4.

**Effect of solution composition on titration curves.** Since the composition of a solution governs the value of the conditional formation constant for the zinc-EDTA complex—because $\alpha_{Zn^{2+}}$ and $\alpha_{Y^{4-}}$ are affected—the success of a complexometric titration with EDTA rests in part on the auxiliary complexing agent.

INFLUENCE OF AMMONIA. How does the concentration of ammonia affect the sharpness of the titration curve? Figure 6–5 shows three titration curves for the reaction of 0.001 $M$ zinc(II) with 0.1 $F$ EDTA in the presence of ammonia-ammonium chloride buffers. In each case, the ammonia-ammonium ion concentration ratio is unity, so that a constant pH of 9.26 is maintained, but the absolute concentrations are different. According to the calculations that we performed in constructing the titration curve depicted in Figure 6–4, values of pZn *before the equivalence point* depend only upon $\alpha_{Zn^{2+}}$ for a specified initial concentration of zinc(II). As seen in Figure 6–5, variations in the ammonia concentration do shift the position of the titration curve before the equivalence point, but have no effect on the curve after the equivalence point. In particular, an increase in the concentration of ammonia causes a decrease in $\alpha_{Zn^{2+}}$ and diminishes the change in pZn in the region of the equivalence point, making the equivalence point more difficult to locate accurately. For an ammonia concentration of 1 $F$, the change in pZn near the equivalence point is inadequate for a direct titration.

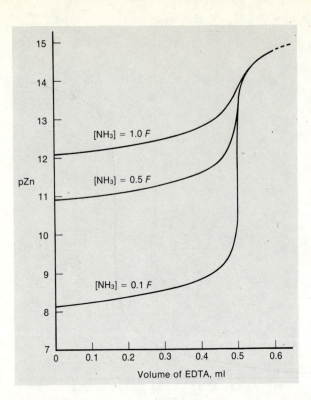

*Figure 6–5.* Effect of ammonia concentration on the titration curve for the titration of 50.00 ml of 0.001000 $M$ zinc(II) with 0.1000 $F$ EDTA in ammonia-ammonium chloride buffers. For each curve, the ammonia concentration is as shown and the ammonium chloride concentration has an identical value so that the pH is 9.26.

A decrease in the ammonia concentration below 0.1 $F$ would improve the quality of the titration curve. However, lowering the concentration of ammonia indiscriminately will lead to the precipitation of zinc hydroxide because, as shown in Figure 6–1, the relative amount of hydrated zinc(II) increases very rapidly for ammonia concentrations less than 0.01 $F$. Therefore, in employing an auxiliary complexing agent, one must seek a compromise—the concentration of the complexing agent must be high enough to prevent precipitation of the metal ion being determined but not so high that a sharp end point cannot be obtained. Sometimes the selection of a buffer with a smaller pH lowers the risk of precipitating a metal hydroxide; however, as revealed in the following discussion, this expedient has its limitations too.

INFLUENCE OF pH. Figure 6–6 illustrates the effect of pH on the titration curve for the zinc(II)-EDTA reaction in various solutions containing the same ammonia concentration of 0.1 $F$. In contrast to the influence of the ammonia concentration (Figure 6–5), the pZn values are independent of pH before the equivalence point. On the other hand, the titration curve *after the equivalence point* is displaced toward higher values of pZn as the pH increases from 7 to 10 because more and more EDTA exists in the form of the $Y^{4-}$ anion.

In the presence of 0.1 $F$ ammonia, some zinc(II) might be precipitated as insoluble $Zn(OH)_2$ at pH values above 10, although the latter compound is converted to the zinc-EDTA complex during the titration. This difficulty can be overcome through the use of a larger initial concentration of ammonia, but a price must be paid because the titration curve before the equivalence point is shifted upward if extra ammonia is added. For any ligand that forms complexes with zinc(II) as stable as the zinc-ammine species, the change of pZn near the equivalence point becomes undesirably small below pH 7. However, zinc(II) can be titrated accurately with EDTA in a solution with a pH as low as 4 if one utilizes a buffer, such as an acetic acid-sodium acetate mixture, which does not interact strongly with zinc(II).

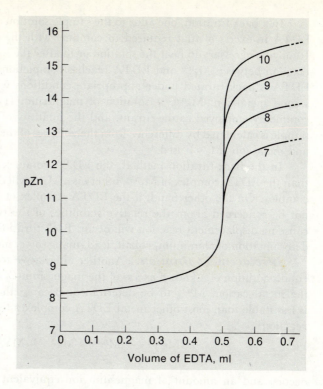

*Figure 6–6.* Effect of pH on the titration curve for the reaction of 0.001000 *M* zinc(II) with 0.1000 *F* EDTA in the presence of 0.100 *F* ammonia. Numbers on the titration curves denote the pH of the ammonia-ammonium chloride buffer. Different pH values are obtained in practice through the use of a range of ammonium chloride concentrations.

It should be mentioned here that, for the titration of zinc(II) with EDTA in an ammonia-ammonium chloride buffer, the value of pZn *at the equivalence point* is controlled by both $\alpha_{Zn^{2+}}$ and $\alpha_{Y^{4-}}$—that is, by both the ammonia concentration and the pH.

## Techniques of Complexometric Titrations

There are four different methods employed to perform complexometric titrations with EDTA—direct titration, back titration, displacement titration, and alkalimetric titration.

**Direct titration.** As the name suggests, one carries out a direct determination of a metal ion by adding standard EDTA titrant to the sample solution until an appropriate end-point signal is observed. Usually, an auxiliary complexing agent is included in the sample medium to control the pH and to prevent precipitation of the metal hydroxide.

Direct titrations are feasible for a large number of metal cations, including aluminum, barium, bismuth, cadmium, calcium, cerium, cobalt, copper, gallium, indium, iron, lead, magnesium, manganese, mercury, nickel, scandium, strontium, thallium, thorium, zinc, and zirconium as well as the lanthanides. Only those cations that react instantaneously with EDTA are suitable candidates for a direct titration.

**Back titration.** Certain metal cations react too slowly with EDTA to be titrated directly, whereas others form precipitates within the pH range normally chosen for a direct titration. Moreover, satisfactory end-point indicators are not available for some metal ions that might be determined by means of a direct-titration procedure. When any of these difficulties arise, the technique of back titration is often successful.

In a back titration, one adds to the sample solution a known volume of standard EDTA in excess of that required to combine with the desired metal cation, $M_1^{n+}$. It may be necessary to heat the solution or to alter the solution conditions so that the reaction between $M_1^{n+}$ and EDTA reaches completion rapidly. Finally, the excess EDTA is back titrated under appropriate conditions with a standard solution of a second metal ion, $M_2^{n+}$. A solution of magnesium(II), zinc(II), or copper(II) is commonly employed as the titrant, and the quantity of metal cation ($M_1^{n+}$) in the sample is calculated by difference from the volumes of the standard solutions of EDTA and metal ion ($M_2^{n+}$) used.*

In the back-titration method, the EDTA complex of $M_2^{n+}$ must be *less* stable than the EDTA complex of $M_1^{n+}$; otherwise, $M_2^{n+}$ will displace $M_1^{n+}$ from its EDTA complex. On the other hand, if the EDTA complex of $M_1^{n+}$ is nonlabile, one need not be concerned about the relative stabilities of the two metal-EDTA species because no displacement reaction will occur. Back titrations may be performed for the determination of aluminum, cobalt, lead, manganese, mercury, nickel, and thallium.

*Displacement titration.* Another alternative to the direct-titration method entails addition of a known excess of the magnesium-EDTA complex to a solution of the metal cation, $M^{n+}$, to be determined. Because the magnesium-EDTA species is less stable than most other metal-EDTA complexes (Table 6–1), the displacement reaction

$$MgY^{2-} + M^{n+} \rightleftharpoons Mg^{2+} + MY^{(n-4)+}$$

occurs, and an amount of magnesium ion equivalent to the quantity of $M^{n+}$ is liberated. Titration of the free magnesium ion with a standard EDTA solution permits the determination of $M^{n+}$. Sometimes the zinc-EDTA complex is used instead of the magnesium-EDTA species.

*Alkalimetric titration.* Since the reaction between a metal cation, $M^{n+}$, and $H_2Y^{2-}$ produces hydrogen ion,

$$M^{n+} + H_2Y^{2-} \rightleftharpoons MY^{(n-4)+} + 2 H^+$$

titration with a standard sodium hydroxide solution provides a means to determine how much metal ion is present. Care must be taken to ensure that the solution of the metal ion contains no buffering agents. In practice, one uses two burets containing standard EDTA and sodium hydroxide solutions, respectively, introduces an acid-base indicator and adjusts the initial pH of the sample solution, and adds first one and then the other titrant back and forth until a final increment of EDTA solution causes no change in the color of the indicator.

## Selectivity in Complexometric Titrations

As mentioned earlier, EDTA forms stable complexes with most cations. Therefore, this reagent frequently lacks selectivity if one is interested in determining a single metal cation in a mixture or in performing a successive titration of several species in the same solution. What can be done to improve the selectivity of complexometric titrations with EDTA?

*Chemical separations.* An obvious answer is to separate the desired species from the other components of the sample. Among the techniques that may be

---

* There need be no restriction on the charges of the two metal cations.

employed are precipitation, liquid-liquid extraction, and ion-exchange chromatography—all of which are discussed in later chapters. These methods have the disadvantages, however, that the composition of the sample solution may be altered, perhaps drastically, and that the various components of the sample, once separated, must be recovered in a suitable form before a complexometric titration can be performed.

*Control of acidity.* Simply by adjusting the pH of a sample solution containing several metal cations, one may cause only a single species to react with EDTA. This is due to the fact that conditional formation constants for metal-EDTA complexes depend in part on the hydrogen ion concentration.

Let us consider the determination of nickel(II). In a solution having an ammonia concentration of $1 F$ and a pH of 10, the conditional formation constant $(K'_{NiY^{2-}})$ for the nickel-EDTA complex is $1.72 \times 10^8$, so that nickel(II) by itself can be accurately titrated with EDTA. However, if the sample solution contains any of the alkaline-earth cations—magnesium, calcium, strontium, and barium—these species (which form no stable ammine complexes) will be co-titrated with nickel(II) because their conditional formation constants at pH 10 are comparable ($1.77 \times 10^8$, $1.77 \times 10^{10}$, $1.41 \times 10^8$, and $2.23 \times 10^7$, respectively) to that for the nickel-EDTA complex. Fortunately, the interference caused by these elements can be prevented if the pH of the sample solution is kept at 5 with an acetic acid-acetate buffer. Ignoring the possible formation of acetate complexes of the various metal cations and recalculating the conditional formation constants for the EDTA complexes of nickel, magnesium, calcium, strontium, and barium at pH 5, we obtain values of $1.41 \times 10^{12}$, 177, $1.77 \times 10^4$, 141, and 22.3, respectively. Thus, the EDTA complexes of the alkaline-earth ions have become too unstable to interfere with the titration of nickel(II).

Similar reasoning shows that aluminum, cadmium, cobalt, copper, lead, manganese, and zinc are other members of a group of elements any one of which can be titrated selectively in the presence of alkaline-earth cations at pH 5. Likewise, at pH 1 to 2 a trivalent or tetravalent species such as bismuth(III), indium(III), iron(III), scandium(III), thorium(IV), uranium(IV), or vanadium(III) may be determined in a solution containing several divalent heavy-metal ions.

*Use of masking agents.* Nickel(II) can be titrated with EDTA in the company of alkaline-earth cations if the pH of the sample solution is properly adjusted. On the other hand, simple pH control yields unsatisfactory results for the determination of manganese(II) in a solution containing other divalent heavy-metal ions such as cadmium, cobalt, copper, mercury, nickel, and zinc. In the latter case, selectivity for the titration of manganese(II) requires addition to the sample solution of a **masking agent**, some substance that reacts with the interfering cations to prevent them from combining with EDTA. Cyanide ion is a widely used masking agent and is an ideal choice for the determination of manganese(II) with EDTA because the other cations form exceedingly stable cyano complexes—$Cd(CN)_4^{2-}$, $Co(CN)_6^{2-}$, $Cu(CN)_3^{2-}$, $Hg(CN)_4^{2-}$, $Ni(CN)_4^{2-}$, and $Zn(CN)_4^{2-}$—which do not react with the titrant. In performing this determination, one adds potassium cyanide (to mask the other divalent cations), potassium sodium tartrate (an auxiliary complexing agent to prevent precipitation of manganese(II) hydroxide), ascorbic acid (a reducing agent to stop atmospheric oxidation of manganese(II) to manganese dioxide), and an ammonia-ammonium chloride buffer of pH 10. Then, at a temperature of 70 to 80°C, manganese(II) is titrated with a standard EDTA solution.

A large number of masking agents have been utilized for complexometric titrations. Most of these reagents function as complexing ligands, although some serve as precipitants or as substances that produce a change in the oxidation state of a likely interferent. Table 6–3 lists four common masking agents and a few of their uses.

**Table 6-3. Some Masking Agents Used in Complexometric Titrations with EDTA**

| Masking Agent | Typical Applications |
|---|---|
| Cyanide, $CN^-$ | (a) Titrate Pb in ammoniacal tartrate medium in presence of Cu masked as $Cu(CN)_3^{2-}$ or Co masked as $Co(CN)_6^{4-}$ <br> (b) Titrate Ca at pH 10 in presence of divalent heavy metals masked as cyanide complexes <br> (c) Titrate In in ammoniacal tartrate medium in presence of Cd, Co, Cu, Hg, Ni, and Zn masked as cyanide complexes |
| Triethanolamine, $N(CH_2CH_2OH)_3$ | (a) Titrate Ni in ammoniacal medium in presence of Al, Fe, and Mn masked as triethanolamine complexes <br> (b) Titrate Mg in ammoniacal medium of pH 10 in presence of Al masked as a triethanolamine complex <br> (c) Titrate Zn or Cd in ammoniacal medium in presence of Al masked as a triethanolamine complex |
| Fluoride, $F^-$ | (a) Titrate Zn in ammoniacal medium in presence of Al masked as $AlF_6^{3-}$ or Mg and Ca masked by precipitation as $MgF_2$ and $CaF_2$, respectively <br> (b) Titrate Ga in glacial acetic acid at pH 2.8 in presence of Al masked as $AlF_6^{3-}$ |
| 2,3-Dimercaptopropanol, $CH_2$—$CH$—$CH_2$ with $OH$ $SH$ $SH$ | (a) Titrate Mg in ammoniacal medium of pH 10 in presence of Bi, Cd, Cu, Hg, and Pb masked as complexes with 2,3-dimercaptopropanol <br> (b) Titrate Th at pH 3 in the presence of Bi and Pb masked as complexes with 2,3-dimercaptopropanol |

Quite frequently, it is possible to **de-mask** a metal cation which has been previously masked so that it can be titrated with EDTA. Sequential masking and de-masking are especially useful for consecutive determinations of several cations in a mixture. De-masking can often be accomplished if the pH of the sample solution is changed. For example, to titrate calcium in the presence of magnesium, one may adjust the pH of the solution to 12 in order to mask magnesium as insoluble $Mg(OH)_2$. Then, if the pH is lowered to approximately 9 after calcium has been determined, magnesium hydroxide dissolves and free magnesium ion can be titrated with EDTA. As pointed out above, cyanide ion is a valuable masking agent for many cations, including zinc(II) and cadmium(II) which form $Zn(CN)_4^{2-}$ and $Cd(CN)_4^{2-}$, respectively. Formaldehyde de-masks zinc(II) from the tetracyano complex by reacting with the cyanide ligands to yield the nitrile of glycolic acid:

$$Zn(CN)_4^{2-} + 4\ HCHO + 4\ H^+ \rightarrow Zn^{2+} + 4\ HO—CH_2—CN$$

Cadmium(II) undergoes a similar de-masking. However, other cyano complexes such as $Co(CN)_6^{4-}$, $Cu(CN)_3^{2-}$, $Hg(CN)_4^{2-}$, and $Ni(CN)_4^{2-}$ remain intact upon treatment with formaldehyde. By taking advantage of this selectivity of the de-masking reaction, one can determine zinc(II) or cadmium(II) in the presence of several other divalent heavy metals.

# Metallochromic Indicators for End-Point Detection

For some complexometric titrations with EDTA, especially those used for the determination of iron(III), a few highly specific end-point indicators are available. Thiocyanate ion which forms blood-red-colored complexes with iron(III)—$FeSCN^{2+}$ and $Fe(SCN)_2^+$—can be employed to signal the end point of a complexometric titration of iron(III) in a strongly acidic medium, a sudden disappearance of color occurring at the equivalence point. Salicylic acid

reacts with iron(III) to yield very stable, reddish-brown complexes having one, two, or three bidentate salicylate anions coordinated to the central atom. When a solution containing iron(III) and a small quantity of salicylic acid is titrated with EDTA at pH 3, the end point is marked by a distinct change from the red-brown color of the iron(III)-salicylate species to the colorlessness of the iron-EDTA complex.

Far more important is the large number of organic dyes, called **metallo-chromic indicators**, which have been utilized for end-point detection in complexo-metric titrations. Metallochromic indicators not only form stable, brightly colored complexes with most metal ions, but are acid-base indicators as well. Table 6–4 summarizes information about five typical metallochromic indicators and lists some of the complexometric titrations for which these substances are used.

*Eriochrome Black T.* Perhaps the most widely employed metallochromic indicator is Eriochrome Black T, which has the structure shown in Table 6–4. We can gain an understanding of metallochromic indicators from a detailed examination of the behavior of Eriochrome Black T.

Eriochrome Black T is actually a triprotic acid which may be abbreviated as $H_3In$, but loss of a proton from the —$SO_3H$ group, like the first acid dissociation of sulfuric acid, is virtually complete in water. However, the second and third steps of dissociation can be represented by the following equilibria:

$$H_2In^- \rightleftharpoons H^+ + HIn^{2-}; \qquad K_{a2} = 5.01 \times 10^{-7}$$

$$HIn^{2-} \rightleftharpoons H^+ + In^{3-}; \qquad K_{a3} = 2.51 \times 10^{-12}$$

In an aqueous medium, the colors of the species $H_2In^-$, $HIn^{2-}$, and $In^{3-}$ are red, blue, and orange, respectively. Because most EDTA titrations utilizing Eriochrome Black T are performed in the presence of a buffer such as ammonia-ammonium chloride whose pH is between 8 and 10, the predominant form of the free indicator is usually the bright-blue-colored $HIn^{2-}$ anion. In addition, Eriochrome Black T combines with most of the metal cations listed in Table 6–1 to yield stable, one-to-one, wine-red-colored complexes.

In the direct titration of zinc(II) with EDTA in an ammonia-ammonium ion buffer containing $10^{-6}$ to $10^{-5}\,F$ Eriochrome Black T, the solution has a wine-red color before the equivalence point due to the zinc-indicator complex. Suddenly, after the first tiny excess of EDTA titrant is added, the color of the solution becomes bright blue because formation of the zinc-EDTA complex liberates the $HIn^{2-}$ indicator ion according to the reaction

$$ZnIn^- + HY^{3-} \rightleftharpoons ZnY^{2-} + HIn^{2-}$$
$$\text{(wine-red)} \hspace{4.5cm} \text{(blue)}$$

Table 6-4. Properties and Uses of Some Selected Metallochromic Indicators

| Indicator | Structure | Dissociation Constants and Colors of Free Indicator Species | Colors of Metal-Indicator Complexes | Applications |
|---|---|---|---|---|
| Eriochrome Black T | ($H_2In^-$) | $H_2In^-$ (red); $pK_{a2} = 6.3$<br>$HIn^{2-}$ (blue); $pK_{a3} = 11.6$<br>$In^{3-}$ (orange) | wine-red | Direct titration of Ba, Ca, Cd, In, Pb, Mg, Mn, Sc, Sr, Tl, Zn, and lanthanides<br>Back titration of Al, Ba, Bi, Ca, Co, Cr, Fe, Ga, Pb, Mn, Hg, Ni, Pd, Sc, Tl, V<br>Displacement titration of Ba, Ca, Cu, Au, Pb, Hg, Pd, Sr |
| Calmagite | ($H_2In^-$) | $H_2In^-$ (red); $pK_{a2} = 8.1$<br>$HIn^{2-}$ (blue); $pK_{a3} = 12.4$<br>$In^{3-}$ (orange) | wine-red | Titrations performed with Eriochrome Black T as indicator may be carried out equally well with Calmagite |
| PAN | (HIn) | HIn (orange-red); $pK_a = 12.3$<br>$In^-$ (pink) | red | Direct titration of Cd, Cu, In, Sc, Tl, Zn<br>Back titration of Cu, Ga, Fe, Pb, Ni, Sc, Sn, Zn<br>Displacement titration of Al, Ca, Co, Ga, In, Fe, Pb, Mg, Mn, Hg, Ni, V, Zn |

| | Indicator | $pK_a$ values | Color | Uses |
|---|---|---|---|---|
| Murexide | (H₄In⁻) | $H_4In^-$ (red-violet); $pK_{a2} = 9.2$; $H_3In^{2-}$ (violet); $pK_{a3} = 10.9$; $H_2In^{3-}$ (blue) | red with $Ca^{2+}$, yellow with $Co^{2+}$, $Ni^{2+}$, and $Cu^{2+}$ | Direct titration of Ca, Co, Cu, Ni; Back titration of Ca, Cr, Ga; Displacement titration of Au, Pd, Ag |
| Pyrocatechol Violet | (H₄In) | $H_4In$ (red); $pK_{a1} = 0.2$; $H_3In^-$ (yellow); $pK_{a2} = 7.8$; $H_2In^{2-}$ (violet); $pK_{a3} = 9.8$; $HIn^{3-}$ (red-purple); $pK_{a4} = 11.7$ | blue, except red with Th(IV) | Direct titration of Al, Bi, Cd, Co, Ga, Fe, Pb, Mg, Mn, Ni, Th, Zn; Back titration of Al, Bi, Ga, In, Fe, Ni, Pd, Sn, Th, Ti |

For direct titrations of other cations, the same color change is observed at the end point, whereas just the opposite change in color takes place in a back titration. Note that successful end-point detection places restrictions on the equilibrium constants for the competing reactions. With reference to the direct titration of zinc(II), the stability of the zinc-Eriochrome Black T complex must be *less* than that of the zinc-EDTA species but *greater* than that of any zinc ammine under the extant solution conditions.

Let us verify quantitatively that the experimental end point for the direct titration of zinc(II) with EDTA in an ammoniacal medium coincides closely with the theoretical equivalence point. For the formation of the zinc-Eriochrome Black T complex, we may write

$$Zn^{2+} + In^{3-} \rightleftharpoons ZnIn^-; \qquad K_{ZnIn^-} = 7.95 \times 10^{12}$$

Combination of this equilibrium with the expression for the third acid dissociation of Eriochrome Black T gives the net reaction

$$Zn^{2+} + HIn^{2-} \rightleftharpoons ZnIn^- + H^+$$

which has the following equilibrium-constant relation:

$$\frac{[ZnIn^-][H^+]}{[Zn^{2+}][HIn^{2-}]} = K_{ZnIn^-} K_{a3} = 20.0$$

For the direct titration of zinc(II) with EDTA described earlier (page 176), the pH of the solution was 9.26 and the hydrogen ion concentration was $5.55 \times 10^{-10}$ $M$. If this value for $[H^+]$ is substituted into the preceding equation, which is then solved for $[Zn^{2+}]$, the result is

$$[Zn^{2+}] = \frac{[ZnIn^-][H^+]}{20.0[HIn^{2-}]} = \frac{[ZnIn^-](5.55 \times 10^{-10})}{20.0[HIn^{2-}]} = 2.77 \times 10^{-11} \frac{[ZnIn^-]}{[HIn^{2-}]}$$

Now, as in the case of acid-base indicators discussed in Chaper 4 (page 116), a visual end point will not be recognized until the concentration of blue-colored $HIn^{2-}$ is at least 10 times larger than the concentration of wine-red-colored $ZnIn^-$. When such a point is reached, an observer will recognize that a color change from red to blue has taken place. Accordingly, at the experimental end point, we will let the ratio of the concentration of $ZnIn^-$ to the concentration of $HIn^{2-}$ be 1/10. Substituting this ratio into the last relation, we have

$$[Zn^{2+}] = 2.77 \times 10^{-11} \frac{[ZnIn^-]}{[HIn^{2-}]}$$

$$= \frac{2.77 \times 10^{-11}}{10} = 2.77 \times 10^{-12} M; \qquad pZn = 11.56$$

This value of pZn agrees well with the theoretical equivalence-point value of 11.78 (Table 6–2), so we can conclude that Eriochrome Black T is an excellent end-point indicator for the titration.

*Some other metallochromic indicators.* One of the chief difficulties in using Eriochrome Black T is that solutions of it decompose rather rapidly. Only with relatively freshly prepared solutions can the proper end-point color change be realized. This instability appears to stem from the fact that the Eriochrome Black T molecule possesses an oxidizing substituent, the nitro group ($-NO_2$), and two different types of reducing groups, azo ($-N{=}N-$) and phenolic hydroxyl, which are probably involved in intermolecular redox reactions leading to destruction of the indicator.

In recent years, Calmagite has tended to replace Eriochrome Black T as an end-point indicator for complexometric titrations. Calmagite has chemical and physical properties that are remarkably similar to those of Eriochrome Black T, but its structural formula (Table 6–4) reveals the absence of a nitro group. A solution of Calmagite is exceedingly stable, having a shelf life of years. More important is that Calmagite may be used routinely as a substitute for Eriochrome Black T in all titrations requiring the latter indicator without any modification of the procedures.

Certain metal cations form such stable complexes with Eriochrome Black T or Calmagite that EDTA cannot displace the indicator from the metal-indicator species. Therefore, no color change is observed at the equivalence point of the titration or, if a color change does occur, it takes place only gradually after an excess of EDTA has been added. Nickel(II), copper(II), and titanium(IV) behave in this manner. However, nickel(II) and copper(II) can be determined by direct titration with EDTA in an ammoniacal medium if murexide is used as the metallochromic indicator, the end point corresponding to conversion of the yellow metal-murexide complex to the free violet-colored indicator anion.

Another indicator for the direct titration of copper(II) is PAN, the properties of which are compiled in Table 6–4. Usually, the determination is performed in an acetic acid-acetate buffer of pH 5 that contains approximately 50 per cent ethanol. This alcoholic solution sharpens the end point by raising the solubility of the copper-PAN complex so that the rate of reaction between EDTA and the copper-PAN species is increased.

Unlike the majority of metallochromic indicators, pyrocatechol violet is not a hydroxyazo dyestuff but a derivative of triphenylmethane. Pyrocatechol violet can be employed in acidic, neutral, and alkaline solutions for the direct titration of numerous metal cations, although the end-point color change in an ammoniacal medium is a somewhat unsatisfactory blue-to-violet transition.

## Instrumental Methods of End-Point Detection

Metallochromic indicators are especially convenient and straightforward to use, but there are occasions when other methods of end-point detection are needed. Visual determination of the equivalence point of a complexometric titration may be impossible if the sample solution is either turbid or intensely colored or if the solution contains an oxidizing agent which destroys the metallochromic indicator. In these situations, one often utilizes spectrophotometry, potentiometry, amperometry, or another instrumental technique to detect the end point. Although the principles and applications of these procedures are discussed later in this book, some features of the methods should be outlined here.

*Spectrophotometric titrations.* Metallochromic indicators are almost invariably intensely colored, and they form brightly colored complexes with most metal ions. Moreover, the complexes formed between metal cations and some auxiliary complexing agents are distinctively colored. Thus, spectrophotometry can be employed as a method for the precise location of end points of complexometric titrations. However, unlike visual end-point detection, the spectrophotometric technique

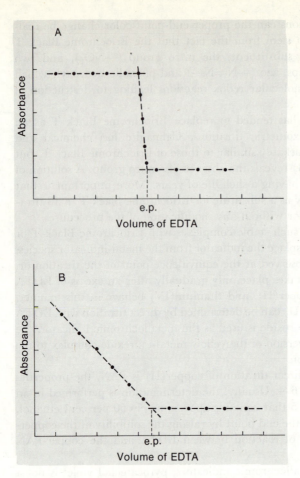

*Figure 6–7.* Spectrophotometric titration curves for complexometric titrations. Both axes have arbitrary units, and only the region of the titration curve near the equivalence point (e.p.) is shown. It is assumed that a correction is made for dilution or that the EDTA titrant is concentrated enough to prevent significant dilution. Curve A: direct titration of calcium in the presence of murexide indicator with EDTA; the equivalence point occurs at the point where the last trace of the calcium-murexide complex is converted to the calcium-EDTA species. Note that the free calcium ion is titrated before the calcium-murexide complex. Curve B: direct titration of copper(II) with EDTA in an ammoniacal solution. (See text for further discussion.)

frequently permits one to monitor changes in the concentration of a single colored species, even if the solution contains appreciable concentrations of other complexes having different colors.

Consider the complexometric determination of calcium in blood serum, urine, or spinal fluid. Classically, the titration of calcium with EDTA in a strongly alkaline medium involves the use of murexide as metallochromic indicator, the end point being taken as the first perceptible change from the red color of the calcium-murexide complex to the blue color of the free indicator ($H_2In^{3-}$). Because the desired color change happens to be somewhat indistinct to the human eye, spectrophotometric end-point detection is useful. If a titration vessel is built into a visible spectrophotometer, and if the absorption of light at 480 nm by the calcium-murexide complex is monitored as EDTA titrant is added, the absorbance of the solution will remain constant prior to the equivalence point, provided a correction is made for the dilution effect. At the equivalence point, when the calcium-murexide species is converted to the calcium-EDTA complex, the absorbance shows a very abrupt decrease. Beyond the equivalence point, the absorbance levels off again. Such behavior is illustrated in Figure 6–7A. Small increments of titrant must be added near the equivalence point so that the sharp change in absorbance can be accurately located.

Another example of spectrophotometric end-point detection is provided by the titration of copper(II) with EDTA in an ammoniacal medium. Since the copper(II)-ammine species are intensely blue, one can observe the absorption of light by these complexes at a wavelength (650 nm) where the copper(II)-EDTA complex does not absorb light to a significant extent. Then, as titrant is added, the absorbance of the solution decreases linearly (if a correction for dilution is made) until the equivalence point is reached, after which no further change in absorbance is seen. As shown in Figure 6–7B, the equivalence point can be located if one extrapolates the two straight-line portions of the spectrophotometric titration curve to their point of intersection.

*Potentiometric and amperometric titrations.* In a potentiometric titration, an indicator electrode and a reference electrode, the latter having a fixed potential, are immersed in the sample solution. Variations in the potential of the indicator electrode with respect to the reference electrode as titrant is added allow the progress of the titration to be followed. An indicator electrode may respond either directly or indirectly to the metal ion being titrated. For example, a mercury indicator electrode placed in an acetic acid solution of mercury(II) exhibits a potential that is linearly related to pHg, the negative logarithm of the concentration (activity) of mercury(II). On the other hand, the potential of a mercury indicator electrode immersed in a solution containing both some metal cation ($M^{n+}$) to be determined and a small concentration ($10^{-4}\,M$) of the mercury(II)-EDTA complex depends linearly, though indirectly, on pM, the negative logarithm of the concentration (activity) of $M^{n+}$. For each situation, a plot of the measured potential of the mercury indicator electrode as a function of the volume of titrant will have the general shape of the complexometric titration curves shown in Figures 6–4, 6–5, and 6–6, and the equivalence point can be located according to one of the procedures outlined in Chapter 11.

As described in Chapter 13, one performs an amperometric titration by placing two electrodes in the solution to be titrated; most commonly, one electrode is a microelectrode (often a dropping mercury electrode) and the other is a suitable (nonpolarizable) reference electrode. Then, the potential of the microelectrode is held at some preselected value versus the reference electrode, and the current (usually in the microampere range) that flows due to the reduction or oxidation of a reactant or product is plotted as a function of the volume of added titrant. Changes in the observed current are proportional to changes in the concentration of a particular species. As shown in Figure 13–9 (page 453), an amperometric titration curve consists of two straight lines whose point of intersection marks the equivalence point. Some typical examples of complexometric titrations that can be monitored amperometrically are listed in Table 13–3 (page 455).

## Scope of Complexometric Titrations with EDTA

Complexometric titrations with EDTA as well as other aminopolycarboxylic acids have been utilized for the determination of practically every metal cation, both singly and in mixtures. In Table 6–5 is a list of some of the individual metal ions that can be determined and of the kinds of samples that can be analyzed. Needless to say, real samples such as alloys or pharmaceutical products often require considerable pretreatment before a final titration with EDTA is performed. When, as usually occurs, mixtures of several metal ions are present in a sample, it may be necessary to separate the species, for example, by precipitation, extraction, or ion-exchange chromatography. Alternatively, appropriate masking agents can be added to the sample solution, or the pH of the sample solution can be adjusted, so that one metal cation can be titrated in the presence of others. Procedures for the analysis of mixtures of metals are contained in several of the references listed at the end of this chapter.

**Table 6-5. Some Representative Analyses Involving EDTA Titrations***

| Species Determined | Procedure | Typical Materials Analyzed |
|---|---|---|
| Al(III) | Add slight excess of standard EDTA; adjust pH to 6.5; dissolve salicylic acid indicator in the solution; back-titrate unreacted EDTA with standard iron-(III) solution | alloys, cryolite ores, cracking catalysts, clays |
| Bi(III) | Adjust pH of sample solution to 2.5; add pyrocatechol violet indicator; titrate with EDTA | alloys, pharmaceuticals |
| Ca(II) | Add NaOH to neutral sample solution to obtain pH > 12; add murexide; titrate with EDTA | pharmaceuticals, phosphate rocks, water, biological fluids |
| Cd(II) or Zn(II) | To neutral sample solution, add $NH_3$–$NH_4^+$ buffer; add Eriochrome Black T indicator and titrate with EDTA | plating baths, alloys |
| Co(II) | Adjust pH of sample solution to 6; add murexide indicator; add ammonia to obtain orange color of cobalt-murexide complex; titrate with EDTA | paint driers, magnet alloys, cemented carbides |
| Cu(II) | Adjust pH of sample solution to 8 with $NH_3$ and $NH_4Cl$; add murexide indicator and titrate with EDTA | alloys, ores, electroplating baths |
| Fe(III) | Adjust pH of sample solution to 4; dissolve salicylic acid indicator in the solution; titrate with EDTA | hemoglobin, limestone, cement, tungsten alloys, paper pulp, boiler scale |
| Ga(III) | To neutral sample solution, add $NH_3$ and $NH_4^+$ to reach pH 9; add Eriochrome Black T indicator and a known excess of standard EDTA; back-titrate the unreacted EDTA with a standard manganese(II) solution | |
| Hf(IV) or Zr(IV) | [same procedure as for Al(III)] | alloys, ores, paint driers, sand |
| Hg(II) | To the sample solution, add a known excess of standard magnesium(II)-EDTA solution; neutralize the solution; add $NH_3$–$NH_4^+$ buffer and Eriochrome Black T indicator; perform substitution titration of free magnesium(II) with EDTA | mercury-containing pharmaceuticals, organomercury compounds, ores |
| In(III) | Add tartaric acid to an acidic sample solution; neutralize the solution; add $NH_3$–$NH_4^+$ buffer and Eriochrome Black T indicator; heat solution to boiling and titrate with EDTA | |
| Mg(II) | [same procedure as for Cd(II)] | pharmaceuticals, aluminum alloys, gun powder, soil, plant materials, water, biological fluids |
| Mn(II) | Add triethanolamine and ascorbic acid to acidic sample solution; neutralize the solution; add $NH_3$–$NH_4^+$ buffer; add Eriochrome Black T and titrate with EDTA | metallurgical slags, silicate rocks, alloys, ferromanganese |
| Ni(II) | Add murexide indicator to neutral sample solution; add $NH_3$ to obtain orange color of nickel-murexide complex; titrate with EDTA | electroplating baths, manganese catalysts, Alnico, ferrites |

| Species Determined | Procedure | Typical Materials Analyzed |
|---|---|---|
| Pb(II) | [same procedure as for In(III), except that a room-temperature titration is performed] | gasoline, ores, paints, alloys, pharmaceuticals |
| Sn(IV) | To strongly acidic sample solution, add known excess of standard EDTA; buffer the solution at pH 5; heat solution to 75°C, add pyrocatechol violet indicator, and titrate unreacted EDTA with standard Zn(II) solution | alloys, electroplating baths |
| Th(IV) | Adjust pH of sample solution to 2; add pyrocatechol violet indicator; warm solution to 40°C and titrate with EDTA | reactor fuels, ores, minerals, glasses, alloys |
| Tl(III) | [same procedure as for Hg(II)] | alloys |

* Information excerpted from G. Schwarzenbach and H. Flaschka: *Complexometric Titrations.* Second English edition, translated by H. M. N. H. Irving, Methuen, London, 1969.

**Determination of water hardness.** Without doubt, the most important early application of titrations with EDTA was the determination of water hardness. In talking about water hardness, we can distinguish between *total* water hardness and *individual* water hardness due separately to calcium and magnesium. To determine total water hardness, one measures the *sum* of the amounts of calcium and magnesium, but usually reports the result in terms of calcium. Generally, the final titration is performed in a solution, buffered at pH 10 with ammonia and ammonium chloride, containing Eriochrome Black T indicator and appropriate masking agents to prevent traces of heavy metals from interfering with the action of the indicator. Furthermore, the EDTA titrant frequently contains a small concentration of the magnesium-EDTA complex, because the latter improves the sharpness of the end-point color change.

To evaluate the hardness of water due to calcium and magnesium individually, it is common practice to follow a procedure with two separate aliquots of the water sample. For one aliquot, the sum of the amounts of calcium and magnesium is established as described in the preceding paragraph. For the second aliquot, only calcium is determined, so that the magnesium content of the water is obtained by difference. In order to determine calcium in the presence of magnesium, the former is titrated with EDTA in a solution having a pH greater than 12; under such conditions, magnesium cannot interfere because it is precipitated as insoluble $Mg(OH)_2$. Murexide is employed as the end-point indicator, and suitable reagents can be added to the solution to mask heavy-metal impurities.

**Determination of organic compounds.** Methods have been devised for the determination of organic compounds and functional groups, particularly those found in pharmaceutical materials. These techniques are based upon the reaction between a known excess of a metal ion and the substance to be determined. Depending upon the nature of the chemical system, one can calculate the quantity of organic compound by measuring either the amount of metal ion that has reacted with the organic substance or the quantity of metal ion that remains unreacted.

A number of compounds form precipitates with mercury(II), after which the excess, unreacted metal in the filtrate is determined; these substances include nicotinamide, theobromine, phenobarbital, and ethyl gallate (an antioxidant). Caffeine, quinine, antihistamines, codeine, strychnine, thiamine, and morphine can each be precipitated from an acidic solution as a salt of the anion $BiI_4^-$, and the excess bismuth(III) can be determined by titration with EDTA. Biological fluids, chocolates, and sweetened and fermented drinks may be analyzed for sugar if one treats the sample with an alkaline solution of copper(II); the copper(II) is reduced to insoluble $Cu_2O$ by the sugar, and one can determine either how much cuprous oxide is formed or how much copper(II) remains.

Carbon-carbon double bonds can be determined if the sample material is treated with an excess of mercury(II) acetate in the presence of a $BF_3$-etherate catalyst; one measures the mercury(II), which does not add to the double bond, by first adding a known excess of EDTA and then back-titrating the unreacted ligand with a standard solution of zinc(II). If a solution containing amino acids or peptides is shaken with freshly precipitated copper(II) phosphate, the copper(II) that becomes complexed by the amino acid group may be found by a titration with EDTA after the solution is centrifuged to remove the excess precipitate. More applications of complexometric titrimetry to the analysis of organic compounds are described in reference 5 of the list of supplemental reading.

## QUESTIONS AND PROBLEMS

1. In view of the fact that ethylenediaminetetraacetate (EDTA) forms stable complexes with so many cations, discuss the approaches by means of which reactions between EDTA and metal ions can be made selective.

2. Account qualitatively for the general stability of metal-EDTA complexes as compared to other types of complexes involving the same cations.

3. Suggest an analytical procedure, involving an EDTA titration, for the direct or indirect determination of sulfate ion.

4. Define or identify each of the following terms: labile complex, nonlabile complex, coordination number, bidentate ligand, chelate, binuclear complex, stepwise formation constant, overall formation constant, ammine complex, auxiliary complexing agent, metallochromic indicator, chelate effect, conditional formation constant, masking agent.

5. Estimate from Figure 6-1 the fractions of the various zinc-ammine complexes in a solution containing a free ammonia concentration of $0.00316\,F$. If the total concentration of zinc-ammine complexes in such a solution is $0.00100\,M$, calculate the concentration of each individual zinc-ammine complex.

6. Write the equilibrium reaction and the equilibrium expression corresponding to the fourth stepwise formation constant for the $Ni(NH_3)_4(H_2O)_2^{2+}$ complex.

7. Write the equilibrium reaction and the equilibrium expression corresponding to the third overall formation constant for the $Cd(NH_3)_3(H_2O)_3^{2+}$ complex.

8. (a) Construct a distribution diagram to show how the fractions of $Ni(H_2O)_6^{2+}$ and each nickel-ammine complex vary as a function of the logarithm of the free (uncomplexed) ammonia concentration.
   (b) List the formulas of the seven nickel(II) species in order of *decreasing* concentration in a solution that contains a free ammonia concentration of $1.000\,F$.
   (c) Calculate the concentration of each of the seven nickel(II) species in a solution containing a total nickel(II) concentration of $0.00250\,F$ and a free ammonia concentration of $0.750\,F$.

9. Calculate the molar concentration of each of the species, $Cu(NH_3)^{2+}$, $Cu(NH_3)_2^{2+}$, $Cu(NH_3)_3^{2+}$, and $Cu(NH_3)_4^{2+}$, in a solution in which the equilibrium concentrations of ammonia and $Cu^{2+}$ ion are $2.50 \times 10^{-3}\,F$ and $1.50 \times 10^{-4}\,M$, respectively.

10. What is the free ammonia concentration in a system containing the nickel-ammine complexes, if the concentration of $Ni(NH_3)_4^{2+}$ is exactly ten times the $Ni(NH_3)_3^{2+}$ concentration?

11. Calculate the equilibrium concentrations of $Zn(NH_3)_2^{2+}$ and $Zn(NH_3)_3^{2+}$ in a solution prepared by the mixing of 10.0 ml of 0.00200 $F$ zinc nitrate solution with 40.0 ml of 0.200 $F$ aqueous ammonia.

12. Calculate the concentration of each of the three silver(I) species in a solution initially containing 0.400 $F$ ammonia and a total silver(I) concentration of 0.00300 $M$.

13. Calculate the equilibrium concentrations of the three species, $Ag^+$, $CN^-$, and $Ag(CN)_2^-$, resulting from the mixing of 35.0 ml of 0.250 $F$ sodium cyanide solution with 30.0 ml of 0.100 $F$ silver nitrate solution.

14. Calculate the equilibrium concentrations of the three species, $Ag^+$, $CN^-$, and $Ag(CN)_2^-$, resulting from the mixing of 50.0 ml of 0.200 $F$ sodium cyanide solution with 70.0 ml of 0.100 $F$ silver nitrate solution.

15. Calculate the concentration of each of the five EDTA species (ignoring $H_6Y^{2+}$ and $H_5Y^+$) in a 0.0250 $F$ EDTA solution of pH 9.50.

16. What is the concentration of uncomplexed magnesium ion in a solution prepared by the mixing of equal volumes of 0.200 $F$ EDTA and 0.100 $F$ magnesium nitrate solutions, assuming that the pH is 9.00?

17. Calculate the concentration of uncomplexed nickel(II) in a solution prepared by the mixing of equal volumes of 0.150 $F$ EDTA solution and 0.100 $F$ nickel(II) nitrate solution. Assume that the solution is buffered at pH 10.50, but that the buffer constituents do not complex nickel(II).

18. A solution was prepared by the addition of 500 ml of a solution 0.01000 $F$ in ethylenediaminetetraacetic acid ($H_4Y$) to 500 ml of a solution 0.02000 $F$ in tetrasodium ethylenediaminetetraacetate ($Na_4Y$) and 0.01500 $F$ in sodium hydroxide. Calculate the pH of the resulting solution and the concentrations of all five EDTA species (ignoring $H_6Y^{2+}$ and $H_5Y^+$).

19. For a solution in which the *total* analytical concentration of silver ion is 0.00500 $M$, what must be the concentrations of ammonia and ammonium ion so that the pH of the solution is 9.70 and the average number of ammonia ligands per silver atom is 1.50? Successive formation constants for the silver-ammine complexes are given in Appendix 3.

20. Nitrilotriacetic acid (NTA)

$$HN^+ \Big\langle \begin{array}{l} CH_2\text{—}COOH \\ CH_2\text{—}COO^- \\ CH_2\text{—}COOH \end{array}$$

is a triprotic acid whose successive dissociation reactions may be represented as follows:

$$H_3Y \rightleftharpoons H^+ + H_2Y^-; \qquad K_{a1} = 1.26 \times 10^{-2}$$

$$H_2Y^- \rightleftharpoons H^+ + HY^{2-}; \qquad K_{a2} = 3.24 \times 10^{-3}$$

$$HY^{2-} \rightleftharpoons H^+ + Y^{3-}; \qquad K_{a3} = 1.86 \times 10^{-10}$$

(a) Construct a distribution diagram to show how the fraction ($\alpha$) of each NTA species varies as a function of pH.

(b) Between what pH values is $H_2Y^-$ the predominant species in an aqueous solution of NTA? Between what pH values is $HY^{2-}$ the predominant species in an aqueous solution of NTA?

(c) Calculate the concentration of each NTA species in a solution with a pH of 4.12 and with a total NTA concentration of 0.01500 $F$.

21. Calculate the conditional formation constant for the nickel(II)-EDTA complex in a solution containing 0.500 $M$ ammonium ion and 0.500 $F$ free (uncomplexed) ammonia.

22. Calculate the conditional formation constant for the mercury(II)-EDTA complex in a solution of pH 11.00 containing 0.0100 $M$ free (uncomplexed) cyanide ion.

23. Construct the titration curve for the titration of 50.00 ml of a 0.001000 $M$ copper(II) solution with a 0.1000 $F$ EDTA solution. Assume that the copper(II) sample solution contains 0.400 $F$ ammonia and 0.200 $F$ ammonium

nitrate. Successive formation constants for the copper(II)-ammine complexes are listed in Appendix 3.

24. Construct the titration curve for the titration of 100.0 ml of a 0.002000 $M$ magnesium ion solution with a 0.1000 $F$ EDTA solution at pH 10.

25. A standard solution of calcium chloride was prepared by dissolution of 0.2000 gm of pure calcium carbonate in hydrochloric acid. Then the solution was boiled to remove carbon dioxide and was diluted to 250.0 ml in a volumetric flask. When a 25.00-ml aliquot of the calcium chloride solution was used to standardize an EDTA solution by titration at pH 10, 22.62 ml of the EDTA solution was required. Calculate the concentration of the EDTA solution.

26. A 1.000-ml aliquot of a nickel(II) solution was diluted with distilled water and an ammonia-ammonium chloride buffer; it was then treated with 15.00 ml of a 0.01000 $F$ EDTA solution. The excess EDTA was back-titrated with a standard 0.01500 $F$ magnesium chloride solution, of which 4.37 ml was required. Calculate the concentration of the original nickel(II) solution.

27. Suppose that the direct EDTA titration of a metal cation, $M^{2+}$, of initial concentration 0.0100 $M$ is performed in an acetic acid-sodium acetate buffer of pH 5.00 with a metallochromic indicator of formula $H_2In$. Assume that the following data are known:

$$M^{2+} + Y^{4-} \rightleftharpoons MY^{2-}; \qquad K_{MY} = 1.00 \times 10^{18}$$
$$H_2In \rightleftharpoons H^+ + HIn^-; \qquad K_{a1} = 1.00 \times 10^{-2}$$
$$HIn^- \rightleftharpoons H^+ + In^{2-}; \qquad K_{a2} = 1.00 \times 10^{-7}$$
$$M^{2+} + In^{2-} \rightleftharpoons MIn; \qquad K_{MIn} = 1.60 \times 10^{11}$$

Also assume that $M^{2+}$ does not form any stable hydroxide or acetate complexes at pH 5.00. If the end-point color change is observed when the concentration of the metal-indicator complex (MIn) and the concentration of the free indicator (in all its forms) are equal, calculate the theoretical titration error in per cent with the proper sign. The original solution volume may be taken to be 50.0 ml, and dilution may be ignored.

28. To a solution containing gallium(III) ion was added an excess of an ammonia-ammonium nitrate buffer of pH 10 and exactly 25.00 ml of a 0.05862 $F$ solution of the magnesium-EDTA complex. Then a few drops of Eriochrome Black T indicator was introduced. Finally, the solution was titrated with 0.07010 $F$ EDTA until a color change from red to blue was observed, 5.91 ml of titrant being used. Calculate the weight in milligrams of gallium present in the original solution.

29. An alkalimetric titration was performed for the determination of zinc(II) in an unbuffered aqueous medium. Two titrants, one a 0.1064 $F$ sodium hydroxide solution and the other a 0.05942 $F$ solution of the disodium salt of nitrilotriacetic acid (Figure 6–2) which we can abbreviate as $Na_2HY$, were utilized. Methyl red indicator was added to the zinc(II) solution, and the pH of the solution was carefully adjusted until the red acid form of methyl red just changed to the yellow base form. Some of the standard $Na_2HY$ solution was introduced; then the acid produced by the complexation reaction between zinc(II) and $HY^{2-}$ was titrated with the sodium hydroxide solution until the yellow color of methyl red reappeared. The successive addition of small portions of the $Na_2HY$ and NaOH was continued until an increment of the $Na_2HY$ titrant produced no change in the yellow color of methyl red. In all, a total of 28.13 ml of the sodium hydroxide solution was used in the titration. Calculate the weight in milligrams of zinc in the sample solution.

30. Total water hardness is usually expressed in terms of the number of milligrams of calcium carbonate ($CaCO_3$) per liter of water (in other words, parts per million, ppm, of calcium carbonate), although the titrimetric procedure actually entails a determination of the sum of the amounts of calcium(II) and magnesium(II) in solution.

A 100.0-ml sample of water was checked for total hardness by means of the procedure described on page 193. In the titration, which was performed in an ammonia-ammonium ion buffer of pH 10 with Calmagite as metallochromic indicator, a volume of 12.58 ml of 0.008826 $F$ EDTA was needed to reach the desired end point. Calculate the total water hardness in terms of parts per million (ppm) of calcium carbonate.

For another 100.0-ml sample of the water, the calcium content was determined by means of a titration with EDTA at pH 13. Murexide was employed as indicator, and 10.11 ml of the previous EDTA titrant was required. Using the titration data, calculate the calcium and magnesium content of the water in terms of parts per million (ppm) of $Ca^{2+}$ and $Mg^{2+}$, respectively.

31. A solution containing iron(III) and aluminum(III) was analyzed by means of consecutive complexometric titrations. The pH of a 50.00-ml aliquot of the sample solution was buffered at approximately 2, and about 200 mg of salicylic acid was dissolved in it. Then, 0.04016 $F$ EDTA titrant was added until the red color of the iron(III)-salicylate complex just vanished, 29.61 ml being required. Then, exactly 50.00 ml of the same EDTA titrant was pipetted into the solution; next, the solution was boiled briefly to ensure the complete complexation of aluminum and the pH was readjusted to a value of 5. Finally, the solution, which contained excess EDTA, was back-titrated with 19.03 ml of a 0.03228 $F$ iron(III) solution, the end point being taken as the first appearance of the red color of the iron(III)-salicylate complex. Using the analytical data, calculate the concentrations of iron(III) and aluminum(III) in the original sample solution.

32. Impure sodium phenobarbital ($NaC_{12}H_{11}N_2O_3$) was assayed by means of a procedure involving a complexometric titration. A 0.2438-gm sample of the powder was dissolved in 100 ml of 0.02 $F$ sodium hydroxide solution at 60°C. After being cooled, the solution was acidified with acetic acid and was transferred to a 250.0-ml volumetric flask. Then a 25.00-ml aliquot of a 0.02031 $F$ mercury(II) perchlorate solution was added to the flask. Next, the solution was diluted to the mark on the volumetric flask and was allowed to stand until the precipitate formed according to the reaction

$$Hg^{2+} + 2\ C_{12}H_{11}N_2O_3{}^- \rightleftharpoons Hg(C_{12}H_{11}N_2O_3)_2$$

Finally, the solution was filtered to remove the precipitate and the filtrate was analyzed for unreacted mercury(II). A 50.00-ml aliquot of the filtrate was treated with 10.00 ml of a 0.01128 $M$ solution of the magnesium-EDTA complex, and the liberated magnesium ion was titrated at pH 10 to an Eriochrome Black T end point with 5.89 ml of 0.01212 $F$ EDTA solution. Calculate the percentage of sodium phenobarbital in the powder.

33. A mixture of caffeine, quinine hydrochloride, and antipyrine (an analgesic and antipyretic substance) was analyzed for its caffeine content. A sample weighing 0.3871 gm was dissolved in 30 ml of water contained in a 50.00-ml volumetric flask. Then approximately 10 ml of a 0.25 $F$ solution of potassium tetraiodomercurate(II), $K_2HgI_4$, and 1 ml of concentrated hydrochloric acid were added, whereupon the tetraiodomercurate(II) salts of quinine and antipyrine precipitated, leaving caffeine in solution. Water was added up to the mark on the volumetric flask, and the mixture was shaken and filtered. A 20.00-ml aliquot of the filtrate was pipetted into a clean, dry flask and exactly 5.00 ml of 0.2507 $F$ $KBiI_4$ solution was added. Reaction between the protonated caffeine molecule and the $BiI_4{}^-$ anion produced a precipitate:

$$(C_8H_{10}N_4O_2)H^+ + BiI_4{}^- \rightleftharpoons (C_8H_{10}N_4O_2)H\ BiI_4$$

After the precipitate was filtered from the solution, a 10.00-ml portion of the filtrate was analyzed for unreacted bismuth(III) by means of a complexometric titration. The titration was performed in an acetic acid-sodium acetate buffer; 5.11 ml of a 0.04919 $F$ EDTA solution was needed to just cause the disappearance of the yellow color due to the presence of the $BiI_4{}^-$ complex. Calculate the percentage of caffeine ($C_8H_{10}N_4O_2$) in the sample material.

34. Acrylonitrile ($H_2C{=}CHCN$), which is a toxic and flammable liquid, polymerizes to form Orlon fibers. A partially polymerized sample of acrylonitrile, weighing 0.2795 gm, was dissolved in 10 ml of a 0.05 $F$ solution of $BF_3 \cdot O(C_2H_5)_2$ in methanol. A 0.1772-gm portion of pure, anhydrous

mercury(II) acetate was dissolved in the methanol solution, and the reaction between monomeric acrylonitrile and mercury(II) acetate

$$H_2C\!=\!CHCN + Hg(CH_3COO)_2 + CH_3OH \longrightarrow \begin{array}{c} H_2C\!-\!CHCN \\ |\qquad | \\ H_3CO\quad Hg(CH_3COO) \end{array} + CH_3COOH$$

was allowed to reach to completion. Then, 10 ml of an ammonia-ammonium nitrate buffer, 5.000 ml of a 0.1016 $M$ solution of the zinc(II)-EDTA complex, 20 ml of water, and a few drops of Eriochrome Black T indicator were added. Zinc(II), released by the reaction between unconsumed mercury(II) and the zinc(II)-EDTA complex, was titrated with 0.05121 $F$ EDTA solution, a volume of 2.743 ml being needed to reach the end point. Calculate the weight per cent of monomeric acrylonitrile in the sample.

## SUGGESTIONS FOR ADDITIONAL READING

1. F. Basolo and R. Johnson: *Coordination Chemistry*. W. A. Benjamin, New York, 1964, pp. 114–140.
2. H. Flaschka: *EDTA Titrations*. Second edition, Pergamon Press, New York, 1964.
3. H. Flaschka and A. J. Barnard, Jr.: Titrations with EDTA and related compounds. *In* C. L. Wilson and D. W. Wilson, eds.: *Comprehensive Analytical Chemistry*. Volume IB, Elsevier, New York, 1960, pp. 288–385.
4. A. Ringbom: Complexation reactions. *In* I. M. Kolthoff and P. J. Elving, eds.: *Treatise on Analytical Chemistry*. Part I, Volume 1, Wiley-Interscience, New York, 1959, pp. 543–628.
5. G. Schwarzenbach and H. Flaschka: *Complexometric Titrations*. Second English edition, translated by H. M. N. H. Irving, Methuen, London, 1969.
6. L. G. Sillén: Graphic presentation of equilibrium data. *In* I. M. Kolthoff and P. J. Elving, eds.: *Treatise on Analytical Chemistry*. Part I, Volume 1, Wiley-Interscience, New York, 1959, pp. 277–317.
7. F. J. Welcher: *The Analytical Uses of Ethylenediaminetetraacetic Acid*. Van Nostrand, Princeton, New Jersey, 1958.
8. T. S. West: *Complexometry*. Third edition, British Drug Houses Chemicals Limited, Poole, England, 1969.

# FORMATION AND DISSOLUTION OF PRECIPITATES

**7**

There are two classes of analytical methods based on the formation of precipitates—gravimetric determinations and precipitation titrations. In the usual gravimetric precipitation method, a solution of the species to be determined is mixed with a second solution containing an excess of some reagent that forms a precipitate with the substance sought. Then the resulting precipitate is separated from the solution, washed free of excess precipitant, dried at an appropriate temperature, and finally weighed so that the quantity of substance sought can be calculated.

In a precipitation titration, a solution containing the species to be determined is titrated with a standard solution of a reagent that forms a precipitate with the desired substance. If some technique is available to signal when an amount of titrant stoichiometrically equivalent to the substance sought has been added, the quantity of that substance can be computed from the concentration and volume of titrant used.

Quite obviously, a precipitation titration differs from a gravimetric determination in that the former does not require separation of a precipitate from the solution in which it is formed. However, regardless of which of these two analytical methods is employed, three requirements must be fulfilled by the precipitation process.

First, *the desired constituent must be precipitated quantitatively*. In other words, the quantity of the desired substance left in solution must be a negligible fraction of the total amount of that species originally present.

Second, *the precipitate must be pure.* In a gravimetric determination, the precipitate must not contain significant amounts of impurities unless these substances are readily removable in the washing and drying steps of the procedure. In a precipitation titration, the precipitate must not contain impurities that significantly alter the stoichiometric relationship between the titrant and the constituent being titrated.

Third, *the precipitate must be in a suitable physical form.* In a gravimetric determination, the precipitate must consist of particles that are large enough to be retained by the medium used for filtration. In a titrimetric procedure, the particles of precipitate must either remain dispersed or settle rapidly as particular circumstances require, so that, for example, the indicator color change marking the end point of the titration is not concealed.

Every precipitation method of analysis must be designed and conducted to meet these three requirements. Decisions concerning such diverse factors as the choice of compound to be precipitated, selection of the precipitating agent, volume and concentration of reagent solutions, presence and concentration ranges of other constituents, choice of solvent, temperature, pH, rate of addition of precipitating reagent, and time and method of digestion and washing must all be based upon the fulfilling of the three requirements. These requirements are closely interrelated, and a condition which may be desirable from the standpoint of one requirement may adversely affect fulfillment of another requirement. Therefore, the analytical procedure ultimately adopted is necessarily the result of a series of compromises arrived at in an effort to optimize the extent to which all three requirements are met.

In the present chapter, we will consider in detail the nature of the reaction between dissolved species which yields a precipitate as well as the physical and chemical interactions between the precipitate and solution phases. Then, in the next chapter, some typical gravimetric determinations and precipitation titrations will be described.

## NUCLEATION AND CRYSTAL GROWTH

Formation of a precipitate involves both a physical process and a chemical process. Generally, the physical reaction consists of two phenomena, **nucleation** and **crystal growth**. Nucleation proceeds through the formation within a supersaturated solution of the smallest particles of a precipitate (nuclei) capable of spontaneous growth. Crystal growth consists of the deposition of ions from the solution upon the surfaces of solid particles which have already been nucleated. As a rule, the number of particles and, therefore, the particle size of a precipitate are determined by the number of nuclei formed in the nucleation step. In turn, the number of nuclei that form is believed to be governed by the interplay between two factors—first, the extent of supersaturation in the immediate environment where nucleation occurs and, second, the number and effectiveness of sites upon which nuclei may form.

*Significance of supersaturation in nucleation.* Supersaturation is a condition in which a solution phase contains more of the dissolved precipitate than can be in equilibrium with the solid phase. It is usually a transient condition, particularly when some crystals of the solid phase are present, although some solutions may remain supersaturated for considerable lengths of time.

It has been found that, to a reasonably good approximation, the number of nuclei formed in the nucleation step is directly proportional to a parameter known as the **relative supersaturation**, defined by the ratio $(Q - S)/S$, where $Q$ is the *actual* concentration of solute at the instant precipitation begins and $S$ is the *equilibrium* concentration of solute in a saturated solution.

Consider the precipitation of barium sulfate when 100 ml of 0.01 $F$ barium chloride solution and 1 ml of 1 $F$ sodium sulfate solution are mixed. If the two solutions were completely mixed before any precipitation occurred, the concentrations of barium ion and sulfate ion would each be very close to 0.01 $M$. Since the equilibrium solubility $S$ of barium sulfate is approximately $1 \times 10^{-5} F$, the $(Q - S)/S$ ratio would be 1000. However, at the spot where the sodium sulfate solution first comes into contact with the barium chloride solution, the momentary concentration of sulfate ion would be much higher than 0.01 $M$, even close to 1 $M$, so the momentary $Q$, calculated as $([Ba^{2+}][SO_4{}^{2-}])^{1/2}$, could approach 0.1 $M$ and the corresponding $(Q - S)/S$ ratio would be 10,000. Although the momentary relative supersaturation is relieved by precipitation of barium sulfate as well as by mixing, the particle size of the final precipitate is markedly influenced by this transient condition. In general, the *larger* the relative supersaturation, the *smaller* will be the size of the individual particles of the precipitate. For this reason, it is desirable to have the solutions quite dilute at the time of mixing, particularly if the precipitate is an extremely insoluble one, in order to promote the formation of large crystals. Furthermore, it is important to form a precipitate under conditions for which it is not so extremely insoluble. For example, the solubility of barium sulfate is greater in an acid medium than in a neutral medium, so the momentary $(Q - S)/S$ value is lower if precipitation occurs from an acid medium.

***Spontaneous and induced nucleation.*** Theoretically, it is possible for a sufficiently large cluster of ions to come together in a supersaturated solution to form a nucleus by the process known as **spontaneous nucleation**. In practical situations, however, it is highly probable that purely spontaneous nucleation is far less frequent than **induced nucleation**, in which the initial clustering of ions is aided by the presence in the solution of sites which can attract and hold ions. Introduction of particles of a precipitate into a solution supersaturated with that solid can initiate further precipitation. In addition, other solid particles or surfaces can serve as nucleation sites. Surfaces of the container in which precipitation occurs provide many sites, as is indicated by the fact that the particle size of a precipitate may be strongly influenced by the type of container, how scratch-free it is, and how it was cleaned prior to use. Insoluble impurities in the reagents and in the solvent used to prepare solutions provide nucleation sites. Reagent-grade chemicals typically contain 0.005 to 0.010 per cent insoluble matter which, when dispersed throughout a solution as tiny particles, can provide extremely large numbers of sites for nucleation.

***Processes of crystal growth.*** Crystal growth, once a nucleus has been formed, consists of two steps—the *diffusion* of ions to the surface of the growing crystal and the *deposition* of these ions on the surface. Either process can be rate-determining. In general, the diffusion rate is influenced by the specific nature of the ions and their concentrations, by the rate of stirring, and by the temperature of the solution, whereas the rate of deposition of ions is affected by concentration, by impurities on the surface, and by the growth characteristics of the particular crystal. Crystals having perfect geometrical shape are rarely formed in analytical precipitation processes because crystal growth usually does not occur uniformly on all faces of a crystal. Preferential growth on certain faces may result, for example, in flat plates or in sticks or rods, whereas preferential growth on corners may result in irregular, branched crystals called dendrites. Foreign ions can markedly influence the shape of precipitated particles, probably because by being adsorbed they can inhibit or encourage growth on certain faces.

Nucleation and crystal growth have been studied in numerous laboratories for many years. Although much knowledge has been obtained, these processes are so complex that much additional research would be required for complete comprehension of the phenomena.

## COMPLETENESS OF PRECIPITATION

Completeness of precipitation of a desired species is generally determined by the equilibrium solubility of that substance under the conditions prevailing either at the time of precipitation or at the time of the subsequent filtration and washing operations. During some gravimetric determinations, however, it is possible for a state of supersaturation to persist, resulting in a less complete separation or recovery of the desired precipitate than would be obtained if supersaturation did not occur. For a moderately soluble substance such as magnesium ammonium phosphate, appreciable solubility loss would occur through supersaturation if the precipitate were filtered immediately after formation, but this source of error is negligible after a reasonable digestion period. Usually, supersaturation is relieved by the presence of some of the precipitate, by stirring of the solution, and by the existence of fine scratches on the inner wall of the vessel used for the precipitation.

### Solubility, Activity Products, and Solubility Products

We shall begin our discussion of equilibrium solubility with a consideration of the dissolution of silver chloride, AgCl, in pure water. In most treatments of the solubility of silver chloride, it is stated that the simple equilibrium

$$AgCl(s) \rightleftharpoons Ag^+ + Cl^-$$

prevails after an aqueous suspension of the solid has been thoroughly shaken. However, one of the interesting features of the behavior of silver chloride is that silver ion and chloride ion may combine to form a soluble, undissociated silver chloride molecule:

$$Ag^+ + Cl^- \rightleftharpoons AgCl(aq)$$

This aqueous, undissociated molecule will be in equilibrium with solid silver chloride, according to the reaction

$$AgCl(s) \rightleftharpoons AgCl(aq)$$

Furthermore, the aqueous, undissociated molecule can react with a chloride ion, originating from silver chloride, to form an anionic complex:

$$AgCl(aq) + Cl^- \rightleftharpoons AgCl_2^-$$

If the chloride ion concentration were high enough, other complexes such as $AgCl_3^{2-}$ or $AgCl_4^{3-}$ could form. Thus, the solubility behavior of even a simple compound can be extremely complicated. Only through a careful investigation can one identify all the reactions which occur. Fortunately, in pure water the concentration of chloride ion is so small that the amounts of the anionic complexes are negligible. However, the solubility of silver chloride should still be considered as the sum of the concentrations of $Ag^+$ and $AgCl(aq)$.

If we omit the anionic complexes just mentioned, the solubility behavior of silver chloride may be represented in the following way:

$$AgCl(s) \rightleftharpoons AgCl(aq) \rightleftharpoons Ag^+ + Cl^-$$

According to this picture, solid silver chloride dissolves to form the aqueous, undissociated molecule which, in turn, dissociates into silver and chloride ions. It is important to recognize that all the various equilibria are interrelated. Equilibrium-constant expressions corresponding to this sequence of reactions must be written in terms of activities rather than analytical concentrations, if we are interested in describing the solubility in a completely rigorous manner.

*Activity and intrinsic solubility of a solid.* For the first step of the solubility reactions discussed in the preceding paragraph, the equilibrium-constant expression can be written as

$$\frac{(AgCl(aq))}{(AgCl(s))} = S^\circ$$

where the symbols in parentheses denote activities of the solid and aqueous forms of silver chloride and where $S^\circ$ is called the **intrinsic solubility** of silver chloride. Let us examine the significance of the terms in this equation. Since the standard state of silver chloride is defined to be the pure solid at 25°C, we may conclude that

$$(AgCl(s)) = 1$$

if the silver chloride is not contaminated in any way with impurities and if the temperature is 25°C. Solids which do contain extraneous substances have activities *less* than unity. Experimental studies have indicated that the activity of exceedingly small particles of a solid is actually *greater* than unity. However, a detailed discussion of these subjects would be extremely lengthy and is beyond the scope of the present treatment of solubility. Suffice it to say that particles of reasonable purity and greater than approximately 2 microns in diameter exhibit unit activity.

Throughout the remainder of this chapter, the activities of solid phases will always be assumed to be unity, so that for the silver chloride system we may write

$$(AgCl(aq)) = S^\circ$$

This relation states that the activity of the soluble molecular form of silver chloride is a constant whenever one has solid AgCl in equilibrium with pure water or any other aqueous solution. Regardless of what other competing reactions occur, the activity of AgCl(aq) remains unchanged, which is why $S^\circ$ is termed the intrinsic solubility of silver chloride. It is exceedingly difficult to determine the value of $S^\circ$ experimentally because the quantity of AgCl(aq) is almost always small compared to the concentrations of other dissolved silver species. Values for $S^\circ$ reported in the chemical literature vary from $1.0 \times 10^{-7}$ to $6.2 \times 10^{-7}$ $M$.

Apparently, a small concentration of the aqueous, undissociated molecular form of a compound always exists in equilibrium with the solid phase. For AgCl, AgBr, AgI, AgSCN, and $AgIO_3$, the concentration of the molecular species is approximately 0.1 to 1 per cent of the total solubility of the parent compound in water. Dissolution of many metallic hydroxides and sulfides, including $Ni(OH)_2$, $Fe(OH)_3$, $Zn(OH)_2$, HgS, CdS, and CuS, is known to yield a small concentration of the corresponding aqueous molecule.

*Activity product.* We may consider the second step of the solubility reactions of silver chloride as the dissociation of the aqueous silver chloride molecule into silver ion and chloride ion:

$$AgCl(aq) \rightleftharpoons Ag^+ + Cl^-$$

As the molecular species is consumed in this reaction, additional solid silver chloride dissolves. If we write the thermodynamic equilibrium-constant expression for the dissociation of AgCl(aq), the result is

$$\frac{(Ag^+)(Cl^-)}{(AgCl(aq))} = \frac{1}{K_1}$$

where $K_1$ is the *first stepwise formation constant* for the silver chloride system (see Chapter 6, page 164). Since the equilibrium-constant expression contains, as one term, the activity of AgCl(aq), the value of the equilibrium constant has the same uncertainty as the intrinsic solubility, $S^\circ$.

If we now add the two equilibria,

$$AgCl(s) \rightleftharpoons AgCl(aq)$$

and

$$AgCl(aq) \rightleftharpoons Ag^+ + Cl^-$$

the result is

$$AgCl(s) \rightleftharpoons Ag^+ + Cl^-$$

When two reactions are *added* together to yield a third reaction, the equilibrium constants for these two reactions are *multiplied* together to give the equilibrium constant for the third reaction. Therefore,

$$(AgCl(aq)) \cdot \frac{(Ag^+)(Cl^-)}{(AgCl(aq))} = S^\circ \cdot \frac{1}{K_1}$$

or

$$(Ag^+)(Cl^-) = \frac{S^\circ}{K_1}$$

with the activity of solid silver chloride still taken to be unity. Although individual values of $S^\circ$ and $K_1$ are not accurately known, the *ratio* of these two quantities can be obtained precisely, because the product of the activities of silver ion and chloride ion is accessible to direct experimental measurement. The product of the activities of silver ion and chloride ion is defined as the **thermodynamic activity product**, $K_{ap}$:

$$(Ag^+)(Cl^-) = K_{ap}$$

This relation is a statement of the fact that, in a solution saturated with solid silver chloride, the product of the activities of silver ion and chloride ion is always constant.

How can the value of the thermodynamic activity product be determined experimentally? Recalling that the activity of an ion is given by the equation

$$a_i = f_i C_i$$

we can formulate the thermodynamic activity-product expression as

$$f_{Ag^+}[Ag^+]f_{Cl^-}[Cl^-] = K_{ap}$$

whereas, if mean activity coefficients are used, we obtain

$$f_{\pm}^{2}[Ag^+][Cl^-] = K_{ap}$$

Suppose that solid silver chloride is shaken with pure water until solubility equilibrium is attained. After the excess solid is separated from its saturated solution, one can determine by means of titrimetry the analytical concentrations of silver ion and chloride ion. Then, because the solubility of silver chloride is small, one may confidently employ the Debye-Hückel limiting law to evaluate the mean activity coefficient of silver chloride. Finally, the mean activity coefficient and the analytical data are combined to calculate the thermodynamic activity product. Numerous measurements of the solubility of silver chloride at 25°C have been performed, and the results are consistent with the conclusion that

$$(Ag^+)(Cl^-) = K_{ap} = 1.78 \times 10^{-10}$$

It should be noted that a correction must be made for the presence of AgCl(aq) when one wishes to determine the analytical concentration of just silver ion, because each of these species behaves identically during a titration or similar procedure.

*Solubility product.* In analytical chemistry, we are often concerned with the concentration of a species instead of its activity, since the former provides information about the quantity of material present in a particular phase or system. For example, it may be important to know the analytical concentration of silver ion in a certain solution. In addition, as discussed in Chapter 3, activity coefficients for ions cannot be reliably calculated for electrolyte solutions other than very dilute ones. Moreover, one generally performs equilibrium calculations in order to predict the feasibility of a method of separation or analysis. For these reasons, it is usually satisfactory to omit the use of activity coefficients and to write relations involving only analytical concentrations, such as

$$[Ag^+][Cl^-] = K_{sp}$$

This latter equation is the familiar **solubility-product expression** for silver chloride, and the equilibrium constant is called the **solubility product** or the **solubility-product constant**. For almost all equilibrium calculations concerned with the solubility of precipitates, we shall use analytical concentrations and we will assume that the activity product $(K_{ap})$ and the solubility product $(K_{sp})$ are synonymous. Appendix 1 contains a compilation of solubility products.

Let us consider two examples of solubility-product calculations. Suppose that the solubility of nickel hydroxide is known to be 0.00011 gm per liter of water and that we wish to calculate the solubility product of this compound. The simplest possible representation of the solubility reaction is

$$Ni(OH)_2(s) \rightleftharpoons Ni^{2+} + 2\,OH^-$$

and the solubility-product expression is

$$[Ni^{2+}][OH^-]^2 = K_{sp}$$

To obtain the molar solubility of nickel hydroxide, we divide the weight solubility by the molecular weight of $Ni(OH)_2$, the latter being 92.7. Thus, the molar solubility, $S$, is given by

$$S = \frac{0.00011 \text{ gm}}{1 \text{ liter}} \frac{1 \text{ mole}}{92.7 \text{ gm}} = 1.2 \times 10^{-6} \, M$$

Since one nickel ion is formed for each nickel hydroxide molecule which dissolves,

$$[Ni^{2+}] = 1.2 \times 10^{-6} \ M$$

whereas the hydroxide ion concentration is twice that of nickel(II):

$$[OH^-] = 2.4 \times 10^{-6} \ M$$

It follows that the solubility product is given by

$$(1.2 \times 10^{-6})(2.4 \times 10^{-6})^2 = K_{sp}$$
$$6.9 \times 10^{-18} = K_{sp}$$

Starting with the solubility-product expression for a compound, we can calculate the molar solubility. For example, the solubility-product expression for cadmium(II) hydroxide is

$$[Cd^{2+}][OH^-]^2 = 5.9 \times 10^{-15}$$

corresponding to the equilibrium

$$Cd(OH)_2(s) \rightleftharpoons Cd^{2+} + 2\ OH^-$$

Since we obtain one $Cd^{2+}$ ion and two $OH^-$ ions from each cadmium(II) hydroxide molecule which dissolves, the concentration of cadmium ion should be equal to the molar solubility ($S$) of cadmium(II) hydroxide

$$[Cd^{2+}] = S$$

whereas the concentration of the hydroxide ion will be twice the molar solubility:

$$[OH^-] = 2\ S$$

Substituting these results into the solubility-product expression, we obtain

$$(S)(2S)^2 = 5.9 \times 10^{-15}$$
$$4S^3 = 5.9 \times 10^{-15}$$
$$S = 1.1 \times 10^{-5} \ M$$

It should be mentioned that the actual solubility behavior of nickel hydroxide and cadmium hydroxide is much more complicated than these simple calculations suggest. We have omitted consideration of the intrinsic solubilities of these compounds as well as formation of the (unhydrated) monohydroxo species, $NiOH^+$ and $CdOH^+$. In addition, we have neglected the fact that the autoprotolysis of water contributes hydroxide ions to the systems.

## Effect of Ionic Strength

Although it is convenient and simple to discuss the solubility of a substance in pure water, the formation and dissolution of a precipitate in analytical procedures invariably occur in the presence of relatively large concentrations of foreign electrolytes. For example, if one prepares pure silver chloride by mixing solutions containing stoichiometric quantities of sodium chloride and silver nitrate, the precipitate will be dispersed in a medium containing dissolved sodium nitrate. Therefore, we must be concerned about the solubility behavior of a precipitate in the presence of a foreign electrolyte. However, we shall restrict this discussion to systems wherein the desired compound is in equilibrium with a solution which contains no ions common to the precipitate—that is, no ions of which the solid is composed, except those produced when the precipitate itself dissolves.

Before attempting to predict quantitatively how the solubility of a precipitate such as silver chloride depends on the concentration of foreign electrolyte, let us consider a qualitative picture of the interaction between silver ion and chloride ion in two different environments—pure water and $0.1 F$ nitric acid. In the nitric acid medium, the positively charged silver ion exhibits a strong attraction for nitrate anions, whereas chloride ions attract hydrogen ions. Obviously, silver ion and chloride ion have a mutual attraction for each other. However, because the concentrations of silver ion and chloride ion are quite small compared to the quantities of hydrogen ion and nitrate ion, the ion atmosphere of a silver ion contains primarily nitrate ions and the ion atmosphere of chloride consists predominantly of hydrogen ions. These ion atmospheres partially neutralize the charges on silver ion and chloride ion, thereby decreasing their attraction for each other. If the force of attraction between $Ag^+$ and $Cl^-$ is less in nitric acid than in pure water, the solubility of silver chloride should be larger in nitric acid. This conclusion is correct; such a simple picture affords a valuable way to predict the effects of foreign electrolytes on chemical equilibria.

Quantitative predictions of the effect of a foreign electrolyte on the solubility of silver chloride can be based upon the equation introduced previously,

$$f_\pm^2 [Ag^+][Cl^-] = K_{ap}$$

where $f_\pm$ is the mean activity coefficient for silver chloride and $[Ag^+]$ and $[Cl^-]$ represent the analytical concentrations of silver ion and chloride ion, respectively. Since the solubility $S$ of silver chloride is measured by the concentration of silver ion (neglecting the presence of aqueous molecular silver chloride) and since the concentrations of silver and chloride ions are equal, we may write

$$f_\pm^2 S^2 = K_{ap}$$

or

$$S = \frac{(K_{ap})^{1/2}}{f_\pm}$$

It is evident from the latter equation that the solubility of silver chloride should increase as the mean activity coefficient decreases; and, of course, the converse is true. By examining the variation of activity coefficients with ionic strength in Figures 3–2, 3–3, and 3–4, we can conclude that a plot of solubility versus ionic strength will exhibit an initial rise followed by a decrease. It should be recalled (see Table 3–2) that activity coefficients for ions with multiple charges are much more sensitive to variations in ionic strength than those for singly charged species. Consequently, the solubility of precipitates consisting of multicharged ions will change much more drastically with ionic strength than, say, the solubility of silver chloride.

Let us compare the solubility of silver chloride in water and in $0.1 F$ nitric acid. For water, the ionic strength $I$ is nearly zero because the concentrations of dissolved silver ion and chloride ion are very small. Thus, we shall assume that the mean activity coefficient is essentially unity. Therefore, the solubility $S$ of silver chloride is given by the simple relation

$$S = (K_{ap})^{1/2} = (1.78 \times 10^{-10})^{1/2}$$

$$S = 1.33 \times 10^{-5} \, M$$

A calculation of the solubility of silver chloride in 0.1 $F$ nitric acid requires that we determine the mean activity coefficient of silver chloride. Although the ionic strength of a solution does depend on the concentrations of all ionic species, the contribution by dissolved silver chloride is negligible. Thus, the ionic strength is essentially equal to the concentration of nitric acid, that is, 0.1. To obtain the mean activity coefficient of silver chloride at this ionic strength, we may consult Table 3–2 (page 52). In the table, the single-ion activity coefficients for $Ag^+$ and $Cl^-$ are both found to be 0.755 at an ionic strength of 0.1, so the mean activity coefficient for silver chloride is 0.755 as well. If we now combine this result with the equation for the solubility of silver chloride, the value of $S$ can be calculated:

$$S = \frac{(K_{ap})^{1/2}}{f_\pm} = \frac{(1.78 \times 10^{-10})^{1/2}}{0.755}$$

$$S = 1.77 \times 10^{-5} \, M$$

Therefore, the solubility of silver chloride is predicted to be about 33 per cent higher in 0.1 $F$ nitric acid than in water.

**Common Ion Effect**

Quite often, the solubility of a precipitate is lower in an aqueous solution containing a common ion—that is, one of the ions comprising the compound—than in pure water alone. We can illustrate this so-called **common ion effect** by calculating and comparing the solubility of lead iodate, $Pb(IO_3)_2$, in water and in 0.03 $F$ potassium iodate solution.

For lead iodate, the solubility reaction and solubility-product constant may be written as

$$Pb(IO_3)_2(s) \rightleftharpoons Pb^{2+} + 2\,IO_3^-; \qquad K_{sp} = 2.6 \times 10^{-13}$$

In water, each dissolved lead iodate molecule yields one lead ion and two iodate ions. If the solubility of lead iodate is denoted by the symbol $S$, we can write

$$[Pb^{2+}] = S \quad \text{and} \quad [IO_3^-] = 2\,S$$

When these mathematical statements are substituted into the solubility-product expression, we obtain

$$[Pb^{2+}][IO_3^-]^2 = (S)(2S)^2 = 2.6 \times 10^{-13}$$

$$4S^3 = 2.6 \times 10^{-13}$$

$$S = 4.0 \times 10^{-5} \, M$$

for the solubility of lead iodate in water.

In the 0.03 $F$ potassium iodate medium, the situation is somewhat more complicated, for there are two sources of iodate ion. At equilibrium, the concentration of iodate ion will be the sum of the contributions from potassium iodate and lead iodate: 0.03 $M$ from potassium iodate and $2S$ from the dissolution of lead iodate, where $S$ represents the equilibrium concentration of lead ion (which is different from that found in the preceding calculation). Thus, the appropriate solubility-product expression is

$$(S)(2S + 0.03)^2 = 2.6 \times 10^{-13}$$

This relation could be solved in a completely rigorous manner, but the manipulations would involve a third-order equation. Therefore, it is worthwhile to consider a suitable simplifying assumption. Since the solubility of lead iodate in water is only $4.0 \times 10^{-5} M$ and since the presence of potassium iodate should repress the dissolution of lead iodate, we will neglect the $2S$ term in the preceding expression:

$$(S)(0.03)^2 = 2.6 \times 10^{-13}$$

This equation may be readily solved:

$$S = \frac{2.6 \times 10^{-13}}{9 \times 10^{-4}} = 2.9 \times 10^{-10} M$$

Certainly our neglect of the $2S$ term was justified, because the solubility of lead iodate is substantially decreased by the presence of 0.03 $F$ potassium iodate.

Although it is important to recognize the usefulness of the common ion effect in the prediction of experimental conditions for a desired gravimetric determination, it is equally important to realize that serious errors can result if one places too much confidence in such calculations. Bad mistakes may arise if one neglects or is unaware of other chemical equilibria that take place in the system of interest. For example, if we examine the solubility-product expression for silver chloride,

$$[Ag^+][Cl^-] = K_{sp} = 1.78 \times 10^{-10}$$

it appears that any desired completeness of precipitation could be achieved through a suitable adjustment of the chloride concentration. If the chloride ion concentration were $1 \times 10^{-3} M$, the concentration of silver ion should be $1.78 \times 10^{-7} M$, whereas the concentration of silver ion ought to be only $1.78 \times 10^{-9} M$ when the chloride concentration is 0.1 $M$. However, as stated previously, a certain concentration, approximately $2 \times 10^{-7} M$, of aqueous molecular silver chloride exists in equilibrium with the solid, and this concentration is unaffected by the occurrence of other chemical reactions. Therefore, the concentration of AgCl(aq) represents the minimum concentration of dissolved silver chloride attainable in an aqueous solution—in spite of the common ion effect.

Earlier, in discussing the solubility of silver chloride in water, we mentioned that anionic complexes such as $AgCl_2^-$, $AgCl_3^{2-}$, and $AgCl_4^{3-}$ may exist to a significant extent when the free chloride concentration is relatively large. Thus, the phenomenon whereby an excess of a common ion represses solubility may, in fact, be reversed in the presence of a very large excess of common ion.

## Complexation of Cation with Foreign Ligand

If a foreign complexing agent or ligand is available which can react with the cation of a precipitate, the solubility of a compound can be markedly enhanced.

As examples, we may cite the increase in the solubility of the silver halides, AgCl, AgBr, and AgI, in ammoniacal solutions.

For silver bromide in aqueous ammonia solutions, the position of equilibrium for the reaction

$$AgBr(s) \rightleftharpoons Ag^+ + Br^-; \qquad K_{sp} = 5.25 \times 10^{-13}$$

will be shifted toward the right because of the formation of stable silver-ammine complexes. Silver ion reacts with ammonia in a stepwise manner, according to the equilibria

$$Ag^+ + NH_3 \rightleftharpoons AgNH_3^+; \qquad K_1 = 2.4 \times 10^3$$

and

$$AgNH_3^+ + NH_3 \rightleftharpoons Ag(NH_3)_2^+; \qquad K_2 = 6.9 \times 10^3$$

Let us calculate the solubility of silver bromide in a $0.1\,F$ ammonia solution. On the basis of the preceding reactions, we can conclude that each silver bromide molecule which dissolves will yield one bromide ion and either one $Ag^+$, one $AgNH_3^+$, or one $Ag(NH_3)_2^+$. Therefore, the solubility $S$ of silver bromide is measured by the concentration of bromide ion or, alternatively, by the *sum* of the concentrations of all soluble silver species, *i.e.*,

$$S = [Ag^+] + [AgNH_3^+] + [Ag(NH_3)_2^+] = [Br^-]$$

This relation may be transformed into one equation with a single unknown if we introduce the equilibrium-constant expressions for the silver-ammine species. For the concentration of the silver-monoammine complex, it can be shown that

$$[AgNH_3^+] = K_1[Ag^+][NH_3]$$

and the concentration of the silver-diammine complex may be written as

$$[Ag(NH_3)_2^+] = K_2[AgNH_3^+][NH_3]$$

or

$$[Ag(NH_3)_2^+] = K_1K_2[Ag^+][NH_3]^2$$

If we substitute the relations for the concentrations of the two silver-ammine complexes into the solubility equation, the result is

$$S = [Ag^+] + K_1[Ag^+][NH_3] + K_1K_2[Ag^+][NH_3]^2 = [Br^-]$$

In a solution saturated with silver bromide, the solubility-product expression must be obeyed; that is,

$$[Ag^+] = \frac{K_{sp}}{[Br^-]}$$

so it follows that

$$S = \frac{K_{sp}}{[Br^-]} + \frac{K_1K_{sp}[NH_3]}{[Br^-]} + \frac{K_1K_2K_{sp}[NH_3]^2}{[Br^-]} = [Br^-]$$

However, since $[Br^-] = S$, we can write

$$S = \frac{K_{sp}}{S} + \frac{K_1K_{sp}[NH_3]}{S} + \frac{K_1K_2K_{sp}[NH_3]^2}{S}$$

Finally, solving for the solubility $S$, we obtain

$$S^2 = K_{sp} + K_1 K_{sp}[NH_3] + K_1 K_2 K_{sp}[NH_3]^2$$

$$S = (K_{sp})^{1/2}(1 + K_1[NH_3] + K_1 K_2[NH_3]^2)^{1/2}$$

Before we insert values for the ammonia concentration and the various constants into the latter equation, note that this relation is a general one for the solubility of *any* sparingly soluble silver salt in an ammonia solution. For the solubility of silver bromide in a 0.1 $F$ ammonia solution, we have

$$S = (5.25 \times 10^{-13})^{1/2}[1 + (2.4 \times 10^3)(0.1) + (2.4 \times 10^3)(6.9 \times 10^3)(0.1)^2]^{1/2}$$

$$S = 2.9 \times 10^{-4}\ M$$

In addition, we can calculate the concentration of each soluble silver species. For $Ag^+$:

$$[Ag^+] = \frac{K_{sp}}{[Br^-]} = \frac{5.25 \times 10^{-13}}{2.9 \times 10^{-4}} = 1.78 \times 10^{-9}\ M$$

For $AgNH_3^+$:

$$[AgNH_3^+] = K_1[Ag^+][NH_3] = (2.4 \times 10^3)(1.78 \times 10^{-9})(0.1)$$

$$[AgNH_3^+] = 4.3 \times 10^{-7}\ M$$

For $Ag(NH_3)_2^+$:

$$[Ag(NH_3)_2^+] = K_2[AgNH_3^+][NH_3] = (6.9 \times 10^3)(4.3 \times 10^{-7})(0.1)$$

$$[Ag(NH_3)_2^+] = 2.9 \times 10^{-4}\ M$$

Therefore, the only species which contributes significantly to the solubility of silver bromide is the silver-diammine complex. One of the possible complications which we overlooked in solving this problem was that the formation of silver-ammine complexes might alter the concentration of ammonia. However, the solubility of silver bromide is small enough that neglect of this effect was justified. If the solubility of silver bromide were of comparable magnitude to the ammonia concentration, it would be necessary to consider the change in the concentration of ammonia.

## Reaction of Anion with Acid

In certain instances, the anion of a slightly soluble compound reacts with one or more of the constituents of a solution phase. For example, if the solubility of lead sulfate in nitric acid media is investigated, one finds that the solubility is greater than in pure water and that the solubility becomes larger as the concentration of nitric acid increases.

In pure water, the solubility of lead sulfate is governed by the equilibrium

$$PbSO_4(s) \rightleftharpoons Pb^{2+} + SO_4^{2-}; \qquad K_{sp} = 1.6 \times 10^{-8}$$

and, if no other equilibria prevail, the solubility of lead sulfate should essentially be the square root of the solubility product, or approximately $1.3 \times 10^{-4}\ M$.

If nitric acid is added to the system, the solubility of lead sulfate increases because sulfate ion reacts with hydrogen ion to form the hydrogen sulfate anion:

$$H^+ + SO_4{}^{2-} \rightleftharpoons HSO_4{}^-$$

However, the addition of another proton to the hydrogen sulfate ion,

$$H^+ + HSO_4{}^- \rightleftharpoons H_2SO_4$$

does not occur to any appreciable extent because sulfuric acid is a very strong acid. Therefore, when a lead sulfate molecule dissolves in nitric acid, one lead ion is formed, but the sulfate ion from the precipitate may be present as either $SO_4{}^{2-}$ or $HSO_4{}^-$. Accordingly, we can express the solubility $S$ of lead sulfate in nitric acid by means of the equation

$$S = [Pb^{2+}] = [SO_4{}^{2-}] + [HSO_4{}^-]$$

A relationship between the concentrations of sulfate ion and hydrogen sulfate ion can be based on the equilibrium expression for the *second* dissociation of sulfuric acid, namely,

$$HSO_4{}^- \rightleftharpoons H^+ + SO_4{}^{2-}; \qquad K_{a2} = 1.2 \times 10^{-2}$$

By substituting the latter information into the solubility equation, we obtain

$$S = [Pb^{2+}] = [SO_4{}^{2-}] + \frac{[H^+][SO_4{}^{2-}]}{K_{a2}}$$

and, by utilizing the solubility-product expression for lead sulfate, we can write

$$S = [Pb^{2+}] = \frac{K_{sp}}{[Pb^{2+}]} + \frac{K_{sp}[H^+]}{K_{a2}[Pb^{2+}]}$$

$$[Pb^{2+}]^2 = K_{sp} + \frac{K_{sp}[H^+]}{K_{a2}}$$

$$[Pb^{2+}] = \left( K_{sp} + \frac{K_{sp}[H^+]}{K_{a2}} \right)^{1/2}$$

If the hydrogen ion concentration is known, we can use the latter relation to calculate the concentration of lead ion in equilibrium with solid lead sulfate and, therefore, the solubility of lead sulfate itself.

Suppose that we wish to predict the solubility of lead sulfate in $0.10\,F$ nitric acid. One question to be answered is whether the formation of the hydrogen sulfate anion, through the reaction of $SO_4{}^{2-}$ with $H^+$, will consume enough hydrogen ion to change significantly the concentration of the latter. We can either assume that the change will be important—in which case the computations are difficult—or that this change may be neglected as a first approximation. If the second alternative is chosen, it will be necessary to justify the validity of the assumption after the calculations are completed. Let us assume that the concentration of hydrogen ion remains essentially constant at $0.1\,M$. Thus, the lead ion concentration is given by the equation

$$[Pb^{2+}] = \left( 1.6 \times 10^{-8} + \frac{(1.6 \times 10^{-8})(0.1)}{(1.2 \times 10^{-2})} \right)^{1/2}$$

which, when solved, yields

$$[Pb^{2+}] = 3.9 \times 10^{-4}\,M$$

Even if all the sulfate derived from the dissolution of lead sulfate were converted to hydrogen sulfate anion, the change in the hydrogen ion concentration would be negligibly small. Therefore, by making a simplifying assumption and later justifying its validity, we obtain the desired result in a straightforward way. If, on the other hand, the lead ion concentration is shown to be comparable to the original concentration of nitric acid, we should immediately suspect that our assumption was false and return to a more rigorous solution to the problem. It should be stressed that the primary reason for making the assumption was the smallness of the solubility product of lead sulfate compared to the initial nitric acid concentration.

## PURITY OF A PRECIPITATE

A desired precipitate may be contaminated with one or more other substances because those substances are themselves insufficiently soluble in the mother liquid. It is not feasible, for example, to separate chloride from bromide by precipitation of the latter with silver nitrate because the equilibrium solubilities of the two silver halides are not sufficiently different from each other. Such situations can be avoided if one resorts to some other means of separation. In addition, a precipitate can become contaminated with substances from its mother liquid even when the equilibrium solubilities of those other substances are not exceeded. To understand this latter phenomenon, let us review some properties of the colloidal state.

### Colloidal State

By the term **colloidal state**, we mean the dispersion of one phase within a second phase. Colloids of most interest in chemical analysis are dispersions of solid particles in a solution phase—in other words, a precipitated substance in its mother liquid. Particles of a colloidal precipitate range from approximately $1 \times 10^{-7}$ to $2 \times 10^{-5}$ cm in diameter.

A colloidal dispersion is not a true solution. In a true solution, the dispersed particles are of ionic or molecular dimensions, whereas they are larger in a colloidal dispersion. A colloidal dispersion is a mixture and a suspension, yet the properties of colloids differ so markedly from simple mixtures and suspensions that it is meaningful to consider colloidal dispersions as a separate subject.

It is undesirable for a precipitate to be colloidal at the time of its filtration, because colloidal particles, although they constitute a distinct phase, are so fine that they pass directly through ordinary filtering media. In addition, two structural features of colloids—the localized electrical character of their surfaces and their tremendously large surface area per unit mass—control and influence the purity of the precipitate.

*Electrical charges on surfaces.* Most inorganic substances of analytical interest are ionic solids consisting of cations and anions arranged in a specific crystal lattice. A portion of a cubic crystal of silver chloride is depicted by the model in Figure 7–1, in which the dark balls represent silver ions and the light balls represent chloride ions. Each chloride ion in the *interior* of the crystal is surrounded by six silver ions which are its closest neighbors, and each silver ion is similarly surrounded by six chloride ions. However, each chloride ion on the *face* of a crystal is associated

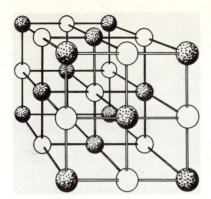

*Figure 7–1.* Model of a portion of the cubic crystal lattice of silver chloride.

with but five silver ions, and each silver ion with only five chloride ions. Under such circumstances, the chloride and silver ions on the faces of a crystal possess partial residual charges—negative and positive, respectively. An ion on an *edge* is surrounded by only four oppositely charged neighbors, and a *corner* ion has only three such neighbors. So the residual charges are even larger for ions on the edges and at the corners of a crystal. Thus, although the overall surface of a crystal is neutral, it is covered with localized centers of positive and negative charge.

*Ratio of surface to mass.* Perhaps the most significant characteristic of the colloidal state of matter is the enormous ratio of surface to mass which it exhibits. For example, a 1-cm cube of a substance has a total surface area of 6 cm². If the same mass of the same material is subdivided into $1 \times 10^{-6}$ cm cubes, there are about $10^{18}$ cubes, and they expose a total surface area of about 6,000,000 cm². In the 1-cm cube, less than one in 10,000,000 ions is on the surface, whereas in the $1 \times 10^{-6}$ cm cube about one in every 12 ions is a surface ion (if we assume an ion diameter of $2 \times 10^{-8}$ cm). Colloidal particles are characterized by very large surface to mass ratios, so surface effects are very important. Any surface features, such as the localized positive and negative charges, are very pronounced when the particles are in the colloidal state.

*Stability of colloids.* Colloidal dispersions can be exceedingly stable, even though gravity should make colloidal particles settle to the bottom of a container. One reason for the stability of colloids is that the particles are in a state of constant motion. Particles continually collide with and rebound from each other as well as the walls of the container. Colloidal particles are doubtless being hit continually by the smaller molecules of the suspending medium and by any ions contained therein. This continual motion of colloidal particles is called **Brownian movement** and tends to overcome the settling influence of gravity.

A second reason for the stability of colloids is even more significant in chemical analysis than is Brownian movement. Ions from the mother liquid which are adsorbed at the localized negative or positive charge centers on the surface of colloidal particles impart an electrical charge to those surfaces. Generally, one particular type of ion in the mother liquid will be attracted more strongly than other species, so this ion will be preferentially adsorbed and will impart its charge, either positive or negative, to all the surfaces. If a cation is more readily adsorbed, all the colloidal particles become positively charged, whereas all surfaces of the particles become negatively charged if some anion is more readily adsorbed. In either case, the adsorbed ion layer imparts stability to the colloidal dispersion because the charged particles repel each other.

*Selectivity of ion adsorption.* Adsorption of ions upon the surface of solid particles in contact with the mother liquid is based upon an electrical attraction.

However, it is not entirely that, or the adsorption would not be so selective. There are four factors that influence the tendency of a colloid to adsorb one ion in preference to another.

PANETH-FAJANS-HAHN LAW.  When two or more types of ions are available for adsorption and when other factors are equal, that ion which forms a compound with the lowest solubility with one of the lattice ions will be adsorbed preferentially. For example, silver chloride adsorbs silver ions in preference to sodium ions because silver chloride is less soluble than sodium chloride. Silver chloride adsorbs silver ion, a cation, in preference to nitrate ion, an anion, because silver ion forms a less soluble compound with the lattice anion than nitrate does with the lattice cation. Above all, if either lattice ion is present in excess in the mother liquid, it will be adsorbed preferentially.

CONCENTRATION EFFECT.  Other factors being equal, the ion present in greater concentration will be adsorbed preferentially. Furthermore, the quantity of any ion which is adsorbed varies directly with its concentration.

IONIC CHARGE EFFECT.  Other factors being equal, a multicharged ion will be adsorbed more readily than a singly charged ion, because the strength of adsorption is governed in part by the electrostatic attraction between the ion and the oppositely charged centers on the crystal surface.

SIZE OF ION.  Other factors being equal, the ion which is more nearly the same size as the lattice ion which it replaces will be adsorbed preferentially. For example, radium ion is adsorbed tightly onto barium sulfate but not onto calcium sulfate; the radium ion is close to the size of a barium ion but is considerably larger than a calcium ion.

These four factors operate simultaneously in each particular situation. Although the first factor is very often dominant, any of the four may predominate.

*Coagulation and peptization.*  The process whereby colloidal particles agglomerate to form larger particles which can settle to the bottom of the container is called **coagulation**. In order for colloidal particles to coagulate, the electrical charge imparted to their surfaces by the adsorbed ion layer must be either removed or neutralized. Removal of this charge is unlikely, because the forces of attraction between the surface and the adsorbed ion layer are quite intense. However, the net charge on the particles can be modified by an adsorbed layer of water and by a second layer of adsorbed ions.

In the lower portion of Figure 7–2 is a row of alternating silver and chloride ions on the surface of a particle of silver chloride. If an excess of silver ions is present in the mother liquid, the primary adsorbed layer will consist of silver ions as shown, and the surface will acquire a net positive charge. Water molecules are polar, so water molecules may be adsorbed onto the primary layer on the particle surface. In addition, there is a tendency for the adsorption of a second layer of ions, called the counter-ion layer, which in Figure 7–2 consists of nitrate ions.

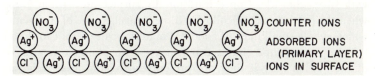

*Figure 7–2.*  Schematic representation of adsorption of nitrate counter ions onto a primary adsorbed layer of silver ions at the surface of a silver chloride crystal.

In large part, the tightness with which counter ions are held in and with the water layer, or the completeness with which they cover the primary adsorbed ion layer, determines the stability of the colloidal dispersion. If this secondary layer is sufficient to neutralize the charge due to the primary adsorbed ion layer, the particles coagulate rather than repel each other.

What factors influence the tightness and completeness of the counter-ion layer? Adsorption of the counter-ion layer is much less selective than is adsorption of the primary layer. Of much less significance are the Paneth-Fajans-Hahn factor and the ionic size factor because the counter-ion layer is more remote from the body of the crystal lattice and because of the presence of adsorbed water molecules. Generally, the concentration factor is most significant in determining both the selectivity and the overall extent of counter-ion adsorption—the greater the ionic content of the mother liquid, the more completely will the secondary layer neutralize the charge imparted to the surface of the particles by the primary layer.

Coagulation of colloidal particles through counter-ion adsorption is a reversible process. The process whereby coagulated particles pass back into the colloidal state is designated **peptization**. Special precaution must be taken during the washing of a coagulated precipitate in a gravimetric determination to prevent the agglomerates from peptizing and passing through the filter. When coagulation is accomplished through charge neutralization, as it is in the precipitation of silver chloride, peptization would occur if the precipitate were washed with pure water. Instead the wash liquid must contain an electrolyte, such as nitric acid, which will be volatile upon subsequent drying or ignition of the precipitate, lest it contribute to lack of purity of the precipitate at the time of weighing.

## Coprecipitation

Coprecipitation is the precipitation of an otherwise soluble substance along with an insoluble precipitate. These two substances may precipitate simultaneously or one may follow the other. Although it is frequently possible to precipitate one component under such conditions that all other components would be predicted to remain completely in solution, coprecipitation always occurs to some extent. Coprecipitation of impurities may be insignificant, even in highly accurate work, as with a properly prepared precipitate of silver chloride, or may be relatively serious, as with hydrous ferric oxide.

There are several mechanisms whereby coprecipitation can occur: surface adsorption, occlusion, post-precipitation, and mechanical entrapment.

*Surface adsorption.* As already described, ions are adsorbed from the mother liquid onto the surfaces of precipitated particles. This adsorption involves a primary adsorbed ion layer, which is held very tightly, and a counter-ion layer, which is held more or less loosely. These ions are carried down with the precipitate, causing it to be impure. In the example illustrated by Figure 7–2, silver nitrate is coprecipitated along with silver chloride. Coprecipitation by surface adsorption is especially significant when the particles are of colloidal dimensions because of the tremendous surface area which even a small mass of colloidal material can present to the mother liquid.

Impurities coprecipitated by surface adsorption cause an error in a gravimetric determination only if they are present during the final weighing of a precipitate. During washing it is sometimes possible to replace initially adsorbed ions with species that will be subsequently volatile in the drying or ignition of the precipitate. Preferably, one or more of the following steps should be taken at the time of precipitation to minimize surface adsorption, except insofar as it is needed to cause the precipitate to coagulate:

1. Ensure that the solution in which the precipitate forms is dilute with respect to all foreign ions.

2. Form the precipitate so that large crystals are obtained by stirring the reagent solutions as they are mixed and by using dilute solutions.

3. Precipitate the substance of interest from a hot solution to increase the solubility of all components, to decrease the tendency toward momentary supersaturation and formation of colloidal particles, and to decrease the selective attractive forces upon which the Paneth-Fajans-Hahn law is based.

4. Replace foreign ions, which form relatively insoluble compounds with the ions of the precipitate, by other ions forming more readily soluble compounds prior to precipitation.

5. Remove from the solution, or convert to forms of lower charge, highly charged ions of substances which show a tendency to coprecipitate.

6. Choose a precipitate of the desired ion such that no other ions in the solution are of the same size as any lattice ion.

Digestion of a precipitate serves to minimize surface adsorption if the individual crystals undergo recrystallization to form larger, and fewer, crystals with correspondingly less surface area. However, digestion may prevent subsequent removal of adsorbed ions if the individual particles coagulate to form tightly packed, nonporous aggregates of the tinier particles.

***Mechanical entrapment.*** Mechanical entrapment is the simple physical enclosure of a small portion of the mother liquid within tiny hollows or flaws which form during the rapid growth and coalescence of the crystals. These pockets remain filled with the mother liquid and eventually become completely enclosed by the precipitate. In a typical case, 0.1 to 0.2 per cent of a precipitate formed from solution may consist of the mother liquid from which it was separated.

Ordinary washing is of no aid in removing entrapped material. When a precipitate is ignited at high temperature, the internal pressure in the pockets may rupture the particles with resultant release of the trapped solvent. However, any nonvolatile solutes present in the trapped mother liquid remain as impurities in the precipitate.

Measures which can be taken to minimize mechanical entrapment include the following:

1. Keep the solution dilute with respect to all components so that the mother liquid which is trapped will not contain much solute.

2. Perform the precipitation under conditions that promote slow growth of crystals, thus minimizing the formation of flaws and voids within the crystals and crystal aggregates.

3. Perform the precipitation under conditions in which the precipitate has appreciable solubility so that smaller particles dissolve and reprecipitate on larger ones.

***Post-precipitation.*** Another type of precipitate contamination closely associated with surface adsorption is post-precipitation. It may best be described by an example, namely, the separation of calcium ion from magnesium ion by precipitation with oxalate. Calcium oxalate is a moderately insoluble compound which may be precipitated quantitatively. Since it tends to precipitate slowly, it is permitted to remain in contact with the mother liquid for some time prior to filtration. Magnesium oxalate is too soluble to precipitate by itself under ordinary conditions. However, if calcium oxalate is precipitated from a solution containing magnesium ions and if the precipitate is allowed to remain in contact with the mother liquid for an excessive time, magnesium oxalate coprecipitates. Apparently, oxalate ion, present in excess in the solution, comprises the primary adsorbed ion layer. This effectively produces a relatively high concentration of oxalate ion localized on the calcium oxalate surface, even to the extent of providing a local state of supersaturation with respect to magnesium oxalate, so that precipitation of some magnesium oxalate ensues.

Post-precipitation is not uncommon, yet it seldom imparts significant errors to the final results of a gravimetric determination. It may be minimized if one brings the desired precipitate to a filterable condition as soon as possible after its first formation.

**Occlusion.** In the process of occlusion, one ion within a crystal is replaced in the crystal lattice by another ion of similar size and structure. Through this mechanism, the impurity becomes incorporated permanently into the crystal lattice, and it cannot be removed by washing. Thus, the only way to eliminate this type of contamination is to remove the offending ion prior to precipitation of the desired compound or to dissolve the precipitate and reform it under more favorable conditions.

Experimentally, it has been found that potassium ion coprecipitates markedly with barium sulfate, whereas sodium ion does not. This is clearly an example of contamination by occlusion, because the radii of potassium ion (1.33 Å) and barium ion (1.35 Å) are almost identical, but sodium ion has a much smaller radius (0.95 Å). Furthermore, replacement of one barium ion by one potassium ion must be accompanied by some other modification of the crystal lattice in order to maintain electrical neutrality. Space considerations would not permit two potassium ions to occupy the lattice site of a single barium ion. However, a hydrogen sulfate ion ($HSO_4^-$) is similar spatially to a sulfate ion, so electrical neutrality is preserved if a hydrogen sulfate anion replaces a sulfate ion every time a potassium ion replaces a barium ion. That this must occur is indicated by the observation that coprecipitation of potassium ion is almost negligible at pH 5, where there are very few hydrogen sulfate ions; but, coprecipitation of potassium ions is appreciable at pH 1, where most of the sulfate does exist as $HSO_4^-$.

## PRECIPITATION FROM HOMOGENEOUS SOLUTION

It has been shown that the supersaturation ratio at the time and place of nucleation plays a major role in determining the particle size of a precipitate. Supersaturation should be held to a minimum in order to obtain a precipitate in the form of relatively large individual crystals. To aid in the accomplishment of this goal, the method of **precipitation from homogeneous solution** has been developed. In this method, the precipitating agent is not added directly but rather is generated slowly by means of a homogeneous chemical reaction within the solution at a rate comparable to the rate of crystal growth. Thus, the supersaturation ratio does not reach as high a value as would exist if the two reagent solutions were simply mixed directly.

This homogeneous precipitation technique is applicable to any precipitation process in which the necessary reagent can be generated slowly by some chemical reaction occurring within the solution of the unknown. A few selected examples will be mentioned here.

**Precipitation by means of urea hydrolysis.** Urea is an especially useful reagent for the homogeneous precipitation or separation of any substance whose solubility is affected by pH. It hydrolyzes slowly in hot aqueous solutions according to the reaction

$$(NH_2)_2CO + H_2O \rightarrow CO_2 + 2\,NH_3$$

Slow generation of ammonia within an initially acidic or neutral solution serves to raise the pH gradually and uniformly. By adjusting the initial concentration of urea and the pH of the sample solution, and by controlling the temperature at which the solution is heated, one can obtain any desired final pH value or rate of increase of pH.

One of the classic gravimetric procedures entails the precipitation of aluminum, chromium, or iron hydrous oxide or hydroxide by the addition of aqueous ammonia to a solution containing one of these species. However, the voluminous and gelatinous character of the resulting precipitate causes numerous complications including difficulty in the filtering and washing steps and serious coprecipitation of other cations and anions. These problems are largely overcome if the homogeneous precipitation technique is employed. Thus, for example, the pH of an aluminum ion solution can be adjusted to a value at which hydrous aluminum oxide is soluble, an appropriate quantity of urea added, and the solution heated to hydrolyze the urea and to raise the pH so that hydrous aluminum oxide will precipitate quantitatively. This precipitate possesses more desirable physical characteristics—high density and crystallinity—than that obtained by the direct mixing of reagents. Upon subsequent ignition, the hydrous aluminum oxide is converted into aluminum oxide, $Al_2O_3$, an excellent weighing form for the gravimetric determination of aluminum.

*Homogeneous generation of sulfate and phosphate ions.* Sulfate ion can be generated homogeneously if one heats a solution containing sulfamic acid

$$HSO_3NH_2 + H_2O \rightarrow H^+ + SO_4^{2-} + NH_4^+$$

or if one causes the slow hydrolysis of diethyl or dimethyl sulfate:

$$(C_2H_5)_2SO_4 + 2\,H_2O \rightarrow 2\,H^+ + SO_4^{2-} + 2\,C_2H_5OH$$

$$(CH_3)_2SO_4 + 2\,H_2O \rightarrow 2\,H^+ + SO_4^{2-} + 2\,CH_3OH$$

Barium ion, or any cation which forms an insoluble sulfate, can be precipitated homogeneously by means of these reactions. By using mixtures of alcohol and water as the solvent along with the homogeneous precipitation technique, one can effect sharper separations of the alkaline earth elements—barium, strontium, and calcium— than is possible by direct addition of a solution containing sulfate ion.

Phosphate ion can be generated homogeneously in solution by the hydrolysis of triethyl phosphate or trimethyl phosphate. This procedure is valuable in the separation and determination of zirconium and a few other elements.

*Precipitation of sulfides.* Many metal ions form insoluble sulfides, the solubilities of which are very much influenced by pH because hydrogen sulfide is an extremely weak acid. When sulfides are precipitated by the bubbling of hydrogen sulfide gas into a solution, the resulting precipitates have undesirable physical characteristics, not to mention the unpleasant and toxic nature of hydrogen sulfide itself. Metal sulfides can be precipitated homogeneously by means of the slow acid-or-base-catalyzed hydrolysis of thioacetamide:

$$CH_3CSNH_2 + H_2O \rightarrow CH_3CONH_2 + H_2S$$

This homogeneous precipitation method yields metal sulfides which are distinctly granular and much easier to handle in subsequent washing and filtering operations. However, application of this method of generating hydrogen sulfide to analytical procedures is complicated by the fact that some metal ions react directly with thioacetamide to form metal sulfides. In addition to simple hydrolysis, thioacetamide can undergo specific sulfide-producing reactions with substances such as ammonia, carbonate, and hydrazine. Nevertheless, thioacetamide is widely used in the qualitative identification of cations, and several quantitative procedures have been developed.

***Method of synthesis of precipitant.*** Nickel ion may be precipitated quantitatively as bis(dimethylglyoximato)nickel(II) by direct addition of dimethylglyoxime to a solution having a pH between 5 and 9:

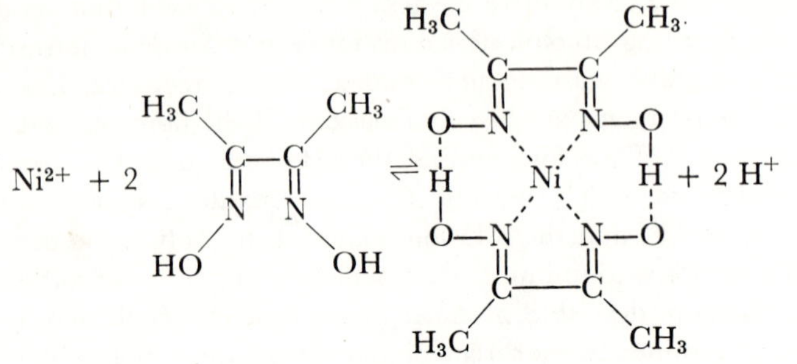


If one replaces the direct addition of dimethylglyoxime with the in situ synthesis of this precipitant, the advantages of precipitation from homogeneous solution can be realized. In practice, the desired synthesis is accomplished by reaction of biacetyl with hydroxylamine, which proceeds in two steps:

(1)

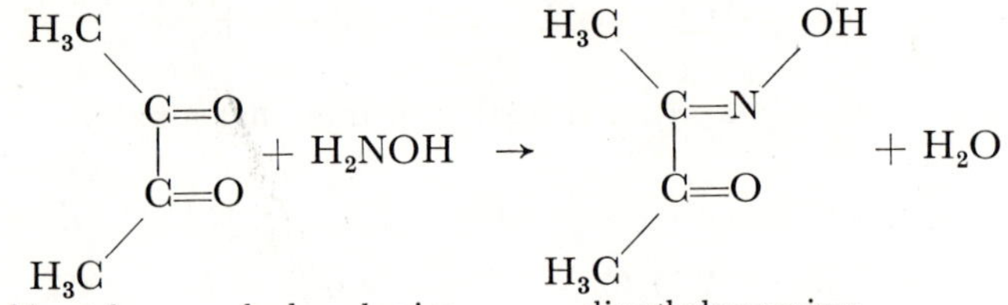


biacetyl hydroxylamine dimethylmonoxime

(2)

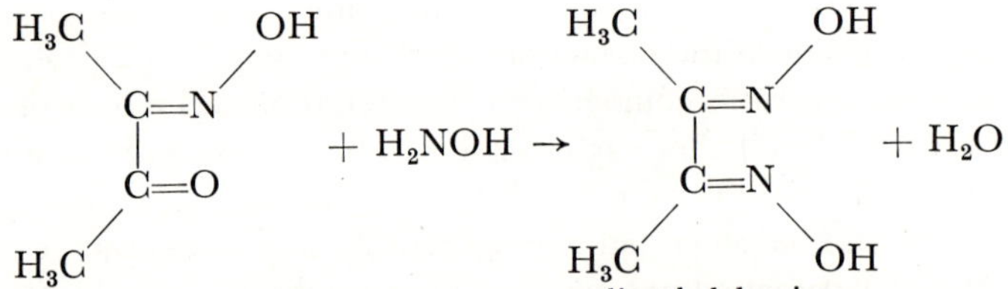


dimethylglyoxime

Thus, dimethylglyoxime is synthesized homogeneously throughout the solution in the presence of nickel ion, and the resulting precipitate does consist of crystals which are appreciably larger and easier to filter than those obtained by the direct addition of the precipitant.

***Nucleation in precipitation from homogeneous solution.*** Ideally, a precipitate formed within a homogeneous solution phase ought to consist of large crystals which can be handled easily in subsequent operations and which possess relatively little surface area for the adsorption of impurities. Although this is often the case, the particle size frequently falls short of what is desired or expected. To understand the reason why this is so, consider the precipitation of barium sulfate by the hydrolysis of sulfamic acid in a solution in which the initial barium ion concentration is 0.01 *M*. Since the solubility product of barium sulfate is $1.08 \times 10^{-10}$, a sulfate ion concentration of $1.08 \times 10^{-8}$ *M* would be sufficient to saturate the solution with barium sulfate. Although the hydrolysis rate of sulfamic acid at room temperature is not known accurately, approximate data indicate that only about 1 second of hydrolysis of a 1 per cent sulfamic acid solution would yield enough sulfate ion to start the precipitation. Of even more significance is the fact that reagent-grade sulfamic acid may contain up to 0.050 per cent free sulfate ion, which means that the sulfate ion concentration in a 1 per cent solution of sulfamic acid is $5 \times 10^{-5} M$, or 5000 times greater than that required to equal the solubility product without any hydrolysis whatsoever. It appears that the initial stage of precipitation, during which all nucleation occurs, is not one of homogeneous precipitation at all but rather one of direct mixing of the barium solution with a solution of sulfamic acid containing sulfate ion. Only the subsequent stages of crystal growth proceed by homogeneous reaction.

Use of reagents other than sulfamic acid presents different situations, but the conclusions are probably the same for most so-called homogeneous precipitation processes. Nevertheless, in spite of these considerations, the concentration of precipitant at the time precipitation starts is much smaller in the homogeneous type of procedure than in the usual direct mixing procedures. Thus, the use of the homogeneous method does result in less supersaturation at the time of nucleation, and thereby in improved particle size, even though the ideal of a completely homogeneous precipitation process is not attained.

## QUESTIONS AND PROBLEMS

1. Distinguish between spontaneous and induced nucleation, and state the relation between each and the supersaturation ratio.
2. Explain why most inorganic substances are more soluble in water than in an organic liquid such as benzene.
3. Why is an indefinitely large excess of a common ion not advisable in a precipitation process?
4. Define or characterize the following terms: colloid, supersaturation, coagulation, peptization, counter ions, coprecipitation, occlusion, Brownian movement, nucleation, crystal growth, homogeneous precipitation, post-precipitation, surface adsorption, mechanical entrapment.
5. To illustrate the importance of surface area in colloids, calculate the surface area of each of the following systems:
   (a) a cube 1 cm on a side
   (b) the cube of (a) cut into eight cubes by being sliced in half in three directions at right angles to each other
   (c) the cubes of (b), each sliced further into eight cubes in the same way as before
6. Explain the role of the counter-ion layer in determining the stability of a colloidal dispersion.
7. Why are the forces which hold counter ions less selective than those which hold primary adsorbed ions?
8. Formulate solubility-product expressions for $CaF_2$, $Fe(OH)_3$, $Ca_3(PO_4)_2$, $MgNH_4PO_4$, and $Ag_2CrO_4$.
9. Two compounds with hypothetical formulas $A(OH)_2$ and $B(OH)_3$ have solubility products of the same numerical value. Which compound is more soluble in a $0.1 F$ sodium hydroxide medium?
10. If the solubility product for silver chloride is $1.78 \times 10^{-10}$ and the solubility product for silver chromate is $2.45 \times 10^{-12}$, which compound is more soluble in water?
11. With the aid of chemical reactions, predict and explain the effect of
    (a) the magnesium ion concentration upon the solubility of magnesium ammonium phosphate, $MgNH_4PO_4$
    (b) the hydrogen ion concentration upon the solubility of calcium carbonate, $CaCO_3$
    (c) the thiosulfate ion concentration upon the solubility of silver chromate, $Ag_2CrO_4$

12. Given the following solubility data in water, evaluate the solubility-product constant for each of the compounds:
    (a) TlCl, 0.32 gram per 100 ml
    (b) $Pb(IO_3)_2$, $3.98 \times 10^{-5}$ mole per liter
    (c) AgI, $1.40 \times 10^{-6}$ gram per 500 ml
    (d) $Mg(OH)_2$, 0.0085 gram per liter
13. From the solubility-product constants tabulated in Appendix 1, calculate the solubility of each of the following compounds in formula weights per liter:
    (a) CdS, (b) $BaF_2$, (c) $BiI_3$, (d) $Cu(IO_3)_2$, (e) $SrSO_4$.
14. Which solution contains a higher concentration of silver ion, pure water saturated with silver iodate or pure water saturated with silver chromate?
15. What is the maximum concentration of calcium ion that can be present in one liter of solution containing 3.00 moles of fluoride ion?
16. What hydroxide ion concentration is needed just to begin precipitation of magnesium hydroxide from a 0.01 $F$ solution of magnesium sulfate?
17. If 25 mg of magnesium chloride is dissolved in a solution formed by dilution of 10 ml of 0.10 $F$ sodium hydroxide solution to 1.0 liter, will a precipitate of magnesium hydroxide appear?
18. To 150 ml of a solution containing 0.50 gram of $Na_2SO_4$ is added 50 ml of a solution containing 1.00 gram of $BaCl_2$. Calculate the weight in milligrams of sulfate ion left unprecipitated.
19. Concentrated potassium iodide (KI) solution is slowly added to a solution that is initially 0.02 $M$ in lead ion and 0.02 $M$ in silver ion. Which cation will precipitate first? What will be its concentration when the second cation starts to be precipitated? Neglect dilution of the sample solution by the potassium iodide medium.
20. Given the following equilibrium data

$$Ag^+ + 2 NH_3 \rightleftharpoons Ag(NH_3)_2{}^+; \qquad \beta_2 = 1.62 \times 10^7$$

$$AgBr(s) \rightleftharpoons Ag^+ + Br^-; \qquad K_{sp} = 5.25 \times 10^{-13}$$

calculate the concentration of ammonia required to prevent precipitation of silver bromide from a solution which is 0.025 $M$ in bromide and 0.045 $M$ in $Ag(NH_3)_2{}^+$.
21. A solution is saturated with respect to a compound of the general formula $AB_2C_3$:

$$AB_2C_3(s) \rightleftharpoons A^+ + 2 B^+ + 3 C^-$$

If this solution is found to contain ion $C^-$ at a concentration of 0.003 $M$, calculate the solubility product of the compound $AB_2C_3$.
22. At 25°C the solubility of barium sulfate ($BaSO_4$) in pure water is $0.96 \times 10^{-5}$ $F$. Using the Debye-Hückel limiting law, calculate the solubility of barium sulfate in a 0.030 $F$ potassium nitrate ($KNO_3$) solution.
23. (a) In 1 liter of a 1.0 $F$ hydrochloric acid solution, $1.3 \times 10^{-3}$ formula weight of lead sulfate dissolves. Calculate the solubility of lead sulfate in pure water.
    (b) Let the solubility of lead sulfate in pure water be $S_0$ (formula weights per liter). Suppose that solid sodium nitrate is dissolved in a saturated lead sulfate solution in equilibrium with excess solid lead sulfate in such a way that the ionic strength of the solution is increased until it becomes saturated with sodium nitrate. Construct a qualitative plot to show how the solubility of lead sulfate varies with ionic strength. How do you interpret this plot?
24. Predict qualitatively and explain briefly how an *increase* in ionic strength affects the following:
    (a) the *solubility* of silver chloride, according to the equilibrium

$$AgCl(s) \rightleftharpoons Ag^+ + Cl^-$$

    (b) the *dissociation* of $AgCl_2{}^-$, according to the equilibrium

$$AgCl_2{}^- \rightleftharpoons AgCl(aq) + Cl^-$$

(c) the *dissociation* of $AgCl_3^{2-}$, according to the equilibrium

$$AgCl_3^{2-} \rightleftharpoons Ag^+ + 3\ Cl^-$$

25. Calculate the solubility of silver iodide (AgI) in a solution containing 0.01 $F$ sodium thiosulfate ($Na_2S_2O_3$) and 0.01 $F$ potassium iodide (KI).

$$
\begin{aligned}
AgI(s) &\rightleftharpoons Ag^+ + I^-; & K_{sp} &= 8.3 \times 10^{-17} \\
AgI(s) &\rightleftharpoons AgI(aq); & S^\circ &= 6.0 \times 10^{-9} \\
Ag^+ + S_2O_3^{2-} &\rightleftharpoons AgS_2O_3^-; & K_1 &= 6.6 \times 10^8 \\
AgS_2O_3^- + S_2O_3^{2-} &\rightleftharpoons Ag(S_2O_3)_2^{3-}; & K_2 &= 4.4 \times 10^4 \\
Ag(S_2O_3)_2^{3-} + S_2O_3^{2-} &\rightleftharpoons Ag(S_2O_3)_3^{5-}; & K_3 &= 4.9
\end{aligned}
$$

26. If the concentration of rubidium ion, $Rb^+$, is decreased by a factor of 111 when 3.00 moles of lithium perchlorate, $LiClO_4$, is added to 1 liter of a saturated rubidium perchlorate solution, what is the solubility product of $RbClO_4$?

27. Calculate the solubility of cuprous iodide in a 0.35 $F$ ammonia solution.

$$
\begin{aligned}
CuI(s) &\rightleftharpoons Cu^+ + I^-; & K_{sp} &= 5.1 \times 10^{-12} \\
Cu^+ + NH_3 &\rightleftharpoons CuNH_3^+; & K_1 &= 8.5 \times 10^5 \\
CuNH_3^+ + NH_3 &\rightleftharpoons Cu(NH_3)_2^+; & K_2 &= 8.5 \times 10^4
\end{aligned}
$$

28. Calcium is frequently determined through its precipitation as $CaC_2O_4 \cdot H_2O$, the dissolution of the precipitate in an acid medium, and the subsequent determination of oxalate by titration with standard potassium permanganate. Suppose that 80 mg of calcium is to be determined by means of this procedure and that for the precipitation of $CaC_2O_4 \cdot H_2O$ the following conditions are to be established:
(a) volume of solution is 100 ml
(b) maximal weight of dissolved calcium ion in solution at equilibrium is 0.80 $\mu g$
(c) the pH of the solution is controlled with an acetic acid-sodium acetate buffer at 4.70.
In order to meet the desired conditions for this precipitation (so that the maximal amount of dissolved calcium ion is not exceeded), what must be the *total concentration* of *all* oxalate species in the solution?

$$
\begin{aligned}
CaC_2O_4(s) &\rightleftharpoons Ca^{2+} + C_2O_4^{2-}; & K_{sp} &= 2.6 \times 10^{-9} \\
CH_3COOH &\rightleftharpoons H^+ + CH_3COO^-; & K_a &= 1.8 \times 10^{-5} \\
H_2C_2O_4 &\rightleftharpoons H^+ + HC_2O_4^-; & K_{a1} &= 6.5 \times 10^{-2} \\
HC_2O_4^- &\rightleftharpoons H^+ + C_2O_4^{2-}; & K_{a2} &= 6.1 \times 10^{-5}
\end{aligned}
$$

29. Calculate the solubility of barium chromate in an acetic acid-sodium acetate buffer of pH 4.26.

$$
\begin{aligned}
BaCrO_4(s) &\rightleftharpoons Ba^{2+} + CrO_4^{2-}; & K_{sp} &= 2.4 \times 10^{-10} \\
HCrO_4^- &\rightleftharpoons H^+ + CrO_4^{2-}; & K_{a2} &= 3.2 \times 10^{-7} \\
2\ HCrO_4^- &\rightleftharpoons Cr_2O_7^{2-} + H_2O; & K &= 33
\end{aligned}
$$

30. Excess solid barium sulfate was shaken with 3.6 $F$ hydrochloric acid until equilibrium was attained. Calculate the equilibrium solubility of barium sulfate and the concentrations of $SO_4^{2-}$ and $HSO_4^-$.

$$
\begin{aligned}
BaSO_4(s) &\rightleftharpoons Ba^{2+} + SO_4^{2-}; & K_{sp} &= 1.1 \times 10^{-10} \\
H_2SO_4 &\rightleftharpoons H^+ + HSO_4^-; & K_{a1} &\gg 1 \\
HSO_4^- &\rightleftharpoons H^+ + SO_4^{2-}; & K_{a2} &= 1.2 \times 10^{-2}
\end{aligned}
$$

31. If 1 liter of a saturated solution of $AgIO_3$ (no solid present) is equilibrated with $3.5 \times 10^{-5}$ mole of solid $Pb(IO_3)_2$, what will be the equilibrium concentrations of silver, lead, and iodate ions, and how many milligrams of $AgIO_3$, if any, will precipitate?

$$AgIO_3(s) \rightleftharpoons Ag^+ + IO_3^-; \qquad K_{sp} = 3.0 \times 10^{-8}$$
$$Pb(IO_3)_2(s) \rightleftharpoons Pb^{2+} + 2\ IO_3^-; \qquad K_{sp} = 2.6 \times 10^{-13}$$

32. Calculate the solubility of zinc hydroxide, $Zn(OH)_2(s)$, in a $0.1\ F$ sodium hydroxide solution, given the following equilibria:

$$Zn(OH)_2(s) \rightleftharpoons Zn^{2+} + 2\ OH^-; K_{sp} = 1.2 \times 10^{-17}$$
$$Zn(OH)_2(s) \rightleftharpoons Zn(OH)_2(aq); \qquad S^\circ = 10^{-6}$$
$$Zn(OH)_2(s) \rightleftharpoons ZnOH^+ + OH^-; \quad K = 2.2 \times 10^{-13}$$
$$Zn(OH)_2(s) + OH^- \rightleftharpoons Zn(OH)_3^-; \qquad K' = 2.0 \times 10^{-3}$$
$$Zn(OH)_2(s) + 2\ OH^- \rightleftharpoons Zn(OH)_4^{2-}; \qquad K'' = 3.6 \times 10^{-2}$$

33. Equilibria involving the solubilities of lead sulfate and strontium sulfate and the dissociation of sulfuric acid are as follows:

$$PbSO_4(s) \rightleftharpoons Pb^{2+} + SO_4^{2-}; \qquad K_{sp} = 1.6 \times 10^{-8}$$
$$SrSO_4(s) \rightleftharpoons Sr^{2+} + SO_4^{2-}; \qquad K_{sp} = 3.8 \times 10^{-7}$$
$$H_2SO_4 \rightleftharpoons H^+ + HSO_4^-; \qquad K_{a1} \gg 1$$
$$HSO_4^- \rightleftharpoons H^+ + SO_4^{2-}; \qquad K_{a2} = 1.2 \times 10^{-2}$$

If an excess of both lead sulfate and strontium sulfate is shaken with a $0.60\ F$ nitric acid solution until equilibrium is attained, what will be the concentrations of $Pb^{2+}$, $Sr^{2+}$, $HSO_4^-$, and $SO_4^{2-}$?

34. Excess barium sulfate was shaken with $0.25\ F$ sulfuric acid until equilibrium was attained. What was the concentration of barium ion in the solution?

$$BaSO_4(s) \rightleftharpoons Ba^{2+} + SO_4^{2-}; \qquad K_{sp} = 1.1 \times 10^{-10}$$
$$H_2SO_4 \rightleftharpoons H^+ + HSO_4^-; \qquad K_{a1} \gg 1$$
$$HSO_4^- \rightleftharpoons H^+ + SO_4^{2-}; \qquad K_{a2} = 1.2 \times 10^{-2}$$

35. A saturated solution of magnesium ammonium phosphate $(MgNH_4PO_4)$ in pure water has a pH of 9.70, and the concentration of magnesium ion in the solution is $5.60 \times 10^{-4}\ M$. Evaluate the solubility product of $MgNH_4PO_4$.

$$H_3PO_4 \rightleftharpoons H^+ + H_2PO_4^-; \qquad K_{a1} = 7.5 \times 10^{-3}$$
$$H_2PO_4^- \rightleftharpoons H^+ + HPO_4^{2-}; \qquad K_{a2} = 6.2 \times 10^{-8}$$
$$HPO_4^{2-} \rightleftharpoons H^+ + PO_4^{3-}; \qquad K_{a3} = 4.8 \times 10^{-13}$$
$$NH_3 + H_2O \rightleftharpoons NH_4^+ + OH^-; \qquad K_b = 1.80 \times 10^{-5}$$
$$H_2O \rightleftharpoons H^+ + OH^-; \qquad K_w = 1.00 \times 10^{-14}$$

36. Pertinent equilibria for the nickel(II)-hydroxide system are as follows:

$$Ni^{2+} + OH^- \rightleftharpoons NiOH^+; \qquad K_1 = 1.2 \times 10^3$$
$$NiOH^+ + OH^- \rightleftharpoons Ni(OH)_2(aq); \qquad K_2 = 6.3 \times 10^8$$
$$Ni(OH)_2(aq) + OH^- \rightleftharpoons Ni(OH)_3^-; \qquad K_3 = 6.3 \times 10^2$$
$$Ni(OH)_2(s) \rightleftharpoons Ni^{2+} + 2\ OH^-; \qquad K_{sp} = 6.5 \times 10^{-18}$$

(a) Calculate the *intrinsic solubility* $(S^\circ)$ of $Ni(OH)_2(s)$.
(b) Calculate the *total* solubility of $Ni(OH)_2(s)$, if excess solid nickel hydroxide

is shaken with a well buffered solution of pH 10.00. You may assume activity coefficients of unity and that the buffering agent does not undergo any chemical reactions with $Ni(OH)_2(s)$.

(c) What is the predominant nickel(II) species in the solution in (b)?

(d) Explain why the calculation of the solubility of solid nickel hydroxide in pure water would be more difficult than the problem in (b).

37. Solubility-product data for cupric iodate, lanthanum iodate, and silver iodate are as follows:

$$Cu(IO_3)_2(s) \rightleftharpoons Cu^{2+} + 2\ IO_3^-; \qquad K_{sp} = 7.4 \times 10^{-8}$$

$$La(IO_3)_3(s) \rightleftharpoons La^{3+} + 3\ IO_3^-; \qquad K_{sp} = 6.0 \times 10^{-10}$$

$$AgIO_3(s) \rightleftharpoons Ag^+ + IO_3^-; \qquad K_{sp} = 3.0 \times 10^{-8}$$

Suppose that you prepare a separate saturated solution of each salt in pure water. Calculate the equilibrium concentration of iodate ion in each of these three solutions. Neglect activity effects.

38. If the solubility of solid barium iodate, $Ba(IO_3)_2$, in a $0.000540\ F$ potassium iodate $(KIO_3)$ solution is 0.000540 formula weight per liter, what is the solubility-product constant for $Ba(IO_3)_2$?

39. Equilibria involving the solubilities of strontium fluoride and calcium fluoride and the dissociation of hydrofluoric acid are as follows:

$$SrF_2(s) \rightleftharpoons Sr^{2+} + 2\ F^-; \qquad K_{sp} = 2.8 \times 10^{-9}$$

$$CaF_2(s) \rightleftharpoons Ca^{2+} + 2\ F^-; \qquad K_{sp} = 4.0 \times 10^{-11}$$

$$HF \rightleftharpoons H^+ + F^-; \qquad K_a = 6.7 \times 10^{-4}$$

(a) If an excess of both strontium fluoride and calcium fluoride is shaken with an innocuous buffer of pH 3.00 until equilibrium is attained, what will be the concentrations of $Sr^{2+}$ and $Ca^{2+}$?

(b) Let the solubility of calcium fluoride (in formula weights per liter) be $S_0$. Suppose that solid potassium nitrate is dissolved in a saturated calcium fluoride solution in equilibrium with excess solid calcium fluoride in such a way that the ionic strength of the solution is slowly increased. Construct a qualitative plot to show how the solubility of calcium fluoride varies with ionic strength. How do you interpret this plot? How do changes in ionic strength affect $K_{sp}$ for calcium fluoride and $K_a$ for hydrofluoric acid?

40. The solubility equilibrium and the solubility-product constant for barium oxalate are as follows:

$$BaC_2O_4(s) \rightleftharpoons Ba^{2+} + C_2O_4^{2-}; \qquad K_{sp} = 2.3 \times 10^{-8}$$

In addition, oxalic acid is a moderately weak diprotic acid which dissociates according to the following equilibria:

$$H_2C_2O_4 \rightleftharpoons H^+ + HC_2O_4^-; \qquad K_{a1} = 6.5 \times 10^{-2}$$

$$HC_2O_4^- \rightleftharpoons H^+ + C_2O_4^{2-}; \qquad K_{a2} = 6.1 \times 10^{-5}$$

Calculate the solubility of barium oxalate in an aqueous solution whose pH is maintained constant at exactly 2.00 by means of an innocuous buffer.

41. Solubility-product data for silver chloride and silver chromate are as follows:

$$AgCl(s) \rightleftharpoons Ag^+ + Cl^-; \qquad K_{sp} = 1.78 \times 10^{-10}$$

$$Ag_2CrO_4(s) \rightleftharpoons 2\ Ag^+ + CrO_4^{2-}; \qquad K_{sp} = 2.45 \times 10^{-12}$$

(a) Suppose that you have an aqueous solution containing $0.0020\ M\ CrO_4^{2-}$ and $0.000010\ M\ Cl^-$. If *concentrated* silver nitrate solution (so that volume changes may be neglected) is added gradually with good stirring to this solution, which precipitate $(Ag_2CrO_4$ or AgCl) will form first? Justify your conclusion with a calculation.

(b) Eventually, the silver ion concentration should increase enough to cause precipitation of the second ion. What will be the concentration of the *first* ion when the second ion just begins to precipitate?

(c) What *percentage* of the *first* ion is already precipitated when the second ion just begins to precipitate?

42. Solubility-product data for lead hydroxide and chromium hydroxide are as follows:

$$Pb(OH)_2(s) \rightleftharpoons Pb^{2+} + 2 OH^-; \qquad K_{sp} = 1.2 \times 10^{-15}$$

$$Cr(OH)_3(s) \rightleftharpoons Cr^{3+} + 3 OH^-; \qquad K_{sp} = 6.0 \times 10^{-31}$$

(a) Suppose that you have an aqueous solution of pH 2 containing 0.030 $M$ $Pb^{2+}$ and 0.020 $M$ $Cr^{3+}$. If the pH of this solution is gradually increased by the addition of a concentrated sodium hydroxide solution (so that volume changes may be neglected), which precipitate, $Pb(OH)_2$ or $Cr(OH)_3$, will form *first*? (A guess is unacceptable; justify your answer by calculations.)

(b) Calculate the pH values at which $Pb(OH)_2$ and $Cr(OH)_3$ will each start to precipitate.

(c) If it is desired to *separate* the first ion to precipitate from the second ion to precipitate by pH control, and if it is desired to have no more than 0.1 per cent of the first ion remaining *unprecipitated* when the second ion begins to precipitate, what *range* of pH values is necessary to accomplish this separation?

43. If 1 liter of a saturated solution (*no* excess solid present) of barium iodate, $Ba(IO_3)_2$, is equilibrated with $3.5 \times 10^{-5}$ mole of pure solid lead iodate, $Pb(IO_3)_2$, what will be the equilibrium concentrations of $Ba^{2+}$, $Pb^{2+}$, and $IO_3^-$?

$$Ba(IO_3)_2(s) \rightleftharpoons Ba^{2+} + 2 IO_3^-; \qquad K_{sp} = 1.5 \times 10^{-9}$$

$$Pb(IO_3)_2(s) \rightleftharpoons Pb^{2+} + 2 IO_3^-; \qquad K_{sp} = 2.6 \times 10^{-13}$$

44. An excess of pure solid calcium sulfate $(CaSO_4)$ was shaken with an aqueous solution whose pH was maintained constant at exactly 2 by means of an innocuous buffer. The solubility of calcium sulfate in the pH 2 solution was determined by measurement of the calcium ion concentration. On the basis of the information

$$[Ca^{2+}] = 1.0 \times 10^{-2} M$$

$$[H^+] = 1.0 \times 10^{-2} M$$

$$HSO_4^- \rightleftharpoons H^+ + SO_4^{2-}; \qquad K_{a2} = 1.2 \times 10^{-2}$$

(a) calculate the solubility-product constant of $CaSO_4$ in water

(b) calculate the solubility (moles per liter) of calcium sulfate in pure water

45. The solubility equilibrium and the solubility-product constant for zinc arsenate are as follows:

$$Zn_3(AsO_4)_2(s) \rightleftharpoons 3 Zn^{2+} + 2 AsO_4^{3-}; \qquad K_{sp} = 1.3 \times 10^{-28}$$

Calculate the final concentration of $AsO_4^{3-}$ required to precipitate all but 0.1 per cent of the $Zn^{2+}$ in 250 ml of a solution that originally contains 0.2 gm of $Zn(NO_3)_2$. Zinc nitrate is very soluble in water.

46. In pure water, the solubility of cerium(III) iodate, $Ce(IO_3)_3$, is 124 mg per 100 ml.

(a) Calculate the solubility-product constant for cerium(III) iodate, which dissolves according to the reaction

$$Ce(IO_3)_3(s) \rightleftharpoons Ce^{3+} + 3 IO_3^-$$

(b) Calculate the solubility (moles per liter) of cerium(III) iodate in a 0.050 $F$ potassium iodate $(KIO_3)$ solution.

47. The solubility product for silver chloride is given by

$$AgCl(s) \rightleftharpoons Ag^+ + Cl^-; \qquad K_{sp} = 1.8 \times 10^{-10}$$

and the dissociation constant for the silver-diammine complex is given by

$$Ag(NH_3)_2{}^+ \rightleftharpoons Ag^+ + 2\,NH_3; \qquad K = 6.2 \times 10^{-8}$$

Suppose that you wish to dissolve completely 0.0035 mole of solid AgCl in 200 ml of an aqueous ammonia solution. What must be the final concentration of $NH_3$ to accomplish this?

48. The solubility of silver azide, $AgN_3$, in water is $5.4 \times 10^{-5}$ mole per liter and the solubility reaction may be written as

$$AgN_3(s) \rightleftharpoons Ag^+ + N_3{}^-$$

Hydrazoic acid is a weak acid as indicated by the equilibrium

$$HN_3 \rightleftharpoons H^+ + N_3{}^-; \qquad K_a = 1.9 \times 10^{-5}$$

If excess solid silver azide is added to 1 liter of 3 $F$ nitric acid, what will be the final equilibrium concentration of $Ag^+$?

## SUGGESTIONS FOR ADDITIONAL READING

1. A. J. Bard: *Chemical Equilibrium*. Harper & Row, New York, 1966, pp. 46–62.
2. T. R. Blackburn: *Equilibrium*. Holt, Rinehart and Winston, New York, 1969, pp. 93–112.
3. J. N. Butler: *Ionic Equilibrium*. Addison-Wesley Publishing Company, Reading, Massachusetts, 1964, pp. 174–205.
4. L. Gordon, M. L. Salutsky, and H. H. Willard: *Precipitation from Homogeneous Solution*. John Wiley and Sons, New York, 1959.
5. I. M. Kolthoff, E. B. Sandell, E. J. Meehan, and S. Bruckenstein: *Quantitative Chemical Analysis*. Fourth edition, The Macmillan Company, New York, 1969, pp. 125–154, 198–247.
6. M. L. Salutsky: Precipitates: Their formation, properties, and purity. *In* I. M. Kolthoff and P. J. Elving, eds.: *Treatise on Analytical Chemistry*. Part I, Volume 1, Wiley-Interscience, New York, 1959, pp. 733–766.
7. C. L. Wilson, F. E. Beamish, W. A. E. McBryde, and L. Gordon: Inorganic gravimetric analysis. *In* C. L. Wilson and D. W. Wilson, eds.: *Comprehensive Analytical Chemistry*. Volume IA, Elsevier, New York, 1959, pp. 430–547.

# ANALYTICAL APPLICATIONS OF PRECIPITATION REACTIONS

# 8

Precipitation processes are useful in several different kinds of quantitative analyses. A gravimetric determination, which is characterized by the measurement of mass or weight, usually requires two weighings, the first being the original sample weight and the second being the final weight of a pure precipitate that contains the desired sample constituent. In a precipitation titration, the analytical result is based on the stoichiometry of the reaction between a titrant species and the substance being titrated. In still other applications, precipitation is utilized to separate one chemical species from another prior to the actual determination of the desired substance. In addition, precipitation is employed in many chemical manufacturing processes.

In this chapter, we shall examine both gravimetric and volumetric methods of analysis based on the formation of precipitates.

## GRAVIMETRIC DETERMINATIONS

To serve as the basis for a gravimetric method of analysis, a precipitation process must satisfy the three requirements stated in the preceding chapter. To review, these requirements are that the unknown species must be precipitated quantitatively, that the resulting precipitate must have a physical form suitable for subsequent handling, and that the precipitate must be pure or of known purity at the time of the final weighing.

In the following sections, we will consider gravimetric determinations that encompass a variety of precipitates, procedures, and principles. Emphasis will be placed on how procedures are designed or modified to fulfill the previously listed requirements for a successful gravimetric analysis.

## Determination of Chloride

Chloride ion in a wide variety of samples can be determined if one adds an excess of silver nitrate solution to a solution of the unknown and collects, dries, and weighs the resulting silver chloride precipitate. This gravimetric method centers on a single chemical reaction:

$$Ag^+ + Cl^- \rightleftharpoons AgCl(s)$$

Silver chloride is a curdy precipitate consisting of very small individual particles that must be coagulated into aggregates before filtration is performed.

*Completeness of precipitation.* In pure water, the solubility of silver chloride is 0.0014 gm per liter at 20°C and 0.022 gm per liter at 100°C. To avoid appreciable solubility losses, the precipitate should be washed and filtered at room temperature or below, even if it has been formed at an elevated temperature. Excess silver nitrate, present in the mother liquid and in the first washings, represses the solubility loss through the common ion effect. Even if the precipitate is washed with as much as 150 ml of cold, dilute nitric acid, the solubility loss seldom reaches 1 mg because the wash liquid does not ordinarily become saturated with silver chloride. If this loss should be intolerable because of an extremely small amount of precipitate or because of a required high accuracy, a wash liquid containing a little silver nitrate may be employed. Through the common ion effect, 150 ml of water containing 0.01 gm of silver nitrate dissolves somewhat less than 0.00001 gm of silver chloride at room temperature. A very dilute nitric acid medium is desirable to hasten and to maintain flocculation of the tiny particles of the precipitate, but the nitric acid does not markedly alter the solubility of silver chloride. Therefore, the precipitation of silver chloride may readily be made quantitatively complete, and no appreciable error need result from solubility losses.

Chloride ion is an almost ubiquitous contaminant, and great care must be exercised to ensure that no undesired chloride is introduced during the determination. In particular, tap water usually contains some chloride ion, which must be removed from all vessels used in the determination with the aid of one or two careful rinsings with distilled water.

*Purity of the precipitate.* Silver ion forms a large number of sparingly soluble salts. However, the specificity of the precipitating agent for chloride ion is much improved in the presence of a small amount of acid, such as nitric acid. Under these conditions, most of the anions of weak acids that form insoluble silver salts will not produce precipitates. For example, silver phosphate precipitates in an alkaline medium. However, in an acidic solution, phosphate ion forms $H_2PO_4^-$ and $H_3PO_4$, so that no precipitate is obtained when silver nitrate solution is added. On the other hand, bromide and iodide ions yield insoluble precipitates with silver ion even in dilute nitric acid, so these anions must be entirely absent or removed before a successful determination of chloride may be undertaken.

A silver chloride precipitate is notable because, when cold-digested, it exists as a porous spongy mass that is readily washed free of practically all common impurities with the aid of a very dilute nitric acid washing fluid.

At the time of weighing, the precipitate must be pure, so any changes in its composition subsequent to precipitation but prior to weighing are significant. One can dry the precipitate quite adequately by heating it for an hour at 110°C. A few hundredths of 1 per cent of water is held very tightly and can be removed only by fusion of the precipitate at about 450°C. However, silver chloride may be extensively decomposed upon fusion by traces of organic matter with which it has become contaminated and, furthermore, the material is slightly volatile at temperatures not much above its melting point. Since silver chloride is readily reduced when strongly heated with organic material, the precipitate should not be collected on filter paper because of the resulting necessity of charring off the paper.

Silver halides are decomposed into their constituent elements on exposure to light. Such decomposition is particularly grave during the early stages of the precipitation when the milky suspension presents an enormous surface for photochemical reaction. Therefore, the precipitation should not be conducted in bright light; and the precipitate should, insofar as reasonably possible, be kept out of sunlight, set inside a desk during digestion, and worked on in a shade-darkened room. Silver chloride is almost pure white, but the precipitate is likely to acquire a purplish hue by the time it is weighed. This is due to tiny grains of metallic silver on the surface of the white silver chloride. Even a small amount of photodecomposition results in an appreciable purplish coloration. Therefore, one should not be alarmed at the coloration if he has exercised reasonable caution to prevent excessive decomposition. This photodecomposition may cause either a positive or negative error in the determination. Initial decomposition yields free silver and free chlorine:

$$2 \; AgCl \xrightarrow{\text{light}} 2 \; Ag + Cl_2$$

If excess silver ions are present, as is true prior to the final washing, the liberated chlorine reacts with silver ion to form more silver chloride:

$$5 \; Ag^+ + 3 \; Cl_2 + 3 \; H_2O \rightarrow 5 \; AgCl + ClO_3^- + 6 \; H^+$$

Stoichiometrically, then, six molecules of silver chloride which decompose result in the presence in the final precipitate of six atoms of elemental silver plus five molecules of silver chloride. Six silver atoms plus five silver chloride molecules weigh more than the six silver chloride molecules which should have been present, so the precipitate is too heavy and the result of the determination is high. If no excess silver ions are present, as is the case after washing, the chlorine liberated through photodecomposition of the precipitate is evolved into the atmosphere, so the result of the determination is low.

**_Physical form of the precipitate._**  As already noted, silver chloride is a curdy precipitate, and the filtered material consists of porous aggregates of tiny particles. Such a flocculated precipitate is almost ideal for filtration and washing. A coarse filter is adequate, so filtration is quite rapid and there is no tendency for the pores of the filtering membrane to become clogged. Wash liquid can readily come in contact with virtually all the initially tiny particles because the larger aggregates are quite porous. However, a serious difficulty arises in that the aggregates can conceivably disperse during washing into separate, tinier particles which pass right through the filter membrane. To suppress this phenomenon, known as peptization, the precipitate must be washed with a liquid containing some electrolyte. This electrolyte must be volatile upon subsequent drying of the precipitate. Nitric acid serves very well, not only in preventing peptization but in preventing precipitation as silver salts of any other substances which were initially kept in solution by the acid medium at the time of precipitation.

*Other applications of the method.* Bromide ion may be determined by means of the same general procedure and with the same order of accuracy. Iodide determinations are possible, but the procedure is less satisfactory for three reasons— silver iodide tends to peptize more readily than silver chloride, silver iodide adsorbs certain impurities quite tenaciously, and silver iodide is extremely photosensitive. Thiocyanate and cyanide ions can also be determined in this fashion.

Several different anions of weak acids can be determined by precipitation of their silver salts from a nearly neutral medium. For a successful determination, the sample solution must contain only one kind of anion which can form an insoluble silver salt. For example, phosphate ion may be determined by precipitation and weighing of silver phosphate. In most instances, however, the weak acid anions are more conveniently determined by means of other procedures.

Several cations may be determined by a reversal of the procedure. Thus, silver may be determined quite accurately by the precipitation and weighing of either silver chloride or silver bromide. Mercury(I) and lead(II) may also be precipitated as their chlorides, but solubility losses are considerable with lead chloride. Thallous ion may likewise be determined by precipitation and weighing of thallous chloride.

## Determination of Sulfate

Sulfur is commonly determined by conversion of all the sulfur-containing material in a weighed sample to sulfate ion, after which barium sulfate is precipitated, washed, filtered, dried, and weighed. This determination centers on the reaction

$$Ba^{2+} + SO_4^{2-} \rightleftharpoons BaSO_4(s)$$

Barium sulfate is a distinctly crystalline precipitate, the individual crystals being large enough so that filtration is easy. Although some coagulation is desirable prior to filtration, the individual crystals are much larger than those in a curdy precipitate such as silver chloride.

*Completeness of precipitation.* Barium sulfate is soluble in pure water at room temperature only to the extent of about 3 mg per liter. In practice, the solubility is sharply diminished by the presence of excess barium ion in the mother liquid at the time of precipitation. Barium sulfate is only slightly more soluble at elevated temperatures. This is of particular importance because it permits the use of hot wash water for more effective removal of impurities.

Barium sulfate is markedly more soluble in acidic media than in pure water. For example, its solubility in a 1 $F$ hydrochloric acid solution is about 30 times greater than in pure water. Such behavior parallels that described in the preceding chapter (page 212) in connection with the solubility of lead sulfate in a nitric acid solution. Thus, the presence of an acid increases the solubility of barium sulfate because sulfate ion combines with a proton to form the hydrogen sulfate anion. If we wish to determine sulfate ion by adding barium ion to an acidic solution of a sample, both $H^+$ and $Ba^{2+}$ will compete for sulfate. Barium ion is the more successful competitor because barium sulfate is quite insoluble,

$$BaSO_4(s) \rightleftharpoons Ba^{2+} + SO_4^{2-}; \qquad K_{sp} = 1.08 \times 10^{-10}$$

but some sulfate remains in solution as the $HSO_4^-$ ion.

How much sulfate does remain unprecipitated under the conditions of a typical gravimetric procedure? Suppose that 500 mg of a water-soluble sulfate salt is taken and that the sample contains 40 per cent sulfate. This corresponds to 200 mg or approximately 2.1 millimoles of sulfate. In the usual procedure, the sample is dissolved in water containing 2 ml of concentrated ($12\,F$) hydrochloric acid, and then to this solution is added a 10 per cent excess of barium ion. We shall assume that 2.3 millimoles of $Ba^{2+}$ is added and that the final solution volume is 200 ml. Now let us calculate the concentrations of $SO_4^{2-}$ and $HSO_4^-$ remaining unprecipitated.

Since barium sulfate is insoluble and since an excess of barium ions is present, nearly all the 2.1 millimoles of sulfate is precipitated as $BaSO_4$, leaving essentially 0.2 millimole of barium ion in excess. Therefore, the concentration of barium ion is given by

$$[Ba^{2+}] = \frac{0.2 \text{ millimole}}{200 \text{ milliliters}} = 1.0 \times 10^{-3}\,M$$

From the solubility-product expression for barium sulfate, we can calculate the sulfate ion concentration in equilibrium with the precipitate:

$$[SO_4^{2-}] = \frac{K_{sp}}{[Ba^{2+}]} = \frac{1.08 \times 10^{-10}}{1.0 \times 10^{-3}} = 1.08 \times 10^{-7}\,M$$

This result does not provide us with the true concentration of *unprecipitated* sulfate because, as yet, we have not considered the presence of $HSO_4^-$. To do this, we must first calculate the hydrogen ion concentration by recalling that 2.0 ml of $12\,F$ hydrochloric acid is present in 200 ml of solution. Thus,

$$[H^+] = \frac{24 \text{ millimoles}}{200 \text{ milliliters}} = 0.12\,M$$

Finally, we can calculate the concentration of $HSO_4^-$ by making use of the expression for the second acid dissociation of sulfuric acid:

$$[HSO_4^-] = \frac{[H^+][SO_4^{2-}]}{K_{a2}} = \frac{(0.12)(1.08 \times 10^{-7})}{(1.2 \times 10^{-2})} = 1.08 \times 10^{-6}\,M$$

Therefore, the *total concentration of unprecipitated sulfate*, expressed as the sum of $[SO_4^{2-}]$ and $[HSO_4^-]$, is close to $1.2 \times 10^{-6}\,M$. In 200 ml of solution, this is $(200)(1.2 \times 10^{-6})$ or $2.4 \times 10^{-4}$ millimole of total sulfate, or approximately 0.024 mg. On the basis of these calculations, we conclude that the loss of total sulfate due to incomplete precipitation is negligible.

For another reason to be encountered shortly, a low pH at the time of precipitation is desirable because the purity of the barium sulfate precipitate is substantially increased.

It is worth mentioning here that the barium sulfate precipitate is finally washed with hot water. If, for predictive purposes, it is assumed that 100 ml of wash water is used and that the solubility of barium sulfate is approximately $10^{-5}\,F$, we estimate that $10^{-3}$ millimole or 0.1 mg of sulfate would be lost *if solubility equilibrium were attained*. Fortunately, solubility equilibrium is undoubtedly not attained, and, besides, there would be a considerable amount of barium ion in contact with the $BaSO_4$ during the early stages of washing. Consequently, the true solubility loss on washing should be much less than this estimated value. Nevertheless, this calculation serves to illustrate the danger of indiscriminate washing of a precipitate.

***Purity of the precipitate.*** Barium ion forms insoluble precipitates with a variety of anions other than sulfate. Yet most of them are anions of weak acids so that their barium salts are soluble in acidic media. In dilute acid solutions, only a trace of the sulfate is lost as hydrogen sulfate ion, but essentially all anions of much weaker acids are effectively removed from the scene of action through formation of their undissociated acids. Fluoride is the only anion that remains troublesome under these conditions. Barium fluoride is quite insoluble in dilute acid solutions, and so fluoride must be removed prior to precipitation of barium sulfate. Such removal may be accomplished readily through volatilization of hydrogen fluoride or through the complexation reaction between boric acid and fluoride.

Of the more common cations other than barium, only lead, calcium, and strontium form nominally insoluble sulfates. Interference from lead may be prevented if one adds acetate to complex the lead. Calcium and strontium must be removed prior to precipitation of barium sulfate.

Precipitated barium sulfate tends to retain many extraneous substances from its mother liquid, and herein lies the chief deterrent to the highly accurate determination of sulfate. Many species can and do coprecipitate with barium sulfate.

Consider first the coprecipitation of anions, such as chloride and nitrate, in a sulfate determination. Negative charges of coprecipitated anions must be compensated electrically by positive ions. Since barium ions are the most abundant cations in the medium in which the precipitate is formed, there will be coprecipitation of $BaCl_2$ or $Ba(NO_3)_2$ along with $BaSO_4$. Insofar as the determination of sulfate is concerned, these coprecipitated substances are merely extra precipitates for weighing. Therefore, when foreign anions coprecipitate with barium sulfate in a gravimetric sulfate determination, the results tend to be high.

Consider next the coprecipitation of a cation, such as ferric ion, in a sulfate determination. In this instance, the positive charge of the cation must be balanced by negative charge from an anion. As the precipitation is normally conducted, this anion is invariably the sulfate ion itself. Thus, the precipitate is partially ferric sulfate, $Fe_2(SO_4)_3$, which is decomposed upon ignition to ferric oxide, $Fe_2O_3$. Because the weight of iron associated with one sulfate ion is less than the weight of barium associated with a sulfate ion, the precipitate is too light, and the result for sulfate comes out low. Generally, coprecipitation of foreign cations leads to results that are too low in the determination of sulfate.

One fortunate aspect of these coprecipitation phenomena is clear—when both foreign cations and foreign anions are coprecipitated, the errors tend to compensate and may yield fairly accurate results. However, the balancing of positive and negative errors is never perfect, of course, unless it be by mere chance.

Barium sulfate must be ignited at a temperature of 500°C or higher to free it of water. Although barium sulfate itself is stable well above this temperature, it may be reduced by the carbon from a filter paper:

$$BaSO_4 + 4\,C \rightarrow BaS + 4\,CO$$

This possible source of error is avoided entirely if a sintered-porcelain filtering crucible is used. However, good results can be obtained with filter paper if the paper is charred off at the lowest possible temperature, if the paper does not inflame, and if there is free access to air during the ignition.

***Physical form of the precipitate.*** Barium sulfate is a crystalline precipitate so there is little danger of its passing through the filter medium. Barium sulfate often exhibits a tendency to "creep"—that is, the fine clumps of precipitate, supported on a liquid surface and moving over it through the action of surface tension, distribute themselves over the entire wetted surface of the containing vessel, even climbing up and over the walls of the container if these surfaces are wet. Thus, the particles of the precipitate can climb up a wetted filter paper onto the side of the glass funnel, if the latter is wet, and may be lost. This source of error can be suppressed if one refrains from filling the filter paper any closer than about 2 cm from the top or if one uses a filtering crucible and keeps the upper part entirely dry.

Conditions at the time of precipitation markedly influence the physical characteristics of barium sulfate. For example, particles precipitated from a relatively dilute solution are better perfected crystals than those formed from a more concentrated solution, and the crystals precipitated at low pH values are smaller but more highly perfected than those obtained at high pH. Even such a seemingly insignificant factor as the age of the barium chloride solution used to precipitate the sulfate ion can influence the size of the resultant crystals of barium sulfate.

*Other applications of the method.* By using the sulfate procedure in reverse, one may determine barium by adding an excess of sulfate ion to a solution of the unknown. Coprecipitation errors may be serious. Reasoning similar to that discussed earlier reveals that, in a barium determination, anion coprecipitation leads to low results and cation coprecipitation to high results—just opposite to the situation in sulfate determinations.

Lead may be determined by means of a similar procedure, but lead sulfate is soluble enough (about 4 mg per 100 ml of water at room temperature) to make solubility losses somewhat more significant than in barium sulfate precipitations. Strontium and calcium sulfates are nominally insoluble, but their solubilities (15 mg of $SrSO_4$ and 100 mg of $CaSO_4$, respectively, per 100 ml of water) are too great for good quantitative results. These compounds are rendered less soluble in a mixed alcohol-water solvent, but even then the determinations are not entirely satisfactory.

## Precipitation of Hydrous Ferric Oxide

Iron(III) can be precipitated from weakly acidic, neutral, or slightly alkaline solutions as hydrous ferric oxide:

$$Fe^{3+} + 3\ OH^- + n\ H_2O \rightarrow Fe(OH)_3 \cdot nH_2O$$

After the filtration and washing steps of the procedure, the precipitate can be ignited to yield anhydrous ferric oxide,

$$2\ Fe(OH)_3 \cdot nH_2O \rightarrow Fe_2O_3 + (2n + 3)\ H_2O$$

which can be weighed. At one time, the gravimetric determination of iron was widely used, but a number of easier, faster, and more accurate analytical methods for iron are now available. However, separation of iron by precipitation is still employed in conjunction with other procedures.

A precipitate of hydrous ferric oxide is gelatinous, meaning that it contains as an intimate part of itself an appreciable but indefinite amount of water. Hydrous ferric oxide is extremely insoluble in water. It is difficult to specify quantitatively just what its solubility is, because the solution in equilibrium with a precipitate of hydrous ferric oxide contains a whole family of soluble species, such as $Fe(H_2O)_6^{3+}$, $Fe(H_2O)_5OH^{2+}$, $Fe(H_2O)_4(OH)_2^{+}$, and $Fe_2(OH)_2(H_2O)_8^{4+}$. Concentrations of these species are related to each other through complicated equilibria, and in most cases the pertinent equilibrium constants are unknown and cannot be easily evaluated.

Iron(III) is complexed by a number of anions, including citrate, tartrate, thiocyanate, ethylenediaminetetraacetate, sulfate, chloride, fluoride, oxalate, cyanide, and phosphate. Therefore, the presence of any of these ions may partially or entirely prevent the precipitation of hydrous ferric oxide. This possibility of forming a complex ion is a disadvantage in the iron precipitation, but it may be an advantage in other situations. For example, aluminum or chromium(III) may be precipitated selectively as a hydroxide in the presence of complexed iron(III). Similarly, in the separation of nickel from iron, nickel(II) can be precipitated as bis(dimethylglyoximato)nickel(II) from an ammonia solution in which iron(III) is complexed with tartrate.

Nearly every metallic ion other than the alkali metal ions can form a hydroxide, hydrous oxide, or basic salt in alkaline solution, and so all may interfere in the gravimetric determination or separation of iron. Careful control of the hydroxide ion concentration often provides considerable selectivity in the precipitation of the hydroxide or hydrous oxide of one cation in the presence of another metal ion. In the case of iron(III), quantitative precipitation and separation of hydrous ferric oxide can be achieved at a pH as low as 3 or 4. This relatively low pH prevents or minimizes the precipitation of most oxides and hydroxides of dipositive metal cations, although coprecipitation may still be a problem. In practice, the pH can conveniently be controlled near a value of 4 through the use of an acetic acid-acetate buffer or a pyridine-pyridinium ion buffer. An advantage of the latter buffer is that pyridine forms stable complexes with many dipositive metal ions, a factor which decreases coprecipitation.

As already noted, hydrous ferric oxide is a gelatinous precipitate and retains within the solid phase an appreciable quantity of the mother liquid. Even though the water retained by the precipitate may be volatilized subsequently during ignition, any nonvolatile components present in the mother liquid will remain through the final weighing operation. Gelatinous precipitates are seldom pure, but retain as impurities at least some of any foreign ion present to an appreciable extent in the mother liquid. In some cases, it is beneficial to redissolve a gelatinous precipitate after filtration and then to precipitate it again; any impurities present in the mother liquid at the time of the first precipitation are present in much less concentration for the reprecipitation.

## Organic Precipitants

A large number of organic reagents react with metal cations to form insoluble precipitates suitable for gravimetric determinations. Most of these precipitants interact with metal ions to produce covalently bonded chelate compounds. Some of these substances combine with cations to yield insoluble salts. In the discussions which follow, we will examine briefly the analytical uses of three representative organic precipitants—dimethylglyoxime, 8-hydroxyquinoline (oxine), and sodium tetraphenylborate. More comprehensive summaries of the applications of organic reagents to gravimetric analysis may be found in the references listed at the end of this chapter.

*Dimethylglyoxime.* One of the most selective of all organic precipitants is dimethylglyoxime. This compound behaves as a very weak monoprotic acid in water, having a dissociation constant of only $2.2 \times 10^{-11}$ at 25°C. Since the solubility of dimethylglyoxime in water is very low, a solution of the compound in ethanol is ordinarily employed as precipitating agent, although one can use an aqueous solution of the sodium salt of dimethylglyoxime.

In weakly acidic or ammoniacal media, two dimethylglyoxime molecules combine with nickel(II) to yield the familiar scarlet-colored bis(dimethylglyoximato)-nickel(II) chelate compound:

$$Ni^{2+} + 2 \quad [dimethylglyoxime] \quad \rightleftharpoons \quad [bis(dimethylglyoximato)nickel(II)] + 2\,H^+$$

Palladium(II) forms a yellow precipitate with dimethylglyoxime in dilute acid solutions, and bismuth(III) reacts with dimethylglyoxime at approximately pH 11 to yield an insoluble compound. However, these are the only species that are precipitated by dimethylglyoxime. Therefore, it has become a routine procedure to separate and determine nickel(II) through the use of dimethylglyoxime as a precipitant in solutions with pH values between approximately 5 and 10.

This method has been applied extensively to the determination of nickel in steel. After a weighed sample of the steel is dissolved in a mixture of hydrochloric and nitric acids, a complexing agent (tartaric acid) is introduced into the solution to prevent precipitation of the hydrous oxides of iron(III) and chromium(III) when the pH is subsequently raised. Then, a modest excess of an alcoholic solution of dimethylglyoxime is added, the solution is heated to about 60°C, and the pH is increased through addition of aqueous ammonia to obtain complete precipitation of nickel(II). If cobalt(II) and copper(II) are present, they form soluble complexes with dimethylglyoxime, so that extra reagent may be needed for quantitative precipitation of nickel(II). However, since dimethylglyoxime is not very soluble in water, too much precipitant must be avoided, because it may contaminate the nickel-(II) compound. Bis(dimethylglyoximato)nickel(II) can be conveniently separated from the solution phase by means of filtration through a previously weighed sintered-glass crucible. Finally, the precipitate is washed with cold water, and the crucible and its contents are dried in an oven at 105°C and then weighed.

Of course, palladium(II) may be determined by means of a similar procedure if the precipitation is performed in a dilute acid medium, whereas bismuth(III) can be separated as a precipitate from an alkaline solution.

Various other dioximes have been recommended for analytical use as precipitants, principally because they are more soluble in water than dimethylglyoxime. Among these compounds are methylbenzoyldioxime and 1,2-cyclohexanedionedioxime.

*8-Hydroxyquinoline (Oxine).* Whereas dimethylglyoxime has unusual selectivity for just a few species, 8-hydroxyquinoline forms insoluble chelate compounds with practically every metal in the periodic table except the alkali metal elements. Trivalent metal cations, including aluminum, bismuth, cerium, iron, scandium, and thallium, react with three oxine ligands according to the generalized equilibrium

$$M^{3+} + 3 \quad [8\text{-hydroxyquinoline}] \quad \rightleftharpoons \quad M[oxine]_3 + 3\,H^+$$

whereas dipositive cations, such as cadmium, copper, magnesium, nickel, lead, strontium, and zinc, combine with oxine in a one-to-two metal-to-oxine ratio. Oxine is amphiprotic in aqueous media because it can either gain or lose a proton, as shown by the following equilibria:

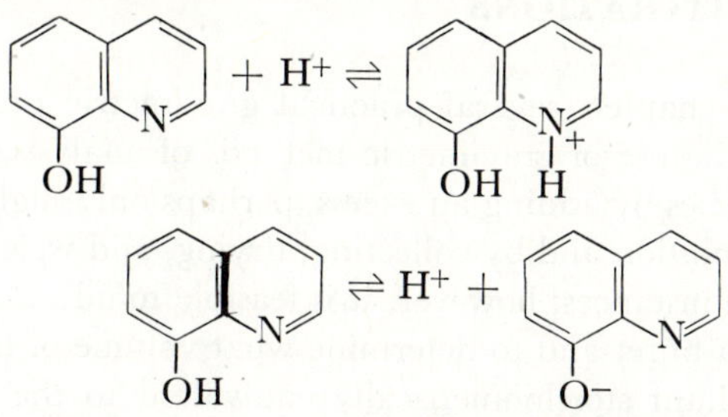

Therefore, depending on the stability of the desired metal oxinate, one can adjust the pH of the solution to achieve a certain degree of selectivity in the precipitation. However, the selectivity of the method can be increased considerably through the use of masking agents. These masking agents are other complexing species, such as cyanide, tartrate, and EDTA, which are added to a sample solution and which prevent the reaction between oxine and interfering metal ions.

As an example, in the gravimetric procedure for the determination of aluminum, many elements including cadmium, cobalt, copper, iron, magnesium, manganese, tin, and zinc do not interfere with the formation of a pure aluminum oxinate precipitate if the precipitation is conducted in the presence of ammonia, cyanide, and EDTA as masking agents. Thereafter, it is straightforward to transfer the precipitate to a previously weighed filter crucible, to wash the precipitate first with warm water containing small amounts of oxine and ammonia and then with cold water, and to heat the crucible and precipitate to constant weight at 140°C. Sometimes the aluminum oxinate can be ignited at 1200°C to yield aluminum oxide ($Al_2O_3$), which is subsequently weighed. Alternatively, the aluminum oxinate can be redissolved in an acidic solution, and the liberated oxine determined by means of a bromination procedure that is described in Chapter 10.

**Sodium tetraphenylborate.** In contrast to the two previous reagents, the precipitates formed by reaction between cations and sodium tetraphenylborate are salts rather than chelate compounds. Sodium tetraphenylborate is especially useful for the quantitative precipitation of ammonium ion and alkali metal cations (potassium, rubidium, and cesium).

For the determination of potassium ion, the sample solution is usually acidified with nitric acid or acetic acid. Then, an appropriate volume of a 3 per cent aqueous solution of sodium tetraphenylborate is added, whereupon potassium tetraphenylborate is immediately precipitated:

$$K^+ + B(C_6H_5)_4^- \rightarrow KB(C_6H_5)_4$$

This compound can be collected on a sintered-glass filter crucible, washed several times with a dilute aqueous solution of sodium tetraphenylborate, washed with cold water, and dried at approximately 100°C until constant weight is attained.

A mixture of ammonium ion and potassium ion can be analyzed. First, both cations are precipitated together as tetraphenylborate salts, and the sum of the weights of the two precipitates is determined. Second, the mixed precipitate is dissolved in acetone, sodium hydroxide solution is added, and the alkaline solution is evaporated to expel ammonia. Third, the residue is dissolved in water, and potassium tetraphenylborate alone is collected, washed, and dried. Using the two weight measurements, one can calculate the individual amounts of ammonium and potassium ions in the original sample.

## PRECIPITATION TITRATIONS

Thus far in this chapter, several practical gravimetric procedures have been examined. Through the use of gravimetric methods of analysis, one can determine a large number of species by adding an excess, perhaps only slight, of a suitable precipitant to a sample solution and by collecting, drying, and weighing the precipitate that forms. In certain instances, however, it is feasible to add a standard solution of the precipitant from a buret and to determine what volume of the solution contains an amount of precipitant stoichiometrically equivalent to the species that is precipitated. Then, without ever collecting, drying, and weighing the precipitate, one can calculate the quantity of species in the original sample solution by knowing the stoichiometry of the precipitation reaction. This latter idea is the basis of a precipitation titration.

### Requirements for Precipitation Titrations

To be suitable for a precipitation titration, a chemical reaction leading to the formation of a slightly soluble compound must satisfy three requirements. First, the rate of reaction between a precipitant and the substance to be precipitated, as well as the attainment of solubility equilibrium, must be rapid in order to permit the progress of a titration, as reflected by the changes in concentrations, to be followed accurately. However, if the rate of precipitation or attainment of solubility equilibrium is slow, the technique of back-titration can sometimes be employed. In this method, a measured excess of the precipitant is added; then, the unreacted precipitant is back-titrated with a standard solution of another reagent.

Second, the precipitation reaction should be quantitative and should proceed according to well-defined stoichiometry. This requirement poses the greatest limitation to the wide application of precipitation reactions in chemical analysis. Few precipitates of analytical interest are formed from a complex aqueous solution with the high purity and definiteness of composition necessary for a successful titration. For example, when a solution containing chloride is titrated with standard silver nitrate, chloride and silver ions must react in an exact one-to-one ratio. In this instance, the reaction stoichiometry is nicely obeyed. As another example, many metallic cations, including zinc, nickel, cobalt, manganese, aluminum, iron, chromium, lead, copper, bismuth, and cadmium, form very insoluble hydroxides. It might be expected that these elements could be determined by precipitation titration with standard sodium hydroxide solution. Unfortunately, the precipitations of the hydroxides of these metals do not proceed according to straightforward stoichiometric relationships. Instead, the hydroxides adsorb hydroxide ions and extraneous cations, and the amounts of adsorbed substances vary greatly, depending upon the temperature and the concentration and composition of the solution. In gravimetric analysis, an impure or contaminated precipitate can be dissolved and reprecipitated under conditions favoring the formation of a pure compound, but this is impractical in titrimetry.

Third, a suitable means must be available for locating or identifying the equivalence point of the titration. A wide variety of end-point detection methods have been developed, several of which will be discussed subsequently. Other methods, based on the measurement of electromotive force, current, or light absorption, are described in later chapters.

In summation, the rate of precipitation and attainment of solubility equilibrium and the stoichiometry of the reaction usually are the important factors governing the feasibility of a precipitation titration. Generally, the silver halides, silver thiocyanate, and a few mercury, lead, and zinc salts are the compounds most frequently involved in precipitation titrations.

## Titration of Chloride with Silver Ion

Perhaps the most classic of all precipitation titrations is the determination of chloride with a standard solution of silver nitrate. In the discussion to follow, we shall consider this titration in some detail by constructing the titration curve and by examining the methods used to locate the equivalence point.

*Construction of the titration curve.* A titration curve for a particular system provides a graphical representation of the variation in the concentration of the substance being determined as the titration progresses. Furthermore, a titration curve gives information about the precision with which the equivalence point of a titration can be determined as well as information which aids selection of a method for endpoint detection. Accordingly, let us construct the titration curve for the titration of 50.00 ml of $0.1000\,F$ sodium chloride solution with $0.1000\,F$ silver nitrate solution. This curve will consist of a plot of the negative base-ten logarithm of the chloride concentration (pCl) versus the volume of titrant.

BEFORE THE EQUIVALENCE POINT. Prior to introduction of any silver nitrate solution, the chloride concentration is $0.1000\,M$, so that pCl is 1.00. However, as soon as the first increment of silver nitrate solution is added, silver chloride is precipitated according to the reaction

$$Ag^+ + Cl^- \rightleftharpoons AgCl(s)$$

To calculate pCl prior to the equivalence point, we must note that the *total* chloride concentration is the *sum* of the chloride ion remaining untitrated plus the chloride ion resulting from dissolution of solid silver chloride:

$$[Cl^-] = [Cl^-]_{untitrated} + [Cl^-]_{from\ AgCl}$$

Now, the concentration of chloride remaining untitrated is given by the total millimoles of chloride originally taken minus the number of millimoles of silver nitrate added, divided by the total volume of the solution in milliliters at the particular point:

$$[Cl^-]_{untitrated} = \frac{millimoles\ Cl^-\ taken - millimoles\ Ag^+\ added}{total\ solution\ volume\ in\ milliliters}$$

To calculate the chloride concentration resulting from dissolution of solid silver chloride, let us first assume that all the added silver nitrate reacts to precipitate silver chloride. Then, we will allow silver chloride to redissolve until solubility equilibrium is established. For each silver chloride molecule that dissolves, one silver ion and one chloride ion are formed. Therefore,

$$[Cl^-]_{from\ AgCl} = [Ag^+]$$

However, the silver ion concentration is, in turn, given by the solubility product of silver chloride divided by the *total* concentration of chloride ion in the solution, which is [Cl⁻]. Thus,

$$[\text{Cl}^-]_{\text{from AgCl}} = [\text{Ag}^+] = \frac{K_{\text{sp}}}{[\text{Cl}^-]}$$

If the preceding equations are combined, we obtain a general relation for the *total* chloride concentration for any point, except the initial point, up to and including the theoretical equivalence point:

$$[\text{Cl}^-] = \frac{\text{millimoles Cl}^- \text{ taken} - \text{millimoles Ag}^+ \text{ added}}{\text{total solution volume in milliliters}} + \frac{K_{\text{sp}}}{[\text{Cl}^-]}$$

Using the above expression, let us calculate pCl values corresponding to the addition of various volumes of the silver nitrate solution. After 10.00 ml of titrant is added,

$$[\text{Cl}^-] = \frac{5.000 \text{ millimoles} - 1.000 \text{ millimole}}{60.00 \text{ milliliters}} + \frac{1.78 \times 10^{-10}}{[\text{Cl}^-]}$$

$$[\text{Cl}^-] = \frac{4.000 \text{ millimoles}}{60.00 \text{ milliliters}} + \frac{1.78 \times 10^{-10}}{[\text{Cl}^-]}$$

$$[\text{Cl}^-] = 0.0667 + \frac{1.78 \times 10^{-10}}{[\text{Cl}^-]}$$

This equation can be solved by means of the quadratic formula. However, the second term on the right-hand side of the equation can be neglected, as shown by the following reasoning. If the value of [Cl⁻] is provisionally taken to be 0.0667 $M$, this second term has a value of only 2.67 × 10⁻⁹ $M$, which is negligibly small compared to 0.0667 $M$. In other words, the value 0.0667 $M$ is an excellent approximation for [Cl⁻], and so pCl = 1.18.

This simplified approach—neglect of the second term (the contribution to [Cl⁻] due to the dissolution of silver chloride)—is valid in the present example until one is very close to the equivalence point. However, if the original chloride concentration is much smaller than 0.1 $M$, say less than 10⁻⁴ $M$, then, even at the beginning of the titration, the dissolution of silver chloride must be considered and the equation solved exactly.

After 20.00 ml of titrant is added, [Cl⁻] = 0.0429 $M$; pCl = 1.37. After 30.00 ml of titrant is added, [Cl⁻] = 0.0250 $M$; pCl = 1.60. After 40.00 ml of titrant is added, [Cl⁻] = 0.0111 $M$; pCl = 1.95. After 49.00 ml of titrant is added, [Cl⁻] = 0.00101 $M$; pCl = 3.00. After 49.90 ml of titrant is added, [Cl⁻] = 1.00 × 10⁻⁴ $M$; pCl = 4.00. Note that the dissolution of silver chloride has been neglected up to this point.

After 49.99 ml of titrant is added, neglect of the dissolution of silver chloride is no longer justifiable, and so the exact equation must be used to calculate [Cl⁻]:

$$[\text{Cl}^-] = \frac{5.000 \text{ millimoles} - 4.999 \text{ millimoles}}{99.99 \text{ milliliters}} + \frac{1.78 \times 10^{-10}}{[\text{Cl}^-]}$$

$$[\text{Cl}^-] = \frac{0.001 \text{ millimole}}{100.0 \text{ milliliters}} + \frac{1.78 \times 10^{-10}}{[\text{Cl}^-]}$$

Using the quadratic formula, one calculates that $[Cl^-] = 1.92 \times 10^{-5} M$; $pCl = 4.72$.

AT THE EQUIVALENCE POINT. When 50.00 ml of silver nitrate solution has been added (the theoretical equivalence point), the exact equation simplifies to

$$[Cl^-] = \frac{K_{sp}}{[Cl^-]}$$

so that $[Cl^-]^2 = K_{sp}$; $[Cl^-] = (K_{sp})^{1/2} = 1.33 \times 10^{-5} M$; $pCl = 4.87$.

AFTER THE EQUIVALENCE POINT. Beyond the equivalence point, an excess of silver ion is present in the solution; the *total* silver ion concentration is the *sum* of the contributions from the excess titrant and the dissolution of the silver chloride precipitate:

$$[Ag^+] = [Ag^+]_{excess} + [Ag^+]_{from\ AgCl}$$

To obtain the concentration of silver ion contributed by the excess titrant, we divide the number of millimoles of silver nitrate added beyond the equivalence point by the volume of the solution in milliliters,

$$[Ag^+]_{excess} = \frac{millimoles\ of\ AgNO_3\ added\ beyond\ equivalence\ point}{total\ volume\ of\ solution\ in\ milliliters}$$

and the silver ion concentration from the dissolution of silver chloride is exactly equal to the chloride ion concentration (since, beyond the equivalence point, dissolved silver chloride is the only source of chloride):

$$[Ag^+]_{from\ AgCl} = [Cl^-]$$

In turn, the chloride concentration, which is the quantity desired, is related to the *total* silver ion concentration through the solubility product:

$$[Ag^+]_{from\ AgCl} = [Cl^-] = \frac{K_{sp}}{[Ag^+]}$$

When these relations are combined, the following equation for the *total* silver ion concentration is obtained:

$$[Ag^+] = \frac{millimoles\ of\ AgNO_3\ added\ beyond\ equivalence\ point}{total\ volume\ of\ solution\ in\ milliliters} + \frac{K_{sp}}{[Ag^+]}$$

Once $[Ag^+]$ is calculated from this relationship, the value of $[Cl^-]$, and hence $pCl$, can be determined from the solubility-product expression. Except for points very close to the equivalence point or for the titration of very dilute solutions, the second term on the right-hand side of this last equation can be neglected.

After 60.00 ml of titrant is added, there is an excess of 1.000 millimole of silver ion in a total volume of 110.0 ml of solution:

$$[Ag^+] = \frac{1.000\ millimole}{110.0\ milliliters} = 9.09 \times 10^{-3} M$$

$$[Cl^-] = \frac{1.78 \times 10^{-10}}{9.09 \times 10^{-3}} = 1.96 \times 10^{-8} M; \qquad pCl = 7.71$$

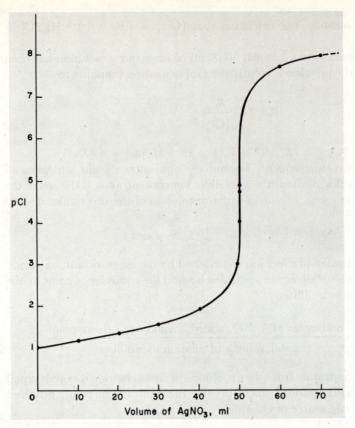

*Figure 8–1.* Titration curve for the titration of 50.00 ml of 0.1000 F sodium chloride solution with 0.1000 F silver nitrate solution.

Note that the silver ion concentration contributed from the dissolution of silver chloride would be virtually identical to $[Cl^-]$; thus, it is negligible compared to $9.09 \times 10^{-3} M$.

After 70.00 ml of titrant is added,

$$[Ag^+] = \frac{2.000 \text{ millimoles}}{120.0 \text{ milliliters}} = 0.0167 M$$

$$[Cl^-] = \frac{1.78 \times 10^{-10}}{1.67 \times 10^{-2}} = 1.07 \times 10^{-8} M; \qquad pCl = 7.97$$

Figure 8–1 shows the titration curve (pCl versus volume of silver nitrate) constructed on the basis of these calculations.

**Factors which govern the shapes of titration curves.** It is the steepness or sharpness of the titration curve that determines the precision with which the equivalence point of a titration can be located. With reference to the titration curve illustrated in Figure 8–1, there are two factors which govern the size of the "break" in the pCl-volume curve in the region of the equivalence point—first, the original concentration of the chloride solution and, second, the equilibrium constant (namely, the reciprocal of the solubility product) for the titration reaction. In general, the higher the concentration of the species being titrated and the more insoluble the precipitate that is formed, the sharper will be the titration curve and the greater should be the precision in the location of the equivalence point. In some cases, the solubility of a precipitate can be markedly decreased through a change in the solvent system. For example, the addition of water-miscible organic solvents, such as acetone or ethanol, to an aqueous medium will significantly lower the solubilities of most inorganic salts and, therefore, will improve the shapes of titration curves.

**End-point detection.** Undoubtedly, the most important characteristic of a titration curve is the large and abrupt change in the concentration of the substance being titrated in the vicinity of the equivalence point. Thus, the purpose of any method of end-point detection is to provide a signal that this point of rapid concentration change has been reached. Ideally, the end-point signal should occur exactly at the theoretical equivalence point. Any difference between the true equivalence point of the titration and the experimentally detected end point constitutes an error in the titration. In the following discussions, we will examine three end-point detection methods useful for the titrimetric determination of chloride.

MOHR TITRATION. In the Mohr method for the determination of chloride, the chloride ion is precipitated by titration with standard silver nitrate solution in the presence of a small concentration of chromate ion, and the end point is signaled by the first detectable and permanent appearance of a precipitate of brick-red-colored silver chromate throughout the solution. Silver chromate forms upon addition of a slight excess of silver nitrate only after virtually all the chloride ion has been precipitated as silver chloride.

For this chemical system, the pertinent equilibria are as follows:

$$AgCl(s) \rightleftharpoons Ag^+ + Cl^-; \qquad K_{sp} = 1.78 \times 10^{-10}$$

$$Ag_2CrO_4(s) \rightleftharpoons 2\ Ag^+ + CrO_4{}^{2-}; \qquad K_{sp} = 2.45 \times 10^{-12}$$

Note that silver chromate, which has a smaller $K_{sp}$, dissolves to form three ions and is actually more soluble than silver chloride. Thus, for a small concentration of chromate ion in the presence of a relatively large amount of chloride, silver chloride is preferentially precipitated upon addition of silver nitrate, and the free silver ion concentration never becomes high enough to cause the solubility product of silver chromate to be exceeded until chloride ion has been almost completely precipitated.

Ideally, it is desirable for red-colored silver chromate just to begin to precipitate at the equivalence point of the titration of chloride ion with silver nitrate, or when $[Cl^-] = [Ag^+] = (K_{sp})^{1/2} = 1.33 \times 10^{-5}\ M$. We can calculate the concentration of chromate ion which must be present to cause precipitation of silver chromate when $[Ag^+]$ reaches $1.33 \times 10^{-5}\ M$ from the solubility-product relationship for silver chromate:

$$[CrO_4{}^{2-}] = \frac{K_{sp}}{[Ag^+]^2} = \frac{2.45 \times 10^{-12}}{(1.33 \times 10^{-5})^2} = 0.014\ M$$

Unfortunately, because chromate ion has a fairly intense canary-yellow color, this concentration of chromate tends to conceal the appearance of the red-colored silver chromate precipitate. Therefore, in practice it is necessary to use a somewhat lower chromate concentration, and 0.005 $M$ is usually chosen to be the value of the chromate concentration at the end point.

With this lower chromate concentration, a slight excess of silver nitrate must be added beyond the equivalence point in order to exceed the solubility product and to cause precipitation of silver chromate. Also, the equivalence point must be surpassed even more because a finite amount of silver chromate must be precipitated before the end point can be seen. Calculations reveal that an excess of about 0.03 ml of 0.1 $F$

silver nitrate, corresponding to an error of $+0.07$ per cent for the titration of 50 ml of a 0.1 $M$ chloride solution, is required to precipitate a detectable amount of silver chromate.

Frequently, one performs a blank titration with a solution containing no chloride ion but the same concentration of chromate as an actual experiment. A suspension of white-colored calcium carbonate is used to simulate the silver chloride precipitate, the mixture is titrated with silver nitrate, and the volume of silver nitrate solution required to produce the same end-point color is measured. This small volume, when subtracted from the total volume of titrant used in an actual titration, provides an end-point correction.

It is essential that the Mohr titration be carried out in a solution which is neither too acidic nor too basic. At high pH values, there is danger that silver hydroxide or oxide may precipitate:

$$Ag^+ + OH^- \rightleftharpoons AgOH$$

$$2\ AgOH \rightleftharpoons Ag_2O + H_2O$$

One form of the solubility-product relationship for silver hydroxide or oxide is given by

$$[Ag^+][OH^-] = 2.6 \times 10^{-8}$$

In an actual titration, the concentration of silver ion at the end point is about $2.2 \times 10^{-5}\ M$. In order that the formation of silver hydroxide not interfere in the titration, the hydroxide ion concentration cannot exceed the value given by

$$[OH^-] = \frac{2.6 \times 10^{-8}}{[Ag^+]} = \frac{2.6 \times 10^{-8}}{2.2 \times 10^{-5}} = 1.2 \times 10^{-3}\ M$$

This corresponds to a pH of 11.1. However, to avoid the possibility of the transient formation of AgOH or $Ag_2O$, the upper pH limit for practical titrations must be lower than approximately 10.

In slightly acidic solutions, chromate ion becomes protonated

$$CrO_4{}^{2-} + H^+ \rightleftharpoons HCrO_4{}^-; \qquad K = 3.2 \times 10^6$$

and the resulting hydrogen chromate anion dimerizes to form dichromate:

$$2\ HCrO_4{}^- \rightleftharpoons Cr_2O_7{}^{2-} + H_2O; \qquad K = 33$$

Using these equilibria, one can calculate the effect of pH on the concentration of chromate ion. Below pH 7 the chromate ion concentration decreases quite rapidly, and the titration error increases correspondingly, because more and more titrant must be added beyond the equivalence point to exceed the solubility product of silver chromate. Therefore, in the Mohr titration of chloride, the solution pH should be no less than approximately 7.

Both bromide and chloride may be determined by means of the Mohr titration. However, iodide and thiocyanate do not give satisfactory results because precipitates of silver iodide and silver thiocyanate adsorb chromate ions, so that indistinct end points are obtained.

VOLHARD TITRATION. In the Volhard titration for the direct determination of silver ion, or for the indirect determination of a number of anions that form insoluble silver salts, the end point is detected by means of the formation of a soluble, colored complex ion.

Advantage is taken of the fact that iron(III) in an acidic medium forms an intense blood-red complex with thiocyanate:

$$Fe^{3+} + SCN^- \rightleftharpoons FeSCN^{2+}; \qquad K_1 = 138$$

Although $FeSCN^{2+}$ is not especially stable, a very sensitive end-point signal can be obtained because this complex is detectable at very low concentration levels. It has been established experimentally that $6.4 \times 10^{-6}\ M$ $FeSCN^{2+}$ imparts a noticeable color to a solution. Iron(III) and thiocyanate ions react to form higher complexes, e.g., $Fe(SCN)_2^+$, but these species are unimportant in the Volhard titration because the thiocyanate concentration is always quite low.

It is important that the Volhard titration be performed in an acidic solution. In aqueous media, the parent form of iron(III) is the colorless hexaaquo species, $Fe(H_2O)_6^{3+}$. However, $Fe(H_2O)_6^{3+}$ is a fairly strong acid and undergoes acid dissociations such as

$$Fe(H_2O)_6^{3+} \rightleftharpoons Fe(H_2O)_5(OH)^{2+} + H^+$$

and

$$Fe(H_2O)_5(OH)^{2+} \rightleftharpoons Fe(H_2O)_4(OH)_2^+ + H^+$$

the first acid dissociation constant for $Fe(H_2O)_6^{3+}$ being about $10^{-3}$. Because the species $Fe(H_2O)_5(OH)^{2+}$ and $Fe(H_2O)_4(OH)_2^+$ are intensely orange and brown, their presence would completely mask the appearance of red-colored $FeSCN^{2+}$. Therefore, to repress the acid dissociation of $Fe(H_2O)_6^{3+}$, the acidity of the solution being titrated must be in the neighborhood of 0.1 to 1 $M$.

*Direct Determination of Silver.* In the Volhard method for the direct determination of silver, an acidic solution containing silver ion (and a small concentration of iron(III) as indicator) is titrated with a standard potassium thiocyanate solution. As noted above, the end point is marked by the appearance of the blood-red-colored $FeSCN^{2+}$ complex upon addition of the first slight excess of thiocyanate into the titration vessel.

We can write the titration reaction as

$$Ag^+ + SCN^- \rightleftharpoons AgSCN(s)$$

where the solubility product for silver thiocyanate is $1.00 \times 10^{-12}$. Usually, the indicator is added to the solution prior to the start of the titration in the form of an acidic solution of ferric ammonium sulfate (ferric alum). However, the concentration of iron(III) must be adjusted so that the first permanent appearance of $FeSCN^{2+}$ occurs at the equivalence point of the titration. At the equivalence point of the titration, $[Ag^+] = [SCN^-] = (K_{sp})^{1/2} = 1.00 \times 10^{-6}\ M$. Therefore, the required ferric ion concentration can be calculated from the expression

$$[Fe^{3+}] = \frac{[FeSCN^{2+}]}{138[SCN^-]}$$

where $[SCN^-]$ is taken to be the equivalence-point value $(1.00 \times 10^{-6}\ M)$ and $[FeSCN^{2+}]$ is the minimum detectable concentration of the iron(III)-thiocyanate complex $(6.4 \times 10^{-6}\ M)$. Thus,

$$[Fe^{3+}] = \frac{(6.4 \times 10^{-6})}{(138)(1.0 \times 10^{-6})} = 0.046\ M$$

This concentration of ferric ion is sufficiently large that, in spite of the high acidity of the solution being titrated, the acid dissociation of $Fe(H_2O)_6^{3+}$ proceeds far enough to cause a distinct orange-brown color in the solution. Consequently, most practical procedures for the Volhard determination of silver call for the use of approximately 0.015 $M$ ferric ion. This compromise does not affect the accuracy of the Volhard titration for silver nearly as much as the similar compromise in the Mohr titration for chloride. For the Volhard titration of 0.1 $M$ silver ion with 0.1 $M$ thiocyanate, the error is somewhat less than +0.02 per cent.

*Indirect Determination of Chloride.* There has been extensive use of the Volhard method for the determination of chloride and other anions. In the classic Volhard titration, a carefully measured excess of standard silver nitrate is added to the chloride sample solution, the proper amounts of acid and ferric alum indicator having been introduced. An amount of silver nitrate stoichiometrically equivalent to the chloride reacts to precipitate silver chloride; then, the excess silver nitrate is back-titrated with standard potassium thiocyanate solution to the appearance of the red $FeSCN^{2+}$ complex. To compute the quantity of chloride (or other anion) in the sample solution, one subtracts the millimoles of silver ion equivalent to the thiocyanate used in the back-titration from the total number of millimoles of silver ion originally added to the sample solution.

A serious difficulty occurs when the excess silver ion is back-titrated with potassium thiocyanate in the presence of the silver chloride precipitate. Because silver chloride is more soluble than silver thiocyanate, the addition of thiocyanate causes *metathesis* of silver chloride according to the reaction

$$AgCl + SCN^- \rightleftharpoons Cl^- + AgSCN$$

Taking $1.00 \times 10^{-12}$ for the solubility product of silver thiocyanate and $1.78 \times 10^{-10}$ for the solubility product of silver chloride, we can calculate the equilibrium constant for this reaction:

$$K = \frac{[Cl^-]}{[SCN^-]} = \frac{[Ag^+][Cl^-]}{[Ag^+][SCN^-]} = \frac{1.78 \times 10^{-10}}{1.00 \times 10^{-12}} = 178$$

In effect, some of the silver nitrate reacts both with chloride and with thiocyanate, so the resulting titration data are useless.

Perhaps the most obvious way to eliminate metathesis is to remove the silver chloride by filtration after the excess of silver nitrate is added. Care must be taken to wash the silver chloride completely free of the excess silver ions. After removal of the silver chloride, the titration of silver ion with thiocyanate is identical to the direct determination of silver outlined earlier.

A second way to overcome the problem of metathesis is based on decreasing the rate of the (heterogeneous) reaction between the solid silver chloride and thiocyanate ion. This can be accomplished if the effective surface area of the silver chloride precipitate is decreased. There are two possibilities—first, boil the solution to coagulate the silver chloride, cool the solution, then titrate it with thiocyanate and, second, coat the silver chloride particles with a water-immiscible substance such as nitrobenzene.

It is possible to perform the indirect Volhard determination of other anions. In particular, for the determinations of bromide and iodide, the metathesis of the silver halide by excess thiocyanate does not occur because silver bromide ($K_{sp} = 5.25 \times 10^{-13}$) and silver iodide ($K_{sp} = 8.31 \times 10^{-17}$) both have smaller solubility products than silver thiocyanate. None of the special precautions used in the determination of chloride has to be employed for bromide or iodide. However, because iodide does react with iron(III), it is necessary to add the excess of silver nitrate before the ferric alum indicator. A successful determination of anions which form silver salts *more soluble* than silver chloride requires that an excess of silver nitrate be added and that the silver salt be removed by filtration before the back-titration with standard potassium thiocyanate solution.

USE OF ADSORPTION INDICATORS. Certain organic dyes are adsorbed by colloidal precipitates more strongly on one side of the equivalence point than on the other. In certain cases, the dye undergoes an abrupt color change in the process of being adsorbed, and it may serve as a sensitive end-point indicator in titrations. Analytical uses of these so-called *adsorption indicators* are quite specific and relatively few in number, but they are of considerable importance nevertheless.

We can understand the behavior of an adsorption indicator in terms of the properties of colloidally dispersed precipitates in an electrolyte solution. As discussed in Chapter 7, a colloidal particle attracts to its surface a primary adsorbed ion layer consisting preferentially of the lattice ion which is present in excess in the surrounding solution phase. In other words, a precipitate particle most strongly adsorbs the ions of which it is composed. For example, in the titration of chloride with a standard silver nitrate solution, the primary ion layer adsorbed on the silver chloride consists of chloride ions prior to the equivalence point. After the equivalence point, the primary ion layer contains adsorbed silver ions. In addition, a secondary (counter) ion layer is attracted more or less strongly, the sign of the charges of the ions in the counter-ion layer being opposite of those of the primary layer.

One important family of adsorption indicators is derived from fluorescein. Quite often, the sodium salt of this compound, sodium fluoresceinate, is used as an adsorption indicator for the titration of chloride with silver nitrate in a neutral or very slightly basic solution. Sodium fluoresceinate is ionized in solution to form sodium ions and fluoresceinate ions, which will be designated as the indicator anions, $In^-$. As soon as the titration is begun, some solid silver chloride is formed in the titration vessel. At all points prior to the equivalence point, chloride ion is in excess, so that the primary ion layer consists of adsorbed chloride ions and the secondary ion layer of any available cations, such as sodium or hydrogen ions. Few negatively charged indicator ions, $In^-$, are adsorbed because chloride displaces them. However, beyond the equivalence point, excess silver nitrate is present, so the primary adsorbed ion layer is composed of silver ions and the secondary ion layer consists of negative ions, an appreciable number of which will be the indicator anions. Sodium fluoresceinate imparts a fluorescent yellow-green color to the solution, but when the indicator anions are adsorbed as counter ions on the precipitate, a dramatic color change occurs and the precipitate particles become bright pink. This color change is believed to be due to a distortion or bending of the fluoresceinate structure when it is attracted to a precipitate particle by the presence of the positively charged, adsorbed silver ions. In effect, the fluoresceinate anion acts as an indicator for adsorbed silver ions. Thus, the end point is marked by the change from a green solution to a pink precipitate. In practice, the particles of precipitate are kept well dispersed so that the observed color change appears to be one of yellow-green to pink throughout the entire solution. This procedure for determining chloride is commonly called the Fajans method.

Fluoresceinate is the anion of the weak acid fluorescein, whose $pK_a$ is approximately 7. Therefore, the titration of chloride with silver nitrate must be performed in neutral or slightly alkaline solutions. Naturally, the pH cannot be so high that silver oxide might be formed. If the titration is attempted below pH 6, the indicator anions are converted to the uncharged parent acid, which displays no tendency to be adsorbed.

Since the color on one side of the equivalence point appears on the surface of the precipitate particles, it is desirable to have as large a precipitate surface area as possible. This means that the individual particles should be kept very small and should be well dispersed throughout the solution. Dextrin is used as a protective colloid in the determination of chloride by means of the adsorption indicator method to help ensure the desired physical characteristics of the precipitate. These characteristics are exactly opposite to those deemed desirable in gravimetric analysis.

Dichlorofluorescein ($pK_a$ about 4) and eosin ($pK_a$ about 2) are stronger acids than fluorescein and serve as adsorption indicators over a wider pH range. Whereas fluorescein is applicable to silver halide precipitations only if the pH is within the approximate limits of 6 to 10, a pH as low as 4 is acceptable with dichlorofluorescein. In practice, the use of dichlorofluorescein rather than fluorescein is preferable for the titration of halides. Eosin is suitable as an indicator for the precipitation titrations of bromide, iodide, and thiocyanate (but not chloride) in solutions as acidic as pH 2. Chloride cannot be determined because the eosinate anion is more strongly adsorbed than chloride even at the start of the titration. Thus, the relative strength of adsorption of an indicator anion is another important criterion in the selection of a suitable adsorption indicator.

### Titration of Halide Mixtures with Silver Ion

If a sample solution contains several ions, each of which forms an insoluble precipitate with a particular titrant solution, and if the solubility products of the

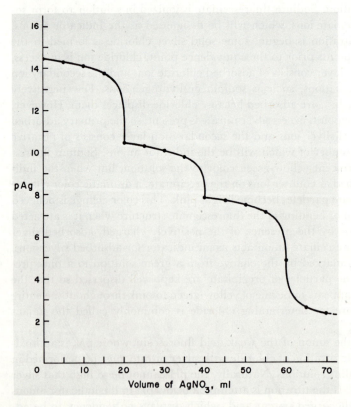

*Figure 8-2.* Titration curve for the titration of 50.00 ml of a solution containing 0.02000 $M$ each of iodide, bromide, and chloride with a 0.05000 $F$ silver nitrate solution.

resulting precipitates differ enough, then it may be possible to titrate the solution and to obtain an end point corresponding to the successive titration of each component. Such is the case for mixtures of iodide, bromide, and chloride. Since the solubility products of the silver halides are $8.31 \times 10^{-17}$ for silver iodide, $5.25 \times 10^{-13}$ for silver bromide, and $1.78 \times 10^{-10}$ for silver chloride, a solution containing a mixture of halides of not too greatly differing concentrations can be titrated with a standard silver nitrate solution, and the amount of each individual halide can be determined.

In Figure 8–2 is shown the theoretical titration curve—a plot of pAg versus the volume of titrant—for the titration of 50.00 ml of a solution which is 0.02000 $M$ each in iodide, bromide, and chloride with a 0.05000 $F$ silver nitrate solution. In accord with the relative magnitudes of the solubility products for the three silver halides, the first step of the curve corresponds to titration of iodide, the second step to titration of bromide, and the third step to titration of chloride.

In practice, the idealized titration curve of Figure 8–2 is not followed precisely. One problem in the titration of halide mixtures is contamination of one silver halide through coprecipitation of a second, more soluble silver halide. This results from the tendency of the silver halides to adsorb other halide ions and to form solid solutions or mixed crystals with each other. Because silver bromide and silver chloride have identical (cubic) crystal structures, the formation of solid solutions of these precipitates is especially serious. Silver iodide and silver bromide are less prone to form solid solutions, and silver iodide and silver chloride form almost no mixed crystals at all. Depending upon the relative concentrations of halides in a mixture, the phenomena of ion adsorption and mixed-crystal formation can frequently cause titration errors on the order of 1 per cent for the determination of individual halides in a mixture.

## QUESTIONS AND PROBLEMS

1. Water in a sample is often determined by loss of weight upon heating. How can the analytical chemist be sure that no other component is being lost at the same time as water?

2. State whether each of the following occurrences would tend to make the results of a determination of water in barium chloride dihydrate by volatilization too high or too low or whether they would be of no effect. Explain each answer.
   (a) The dehydration temperature is too low.
   (b) The dehydration temperature is too high.
   (c) The crucible and anhydrous barium chloride are placed on the balance pan and weighed while they are still hot.
   (d) The anhydrous barium chloride is allowed to stand uncovered in air between the time it is heated and subsequently weighed.
   (e) A bit of plaster unsuspectedly falls from the wall into the crucible before the final weighing of anhydrous barium chloride.

3. Will the results of a determination of chloride by the usual gravimetric procedure be made too high or too low, or will there be no effect, if each of the following occurs? Explain each answer.
   (a) Some carbonate is present in the original sample.
   (b) The water used in preparing the wash liquid contains some chloride ion.
   (c) The precipitate is exposed to bright sunlight while it is in the mother liquid.
   (d) The precipitate is exposed to bright sunlight after it is washed but prior to its weighing.
   (e) The precipitate is washed with pure water rather than with water containing an electrolyte.

4. Will the results of a determination of sulfate by the usual gravimetric procedure be made too high or too low, or will there be no effect, if each of the following occurs? Explain each answer.
   (a) An excessive amount of acid is present in the mother liquid.
   (b) Fluoride ion is present at the time of precipitation of barium sulfate.
   (c) Nitrate ion is coprecipitated.
   (d) Aluminum ion is coprecipitated.
   (e) The ignition temperature becomes too high before the combustion of the filter paper is complete.

5. A 0.8046-gm sample of impure barium chloride dihydrate weighed 0.7082 gm after it was dried at 200°C for two hours. Calculate the percentage of water in the sample.

6. A silver chloride precipitate weighing 0.3221 gm was obtained from 0.4926 gm of a soluble salt mixture. Calculate the percentage of chloride in the sample.

7. The sulfur content of a sample is usually expressed in terms of the percentage of $SO_3$ in that sample. Calculate the percentage of $SO_3$ in a 0.3232-gm sample of a soluble salt mixture which yielded a barium sulfate precipitate weighing 0.2982 gm.

8. What is the percentage of iron in a 0.9291-gm sample of an acid-soluble iron ore which yields a ferric oxide ($Fe_2O_3$) precipitate weighing 0.6216 gm?

9. From a 0.6980-gm sample of an impure, acid-soluble magnesium compound, a precipitate of magnesium pyrophosphate ($Mg_2P_2O_7$) weighing 0.4961 gm was obtained. Calculate the magnesium content of the sample in terms of the percentage of magnesium oxide (MgO) present in that sample.

10. A 1.0000-gm sample of a mixture of $K_2CO_3$ and $KHCO_3$ yielded 0.4000 gm of carbon dioxide ($CO_2$) upon ignition. What weight of each compound was present in the mixture?

11. A 1.1374-gm sample contains only sodium chloride and potassium chloride. Upon dissolution of the sample and precipitation of the chloride as silver chloride, a precipitate weighing 2.3744 gm was obtained. Calculate the percentage of sodium chloride in the sample.

12. A 0.5000-gm sample of a clay was analyzed for sodium and potassium by precipitation of both as chlorides. The combined weight of the sodium and potassium chlorides was found to be 0.0361 gm. The potassium in the chloride precipitate was reprecipitated as $K_2PtCl_6$, which weighed 0.0356 gm. Calculate the sodium and potassium content of the clay in terms of the percentages of hypothetical $Na_2O$ and $K_2O$ present in the clay.

13. A 0.2000-gm sample of an alloy containing only silver and lead was dissolved in nitric acid. Treatment of the resulting solution with cold hydrochloric acid gave a mixed chloride precipitate (AgCl and $PbCl_2$) weighing 0.2466 gm. When this mixed chloride precipitate was treated with hot water to dissolve all the lead chloride, 0.2067 gm of silver chloride remained. Calculate the percentage of silver in the alloy and calculate what weight of lead chloride was not precipitated by the addition of cold hydrochloric acid.

14. In the determination of chloride by the Mohr procedure, a chemist added the proper amount of potassium chromate indicator, made the solution distinctly acidic, and titrated the solution with silver nitrate. Was the quantity of chloride found by this technique high, low, or correct? Explain.

15. Make sketches of the silver chloride precipitate during a chloride determination according to the Fajans method, showing the relative positions and composition of the primary adsorbed ion layer and the counter-ion layer.

16. What volume of 0.2000 $F$ potassium thiocyanate solution is required to precipitate the silver from a solution containing 0.4623 gm of silver nitrate?

17. If 25.00 ml of a sodium chloride solution was required to precipitate the silver in the solution obtained by dissolution of 0.2365 gm of 98.00 per cent pure silver metal, what was the concentration of the sodium chloride solution?

18. Calculate the percentage of potassium iodide (KI) in a 2.145-gm sample of a mixture of potassium iodide and potassium carbonate that, when analyzed according to the Volhard procedure, required 3.32 ml of a 0.1212 $F$ potassium thiocyanate solution after the addition of 50.00 ml of a 0.2429 $F$ silver nitrate solution.

19. For the analysis of a commercial solution of silver nitrate, a 2.075-gm sample of the solution was weighed out and diluted to 100.0 ml in a volumetric flask.

A 50.00-ml aliquot of the solution was then titrated with 35.55 ml of a potassium thiocyanate solution, of which 1.000 ml corresponds to exactly 5.000 mg of silver. Calculate the weight percentage of silver nitrate in the original solution.

20. An analytical chemist analyzed 0.5000 gm of an arsenic-containing sample by oxidizing the arsenic to arsenate, precipitating silver arsenate ($Ag_3AsO_4$), dissolving the precipitate in acid, and titrating the silver with a $0.1000 F$ potassium thiocyanate solution—45.45 ml being required. Calculate the percentage of arsenic in the sample.

21. A 25.00-ml portion of a silver nitrate solution was treated with an excess of sodium chloride solution, and the resulting silver chloride precipitate was coagulated, filtered, washed, and found to weigh 0.3520 gm. In a subsequent experiment, 22.00 ml of the same original silver nitrate solution was found to react with 16.25 ml of a potassium thiocyanate solution. Calculate the concentrations of the silver nitrate and potassium thiocyanate solutions.

22. A solid sample is known to contain only sodium hydroxide, sodium chloride, and water. A 6.700-gm sample was dissolved in distilled water and diluted to 250.0 ml in a volumetric flask. A one-tenth aliquot required 22.22 ml of $0.4976 F$ hydrochloric acid for titration to a phenolphthalein end point. When another one-tenth aliquot of the sample solution was titrated according to the Volhard procedure, 35.00 ml of a $0.1117 F$ silver nitrate solution was added and 4.63 ml of $0.0962 F$ potassium thiocyanate solution was required for the back-titration. The water content of the sample was determined by difference. What was the percentage composition of the original sample?

23. Suppose that you desire to determine iodate ion according to the Mohr procedure. Calculate the concentration of chromate ion which must be present so that precipitation of silver chromate just begins at the equivalence point.

24. Construct the titration curve for the titration, with a $0.05000 F$ silver nitrate solution, of 50.00 ml of a solution containing 0.03000 $M$ iodide ion, 0.01500 $M$ bromide ion, and 0.02500 $M$ chloride ion.

25. A method for the determination of low concentrations of chloride ion takes advantage of the common ion effect and the solubility-product principle. The "concentration solubility product" of AgCl(s) in $0.0250 F$ nitric acid at 25°C is $2.277 \times 10^{-10}$. The concentration of chloride in an unknown was determined as follows: A $0.0250 F$ nitric acid solution containing chloride ion at an unknown concentration was shaken with excess pure AgCl(s) at 25°C until solubility equilibrium was attained. A 100.0-ml sample of the equilibrated solution was withdrawn by suction through a sintered-glass filter stick in order to remove any solid AgCl which might have been dispersed in the solution. The 100.0-ml sample was then titrated with a standard $1.020 \times 10^{-4} F$ potassium iodide (KI) solution. The end point was reached when 7.441 ml of the standard KI had been added. Calculate the concentration of chloride ion in the original solution to four significant figures. Neglect activity effects, the presence of molecular AgCl(aq), and the possible formation of anionic complexes between silver and chloride ions.

26. To each of three flasks, labeled A, B, and C, was added 50 ml of $0.05 F$ silver nitrate solution and three drops of a 0.10 per cent aqueous solution of sodium fluoresceinate. The solution in each flask was treated separately as follows:

    Flask A: When 100 mg of dextrin and 1.0 ml of a $1.0 F$ potassium chloride solution were added, the mixture became bright pink.

    Flask B: When 100 mg of dextrin and 50 ml of a $1.0 F$ potassium chloride solution were added, white AgCl was observed to be in suspension and the solution was colored a fluorescent yellow-green.

    Flask C: When 100 mg of dextrin and 1.0 ml of a $1.0 F$ potassium chloride solution were added, the mixture became bright pink. However, upon the addition of 5.0 ml of concentrated nitric acid, white AgCl was observed to be in suspension and the solution was colored a fluorescent yellow-green.

    Explain the phenomena which occurred in each of these three experiments.

27. Construct the titration curve for the titration of 50.00 ml of a $0.01000 F$ silver nitrate solution with $0.1000 F$ potassium thiocyanate solution in a $1 F$ nitric acid medium.

28. Construct the titration curve for the titration of 50.00 ml of a 0.02000 $F$ mercuric nitrate solution with a 0.1000 $F$ potassium thiocyanate solution.

29. For the calculation and construction of the titration curve for the titration of 50.00 ml of 0.1000 $F$ sodium chloride with 0.1000 $F$ silver nitatte discussed in this chapter, it was necessary to consider the dissolution of solid AgCl in order to calculate the chloride ion concentration and the value of pCl after the addition of 49.99 ml of titrant. What would have been the values of [Cl⁻] and pCl if the dissolution of AgCl had been ignored? What are the magnitudes of the errors in [Cl⁻] and pCl when the dissolution of AgCl is ignored?

30. It might be possible to determine $Pb^{2+}$ by titration with a standard solution of potassium chromate. For the end-point indicator one could conceivably use silver ion, which would form a precipitate of silver chromate. Lead chromate is bright yellow, whereas silver chromate is brick-red.

    (a) Calculate what concentration of silver ion would be required so that silver chromate would just begin to precipitate at the equivalence point of the titration. (This should be a theoretical calculation, uncomplicated by solid-solution formation, activities, adsorption, or the requirement that a finite amount of silver chromate would be needed to see the end point.)

    In practice, it seems evident that an unusually large amount of red-colored silver chromate must be formed in order to observe the end point in the presence of yellow-colored lead chromate. Therefore, to compensate for this, one might arrange conditions so that silver chromate would start to precipitate a little *before* the equivalence point, and thus obtain enough silver chromate to see the end point just at the equivalence point. In other words, you would probably want to make the silver ion concentration larger than the theoretical value, so that, when you see the end point, the amount of chromate added in the titration would really equal the amount of lead in the sample. Thus, although some lead remains unprecipitated, this should be exactly compensated by the chromate consumed in the precipitation of silver chromate.

    (b) Suppose that the minimum detectable concentration of silver chromate is $1 \times 10^{-4}$ $M$ in the presence of yellow-colored lead chromate. Calculate what concentration of $Ag^+$ should be present to make the titration error virtually *zero* for the titration of 50.00 ml of 0.1000 $M$ $Pb^{2+}$ with 0.1000 $F$ potassium chromate.

31. An attempt was made to analyze a solution of potassium iodate according to the Volhard method. To a 50.00-ml sample of the potassium iodate solution containing nitric acid, 50.00 ml of 0.1000 $F$ silver nitrate was added to precipitate silver iodate. Then 5.00 ml of a 2.00 $F$ ferric nitrate solution was added, and, without removal of the silver iodate precipitate, the mixture was back-titrated with a 0.1000 $F$ potassium thiocyanate solution. An end point was reached after the addition of 34.80 ml of potassium thiocyanate solution.

    (a) Calculate the concentrations of thiocyanate and iodate ions at the end point.

    (b) Calculate the titration error in per cent with the proper sign.

    (c) Calculate the concentration of the original potassium iodate solution.

## SUGGESTIONS FOR ADDITIONAL READING

1. C. Ayers: Argentometric methods. *In* C. L. Wilson and D. W. Wilson, eds.: *Comprehensive Analytical Chemistry.* Volume IB, Elsevier, New York, 1960, pp. 222–237.

2. J. F. Coetzee: Equilibria in precipitation reactions. *In* I. M. Kolthoff and P. J. Elving, eds.: *Treatise on Analytical Chemistry.* Part I, Volume 1, Wiley-Interscience, New York, 1959, pp. 767–809.

3. W. F. Hillebrand, G. E. F. Lundell, H. A. Bright, and J. I. Hoffman: *Applied Inorganic Analysis.* Second edition, John Wiley and Sons, New York, 1953, pp. 44–46, 711–723, 822–835.

4. I. M. Kolthoff, E. B. Sandell, E. J. Meehan, and S. Bruckenstein: *Quantitative Chemical Analysis.* Fourth edition, The Macmillan Company, New York, 1969, pp. 248–285, 565–677, 716–725, 795–802.
5. I. M. Kolthoff and V. A. Stenger: *Volumetric Analysis.* Second edition, Volume II, Wiley-Interscience, New York, 1947, pp. 239–344.
6. H. F. Walton: *Principles and Methods of Chemical Analysis.* Second edition, Prentice-Hall, Englewood Cliffs, New Jersey, 1964, pp. 80–114, 373–389.

# PRINCIPLES AND THEORY OF OXIDATION-REDUCTION METHODS

# 9

Oxidation-reduction methods of analysis, commonly called **redox methods**, probably rank as the most widely used volumetric procedures. Although acid-base, precipitation, and complexometric titrations have usefulness, many substances cannot be determined satisfactorily by means of these kinds of titrations. Redox methods are, however, of much broader applicability for the determination of a wide variety of chemical species, and additional emphasis has been placed on these techniques through the large amount of research in the field of electroanalytical chemistry.

## ELECTROCHEMICAL CELLS

Consideration of the behavior and properties of electrochemical cells provides an introduction to the subject of oxidation-reduction processes. An **electrochemical cell** consists essentially of two electrodes which are immersed either into the same solution or into two different solutions in electrolytic contact with one another. Across the interface where an electrode meets a solution, a difference in electrical potential

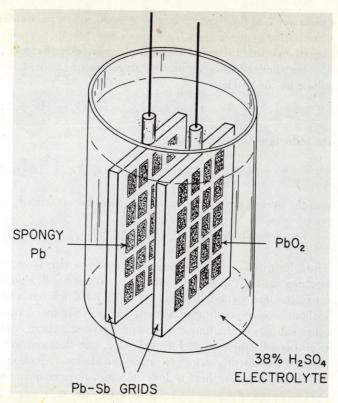

SPONGY Pb

PbO$_2$

Pb-Sb GRIDS

38% H$_2$SO$_4$ ELECTROLYTE

*Figure 9–1.* A simplified version of the lead storage cell. This galvanic cell consists of two electrodes immersed in 38 per cent sulfuric acid. The anode is a lead-antimony support or grid impregnated with spongy lead. The cathode is a second lead-antimony grid impregnated with solid lead dioxide. A fully charged lead storage cell has an emf close to 2 v.

develops—which we shall refer to simply as a **potential**. Likewise, across the interface between two solutions of differing composition, a so-called **liquid-junction potential** arises, although it really is a potential difference across the boundary where the two media come into contact. In addition, every electrochemical cell exhibits an **electromotive force**\* or **emf**, expressed in volts, which is the *sum* of various potential differences within the cell, including the two electrode potentials and perhaps one or more liquid-junction potentials.

A familiar electrochemical cell is the lead storage cell, shown schematically in Figure 9–1. One electrode comprises a lead-antimony support or grid which is impregnated with spongy metallic lead, whereas the other electrode is a similar lead-antimony grid impregnated with solid lead dioxide (PbO$_2$). Both electrodes are immersed in an aqueous sulfuric acid medium. If we short-circuit this cell, that is, if we connect the two electrodes to each other by a conducting wire, electrolysis occurs. Just after electrolysis begins, microscopic examination of each electrode reveals that the spongy lead is *oxidized* to solid lead sulfate, which adheres to the electrode surface, and that the lead dioxide is *reduced*, also to lead sulfate.

*Oxidation and reduction.* Oxidation is the loss of electrons, and reduction is the gain of electrons. A substance which is reduced, that is, a substance which

---

\* Instead of electromotive force, one may refer to the **voltage** of an electrochemical cell. Other important aspects of nomenclature are emphasized in the footnotes on pages 264, 267, and 273.

causes another chemical species to be oxidized, is an **oxidizing agent** or **oxidant**. Conversely, a substance which causes another species to be reduced, thereby becoming oxidized itself, is a **reducing agent** or **reductant**.

For the lead storage cell, we can represent the oxidation of spongy lead to lead sulfate by the reaction

$$Pb + HSO_4^- \rightleftharpoons PbSO_4 + H^+ + 2 e$$

the reduction of lead dioxide as

$$PbO_2 + 3 H^+ + HSO_4^- + 2 e \rightleftharpoons PbSO_4 + 2 H_2O$$

and the overall process as the sum of these two reactions:

$$Pb + PbO_2 + 2 H^+ + 2 HSO_4^- \rightleftharpoons 2 PbSO_4 + 2 H_2O$$

Thus, lead metal is the reductant in this reaction; it reduces $PbO_2$ to $PbSO_4$ and becomes oxidized in the process. On the other hand, lead dioxide is an oxidant, for it oxidizes lead metal to lead sulfate and is itself reduced. Oxidation cannot take place without a corresponding reduction, and no substance can be reduced without some other substance simultaneously being oxidized. There is no direct transfer of electrons between the lead and lead dioxide, but electrons are transferred through the external conducting wire. In many oxidation-reduction reactions employed for analytical purposes, there is direct electron-transfer from one reactant to another.

In an electrochemical cell, the electrode at which oxidation occurs is always called the **anode**, whereas the **cathode** is the electrode at which the reduction process takes place. Therefore, the spongy lead electrode is the *anode* because it undergoes oxidation to lead sulfate, and the lead dioxide electrode functions as the *cathode* because it is reduced to lead sulfate.

*Galvanic and electrolytic cells.* There are two kinds of electrochemical cells. A **galvanic cell** is an electrochemical cell in which the *spontaneous* occurrence of electrode reactions produces electrical energy which can be converted into useful work. Thus, the "discharge" of a lead storage cell as described earlier is an example of a spontaneous process which, in turn, causes a flow of electrons in an external circuit and makes useful electrical energy available. Three or six lead storage cells in series comprise the familiar 6- or 12-v lead storage battery used in automobile ignition systems.

In an **electrolytic cell**, nonspontaneous electrode reactions are forced to proceed when an external source of emf is connected across the two electrodes. In the operation of an electrolytic cell, electrical energy or work must be expended in causing the electrode reactions to occur. As a lead storage cell continues to discharge spontaneously for a time, according to the reaction

$$Pb + PbO_2 + 2 H^+ + 2 HSO_4^- \rightleftharpoons 2 PbSO_4 + 2 H_2O$$

the reactants—lead, lead dioxide, and sulfuric acid—are gradually exhausted, and the electrical energy obtainable from the cell decreases. To restore the cell to its original condition, we must recharge it by placing across the two electrodes an external emf which both opposes the emf of the cell itself and is larger than the cell emf. In this way the reaction occurring within the cell is reversed—lead sulfate is reduced back to spongy lead at one electrode, lead sulfate is oxidized to lead dioxide, and more sulfuric acid is produced.

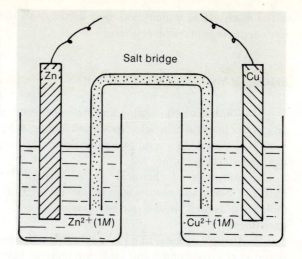

*Figure 9–2.* A simple galvanic cell involving a zinc metal–zinc ion half-cell and a cupric ion–copper metal half-cell connected by a potassium chloride salt bridge. For this cell the concentration (activity) of each ionic species, $Zn^{2+}$ and $Cu^{2+}$, is unity, although in general any values are possible.

Another example of a galvanic cell is the system depicted in Figure 9–2. One beaker contains a solution of some zinc salt into which a metallic zinc electrode is immersed, and the other beaker contains a solution of a cupric salt into which an electrode of metallic copper is placed. Connecting the two beakers is a **salt bridge**, which provides a pathway for the migration of ions (flow of current) from one beaker to the other when an electrical circuit is completed, yet prevents mixing of the solutions in the two beakers as well as any direct electron-transfer reaction between one electrode and the solution in the opposite beaker. In its simplest form, a salt bridge consists of an inverted U-tube filled with a mixture of potassium chloride solution and agar which forms a gel to minimize the leakage of potassium chloride into the two beakers.

If the two metal electrodes are connected by means of a conducting wire, the zinc electrode dissolves according to the reaction

$$Zn \rightleftharpoons Zn^{2+} + 2\ e$$

and cupric ions are reduced to copper atoms which deposit upon the copper electrode:

$$Cu^{2+} + 2\ e \rightleftharpoons Cu$$

Again, the overall process occurring in this galvanic cell may be represented as the combination of the two individual electrode reactions; that is,

$$Zn + Cu^{2+} \rightleftharpoons Zn^{2+} + Cu$$

Electrons produced at the zinc electrode, as zinc atoms are oxidized to zinc ions, flow through the external wire to the copper electrode, where they are available to combine with incoming cupric ions to form more copper metal. Along with the flow of electrons in the external circuit, negatively charged ions in the solutions migrate from the copper half-cell through the salt bridge in the direction of the zinc half-cell. Migration of ions in solution is not all one way, for the production of zinc ions and the consumption of cupric ions cause the movement of cations from the zinc half-cell toward the copper half-cell.

In keeping with previous definitions, the zinc electrode is the *anode*, because it is the electrode at which oxidation occurs, and the copper electrode is the *cathode* since reduction takes place there.

## SHORTHAND REPRESENTATION OF CELLS

It is both time-consuming and space-consuming to draw a complete diagram of an electrochemical cell—electrodes, salt bridge, solutions, and containers—each time a cell is discussed. As a result, a set of rules for representing cells in a shorthand fashion has been developed.

1. Conventional chemical symbols are used to indicate the ions, molecules, elements, gases, and electrode materials involved in a cell. Concentrations of ions or molecules, as well as the partial pressure of each gaseous species, are enclosed in parentheses.

2. A single vertical line | is employed to designate the fact that a boundary between an electrode phase and a solution phase or between two different solution phases exists and that the potential difference developed across this interface is included in the total emf of the cell.

3. A double vertical line ‖ indicates that the liquid-junction potential developed across the interface between two different solutions is ignored or that it is minimized or eliminated by placing a suitable salt bridge between the two solutions. A liquid-junction potential originates at the interface between two nonidentical solutions because anions and cations diffuse across this interface at different rates. A salt bridge containing a saturated solution of potassium chloride *almost* eliminates the liquid-junction potential because of the nearly equal mobilities of potassium and chloride ions.

Several examples will demonstrate the application of these rules.

*Example 9–1.* We may represent the zinc-copper cell depicted in Figure 9–2 as

$$Zn \mid Zn^{2+} \ (1 \ M) \ \| \ Cu^{2+} \ (1 \ M) \mid Cu$$

This shorthand abbreviation indicates that a zinc electrode in contact with a 1 $M$ solution of zinc ion gives rise to a potential and, similarly, that a 1 $M$ cupric ion solution in contact with a copper electrode produces another potential—the sum of these two potentials being the emf of the galvanic cell. A double vertical line indicates the presence of a salt bridge and reminds us that any liquid-junction potential is neglected in considering the emf of the cell.

*Example 9–2.* If, for the zinc-copper cell, we wished to specify 1 $F$ solutions of particular salts, such as $Zn(NO_3)_2$ and $CuSO_4$, and if instead of a salt bridge we allowed the two solutions to contact each other directly through a thin porous membrane (to prevent mixing of the two solutions), the cell would be correctly indicated as

$$Zn \mid Zn(NO_3)_2 \ (1 \ F) \mid CuSO_4 \ (1 \ F) \mid Cu$$

*Example 9–3.* Suppose we construct an entirely new galvanic cell. Let one electrode consist of a platinum wire in a mixture of 0.2 $M$ ferric ion and 0.05 $M$ ferrous ion in a 1 $F$ hydrochloric acid medium, and let the other electrode be a second platinum wire immersed in a 2 $F$ hydrochloric acid solution saturated with chlorine gas at a pressure of 0.1 atm. A salt bridge can be used to prevent mixing of the two solutions. We can represent this cell as

$$Pt \mid Fe^{3+} \ (0.2 \ M), \ Fe^{2+} \ (0.05 \ M), \ HCl \ (1 \ F) \ \| \ HCl \ (2 \ F) \mid Cl_2 \ (0.1 \ atm), \ Pt$$

This example brings up two other significant points. First, when several soluble species are present in the same solution, no special order of listing these is needed. Second, although electron transfer between chloride ion and chlorine molecules involves dissolved species, it is customary to indicate gaseous substances as part of the electrode phase along with the electrode material, *e.g.*, platinum metal.

## FREE-ENERGY CHANGE AND ELECTROMOTIVE FORCE

As discussed in Chapter 3, the free-energy change ($\Delta G$) for a chemical process measures the driving force or tendency for reaction to occur. In addition, $\Delta G$ represents the *maximal amount of useful energy or work* (aside from work due to expansion) that can be obtained from the process. However, in order for this maximal energy to be realized, the process must occur infinitely slowly or with thermodynamic reversibility. A reaction is said to be thermodynamically reversible if we can cause it to shift back and forth about some equilibrium position by applying a succession of appropriate infinitesimal forces or stresses. If a process occurs rapidly, at a finite or observable rate, the maximal quantity of useful energy cannot be obtained.

To explore these points further, let us consider the behavior of the lead storage cell. If a lead storage cell is discharged infinitely slowly, the current flowing through the external circuit will be infinitesimally small, and the theoretical, maximal amount of useful energy will be obtained (although the *rate* of production of energy would be intolerably low). On the other hand, if we allow a lead storage cell to discharge rapidly, the quantity of useful energy will be less than before, because the flow of a sizable current through the external circuit dissipates some of the energy as heat. No *real* process occurs infinitely slowly, and so the maximal amount of useful energy, $\Delta G$, can never be obtained. Yet a galvanic cell provides a means for studying chemical reactions under conditions which come remarkably close to thermodynamic reversibility.

*Electromotive force.* A quantity which is characteristic of any galvanic cell is $E$, the emf expressed in volts. In turn, the free-energy change for a process which occurs or can be made to occur in a galvanic cell is related to the emf by the equation

$$\Delta G = -nFE$$

where $n$ is the number of faradays of electricity generated by the cell reaction and $F$ is the Faraday constant (96,487 coulombs or 23,060 calories per faraday).

By measuring the emf of a galvanic cell, we can determine the free-energy change $\Delta G$ for a given process when the reaction virtually does not occur at all, that is, under conditions very close to thermodynamic reversibility. Let us see how this can be done for the reaction

$$Zn + Cu^{2+} \rightleftharpoons Zn^{2+} + Cu$$

A galvanic cell in which the desired reaction can occur is shown in Figure 9–2. In setting up a galvanic cell for the measurement of emf, one must specify the concentrations (activities) of all ionic and molecular species in solution as well as the temperature. If gaseous reactants or products are involved, their partial pressures should be stated. In the cell of Figure 9–2, the concentrations of $Zn^{2+}$ and $Cu^{2+}$ are both 1 $M$, but the concentration (activity) of each ion could have any value.

When the zinc and copper electrodes are connected by a conducting wire, there is a flow of electrons from the zinc to the copper electrode. Simultaneously, the zinc electrode dissolves to form zinc ions in the left-hand solution, and cupric ions in the right-hand solution are reduced and plated upon the copper electrode. At the instant the external wire is first connected between the two electrodes, there will be a certain initial value of the emf $E$. However, the production of zinc ions in the left cell and the consumption of cupric ions in the right cell will change the original concentrations of these ions, thereby diminishing the emf $E$ and the value of $\Delta G$.

It is known that the concentrations of the ions immediately adjacent to the surfaces of the electrodes govern the values of $E$ and $\Delta G$ for the reaction, so these surface concentrations must be accurately established. Unfortunately, it is extremely difficult to determine what the surface concentrations of zinc and cupric ions are because they keep changing as current continues to flow through the cell. Why not prevent the flow of current in order to avoid altering the surface concentrations of zinc and cupric ions? Then, the surface concentrations would be invariant and equal to the original concentrations, *e.g.*, 1 $M$. This can be done, and, in fact, this is exactly what is done in the measurement of the emf of a galvanic cell.

*Potentiometric measurement of electromotive force.* A **potentiometer** is an instrument used to determine the emf of a galvanic cell. As a brief introduction to the principles of operation of a potentiometer, let us consider the following simple concepts. Figure 9–3A shows the zinc-copper cell with an external conducting wire connecting the zinc and copper electrodes. In addition, a current-measuring galvanometer, $G$, has been inserted into the external circuit.

Suppose we introduce into the external circuit a source of known emf, as shown in Figure 9–3B, in such a way that it opposes the zinc-copper cell. If the known emf is now varied until it becomes just equal to the emf of the zinc-copper cell, the galvanometer will register a current of zero. When no current flows in the circuit, it

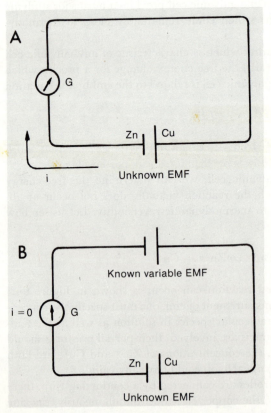

*Figure 9–3.* Diagram to illustrate the principles of the potentiometric measurement of electromotive force. It should be noted that the direction of current flow shown in A is the direction of electron flow. (See text for discussion.)

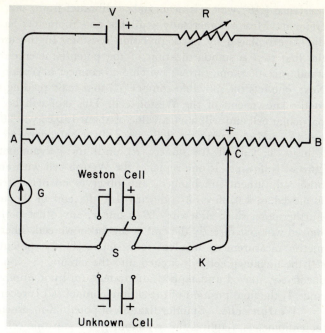

*Figure 9–4.* Circuit diagram of a simple potentiometer with a linear voltage divider.

follows that the emf values of the zinc-copper cell and the known source are identical, and hence the emf of the zinc-copper cell is determined. In practice, an actual potentiometric measurement consists of varying the known emf until the galvanometer indicates zero current.

If the known emf is *less* than that of the zinc-copper cell, the latter will discharge spontaneously as a normal *galvanic* cell according to the reaction

$$Zn + Cu^{2+} \rightleftharpoons Zn^{2+} + Cu$$

at a rate proportional to the difference in the emf values. On the other hand, if the known emf is *greater* than that of the zinc-copper cell, the latter will behave as an *electrolytic* cell and the reverse reaction

$$Zn^{2+} + Cu \rightleftharpoons Zn + Cu^{2+}$$

will occur at a rate governed again by the difference in the emf values. However, at the point of potentiometric balance, where virtually no current flows, the unknown and the known emf values are equal and no net reaction occurs in the zinc-copper cell. Therefore, the major advantages of a potentiometric measurement are that we prevent any electrochemical reaction from occurring in the zinc-copper cell, we avoid disturbing or changing the concentrations of ions at the electrode surfaces, and we obtain a highly accurate value for the emf of the zinc-copper cell.

Unfortunately, the circuit discussed above lacks the sophistication necessary for the accurate measurement of emf. Introduction of several improvements provides the precision potentiometer shown in Figure 9–4. There must be a source of emf (*V*), which is typically a 3-v battery, but the value of which need not be known exactly. This emf source is connected across the ends of a uniform or linear slidewire resistance (*AB*) which is accurately calibrated from, say, 0 to 1.5 v in 0.1-mv increments. Most commercially available potentiometers have a large fraction of the slidewire replaced by fixed, precision resistors. In the use of such an instrument, the first step is standardization of the potentiometer circuit. To accomplish this standardization, one throws switch *S* to connect a Weston cell across the slidewire. Next, one sets the movable contact (*C*) to a scale reading along slidewire *AB* equal to the known emf of the Weston cell. This emf will be precisely known for any particular cell and will have a value of about 1.0186 v at 20°C. Then, by tapping the key (*K*) and observing the deflections of the galvanometer needle (*G*), one carefully adjusts the variable resistance (*R*) until the galvanometer indicates that no *net* current is flowing. If one replaces the Weston cell with an unknown cell, keeps all other adjustments unchanged, and observes no net current-flow when contact *C* is moved to 1.0186 v on slidewire *AB*, the emf of the unknown cell is 1.0186 v. Furthermore, since slidewire *AB* is linear, any other position to which *C* must be moved will give directly the emf of the unknown cell. Therefore, an actual potentiometric measurement involves four steps—(1) the potentiometer is standardized, (2) the unknown cell is switched into the circuit, (3) contact *C* and key *K* are, respectively, moved and tapped alternately until potentiometric balance is achieved, and (4) the final position of the sliding contact (*C*) is recorded.

**Weston cells.** Standardization of potential-measuring instruments, including potentiometers, requires the availability of a galvanic cell whose emf is known very accurately. An internationally accepted standard for calibration purposes is the *saturated* Weston cell, whose shorthand representation is

$$-\mathrm{Cd(Hg)} \mid \mathrm{CdSO_4} \cdot \tfrac{8}{3}\mathrm{H_2O(s)}, \mathrm{Hg_2SO_4(s)} \mid \mathrm{Hg}+$$

As shown in Figure 9–5, the Weston cell is normally fabricated in the form of an H-shaped glass vessel; the anode is a saturated cadmium amalgam in equilibrium with an aqueous phase which is saturated with both hydrated cadmium sulfate and mercurous sulfate, whereas the cathode consists of a pool of pure mercury in equilibrium with the same aqueous phase. Excess cadmium sulfate crystals and solid mercurous sulfate are present to ensure that the aqueous phase will be saturated at all temperatures. At 20°C the emf of the saturated Weston cell is 1.0186 v. Weston cells encountered in a typical laboratory and employed for routine electrochemical measurements are almost always *unsaturated* with respect to cadmium sulfate. An

*Figure 9–5.* Saturated Weston standard cell.

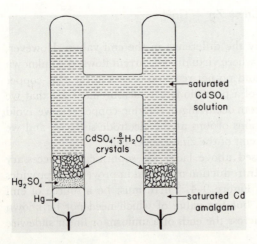

unsaturated Weston cell usually exhibits an emf between 1.0185 and 1.0195 v—the exact value being determined by potentiometric comparison to a saturated Weston cell.

## THERMODYNAMIC AND ELECTROCHEMICAL SIGN CONVENTIONS

We shall now explore some of the relationships between thermodynamics and electrochemistry. In the galvanic cell shown in Figure 9–6, the left-hand compartment contains a platinum electrode dipped into a hydrochloric acid solution of *unit activity* and bathed with hydrogen gas at a pressure of 1 atm. For convenience, the platinum electrode can be sealed into an inverted tube, with an enlarged open bottom to admit the hydrochloric acid solution and with exit ports for the hydrogen to escape. Hydrogen gas at 1 atm pressure enters through a side-arm tube. Hydrogen gas must come into intimate contact with the platinum electrode. This requirement can be fulfilled if the platinum electrode is coated prior to its use with a thin layer of finely divided, spongy platinum metal (so-called **platinum black**). A platinum-black electrode is easily prepared if one cathodically polarizes a shiny platinum electrode in a dilute solution of $PtCl_6^{2-}$. Hydrogen gas can permeate the platinum black, and electron transfer between hydrogen ions and hydrogen molecules (or atoms) is greatly facilitated. This special combination of a platinum-black electrode, hydrochloric acid of unit activity, and hydrogen gas at 1 atm pressure is called the **standard hydrogen electrode** (SHE) or the **normal hydrogen electrode** (NHE).

In the right-hand compartment is a silver electrode coated with a layer of silver chloride and placed into a hydrochloric acid solution of unit activity which is saturated with silver chloride. A porous membrane can be inserted between the two compartments to prevent direct mixing of the solutions. However, the liquid-junction potential is virtually zero because the solutions on each side of the membrane are nearly identical in terms of their ionic compositions; the low solubility of silver chloride in hydrochloric acid means that both compartments contain essentially hydrochloric acid at unit activity.

This word and picture description of the cell can be portrayed by means of the

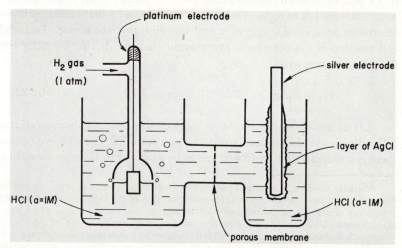

*Figure 9–6.* A galvanic cell consisting of a standard or normal hydrogen electrode (NHE) and a silver-silver chloride electrode.

shorthand cell representation

$$\text{Pt, H}_2 \text{ (1 atm)} \mid \text{HCl } (a = 1 \text{ } M), \text{AgCl(s)} \mid \text{Ag}$$

Associated with every shorthand cell representation is a so-called **conventional cell reaction**, which may or may not proceed as written when the electrodes are short-circuited. To write this cell reaction, we must adopt the following rule:

**Looking at the shorthand cell representation, we combine as *reactants* the reductant (reducing agent) of the left-hand electrode and the oxidant (oxidizing agent) of the right-hand electrode. As *products* we obtain the oxidant of the left-hand electrode and the reductant of the right-hand electrode.**

If we now apply this rule, hydrogen gas (reductant of the left-hand electrode) reacts with silver chloride (oxidant of the right-hand electrode) to yield hydrogen ions (oxidant of the left-hand electrode) and elemental silver (reductant of the right-hand electrode). Chloride ions, freed by the reduction of silver chloride, are also formed. Thus, the cell reaction is

$$\text{H}_2 + 2 \text{ AgCl} \rightleftharpoons 2 \text{ H}^+ + 2 \text{ Ag} + 2 \text{ Cl}^-$$

Information about the tendency for this cell reaction to occur as written (when the electrodes are short-circuited) is furnished by the free-energy change ($\Delta G$) for the process. According to thermodynamic conventions, the cell reaction will be spontaneous if $\Delta G$ is a negative quantity. If $\Delta G$ is positive, the reaction is nonspontaneous and, in fact, will proceed in the reverse direction. In the event that $\Delta G$ is zero, the system is at equilibrium and no net cell reaction will occur. From the relationship between free-energy change $\Delta G$ and electromotive force $E$,

$$\Delta G = -nFE$$

it can be seen that $E$ is positive for spontaneous processes, negative for nonspontaneous reactions, and zero for systems at equilibrium.*

If the emf for the cell is measured potentiometrically as described previously, it is found to have an absolute magnitude of 0.2222 v. In addition, the cell reaction proceeds from left to right as written; this fact can be demonstrated if we connect the electrodes by a conducting wire and examine their behavior. To indicate that the cell reaction is a spontaneous process and that the driving force for this reaction is 0.2222 v, we can write

$$\text{H}_2 + 2 \text{ AgCl} \rightleftharpoons 2 \text{ H}^+ + 2 \text{ Ag} + 2 \text{ Cl}^-; \qquad E = +0.2222 \text{ v}$$

Let us now focus our attention more closely on the galvanic cell itself. Since the cell reaction should proceed spontaneously from left to right, we can conclude that there is a tendency for hydrogen gas to be oxidized to hydrogen ions and a tendency

---

* Strictly speaking, a chemical process can have a free-energy change, but not an emf; only an electrochemical cell possesses an emf. However, because of the fundamental relationship between the free-energy change for an overall cell reaction and the emf for the cell in which that reaction takes place, namely $\Delta G = -nFE$, it is common practice to speak of the emf of the overall cell reaction. When we talk about the emf of an overall cell reaction, we will actually mean the free-energy change for the reaction computed from the equation in the preceding sentence. In addition, the emf of a cell properly has no sign, whereas the emf of a cell reaction is a thermodynamic quantity and can have a + or − sign in accordance with the direction in which the process is written.

for silver chloride to be reduced to silver metal. Therefore, the hydrogen gas-hydrogen ion electrode is the *anode*, and the silver-silver chloride electrode functions as the *cathode*. Likewise, there is a tendency for an excess of electrons to accumulate at the left-hand electrode and for a deficiency of electrons to exist at the right-hand electrode. We can speak only of a *tendency* because, under the conditions of an ideal potentiometric emf measurement, there is no flow of electrons. An excess of electrons at the hydrogen gas-hydrogen ion electrode causes the anode to be the *negative* electrode of the galvanic cell, and a deficiency of electrons at the silver-silver chloride electrode means that the cathode is the *positive* electrode. On the shorthand cell representation, the signs of the electrodes are indicated as follows:

$$-\text{Pt}, \text{H}_2 \ (1 \text{ atm}) \mid \text{HCl} \ (a = 1 \ M), \text{AgCl(s)} \mid \text{Ag} +$$

We come now to one of the most important and fundamental conclusions of this discussion:

**There is no direct relationship between the sign of the emf for a cell reaction and the signs (+ or −) of the electrodes of the cell. An emf for a cell reaction is a *thermodynamic* concept; its sign may be positive (+) when a process occurs spontaneously or negative (−) when a process is nonspontaneous. On the other hand, the sign of any particular electrode of a galvanic cell is an *electrochemical* concept and is invariant.**

To illustrate the significance of these statements, the following galvanic cell may be considered:

$$\text{Ag} \mid \text{AgCl(s)}, \text{HCl} \ (a = 1 \ M) \mid \text{H}_2 \ (1 \text{ atm}), \text{Pt}$$

This is the same galvanic cell we have been discussing, except for the fact that we are viewing it from the other side of the bench top. According to our rule for writing the reaction corresponding to a shorthand cell representation, we obtain

$$2 \text{ H}^+ + 2 \text{ Ag} + 2 \text{ Cl}^- \rightleftharpoons \text{H}_2 + 2 \text{ AgCl}$$

This reaction is exactly the reverse of the process we considered earlier, so it follows that the present reaction is nonspontaneous. Since the galvanic cell is the same as before, the absolute magnitude of the emf is identical to the previous value. To indicate these facts, we may write

$$2 \text{ H}^+ + 2 \text{ Ag} + 2 \text{ Cl}^- \rightleftharpoons \text{H}_2 + 2 \text{ AgCl}; \qquad E = -0.2222 \text{ v}$$

From a *thermodynamic* viewpoint, whenever we reverse the direction of a process, the magnitude of the emf remains the same, whereas the sign of the emf is changed to denote whether the process is spontaneous or nonspontaneous. Thus, the reaction *actually would proceed from right to left if the electrodes were short-circuited.* There is a tendency for hydrogen gas to be oxidized to hydrogen ions and for silver chloride to be reduced to silver metal, so the *electrochemical* conclusions made previously about the galvanic cell,

$$-\text{Pt}, \text{H}_2 \ (1 \text{ atm}) \mid \text{HCl} \ (a = 1 \ M), \text{AgCl(s)} \mid \text{Ag} +$$

are completely valid for the cell

$$+\text{Ag} \mid \text{AgCl(s)}, \text{HCl} \ (a = 1 \ M) \mid \text{H}_2 \ (1 \text{ atm}), \text{Pt} -$$

Accordingly, the hydrogen gas-hydrogen ion electrode is the *anode* and the *negative* electrode, whereas the silver-silver chloride electrode is the *cathode* and the *positive* electrode.

## STANDARD POTENTIALS AND HALF-REACTIONS

We may consider every overall cell reaction as being composed of two separate half-reactions. A **half-reaction** is a balanced chemical equation showing the transfer of electrons between two different oxidation states of the same element. Each half-reaction must represent as accurately as possible one of the two individual electrode processes which occur in a cell. For the cell

$$-Pt, H_2 \text{ (1 atm)} \,|\, HCl \,(a = 1 \, M), AgCl(s) \,|\, Ag+$$

the overall reaction is

$$H_2 + 2 \, AgCl \rightleftharpoons 2 \, H^+ + 2 \, Ag + 2 \, Cl^-$$

and the two half-reactions are
$$H_2 \rightleftharpoons 2 \, H^+ + 2 \, e$$
and
$$2 \, AgCl + 2 \, e \rightleftharpoons 2 \, Ag + 2 \, Cl^-$$

Each half-reaction contains the same number of electrons, so when these half-reactions are added together the electrons cancel and the overall reaction is obtained. Alternatively, the overall reaction could be written as

$$\tfrac{1}{2} H_2 + AgCl \rightleftharpoons H^+ + Ag + Cl^-$$

and the half-reactions as
$$\tfrac{1}{2} H_2 \rightleftharpoons H^+ + e$$
and
$$AgCl + e \rightleftharpoons Ag + Cl^-.$$

It is noteworthy that the emf for a cell or for an overall reaction is *independent* of how much reaction occurs. For example, a common 1.5-v dry cell has the same emf regardless of its physical size. A large dry cell can supply more electrical energy than a small cell, simply because it contains more chemical ingredients, but the emf values are identical. Furthermore, the emf for the reaction

$$H_2 + 2 \, AgCl \rightleftharpoons 2 \, H^+ + 2 \, Ag + 2 \, Cl^-$$

has exactly the same sign and magnitude as that for the process

$$\tfrac{1}{2} H_2 + AgCl \rightleftharpoons H^+ + Ag + Cl^-$$

although the former reaction involves twice as many units of the reactant and product species as the latter.

**Definition and determination of standard potentials.** We can regard the emf for the cell

$$-Pt, H_2 \text{ (1 atm)} \,|\, HCl \,(a = 1 \, M), AgCl(s) \,|\, Ag+$$

as the sum of two potentials, one being associated with the half-reaction occurring at each electrode of the cell.* Unfortunately, it is impossible to determine the *absolute* value for the potential of any individual electrode. All that can be measured is the *sum* of no less than two potentials (actually, potential differences) within a cell; from this sum, one cannot extract the potential of just one electrode. Therefore, it is necessary to define arbitrarily the value for the potential of some electrode chosen as a reference; then, the potentials for all other electrodes can be quoted *relative* to this standard.

Accordingly, the **normal hydrogen electrode** (NHE) has been adopted as the standard. This electrode (Figure 9–6) involves the hydrogen ion-hydrogen gas half-reaction with hydrogen ion and hydrogen gas in their standard states of unit activity and 1 atm pressure, respectively. Under these conditions, the potential for this system is assigned the value of *zero* at all temperatures and is called the *standard potential, $E^0$, for the hydrogen ion-hydrogen gas electrode*:

$$2\,H^+ + 2\,e \rightleftharpoons H_2; \quad E^0 = 0.0000\ v$$

One can determine the standard potential ($E^0$), *relative to the standard hydrogen electrode*, for an electrode at which any other half-reaction occurs, if a suitable galvanic cell is constructed in which one electrode is the standard hydrogen electrode and the second electrode involves the half-reaction of interest. Although $E^0$ for the standard hydrogen electrode is *defined* to be independent of temperature, standard potentials for electrodes involving all other half-reactions do vary with temperature. By convention, standard potentials are always determined at 25°C. Conditions necessary for the successful measurement of a standard potential are included in the definition of this quantity—**the *standard potential* for any redox couple is the potential (sign and magnitude) of an electrode consisting of that redox couple under standard-state conditions measured in a galvanic cell against the standard hydrogen electrode at 25°C.**

Notice that the cell

$$-Pt, H_2\ (1\ atm)\ |\ HCl\ (a = 1\ M),\ AgCl(s)\ |\ Ag+$$

fulfills all requirements of the definition of standard potential, inasmuch as all species are present in their standard states. Since the magnitude of the emf of the galvanic cell is 0.2222 v, and since the silver chloride-silver metal electrode is always the positive electrode (regardless of the bench-top orientation of the cell), we can conclude that *the standard potential, $E^0$, of the silver chloride-silver metal electrode is* $+0.2222\ v$ (relative to the normal hydrogen electrode).

*Table of standard potentials.* Table 9–1 presents a short list of standard potentials along with the half-reaction to which each pertains, whereas a more complete compilation is provided in Appendix 4. All of the half-reactions are written as

---

* Recall that we customarily ascribe an emf to an electrochemical cell as well as to the corresponding overall cell reaction. A similar practice is followed to describe the properties of individual electrodes and half-reactions. An electrode in contact with a solution has a potential, but the half-reaction (or electron-transfer process) which takes place at that electrode cannot truly be said to possess a potential. Instead, the half-reaction has a free-energy change ($\Delta G$) associated with it, and this free-energy change may be related mathematically through the expression, $\Delta G = -nFE$, to the potential ($E$) of the electrode at which that half-reaction occurs. Provided that we recognize these distinctions (and the fact that the potentials and free-energy changes to which we refer are relative values with respect to the normal [or standard] hydrogen electrode), *it is convenient to discuss the potential of either an electrode or a half-reaction.* However, the potential of any individual electrode is a sign-invariant quantity, whereas the potential of a half-reaction is a thermodynamic parameter and may have either a + or − sign, depending upon the direction in which the half-reaction is written. Finally, note that $E$ can symbolize either emf or potential, the context of usage indicating which meaning is intended.

## Table 9-1. Standard and Formal Potentials*

| $E^0$ | Half-reaction |
|---|---|
| 1.685 | $PbO_2 + SO_4^{2-} + 4\,H^+ + 2\,e \rightleftharpoons PbSO_4 + 2\,H_2O$ |
| 1.51 | $MnO_4^- + 8\,H^+ + 5\,e \rightleftharpoons Mn^{2+} + 4\,H_2O$ |
| *1.44* | $Ce^{4+} + e \rightleftharpoons Ce^{3+}\ (1\,F\,H_2SO_4)$ |
| 1.3595 | $Cl_2 + 2\,e \rightleftharpoons 2\,Cl^-$ |
| 1.33 | $Cr_2O_7^{2-} + 14\,H^+ + 6\,e \rightleftharpoons 2\,Cr^{3+} + 7\,H_2O$ |
| 1.15 | $SeO_4^{2-} + 4\,H^+ + 2\,e \rightleftharpoons H_2SeO_3 + H_2O$ |
| 1.000 | $VO_2^+ + 2\,H^+ + e \rightleftharpoons VO^{2+} + H_2O$ |
| 0.7995 | $Ag^+ + e \rightleftharpoons Ag$ |
| 0.771 | $Fe^{3+} + e \rightleftharpoons Fe^{2+}$ |
| 0.740 | $H_2SeO_3 + 4\,H^+ + 4\,e \rightleftharpoons Se + 3\,H_2O$ |
| *0.70* | $Fe^{3+} + e \rightleftharpoons Fe^{2+}\ (1\,F\,HCl)$ |
| *0.68* | $Fe^{3+} + e \rightleftharpoons Fe^{2+}\ (1\,F\,H_2SO_4)$ |
| 0.559 | $H_3AsO_4 + 2\,H^+ + 2\,e \rightleftharpoons HAsO_2 + 2\,H_2O$ |
| 0.5355 | $I_3^- + 2\,e \rightleftharpoons 3\,I^-$ |
| 0.361 | $VO^{2+} + 2\,H^+ + e \rightleftharpoons V^{3+} + H_2O$ |
| 0.337 | $Cu^{2+} + 2\,e \rightleftharpoons Cu$ |
| 0.334 | $UO_2^{2+} + 4\,H^+ + 2\,e \rightleftharpoons U^{4+} + 2\,H_2O$ |
| 0.2415 | $Hg_2Cl_2 + 2\,K^+ + 2\,e \rightleftharpoons 2\,Hg + 2\,KCl(s)$ |
| | (saturated calomel electrode) |
| 0.2222 | $AgCl + e \rightleftharpoons Ag + Cl^-$ |
| 0.153 | $Cu^{2+} + e \rightleftharpoons Cu^+$ |
| 0.10 | $TiO^{2+} + 2\,H^+ + e \rightleftharpoons Ti^{3+} + H_2O$ |
| 0.0000 | $2\,H^+ + 2\,e \rightleftharpoons H_2$ |
| −0.126 | $Pb^{2+} + 2\,e \rightleftharpoons Pb$ |
| −0.255 | $V^{3+} + e \rightleftharpoons V^{2+}$ |
| −0.3563 | $PbSO_4 + 2\,e \rightleftharpoons Pb + SO_4^{2-}$ |
| −0.41 | $Cr^{3+} + e \rightleftharpoons Cr^{2+}$ |
| −0.763 | $Zn^{2+} + 2\,e \rightleftharpoons Zn$ |

* Formal potentials are indicated by italicized entries. Only those standard and formal potentials utilized in this chapter are listed here; a much more complete tabulation is presented in Appendix 4.

*reductions*. This practice follows the recommendation set down by the International Union of Pure and Applied Chemistry (IUPAC) in 1953. Each half-reaction in a table of standard potentials automatically implies an overall cell reaction in which the half-reaction of interest (written as a reduction) is combined with the *hydrogen gas-hydrogen ion* half-reaction. Although only half-reactions appear in the table, standard potentials are always derived from a galvanic cell which consists of a standard hydrogen electrode and an electrode involving the desired half-reaction under standard-state conditions. For example, the half-reaction

$$AgCl + e \rightleftharpoons Ag + Cl^-$$

should be thought of as the overall reaction

$$\tfrac{1}{2}\,H_2 + AgCl \rightleftharpoons H^+ + Ag + Cl^-$$

Similarly, the entry

$$Zn^{2+} + 2\,e \rightleftharpoons Zn$$

may be considered as

$$H_2 + Zn^{2+} \rightleftharpoons 2\,H^+ + Zn$$

In turn, the first of these overall reactions pertains to the galvanic cell

$$-\text{Pt, H}_2 \text{ (1 atm)} \,\big|\, \text{HCl } (a = 1\ M), \text{AgCl(s)} \,\big|\, \text{Ag}+$$

and the second reaction is correlated with the galvanic cell

$$+\text{Pt, H}_2 \text{ (1 atm)} \,\big|\, \text{H}^+ (a = 1\ M) \,\big\|\, \text{Zn}^{2+} (a = 1\ M) \,\big|\, \text{Zn}-$$

When any half-reaction is written as a reduction (either by itself or combined in an overall reaction with the hydrogen gas-hydrogen ion half-reaction under standard-state conditions), the potential of this half-reaction (or the emf of the overall reaction, since the potential for the standard hydrogen electrode is zero) is identical in *sign* and *magnitude* to the potential for the actual physical electrode at which this half-reaction occurs. Specifically, if the reactants and products are present in their standard states, the potential of a reduction half-reaction is the same as the standard potential for the electrode at which the half-reaction takes place. For example, under standard-state conditions, the potential of the lead(II)-lead metal half-reaction written as a reduction

$$\text{Pb}^{2+} + 2\ \text{e} \rightleftharpoons \text{Pb}$$

and the standard potential for the lead(II)-lead metal electrode are both $-0.126$ v.

Standard-potential data are utilized for many types of thermodynamic calculations. In cases where reduction processes are involved, one may employ the data exactly as they appear in the table of standard potentials. Thus, for the reduction of dichromate to chromic ion, we obtain from the table

$$\text{Cr}_2\text{O}_7{}^{2-} + 14\ \text{H}^+ + 6\ \text{e} \rightleftharpoons 2\ \text{Cr}^{3+} + 7\ \text{H}_2\text{O}; \qquad E^0 = +1.33\ \text{v}$$

A positive sign for the potential indicates that, relative to the hydrogen ion-hydrogen gas half-reaction, the reduction is spontaneous. However, if we wish to consider the oxidation of chromic ion to dichromate, both the direction of the half-reaction and the sign of the potential must be reversed; that is,

$$2\ \text{Cr}^{3+} + 7\ \text{H}_2\text{O} \rightleftharpoons \text{Cr}_2\text{O}_7{}^{2-} + 14\ \text{H}^+ + 6\ \text{e}; \qquad E^0 = -1.33\ \text{v}$$

Because the potential of a half-reaction is a thermodynamic quantity, we shall use subscripts to indicate the direction of the half-reaction to which each potential corresponds, particularly when we write the potential value separately from the half-reaction. Thus, $E^0_{\text{Cr}_2\text{O}_7{}^{2-},\text{Cr}^{3+}}$ or, more generally, $E_{\text{Cr}_2\text{O}_7{}^{2-},\text{Cr}^{3+}}$ denotes the potential for the half-reaction

$$\text{Cr}_2\text{O}_7{}^{2-} + 14\ \text{H}^+ + 6\ \text{e} \rightleftharpoons 2\ \text{Cr}^{3+} + 7\ \text{H}_2\text{O}$$

$E_{\text{Zn,Zn}^{2+}}$ would represent the potential for the half-reaction

$$\text{Zn} \rightleftharpoons \text{Zn}^{2+} + 2\ \text{e}$$

and $E_{\text{I}_3{}^-,\text{I}^-}$ would stand for the potential of the half-reaction

$$\text{I}_3{}^- + 2\ \text{e} \rightleftharpoons 3\ \text{I}^-$$

Although the normal hydrogen electrode is the international standard, it is seldom used in practice because it is cumbersome as well as hazardous. Instead, secondary standard electrodes (whose potentials have been previously measured against the hydrogen electrode) are employed. Two especially good secondary standards are the saturated calomel electrode and the silver chloride-silver metal electrode, because they are easy to prepare and handle and because they are exceptionally stable and reproducible. It is a simple matter to relate the potential of an electrode, measured against a secondary standard electrode, to the standard hydrogen electrode itself.

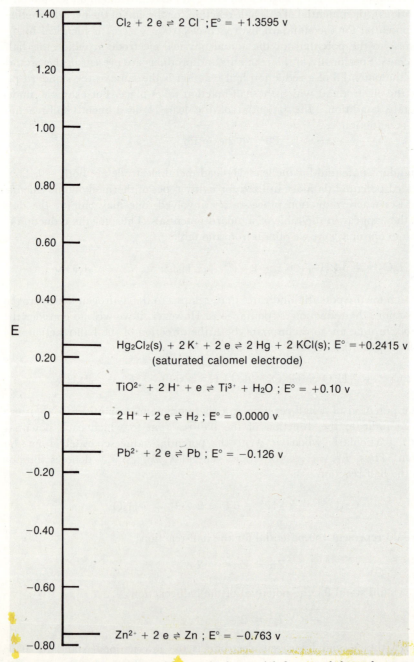

*Figure 9–7.* Relative positions of the standard potentials for several electrode systems.

Figure 9–7 shows the relative positions on the scale of standard potentials of the chlorine gas-chloride ion, saturated calomel, titanium(IV)-titanium(III), hydrogen ion-hydrogen gas, lead(II)-lead metal, and zinc(II)-zinc metal electrodes. It can be seen that the standard potentials for the chlorine gas-chloride ion, titanium(IV)-titanium(III), lead(II)-lead metal, and zinc(II)-zinc metal electrodes are +1.3595, +0.10, −0.126, and −0.763 v, respectively, relative to the normal hydrogen electrode. However, the same four electrode systems have respective standard potentials of +1.1180, −0.14, −0.368, and −1.005 v with respect to the saturated calomel electrode—these values being 0.2415 v *more negative* than the standard potentials on the hydrogen-electrode scale. Thus, if one measures the potential of some electrode against the saturated calomel electrode, it is only necessary to *add* 0.2415 v to obtain the electrode potential on the hydrogen scale.

Note that the emf of a galvanic cell composed of any pair of electrode systems included in Figure 9–7 is the vertical distance expressed in volts between the two electrodes of interest. A cell consisting of chlorine gas-chloride ion and zinc(II)-zinc metal electrodes has an emf under standard-state conditions of 1.3595 − (−0.763) or 2.123 v; the cell constructed from saturated calomel and titanium(IV)-titanium-(III) electrodes should exhibit an emf of 0.2415 − (0.10) or 0.14 v; and a cell composed of the lead(II)-lead metal and zinc(II)-zinc metal electrodes would have an emf of −0.126 − (−0.763) or 0.637 v. In addition, the positive electrode in each of these galvanic cells is the one for which the corresponding reduction half-reaction has the more positive potential. Thus, lead metal is the positive electrode in a cell consisting of lead(II)-lead and zinc(II)-zinc metal electrodes.

*Formal potentials.* Frequently, the use of standard potentials to make chemical and thermodynamic predictions is unjustified and may even lead to serious error. This situation arises because the nature of solutions is very complicated. One difficulty is that aqueous solutions of analytical interest are usually so concentrated that activity coefficients cannot be reliably calculated. In turn, if activity coefficients are unknown, the activities of ions and molecules involved in half-reactions cannot be determined. Another problem is that the simple ionic and molecular species written in half-reactions do not necessarily represent the true state of affairs in aqueous solution. For example, the entry

$$Fe^{3+} + e \rightleftharpoons Fe^{2+}; \qquad E^0 = +0.771 \text{ v}$$

in Table 9–1 applies to an aqueous solution in which the activities of the species $Fe(H_2O)_6^{3+}$ and $Fe(H_2O)_6^{2+}$ are both unity. However, it is virtually impossible to prepare an aqueous solution containing just these species, so the validity of using this standard potential may be questioned. Most cations form complexes with anions present, and some highly charged cations also form polymeric species. In a $1\,F$ hydrochloric acid medium, the species of iron(III) include $Fe(H_2O)_6^{3+}$, $Fe(H_2O)_5OH^{2+}$, $Fe(H_2O)_3Cl_3$, $Fe(H_2O)_2Cl_4^{-}$, $Fe(H_2O)_4Cl_2^{+}$, and $Fe_2(H_2O)_8$-$(OH)_2^{4+}$, whereas $Fe(H_2O)_6^{2+}$, $Fe(H_2O)_5Cl^{+}$, and $Fe(H_2O)_4Cl_2$ are likely forms of iron(II). Unfortunately, the identities of the various species and the equilibria and equilibrium constants which describe this and most other systems are difficult to ascertain.

To overcome the uncertainties resulting from the variation of activity coefficients, from the formation of complexes, and from other chemical interactions, the use of formal potentials has been suggested. A **formal potential** is defined as the potential (relative to the standard hydrogen electrode) of an electrode involving the half-reaction of interest in a specified electrolyte solution when the *formal concentrations* of the oxidant and reductant are both unity. A formal potential is usually denoted by the symbol $E^{0\prime}$. As an example, the formal potential for the iron(III)-iron(II) half-reaction is $+0.70$ v in $1 F$ hydrochloric acid, which differs significantly from the standard potential of $+0.771$ v for the iron(III)-iron(II) couple.

Some formal potentials are included in Table 9-1, as well as in Appendix 4; the specific electrolyte solutions to which these data pertain, *e.g.*, $2 F$ hydrochloric acid or $0.1 F$ potassium chloride, are indicated in parentheses following the half-reactions. When available, a formal potential should be used for thermodynamic calculations *if experimental conditions correspond closely* to those used in the original measurement of the formal potential.

## THE NERNST EQUATION

For a chemical reaction, the driving force, and thus the emf, depends on the activities (concentrations) of reactants and products. We can derive an equation which relates emf to the activities of the reactants and products by starting with the fundamental expression

$$\Delta G = \Delta G^0 + RT \ln \frac{(a_C)^c (a_D)^d}{(a_A)^a (a_B)^b}$$

This equation gives the change in free energy $\Delta G$ for any process, such as

$$aA + bB \rightleftharpoons cC + dD$$

in terms of the *standard free-energy change*, $\Delta G^0$, for the process and the activities of the reactants and products raised to appropriate powers.

It has already been indicated that the relation between emf and free-energy change is

$$\Delta G = -nFE$$

and we may write the specific equation

$$\Delta G^0 = -nFE^0$$

If we substitute the two latter expressions into the fundamental relation, the result is

$$-nFE = -nFE^0 + RT \ln \frac{(a_C)^c (a_D)^d}{(a_A)^a (a_B)^b}$$

If we divide each member of this expression by the term $-nF$, the familiar **Nernst equation**\* is obtained:

$$E = E^0 - \frac{RT}{nF} \ln \frac{(a_C)^c (a_D)^d}{(a_A)^a (a_B)^b}$$

---

\* As derived above, the Nernst equation is applicable specifically to the calculation of the emf for an overall cell reaction; but, it can be employed to compute the emf of an electrochemical cell, the potential of a half-reaction, or the potential of an electrode.

Since the *emf for a cell reaction* is a thermodynamic quantity, care must be taken that the way in which the Nernst equation is formulated corresponds to the direction in which the overall reaction is written, and that thermodynamic sign conventions are obeyed.

Since activity coefficients cannot be accurately determined for the concentrated solutions normally encountered, we will replace activities with concentrations. Furthermore, if we convert to base-ten logarithms, if the temperature $T$ is taken to be 298°K (25°C), and if the values used for $R$ and $F$ are 1.987 calories mole$^{-1}$ degree$^{-1}$ and 23,060 calories per faraday, respectively, the Nernst equation becomes

$$E = E^0 - \frac{0.05915}{n} \log \frac{[C]^c[D]^d}{[A]^a[B]^b}$$

For most calculations the numerical constant 0.05915, valid only for 25°C, is shortened to 0.059.

When we utilize the latter form of the Nernst equation, molar concentrations of dissolved ionic and molecular species are employed, and pressure in atmospheres is used for gases. In addition, the concentration of water is taken to be unity, and concentrations of pure solids are likewise assumed to be unity.*

*Example 9–4.* Write the Nernst equation corresponding to the half-reaction

$$UO_2^{2+} + 4\,H^+ + 2\,e \rightleftharpoons U^{4+} + 2\,H_2O$$

We may formulate the expression as

$$E_{UO_2^{2+}, U^{4+}} = E^0_{UO_2^{2+}, U^{4+}} - \frac{0.059}{n} \log \frac{[U^{4+}]}{[UO_2^{2+}][H^+]^4}$$

$$= +0.334 - \frac{0.059}{2} \log \frac{[U^{4+}]}{[UO_2^{2+}][H^+]^4}$$

Several features of this equation should be noted—the order of the subscripts of $E$ and $E^0$ indicates the direction of reaction, the logarithmic term contains products in the numerator and reactants in the denominator, and the value for $n$ of 2 is readily determined by inspection of the half-reaction.

---

* Although it may not be apparent, the logarithmic term of the Nernst equation contains no units (such as moles per liter or atmospheres). This is so because each individual term in the numerator and denominator is really a *ratio* of the actual concentration (activity) of a species to the concentration (activity) of the same species in its standard state, the latter being defined as unity; thus, all units cancel out.

---

If we wish to evaluate the *emf for an actual cell*, the procedure is the same as in the calculation of the emf for a cell reaction, except that the emf for a physically real cell has no sign, only a magnitude, so the directionality of the Nernst equation is immaterial.

A half-reaction cannot exist by itself, but only in combination with a second half-reaction. Thus, when the Nernst equation is used to compute the *potential of a half-reaction*, we are actually calculating the emf of an overall reaction composed of the half-reaction of interest and the hydrogen gas-hydrogen ion half-reaction under standard-state conditions. However, the potential of the latter half-reaction is defined to be zero, so the Nernst equation written for a half-reaction is an acceptable abbreviation. Since the potential of a half-reaction is a thermodynamic parameter, the Nernst equation must be formulated in the same direction as the desired half-reaction, with proper attention given to signs.

To calculate the *potential of an electrode* at which a certain half-reaction occurs, the Nernst equation for the appropriate half-reaction should be written as a *reduction*. Then, the potential (sign and magnitude) determined from the Nernst equation will be the actual potential of the electrode with respect to the standard hydrogen electrode.

In almost all discussions in this textbook, we will work with the Nernst equation written for half-reactions.

*Example 9–5.* Write Nernst equations for each of the following half-reactions:

(1)
$$2\,H^+ + 2\,e \rightleftharpoons H_2$$

$$E_{H^+,\,H_2} = E^0_{H^+,\,H_2} - \frac{0.059}{n} \log \frac{p_{H_2}}{[H^+]^2} = 0.0000 - \frac{0.059}{2} \log \frac{p_{H_2}}{[H^+]^2}$$

(2)
$$Cr_2O_7^{2-} + 14\,H^+ + 6\,e \rightleftharpoons 2\,Cr^{3+} + 7\,H_2O$$

$$E_{Cr_2O_7^{2-},\,Cr^{3+}} = E^0_{Cr_2O_7^{2-},\,Cr^{3+}} - \frac{0.059}{n} \log \frac{[Cr^{3+}]^2}{[Cr_2O_7^{2-}][H^+]^{14}}$$

$$= +1.33 - \frac{0.059}{6} \log \frac{[Cr^{3+}]^2}{[Cr_2O_7^{2-}][H^+]^{14}}$$

(3)
$$MnO_4^- + 8\,H^+ + 5\,e \rightleftharpoons Mn^{2+} + 4\,H_2O$$

$$E_{MnO_4^-,\,Mn^{2+}} = E^0_{MnO_4^-,\,Mn^{2+}} - \frac{0.059}{n} \log \frac{[Mn^{2+}]}{[MnO_4^-][H^+]^8}$$

$$= +1.51 - \frac{0.059}{5} \log \frac{[Mn^{2+}]}{[MnO_4^-][H^+]^8}$$

Since the potential of a half-reaction is independent of the quantity of chemical reaction, each of the preceding relations could be multiplied or divided by an integer without altering the value of $E$ obtained from the Nernst equation. For example, the dichromate-chromic ion half-reaction could be written as

$$\tfrac{1}{2}\,Cr_2O_7^{2-} + 7\,H^+ + 3\,e \rightleftharpoons Cr^{3+} + 3\tfrac{1}{2}\,H_2O$$

and the corresponding Nernst equation would be

$$E_{Cr_2O_7^{2-},\,Cr^{3+}} = E^0_{Cr_2O_7^{2-},\,Cr^{3+}} - \frac{0.059}{n} \log \frac{[Cr^{3+}]}{[Cr_2O_7^{2-}]^{1/2}[H^+]^7}$$

$$= +1.33 - \frac{0.059}{3} \log \frac{[Cr^{3+}]}{[Cr_2O_7^{2-}]^{1/2}[H^+]^7}$$

Sometimes it is necessary or desirable to write a half-reaction as an oxidation rather than a reduction. If we are interested in calculating $E$ for the half-reaction

$$Mn^{2+} + 4\,H_2O \rightleftharpoons MnO_4^- + 8\,H^+ + 5\,e$$

the proper form for the Nernst equation is

$$E_{Mn^{2+},\,MnO_4^-} = E^0_{Mn^{2+},\,MnO_4^-} - \frac{0.059}{n} \log \frac{[MnO_4^-][H^+]^8}{[Mn^{2+}]}$$

$$= -1.51 - \frac{0.059}{5} \log \frac{[MnO_4^-][H^+]^8}{[Mn^{2+}]}$$

Notice that the order of the subscripts for $E$ and $E^0$ has been appropriately reversed to show the direction of the half-reaction being considered. Also, the sign of $E^0$ has been changed and the concentration terms in the logarithmic part of the equation have been inverted.

# APPLICATIONS OF STANDARD POTENTIALS AND THE NERNST EQUATION

This section of the chapter is concerned with specific examples of the various types of thermodynamic calculations which make use of standard-potential data and the Nernst equation.

*Calculation of the electromotive force for an overall cell reaction.* Since knowledge of the emf for an overall cell reaction is often desirable, we shall first consider how to calculate this quantity.

*Example 9-6.* Calculate the emf for the overall reaction corresponding to the galvanic cell

$$Pb \mid Pb^{2+} \ (a = 1 \ M) \parallel Cr^{2+} \ (a = 1 \ M), Cr^{3+} \ (a = 1 \ M) \mid Pt$$

First, we should write the overall cell reaction according to the rule established earlier:

$$Pb + 2 \ Cr^{3+} \rightleftharpoons Pb^{2+} + 2 \ Cr^{2+}$$

This overall reaction can be viewed in terms of its component half-reactions,

$$Pb \rightleftharpoons Pb^{2+} + 2 \ e$$

and

$$2 \ Cr^{3+} + 2 \ e \rightleftharpoons 2 \ Cr^{2+}$$

Since the activities of all species which comprise the galvanic cell are unity, the value for the potential of each half-reaction may be obtained directly from Table 9–1. Thus, we can write

$$E_{Pb,Pb^{2+}} = E^0_{Pb,Pb^{2+}} = +0.126 \ v$$

and

$$E_{Cr^{3+},Cr^{2+}} = E^0_{Cr^{3+},Cr^{2+}} = -0.41 \ v$$

Although we are considering two units of the chromium(III)-chromium(II) half-reaction, we do *not* multiply $E^0_{Cr^{3+},Cr^{2+}}$ by a factor of two—emf or potential is an intensity factor and does not depend on the quantity of reaction.

Now, if we combine two half-reactions to obtain an overall reaction, what relationships exist between the properties of the half-reactions and the properties of the net reaction? It is rigorously true that the values of an *extensive* thermodynamic property for the individual half-reactions can be *added* to yield the value of the same *extensive* thermodynamic property for the net reaction. An **extensive thermodynamic property**, such as free-energy change ($\Delta G$), enthalpy change ($\Delta H$), or entropy change ($\Delta S$), is one whose value is proportional to the amount of reaction being considered. Thus, the free-energy change for the overall reaction is equal to the *sum* of the free-energy changes for the two individual half-reactions:

$$\Delta G_{overall} = \Delta G_{Pb,Pb^{2+}} + \Delta G_{Cr^{3+},Cr^{2+}}$$

However, since

$$\Delta G = -nFE$$

we can write

$$-nFE_{overall} = -nFE_{Pb,Pb^{2+}} - nFE_{Cr^{3+},Cr^{2+}}$$

or

$$-nFE_{overall} = -nFE^0_{Pb,Pb^{2+}} - nFE^0_{Cr^{3+},Cr^{2+}}$$

Inspection of the two half-reactions, which when added give the overall cell reaction, shows that each involves *two electrons*, since the electrons must cancel to produce the net reaction. In addition, the overall cell reaction must itself involve *two electrons*. Accordingly, the value of $n$ is the *same* for all three terms of the preceding equation. Dividing each term by the common factor $-nF$, we obtain

$$E_{\text{overall}} = E^0_{\text{Pb,Pb}^{2+}} + E^0_{\text{Cr}^{3+},\text{Cr}^{2+}} = +0.126 - 0.41 = -0.28 \text{ v}$$

Although this result is indistinguishable from that obtained by straightforward addition of the potentials for the two half-reactions, in general it is not permissible to add *intensive* thermodynamic properties such as potential, emf, or pressure. Only because $n$ happens to be identical for the overall cell reaction and the two half-reactions does the sum of the potentials yield the same result as the sum of the free-energy changes.

*Example 9–7.* Calculate the emf for the overall reaction pertaining to the galvanic cell

$$\text{Cu} \mid \text{Cu}^{2+} \ (0.01 \ M) \parallel \text{Fe}^{2+} \ (0.1 \ M), \ \text{Fe}^{3+} \ (0.02 \ M) \mid \text{Pt}$$

First, the overall cell reaction and the two half-reactions are as follows:

$$\text{Cu} + 2 \ \text{Fe}^{3+} \rightleftharpoons \text{Cu}^{2+} + 2 \ \text{Fe}^{2+}$$

$$\text{Cu} \rightleftharpoons \text{Cu}^{2+} + 2 \ \text{e}$$

$$2 \ \text{Fe}^{3+} + 2 \ \text{e} \rightleftharpoons 2 \ \text{Fe}^{2+}$$

Since the free-energy change for the overall reaction equals the sum of the free-energy changes for the two half-reactions, we have

$$\Delta G_{\text{overall}} = \Delta G_{\text{Cu,Cu}^{2+}} + \Delta G_{\text{Fe}^{3+},\text{Fe}^{2+}}$$

or

$$-nFE_{\text{overall}} = -nFE_{\text{Cu,Cu}^{2+}} - nFE_{\text{Fe}^{3+},\text{Fe}^{2+}}$$

Again, the overall reaction and the half-reactions involve the same number of electrons, so

$$E_{\text{overall}} = E_{\text{Cu,Cu}^{2+}} + E_{\text{Fe}^{3+},\text{Fe}^{2+}}$$

For the present example, the concentrations of the various ions differ from unity, and so it is necessary to use the Nernst equation to calculate the potential for each half-reaction:

$$E_{\text{Cu,Cu}^{2+}} = E^0_{\text{Cu,Cu}^{2+}} - \frac{0.059}{n} \log [\text{Cu}^{2+}] = -0.337 - \frac{0.059}{2} \log (0.01)$$

$$= -0.337 - \frac{0.059}{2} (-2) = -0.278 \text{ v}$$

$$E_{\text{Fe}^{3+},\text{Fe}^{2+}} = E^0_{\text{Fe}^{3+},\text{Fe}^{2+}} - \frac{0.059}{n} \log \frac{[\text{Fe}^{2+}]}{[\text{Fe}^{3+}]}$$

$$= +0.771 - \frac{0.059}{1} \log \frac{(0.1)}{(0.02)} = +0.771 - 0.059 \log 5$$

$$= +0.771 - 0.059 (0.70) = +0.730 \text{ v}$$

$$E_{\text{overall}} = -0.278 + 0.730 = +0.452 \text{ v}$$

It is not necessary that an overall chemical reaction be related to a galvanic cell in which that reaction will occur. Thus, Example 9–7 could have been phrased as follows: Calculate the emf for the reaction

$$Cu + 2\ Fe^{3+} \rightleftharpoons Cu^{2+} + 2\ Fe^{2+}$$

when the initial concentrations of cupric, ferric, and ferrous ions are 0.01, 0.02, and 0.1 $M$, respectively. Our method of solving the problem as well as the numerical result would be identical to that just presented, but mention of a galvanic cell is omitted.

**Calculation of equilibrium constants.** One of the most important applications of emf measurements is the evaluation of equilibrium constants for chemical reactions.

A relation between the equilibrium constant and the emf for an overall reaction can be derived. Recall from Chapter 3 that the standard free-energy change for a reaction is related to the equilibrium constant (written in terms of *concentrations*) by the equation

$$\Delta G^0 = -RT \ln K$$

In addition, the standard free-energy change and the emf for a reaction under standard-state conditions may be equated through the expression

$$\Delta G^0 = -nFE^0$$

By combining these two equations, we obtain

$$RT \ln K = nFE^0$$

If we convert this relation for use with base-ten logarithms and substitute appropriate values for the various physical constants ($F = 23,060$ calories per faraday, $R = 1.987$ calories mole$^{-1}$ degree$^{-1}$, and $T = 298°K$), the following result is obtained:

$$\log K = \frac{nE^0}{0.059}$$

This equation is valid only for 25°C or 298°K. It should be noted that $E^0$ is the emf for the reaction of interest under *standard-state conditions*; this value can be easily determined according to the procedure developed in the two preceding examples.

*Example 9–8.* Compute the equilibrium constant for the reaction

$$HAsO_2 + I_3^- + 2\ H_2O \rightleftharpoons H_3AsO_4 + 2\ H^+ + 3\ I^-$$

We may express the equilibrium constant as

$$K = \frac{[H_3AsO_4][H^+]^2[I^-]^3}{[HAsO_2][I_3^-]}$$

Next, we can show that the overall reaction is composed of the half-reactions

$$HAsO_2 + 2\ H_2O \rightleftharpoons H_3AsO_4 + 2\ H^+ + 2\ e$$

and

$$I_3^- + 2\ e \rightleftharpoons 3\ I^-$$

which appear side by side in Table 9–1. In accord with previous calculations, we may write

$$\Delta G^0_{\text{overall}} = \Delta G^0_{\text{HAsO}_2, \text{H}_3\text{AsO}_4} + \Delta G^0_{\text{I}_3^-, \text{I}^-}$$

or

$$-nFE^0_{\text{overall}} = -nFE^0_{\text{HAsO}_2, \text{H}_3\text{AsO}_4} - nFE^0_{\text{I}_3^-, \text{I}^-}$$

which, because the $n$ values are identical, reduces to

$$E^0_{\text{overall}} = E^0_{\text{HAsO}_2, \text{H}_3\text{AsO}_4} + E^0_{\text{I}_3^-, \text{I}^-}$$

From data in Table 9–1, it follows that

$$E^0_{\text{overall}} = -0.559 + 0.536 = -0.023 \text{ v}$$

If this value is substituted into the expression for log $K$ along with $n = 2$, we find that

$$\log K = \frac{nE^0}{0.059} = \frac{(2)(-0.023)}{0.059} = -0.78; \qquad K = 0.17$$

*Example 9–9.* From standard-potential data, calculate the solubility-product constant for silver chloride, which dissolves according to the reaction

$$\text{AgCl} \rightleftharpoons \text{Ag}^+ + \text{Cl}^-$$

A search of Table 9–1 shows that the two half-reactions

$$\text{Ag} \rightleftharpoons \text{Ag}^+ + \text{e}$$

and

$$\text{AgCl} + \text{e} \rightleftharpoons \text{Ag} + \text{Cl}^-$$

may be added together to yield the desired overall reaction. It follows that

$$\Delta G^0_{\text{overall}} = \Delta G^0_{\text{Ag}, \text{Ag}^+} + \Delta G^0_{\text{AgCl}, \text{Ag}}$$

and

$$-nFE^0_{\text{overall}} = -nFE^0_{\text{Ag}, \text{Ag}^+} - nFE^0_{\text{AgCl}, \text{Ag}}$$

but, since $n = 1$, we obtain

$$E^0_{\text{overall}} = E^0_{\text{Ag}, \text{Ag}^+} + E^0_{\text{AgCl}, \text{Ag}} = -0.7995 + 0.2222 = -0.5773 \text{ v}$$

Substituting this latter value for $E^0_{\text{overall}}$ into our equation for log $K$, we obtain

$$\log K = \frac{nE^0}{0.05915} = \frac{(1)(-0.5773)}{0.05915} = -9.760$$

$$K = K_{\text{sp}} = 1.74 \times 10^{-10}$$

Using this approach, we can calculate equilibrium constants for redox reactions, solubility products, formation constants for complex ions, and dissociation constants for acids and bases. All we need do is represent the desired reaction as the sum of appropriate half-reactions. If standard-potential data are available, the calculation may be performed as in the preceding two examples. However, if the pertinent standard potentials are unknown, it may be possible to set up a galvanic cell suitable for the experimental determination of the equilibrium constant.

***Calculation of the standard potential for an unknown half-reaction.*** Another application of standard-potential data is the evaluation of standard potentials for half-reactions which do not appear in a table.

*Example 9–10.* Calculate the standard potential which pertains to the half-reaction

$$Cu^+ + e \rightleftharpoons Cu$$

Note that the potential for this reduction half-reaction under standard-state conditions is identical in sign and magnitude to the standard potential for the copper(I)-copper metal electrode. However, the standard potential is not to be found in either Table 9–1 or Appendix 4, so it must be calculated from other information. Two entries which do appear in Table 9–1 are

$$Cu^+ \rightleftharpoons Cu^{2+} + e$$

and

$$Cu^{2+} + 2 e \rightleftharpoons Cu$$

which, if added together in the manner just written, will yield the desired half-reaction.

Following the pattern of earlier examples, we can state that the standard free-energy change for the copper(I)-copper metal half-reaction is the sum of the standard free-energy changes for the other two half-reactions:

$$\Delta G^0_{Cu^+, Cu} = \Delta G^0_{Cu^+, Cu^{2+}} + \Delta G^0_{Cu^{2+}, Cu}$$

Each free-energy term can be related to the potential for the appropriate half-reaction. Thus,

$$- nFE^0_{Cu^+, Cu} = - nFE^0_{Cu^+, Cu^{2+}} - nFE^0_{Cu^{2+}, Cu}$$

If we substitute values for $n$ from each half-reaction written above and values for $E^0_{Cu^+, Cu^{2+}}$ and $E^0_{Cu^{2+}, Cu}$ from Table 9–1, the equation becomes

$$-1 \ FE^0_{Cu^+, Cu} = -1 \ F(-0.153) - 2 \ F(+0.337)$$

By dividing each term of this expression by $-F$, we can simplify the relation to

$$E^0_{Cu^+, Cu} = -0.153 + 0.674 = +0.521 \text{ v}$$

Thus, the standard potential for the copper(I)-copper metal electrode is $+0.521$ v versus the standard hydrogen electrode.

*Example 9–11.* Calculate the standard potential for the half-reaction

$$SeO_4^{2-} + 8 H^+ + 6 e \rightleftharpoons Se + 4 H_2O$$

Two half-reactions in Table 9–1 from which the desired standard potential can be computed are as follows:

$$SeO_4^{2-} + 4 H^+ + 2 e \rightleftharpoons H_2SeO_3 + H_2O; \qquad E^0 = +1.15 \text{ v}$$

$$H_2SeO_3 + 4 H^+ + 4 e \rightleftharpoons Se + 3 H_2O; \qquad E^0 = +0.740 \text{ v}$$

To begin, we may write

$$\Delta G^0_{SeO_4^{2-},Se} = \Delta G^0_{SeO_4^{2-},H_2SeO_3} + \Delta G^0_{H_2SeO_3,Se}$$

and

$$-nFE^0_{SeO_4^{2-},Se} = -nFE^0_{SeO_4^{2-},H_2SeO_3} - nFE^0_{H_2SeO_3,Se}$$

From the above half-reactions, it follows that

$$-6\ FE^0_{SeO_4^{2-},Se} = -2\ F(+1.15) - 4\ F(+0.740)$$

or

$$6\ E^0_{SeO_4^{2-},Se} = 2(+1.15) + 4(+0.740)$$

Solution of the preceding expression yields

$$E^0_{SeO_4^{2-},Se} = \tfrac{1}{3}(+1.15) + \tfrac{2}{3}(+0.740) = +0.383 + 0.493 = +0.876\ v$$

This result is the standard potential for the selenate-selenium metal electrode relative to the standard hydrogen electrode.

## THE CERIUM(IV)-IRON(II) TITRATION

One of the classic oxidation-reduction or redox procedures is the determination of iron(II) by titration with a standard solution of cerium(IV):

$$Fe^{2+} + Ce^{4+} \rightleftharpoons Fe^{3+} + Ce^{3+}$$

This simple representation implies the reaction between $Fe(H_2O)_6^{2+}$ and $Ce(H_2O)_6^{4+}$ to form $Fe(H_2O)_6^{3+}$ and $Ce(H_2O)_6^{3+}$. Although the true nature of the reactants and products is complicated and unknown, these simple aquated ions do not predominate in aqueous media. Cerium(IV) is so strong an acid that proton-transfer reactions such as

$$Ce(H_2O)_6^{4+} + H_2O \rightleftharpoons Ce(H_2O)_5(OH)^{3+} + H_3O^+$$

$$Ce(H_2O)_5(OH)^{3+} + H_2O \rightleftharpoons Ce(H_2O)_4(OH)_2^{2+} + H_3O^+$$

and

$$2\ Ce(H_2O)_5(OH)^{3+} + H_2O \rightleftharpoons [(OH)(H_2O)_4CeOCe(H_2O)_4(OH)]^{4+} + 2\ H_3O^+$$

occur to a significant extent even in $1\ F$ acid media. Species produced by these reactions, especially dimeric $[(OH)(H_2O)_4CeOCe(H_2O)_4(OH)]^{4+}$, tend to interact somewhat slowly with iron(II). Iron(II) is not nearly as acidic as cerium(IV), and reactions of the types just written do not occur until the pH of a solution exceeds 7.

Usually, the titration of iron(II) is performed in a 0.5 to $2\ F$ sulfuric acid medium, the presence of a large concentration of acid helping to repress the acid dissociation of cerium(IV) and to promote the rapid reaction between iron(II) and cerium(IV), and with the titrant consisting of a solution of cerium(IV) in $1\ F$ sulfuric acid. In sulfuric acid the chemistry of both cerium(IV) and iron(II) is modified by formation of complexes such as $Ce(H_2O)_5SO_4^{2+}$ and $Fe(H_2O)_4(SO_4)_2^{2-}$.

In the present discussion, we shall consider the cerium(IV)-iron(II) titration in a $1\ F$ sulfuric acid medium. Therefore, it is appropriate to utilize the *formal* potentials for the iron(III)-iron(II) and cerium(IV)-cerium(III) half-reactions in $1\ F$ sulfuric acid, which are listed in Table 9–1 as

$$Ce^{4+} + e \rightleftharpoons Ce^{3+}; \quad E^{0\prime} = +1.44 \text{ v}$$

and

$$Fe^{3+} + e \rightleftharpoons Fe^{2+}; \quad E^{0\prime} = +0.68 \text{ v}$$

Such data account for the combined effects of the acidic characters of the reactants and products, the formation of sulfate complexes, and the ionic strength. If the equilibrium constant for the titration reaction is computed according to the method discussed in the preceding section, the result is

$$\log K = \frac{nE^{0\prime}}{0.059} = \frac{(1)(-0.68 + 1.44)}{0.059} = +12.9$$

$$K = 8 \times 10^{12}$$

Such a relatively large equilibrium constant indicates that the reaction between iron(II) and cerium(IV) is strongly favored and, therefore, that the quantitative determination of iron(II) should be successful. However, a thermodynamically favorable reaction may not proceed at a conveniently rapid rate. For some redox reactions, the rate of electron exchange between the reductant and oxidant is so slow that the reaction cannot be usefully employed as a titrimetric method of analysis. We shall mention some features of slow redox reactions in subsequent chapters. Fortunately, the iron(II)-cerium(IV) reaction is fast, partly because it involves the transfer of only one electron.

*Construction of the titration curve.* As for all types of titrations, calculation and construction of a titration curve provides insight into the probable success of the proposed titration and gives information which aids in the selection of either a colored end-point indicator or a physico-chemical method of end-point detection. We shall discuss the titration of 25.00 ml of 0.1000 $M$ iron(II) with 0.1000 $M$ cerium(IV) in a 1 $F$ sulfuric acid medium.

According to the Nernst equation, the potential of an electrode depends on the concentrations (activities) of reducing and oxidizing agents in solution. Therefore, the variation of potential as a function of the volume of added titrant serves to show the progress of a redox titration. A key point of this discussion centers on the technique used to monitor the progress of the titration. In order to observe continuously the variation of potential during the course of the titration, it is necessary to make potentiometric measurements of the emf of a galvanic cell which consists of a reference electrode and an indicator electrode. A **reference electrode** is one whose potential is fixed with respect to the standard hydrogen electrode. In principle, the reference electrode may be the standard hydrogen electrode, but the saturated calomel electrode is almost universally employed. An **indicator electrode** is one whose potential varies in accord with the Nernst equation as the concentrations of reactant and product change during a titration. For example, a platinum wire dipped into an iron(II)-iron(III) mixture in 1 $F$ sulfuric acid exhibits a potential which depends on the ratio of concentrations (activities) of iron(II) and iron(III). When a reference electrode and indicator electrode are combined to make a galvanic cell, any change in the overall cell emf is due solely to a change in the potential of the indicator electrode, which reflects in turn a variation in the concentrations of oxidized and reduced species. For convenience, we shall take the standard hydrogen electrode (with a potential defined to be zero) as the reference electrode and a platinum wire electrode as the indicator electrode. Then the emf of the complete galvanic cell consisting of these two electrodes will be identical in magnitude to the potential of the indicator electrode alone.

It should be recalled that the *sign* and *magnitude* of the potential of an electrode in a galvanic cell do not vary if the bench-top orientation of a cell is altered, whereas the potential of a half-reaction as computed from the Nernst equation is a thermo-dynamic quantity and does depend on the direction in which the half-reaction is written. For the purpose of constructing a redox titration curve, *the potential for the half-reaction of interest, when it is written as a reduction process, is always identical in both sign and magnitude to the potential of the indicator electrode versus the standard hydrogen electrode.* Therefore, as we construct the titration curve, we shall apply the Nernst equation specifically to half-reactions written as reductions.

AT THE START OF THE TITRATION. It is not possible to predict what the potential of the indicator electrode is at the beginning of the titration, since the iron(II)-iron(III) concentration ratio is unknown; thus, the Nernst equation cannot be applied. Although the sample solution is specified to contain only iron(II), a very tiny amount of iron(III) will inevitably exist. There are three reasons why this is so. First, a small quantity of iron(III) is present because dissolved atmospheric oxygen slowly oxidizes iron(II). Second, in the determination of iron(II) by titration with a standard solution of cerium(IV) or some other oxidant, the original sample invariably consists either partly or wholly of iron(III). Prior to the actual titration, the iron(III) must be converted quantitatively to iron(II) with the aid of some reducing agent. Regardless of the strength of this reducing agent, at least a trace of unreduced iron(III) will remain. Third, an absolutely pure solution of iron(II) in aqueous sulfuric acid would be unstable; iron(II) would reduce hydrogen ion to hydrogen gas, forming a very small amount of iron(III), because a hypothetical solution initially containing only iron(II) would possess an infinitely great reducing strength. Under the usual titration conditions, the quantity of iron(III) formed or remaining unreduced is too small to cause a significant error in the determination of iron.

PRIOR TO THE EQUIVALENCE POINT. As the first definite point on the titration curve, let us calculate the potential of the platinum indicator electrode after the addition of 1.00 ml of 0.1000 $M$ cerium(IV) titrant to the 25.00-ml sample of 0.1000 $M$ iron(II) in 1 $F$ sulfuric acid.

Initially, 2.500 millimoles of iron(II) is present. Addition of 1.00 ml of 0.1000 $M$ cerium(IV), or 0.100 millimole, results in the formation of 0.100 millimole of iron(III) and 0.100 millimole of cerium(III) as reaction products, and 2.400 millimoles of iron(II) remains. These conclusions follow from the stoichiometry of the titration reaction, because the equilibrium constant is so large ($K = 8 \times 10^{12}$) that the reaction between cerium(IV) and iron(II) is essentially complete. For the same reason, the quantity of cerium(IV) titrant remaining unreacted is extremely small, but may be computed from the known quantities of cerium(III), iron(II), and iron(III) and the equilibrium constant. It should be emphasized that the potential of the platinum indicator electrode can have only one value which is governed in principle by either the iron(II)-iron(III) concentration ratio or the cerium(III)-cerium(IV) concentration ratio. However, the iron(II)-iron(III) ratio prior to the equivalence point can be easily determined from the reaction stoichiometry, whereas the determination of the cerium(III)-cerium(IV) ratio requires a calculation involving the use of the equilibrium expression. Therefore, it is more straightforward to employ the Nernst equation for the iron(III)-iron(II) half-reaction

$$E_{\mathrm{Fe^{3+},\,Fe^{2+}}} = E^{0'}_{\mathrm{Fe^{3+},\,Fe^{2+}}} - \frac{0.059}{n} \log \frac{[\mathrm{Fe^{2+}}]}{[\mathrm{Fe^{3+}}]}$$

for all points on the titration curve *prior* to the equivalence point. Since the logarithmic term involves the ratio of two concentrations, each raised to the first power, it is not necessary to convert the quantities of iron(II) and iron(III) expressed in millimoles into concentration units. After the addition of 1.00 ml of cerium(IV) titrant, there are 0.100 millimole of iron(III) and 2.400 millimoles of iron(II) present. Thus,

$$E_{Fe^{3+}, Fe^{2+}} = +0.68 - 0.059 \log \frac{(2.400)}{(0.100)} = +0.68 - 0.059 \log 24$$

$$= +0.68 - 0.08 = +0.60 \text{ v}$$

which is the potential of the platinum indicator electrode with respect to the standard hydrogen electrode.

Similar calculations were performed for other titrant volumes prior to the equivalence point up to 24.99 ml, the results of which are listed in Table 9–2. Even at a point on the titration curve as close to the equivalence point as 24.99 ml, it is valid to determine the quantities of iron(III) and iron(II) from the simple reaction stoichiometry. For points much closer to the equivalence point, it is necessary to use the equilibrium expression for the titration reaction to obtain accurate values for the concentrations of iron(III) and iron(II).

AT THE EQUIVALENCE POINT. To calculate the potential of the platinum indicator electrode at the equivalence point, recall that the potential can be viewed as being determined by either the iron(III)-iron(II) or the cerium(IV)-cerium(III) half-reaction. Let us symbolize the potential for the indicator electrode as well as for each half-reaction (written as a reduction) by $E_{ep}$ at the equivalence point. Then the Nernst equation for each half-reaction may be written:

$$E_{ep} = E^{0\prime}_{Fe^{3+}, Fe^{2+}} - 0.059 \log \frac{[Fe^{2+}]}{[Fe^{3+}]}$$

$$E_{ep} = E^{0\prime}_{Ce^{4+}, Ce^{3+}} - 0.059 \log \frac{[Ce^{3+}]}{[Ce^{4+}]}$$

Table 9–2. Variation of Potential of a Platinum Indicator Electrode During Titration of 25.00 ml of 0.1000 *M* Iron(II) with 0.1000 *M* Cerium(IV)

| Volume of Cerium(IV), ml | E *vs.* NHE |
| --- | --- |
| 0 | — |
| 1.00 | +0.60 |
| 2.00 | 0.62 |
| 5.00 | 0.64 |
| 10.00 | 0.67 |
| 12.50 | 0.68 |
| 15.00 | 0.69 |
| 20.00 | 0.72 |
| 24.00 | 0.76 |
| 24.50 | 0.78 |
| 24.90 | 0.82 |
| 24.99 | 0.88 |
| 25.00 | 1.06 |
| 26.00 | 1.36 |
| 30.00 | 1.40 |
| 40.00 | 1.43 |
| 50.00 | 1.44 |

An inspection of the titration reaction

$$Fe^{2+} + Ce^{4+} \rightleftharpoons Fe^{3+} + Ce^{3+}$$

provides us with two useful relations. By definition, the *equivalence point* of any titration is the point at which the quantity of added titrant is stoichiometrically equivalent to the substance being titrated. Since we are titrating 2.500 millimoles of iron(II), the equivalence point corresponds to the addition of 2.500 millimoles of cerium(IV). No reaction is ever 100 per cent complete, but the iron(II)-cerium(IV) reaction has a large equilibrium constant and does attain a high degree of completion. At the equivalence point, the small concentration of unreacted iron(II) is equal to the small concentration of unreacted cerium(IV):

$$[Fe^{2+}] = [Ce^{4+}]$$

Similarly, the concentrations of the products, iron(III) and cerium(III), are equal:

$$[Fe^{3+}] = [Ce^{3+}]$$

If we rewrite the first of the preceding two Nernst equations,

$$E_{ep} = E^{0\prime}_{Fe^{3+},Fe^{2+}} - 0.059 \log \frac{[Fe^{2+}]}{[Fe^{3+}]}$$

substitute the concentration relationships just derived into the second Nernst equation,

$$E_{ep} = E^{0\prime}_{Ce^{4+},Ce^{3+}} - 0.059 \log \frac{[Fe^{3+}]}{[Fe^{2+}]}$$

and add these two equations together, the result is

$$2E_{ep} = E^{0\prime}_{Fe^{3+},Fe^{2+}} + E^{0\prime}_{Ce^{4+},Ce^{3+}} - 0.059 \log \frac{[Fe^{2+}]}{[Fe^{3+}]} - 0.059 \log \frac{[Fe^{3+}]}{[Fe^{2+}]}$$

However, the two logarithmic terms cancel, so at the equivalence point the potential of the platinum indicator electrode with respect to the standard hydrogen electrode is

$$E_{ep} = \frac{E^{0\prime}_{Fe^{3+},Fe^{2+}} + E^{0\prime}_{Ce^{4+},Ce^{3+}}}{2} = \frac{0.68 + 1.44}{2} = +1.06 \text{ v}$$

In the present case, the potential at the equivalence point is the simple average of the formal potentials for the two half-reactions. However, in general the equivalence-point potential is a *weighted* average of the formal or standard potentials for the two half-reactions involved in the titration. In some situations the potential at the equivalence point may depend on the pH and on the concentrations of one or more of the reactants or products.

It is of interest to use the equivalence-point potential to calculate the concentration of iron(II) and thereby determine how complete the oxidation of iron(II) to iron(III) is. If we assume that the iron(III) concentration is essentially 0.05 $M$, which because of dilution is exactly one half of the original concentration of iron(II), and substitute this value along with $E_{ep}$ and $E^{0'}_{Fe^{3+}, Fe^{2+}}$ into the Nernst equation,

$$E_{ep} = E^{0'}_{Fe^{3+}, Fe^{2+}} - 0.059 \log \frac{[Fe^{2+}]}{[Fe^{3+}]}$$

we get

$$+1.06 = +0.68 - 0.059 \log \frac{[Fe^{2+}]}{(0.05)}$$

$$+0.38 = -0.059 \log \frac{[Fe^{2+}]}{(0.05)}$$

$$-\frac{0.38}{0.059} = -6.4 = \log \frac{[Fe^{2+}]}{(0.05)}$$

$$\frac{[Fe^{2+}]}{(0.05)} = 4 \times 10^{-7}$$

$$[Fe^{2+}] = 2 \times 10^{-8} \, M$$

BEYOND THE EQUIVALENCE POINT.   Points on the titration curve after the equivalence point can be most readily determined from the Nernst equation for the cerium(IV)-cerium(III) reaction. It is reasonable to assert that 2.500 millimoles of cerium(III) is present at and beyond the equivalence point, because this is the quantity formed upon oxidation of 2.500 millimoles of iron(II) by cerium(IV). Addition of cerium(IV) after the equivalence point simply introduces excess unreacted cerium(IV). For example, after the addition of a *total* of 26.00 ml of 0.1000 $M$ cerium(IV) titrant, or 1.00 ml in excess of the volume needed to reach the equivalence point, the solution contains 2.500 millimoles of cerium(III) and 0.100 millimole of cerium(IV). Thus we can write

$$E_{Ce^{4+}, Ce^{3+}} = E^{0'}_{Ce^{4+}, Ce^{3+}} - \frac{0.059}{n} \log \frac{[Ce^{3+}]}{[Ce^{4+}]}$$

$$E_{Ce^{4+}, Ce^{3+}} = +1.44 - 0.059 \log \frac{(2.500)}{(0.100)} = +1.44 - 0.059 \log 25$$

$$= +1.44 - 0.08 = +1.36 \, v$$

Results of other calculations of the potential of the indicator electrode beyond the equivalence point are included in Table 9–2.

After the addition of 50.00 ml of cerium(IV) titrant, $E_{Ce^{4+}, Ce^{3+}} = +1.44$ v. This is a special point on the titration curve, for the quantities of cerium(III) and cerium(IV) present are equal. Thus, the potential of the indicator electrode is identical to the formal potential for the cerium(IV)-cerium(III) half-reaction. There is another special point on the titration curve which occurs halfway to the equivalence point. After the addition of exactly 12.50 ml of cerium(IV) titrant, the quantity of iron(III) produced and the amount of untitrated iron(II) are the same, so the potential of the indicator electrode is identical to the formal potential ($+0.68$ v) for the iron(III)-iron(II) couple in 1 $F$ sulfuric acid. In Figure 9–8 is plotted the complete titration curve for the iron(II)-cerium(IV) reaction.

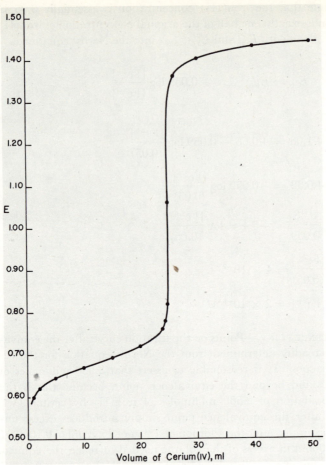

*Figure 9–8.* Titration curve for the titration of 25.00 ml of 0.1000 *M* iron(II) with 0.1000 *M* cerium(IV) in a 1 *F* sulfuric acid medium.

## OXIDATION-REDUCTION INDICATORS

There are many ways in which the end point of an oxidation-reduction titration can be detected. Most of these techniques are common to all types of titrations. As the well-defined titration curve of Figure 9–8 suggests, one good approach is to plot or record the complete titration curve. Then the end point can simply be taken as the midpoint of the steeply rising portion of the titration curve. Alternatively, if the theoretical equivalence-point potential is known, it may be feasible to titrate carefully just to or very close to that potential. Any titration which is followed by measurement of the potential of an indicator electrode as a function of added titrant is termed a **potentiometric titration**. Potentiometric titrations will be discussed in detail in Chapter 11.

A few titrants have such intense colors that they may serve as their own end-point indicators. For example, even a small amount of potassium permanganate imparts a distinct purple tinge to an otherwise colorless solution, so the appearance of the color due to the first slight excess of permanganate in a titration vessel marks the end point of the titration. However, this method of end-point detection may not be applicable if the sample solution contains other highly colored substances.

For some titrations, it is possible to add a species to the sample solution which reacts either with the substance being titrated or with the titrant to produce a sharp color change at the end point. Thus, starch is frequently used as an indicator in titrations involving iodine. Starch forms an intensely colored, dark blue complex with iodine which can serve to indicate the disappearance of the last trace of iodine in the titration of iodine with thiosulfate or the appearance of the first small excess of iodine in the titration of arsenic(III) with triiodide solution.

Redox indicators comprise another class of end-point indicators for oxidation-reduction titrations. A **redox indicator** is a substance which can undergo an oxidation-reduction reaction and whose oxidized and reduced forms differ in color. We may represent the behavior of a redox indicator by the half-reaction

$$Ox + ne \rightleftharpoons Red$$

where Ox and Red are the oxidized and reduced forms of the indicator, respectively. This half-reaction implies that no hydrogen ions are involved in the conversion of the oxidized form of the indicator to the reduced form. Although such a statement is valid for some redox indicators, hydrogen ions do enter into the oxidation-reduction reactions of most indicators. However, the concentration of indicator is always small and, therefore, the quantity of hydrogen ion produced or consumed by oxidation or reduction of the indicator does not alter appreciably the pH of the solution.

We can write the Nernst equation corresponding to the above half-reaction as

$$E_{Ox,Red} = E^0_{Ox,Red} - \frac{0.059}{n} \log \frac{[Red]}{[Ox]}$$

where $E^0_{Ox,Red}$ is the standard potential for the indicator half-reaction *at a specified hydrogen ion concentration*. As in the case of acid-base or metallochromic indicators, a redox indicator will appear to be the color of the *oxidized* form if [Ox] is at least 10 times greater than [Red]. However, the indicator will have the color of the *reduced* form if [Red] is 10 or more times greater than [Ox]. From the preceding Nernst equation, we can conclude that the color of the *oxidized* form will be seen when

$$E_{Ox,Red} = E^0_{Ox,Red} - \frac{0.059}{n} \log \frac{1}{10} = E^0_{Ox,Red} + \frac{0.059}{n}$$

but that the color of the *reduced* form will be observed if

$$E_{Ox,Red} = E^0_{Ox,Red} - \frac{0.059}{n} \log 10 = E^0_{Ox,Red} - \frac{0.059}{n}$$

At intermediate potentials, the indicator will appear to be a mixture of colors. Thus, a redox indicator requires a potential change of $2 \times (0.059/n)$ or a maximum of about 0.12 v to change from one color form to the other. Theoretically, the range over which this color change occurs is centered on the standard-potential value $(E^0_{Ox,Red})$ of the indicator. For many redox indicators, the number of electrons $(n)$ involved in the half-reaction is 2, so a potential change of only 0.059 v causes the indicator to change color.

In Table 9–3 are listed some common oxidation-reduction indicators along with their standard potentials and the colors of the oxidized and reduced forms. As pointed out previously, the oxidation-reduction reactions of most redox indicators

### Table 9–3. Selected Oxidation-Reduction Indicators

| Indicator | Color of Reduced Form | Color of Oxidized Form | Standard Potential, $E^0$ |
|---|---|---|---|
| Indigo monosulfate | colorless | blue | 0.26 |
| Methylene blue | colorless | blue | 0.36 |
| 1-naphthol-2-sulfonic acid indophenol | colorless | red | 0.54 |
| Diphenylamine | colorless | violet | 0.76 |
| Diphenylbenzidine | colorless | violet | 0.76 |
| Barium diphenylamine sulfonate | colorless | red-violet | 0.84 |
| Sodium diphenylbenzidine sulfonate | colorless | violet | 0.87 |
| Tris(2,2′-bipyridine)-iron(II) sulfate | red | pale blue | 0.97 |
| Erioglaucine A | green | red | 1.00 |
| Tris(5-methyl-1,10-phenanthroline)iron(II) sulfate (methyl ferroin) | red | pale blue | 1.02 |
| Tris(1,10-phenanthroline)-iron(II) sulfate (ferroin) | red | pale blue | 1.06 |
| N-phenylanthranilic acid | colorless | pink | 1.08 |
| Tris(5-nitro-1,10-phenanthroline)iron(II) sulfate (nitroferroin) | red | pale blue | 1.25 |
| Tris(2,2′-bipyridine)-ruthenium(II) nitrate | yellow | pale blue | 1.25 |

involve hydrogen ions, and so the potentials at which these indicators change color depend on pH. Accordingly, the standard potentials listed in Table 9–3 pertain to a hydrogen ion concentration (activity) of 1 $M$.

With certain redox indicators the oxidation-reduction process is not reversible. For instance, in the presence of excess oxidizing titrant (permanganate, cerium(IV), or dichromate), indicators such as diphenylamine, diphenylbenzidine, and erioglaucine are destructively oxidized and may not be converted back to their reduced forms. On the other hand, some redox indicators are metal complexes—a central metal ion surrounded by organic ligands—and appear to function by the simple oxidation and reduction of the metal cation. Generally, these latter indicators behave in a reversible manner and may be oxidized and reduced repeatedly.

Most frequently used as redox indicator for the iron(II)-cerium(IV) titration in 1 $F$ sulfuric acid is tris(1,10-phenanthroline)iron(II) sulfate or **ferroin**. As shown in Figure 9–9, three 1,10-phenanthroline (or $o$-phenanthroline) ligands are coordinated to one iron(II) cation through the lone electron pairs on the nitrogen atoms to form the octahedral indicator complex, which is frequently abbreviated $Fe(phen)_3^{2+}$. In its reduced iron(II) form, the indicator has an intense red color, whereas the oxidized iron(III) indicator complex is pale blue; the indicator half-reaction may be written as

$$Fe(phen)_3^{3+} + e \rightleftharpoons Fe(phen)_3^{2+}; \qquad E^0 = +1.06 \text{ v}$$
$$\text{(pale blue)} \qquad\qquad \text{(red)}$$

In the titration of iron(II) with cerium(IV) in a sulfuric acid medium, the ferroin indicator is initially in its reduced, red form, and the end point is signaled by the

*Figure 9–9.* Structure of the tris(1,10-phenanthroline)iron(II) or ferroin redox indicator.

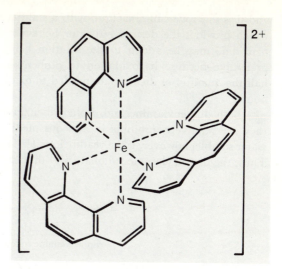

sudden change to the pale blue color of the oxidized form. Actually, the combination of the blue color of the oxidized indicator and the yellow color of the first slight excess of cerium(IV) titrant imparts a green tinge to the solution at the end point. Conversely, the pale-blue, oxidized form of ferroin is an excellent end-point indicator for the titrimetric determination of cerium(IV) with a standard iron(II) solution. It is very stable in the presence of cerium(IV) prior to the end point and undergoes a sharp color change at the end point to the intensely colored red form.

A number of substituted 1,10-phenanthroline ligands have been employed for the preparation of other ferroin-like indicators. Two of these indicators, tris(5-methyl-1,10-phenanthroline)iron(II) and tris(5-nitro-1,10-phenanthroline)iron(II), are included in Table 9–3. Addition of methyl ($—CH_3$) and nitro ($—NO_2$) groups to the 1,10-phenanthroline molecule alters the electronic properties of the organic ligand, so that redox indicators with different standard potentials may be synthesized.

Another family of redox indicators is derived from the colorless compound diphenylamine. In the presence of an oxidant, such as potassium dichromate, diphenylamine undergoes an irreversible oxidative coupling reaction to yield diphenylbenzidine, which is also colorless:

$$2 \quad \text{diphenylamine} \rightarrow \text{diphenylbenzidine} \quad + \; 2\,H^+ + 2\,e$$

diphenylamine            diphenylbenzidine

However, upon further oxidation, diphenylbenzidine is converted to a bright purple compound, known as diphenylbenzidine violet, which owes its deep color to a system

diphenylbenzidine violet

of conjugated double bonds. This color is sufficiently intense to be easily discernible even in the presence of green-colored chromium(III), the reduction product of dichromate ion. Diphenylbenzidine and diphenylbenzidine violet are interconvertible, and it is this reversible reaction, with a standard potential of +0.76 v versus NHE, which provides the desired indicator behavior. Either diphenylamine or diphenylbenzidine may be added to the solution being titrated, but diphenylbenzidine is preferable because its oxidation to diphenylbenzidine violet consumes only one-half the quantity of dichromate needed to oxidize diphenylamine to the violet compound.

Both diphenylamine and diphenylbenzidine suffer from the disadvantage that they are relatively insoluble in aqueous media. These substances can be rendered water soluble, however, if the readily ionizable sulfonic acid derivatives are prepared. Thus, the diphenylamine sulfonate and diphenylbenzidine sulfonate ions, usually in

diphenylamine sulfonate

diphenylbenzidine sulfonate

the form of their barium or sodium salts, are soluble in acidic media and undergo exactly the same color change as the unsulfonated parent compounds. In fact, the only difference between the unsulfonated and sulfonated indicator species is that the standard potentials of the latter are approximately 80 to 110 mv more positive (Table 9–3).

## FEASIBILITY OF TITRATIONS

A titration curve provides direct information concerning whether a titration is feasible and, if so, what method of end-point detection should be used. An important factor which determines the probable success of any proposed redox titration is the difference between the standard or formal potentials for the titrant and the substance being titrated. For example, we have already seen that the difference between the formal potentials in $1\ F$ sulfuric acid for the cerium(IV)-cerium(III) couple ($E^{0\prime} = +1.44$ v) and the iron(III)-iron(II) couple ($E^{0\prime} = +0.68$ v) is large enough to yield a very well defined titration curve (Figure 9–8).

Let us ask the following question: In order to obtain a sufficiently well-defined titration curve, what is the minimum allowable difference in the standard or formal potentials for the titrant and the species to be titrated?

We can gain some insight into this matter by inspecting the titration curves shown in Figure 9–10. These curves show what happens when cerium(IV) is used to titrate reducing agents of various strengths. For our discussion let us assume, first, that the titrant is 0.1 $M$ cerium(IV) and, second, that each of the different reductants is at an initial concentration of 0.1 $M$ and also that each undergoes a one-electron oxidation. If we start with 50 ml of 0.1 $M$ reducing agent, exactly 50 ml of 0.1 $M$ cerium-(IV) solution will be needed in each titration. We have calculated and plotted the titration curves of Figure 9–10 on the assumption that the solutions involved in these titrations are $1\ F$ in sulfuric acid, and so the formal potential for the cerium(IV)-cerium(III) half-reaction ($E^{0\prime} = +1.44$ v versus NHE) is applicable. Standard potentials for the reductants being titrated are printed on the titration curves.

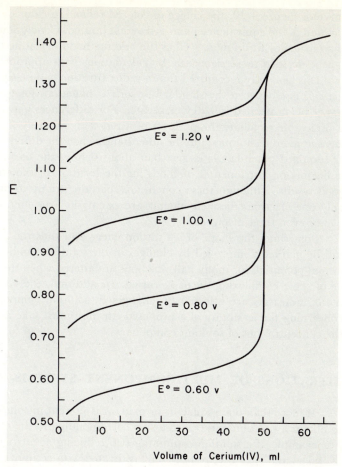

*Figure 9–10.* Titration curves for reducing agents of various strengths with 0.10 $M$ cerium(IV) solution in a 1 $F$ sulfuric acid medium. In each case, it is assumed that 50 ml of 0.10 $M$ reducing agent is titrated and that each reducing agent undergoes a one-electron oxidation. The standard potential for each species titrated is printed on the titration curve. The formal potential for the cerium-(IV)-cerium(III) couple is +1.44 v versus NHE in a 1 $F$ sulfuric acid solution.

One fact emerges immediately from an inspection of Figure 9–10. That is, the magnitude of the emf change at the equivalence point of a titration is proportional to the *difference* between the formal potential for the cerium(IV)-cerium(III) half-reaction and the standard potential for the substance being titrated. If a species with a standard potential of +0.60 v (versus the hydrogen electrode) is titrated, the emf jumps at least 500 mv near the equivalence point. In the case of a substance with a standard potential of +1.00 v, the emf changes only about 150 mv near the equivalence point. With a species having a standard potential of +1.20 v, there is no very steep portion of the titration curve. Obviously, the size and steepness of the vertical portion of a titration curve govern the accuracy and the precision with which the equivalence point can be located, and thereby govern the feasibility of a titration itself. When the vertical portion is large and steep, as it is for the two bottom titration curves in Figure 9–10, the equivalence point can be easily determined. On the other hand, if the vertical portion is small and not sharply defined, as in the uppermost curve of Figure 9–10, the choice of colored redox indicator or other method for the location of the equivalence point is severely limited. Conclusions about the feasibility of a particular titration, as well as the best method of locating the equivalence point, can be decided more rigorously by calculations with appropriate Nernst equations.

It is generally recognized that a redox titration can be successfully performed if the *difference* between the standard or formal potentials for the titrant and the substance being titrated is 0.20 v or greater. For differences between 0.20 and 0.40 v, it is preferable to follow the progress of a redox titration by means of potentiometric measurement techniques. On the other hand, when the difference between the formal or standard potentials is larger than about 0.40 v, the use of either colored redox indicators or instrumental methods for the location of the equivalence point gives good results. Although these conclusions pertain strictly to half-reactions involving only one electron, these statements are essentially true for half-reactions involving two or even three electrons.

Sometimes the shape of a titration curve and thus the feasibility of a titration can be markedly improved by changes in the solution conditions. For example, the formal potentials for many half-reactions are altered when the concentration of acid or of some complexing agent is varied. In addition, the color-change intervals of redox indicators are affected by changes in solution composition. Any prediction concerning the feasibility of a proposed titration must take account of the effects of the concentrations of solution components.

## TITRATIONS OF MULTICOMPONENT SYSTEMS

If more than one reductant or oxidant is present in solution, it may be possible to obtain a titration curve with a step corresponding to the titration of each species. For example, if a solution containing two reducing agents with standard potentials of +0.60 and +1.00 v, respectively, is titrated with cerium(IV), it can be seen from Figure 9–10 that the first substance will be completely titrated before the potential rises high enough to begin the titration of the second reductant. If, however, the two reducing agents had the same standard potential, they would be titrated simultaneously rather than successively. In order for one reducing agent to be titrated individually in the presence of another reducing agent, the standard potentials for these two must differ by at least 0.20 v.

In addition, a solution containing only a single element may give a titration curve with two or more steps if that element exhibits several stable oxidation states. Uranium can exist in four different oxidation states, +3, +4, +5, and +6. If one titrates an acidic solution of uranium(III) with a moderately strong oxidizing agent, the titration curve will show two steps, representing the oxidation of uranium(III) to uranium(IV) and of uranium(IV) to uranium(VI), respectively. Both molybdenum(III) and tungsten(III) can be oxidized in two steps, first to the corresponding +5 state and then to the +6 state. From an inspection of Table 9–1, the following three entries, involving four different oxidation states of vanadium, may be found:

$$V^{3+} + e \rightleftharpoons V^{2+}; \qquad E^0 = -0.255 \text{ v}$$

$$VO^{2+} + 2\,H^+ + e \rightleftharpoons V^{3+} + H_2O; \qquad E^0 = +0.361 \text{ v}$$

$$VO_2{}^+ + 2\,H^+ + e \rightleftharpoons VO^{2+} + H_2O; \qquad E^0 = +1.000 \text{ v}$$

As shown in Figure 9–11, if a solution originally containing only vanadium(II) is titrated with a strong oxidizing agent, three distinct steps will appear in the titration

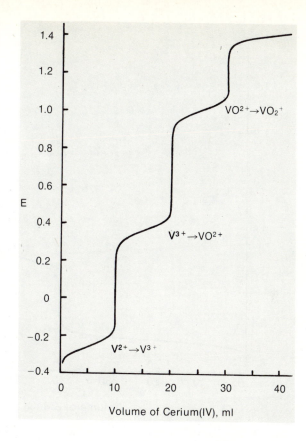

*Figure 9–11.* Stepwise titration of a 1 $F$ sulfuric acid solution containing 1.0 millimole of vanadium(II) with 0.10 $M$ cerium(IV). The oxidation reaction occurring at each step of the titration is shown on the curve.

curve, corresponding in order to the oxidation of vanadium(II) to vanadium(III), of vanadium(III) to vanadium(IV), and of vanadium(IV) to vanadium(V).

**Mixture of uranium(IV) and iron(II).** Titration curves for systems consisting of either several species or a single substance which undergoes stepwise reaction can be constructed on the basis of principles discussed earlier. Figure 9–12 depicts a titration curve for the titration of a mixture of 1 millimole of uranium(IV) and 1 millimole of iron(II) with 0.01 $F$ potassium permanganate in 1 $F$ sulfuric acid. Uranium(IV) undergoes a two-electron oxidation to uranium(VI), the standard potential for the uranium(VI)-uranium(IV) couple being given by

$$UO_2^{2+} + 4\,H^+ + 2\,e \rightleftharpoons U^{4+} + 2\,H_2O; \qquad E^0 = +0.334\,v$$

Iron(II) is oxidized to iron(III), and the pertinent formal potential is

$$Fe^{3+} + e \rightleftharpoons Fe^{2+}; \qquad E^{0\prime} = +0.68\,v$$

In 1 $F$ sulfuric acid, permanganate is a somewhat better oxidizing agent than cerium(IV):

$$MnO_4^- + 8\,H^+ + 5\,e \rightleftharpoons Mn^{2+} + 4\,H_2O; \qquad E^0 = +1.51\,v$$

Since uranium(IV) is the stronger of the two reducing agents, it is oxidized first. In the region of the first equivalence point, the titration curve is well defined since the

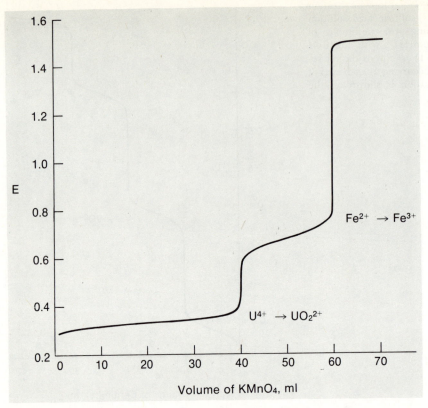

*Figure 9–12.* Titration curve for the titration of a $1 \, F$ sulfuric acid solution containing 1 millimole of uranium(IV) and 1 millimole of iron(II) with $0.01 \, F$ potassium permanganate solution.

uranium and iron systems differ in their respective standard and formal potentials by 0.35 v. However, the second equivalence point is even sharper because the potentials for the iron and permanganate half-reactions differ by 0.83 v.

Up to the first equivalence point, the shape of the titration curve is governed solely by the Nernst equation for the uranium(VI)-uranium(IV) half-reaction,

$$E_{UO_2^{2+},U^{4+}} = E^0_{UO_2^{2+},U^{4+}} - \frac{0.059}{2} \log \frac{[U^{4+}]}{[UO_2^{2+}][H^+]^4}$$

but calculation of the potential at the *first* equivalence point requires some new considerations. At the first equivalence point, the solution is identical to one obtained by mixing 1 millimole each of uranium(VI) and iron(II), plus 0.4 millimole of manganese(II). Two reactions which may occur in such a solution are

$$UO_2^{2+} + 2 \, Fe^{2+} + 4 \, H^+ \rightleftharpoons U^{4+} + 2 \, Fe^{3+} + 2 \, H_2O; \qquad E^{0\prime} = -0.35 \text{ v}$$

and

$$5 \, UO_2^{2+} + 2 \, Mn^{2+} + 4 \, H^+ \rightleftharpoons 5 \, U^{4+} + 2 \, MnO_4^- + 2 \, H_2O; \qquad E^0 = -1.18 \text{ v}$$

Neither reaction is thermodynamically favored, but the *first* reaction is at least *less unfavorable*. Hence, at equilibrium there will be minute concentrations of uranium(IV) and iron(III) and, from the reaction stoichiometry, we can write

$$2\,[U^{4+}] = [Fe^{3+}]$$

Since the extent of the uranium(VI)-iron(II) reaction is so slight, we will still have essentially 1 millimole of uranium(VI) and 1 millimole of iron(II). Thus,

$$[UO_2^{2+}] = [Fe^{2+}]$$

At the first equivalence point, the Nernst equations for the uranium and iron half-reactions are

$$E_{ep} = E^0_{UO_2^{2+},U^{4+}} - \frac{0.059}{2} \log \frac{[U^{4+}]}{[UO_2^{2+}][H^+]^4}$$

and

$$E_{ep} = E^{0\prime}_{Fe^{3+},Fe^{2+}} - 0.059 \log \frac{[Fe^{2+}]}{[Fe^{3+}]}$$

If the last four equations are combined and manipulated as in the discussion of the cerium(IV)-iron(II) titration earlier in this chapter, the result is

$$3\,E_{ep} = 2\,E^0_{UO_2^{2+},U^{4+}} + E^{0\prime}_{Fe^{3+},Fe^{2+}} + 0.059 \log\,(2[H^+]^4)$$

and, if $[H^+]$ is taken to be 1 $M$,

$$3\,E_{ep} = 2(0.334) + 0.68 + 0.059(0.30) = 1.37$$

$$E_{ep} = +0.46 \text{ v versus NHE}$$

Between the first and second equivalence points, the titration curve can be calculated from the Nernst equation for the iron(III)-iron(II) couple, whereas the permanganate-manganous ion half-reaction governs the curve *after* the second equivalence point.

At the second equivalence point, the solution contains 1 millimole of iron(III) and 0.6 millimole of manganese(II), as well as 1 millimole of uranium(VI), so

$$[Mn^{2+}] = 0.6\,[Fe^{3+}]$$

From the stoichiometry of the titration reaction,

$$5\,Fe^{2+} + MnO_4^- + 8\,H^+ \rightleftharpoons 5\,Fe^{3+} + Mn^{2+} + 4\,H_2O$$

it follows that

$$[MnO_4^-] = 0.2\,[Fe^{2+}]$$

In addition, we can write the following two Nernst equations:

$$E_{ep} = E^{0\prime}_{Fe^{3+},Fe^{2+}} - 0.059 \log \frac{[Fe^{2+}]}{[Fe^{3+}]}$$

$$E_{ep} = E^{0}_{MnO_4^-,Mn^{2+}} - \frac{0.059}{5} \log \frac{[Mn^{2+}]}{[MnO_4^-][H^+]^8}$$

As before, proper combination of the preceding four mathematical relations leads to an expression for the potential at the *second* equivalence point:

$$6\,E_{ep} = E^{0\prime}_{Fe^{3+},Fe^{2+}} + 5\,E^{0}_{MnO_4^-,Mn^{2+}} - 0.059 \log \frac{3}{[H^+]^8}$$

Finally, if the value of 1 $M$ for the hydrogen ion concentration is inserted into the latter relation, we have

$$6\,E_{ep} = +0.68 + 5(1.51) - 0.059(0.477) = 8.26$$

$$E_{ep} = +1.38 \text{ v versus NHE}$$

*Uranium-vanadium mixtures.*  Suppose it is necessary to determine the individual concentrations of both uranium and vanadium in an acidic solution of these two elements. Very likely, in the preparation of the sample solution, an oxidizing agent would be employed, so let us assume that uranium and vanadium are both present in their highest oxidation states—$UO_2^{2+}$ and $VO_2^+$, respectively. If this sample were titrated with a standard titanium(III) solution—a very good reducing agent—vanadium(V) would be quantitatively reduced to vanadium(IV):

$$VO_2^+ + Ti^{3+} \rightleftharpoons VO^{2+} + TiO^{2+}; \qquad E^0 = +0.90 \text{ v}$$

From the volume of titanium(III) titrant required to reach the first equivalence point $(v_1)$, the amount of vanadium present could be calculated directly. If the addition of titanium(III) titrant were continued, a second equivalence point could be obtained, corresponding to the *combined* reductions of vanadium(IV) to vanadium(III) and uranium(VI) to uranium(IV):

$$VO^{2+} + Ti^{3+} \rightleftharpoons V^{3+} + TiO^{2+}; \qquad E^0 = +0.26 \text{ v}$$

$$UO_2^{2+} + 2\,Ti^{3+} \rightleftharpoons U^{4+} + 2\,TiO^{2+}; \qquad E^0 = +0.23 \text{ v}$$

Vanadium(IV) and uranium(VI) are titrated simultaneously because the standard potentials for the vanadium(IV)-vanadium(III) and uranium(VI)-uranium(IV) couples are very close together, being +0.361 and +0.334 v versus NHE, respectively. If the volume of titanium(III) solution required to go from the first to the second equivalence point is designated as $v_2$, the concentration of uranium(VI) originally present is proportional to $v_2 - v_1$.

Another approach to the same analytical problem involves the preliminary reduction of uranium and vanadium to $U^{4+}$ and $V^{3+}$, respectively, with a reductant such as metallic lead. Then the solution is titrated with a suitable oxidizing agent—permanganate, cerium(IV), or dichromate. As before, the resulting titration curve exhibits two steps, and the oxidation processes are the reverse of those just described.

QUESTIONS AND PROBLEMS

1. Balance the following oxidation-reduction equations:
    (a) $Al + H^+ \rightleftharpoons Al^{3+} + H_2$
    (b) $HAsO_2 + Ce^{4+} + H_2O \rightleftharpoons H_2AsO_4^- + Ce^{3+} + H^+$
    (c) $SCN^- + MnO_4^- + H^+ \rightleftharpoons SO_4^{2-} + CN^- + Mn^{2+} + H_2O$
    (d) $ClO_3^- + Cl^- + H^+ \rightleftharpoons Cl_2 + H_2O$
    (e) $ClO_3^- + H_2S \rightleftharpoons S + Cl^- + H_2O$
    (f) $H_2S + Br_2 + H_2O \rightleftharpoons SO_4^{2-} + H^+ + Br^-$
    (g) $Cr^{3+} + MnO_2 + H_2O \rightleftharpoons Mn^{2+} + CrO_4^{2-} + H^+$
    (h) $Zn + OH^- \rightleftharpoons Zn(OH)_4^{2-} + H_2$
    (i) $BrO_3^- + I^- + H^+ \rightleftharpoons Br^- + I_2 + H_2O$
    (j) $NO_2^- + Al + OH^- + H_2O \rightleftharpoons NH_3 + Al(OH)_4^-$
    (k) $Cu(NH_3)_4^{2+} + CN^- + OH^- \rightleftharpoons Cu(CN)_3^{2-} + CNO^- + NH_3 + H_2O$
    (l) $NO_3^- + Bi_2S_3 + H^+ \rightleftharpoons NO + Bi^{3+} + S + H_2O$
    (m) $HOI + OH^- \rightleftharpoons IO_3^- + I^- + H_2O$
    (n) $Cu^{2+} + I^- \rightleftharpoons CuI + I_3^-$

2. Write the shorthand cell representation for each of the following galvanic cells:
    (a) A metallic silver electrode in a $0.015\ F$ silver nitrate solution connected, through a potassium nitrate salt bridge to eliminate the liquid-junction potential, to a $0.028\ F$ nickel chloride solution into which a nickel metal rod is immersed.
    (b) A platinum wire in a mixture of $0.10\ M$ cerium(IV) and $0.05\ M$ cerium(III) in a $1\ F$ sulfuric acid solution which is in contact through a permeable membrane with a $10\ F$ sodium hydroxide solution containing $0.05\ M$ permanganate ion $(MnO_4^-)$ and $0.001\ M$ manganate ion $(MnO_4^{2-})$ in which a gold wire electrode is immersed.
    (c) A palladium wire in a $0.025\ F$ hydrochloric acid solution which is saturated with hydrogen gas at a pressure of $0.5$ atm and which is connected, through a potassium chloride salt bridge to minimize the liquid-junction potential, to another half-cell consisting of a zinc metal rod immersed in a $0.04\ F$ zinc nitrate solution.

3. Write the shorthand cell representation of a galvanic cell which could be employed to determine the solubility-product constant of silver chloride. In other words, construct a suitable cell in which the cell reaction is

$$AgCl(s) \rightleftharpoons Ag^+ + Cl^-$$

4. Write the shorthand cell representation of a galvanic cell which can be used to measure the ion-product constant for water. In other words, construct a cell for which the overall reaction is

$$H_2O \rightleftharpoons H^+ + OH^-$$

5. Write the shorthand cell representation for a suitable galvanic cell which might be used in the evaluation of the formation constant for the mercury(II)-EDTA complex. In other words, construct a cell to study the reaction

$$Hg^{2+} + Y^{4-} \rightleftharpoons HgY^{2-}$$

where $Y^{4-}$ represents the ethylenediaminetetraacetate ion.

6. Write the Nernst equation which corresponds to each of the following half-reactions:
    (a) $Cd + 4\ CN^- \rightleftharpoons Cd(CN)_4^{2-} + 2\ e$
    (b) $Cu^{2+} + I^- + e \rightleftharpoons CuI(s)$
    (c) $TiO^{2+} + 2\ H^+ + e \rightleftharpoons Ti^{3+} + H_2O$
    (d) $Mn^{2+} + 2\ H_2O \rightleftharpoons MnO_2 + 4\ H^+ + 2\ e$
    (e) $2\ Hg^{2+} + 2\ e \rightleftharpoons Hg_2^{2+}$
    (f) $Hg_2SO_4(s) + 2\ e \rightleftharpoons 2\ Hg + SO_4^{2-}$

7. Calculate the actual emf for the overall reaction occurring in each of the following galvanic cells:

    (a) $Ni \mid Ni^{2+}$ (0.200 $M$) $\parallel Ag^+$ (0.00500 $M$) $\mid Ag$

    (b) $Pt \mid I^-$ (0.500 $M$), $I_3^-$ (0.0300 $M$) $\parallel Cd^{2+}$ (0.100 $M$) $\mid Cd$

    (c) $Pb \mid Pb^{2+}$ (0.0250 $M$) $\parallel Cu^{2+}$ (0.300 $M$) $\mid Cu$

    (d) $Ag \mid AgCl(s)$, $Cl^-$ (0.100 $M$) $\parallel Hg_2SO_4(s)$, $SO_4^{2-}$ (2.00 $M$) $\mid Hg$

    (e) $Pt \mid VO_2^+$ (0.0200 $M$), $VO^{2+}$ (0.100 $M$),
$$H^+ \text{ (0.500 } M) \parallel Tl^+ \text{ (0.0400 } M) \mid Tl$$

8. For each of the following galvanic cells: (a) Write the two half-reactions and the overall cell reaction, (b) calculate the actual emf for each cell from the Nernst equation, (c) calculate the equilibrium constant for each cell reaction, and (d) identify the anode and cathode and label the electrodes appropriately with either a $+$ or $-$ sign.

    $Cu \mid CuSO_4$ (0.02 $F$) $\parallel Fe^{2+}$ (0.2 $M$), $Fe^{3+}$ (0.01 $M$), $HCl$ (1 $F$) $\mid Pt$

    $Pt \mid Pu^{4+}$ (0.1 $M$), $Pu^{3+}$ (0.2 $M$) $\parallel AgCl(s)$, $HCl$ (0.03 $F$) $\mid Ag$

    $Pt \mid KBr$ (0.2 $F$), $Br_3^-$ (0.03 $M$) $\parallel Ce^{4+}$ (0.01 $M$), $Ce^{3+}$ (0.002 $M$),
$$H_2SO_4 \text{ (1 } F) \mid Pt$$

    $Pt, Cl_2$ (0.1 atm) $\mid HCl$ (2.0 $F$) $\parallel HCl$ (0.1 $F$) $\mid H_2$ (0.5 atm), $Pt$

    $Zn \mid ZnCl_2$ (0.02 $F$) $\parallel Na_2SO_4$ (0.1 $F$), $PbSO_4(s)$ $\mid Pb$

    $Pt \mid UO_2^{2+}$ (0.005 $M$), $U^{4+}$ (0.1 $M$), $HClO_4$ (0.2 $F$) $\parallel HBr$ (0.3 $F$),
$$AgBr(s) \mid Ag$$

9. Evaluate the equilibrium constant for each of the following reactions:

    (a) $Ag + Fe^{3+} \rightleftharpoons Ag^+ + Fe^{2+}$

    (b) $2 Cr^{2+} + Co^{2+} \rightleftharpoons 2 Cr^{3+} + Co$

    (c) $Br_2(aq) + 3 I^- \rightleftharpoons 2 Br^- + I_3^-$

    (d) $2 V^{3+} + UO_2^{2+} \rightleftharpoons 2 VO^{2+} + U^{4+}$

    (e) $2 Co^{3+} + Co \rightleftharpoons 3 Co^{2+}$

10. Compute the equilibrium constant for the reaction

$$AuCl_4^- + 2 Au + 2 Cl^- \rightleftharpoons 3 AuCl_2^-$$

from the standard potentials for the following half-reactions:

$$AuCl_2^- + e \rightleftharpoons Au + 2 Cl^-; \qquad E^0 = +1.154 \text{ v}$$
$$AuCl_4^- + 2 e \rightleftharpoons AuCl_2^- + 2 Cl^-; \qquad E^0 = +0.926 \text{ v}$$

11. Given the information that

$$I_2(aq) + 2 e \rightleftharpoons 2 I^-; \qquad E^0 = +0.6197 \text{ v}$$

and

$$I_2(s) + 2 e \rightleftharpoons 2 I^-; \qquad E^0 = +0.5345 \text{ v}$$

where the symbol (aq) indicates unit activity of iodine in the aqueous phase and (s) denotes pure solid iodine at unit activity, calculate the solubility (moles/liter) of solid iodine in water.

12. Evaluate the equilibrium constant for the formation of the triiodide ion,

$$I_2(aq) + I^- \rightleftharpoons I_3^-$$

from the knowledge that

$$I_2(aq) + 2 e \rightleftharpoons 2 I^-; \qquad E^0 = +0.6197 \text{ v}$$

and

$$I_3^- + 2 e \rightleftharpoons 3 I^-; \qquad E^0 = +0.5355 \text{ v}$$

13. Calculate the solubility-product constant for copper(I) iodide,

$$CuI(s) \rightleftharpoons Cu^+ + I^-$$

given the following information:

$$Cu^{2+} + e \rightleftharpoons Cu^+; \qquad E^0 = +0.153 \text{ v}$$

and

$$Cu^{2+} + I^- + e \rightleftharpoons CuI; \qquad E^0 = +0.86 \text{ v}$$

14. Given that

$$Ag + 2\ CN^- \rightleftharpoons Ag(CN)_2^- + e; \qquad E^0 = +0.31\ v$$

and

$$Ag \rightleftharpoons Ag^+ + e; \qquad E^0 = -0.80\ v$$

calculate the equilibrium (dissociation) constant for the following reaction:

$$Ag(CN)_2^- \rightleftharpoons Ag^+ + 2\ CN^-$$

15. Given the information that

$$Cu^{2+} + 2\ e \rightleftharpoons Cu; \qquad E^0 = +0.34\ v$$

$$Cu^+ + e \rightleftharpoons Cu; \qquad E^0 = +0.52\ v$$

$$Cu^{2+} + Cl^- + e \rightleftharpoons CuCl(s); \qquad E^0 = +0.54\ v$$

calculate the solubility-product constant for $CuCl(s)$.

16. Calculate the standard potential ($E^0$) for the half-reaction

$$Au^+ + e \rightleftharpoons Au$$

from the following data:

$$Au^{3+} + 2\ e \rightleftharpoons Au^+; \qquad E^0 = +1.41\ v$$

$$Au^{3+} + 3\ e \rightleftharpoons Au; \qquad E^0 = +1.50\ v$$

17. Given the following data

$$MnO_4^- + 8\ H^+ + 5\ e \rightleftharpoons Mn^{2+} + 4\ H_2O; \qquad E^0 = +1.51\ v$$

$$Mn^{3+} + e \rightleftharpoons Mn^{2+}; \qquad E^0 = +1.51\ v$$

calculate the standard potential ($E^0$) for

$$MnO_4^- + 8\ H^+ + 4\ e \rightleftharpoons Mn^{3+} + 4\ H_2O$$

18. Given that

$$Br_2 + 2\ H_2O \rightleftharpoons 2\ HOBr + 2\ H^+ + 2\ e; \qquad E^0 = -1.59\ v$$

and

$$Br_2 + 6\ H_2O \rightleftharpoons 2\ BrO_3^- + 12\ H^+ + 10\ e; \qquad E^0 = -1.52\ v$$

calculate $E^0$ for the half-reaction

$$BrO_3^- + 5\ H^+ + 4\ e \rightleftharpoons HOBr + 2\ H_2O$$

19. Consider the following galvanic cell:

$$Pt\ |\ PuO_2^{2+}\ (0.01\ M),\ Pu^{4+}\ (0.001\ M),\ H^+\ (0.1\ M)\ \|\ Cu^{2+}\ (0.001\ M)\ |\ Cu$$

(a) Write the two pertinent half-reactions and the overall cell reaction.
(b) Calculate the $E^0$ and the equilibrium constant for the overall cell reaction.
(c) Calculate the actual emf of the galvanic cell.
(d) Which electrode is the negative electrode?
(e) A bar of pure copper metal weighing 127 gm was placed into 1 liter of a 0.0500 $F$ $PuO_2(ClO_4)_2$ solution. The pH of this solution was maintained constant at exactly 3.00. What was the concentration of $PuO_2^{2+}$ at equilibrium?

20. Given the following data:

$$PtCl_4^{2-} + 2\ e \rightleftharpoons Pt + 4\ Cl^-; \qquad E^0 = +0.73\ v$$

$$PtCl_6^{2-} + 4\ e \rightleftharpoons Pt + 6\ Cl^-; \qquad E^0 = +0.72\ v$$

(a) Calculate the $E^0$ and the equilibrium constant (to two significant figures) for the reaction

$$2\ PtCl_4^{2-} \rightleftharpoons Pt + PtCl_6^{2-} + 2\ Cl^-$$

(b) Suppose that a $1.00\ F$ hydrochloric acid solution, which is initially $1.00 \times 10^{-6}\ M$ in $PtCl_4^{2-}$ and $1.00 \times 10^{-6}\ M$ in $PtCl_6^{2-}$, is equilibrated with a large piece of platinum foil. What will be the final equilibrium concentrations of $PtCl_4^{2-}$ and $PtCl_6^{2-}$?

(c) Calculate the $E^0$ for the half-reaction

$$PtCl_4^{2-} + 2\ Cl^- \rightleftharpoons PtCl_6^{2-} + 2\ e$$

(d) If, in the experiment of part (b), electrical contact is made to the platinum foil and its potential measured versus a saturated calomel electrode (SCE) with a potentiometer, what will be the observed potential versus SCE?

21. Consider the following galvanic cell:

$$Pt\ |\ Cr^{3+}\ (0.30\ M),\ Cr^{2+}\ (0.010\ M)\ \|\ AgCl(s),\ Cl^-\ (0.025\ M)\ |\ Ag$$

What will be the qualitative effect of each of the following upon the emf of the galvanic cell:
(a) The addition of 10 grams of solid silver chloride to the right half-cell?
(b) The dissolution of 10 grams of sodium chloride in the right half-cell?
(c) The addition of 10 ml of water to the left half-cell?
(d) The addition of some soluble chromium(III) salt to the left half-cell?
(e) The dissolution of some silver nitrate in the right half-cell?

22. Consider the following galvanic cell:

$$Pt\ |\ Fe^{3+}\ (0.01\ M),\ Fe^{2+}\ (0.02\ M),\ H^+\ (1\ M)\ \|\ H^+\ (1\ M)\ |\ H_2\ (1\ atm),\ Pt$$

What will be the qualitative effect of each of the following upon the emf of this galvanic cell:
(a) The addition of 50 ml of 1 $M$ acid to the left half-cell?
(b) The addition of some soluble iron(II) salt to the left half-cell?
(c) The addition of a small amount of solid potassium permanganate to the left half-cell?
(d) The decrease of the pressure of hydrogen gas in the right half-cell?

23. Calculate the equilibrium constant for each of the following reactions:
(a) $Ce^{4+} + Fe(CN)_6^{4-} \rightleftharpoons Ce^{3+} + Fe(CN)_6^{3-}$ ($1\ F\ HClO_4$ medium)
(b) $Pb + SnCl_6^{2-} \rightleftharpoons Pb^{2+} + SnCl_4^{2-} + 2\ Cl^-$ ($1\ F\ HCl$ medium)
(c) $5\ Fe^{2+} + MnO_4^- + 8\ H^+ \rightleftharpoons 5\ Fe^{3+} + Mn^{2+} + 4\ H_2O$
(d) $Fe^{2+} + Cu^{2+} \rightleftharpoons Fe^{3+} + Cu^+$

24. The following galvanic cell was constructed for the measurement of the dissociation constant of a weak monoprotic acid, HA:

$$-Pt,\ H_2\ (0.8\ atm)\ |\ HA\ (0.5\ F),\ NaCl\ (1.0\ F),\ AgCl(s)\ |\ Ag+$$

If the observed electromotive force of this cell was 0.568 v, what is the dissociation constant for the weak acid?

25. Calculate the equilibrium concentrations of cobalt(II) and thallium(I) ions resulting when a $0.250\ F$ cobalt(II) sulfate, $CoSO_4$, solution is treated with an excess of pure thallium metal.

26. Titanium(III) may be oxidized by iron(III) in an acidic solution, as shown by the reaction

$$Ti^{3+} + Fe^{3+} + H_2O \rightleftharpoons TiO^{2+} + Fe^{2+} + 2\ H^+$$

Calculate the final concentration of $Ti^{3+}$ ion in a solution prepared by mixing 25.0 ml each of 0.0200 $M$ $Ti^{3+}$ and $Fe^{3+}$ solutions, both being 1 $F$ in sulfuric acid.

27. As described earlier in this chapter, cerium(IV) and iron(II) react to form cerium(III) and iron(III):

$$Ce^{4+} + Fe^{2+} \rightleftharpoons Ce^{3+} + Fe^{3+}$$

(a) Calculate the equilibrium concentrations of all four species in a solution prepared by mixing 5.0 ml of 0.0500 $F$ cerium(IV) sulfate with 25.0 ml of 0.0150 $F$ iron(II) sulfate. Assume that the solutions contain 1 $F$ sulfuric acid.

(b) Calculate the equilibrium concentrations of all four species in a solution prepared by mixing 25.0 ml of 0.0200 $F$ cerium(IV) sulfate with 17.0 ml of 0.0200 $F$ iron(II) sulfate. Assume that the solutions contain 1 $F$ sulfuric acid.

28. The hypothetical galvanic cell

$$-A \mid A^{2+} \parallel B^{2+} \mid B+$$

has an emf of 0.360 v when the concentrations of $A^{2+}$ and $B^{2+}$ are equal. What will be the observed emf of the cell if the concentration of $A^{2+}$ is 0.100 $M$ and the concentration of $B^{2+}$ is $1.00 \times 10^{-4}$ $M$?

29. Given the following standard-potential data,

$$Cd^{2+} + 2 e \rightleftharpoons Cd; \qquad E^0 = -0.403 \text{ v}$$
$$Fe^{2+} + 2 e \rightleftharpoons Fe; \qquad E^0 = -0.440 \text{ v}$$

calculate the equilibrium concentration of cadmium ion when a 0.0500 $M$ cadmium ion solution is shaken with an excess of pure iron filings.

30. Consider the following galvanic cell:

$$Cd \mid Cd^{2+} (0.100 \ M) \parallel Hg^{2+} (0.000300 \ M) \mid Hg$$

(a) Write the two half-reactions which occur in this cell when the electrodes are connected with a conducting wire.
(b) Write the overall cell reaction.
(c) Calculate the $E^0$ for the overall cell reaction.
(d) Calculate the actual emf of the galvanic cell.

31. Consider the following two galvanic cells and assume that the copper electrode of the first cell is permanently connected to the iron electrode of the second cell, as shown below:

$$Cu \mid Cu^{2+} (0.100 \ M) \parallel Pb^{2+} (0.200 \ M) \mid Pb$$
$$Fe \mid Fe^{2+} (0.300 \ M) \parallel Ag^+ (0.500 \ M) \mid Ag$$

(a) Calculate what emf would be observed or measured if a potentiometer were connected across the Pb and Ag electrodes.
(b) Which electrode is the negative electrode?
(c) Suppose that you connect the Ag and Pb electrodes with a conducting wire. Write the half-reaction which would occur at each of the four electrodes, being sure to *write each half-reaction in the direction in which it proceeds spontaneously.*

32. Cadmium metal reacts with vanadium(III) according to the reaction

$$Cd + 2 V^{3+} \rightleftharpoons Cd^{2+} + 2 V^{2+}$$

If a 0.0750 $M$ vanadium(III) solution is shaken with an excess of cadmium metal, what will be the concentrations of $V^{3+}$, $V^{2+}$, and $Cd^{2+}$ at equilibrium?

33. Tin(II) is often employed to reduce iron(III) to iron(II) in a hydrochloric acid medium:

$$SnCl_4{}^{2-} + 2 Fe^{3+} + 2 Cl^- \rightleftharpoons SnCl_6{}^{2-} + 2 Fe^{2+}$$

If 1.00 ml of 0.0500 $M$ tin(II) is added to 22.00 ml of 0.00450 $M$ iron(III), what will be the equilibrium concentrations of $SnCl_4^{2-}$, $SnCl_6^{2-}$, $Fe^{2+}$, and $Fe^{3+}$? Assuming that the solution is 1 $F$ in hydrochloric acid, use the appropriate formal-potential data in Appendix 4.

34. If the electromotive force of the galvanic cell

$$+Pb \mid Pb^{2+} \ (0.0860 \ M) \parallel Tl^+ \mid Tl -$$

is 0.280 v, what must be the concentration of thallium ion in the right half-cell?

35. If the electromotive force of the galvanic cell

$$-Ag \mid AgCl(s), \ Cl^- \ (0.100 \ M) \parallel Fe^{3+} \ (0.0200 \ M), \ Fe^{2+} \mid Pt +$$

is 0.319 v, what must be the concentration of $Fe^{2+}$ in the right half-cell? Assume that the solution in the right half-cell is 1 $F$ in sulfuric acid.

36. If the electromotive force of the galvanic cell

$$-Pt, \ H_2 \ (0.250 \ atm) \mid \text{solution of unknown pH} \parallel AgCl(s), \ Cl^- \ (1.00 \ M) \mid Ag +$$

is 0.621 v, what is the pH of the unknown sample?

37. Derive a general relation for the potential of a platinum indicator electrode versus the normal hydrogen electrode at the equivalence point of the titration of iron(II) with permanganate, according to the reaction

$$5 \ Fe^{2+} + MnO_4^- + 8 \ H^+ \rightleftharpoons 5 \ Fe^{3+} + Mn^{2+} + 4 \ H_2O$$

38. The determination of iron is frequently accomplished by means of the potentiometric titration of an iron(II) solution with a standard potassium dichromate solution:

$$Cr_2O_7^{2-} + 6 \ Fe^{2+} + 14 \ H^+ \rightleftharpoons 2 \ Cr^{3+} + 6 \ Fe^{3+} + 7 \ H_2O$$

If it is assumed that the hydrogen ion concentration is exactly 1 $M$, prove that the equivalence-point potential is given by the expression

$$E_{ep} = \frac{E^0_{Fe^{3+},Fe^{2+}} + 6 \ E^0_{Cr_2O_7^{2-},Cr^{3+}}}{7} + \frac{0.059}{7} \log \ (\tfrac{2}{3}[Fe^{3+}])$$

39. A 20.00-ml sample of 0.1000 $M$ uranium(IV) in a 1 $F$ sulfuric acid medium is titrated with 0.1000 $M$ cerium(IV) in 1 $F$ sulfuric acid, according to the reaction

$$2 \ Ce^{4+} + U^{4+} + 2 \ H_2O \rightleftharpoons 2 \ Ce^{3+} + UO_2^{2+} + 4 \ H^!$$

Suppose that the progress of the titration is followed by means of the potentiometric measurement of the potential of a platinum indicator electrode versus a saturated calomel electrode (SCE).
(a) Calculate the potential of the platinum indicator electrode versus SCE at the equivalence point of the titration.
(b) What concentration of uranium(IV) remains unoxidized at the equivalence point?
(c) Calculate what volume of cerium(IV) titrant solution must be added to cause the potential of the platinum indicator electrode to become +0.334 v versus SCE.
(d) Calculate the potential of the platinum electrode versus SCE after the addition of 50.00 ml of cerium(IV) titrant.
(e) What would be the titration error, in per cent with the proper sign, if the titration were mistakenly terminated at an observed potential of +0.150 v versus SCE?

40. A sample of pure vanadium metal weighing 5.10 gm was treated with 500 ml of a 0.500 $M$ $VO_2^+$ solution. Determine what vanadium species are present at equilibrium and calculate their concentrations. Assume that the solution is

air free to prevent the extraneous oxidation of vanadium species, that the solution is maintained 1.00 $F$ in perchloric acid, and that any reaction between hydrogen ion and vanadium species may be neglected.

41. Consider the titration of 20.00 ml of 0.05000 $M$ iron(III) in a 1 $F$ sulfuric acid medium with 0.02000 $M$ titanium(III), also in 1 $F$ sulfuric acid:

$$Fe^{3+} + Ti^{3+} + H_2O \rightleftharpoons Fe^{2+} + TiO^{2+} + 2\,H^+$$

(a) Calculate and plot the complete titration curve, taking enough points to define the titration curve smoothly.
(b) Calculate the concentration of iron(III) unreduced at the equivalence point of the titration.
(c) If methylene blue indicator (see Table 9–3), which undergoes a sharp color change from blue to colorless at a potential of $+0.33$ v versus the normal hydrogen electrode (NHE), is employed to signal the end point of this titration, what volume of titanium(III) titrant would be required to reach this end point?
(d) What would be the titration error, in per cent with the proper sign, if the methylene blue end point in part (c) was assumed to be the equivalence point?

42. Suppose that you perform a potentiometric study to evaluate the formation constants for the one-to-one complexes which mercury(II) and lead(II) form with a ligand $Y^{4-}$. Assume that the following three galvanic cells are constructed and that you measure the emf value for each cell, with the results indicated below.

$$-Pt, H_2 \text{ (1 atm)} \mid H^+ \text{ (1 } M) \parallel Hg^{2+} \text{ (0.001 } M) \mid Hg+; \qquad E_{cell} = 0.765 \text{ v}$$

$$-Pt, H_2 \text{ (1 atm)} \mid H^+ \text{ (1 } M) \parallel HgY^{2-} \text{ (0.001 } M), Y^{4-} \text{ (0.001 } M) \mid Hg+;$$
$$E_{cell} = 0.280 \text{ v}$$

$$-Pt, H_2 \text{ (1 atm)} \mid H^+ \text{ (1 } M) \parallel HgY^{2-} \text{ (0.001 } M), Pb^{2+} \text{ (0.001 } M),$$
$$PbY^{2-} \text{ (0.001 } M) \mid Hg+; \qquad E_{cell} = 0.581 \text{ v}$$

From the results of these experiments, calculate values for the formation constants represented by the expressions

$$K_{HgY} = \frac{[HgY^{2-}]}{[Hg^{2+}][Y^{4-}]} \qquad \text{and} \qquad K_{PbY} = \frac{[PbY^{2-}]}{[Pb^{2+}][Y^{4-}]}$$

43. Calculate and plot the titration curves for each of the following:
(a) The titration of 20.00 ml of 0.02000 $M$ tin(II) in a 1 $F$ hydrochloric acid medium with 0.02000 $M$ vanadium(V) in 1 $F$ hydrochloric acid:

$$SnCl_4{}^{2-} + 2\,VO_2{}^+ + 4\,H^+ + 2\,Cl^- \rightleftharpoons SnCl_6{}^{2-} + 2\,VO^{2+} + 2\,H_2O$$

(b) The titration of 25.00 ml of 0.1000 $M$ iron(III) in 1 $F$ sulfuric acid with 0.05000 $M$ chromium(II) in a 1 $F$ sulfuric acid medium:

$$Fe^{3+} + Cr^{2+} \rightleftharpoons Fe^{2+} + Cr^{3+}$$

44. A 25.00-ml sample of 0.02000 $M$ $VO_2{}^+$ in 1 $F$ perchloric acid was titrated with 0.05000 $M$ titanium(III) in 1 $F$ perchloric acid according to the following reaction:

$$VO_2{}^+ + Ti^{3+} \rightleftharpoons VO^{2+} + TiO^{2+}$$

The titration was followed potentiometrically by measuring the potential between a platinum indicator electrode and a saturated calomel reference electrode (SCE). Determine the following information:
(a) The equilibrium constant for the titration reaction.

(b) The potential of the platinum indicator electrode versus the normal hydrogen electrode (NHE) at the equivalence point of the titration.

(c) The concentration of $VO_2^+$ at the equivalence point.

(d) The volume of titanium(III) titrant added at the point at which the potential of the platinum indicator electrode is $+1.00$ v versus the normal hydrogen electrode (NHE).

(e) The potential of the platinum indicator electrode versus the saturated calomel electrode (SCE) after 8.00 ml of the titanium(III) had been added.

45. Calculate and plot the titration curves for each of the following:

(a) The two-step titration of a 20.00-ml sample of 0.1000 $M$ vanadium(III) with 0.02000 $M$ permanganate in a 1 $F$ sulfuric acid medium:

$$5 \text{ V}^{3+} + \text{MnO}_4^- + \text{H}_2\text{O} \rightleftharpoons 5 \text{ VO}^{2+} + \text{Mn}^{2+} + 2 \text{ H}^+$$

and

$$5 \text{ VO}^{2+} + \text{MnO}_4^- + \text{H}_2\text{O} \rightleftharpoons 5 \text{ VO}_2^+ + \text{Mn}^{2+} + 2 \text{ H}^+$$

(b) The titration of a mixture of 1.000 millimole of permanganate and 2.000 millimoles of plutonium(IV) in a 1 $F$ sulfuric acid medium with 0.05000 $M$ titanium(III) in 1 $F$ sulfuric acid:

$$\text{MnO}_4^- + 5 \text{ Ti}^{3+} + \text{H}_2\text{O} \rightleftharpoons \text{Mn}^{2+} + 5 \text{ TiO}^{2+} + 2 \text{ H}^+$$

and

$$\text{Pu}^{4+} + \text{Ti}^{3+} + \text{H}_2\text{O} \rightleftharpoons \text{Pu}^{3+} + \text{TiO}^{2+} + 2 \text{ H}^+$$

## SUGGESTIONS FOR ADDITIONAL READING

1. R. G. Bates: Electrode potentials. *In* I. M. Kolthoff and P. J. Elving, eds.: *Treatise on Analytical Chemistry*. Part I, Volume 1, Wiley-Interscience, New York, 1959, pp. 319–359.

2. H. A. Laitinen: *Chemical Analysis*. McGraw-Hill Book Company, New York, 1960, pp. 276–297, 326–341.

3. W. M. Latimer: *Oxidation Potentials*. Second edition, Prentice-Hall, Englewood Cliffs, New Jersey, 1952.

4. J. J. Lingane: *Electroanalytical Chemistry*. Second edition, Wiley-Interscience, New York, 1958, pp. 1–70, 129–157.

5. S. Wawzonek: Potentiometry: Oxidation-reduction potentials. *In* A. Weissberger and B. W. Rossiter, eds.: *Physical Methods of Chemistry*. Volume I, Part IIA, Wiley-Interscience, New York, 1971, pp. 1–60.

# ANALYTICAL APPLICATIONS OF OXIDATION-REDUCTION REACTIONS

# 10

Because oxidation-reduction reactions provide a foundation for innumerable methods of chemical analysis, it is appropriate to explore the properties and uses of some of the most popular titrants. Three of the strongest oxidizing agents used in redox titrimetry—potassium permanganate, potassium dichromate, and cerium(IV) —will be described first. Then, the great versatility of the triiodide-iodide system will be demonstrated through a discussion of reactions in which triiodide ion serves as an oxidizing agent and of processes in which iodide ion acts as a reductant for many oxidizing agents. In addition, the analytical uses of iodate, periodate, and bromate—especially for the determination of organic substances—will be considered. Finally, iron(II), titanium(III), and chromium(II), three valuable reducing titrants, will be examined briefly.

Before the individual reagents are discussed, it is worthwhile to outline what characteristics one seeks in choosing a titrant and what preliminary steps must be taken to prepare the substance being determined for reaction with the titrant.

# REQUIREMENTS FOR SELECTION OF TITRANTS

Although examination of a table of standard potentials reveals the existence of species covering a wide range of oxidizing and reducing strengths, relatively few substances qualify for selection as titrants for redox methods of analysis. For any particular titration, the titrant must satisfy several requirements. First, the titrant must be strong enough to react to practical completion with the substance being titrated. As discussed in Chapter 9, this requirement means that the standard potential for the half-reaction involving an oxidizing titrant must be at least 0.2 v more positive, whereas the standard potential for the half-reaction involving a reducing titrant must be at least 0.2 v more negative, than the standard potential for the half-reaction pertaining to the substance being titrated. Second, the titrant must not be so powerful that it is able to react with some component of the solution being titrated other than the desired species. In other words, the redox reaction must proceed stoichiometrically. For example, strong oxidants such as silver(II) and cobalt(III) easily satisfy the first requirement, but fail to meet the second requirement because they quickly oxidize the solvent (water) in which they are dissolved. Third, the titrant must react rapidly with the substance being determined. A certain reaction may appear to be favorable from a thermodynamic point of view—that is, the emf for the overall reaction may be large in magnitude and positive in sign—but the reaction may not occur at a convenient rate. This is often true if the redox reaction involves multiple electron transfer, or the formation or rupture of chemical bonds. Finally, a simple and precise method for location of the equivalence point must be available. Most of the titrants encountered in this chapter fulfill these requirements only insofar as careful attention is given to control of such experimental variables as pH, temperature, use of catalysts, and reaction time.

# PRELIMINARY ADJUSTMENT OF OXIDATION STATES OF TITRATED SUBSTANCES

For the performance of a successful redox titration, the species to be determined must exist in a single oxidation state which can react stoichiometrically and rapidly with the titrant. However, preparation of a sample solution often leaves the desired substance in an oxidation state which is unreactive toward the titrant or perhaps in a mixture of oxidation states. In these instances, one must treat the sample solution with a suitable oxidizing or reducing agent to adjust the oxidation state of the substance prior to the final titration. What characteristics should this oxidizing or reducing agent possess? Clearly, the reagent must convert the desired substance quickly and quantitatively to the proper oxidation state. In addition, it must be possible, as well as convenient, to separate or remove the excess oxidant or reductant from the sample solution, lest the reagent interact later with the titrant. Finally, the oxidizing or reducing agent should display a certain degree of selectivity. Let us consider some of the more widely employed oxidizing and reducing agents used to adjust the oxidation state of a species.

## Oxidizing Agents

*Sodium bismuthate.* A powerful oxidizing agent, this sparingly soluble compound has an indefinite composition, but consists largely of $NaBiO_3$. In the presence of excess sodium bismuthate in a nitric acid medium, manganese(II) is converted to permanganate, chromium(III) becomes dichromate, and cerium(III) is oxidized to cerium(IV). Simultaneously, bismuth(V) is reduced to soluble and innocuous bismuth(III), and the unreacted sodium bismuthate may be separated from the solution by filtration.

*Potassium peroxodisulfate.* In boiling acidic solutions containing a trace of silver(I) ion as catalyst, potassium peroxodisulfate ($K_2S_2O_8$) quantitatively oxidizes cerium(III) to cerium(IV), manganese(II) to permanganate, chromium(III) to dichromate, and vanadium(IV) to vanadium(V). Ammonium peroxodisulfate may be similarly employed as a strong oxidant. After the oxidation reactions are complete, continued boiling of the solution for approximately 10 minutes decomposes the excess peroxodisulfate:

$$2\,S_2O_8{}^{2-} + 2\,H_2O \rightleftharpoons 4\,SO_4{}^{2-} + O_2 + 4\,H^+$$

*Silver(II) oxide.* Commercially available, dark brown silver(II) oxide, AgO, is an exceptionally strong oxidant which smoothly converts manganese(II) to permanganate, chromium(III) to dichromate, cerium(III) to cerium(IV), and vanadium(IV) to vanadium(V) as small portions of the reagent are dissolved in a cold perchloric acid or nitric acid medium. It is easy to recognize when enough silver(II) oxide has been added due to the chocolate-brown color of unreacted $Ag^{2+}$ ion; the excess silver(II) is easily reduced by water if the solution is warmed:

$$4\,Ag^{2+} + 2\,H_2O \rightleftharpoons 4\,Ag^+ + O_2 + 4\,H^+$$

*Hydrogen peroxide.* In an alkaline medium, hydrogen peroxide can be employed to oxidize chromium(III) to chromate, manganese(II) to manganese dioxide, arsenic(III) to arsenic(V), antimony(III) to antimony(V), and vanadium(IV) to vanadium(V). On the other hand, in acidic solutions, this reagent quantitatively converts iron(II) to iron(III) and iodide ion to molecular iodine, but reduces dichromate to chromium(III) and permanganate to manganese(II). Excess hydrogen peroxide decomposes by disproportionation when the acidic or alkaline solution is boiled for a few minutes:

$$2\,H_2O_2 \rightleftharpoons O_2 + 2\,H_2O$$

## Reducing Agents

*Hydrogen sulfide and sulfur dioxide.* These gases are readily soluble in aqueous media and are relatively mild reducing agents, being widely employed for the reduction of iron(III) to iron(II) in acid solutions prior to titrations with standard solutions of oxidants. In addition, hydrogen sulfide and sulfur dioxide reduce vanadium(V) to vanadium(IV), as well as stronger oxidizing agents such as permanganate, cerium(IV), and dichromate, whereas titanium(IV) and chromium(III) are unaffected. One need only boil the solution if acidic to remove the excess of either gas, although the operation is time-consuming. Besides the unpleasant and toxic nature of these gases, reductions with sulfur dioxide tend to be rather slow, and the use of hydrogen sulfide leads to the formation of colloidal sulfur which may react with strongly oxidizing titrants.

*Tin(II) chloride.* Hydrochloric acid solutions of this reagent are utilized almost exclusively for the quantitative reduction of iron(III) to iron(II) prior to titration of the latter with a standard solution of permanganate, dichromate, or cerium(IV). Further discussion of the use of tin(II) chloride is presented later.

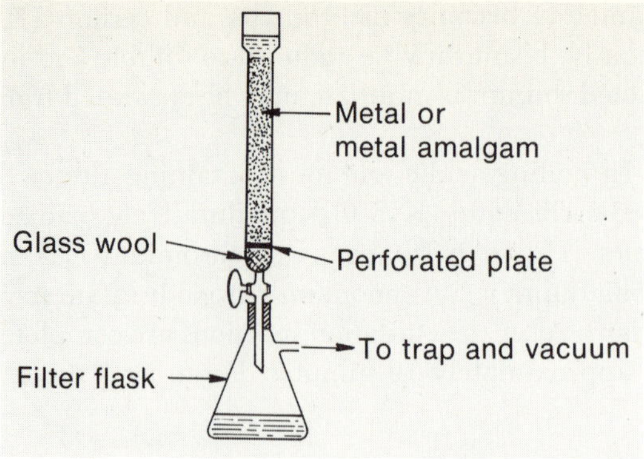

*Figure 10–1.* A metal or metal amalgam reductor.

Metal or metal amalgam

Glass wool

Perforated plate

To trap and vacuum

Filter flask

***Metals and metal amalgams.*** Perhaps the most versatile method to accomplish reduction of a species to a definite oxidation state is treatment of the sample solution with a metal. Among the metallic elements employed as reductants are zinc, aluminum, cadmium, silver, mercury, copper, nickel, bismuth, lead, tin, and iron.

In the form of wire or rod, the metal can be inserted directly into the solution and, when the reduction is judged to be complete, the unused reductant may be washed carefully and withdrawn. To enhance the rate of reduction, it is often desirable to use metallic powder or granules which present a large surface area to the solution; if the metallic reductant is simply added to the solution, filtration suffices to separate the unreacted metal from the sample. However, one usually finds it more convenient to prepare a so-called *reductor*, as shown in Figure 10–1, by packing granules of the appropriate metal reductant into a glass column; then the sample solution is percolated through the column and is collected in a receiving flask. Generally, a solution treated with a metal reductant must be blanketed with an inert gas such as nitrogen or carbon dioxide to prevent exposure to the atmosphere before the final titration, since lower oxidation states of most species are quickly oxidized by oxygen.

Inspection of a table of standard potentials reveals that aluminum, zinc, and cadmium are especially potent reductants, which means, incidentally, that they lack specificity. As a consequence of their reducing power, these substances react vigorously with hydrogen ion to yield hydrogen gas, *e.g.,*

$$2 \, Al + 6 \, H^+ \rightarrow 2 \, Al^{3+} + 3 \, H_2$$

This side-reaction wastes the metallic reductant and introduces unwanted amounts of the corresponding metal ion into the sample solution. Moreover, if a reductor column is employed, evolution of hydrogen gas, as the sample solution passes through the column, may render reduction of the desired species less than quantitative. To eliminate this source of difficulty, zinc and cadmium may be amalgamated with mercury. In the case of zinc, this process involves thorough agitation of a solution of mercury(II) nitrate in contact with zinc granules for a few minutes until the reaction

$$Zn + Hg^{2+} \rightleftharpoons Zn^{2+} + Hg$$

causes each metal particle to become coated with a thin film of elemental mercury. Amalgamated zinc and cadmium have little tendency to reduce hydrogen ion in acidic solutions and, therefore, make ideal packings for reductor columns.

Without doubt, the **Jones reductor**—a glass tube approximately 2 cm in diameter packed with a 30- to 40-cm column of amalgamated zinc (Figure 10–1)—is more extensively used than any other amalgam reductor. As a general rule, the reducing strength of an amalgamated metal is such that its potential in contact with a 1 $M$ solution of the corresponding metal ion is about 0.05 v *more positive* than the potential for a system involving the pure metal and the metal ion. Thus, for pure zinc metal, we obtain from a table of standard potentials

$$Zn^{2+} + 2\ e \rightleftharpoons Zn; \qquad E^0 = -0.763 \text{ v}$$

whereas, for zinc metal amalgamated with 1 per cent of mercury, we have

$$Zn^{2+} + Hg + 2\ e \rightleftharpoons Zn(Hg); \qquad E \sim -0.71 \text{ v}$$

A list of half-reactions for some commonly reduced species is presented in Table 10–1.

**Table 10–1.** **Comparison of Reductions Accomplished with the Jones Reductor and the Silver Reductor***

| Jones Reductor (1 $F$ $H_2SO_4$ Medium) | Silver Reductor (1 $F$ HCl Medium) |
|---|---|
| $Cr^{3+} + e \rightleftharpoons Cr^{2+}$ | $Cr^{3+}$ is not reduced |
| $Cu^{2+} + 2\ e \rightleftharpoons Cu$ | $Cu^{2+} + 2\ Cl^- + e \rightleftharpoons CuCl_2^-$ |
| $Fe^{3+} + e \rightleftharpoons Fe^{2+}$ | $Fe^{3+} + e \rightleftharpoons Fe^{2+}$ |
| $MoO_2^{2+} + 4\ H^+ + 3\ e \rightleftharpoons Mo^{3+} + 2\ H_2O$ | $MoO_2^{2+} + e \rightleftharpoons MoO_2^+$ |
| $TiO^{2+} + 2\ H^+ + e \rightleftharpoons Ti^{3+} + H_2O$ | $TiO^{2+}$ is not reduced |
| $\left\{\begin{array}{l} UO_2^{2+} + 4\ H^+ + 2\ e \rightleftharpoons U^{4+} + 2\ H_2O \\ UO_2^{2+} + 4\ H^+ + 3\ e \rightleftharpoons U^{3+} + 2\ H_2O \end{array}\right\}$ † | $UO_2^{2+} + 4\ H^+ + 2\ e \rightleftharpoons U^{4+} + 2\ H_2O$ |
| $VO_2^+ + 4\ H^+ + 3\ e \rightleftharpoons V^{2+} + 2\ H_2O$ | $VO_2^+ + 2\ H^+ + e \rightleftharpoons VO^{2+} + H_2O$ |

* Taken with permission from I. M. Kolthoff and R. Belcher: *Volumetric Analysis*. Volume 3, Interscience Publishers, Inc., New York, 1957, p. 12.

† A mixture of uranium(III) and uranium(IV) is obtained from the Jones reductor; however, the uranium(III) can be converted to uranium(IV) if the solution is shaken in contact with oxygen for several minutes.

If a column similar to that employed for the Jones reductor is filled with pure silver granules, one obtains the well known **silver reductor**. By itself, elemental silver is a poor reducing agent. However, in the presence of hydrochloric acid—with which medium the silver reductor is almost always utilized—metallic silver becomes an effective, though mild, reductant. For example, when a 1 $F$ hydrochloric acid solution of uranium(VI) is passed through the reductor, uranium(IV) is formed quantitatively while metallic silver is oxidized to silver chloride which adheres to the particles of silver:

$$UO_2^{2+} + 4\ H^+ + 2\ Ag + 2\ Cl^- \rightleftharpoons U^{4+} + 2\ AgCl + 2\ H_2O; \qquad E^0 = +0.112 \text{ v}$$

Since silver metal in hydrochloric acid is not as strong a reductant as amalgamated zinc, use of a silver reductor may permit greater selectivity in the preferential adjustment of the oxidation state of one substance in a mixture of several species. A comparison of the capabilities of the Jones reductor and the silver reductor is presented in Table 10–1. According to the Nernst equation, the reducing strength of a silver reductor depends on the concentration of chloride ion in the solution passed through the column. It can be shown that the effective emf for the preceding reaction is $-0.006$ v in the presence of 0.01 $M$ chloride ion, so uranium(VI) would not be completely reduced. Thus, variation of the chloride concentration provides additional selectivity in the use of a silver reductor. Since the efficiency of a silver reductor is impaired as the metallic silver becomes coated with silver chloride, it is necessary to regenerate the column periodically by treatment with a strong reductant.

A number of liquid amalgams have been employed to perform reductions in a manner analogous to the use of solid reductants. Thus, a solution of elemental lead, bismuth, or tin in mercury may be shaken in an ordinary separatory funnel with the sample solution until the reduction is complete, the amalgam and solution phases separated, and the aqueous medium subjected to appropriate titration. In addition, pure mercury in contact with 1 $F$ hydrochloric acid has nearly the same reducing power as silver metal in that medium.

## POTASSIUM PERMANGANATE

### Half-Reactions Involving Permanganate Ion

Manganese may exist in a number of stable oxidation states. However, for the majority of oxidation-reduction titrations, the important states are manganese(VII), manganese(IV), and manganese(II). In Appendix 4, one finds the two half-reactions

$$MnO_4^- + 8\,H^+ + 5\,e \rightleftharpoons Mn^{2+} + 4\,H_2O; \qquad E^0 = +1.51\text{ v}$$

and

$$MnO_4^- + 4\,H^+ + 3\,e \rightleftharpoons MnO_2 + 2\,H_2O; \qquad E^0 = +1.695\text{ v}$$

Factors that govern which of the two reduction products, $Mn^{2+}$ or $MnO_2$, is actually formed are quite complex and involve consideration of both kinetics and thermodynamics. Suffice it to say that, in neutral and alkaline media as well as in weakly acidic solutions, manganese dioxide is the product. Except for some reactions with organic compounds, potassium permanganate is usually employed for titrations in acidic media having hydrogen ion concentrations of 0.1 $M$ or greater, in which manganous ion is the reduction product.

Two other oxidation states of manganese are involved to a limited extent in analytical applications of permanganate solutions. Although manganese(III) normally disproportionates into manganous ion and manganese dioxide,

$$2\,Mn^{3+} + 2\,H_2O \rightleftharpoons Mn^{2+} + MnO_2 + 4\,H^+$$

the $Mn^{3+}$ species can be stabilized by pyrophosphate and by fluoride. For example, permanganate may be reduced to the manganese(III)-pyrophosphate complex in a solution of pH 4 to 7 containing sodium pyrophosphate:

$$MnO_4^- + 3\,H_2P_2O_7^{2-} + 8\,H^+ + 4\,e \rightleftharpoons Mn(H_2P_2O_7)_3^{3-} + 4\,H_2O$$

Unfortunately, the wine-red color of the $Mn(H_2P_2O_7)_3^{3-}$ complex resembles the color of permanganate so closely that visual end-point detection is impossible.

Bright green manganate ion, $MnO_4^{2-}$, is the reduction product of permanganate in strongly alkaline media:

$$MnO_4^- + e \rightleftharpoons MnO_4^{2-}; \qquad E^0 = +0.564\text{ v}$$

In sodium hydroxide solutions more concentrated than $2 F$, organic compounds reduce permanganate according to this half-reaction.

## Preparation and Stability of Permanganate Solutions

Potassium permanganate is rarely available in a sufficiently high state of purity to permit its direct use as a primary standard substance. Even reagent-grade potassium permanganate is invariably contaminated with small quantities of manganese dioxide. In addition, ordinary distilled water, from which potassium permanganate solutions are prepared, contains organic matter which can reduce permanganate to manganese dioxide.

Permanganate solutions are inherently unstable because $MnO_4^-$ is capable of oxidizing water spontaneously:

$$4 \, MnO_4^- + 2 \, H_2O \rightleftharpoons 4 \, MnO_2 + 3 \, O_2 + 4 \, OH^-; \qquad E^0 = +0.187 \, v$$

Fortunately, the rate of this reaction is exceedingly slow if proper precautions are taken in the original preparation of the solution. Heat, light, acids, bases, manganese(II) salts, and especially manganese dioxide catalyze the permanganate-water reaction. Thus, any manganese dioxide initially present in the potassium permanganate solution must be removed if a stable titrant is to be obtained. Since the reaction between permanganate and the organic matter in distilled water cannot conveniently be eliminated, it should be allowed to proceed to completion, or nearly so, prior to standardization of the solution. After the potassium permanganate is dissolved, the solution should be heated to hasten this decomposition reaction and allowed to stand so that the initially colloidal manganese dioxide can coagulate. Next, the manganese dioxide precipitate must be removed from the solution by filtration. Solutions of potassium permanganate should be stored in dark bottles and kept out of bright light and away from dust as much as possible. Potassium permanganate solutions of a concentration not less than $0.02 \, F$, when prepared and stored in the manner just described, are stable for many months.

## Standardization of Permanganate Solutions

Among the substances suitable for the standardization of potassium permanganate solutions are arsenious oxide ($As_2O_3$), sodium oxalate ($Na_2C_2O_4$), and pure iron wire.

Arsenious oxide, commercially available as a primary-standard-grade solid, is initially dissolved in a sodium hydroxide solution,

$$As_2O_3 + 2 \, OH^- \rightarrow 2 \, AsO_2^- + H_2O$$

which is subsequently acidified with hydrochloric acid and titrated with potassium permanganate according to the reaction*

$$2 \, MnO_4^- + 5 \, HAsO_2 + 6 \, H^+ + 2 \, H_2O \rightleftharpoons 2 \, Mn^{2+} + 5 \, H_3AsO_4$$

---

* Some authors use $H_3AsO_3$ as the formula of arsenious acid. We prefer to write $HAsO_2$ as the predominant form of arsenic(III) because arsenious acid behaves as a weak monoprotic acid in aqueous media.

However, this reaction does not proceed rapidly without a catalyst. One possible explanation for its slowness is that permanganate is partially reduced to manganese(III) and manganese(IV) species, which are stabilized as arsenate complexes. Iodine monochloride is an excellent catalyst for this titration although, in a hydrochloric acid solution, iodine monochloride actually exists as the species $ICl_2^-$.

Sodium oxalate is another substance used for the standardization of potassium permanganate solutions. It is available commercially in a very pure state, and it dissolves in sulfuric acid media with the resultant formation of oxalic acid. We may represent the stoichiometry of the permanganate-oxalic acid reaction as

$$2 \, MnO_4^- + 5 \, H_2C_2O_4 + 6 \, H^+ \rightleftharpoons 2 \, Mn^{2+} + 10 \, CO_2 + 8 \, H_2O$$

However, the mechanism by which this reaction proceeds is exceedingly complex, and reproducible and stoichiometric results are obtained only when certain empirical conditions are fulfilled. Temperature plays a very important role in the success of the titration; usually, the titration is carried out at solution temperatures near 70°C so that the rate of reaction is conveniently fast. Another interesting feature of this redox process is that the first few drops of permanganate react very slowly, as evidenced by the fact that the permanganate color does not disappear for many seconds, but succeeding portions of titrant react more and more rapidly until the reaction becomes essentially instantaneous. This behavior is typical of an *autocatalytic process*, in which one of the reaction products functions as a catalyst. In the present situation, manganese(II) is this catalyst.

Specially purified iron wire may be dissolved in acid, converted to iron(II), and the iron(II) titrated to iron(III) with the permanganate solution to be standardized.

### End-Point Detection for Permanganate Titrations

Solutions of potassium permanganate are so intensely colored that a single drop of a $0.02 \, F$ titrant imparts a perceptible color to 100 ml of water, the reduction product of permanganate in an acid medium, $Mn^{2+}$, being practically colorless. Therefore, if the solution being titrated is colorless, the appearance of the pale pink color due to the first trace of excess titrant may be taken as the equivalence point. If the titrant is too dilute to serve as its own indicator, tris(1,10-phenanthroline)iron(II), or ferroin, plus other substituted ferroin-type indicators discussed in Chapter 9 may be used.

With $0.02 \, F$ potassium permanganate in a buret, the meniscus of the solution is not visible, so the position of the top of the solution is usually observed. However, if a flashlight is held behind the buret, the meniscus can be seen.

### Analytical Uses of Potassium Permanganate in Acidic Solutions

As already mentioned, except for the oxidation of some organic compounds with permanganate in sodium hydroxide media, most applications involve titrimetry under strongly acidic conditions, for which permanganate is reduced to manganous ion. Information about a number of useful direct titrations with a standard potassium permanganate solution is summarized in Table 10–2. In the following paragraphs, we shall examine two of these applications in more detail.

*Determination of iron in an ore.* Hematite ($Fe_2O_3$), limonite ($2 \, Fe_2O_3 \cdot 3 \, H_2O$), and magnetite ($Fe_3O_4$), which are the important iron ores, can usually be

**Table 10–2. Some Titrations Performed with Standard Potassium Permanganate in Acidic Media**

| Substance Determined | Half-Reaction for Substance Titrated | Procedure and Conditions for Titration |
|---|---|---|
| Sb(III) | $SbCl_4^- + 2\ Cl^- \rightleftharpoons SbCl_6^- + 2\ e$ | Titrate in $2\ F$ HCl |
| As(III) | $HAsO_2 + 2\ H_2O \rightleftharpoons H_3AsO_4 + 2\ H^+ + 2\ e$ | Titrate in $1\ F$ HCl with ICl as catalyst |
| Br⁻ | $2\ Br^- \rightleftharpoons Br_2 + 2\ e$ | Titrate in boiling $2\ F$ $H_2SO_4$ to expel $Br_2$ |
| Fe(CN)₆⁴⁻ | $Fe(CN)_6^{4-} \rightleftharpoons Fe(CN)_6^{3-} + e$ | Titrate in $0.2\ F$ $H_2SO_4$ with erioglaucine A as indicator |
| H₂O₂ | $H_2O_2 \rightleftharpoons O_2 + 2\ H^+ + 2\ e$ | Titrate in $1\ F$ $H_2SO_4$ |
| I⁻ | $I^- + HCN \rightleftharpoons ICN + H^+ + 2\ e$ | Titrate in presence of $0.1\ F$ HCN and $1\ F$ $H_2SO_4$; ferroin indicator |
| Fe(II) | $Fe^{2+} \rightleftharpoons Fe^{3+} + e$ | Reduce iron(III) with tin(II) or in Jones reductor; titrate in $1\ F$ $H_2SO_4$, or in $1\ F$ HCl with Zimmermann-Reinhardt reagent added |
| Mo(III) | $Mo^{3+} + 2\ H_2O \rightleftharpoons MoO_2^{2+} + 4\ H^+ + 3\ e$ | Reduce molybdenum to molybdenum(III) in Jones reductor; treat with excess iron(III) in $1\ F$ $H_2SO_4$ and titrate iron(II) formed |
| H₂C₂O₄ | $H_2C_2O_4 \rightleftharpoons 2\ CO_2 + 2\ H^+ + 2\ e$ | Titrate in $1\ F$ $H_2SO_4$ at 70°C |
| Ca(II), Mg(II), Zn(II), Co(II), La(III), Th(IV), Ba(II), Sr(II), Pb(II) | $H_2C_2O_4 \rightleftharpoons 2\ CO_2 + 2\ H^+ + 2\ e$ | Metal oxalates are precipitated, washed, and redissolved in $1\ F$ $H_2SO_4$; liberated oxalic acid titrated at 70°C |
| Sn(II) | $SnCl_4^{2-} + 2\ Cl^- \rightleftharpoons SnCl_6^{2-} + 2\ e$ | Reduce tin(IV) to tin(II) with bismuth amalgam in 5 to 12 $F$ HCl; exclude oxygen during titration |
| Ti(III) | $Ti^{3+} + H_2O \rightleftharpoons TiO^{2+} + 2\ H^+ + e$ | Reduce titanium(IV) to titanium(III) with zinc metal, lead amalgam, or in Jones reductor; exclude oxygen; titrate in $1\ F$ $H_2SO_4$ or HCl |
| W(III) | $W^{3+} + 2\ H_2O \rightleftharpoons WO_2^{2+} + 4\ H^+ + 3\ e$ | Reduce tungsten to tungsten(III) with lead amalgam at 50°C; titrate in $2\ F$ HCl |
| U(IV) | $U^{4+} + 2\ H_2O \rightleftharpoons UO_2^{2+} + 4\ H^+ + 2\ e$ | Reduce uranium to uranium(III) in Jones reductor; expose solution to air to obtain uranium(IV); titrate in $1\ F$ $H_2SO_4$ |
| V(IV) | $VO^{2+} + H_2O \rightleftharpoons VO_2^+ + 2\ H^+ + e$ | Reduce vanadium to vanadium(IV) with bismuth amalgam; titrate in $1\ F$ $H_2SO_4$ |

dissolved in hot hydrochloric acid. If the iron ore is highly refractory, fusion of a sample with a flux of sodium carbonate or potassium acid sulfate may be necessary to render the ore soluble. Most iron ores contain silica, which will not dissolve in hydrochloric acid. However, the white residue of hydrated silica is not harmful to the determination of iron, and it may be distinguished easily from particles of undissolved and dark-colored iron oxide.

After the oxide is dissolved, the iron will exist wholly or partly as iron(III). Since the titration with standard potassium permanganate solution requires that all iron be present as iron(II), the iron(III) formed during dissolution of the sample must be quantitatively reduced. Any of the procedures described earlier for the preliminary adjustment of the oxidation state of a substance may be employed. Hydrogen sulfide and sulfur dioxide have both been used to reduce iron(III). If one simply boils the solution, the excess of either gaseous reductant may be removed, although care must be taken not to reoxidize the iron(II). Alternatively, one may utilize the Jones reductor, but due consideration must be given to the fact that elements which often accompany iron in an ore or alloy—titanium, vanadium, chromium, uranium, tungsten, arsenic, and molybdenum—will be reduced to their lower oxidation states and will consume some of the permanganate titrant. On the other hand, use of a silver reductor permits preferential reduction of iron(III) in the presence of both titanium(IV) and chromium(III).

A preferred method for the reduction of iron(III) entails addition of a very small excess of tin(II) chloride to the hot hydrochloric acid solution of the iron sample,

$$2\ Fe^{3+} + SnCl_4^{2-} + 2\ Cl^- \rightleftharpoons 2\ Fe^{2+} + SnCl_6^{2-}$$

followed by destruction of the excess tin(II) with mercury(II):

$$SnCl_4^{2-} + 2\ HgCl_4^{2-} \rightleftharpoons SnCl_6^{2-} + Hg_2Cl_2(s) + 4\ Cl^-$$

Enough tin(II) must be introduced to cause complete reduction of iron(III). However, if the excess of tin(II) is too great, elemental mercury rather than mercurous chloride (calomel) may form:

$$SnCl_4^{2-} + HgCl_4^{2-} \rightleftharpoons SnCl_6^{2-} + Hg(1) + 2\ Cl^-$$

Although solid mercurous chloride does not interfere with a successful titration, metallic mercury ruins the determination because, being present in a finely divided colloidal state, it reacts to a significant extent with permanganate.

When iron(II) in hydrochloric acid is titrated with permanganate, significant positive errors may arise, because the iron(II)-permanganate reaction *induces* some oxidation of chloride to hypochlorous acid (HOCl). Two methods have been devised to eliminate this source of error—removal of hydrochloric acid prior to the titration by evaporation with sulfuric acid, or addition of Zimmermann-Reinhardt reagent followed by titration in the presence of the hydrochloric acid. Zimmermann-Reinhardt reagent consists of manganous sulfate, phosphoric acid, and sulfuric acid—manganese(II) prevents the accumulation of local excesses of the permanganate titrant, and phosphoric acid forms colorless complexes with iron(III) which are more stable than yellow-colored iron(III)-chloride complexes and which do not obscure the appearance of the permanganate end point. More importantly, it appears that phosphoric acid weakens the strength of the active oxidizing agent in this system, manganese(III), by forming complexes with it, so that chloride can no longer be oxidized.

*Determination of calcium in limestone.* A useful indirect oxidation-reduction method of analysis is the determination of calcium in limestone. Although the principal components of dolomitic limestone are calcium carbonate and magnesium carbonate, smaller percentages of calcium and magnesium silicates, plus carbonates and silicates of such elements as aluminum, iron, and manganese, are usually present. In addition, minor amounts of titanium, sodium, and potassium exist in most specimens.

Preparation of the sample solution usually entails dissolution of a weighed portion of powdered limestone in hydrochloric acid:

$$CaCO_3 + 2 H^+ \rightarrow Ca^{2+} + CO_2 + H_2O$$

Next, the calcium is precipitated as calcium oxalate monohydrate under carefully adjusted conditions:

$$Ca^{2+} + C_2O_4^{2-} + H_2O \rightleftharpoons CaC_2O_4 \cdot H_2O$$

Then the precipitate is separated from its mother solution by filtration and is washed free of excess oxalate. Finally, the calcium oxalate monohydrate is dissolved in sulfuric acid, the oxalate species being converted into oxalic acid,

$$CaC_2O_4 \cdot H_2O + 2 H^+ \rightleftharpoons Ca^{2+} + H_2C_2O_4 + H_2O$$

and the oxalic acid solution is titrated with a standard potassium permanganate solution. Although calcium ion is not involved directly in the final titration, it is stoichiometrically related to the amount of oxalic acid titrated. A number of other elements which form insoluble oxalates—zinc, cadmium, lead, cobalt, and nickel—may be determined according to the procedure. These elements constitute potential interferences in the successful determination of calcium.

## Determination of Organic Compounds with Alkaline Permanganate

A major difficulty encountered in the development of procedures for the determination of organic substances is their inherently slow rate of reaction with permanganate. Such behavior is understandable, because most organic compounds are eventually degraded to carbon dioxide and water, a process which usually involves the rupture of at least several carbon-carbon and carbon-hydrogen bonds. Furthermore, many organic compounds exhibit little or no specificity in the oxidation processes which they undergo, and so mixtures of organic substances may not always be analyzed. Slow reactions can be overcome if the desired organic compound is treated with an excess of permanganate and if the reaction is allowed to proceed at an elevated temperature for extended periods of time. Unfortunately, these extreme conditions may lead to undesired side reactions and, therefore, to oxidation processes which appear to be nonstoichiometric. Nevertheless, if reaction conditions are carefully controlled and reproduced, numerous valuable and practical determinations can be performed.

Earlier, we mentioned that permanganate is reduced to the green-colored manganate ion by organic substances in concentrated sodium hydroxide solutions.

This reaction serves as the basis of a quantitative method for the determination of a number of organic compounds, if the desired substance is dissolved in a known excess of strongly alkaline standard permanganate solution and allowed to react at room temperature for approximately 30 minutes. For example, glycerol which is oxidized to carbonate

$$H_2C\text{----}CH\text{----}CH_2 + 14\ MnO_4^- + 20\ OH^- \rightarrow$$
$$\underset{OH}{\phantom{|}}\ \underset{OH}{\phantom{|}}\ \underset{OH}{\phantom{|}}$$
$$3\ CO_3^{2-} + 14\ MnO_4^{2-} + 14\ H_2O$$

may be determined by acidification of the solution after the alkaline oxidation, followed by titrimetric reduction of all higher oxidation states of manganese to manganese(II); the glycerol content of the sample is computed from the *difference* between the amount of permanganate originally taken and the quantity of higher oxidation states of manganese found in the final titration.

If an aqueous formic acid-acetic acid mixture is treated with a known volume of alkaline standard permanganate solution, only the formate ion undergoes oxidation,

$$HCOO^- + 2\ MnO_4^- + 3\ OH^- \rightarrow CO_3^{2-} + 2\ MnO_4^{2-} + 2\ H_2O$$

whereas acetate remains unattacked. Therefore, subsequent acidification of the alkaline medium and titration of the higher oxidation states of manganese permit one to determine directly the quantity of formic acid in the original sample. If the result of the oxidation-reduction procedure is combined with a determination of the total acid content of a separate sample of the mixture, the individual concentrations of both formic acid and acetic acid may be ascertained.

One other application of the alkaline permanganate procedure for organic compounds is the determination of methanol. In the presence of excess permanganate-sodium hydroxide solution, methanol is oxidized to carbonate according to the reaction

$$CH_3OH + 6\ MnO_4^- + 8\ OH^- \rightarrow CO_3^{2-} + 6\ MnO_4^{2-} + 6\ H_2O$$

Instead of acidifying the solution, one can determine the excess, unreacted permanganate by adding barium ion and titrating the mixture with a standard solution of sodium formate. Formate reduces the permanganate only to manganate in an alkaline medium as shown previously, and the titration reaction is usually formulated as

$$2\ MnO_4^- + HCOO^- + 3\ Ba^{2+} + 3\ OH^- \rightarrow 2\ BaMnO_4(s) + BaCO_3(s) + 2\ H_2O$$

Notice that barium ion reacts with both manganate and carbonate ion to form insoluble precipitates and that, in particular, removal of the green-colored manganate ion from the solution phase aids considerably in helping one recognize the final disappearance of the permanganate color. If one subtracts the equivalents of formate needed for the titration from the total equivalents of permanganate used originally, the quantity of methanol in the sample is obtained.

Other organic compounds which have been determined by means of the permanganate procedure are glycolic acid, tartaric acid, citric acid, ethylene glycol, phenol, salicylic acid, formaldehyde, glucose, and sucrose.

## POTASSIUM DICHROMATE

### Reduction of Dichromate Ion

When we turn to the dichromate ion, we find that the only important reduction involving this species is

$$Cr_2O_7^{2-} + 14\,H^+ + 6\,e \rightleftharpoons 2\,Cr^{3+} + 7\,H_2O; \qquad E^0 = +1.33\,v$$

However, the formal potential for this half-reaction is $+1.00$ v in $1\,F$ hydrochloric acid and $+1.11$ v in $2\,F$ sulfuric acid. Potassium dichromate is invariably employed as a titrant in acidic media, for in neutral and alkaline solutions chromic ion, $Cr^{3+}$ or $Cr(H_2O)_6^{3+}$, forms an insoluble hydrous oxide, and dichromate is converted to the chromate ion:

$$Cr_2O_7^{2-} + 2\,OH^- \rightleftharpoons 2\,CrO_4^{2-} + H_2O$$

### Preparation and Stability of Standard Dichromate Solutions

Potassium dichromate may be purchased as a high-purity, reagent-grade solid which is suitable for the direct preparation, by weight, of standard solutions. Furthermore, potassium dichromate is readily soluble in water, and the resulting solutions are stable for many years if protected against evaporation. In addition, dichromate solutions can be boiled without the occurrence of any detectable decomposition.

### End-Point Detection for Titrations with Dichromate

Dichromate solutions have an orange-yellow color, whereas chromium(III) may be either green or violet, depending upon the composition of the medium in which it is formed as a reduction product. Not surprisingly, the color due to excess dichromate is masked by that of chromium(III), so dichromate cannot serve as its own end-point indicator. Any indicator which is selected must be one whose color change is so distinct that it can be readily recognized in the presence of the other colored species in solution at and near the equivalence point. Sodium diphenylbenzidine sulfonate and barium diphenylamine sulfonate (Table 9–3) are well suited as end-point indicators for titrations with potassium dichromate.

### Analytical Uses of Potassium Dichromate

Because potassium dichromate is a less potent oxidant than potassium permanganate, it has fewer analytical applications. Undoubtedly, the most important use of dichromate is the titration of iron(II),

$$Cr_2O_7^{2-} + 6\,Fe^{2+} + 14\,H^+ \rightleftharpoons 2\,Cr^{3+} + 6\,Fe^{3+} + 7\,H_2O$$

since the method is highly precise in the presence of hydrochloric acid, the most common solvent for iron-containing alloys and ores.

Species which react quantitatively with iron(III) to produce a stoichiometric amount of iron(II) can be determined indirectly by means of the preceding reaction.

Copper(I) and uranium(IV) fall into this category. In addition, if an oxidizing agent such as chlorate, nitrate, permanganate, or trisoxalatocobaltate(III) is treated with a known excess of iron(II), and if the unreacted iron(II) is titrated with dichromate, one can determine by difference the amount of the original oxidant.

Although hydroquinone in $2\ F$ hydrochloric acid at 50°C may be titrated directly with a standard potassium dichromate solution, the oxidation of most organic compounds is much too slow to be of practical value.

## CERIUM(IV)

### Reactions of Cerium(IV) and Cerium(III)

Although the half-reaction

$$Ce^{4+} + e \rightleftharpoons Ce^{3+}$$

implies the one-electron reduction of $Cc(H_2O)_6^{4+}$ to $Ce(H_2O)_6^{3+}$, these simple aquo ions do not predominate in an aqueous medium. As pointed out in the discussion of the cerium(IV)-iron(II) titration in Chapter 9, $Ce(H_2O)_6^{4+}$ is a strong acid, and proton-transfer reactions such as

$$Ce(H_2O)_6^{4+} + H_2O \rightleftharpoons Ce(H_2O)_5(OH)^{3+} + H_3O^+$$

and

$$Ce(H_2O)_5(OH)^{3+} + H_2O \rightleftharpoons Ce(H_2O)_4(OH)_2^{2+} + H_3O^+$$

occur regardless of what anions may be present in solution. In addition, the formation of dimeric species and higher polymers is known to take place. Analogous reactions prevail with cerium(III), although the extent of such processes is less, owing to the smaller positive charge of cerium(III). These reactions largely depend on the hydrogen ion concentration of the particular solution and on the total concentrations of cerium(IV) and cerium(III).

In addition, the appearance of three different *formal* potentials in Appendix 4 for the cerium(IV)-cerium(III) couple ($+1.70$ v in $1\ F$ perchloric acid, $+1.61$ v in $1\ F$ nitric acid, and $+1.44$ v in $1\ F$ sulfuric acid) indicates that both cerium(IV) and cerium(III) form complexes with anions of the mineral acids. Even the perchlorate ion, considered to be one of the poorest ligands with regard to complex formation, apparently coordinates with cerium(III) and cerium(IV) to yield, respectively, $Ce(H_2O)_5(ClO_4)^{2+}$ and $Ce(H_2O)_5(ClO_4)^{3+}$. In nitric acid solutions there seems to be definite interaction between nitrate and cerium(IV) leading to the formation of $Ce(H_2O)(NO_3)_5^-$ and $Ce(NO_3)_6^{2-}$; and, in sulfuric acid, species such as $Ce(H_2O)_5(SO_4)^{2+}$, $Ce(H_2O)_3(SO_4)_3^{2-}$, and $Ce(H_2O)_5(SO_4)^+$ have been identified. Finally, cerium(IV) complexes such as $Ce(H_2O)_5Cl^{3+}$ and $Ce(H_2O)_4Cl_2^{2+}$ are known to exist in chloride-containing solutions.

### Preparation and Stability of Cerium(IV) Titrants

Ammonium hexanitratocerate(IV), $(NH_4)_2Ce(NO_3)_6$, is a good primary standard for the direct preparation of standard cerium(IV) solutions. However, cerium(IV) solutions are more commonly prepared from cerium(IV) bisulfate, $Ce(HSO_4)_4$;

from ammonium sulfatocerate(IV), $(NH_4)_4Ce(SO_4)_4 \cdot 2 H_2O$; or from hydrous ceric oxide, $CeO_2 \cdot xH_2O$—all of which are commercially available. Moreover, the solution must contain a high concentration of acid, usually 1 $M$ or greater, in order to prevent the precipitation of hydrous ceric oxide.

Sulfuric acid solutions of cerium(IV) are stable indefinitely and, if necessary, they may even be heated for short periods of time. On the other hand, although solutions of cerium(IV) in nitric acid and in perchloric acid possess greater oxidizing strength, they undergo slow decomposition because of the reduction of cerium(IV) by water, a process which is light-induced. Solutions of cerium(IV) in hydrochloric acid are decidedly unstable, owing to the oxidation of chloride ion to chlorine gas by cerium(IV). However, no particular difficulty is experienced when one titrates species dissolved in hydrochloric acid with a sulfuric acid solution of cerium(IV), because cerium(IV) reacts more readily with these other substances than with chloride prior to the equivalence point.

## Standardization of Cerium(IV) Solutions

Procedures used in the standardization of cerium(IV) titrants differ little from those used to standardize potassium permanganate solutions. Substances suitable for standardizing cerium(IV) solutions include the three already mentioned for permanganate—namely, arsenious oxide, sodium oxalate, and iron wire.

As in the standardization of potassium permanganate described previously, a catalyst is required for rapid reaction between cerium(IV) and arsenic(III). If arsenic(III) is titrated in a hydrochloric acid medium, iodine monochloride may serve as the catalytic species. More commonly, after the solid arsenious oxide is initially dissolved in a sodium hydroxide medium, sulfuric acid rather than hydrochloric acid is used to acidify the solution, osmium tetroxide ($OsO_4$) is added as the catalyst, and the mixture is titrated with the cerium(IV) solution to be standardized.

Pure sodium oxalate may be dissolved in perchloric acid and employed for the standardization at room temperature of cerium(IV) solutions in either nitric acid or perchloric acid. If one desires to standardize a cerium(IV) sulfate titrant by an analogous procedure, it is necessary to heat the oxalic acid solution to approximately 70°C so that the rate of the cerium(IV)-oxalic acid reaction will be adequately fast. At room temperature the presence of relatively stable and unreactive sulfate complexes of cerium(IV) inhibits the titration reaction. However, the use of iodine monochloride as a catalyst permits the room-temperature standardization of cerium(IV) sulfate solutions against oxalic acid in a hydrochloric acid medium.

In addition to pure iron wire, other substances for the standardization of cerium(IV) titrants include Oesper's salt, $FeC_2H_4(NH_3)_2(SO_4)_2 \cdot 2 H_2O$, and Mohr's salt, $Fe(NH_4)_2(SO_4)_2 \cdot 6 H_2O$. These compounds are available in exceptionally pure form and are readily soluble in water. They are suitable for the standardization of potassium permanganate and potassium dichromate solutions as well.

## End-Point Detection for Cerium(IV) Titrations

Although cerium(IV) is yellow in all mineral acid solutions, whereas its reduced form, cerium(III), is colorless, the color of the former is not sufficiently intense to serve as an end-point signal, nor does it interfere with recognition of the color changes of redox indicators.

Some common indicators, listed in Table 9–3, for cerium(IV) titrations are tris(1,10-phenanthroline)iron(II), tris(5-nitro-1,10-phenanthroline)iron(II), and tris-(5-methyl-1,10-phenanthroline)iron(II). Standard-potential values for these three indicators range from $+1.02$ to $+1.25$ v versus NHE, making one or more of them suitable for the titrations with cerium(IV) of iron(II), oxalic acid, arsenic(III), hydrogen peroxide, antimony(III), molybdenum(V), plutonium(III), tin(II), and numerous other species.

### Analytical Uses of Cerium(IV) Solutions

Many volumetric analyses involving direct titration with a standard potassium permanganate solution can be performed equally well with cerium(IV). As a matter of fact, all of the procedures listed in Table 10–2 are adaptable with little or no modification for use with cerium(IV) titrants. There are, however, some unique applications of cerium(IV) for the determination of organic substances.

*Determination of organic compounds.* Cerium(IV) in $4F$ perchloric acid quantitatively oxidizes many organic substances, frequently within about 15 minutes at room temperature. Perhaps the most striking feature about the reactions between cerium(IV) and organic compounds is that the oxidation products depend on the nature of the organic substance. Some of the typical reactions may be summarized as follows:

1. A compound with *hydroxyl* groups on two or more adjacent carbon atoms has the bond between each of these carbon atoms cleaved, and each fragment is oxidized to a saturated carboxylic acid:

$$
\underset{\underset{\text{glycerol}}{}}{\underset{\overset{|}{OH} \quad \overset{|}{OH} \quad \overset{|}{OH}}{H_2C\!\!-\!\!-\!\!CH\!\!-\!\!-\!\!CH_2}} + 3\,H_2O \rightarrow 3\,HCOOH + 8\,H^+ + 8\,e
$$

2. A compound with an *aldehyde* group bound to a carbon atom also having a ketone group or a hydroxyl group has the bond between the two carbon atoms cleaved, and each fragment is oxidized to a saturated carboxylic acid:

$$
\underset{\underset{\text{glyceraldehyde}}{}}{\underset{\overset{\|}{O} \quad \overset{|}{OH} \quad \overset{|}{OH}}{H\!\!-\!\!-\!\!C\!\!-\!\!-\!\!CH\!\!-\!\!-\!\!CH_2}} + 3\,H_2O \rightarrow 3\,HCOOH + 6\,H^+ + 6\,e
$$

3. A compound with a *carboxyl* group bound to a carbon atom which also has a hydroxyl group, or a *carboxyl* group bound to an aldehyde group or another carboxyl group, has the bond between the two carbon atoms broken; each carboxyl group is oxidized to carbon dioxide, and the other fragments are oxidized to saturated carboxylic acids:

$$
\underset{\underset{\text{tartaric acid}}{}}{\underset{\overset{\phantom{HO}}{HO}}{\overset{\overset{\displaystyle O}{\big\|}}{\underset{\overset{|}{OH}\ \overset{|}{OH}}{C\!\!-\!\!CH\!\!-\!\!CH\!\!-\!\!C}}}\overset{\overset{\displaystyle O}{\big\|}}{\underset{\overset{|}{OH}}{\phantom{C}}}} + 2\,H_2O \rightarrow 2\,CO_2 + 2\,HCOOH + 6\,H^+ + 6\,e
$$

4. A compound with a *ketone* group adjacent to a carboxyl group, an aldehyde group, or another ketone group—or a *ketone* group bound to a carbon atom with a

hydroxyl group—has the bond between the two carbon atoms cleaved; each fragment with a ketone group is oxidized to a saturated carboxylic acid, and the other fragments are oxidized in accordance with previous rules:

$$\underset{\substack{\| \quad \| \\ \text{O} \quad \text{O}}}{\text{H—C—C—CH}_2\text{—CH}_3} + 2\ H_2O \rightarrow HCOOH + CH_3CH_2COOH + 2\ H^+ + 2\ e$$

1,2-butanedione

5. A compound with an active *methylene* group (—CH$_2$—), that is, a methylene group bound to two carboxyl groups, two aldehyde groups, two ketone groups, or combinations of these groups, has each of the two carbon-carbon bonds broken; each methylene fragment is oxidized to formic acid, and each other fragment is oxidized in accordance with the preceding rules:

$$\underset{\text{HO}}{\overset{\text{O}}{\diagdown}}\text{C—CH}_2\text{—C}\underset{\text{OH}}{\overset{\text{O}}{\diagup}} + 2\ H_2O \rightarrow 2\ CO_2 + HCOOH + 6\ H^+ + 6\ e$$

malonic acid

In order to determine an organic compound, such as one of those just mentioned, the sample is added to a known excess of cerium(IV) in 4 $F$ perchloric acid and the reaction mixture is allowed to stand (or perhaps is heated at 50 to 60°C) for up to 20 minutes. Unreacted cerium(IV) can be back-titrated with a standard oxalic acid solution, with tris(5-nitro-1,10-phenanthroline)iron(II) as the end-point indicator.

Organic compounds may be oxidized and determined quantitatively through the use of excess cerium(IV) in 10 to 12 $F$ sulfuric acid, although the reaction mixture must be boiled and refluxed for at least one hour. After oxidation of the desired organic substance has reached completion, the mixture is cooled and the excess, unreacted cerium(IV) is titrated with a standard solution of a reducing agent, such as iron(II) or oxalic acid.

Tartaric acid, which is converted to two molecules of carbon dioxide and two molecules of formic acid by cerium(IV) in 4 $F$ perchloric acid, undergoes complete oxidation to four molecules of carbon dioxide in 12 $F$ sulfuric acid:

$$\underset{\text{HO}\quad\text{OH}\ \text{OH}\quad\text{OH}}{\overset{\text{O}\qquad\qquad\text{O}}{\diagdown\qquad\qquad\diagup}}\text{C—CH—CH—C} + 10\ Ce^{4+} + 2\ H_2O \rightarrow 4\ CO_2 + 10\ Ce^{3+} + 10\ H^+$$

Formic acid is unreactive toward cerium(IV) in perchloric acid, but is oxidized to carbon dioxide in a concentrated sulfuric acid medium:

$$HCOOH + 2\ Ce^{4+} \rightarrow CO_2 + 2\ Ce^{3+} + 2\ H^+$$

Some of the other organic compounds which can be oxidized, all of them to carbon dioxide, in 12 $F$ sulfuric acid with cerium(IV) are malonic acid, citric acid, benzoic acid, salicylic acid, fumaric acid, and phthalic acid.

## IODINE (TRIIODIDE)

Iodine exists in a number of analytically important oxidation states, represented by such familiar species as iodide, molecular iodine (or triiodide ion), iodine monochloride, iodate, and periodate. Of particular interest are oxidation-reduction processes involving the two lowest oxidation states, namely, iodide and iodine (triiodide ion). Although three different half-reactions listed in Appendix 4 pertain to these substances

$$I_2(aq) + 2\,e \rightleftharpoons 2\,I^-; \qquad E^0 = +0.6197\ v$$

$$I_2(s) + 2\,e \rightleftharpoons 2\,I^-; \qquad E^0 = +0.5345\ v$$

$$I_3^- + 2\,e \rightleftharpoons 3\,I^-; \qquad E^0 = +0.5355\ v$$

the third half-reaction provides the most realistic picture of the redox behavior of the iodine-iodide system because it includes the two predominant species, triiodide and iodide ions, encountered in practical situations.

Triiodide ion is a sufficiently good oxidizing agent to react quantitatively with a number of reductants. On the other hand, iodide ion is easily enough oxidized to permit its quantitative reaction with many strong oxidizing agents. Accordingly, two major classifications of redox methods involving the use of the triiodide-iodide half-reaction have been developed. First, there are the *direct* methods in which a standard solution of triiodide, or iodine dissolved in a potassium iodide medium, is utilized for the direct titration of the substance to be determined. Second, many *indirect* procedures are employed in which triiodide is formed through the reaction of excess iodide ion with some oxidizing agent, followed by the titration of the triiodide ion with a standard solution of sodium thiosulfate. In a direct method, the end point is marked by the first permanent appearance of free iodine (triiodide) in the titration vessel; in an indirect procedure, the final disappearance of triiodide ion signals the end point.

### Preparation and Stability of Iodine (Triiodide) Solutions

Although solid iodine is sparingly soluble in water (0.00133 $M$), its solubility is considerably enhanced in an aqueous potassium iodide medium, owing to the formation of triiodide ion:

$$I_2(aq) + I^- \rightleftharpoons I_3^-; \qquad K = 708$$

Thus, a so-called *iodine* titrant is actually a solution of potassium iodide in which the requisite weight of solid molecular iodine has been dissolved; typically, such a solution contains 0.05 $M$ $I_3^-$, 0.20 $M$ $I^-$, and only $3.5 \times 10^{-4}$ $M$ $I_2$, and might be more appropriately termed a *triiodide* titrant. Nevertheless, triiodide ion and molecular iodine behave identically in oxidation-reduction processes.

Standard triiodide solutions may be prepared directly from an accurately weighed portion of pure solid iodine. However, since solid iodine has an appreciable vapor pressure (0.31 mm at room temperature), special precautions are necessary to prevent the loss of iodine during weighing and handling operations. Therefore, it is simpler to prepare an iodine (triiodide) solution of approximately the desired concentration and to standardize it against pure arsenious oxide.

Triiodide solutions are inherently unstable for at least two reasons. One cause of this instability is the volatility of iodine. Loss of iodine can occur in spite of the facts that excess potassium iodide is present and that most of the dissolved iodine really exists as the triiodide ion. However, if the standard solution is stored in a tightly stoppered bottle, there should be no significant loss of iodine for several weeks.

Another reason why triiodide solutions undergo changes in concentration is that dissolved atmospheric oxygen causes the oxidation of iodide ion to iodine (triiodide):

$$6\,I^- + O_2 + 4\,H^+ \rightleftharpoons 2\,I_3^- + 2\,H_2O; \qquad E^0 = +0.693\,v$$

Fortunately, this oxidation proceeds very slowly, even though the forward reaction is strongly favored. Since hydrogen ion is a reactant, the oxidation of iodide becomes more significant as the pH decreases, but the problem is not at all serious for essentially neutral triiodide solutions. Heat and light both induce the air oxidation of iodide, as does the presence of heavy metal impurities; so care should be taken in the preparation, handling, and storage of triiodide titrants.

### End-Point Detection for Titrations Involving Iodine

In our classification of direct and indirect methods of analysis involving iodine, we pointed out that the end point for the former type of titration is marked by the first detectable excess of iodine in the titration vessel, whereas the disappearance of the last discernible quantity of iodine signals the end point in the latter category of titrations. Three different techniques are commonly employed for end-point detection in titrations involving iodine.

Provided that iodine (triiodide) is the only colored substance in the system, the appearance or disappearance of the yellow-brown triiodide color is itself an extremely sensitive way to locate the end point of a titration. In a perfectly colorless solution, it is possible to detect visually a concentration of triiodide as low as $5 \times 10^{-6}\,M$ which, in a solution volume of 100 ml, corresponds to only 0.01 ml of $0.05\,F$ iodine (triiodide) titrant.

Triiodide ion forms a very intense blue-colored complex with colloidally dispersed starch, which serves well as an indication of the presence of iodine. Thus, the end point of a titration with a triiodide solution is marked by the appearance of the blue starch-iodine color, whereas the blue color disappears at the end point of an indirect iodine procedure. As little as $2 \times 10^{-7}\,M$ iodine (triiodide) gives a detectable blue color with starch under optimum conditions.

Starch indicator solution should not be added until just prior to the end point of an indirect iodine method, because large amounts of iodine cause a starch suspension to coagulate, and promote the decomposition of starch as well. Starch suspensions must not be exposed unnecessarily to air and to any possible source of bacteria, because the decomposition of starch is promoted by both oxygen and microorganisms. Various preservatives, such as mercuric iodide, thymol, and glycerol, help to minimize bacterially induced decomposition. It is preferable to prepare a fresh starch suspension at least within a day or two of the time it is to be used. Starch indicator should be employed only at or near room temperature, as the sensitivity of the starch-iodine end point is markedly decreased at higher temperatures. Acid decomposes starch through a hydrolysis reaction, so starch should not be used in strongly acidic media.

Another form of end-point detection involves the addition of a few milliliters of a water-immiscible liquid, usually carbon tetrachloride or chloroform, to the titration vessel. Since molecular iodine is much more soluble in the nonpolar organic layer than in the aqueous phase, iodine will concentrate itself in the heavier, lower layer and will impart its violet color to this layer. In practice, one performs the titration with the aid of a glass-stoppered flask. As the end point is approached, the flask is stoppered and the solution is thoroughly shaken after each addition of titrant. Then the flask is inverted so that the organic liquid collects in the narrow neck of the flask, and the organic phase is examined for the presence or absence of the iodine color.

## Standardization of Iodine (Triiodide) Solutions

Iodine or triiodide solutions are almost always standardized by titration against a solution of arsenious oxide. We may write the pertinent reaction* as

$$HAsO_2 + I_3^- + 2\,H_2O \rightleftharpoons H_3AsO_4 + 3\,I^- + 2\,H^+$$

whereas the corresponding equilibrium-constant expression† is

$$\frac{[H_3AsO_4][I^-]^3[H^+]^2}{[HAsO_2][I_3^-]} = 0.166$$

Due to the small magnitude of the equilibrium constant, the titration reaction will be adequately complete only if the concentrations of the product species remain low. In particular, the hydrogen ion concentration can be adjusted to and controlled at almost any suitably small value through the use of an appropriate buffer.

Several facts about the arsenic(III)-triiodide reaction have been confirmed experimentally. First, this reaction does proceed quantitatively to completion at pH values as low as 3.5; but below pH 5 the *rate* of the reaction is so slow that fading, premature end points are obtained. Second, pH 7 to 9 is the optimum range for the arsenic(III)-triiodide reaction. Third, the upper pH limit is approximately 11; above this latter pH, disproportionation of iodine (triiodide) into hypoiodous acid, iodate, and iodide occurs. Finally, it is not sufficient that the initial pH merely lie between 7 and 9; the pH must be buffered within this range. If hydrogen ion produced during the titration is not neutralized by the buffer, the marked increase in the acidity of the solution can bring the arsenic(III)-triiodide reaction to a halt long before the true equivalence point is reached.

Accordingly, after a weighed portion of arsenious oxide is dissolved in sodium hydroxide medium, enough hydrochloric acid is added to neutralize or slightly acidify the solution. Next, several grams of sodium bicarbonate are added to provide a carbonic acid-bicarbonate buffer of pH 7 to 8; then starch indicator is added and the solution is titrated with the iodine (triiodide) solution to be standardized.

## Reaction between Iodine (Triiodide) and Thiosulfate

Whereas most other oxidizing agents convert thiosulfate to sulfite or sulfate, the reaction between thiosulfate and iodine (or triiodide) to yield the tetrathionate ion

$$I_3^- + 2\,S_2O_3^{2-} \rightleftharpoons 3\,I^- + S_4O_6^{2-}$$

is unusual. In general, the iodine-thiosulfate reaction proceeds rapidly according to the preceding well-defined stoichiometry at all pH values between 0 and 7. In mildly alkaline media, however, even iodine oxidizes thiosulfate to sulfate,

$$4\,I_3^- + S_2O_3^{2-} + 10\,OH^- \rightleftharpoons 2\,SO_4^{2-} + 12\,I^- + 5\,H_2O$$

---

* See footnote at the bottom of page 311.
† See page 277 for a discussion of the formulation of this equilibrium-constant expression.

although the reaction is by no means quantitative until the pH becomes very high. In alkaline solutions, the nature of the iodine-thiosulfate reaction changes because iodine or triiodide is no longer the active oxidant. Above a pH of approximately 8 or 9, triiodide disproportionates into iodide and hypoiodous acid,

$$I_3^- + OH^- \rightleftharpoons 2\,I^- + HOI$$

the latter appearing to be the substance which most readily oxidizes thiosulfate to sulfate:

$$4\,HOI + S_2O_3^{2-} + 6\,OH^- \rightleftharpoons 4\,I^- + 2\,SO_4^{2-} + 5\,H_2O$$

However, the chemistry of iodine in alkaline media is complicated further by the disproportionation of hypoiodous acid into iodate and iodide:

$$3\,HIO + 3\,OH^- \rightleftharpoons IO_3^- + 2\,I^- + 3\,H_2O$$

Since such side reactions are not unique to the iodine-thiosulfate system, the successful use of iodine or triiodide titrants is generally restricted to solutions having pH values less than about 8.

## Preparation and Stability of Sodium Thiosulfate Solutions

Pure sodium thiosulfate pentahydrate, $Na_2S_2O_3 \cdot 5\,H_2O$, can be crystallized and stored under carefully controlled conditions, and an accurately weighed portion of the solid may be dissolved in water for the direct preparation of a standard solution. However, it is more straightforward to prepare a solution of approximately the desired strength and to standardize it against potassium dichromate, potassium iodate, or a previously standardized iodine (triiodide) solution.

There are several factors which influence the stability of a thiosulfate solution—the pH of the solution, the presence of heavy metal impurities, and the presence of sulfur-consuming bacteria.

In a very dilute acid medium, such as might arise in an aqueous solution saturated with carbon dioxide, thiosulfate slowly decomposes with the resultant formation of elemental sulfur and hydrogen sulfite ion,

$$S_2O_3^{2-} + H^+ \rightleftharpoons HS_2O_3^- \rightarrow S + HSO_3^-$$

but, in a $1\,F$ hydrochloric acid medium, decomposition of thiosulfate is extensive within only a minute or two. This decomposition changes the effective concentration of thiosulfate titrants because the $HSO_3^-$ ion reduces twice the amount of iodine (triiodide) that thiosulfate does, as can be seen from the equilibria

$$HSO_3^- + I_3^- + H_2O \rightleftharpoons HSO_4^- + 3\,I^- + 2\,H^+$$

and

$$2\,S_2O_3^{2-} + I_3^- \rightleftharpoons S_4O_6^{2-} + 3\,I^-$$

Although thiosulfate is unstable in acidic media, nothing prevents its use as a titrant for iodine or triiodide in even relatively concentrated acid solutions, if the titration is carried out in such a manner that no significant local excess of thiosulfate is present. Traces of heavy metal impurities, in the presence of oxygen, cause the gradual oxidation of thiosulfate to tetrathionate ion. Copper(II) oxidizes thiosulfate to tetrathionate,

$$2\,Cu^{2+} + 2\,S_2O_3^{2-} \rightarrow 2\,Cu^+ + S_4O_6^{2-}$$

and then dissolved oxygen converts copper(I) back to copper(II) to reinitiate the sequence of reactions.

Sulfur-metabolizing bacteria present in distilled water can convert thiosulfate to a variety of products, including elemental sulfur, sulfite, and sulfate. One usually boils the distilled water in which the solid reagent is to be dissolved as a means of destroying the bacteria. Furthermore, it is common to introduce 50 to 100 mg of sodium bicarbonate into a liter of sodium thiosulfate titrant because bacterial action is minimal near pH 9 or 10. Finally, by boiling the distilled water and adding a small amount of sodium bicarbonate to the titrant solution, one eliminates carbon dioxide and traces of other acidic substances which promote the acid-catalyzed decomposition of thiosulfate.

## Standardization of Sodium Thiosulfate Solutions

All procedures employed for the standardization of thiosulfate solutions ultimately involve the reaction between iodine (triiodide) and thiosulfate. A previously standardized iodine (triiodide) solution suffices well for the standardization of a sodium thiosulfate titrant. In addition, strong oxidizing agents, such as potassium dichromate and potassium iodate, may serve for the standardization through the indirect method of analysis involving iodine.

Potassium dichromate is an excellent reagent for the standardization of thiosulfate solutions. If excess potassium iodide is dissolved in an acidic solution containing an accurately known amount of dichromate, the latter oxidizes iodide to triiodide

$$Cr_2O_7^{2-} + 9\,I^- + 14\,H^+ \rightleftharpoons 2\,Cr^{3+} + 3\,I_3^- + 7\,H_2O$$

and the liberated iodine may be titrated with the sodium thiosulfate solution to a starch-iodine end point. However, the dichromate-iodide reaction is relatively slow and, in order for the oxidation of iodide to proceed quantitatively, careful control must be maintained over the experimental variables—the concentrations of iodide and acid, the time of reaction, and even the order in which the reagents are mixed.

Another substance which quantitatively oxidizes iodide to triiodide in an acid medium is potassium iodate:

$$IO_3^- + 8\,I^- + 6\,H^+ \rightleftharpoons 3\,I_3^- + 3\,H_2O$$

One may titrate the iodine (triiodide) produced by this reaction with the sodium thiosulfate solution. When compared to the dichromate-iodide reaction, the iodate-iodide system possesses at least two advantages. First, the iodate-iodide reaction occurs almost instantaneously, even in very dilute acid media. Second, whereas the disappearance of the blue starch-iodine color is partially obscured by the green-colored chromic ion in the standardization of thiosulfate against potassium dichromate, the end point in the iodate procedure is signaled by a sharp change from blue to colorless. One disadvantage of potassium iodate stems from the fact that, for each standardization of a 0.1 F sodium thiosulfate solution, the required weight of potassium iodate is only about 0.12 gm, so a typical weighing uncertainty of 0.1 mg exceeds the limit of error normally desired in a standardization procedure.

## Analytical Uses of the Triiodide-Iodide System

*Direct methods.* Because triiodide ion is a relatively mild oxidant, it can react quantitatively only with substances which are easily oxidizable. Some of the substances which can be determined by direct titration with a standard iodine (triiodide) solution are listed in Table 10–3, along with the half-reaction which each species undergoes, plus information about the experimental procedure. A mixture of arsenic, antimony, and tin provides a specific example for more detailed discussion.

**Separation and determination of arsenic, antimony, and tin.** Arsenic, antimony, and tin are commonly encountered together as sulfides—$As_2S_3$, $As_2S_5$, $Sb_2S_3$, $Sb_2S_5$, $SnS$, and $SnS_2$—in the analysis of complex samples. Therefore, it is of interest to consider, first, the separation of these elements and, second, the individual determination of each element.

Separation of these elements can be accomplished by means of distillation. Prior to the distillation, however, one must obtain a solution of only arsenic(III), antimony(III), and tin(IV) by dissolving the sulfides in hot, concentrated sulfuric acid to which is added hydrazine sulfate. Next, the resulting solution is transferred to an all-glass distillation apparatus, concentrated hydrochloric acid is introduced, and $AsCl_3$ is distilled into a receiving flask containing water. After the arsenic is removed, more hydrochloric acid and some phosphoric acid are added to the distilling flask and $SbCl_3$ is distilled into another receiving vessel; tin(IV) remains in the distilling flask as a phosphate complex. Finally, hydrobromic acid is introduced into the distilling flask and $SnBr_4$ is distilled into a third receiving vessel.

Arsenic(III) in the first portion of distillate may be determined by direct titration at pH 7 to 8 with a standard triiodide solution.

With one exception, the procedure for the determination of antimony(III) is identical to that used for arsenic(III). Unlike arsenic, antimony(III) and antimony-(V) both form insoluble basic salts, such as $SbOCl$ and $SbO_2Cl$, in slightly acidic and neutral solutions. Antimony(III) and antimony(V) can be kept in solution as complex ions, probably $SbOC_4H_4O_6^-$ and $SbO_2C_4H_4O_6^-$, by the addition of tartaric acid, so that the reaction between antimony(III) and triiodide may be formulated as

$$SbOC_4H_4O_6^- + I_3^- + H_2O \rightleftharpoons SbO_2C_4H_4O_6^- + 3\,I^- + 2\,H^+$$

Tin(IV) in the third portion of the distillate must be reduced to tin(II) prior to titration with a standard triiodide solution, metallic lead or nickel being commonly employed as reductants. Both the reduction of tin(IV) and the subsequent titration of tin(II) are performed in a moderately strong acid medium. Because of the ease with which tin(II) is reoxidized by atmospheric oxygen, the solution must be blanketed with carbon dioxide or nitrogen during the prereduction and titration steps.

**Determination of organic compounds.** A number of useful applications have been developed in which iodine (triiodide) is employed to oxidize organic substances. Usually, a known excess of standard iodine solution is added to the sample; then, after the reaction between iodine and the organic compound is complete, the unreacted iodine is back-titrated with a standard sodium thiosulfate solution. For some organic molecules, the reaction mixture must be strongly alkaline because hypoiodite ($IO^-$), not iodine, is the active oxidant; however, prior to the thiosulfate titration, the solution must be acidified. Table 10–4 outlines some of these determinations.

**Indirect methods.** Many oxidizing agents are capable of converting iodide quantitatively to free iodine, which in the presence of excess iodide forms the triiodide ion. Since a stoichiometric relationship exists between the original quantity of oxidant and the amount of triiodide produced, a determination of triiodide, by titration with a standard solution of sodium thiosulfate, provides data from which the quantity of oxidizing agent may be calculated. In applying the indirect method of analysis, one must always be sure that the reaction between the strong oxidant and excess iodide has reached completion before beginning the titration with thiosulfate.

Table 10-3. Some Titrations Performed with Standard Iodine (Triiodide) Solution

| Substance Determined | Half-Reaction for Substance Determined | Procedure and Conditions for Titration |
| --- | --- | --- |
| Sb(III) | $SbOC_4H_4O_6^- + H_2O \rightleftharpoons SbO_2C_4H_4O_6^- + 2 H^+ + 2 e$ | Titrate in NaHCO$_3$ medium containing tartrate to complex antimony species |
| As(III) | $HAsO_2 + 2 H_2O \rightleftharpoons H_3AsO_4 + 2 H^+ + 2 e$ | Titrate in NaHCO$_3$ medium |
| N$_2$H$_4$ | $N_2H_4 \rightleftharpoons N_2 + 4 H^+ + 4 e$ | Titrate in NaHCO$_3$ medium |
| H$_2$S | $H_2S \rightleftharpoons S + 2 H^+ + 2 e$ | Add sample to excess of standard I$_3^-$ in 1 $F$ HCl; back-titrate unreacted I$_3^-$ with standard thiosulfate |
| Cd(II), Zn(II) | $H_2S \rightleftharpoons S + 2 H^+ + 2 e$ | Precipitate and wash CdS or ZnS; dissolve precipitate in 3 $F$ HCl containing known excess of standard I$_3^-$; back-titrate unreacted I$_3^-$ with standard thiosulfate |
| H$_2$SO$_3$ | $H_2SO_3 + H_2O \rightleftharpoons SO_4^{2-} + 4 H^+ + 2 e$ | Add sample to excess standard I$_3^-$ in dilute acid; back-titrate unreacted I$_3^-$ with standard thiosulfate |
| S$_2$O$_3{}^{2-}$ | $2 S_2O_3^{2-} \rightleftharpoons S_4O_6^{2-} + 2 e$ | Titrate in neutral medium |
| Sn(II) | $SnCl_4^{2-} + 2 Cl^- \rightleftharpoons SnCl_6^{2-} + 2 e$ | Reduce tin(IV) to tin(II) with metallic Pb or Ni in 1 $F$ HCl; exclude oxygen; titrate with air-free I$_3^-$ |

**Table 10-4. Determination of Organic Compounds with Standard Iodine (Triiodide)**

| Substance Determined | Half-Reaction for Substance Determined | Procedure and Conditions for Determination |
|---|---|---|
| Formaldehyde | $HCHO + 3 OH^- \rightleftharpoons HCOO^- + 2 H_2O + 2 e$ | Add NaOH plus known excess of $I_3^-$ to sample; wait 5 minutes; add HCl and back-titrate unreacted $I_3^-$ with $S_2O_3^{2-}$ |
| Glucose, galactose, maltose | $RCOH + 3 OH^- \rightleftharpoons RCOO^- + 2 H_2O + 2 e$ | (same procedure as above) |
| Acetone | $(CH_3)_2CO + 3 I^- + 4 OH^- \rightleftharpoons CHI_3 + CH_3COO^- + 3 H_2O + 6 e$ | (same procedure as above) |
| Thiourea | $CS(NH_2)_2 + 10 OH^- \rightleftharpoons CO(NH_2)_2 + SO_4^{2-} + 5 H_2O + 8 e$ | (same procedure as above) |
| Semicarbazide | $NH_2NHCONH_2 + 3 OH^- \rightleftharpoons N_2 + CO_2 + NH_4^+ + 2 H_2O + 4 e$ | Add known excess of $I_3^-$ to sample at pH 7; back-titrate unreacted $I_3^-$ with $S_2O_3^{2-}$ |
| Resorcinol | $C_6H_4(OH)_2 + 3 I^- \rightleftharpoons C_6HI_3(OH)_2 + 3 H^+ + 6 e$ | Add known excess of $I_3^-$ to sample at pH 5; wait 2 minutes; back-titrate unreacted $I_3^-$ with $S_2O_3^{2-}$ |
| Hydroquinone | $C_6H_4(OH)_2 \rightleftharpoons C_6H_4O_2 + 2 H^+ + 2 e$ | (same procedure as above) |
| Tetraethyl lead | $Pb(C_2H_5)_4 + 2 I^- \rightleftharpoons PbI(C_2H_5)_3 + C_2H_5I + 2 e$ | Titrate methanol solution of sample directly with $I_3^-$ |
| Thioglycolic acid, cysteine, glutathione | $2 RSH \rightleftharpoons RSSR + 2 H^+ + 2 e$ | Direct titration with $I_3^-$ at pH 4–5 |

Otherwise, any strong oxidizing agent present at the start of the titration will probably oxidize thiosulfate to sulfur and sulfate as well as tetrathionate, thereby upsetting the stoichiometry of the triiodide-thiosulfate reaction.

Included in Table 10–5 are many of the species which can be determined by means of the indirect method. One of these procedures will now be described in some detail.

**Determination of copper.** When an excess of potassium iodide is added to an acidic solution of copper(II), the following reaction occurs:

$$2 \, Cu^{2+} + 5 \, I^- \rightleftharpoons 2 \, CuI(s) + I_3^-$$

By titrating the triiodide with standard sodium thiosulfate solution, one can achieve accurate determinations of copper in a variety of samples.

Although hydrogen ion does not appear as a reactant or product in the preceding reaction, pH does exert an influence on the rate of the process as well as on the accuracy of the determination. To repress the acid dissociation of $Cu(H_2O)_4^{2+}$ which results in the formation of $Cu(H_2O)_3(OH)^+$ and $Cu(H_2O)_2(OH)_2$—species that react slowly with iodide—the pH of the solution must be less than approximately 4. On the other hand, if the concentration of hydrogen ion exceeds 0.3 $M$, the air oxidation of iodide is induced by the copper(II)-iodide reaction.

A significant source of negative error in the titration arises from adsorption of triiodide ion by the CuI precipitate. Triiodide ion is held so tenaciously on the surface of the precipitate that the reaction between thiosulfate and triiodide is incomplete. This problem may be overcome if a small excess of potassium thiocyanate is added to the solution as the end point is approached, for copper(I) iodide is metathesized to copper(I) thiocyanate

$$CuI(s) + SCN^- \rightleftharpoons CuSCN(s) + I^-$$

and the latter compound has almost no tendency to adsorb triiodide. However, thiocyanate may not be added at the beginning of the titration, for triiodide slowly oxidizes it.

Certain elements which accompany copper in brass, other alloys, and ores may interfere with the iodometric determination. Copper alloys contain zinc, lead, and tin as well as smaller amounts of iron and nickel, whereas iron, arsenic, and antimony are often found in copper-containing ores.

Hot, concentrated nitric acid is invariably employed to dissolve brass and ore samples. Copper, zinc, lead, and nickel enter the solution in their dipositive oxidation states, but the oxidizing properties of nitric acid cause the formation of iron(III), arsenic(V), and antimony(V) as well as the precipitation of hydrated tin(IV) oxide, $SnO_2 \cdot xH_2O$. Since zinc, nickel, and tin are innocuous to the determination of copper, we may disregard them in the remainder of this discussion.

After the sample has been dissolved in nitric acid, a small volume of concentrated sulfuric acid is added and the solution is boiled to remove nitrate and other oxides of nitrogen which, if allowed to remain, would later oxidize iodide. When the sulfuric acid solution is diluted with water and cooled, lead sulfate precipitates almost quantitatively and should be removed by filtration. Otherwise, when iodide is subsequently added for the determination of copper, the formation of canary-yellow lead iodide ($PbI_2$) will obscure the end point of the titration.

In strongly acidic media, iron(III), antimony(V), and arsenic(V) all oxidize iodide to triiodide. Therefore, the procedure must be modified to prevent the interference of these elements. If the pH of the solution is greater than approximately 3,

Table 10-5. Some Indirect Iodine Methods

| Substance Determined | Half-Reaction for Substance Determined | Experimental Conditions |
|---|---|---|
| $IO_4^-$ | $IO_4^- + 8\,H^+ + 7\,e \rightleftharpoons \tfrac{1}{2}\,I_2 + 4\,H_2O$ | $0.5\,F$ HCl |
| $IO_3^-$ | $IO_3^- + 6\,H^+ + 5\,e \rightleftharpoons \tfrac{1}{2}\,I_2 + 3\,H_2O$ | $0.5\,F$ HCl |
| $BrO_3^-$ | $BrO_3^- + 6\,H^+ + 6\,e \rightleftharpoons Br^- + 3\,H_2O$ | $0.5\,F\ H_2SO_4$ or HCl |
| $Br_2$ | $Br_2 + 2\,e \rightleftharpoons 2\,Br^-$ | Dilute acid |
| $Cl_2$ | $Cl_2 + 2\,e \rightleftharpoons 2\,Cl^-$ | Dilute acid |
| $HOCl$ | $HOCl + H^+ + 2\,e \rightleftharpoons Cl^- + H_2O$ | $0.5\,F\ H_2SO_4$ |
| $O_3$ | $O_3 + 2\,H^+ + 2\,e \rightleftharpoons O_2 + H_2O$ | Absorb $O_3$ in slightly alkaline KI solution; add $H_2SO_4$ and titrate liberated $I_3^-$ |
| $O_2$ | $(O_2 + 4\,Mn(OH)_2 + 2\,H_2O \rightleftharpoons 4\,Mn(OH)_3)$ $Mn(OH)_3 + 3\,H^+ + e \rightleftharpoons Mn^{2+} + 3\,H_2O$ | To $O_2$ containing sample, add solutions of $Mn^{2+}$ and of NaOH–KI; after 1 minute, acidify with $H_2SO_4$ and titrate liberated $I_3^-$ |
| $H_2O_2$ | $H_2O_2 + 2\,H^+ + 2\,e \rightleftharpoons 2\,H_2O$ | $1\,F\ H_2SO_4$ with $NH_4MoO_3$ catalyst |
| $MnO_4^-$ | $MnO_4^- + 8\,H^+ + 5\,e \rightleftharpoons Mn^{2+} + 4\,H_2O$ | $0.1\,F$ HCl or $H_2SO_4$ |
| $MnO_2$ | $MnO_2 + 4\,H^+ + 2\,e \rightleftharpoons Mn^{2+} + 2\,H_2O$ | $0.5\,F\ H_3PO_4$ or HCl |
| $PbO_2$ | $PbO_2 + 4\,H^+ + 2\,e \rightleftharpoons Pb^{2+} + 2\,H_2O$ | $2\,F$ HCl |
| $Cr_2O_7^{2-}$ | $Cr_2O_7^{2-} + 14\,H^+ + 6\,e \rightleftharpoons 2\,Cr^{3+} + 7\,H_2O$ | $0.4\,F$ HCl; wait 5 minutes before titration |
| $Ba(II),\ Pb(II)$ | $Cr_2O_7^{2-} + 14\,H^+ + 6\,e \rightleftharpoons 2\,Cr^{3+} + 7\,H_2O$ | Precipitate $BaCrO_4$ or $PbCrO_4$; dissolve washed precipitate in $1\,F$ HCl |
| $Ce(IV)$ | $Ce^{4+} + e \rightleftharpoons Ce^{3+}$ | $1\,F\ H_2SO_4$ |
| $S_2O_8^{2-}$ | $S_2O_8^{2-} + 2\,e \rightleftharpoons 2\,SO_4^{2-}$ | Mix $S_2O_8^{2-}$ and $I^-$ in neutral medium; after 15 minutes, acidify solution and titrate $I_3^-$ |
| $Cu(II)$ | $Cu^{2+} + I^- + e \rightleftharpoons CuI(s)$ | pH 1–2 |
| $Fe(CN)_6^{3-}$ | $Fe(CN)_6^{3-} + e \rightleftharpoons Fe(CN)_6^{4-}$ | $1\,F$ HCl |
| $As(V)$ | $H_3AsO_4 + 2\,H^+ + 2\,e \rightleftharpoons HAsO_2 + 2\,H_2O$ | $5\,F$ HCl |
| $Sb(V)$ | $SbCl_6^- + 2\,e \rightleftharpoons SbCl_4^- + 2\,Cl^-$ | $5\,F$ HCl |

the tendencies for arsenic(V) and antimony(V) to oxidize iodide can be nullified, but the solution pH must not exceed 4 for the reason stated earlier. If one adds a reagent which strongly complexes iron(III), it cannot oxidize iodide ion. When antimony and arsenic are absent, and acidity is not so critical, phosphoric acid may be used to complex iron(III). When antimony, arsenic, and iron are all present, a fluoride-hydrofluoric acid buffer is employed to adjust the pH to 3.2 and to complex iron as $FeF_4^-$.

## POTASSIUM IODATE

Because of the ideal physical properties of potassium iodate and its availability in a high state of purity, and because exceedingly stable standard solutions can be prepared directly by dissolution of a weighed portion of the compound in water, many analytical uses of this reagent have been developed.

### Reactions of Iodate

As described earlier, potassium iodate reacts quantitatively with iodide in an acid medium:

$$IO_3^- + 8\,I^- + 6\,H^+ \rightleftharpoons 3\,I_3^- + 3\,H_2O$$

However, the position of equilibrium is strongly influenced by the pH of the solution and, in a neutral medium, iodate exhibits no tendency to oxidize iodide. On the other hand, addition of a known volume of standard potassium iodate solution to an excess of an acidic iodide medium results in the immediate formation of a stoichiometric quantity of triiodide. This liberated triiodide may be employed for the standardization of a thiosulfate solution, for the determination of hydrogen sulfide, or for the determination of tin(II), to name just three representative uses. In the two latter situations, an excess of triiodide remains after the reaction of iodine with hydrogen sulfide or tin(II), so the complete analytical procedure entails a final titration of the unreacted iodine with standard sodium thiosulfate solution.

A number of applications of potassium iodate as an oxidizing agent have been devised which do not fit into the normal classification of direct or indirect iodine methods. If a suitable reductant is titrated with a standard potassium iodate solution in an acid medium, the initial portions of the iodate are reduced to free iodine,

$$2\,IO_3^- + 12\,H^+ + 10\,e \rightleftharpoons I_2 + 6\,H_2O; \qquad E^0 = +1.20\ v$$

which, upon reaction with additional iodate, is oxidized to unipositive iodine. Although the species $I^+$ is unstable, iodine in the unipositive state is stabilized through the formation of complexes with chloride, bromide, and cyanide ions. For example, in hydrochloric acid solutions ranging in concentration from 3 to 9 $F$, iodate oxidizes molecular iodine to the iodine dichloride complex:

$$IO_3^- + 2\,I_2 + 10\,Cl^- + 6\,H^+ \rightleftharpoons 5\,ICl_2^- + 3\,H_2O$$

In concentrated hydrobromic acid solutions, the species $IBr_2^-$ may be produced and, when hydrogen cyanide is present in 1 $F$ hydrochloric acid or 2 $F$ sulfuric acid, iodate can oxidize iodine to the iodine monocyanide molecule.

## Analytical Uses of Potassium Iodate Solutions

Antimony(III) may be titrated with a standard solution of potassium iodate if the titration medium contains $3\,F$ hydrochloric acid. During the early stages of the titration, the reduction of iodate to molecular iodine occurs,

$$2\,IO_3^- + 5\,SbCl_4^- + 12\,H^+ + 10\,Cl^- \rightleftharpoons I_2 + 5\,SbCl_6^- + 6\,H_2O$$

as evidenced by the accumulation of iodine in the titration vessel. However, after all of the antimony(III) has been oxidized, further additions of titrant cause the disappearance of iodine, owing to the formation of iodine dichloride:

$$IO_3^- + 2\,I_2 + 10\,Cl^- + 6\,H^+ \rightleftharpoons 5\,ICl_2^- + 3\,H_2O$$

Thus, the end point of the titration is signaled when the last detectable trace of molecular iodine is oxidized to the iodine dichloride complex, and the stoichiometric relationship between iodate and antimony(III) is readily obtained if the two preceding reactions are combined:

$$IO_3^- + 2\,SbCl_4^- + 6\,H^+ + 6\,Cl^- \rightleftharpoons ICl_2^- + 2\,SbCl_6^- + 3\,H_2O$$

Determination of tin(II) in concentrated hydrochloric acid by direct titration with standard potassium iodate solution is interesting because two different end points may be obtained. In contrast to the antimony(III)-iodate system, no molecular iodine is formed in the first step of the iodate-tin(II) titration,

$$IO_3^- + 3\,SnCl_4^{2-} + 6\,H^+ + 6\,Cl^- \rightleftharpoons I^- + 3\,SnCl_6^{2-} + 3\,H_2O$$

because tin(II) can reduce any transiently formed iodine to iodide,

$$SnCl_4^{2-} + I_2 + 2\,Cl^- \rightleftharpoons SnCl_6^{2-} + 2\,I^-$$

whereas antimony(III) is not a sufficiently strong reductant to have a similar capability. When tin(II) has been quantitatively oxidized, the next drop of titrant oxidizes iodide to iodine (triiodide), this end point being recognizable by the appearance of the yellow color of the latter. However, if one continues to add the iodate solution, all of the iodine is eventually oxidized to iodine dichloride; so the disappearance of the iodine color constitutes a second end point. At this second end point, the net titration reaction is represented as

$$IO_3^- + 2\,SnCl_4^{2-} + 6\,H^+ + 6\,Cl^- \rightleftharpoons ICl_2^- + 2\,SnCl_6^{2-} + 3\,H_2O$$

Among the other species which have been successfully titrated with a standard potassium iodate solution in concentrated hydrochloric acid medium are arsenic(III), iron(II), thallium(I), iodide, sulfurous acid, hydrazine, thiocyanate, and thiosulfate.

## End-Point Detection for Iodate Titrations

Although iodate is ultimately reduced only to the unipositive state of iodine in the present family of oxidation-reduction methods, molecular iodine is almost always formed as an intermediate substance during these titrations. Consequently, the disappearance of the last detectable quantity of iodine is usually employed to signal the end point of an iodate titration. In general, the extraction method of end-point detection is utilized in titrations with potassium iodate, because the high acidity of the solutions precludes the use of starch.

Several organic dye molecules—amaranth, brilliant ponceau 5R, and naphthol blue black—have been proposed as end-point indicators for iodate titrimetry, the color of each substance undergoing a marked change near the equivalence point of the titration. Unfortunately, all of these compounds are destructively oxidized by iodate, so the indicator action is irreversible.

## PERIODIC ACID

To the chemist interested in the determination of organic compounds, the oxidizing properties of periodic acid have special significance because this reagent undergoes a number of highly selective reactions with organic functional groups.

### Nature of Periodic Acid Solutions

Detailed studies have demonstrated that **paraperiodic acid**, $H_5IO_6$, is the parent compound in aqueous periodic acid solutions. In an aqueous medium, paraperiodic acid behaves as a moderately weak diprotic acid:

$$H_5IO_6 \rightleftharpoons H^+ + H_4IO_6^-; \qquad K_{a1} = 2.3 \times 10^{-2}$$

$$H_4IO_6^- \rightleftharpoons H^+ + H_3IO_6^{2-}; \qquad K_{a2} = 4.4 \times 10^{-9}$$

Furthermore, the product of the first acid dissociation, $H_4IO_6^-$, may dehydrate to form the **metaperiodate ion**, $IO_4^-$:

$$H_4IO_6^- \rightleftharpoons IO_4^- + 2\,H_2O; \qquad K = 40$$

Paraperiodic acid is a valuable reagent for the oxidation of organic compounds because it is an extremely powerful oxidant, the standard potential for the half-reaction

$$H_5IO_6 + H^+ + 2\,e \rightleftharpoons IO_3^- + 3\,H_2O$$

being in the neighborhood of $+1.6$ to $+1.7$ v versus NHE, although the precise value is unknown.

### Preparation and Standardization of Periodate Solutions

Periodate solutions may be prepared from paraperiodic acid, potassium metaperiodate, or sodium metaperiodate, but the latter compound is preferred because it exhibits high solubility and is easily purified. Despite the oxidizing strength of periodic acid and periodate salts, aqueous solutions containing these species are remarkably stable, and water is only slowly oxidized to oxygen and ozone. All organic matter must be rigorously excluded from periodate solutions.

At least two methods are useful for the standardization of periodate solutions. If an aliquot of the periodate solution is added to an excess of potassium iodide dissolved in a bicarbonate or borate buffer of pH 8 to 9, periodate is reduced to iodate and an equivalent quantity of triiodide is liberated

$$IO_4^- + 3\ I^- + H_2O \rightleftharpoons IO_3^- + I_3^- + 2\ OH^-$$

and a standard solution of arsenic(III) can be employed to titrate the triiodide to a starch-iodine end point.

Alternatively, a known volume of the periodate solution is pipetted into an acidic iodide medium, the triiodide produced in the reaction

$$H_5IO_6 + 11\ I^- + 7\ H^+ \rightleftharpoons 4\ I_3^- + 6\ H_2O$$

being titrated with a standard sodium thiosulfate solution.

## Reactions of Periodate with Organic Compounds

Since oxidations of organic molecules by periodic acid are slow, an excess of this reagent, as well as a reaction time ranging from a few minutes to several hours, is needed to achieve quantitative results. Most periodate oxidations are performed in an aqueous medium, although the limited solubility of the organic compound in water may require that ethanol or glacial acetic acid be used as solvent. Oxidations are usually carried out at or below room temperature because undesired side reactions occur at elevated temperatures. In addition, the rate of reaction of periodic acid with organic compounds is most rapid near pH 4.

Organic compounds with hydroxyl (—OH) groups on two adjacent carbon atoms are oxidized by periodic acid at room temperature. This oxidation process, known as the **Malaprade reaction**, involves scission of the bond between these carbon atoms followed by conversion of each hydroxyl group to a carbonyl group. For example, 1,2-butanediol is oxidized to formaldehyde and propionaldehyde:

$$\begin{array}{c} H_2C\!-\!-\!CH\!-\!-\!CH_2\!-\!-\!CH_3 \\ |\quad\ | \\ OH\ \ OH \end{array} + H_4IO_6^- \rightarrow$$

$$\begin{array}{c} H_2C \\ \| \\ O \end{array} + \begin{array}{c} HC\!-\!-\!CH_2\!-\!-\!CH_3 \\ \| \\ O \end{array} + IO_3^- + 3\ H_2O$$

Many organic substances have a carbonyl group ($>C\!=\!O$) attached either to another carbonyl group or to a carbon atom with a hydroxyl group. Periodic acid reacts with these compounds to cleave the carbon-carbon bond and to oxidize a carbonyl group to a carboxyl group (—COOH) and a hydroxyl group to a carbonyl group. Thus, the oxidation of one mole of glyoxal yields two moles of formic acid,

$$\begin{array}{c} H\!-\!C\!-\!C\!-\!H \\ \|\ \ \| \\ O\ \ O \end{array} + H_4IO_6^- \rightarrow \begin{array}{c} H\!-\!C\!-\!OH \\ \| \\ O \end{array} + \begin{array}{c} HO\!-\!C\!-\!H \\ \| \\ O \end{array} + IO_3^- + H_2O$$

and the oxidation of 2-hydroxy-3-pentanone results in the formation of propanoic acid and acetaldehyde:

$$\begin{array}{c} H_3C\!-\!CH_2\!-\!C\!-\!CH\!-\!CH_3 \\ \|\ \ | \\ O\ \ OH \end{array} + H_4IO_6^- \rightarrow$$

$$\begin{array}{c} H_3C\!-\!CH_2\!-\!C\!-\!OH \\ \| \\ O \end{array} + \begin{array}{c} HC\!-\!CH_3 \\ \| \\ O \end{array} + IO_3^- + 2\ H_2O$$

A class of organic compounds which we have not previously discussed comprises those molecules in which a carbon atom with a primary or secondary amine group is adjacent either to a carbonyl group or to a carbon atom with a hydroxyl group. As in other periodate oxidations, the bond between the adjacent carbon atoms is broken, a carbonyl group is oxidized to a carboxyl group, and a hydroxyl group is converted to a carbonyl group. However, the fate of an amine group differs from that of other groups. A *primary* amine group is lost as ammonia, and the carbon atom is transformed into an aldehyde. A *secondary* amine group is lost as a primary amine, and the carbon atom is converted to an aldehyde. Two typical reactions are

$$\underset{\substack{| \quad | \\ OH \quad NH(CH_3)}}{H_2C\text{——}CH_2} + H_4IO_6^- \rightarrow \underset{\substack{\| \quad \| \\ O \quad O}}{H_2C \quad CH_2} + CH_3NH_2 + IO_3^- + 2\,H_2O$$

N-methyl-ethanolamine

and

$$\underset{\substack{| \quad | \\ OH \quad NH_2}}{H_3C\text{—}CH\text{—}CH\text{—}COOH} + H_4IO_6^- \rightarrow$$

threonine

$$\underset{\substack{\| \\ O}}{H_3C\text{—}CH} + \underset{\substack{\| \\ O}}{HC\text{—}COOH} + NH_3 + IO_3^- + 2\,H_2O$$

Notice that glyoxylic acid, CHO—COOH, is a stable product of the threonine-periodate reaction; a carboxyl group adjacent to any of the functional groups mentioned so far displays little, if any, reactivity toward periodate. Moreover, a molecule such as ethylenediamine, $H_2N\text{—}CH_2CH_2\text{—}NH_2$, with primary amine groups on two adjacent carbon atoms, reacts very slowly with periodic acid. Finally, it is of considerable significance to the success of periodate oxidations that organic compounds with single or isolated hydroxyl, carbonyl, or amine groups are unreactive *at room temperature*. At elevated temperatures, the selectivity of periodate oxidations is lost because carboxyl groups as well as the isolated functional groups will react, and the rules for predicting the course of these reactions become invalid.

Situations may be encountered in which three or more adjacent functional groups are present in the same molecule, as, for example, in glycerol. To establish what products are formed in the oxidation of glycerol by periodate, we may apply two rules: (1) assume that chemical attack begins at two adjacent functional groups on one end of the molecule and progresses one carbon-carbon bond at a time toward the other end of the molecule, and (2) use information from previous examples to determine the nature of the products. Accordingly, the first step of the oxidation,

$$\underset{\substack{| \quad | \quad | \\ OH \quad OH \quad OH}}{H_2C\text{——}CH\text{——}CH_2} + H_4IO_6^- \rightarrow \underset{\substack{\| \\ O}}{H_2C} + \underset{\substack{\| \quad | \\ O \quad OH}}{HC\text{——}CH_2} + IO_3^- + 3\,H_2O$$

glycerol

is followed by the reaction

$$\underset{\substack{\| \quad | \\ O \quad OH}}{HC\text{——}CH_2} + H_4IO_6^- \rightarrow \underset{\substack{\| \\ O}}{HC\text{—}OH} + \underset{\substack{\| \\ O}}{CH_2} + IO_3^- + 2\,H_2O$$

so the overall oxidation of 1 mole of glycerol yields 1 mole of formic acid plus 2 moles of formaldehyde.

## Analytical Uses of Periodate Solutions

Two procedures are employed in the application of the Malaprade reaction to the determination of organic compounds. In the first method, a titrimetric measurement of the quantity of periodate consumed in the oxidation process is performed and, in the second, more elaborate technique, a detailed analysis of the identities and quantities of the reaction products is involved.

*Determination of ethylene glycol.* Consider the analysis of an aqueous ethylene glycol solution. If a known excess of a standard sodium metaperiodate solution is added to the sample and the mixture set aside for one hour at room temperature, the ethylene glycol will be quantitatively oxidized to formaldehyde, and an equivalent amount of periodate will be reduced to iodate. As described earlier, one may adjust the pH of the reaction mixture to 7 (employing a carbonic acid-bicarbonate buffer), add an excess of potassium iodide, and titrate the resulting triiodide with a standard arsenic(III) solution; the *difference* between the amount of periodate originally used in the reaction mixture and that found by means of the arsenic(III) titration is proportional to the quantity of ethylene glycol in the sample.

*Analysis of a mixture of ethylene glycol, 1,2-propylene glycol, and glycerol.* Treatment of a solution containing ethylene glycol, 1,2-propylene glycol, and glycerol with excess periodic acid causes the following reactions to occur:

$$\underset{\text{ethylene glycol}}{\overset{\displaystyle H_2C\text{----}CH_2}{\underset{\displaystyle OH \quad OH}{|\qquad|}}} + H_5IO_6 \rightarrow 2\ HCHO + IO_3^- + H^+ + 3\ H_2O$$

$$\underset{\text{1,2-propylene glycol}}{\overset{\displaystyle H_2C\text{----}CH\text{----}CH_3}{\underset{\displaystyle OH \quad OH}{|\qquad|}}} + H_5IO_6 \rightarrow HCHO + CH_3CHO + IO_3^- + H^+ + 3\ H_2O$$

$$\underset{\text{glycerol}}{\overset{\displaystyle H_2C\text{----}CH\text{----}CH_2}{\underset{\displaystyle OH \quad OH \quad OH}{|\qquad|\qquad|}}} + 2\ H_5IO_6 \rightarrow 2\ HCHO + HCOOH + 2\ IO_3^- + 2\ H^+ + 5\ H_2O$$

Formic acid arises only from oxidation of glycerol. Therefore, if the reaction mixture is titrated with a standard sodium hydroxide solution after the periodate oxidation is complete, the quantity of formic acid may be related to the concentration of glycerol in the sample. However, one must correct for the amount of periodic acid in the reaction mixture by titration with sodium hydroxide of a blank solution containing all of the reagents except the three unknowns.

That acetaldehyde originates only from 1,2-propylene glycol is utilized in the determination of the latter substance. However, it is essential to separate acetaldehyde from formaldehyde, which is an oxidation product of all three compounds in the mixture. To perform this separation, one passes a stream of carbon dioxide through the solution resulting from the periodic acid treatment. Highly volatile formaldehyde and acetaldehyde are both carried by the carbon dioxide gas through a closed system which includes specific collecting solutions for the aldehydes. Formaldehyde is preferentially absorbed by a glycine solution,

$$HCHO + H_2NCH_2COOH \rightarrow H_2C{=}NCH_2COOH + H_2O$$

and acetaldehyde is trapped in a sodium bisulfite medium:

$$CH_3CHO + NaHSO_3 \rightarrow H_3C-\underset{\underset{H}{|}}{\overset{\overset{SO_3H}{|}}{C}}-O^-Na^+$$

Then, the excess of the sodium bisulfite collecting solution is destroyed, the acetaldehyde-bisulfite addition compound is decomposed, and the liberated bisulfite (which is stoichiometrically equivalent to acetaldehyde) is titrated with a standard iodine solution.

Since oxidation of ethylene glycol does not lead to any unique products, it is necessary to establish the ethylene glycol content of the mixture indirectly. Accordingly, the sum of all three components in the sample can be determined from the periodic acid consumed in the oxidation reactions; then the previously measured amounts of glycerol and 1,2-propylene glycol are subtracted from this sum to obtain the quantity of ethylene glycol.

## POTASSIUM BROMATE

A standard solution of potassium bromate can be readily prepared by weight from the reagent-grade solid and may be employed as an oxidant for the direct titration of numerous reducing agents in strongly acidic media. More importantly, a neutral potassium bromate medium containing an excess of potassium bromide serves, upon introduction of acid, as a convenient source of molecular bromine for use in the analysis and determination of organic compounds.

### Reactions of Bromate

In titrations involving a variety of reductants in acid solutions, bromate is converted to bromide ion, according to the half-reaction

$$BrO_3^- + 6 H^+ + 6 e \rightleftharpoons Br^- + 3 H_2O; \qquad E^0 = +1.45 \text{ v}$$

which indicates the substantial oxidizing strength of this species. Beyond the equivalence point, when the first excess of titrant is added, bromide ion is oxidized by bromate to elemental bromine

$$BrO_3^- + 5 Br^- + 6 H^+ \rightleftharpoons 3 Br_2(aq) + 3 H_2O$$

whose appearance can be utilized to signal the end point of the titration.

Many organic compounds are not oxidized by bromate, but do react quantitatively with molecular bromine, a somewhat weaker oxidant:

$$Br_2(aq) + 2 e \rightleftharpoons 2 Br^-; \qquad E^0 = +1.087 \text{ v}$$

Unfortunately, due to the high volatility of bromine, standard solutions of this reagent in water exhibit a rapid decrease in concentration—an undesirable characteristic. On the other hand, a standard solution of potassium bromate in which potassium bromide is dissolved at pH 7 is a stable source of bromine; when an aliquot of this solution is acidified, the occurrence of the reaction

$$BrO_3^- + 5\,Br^- + 6\,H^+ \rightleftharpoons 3\,Br_2(aq) + 3\,H_2O$$

releases a precisely known amount of bromine for the oxidation of certain organic molecules. From the above standard-potential data, we can obtain the equilibrium constant for the latter process:

$$\frac{[Br_2(aq)]^3}{[BrO_3^-][Br^-]^5[H^+]^6} = 8 \times 10^{36}$$

Because of the sixth-order dependence on the hydrogen ion concentration, the position of equilibrium for the bromate-bromide reaction is extremely sensitive to pH. At a pH of 7, $[H^+] = 1 \times 10^{-7}\,M$ and $[H^+]^6 = 1 \times 10^{-42}$, so

$$\frac{[Br_2(aq)]^3}{[BrO_3^-][Br^-]^5} = 8 \times 10^{-6}$$

whereas, in a solution whose hydrogen ion concentration is $1\,M$,

$$\frac{[Br_2(aq)]^3}{[BrO_3^-][Br^-]^5} = 8 \times 10^{36}$$

Thus, bromate and bromide have little tendency to interact in a *neutral* medium; but, in an *acidic* solution, bromate ion is quickly and quantitatively reduced by bromide ion to yield bromine.

## End-Point Detection for Titrations with Bromate

For determinations in which a reducing agent reacts directly and instantaneously with bromate in an acid solution, the end point of the titration is marked by the formation of elemental bromine, as mentioned previously. Similarly, if a neutral bromate-bromide mixture is titrated into an acidic medium containing an organic compound, and if the molecular bromine generated by the bromate-bromide reaction combines rapidly with the organic substance, the appearance of unreacted bromine will indicate the end point.

Although the color of elemental bromine is sufficiently intense to serve as an end-point indicator in some instances, several organic substances provide a more sensitive means to detect the first slight excess of bromine. Quinoline yellow, α-naphthoflavone, and p-ethoxychrysoidine undergo reversible reactions with bromine, their brominated and unbrominated forms having intense and distinctly different colors.

## Analytical Uses of Bromate Solutions

*Titrations with potassium bromate.* Table 10–6 lists some of the inorganic and organic species which can be determined by direct titration with a standard potassium bromate solution. In each application, bromate undergoes a six-electron reduction to bromide ion, whereas the half-reaction (written as an oxidation) for the reductant is indicated in the table; for the majority of titrations, the sample solution must be at least $1\,F$ in hydrochloric acid.

**Table 10–6. Substances Determined by Direct Titration with Potassium Bromate Solution**

| Substance Determined | Half-Reaction |
|---|---|
| Sb(III) | $SbCl_4^- + 2\ Cl^- \rightleftharpoons SbCl_6^- + 2\ e$ |
| As(III) | $HAsO_2 + 2\ H_2O \rightleftharpoons H_3AsO_4 + 2\ H^+ + 2\ e$ |
| Cu(I) | $CuCl_2^- \rightleftharpoons Cu^{2+} + 2\ Cl^- + e$ |
| $N_2H_4$ | $N_2H_5^+ \rightleftharpoons N_2 + 5\ H^+ + 4\ e$ |
| $H_2O_2$ | $H_2O_2 \rightleftharpoons O_2 + 2\ H^+ + 2\ e$ |
| Fe(II) | $Fe^{2+} \rightleftharpoons Fe^{3+} + e$ |
| Se(0) | $Se + 3\ H_2O \rightleftharpoons H_2SeO_3 + 4\ H^+ + 4\ e$ |
| Tl(I) | $Tl^+ \rightleftharpoons Tl^{3+} + 2\ e$ |
| Sn(II) | $SnCl_4^{2-} + 2\ Cl^- \rightleftharpoons SnCl_6^{2-} + 2\ e$ |
| $H_2C_2O_4$ | $H_2C_2O_4 \rightleftharpoons 2\ CO_2 + 2\ H^+ + 2\ e$ |
| Alkyl disulfides | $RSSR + 4\ H_2O + 2\ Br^- \rightleftharpoons 2\ RSO_2Br + 8\ H^+ + 10\ e$ |
| Dialkyl sulfides | $R_2S + H_2O \rightleftharpoons R_2SO + 2\ H^+ + 2\ e$ |
| Ascorbic acid (vitamin C) | $C_6H_8O_5 \rightleftharpoons C_6H_6O_5 + 2\ H^+ + 2\ e$ |

***Bromination of organic compounds.*** Most of the analytically important organic bromination reactions involve replacement of one or more hydrogen atoms on a phenol or a primary aromatic amine with bromine atoms. For example, in the presence of excess bromine, phenol is converted quantitatively to tribromophenol

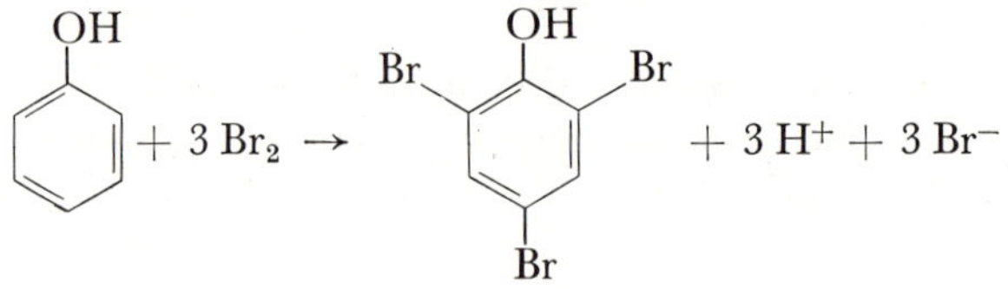

whereas *o*-nitroaniline forms a dibromo derivative:

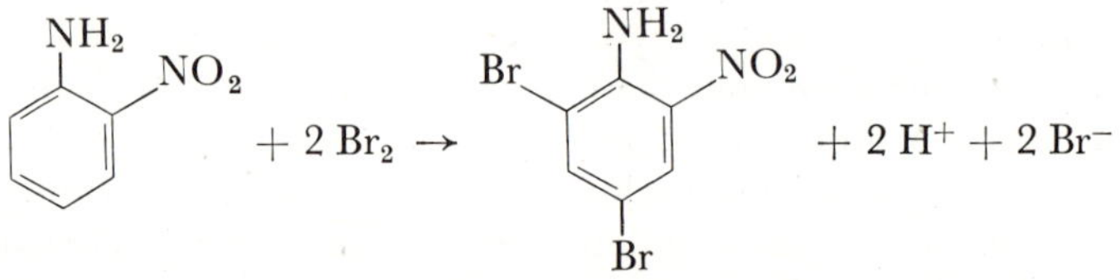

By using the bromate-bromide reagent as a source of molecular bromine, one can determine a large number of organic compounds.

Depending upon the reactivity of a substance toward bromine, either of two procedures may be employed. If the organic compound undergoes rapid bromine substitution, an acidified sample can be titrated directly with a standard potassium bromate solution containing excess potassium bromide to the first permanent appearance of bromine. For the more common situation in which the organic compound reacts slowly with bromine, one adds a known excess of the neutral bromate-bromide reagent to the sample, followed by acid to release molecular bromine, and the flask containing the mixture is immediately stoppered and set aside until the bromination reaction is complete, reaction times ranging from a few minutes to more than an hour. Finally, the unreacted bromine is back-titrated with a standard arsenic(III) solution, and the amount of organic compound is computed by difference. Alternatively, to the solution containing unreacted bromine is added a small excess of potassium iodide—which is quickly oxidized by bromine to triiodide—and the liberated triiodide is titrated with standard sodium thiosulfate solution.

Among the compounds which may be accurately determined by means of the bromination method are phenol, *m*-nitrophenol, salicylic acid, *m*-cresol, resorcinol, aniline, *m*-nitroaniline, acetanilide, and sulfathiazole, all of which form tribromo derivatives; *p*-chlorophenol, *o*-nitrophenol, *p*-nitrophenol, thymol, *p*-chloroaniline, *o*-nitroaniline, and sulfanilamide, which yield dibromo products; and *β*-naphthol and 2,4-dinitrophenol, both of which give monobromo species.

Bromination of 8-hydroxyquinoline, commonly known as oxine,

provides an analytical method for the determination of many metal cations, including aluminum(III), cadmium(II), calcium(II), cerium(III), cobalt(II), gallium(III), indium(III), iron(III), lanthanum(III), magnesium(II), manganese(II), molybdenum(VI), nickel(II), thorium(IV), titanium(IV), tungsten(VI), uranium(VI), vanadium(V), zinc(II), and zirconium(IV). As an illustration of the procedure, let us consider the determination of magnesium. If an aqueous solution of 8-hydroxyquinoline is added to an ammoniacal medium (pH 9) containing magnesium ion, a crystalline precipitate of magnesium oxinate is obtained:

$$Mg^{2+} + 2\ HOC_9H_6N \rightleftharpoons Mg(OC_9H_6N)_2 + 2\ H^+$$

After the precipitate is separated from the solution by filtration and washed, it is redissolved in approximately $2\ F$ hydrochloric acid. Then the 8-hydroxyquinoline is determined by bromination, according to either of the two techniques described earlier. From the quantity of 8-hydroxyquinoline found, one can compute how much magnesium was present in the original sample.

All of the above-named cations form insoluble compounds with 8-hydroxy-quinoline, although the solution conditions, especially pH, required for complete precipitation vary with the solubility product of the desired metal oxinate. For example, quantitative precipitation of aluminum oxinate can be achieved at pH 5 (acetic acid-acetate buffer), whereas a pH of at least 9 (ammonium ion-ammonia buffer) is needed for the magnesium salt. When several metal ions are encountered together, it may be necessary to effect their separation before 8-hydroxyquinoline is added; sometimes, control of the solution pH during the precipitation accomplishes the desired separation.

## REDUCING AGENTS AS TITRANTS

Besides sodium thiosulfate, a number of other reducing agents have been utilized as titrants, including ascorbic acid, chromium(II), iron(II), tin(II), titanium(III), and vanadium(II). Compared to the common oxidizing agents, however, these reductants are relatively little used in volumetric analysis because their ease of oxidation by atmospheric oxygen requires special procedures for storing and dispensing the solutions under air-free conditions as well as frequent restandardizations. Moreover, chromium(II) and vanadium(II) titrants, which normally contain hydrochloric acid or sulfuric acid, spontaneously reduce hydrogen ion, a process responsible for further instability of these reagents. Another obstacle to titrimetry with standard solutions of reductants is the unavailability of a wide selection of end-point indicators; in most instances, the progress of a titration must be followed potentiometrically.

Despite these problems, valuable analytical applications have been developed. Direct titration with a standard solution of an individual reducing agent is possible if the reaction between the reducing agent and the substance to be determined is fast. A sluggish reaction can be overcome if a known excess of the reductant is added to the sample to drive the reaction to completion, followed by back-titration of the unreacted reductant with a standard solution of an oxidant.

We shall now explore briefly the properties and analytical applications of iron-(II), titanium(III), and chromium(II).

## Iron(II)

Standard solutions of iron(II) in 1 $F$ sulfuric acid are best prepared by weight from Mohr's salt, $Fe(NH_4)_2(SO_4)_2 \cdot 6 H_2O$, which is commercially available as a reagent-grade material. Although solutions of moderate acidity are reasonably stable, oxidation of iron(II) by atmospheric oxygen is relatively rapid in neutral and in very strongly acid media. It is good practice to restandardize an iron(II) solution at the time of use by titration with standard potassium dichromate.

With a standard solution of iron(II), one can perform direct titrations of vanadium(V) to vanadium(IV), dichromate to chromium(III), gold(III) to elemental gold, cerium(IV) to cerium(III), and permanganate to manganese(II).

## Titanium(III)

Solutions are usually prepared by dissolution of an appropriate titanium(III) salt in approximately 0.3 $F$ hydrochloric or sulfuric acid. These titrants are extremely air-sensitive and must be stored under an inert gas, and titrations involving titanium-(III) should be performed in an oxygen-free environment. Reduction of hydrogen ion by titanium(III) is quite slow in an acidic medium but increases markedly in neutral and alkaline solutions, especially at elevated temperatures. Standardization of a titanium(III) solution should be done regularly by direct titration against a known quantity of potassium dichromate.

By means of direct titration with a standard titanium(III) solution, the following quantitative reductions can be accomplished: iron(III) to iron(II); copper(II) to copper(I), in the presence of excess thiocyanate which forms an insoluble compound with cuprous ion; antimony(V) to antimony(III); uranium(VI) to uranium(IV); bismuth(III) to elemental bismuth; ruthenium(IV) to ruthenium(III); osmium-(VIII) to osmium(IV); iridium(IV) to iridium(III); vanadium(V) to vanadium-(IV); and selenium(IV) to elemental selenium.

Some interesting uses of titanium(III) for the determination of organic substances have been devised. If a nitro ($R\!-\!NO_2$), nitroso ($R\!-\!NO$), or azo ($R\!-\!N\!=\!N\!-\!R'$) compound is treated with a slight excess of titanium(III) in a sodium citrate buffer (pH 7), it is quantitatively reduced within a few minutes at room temperature to a primary amine, $e.g.$,

$$\underset{NO_2}{\bigcirc} + 6 Ti^{3+} + 4 H_2O \rightleftharpoons \underset{NH_2}{\bigcirc} + 6 TiO^{2+} + 6 H^+$$

After the reduction is complete, the solution is acidified with hydrochloric acid, the unreacted titanium(III) is titrated with a standard solution of iron(III), and the quantity of organic compound is calculated by difference. Quinone can be quantitatively converted to hydroquinone by direct titration with a standard solution of titanium(III) in hydrochloric acid. In the presence of a known excess of titanium-(III), a saturated sulfoxide is reduced to the corresponding sulfide, *e.g.*,

$$(CH_3)_2SO + 2\ Ti^{3+} + H_2O \rightleftharpoons (CH_3)_2S + 2\ TiO^{2+} + 2\ H^+$$

If iron(III) is added to the solution containing unreacted titanium(III), the iron(II) so produced can be titrated with standard potassium dichromate, and the amount of the sulfoxide can be computed from the quantities of titanium(III) and dichromate employed.

## Chromium(II)

Chromium(II) is one of the most potent reducing agents available in an acidic medium. Solutions of this reagent may be prepared by preliminary reduction of potassium dichromate in 0.1 $F$ hydrochloric acid to chromium(III) with hydrogen peroxide. After the excess hydrogen peroxide is decomposed, the chromium(III) solution is transferred to a storage vessel containing amalgamated zinc, wherein chromium(III) is converted quantitatively to chromium(II). If the chromium(II) solution is kept in contact with amalgamated zinc, and if atmospheric oxygen is rigorously excluded, a stable titrant results. For convenience in titrimetry, including standardization of the chromium(II) solution against copper(II) in hydrochloric acid, the storage vessel is usually connected directly to a buret, so that the reagent can be dispensed and used in the complete absence of oxygen.

Included in the list of inorganic species which can be determined by potentiometric titration with a standard chromium(II) solution at elevated temperatures are iron(III), copper(II), titanium(IV), vanadium(V), chromium(VI), molybdenum-(VI), tungsten(VI), mercury(II), gold(III), silver(I), antimony(V), tin(IV), bismuth(III), selenium(IV), tellurium(IV), nitrate, peroxodisulfate, and ferricyanide.

Applications of chromium(II) to the determination of organic substances resemble those performed with titanium(III), except that chromium(II) is more reactive. Nitro, nitroso, and azo compounds are quantitatively reduced to primary amines, and acetylene dicarboxylic acid undergoes reduction to the ethylenic derivative, in the presence of excess chromium(II) at room temperature in only a minute or two; these compounds can be determined by difference if the unreacted chromium(II) is back-titrated with a standard iron(III) solution. Anthraquinone-2,7-disodium sulfonate, which is reduced to the corresponding anthrahydroquinone,

has been determined by direct potentiometric titration with a standard chromium(II) solution.

## QUESTIONS AND PROBLEMS

1. Formal potentials for the cerium(IV)-cerium(III) half-reaction (versus NHE) in various concentrations of perchloric acid, nitric acid, and sulfuric acid are listed in the following table:

| Acid Concentration, $F$ | $HClO_4$ | $HNO_3$ | $H_2SO_4$ |
|---|---|---|---|
| 1 | +1.70 | +1.61 | +1.44 |
| 2 | 1.71 | 1.62 | 1.44 |
| 4 | 1.75 | 1.61 | 1.43 |
| 6 | 1.82 | 1.56 | — |
| 8 | 1.87 | — | 1.42 |

Discuss the significance of these formal potentials in terms of the possible chemical interactions of cerium(IV) and cerium(III) with the acid media.

2. Calculate the formality of a potassium dichromate solution 1.000 ml of which is equivalent to 0.005000 gm of iron.

3. Given the following information,

$$Cr_2O_7^{2-} + 14 H^+ + 6 e \rightleftharpoons 2 Cr^{3+} + 7 H_2O; \quad E^0 = +1.33 \text{ v}$$

$$Cr(OH)_3 \rightleftharpoons Cr^{3+} + 3 OH^-; \quad K_{sp} = 6 \times 10^{-31}$$

$$CrO_4^{2-} + H^+ \rightleftharpoons HCrO_4^-; \quad K = 3.2 \times 10^6$$

$$2 HCrO_4^- \rightleftharpoons Cr_2O_7^{2-} + H_2O; \quad K' = 33$$

evaluate the standard potential for the half-reaction

$$CrO_4^{2-} + 4 H_2O + 3 e \rightleftharpoons Cr(OH)_3 + 5 OH^-$$

4. From standard-potential data for the half-reactions

$$MnO_4^- + 4 H^+ + 3 e \rightleftharpoons MnO_2 + 2 H_2O$$

and

$$MnO_4^- + e \rightleftharpoons MnO_4^{2-}$$

calculate the equilibrium constant for the disproportionation of the manganate ion

$$3 MnO_4^{2-} + 2 H_2O \rightleftharpoons 2 MnO_4^- + MnO_2 + 4 OH^-$$

and calculate what must be the minimum concentration of hydroxide ion required to prevent more than 1 per cent disproportionation of 0.01 $M$ manganate ion.

5. Calculate the formality of a potassium permanganate solution, 35.00 ml of which is equivalent to 0.2500 gm of 98.00 per cent pure calcium oxalate.

6. If 40.32 ml of a solution of oxalic acid can be titrated with 37.92 ml of a 0.4736 $F$ sodium hydroxide solution and if 30.50 ml of the same oxalic acid solution reacts with 47.89 ml of a potassium permanganate solution, what is the formality of the permanganate solution?

7. Barium ion is frequently added to the strongly alkaline reaction medium during the oxidation of organic compounds with potassium permanganate.

How does barium ion influence the effective oxidizing strength of permanganate under these conditions? Explain your conclusion.

8. Calculate the per cent purity of a sample of impure $H_2C_2O_4 \cdot 2\,H_2O$ if 0.4006 gm of this material requires a titration with 28.62 ml of a potassium permanganate solution, 1.000 ml of which contains 5.980 mg of potassium permanganate.

9. A slag sample is known to contain all of its iron in the forms FeO and $Fe_2O_3$. A 1.000-gm sample of the slag was dissolved in hydrochloric acid according to the usual procedure, reduced with stannous chloride, and eventually titrated with 28.59 ml of a 0.02237 $F$ potassium permanganate solution. A second slag sample, weighing 1.500 gm, was dissolved in a nitrogen atmosphere in order to prevent atmospheric oxidation of iron(II) during the dissolution process and, without any further adjustment of the oxidation state of iron, it was immediately titrated with the same potassium permanganate solution. If 15.60 ml of the permanganate solution was required in the second experiment, calculate (a) the total percentage of iron in the slag and (b) the individual percentages of FeO and $Fe_2O_3$.

10. A 20.00-ml aliquot of a mixture of formic acid and acetic acid was titrated with a 0.1209 $F$ sodium hydroxide solution, 23.12 ml being required to reach a phenolphthalein end point. A second 20.00-ml sample of the acid mixture was pipetted into 50.00 ml of a strongly alkaline 0.02832 $F$ potassium permanganate solution, and the reaction mixture was allowed to stand for 30 minutes at room temperature. Then the solution was acidified with sulfuric acid and titrated with a standard 0.1976 $F$ iron(II) solution. The titration required 28.51 ml of the iron(II) solution. Calculate the molar concentrations of formic acid and acetic acid in the original acid mixture.

11. The quantity of glycerol in a sample solution was determined by reacting the glycerol with 50.00 ml of a strongly alkaline 0.03527 $F$ potassium permanganate solution at room temperature for 25 minutes. After the reaction was complete, the mixture was acidified with sulfuric acid, and the excess unreacted permanganate was reduced with 10.00 ml of a 0.2500 $F$ oxalic acid solution. Then the excess oxalic acid was back-titrated with a 0.01488 $F$ potassium permanganate solution, 26.18 ml being required. Calculate the weight of glycerol in the original sample.

12. In the determination of the manganese dioxide content of a pyrolusite ore, a 0.5261-gm sample of the pyrolusite was treated with 0.7049 gm of pure sodium oxalate, $Na_2C_2O_4$, in an acid medium. After the reaction had gone to completion, 30.47 ml of 0.02160 $F$ potassium permanganate solution was needed to titrate the excess, unreacted oxalic acid. Calculate the percentage of manganese dioxide in the pyrolusite.

13. A 25.00-ml aliquot of a stock hydrogen peroxide solution was transferred to a 250.0-ml volumetric flask, and the solution was diluted and mixed. A 25.00-ml sample of the diluted hydrogen peroxide solution was acidified with sulfuric acid and titrated with 0.02732 $F$ potassium permanganate, 35.86 ml being required. Calculate the number of grams of hydrogen peroxide per 100.0 ml of the original stock solution.

14. What weight of an iron ore should be taken for analysis so that the volume of 0.1046 $F$ cerium(IV) sulfate solution used in the subsequent titration will be numerically the same as the percentage of iron in the ore?

15. Calculate the weight of potassium dichromate which must be used to prepare 1.000 liter of solution such that, when the solution is employed in iron determinations with 1.000-gm samples, the buret reading in milliliters will equal the percentage of iron in the sample.

16. A silver reductor was employed to reduce 0.01000 $M$ uranium(VI) to uranium-(IV) in a 2.00 $F$ hydrochloric acid medium according to the reaction

$$UO_2{}^{2+} + 2\,Ag + 2\,Cl^- + 4\,H^+ \rightleftharpoons U^{4+} + 2\,AgCl + 2\,H_2O$$

Calculate the fraction of uranium(VI) which remains unreduced after this treatment.

17. What procedure would you employ to regenerate a silver reductor?

18. Cerium(IV) in 4 $F$ perchloric acid is an effective oxidant for many organic

compounds. Write balanced half-reactions for the oxidation of each of the following organic substances:

(a) ethylene glycol

$$\begin{array}{cc} H_2C\text{———}CH_2 \\ | \quad\quad | \\ OH \quad OH \end{array}$$

(b) formaldehyde

$$\begin{array}{c} O \\ \| \\ H\text{—}C \\ \diagdown \\ H \end{array}$$

(c) citric acid

$$\begin{array}{c} OH \\ | \\ HOOC\text{—}CH_2\text{—}C\text{—}CH_2\text{—}COOH \\ | \\ COOH \end{array}$$

(d) malic acid

$$\begin{array}{c} HOOC\text{—}CH\text{—}CH_2\text{—}COOH \\ | \\ OH \end{array}$$

(e) 2,3-butanediol

$$\begin{array}{c} H_3C\text{—}CH\text{—}CH\text{—}CH_3 \\ | \quad\quad | \\ OH \quad OH \end{array}$$

(f) 2,3,5-hexanetrione

$$\begin{array}{c} H_3C\text{—}C\text{—}C\text{—}CH_2\text{—}C\text{—}CH_3 \\ \| \quad \| \quad\quad\quad \| \\ O \quad O \quad\quad\quad O \end{array}$$

19. A 20.00-ml aliquot of a 1 $F$ hydrochloric acid solution containing iron(III) and vanadium(V) was passed through a silver reductor. The resulting solution was titrated with a 0.01020 $F$ potassium permanganate solution, 26.24 ml being required to reach a visual end point. A second 20.00-ml aliquot of the sample solution was passed through a Jones reductor, proper precautions being taken to ensure that air oxidation of the reduced species did not occur. When the solution was titrated with the 0.01020 $F$ permanganate solution, 31.80 ml was required for the titration. Calculate the concentrations of iron(III) and vanadium(V) in the original sample solution.

20. A potassium permanganate solution was standardized against 0.2643 gm of pure arsenious oxide. The arsenious oxide was dissolved in a sodium hydroxide solution; the resulting solution was acidified and finally titrated with the permanganate solution, of which 40.46 ml was used. Calculate the formality of the potassium permanganate solution.

21. A cerium(IV) sulfate solution was standardized against 0.2023 gm of 97.89 per cent pure iron wire. If 43.61 ml of the cerium(IV) solution was used in the final titration, calculate its formality.

22. A 10.00-ml sample of a mixture of tartaric acid and formic acid was boiled with 25.00 ml of 0.1625 $F$ cerium(IV) in 10 $F$ sulfuric acid for approximately one hour. The back-titration of the unreacted cerium(IV) required 3.41 ml of a 0.1009 $F$ iron(II) solution. A second 10.00-ml sample of the acid mixture was treated for 15 minutes at room temperature with 20.00 ml of 0.1119 $F$ cerium-(IV) in 4 $F$ perchloric acid. After the reaction was complete, the unreacted cerium(IV) was back-titrated with a 0.1031 $F$ oxalic acid solution, 8.12 ml being required. Calculate the concentrations of tartaric acid and formic acid in the original mixture.

23. If 36.81 ml of a 0.1206 $F$ sodium hydroxide solution is needed to neutralize the acid formed when 34.76 ml of a potassium permanganate solution is treated with sulfur dioxide,

$$2\ MnO_4^- + 5\ H_2SO_3 \rightleftharpoons 2\ Mn^{2+} + 4\ H^+ + 5\ SO_4^{2-} + 3\ H_2O$$

the excess sulfur dioxide being removed by boiling, what volume of 0.1037 $M$ iron(II) solution would be oxidized by 38.41 ml of the potassium permanganate solution?

24. If 5.000 ml of a commercial hydrogen peroxide solution, having a density of 1.010 gm/ml, requires 18.70 ml of a 0.1208 $F$ cerium(IV) solution for titration in an acid medium, what is the percentage by weight of hydrogen peroxide in the commercial solution?

25. Consider the titration of 50.00 ml of 0.06000 $M$ iron(II) with 0.01000 $F$ potassium dichromate in 1 $F$ hydrochloric acid, assuming the validity of the following formal-potential data:

$$Fe^{3+} + e \rightleftharpoons Fe^{2+}; \qquad E^{0\prime} = +0.70 \text{ v versus NHE}$$

$$Cr_2O_7^{2-} + 14 H^+ + 6 e \rightleftharpoons 2 Cr^{3+} + 7 H_2O;$$

$$E^{0\prime} = +1.05 \text{ v versus NHE}$$

Determine the following information:

(a) The potential of a platinum indicator electrode versus a normal hydrogen electrode at the theoretical equivalence point of the titration.

(b) The potential of a platinum indicator electrode versus a saturated calomel electrode (SCE) after the addition of 37.60 ml of the potassium dichromate titrant.

(c) The per cent titration error, with the proper sign, if barium diphenylamine sulfonate, which changes to a violet color at +0.84 v versus NHE, is used as the end-point indicator.

(d) Repeat the calculations of parts (a), (b), and (c), assuming that the titration is performed in a 1 $F$ hydrochloric acid-0.25 $F$ phosphoric acid mixture in which the formal potential for the iron(III)-iron(II) couple is shifted to +0.51 v versus NHE.

26. Calculate the equilibrium concentrations of $H_5IO_6$, $H_4IO_6^-$, $H_3IO_6^{2-}$, and $IO_4^-$ in a solution prepared by dissolving 0.0100 mole of sodium metaperiodate ($NaIO_4$) in 1 liter of a pH 7.00 buffer.

27. In the titration of 50.00 ml of 0.05000 $M$ iodide ion with a 0.05000 $M$ cerium(IV) solution, calculate what volume of cerium(IV) will be added before solid iodine just begins to precipitate. Take dilution into account.

28. What can be the maximal initial iodide concentration in 50.00 ml of solution such that solid iodine will not precipitate during a titration of the iodide with a 0.05000 $M$ cerium(IV) solution?

29. From standard-potential data given in Appendix 4, calculate the solubility of bromine in water at 25°C and calculate the equilibrium constant for the reaction

$$Br_2(aq) + Br^- \rightleftharpoons Br_3^-$$

30. In order to gain some idea of the importance of pH control in the titration of arsenic(III) with a triiodide solution, consider the following situation: Assume that 50.00 ml of 0.04000 $M$ arsenic(III) solution is mixed with 50.00 ml of 0.04000 $M$ iodine (in 0.25 $F$ potassium iodide) solution and that both solutions are initially unbuffered at pH 7.00. Using appropriate equilibrium considerations, calculate what fraction of arsenic(III) is oxidized when equilibrium is finally established.

31. From data given in Appendix 4, evaluate the standard potentials for each of the following half-reactions:

(a) $3 IO_3^- + 18 H^+ + 16 e \rightleftharpoons I_3^- + 9 H_2O$
(b) $IO_3^- + 6 H^+ + 2 Cl^- + 4 e \rightleftharpoons ICl_2^- + 3 H_2O$
(c) $BrO_3^- + 6 H^+ + 6 e \rightleftharpoons Br^- + 3 H_2O$

32. Determine the equilibrium constant for the reaction

$$BrO_3^- + 5 Br^- + 6 H^+ \rightleftharpoons 3 Br_2 + 3 H_2O$$

and compute the concentration of free bromine in a solution of pH 7.00 containing 0.1000 $F$ potassium bromate and 0.70 $F$ potassium bromide.

33. If 37.12 ml of an iodine (triiodide) solution is required to titrate the sample solution prepared from 0.5078 gm of pure arsenious oxide, calculate the formality of the iodine (triiodide) solution.

34. A 0.1936-gm sample of primary-standard potassium dichromate was dissolved in water, the solution was acidified, and, after the addition of potassium iodide, the titration of the liberated iodine required 33.61 ml of a sodium thiosulfate solution. Calculate the formality of the thiosulfate solution.

35. The sulfur from a 5.141-gm steel sample was evolved as hydrogen sulfide and collected in an excess of an ammoniacal cadmium solution. The resulting cadmium sulfide precipitate was washed and suspended in water to which a few drops of acetic acid had been added. Then 25.00 ml of a 0.002027 $F$ potassium iodate solution was added to the mixture, followed by 3 gm of potassium iodide and 10 ml of concentrated hydrochloric acid. The liberated iodine oxidized the hydrogen sulfide gas to sulfur, and the excess unreacted iodine was titrated with 0.1127 $F$ sodium thiosulfate solution from a micro-buret, 1.085 ml being used. Calculate the percentage of sulfur in the steel.

36. What weight of a copper ore must be taken for analysis so that each 1.00 ml of a 0.1050 $F$ sodium thiosulfate solution represents exactly 1.00 per cent copper in the sample?

37. From a 0.5635-gm sample of a copper ore, the copper was precipitated as copper(I) thiocyanate, CuSCN. The precipitate was treated with a 4 $F$ hydrochloric acid solution and then titrated with a standard potassium iodate solution according to the reaction

$$4 \text{ CuSCN} + 7 \text{ IO}_3^- + 18 \text{ H}^+ + 14 \text{ Cl}^-$$
$$\rightleftharpoons 4 \text{ HCN} + 4 \text{ Cu}^{2+} + 4 \text{ HSO}_4^- + 7 \text{ ICl}_2^- + 5 \text{ H}_2\text{O}$$

If the standard potassium iodate solution contained 10.71 gm of $KIO_3$ per liter of solution and if 39.57 ml of this solution was required for the titration, calculate the percentage of copper in the ore.

38. A certain oxidizing agent has a molecular weight of 250.0. A 0.3125-gm sample of this compound was treated with excess potassium iodide in an acidic medium, and the liberated iodine (triiodide) was titrated with 20.00 ml of a 0.1250 $F$ sodium thiosulfate solution. How many electrons per molecule are gained by the oxidizing agent in its reaction with iodide?

39. Write balanced equations showing all reactants and products for the oxidation of each of the following compounds with periodic acid at room temperature:

    (a) acetoin

$$\begin{array}{c} \text{H}_3\text{C—C—CH—CH}_3 \\ \quad\ \| \quad\ | \\ \quad\ \text{O} \quad \text{OH} \end{array}$$

    (b) biacetyl

$$\begin{array}{c} \text{H}_3\text{C—C—C—CH}_3 \\ \quad\ \| \ \ \| \\ \quad\ \text{O} \ \ \text{O} \end{array}$$

    (c) glycolic aldehyde

$$\begin{array}{c} \text{H}_2\text{C———C—H} \\ | \quad\ \ \| \\ \text{OH} \ \ \text{O} \end{array}$$

    (d) gluconic acid

$$\begin{array}{c} \text{H}_2\text{C———CH—CH—CH—CH—COOH} \\ | \qquad | \quad\ | \quad\ | \quad\ | \\ \text{OH} \ \ \text{OH} \ \text{OH} \ \text{OH} \ \text{OH} \end{array}$$

    (e) fructose

$$\begin{array}{c} \text{H}_2\text{C———CH—CH—CH—C—CH}_2 \\ | \qquad | \quad\ | \quad\ | \quad\ \| \quad\ | \\ \text{OH} \ \ \text{OH} \ \text{OH} \ \text{OH} \ \text{O} \ \text{OH} \end{array}$$

    (f) dihydroxyacetone

$$\begin{array}{c} \text{H}_2\text{C———C—CH}_2 \\ | \quad\ \ \| \quad\ | \\ \text{OH} \ \ \text{O} \ \ \text{OH} \end{array}$$

    (g) serine

$$\begin{array}{c} \text{H}_2\text{C———CH—COOH} \\ | \qquad\ | \\ \text{OH} \ \ \text{NH}_2 \end{array}$$

    (h) $\beta$-hydroxyglutamic acid

$$\begin{array}{c} \text{HOOC—CH———CH—CH}_2\text{—COOH} \\ | \qquad\quad | \\ \text{NH}_2 \quad\ \ \text{OH} \end{array}$$

40. A sample solution containing glycerol and ethylene glycol was transferred to a 100-ml volumetric flask, 50.00 ml of a standard $0.01913 F$ paraperiodic acid ($H_5IO_6$) solution was added, and the mixture was diluted to the calibration mark with distilled water. The reaction solution was allowed to stand at room temperature for one hour; then the following analyses were performed on aliquots of this reaction mixture:

A 25.00-ml aliquot was titrated with a $0.02041 F$ sodium hydroxide solution, 14.39 ml being required to reach a phenolphthalein end point. In a blank titration, 24.90 ml of the sodium hydroxide solution was required to titrate a 25.00-ml portion of the *original* $0.01913 F$ paraperiodic acid to the same end point.

A second 25.00-ml aliquot of the reaction mixture was neutralized with sodium bicarbonate to a pH of approximately 7, and potassium iodide was added. The liberated iodine was titrated with a $0.01029 F$ arsenic(III) solution to a starch-iodine end point, 5.47 ml being needed.

Calculate the total number of milligrams of glycerol and ethylene glycol in the original sample.

41. A 0.4191-gm sample of Paris green, an arsenic-containing insecticide, was treated with hydrochloric acid and a reducing agent; and the arsenic was distilled as arsenic trichloride ($AsCl_3$) into a receiving vessel which contained distilled water. The hydrogen chloride which accompanied the arsenic trichloride was neutralized with an excess of solid sodium bicarbonate and the solution was titrated with a $0.04489 F$ iodine (triiodide) solution, 37.06 ml being required. Calculate the percentage of arsenious oxide, $As_2O_3$, in the sample.

42. Ascorbic acid (vitamin C) in a juice drink was determined by means of a redox procedure involving iodine. A 100.0-ml sample of the drink was acidified with 20 ml of $6 F$ sulfuric acid, and exactly 25.00 ml of $0.005204 F$ iodine (triiodide) solution was pipetted into the sample. During a one-minute waiting period, ascorbic acid was oxidized to dehydroascorbic acid according to the half-reaction

Next, 3 ml of 0.5 per cent starch indicator was added. Excess, unreacted iodine (triiodide) was reduced through the addition of a known small excess of standard sodium thiosulfate solution; 5.00 ml of $0.009744 F$ thiosulfate solution was used. Finally, the sample was titrated with the $0.005204 F$ iodine solution to the first permanent appearance of the starch-iodine color, 2.23 ml being required. Calculate the weight of ascorbic acid in 100 ml of the juice drink. How many fluid ounces of the drink must be consumed to provide the minimum daily requirement (30 mg) of vitamin C?

43. An aqueous solution of hydrazine was analyzed by means of a direct titration with a standard $0.02722 F$ potassium iodate solution. A 25.00-ml aliquot of the hydrazine solution was pipetted into a 250-ml iodine flask, approximately 25 ml of concentrated ($12 F$) hydrochloric acid and 5 ml of carbon tetrachloride were added to the flask, and the titration with potassium iodate solution was begun. During the early stages of the titration, the brown color of molecular iodine appeared. As the titration progressed, the iodine was oxidized to the colorless iodine dichloride complex. Near the end point, the flask was stoppered and shaken after each drop of titrant was added, and the carbon tetrachloride phase was examined for the presence or absence of the violet color of dissolved iodine. The last trace of iodine disappeared from the carbon tetrachloride phase when 29.18 ml of the potassium iodate solution had been used. Calculate the formal concentration of the hydrazine solution, utilizing information about the redox chemistry of hydrazine available in Appendix 4.

44. Of the amino acids generally present in protein materials, threonine ($\alpha$-amino-$\beta$-hydroxybutyric acid) is the only one which yields acetaldehyde when treated with sodium metaperiodate:

$$CH_3CH(OH)CH(NH_2)COOH + H_4IO_6^- \rightarrow$$

$$CH_3CHO + CHOCOOH + NH_3 + IO_3^- + 2 H_2O$$

A protein food tablet, weighing 660 mg, was subjected to hydrolysis in the presence of 6 $F$ hydrochloric acid, after which the excess acid was removed by means of vacuum evaporation. The protein hydrolysate was treated at room temperature with sodium metaperiodate in a solution containing excess sodium bicarbonate. A gas train was set up so that acetaldehyde could be separated from formaldehyde, the latter resulting from the periodate oxidation of serine and hydroxylysine. Acetaldehyde was collected in a solution of sodium bisulfite, in which a stable acetaldehyde-bisulfite addition compound is formed. After the excess sodium bisulfite had been decomposed (by oxidation of $HSO_3^-$ with $I_3^-$), borax and sodium carbonate were added to the solution to decompose the acetaldehyde-bisulfite addition compound. Then the liberated bisulfite was titrated with a standard 0.01125 $F$ iodine (triiodide) solution to a starch-iodine end point:

$$HSO_3^- + I_3^- + 3\, OH^- \rightleftharpoons SO_4^{2-} + 3\, I^- + 2\, H_2O$$

If 17.92 ml of the iodine titrant was used, calculate the percentage by weight of threonine in the protein tablet.

45. A 0.2536-gm sample of a mixture of only diethylsulfide and di-$n$-butylsulfide was dissolved in approximately 40 ml of glacial acetic acid contained in a 250-ml flask. Then, 3 ml of concentrated hydrochloric acid was added, and the mixture was titrated with a 0.01702 $F$ potassium bromate solution containing excess potassium bromide until the first permanent pale yellow color of molecular bromine appeared. A volume of 53.50 ml of titrant was used; however, in a blank titration performed without the dialkyl sulfides, 0.24 ml of the potassium bromate reagent was consumed before a stable end point was obtained. Calculate the percentages of diethylsulfide and di-$n$-butylsulfide in the mixture.

46. An aqueous solution containing only aluminum(III) was buffered with ammonia and ammonium acetate at pH 9. To this solution was added a slight excess of 8-hydroxyquinoline (oxine), which caused the quantitative precipitation of aluminum oxinate:

$$Al^{3+} + 3\, HOC_9H_6N \rightleftharpoons Al(OC_9H_6N)_3 + 3\, H^+$$

The precipitate was collected with suction on a sintered-glass filter crucible and was washed free of excess oxine. Then, the precipitate was dissolved in 2 $F$ hydrochloric acid, and the resulting acidic solution of 8-hydroxyquinoline was treated with 15.00 ml of 0.1238 $F$ potassium bromate-1 $F$ potassium bromide reagent. After the bromination of oxine was complete, an excess of potassium iodide was added, which was immediately oxidized to triiodide by the unconsumed bromine:

$$Br_2 + 3\, I^- \rightleftharpoons 2\, Br^- + I_3^-$$

Finally, the liberated iodine (triiodide) was titrated with a 0.1028 $F$ sodium thiosulfate solution to a starch-iodine end point, 5.45 ml being needed. From the experimental data, calculate the number of milligrams of aluminum present in the original sample.

47. Exactly 250.0 ml of a standard solution of iron(II) was freshly prepared by dissolution of 9.932 gm of $Fe(NH_4)_2(SO_4)_2 \cdot 6\, H_2O$, Mohr's salt, in 1 $F$ sulfuric acid. This titrant was employed in the standardization of an ammonium vanadate ($NH_4VO_3$) solution. To a 25.00-ml aliquot of the ammonium vanadate solution was added 125 ml of water and 50 ml of concentrated sulfuric acid. The solution was cooled, 1 drop of ferroin indicator was introduced, and the mixture was titrated with the iron(II) solution until a color change from greenish blue to reddish green was observed. If 28.59 ml of the iron(II)

titrant was used, calculate the formal concentration of the ammonium vanadate solution.

48. To determine the quantity of nitroglycerin

$$
\begin{array}{ccc}
\text{CH}_2 & \text{—CH—} & \text{CH}_2 \\
| & | & | \\
\text{ONO}_2 & \text{ONO}_2 & \text{ONO}_2
\end{array}
$$

in dynamite, a 0.01982-gm sample of the material was treated with 25 ml of methanol. The resulting solution was placed under a carbon dioxide atmosphere to exclude oxygen, and 25.00 ml of $0.04919\,F$ titanium(III) in $0.3\,F$ hydrochloric acid was introduced. The reaction mixture was allowed to stand for 10 minutes, and was then boiled for 10 minutes. Finally, the solution was cooled, some potassium thiocyanate indicator was added, and a standard $0.02441\,F$ iron(III)-$1\,F$ sulfuric acid titrant was introduced until the appearance of the red color of the $\text{FeSCN}^{2+}$ complex signaled the end point; 5.77 ml of the iron(III) solution was required. Calculate the percentage by weight of nitroglycerin in the dynamite.

49. A sample of anthraquinone-2,7-disodium sulfonate, suspected to be contaminated with anthracene-2,7-disodium sulfonate, was analyzed by means of a titration with chromium(II). Exactly 0.2250 gm of the solid material was dissolved in 100 ml of dilute hydrochloric acid, and the resulting solution was titrated potentiometrically with $0.05741\,F$ chromium(II) in $0.1\,F$ hydrochloric acid. A volume of 18.53 ml of the titrant was required. What can be concluded about the purity of the sample?

50. The reaction between hypochlorite and iodide ions in an acidic medium

$$\text{OCl}^- + 3\,\text{I}^- + 2\,\text{H}^+ \rightleftharpoons \text{Cl}^- + \text{I}_3^- + \text{H}_2\text{O}$$

is the basis of a method for the determination of "available chlorine" in bleaching powders and bleach solutions. A 2.622-gm sample of a household bleach (sodium hypochlorite) solution was treated with excess potassium iodide, and the solution was acidified with 15 ml of $2\,F$ sulfuric acid. The liberated iodine (triiodide) was immediately titrated in the presence of starch indicator with $0.1109\,F$ sodium thiosulfate solution, 35.58 ml being required. Calculate the "available chlorine" in the bleach solution.

The strength of a chlorine-containing bleaching agent is expressed in terms of "available chlorine." Available chlorine is a measure of the oxidizing power of the chlorine present in the bleaching agent. For example, pure, solid sodium hypochlorite (NaOCl) contains 47.62 weight per cent chlorine. However, the oxidizing strength of chlorine in $\text{OCl}^-$ is twice that of chlorine in $\text{Cl}_2$ because the reduction of $\text{OCl}^-$ to $\text{Cl}^-$ is a two-electron reaction whereas the reduction of $\text{Cl}_2$ to $\text{Cl}^-$ is a one-electron process *per* chlorine. Thus, the "available chlorine" in pure NaOCl is said to be $(2)(47.62)$ or 95.24 per cent.

## SUGGESTIONS FOR ADDITIONAL READING

1. A. Berka, J. Vulterin, and J. Zýka: *Newer Redox Titrants*. Pergamon Press, London, 1965.
2. I. M. Kolthoff and R. Belcher: *Volumetric Analysis*. Volume 3, Wiley-Interscience, New York, 1957.
3. B. Kratochvil: Analytical oxidation-reduction reactions in organic solvents. In *Critical Reviews in Analytical Chemistry*. Volume 1, Chemical Rubber Company, Cleveland, Ohio, 1971, pp. 415–454.
4. H. A. Laitinen: *Chemical Analysis*. McGraw-Hill Book Company, New York, 1960, pp. 326–451.
5. G. F. Smith: *Analytical Applications of Periodic Acid and Iodic Acid*. G. F. Smith Chemical Company, Columbus, Ohio, 1950.
6. G. F. Smith: *Cerate Oxidimetry*. G. F. Smith Chemical Company, Columbus, Ohio, 1942.

# DIRECT POTENTIOMETRY AND POTENTIOMETRIC TITRATIONS

**11**

We observed in Chapter 9 that the emf of a galvanic cell depends on the activities of the reducing and oxidizing agents in equilibrium with the electrodes. Such behavior may be exploited analytically in at least two distinct ways. First, in certain well-defined situations, the magnitude of the emf of a galvanic cell can be related by means of the Nernst equation to the activity of a single species, providing us with the procedure known as **direct potentiometry**. Second, the variation of the emf of a galvanic cell in response to changes in the activities of chemical species can serve as a valuable method for following the progress of a titration, commonly referred to as the technique of **potentiometric titration**.

In the past, the use of direct potentiometry in chemical analysis had been largely restricted to pH measurements. However, within the last decade or so, the development of numerous ion-selective electrodes has revived interest in the subject of direct potentiometry. Potentiometric titrimetry has enjoyed widespread popularity because it is applicable to every kind of chemical reaction and because the required instrumentation is relatively simple and inexpensive.

# INSTRUMENTATION FOR POTENTIOMETRIC
# METHODS OF ANALYSIS

Apparatus required for the performance of either direct potentiometry or potentiometric titrations involves the same components. Every potentiometric measurement requires that a suitable galvanic cell, consisting of a sensitive indicator electrode and a stable reference electrode, be constructed with the solution to be studied in direct electrolytic contact with the indicator electrode. A reference electrode may be inserted into the sample solution or it can be brought into contact with the sample solution through an appropriate salt bridge. To determine the absolute magnitude of the emf of the galvanic cell or to follow the changes which occur in emf as titrant is added during a potentiometric titration, a reliable potential-measuring instrument is needed.

## Indicator Electrodes

Although we shall consider specific types of indicator electrodes in detail throughout this chapter, three general characteristics must be sought when one selects an indicator electrode for direct potentiometry or for a potentiometric titration. First, the potential of an indicator electrode should be related through the Nernst equation to the activity of the species being determined. This requirement is especially critical in the technique of direct potentiometry. Second, it is desirable that an indicator electrode respond rapidly and reproducibly to variations in the activity of the substance of interest. Third, an indicator electrode should possess a physical form which permits one to perform measurements conveniently.

## Reference Electrodes

A reference electrode is one whose potential remains constant during a single potentiometric measurement or throughout the course of a potentiometric titration. Actually, the roster of suitable reference electrodes is very small, the selection being limited by several practical considerations—preparation and maintenance of a reference electrode should be simple; the potential of a reference electrode should be reproducible; and the electrode must undergo no significant change in potential when a small current passes through the electrode during potentiometric measurements.

*Mercury-mercurous chloride (calomel) electrodes.* Without doubt, the most commonly used reference electrode is the **saturated calomel electrode (SCE)**, composed of metallic mercury and solid mercurous chloride (calomel) in contact with, and in equilibrium with, a saturated aqueous solution of potassium chloride. We may represent the electrochemical equilibrium which characterizes the behavior of a calomel electrode by the half-reaction

$$Hg_2Cl_2(s) + 2\,e \rightleftharpoons 2\,Hg(l) + 2\,Cl^-$$

and the corresponding Nernst equation can be written as

$$E_{Hg_2Cl_2,Hg} = E^0_{Hg_2Cl_2,Hg} - \frac{0.059}{2} \log [Cl^-]^2$$

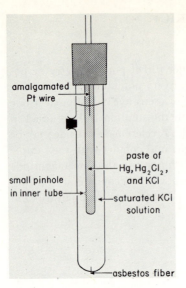

amalgamated
Pt wire

small pinhole
in inner tube

paste of
Hg, Hg₂Cl₂,
and KCl

saturated KCl
solution

asbestos fiber

*Figure 11–1.* Saturated calomel reference electrode. An inner tube—the electrode proper—contains a paste of metallic mercury, mercurous chloride, and potassium chloride in contact with an amalgamated platinum wire which leads to a potential-measuring instrument. An outer tube—actually a salt bridge—is filled with a saturated potassium chloride solution. Electrolytic contact between the inner and outer compartments is obtained by means of a pinhole in the side of the inner tube. A porous asbestos fiber, sealed into the tip of the outer (salt bridge) tube, provides contact to the sample solution.

Thus, the potential of a calomel reference electrode depends only on the chloride ion concentration (activity). Since mercurous chloride is very insoluble ($K_{sp} = 1.3 \times 10^{-18}$), the concentration of chloride ion is governed essentially by the amount of potassium chloride used in the preparation of the electrode. At 25°C, the potential of the saturated calomel electrode is $+0.2415$ v versus NHE. Saturated calomel electrodes are commercially available, one of the most familiar forms appearing in Figure 11–1.

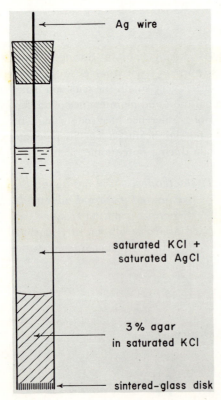

Ag wire

saturated KCl +
saturated AgCl

3 % agar
in saturated KCl

sintered–glass disk

*Figure 11–2.* Silver-silver chloride reference electrode.

Two other calomel electrodes are occasionally utilized for electrochemical measurements—the **normal calomel electrode**, in which the aqueous medium is $1\,F$ potassium chloride solution; and the **decinormal calomel electrode**, in which $0.1\,F$ potassium chloride solution is employed. These reference electrodes have potentials of $+0.280$ and $+0.334$ v versus NHE, respectively.

*Silver-silver chloride electrode.* Another useful reference electrode is the silver-silver chloride electrode, shown in Figure 11–2. It is analogous to the calomel electrode, except that silver metal and silver chloride replace the mercury and mercurous chloride, respectively, and the pertinent half-reaction is

$$AgCl(s) + e \rightleftharpoons Ag(s) + Cl^-$$

Once again, the potential of this reference electrode depends solely on the concentration (activity) of chloride ion in equilibrium with solid silver chloride. Unlike the family of calomel reference electrodes, the silver-silver chloride electrode is usually prepared with saturated potassium chloride solution, its potential being $+0.197$ v versus NHE at $25°C$.

## Potential-Measuring Instruments

As discussed in Chapter 9, a meaningful value for the emf of a galvanic cell can be obtained only if virtually no current flows through the cell during the experimental measurement, because the passage of current through a galvanic cell has two adverse effects upon the magnitude of the observed emf. First, the concentrations (activities) of the potential-determining species at the surface of each electrode change as a consequence of current flow, thereby leading to an erroneous emf. Second, every cell possesses an internal resistance, $R$, and so the flow of a current, $i$, through the cell produces a voltage, $iR$, which, depending upon the direction of current flow, either opposes or reinforces the emf of the cell, making the observed reading either less than or greater than the true value.

A high input impedance vacuum-tube or field-effect-transistor voltmeter can be employed to follow the progress of a potentiometric titration, for which the change of emf of a galvanic cell is of primary concern and for which continuous observation of emf is desirable. Although we shall not discuss the behavior or circuitry of these devices, such instruments do allow accurate emf measurements without causing or permitting a significant current to flow through the cell. In fact, the current need not exceed $10^{-10}$ amp and may be much smaller. High input impedance voltmeters (with resistances typically of $10^{14}$ ohms) are usually incorporated into direct-reading pH meters.

Most reliable values for the emf of a galvanic cell are derived from measurements with a potentiometer, which was discussed in Chapter 9. Fundamentally, the precision of any potentiometric measurement depends on the sensitivity with which one can balance the known, variable emf against the emf of the galvanic cell. This sensitivity is governed by the device used to detect the tiny current that flows through the circuit when the two opposing emf values are not perfectly balanced. In Figures 9–3 and 9–4, the use of a galvanometer as a current-measuring instrument is depicted. Provided that the internal resistance of the galvanic cell and its associated circuitry does not exceed 1 megohm, it is feasible to employ a galvanometer for this purpose. However, for pH measurements, where the resistance of the glass membrane electrode itself may be 100 megohms or more, one must replace the galvanometer with a high-impedance voltage amplifier in order to detect very small differences between the two opposing emf values. Additional information about pH meters is presented later in this chapter.

## DIRECT POTENTIOMETRY

### PRINCIPLES AND TECHNIQUES OF DIRECT POTENTIOMETRY

Ideally, the direct potentiometric measurement of the emf of a galvanic cell can be related through the Nernst equation to the activity* of a single desired species, if the activities of all other substances in the system are known. Unfortunately, this ostensibly simple procedure is complicated by both practical and theoretical problems.

Suppose that we wish to analyze a dilute aqueous solution of silver nitrate by means of direct potentiometry. One approach is to construct a galvanic cell consisting of a saturated calomel reference electrode and a silver wire indicator electrode:

$$\text{Hg} \mid \text{Hg}_2\text{Cl}_2(s), \text{KCl}(s) \mid \text{KNO}_3 \ (1 \ F) \mid \text{AgNO}_3 \mid \text{Ag}$$

$$E_{\text{Hg},\text{Hg}_2\text{Cl}_2} \qquad E_\text{j} \qquad E_\text{j}' \qquad E_{\text{Ag}^+,\text{Ag}}$$

In addition, to prevent the precipitation of silver chloride, we may separate the calomel electrode from the silver nitrate solution by a salt bridge containing 1 $F$ potassium nitrate solution.

Now, the overall emf of the galvanic cell is the sum of the potentials of the reference and indicator electrodes, plus two liquid-junction potentials—one at the boundary between the saturated calomel electrode and the potassium nitrate salt bridge, and the other at the interface between the salt bridge and the silver nitrate solution:†

$$E_{\text{cell}} = E_{\text{Hg},\text{Hg}_2\text{Cl}_2} + E_\text{j} + E_\text{j}' + E_{\text{Ag}^+,\text{Ag}}$$

By utilizing a potentiometer, we can determine the overall emf of the galvanic cell with an error not exceeding 1 mv. If we extract from this measurement the value for the potential of the silver indicator electrode, the activity of silver ion may be computed from the relation

$$E_{\text{Ag}^+,\text{Ag}} = E^0_{\text{Ag}^+,\text{Ag}} - 0.059 \log \frac{1}{(\text{Ag}^+)}$$

To obtain $E_{\text{Ag}^+,\text{Ag}}$, it is necessary to subtract individual values for $E_{\text{Hg},\text{Hg}_2\text{Cl}_2}$, $E_\text{j}$, and $E_\text{j}'$ from $E_{\text{cell}}$. Obviously, any uncertainties or variations in the potential of the

---

* Although the Nernst equation involves activities of chemical species, concentrations can be computed if the pertinent activity coefficients are known.

† This four-term summation actually gives the emf (both sign and magnitude) for the cell reaction

$$2 \ \text{Hg} + 2 \ \text{Ag}^+ + 2 \ \text{Cl}^- \rightleftharpoons \text{Hg}_2\text{Cl}_2(s) + 2 \ \text{Ag}$$

However, recall (footnote on page 264) that the emf for a physically real galvanic cell has no sign, only a magnitude; therefore, if we are interested in the emf of this cell, it is correct to report only the *absolute* value obtained from the summation of the four terms. But the way we must connect a potentiometer to the cell to achieve a reading tells us the polarity of the indicator electrode; for example, the silver indicator electrode is positive with respect to the reference electrode if the positive lead of the potentiometer is attached to the indicator electrode at potentiometric balance.

reference electrode or the liquid-junction potentials will produce corresponding errors in the values of $E_{Ag^+,Ag}$ and the silver ion activity calculated from the Nernst equation. It is relatively easy to obtain a saturated calomel reference electrode whose potential is known to within 1 mv or less. What do we know about the liquid-junction potentials?

## Liquid-Junction Potentials

When two solutions containing dissimilar ions—or the same ions at different concentrations—are brought into contact with each other, a potential difference develops at the boundary between these two media. This **liquid-junction potential** arises because the various ionic species diffuse or migrate across the interface at characteristically different rates.

As a simple illustration of the phenomenon of a liquid-junction potential, let us examine what happens at the interface between a saturated $(4.2\,F)$ potassium chloride solution and a $1\,F$ potassium chloride medium. Schematically, this phase boundary may be depicted as

$$
\begin{array}{l}
\text{K}^+ \longrightarrow \\
\text{Cl}^- \longrightarrow \\
\phantom{xxxxxxxxxx} \text{K}^+ \\
\phantom{xxxxxxxxxx} \text{Cl}^- \\
(4.2\,F\,\text{KCl}) + \mid - \phantom{xx} (1\,F\,\text{KCl})
\end{array}
$$

Under the influence of a concentration gradient, potassium ions and chloride ions both diffuse from left to right across the boundary, but chloride ion is approximately 4 per cent more mobile than potassium ion. Therefore, the former outraces the latter, and the right side of the boundary acquires a small negative charge, whereas the left side of the liquid junction becomes slightly positive due to an accumulation of slower-moving potassium ions. Thus, it is the separation of charge which causes the liquid-junction potential.* However, this charge separation cannot increase beyond a certain equilibrium state. On the right side of the phase boundary, the tiny excess of negatively charged ions retards further migration of chloride ions and, simultaneously, accelerates the transport of potassium ion from left toward right to restore neutrality. Similarly, for the left side of the liquid junction, the positive charge hinders the movement of chloride ions and enhances the migration of potassium ions from left to right.

Liquid-junction potentials encompass a large range of values, depending upon the charges, mobilities, and concentrations of the ionic species and the nature of the solvent on each side of the boundary. Moreover, in many galvanic cells of analytical usefulness, the compositions of the solutions on either side of a liquid junction differ so much that the resulting liquid-junction potential cannot be predicted. At best, when the two opposing solutions are quite similar, the liquid-junction potential may be only 1 or 2 mv—a modest uncertainty which, as we shall see, seriously affects the accuracy of direct potentiometry. More often than not, liquid-junction potentials reach 10 mv, if not higher values. Unavoidable and unknown liquid-junction potentials constitute one of the major drawbacks to the analytical use of direct potentiometry.

---

* By convention, a liquid-junction potential is given a positive sign when the left-hand side of the junction is negative, and vice versa. For the present example, the liquid-junction potential is approximately $-1$ mv.

To reduce the liquid-junction potential between two solutions to a few milli-volts or less, it is common practice to interpose between these media a salt bridge containing a saturated $(4.2\,F)$ potassium chloride solution, because potassium and chloride ions have similar mobilities and tend to migrate across the liquid junctions at nearly equal rates. This procedure is successful only when the concentrations of all other ionic species at the solution boundaries are small compared to that of potassium chloride; otherwise, the liquid-junction potentials will remain relatively large.

## Calibration Techniques in Direct Potentiometry

Most analytical applications of direct potentiometry rely upon an empirical calibration technique which minimizes the effects of uncertainties in the potential of the reference electrode and in the liquid-junction potentials. Assume we are interested in the determination of silver ion in unknown silver nitrate solutions, and let us prepare a series of solutions containing known concentrations of silver ion. Each known solution must have the same composition and ionic strength as the anticipated unknown samples so that activity coefficients will be constant. Next, the known solutions are transferred, one by one, into a galvanic cell such as that described earlier, and the overall emf of the cell is measured potentiometrically for each solution. Finally, we construct a calibration curve, which is essentially a straight line, by plotting the overall emf of the galvanic cell as a function of the logarithm of the silver ion concentration.

When an unknown silver ion solution is subsequently placed in the galvanic cell, the observed emf can be translated into the concentration of silver ion by interpolation of the calibration curve. Although the reference-electrode potential and the liquid-junction potentials need not be known in this procedure, the values of these parameters must remain constant during the establishment of the calibration curve and the measurement of the emf of the galvanic cell containing the unknown solution. Note that the calibration technique permits direct determination of the analytical concentration of a species, whereas the potential of an indicator electrode is governed by the activity of that species.

## Method of Standard Addition

Another way to overcome the effects of uncertain liquid-junction potentials and changing activity coefficients on direct potentiometric measurements is the method of standard addition.

Suppose that we wish to determine the concentration of silver ion in a solution by using the electrochemical cell described earlier. First, we transfer an accurately known volume of the sample solution into the cell. Second, we measure the potential of the silver indicator electrode $(E_1)$ relative to the saturated calomel reference electrode. Third, we pipet a known volume of a standard silver ion solution into the sample solution contained in the cell. This standard solution should be relatively concentrated so that only a tiny volume needs to be added; in this way, there is almost no change at all in the composition and ionic strength of the original sample medium, so that the liquid-junction potential and activity coefficients remain essentially constant. Fourth, we measure the new potential of the silver indicator electrode $(E_2)$ versus the reference electrode. Then, knowing the change in the potential of the indicator electrode $(\Delta E = E_2 - E_1)$, the volume $(V_x)$ of the original sample solution, and the volume ($V_s$) and concentration ($C_s$) of the standard silver ion solution, we can calculate the concentration ($C_x$) of silver ion in the sample by means of the relation

$$C_x = C_s \left( \frac{V_s}{V_x + V_s} \right) \left[ 10^{n(\Delta E)/0.059} - \left( \frac{V_x}{V_x + V_s} \right) \right]^{-1}$$

Of the various terms in the preceding expression, it is $\Delta E$ (or the difference between $E_2$ and $E_1$) that is subject to the most uncertainty, because $C_s$, $V_s$, and $V_x$ can be determined very accurately. Errors in the measurement of $E_1$ and $E_2$ become more serious as $\Delta E$ decreases. Therefore, it is desirable to make $\Delta E$ as large as possible by the addition of a reasonably large volume ($V_s$) of the standard solution. However, if too great a volume of standard solution is added, the composition of the unknown medium may be altered and there may be uncertain changes in the liquid-junction potential and activity coefficients. To obtain the best results, one must seek a compromise. In the exercises at the end of this chapter, the theory and application of the method of standard addition will be treated.

## A Fundamental Limitation of Direct Potentiometry

Even when the greatest possible care is exercised, it is practically impossible to measure the potential of an indicator electrode (or the emf of a galvanic cell) with an accuracy better than 1 mv, although the measurement may be precise to within 0.1 mv or less. An uncertainty of 1 mv or more may arise from a number of sources, including a change in the potential of the reference electrode, an unknown liquid-junction potential, a variation in temperature, an undetected shift in the standardization of the potentiometer circuit, or an absence of electrochemical equilibrium at the surface of the indicator electrode.

If we wish to determine the activity (concentration) of a species by means of direct potentiometry, how seriously will the result be affected by a 1-mv error in the potential of the indicator electrode or in the overall emf of a galvanic cell? To answer this question, let us perform two calculations. Suppose that the *true* potential of a silver indicator electrode immersed in a silver ion solution at 25°C should be +0.592 v versus NHE. Taking the standard potential for the silver ion-silver metal half-reaction to be +0.800 v versus NHE, we may write

$$E_{Ag^+, Ag} = E^0_{Ag^+, Ag} - 0.05915 \log \frac{1}{(Ag^+)}$$

$$+0.592 = +0.800 - 0.05915 \log \frac{1}{(Ag^+)}$$

$$\log \frac{1}{(Ag^+)} = \frac{0.208}{0.05915} = 3.51_6$$

$$\log (Ag^+) = -3.51_6 = 0.48_4 - 4.000$$

$$(Ag^+) = 3.05 \times 10^{-4} \, M$$

Now let us assume that the potential of the silver indicator electrode immersed in the same sample solution was *observed* to be +0.593 v versus NHE, an error of only 1 mv. If the silver-ion activity is calculated as before, we obtain $(Ag^+) = 3.16 \times 10^{-4}$ $M$. Thus, a 1-mv uncertainty in the indicator-electrode potential leads to an error of just under 4 per cent in the activity of silver ion. Since possible uncertainties due to unknown liquid-junction potentials may easily exceed 1 mv, the attainment of accurate analytical results from direct potentiometry usually requires the use of an empirical calibration technique or the method of standard addition.

## APPLICATIONS OF DIRECT POTENTIOMETRY

During the last ten years, the commercial availability of a family of ion-selective electrodes responding to more than twenty different cations and anions—and the hundreds of novel and diverse applications of these ion sensors to many scientific and technological disciplines—has created a spectacular renaissance in the field of direct potentiometry. Among the virtues of direct potentiometry are that measurements can be performed on exceedingly small samples—much less than a milliliter—and that the sample solution is not altered or destroyed during the experimental measurements. In addition, potentiometry is suitable for the analysis of relatively dilute solutions.

Two of the more exciting applications of potentiometry lie in the areas of continuous, automated analysis and bioanalytical chemistry. It is possible to insert suitable indicator and reference electrodes directly into chemical systems which might otherwise be inaccessible. Thus, one can obtain information concerning the composition of these systems in the form of a readily recorded and easily utilized electrical signal—the emf of a galvanic cell—which can be employed to monitor pollution levels in industrial processes, or nuclear reactors, or other remote or dangerous facilities, and which, in conjunction with computer-controlled instrumentation, can be used to regulate many processes. Relatively recent development of reliable indicator electrodes sensitive to sodium, potassium, and calcium ions permits the study of such phenomena as binding of sodium ion in brain tissue, diffusion of potassium and sodium ions through nerve and muscle membranes, composition of cerebrospinal fluid, blood coagulation, and intestinal secretion.

### Determination of pH

In biochemical processes, in the rates of many organic and inorganic reactions, and in a wide variety of separations and measurements in analytical chemistry, the activity or concentration of hydrogen ion in a solution not only plays a critical role but also may vary over an enormous range. Consequently, the determination of pH remains one of the most important practical applications of direct potentiometry.

*Hydrogen electrode.* A logical indicator electrode for the measurement of pH is the hydrogen electrode, a convenient form of which is incorporated into the galvanic cell shown in Figure 9–6 (page 263). Although the potential of the hydrogen electrode exhibits a dependence on the hydrogen ion activity in accord with the Nernst equation, a hydrogen electrode can be used only in media which are free from extraneous oxidizing and reducing agents, inasmuch as these species affect the potential of the electrode. Furthermore, traces of numerous substances, including organic molecules, arsine, and hydrogen sulfide, tend to be adsorbed upon the electrode, causing it to perform sluggishly and erratically. These interferences, together with the hazard associated with the use of hydrogen gas, severely limit the applicability of the hydrogen electrode in analytical chemistry.

*Glass membrane electrode.* Of all the electrodes sensitive to hydrogen ions, the glass membrane electrode, or simply the glass electrode, is unique, because the mechanism of its response to hydrogen ion is totally different, involving an ion-exchange process rather than an electron-transfer reaction. Consequently, the glass electrode is not subject to interferences by oxidizing and reducing agents in the sample solution. Furthermore, the rapidity and accuracy with which the glass electrode responds to sudden changes in hydrogen ion activity make it the universal choice of indicator electrode for potentiometric acid-base titrations in both aqueous and non-aqueous media.

Glass electrodes may be purchased in a wide variety of sizes and shapes for such diverse purposes as the determination of the pH of blood and other biological fluids, the continuous measurement and recording of the hydrogen ion activity in flowing solutions, or the evaluation of the pH of a single drop or less of solution. However, the most familiar form of the glass membrane electrode is depicted in Figure 11–3. A thin-walled bulb, fabricated from a special glass highly sensitive to the hydrogen ion activity of a solution, is sealed to the bottom of an ordinary glass tube. Inside the glass bulb is a dilute aqueous hydrochloric acid solution, usually 0.1 $F$ in concentration. A silver wire coated with a layer of silver chloride is immersed into the hydrochloric acid medium, and the silver wire is extended upward through the resin-filled tube to provide electrical contact to the external circuit. Thus, the glass electrode comprises an inner silver-silver chloride reference electrode dipping into a dilute hydrochloric acid solution confined within a pH-sensitive glass bulb. Other inner reference electrodes, including the calomel electrode, are sometimes employed.

Most glasses employed for the construction of durable pH-sensitive membrane electrodes contain at least 60 weight per cent $SiO_2$, along with smaller amounts of oxides of the alkali metal and alkaline earth elements. For many years soda-lime glass, consisting by weight of approximately 72 per cent $SiO_2$, 22 per cent $Na_2O$, and 6 per cent $CaO$, was widely used. However, this formulation has been almost entirely replaced by several so-called lithia glasses containing $SiO_2$, $Li_2O$, and $CaO$, or $SiO_2$, $Li_2O$, and $BaO$, whose advantages will be discussed later.

By means of x-ray studies, the structure of a silicate glass has been characterized as an irregular, three-dimensional network in which each silicon atom is tetrahedrally bonded to four oxygen atoms, with each oxygen being shared by two silicon atoms. This framework contains interstitial holes which, in the case of lithia glass, are occupied by rather loosely held lithium ions and by more strongly bound caclium or barium ions.

*Figure 11–3.* Glass membrane electrode.

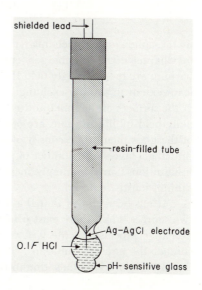

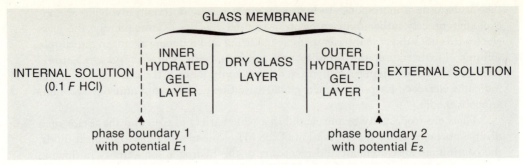

Figure 11–4.   Cross section of hydrated glass membrane electrode. (See text for discussion.)

In order to function usefully as a pH-sensitive electrode, the surfaces of a glass membrane must be hydrated, typically to the extent that 50 to 100 mg of water is absorbed per cubic centimeter of glass. When a glass electrode is exposed to water, solvent molecules penetrate the surface of the silicate lattice, causing it to swell, and after a time the cross-section of the glass membrane, including the internal and external solutions, is believed to have the general structure depicted in Figure 11–4. According to this picture, a well-soaked glass membrane consists of a middle layer of dry glass, perhaps 0.5 mm in thickness, sandwiched between inner and outer hydrated layers which are typically $10^{-4}$ mm thick. Indeed, the presence of water in these inner and outer layers is essential, as shown by the fact that a glass electrode begins to malfunction soon after it is immersed in a dehydrating medium such as concentrated sulfuric acid or anhydrous ethanol; however, the electrode is restored to normal behavior by being returned to water for a short time. Nonhygroscopic glasses such as Pyrex and Vycor, which are used to manufacture laboratory glassware, cannot be employed as pH-sensitive membranes.

As hydration of a pH-sensitive membrane takes place, lithium ions (or other singly charged cations) in the surface layers of the glass become aquated and can be exchanged for protons from the adjacent solution

$$Li^+_{glass} + H^+_{solution} \rightleftharpoons Li^+_{solution} + H^+_{glass}$$

by the same kind of mechanism that occurs with the cation-exchange resins discussed in Chapter 17. Multicharged cations in the silicate lattice, such as calcium and barium, are more strongly held than lithium ions and do not participate to any significant extent in the ion-exchange reaction. So complete is the ion-exchange process at the surfaces of the inner and outer hydrated layers of the glass that all sites originally held by lithium ions are occupied by hydrogen ions; thus, the exterior surfaces of the glass membrane consist of a gel of hydrated silicic acid. Deeper into the hydrated gel layers, the number of protons decreases progressively, while the number of lithium ions correspondingly increases. Within the dry-glass layer itself, none of the lithium ions is replaced.

It has been shown that the total potential across a glass membrane arises from two sources. First, there exist **phase-boundary potentials** associated with the ion-exchange processes at the *surfaces* of the inner and outer hydrated gel layers which contact the aqueous solutions. Second, there are **diffusion potentials**, similar to liquid-junction potentials, due to the differing mobilities of protons and lithium ions (or other alkali metal cations) *within* the inner and outer hydrated gel layers.* However, if protons fully saturate all ion-exchange sites on the exterior surfaces of the two hydrated gel layers, as occurs for a properly responding pH-sensitive glass electrode, and if the two surfaces of the gel layers are identical in their physical characteristics, the two diffusion potentials cancel each other. Therefore, the net potential across the glass membrane is the sum of two phase-boundary potentials, $E_1$ and $E_2$, shown in Figure 11–4; that is,

$$E_{\text{membrane}} = E_1 + E_2$$

Thermodynamic arguments, which will not be elaborated here, lead to the conclusion that each phase-boundary potential is governed in Nernstian fashion by the *ratio* of the hydrogen ion activity on the *right* side of the boundary to the hydrogen ion activity on the *left* side of the boundary. Accordingly,

$$E_{\text{membrane}} = -\frac{RT}{F} \ln \frac{(a_{\text{H}^+})_1'}{(a_{\text{H}^+})_1} - \frac{RT}{F} \ln \frac{(a_{\text{H}^+})_2}{(a_{\text{H}^+})_2'} = -\frac{RT}{F} \ln \frac{(a_{\text{H}^+})_1'(a_{\text{H}^+})_2}{(a_{\text{H}^+})_1(a_{\text{H}^+})_2'}$$

where $(a_{\text{H}^+})_1$ and $(a_{\text{H}^+})_2$ are the hydrogen ion activities of the *solutions* at phase boundaries 1 and 2, respectively, and where $(a_{\text{H}^+})_1'$ and $(a_{\text{H}^+})_2'$ correspond to the *surface* hydrogen ion activities in the hydrated gel layers at phase boundaries 1 and 2. Now, if the surfaces of the two hydrated gel layers are identical in the number of sites available to hydrogen ions, and if all sites are occupied by protons, $(a_{\text{H}^+})_1' = (a_{\text{H}^+})_2'$ and the latter expression assumes the simple form

$$E_{\text{membrane}} = -\frac{RT}{F} \ln \frac{(a_{\text{H}^+})_2}{(a_{\text{H}^+})_1}$$

or

$$E_{\text{membrane}} = -0.059 \log \frac{(a_{\text{H}^+})_2}{(a_{\text{H}^+})_1}$$

Thus, for these experimentally realizable conditions, the total phase-boundary potential depends only on the hydrogen ion activities of the two aqueous solutions in contact with the glass membrane.

If we let the activity of hydrogen ion on the left side of the glass membrane, $(a_{\text{H}^+})_1$, be constant—as is true for the inner compartment of the glass membrane electrode shown in Figure 11–3—the preceding equation may be rewritten as

$$E_{\text{membrane}} = k - 0.059 \log (a_{\text{H}^+})_2$$

where $k$ incorporates the hydrogen ion activity, $(a_{\text{H}^+})_1$, inside the bulb of the glass electrode. Finally, the activity of hydrogen ion, $(a_{\text{H}^+})_2$, in the solution on the right side of the membrane—which may be regarded as the sample solution to be analyzed—can be expressed in terms of the pH of that medium through the relation

$$E_{\text{membrane}} = k + 0.059 \text{ pH}$$

---

*  Diffusion occurs only within the hydrated gel layers themselves. Most definitely, the mode of functioning of a pH-sensitive electrode does *not* involve selective penetration of hydrogen ions all the way through the glass membrane, as was erroneously believed for many years until it was demonstrated that prolonged electrolysis in a cell containing (radioactive) tritium-labeled water on one side of a glass membrane did not result in any detectable transport of tritium through the membrane.

Thus, we can attribute any variation in the potential across a glass membrane electrode to a change in the phase-boundary potential at the interface between the outer hydrated gel layer and the adjacent solution. Accordingly, when a sample solution of unknown pH is brought into contact with the outer hydrated gel layer of a glass membrane electrode, the magnitude of the resulting phase-boundary potential can be correlated with a numerical value for the pH of that sample. We must now explore how measurements of pH are accomplished in a practical way.

*Practical pH measurements.* A glass membrane electrode in combination with the saturated calomel reference electrode provides the galvanic cell usually employed for practical pH measurements. We can describe such a cell in shorthand fashion as

$$\text{Ag} \left| \text{AgCl(s), HCl (0.1 } F\text{)} \left| \begin{array}{c} \text{glass} \\ \text{mem-} \\ \text{brane} \end{array} \right| \begin{array}{c} \text{sample} \\ \text{solution of} \\ \text{unknown pH} \end{array} \right| \text{Hg}_2\text{Cl}_2\text{(s), KCl(s)} \left| \text{Hg} \right.$$

Each vertical line in the cell representation denotes a phase boundary across which a potential develops. Therefore, the emf for this galvanic cell is composed of five parts: (1) the potential of the silver-silver chloride electrode, (2) the phase-boundary potential between the hydrochloric acid solution inside the glass electrode and the inner wall of the glass membrane, (3) the phase-boundary potential between the outer wall of the glass membrane and the solution of unknown pH, (4) the liquid-junction potential between the solution of unknown pH and the saturated potassium chloride solution of the calomel electrode, and (5) the potential of the saturated calomel electrode.

Included in the overall emf of the galvanic cell is the so-called **asymmetry potential** of the glass membrane. This asymmetry potential, perhaps 1 or 2 mv in magnitude, appears because the inner and outer hydrated gel layers differ in their absorptive power for water and in their ion-exchange capacity for hydrogen ions. These differences, which are often time-dependent, result from unequal strains in the two surfaces of the glass bulb due to nonuniform heating when the electrode is fabricated, from chemical etching and mechanical abrasion during electrode use, from coating of the outer surface of the membrane with grease films and adsorbable materials, and from severe dehydration of the glass.

Since the composition of the solution inside the glass electrode remains constant, and since the potential of the saturated calomel electrode is fixed, the first, second, and fifth sources of potential named above are invariant. Any shift in the observed emf of the galvanic cell, as solutions of different pH are transferred into the sample compartment, can be attributed to three effects—small variations in the liquid-junction and asymmetry potentials and, most importantly, a change in the phase-boundary potential at the interface between the outer wall of the glass membrane and the unknown sample solution.

Many careful experimental studies have led to the conclusion that the overall emf, $E_{cell}$, for such a cell is related to the pH of the sample solution by the expression

$$E_{cell} = K + 0.059 \text{ pH}$$

In this equation, $K$ includes the first, second, and fifth sources of potential listed earlier, plus the liquid-junction and asymmetry potentials. However the value of $K$ can never be precisely known because the liquid-junction potential and the asymmetry potential are both uncertain. *Therefore, all practical* pH *determinations necessarily involve a calibration procedure in which the pH of an unknown solution is compared to the pH of a standard buffer.* If we transfer a standard buffer into the galvanic cell and measure the emf of the cell, it follows that

$$(E_{cell})_s = K + 0.059 \; (pH)_s$$

where the subscript *s* pertains to the standard buffer. Similarly, if an unknown solution is present in the cell, we obtain

$$(E_{cell})_x = K + 0.059 \; (pH)_x$$

where the subscript *x* refers to the unknown solution. When the latter two relationships are combined, the result is

$$(pH)_x = (pH)_s + \frac{(E_{cell})_x - (E_{cell})_s}{0.059}$$

which has been adopted at the National Bureau of Standards as the *operational definition* of pH. Implicit in this definition are two assumptions, namely, that the pH of the standard buffer is accurately known and that $K$ has the same value when either the standard buffer or the unknown sample is present in the galvanic cell.

At the National Bureau of Standards, the pH values for standard buffers have been determined from emf measurements of galvanic cells *without* liquid-junction potentials. Although details of this procedure can be found in the specialized references at the end of this chapter, it can be stated here that the pH values of standard buffers are known with an accuracy of $\pm 0.005$ pH unit at 25°C. A list of accepted pH values for several standard buffer solutions is presented in Table 11–1. As stated, the value of $K$ must remain constant if the operational definition of pH is to be truly valid. However, liquid-junction potentials vary considerably for even small changes in solution composition. In practice, one attempts to overcome this difficulty by selecting a standard buffer with a pH as close as possible to that of the unknown sample. For example, if the unknown solution has a pH close to 4, it is preferable that the standard buffer be potassium acid phthalate, the latter having a pH of 4.01 at 25°C (Table 11–1).

**Table 11–1.  pH Values of Standard Buffers\***

| Temperature, °C | Potassium Tetroxalate, 0.05 $M$ | Potassium Hydrogen Tartrate, Saturated at 25°C | Potassium Acid Phthalate, 0.05 $M$ | $KH_2PO_4$, 0.025 $M$; $Na_2HPO_4$, 0.025 $M$ | Borax, 0.01 $M$ |
|---|---|---|---|---|---|
| 0 | 1.666 | — | 4.003 | 6.984 | 9.464 |
| 10 | 1.670 | — | 3.998 | 6.923 | 9.332 |
| 20 | 1.675 | — | 4.002 | 6.881 | 9.225 |
| 25 | 1.679 | 3.557 | 4.008 | 6.865 | 9.180 |
| 30 | 1.683 | 3.552 | 4.015 | 6.853 | 9.139 |
| 40 | 1.694 | 3.547 | 4.035 | 6.838 | 9.068 |
| 50 | 1.707 | 3.549 | 4.060 | 6.833 | 9.011 |

\* Values taken with permission from R. G. Bates: *Determination of pH*. John Wiley and Sons, New York, 1964, p. 76. The uncertainties in these values are about $\pm 0.005$ unit at 25°C, but somewhat larger at other temperatures.

Ultimately, the accuracy of pH measurements is dictated by the constancy of the parameter $K$ in the preceding equations. For optimum conditions, when the pH values of the standard buffer and unknown sample are essentially identical and when any variations in the liquid-junction potential are minimized as much as possible, the uncertainty in the emf of an appropriate galvanic cell can be reduced to approximately 1 mv. In turn, this corresponds to an error in the pH of the unknown solution of about $\pm 0.02$ pH unit. It is noteworthy that pH meters are available commercially from which one can read pH values with a *precision* of $\pm 0.003$ unit or better, a feature which is frequently useful for studying *changes* in the pH of a system under carefully controlled conditions. However, such precision must not be misconstrued, for the *accuracy* of the measurements is still no better than one or two hundredths of a pH unit.

**Errors in glass-electrode pH measurements.** Although the glass membrane electrode does not usually suffer from the interferences seen with most pH-sensitive electrodes, such as the hydrogen gas electrode, the former exhibits several peculiarities which limit its application in certain kinds of media.

In strongly basic media, or in moderately alkaline solutions containing high concentrations of alkali metal ions, glass membrane electrodes show a so-called **alkaline error**, in which the observed pH is *lower* than the true value by an amount which increases as the pH becomes greater. This error arises because cations other than hydrogen ion—particularly, sodium ion and, to a lesser extent, lithium and potassium ions—compete for the ion-exchange sites at the surface of the outer hydrated gel layer, thereby altering the phase-boundary potential as well as the diffusion potential. Careful investigations have demonstrated that the size of the alkaline error varies with the identity and concentration of the extraneous cation, with the temperature, and with the glass composition.

Soda-lime glass (22 per cent $Na_2O$, 6 per cent $CaO$, and 72 per cent $SiO_2$), for a long time the most popular material for the construction of pH-sensitive membrane electrodes, exhibits a correct hydrogen-ion response at room temperature between pH 1 and 9. However, in the presence of 1 $M$ sodium ion at $25°C$, the observed pH is *low* by 0.2 unit at pH 10, by 1.0 unit at pH 12, and by 2.5 units at pH 14. In general, the corresponding alkaline errors are less when 1 $M$ lithium ion is present, and smaller still for solutions containing 1 $M$ potassium ion. It has been discovered that the alkaline error is greatly reduced if sodium oxide in a glass is replaced with lithium oxide. For example, a glass composed of 10 per cent $Li_2O$, 10 per cent $CaO$, and 80 per cent $SiO_2$ by weight gives an alkaline error of less than 0.2 pH unit in the presence of 1 $M$ sodium ion at pH 13. Nowadays, almost all commercially available glass electrodes for pH measurements contain lithium oxide, providing quite accurate hydrogen-ion response up to nearly pH 13.*

At the other end of the pH scale, the glass electrode exhibits an **acid error** which is opposite from that observed in alkaline solutions. In very strongly acidic media, the observed pH is *higher* than the true value, but the cause of this behavior is unknown. Fortunately, the magnitude of the acid error is negligibly small down to at least pH 1, so this source of uncertainty is not often important in practical pH determinations.

Finally, a glass electrode does not respond rapidly when inserted into solutions which are poorly buffered, presumably because the rate of attainment of ion-exchange equilibrium in the outer hydrated gel layer is slow.

---

* Lithium-oxide glasses do exhibit a significant alkaline error in the presence of lithium ion, but the latter is a very uncommon constituent of aqueous solutions compared to either sodium or potassium ion.

***pH meters.*** One of the characteristics of a glass membrane is its extra-ordinarily high resistance, which ranges from 15 to 200 megohms for a typical electrode. This property adds another problem to the experimental measurement of pH.

Recall that all galvanic cells have a certain internal resistance, $R$, that a tiny current, $i$, inevitably flows through the cell during an emf measurement, and that the product of these two quantities, $iR$, is a voltage which opposes or adds to the emf of a cell. Thus, if one does not wish to exceed a 1-mv error in potentiometric measurements, the quantity $iR$ must be kept at or below this value. For galvanic cells employed in pH measurements, virtually all the internal resistance of the cell is provided by the glass membrane. If the largest permissible value of $iR$ is 1 mv and if $R$ is taken to be 50 megohms, it follows that the maximal current which can flow through a galvanic cell during the determination of pH is $2 \times 10^{-11}$ amp.* Unfortunately, the minimal detectable deflection of an ordinary galvanometer, such as that in the potentiometer circuit of Figure 9–4 (page 261), corresponds to a current which is perhaps $10^{-9}$ amp, or 50 times greater than the acceptable value. If one balances a potentiometric circuit but cannot detect any difference in the final potentiometer setting while a current of $10^{-9}$ amp flows through the cell, the error in the emf measurement may be as large as 50 mv, and the pH of the solution can be uncertain by nearly one pH unit. For these reasons, an ordinary potentiometer cannot be used for pH determinations; however, two kinds of pH meters are employed.

A *direct-reading* pH meter is essentially a high input impedance voltmeter in which the emf of the galvanic cell is measured and converted to a proportional current which is passed through an ammeter whose dial face is calibrated directly in pH units. A *potentiometric* or *null-detector* pH meter incorporates a relatively straightforward potentiometer circuit, except that the usual galvanometer is replaced with a vacuum-tube or solid-state amplifier so that currents as small as $10^{-12}$ amp can be readily detected at potentiometric balance. A direct-reading pH meter is especially convenient for use in potentiometric acid-base titrations, where the variation of pH as a function of added titrant is more important than any individual value of pH. Direct-reading pH instruments are usually accurate to $\pm 0.1$ pH unit. On the other hand, null-detector meters yield pH values with accuracies of 0.01 to 0.02 pH unit, and consequently they are preferable for precise measurements.

## Ion-Selective Electrodes for the Determination of Other Cations and Anions

For a long time, the analytical uses of direct potentiometry were limited to the determination of pH, primarily because of the lack of sensitive and selective indicator electrodes for other ions. However, recent years have witnessed the development of a number of interesting ion-selective electrodes. Such electrodes are finding important applications in the monitoring of industrial operations, in water analysis, in oceanography, in medical diagnosis, in the measurement of pollution, and in the study of biochemical systems.

Currently, there are perhaps four broad classes of ion-selective indicator electrodes—glass membrane electrodes (for cations other than hydrogen ion), liquid ion-exchange membrane electrodes, solid-state electrodes, and heterogeneous electrodes. All these electrodes respond, as does the pH-sensitive glass electrode, to changes in the activity of a species in the same oxidation state but in two different

---

* Conduction of an electrical current through the glass membrane itself occurs by the movement (jumping) of lithium ions within the silicate network.

phases separated by a membrane, although the physical form of the membrane may vary. Contrast this behavior to that of a platinum indicator electrode responding to changes in the ratio of activities of two different oxidation states of some element in a single phase (*e.g.*, a solution containing $Fe^{2+}$ and $Fe^{3+}$) or to that of a silver indicator electrode responding to changes in the activity of silver ion in solution; these systems involve electron transfer between two oxidation states of the same element.

**Glass electrodes for the determination of cations.** From early research aimed at minimizing the alkaline error in pH measurements through the development of lithium-oxide glass came knowledge that has led to the fabrication of glass membrane electrodes which respond selectively to a number of cations. As long ago as

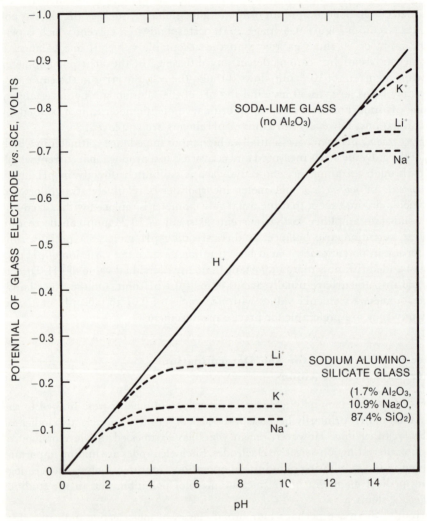

*Figure 11–5.* Effect of aluminum oxide on the cation response of glass electrodes. Solid, diagonal line represents behavior of a glass electrode responding solely to protons. Dashed lines in upper right corner pertain to behavior of ordinary soda-lime glass with no added $Al_2O_3$ in solutions containing 0.1 $M$ concentrations of indicated cations. Dashed lines in lower left corner show behavior of sodium aluminosilicate glass in solutions containing 0.1 $M$ concentrations of specified ions. (Redrawn, with permission, from C. N. Reilley, ed.: *Advances in Analytical Chemistry and Instrumentation.* Volume 4, John Wiley and Sons, New York, 1965, p. 219.)

1925, it was discovered that the addition of small percentages of $Al_2O_3$ or $B_2O_3$ to pH-sensitive glass results in much larger alkaline errors—in other words, enhanced response to alkali metal ions—than are observed with ordinary soda-lime glass. Yet, thirty years elapsed before careful investigations confirmed that electrodes made from glass composed of $Na_2O$, $Al_2O_3$, and $SiO_2$ are sensitive to sodium ion as well as the other alkali metal cations.

Figure 11–5 reveals the effect of aluminum oxide on the response of a glass membrane electrode. If a glass electrode responded perfectly to hydrogen ions throughout the usual pH range, the potential of the electrode would vary linearly with pH, as indicated by the diagonal solid line in Figure 11–5. An electrode manufactured from conventional soda-lime glass exhibits the expected linear hydrogen-ion response up to almost pH 10, beyond which point there occur deviations, or alkaline errors, due to the interference of alkali metal cations; sodium ion is the greatest offender, followed by lithium and potassium ions. However, a glass membrane electrode composed of 1.7 mole per cent $Al_2O_3$, 10.9 mole per cent $Na_2O$, and 87.4 mole per cent $SiO_2$ behaves quite differently; in very strongly acidic media, a normal pH response is observed, but the electrode becomes noticeably sensitive to 0.1 $M$ sodium or potassium ion above pH 2 and to 0.1 $M$ lithium ion above pH 4. For equal concentrations of hydrogen ion and an alkali metal cation, the $Al_2O_3$-containing glass electrode is inherently more sensitive to the former species, but by raising the pH one can enhance the selectivity of such an electrode toward the alkali metal ion. Between pH 5 and 6, the dashed lines near the bottom of Figure 11–5 become horizontal, signifying that the sodium aluminosilicate glass no longer responds to protons, but only to the alkali metal ions. Although the properties of the particular sodium aluminosilicate glass depicted in Figure 11–5 are not optimal, the ion-exchange sites in the outer hydrated gel layer of the electrode do exhibit a preference for sodium over lithium ions.

It has been established that sodium silicate glasses, such as the soda-lime variety, contain so-called $SiO^-$ exchange sites which have a high electrostatic field strength and a strong affinity for hydrogen ions. On the other hand, sodium aluminosilicate glasses possess what may be termed $AlOSi^-$ sites with a weaker electrostatic field strength and a marked preference for cations other than protons.

By suitable variations in the composition of sodium aluminosilicate glasses and related materials, glass membrane electrodes have been developed for the determination of $Li^+$, $Rb^+$, $Cs^+$, $Tl^+$, $NH_4^+$, $Na^+$, $K^+$, and $Ag^+$, the latter three being available commercially. Table 11–2 presents information about some of the most useful glass membrane electrodes. Included in the data are selectivity ratios,* which measure the seriousness of interferences. For example, the electrode for the determination of lithium ion has a selectivity constant $K_{Li^+/Na^+}$ of approximately 3, so that lithium ion is only three times more effective than sodium ion in governing the electrode potential. However, the same glass electrode has better than a 1000-fold preference for lithium ion over potassium ion; that is, $K_{Li^+/K^+} > 1000$.

Analytical uses of cation-selective glass electrodes are impressive in both number and scope. Such electrodes have been employed for potentiometric titrations, for investigations of activity coefficients, for measurements of equilibrium constants, for

---

* Selectivity ratios depend on the nature and composition of the solution being investigated. Values quoted in this chapter are primarily intended to give semiquantitative information about the preferential response of an electrode to equal concentrations of two different ions. Some authors prefer to report the selectivity ratio as a number less than 1; for example, the selectivity ratio ($K_{Li^+/Na^+}$) for the lithium electrode in Table 11–2 might be written as 1/3 or 0.33. For any particular situation, it is necessary to identify which form of expression is meant. Most likely, this source of ambiguity will be resolved in the near future by one of the committees of the International Union of Pure and Applied Chemistry (IUPAC).

**Table 11–2. Compositions and Characteristics of Some Cation-Selective Glasses***

| Cation of Interest | Glass Composition (mole per cent) | Selectivity Ratio** | Characteristics |
|---|---|---|---|
| Li+ | 15% $Li_2O$, 25% $Al_2O_3$, 60% $SiO_2$ | $K_{Li^+/Na^+} \approx 3$ <br><br> $K_{Li^+/K^+} > 1000$ | Best for Li+ in presence of H+ and Na+ |
| Na+ | 11% $Na_2O$, 18% $Al_2O_3$, 71% $SiO_2$ | $K_{Na^+/K^+} \approx 2800$ at pH 11 <br><br> $K_{Na^+/K^+} \approx 300$ at pH 7 | Nernstian response down to approximately $10^{-5} M$ Na+ |
| K+ | 27% $Na_2O$, 5% $Al_2O_3$, 68% $SiO_2$ | $K_{K^+/Na^+} \approx 20$ | Nernstian response down to somewhat less than $10^{-4} M$ K+ |
| Ag+ | 11% $Na_2O$, 18% $Al_2O_3$, 71% $SiO_2$ | $K_{Ag^+/Na^+} > 1000$ | Less selective but more reproducible than glass containing 28.8% $Na_2O$, 19.1% $Al_2O_3$, and 52.1% $SiO_2$ with $K_{Ag^+/H^+} \approx 100,000$ |

\* Data taken with permission from G. A. Rechnitz: Chem. Eng. News *45*:146, June 12, 1967.

\*\* See footnote on page 369.

continuous analysis, and for kinetics studies. Due to the availability of glass electrodes sensitive to sodium and potassium ions, the perfection of special miniature and flow-through electrodes for the determination of these two cations, and the physiological importance of these two species, some of the most interesting applications are in the field of biomedical analysis. Thus, it is feasible to measure the activities of sodium and potassium ions in urine, serum, cerebrospinal fluid, whole blood, plasma, bile, brain cortex, kidney tubules, and muscle fibers. In many situations, the accuracy of the results is comparable to, if not better than, that obtainable by means of flame photometry; often, measurements can be accomplished more quickly with glass electrodes. Cystic fibrosis, characterized by abnormally high sodium levels in sweat, can be diagnosed through determination of the sodium ion activity on the skin surface. There are numerous practical uses of sodium- and potassium-selective glass electrodes for the analysis of water and soil extracts. Since the roster of cation-selective glass electrodes will undoubtedly expand in the future, many new applications should appear.

One recent development entails the use of a glass electrode (27 mole per cent $Na_2O$, 4 mole per cent $Al_2O_3$, and 69 mole per cent $SiO_2$) responsive to ammonium ion for the determination of enzymes and their corresponding substrates. For example, urease is an enzyme which catalyzes the hydrolysis of urea to ammonium ion,

$$CO(NH_2)_2 + H_2O + 2\ H^+ \xrightarrow{\text{urease}} CO_2 + 2\ NH_4^+$$

the latter being measured potentiometrically to obtain the original quantity of urea. A similar procedure may be followed for the analysis of solutions containing glutamine, asparagine, glutamic acid, as well as various amines. Alternatively, the

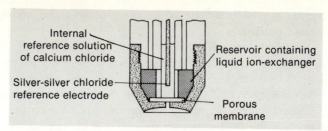

*Figure 11–6.* Schematic diagram of calcium-selective liquid ion-exchange membrane electrode.

ammonium-ion electrode may be employed as an enzyme-sensing device. Suppose that one seals a cellophane membrane around the glass bulb of the electrode, and injects a solution of urea into the annular space between the cellophane and glass bulb. If this electrode system is immersed into an unknown urease solution, urea diffuses outward through the cellophane and reacts with urease to form ammonium ion. Then, some of the ammonium ions diffuse through the cellophane membrane and are detected by the electrode. A calibration curve can be constructed so that the response of the electrode may be translated into the amount of urease present.

*Liquid ion-exchange membrane electrodes.* If a porous membrane is impregnated with a liquid ion-exchange substance, and if this membrane is placed between two different solutions containing a species which can bond with the ion exchanger, a phase-boundary potential develops at each solution-membrane interface, and diffusion potentials arise inside the membrane. Most significant from the viewpoint of chemical analysis is that the potential across the membrane depends on the activities of the desired ion in the solutions on either side of the membrane. If the activity of the ion on one side of the membrane remains constant, variations in the membrane potential reflect changes in the ion activity in the second solution. This behavior, which mimics that of glass membrane electrodes, has permitted useful liquid ion-exchange membrane electrodes to be devised.

A liquid ion exchanger consists of a water-immiscible solvent containing a high-molecular-weight organic solute with acidic or basic functional groups, or perhaps chelating properties, which interacts strongly and rather selectively with the ion to be determined. A number of liquid ion-exchange materials are known, some which combine with cations and others which react with anions. Several of these substances have been successfully incorporated into ion-selective electrodes.

Figure 11–6 illustrates a commercially available electrode designed for the determination of calcium ion. A porous, plastic filter-membrane is held in contact with a reservoir filled with the liquid ion exchanger—a solution of calcium didecylphosphate in di-*n*-octylphenyl phosphonate—so that the membrane becomes permeated by the liquid cation exchanger. This membrane separates the solution to be analyzed from the internal compartment of the electrode. In the internal compartment is a calcium chloride solution of fixed concentration into which dips a silver-silver chloride reference electrode. It can be shown that the overall emf of a galvanic cell in which the calcium-selective electrode and a saturated calomel reference electrode are in electrolytic contact with a calcium-ion solution follows the equation

$$E_{\text{cell}} = K + \frac{0.059}{2} \, \text{pCa}$$

where $K$ includes the potentials of the calomel electrode and the inner silver-silver chloride electrode as well as the inner phase-boundary potential and any liquid-junction potentials.

A calcium-selective electrode obeys the preceding relation for calcium-ion activities as low as $10^{-5}$ $M$ and exhibits marked specificity for calcium ion in the presence of strontium, magnesium, barium, sodium, and potassium ions. However, this electrode cannot be employed in solutions of pH 11 or higher, owing to the precipitation of calcium hydroxide, nor in media with pH values much below 6 because protons compete with calcium ions for the cation-exchange substance.

Calcium-ion activity in living organisms is known to affect many physiological processes, including blood coagulation, enzyme function, nerve conduction, bone formation, intestinal secretion and absorption, hormonal release from endocrine glands, and cerebral function. Thus, it is not surprising that some of the most important applications of the calcium-selective electrode are in the areas of biomedical research and clinical medicine.

A liquid ion-exchange membrane electrode has been developed for the measurement of potassium-ion activity. Closely resembling the calcium-selective electrode in design, this electrode utilizes a dilute solution of valinomycin in diphenyl ether as the liquid ion exchanger. As shown in Figure 11–7, valinomycin (an antibiotic) is an uncharged, cyclic macromolecule with a high affinity for potassium ion, but not sodium ion. This sensor exhibits a selectivity for $K^+$ over $Na^+$ of approximately 13,000 to 1, and for $K^+$ over either $Ca^{2+}$ or $Mg^{2+}$ of better than 5,000 to 1. In addition, a Nernstian response to potassium-ion activities ranging from $10^{-6}$ to 0.1 $M$ is observed. Thus, the valinomycin electrode is vastly superior to any available glass membrane electrode for the determination of potassium in urine, serum, kidney dialyzates, or any other sample in which appreciable amounts of sodium ion are present.

An electrode similar to the potassium-selective valinomycin sensor has been prepared for the measurement of ammonium ion in the presence of calcium ion and the alkali metal cations. It employs a mixture of nonactin and monactin (which are macrocyclic antibiotics) dissolved in tris(2-ethylhexyl)phosphate as the liquid ion exchanger.

A liquid ion-exchange membrane electrode, responding almost equally well to calcium and magnesium ions, has been devised for the monitoring of water hardness.

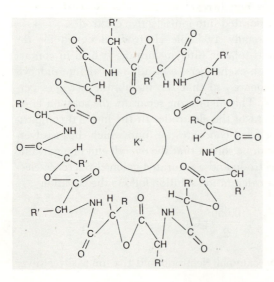

*Figure 11–7.* Structure of the valinomycin-potassium ion complex (R = —CH$_3$, R′ = —CH(CH$_3$)$_2$). Potassium ion fits into the central opening of the valinomycin molecule, the six carbonyl oxygen atoms which coordinate with $K^+$ being at the corners of a regular octahedron. Further details, including drawings of the conformation of the complex, may be found in V. T. Ivanov, *et al.*: Biochem. Biophys. Res. Communs., *34*:803, 1969, and in D. F. Mayers and D. W. Urry: J. Amer. Chem. Soc., *94*:77, 1972.

Other electrodes, which contain chelating ion-exchange substances with the general formulas $(R—S—CH_2COO)_2Cu$ and $(R—S—CH_2COO)_2Pb$, are sensitive to copper(II) and lead(II), respectively. Liquid membrane electrodes employing substituted 1,10-phenanthroline complexes of iron(III) and nickel(II) as anion exchangers can be purchased for the determination of $BF_4^-$, $ClO_4^-$, and $NO_3^-$. An electrode utilizing the salt of a high-molecular-weight tetraalkylammonium cation serves as a chloride sensor. Properties of these liquid ion-exchange membrane electrodes are compiled in Table 11–3.

Table 11–3.  Characteristics of Liquid Ion-Exchange Membrane Electrodes*

| Ion Determined | Measurement Range | pH Range for Use | Interferences | Applications |
|---|---|---|---|---|
| $Ca^{2+}$ | pCa = 0 to 5 | 5.5 to 11 | $Zn^{2+}$, $Fe^{2+}$, $Pb^{2+}$ | biological fluids, EDTA titrations, water analysis |
| $Cu^{2+}$ | pCu = 1 to 5 | 4 to 7 | $Fe^{2+}$, $Ni^{2+}$, $Zn^{2+}$ | |
| $Pb^{2+}$ | pPb = 2 to 5 | 3.5 to 7.5 | $Cu^{2+}$, $Fe^{2+}$, $Zn^{2+}$, $Ni^{2+}$ | |
| $BF_4^-$ | pBF$_4$ = 1 to 5 | 2 to 12 | $I^-$, $NO_3^-$, $Br^-$ | analysis of alloys, semiconductors, paints, and pigments after conversion of boron to $BF_4^-$ by treatment with HF |
| $Cl^-$ | pCl = 0 to 5 | 2 to 10 | $ClO_4^-$, $I^-$, $NO_3^-$, $Br^-$, $OH^-$, $HCO_3^-$, $SO_4^{2-}$, $CH_3COO^-$ | foods, pharmaceuticals, water analysis, soaps, especially useful in solutions of sulfide or strong reductants |
| $ClO_4^-$ | pClO$_4$ = 1 to 5 | 4 to 10 | $OH^-$, $I^-$ | potentiometric titration of $ClO_4^-$ with tetraphenylarsonium chloride |
| $NO_3^-$ | pNO$_3$ = 1 to 5 | 2 to 12 | $ClO_4^-$, $I^-$, $ClO_3^-$, $Br^-$, $S^{2-}$ | fertilizers, soils, foods, plants, explosives, water supplies |

* Data taken in part from Orion Research, Inc., *Analytical Methods Guide*, fourth edition, October, 1972.

*Solid-state membrane electrodes.*   Ionically conducting materials—single crystals, mixed crystals, and polycrystalline solids—have been employed for the construction of a variety of solid-state membrane electrodes for the direct potentiometric determination of cations and anions. A comparison of the design features of these solid-state sensors with the more familiar glass membrane electrode is provided in Figure 11–8.

Crystalline lanthanum fluoride $(LaF_3)$ has a high electrical conductivity, attributable to the exceptional mobility of fluoride ion in the solid lattice. Thus, a synthetic, *single crystal* of lanthanum fluoride, doped with europium(II) to enhance its conductivity, makes an excellent solid-state membrane electrode which responds selectively to fluoride ion over a range of activity from $10^{-6}$ to 1 $M$. Within the internal compartment of the commercially available model is a solution containing 0.1 $F$

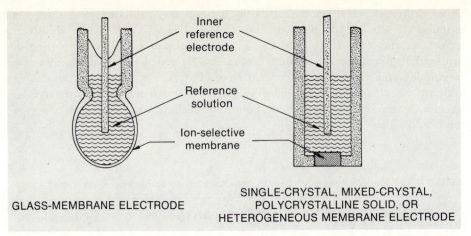

*Figure 11–8.* Comparison between the construction of a glass-membrane electrode and the various solid-state and heterogeneous membrane electrodes.

sodium fluoride and $0.1\,F$ sodium chloride along with a silver-silver chloride reference electrode. Hydroxide ion is the only significant interferent to the use of the electrode for fluoride measurements, whereas the electrode exhibits at least a thousand-to-one preference for fluoride over chloride, bromide, iodide, nitrate, bicarbonate, and sulfate ions. In acidic media, fluoride is converted to hydrofluoric acid, to which the electrode is insensitive. There are a number of applications for the fluoride-selective electrode. It has been utilized for continuous monitoring of fluoride levels in drinking water. In pollution control, the electrode is useful for the determination of fluoride in air and stack-gas samples. Fluoride in toothpaste, pharmaceuticals, urine, saliva, dissolved bone and teeth, and multiple vitamins is measurable. Among the industrial uses of the electrode are fluoride analyses of chromium electroplating baths, plastics, pesticides, and fertilizers.

Solid-state electrodes responsive to cadmium, copper, and lead ions have been prepared from *mixed-crystal* membranes consisting of silver sulfide to which is added CdS, CuS, or PbS, respectively. Electrodes selective toward thiocyanate, chloride, bromide, and iodide can be fabricated if silver sulfide containing finely divided and well dispersed AgSCN, AgCl, AgBr, or AgI is pressed into the form of a disc or pellet, and mounted into the bottom of a glass tube as shown in Figure 11–8. Moreover, the silver iodide-silver sulfide mixture provides a solid-state membrane electrode suitable for the measurement of cyanide ion. By itself, *polycrystalline* silver sulfide may be subjected to conventional pellet-pressing techniques for the preparation of a solid-state electrode which responds to both sulfide and silver ions. In addition, the latter is a useful indicator electrode for potentiometric titrations of halide mixtures or of cyanide with standard silver nitrate solution. Some analytical applications of these various electrodes, as well as substances that interfere with their response, are listed in Table 11–4.

One interesting use of the cyanide-selective electrode involves the indirect determination of amygdalin,

$$\text{CN}$$
$$|$$
$$\text{CH}$$
$$|$$
$$\text{OC}_{12}\text{H}_{21}\text{O}_{10}$$

**Table 11–4. Characteristics of Solid-State and Heterogeneous Membrane Electrodes***

| Ion Determined | Measurement Range | Interferences | Applications |
|---|---|---|---|
| | | SOLID-STATE ELECTRODES | |
| $F^-$ | $pF = 0$ to $6$ | $OH^-$ ($pH < 8.5$ for $pF = 6$, $pH < 11$ for $pF = 2$) | water supplies, plating baths, toothpaste, pharmaceuticals, teeth and bone |
| $Cd^{2+}$ | $pCd = 1$ to $7$ | $Fe^{3+}$, $Pb^{2+}$, $Ag^+$, $Hg^{2+}$, $Cu^{2+}$ | water, industrial wastes, plating baths, paper, and pigments |
| $Cu^{2+}$ | $pCu = 0$ to $8$ | $S^{2-}$, $Ag^+$, $Hg^{2+}$, $Cl^-$, $Br^-$, $Fe^{3+}$, $Cd^{2+}$ | plating baths, water supplies, sewage, ore refining, fungicides |
| $Pb^{2+}$ | $pPb = 1$ to $7$ | $Cd^{2+}$, $Ag^+$, $Hg^{2+}$, $Cu^{2+}$, $Fe^{3+}$ | petroleum products, foods, air pollution, EDTA titration of lead, titration of $SO_4^{2-}$ with lead |
| $CN^-$ | $pCN = 2$ to $6$ | $S^{2-}$, $I^-$ | industrial wastes, plating baths, determination of amygdalin |
| $SCN^-$ | $pSCN = 0$ to $5$ | $I^-$, $Cl^-$, strong reductants, species that form insoluble silver salts or stable silver complexes | |
| $Cl^-$ | $pCl = 0$ to $4.3$ | $S^{2-}$, $I^-$, $Br^-$, $CN^-$ | foods, pharmaceuticals, soaps, plastics, glass, water analysis |
| $Br^-$ | $pBr = 0$ to $5.3$ | $S^{2-}$, $I^-$, $CN^-$ | titration of epoxy groups with HBr |
| $I^-$ | $pI = 0$ to $7.3$ | $S^{2-}$, $CN^-$ | |
| $S^{2-}$ | $pS = 0$ to $17$ | (none) | industrial wastes, paper and pulp industries |
| $Ag^+$ | $pAg = 0$ to $17$ | $Hg^{2+}$ | titration of mixed halides or of $CN^-$ with silver ion |
| | | HETEROGENEOUS ELECTRODES | |
| $I^-$ | $pI = 1$ to $7$ | $pCl$, $pBr$, $pSO_4$, and $pNO_3$ must be greater than 1 | titration of mixed halides; electrode can be used to determine silver ion ($pAg = 1$ to $5$) and cyanide ($pCN = 1$ to $5$) |
| $Br^-$ | $pBr = 1$ to $6$ | $pCl$, $pSO_4$, and $pNO_3$ must be greater than 1 | |
| $Cl^-$ | $pCl = 1$ to $5$ | $pSO_4$ and $pNO_3$ must be greater than 1 | urine and serum analyses |
| $S^{2-}$ | $pS = 1$ to $17$ | $CN^-$ | wastes from paper pulp production, leather industry |

* Data taken from Orion Research, Inc., *Analytical Methods Guide*, fourth edition, October, 1972, and from A. K. Covington: Chem. Brit., *5*:388, 1969.

a glucoside found in the seed of the bitter almond. In the presence of the enzyme β-glucosidase, hydrolysis of amygdalin to hydrogen cyanide, benzaldehyde, and glucose occurs according to the reaction

$$C_6H_5CH(CN)OC_{12}H_{21}O_{10} + 2 H_2O \xrightarrow{\beta-glucosidase} HCN + C_6H_5CHO + 2 C_6H_{12}O_6$$

If a solid-state, cyanide-selective electrode is coated with a film of acrylamide gel containing β-glucosidase, and if the electrode is immersed into a solution of amygdalin, the latter reacts with the enzyme to produce HCN, which can be measured potentiometrically as cyanide.

**Heterogeneous membrane electrodes.** Anion-selective electrodes have been developed in which an insoluble precipitate containing the desired anion is imbedded in an inert, solid binder material. For example, one can prepare an electrode for iodide ion by causing monomeric silicone rubber to polymerize in the presence of an equal weight of silver iodide particles. Silicone rubber provides a flexible, heterogeneous membrane with resistance to cracking and swelling; however, the individual particles of precipitate must touch each other so that the membrane is ionically conducting. After the silicone rubber-precipitate mixture solidifies, it is cut into discs, approximately 0.5 mm in thickness. Each disc is sealed to the bottom of a glass tube; then a potassium iodide solution and a silver wire are placed in the tube. Similar electrodes have been fabricated from silver chloride, silver bromide, and silver sulfide, and properties of these heterogeneous membrane electrodes are included in Table 11–4.

There has been some success in the preparation of other heterogeneous membrane electrodes. A barium sulfate-impregnated electrode responds to changes in sulfate-ion activity over the range from $10^{-6}$ to 0.1 $M$. Efforts have been directed toward the development of a phosphate-selective electrode which incorporates bismuth phosphate. Both lanthanum fluoride and calcium fluoride appear to be suitable for the fabrication of electrodes sensitive to fluoride ion. Cation sensors made from heterogeneous membranes containing ion-exchange resins can be prepared. Several investigators have tested a nickel-selective electrode consisting of a membrane imbedded with bis(dimethylglyoximato)nickel(II).

## POTENTIOMETRIC TITRATIONS

We may characterize a potentiometric titration as one in which the change of the emf of a galvanic cell during a titration is recorded as a function of added titrant. Although the major aim of this procedure is the precise location of the equivalence point, thermodynamic information, including dissociation constants for weak acids and formation constants for complex ions, may be obtained from potentiometric titration curves.

Compared to other methods for the location of equivalence points, the technique of potentiometric titrimetry offers a number of advantages. It is applicable to systems which are so brightly colored that visual methods of end-point detection are useless, and it is especially valuable when no internal chemical indicator is available. Moreover, it eliminates subjective decisions concerning color changes of end-point indicators as well as the need for indicator blank corrections.

### TECHNIQUES OF POTENTIOMETRIC TITRIMETRY

Equipment required for the performance of potentiometric titrations is relatively simple, as shown in Figure 11–9. A titration vessel, a buret containing standard

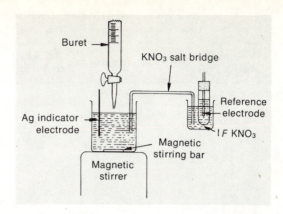

*Figure 11–9.* Apparatus for potentiometric titration of chloride with silver nitrate solution.

Buret

KNO$_3$ salt bridge

Ag indicator electrode

Reference electrode

1 F KNO$_3$

Magnetic stirring bar

Magnetic stirrer

titrant, and suitable indicator and reference electrodes are the necessities, and there should be provision for efficient stirring of the solution. Usually, the titrant can be added rather rapidly at the beginning of the titration until the equivalence point is approached. Thereafter, until the equivalence point is passed, one should introduce the titrant in small increments and should wait for equilibrium to be established before adding more reagent. After the addition of each increment, sufficient time must be allowed for the indicator-electrode potential to become reasonably steady, so that it exhibits a drift of perhaps not more than a few millivolts per minute.

In contrast to direct potentiometry, the measurement of the emf of a galvanic cell does not require the highest possible accuracy, because only the rather large *changes* in the potential of an indicator electrode near the equivalence point of a titration have significance. It is unnecessary to know the potential of the reference electrode or the magnitudes of any liquid-junction potentials.

## Determination of Equivalence Points

A typical set of potentiometric-titration data for the precipitation reaction involving chloride and silver ions is presented in Table 11–5. In the first column is a list of buret readings, whereas the second column shows the potential of a silver indicator electrode, measured with respect to the saturated calomel electrode (SCE), corresponding to each volume reading. Notice that the volume increments are large near the beginning of the titration when the potential of the indicator electrode varies only slightly. However, in the region of the equivalence point, *small* and *equal* volume increments are introduced. Previous experience with a particular titration alerts one to the fact that the equivalence point is being approached. Any of several methods may be employed for determination of the equivalence point of a potentiometric titration.

*Graphical methods.* One procedure involves visual inspection of the complete titration curve. If one plots the potential of the indicator electrode versus the volume of titrant, the resulting titration curve exhibits a maximal slope—that is, a maximal value of $\Delta E/\Delta V$—which may be taken as the equivalence point. Figure 11–10A illustrates this approach with the experimental data in Table 11–5, although only that portion of the titration curve near the equivalence point is pictured. For sharply defined titration curves, the uncertainty in the method is small, but considerable trouble may be encountered in less favorable situations.

An extension of the preceding technique entails preparation of a plot of $\Delta E/\Delta V$, the change in potential per volume increment of titrant, as a function of the volume of

Table 11–5. Potentiometric Titration Data for Titration of 3.737 Millimoles of Chloride with 0.2314 $F$ Silver Nitrate

| Volume of AgNO$_3$, ml | $E$ vs. SCE, volts | $\Delta E/\Delta V$, mv/0.1 ml | $\Delta^2 E/\Delta V^2$ | $\Delta V/\Delta E$ |
|---|---|---|---|---|
| 0 | +0.063 | | | |
| 5.00 | 0.071 | | | |
| 10.00 | 0.086 | | | |
| 15.00 | 0.138 | | | |
| 15.20 | 0.145 | | | |
| 15.40 | 0.153 | | | |
| 15.60 | 0.161 | | | |
| | | 5 | | 0.200 |
| 15.70 | 0.166 | | | |
| | | 7 | | 0.143 |
| 15.80 | 0.173 | | | |
| | | 9 | | 0.111 |
| 15.90 | 0.182 | | +5 | |
| | | 14 | | 0.071 |
| 16.00 | 0.196 | | +15 | |
| | | 29 | | 0.034 |
| 16.10 | 0.225 | | +35 | |
| | | 64 | | 0.016 |
| 16.20 | 0.289 | | −34 | |
| | | 30 | | 0.033 |
| 16.30 | 0.319 | | −14 | |
| | | 16 | | 0.063 |
| 16.40 | 0.335 | | −5 | |
| | | 11 | | 0.091 |
| 16.50 | 0.346 | | | |

titrant. Such a graph, derived from the titration data of Table 11–5, is shown in Figure 11–10B. If the volume increments near the equivalence point are equal, one may simply take the difference in millivolts between each successive pair of potential readings as $\Delta E/\Delta V$; however, if the volume increments are unequal, it is necessary to normalize the data. To obtain the volume, $V$, pertaining to each value of $\Delta E/\Delta V$, one must *average* the two volumes corresponding to the successive potential readings. For example, in Table 11–5 the change in potential, $\Delta E$, is 29 mv as one proceeds from 16.00 to 16.10 ml of titrant added. Accordingly, $\Delta E/\Delta V$ is 29 mv per 0.1 ml, whereas $V$ should be taken as 16.05 ml. Figure 11–10B consists of two curves extrapolated to a point of intersection whose position along the abscissa indicates the equivalence-point volume. It is only a happenstance that one of the experimental points lies so close to this point of intersection. Because of the tediousness and arbitrariness of the extrapolation method, it has little advantage over the previous approach.

Figure 11–11, called a **Gran plot**, illustrates an especially simple and precise way to locate an equivalence point. As an example, we have plotted $\Delta V/\Delta E$, the reciprocal of each value of $\Delta E/\Delta V$ listed in Table 11–5, against the *average* volume of titrant. Just before and after the equivalence point of a potentiometric titration, $\Delta V/\Delta E$ varies linearly with volume, so the experimental data fall upon two straight lines which intersect at the volume axis, the point of intersection being the equivalence-point volume. Due to the slight, though inherent, incompleteness of any titration reaction, the experimental points near the apex of the V-shaped plot deviate from the straight-line relationship. Besides offering the convenience and accuracy of a straight-line graph, a Gran plot possesses the important advantage that it does not

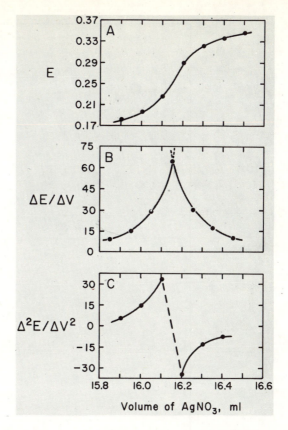

*Figure 11–10.* Titration curves for the titration of 3.737 millimoles of chloride with 0.2314 $F$ silver nitrate. Curve A: normal titration curve, showing the region near the equivalence point. Curve B: first derivative titration curve. Curve C: second derivative titration curve. All data points are taken from Table 11–5.

require titration data *very* close to the equivalence point, where drifting potentials are often encountered because of the slow attainment of chemical equilibrium.

**Analytical method.** A rapid, accurate, and convenient procedure for the location of an equivalence point relies upon the mathematical definition that the second derivative ($\Delta^2 E/\Delta V^2$) of the titration curve is zero at the point where the first derivative ($\Delta E/\Delta V$) is a maximum. In the fourth column of Table 11–5 are listed the individual values of $\Delta^2 E/\Delta V^2$—the difference between each pair of $\Delta E/\Delta V$ values—and Figure 11–10C presents a graph of these results as a function of the volume of titrant. Since the second derivative ($\Delta^2 E/\Delta V^2$) changes sign and passes through zero between volume readings of 16.10 and 16.20 ml, the equivalence-point volume must lie within this interval. If the volume increments are small enough, one may assume that the ends of the upper and lower curves in Figure 11–10C can be joined by a straight line. Provided that this approach is valid, one can show by mathematical interpolation that the equivalence-point volume should be

$$16.10 + 0.10\left(\frac{35}{35 + 34}\right) = 16.15 \text{ ml}$$

Note that the *analytical method* requires only experimental data similar to those listed in Table 11–5; no graphical constructions are necessary.

Implicit in the graphical and analytical methods just described is the assumption that the theoretical equivalence point for a titration coincides with the maximal value of $\Delta E/\Delta V$ for the titration curve. However, this requirement is fulfilled only for processes in which the reactants combine in a one-to-one stoichiometric ratio, that

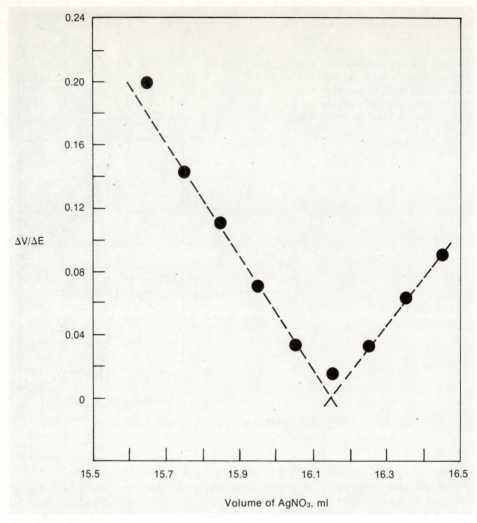

*Figure 11–11.* A Gran plot constructed from potentiometric-titration data presented in Table 11–5. (See text for discussion as well as the original paper by G. Gran: Acta Chim. Scand., *4*:599, 1950.)

is, for equilibria such as

$$Ag^+ + Cl^- \rightleftharpoons AgCl(s)$$

and

$$Fe^{2+} + Ce^{4+} \rightleftharpoons Fe^{3+} + Ce^{3+}$$

which are *symmetrical*. When the reaction between the titrant and the substance being titrated involves unequal numbers of species, as in the titration of iron(II) with dichromate ion,

$$6\,Fe^{2+} + Cr_2O_7^{2-} + 14\,H^+ \rightleftharpoons 6\,Fe^{3+} + 2\,Cr^{3+} + 7\,H_2O$$

an *asymmetric* titration curve results, and the theoretical equivalence point deviates from the region of maximal slope. A significant error can arise in especially unfavorable situations if the analytical or graphical methods are used, particularly if exceptional titration accuracy is sought. However, the difference between the potential at which $\Delta E/\Delta V$ is a maximum and the theoretical equivalence-point potential is small enough to be neglected for most titrations.

*Titration to the equivalence-point potential.* If the equivalence-point potential has been established from previous titrations or by means of theoretical calculations, one can simply introduce titrant—slowly as the equivalence point is approached—until the potential of the indicator electrode attains the desired value. Obviously, conditions must be controlled sufficiently well to ensure that the equivalence-point potential is reasonably reproducible from one titration to another. In many instances, the titration curve rises so steeply near the equivalence point that a considerable latitude of potential values still guarantees highly accurate results.

## Automatic Potentiometric Titrations

Several automatic titrators have been manufactured by means of which titrant can be delivered from a buret whose stopcock is replaced with a solenoid-operated valve. When the potential of the indicator electrode reaches a certain preselected value, the flow of titrant is interrupted automatically. Some instruments can anticipate the approach of the equivalence point through circuitry which essentially monitors $\Delta E/\Delta V$. Other automatic titrators can perform overnight and unattended a series of identical titrations; such instruments have provision for permanent recording of the titrant volume used for each sample. With some devices, it is possible to obtain completely recorded potentiometric titration curves. It should be emphasized that the possible success of an automatic potentiometric titration is governed by the response of the indicator electrode to species in solution and not by the action of the electronic sensing system of the titrator.

## APPLICATIONS OF POTENTIOMETRIC TITRIMETRY

Potentiometric titrations may be performed for precipitation, complex-formation, acid-base, and oxidation-reduction reactions. Titration accuracies are fully as good as with conventional end-point indicators and may be even better.

## Precipitation Titrations

Potentiometric end-point detection is applicable to precipitation titrations involving silver(I), mercury(II), lead(II), and zinc(II), as well as anions such as chloride, bromide, iodide, thiocyanate, and ferrocyanide.

Indicator electrodes for potentiometric precipitation titrations may be fabricated from the metal whose cation comprises the precipitate. Alternatively, a membrane electrode which responds to an appropriate ion is often useful. In some instances, the indicator electrode may be one whose potential is governed by the anion of the precipitate. For example, the progress of a titration of chloride ion with standard silver nitrate solution can be monitored if the potential of a silver indicator electrode versus a saturated calomel reference electrode is measured. Figure 11–9 illustrates the physical appearance of the titration apparatus, whereas the shorthand representation of the galvanic cell shown in this diagram is

$$\text{Hg} \,|\, \text{Hg}_2\text{Cl}_2(s), \text{KCl}(s) \,|\, \text{KNO}_3 \,(1\,F) \,|\, \text{chloride sample solution} \,|\, \text{Ag}$$

A potassium nitrate salt bridge is inserted between the sample solution and the saturated calomel electrode to prevent diffusion of potassium chloride solution into the sample.

Once a small volume of silver nitrate titrant is added, the solution becomes saturated with silver chloride and the indicator electrode behaves as a silver-silver chloride electrode,

$$AgCl(s) + e \rightleftharpoons Ag + Cl^-; \qquad E^0 = +0.2222 \text{ v}$$

its potential being governed by the Nernst equation

$$E_{AgCl,Ag} = E^0_{AgCl,Ag} - 0.059 \log (Cl^-)$$

If we neglect liquid-junction potentials, the overall emf of the galvanic cell can be written as

$$E_{cell} = E_{Hg,Hg_2Cl_2} + E_{AgCl,Ag}$$

and, if the Nernst equation for the silver chloride-silver metal half-reaction is substituted into the latter relation, it becomes

$$E_{cell} = E_{Hg,Hg_2Cl_2} + E^0_{AgCl,Ag} - 0.059 \log (Cl^-)$$

When the respective values for $E_{Hg,Hg_2Cl_2}$ and $E^0_{AgCl,Ag}$ (namely, $-0.2415$ and $+0.2222$ v versus NHE) are substituted into this equation, the result is

$$E_{cell} = -0.0193 - 0.059 \log (Cl^-)$$

or

$$E_{cell} = -0.0193 + 0.059 \text{ pCl}$$

Thus, a linear relationship exists between the value of pCl and the emf of the galvanic cell. Accordingly, if one performs a potentiometric titration of chloride with silver nitrate solution and constructs a plot of emf as a function of the volume of titrant, the titration curve would be similar to that shown in Figure 8–1. Equations analogous to those just derived can be obtained for the precipitation titrations of other anions which form insoluble silver salts, as well as for sparingly soluble salts of other cations.

## Complexometric Titrations

Most present applications of complexometric titrations center on the reactions of metal ions with various members of the family of aminopolycarboxylic acids, notably ethylenediaminetetraacetic acid (EDTA). A large number of metal ions form stable one-to-one complexes with the ethylenediaminetetraacetate anion, so there exists considerable interest in the use of potentiometric methods to monitor the progress of metal-EDTA titrations.

In principle, one could follow the titration of a metal ion with EDTA by utilizing an indicator electrode of the parent metal. Unfortunately, most metallic electrodes do not respond in accordance with the Nernst equation over a wide range of metal ion activity (concentration). Moreover, many metal electrodes are sensitive to dissolved oxygen and other extraneous species. Substantial improvement in specificity and response is gained through the use of a glass, liquid ion-exchange, or solid-state membrane electrode which is sensitive to the desired metal ion, but the list of available electrodes is insufficient at present.

Elemental mercury in the presence of a small concentration of the very stable mercury(II)-EDTA complex provides a versatile and unique indicator-electrode system for many metal-EDTA titrations. In practice, the indicator electrode may consist of a pool of mercury in a small cup positioned inside the titration vessel, with some means for connecting the electrode to an external potentiometer circuit. A more convenient form of mercury indicator electrode can be prepared if a gold wire is amalgamated by immersion into mercury; the mercury-coated gold wire is especially easy to manipulate.

Suppose we wish to determine calcium ion in an unknown sample solution by means of a potentiometric EDTA titration. Into the titration vessel containing the sample, let us insert a mercury indicator electrode and a saturated calomel reference electrode. We shall then introduce a small concentration, approximately $10^{-4}$ $M$, of the mercury(II)-EDTA complex. Finally, let us titrate the mixture with a standard solution of EDTA, recording the potential of the mercury indicator electrode as a function of the titrant volume. If the potential of the indicator electrode is plotted versus the volume of titrant, a titration curve similar to those shown in Figures 6–4, 6–5, and 6–6 will be obtained. Let us see why the mercury indicator electrode exhibits a linear response to pCa, the negative logarithm of the calcium ion concentration (activity).

Once addition of EDTA titrant has commenced, the sample solution will contain both calcium ion and the calcium-EDTA complex, as well as the mercury(II)-EDTA species. Thus, the system comprises a galvanic cell which may be represented by the shorthand notation

$$Hg \mid Hg_2Cl_2(s), KCl(s) \parallel Ca^{2+}, CaY^{2-}, HgY^{2-} \mid Hg$$

where $Y^{4-}$ denotes the ethylenediaminetetraacetate ion. A small concentration of this latter anion will, of course, be in equilibrium with the metal-EDTA complexes. Of particular interest is the electrode involving mercury and the mercury(II)-EDTA complex. Throughout the titration this electrode ultimately responds to the concentration (activity) of $Hg^{2+}$, since the potential-determining process is the electron transfer between mercury and dissolved mercury ions. Therefore, the Nernst equation

$$E_{Hg^{2+},Hg} = E^0_{Hg^{2+},Hg} - \frac{0.059}{2} \log \frac{1}{[Hg^{2+}]}$$

is always obeyed. Dissociation of the mercury(II)-EDTA complex, present in solution during the titration, is the primary source of mercuric ion,

$$HgY^{2-} \rightleftharpoons Hg^{2+} + Y^{4-}$$

and the concentrations of these two species are related through the equilibrium expression for the formation of the mercury(II)-EDTA complex:

$$K_{HgY^{2-}} = \frac{[HgY^{2-}]}{[Hg^{2+}][Y^{4-}]} = 6.3 \times 10^{21}$$

If we solve this equilibrium expression for the mercuric ion concentration and substitute the result into the Nernst equation for the mercury indicator electrode, it follows that

$$E_{Hg^{2+},Hg} = E^0_{Hg^{2+},Hg} - \frac{0.059}{2} \log \frac{K_{HgY^{2-}}[Y^{4-}]}{[HgY^{2-}]}$$

For the reaction between calcium ion and EDTA,

$$Ca^{2+} + Y^{4-} \rightleftharpoons CaY^{2-}$$

the equilibrium expression has the form

$$K_{CaY^{2-}} = \frac{[CaY^{2-}]}{[Ca^{2+}][Y^{4-}]} = 5.0 \times 10^{10}$$

If this relation is rearranged,

$$[Y^{4-}] = \frac{[CaY^{2-}]}{[Ca^{2+}]K_{CaY^{2-}}}$$

and substituted into the previous modified Nernst equation, we have

$$E_{Hg^{2+},Hg} = E^0_{Hg^{2+},Hg} - \frac{0.059}{2} \log \frac{K_{HgY^{2-}}[CaY^{2-}]}{K_{CaY^{2-}}[Ca^{2+}][HgY^{2-}]}$$

which becomes

$$E_{Hg^{2+},Hg} = E^0_{Hg^{2+},Hg} - \frac{0.059}{2} \log \frac{K_{HgY}[CaY^{2-}]}{K_{CaY^{2-}}[HgY^{2-}]} + \frac{0.059}{2} \log [Ca^{2+}]$$

or

$$E_{Hg^{2+},Hg} = E^0_{Hg^{2+},Hg} - \frac{0.059}{2} \log \frac{K_{HgY^{2-}}[CaY^{2-}]}{K_{CaY^{2-}}[HgY^{2-}]} - \frac{0.059}{2} pCa$$

In the second term of the right-hand member of the preceding equation are the formation constants for the calcium-EDTA and mercury(II)-EDTA complexes together with the concentrations of these two species. Since the mercury(II)-EDTA ion is very stable, its concentration remains virtually constant during the titration, if dilution effects are minimized through the use of a relatively concentrated titrant. Near the equivalence point, when most of the calcium ion has already been converted to its EDTA complex, the concentration of $CaY^{2-}$ will undergo so little additional increase that it may be assumed to be constant. Accordingly, the first and second terms on the right-hand side of the latter relation both appear to be constant as the equivalence point is approached, with the result that

$$E_{Hg^{2+},Hg} = K - \frac{0.059}{2} pCa$$

Since the value of pCa changes abruptly at the equivalence point of an EDTA titration, the potential of the mercury indicator electrode will exhibit a similarly large variation. It should be noted that the potential of the indicator electrode is actually measured against a saturated calomel reference electrode. This means that the value of $K$ in the previous equation is different and that the entire titration curve is shifted along the potential axis.

Wide applicability of the potentiometric procedure stems from the fact that the mercury(II)-EDTA complex is unusually stable. Indeed, only metal cations whose EDTA complexes are *less* stable than the mercury(II)-EDTA species can be determined satisfactorily. However, Table 6–1 on page 173 provides evidence that this requirement is met by many cations of analytical importance and, in accord with this observation, several dozen elements may be titrated potentiometrically with EDTA.

## Acid-Base Titrations

Availability of glass electrodes, saturated calomel electrodes, and pH meters has rendered potentiometric acid-base titrations more convenient to perform than practically any other method of volumetric analysis. Particularly impressive is the increasing number of applications in the field of nonaqueous acid-base titrimetry, some of which were described in Chapter 5. Furthermore, potentiometric acid-base titrimetry is especially useful in situations where visual end-point detection is precluded by the presence of colored species or turbidity.

Earlier in this chapter, we found that a linear relationship exists between the pH of a solution and the potential of a glass membrane electrode. Since most pH-measuring instruments are calibrated directly in pH units, it is more straightforward for potentiometric acid-base titrations to record pH versus the volume of titrant rather than to convert pH readings into values of potential. In addition, experimental pH data can be readily correlated with theoretical titration curves such as those calculated in Chapter 4. In applying *graphical* or *analytical* methods for the location of equivalence points in potentiometric acid-base titrimetry, one may employ $\Delta pH/\Delta V$ and $\Delta^2 pH/\Delta V^2$, respectively, or $\Delta V/\Delta pH$ if a Gran plot is constructed.

Potentiometric acid-base titrimetry offers several additional advantages which other neutralization methods of analysis lack. First, the former technique allows the recording of complete titration curves, which is especially valuable for the study of mixtures of acids or bases in either aqueous or nonaqueous solvents. Second, the potentiometric procedure permits one to apply graphical or analytical methods for the location of equivalence points, a necessity in the analysis of multicomponent systems. Third, quantitative information concerning the relative acid or base strengths of solute species can be obtained. For example, as shown in Figure 4–9, the titration of phosphoric acid with a standard sodium hydroxide solution produces a titration curve with two steps, corresponding to the consecutive reactions

$$H_3PO_4 + OH^- \rightleftharpoons H_2PO_4^- + H_2O$$

and

$$H_2PO_4^- + OH^- \rightleftharpoons HPO_4^{2-} + H_2O$$

Halfway to the first equivalence point, the concentrations of phosphoric acid and dihydrogen phosphate are essentially equal. Substitution of the latter condition—namely, $[H_3PO_4] = [H_2PO_4^-]$—into the equilibrium expression

$$K_{a1} = \frac{[H^+][H_2PO_4^-]}{[H_3PO_4]}$$

leads to the conclusion that the measured pH is numerically equal to $pK_{a1}$. In a similar fashion, the value of $pK_{a2}$ may be obtained directly from the pH reading midway between the first and second equivalence points.

## Oxidation-Reduction Titrations

Principles and applications of oxidation-reduction methods of analysis have been considered in detail in earlier chapters, so it suffices to mention here that a large majority of the determinations outlined previously can be accomplished by means of potentiometric titrimetry.

Most potentiometric redox titrations involve the use of a platinum indicator electrode, a saturated calomel reference electrode, and a suitable potential-measuring instrument. Unless chloride ion has a deleterious effect upon the desired oxidation-reduction reaction, the calomel reference electrode can be immersed directly into the titration vessel; otherwise, an appropriate salt bridge connecting the reference electrode to the solution in the titration vessel is required. Although a platinum wire or foil is almost invariably employed as the indicator electrode, gold, palladium, and carbon are occasionally used.

If an indicator electrode responds reversibly to the various oxidants and reductants in the chemical system, titration curves recorded in practice exhibit excellent agreement with theoretical curves derived according to procedures discussed in Chapter 9.

## QUESTIONS AND PROBLEMS

1. Suppose that the activity of copper(II) in an aqueous medium is to be determined through a direct potentiometric measurement with a copper metal indicator electrode, the pertinent half-reaction being

$$Cu^{2+} + 2\ e \rightleftharpoons Cu$$

   (a) If an error of exactly 1 mv in the observed potential of the copper electrode is incurred, calculate the corresponding per cent uncertainty in the copper(II) activity.
   (b) If the error in the copper(II) activity is not to exceed 1 per cent, calculate the maximal tolerable uncertainty in the potential of the copper indicator electrode.

2. The following data were observed during the potentiometric pH titration of a 25.00-ml aliquot of a solution containing a weak monoprotic acid with 0.1165 $F$ sodium hydroxide solution:

| Volume, ml | pH | Volume, ml | pH |
|---|---|---|---|
| 0 | 2.89 | 15.60 | 8.40 |
| 2.00 | 4.52 | 15.70 | 9.29 |
| 4.00 | 5.06 | 15.80 | 10.07 |
| 10.00 | 5.89 | 16.00 | 10.65 |
| 12.00 | 6.15 | 17.00 | 11.34 |
| 14.00 | 6.63 | 18.00 | 11.63 |
| 15.00 | 7.08 | 20.00 | 12.00 |
| 15.50 | 7.75 | 24.00 | 12.41 |

   (a) Construct a plot of the experimental titration curve.
   (b) Construct the first derivative titration curve.
   (c) Construct the second derivative titration curve.
   (d) Calculate the concentration of the original acid solution.
   (e) Evaluate the dissociation constant for the weak acid and the pH at the equivalence point of the titration.

3. Consider the determination of zinc by titration with a standard potassium ferrocyanide solution, $K_4Fe(CN)_6$, according to the precipitation reaction

$$3\ Zn^{2+} + 2\ K^+ + 2\ Fe(CN)_6{}^{4-} \rightleftharpoons K_2Zn_3[Fe(CN)_6]_2(s)$$

The solubility product for the precipitate is $1.0 \times 10^{-17}$. Assume that 50.00 ml of a 0.001000 $M$ zinc ion solution, containing 0.001000 $M$ ferricyanide ion, is titrated with 0.2000 $F$ potassium ferrocyanide; the effect of dilution may be ignored. The progress of such a titration may be followed potentiometrically with the aid of a platinum indicator electrode and a saturated calomel reference electrode, the potential of the platinum electrode being governed by the ferricyanide-ferrocyanide half-reaction. Construct a theoretical plot of the potential of the platinum indicator electrode versus the saturated calomel electrode, on the assumption that the formal potential for the ferricyanide-ferrocyanide couple is +0.70 v versus NHE under the conditions of the titration. Neglect the existence of any liquid-junction potentials.

4. Describe in detail how the iodine-iodide couple, in conjunction with a platinum indicator electrode, might be used to follow the progress of the titration of silver ion with a standard potassium iodide solution.

5. Silver ion in 100.0 ml of a dilute nitric acid medium was titrated potentiometrically with a standard $2.068 \times 10^{-4} F$ potassium iodide solution, according to the reaction

$$Ag^+ + I^- \rightleftharpoons AgI(s)$$

A silver iodide indicator electrode and a saturated calomel reference electrode were used, and the following titration data resulted:

| Volume of KI, ml | $E$ versus SCE, volts |
|---|---|
| 0 | +0.2623 |
| 1.00 | 0.2550 |
| 2.00 | 0.2451 |
| 3.00 | 0.2309 |
| 3.50 | 0.2193 |
| 3.75 | 0.2105 |
| 4.00 | 0.1979 |
| 4.25 | 0.1730 |
| 4.50 | 0.0316 |
| 4.75 | −0.0066 |

(a) Calculate the concentration of silver ion in the original sample solution.

(b) Compute the titration error, in per cent with the proper sign, if potassium iodide titrant is added until the potential of the indicator electrode reaches +0.0600 v versus SCE.

6. A 0.2479-gm sample of anhydrous sodium hexachloroplatinate(IV), $Na_2PtCl_6$, was analyzed for its chloride content by means of a potentiometric titration with standard silver nitrate solution. The weighed sample was decomposed in the presence of hydrazine sulfate, the platinum(IV) being reduced to the metal, and the liberated chloride ion was titrated with 0.2314 $F$ silver nitrate. A silver indicator electrode, a saturated calomel reference electrode, and a nitric acid salt bridge were used in the performance of the titration. The resulting experimental data were as follows:

| Volume of $AgNO_3$, ml | $E$ versus SCE, volts |
|---|---|
| 0 | +0.072 |
| 13.00 | 0.140 |
| 13.20 | 0.145 |
| 13.40 | 0.152 |
| 13.60 | 0.160 |
| 13.80 | 0.172 |
| 14.00 | 0.196 |
| 14.20 | 0.290 |
| 14.40 | 0.326 |
| 14.60 | 0.340 |

Calculate the apparent percentage of chloride in the sample, the true percentage of chloride in sodium hexachloroplatinate(IV), and the per cent relative error between these two results.

7. Under what conditions may a glass membrane electrode be employed as a reference electrode in a potentiometric titration?

8. In the potentiometric titration of an aqueous potassium chloride solution with 0.09603 $F$ silver nitrate, the following data were obtained:

| Volume of AgNO$_3$, ml | $E$ versus SCE, volts |
|:---:|:---:|
| 10.00 | +0.062 |
| 20.00 | 0.085 |
| 25.00 | 0.107 |
| 27.00 | 0.123 |
| 28.00 | 0.138 |
| 28.50 | 0.146 |
| 28.80 | 0.161 |
| 29.00 | 0.173 |
| 29.10 | 0.180 |
| 29.20 | 0.192 |
| 29.30 | 0.221 |
| 29.40 | 0.288 |
| 29.50 | 0.304 |
| 29.60 | 0.314 |
| 29.70 | 0.321 |
| 30.00 | 0.336 |
| 30.50 | 0.348 |
| 31.00 | 0.359 |
| 33.00 | 0.389 |

A silver wire indicator electrode and a saturated calomel reference electrode (SCE) were employed.

(a) Construct a Gran plot, locate the equivalence point, and calculate the millimoles of chloride ion in the sample.

(b) Treat the titration data by means of the so-called *analytical method* to locate the equivalence point, and compare the result with that obtained from the Gran plot.

9. Derive the equation, shown on page 359, pertaining to the method of standard addition as employed in direct potentiometry.

10. Calcium ion in a water sample was determined by means of a direct potentiometric measurement. Exactly 100.0 ml of the water was placed in a beaker, and a saturated calomel reference electrode (SCE) and a calcium-selective membrane electrode were immersed in the solution. The potential of the calcium electrode was found to be $-0.0619$ volt versus SCE. When a 10.00-ml aliquot of 0.00731 $F$ calcium nitrate was pipetted into the beaker and mixed thoroughly with the water sample, the new potential of the calcium electrode was observed to be $-0.0483$ volt versus SCE. Calculate the molar concentration of calcium ion in the original water sample.

11. When the emf of the galvanic cell

$$-Ag \mid Ag(S_2O_3)_2{}^{3-} \ (0.001000 \ M), \ S_2O_3{}^{2-} \ (2.00 \ M) \parallel Ag^+ \ (0.0500 \ M) \mid Ag+$$

was measured, a value of 0.903 volt was obtained. Calculate the overall formation constant, $\beta_2$, for the $Ag(S_2O_3)_2{}^{3-}$ complex, which is formed according to the reaction

$$Ag^+ + 2 \ S_2O_3{}^{2-} \rightleftharpoons Ag(S_2O_3)_2{}^{3-}$$

12. If the electromotive force of the galvanic cell

$$-Pt, \ H_2 \ (0.500 \ atm) \mid \text{solution of unknown pH} \mid Hg_2Cl_2(s), \ KCl(s) \mid Hg+$$

is 0.366 volt, calculate the pH of the unknown solution, neglecting the liquid-junction potential.

13. A liquid ion-exchange membrane electrode useful for the determination of ammonium ion ($NH_4^+$) was prepared by impregnation of a Millipore filter with a saturated solution of 72 per cent nonactin and 28 per cent monactin in tris(2-ethylhexyl)phosphate. A silver-silver chloride electrode was placed behind the ammonium-selective membrane; a saturated calomel reference electrode was combined with the ammonium ion electrode to obtain a complete electrochemical cell:

$$-Ag \left| \begin{array}{c} AgCl(s), \\ LiCl\ (0.0100\ F) \end{array} \right| \begin{array}{c} nonactin- \\ monactin \\ membrane \end{array} \left| \begin{array}{c} solution \\ of\ unknown \\ NH_4^+ \\ activity \end{array} \right\| Hg_2Cl_2(s), KCl(s) \left| Hg+ \right.$$

When two different solutions having ammonium ion activities of $3.19 \times 10^{-3}$ and $1.08 \times 10^{-5}\ M$ were placed in the cell, the observed emf values were 0.2105 and 0.0644 volt, respectively, at 25.00°C. What is the ammonium ion activity of a sample solution that yields an emf of 0.1128 volt? Assume that the liquid-junction potentials remain the same throughout the series of measurements.

14. Null-point potentiometry can be used for the determination of chloride ion at low concentration levels. To perform the measurements, one constructs an electrochemical cell consisting of a solution of known chloride concentration and a solution of unknown chloride concentration separated by a porous membrane (to prevent mixing of the two solutions). Into each solution is dipped an identical silver-silver chloride electrode. Thus, the cell may be designated as follows:

$$Ag \left| \begin{array}{c} \\ AgCl(s), \end{array} \begin{array}{c} unknown \\ chloride \\ solution \end{array} \right\| \begin{array}{c} known \\ chloride \\ solution, \end{array} AgCl(s) \left| Ag \right.$$

What one does is change (increase or decrease) the concentration of chloride ion in the left-hand compartment either by adding a known volume of a standard sodium chloride solution or by adding a known volume of a chloride-free solution until the emf of the cell becomes zero. Under ideal conditions, when the cell emf is zero, the concentrations of chloride ion in the solutions on each side of the porous membrane must be equal. To ensure ideal conditions as nearly as possible, the ionic strength and composition of the solutions on each side of the membrane must be the same. In practice, a fixed high concentration of some electrolyte, *e.g.*, 1 F sulfuric acid, is added to all solutions in order to swamp out any minor changes in ionic strength and composition of the unknown samples.

In the determination of chloride ion by means of null-point potentiometry, a solution containing $2.603 \times 10^{-4}\ M$ chloride ion was placed in the right-hand compartment of the cell. A 25.00-ml aliquot of an unknown chloride solution was pipetted into the left-hand compartment. A preliminary potentiometric measurement revealed that the concentration of chloride in the unknown was less than in the known solution on the right. Next, a standard solution containing $5.206 \times 10^{-4}\ M$ chloride ion was slowly added to the left-hand compartment. Each portion of the standard chloride solution was thoroughly mixed with the unknown solution. After exactly 21.39 ml of the standard chloride solution had been added, the emf of the cell was found to be zero. All solutions used for the measurements contained 1.000 F sulfuric acid. Calculate the concentration of chloride ion in the original unknown.

15. A lanthanum fluoride ($LaF_3$) solid-state electrode can be used to follow the progress of potentiometric titrations of fluoride ion with standard lanthanum nitrate solution and can be employed for the direct measurement of the fluoride ion concentration in a solution. Exactly 100.0 ml of a 0.03095 F sodium fluoride (NaF) solution was titrated potentiometrically with 0.03318 F lanthanum nitrate solution, the titration reaction being

$$La^{3+} + 3\ F^- \rightleftharpoons LaF_3(s)$$

A solid-state LaF$_3$ membrane electrode, a saturated potassium chloride salt bridge, and a saturated calomel reference electrode (SCE) were employed, and the following titration data were obtained:

| Volume of La(NO$_3$)$_3$, ml | E versus SCE, volts |
|---|---|
| 0 | −0.1046 |
| 29.00 | −0.0249 |
| 30.00 | −0.0047 |
| 30.30 | +0.0041 |
| 30.60 | 0.0179 |
| 30.90 | 0.0410 |
| 31.20 | 0.0656 |
| 31.50 | 0.0769 |
| 32.50 | 0.0888 |
| 36.00 | 0.1007 |
| 41.00 | 0.1069 |
| 50.00 | 0.1118 |

(a) Using the known concentrations of the sodium fluoride and lanthanum nitrate solutions, calculate the theoretical volume of titrant needed. How does this value compare to that obtained by means of the so-called analytical method for locating the equivalence point?

(b) It has been established that the emf of a cell consisting of a lanthanum fluoride membrane electrode and a saturated calomel reference electrode in contact with a sodium fluoride solution follows the equation

$$E = K + 0.05915 \text{ pF}$$

where $K$ is a constant for the system and where pF denotes the negative logarithm of the fluoride ion concentration. Using the initial point of the titration data, evaluate the constant in the above relation.

(c) Using the result of part (b), determine the concentration of fluoride ion after the addition of 50.00 ml of titrant. Ignore changes in activity coefficients due to variations in solution composition.

(d) Calculate the concentration of free lanthanum ion after the addition of 50.00 ml of titrant.

(e) Using the results of parts (c) and (d), evaluate the solubility-product constant for LaF$_3$(s).

(f) Would you expect the thermodynamic activity product for LaF$_3$(s) to be larger or smaller than the simple solubility-product constant? Why?

16. Quinhydrone is a sparingly soluble one-to-one addition compound formed from hydroquinone and quinone. When solid quinhydrone is dissolved in an aqueous medium, equal concentrations of hydroquinone and quinone result. If a platinum wire is dipped into the solution, the potential of the electrode is governed by the reversible reaction

$$+ 2 \text{ H}^+ + 2 \text{ e} \rightleftharpoons \qquad ; \quad E^0 = +0.6994 \text{ v versus NHE}$$

Because the potential of the quinone-hydroquinone half-reaction depends on the concentration of hydrogen ion, it is possible to utilize this system for the measurement of pH.

A solution of unknown pH was saturated with quinhydrone and a platinum electrode was dipped into it. Then electrolytic contact was made between the sample solution and a saturated calomel reference electrode

(SCE) by means of a potassium chloride salt bridge. Schematically, the electrochemical cell may be represented as

$$\text{Pt} \left| \begin{array}{l} \text{solution of unknown pH} \\ \text{saturated with quinhydrone} \end{array} \right\| \text{Hg}_2\text{Cl}_2(s), \quad \text{KCl}(s) \right|$$

If the potential of the platinum electrode was found to be $+0.2183$ volt versus SCE, what was the pH of the unknown solution?

17. In order to determine the equilibrium constant for the proton-transfer reaction between pyridine and water,

$$\text{C}_5\text{H}_5\text{N} + \text{H}_2\text{O} \rightleftharpoons \text{C}_5\text{H}_5\text{NH}^+ + \text{OH}^-$$

the following electrochemical cell was constructed:

$$-\text{Pt}, \quad \text{H}_2 \text{ (0.200 atm)} \left| \begin{array}{l} \text{C}_5\text{H}_5\text{N (0.189 } M), \\ \text{C}_5\text{H}_5\text{NH}^+\text{Cl}^- \text{ (0.0536 } M) \end{array} \right\| \begin{array}{l} \text{Hg}_2\text{Cl}_2(s), \\ \text{KCl}(s) \end{array} \right| \text{Hg} +$$

If the electromotive force of the cell is 0.563 volt at 25°C, what is the equilibrium constant ($K_b$) for the above reaction?

18. Suppose that you wish to utilize the following electrochemical cell for the direct potentiometric determination of oxalate ion ($\text{C}_2\text{O}_4^{2-}$):

$$\text{Ag} \mid \text{Ag}_2\text{C}_2\text{O}_4(s), \quad \text{C}_2\text{O}_4^{2-} \text{ (unknown activity)} \parallel \text{AgCl}(s), \quad \text{KCl}(s) \mid \text{Ag}$$

(a) Derive an equation showing how the overall emf of the cell is related to $p\text{C}_2\text{O}_4$, the negative logarithm of the oxalate ion activity.

(b) When an unknown sodium oxalate solution was transferred into the electrochemical cell, the emf was found to be 0.402 volt at 25°C, with the silver-silver chloride reference electrode being negative. Determine the value of $p\text{C}_2\text{O}_4$ in the unknown solution.

19. Three electrochemical cells were constructed and their emf values were determined potentiometrically at 25°C with the following results:

$$+\text{Ag} \mid \text{Ag}^+ \text{ (0.0100 } M) \parallel \text{Hg}_2\text{Cl}_2(s), \quad \text{KCl}(s) \mid \text{Hg} -; \quad E_{\text{cell}} = 0.4397 \text{ v}$$

$$+\text{Ag} \mid \text{AgI}(s) \parallel \text{Hg}_2\text{Cl}_2(s), \quad \text{KCl}(s) \mid \text{Hg} -; \quad E_{\text{cell}} = 0.0824 \text{ v}$$

$$-\text{Ag} \mid \text{PbI}_2(s), \quad \text{AgI}(s) \parallel \text{Hg}_2\text{Cl}_2(s), \quad \text{KCl}(s) \mid \text{Hg} +; \quad E_{\text{cell}} = 0.2382 \text{ v}$$

Using only the experimental data, evaluate the solubility products for silver iodide (AgI) and lead iodide ($\text{PbI}_2$).

20. A sodium-sensitive membrane electrode was fabricated from a glass composed of 11 per cent $\text{Na}_2\text{O}$, 18 per cent $\text{Al}_2\text{O}_3$, and 71 per cent $\text{SiO}_2$. Then, the sodium-selective electrode and a saturated calomel reference electrode were dipped into an aqueous sodium chloride solution of unknown concentration to produce the following electrochemical cell:

$$-\text{Ag} \left| \begin{array}{l} \text{AgCl}(s), \\ \text{NaCl (1.00 } F) \end{array} \right| \begin{array}{l} \text{sodium-sensitive} \\ \text{glass membrane} \end{array} \left| \begin{array}{l} \text{sample solution} \\ \text{of unknown pNa} \end{array} \right\| \begin{array}{l} \text{Hg}_2\text{Cl}_2(s), \\ \text{KCl}(s) \end{array} \right| \text{Hg} +$$

When the emf of the cell was measured at 25°C with a potentiometer, a value of 0.2033 volt was obtained. When the unknown solution was replaced with a sodium chloride solution containing a sodium ion activity of $3.23 \times 10^{-3} M$, the emf of the cell was observed to be 0.1667 volt at 25°C.

(a) Assuming no changes in the liquid-junction potential or the asymmetry potential of the glass membrane, and assuming a perfect Nernstian response of the glass membrane to sodium ion, calculate the sodium ion activity in the unknown solution.

(b) If the accuracy of each emf measurement is no better than $\pm 1$ mv, what is the range of sodium ion activities within which the true value must fall?

# SUGGESTIONS FOR ADDITIONAL READING

1. R. G. Bates: *Determination of pH*. John Wiley and Sons, New York, 1964.
2. R. G. Bates: Electrometric methods of pH determination. *In* F. J. Welcher, ed.: *Standard Methods of Chemical Analysis*. Sixth edition, Volume III-A, Van Nostrand, Princeton, New Jersey, 1966, pp. 521–532.
3. R. P. Buck: Potentiometry: pH measurements and ion selective electrodes. *In* A. Weissberger and B. W. Rossiter, eds.: *Physical Methods of Chemistry*. Volume I, Part IIA, Wiley-Interscience, New York, 1971, pp. 61–162.
4. A. K. Covington: Ion-selective electrodes. Chem. Brit., *5*:388, 1969.
5. D. G. Davis: Potentiometric titrations. *In* F. J. Welcher, ed.: *Standard Methods of Chemical Analysis*. Sixth edition, Volume III-A, Van Nostrand, Princeton, New Jersey, 1966, pp. 283–296.
6. R. A. Durst, ed.: *Ion-Selective Electrodes*. National Bureau of Standards, Special Publication No. 314, Washington, 1969.
7. R. A. Durst: Ion-selective electrodes in science, medicine, and technology. Amer. Sci., *59*:353, 1971.
8. G. Eisenman, ed.: *Glass Electrodes for Hydrogen and Other Cations*. Dekker, New York, 1967.
9. G. Eisenman: The electrochemistry of cation-sensitive glass electrodes. *In* C. N. Reilley, ed.: *Advances in Analytical Chemistry and Instrumentation*. Volume 4, Wiley-Interscience, New York, 1965, pp. 213–369.
10. G. Eisenman, R. Bates, G. Mattock, and S. M. Friedman: *The Glass Electrode*. Wiley-Interscience, New York, 1966.
11. N. H. Furman: Potentiometry. *In* I. M. Kolthoff and P. J. Elving, eds.: *Treatise on Analytical Chemistry*. Part I, Volume 4, Wiley-Interscience, New York, 1959, pp. 2269–2302.
12. J. Koryta: Theory and applications of ion-selective electrodes. Anal. Chim. Acta *61*:329, 1972.
13. H. A. Laitinen: *Chemical Analysis*. McGraw-Hill Book Company, New York, 1960, pp. 22–27, 326–341.
14. J. J. Lingane: *Electroanalytical Chemistry*. Second edition, Wiley-Interscience, New York, 1958, pp. 9–35, 71–167.
15. L. Meites and H. C. Thomas: *Advanced Analytical Chemistry*. McGraw-Hill Book Company, New York, 1958, pp. 17–104.
16. G. A. Rechnitz: Ion-selective electrodes. Chem. Eng. News *45*:146, June 12, 1967.
17. G. A. Rechnitz: Chemical studies at ion-selective membrane electrodes. Accts. Chem. Res., *3*:69, 1970.

# ELECTROGRAVIMETRIC AND COULOMETRIC METHODS OF ANALYSIS

# 12

A number of methods of chemical analysis are based upon the principle that one may cause a desired reaction to take place when an external voltage of appropriate magnitude and polarity is impressed across the electrodes of an electrochemical cell. Some of these techniques are described in this chapter. Using **electrogravimetry**, one can accomplish a number of useful analyses by depositing a metal quantitatively upon a previously weighed electrode and by reweighing the electrode to obtain the amount of the metal. In **controlled-potential coulometry**, the potential of an anode or cathode is maintained constant so that only a single reaction occurs at that electrode. By integrating the current which flows as a function of time, one can determine the total quantity of electricity produced by the desired reaction and can calculate the amount of the species of interest according to Faraday's law. Another technique, **coulometric titration**, is a method in which a titrant, electrochemically generated at constant current, reacts with the substance to be determined. Since the magnitude of the constant current is analogous to the concentration of a standard titrant solution, and the time required to complete the titration is equivalent to the

In the present chapter, we shall investigate what happens when current passes through an electrochemical cell. With this knowledge as background, electrogravimetric analysis, controlled-potential coulometry, and coulometric titrimetry will be discussed.

## BEHAVIOR OF ELECTROLYTIC CELLS

At least three phenomena occur when current passes through an electrochemical cell. First, changes in the concentrations of chemical species arise at one or both electrodes. Second, there is an ohmic potential drop which develops because current flows against the resistance of the cell. Third, effects related to the kinetics of electron-transfer processes influence the behavior of a cell. In addition, if the cell contains different solutions which share a common interface, liquid-junction potentials exist.

### Current-Voltage Curve for a Reversible Cell

Let us examine the behavior of the zinc-copper cell represented by the short-hand notation

$$Zn \mid Zn(NO_3)_2 \,(1\,F) \parallel CuSO_4 \,(1\,F) \mid Cu$$

If, for simplicity, the activity coefficients of zinc and cupric ions are taken to be unity, we can calculate an emf of 1.100 v for this cell under the condition of zero current-flow, the copper electrode being positive with respect to the zinc electrode.

*Ohmic potential drop.* Let us connect an external voltage source to this cell—the positive terminal of the source to the copper electrode, the negative terminal to the zinc electrode—so that the external source opposes the zinc-copper cell. *In addition, let us assume provisionally that in each compartment of the cell perfect homogeneity of the solution is maintained, even in the vicinity of the electrode, by means of highly efficient stirring.* If the magnitude of the external applied voltage is precisely equal to the emf of the zinc-copper cell—that is, 1.100 v—no current passes through the cell and no net reaction occurs. When the applied voltage is increased above 1.100 v, the zinc-copper cell behaves as an electrolytic cell, the reaction

$$Zn^{2+} + Cu \rightarrow Zn + Cu^{2+}$$

is forced to proceed, and the relationship between the applied voltage and the current which flows through the cell is depicted by line $A$ in the *upper* half of Figure 12–1. If the applied voltage is less than 1.100 v, the zinc-copper cell acts as a galvanic cell, the overall reaction becomes

$$Zn + Cu^{2+} \rightarrow Zn^{2+} + Cu$$

and the current-voltage relation is shown by line $A$ in the *lower* half of Figure 12–1.*

---

* Since electrochemical reactions cause changes in the concentrations (activities) of zinc and cupric ions as well as the emf and resistance of the cell, Figure 12–1 is strictly applicable only during the first few moments of current flow.

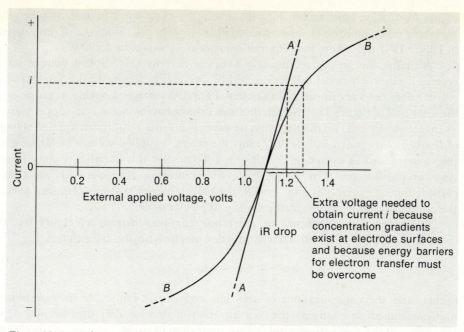

*Figure 12–1.* Current-applied voltage curves for the reversible electrochemical cell

$$\text{Zn} \,|\, \text{Zn (NO}_3)_2\, (1\,F)\,||\,\text{CuSO}_4\,(1\,F)\,|\,\text{Cu}$$

Curve $A$: schematic variation of current with applied voltage for ideal situation in which the concentrations (activities) of zinc(II) and copper(II) are uniform throughout the two half-cells and in which the electron-transfer processes are infinitely fast. Curve $B$: schematic variation of current with applied voltage when concentration gradients exist at the electrode surfaces and when the electron-transfer reactions occur at finite rates. At applied voltages less than 1.100 v, the system behaves as a galvanic cell; at applied voltages greater than 1.100 v, the system behaves as an electrolytic cell.

By convention, when an electrochemical cell operates nonspontaneously, the resulting current is said to be *positive*, whereas the current is *negative* if an electrochemical cell functions spontaneously as a galvanic cell.

For the remainder of this discussion, we shall focus our attention mainly on the behavior of the zinc-copper system as an electrolytic cell. If the zinc nitrate and cupric sulfate solutions remained homogeneous as electrolysis began, the current passing through the cell would exhibit a linear dependence on the quantity $(E_{\text{app}} - E_{\text{cell}})$, where $E_{\text{app}}$ is the magnitude of the external voltage impressed across the two electrodes and $E_{\text{cell}}$ is the zero-current emf of the cell calculated from the concentrations (activities) of zinc and cupric ions. In other words, the electrochemical cell would act as a simple resistance, and the current, $i$, would be governed by Ohm's law,

$$i = \frac{E_{\text{app}} - E_{\text{cell}}}{R}$$

where $R$ is the total cell resistance. Alternatively, this expression may be written as

$$E_{\text{app}} = E_{\text{cell}} + iR$$

If, in the present hypothetical situation, we wished to begin operating the zinc-copper cell electrolytically at a rate corresponding to the passage of current $i$ in

Since the $iR$ drop shown in Figure 12–1 is 0.100 v, the required value of $E_{app}$ would be 1.200 v.

Actually, the proper relationship between current and applied voltage for a reversible electrochemical cell is depicted by curve $B$ in Figure 12–1. Note that the current does not vary linearly with external applied voltage. Looking at curve $B$ in the *upper* half of Figure 12–1, which illustrates the behavior of the zinc-copper system as an electrolytic cell, we discover that passage of current $i$ requires an applied voltage of nearly 1.300 v, not 1.200 v. Why must the external applied voltage be larger than the value needed to overcome the $iR$ drop? To answer this question, we must consider two phenomena—first, the existence of concentration gradients at the anode and cathode during electrolysis and, second, the energy barrier that must be overcome in the electron-transfer process at the surface of an electrode.

***Concentration gradients at electrode surfaces during electrolysis.*** As soon as the zinc-copper cell starts to function electrolytically, the reaction

$$Zn^{2+} + Cu \rightarrow Zn + Cu^{2+}$$

occurs, and the concentrations of zinc ion and cupric ion *at the electrode surfaces* undergo immediate changes—the concentration of zinc ion decreases below 1 $M$ at the surface of the zinc cathode and the concentration of cupric ion around the copper anode becomes larger than 1 $M$—although the bulk concentration of each ion remains at 1 $M$. However, even as electrolysis causes a concentration gradient to develop at the surface of each electrode, there are three mass-transport processes—diffusion, electrical migration, and mechanical stirring—which act simultaneously to diminish these concentration gradients. First, because a chemical species undergoes *diffusion* from a region of higher concentration to one of lower concentration, zinc ions will diffuse from the bulk of the zinc nitrate solution toward the zinc cathode, whereas cupric ions will diffuse from the copper anode into the bulk of the cupric sulfate solution. Second, the direction of current flow through the cell is such that there is *electrical migration* of zinc ions toward the zinc cathode and of cupric ions away from the copper anode. Third, the use of *mechanical stirring* helps decrease the differences between the concentrations of zinc ion and cupric ion at the electrode surfaces and in the bulk of the solutions.

How effective are diffusion, electrical migration, and mechanical stirring in eliminating the concentration gradients caused by electrolysis? Mechanical stirring is most efficient for the kind of electrolysis experiment we are describing now. However, even if the solution in contact with each electrode is very vigorously stirred, the theory of hydrodynamics tells us that a thin layer of stationary liquid always surrounds an electrode immersed in a moving solution. Consequently, movement of zinc ions to the surface of the cathode proceeds in two steps; zinc ions are transported by *stirring* of the zinc nitrate solution up to the edge of the thin film of immobile liquid, and then *diffuse* and *migrate* through the film to the electrode surface at which they are reduced. At the copper anode, the cupric ions *diffuse* and *migrate* away from the electrode through the thin film of stationary liquid and, upon reaching the stirred solution, are *swept* into the main body of the copper sulfate medium.* *In spite of diffusion, migration, and stirring, if the zinc-copper cell operates at some usefully large current level, it is inevitable that the concentration gradient produced at each electrode by electrolysis will persist to some extent.*

---

* Nitrate and sulfate ions also undergo mass transport by means of diffusion, electrical migration, and mechanical stirring. It would be instructive for the reader to chart the movement of these anions during the electrolysis.

Just as the voltage impressed across the anode and cathode of an electrolytic cell must be increased to compensate for the ohmic potential drop, so the existence of a concentration gradient at the surface of each electrode creates a similar need— the applied voltage must be increased still more to cause the desired rate of electrolysis. With reference to Figure 12–1, suppose that we wish to obtain a current flow of $i$ through the zinc-copper cell. Let us assume arbitrarily that, in the attainment of this rate of electrolysis, the surface concentration of zinc ion is lowered to 0.1 $M$ and the surface concentration of cupric ion becomes 1.9 $M$. By using the Nernst equation, we can calculate that the emf of an electrochemical cell consisting of a zinc electrode in contact with a 0.1 $M$ zinc ion solution and of a copper electrode immersed in a 1.9 $M$ cupric ion solution is 1.138 v, which is 38 mv larger than that of a similar cell with 1 $M$ solutions of the two metal ions in contact with the electrodes. Therefore, in this hypothetical situation, if we wish current $i$ to flow initially through the zinc-copper cell, the applied voltage must not only be large enough to overcome the ohmic potential drop but must further exceed the emf of the cell at zero current flow by at least 38 mv.*

*Energy barriers for electron-transfer reactions.* Like other chemical processes, the transfer of electrons from a chemical species to an electrode (or vice versa) involves a reaction pathway that includes an energy barrier with an associated free energy of activation ($\Delta G^{\ddagger}$). In contrast to an ordinary chemical reaction, however, the height of the energy barrier for such an electron-transfer process depends upon the potential of the electrode. We can use this simple picture to examine the behavior of the zinc electrode of the zinc-copper electrolytic cell.

*When no net current flows through the zinc-copper cell*, the reversible electron-transfer reaction

$$Zn^{2+} + 2 e \rightleftharpoons Zn$$

exists in a state of dynamic equilibrium. This means that the rate of the forward reaction (reduction of zinc ion to zinc metal) equals the rate of the backward reaction (oxidation of zinc metal to zinc ion). In other words, equal cathodic and anodic currents† are flowing across the interface between the zinc electrode and the zinc nitrate solution. If the activities of zinc(II) and zinc metal are taken to be unity at the electrode-solution interface, the equality of the cathodic and anodic currents requires that the energy barriers for the forward and backward electron-transfer reactions be identical.

*When a net current does flow through the zinc-copper cell*, the zinc electrode functions as the cathode and net reduction of zinc(II) to zinc metal occurs. Consider the following hypothetical question: if the activities of zinc(II) and zinc metal at the electrode-solution interface are kept at unity, what must be done to cause net reduction of zinc(II) to zinc metal? To obtain the desired result, we need to make the rate of reduction of zinc(II) to zinc metal greater than the rate of oxidation of zinc

---

* As stated in Chapter 9 (page 260), the emf of the zinc-copper cell is governed by the concentrations (activities) of species *at the surfaces of the electrodes*. However, when no current flows through the cell, it is permissible to substitute *bulk concentrations* into the Nernst equation for calculation of the cell emf, because there are no concentration gradients in the solution. When current does flow through the cell, the concentrations of species at the electrode surfaces deviate from the bulk concentrations, so the cell emf will differ from the value based on the bulk concentrations. In many books, this difference is referred to as **concentration overpotential**, a term which implies some departure of the cell from normal behavior. If the true concentrations of species at the electrode surfaces during current flow are substituted into the Nernst equation, the calculated cell emf should agree with the observed emf (if there are no other complications) and concentration overpotential becomes nonexistent. Therefore, we prefer to avoid use of the artificial term concentration overpotential and to focus attention instead on the real phenomena that occur in a cell during electrolysis.

† This current, either anodic or cathodic, is known as the **exchange current**.

metal to zinc(II). This can be accomplished if the energy barrier for the (forward) reaction

$$Zn^{2+} + 2\,e \rightarrow Zn$$

becomes smaller than the energy barrier for the (backward) reaction

$$Zn \rightarrow Zn^{2+} + 2\,e$$

It can be shown, although we will not offer the proof here,* that a shift in the potential of the zinc electrode in the *negative* direction lowers the energy barrier for the forward reaction and, simultaneously, raises the energy barrier for the backward reaction. Therefore, a negative shift in the potential of the zinc cathode from its zero-current, equilibrium value favors the reduction of zinc(II) to zinc metal and causes a net cathodic current to flow. This shift in the potential of an electrode that produces a certain net current (in the absence of any concentration gradients) is called the **activation overpotential**. If we want the net current for the reduction of zinc(II) to zinc metal to increase, the potential of the cathode must be shifted to a greater extent in the negative direction—a bigger current requires a larger activation overpotential.

Similar reasoning leads to the conclusion that (even in the absence of a concentration gradient) the potential of the copper electrode of the zinc-copper cell must be shifted in a *positive* direction from its value when no current is flowing to cause net oxidation of copper metal to cupric ion. Thus, there is an activation overpotential associated with the electron-transfer process at each electrode of an electrolytic cell.

For some reactions, such as the reduction of a metal cation at a mercury surface to form an amalgam, the activation overpotential needed for rapid electron transfer is small. For other processes, such as those involving the formation or rupture of chemical bonds, the activation overpotential required for fast electron transfer is considerably larger. Although it is not possible to predict the magnitude of the activation overpotential needed to cause an electron-transfer reaction to proceed at a specified rate, there are some qualitative guides. First, activation overpotential increases, sometimes quite drastically, as the current density (current per unit electrode area) becomes larger. Second, an elevation of temperature causes the activation overpotential to decrease, because part of the free energy of activation for the electron-transfer process is provided thermally. Third, activation overpotentials are characteristically large for reactions resulting in liberation of gases, for the oxidation or reduction of organic molecules, and for processes involving multiple electron transfers. Fourth, the deposition of one metal upon the surface of a different metal often takes place with some activation overpotential until a complete layer of the desired metal has been plated out. Fifth, electrode processes occurring at the so-called *soft* metals, including lead, zinc, and mercury, usually show much larger activation overpotentials than at *noble* metals such as platinum, palladium, and gold. Table 12–1 lists some values of activation overpotentials for the evolution of hydrogen and oxygen gases at different electrodes and current densities. Notice that the data nicely illustrate several of the qualitative rules just mentioned.

Let us recall the various phenomena that occur during operation of an electrolytic cell. It should now be evident that, to make a definite current flow through a cell, the external applied voltage must be greater than the zero-current emf of the cell by the amounts needed to overcome the ohmic potential drop, to compensate for

---

* A thorough presentation of the subject of electrochemical kinetics is given by B. E. Conway: *Theory and Principles of Electrode Processes*. Ronald Press, New York, 1965, pp. 92–135.

| Electrode Material | Overpotential, Volts | | | | | | | |
|---|---|---|---|---|---|---|---|---|
| | 0.001 amp/cm² | | 0.01 amp/cm² | | 0.1 amp/cm² | | 1.0 amp/cm² | |
| | H₂ | O₂ | H₂ | O₂ | H₂ | O₂ | H₂ | O₂ |
| Pt, smooth | 0.024 | 0.72 | 0.068 | 0.85 | 0.29 | 1.3 | 0.68 | 1.5 |
| Pt, platinized | 0.015 | 0.35 | 0.030 | 0.52 | 0.041 | 0.64 | 0.048 | 0.76 |
| Au | 0.24 | 0.67 | 0.39 | 0.96 | — | — | 0.80 | 1.6 |
| Cu | 0.48 | 0.42 | 0.58 | 0.58 | 0.80 | 0.66 | 1.3 | 0.79 |
| Ni | 0.56 | 0.35 | 0.75 | 0.52 | 1.1 | 0.64 | 1.2 | 0.85 |
| Hg | 0.9 | — | 1.0 | — | 1.1 | — | 1.2 | — |
| Zn | 0.72 | — | 0.75 | — | 1.1 | — | 1.2 | — |
| Fe | 0.40 | — | 0.56 | — | 0.82 | — | 1.3 | — |
| Pb | 0.52 | — | 1.0 | — | 1.2 | — | 1.3 | — |

\* Data from International Critical Tables 6:339,1929.

the existence of concentration gradients, and to supply the activation overpotentials for rapid electron transfer at each electrode.

We have not discussed the detailed behavior of the zinc-copper system as a galvanic cell. It is sufficient to say here that, since the cell behaves reversibly—the overall cell reaction can be made to go in either direction—the upper and lower branches of curve B in Figure 12–1 are symmetrical about the point at which the external applied voltage is 1.100 v. When the cell acts galvanically, the phenomena responsible for the ohmic potential drop, the concentration gradients, and the activation overpotentials are still operative.

## Current-Voltage Curve for an Irreversible Cell

Some electrochemical cells behave irreversibly. As an example, consider the cell

$$\text{Pt} \mid \text{Zn(NO}_3)_2 \ (1 \ F) \parallel \text{CuSO}_4 \ (1 \ F) \mid \text{Cu}$$

in which the zinc electrode of the previous zinc-copper cell has been replaced by a platinum electrode. Assignment of an emf to the present cell is impossible; since the platinum electrode initially has no deposit of zinc metal on its surface, the potential of this electrode is undefined. Furthermore, the absence of zinc metal precludes the spontaneous discharge of this electrochemical cell. What happens when a variable voltage is impressed across the two electrodes of this cell, the negative side of the source being connected to the platinum electrode and the positive side of the source to the copper electrode? Figure 12-2 shows the qualitative dependence of the current upon the value of the applied voltage. As long as the applied voltage remains below 1.100 v, the current is relatively small, although it does increase more or less linearly as the applied voltage becomes larger. However, when the applied voltage reaches 1.100 v, the current rises abruptly, and the current-voltage curve mimics curve B in Figure 12–1.

*Residual current.* What is the origin of the small current—the **residual current**—that flows through the cell at applied voltages less than 1.100 v? At the copper anode, oxidation of a tiny amount of metallic copper to cupric ion occurs.

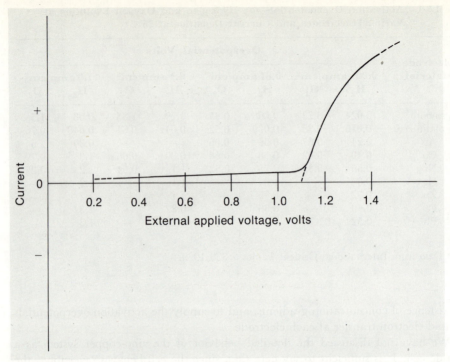

*Figure 12–2.* Current-applied voltage curve for the irreversible electrochemical cell

$$Pt \mid Zn(NO_3)_2 \ (1 \ F) \mid\mid CuSO_4 \ (1 \ F) \mid Cu$$

At any applied voltage, this system can function only as an electrolytic cell. Extrapolation of the rising portion of the curve down to the zero-current line gives the so-called decomposition potential.

On the other hand, at least two current-producing reactions take place at the platinum cathode. First, traces of dissolved oxygen are reduced to hydrogen peroxide and water; other impurities may undergo reduction as well. Second, there can be deposition of a small quantity of zinc metal upon the platinum electrode. Of course, the total anodic residual current is equal to the total cathodic residual current.

It might be expected that no metallic zinc would be deposited until the external applied voltage exceeds 1.100 v, as is true for the reversible zinc-copper cell described earlier (Figure 12–1). However, this statement is correct only if zinc metal is deposited in a state of unit activity (or if a pure zinc electrode is initially in contact with the zinc nitrate solution). When only a minute quantity of zinc metal is plated upon a platinum surface, the activity of zinc metal will be far less than unity—a value such as $10^{-10} \ M$ is not unreasonable. Through use of the Nernst equation, it can be shown for the present cell that deposition of zinc metal at an activity of $10^{-10} \ M$ upon the platinum electrode occurs at an external applied voltage of approximately 0.800 v. Furthermore, deposition of zinc metal at even smaller activities may proceed at lower applied voltages. Thus, part of the cathodic residual current can be attributed to deposition of zinc metal at less than unit activity upon the platinum surface.

***Decomposition potential.*** Commencement of *rapid* deposition of metallic zinc upon the platinum electrode accounts for the sudden rise in current at approximately 1.100 v, because sufficient zinc is plated on the platinum cathode to transform the latter into a zinc electrode. Some electrochemists designate the applied voltage at which zinc ion begins to be reduced as the **decomposition potential**. However, the concept of a decomposition potential lacks theoretical significance for the irreversible cell

$$Pt \mid Zn(NO_3)_2 \ (1 \ F) \parallel CuSO_4 \ (1 \ F) \mid Cu$$

because, as we have seen, some zinc metal is deposited at all values of the applied voltage. Thus, we may only regard the decomposition potential as the approximate applied voltage at which the cell current exhibits a rapid increase and above which reduction of zinc ion occurs at a convenient rate for practical applications. On the other hand, the decomposition potential for the reversible cell

$$Zn \mid Zn(NO_3)_2 \ (1 \ F) \parallel CuSO_4 \ (1 \ F) \mid Cu$$

is identical to the emf of the cell at zero current flow; deposition of zinc metal begins as soon as the applied voltage exceeds 1.100 v.

## Relationship between Applied Voltage, Individual Electrode Potentials, and Ohmic Potential Drop

If we consider a practical electrolysis for the reversible zinc-copper cell described earlier,

$$Zn \mid Zn(NO_3)_2 \ (1 \ F) \parallel CuSO_4 \ (1 \ F) \mid Cu$$

the total voltage ($E_{app}$) impressed across the anode and cathode is the sum of two terms

$$E_{app} = E_{cell} + iR$$

where $E_{cell}$ is the cell emf based on the surface concentrations of zinc ion and cupric ion, plus the sum of the activation overpotentials for the reduction of zinc(II) and the oxidation of copper metal, and where $iR$ represents the ohmic potential drop. In turn, the emf of the cell consists of two contributions

$$E_{cell} = E_{Cu^{2+},Cu} + E_{Zn,Zn^{2+}}$$

if liquid-junction potentials are ignored. Another form of the last equation is

$$E_{cell} = E_{Cu^{2+},Cu} - E_{Zn^{2+},Zn}$$

in which $E_{Cu^{2+},Cu}$ denotes the *actual* potential of the copper electrode and $E_{Zn^{2+},Zn}$ the *actual* potential of the zinc electrode, each measured with respect to the same reference electrode. For each electrode the actual potential is governed by the surface concentration of the pertinent metal cation and by the activation overpotential for the electron-transfer reaction occurring at that electrode. Combining the preceding expressions, we obtain

$$E_{app} = E_{Cu^{2+},Cu} - E_{Zn^{2+},Zn} + iR$$

Noting that the copper electrode is the anode of the zinc-copper electrolytic cell and that the zinc electrode is the cathode, we can write a general equation for the applied voltage in any electrolysis experiment

$$E_{app} = E_a - E_c + iR$$

where $E_a$ and $E_c$ are actual anode and cathode potentials, respectively.

During an electrolysis, the various terms in the preceding equation may change with time. For example, when $E_{app}$ is held constant, the cell resistance ($R$) probably changes as electrolysis proceeds, whereas the current ($i$) decreases as the reacting species are consumed. If the latter two quantities change unequally, the ohmic potential drop ($iR$) will vary, and so must the potentials of the anode and cathode. Similarly, if one performs a constant-current electrolysis, it is necessary to readjust the applied voltage continuously as the resistance varies; but any change in the applied voltage alters the anode and cathode potentials.

## ELECTROGRAVIMETRY

Electrogravimetric methods of analysis usually involve deposition of the desired metallic element upon a previously weighed cathode, followed by subsequent re-weighing of the electrode plus deposit to obtain by difference the quantity of the metal. Among the elements that have been determined in this manner are cadmium, copper, nickel, silver, tin, and zinc. A few substances may be oxidized at a platinum anode to form an insoluble and adherent precipitate suitable for gravimetric measurement. An example of this procedure is the oxidation of lead(II) to lead dioxide in a nitric acid medium. In addition, certain analytical separations can be accomplished through a variation of electrogravimetry, in which easily reducible metallic ions are deposited into a mercury pool cathode, while difficult-to-reduce cations remain in solution. Thus, aluminum, vanadium, titanium, tungsten, and the alkali and alkaline

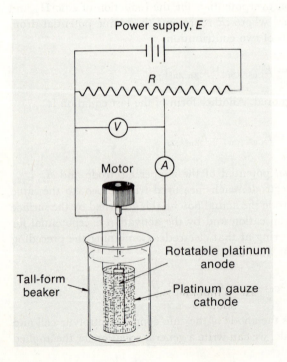

*Figure 12–3.* Simple apparatus for electrodeposition of metals.

Power supply, *E*

*R*

*V*

Motor

*A*

Tall-form beaker

Rotatable platinum anode

Platinum gauze cathode

earth metals may be separated from iron, silver, copper, cadmium, cobalt, and nickel by deposition of the latter group of elements into mercury.

Representative of most electrogravimetric procedures is the determination of copper. Suppose that we desire to measure the quantity of copper(II) in a 1 $F$ sulfuric acid medium by depositing the metal upon a previously weighed platinum electrode. Typical apparatus, consisting of a large platinum gauze cathode and a smaller anode, is depicted in Figure 12–3. Both electrodes are immersed in the solution to be electrolyzed, although some applications may require that the anode and cathode be in different compartments separated by a porous membrane to prevent interaction of the products formed at the anode with the metal deposit on the cathode. A direct-current power supply ($E$) and a rheostat ($R$) enabling one to vary the voltage impressed across the cell constitute the essential electrical circuitry. It is customary to incorporate both a voltmeter ($V$) and an ammeter ($A$) so that the magnitudes of the applied voltage and current can be monitored continuously. Efficient stirring of the solution is mandatory. A magnetic stirrer or a motor-driven propeller is often employed; sometimes, the anode has a design which permits it to be rotated with the aid of a motor.

There are two techniques by which the reduction of copper(II) and deposition of copper metal can be accomplished electrolytically. In the first method the electrolysis is performed at constant applied voltage, whereas the second procedure requires that the current passing through the cell be kept constant.

## Electrolysis at Constant Applied Voltage

In beginning our discussion of electrolysis at constant applied voltage, we shall assume that the original concentration of copper(II) is 0.01 $M$ and that 2.0 to 2.5 v—a voltage sufficient to cause rapid and quantitative deposition of copper—is impressed across the electrochemical cell. As soon as the deposition of copper metal has commenced, the cathodic process is the reduction of cupric ion at a copper-plated platinum electrode

$$Cu^{2+} + 2 e \rightleftharpoons Cu$$

whereas oxidation of water

$$2 H_2O \rightleftharpoons O_2 + 4 H^+ + 4 e$$

occurs at the anode.

Let us examine how the potential of the cathode varies with time. If it is assumed that the surface concentration of cupric ion is essentially 0.01 $M$ at the start of electrolysis because the solution is well stirred, and that the activation overpotential is negligible, we can calculate the potential of the cathode at the moment deposition of copper metal begins:

$$E_{Cu^{2+},Cu} = E^0_{Cu^{2+},Cu} - \frac{0.059}{2} \log \frac{1}{[Cu^{2+}]}$$

$$E_{Cu^{2+},Cu} = 0.337 - \frac{0.059}{2} \log \frac{1}{(0.01)} = +0.278 \text{ v versus NHE}$$

During the first minute of the electrolysis, the potential of the copper cathode changes only slightly from its original value of approximately +0.28 v. However, within only a few minutes, most of the cupric ion is reduced to the elemental state, and the concentration of cupric ion at the electrode surface and in the bulk of the

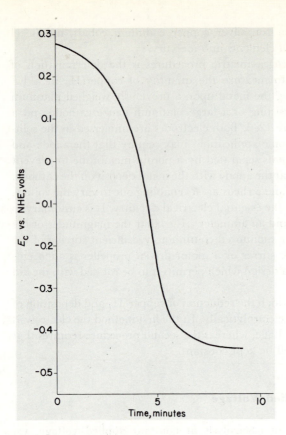

*Figure 12–4.* Schematic variation of potential of copper-plated platinum cathode during deposition of copper by means of electrolysis at constant applied voltage or at constant current from a solution initially 0.01 $M$ in copper(II) and 1 $F$ in sulfuric acid.

solution begins to decrease rapidly. Therefore, the potential of the cathode shifts abruptly toward a value at which another reaction can maintain the current.

Figure 12–4 depicts the experimentally observed potential-time behavior of the copper cathode. It can be seen that the potential of the cathode decreases rapidly after about five minutes to approximately −0.3 v and undergoes a slower drop to a final value near −0.44 v. Any further shift of the cathode potential is halted by the steady evolution of hydrogen gas at the copper-plated platinum electrode. Although reduction of hydrogen ion to hydrogen gas in 1 $F$ sulfuric acid might be expected to occur close to 0 v versus NHE, there is an activation overpotential of 0.4 to 0.5 v for evolution of hydrogen gas at copper.

### Electrolysis at Constant Current

A second method by which electrogravimetric determinations are performed involves the use of constant-current electrolysis. In this technique one varies the external voltage impressed across an electrolytic cell so that the current remains constant during the course of the deposition.

For the copper(II)-sulfuric acid system described above, the potential-time behavior of the copper cathode during a constant-current electrolysis will resemble the curve shown in Figure 12–4. Of course, the larger the constant current, the shorter will be the time required to complete the deposition of copper. However, if an unusually high current is employed, it is impossible for cupric ions alone to sustain the desired rate of electron transfer and, almost as soon as electrolysis begins, a significant fraction of the total current will be due to evolution of hydrogen gas. As a result of vigorous gas evolution, metal deposits obtained by means of constant-current electrolysis frequently have poorer physical properties than those resulting from electrolysis at constant applied voltage. Formation of hydrogen gas can be eliminated through the use of a **cathodic depolarizer**, some substance more easily reduced than hydrogen ion but one that does not interfere with the smooth deposition of copper or any other metal. Nitric acid is a common cathodic depolarizer, because nitrate ion undergoes an eight-electron reduction to ammonium ion

$$NO_3^- + 10\,H^+ + 8\,e \rightleftharpoons NH_4^+ + 3\,H_2O$$

at a more positive potential than the reduction of hydrogen ion to molecular hydrogen.

### Feasibility of Determinations and Separations

That the deposition of copper described in the preceding sections permits the attainment of accurate analytical results may be shown through a simple calculation. Suppose that no more than one part in 10,000 of the original quantity of copper(II) is to remain in solution at the end of the electrolysis. Since the sulfuric acid medium initially contained cupric ion at a concentration of 0.01 $M$, the copper(II) concentration must be decreased to $10^{-6}$ $M$. Using the Nernst equation

$$E_{Cu^{2+},Cu} = E^0_{Cu^{2+},Cu} - \frac{0.059}{2} \log \frac{1}{[Cu^{2+}]}$$

we can conclude that, if the potential of the cathode has a value of

$$E_{Cu^{2+},Cu} = 0.337 - \frac{0.059}{2} \log \frac{1}{(10^{-6})} = +0.160 \text{ v versus NHE}$$

or less when the electrolysis is interrupted, copper metal will have been quantitatively deposited. Figure 12–4 reveals that this requirement is easily met, since the cathode potential reaches $-0.44$ v.

Although the large negative shift of the cathode potential does ensure the quantitative deposition of copper, an uncontrolled change in the potential is undesirable if separation of copper from other elements is to be achieved. Any metal cations which happen to be present in the sulfuric acid solution and which undergo even partial deposition at potentials more positive than $-0.44$ v versus NHE will be plated upon the cathode along with copper. Among the species that could interfere with the determination of copper are bismuth(III), antimony(III), tin(II), and cobalt(II), the pertinent half-reactions and standard potentials being:

$$BiO^+ + 2\,H^+ + 3\,e \rightleftharpoons Bi + H_2O; \qquad E^0 = +0.32 \text{ v}$$
$$SbO^+ + 2\,H^+ + 3\,e \rightleftharpoons Sb + H_2O; \qquad E^0 = +0.212 \text{ v}$$
$$Sn^{2+} + 2\,e \rightleftharpoons Sn; \qquad E^0 = -0.136 \text{ v}$$
$$Co^{2+} + 2\,e \rightleftharpoons Co; \qquad E^0 = -0.277 \text{ v}$$

Codeposition of tin and cobalt could be avoided if the electrolysis were stopped before the cathode potential became too negative. However, as commonly performed, electrogravimetry is a non-selective, "brute-force" technique in which one merely separates metallic elements into two groups—those species more easily reduced than hydrogen ion or water and those elements more difficult to reduce than hydrogen ion or the solvent under the prevailing experimental conditions. Therefore, electrogravimetric analysis is usually applicable to systems in which the metal ion of interest is the only reducible substance other than hydrogen ion or water. In some instances, one may improve the selectivity of electrogravimetry by changing the pH of the sample solution or by adding appropriate complexing agents to alter the reduction potentials for the species of interest.

When a mercury pool cathode is employed in conjunction with electrolysis at either constant applied voltage or constant current, a number of useful separations can be accomplished. This approach may be used as a method of separation to precede another kind of physical or chemical measurement. Due to the unusually high activation overpotential for evolution of hydrogen gas at mercury (Table 12–1), the reduction of hydrogen ion in $1\,F$ hydrochloric acid or perchloric acid does not occur until the potential of the mercury cathode reaches approximately $-1.0$ v versus NHE. Therefore, in a $1\,M$ acid medium, all metal ions except aluminum(III), uranium(III), titanium(III), vanadium(II), molybdenum(III), tungsten(III), the tripositive lanthanide and actinide cations, and the alkaline earth and alkali metal ions are reduced to their elemental states and dissolved in mercury. Manganese(II), not expected to undergo significant reduction even at $-1.0$ v, may be deposited as the metal in mercury under some conditions.

### Influence of Experimental Conditions on Electrolytic Deposits

Analyses based on electrodeposition must fulfill the same requirements as other gravimetric methods. First, the desired substance must be deposited quantitatively. Fundamentally, the completeness of any deposition is governed by the potential of the cathode (or the anode when $PbO_2$ or $Co_2O_3$ is formed) at the time the electrolysis is terminated relative to the initial potential of that electrode. Second, the deposit must be pure; therefore, the composition of the sample solution must be adjusted to prevent codeposition of possible interferents. Third, the deposit must possess physical properties such that subsequent washing, drying, and weighing operations will not cause significant mechanical loss or chemical alteration of the substance. A lustrous, fine-grained, tightly adherent deposit must be obtained; conditions leading to the formation of spongy, porous, or fragile deposits should be avoided if possible. Let us examine some of the factors that influence the properties of electrolytic deposits.

*Current density.* Generally, the variable with the greatest effect upon the physical state of an electrolytic deposit is current density, since increases in the latter reduce the size of individual metal crystals. As a rule, current densities ranging from 0.005 to 0.05 amp/cm² yield metal deposits which are smooth and lustrous. Excessively high current densities are undesirable because the resulting deposit has a tendency to consist of fragile whiskers or dendrites and because simultaneous evolution of hydrogen gas will cause spongy deposits to form.

*Evolution of gas.* Liberation of a gas during the deposition of a metal invariably produces a rough, spongy solid. In addition, vigorous evolution of a gas may detach fragments of the metal deposit from the electrode. Evolution of gas—nearly always hydrogen, since most deposition processes are cathodic reactions—can be minimized or eliminated through the use of a depolarizer.

*Solution composition.* It is known empirically that reduction of a stable metal ion complex almost always yields a much smoother and more tightly adherent deposit than is obtainable from a solution of the aquated cation. For example, metallic silver plated from an alkaline cyanide medium containing the species $Ag(CN)_2^-$, or from an ammonia solution in which $Ag(NH_3)_2^+$ is predominant, is characteristically smooth and shiny. On the other hand, a loosely adherent, dendritic deposit of elemental silver is obtained from an aqueous silver nitrate solution in which the simple hydrated argentous ion exists.

*Nature of electrode surface.* A smooth, scrupulously clean electrode surface is usually preferable for production of a dense, fine-grained deposit. Certain deposits adhere better to one electrode material than another. In some instances, the plated metal forms a surface alloy with the electrode material, which improves the adherence of the deposit.

## Apparatus and Technique for Electrogravimetric Determinations and Separations

Although electrodes and other items of equipment needed for electrogravimetry have been identified earlier (Figure 12–3), additional details concerning experimental techniques and apparatus should be mentioned.

*Power supplies and electrolytic cells.* Power supplies for the performance of an electrolysis at either constant applied voltage or constant current are available commercially in various degrees of sophistication. However, a suitable direct-current power supply can be constructed from a 6- or 12-v storage battery connected in series with a variable-load resistor for adjustment of the applied voltage. A cell vessel for an electrogravimetric analysis normally consists of a tall-form beaker which is covered with a watch glass, split in half to allow the stems of the cathode and anode to protrude above the top of the cell. Since vigorous gas evolution causes the solution to spray, the watch glass minimizes loss of the sample.

*Electrodes for electrogravimetric determinations.* Platinum cathodes and anodes are preferred for electrogravimetry because they are not readily attacked by acids and bases, can be easily cleaned with appropriate solvents to dissolve old metal deposits, and, if necessary, may be heated in a flame to remove organic matter which might cause a poor or nonuniform deposit. Copper, lead, and nickel cathodes are occasionally employed for some electrogravimetric determinations.

To meet the requirements of maximal surface area and minimal weight and to permit efficient circulation of the sample solution around the electrode, a platinum cathode is constructed in the shape of a gauze cylinder, ranging from 3 to 4 cm in diameter and from 4 to 6 cm in height (Figure 12–3). A stout platinum wire is welded to the gauze cylinder for convenience in the handling and weighing operations and for making electrical connections to the electrode. Frequently, the anode is another platinum gauze electrode with dimensions somewhat smaller than the cathode. However, the anode may simply be a coil of platinum wire.

*Cell with mercury pool cathode for electrolytic separations.* As already described, metal ions more easily reduced than hydrogen ion can be deposited into a mercury cathode and thereby separated from metal cations which do not undergo reduction. Figure 12–5 illustrates an electrolysis cell which is suitable for electrolytic separations; the area of the mercury pool cathode may vary from 10 to 50 cm². After the desired separation has been completed, the leveling bulb is lowered without interruption of the electrolysis until the level of mercury in the main vessel drops to the stopcock. This procedure is followed so that metals in the mercury phase cannot be reoxidized by oxygen present in the aqueous medium. Then the stopcock is closed

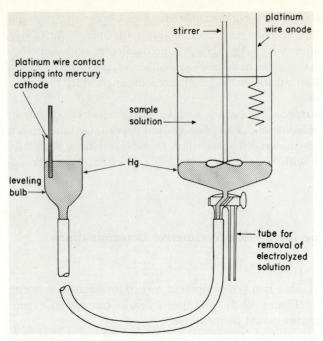

*Figure 12–5.* Electrolysis cell incorporating a mercury pool cathode for separation of metal cations.

to terminate the electrolysis, and the solution is withdrawn from the cell for subsequent analysis. Metals dissolved in the mercury are not recovered, but the impure mercury is saved for repurification and reuse.

***Termination of an electrolysis.*** As soon as the desired metal or metals have been quantitatively deposited upon or into an appropriate electrode, the electrolysis may be terminated. How can one determine when to stop an electrolysis? There are at least three methods. First, if a metal cation is intensely and distinctively colored, such as copper(II), one can continue the electrolysis for just a few minutes beyond the time at which the color of the ion disappears from solution. Second, if the metal ion is not intensely colored, or is perhaps colorless, a few drops of the sample solution may be withdrawn from the cell periodically and tested with a reagent that forms a colored complex or precipitate with the metal cation. Third, one may add a little water to the sample solution after the electrolysis has progressed for awhile; since a clean area of the electrode will become exposed to the solution, one can watch to see if any metal deposit forms on that fresh surface.

## CONTROLLED-POTENTIAL COULOMETRY

In the preceding section of this chapter, we concluded that electrolysis at either constant applied voltage or constant current lacks selectivity for analytical separations and determinations. This is because variations in the potential of the cathode or anode at which the desired reaction is supposed to occur may cause extraneous processes to take place. On the other hand, by controlling the potential of an anode or cathode so that only the desired reaction will proceed and by taking advantage of the influence of solution composition on the standard or formal potentials for half-reactions, one may achieve utmost specificity in electrochemical determinations and separations. This is the principle of controlled-potential coulometry.

## Current-Potential Curves

In controlled-potential coulometry, it is essential to know the behavior of the **working electrode**—that is, the anode or cathode at which the reaction of interest proceeds. Information of this kind may be obtained from current-potential curves such as those illustrated in Figure 12–6. A current-potential curve is a plot of the current, which flows as a consequence of reactions at the working electrode, versus the potential of that electrode, measured against an appropriate reference electrode. Such a curve depicts only the characteristics, including any activation overpotential, of processes which occur at the working electrode. One need not be concerned about the ohmic potential drop or about the reactions that take place at the other electrode

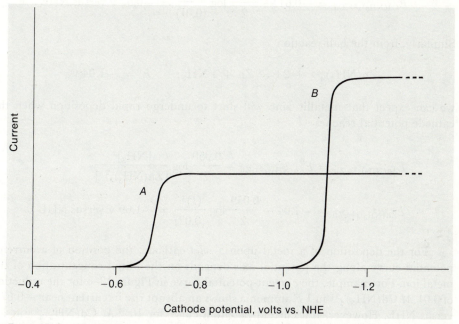

*Figure 12–6.* Idealized current-potential curves. Curve $A$: reduction of $0.01\ M\ \mathrm{Cd(NH_3)_4^{2+}}$ at a platinum cathode in a $1\ F$ aqueous ammonia solution. Curve $B$: reduction of $0.02\ M$ $\mathrm{Zn(NH_3)_4^{2+}}$ at a platinum cathode in a $1\ F$ aqueous ammonia medium.

of the electrolytic cell. To be meaningful, a current-potential curve must be recorded under the same conditions as those to be used in an actual controlled-potential coulometric determination.

Figure 12–6 presents two idealized current-potential curves, one for the reduction of a $0.01\ M$ solution of the $\mathrm{Cd(NH_3)_4^{2+}}$ complex in $1\ F$ aqueous ammonia at a platinum cathode (curve $A$) and the other for reduction of a $0.02\ M$ solution of the $\mathrm{Zn(NH_3)_4^{2+}}$ complex under identical conditions (curve $B$).* Each curve exhibits a small residual current until a potential is reached at which *rapid* deposition of either cadmium or zinc metal commences, whereupon a sudden rise in current is observed.

---

* These curves pertain only to the initial conditions cited. For each example, the position of the curve along the potential axis and the size of the limiting current both change as the concentration of the metal ion complex is varied; the position of each curve along the potential axis also depends on the ammonia concentration.

Since the half-reaction and standard potential for the two-electron reduction of $Cd(NH_3)_4^{2+}$ are

$$Cd(NH_3)_4^{2+} + 2\,e \rightleftharpoons Cd + 4\,NH_3; \qquad E^0 = -0.61\ v$$

we can predict from the Nernst equation

$$E_{Cd(NH_3)_4^{2+},Cd} = E^0_{Cd(NH_3)_4^{2+},Cd} - \frac{0.059}{2} \log \frac{[NH_3]^4}{[Cd(NH_3)_4^{2+}]}$$

that cadmium metal should begin to deposit rapidly when the cathode attains a potential of

$$E_{Cd(NH_3)_4^{2+},Cd} = -0.61 - \frac{0.059}{2} \log \frac{(1)^4}{(0.01)} = -0.67\ v \text{ versus NHE}$$

Similarly, from the half-reaction

$$Zn(NH_3)_4^{2+} + 2\,e \rightleftharpoons Zn + 4\,NH_3; \qquad E^0 = -1.04\ v$$

we can expect that metallic zinc will start to undergo rapid deposition when the cathode potential reaches

$$E_{Zn(NH_3)_4^{2+},Zn} = E^0_{Zn(NH_3)_4^{2+},Zn} - \frac{0.059}{2} \log \frac{[NH_3]^4}{[Zn(NH_3)_4^{2+}]}$$

$$E_{Zn(NH_3)_4^{2+},Zn} = -1.04 - \frac{0.059}{2} \log \frac{(1)^4}{(0.02)} = -1.09\ v \text{ versus NHE}$$

For the deposition of a metal upon a *solid* cathode, the position of a current-potential curve along the potential axis depends on the initial concentration of the metal ion. For example, the current-potential curve in Figure 12–6 for the reduction of 0.01 $M$ $Cd(NH_3)_4^{2+}$ in 1 $F$ ammonia shows an abrupt rise in current near $-0.67$ v versus NHE. However, if a current-potential curve for $10^{-4}$ $M$ $Cd(NH_3)_4^{2+}$ in 1 $F$ ammonia were recorded, one would not observe the characteristic rise in current until the cathode potential reached $-0.73$ v. When both the reactant and product species are soluble, no shift of the current-potential curve along the abscissa is observed as the initial concentration of reactant is varied. A reaction for which the oxidant and reductant are both soluble in the aqueous phase is the one-electron reduction of iron(III) at a platinum or gold cathode:

$$Fe^{3+} + e \rightleftharpoons Fe^{2+}$$

Reduction of lead ion at a mercury cathode

$$Pb^{2+} + Hg + 2\,e \rightleftharpoons Pb(Hg)$$

results in the formation of a lead amalgam. Again, the oxidized and reduced substances are soluble, except that the metallic lead is dissolved in the electrode phase.

Current-potential curves may be recorded for mixtures of substances. If the various species undergo reduction or oxidation at sufficiently different potentials, a separate step or *wave* appears for each reactant.

*Limiting current.* There is another important property of current-potential curves. Observe in Figure 12–6, for either curve $A$ or curve $B$, that as the cathode potential becomes more and more negative a point is reached beyond which the current no longer increases, but remains essentially constant. For the reduction of $Cd(NH_3)_4^{2+}$, this so-called **limiting current** results because, at cathode potentials more negative than $-0.76$ v, virtually all $Cd(NH_3)_4^{2+}$ ions transported to the electrode surface by the processes of mechanical stirring, diffusion, and electrical migration undergo reduction. To a good approximation, the rate of transport of $Cd(NH_3)_4^{2+}$ to the cathode surface—and, therefore, the size of the limiting current—is directly proportional to the *difference* between the concentration of the metal ion complex in the bulk of the solution and at the surface of the electrode:

$$i_{limiting} = k\{[Cd(NH_3)_4^{2+}]_{bulk} - [Cd(NH_3)_4^{2+}]_{surface}\}$$

However, according to the Nernst equation, if the potential of the cathode is $-0.76$ v or more negative, the concentration of $Cd(NH_3)_4^{2+}$ *at the surface of the electrode* is $10^{-5}$ $M$ or less, which is negligibly small compared to the bulk concentration ($0.01$ $M$). Thus, *the limiting current is proportional to the bulk concentration* of $Cd(NH_3)_4^{2+}$:

$$i_{limiting} = k[Cd(NH_3)_4^{2+}]_{bulk}$$

A similar relation holds for all current-potential curves that exhibit a limiting current.

## Feasibility of Separations and Determinations

Suppose that we wish to separate and determine by means of controlled-potential coulometry with a platinum gauze cathode both cadmium(II) and zinc(II) in a 1 $F$ ammonia solution initially containing $0.01$ and $0.02$ $M$ concentrations of the respective metal-ammine complexes. An inspection of Figure 12–6 as well as a review of the calculations performed previously reveals that the potential of the working electrode must not be more positive than $-0.67$ v if deposition of cadmium is to occur. Furthermore, the cathode potential cannot be more negative than $-1.09$ v if codeposition of zinc is to be avoided. Let us control the working-electrode potential at $-0.88$ v, midway between the preceding two values. What concentration of $Cd(NH_3)_4^{2+}$ remains after the solution has been exhaustively electrolyzed? Using the Nernst equation, we can write

$$E_{Cd(NH_3)_4^{2+},Cd} = E^0_{Cd(NH_3)_4^{2+},Cd} - \frac{0.059}{2} \log \frac{[NH_3]^4}{[Cd(NH_3)_4^{2+}]}$$

$$-0.88 = -0.61 - \frac{0.059}{2} \log \frac{(1)^4}{[Cd(NH_3)_4^{2+}]}$$

$$\log [Cd(NH_3)_4^{2+}] = \frac{2(-0.88 + 0.61)}{0.059} = -9.15; \quad [Cd(NH_3)_4^{2+}] = 7.1 \times 10^{-10} \ M$$

This result demonstrates that very complete deposition of cadmium is obtainable.

At the conclusion of the first part of this procedure—signaled by the fact that the current falls practically to zero—the cathode may be removed from the electrolytic cell, and washed, dried, and weighed for the determination of cadmium. Before the deposition of zinc, the cadmium might be dissolved from the platinum cathode with nitric acid. Alternatively, the cadmium could be left on the electrode, and the latter returned to the electrolytic cell. Then the cathode potential is adjusted so that zinc(II) is reduced to the elemental state.

A prediction of the potential needed to ensure the quantitative deposition of zinc can be based on the Nernst equation:

$$E_{Zn(NH_3)_4{}^{2+},Zn} = E^0_{Zn(NH_3)_4{}^{2+},Zn} - \frac{0.059}{2} \log \frac{[NH_3]^4}{[Zn(NH_3)_4{}^{2+}]}$$

To deposit 99.9 per cent of the zinc, corresponding to a final $Zn(NH_3)_4{}^{2+}$ concentration of $2 \times 10^{-5}$ $M$, the required cathode potential is

$$E_{Zn(NH_3)_4{}^{2+},Zn} = -1.04 - \frac{0.059}{2} \log \frac{(1)^4}{(2 \times 10^{-5})} = -1.18 \text{ v versus NHE}$$

If the cathode potential is maintained at $-1.18$ v, the bulk concentration of $Zn(NH_3)_4{}^{2+}$ decreases with time and the limiting current diminishes until it becomes almost zero—a very small residual current will continue to flow. At this point, the controlled-potential deposition of zinc may be regarded as complete, the cathode removed from the cell, and the plated zinc washed, dried, and weighed.

However, the controlled-potential deposition of metals does not require any weighing operations. If the working-electrode potential is controlled and only one electrode reaction occurs, the total quantity of electricity produced by that reaction can be related to the weight or concentration of the original electroactive species through Faraday's law, as will be discussed subsequently.

*Minimal potential difference for separation of two species.* What must be the minimal difference in the standard or formal potentials for half-reactions involving two substances to permit their quantitative separation and determination by means of controlled-potential electrolysis? Suppose that we desire to separate two substances, $Ox_1$ and $Ox_2$, which are reduced, respectively, to $Red_1$ and $Red_2$ according to the half-reactions

$$Ox_1 + n_1e \rightleftharpoons Red_1$$

and

$$Ox_2 + n_2e \rightleftharpoons Red_2$$

Let us assume that the initial concentrations of $Ox_1$ and $Ox_2$ are identical, that $Ox_1$ is more easily reduced than $Ox_2$, that 99.9 per cent of $Ox_1$ should be reduced, but that no more than 0.1 per cent of $Ox_2$ should undergo reduction.

When all of the reactants and products are soluble in an aqueous phase and when both $n_1$ and $n_2$ are equal to one, the standard potentials for the two half-reactions must differ by at least 354 mv to accomplish the desired separation. Similarly, if $n_1$ and $n_2$ are both 2, the difference in standard potentials need be only 177 mv; and, if $n_1$ and $n_2$ are both 3, the two standard potentials may differ by as little as 118 mv.*

When both $Ox_1$ and $Ox_2$ are metal cations, are reduced to the elemental state, and are deposited at unit activity upon a solid electrode, the standard potentials for the reduction of two univalent metal ions should differ by a minimum of 177 mv for the desired separation. In addition, the permissible difference is 89 mv for separation of two divalent metal ions and 59 mv for the separation of a pair of trivalent cations.

---

* These results are also valid for controlled-potential separations in which metal ions are deposited into a mercury pool cathode *if the volumes of the solution and the mercury are equal.*

In view of our ability to make predictions based on the Nernst equation, why is it necessary to record current-potential curves? Real chemical systems are much more complicated than the idealized situations we have discussed. Data are rarely available concerning the effects of complexation, pH, temperature, ionic strength, and other variables on standard potentials. Furthermore, it is impossible to predict the magnitudes of activation overpotentials for electrode reactions. However, current-potential curves, if recorded under conditions identical to those used in a practical controlled-potential separation and determination, do provide the information needed to help one choose the correct control potentials.

## Current-Time Behavior and Faraday's Law

A useful characteristic of controlled-potential electrolysis is the variation of current with time. Let us consider the deposition of cadmium metal upon a platinum cathode from a well-stirred 0.01 $M$ solution of $Cd(NH_3)_4{}^{2+}$ in 1 $F$ aqueous ammonia. Suppose that the cathode potential is controlled at a value on the limiting-current plateau of the current-potential curve for $Cd(NH_3)_4{}^{2+}$ reduction—for example, $-0.88$ v versus NHE on curve $A$ of Figure 12–6. Furthermore, aside from a negligibly small residual current, we will assume that the only electrode process occurring at $-0.88$ v is the reduction of $Cd(NH_3)_4{}^{2+}$. In other words, the reaction of interest is presumed to take place with 100 per cent current efficiency. In fact, the success of controlled-potential coulometry demands that only the species being determined react at the working electrode. At the moment the controlled-potential deposition begins, the current due to reduction of $Cd(NH_3)_4{}^{2+}$ ion has some initial value, $i_0$. When one half of the cadmium has been deposited, the current, being proportional to the bulk concentration of $Cd(NH_3)_4{}^{2+}$, will have diminished to 0.5 $i_0$. After 90 per cent of the $Cd(NH_3)_4{}^{2+}$ ion has undergone reduction, the current would be $0.1i_0$, and the current after 99 and 99.9 per cent of the cadmium is deposited should have a value of $0.01i_0$ and $0.001i_0$, respectively.

For the ideal situation described in the previous paragraph, the current decays in accordance with the exponential expression

$$i_t = i_0 e^{-kt}$$

where $i_0$ is the initial limiting current and $i_t$ is the limiting current at any time $t$. In this relation, the constant $k$ increases if the electrode area or the rate of mechanical stirring or the temperature becomes greater, whereas $k$ decreases if the volume of the solution is increased. To minimize the time required to complete a controlled-potential electrolysis, these experimental parameters should be manipulated to make the value of $k$ as large as possible. Although the form of the exponential decay law prevents the current from ever reaching zero, the magnitude of the final current relative to the initial current does provide the most straightforward means to determine when an electrolysis is quantitatively complete. For example, if an error not exceeding one part per thousand is desired, the electrolysis should be continued until the current decays to 0.1 per cent of its initial value.

Fundamentally, the quantity of electricity, $Q$, in coulombs passed in any electrolysis is given by the relation

$$Q = \int_0^t i_t \, dt$$

where $i_t$ is the current in amperes as a function of time and $t$ is the duration of the electrolysis in seconds. Integration of the current-time curve for a controlled-potential electrolysis may be accomplished in either of two ways. First, a mechanical or electronic *current-time integrator*, with provision for direct readout of the number of coulombs of electricity, can be incorporated into the electrical circuit. Alternatively, the quantity of electricity corresponding to the desired reaction can be determined by means of a *chemical coulometer* placed in series with the electrolysis cell. Some of these devices are described in the next section of this chapter.

Once the quantity of electricity, $Q$, has been measured, the weight, $W$, in grams of the substance being determined is obtainable from Faraday's law of electrolysis

$$W = \frac{QM}{nF}$$

where $M$ is the formula weight of the species that is oxidized or reduced, $n$ is the number of faradays of electricity involved in the oxidation or reduction of one formula weight of substance, and $F$ is the Faraday constant (96,487 coulombs).

## Equipment for Controlled-Potential Coulometry

Figure 12–7 shows the apparatus needed for controlled-potential coulometry—a cell with appropriate working, auxiliary, and reference electrodes; a variable voltage source-potentiostat combination; and a coulometer.

***Electrodes and cells for controlled-potential coulometry.*** A coulometric cell contains three electrodes. First, there is the *working electrode*, whose potential is controlled to permit only one reaction to occur. Second, the *auxiliary electrode* serves, along with the working electrode, to complete the electrolysis circuit. Third, there must be a *reference electrode* against which the potential of the working electrode is measured and controlled.

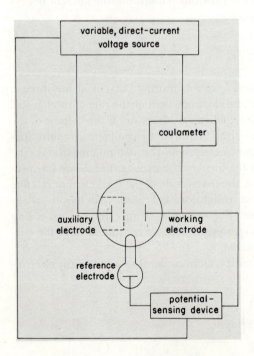

*Figure 12–7.* Apparatus for controlled-potential coulometry.

Platinum and mercury working electrodes are commonly employed, but gold, carbon, and silver are occasionally useful. Mercury is so easily oxidized that it can be used as an anode only for reactions of readily oxidizable substances; platinum, gold, and carbon are much better anode materials. On the other hand, the high activation overpotential for hydrogen gas evolution on mercury renders this electrode superior for cathodic reactions.

An auxiliary electrode is quite often fabricated from the same material as the working electrode. There is a danger that substances formed at the auxiliary electrode may be stirred to the working electrode at which they would react, or vice versa. To prevent this, the auxiliary electrode is situated in a separate compartment of the cell that makes electrolytic contact with the working-electrode compartment through a sintered-glass disk. A saturated calomel or a silver-silver chloride half-cell is usually chosen as the reference electrode; it is brought into electrolytic contact with the sample solution through a salt bridge.

Two cells described previously in this chapter require only slight modification for controlled-potential coulometry. For oxidation or reduction processes at a platinum gauze working electrode, the cell illustrated in Figure 12–3 is suitable. A length of platinum wire placed in a separate compartment of the cell could be employed as the auxiliary electrode, and an appropriate reference electrode must be added. It would be necessary to close the cell with a stopper containing holes for insertion of the electrodes and other accessories used for stirring the solution and removing dissolved oxygen.

Separations and determinations with a mercury pool cathode can be performed in the cell depicted in Figure 12–5, if the auxiliary electrode is placed in a separate compartment, and if the cell is fitted with an air-tight cover and equipped with inlet and outlet tubes so that nitrogen can be bubbled through the solution to remove oxygen. In addition, a saturated calomel reference electrode with a salt-bridge tube long enough to nearly touch the surface of the stirred mercury pool is needed.

**Voltage source and potentiostat.** As discussed earlier, the total voltage, $E_{app}$, impressed across an electrolytic cell is

$$E_{app} = E_a - E_c + iR$$

where $E_a$ and $E_c$ denote the actual potentials (versus a reference electrode) of the anode and cathode, respectively, and where $iR$ is the ohmic potential drop of the cell.

Suppose that we wish to determine the quantity of $Cd(NH_3)_4^{2+}$ ion in a $1\ F$ ammonia solution by depositing cadmium metal on a platinum cathode held at a potential ($E_c$) of $-0.88$ v versus NHE (curve $A$, Figure 12–6). During the reduction of $Cd(NH_3)_4^{2+}$, the ohmic potential drop ($iR$) decreases almost to zero due to the exponential decay of the electrolysis current. Moreover, a decrease in the current causes the activation overpotential for evolution of oxygen at the anode to diminish so that $E_a$ becomes less positive. Therefore, the total applied voltage must be continuously decreased throughout the electrolysis to keep the cathode potential constant. For controlled-potential coulometric determinations, one can now choose any of a number of commercially available electronic *potentiostats* which automatically monitor the potential of the working electrode and adjust the applied voltage to maintain that potential at the preselected value. Because potentiostats are relatively complex instruments, most commercial models are priced between one and two thousand dollars.

**Coulometers.** Measurement of the quantity of electricity corresponding to oxidation or reduction of the desired substance can be accomplished with the aid of either a chemical coulometer or a current-time integrator connected in series with the coulometric cell. Chemical coulometers are electrochemical cells in which specified reactions occur with 100 per cent current efficiency. In a silver coulometer, metallic silver is deposited quantitatively from a silver nitrate solution upon a platinum cathode which is weighed to determine the amount of silver metal and, therefore, the number of coulombs passed during the electrolysis. A hydrogen-oxygen coulometer involves electrolysis of a dilute aqueous potassium sulfate solution between two platinum electrodes. Water is oxidized to oxygen at the anode, whereas hydrogen gas is formed at the cathode. These evolved gases are collected together in a closed chamber above the potassium sulfate solution, and their volume is measured at a known temperature and pressure. From the observed volume, temperature, and pressure, one can compute the quantity of electricity. Sometimes a hydrazine sulfate solution is used instead of the potassium sulfate medium, in which case nitrogen gas is produced at the anode.

One of the most simple current-time integrators employs a precision capacitor which is charged by a known fraction of the current passing through the coulometric cell. Within broad limits, the quantity of electricity, $Q$, in coulombs on a capacitor is the product of the capacitance, $C$, of the capacitor in coulombs per volt and the voltage, $V$, in volts across the plates of the capacitor:

$$Q = CV$$

During a controlled-potential electrolysis, a charge proportional to the number of coulombs of electricity passed through the cell accumulates on the capacitor. Then $V$ is measured with a potentiometer and $Q$ is calculated from the preceding relation. Finally, $Q$ for the capacitor is multiplied by the proportionality constant (which one determines in a separate experiment by finding how many coulombs of charge accumulate on the capacitor when a known number of coulombs of electricity pass through the cell) to obtain the quantity of electricity corresponding to oxidation or reduction of the desired substance.

### Applications of Controlled-Potential Coulometry

Controlled-potential coulometry has been applied to the determination of both inorganic and organic substances. In addition, this technique is valuable for the electrochemical generation of species that cannot be conveniently prepared by means of ordinary chemical methods. In general, analyses based on controlled-potential coulometry are accurate to within 0.5 to 1 per cent, although errors may be as small as 0.1 per cent. Controlled-potential coulometry is best suited for the determination of 0.05 to 1 milliequivalent of the desired substance in 50 to 100 ml of solution.

*Cathodic separation and determination of metals.* An elegant example of the use of controlled-potential coulometry for the separation and determination of metal cations is the analysis of alloys containing copper, bismuth, lead, and tin. A weighed sample is dissolved in a mixture of nitric acid and hydrochloric acid; sodium tartrate, succinic acid, and hydrazine are added to the solution and the pH is adjusted to 6.0. From this solution, copper can be deposited quantitatively upon a platinum cathode whose potential is maintained at −0.30 v versus SCE. Then the cathode is weighed to establish the quantity of copper, the copper-coated cathode is returned to the cell, and bismuth is plated upon the electrode at a potential of −0.40 v. After the gravimetric determination of bismuth, the cathode is again returned to the cell for the deposition of lead at a potential of −0.60 v. Finally, when the weight of the lead deposit has been determined, the cathode is replaced in the cell, the solution is acidified to decompose the tin(IV)-tartrate complex, the tin(IV) is reduced at a potential of $-0.60$ v, and the weight of tin metal is obtained.

Procedures such as the one just outlined are available for the separation and determination of one or more components of the following mixtures: antimony, lead, and tin; lead, cadmium, and zinc; silver and copper; nickel, zinc, aluminum, and iron; and rhodium and iridium.

In the analyses performed with a platinum cathode, the techniques of electrogravimetry and controlled-potential separation are combined; there is no need to measure the quantity of electricity corresponding to deposition of each metal. On the other hand, in controlled-potential coulometry with a mercury pool cathode, measurement of the number of coulombs passed during the electrolysis is essential; weighing of the electrode is unnecessary, and metal ions can be determined that do not form satisfactory deposits on platinum or that do not undergo reduction to the elemental state. A number of separations and determinations have been accomplished with the aid of the mercury pool cathode. Lead(II) can be separated from cadmium-(II) by deposition of lead into a mercury electrode whose potential is controlled at $-0.50$ v versus SCE in a $0.5\,F$ potassium chloride medium. Copper(II) and bismuth(III) in an acidic tartrate solution can be separated and determined by means of controlled-potential coulometry with a mercury pool cathode. A method for the analysis of a mixture of nickel(II) and cobalt(II) entails the selective deposition of nickel from an aqueous pyridine solution of pH 6.5 into mercury at a potential of $-0.95$ v versus SCE. After the quantity of electricity corresponding to reduction of nickel(II) has been obtained, the cathode potential is reset to $-1.20$ v for the reduction of cobalt(II). Uranium(VI) may be determined through controlled-potential reduction to uranium(IV) in a $1\,F$ sulfuric acid medium at a mercury pool cathode.

***Anodic reactions at platinum and silver electrodes.*** Controlled-potential determinations involving anodic processes at platinum electrodes in $1\,F$ sulfuric acid include the oxidation of iron(II) to iron(III), of arsenic(III) to arsenic(V), and of thallium(I) to insoluble thallium(III) oxide.

By using silver electrodes one can carry out the controlled-potential coulometric determination of individual halides as well as the analysis of halide mixtures, the reaction being

$$Ag + Br^- \rightleftharpoons AgBr + e$$

when bromide ion is present.

***Determination of organic compounds.*** A variety of organic substances have been determined by means of controlled-potential coulometry. Reduction of picric acid to triaminophenol at a mercury pool cathode

permits solutions of the original acid to be analyzed. Other nitro compounds (including o-nitrophenol, m-nitrophenol, and p-nitrobenzoic acid) as well as azo dyes and nitroso compounds can be quantitatively reduced at mercury. Trichloroacetate ion undergoes a pair of two-electron reductions at a mercury electrode, first to dichloroacetate

$$Cl_3CCOO^- + H^+ + 2\,e \rightarrow Cl_2HCCOO^- + Cl^-$$

and then to monochloroacetate:

$$Cl_2HCCOO^- + H^+ + 2\,e \rightarrow ClH_2CCOO^- + Cl^-$$

Since the potentials for the two reactions differ by 0.8 v, it is possible to measure trichloroacetate in the presence of much larger or much smaller quantities of dichloroacetate and monochloroacetate. Primary, secondary, and tertiary aliphatic amides such as acetamide, N-methylacetamide, and N,N-dipropylpropionamide, respectively, can be determined by oxidation at a platinum electrode in an acetonitrile medium. Substituted hydroquinones undergo a two-electron reaction at platinum anodes which serves as the basis of an analytical method.

## COULOMETRIC TITRATIONS

Coulometric titrimetry is based upon the constant-current electrolytic generation of a titrant which reacts quantitatively with the substance to be determined. By multiplying together the magnitude of the constant current and the time required to generate an amount of titrant stoichiometrically equivalent to the desired species, one obtains the quantity of the desired substance.

### Principles of Coulometric Titrimetry

Several aspects of coulometric titrimetry can be illustrated if we consider the determination of cerium(IV) by reduction to cerium(III) in a 1 $F$ sulfuric acid medium. Assume that an oxygen-free cerium(IV) solution is contained in a coulometric cell equipped with a stirrer, that a platinum working electrode is immersed in the sample solution, and that a platinum auxiliary electrode is situated in a separate compartment filled with 1 $F$ sulfuric acid.

By referring to a schematic current-potential curve for cerium(IV) reduction at platinum (curve $A$, Figure 12–8), let us see what happens when a constant current $i_1$ is passed through this electrolytic cell such that the working electrode is the cathode. Initially, this current would be sustained completely by reduction of cerium(IV), since the limiting current of curve $A$ exceeds the selected current. As long as the limiting current for cerium(IV) reduction is equal to or greater than $i_1$, the potential of the cathode remains close to 1.44 v and the current efficiency for this process is 100 per cent. As electrolysis continues, the limiting current for reduction of cerium-(IV) drops below $i_1$, the current efficiency becomes less than 100 per cent, and the cathode potential shifts abruptly to a value at which another electrode reaction—hydrogen ion reduction at about 0 v—occurs to maintain the constant current. If the current is originally chosen to be as large as $i_2$, the cathode immediately adopts a potential at which hydrogen gas is evolved and the current efficiency for cerium(IV) reduction is less than 100 per cent, even at the beginning of the electrolysis.

If, as quickly as it were produced, hydrogen gas reacted stoichiometrically and rapidly with cerium (IV), all the electricity passing through the cell would result in reduction of cerium(IV) to cerium(III). Even though cerium(IV) might not undergo direct reaction at the platinum cathode, cerium(IV) would be reduced chemically by hydrogen gas, so the net result would be the same as if cerium(IV) were reduced with 100 per cent current efficiency. However, since hydrogen gas does not react with cerium(IV) at a measurable rate, the constant-current coulometric titration of cerium(IV) fails under these conditions.

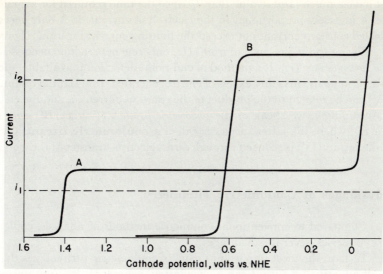

*Figure 12–8.* Schematic current-potential curves for one-electron reductions of cerium(IV), curve *A*, and iron(III), curve *B*, at a platinum cathode in 1 *F* sulfuric acid. At approximately 0 v for each curve, the current rises without apparent limit due to reduction of the high concentration of hydrogen ion. Further discussion of current-potential curves is presented on pages 409–411.

   To overcome this difficulty, one must add to the original cerium(IV) solution some substance which is reduced at a platinum cathode more easily than hydrogen ion but less readily than cerium(IV). Furthermore, the *reduced* form of this substance must undergo a quantitative and instantaneous reaction with cerium(IV). A reagent which fulfills these requirements excellently is iron(III), since its reduction product, iron(II), immediately reduces cerium(IV).

   To understand how the presence of iron(III) ensures the success of the constant-current coulometric titration of cerium(IV), we must refer to a current-potential curve for reduction of iron(III) at a platinum electrode in a 1 *F* sulfuric acid medium (curve *B*, Figure 12–8). We have seen that, if a constant current $i_1$ is passed through the cell, the limiting current for reduction of cerium(IV) eventually decreases below $i_1$, at which time the working-electrode potential must shift in a cathodic direction until some other current-sustaining process occurs. When a relatively large concentration of iron(III) is present, the cathode potential changes only to the value where the half-reaction

$$Fe^{3+} + e \rightleftharpoons Fe^{2+}$$

takes place. However, the iron(II) formed at the electrode surface is stirred into the body of the sample solution and chemically reduces cerium(IV) to cerium(III):

$$Ce^{4+} + Fe^{2+} \rightleftharpoons Ce^{3+} + Fe^{3+}$$

Since iron(II) is converted back to iron(III), the concentration of the latter stays constant. Thus, the limiting current for iron(III) reduction is the same throughout the titration, and there is no danger that the cathode potential will shift to the value at which hydrogen ion is reduced. There is no reason why the constant current should not have a much larger value than $i_1$, since the time required for a coulometric titration is inversely proportional to the electrolysis current. It is only necessary that the selected constant current not exceed the limiting current for iron(III) reduction.

Owing to the presence of iron(III), only one net reaction occurs—*the quantitative reduction of cerium(IV)*. If a method of end-point detection is available, such as a means to determine the first tiny excess of electrogenerated iron(II), the quantity of cerium(IV) can be related to the product of the constant current, $i$, and the electrolysis time, $t$. Accordingly, we speak of the *coulometric titration of cerium(IV) with electrogenerated iron(II)*, where the latter species is called a **coulometric titrant**. Of course, some of the cerium(IV) is reduced through direct electron-transfer at the working electrode.

## Advantages of Coulometric Titrimetry

Compared to conventional volumetric methods of analysis, coulometric titrimetry possesses several advantages.

Perhaps the most important virtue of coulometric titrimetry is that exceedingly small quantities of titrant can be accurately generated. It is not uncommon to perform coulometric titrations in which increments of titrant are generated with a constant current of 10 ma flowing for 0.1 second—which corresponds to only 1 millicoulomb or approximately $1 \times 10^{-5}$ milliequivalent. Such an amount of reagent would require the addition of 1 $\mu$l of 0.01 $N$ titrant from a microburet.

Coulometric titrimetry is inherently capable of much higher accuracy than ordinary volumetric analysis because the two parameters of interest—current and time—can be determined experimentally with exceptional precision. In addition, the preparation, storage, and handling of standard titrant solutions are eliminated. Moreover, titrants which are unstable or otherwise troublesome to preserve can be electrochemically generated in situ. Such uncommon reagents as silver(II), manganese(III), titanium(III), copper(I), tin(II), bromine, and chlorine can be conveniently prepared as coulometric titration.

## Apparatus and Techniques of Coulometric Titrimetry

Basic equipment needed for the performance of coulometric titrations includes a suitable source of constant current, an accurate timing device, and a coulometric cell, as indicated schematically in Figure 12–9.

*Constant-current sources.* For the majority of coulometric titrations, currents ranging from 5 to 20 ma are employed, although the use of constant currents as large as 100 ma or as small as 1 ma is occasionally desirable. Three or four 45-v batteries connected in series with a large current-limiting resistance provide a voltage source which can deliver currents up to 10 ma with a constancy of $\pm0.5$ per cent. However, one must establish the magnitude of the current experimentally by measuring the $iR$ drop across a calibrated resistor ($R$) with a potentiometer (Figure 12–9).

For coulometric titrations of the highest accuracy, electronic instruments, permitting current regulation to within $\pm0.01$ per cent and delivering currents of 200 ma or more, are commercially available. Some of these constant-current sources provide for direct readout of the number of equivalents of the substance being titrated.

*Measurement of time.* In the performance of a coulometric titration, it is invariably necessary to open and close switch $S$ a number of times as the equivalence point is approached so that small increments of titrant can be generated and permitted to react before the titration is either continued or terminated. This procedure

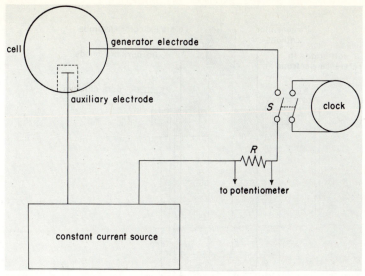

*Figure 12–9.* Diagram of essential apparatus for performance of constant current coulometric titrations.

parallels the manual closing and opening of a buret stopcock in classic volumetric analysis.

Ordinarily, time is measured with the aid of an electric stopclock powered by a synchronous motor which, in turn, is actuated through the same switch $(S)$ used to start and stop the coulometric titration. Some timing devices exhibit a start-stop error, because the synchronous motor lags in starting and coasts in stopping rather than starting and stopping abruptly. Therefore, one should employ a stopclock equipped with a solenoid brake to reduce the start-stop error to about 0.01 second for each separate operation of the switch. Nevertheless, the magnitude of the constant current should be selected to give a titration time of at least several hundred seconds so that the cumulative timing error is negligible.

*Coulometric cells.* Figure 12–10 illustrates a cell that has been employed for the coulometric titration of acids with electrogenerated hydroxide ion. Every coulometric cell contains a **generator electrode,** at which the titrant is produced, and an auxiliary electrode to complete the electrolysis circuit. Generator electrodes typically range from 10 to 25 cm² in area and are usually fabricated from platinum, gold, silver, or mercury. In most instances, a length of platinum wire serves as the auxiliary electrode. Frequently, the reaction at the auxiliary electrode yields a substance that interferes with the coulometric process. Therefore, one customarily isolates the auxiliary electrode from the main sample solution—although not for the cell shown in Figure 12–10—by placing the auxiliary electrode in a tube containing electrolyte solution but closed at the bottom with a porous glass disk.

As for other coulometric methods of analysis, the sample solution must be well stirred. In addition, the cell may be closed with a fitted cover having an inlet tube for the continuous introduction of nitrogen or sometimes carbon dioxide for removal and exclusion of oxygen from the sample solution. Depending upon the technique used to follow the progress of the titration, some cells are equipped with other electrodes. Thus, the cell pictured in Figure 12–10 contains a glass membrane indicator electrode and a salt-bridge tube leading to a reference electrode for potentiometric end-point detection of a coulometric acid-base titration.

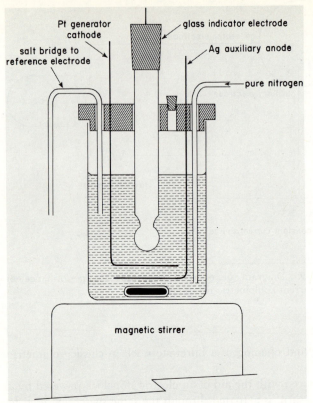

*Figure 12–10.* Coulometric cell employed for titration of acids. (Redrawn, with permission, from a paper by J. J. Lingane: Anal. Chim. Acta *11*, 283, 1954.)

***End-point detection.*** Other than the mode by which titrant is added to the sample solution, there is little difference between coulometric titrimetry and the more familiar methods of classic volumetric analysis. It is not surprising that all of the techniques discussed earlier in the text for end-point detection are applicable. Visual methods for the location of equivalence points, including the use of colored acid-base and redox indicators, have been widely employed in coulometric titrimetry. Among the instrumental techniques of end-point detection are potentiometric, amperometric, and spectrophotometric methods.

Potentiometry is an obvious method of end-point detection. In coulometric acid-base titrimetry, a glass membrane-calomel electrode pair is utilized, whereas a platinum or gold indicator electrode is the logical choice for redox reactions. Complexometric titrations involving the electrogeneration of ethylenediaminetetraacetate (EDTA) can be monitored with the aid of the mercury indicator electrode described in the preceding chapter (page 383), and the equivalence point of a precipitation titration is located through the use of a silver or mercury indicator electrode or one of the family of membrane electrodes.

Spectrophotometry has obvious applicability in conjunction with the use of visual indicators and, in addition, serves well for the location of equivalence points in coulometric titrations involving light-absorbing species such as iodine, bromine, cerium(IV), manganese(III), and titanium(III).

Every kind of titration—acid-base, precipitation, complex-formation, and oxidation-reduction—has been successfully performed coulometrically. Coulometric titrations are best suited to the determination of quantities of a substance ranging from 1 $\mu$g or less up to about 100 mg, errors characteristically being 0.1 to 0.3 per cent as compared to the 1 per cent uncertainties encountered in controlled-potential coulometry.

*Acid-base titrations.* For the determination of both strong and weak acids, advantage may be taken of the fact that electroreduction of water at a platinum cathode yields hydroxide ion:

$$2\,H_2O + 2\,e \rightleftharpoons H_2 + 2\,OH^-$$

If a platinum auxiliary anode is employed, it must be isolated from the main sample solution because oxidation of water,

$$2\,H_2O \rightleftharpoons O_2 + 4\,H^+ + 4\,e$$

produces hydrogen ion which would ruin the determination. Alternatively, the anodic generation of hydrogen ion can be eliminated through the use of the cell shown in Figure 12–10. In this cell, a silver auxiliary anode is situated directly in the sample solution, the latter containing excess bromide ion in addition to the acid to be titrated; the anode reaction

$$Ag + Br^- \rightleftharpoons AgBr + e$$

leads to formation of an innocuous coating of silver bromide upon the silver electrode. An advantage of the constant-current coulometric titration of acids with electrogenerated hydroxide ion is that one may avoid the nuisance of having to protect a standard base solution from atmospheric carbon dioxide.

Hydrogen ion, for the coulometric titration of bases, can be generated from the oxidation of water at a platinum anode. Usually, it is necessary to isolate the auxiliary cathode so that the anodically formed hydrogen ion does not undergo subsequent reduction.

By adding a small amount of water to a nonaqueous solvent, one can electrogenerate hydrogen ions at a platinum anode for the titration of weak bases without seriously modifying the properties of the solvent system. One-milligram samples of pyridine, benzylamine, and triethylamine have been determined with an error of less than 2 per cent in acetonitrile containing lithium perchlorate trihydrate as both the supporting electrolyte and the source of water.

*Precipitation titrations.* If the glass membrane electrode is replaced with a silver indicator electrode for potentiometric end-point detection, the cell of Figure 12–10 serves well for titrations with anodically generated silver ion. Individual halide ions can be accurately determined, and it is possible to analyze mixtures of these species as well. Organic chlorides or bromides may be analyzed if, after combustion of the desired compound, the liberated halide is titrated coulometrically. In addition, mercaptans are determinable with anodically formed silver ion, 3-$\mu$g quantities of mercaptans having been determined with an error of $\pm$0.2 $\mu$g and larger samples with an uncertainty of 1 per cent or less.

Electrogenerated mercury(I) can be employed to titrate chloride, bromide, and iodide individually and in mixtures; chromate, oxalate, molybdate, and tungstate form precipitates with electrogenerated lead(II).

*Complexometric titrations.* Ethylenediaminetetraacetate is generated through reduction of the stable mercury(II)-EDTA-ammine complex at a mercury pool cathode in an ammonia-ammonium nitrate buffer:

$$HgNH_3Y^{2-} + NH_4^+ + 2\,e \rightleftharpoons Hg + 2\,NH_3 + HY^{3-}$$

As the EDTA species ($HY^{3-}$) is released, it reacts rapidly with a metal ion to form the corresponding metal-EDTA complex. This procedure has been applied to the titration of calcium(II), copper(II), lead(II), and zinc(II), but can be extended to the determination of other metal ions which form EDTA complexes.

Table 12–2. Coulometric Oxidation-Reduction Titrations

| Coulometric Titrant | Generator Electrode Reaction | Applications |
|---|---|---|
| $Cl_2$ | $2\,Cl^- \rightleftharpoons Cl_2 + 2\,e$ | chlorination of unsaturated fatty acids; oxidation of As(III) to As(V), of Tl(I) to Tl(III), and of iodide to iodine |
| $Br_2$ | $2\,Br^- \rightleftharpoons Br_2 + 2\,e$ | continuous monitoring of $H_2S$ and $SO_2$ in gases and air; bromination of olefins, phenols, and aromatic amines (see p. 340); bromination of oxine and metal oxinates (see p. 341); determination of mustard gas |
| $OBr^-$ | $Br^- + 2\,OH^- \rightleftharpoons OBr^- + H_2O + 2\,e$ | determination of protein in serum; oxidation of ammonia to nitrogen |
| $I_3^-$ | $3\,I^- \rightleftharpoons I_3^- + 2\,e$ | continuous monitoring of $H_2S$ and $SO_2$ in gases and air; oxidation of As(III) to As(V), of Sb(III) to Sb(V), of hydrogen sulfide to sulfur, and of sulfur dioxide to sulfate |
| $Ce^{4+}$ | $Ce^{3+} \rightleftharpoons Ce^{4+} + e$ | oxidation of Fe(II) to Fe(III), of ferrocyanide to ferricyanide, of Ti(III) to Ti(IV), of As(III) to As(V), of U(IV) to U(VI), of iodide to iodine, and of hydroquinone to quinone |
| $Mn^{3+}$ | $Mn^{2+} \rightleftharpoons Mn^{3+} + e$ | oxidation of As(III) to As(V), of Fe(II) to Fe(III), and of $H_2C_2O_4$ to $CO_2$ |
| $Ag^{2+}$ | $Ag^+ \rightleftharpoons Ag^{2+} + e$ | oxidation of Ce(III) to Ce(IV), of V(IV) to V(V), of As(III) to As(V), and of $H_2C_2O_4$ to $CO_2$ |
| $CuCl_2^-$ | $Cu^{2+} + 2\,Cl^- + e \rightleftharpoons CuCl_2^-$ | reduction of $Cr_2O_7^{2-}$ to Cr(III), of V(V) to V(IV), of Fe(III) to Fe(II), of $IO_3^-$ to $ICl_2^-$, and of Au(III) to Au(0) |
| $Fe^{2+}$ | $Fe^{3+} + e \rightleftharpoons Fe^{2+}$ | reduction of Ce(IV) to Ce(III), of V(V) to V(IV), of $Cr_2O_7^{2-}$ to Cr(III), and of $MnO_4^-$ to Mn(II) |
| $Ti^{3+}$ | $TiO^{2+} + 2\,H^+ + e \rightleftharpoons Ti^{3+} + H_2O$ | reduction of Fe(III) to Fe(II), of Ce(IV) to Ce(III), of U(VI) to U(IV), of Mo(VI) to Mo(V), and of V(V) to either V(IV) or V(III) |

***Oxidation-reduction titrations.*** Coulometric titrimetry has been applied more extensively to redox processes than to all other kinds of reactions combined. Table 12–2 lists some of the titrants and applications which have been investigated. One of the interesting uses of coulometric redox titrimetry is the continuous monitoring of hydrogen sulfide and sulfur dioxide in air and gas samples with electrogenerated bromine or triiodide. In addition, the protein content of serum has been determined with anodically formed hypobromite ion.

## QUESTIONS AND PROBLEMS

1. When a silver electrode is polarized anodically in a solution containing chloride, bromide, or iodide ion, the following reaction occurs

$$Ag + X^- \rightleftharpoons AgX + e$$

where $X^-$ denotes the halide ion of interest.
   (a) Suppose that a silver electrode is to be polarized anodically in the controlled-potential electrolysis of an aqueous solution initially $0.0200\,F$ in sodium bromide (NaBr) and $1\,F$ in perchloric acid. What is the minimal potential versus NHE (normal hydrogen electrode) at which the silver anode must be controlled in order to ensure an accuracy of at least 99.9 per cent in the determination of bromide?
   (b) Suppose that controlled-potential coulometry with a silver anode is to be utilized for the separation and analysis of a mixture of $0.0500\,M$ chloride and $0.0250\,M$ iodide in a $1\,F$ perchloric acid medium. Within what range of potentials (versus NHE) must the silver anode be maintained to ensure the deposition of at least 99.9 per cent of the iodide as silver iodide without removal of any chloride ion from the sample solution? What is the minimal potential (versus NHE) at which the silver anode must be kept to ensure that at least 99.9 per cent of the chloride ion is deposited as silver chloride?

2. Imagine that you wish to separate copper(II) from antimony(III) by controlled-potential deposition of copper upon a platinum cathode from a $1\,F$ perchloric acid medium originally containing $0.0100\,M$ copper(II) and $0.0750\,M$ antimony(III). At what potential versus SCE (saturated calomel electrode) must the platinum cathode be controlled to cause 99.9 per cent deposition of copper? What fraction of the antimony(III) will be reduced to the metallic state? If the volume of the solution is 100 ml, what will be the total gain in weight of the cathode?

$$Cu^{2+} + 2\,e \rightleftharpoons Cu; \qquad\qquad E^0 = +0.337\text{ v versus NHE}$$

$$SbO^+ + 2\,H^+ + 3\,e \rightleftharpoons Sb + H_2O; \quad E^0 = +0.212\text{ v versus NHE}$$

3. If a $1\,F$ perchloric acid solution which is $0.0300\,M$ in antimony(III) and $0.0200\,M$ in bismuth(III) is subjected to electrolysis at constant applied voltage in a cell containing two platinum electrodes, what will be the equilibrium concentration of bismuth(III) when the cathode potential shifts to the value at which antimony(III) starts to undergo reduction? Assume that the stirring of the solution is highly efficient.

$$SbO^+ + 2\,H^+ + 3\,e \rightleftharpoons Sb + H_2O; \quad E^0 = +0.212\text{ v versus NHE}$$

$$BiO^+ + 2\,H^+ + 3\,e \rightleftharpoons Bi + H_2O; \quad E^0 = +0.32\text{ v versus NHE}$$

4. Assume that the half-reactions pertaining to the reduction of two substances, $Ox_1$ and $Ox_2$, are as follows:

$$Ox_1 + 3\,e \rightleftharpoons Red_1; \quad E^0_{Ox_1,Red_1}$$

$$Ox_2 + 2\,e \rightleftharpoons Red_2; \quad E^0_{Ox_2,Red_2}$$

If all oxidized and reduced species are soluble, if $Ox_1$ is more easily reducible than $Ox_2$, and if $Ox_1$ and $Ox_2$ are both initially present at the same concentration, what is the minimal permissible difference between the standard potentials for the two half-reactions if one desires to reduce 99.9 per cent of $Ox_1$, yet not reduce more than 0.1 per cent of $Ox_2$, during a controlled-potential electrolysis?

5. In an aqueous ammonia solution, copper(II) can be reduced at a mercury cathode in a stepwise fashion according to the following half-reactions:

$$Cu(NH_3)_4^{2+} + e \rightleftharpoons Cu(NH_3)_2^+ + 2\,NH_3; \quad E^0 = -0.22 \text{ v versus SCE}$$

$$Cu(NH_3)_2^+ + e + Hg \rightleftharpoons Cu(Hg) + 2\,NH_3; \quad E^0 = -0.51 \text{ v versus SCE}$$

(a) Suppose that you are interested in preparing a solution of copper(I) in $1\,F$ aqueous ammonia by the controlled-potential reduction of copper(II) to copper(I) at a large mercury pool cathode. Furthermore, suppose that you begin with $0.0100\,M$ copper(II) and that you desire to minimize the concentration of copper(II) unreduced as well as the concentration of metallic copper in the mercury (amalgam) phase. What potential would you employ to optimize the generation of copper(I)? What would be the ratio of copper(I) to copper(II) at the termination of the electrolysis? *Hint:* Assume that the species present in solution are $Cu(NH_3)_4^{2+}$ and $Cu(NH_3)_2^+$, that the free ammonia concentration is $1\,M$, and that concentrations (rather than activities) are applicable. Moreover, assume that the volume of mercury is equal to that of the aqueous phase.

(b) Suppose that, in an experiment of the preceding type, you had wanted to reduce at least 99.9 per cent of the copper(II) to copper(I), while at the same time you wished to avoid the reduction of more than 0.1 per cent of the copper(I) to the copper amalgam. What is the *minimal difference* (in millivolts) between the standard potentials for the copper(II)-copper(I) and copper(I)-copper(0) half-reactions necessary to achieve this goal?

(c) What possible advantage(s) does controlled-potential electrolysis possess as a preparative technique over conventional chemical preparations? What disadvantage(s) can you name?

6. An unknown sample of hydrochloric acid was titrated coulometrically with hydroxide ion electrogenerated at a constant current of 20.34 ma, a total generation time of 645.3 seconds being required to reach the equivalence point. Calculate the number of millimoles of hydrochloric acid in the sample.

7. Copper(II) may be titrated coulometrically with electrogenerated tin(II) in a $4\,F$ sodium bromide medium according to the following reaction:

$$2\,CuBr_3^- + SnBr_4^{2-} \rightleftharpoons 2\,CuBr_2^- + SnBr_6^{2-}$$

In the titration of one sample, a constant generating current of 67.63 ma was employed and the titration time was 291.7 seconds. Calculate the weight of copper present in the unknown sample.

8. A 9.14-mg sample of pure picric acid was dissolved in $0.1\,F$ hydrochloric acid and subjected to a controlled-potential coulometric reduction at $-0.65$ v versus SCE. A coulometer in series with the electrolysis cell registered 65.7 coulombs of electricity. Calculate the number of electrons involved in the reduction of picric acid, and write a plausible half-reaction for the process.

9. To determine the quantity of chlorobenzene impurity in benzene, a $1.000$-$\mu l$ sample of the liquid was injected from a microliter syringe into a vaporization chamber at $200°C$. Immediately, the vaporized sample was swept with a stream of nitrogen into an adjoining chamber at $800°C$ and was mixed with oxygen, whereupon rapid combustion of the organic compounds occurred to yield carbon dioxide, water, and hydrogen chloride. These gaseous products were carried by the nitrogen stream into a coulometric titration cell (equipped with a silver generator anode and a platinum auxiliary cathode) and were trapped with a 70 per cent solution of acetic acid in water. Next, a constant current of $64.2\,\mu a$ was passed through the electrolysis cell until the potential of a separate silver indicator electrode versus a silver-silver

acetate reference electrode showed that the equivalence point had been reached. If the coulometric titration required 48.4 seconds, calculate the weight percentage of chlorobenzene in the benzene sample. Assume that the density of the liquid sample is equal to that of pure benzene.

10. In anhydrous acetonitrile containing $0.2 F$ lithium perchlorate as supporting electrolyte, primary, secondary, and tertiary aliphatic amides undergo a one-electron oxidation at platinum, as exemplified by the following reaction scheme for N,N-dimethylacetamide:

$$CH_3CON(CH_3)_2 \rightarrow CH_3C\dot{O}N(CH_3)_2{}^+ + e$$

$$CH_3C\dot{O}N(CH_3)_2{}^+ + CH_3CN \rightarrow CH_3CONH(CH_3)_2{}^+ + \cdot CH_2CN$$

$$2 \cdot CH_2CN \rightarrow NC\!-\!CH_2CH_2\!-\!CN$$

A mixture of acetamide and N,N-dimethylacetamide dissolved in approximately 65 ml of acetonitrile containing $0.2 F$ lithium perchlorate was analyzed by means of controlled-potential coulometry. First, the potential of a platinum gauze working electrode was held at $+1.35$ v versus SCE to accomplish the selective oxidation of the tertiary amide. Then, the potential of the anode was readjusted to a value of $+1.86$ v versus SCE to oxidize the primary amide. If the readings of a coulometer in series with the electrolysis cell were 80.72 coulombs for the first oxidation and 36.98 coulombs for the second oxidation, calculate the weight in milligrams of acetamide and N,N-dimethylacetamide present in the original solution.

11. A weighed platinum gauze electrode is placed in an electrolysis cell containing 100 ml of a solution which is $0.0500 F$ in lead chloride ($PbCl_2$) and $0.0200 F$ in tin(IV) chloride ($SnCl_4$). The pH of the solution is buffered at exactly 3.00. If the solution is well stirred and if a current of 1.00 amp is passed through the cell for exactly 25.0 minutes, what will be the gain in weight of the platinum electrode? (Assume that the platinum electrode is the cathode and that the anode is isolated by a salt bridge so that the cathode and anode products do not mix. In addition, assume that the activation overpotential for evolution of hydrogen gas at the cathode is 0.22 v.)

12. A mixture of nickel(II) and cobalt(II) was analyzed by means of controlled-potential coulometry. An electrolysis cell was assembled having a well-stirred mercury pool cathode and a platinum auxiliary anode. One hundred milliliters of supporting electrolyte solution ($1.00 F$ in pyridine, $0.30 M$ in chloride ion, and $0.20 F$ in hydrazine and with a pH of 6.89) was deaerated with nitrogen gas, and exactly 5.000 ml of the nickel-cobalt sample solution was pipetted into the cell. Quantitative reduction of nickel(II) to the elemental (amalgam) state was accomplished with the potential of the mercury pool held at $-0.95$ v versus SCE; an electromechanical current-time integrator in series with the cell read 60.14 coulombs when the current had decayed. Then the potential of the mercury cathode was adjusted to $-1.20$ v versus SCE for the reduction of cobalt(II) to the elemental (amalgam) state; the final reading of the current-time integrator, corresponding to the sum of the quantities of nickel(II) and cobalt(II) was 351.67 coulombs. Calculate the molar concentrations of nickel(II) and cobalt(II) in the original sample solution.

13. An indium(III)-containing sample solution was buffered with a mixture of acetic acid and sodium acetate, was heated to approximately 75°C, and was treated with a slight excess of 8-hydroxyquinoline (oxine) to precipitate quantitatively the insoluble indium(III) oxinate salt:

$$In^{3+} + 3 HOC_9H_6N \rightleftharpoons In(OC_9H_6N)_3 + 3 H^+$$

After the precipitate was washed, it was redissolved in a minimal volume of concentrated hydrochloric acid, and the resulting solution was diluted with water to exactly 50.00 ml in a volumetric flask. A 200.0-$\mu$l aliquot of this solution was pipetted into a titration vessel containing 50 ml of $0.2 F$ sodium bromide solution and equipped with a platinum generator anode and a platinum auxiliary cathode. Bromine, electrogenerated by oxidation of bromide ion at a constant current of 12.53 ma, reacted with oxine according to the reaction

$$HOC_9H_6N + 2\,Br_2 \rightarrow HOC_9H_4NBr_2 + 2\,HBr$$

which was discussed on page 341. An end point was reached after a titration time of 186.6 seconds. Calculate the weight of indium(III) in milligrams in the original sample solution.

14. Monoprotonated ethylenediaminetetraacetate ($HY^{3-}$) can be electro-generated by means of the constant-current reduction of the mercury(II)-EDTA-ammine complex

$$HgNH_3Y^{2-} + NH_4^+ + 2\,e \rightleftharpoons Hg + 2\,NH_3 + HY^{3-}$$

at a mercury pool cathode (page 424), whereupon the free ligand reacts rapidly and quantitatively with many metal ions, including lead(II):

$$Pb^{2+} + HY^{3-} \rightleftharpoons PbY^{2-} + H^+$$

A 50.00-ml sample of gasoline, containing tetraethyllead, was refluxed with concentrated hydrochloric acid, and the resulting solution was evaporated almost to dryness. Enough water was added to dissolve the slightly soluble precipitate of lead(II) chloride. Then the lead(II) solution was added to a titration vessel containing an ammoniacal solution of $HgNH_3Y^{2-}$ and equipped with a mercury pool cathode and a platinum auxiliary anode. A constant current of 93.44 ma was passed through the cell. A mercury indicator electrode (page 383) was employed to follow the progress of the titration potentiometrically, and the end point was reached after a titration time of 578.3 seconds. If the density of tetraethyllead is 1.65 gm/ml, calculate the number of milliliters of tetraethyllead per gallon of gasoline.

15. MacNevin, Baker, and McIver (Anal. Chem., 25:274, 1953) devised a procedure called *coulogravimetric analysis*, which combines a weight determination with a measurement of the quantity of electricity in a controlled-potential electrolysis, to analyze bromide-chloride mixtures.

Exactly 25.00 ml of an aqueous solution containing bromide and chloride was pipetted into an electrolysis cell. Approximately 75 ml of a solution consisting of 0.2 F concentrations of both acetic acid and sodium acetate was added. A silver-plated anode and a platinum cathode were placed in the cell. Then the potential of the silver electrode was set at +0.25 v versus SCE, and the electrolysis was continued until the current decayed to a very small value. A current-time integrator in series with the cell registered 425.3 coulombs of electricity. When the silver anode (with its deposit of silver bromide and silver chloride) was washed, dried, and placed on an analytical balance, a weight of 7.4228 gm was obtained, whereas the weight of the silver anode before the electrolysis was 7.1994 gm. Calculate the molar concentrations of bromide and chloride ions in the original sample solution.

16. In a solution containing 2.5 F ammonia, 1 F ammonium chloride, and 2 F potassium chloride, trichloroacetate undergoes reduction to dichloroacetate

$$Cl_3CCOO^- + H_2O + 2\,e \rightarrow Cl_2CHCOO^- + OH^- + Cl^-$$

at a mercury pool cathode held at a potential of −0.90 v versus SCE. Dichloroacetate is reduced to monochloroacetate at potentials near −1.65 v versus SCE, whereas monochloroacetate is electroinactive under such conditions. A 0.3339-gm sample of impure trichloroacetic acid, known to be contaminated with dichloroacetic and monochloroacetic acids, was assayed by means of controlled-potential coulometry. An electrolysis cell consisting of a mercury pool cathode and a platinum auxiliary anode was employed, the trichloroacetic acid sample was dissolved in the supporting electrolyte solution specified above, and the potential of the cathode was maintained at −0.90 v versus SCE throughout the electrolysis. A current-time integrator in series with the coulometric cell read 352.8 coulombs when the electrolysis current had dropped essentially to zero. Calculate the weight percentage of trichloroacetic acid in the sample material.

17. For an *ideal* controlled-potential electrolysis, as discussed on page 413, the limiting current ($i_t$) at any time $t$ is related to the initial limiting current ($i_0$) through the expression

$$i_t = i_0 e^{-kt}$$

where $k$ is a constant which is unique to a specific set of experimental conditions, *viz.*, electrode area, rate of stirring of solution, volume of solution, and temperature. In addition, the total quantity of electricity ($Q$) passed during a controlled-potential electrolysis is defined by the equation

$$Q = \int_0^t i_t \, dt$$

(a) Show that the total quantity of electricity passed between zero time and infinite time in an *ideal* controlled-potential electrolysis is given by

$$Q = \frac{i_0}{k}$$

if natural logarithms are used, or by

$$Q = \frac{i_0}{2.303k}$$

if base-ten logarithms are employed.

(b) For the controlled-potential electrolytic oxidation of iron(II) to iron(III) in 1 $F$ sulfuric acid at a platinum gauze anode whose potential was held at $+0.80$ v versus SCE, the following data were obtained:

| Electrolysis time, minutes | Current, milliamperes |
|---|---|
| 1.00 | 188 |
| 2.00 | 164 |
| 4.00 | 123 |
| 6.00 | 92.8 |
| 8.00 | 69.5 |
| 10.00 | 52.5 |
| 12.00 | 39.4 |
| 14.00 | 29.7 |
| 16.00 | 22.3 |

Construct a plot of the base-ten logarithm of the current as a function of time; determine, using the method of least squares, the number of milligrams of iron in the sample solution.

18. Vitamin C tablets, advertised to each contain 120 mg of ascorbic acid, were analyzed by means of controlled-potential coulometry. A single tablet was dissolved in approximately 100 ml of a solution consisting of a 0.1 $F$ biphthalate-phthalate buffer of pH 6.03 along with 0.25 per cent oxalic acid (to inhibit the autoxidation of ascorbic acid). This solution was transferred to an electrolysis cell equipped with a platinum gauze anode and a platinum auxiliary cathode, the potential of the anode was set at $+1.09$ v versus SCE, and the electro-oxidation of ascorbic acid

$$C_6H_8O_6 \rightleftharpoons C_6H_6O_6 + 2\,H^+ + 2\,e$$

was allowed to reach completion. A current-time integrator in series with the electrolysis cell had a final reading (corrected for the oxidation of oxalic acid) of 128.9 coulombs. Calculate the weight of ascorbic acid in the tablet.

19. Jolly and Boyle (Anal. Chem., *43*:514, 1971) have devised a method for the constant-current coulometric titration of weak acids dissolved in liquid ammonia. At a platinum cathode immersed in liquid ammonia containing potassium bromide as a supporting electrolyte, electrons are produced which become solvated by the ammonia:

$$e(\text{platinum}) \rightarrow e(\text{ammonia})$$

These ammoniated electrons can react with any weak acid (HA) to yield hydrogen gas and the conjugate base of the acid:

$$e(\text{ammonia}) + HA \rightarrow \tfrac{1}{2} H_2 + A^-$$

For coulometric titrations in liquid ammonia, the cell must be cooled with a dry ice–alcohol mixture; a platinum auxiliary anode is employed, and potentiometric end-point detection is utilized.

A *moist* sample of ammonium chloride dissolved in liquid ammonia containing potassium bromide was titrated with ammoniated electrons generated with a constant current of 50.99 ma. One end point, corresponding to the conversion of ammonium ion to ammonia, was detected at a titration time of 811.3 seconds. A second end point, due to the conversion of water to hydroxide ion, was observed at a *total* titration time of 948.6 seconds. Calculate the weight in milligrams of ammonium chloride in the sample, and determine the weight percentage of water in the moist ammonium chloride.

20. A 100.0-mg tablet containing sodium phenobarbital, a long-acting sedative with the formula $NaC_{12}H_{11}N_2O_3$, was dissolved in approximately 5 ml of a $0.5\,F$ solution of sodium perchlorate in 20 per cent aqueous acetone. This solution was transferred quantitatively to a coulometric titration vessel equipped with a mercury pool anode and a platinum auxiliary cathode (in an isolated compartment) and thermostated at 37°C. Mercury(II) was generated by anodic polarization of the well-agitated mercury electrode by passage of a constant current of 200.2 ma through the cell. As mercury(II) formed, it reacted rapidly and quantitatively with the barbiturate to yield a one-to-one complex (which may be polymeric). An end point was observed (by means of two-electrode amperometry) at a titration time of 210.3 seconds. Calculate the weight percentage of sodium phenobarbital in the tablet.

## SUGGESTIONS FOR ADDITIONAL READING

1. G. Charlot, J. Badoz-Lambling, and B. Trémillon: *Electrochemical Reactions*. Elsevier, Amsterdam, 1962.
2. D. D. DeFord and J. W. Miller: Coulometric analysis. *In* I. M. Kolthoff and P. J. Elving, eds.: *Treatise on Analytical Chemistry*. Part I, Volume 4, Wiley-Interscience, New York, 1959, pp. 2475–2531.
3. A. J. Lindsey: Electrodeposition. *In* C. L. Wilson and D. W. Wilson, eds.: *Comprehensive Analytical Chemistry*. Volume II-A, Elsevier, Amsterdam, 1964, pp. 7–64.
4. J. J. Lingane: *Electroanalytical Chemistry*. Second edition, Wiley-Interscience, New York, 1958.
5. L. Meites: *Polarographic Techniques*. Second edition, Wiley-Interscience, New York, 1965.
6. L. Meites and H. C. Thomas: *Advanced Analytical Chemistry*. McGraw-Hill Book Company, New York, 1958.
7. G. W. C. Milner and G. Phillips: *Coulometry in Analytical Chemistry*. Pergamon Press, New York, 1967.
8. W. C. Purdy: *Electroanalytical Methods in Biochemistry*. McGraw-Hill Book Company, New York, 1965.
9. G. A. Rechnitz: *Controlled-Potential Analysis*. The Macmillan Company, New York, 1963.
10. W. D. Shults: Coulometric methods. *In* F. J. Welcher, ed.: *Standard Methods of Chemical Analysis*. Sixth edition, Volume III-A, Van Nostrand, Princeton, New Jersey, 1966, pp. 323–376.
11. N. Tanaka: Electrodeposition. *In* I. M. Kolthoff and P. J. Elving, eds.: *Treatise on Analytical Chemistry*. Part I, Volume 4, Wiley-Interscience, New York, 1959, pp. 2417–2473.

# POLAROGRAPHY AND AMPEROMETRIC TITRATIONS

# 13

In 1922, Jaroslav Heyrovský published a paper* describing a new electro-analytical technique called **polarography**, a discovery that eventually led to his receiving the Nobel prize in chemistry for 1959. In the half-century following its inception, polarography has provided a foundation for development of many electro-chemical methods of analysis.

In polarography, one is interested in the current-potential curves resulting from electron-transfer processes that occur at the surface of a dropping mercury electrode only a few square millimeters in area. From such measurements the identity and concentration of the reactant can be determined. In his Nobel lecture,† Heyrovský reviewed some of the virtues of polarography—that solutions as dilute as $10^{-5}\ M$ can be analyzed, that samples as small as 0.05 ml may be handled routinely, and that almost every element, in one form or another, as well as many hundreds of organic compounds can be determined.

Seven years after the first report of polarography, Heyrovský and a co-worker‡

---

* J. Heyrovský: Chem. listy *16*:256, 1922.

† J. Heyrovský: The trends of polarography. In *Nobel Lectures in Chemistry for 1942–1962*, Elsevier Publishing Company, Amsterdam, pp. 564–584.

‡ J. Heyrovský and S. Berezický: Collection Czechoslov. Chem. Commun., *1*:19, 1929.

employed polarographic techniques to locate the equivalence point for the titration of barium ion with sulfate. This method, now known as **amperometric titration**, has many useful applications, some of which we will examine in the second part of this chapter.

## POLAROGRAPHY

Polarography derives its analytical importance from two characteristics of current-potential curves or polarograms obtained with a dropping mercury electrode. First, the position of a polarogram along the potential axis may indicate the identity of the substance which undergoes electron transfer. Second, under experimental conditions that are easily achieved, a polarogram exhibits a diffusion-controlled limiting current whose magnitude is governed by the concentration of the electroactive substance.

### A Polarographic Cell and the Dropping Mercury Electrode

Some of the apparatus used to perform polarographic measurements is pictured in Figure 13–1. A so-called H-cell, comprising two compartments separated by a

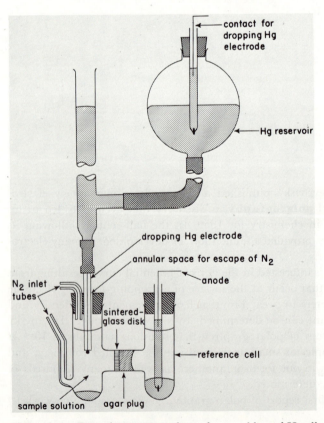

*Figure 13–1.* Dropping mercury electrode assembly and H-cell for polarographic measurements. (Redrawn, with permission, from a paper by J. J. Lingane and H. A. Laitinen: Ind. Eng. Chem., Anal. Ed., *11*: 504, 1939.)

sintered-glass disk and an agar plug saturated with potassium chloride, is commonly employed. A dropping mercury electrode consists of a length of thin-bore capillary tubing attached to the bottom of a stand tube and a mercury reservoir. Under the influence of gravity, mercury issues from the orifice of the capillary in a series of identical droplets, approximately one-half millimeter in diameter and a few square millimeters in area. By adjusting the height of the mercury column, one may vary the **drop time**—that is, the time required for a fresh drop of mercury to emerge, grow, and fall from the tip of the capillary. Drop times range from three to six seconds with mercury heights of 30 cm or more.

Since oxygen undergoes stepwise reduction at a mercury cathode (as shown in Figure 13–4),

$$O_2 + 2\ H^+ + 2\ e \rightleftharpoons H_2O_2$$

and

$$H_2O_2 + 2\ H^+ + 2\ e \rightleftharpoons 2\ H_2O$$

and masks the reactions of other substances, nitrogen is bubbled through the sample solution to remove oxygen prior to the recording of a polarogram and is passed over the surface of the solution during the measurements.

A reference electrode, most often a saturated calomel electrode, is situated in the second compartment of the H-cell. Because the reference electrode serves simultaneously as the auxiliary electrode, it must be large compared to the area of a mercury drop so that the potential of the reference electrode remains essentially constant during the passage of microampere currents through the polarographic cell.

***Advantages and limitations of the dropping mercury electrode.*** Other microelectrodes, particularly those fabricated from platinum, have been used to record current-potential curves, but the dropping mercury electrode possesses three advantages. First, the continuous growth and fall of mercury droplets means that a fresh electrode surface is exposed to the sample solution every few seconds; consequently, the reaction being studied is not affected by the course of electrolysis at previous drops. Second, the large activation overpotential for hydrogen gas evolution at mercury makes the dropping mercury electrode valuable for the study of cathodic processes—most notably, the reduction of metal ions. Third, with the dropping mercury electrode it is possible to obtain a reproducible diffusion-controlled limiting current which is quantitatively related to the concentration of the reacting species.

Mercury has one serious drawback as an electrode because of its relative ease of oxidation. A dropping mercury electrode begins to undergo anodic dissolution in a noncomplexing medium, such as perchloric acid, at a potential of approximately +0.25 v versus SCE. In the presence of 1 $M$ chloride ion, mercury is oxidized to insoluble mercury(I) chloride at about 0 v versus SCE. Thus, the dropping mercury electrode is largely restricted to the study of cathodic processes, although some easily oxidized species can be determined polarographically.

## Nature of a Polarogram

Suppose that an oxygen-free 0.1 $F$ potassium nitrate solution containing $2.00 \times 10^{-3} F$ thallium(I) nitrate is transferred into a polarographic cell. If the potential of the dropping mercury electrode is scanned in a cathodic direction from 0 to −1.8 v versus SCE, the polarogram shown as curve $A$ in Figure 13–2 is observed. This current-potential curve was obtained with the aid of a **polarograph**, an instrument

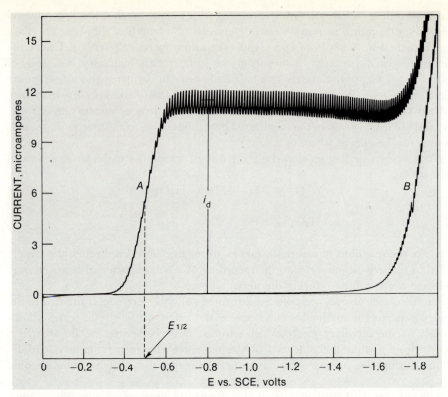

*Figure 13–2.* A representative polarogram. Curve *A*: polarogram for reduction of $2.00 \times 10^{-3} F$ thallium(I) nitrate in a $0.1 F$ potassium nitrate medium containing $0.001\%$ Triton X-100 as maximum suppressor. Curve *B*: residual-current curve for a $0.1 F$ potassium nitrate solution.

which automatically increases the voltage impressed across the polarographic cell and simultaneously records the current-potential curve on moving chart paper. According to polarographic convention, currents resulting from cathodic processes are positive, whereas anodic currents are negative.

A complete plot of current as a function of the dropping-electrode potential is termed a **polarogram**, and the rapidly ascending portion of the curve, as appears around −0.45 v versus SCE in Figure 13–2, is known as a **polarographic wave**. Reduction of thallium(I) with the resultant formation of a thallium amalgam

$$\text{Tl}^+ + e + \text{Hg} \rightleftharpoons \text{Tl(Hg)}$$

is responsible for this polarographic wave. A well-defined limiting-current region for reduction of thallium(I) appears between −0.7 and −1.6 v; however, the abrupt rise in current beyond the latter potential is caused by reduction of potassium ions to form a potassium amalgam.

*Current oscillations.* Variations in current during the lifetime of a single drop are much larger than suggested by the fluctuations shown in Figure 13–2. At the beginning of each drop lifetime, the area of the electrode is equal to the tiny cross section of the capillary orifice. Thus, the current at this moment is practically zero because the quantity of electroactive substance reaching the electrode surface is exceedingly small. As a mercury drop expands, greater surface area is exposed to the solution and the current increases. Just before the drop detaches from the tip of the

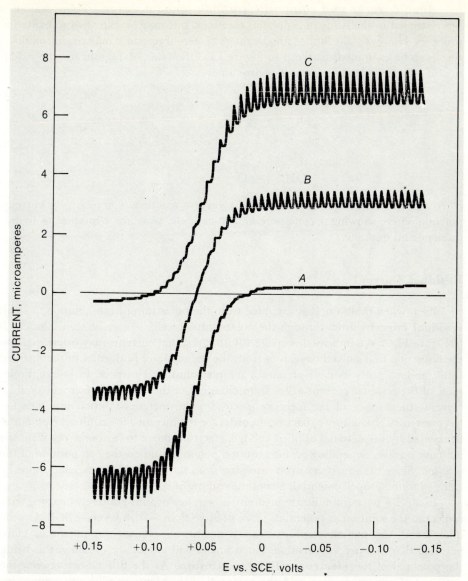

*Figure 13–3.* Polarograms for the quinone-hydroquinone system in a 0.1 $F$ phosphate buffer at pH 7.0. Curve $A$: anodic polarogram for 0.001 $M$ hydroquinone. Curve $B$: composite anodic-cathodic polarogram for a mixture of 0.0005 $M$ hydroquinone and 0.0005 $M$ quinone. Curve $C$: cathodic polarogram for 0.001 $M$ quinone.

capillary, the current attains its maximal value. Therefore, the current oscillations actually range from almost zero up to the maximal current value. Because such large and rapid fluctuations in current are not easy to observe in practice, the current is usually recorded with the aid of a damped galvanometer or other current-measuring instrument. Although the oscillations of the galvanometer are less than the real current fluctuations at the dropping mercury electrode, the average of the galvanometer excursions is virtually identical to the mean current flowing through the cell.

*Anodic waves and composite cathodic-anodic waves.* Polarography is not restricted to the study of reduction reactions, but may be extended to oxidation processes. However, the limited anodic range of mercury permits only readily oxidized species to be examined. Curve *A* in Figure 13–3 shows a polarogram for the oxidation of *p*-benzohydroquinone to *p*-benzoquinone,

$$\text{[OH-benzene-OH]} \rightleftharpoons \text{[O=benzene=O]} + 2\,H^+ + 2\,e$$

whereas curve *C* is a polarogram for the reverse reaction. Curve *B* is a current-potential curve showing a composite anodic-cathodic wave for a mixture of hydroquinone and quinone.

## Residual Current

Even when thallium(I) is excluded from the potassium nitrate solution, a small **residual current** flows through the polarographic cell, as may be seen in curve *B* of Figure 13–2. We indicated on page 400 that residual currents may originate from the reduction of dissolved oxygen or from the reduction of impurities in the distilled water and reagents used to prepare a sample solution. However, in polarography most of the residual current arises from charging of the so-called electrical double-layer at the surface of the mercury electrode; the mercury-solution interface behaves essentially as a tiny capacitor. In order for the mercury to acquire the potential demanded by the external applied voltage, electrons move toward or away from the mercury surface, depending on the required potential and on the composition of the solution. Since each mercury drop emerging from the orifice of the capillary must be charged to the proper potential, a continuous flow of current results.

In a 0.1 *F* potassium nitrate medium, a mercury drop is positively charged with respect to the solution at potentials more positive than −0.53 v versus SCE. On the other hand, the mercury surface bears a net negative charge at potentials more negative than the preceding value, but is uncharged at exactly −0.53 v, this being the potential of the **electrocapillary maximum**. As the difference between the potential of a mercury drop and the potential of the electrocapillary maximum increases, the charging current becomes greater. Note that the charging current is positive when the potential of the dropping mercury electrode is negative with respect to the electrocapillary maximum, and vice versa.

## Polarographic Diffusion Current

There are three mechanisms by which an electroactive substance, such as thallium(I), may reach the surface of an electrode—*migration* of a charged species due to the flow of current through a cell, *convection* caused by stirring or agitation of the solution and by density or thermal gradients, and *diffusion* of a substance from a region of high concentration to one of lower concentration. Successful polarographic measurements demand that the electroactive species reach the surface of a mercury drop solely by diffusion, this being the only process amenable to straightforward mathematical treatment. To eliminate electrical migration of an ion, one introduces into the sample solution a 50- to 100-fold excess of an innocuous **supporting electrolyte**—for example, potassium chloride—which lowers the transference number of the reacting species virtually to zero. Convection can be prevented if the sample solution is unstirred and if the polarographic cell is mounted in a vibration-free location.

Of particular interest in polarography is the **diffusion layer**, a film of solution perhaps 0.05 mm in thickness at the surface of the electrode. By referring to the reduction of thallium(I) in a potassium nitrate medium (curve $A$, Figure 13–2), let us see what happens at the surface of the dropping mercury electrode as its potential is varied. At approximately −0.1 v versus SCE, little, if any, thallium(I) is reduced. If the potential of the dropping mercury electrode is shifted to about −0.45 v, a significant fraction of the thallium(I) at the electrode surface is reduced to yield a thallium amalgam, and a finite current flows through the polarographic cell. As a consequence of this electrolysis, a concentration gradient is established, causing thallium(I) to diffuse from the bulk of the solution toward the electrode surface. If thallium(I) did not diffuse toward the electrode, the current would fall to zero almost instantaneously because the surface concentration of thallium(I) would drop to the value governed by the electrode potential and would change no more. However, the current is sustained by diffusion of additional thallium(I) toward the electrode under the influence of the concentration gradient.

According to the simplest model for a diffusion-controlled process, the observed current is directly related to the difference between the concentration of the electroactive species in the bulk of the solution, $C$, and at the surface of the electrode, $C_0$. Thus, we may write

$$i = k(C - C_0)$$

where $i$ is the current and $k$ is a proportionality constant. Fundamentally, it is the potential of the dropping mercury electrode that fixes the surface concentration ($C_0$) of the electroactive substance. At potentials more negative than −0.70 v versus SCE for the reduction of thallium(I), essentially all of the reactant reaching the electrode surface is immediately reduced. Thus, the surface concentration is so small compared to the bulk concentration that the term $(C - C_0)$ becomes virtually equal to $C$, and the current attains a limiting value called the **diffusion current**, which we denote by the symbol $i_d$ (Figure 13–2):

$$i_d = kC$$

Analytical applications of polarography rely upon the direct proportionality between the diffusion current and the bulk concentration of the electroactive substance.

*Ilkovič equation.* Because diffusion of an electroactive substance to an expanding mercury drop under polarographic conditions is a complicated process, the derivation of a complete and exact expression for the diffusion current has not been achieved. However, several simplifying assumptions lead to the Ilkovič equation

$$i_d = 607nD^{1/2} Cm^{2/3} t^{1/6}$$

which accounts remarkably well for the effect of all important variables on the diffusion current. In this relation $i_d$ is the time-average diffusion current in microamperes that flows during the lifetime of a single mercury drop, $n$ is the number of faradays of electricity per mole of electroactive substance reduced or oxidized, $D$ is the diffusion coefficient (cm²/second) of the reactant, $C$ is the bulk concentration of the electroactive species in millimoles per liter, $m$ is the rate of flow of mercury in milligrams per second, and $t$ is the drop time in seconds. Among other terms, the numerical constant 607 includes the value of the faraday (96,487 coulombs).

*Factors affecting the diffusion current.* Several parameters in the Ilkovič equation are influenced by the nature of the sample solution, the potential of the dropping mercury electrode, and the temperature.

Diffusion coefficients are sensitive to variations in the composition and viscosity of the sample medium, to the temperature, and to the size of the diffusing species. Since the diffusion coefficient of a substance increases approximately 2 per cent for every one degree rise in temperature, the polarographic cell should be thermostated to within a few tenths of a degree for accurate work. Diffusion coefficients of hydrated metal ions are larger than those of the corresponding metal ion complexes formed with bulky ligands such as citrate, tartrate, and ethylenediaminetetraacetate (EDTA).

Since the rate at which mercury flows through the orifice of the capillary ($m$) is directly proportional to the height of the mercury column in the stand tube, whereas the drop time ($t$) is inversely proportional to the mercury height, it can be shown that $m^{2/3} t^{1/6}$, the so-called **capillary constant**, depends on the square root of the mercury column height. Another factor that influences the characteristics of the capillary is the potential of the dropping mercury electrode. At the potential of the electro-capillary maximum, the drop time is longest but decreases to as little as one half of this value at more positive and more negative potentials. Only the one-sixth power of the drop time appears in the Ilkovič equation, so that the diffusion current is not too seriously affected. Variations in potential influence the mercury flow rate less than 1 per cent between 0 and $-1.5$ v versus SCE. For highly accurate polarographic measurements, it is essential to determine the drop time, the mercury flow rate, and the diffusion current at the same potential.

## Polarographic Maxima

If the polarogram for an air-saturated $0.05\,F$ potassium chloride solution is recorded, one obtains curve $A$ in Figure 13–4, showing a **polarographic maximum** on the rising portion of the first wave for reduction of oxygen. Polarographic maxima are not unique to the oxygen system, and occur quite generally. High-speed motion picture photography has demonstrated that the occurrence of a polarographic maximum is associated with streaming or swirling of solution around the dropping mercury electrode. Such convective motion of the solution causes more electroactive substance to reach the electrode surface than would be expected on the basis of diffusion alone. However, the origins of polarographic maxima are not fully understood.

Fortunately, a polarographic maximum can be eliminated through addition to the sample medium of a small concentration of a **maximum suppressor**, some substance which is adsorbed upon the surface of the mercury electrode. Among the species used for the suppression of polarographic maxima are gelatin, methyl red, and Triton X-100, the latter being a commercially available surface-active substance. Curve $B$ in Figure 13–4 shows the polarogram for reduction of oxygen in the presence of a trace of methyl red. In general, the concentration of maximum suppressor should be as small as possible—a concentration of 0.005 per cent is usually sufficient—because too much surface-active substance diminishes the diffusion current.

## Equation of the Polarographic Wave

If we assume that a species originally present in the sample solution undergoes reversible polarographic reduction to some substance which is soluble either in the

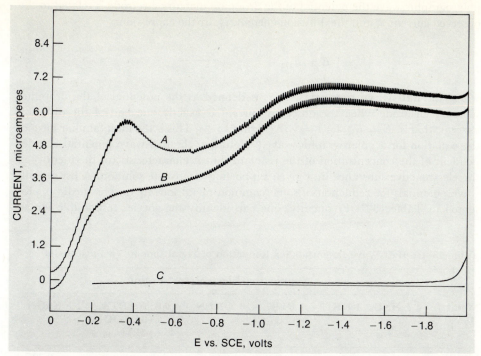

*Figure 13–4.* Polarograms for reduction of oxygen in a 0.05 $F$ potassium chloride medium showing effect of absence and presence of a maximum suppressor. Curve $A$: polarogram for air-saturated solution without maximum suppressor. Curve $B$: polarogram for air-saturated solution containing a trace of methyl red. Curve $C$: residual-current curve after removal of dissolved oxygen with nitrogen. Note that curve $A$ has been shifted upward 0.6 microampere to prevent overlap with curve $B$.

solution or in the mercury phase,

$$Ox + n\,e \rightleftharpoons Red$$

the concentrations (activities) of the reactant and product *at the solution-electrode interface* are related to the potential of the dropping mercury electrode through the Nernst equation:

$$E = E^0 - \frac{0.059}{n} \log \frac{[Red]_{surface}}{[Ox]_{surface}}$$

In this expression, $E$ and $E^0$ represent, respectively, the potential of the dropping mercury electrode and the standard potential for the pertinent half-reaction, each measured with respect to the same reference electrode. In addition, the observed potential of the dropping mercury electrode and the surface concentrations of the oxidized and reduced species are average values during the lifetime of a mercury drop.

*Half-wave potential.* Provided that the reactant and product species are completely soluble in either the solution or the mercury drop, it can be demonstrated, although we shall not present the derivation in this text,* that in the region of the

---

* Details of the derivation may be found in J. J. Lingane: *Electroanalytical Chemistry.* Second edition, Wiley-Interscience, New York, 1958, p. 261.

polarographic wave the potential of the dropping mercury electrode is related to the observed current, $i$, and the diffusion current, $i_d$, by the expression

$$E = E_{1/2} - \frac{0.059}{n} \log \frac{i}{i_d - i}$$

where the **half-wave potential**, $E_{1/2}$, is defined as the potential of the dropping mercury electrode when the observed current is exactly one half of the diffusion current; that is, $E = E_{1/2}$ for $i = i_d/2$ (Figure 13–2). This is the most familiar form of the equation for a polarographic wave.* Note that the half-wave potential is independent of the concentration of the reactant, but is characteristic of the identity of the electroactive substance in a given supporting electrolyte solution. A list of half-wave potentials for reduction of some common metal ions in various media is presented in Table 13–1. Frequently, one can identify the species responsible for an

**Table 13–1. Half-Wave Potentials for Reduction of Metal Ions in Various Media**[a]

| Metal Ion[b] | Supporting Electrolyte Solution | | | | | |
|---|---|---|---|---|---|---|
| | 1 $F$ HNO$_3$ | 1 $F$ KCl | 1 $F$ NH$_3$, 1 $F$ NH$_4$Cl | 1 $F$ KCN | 0.05 $F$ EDTA, 0.8 $F$ CH$_3$COONa pH 12 | 1 $F$ NaOH |
| Cd(II) | −0.59 | −0.64 | −0.81 | −1.18 | −1.28 | −0.78 |
| Co(II) | — | −1.20 | −1.29 | −1.45 | NR[d] | −1.46 |
| Cu(II) | −0.01 | +0.04; −0.22[c] | −0.24; −0.51[c] | NR[d] | −0.51 | −0.41 |
| Ni(II) | — | — | −1.10 | −1.36 | NR[d] | — |
| Pb(II) | −0.41 | −0.44 | — | −0.72 | −1.32 | −0.76 |
| Tl(I) | −0.48 | −0.48 | −0.48 | — | −0.60 | −0.48 |
| Zn(II) | −1.00 | −1.00 | −1.35 | NR[d] | NR[d] | −1.53 |

[a] All half-wave potentials are quoted with respect to the saturated calomel reference electrode (SCE); data excerpted from L. Meites: *Polarographic Techniques.* Second edition, Wiley-Interscience, New York, 1965, pp. 615–670.
[b] All species are reduced to form a metal amalgam aside from exceptions noted for copper(II).
[c] Copper(II) gives two waves, corresponding to stepwise reduction to copper(I) and copper (0).
[d] No wave is observed before reduction of the supporting electrolyte-solvent system itself.

unknown polarographic wave by searching a table of half-wave potentials.

*Polarographic study of metal ion complexes.* As Table 13–1 reveals, a metal ion complex is more difficult to reduce than the corresponding hydrated cation. Compared to the half-wave potential for reduction of the simple aquo ion, *the half-wave potential for reduction of a metal ion complex becomes more negative as the stability*

---

* We have considered only the reversible reduction of a substance to yield a product soluble in the solution or in mercury. For the reversible oxidation of a species to give a soluble product, the equation for the polarographic wave has the same form as that for a cathodic process except that a plus (+) sign precedes the logarithmic term. For a composite cathodic-anodic wave (see curve $B$, Figure 13–3), the pertinent expression is

$$E = E_{1/2} - \frac{0.059}{n} \log \frac{i - (i_d)_a}{(i_d)_c - i}$$

where the subscripts $a$ and $c$ refer, respectively, to the anodic and cathodic diffusion currents. See J. J. Lingane: *Electroanalytical Chemistry.* Second edition, Wiley-Interscience, New York, 1958, p. 263 for the derivation of these relations.

*of the complex increases.* For example, the half-wave potential for the reduction of hydrated cadmium(II), presumably $Cd(H_2O)_4{}^{2+}$, is $-0.59$ v versus SCE, but the half-wave potential is $-0.81$ v in a $1\ F$ ammonia solution in which the predominant species is $Cd(NH_3)_4{}^{2+}$. Even more stable is the one-to-one cadmium(II)-EDTA complex, a fact reflected by its half-wave potential of $-1.28$ v. *For the reduction of any one metal ion complex, the half-wave potential is dependent upon the concentration (activity) of the complexing ligand.* By measuring how much the half-wave potential shifts as the ligand concentration is varied, one can determine both the formula and the formation constant of the metal ion complex.

Let us consider the reduction of a hypothetical metal ion complex at the dropping mercury electrode

$$ML_p{}^{n-pb} + ne + Hg \rightleftharpoons M(Hg) + pL^{b-}$$

where $ML_p{}^{n-pb}$ is the complex, $L^{b-}$ is the free ligand, $n$ is the formal charge of the cation, and $p$ is the number of ligands coordinated to the metal ion. It can be shown[*] that the difference between the half-wave potentials for reduction of the complexed and uncomplexed metal ion is

$$(E_{1/2})_c - E_{1/2} = -\frac{0.059}{n} \log \beta_p - \frac{0.059p}{n} \log [L^{b-}]$$

where $(E_{1/2})_c$ is the half-wave potential for the complex in the presence of a specified free-ligand concentration, $[L^{b-}]$; $E_{1/2}$ is the half-wave potential for the aquated cation; and $\beta_p$ is the overall formation constant for the complex, that is

$$\beta_p = \frac{[ML_p{}^{n-pb}]}{[M^{n+}][L^{b-}]^p}$$

This treatment requires that the concentration of free ligand be strictly constant throughout the solution.

In applying the above equation, one measures the half-wave potential $(E_{1/2})_c$ for reduction of the metal ion complex as a function of the free-ligand concentration, and constructs a plot of $(E_{1/2})_c$ versus $\log [L^{b-}]$. If only a single metal ion complex is predominant, a straight line with a slope of $-0.059\,p/n$ is obtained. Figure 13–5 shows a plot of some actual data for the lead(II)-sodium hydroxide system. From the slope of the line in Figure 13–5, the value of $p$ can be computed as follows:

$$\text{slope} = -0.083 = -\frac{0.059p}{n}$$

$$p = \frac{0.083n}{0.059} = \frac{(0.083)(2)}{0.059} = 2.8 \approx 3$$

Thus, three hydroxide ligands are associated with lead(II), and the formula of the complex is deduced to be $Pb(OH)_3{}^-$.[†] Once the value of $p$ is known, we can calculate

---

[*] A derivation is presented by J. J. Lingane: Chem. Rev., *29*:1, 1941.

[†] Actually, the $Pb(OH)_3{}^-$ ion undergoes a dehydration reaction

$$Pb(OH)_3{}^- \rightleftharpoons HPbO_2{}^- + H_2O$$

but three hydroxide ions are ultimately involved in the polarographic reduction process:

$$HPbO_2{}^- + H_2O + 2\,e + Hg \rightleftharpoons Pb(Hg) + 3\,OH^-$$

Failure to include activity coefficients explains why the calculated value of $p$ is not precisely three.

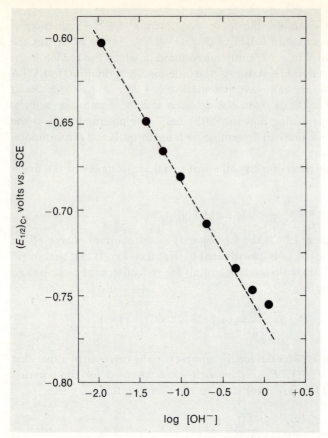

*Figure 13–5.* Plot of half-wave potential for reduction of the lead(II)-hydroxide complex versus the logarithm of the concentration of hydroxide ion. Experimental data were taken from a paper by J. J. Lingane: Chem. Rev., 29: 1, 1941.

the overall formation constant $(\beta_3)$ for the complex by using the value for $(E_{1/2})_c$ of $-0.68$ v at log $[OH^-] = -1.00$ (Figure 13–5) and by taking $E_{1/2}$ to be $-0.41$ v (Table 13–1):

$$(E_{1/2})_c - E_{1/2} = -\frac{0.059}{n}\log\beta_p - \frac{0.059p}{n}\log[L^{b-}]$$

$$-0.68 - (-0.41) = -\frac{0.059}{2}\log\beta_3 - \frac{(0.059)(3)}{2}(-1.00)$$

$$\beta_3 = 1.6\times10^{12}$$

This procedure has been employed for the characterization of many metal ion complexes. Theoretical equations are also available which permit one to study the successive formation of a family of complexes.

## Experimental Techniques of Polarography

Although polarographic measurements are usually accomplished with the aid of instruments that automatically record complete current-potential curves, data of excellent accuracy can be obtained manually.

***Polarographic cells and dropping mercury electrodes.*** Figure 13–1 shows the popular H-cell employed for polarography. Generally, the compartment containing the sample solution is designed to accommodate volumes of 15 to 25 ml, but special cells have been constructed for the analysis of a single drop of solution.

Capillary tubing for the fabrication of a dropping mercury electrode is available commercially. A 10-cm length of marine-barometer tubing, with an internal diameter of 0.06 mm, provides an electrode having a drop time of 3 to 5 seconds for a mercury reservoir height of 40 cm. To ensure capillary characteristics that are both uniform and reproducible, the end of the marine-barometer tubing should be perfectly flat and horizontal, and the tubing must be mounted vertically. A determination of the rate of mercury flow $(m)$ involves collection of mercury for a specified length of time followed by a weight measurement. To obtain the drop time $(t)$, one employs a stop watch and measures the total time required for a certain number of drops to fall. Since the capillary characteristics vary with the potential of the dropping mercury electrode, the mercury flow rate and the drop time should be evaluated for the potential at which one intends to measure the diffusion current.

With proper handling, a dropping mercury electrode will behave reproducibly for many months. Above all, the mercury column should always be high enough so that a positive flow of mercury results. At the conclusion of an experiment, the tip of the capillary should be washed thoroughly with distilled water; then, after the mercury reservoir is lowered to just stop the mercury flow, the capillary may be allowed to stand in air.

***Electrical apparatus.*** To perform polarographic measurements, one should be able to impress voltages ranging from 0 to 3 v across the cell, and one must know the potential of the dropping mercury electrode versus the reference electrode with a precision of 10 mv. In addition, it is necessary to measure the cell current, which usually encompasses a range from 0.1 to 100 $\mu a$, with an accuracy of $\pm 0.01$ $\mu a$.

A simple electrical circuit for polarographic experiments is illustrated in Figure 13–6. Two 1.5-v batteries connected in series provide a source of voltage which can

*Figure 13–6.* Schematic diagram of circuitry for manual performance of polarographic measurements. (Redrawn, with permission, from a paper by J. J. Lingane: Anal. Chem., *21*: 45, 1949.)

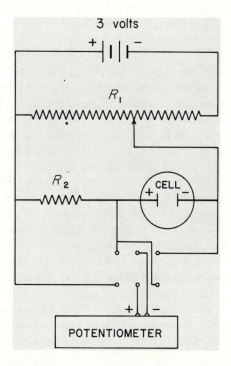

be varied through adjustment of a movable contact along the slidewire $R_1$. If the double-pole double-throw switch is moved to the right-hand position, the potential of the dropping mercury electrode versus the reference electrode can be determined with the aid of a potentiometer. For measurement of the current, the switch is thrown to the left-hand position and the $iR$ drop across a precision 10,000-ohm resistor ($R_2$) is obtained by means of the potentiometer. A galvanometer is often used as a null-point detector in the potentiometer circuit; if so, the galvanometer must be damped by placement of a resistor across its terminals so that the potentiometer can be balanced in spite of the current oscillations caused by the growth and fall of the dropping mercury electrode.

Although the circuitry depicted in Figure 13–6 is excellent insofar as the accuracy of the potential and current measurements is concerned, manual point-by-point determination of a polarogram is tedious. With only slight modification, the circuit of Figure 13–6 can be converted into an automatic recording polarograph. First, the slidewire may be replaced by a motor-driven, 100-ohm, 10-turn potentiometer which varies the applied voltage in a known and linear fashion. Second, the current may be recorded in synchronization with the change in applied voltage, if a strip-chart potentiometer recorder is connected across resistor $R_2$. A large number of all-electronic, recording polarographs are available commercially.

***Measurement of diffusion currents.*** To determine the diffusion current—which is directly related to the concentration of the electroactive species—it is necessary to subtract the residual current from the observed limiting current. Correcting for the residual current becomes more important and more difficult as the concentration of desired substance decreases; at a concentration of $10^{-4} M$, the residual current may be as large as one-tenth of the diffusion current. Ultimately, our inability to obtain a precise value for the diffusion current in the presence of a residual current of comparable magnitude sets the lowest concentration for accurate ($\pm 2$ per cent) determination by means of conventional polarography at approximately $5 \times 10^{-5} M$. Several sophisticated modifications of polarography have been developed in which the residual current can be almost eliminated or cancelled. These techniques permit concentrations as low as $10^{-7} M$ to be determined.

Undoubtedly, the most reliable procedure for measurement of the diffusion current entails the recording of separate current-potential curves for the sample solution and for a blank solution containing all components except the electroactive substance. Next, one selects a potential at which the diffusion current is to be evaluated and measures the difference in current between the two curves. Since the limiting-current region of the polarogram for the desired substance and the residual-current curve often have different slopes, the diffusion current may vary slightly as a function of potential. However, the accuracy of the measurement is not impaired if a self-consistent procedure is followed.

Another method for determining the diffusion current requires that only the polarogram for the sample solution be recorded. One simply extrapolates the linear portion of the current-potential curve preceding the polarographic wave to some potential in the limiting-current region of the polarogram at which the diffusion current is to be measured; the difference between the observed limiting current and the straight-line extension of the residual-current curve is taken as the diffusion current. This procedure is valid only insofar as the charging current increases linearly with applied voltage. In fact, the slope of the residual-current curve changes at the potential of the electrocapillary maximum, so that there is an inherent error if the extrapolated line passes by the latter potential.

***Determination of concentration.*** We shall now examine some procedures used to evaluate the concentration of an electroactive substance.

A simple and straightforward approach involves the preparation of a series of standard solutions, the measurement of the diffusion current for each sample, and the construction of a plot of observed diffusion current versus concentration. With the aid of this calibration curve, the concentration of the desired substance in an unknown solution is determined from a single measurement of the diffusion current. Obviously, the supporting electrolyte solution must be the same for both the standards and the unknown.

A second procedure for determination of the concentration of an electroactive species is called the **method of standard addition**. Suppose that a polarogram is recorded for a solution containing an unknown concentration, $C_u$, of the desired substance and the observed diffusion current is $(i_d)_1$. If an aliquot $V_s$ of a solution of the electroactive substance of known concentration $C_s$ is added to a volume $V_u$ of the unknown solution, the diffusion current will attain a new value $(i_d)_2$. Using data collected from the two experiments, one can compute the concentration of the unknown sample from the relation

$$C_u = \frac{C_s V_s (i_d)_1}{(V_u + V_s)(i_d)_2 - V_u (i_d)_1}$$

This method requires preparation of one standard solution whose supporting electrolyte concentration, pH, and ionic strength are identical to those of the unknown solution. Moreover, this procedure can succeed only if a linear relation exists between the diffusion current and the concentration of the electroactive substance. Best results are achieved if the diffusion currents, $(i_d)_1$ and $(i_d)_2$, differ by at least a factor of two.

## Analytical Applications of Polarography

Hundreds of inorganic and organic substances have been subjected to polarographic investigation, and many procedures for the analysis of individual species and mixtures of compounds have been developed. For routine determinations, the optimum concentration range for polarography is $10^{-4}$ to $10^{-2}$ $M$, and analytical results accurate to within $\pm 2$ per cent are usually attainable. With special care, the uncertainty of a polarographic analysis may be as small as a few tenths of 1 per cent.

*Determination of metal ions.* Through an appropriate choice of supporting electrolyte, pH, and complexing ligands, practically any metal ion can be reduced at the dropping mercury electrode to an amalgam or to a soluble lower oxidation state. In many situations, one obtains polarographic waves suitable for quantitative determination of these species. Divalent cations such as cadmium, cobalt, copper, lead, manganese, nickel, tin, and zinc can be determined in many different complexing and noncomplexing media. Alkaline earth ions—barium, calcium, magnesium, and strontium—exhibit well-defined polarographic waves at approximately $-2.0$ v versus SCE in solutions containing tetraethylammonium iodide as the supporting electrolyte. Cesium, lithium, potassium, rubidium, and sodium undergo reduction between $-2.1$ and $-2.3$ v versus SCE in aqueous and alcoholic tetraalkylammonium hydroxide media. Polarographic characteristics of the tripositive states of aluminum, bismuth, chromium, europium, gallium, gold, indium, iron, samarium, uranium, vanadium, and ytterbium in a number of supporting electrolyte solutions have been reported.

*Determination of inorganic anions.* Polarographic reductions of chromate, iodate, molybdate, selenite, tellurite, and vanadate, as well as the anionic chloride complexes of tungsten(VI), tin(IV), and molybdenum(VI), are included in the list of analytically useful procedures.

In the presence of bromide ion, the dropping mercury electrode is oxidized to insoluble mercurous bromide

$$2 \text{ Hg} + 2 \text{ Br}^- \rightleftharpoons \text{Hg}_2\text{Br}_2(\text{s}) + 2 \text{ e}$$

and a diffusion current proportional to the bulk concentration of the anion is observed. Other species, including chloride, iodide, sulfate, and thiocyanate, exhibit analogous behavior. Cyanide and thiosulfate form soluble complex ions with mercury(II):

$$\text{Hg} + 4 \text{ CN}^- \rightleftharpoons \text{Hg(CN)}_4{}^{2-} + 2 \text{ e}$$

$$\text{Hg} + 2 \text{ S}_2\text{O}_3{}^{2-} \rightleftharpoons \text{Hg(S}_2\text{O}_3)_2{}^{2-} + 2 \text{ e}$$

An anodic polarogram for the oxidation of mercury in a dilute sodium cyanide medium shows a limiting current governed by the diffusion of cyanide to the surface of the dropping mercury electrode.

*Analysis of mixtures.* Solutions containing two or more electroactive substances may be analyzed polarographically if certain requirements are satisfied. Clearly, the half-wave potentials for the various species must differ significantly if the resulting polarogram is to exhibit a distinct wave and limiting-current region for each component. In general, the separation in the half-wave potentials for successive reactions should be at least 0.2 v, but the minimal difference can be calculated as was done previously for controlled-potential coulometry (page 411).

As shown in Figure 13–7, a mixture of thallium(I), cadmium(II), and nickel(II) in an ammonia-ammonium chloride buffer can be resolved and analyzed by means

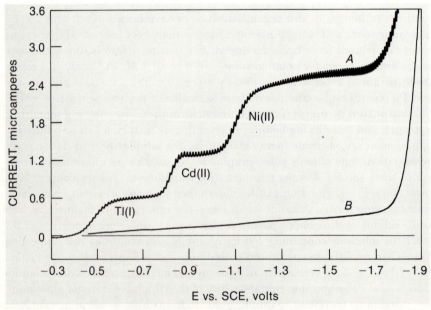

*Figure 13–7.* Curve *A*: polarogram for a mixture of thallium(I), cadmium(II), and nickel(II), each at a concentration of $1.00 \times 10^{-4} M$, in a buffer consisting of 1 *F* ammonia and 1 *F* ammonium chloride with 0.0002 per cent methyl red as maximum suppressor. Curve *B*: residual-current curve for the supporting electrolyte solution alone.

of polarography. Table 13–1 indicates that the respective half-wave potentials for these elements are $-0.48$, $-0.81$, and $-1.10$ v versus SCE. However, difficulty in the analysis of such a mixture can be encountered if the relative concentrations of the three metal ions are greatly disparate. For example, if a small quantity of cadmium is present in a solution containing relatively large amounts of thallium and nickel, the polarographic wave for reduction of cadmium will appear as a tiny plateau on top of the thallium wave and will be followed by another large rise in current because of nickel(II) reduction. One remedy for this problem involves polarographic measurement of thallium(I) and subsequent removal of most, if not all, of this ion by controlled-potential separation. Then, cadmium may be more accurately determined because the interference from thallium is eliminated.

Another approach to the polarographic analysis of mixtures takes advantage of the fact that half-wave potentials change in the presence of complexing agents. As an example, the individual determinations of lead(II) and thallium(I) in a mixture cannot be accomplished if the supporting electrolyte solution is 1 $F$ potassium chloride because the half-wave potentials are too close together, being $-0.44$ and $-0.48$ v versus SCE, respectively (see Table 13–1). However, in a 1 $F$ sodium hydroxide medium, the reduction of thallium(I) yields a well-developed polarographic wave at $-0.48$ v, whereas reduction of lead(II) occurs at a half-wave potential of $-0.76$ v.

*Polarography of organic compounds.*   In contrast to inorganic substances, the reactions of organic compounds at the dropping mercury electrode are usually irreversible and often proceed in several steps. Polarographic data pertaining to organic molecules are frequently difficult to interpret, because of the effect of pH on the rate, mechanism, and products of an organic electrode process, the influence of solvent on the course of the reaction, the importance of the nature and concentration of the supporting electrolyte, and the occurrence of side reactions involving partially oxidized or reduced species.

Despite the just-named problems, there are a number of useful applications of polarography to the determination of organic substances. We have already seen in Figure 13–3 that $p$-benzoquinone and $p$-benzohydroquinone yield well-defined polarograms suitable for analysis. In Table 13–2 are listed some of the functional groups found in organic molecules that are electroactive at the dropping mercury electrode and that give diffusion currents proportional to concentration.

Typical of the complexity of organic polarography is the reduction of benzaldehyde in an aqueous medium illustrated in Figure 13–8. In a phosphate buffer of pH 3, the first step near $-1.00$ v versus SCE corresponds to the one-electron reduction to a free radical

that immediately dimerizes to form hydrobenzoin:

At $-1.30$ v, benzaldehyde undergoes a two-electron reduction to benzyl alcohol:

**Table 13-2. Polarographic Behavior of Organic Functional Groups**

| Functional Group | Class of Compounds | Exemplary Reaction |
|---|---|---|
| C—I<br>C—Br | organic halides | $\text{Ph—CH}_2\text{Br} + \text{H}^+ + 2\,e \longrightarrow \text{Ph—CH}_3 + \text{Br}^-$ |
| C=C | carbon-carbon double bond conjugated with another C=C bond or with an aromatic ring | $\text{Ph—CH=CH}_2 + 2\,\text{H}^+ + 2\,e \longrightarrow \text{Ph—CH}_2\text{CH}_3$ |
| C≡C | phenyl-substituted acetylenes | $\text{Ph—C}\equiv\text{C—Ph} + 4\,\text{H}^+ + 4\,e \longrightarrow \text{Ph—CH}_2\text{CH}_2\text{—Ph}$ |
| C=O | quinones | $\text{O=}\langle\text{ring}\rangle\text{=O} + 2\,\text{H}^+ + 2\,e \longrightarrow \text{HO—}\langle\text{ring}\rangle\text{—OH}$ |
| | aldehydes, ketones | $2\ \text{Ph—C(=O)—CH}_3 + 2\,\text{H}^+ + 2\,e \longrightarrow 2\ \text{Ph—C(OH)·—CH}_3 \longrightarrow \text{dimer}$ |
| C—OH | hydroquinones | $\text{HO—}\langle\text{ring}\rangle\text{—OH} \longrightarrow \text{O=}\langle\text{ring}\rangle\text{=O} + 2\,\text{H}^+ + 2\,e$ |

Table 13-2. Polarographic Behavior of Organic Functional Groups  (continued)

| Functional Group | Class of Compounds | Exemplary Reaction |
|---|---|---|
| C—SH | mercaptans (thiols) | $HS-CH_2COO^- + Hg \longrightarrow HgS-CH_2COO^- + H^+ + e$ |
| S—S | disulfides | $H_5C_2-S-S-C_2H_5 + 2\,H^+ + 2\,e \longrightarrow 2\,C_2H_5-SH$ |
| S=O | sulfoxides | $H_3C-\overset{O}{\underset{\|}{S}}-CH_3 + 2\,H^+ + 2\,e \longrightarrow H_3C-S-CH_3 + H_2O$ |
| O—O | peroxides, hydroperoxides | (benzoyl peroxide structure) $+ 2\,H^+ + 2\,e \longrightarrow 2$ (benzene)—COOH |
| N=O | nitro and nitroso compounds | (phenyl)—C(=O)—$CH_2NO_2 + 4\,H^+ + 4\,e \longrightarrow$ (phenyl)—C(=O)—$CH_2NHOH + H_2O$ |
| N=N | azo compounds | (phenyl)—N=N—(phenyl) $+ 2\,H^+ + 2\,e \longrightarrow$ (phenyl)—NH—NH—(phenyl) |
| C≡N | nitriles | $CH_3CO$—(phenyl)—$CN + 4\,H^+ + 4\,e \longrightarrow CH_3CO$—(phenyl)—$CH_2NH_2$ |
| C=N | imines, oximes | (phenyl)—$CH=CH-\underset{CH_3}{C}=NH + 2\,H^+ + 2\,e \longrightarrow$ (phenyl)—$CH=CH-\underset{CH_3}{CH}-NH_2$ |

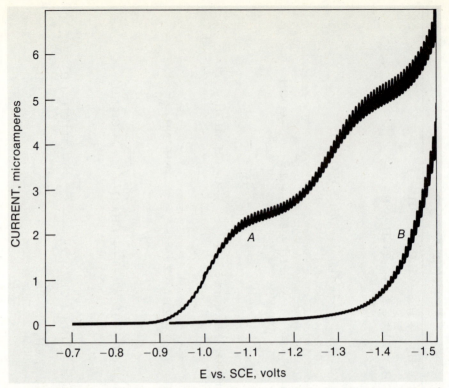

*Figure 13–8.* Curve *A*: polarogram for the stepwise reduction of 0.001 *M* benzaldehyde in an aqueous 0.1 *F* phosphate buffer of pH 2.75 containing 0.002 per cent Triton X-100 as maximum suppressor (see text for further discussion). Curve *B*: residual-current curve for supporting electrolyte solution alone.

Half-wave potentials for the two steps are pH dependent. Furthermore, only the first wave is observed at pH values less than approximately 2, whereas the second wave appears alone above pH 11; between these two pH extremes both processes occur. Thus, the positions of the waves along the potential axis as well as the course of the reaction are affected by the hydrogen ion concentration. Such behavior emphasizes the importance of controlling the pH of the sample solution in the polarographic determination of organic compounds.

## AMPEROMETRIC TITRATIONS

From the viewpoint of chemical analysis, the most significant feature of polarography is the direct proportionality between the diffusion current and the concentration of the electroactive substance. In addition, we have seen earlier how the linear relationship between concentration and limiting current in a well-stirred solution is utilized to follow the course of a controlled-potential electrolysis. It is not surprising that the measurement of current—often termed **amperometry**—can be employed to monitor the concentration changes which occur during a titration. If at least one of the reactants or products involved in a titration is electroactive, a plot of the observed current versus the volume of added titrant may permit the equivalence point to be located. This method is known as the technique of **amperometric titration**. Amperometric titrations are performed in two ways. In one procedure, the potential of a single microelectrode versus a stable reference electrode is adjusted to be in the limiting-current region for the desired electroactive species; then the current flowing at the microelectrode is recorded as increments of titrant are added. Alternatively, the potential difference between a pair of identical microelectrodes is held constant, and the current that flows between them is measured as a function of the volume of titrant.

## AMPEROMETRIC TITRATIONS WITH ONE POLARIZABLE ELECTRODE

Amperometric titrations with a single polarizable electrode* are based upon the polarographic techniques discussed previously in this chapter. Many titrations can be performed in a polarographic H-cell, and changes in current as titrant is added are followed with the aid of a dropping mercury electrode whose potential is fixed with respect to a saturated calomel reference electrode. For amperometric titrations involving species that are oxidized at potentials beyond the anodic range of mercury, a rotating platinum microelectrode may replace the dropping electrode. In this section, we will survey the theory, experimental apparatus, and applications of amperometric titrations with one polarizable electrode.

### Principles and Titration Curves

Suppose that we wish to perform a precipitation titration of thallium(I) with a standard potassium iodide solution, the pertinent reaction being

$$Tl^+ + I^- \rightleftharpoons TlI(s)$$

This determination can be conducted amperometrically if the thallium(I) solution is placed in a polarographic cell equipped with a dropping mercury electrode and a saturated calomel reference electrode. We have seen in Figure 13–2 that the reduction of thallium(I) yields a well-defined polarogram with a limiting-current region extending from $-0.7$ to $-1.6$ v versus SCE. Iodide ion produces an anodic wave due to the oxidation of mercury to mercurous iodide; however, this wave appears at potentials more positive than $-0.1$ v and no significant current flows at potentials from $-0.3$ to $-1.8$ v. These polarographic characteristics of thallium(I) and iodide ion are depicted qualitatively in the left-hand panel of Figure 13–9A by the curves labeled $X$ and $T$, respectively. If the potential of the dropping mercury electrode is held at $-0.8$ v versus SCE, the only current that flows will be caused by reduction of thallium(I). Now, as the potassium iodide titrant is added incrementally, thallium-(I) will be precipitated as insoluble thallous iodide and, with each portion of titrant, the current will decrease in proportion to the quantity of thallium(I) removed from the solution. When thallium(I) has been completely titrated, the current drops essentially to zero and remains unchanged as excess titrant is introduced.

If we plot the current corresponding to each increment of titrant near the equivalence point versus the volume of potassium iodide solution, the result shown in the right-hand panel of Figure 13–9A is obtained. Provided that the titrant is sufficiently concentrated to make dilution negligible, the amperometric titration curve consists

---

\* A polarizable electrode is one whose potential is easily changed upon passage of only a tiny current. On the other hand, the potential of a nonpolarizable electrode, such as a saturated calomel reference electrode, remains essentially constant even when an appreciable current flows.

of two straight-line portions—one on either side of the equivalence point—which can be extrapolated to their point of intersection to locate the end point. Because the reaction between thallium(I) and iodide is not 100 per cent complete, the curve is rounded in the vicinity of the equivalence point. For this reason, experimental data close to the equivalence point are always ignored.

Three other types of amperometric titration curves are illustrated in Figure 13–9, along with current-potential curves for the substance being titrated ($X$) and for the titrant ($T$). If the potential of the polarizable microelectrode is kept at any value within the shaded area of the current-potential diagram shown in a left-hand panel, the equivalence-point region of the resulting titration curve will appear as depicted in the adjacent right-hand panel. Titration curves for mixtures of species consist of combinations of the plots in Figure 13–9.

Figure 13–9B is characteristic of the behavior of a system in which the titrant is electroreducible but the substance to be determined is not reactive. An example is the titration of barium ion with standard potassium chromate solution in a neutral medium, the potential of the dropping mercury electrode being adjusted to a value ($-1.0$ v versus SCE) at which only chromate ion is reduced. When the substance to be determined and the titrant are both reducible within the same range of potentials, a V-shaped titration curve results (Figure 13–9C). A titration of nickel(II) with dimethylglyoxime in an ammoniacal solution yields such a curve if the potential of the dropping mercury electrode is set at approximately $-1.85$ v versus SCE. Finally, Figure 13–9D illustrates what happens if the substance being titrated is not electro-active but the titrant undergoes oxidation at a rotating platinum microelectrode. Quantitative reduction of cerium(IV) to cerium(III) in a sulfuric acid medium with a standard solution of hydroquinone yields the latter type of titration curve with a rotating platinum electrode at a potential of $+0.8$ v versus SCE.

## Experimental Techniques and Apparatus

*Polarizable electrodes.* Usually, either a dropping mercury electrode or a rotating platinum electrode is used as the polarizable microelectrode. Although a dropping mercury electrode is superior for the measurement of currents resulting from cathodic processes, it has a limited anodic range. On the other hand, the rotating platinum electrode is preferable for anodic reactions and is especially useful for amperometric redox titrations.

One constructs a rotating platinum electrode by sealing a small-gauge platinum wire, approximately 6 to 8 mm in length, into the side of a closed glass tube, as shown in Figure 13–10. To make electrical contact with the electrode, the glass tube is filled with mercury and another short length of platinum wire is dipped into the upper, open end of the tube. Then the tube is mounted into the hollow chuck of a synchro-nous motor and is rotated at a constant speed of, say, 600 rpm. As a consequence of mechanical stirring, the limiting current densities obtained with the rotating electrode are 20 to 30 times larger than those for a dropping mercury electrode, so that the former provides higher sensitivity. Otherwise, the shapes of current-potential curves are similar for both electrodes. Among the disadvantages of the rotating platinum electrode are that it has a restricted cathodic range and that the size of the limiting current from one titration to another is affected by how the electrode surface is cleaned between experiments.

*Cells.* An ordinary H-cell may be employed for amperometric titrations with the dropping mercury electrode. Since oxygen is reducible at a mercury cathode, it is often necessary to deaerate the sample solution as well as the titrant. Mixing of the

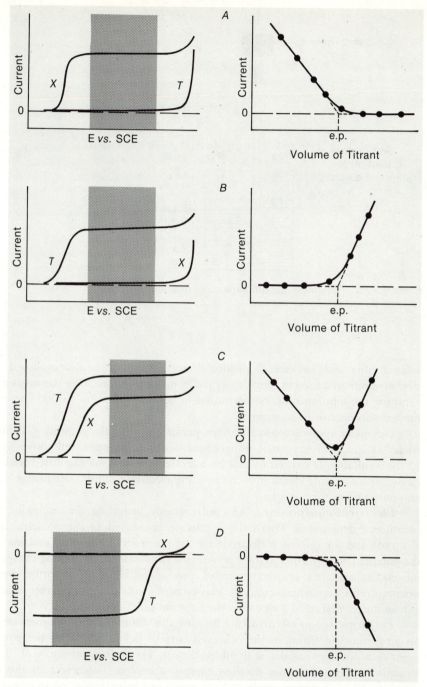

*Figure 13–9.* Current-potential curves (smoothed polarograms) for various titration systems (left-hand panels) and corresponding amperometric titration curves (right-hand panels). In each left-hand panel, $X$ denotes the curve for the substance which is titrated and $T$ represents the curve for the titrant; in each case, the shaded region shows the range of potentials within which the polarizable electrode is maintained. In each right-hand panel, only the portion of the titration curve near the equivalence point is depicted; the solid points represent hypothetical data, whereas the intersection of the extrapolated straight lines indicates the experimental end point. See text for further discussion.

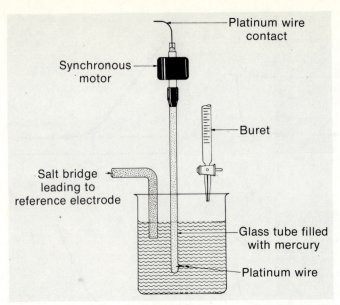

*Figure 13–10.* Apparatus for amperometric titrations with the rotating platinum microelectrode.

solution after addition of each portion of the titrant can be accomplished with the stream of nitrogen used to remove oxygen from the solution. After the mixing and de-aeration, the solution must be permitted to become quiescent so that the diffusion-limited current can be measured.

For many amperometric titrations performed with the rotating platinum electrode, exclusion of oxygen is unnecessary, so that the open cell shown in Figure 13–10 is suitable. If oxygen cannot be tolerated, the cell may be fitted with a cover having holes for introduction of the rotating electrode, buret, salt-bridge, and inlet and outlet tubes for nitrogen.

***Electrical equipment.*** Any polarograph, including the manually operated instrument depicted in Figure 13–6, can be connected to the polarizable micro-electrode and the reference electrode for the purposes of adjusting and maintaining the potential of the former at some preselected value. Since the limiting-current region for most substances is reasonably broad, precise control of the potential is generally not required. Current measurements may be made with a microammeter or a damped galvanometer placed in series with the titration cell.

***Performance of titrations.*** Because the volume of solution increases during an amperometric titration, the observed current before and after the equivalence point does not vary linearly with added titrant. Thus, some curvature of the desired straight-line portions of the titration curve is observed. To correct for this effect of dilution and to obtain two straight lines which can be extrapolated to locate the end point, each current reading must be multiplied by the factor $(V + v)/V$. Before the equivalence point, $V$ is the original volume of the sample solution and $v$ is the volume of titrant added. After the equivalence point, $V$ is the equivalence-point volume and $v$ is the amount of titrant added beyond the equivalence point. An excellent way to avoid this nuisance is to employ a titrant (delivered from a microburet) that is at least 50 times more concentrated than the sample; then the effect of dilution can be ignored.

## Applications

Amperometry is particularly useful as a method of end-point detection for titrations of samples at the $10^{-3}$ $M$ concentration level, analytical errors commonly being as low as a few tenths of 1 per cent. Even for solutions as dilute as $10^{-5}$ $M$, it is possible to obtain results accurate to $\pm 1$ per cent. In general, the accuracy of amperometric titrations surpasses that of polarography and is comparable to that of either potentiometric or spectrophotometric titration. Amperometric end-point detection has been used extensively for constant-current coulometric titrations as well as for a variety of ordinary volumetric analyses. Some of these applications will now be explored.

*Amperometric titrations of inorganic species.* Most of the systems for which amperometric end-point detection has been employed involve the formation of precipitates, although complexometric and redox titrations can be followed amperometrically. Table 13–3 lists a number of these titrations.

**Table 13–3.** Amperometric Titrations of Inorganic Species[a]

| Species Determined | Titrant and Conditions | Polarizable Electrode and Potential versus SCE | Shape of Titration Curve[b] |
|---|---|---|---|
| (a) Precipitation titrations | | | |
| $Ag^+$ | KI in neutral medium | rotating Pt at 0 v | A |
| $Ba^{2+}$ | $K_2CrO_4$ | dropping Hg at $-1.0$ v | B |
| $Br^-$, $Cl^-$, $I^-$ | $AgNO_3$ in 1 $F$ $HNO_3$ | rotating Pt at 0 v | B |
| $K^+$ | add excess $B(C_6H_5)_4^-$; back-titrate with $AgNO_3$ at pH 5 | dropping Hg at 0 v | B |
| $Mg^{2+}$ | 8-hydroxyquinoline in $NH_3$-$NH_4Cl$ buffer | dropping Hg at $-1.7$ v | B |
| $MoO_4^{2-}$ | $Pb(NO_3)_2$ at pH 8 | dropping Hg at $-0.8$ v | C |
| $Pd^{2+}$ | dimethylglyoxime at pH 5 | dropping Hg at 0 v | A |
| $SO_4^{2-}$ | $Pb(NO_3)_2$ in neutral ethanol-water mixture | dropping Hg at $-0.7$ v | B |
| $WO_4^{2-}$ | 8-hydroxyquinoline in acetic acid-acetate buffer | dropping Hg at $-1.4$ v | B |
| $Zn^{2+}$ | $K_4Fe(CN)_6$ in 0.2 $F$ $H_2SO_4$ | rotating Pt at $+0.7$ v | D |
| (b) Complexometric titrations | | | |
| $Bi^{3+}$ | EDTA in dilute $HNO_3$ | dropping Hg at $-0.18$ v | A |
| $Cd^{2+}$ | EDTA in neutral medium | dropping Hg at $-0.80$ v | A |
| $F^-$ | $Fe^{3+}$ in neutral ethanolic NaCl solution | dropping Hg at 0 v | B |
| $In^{3+}$ | EDTA in weakly acidic medium | dropping Hg at $-0.90$ v | A |
| $Tl^{3+}$ | EDTA in pH 2 medium | rotating Pt at 0 v | A |
| (c) Redox titrations | | | |
| $HAsO_2$, $HSbO_2$ | $KBrO_3$ in 1 $F$ HCl and 0.05 $F$ KBr | rotating Pt at 0 v | B |
| $Au^{3+}$ | hydroquinone in 1 $F$ $H_2SO_4$ | rotating Pt at $+1.0$ v | D |
| $Cr_2O_7^{2-}$ | $Fe^{2+}$ in 1 $F$ $H_2SO_4$ | rotating Pt at $+0.8$ v | D |
| $I_2$ | $S_2O_3^{2-}$ in neutral medium | rotating Pt at 0 v | A |

[a] Information excerpted from an article by J. Doležal and J. Zýka: Amperometric titrations. *In* F. J. Welcher, ed.: *Standard Methods of Chemical Analysis.* Sixth edition, Volume III-A, Van Nostrand, Princeton, New Jersey, 1966, pp. 377–403.
[b] Letter designations refer to the shapes of the titration curves shown in Figure 13–9.

*Amperometric titrations of organic compounds.* Several classes of organic molecules may be conveniently determined by means of amperometric titration. For all of the systems to be discussed, the amperometric titration curves appear as shown in Figure 13–9B.

Aldehydes and ketones—substances having a carbonyl group—may be titrated amperometrically in a hydrochloric acid medium with a solution of 2,4-dinitro-phenylhydrazine, which interacts with a carbonyl moiety to form a hydrazone. Under the usual experimental conditions, only the titrant is reducible at a dropping mercury electrode.

Amperometry can be used to follow the progress of the bromination of phenol, aniline, and their derivatives as well as some compounds with carbon-carbon double bonds. A standard potassium bromate-potassium bromide titrant is added to a solution of the sample in hydrochloric acid, and the liberated bromine reacts with the organic molecule as described on page 340. A rotating platinum microcathode is utilized to detect the current due to the first excess of molecular bromine.

One of the most useful applications of amperometric titrimetry is the highly accurate determination of sulfhydryl (—SH) groups in amino acids, peptides, and native and denatured proteins. Silver nitrate may be used as the titrant, the pertinent reaction being

$$Ag^+ + RSH \rightarrow RSAg + H^+$$

With a rotating platinum electrode, the current is essentially zero before the equivalence point but increases linearly after the equivalence point because of the reduction of excess silver ion. However, titrations with silver nitrate often provide large positive errors, apparently because silver ion interacts with other groups on the molecule to be determined. A better titrant is sodium *p*-chloromercuribenzoate, whose reaction with a compound containing a sulfhydryl group may be written as

$$^-OOC\!-\!\langle\bigcirc\rangle\!-\!HgCl + RSH \rightarrow HOOC\!-\!\langle\bigcirc\rangle\!-\!HgSR + Cl^-$$

Although both the titrant and the product are electroactive at the dropping mercury electrode, the potential can be adjusted so that only the titrant is reduced. In the presence of guanidine, disulfide bridges in polypeptides such as insulin are reduced by sulfite ion to sulfhydryl groups which can be determined by means of an amperometric titration.

*Amperometric determination of oxygen in biological systems.* Much effort has been expended in the development of an amperometric method for the continuous measurement of oxygen in blood and other biological fluids or extracts as well as in gases. In some instances, the dropping mercury electrode may be used as a sensor for oxygen. More popular and versatile is the so-called **Clark electrode**, a platinum cathode which is covered with an oxygen-permeable membrane (polyethylene, Mylar, or Teflon) to prevent the electrode surface from becoming contaminated with other constituents of the sample, *e.g.*, erythrocytes in blood. Oxygen undergoes stepwise reaction at a mercury cathode (page 433), but only the first stage of reduction

$$O_2 + 2\,H^+ + 2\,e \rightleftharpoons H_2O_2$$

normally occurs at a platinum surface in acidic and neutral media. In use, the potential of the platinum electrode is held at $-0.5$ to $-0.6$ v with respect to a silver-silver chloride reference electrode, and the limiting current arising from oxygen reduction is obtained. For many measurements, the two electrodes are mounted in a microcell through which the sample is pumped. Some determinations of oxygen are performed with miniature electrodes placed in the tip of a catheter which is inserted directly into the blood stream. Reproducible and accurate results require careful calibration of the electrode system with solutions or gases containing known amounts of oxygen. In addition, there must be strict control of the temperature and the rate of stirring or pumping of sample past the Clark electrode.

## AMPEROMETRIC TITRATIONS WITH TWO POLARIZABLE ELECTRODES

Another form of amperometric titration requires two identical polarizable microelectrodes—usually made of platinum—with areas from 0.1 to 2 cm². In this technique, the sample solution is vigorously stirred and the potential difference between the two electrodes is kept at a specified value, perhaps as small as 10 mv or as large as 500 mv, by means of the same electrical circuitry employed for amperometry with one polarizable electrode. However, although the potential difference between the two microelectrodes does remain constant, the individual electrode potentials shift in unison during the course of an amperometric titration and may not lie within the limiting-current regions for the electroactive species. Nevertheless, if experimental conditions are controlled, a plot of the observed current versus the volume of titrant exhibits an abrupt change at the equivalence point of the titration. Amperometry with dual microelectrodes has been used most frequently to locate equivalence points for redox titrations.

### Principles and Titration Curves

Dual-electrode amperometry may be illustrated by the titration of iron(II) with cerium(IV) in a 4 F sulfuric acid medium. Suppose that we wish to follow the course of this titration amperometrically with two identical platinum microelectrodes whose potentials differ by 200 mv. Let us see how the current varies with added titrant by considering the schematic current-potential curves in Figure 13–11.

At the beginning of the titration, before any cerium(IV) has been introduced, the current-potential curve for a solution of iron(II) in 4 F sulfuric acid appears qualitatively as shown in Figure 13–11A. According to convention, positive currents correspond to cathodic processes, whereas negative currents pertain to anodic reactions. Only one cathodic process occurs, namely, the reduction of hydrogen ion to hydrogen gas. However, there are two anodic reactions—the oxidation of iron(II) to iron(III) and, at a more positive potential, the oxidation of water. If two platinum microelectrodes whose potentials differ by 200 mv are immersed into the solution, how much current flows between them? To answer this question, we must recognize that one electrode functions as a cathode and the other as an anode and that *the cathodic and anodic currents must be equal*. Let us adjust a pair of dividers to a span of 200 mv; then, let us slide the dividers along the abscissa until the points of the dividers lie at two potentials corresponding to identical anodic and cathodic currents. When this is done, we arrive at the position represented by the shaded region of Figure 13–11A; the potentials of the anode and cathode are +0.4 and +0.2 v versus NHE, respectively. It is evident that only a tiny residual current flows at each electrode. For practical purposes the current at the start of the titration is zero. This fact is indicated by the amperometric titration curve in Figure 13–12A.

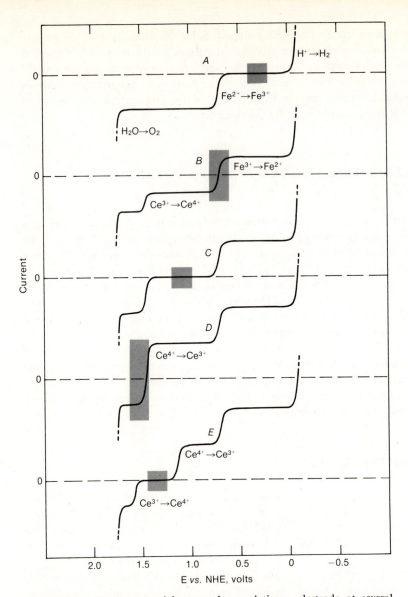

*Figure 13-11* Current-potential curves for a platinum electrode at several stages during the titration of iron(II) with cerium(IV). Curve *A*: at the start of the titration Curve *B*: at the midpoint of the titration. Curve *C*: at the equivalence point. Curves *A* through *D* pertain to a 4 *F* sulfuric acid medium in which the iron(III)-iron(II) and cerium (IV)-cerium(III) couples behave reversibly. Curve *E* shows qualitatively what happens beyond the equivalence point if the titration is performed in a 0.5 *F* sulfuric acid solution in which the cerium(IV)-cerium(III) half-reaction is irreversible; observe the separation of the waves for the reduction of cerium(IV) and the oxidation of cerium(III) caused by this irreversibility. See the text for a discussion of how these current-potential curves are used to determine the amperometric titration curve. Each abbreviated half-reaction printed on the current-potential curves identifies the wave immediately to the left. Note that cathodic processes give rise to positive currents and that anodic processes cause negative currents.

In Figure 13–11*B* is shown the current-potential curve after one-half of the required amount of cerium(IV) titrant is added. Since iron(III) and cerium(III) are formed as products of the titration, two new reactions can occur—the reduction of iron(III) and the oxidation of cerium(III). If, as before, a pair of dividers is used to determine the current, we see from the shaded area in Figure 13–11*B* that the current is much larger than at the beginning of the titration. In fact, the current

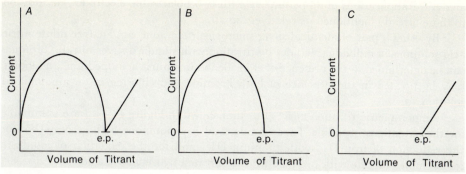

*Figure 13–12.* Amperometric titration curves for redox reactions obtained with two polarizable microelectrodes. Curve *A*: half-reactions for titrant and substance titrated are both reversible. Curve *B*: half-reaction for substance titrated is reversible, but half-reaction for titrant is irreversible. Curve *C*: half-reaction for titrant is reversible, but half-reaction for substance titrated is irreversible. See text for further discussion.

attains its maximal value halfway to the equivalence point of the titration, as Figure 13–12*A* reveals. Moreover, notice that the potentials of the two microelectrodes have shifted to more positive values.

Figures 13–11*C* and 13–11*D* show current-potential curves for the iron(II)-cerium(IV) system at and beyond the equivalence point, respectively. An examination of these two curves indicates that the current should decrease almost to zero at the equivalence point but increase again after the equivalence point. These expectations are borne out by the amperometric titration curve in Figure 13–12*A*.

Thus far, we have considered only a system in which the half-reactions for both reactants are reversible. In other words, the half-wave potentials for the oxidation of iron(II) and reduction of iron(III) are identical, and a similar relationship holds for the cerium(IV)-cerium(III) couple. This explains why we see a composite anodic-cathodic wave for a mixture of iron(III) and iron(II) in Figure 13–11*B* and for a mixture of cerium(IV) and cerium(III) in Figure 13–11*D*. However, if the cerium-(IV)-cerium(III) half-reaction behaved irreversibly, as it does in 0.5 *F* sulfuric acid, the current-potential curve beyond the equivalence point would look like Figure 13–11*E*, in which the anodic wave for cerium(III) and the cathodic wave for cerium(IV) are separated. Then, as indicated by the shaded region of Figure 13–11*E*, the current after the equivalence point would remain very small. As a consequence, the amperometric titration curve would appear as in Figure 13–12*B*. Other systems which yield such an amperometric titration curve are the titration of cerium(IV) with oxalic acid in 3 *F* sulfuric acid and the titration of iodine (triiodide) with thiosulfate.

If the half-reaction for the species being titrated is irreversible, whereas the titrant redox couple behaves reversibly, the amperometric titration curve will resemble Figure 13–12*C*. Examples are the titration of vanadium(IV) to vanadium-(III) with titanium(III) and the oxidation of arsenic(III) to arsenic(V) with iodine (triiodide).

## Applications

Although experimental techniques are very similar to those described earlier for amperometric titrations with a single polarizable electrode, the number of analytical applications of amperometry with two polarizable electrodes, in addition to those already mentioned, is relatively small.

By using a pair of identical silver microelectrodes, one can analyze dilute nitric acid solutions of individual halides by titration with standard silver nitrate. Bromide can be determined in the presence of chloride ion in a nitric acid medium, and iodide may be titrated in the presence of both bromide and chloride in an ammoniacal solution.

Amperometric titration curves like that shown in Figure 13–12A are obtainable for the oxidations of iodide, ferrocyanide, and hydroquinone with cerium(IV), for the reduction of iron(III) with titanium(III), and for the oxidation of iodide to iodine monochloride with bromine in concentrated hydrochloric acid.

## QUESTIONS AND PROBLEMS

1. In a $1\,F$ sodium hydroxide medium, the reduction of the tellurite anion, $TeO_3{}^{2-}$, yields a single, well-developed polarographic wave. With a dropping mercury electrode whose $m$ value was 1.50 mg/second and whose drop time was 3.15 seconds, the observed diffusion current was 61.9 $\mu$a for a 0.00400 $M$ solution of the tellurite ion. If the diffusion coefficient of the tellurite anion is $0.75 \times 10^{-5}$ cm²/second, to what oxidation state is tellurium reduced under these conditions?

2. A $2.23 \times 10^{-3}\,F$ solution of osmium tetroxide $(OsO_4)$ in $1\,F$ hydrochloric acid was examined polarographically. For a dropping mercury electrode with a drop time of 4.27 seconds and a mercury flow rate of 3.29 mg/second, a cathodic diffusion current of 62.4 $\mu$a was observed. Although the diffusion coefficient of $OsO_4$ is unknown, it is reasonable to assume that the value is close to that of the chromate ion $(CrO_4{}^{2-})$, the latter species having a diffusion coefficient of $1.07 \times 10^{-5}$ cm²/second. On the basis of this information, write the half-reaction for the reduction of osmium tetroxide in $1\,F$ hydrochloric acid.

3. Iodate ion undergoes a six-electron reduction to iodide ion at the dropping mercury electrode:

$$IO_3{}^- + 6\,H^+ + 6\,e \rightleftharpoons I^- + 3\,H_2O$$

When a $1.41 \times 10^{-3}\,F$ solution of potassium iodate in $0.1\,F$ perchloric acid was reduced polarographically at a dropping mercury electrode with a drop time of 2.18 seconds and a mercury flow rate of 2.67 mg/second, the diffusion current was observed to be 37.1 $\mu$a. What is the diffusion coefficient of iodate ion in $0.1\,F$ perchloric acid?

4. Derive the relationship shown on page 445 which pertains to polarographic analysis by means of the method of standard addition.

5. An unknown cadmium(II) solution was analyzed polarographically according to the method of standard addition. A 25.00-ml sample of the unknown solution was found to yield a diffusion current of 1.86 $\mu$a. Next, a 5.00-ml aliquot of a standard $2.12 \times 10^{-3}\,M$ cadmium(II) solution was added to the unknown sample, and the resulting mixture gave a diffusion current of 5.27 $\mu$a. Calculate the concentration of cadmium(II) in the unknown solution.

6. A trace impurity of nickel(II) in cobalt(II) salts can be measured polarographically. A 3.000-gm sample of reagent-grade $CoSO_4 \cdot 7\,H_2O$ was transferred to a 100.0-ml volumetric flask and was dissolved in a small amount of water. Then, 2 ml of concentrated hydrochloric acid, 5 ml of pyridine, and 5 ml of 0.2 per cent gelatin were added to the volumetric flask, and the solution was diluted with water to the calibration mark and was mixed well. Exactly 75.00 ml of this solution was placed in a polarographic cell, and the resulting polarogram exhibited a diffusion current of 1.97 $\mu$a at $-0.96$ v

versus SCE. Next, 4.00 ml of a $9.24 \times 10^{-3}\,F$ nickel(II) chloride solution was pipetted into the polarographic cell. Subsequently, the diffusion current at $-0.96$ v versus SCE was increased to 3.95 $\mu$a. Assuming that the nickel(II) impurity was present as $NiSO_4 \cdot 7\,H_2O$, calculate the weight per cent of this compound in the $CoSO_4 \cdot 7\,H_2O$ reagent.

7. In a $1\,F$ potassium chloride medium, the polarographic half-wave potential for the reduction of zinc(II) to a zinc amalgam is $-1.00$ v versus SCE. However, the half-wave potential for the reduction of the tetraamminezinc(II) complex, $Zn(NH_3)_4{}^{2+}$, to a zinc amalgam is $-1.18$ v versus SCE in a $0.1\,F$ ammonia-$1.0\,F$ ammonium chloride buffer. In each instance, the reduction proceeds reversibly. Calculate the overall formation constant $(\beta_4)$ for the tetraamminezinc(II) complex.

8. In a buffer solution consisting of $1\,F$ ammonia and $1\,F$ ammonium chloride, copper(II) exists predominantly as the tetraammine complex, $Cu(NH_3)_4{}^{2+}$. This species undergoes stepwise reduction at the dropping mercury electrode:

$$Cu(NH_3)_4{}^{2+} + e \rightleftharpoons Cu(NH_3)_2{}^+ + 2\,NH_3; \qquad E_{1/2} = -0.24 \text{ v versus SCE}$$

$$Cu(NH_3)_2{}^+ + e + Hg \rightleftharpoons Cu(Hg) + 2\,NH_3; \qquad E_{1/2} = -0.51 \text{ v versus SCE}$$

A series of polarograms was recorded for different concentrations of copper(II) in the ammonia-ammonium chloride buffer, and the limiting current was measured at $-0.70$ v versus SCE (on the plateau of the polarographic wave) for each concentration.

| Concentration of copper(II), millimoles/liter | Limiting current, $\mu$a |
|---|---|
| 0 | 0.18 |
| 0.489 | 3.24 |
| 0.990 | 6.55 |
| 1.97 | 13.18 |
| 3.83 | 25.2 |
| 8.43 | 56.0 |

When the polarogram for a buffer solution containing an unknown concentration of copper(II) was recorded, a limiting current of 7.49 $\mu$a was observed. Calculate the concentration in moles/liter of copper(II) in the unknown solution.

9. In a $0.1\,F$ potassium nitrate medium, the two-electron reduction of cadmium-(II) has a half-wave potential of $-0.578$ v versus SCE. When the polarographic behavior of cadmium(II) in $0.1\,F$ potassium nitrate solutions containing increasing concentrations of ammonia is examined, the half-wave potential for the reduction process appears at progressively more negative values, as the following data reveal:

| Concentration of free $NH_3$, $M$ | $E_{1/2}$, volts versus SCE |
|---|---|
| 0 | $-0.578$ |
| 0.126 | $-0.679$ |
| 0.282 | $-0.719$ |
| 0.446 | $-0.743$ |
| 0.562 | $-0.754$ |
| 0.828 | $-0.774$ |
| 1.23 | $-0.794$ |
| 1.68 | $-0.801$ |
| 2.00 | $-0.819$ |

Using the above information, determine the formula of the predominant cadmium(II)-ammonia complex and its overall formation constant. Neglect any consideration of activity coefficients or changes in the liquid-junction potential.

10. In a polarographic study of the two-electron reduction of $5.14 \times 10^{-4} M$ copper(II) in $0.1 F$ potassium nitrate solutions containing various concentrations of diethylenetriamine, $H_2NCH_2CH_2NHCH_2CH_2NH_2$, the following data have been obtained:

| Concentration of free diethylenetriamine, $M$ | $E_{1/2}$, volts versus SCE |
|:---:|:---:|
| 0 | +0.016 |
| 0.020 | −0.504 |
| 0.040 | −0.520 |
| 0.100 | −0.542 |
| 0.200 | −0.559 |
| 0.400 | −0.575 |
| 1.00 | −0.602 |

Determine the formula of the complex formed between copper(II) and diethylenetriamine, and evaluate the overall formation constant for the species. Suggest a structure for the complex. In what way is this complex unique among the majority of copper(II) coordination compounds? Why?

11. Nickel(II) in an ammoniacal medium can be titrated with a standard solution of dimethylglyoxime dissolved in ethanol, a precipitate being formed according to the reaction

$$Ni(NH_3)_4{}^{2+} + 2 \begin{matrix} HON{=}C{-}CH_3 \\ | \\ HON{=}C{-}CH_3 \end{matrix} \rightleftharpoons Ni \left[ \begin{matrix} ON{=}C{-}CH_3 \\ \| \\ ON{=}C{-}CH_3 \\ H \end{matrix} \right]_2 (s)$$

$$+\, 2\, NH_3 + 2\, NH_4{}^+$$

A 50.00-ml aliquot of a sample solution containing $0.5 F$ ammonia, $0.1 F$ ammonium chloride, and an unknown concentration of nickel(II) was transferred to a polarographic cell, and the potential of the dropping mercury electrode was adjusted to $-1.85$ v versus SCE, at which value both the tetraamminenickel(II) complex and dimethylglyoxime are reducible. Portions of a $0.1076 F$ solution of dimethylglyoxime in 96 per cent ethanol were introduced from a microburet, and current readings were recorded after the precipitation reaction reached equilibrium. In the table below are the titration data:

| Volume of dimethylglyoxime solution, ml | Diffusion current, $\mu a$ |
|:---:|:---:|
| 0 | 18.90 |
| 1.00 | 13.96 |
| 2.00 | 9.66 |
| 3.00 | 5.52 |
| 3.20 | 4.58 |
| 3.40 | 3.64 |
| 3.80 | 2.01 |
| 3.90 | 1.64 |
| 4.00 | 1.50 |
| 4.10 | 2.82 |
| 4.20 | 4.56 |
| 4.40 | 6.44 |

Plot the titration data and determine the concentration of nickel(II) in the original sample solution.

12. An ore was analyzed for gold by means of an amperometric titration with hydroquinone. A 100.0-gm sample of the ore was decomposed with concentrated sulfuric acid, and the mixture was then evaporated and ignited. Aqua

regia was used to redissolve the solid; repeated evaporations with hydro-chloric acid served to expel the nitric acid. Next, 200 ml of 1 $F$ sulfuric acid was added to the hydrochloric acid solution of $AuCl_4^-$, and the solution was transferred to the titration cell. A rotating platinum electrode and a saturated calomel reference electrode were placed in the cell; the potential of the former was set at $+1.0$ v versus SCE; and the solution was titrated at $60°C$ with a $0.002500\ F$ solution of hydroquinone in 1 $F$ sulfuric acid. At $+1.0$ v versus SCE, the limiting current for the rotating platinum anode is due to the oxidation of hydroquinone to quinone:

$$C_6H_4(OH)_2 \rightleftharpoons C_6H_4O_2 + 2\ H^+ + 2\ e$$

The titration reaction itself may be written as

$$2\ AuCl_4^- + 3\ C_6H_4(OH)_2 \rightleftharpoons 2\ Au + 3\ C_6H_4O_2 + 6\ H^+ + 8\ Cl^-$$

For various volumes of the hydroquinone solution, the following values of the limiting current were observed:

| Volume of hydroquinone solution, ml | Limiting current, $\mu a$ |
|:---:|:---:|
| 0 | 2.1 |
| 1.00 | 2.1 |
| 2.00 | 2.1 |
| 3.00 | 2.0 |
| 4.00 | 2.1 |
| 5.00 | 2.2 |
| 5.50 | 2.1 |
| 6.00 | 2.1 |
| 6.50 | 4.4 |
| 7.00 | 10.1 |
| 7.50 | 18.4 |
| 8.00 | 27.1 |
| 9.00 | 43.6 |
| 10.00 | 60.3 |

Determine the weight per cent of gold in the ore sample.

13. Vitamin E ($\alpha$-tocopherol) can be determined by means of an amperometric titration with gold(III) chloride, the actual reaction being

A vitamin E capsule, reputed to contain 30 milligrams of $\alpha$-tocopherol, was analyzed. The contents of the capsule were dissolved in approximately 90 ml of 75 per cent ethanol containing $0.1\ F$ benzoic acid, $0.1\ F$ sodium benzoate, and $0.1\ F$ sodium chloride. This solution was transferred quan-titatively to a polarographic cell, and the potential of the dropping mercury electrode was adjusted to −0.075 v versus SCE, a potential at which only gold(III) is reducible. A 0.00979 $F$ solution of $HAuCl_4$ in 75 per cent ethanol was added from a microburet, and the diffusion current corresponding to each increment of titrant was measured. Titration data were as follows:

| Volume of $HAuCl_4$ solution, ml | Diffusion current, $\mu$a |
|---|---|
| 0 | −0.05 |
| 1.00 | −0.05 |
| 2.00 | −0.05 |
| 3.00 | −0.05 |
| 4.00 | −0.05 |
| 5.00 | +0.43 |
| 5.50 | +0.96 |
| 6.00 | +1.47 |
| 6.50 | +1.98 |

Calculate the weight in milligrams of α-tocopherol in the capsule.

14. Complexometric titrations of mixtures of metal cations with EDTA can be followed amperometrically. Exactly 50.00 ml of a solution containing 0.1 $F$ ammonia, 0.1 $F$ ammonium nitrate, and unknown amounts of copper(II) and calcium(II) was transferred into a polarographic cell, and the potential of the dropping mercury electrode was set at −0.25 v versus SCE, where only the copper(II)-ammine complex is reducible. A standard 0.09071 $F$ EDTA solution was introduced from a microburet, and the diffusion current was measured after the addition of each increment of titrant. After it was evident that the first equivalence point had been exceeded, the potential of the dropping mercury electrode was readjusted to 0 v versus SCE; at this potential, EDTA yields an anodic wave. Then, the addition of titrant and the measurement of the diffusion current were continued. Complete titration data are shown in the following table:

| Volume of EDTA, ml | Diffusion current, $\mu$a |
|---|---|
| 0 | 23.8 |
| 0.50 | 18.9 |
| 0.90 | 15.0 |
| 1.25 | 11.6 |
| 1.70 | 7.2 |
| 2.20 | 2.3 |
| 2.55 | 0.3 |
| 2.95 | 0.3 |
| 3.35 | 0.3 |
| 3.70 | 0.3 |
| (potential changed to 0 v versus SCE) | |
| 3.70 | −0.5 |
| 4.10 | −0.5 |
| 4.45 | −0.5 |
| 4.85 | −0.5 |
| 5.25 | −0.5 |
| 5.90 | −1.8 |
| 6.15 | −3.1 |
| 6.55 | −5.1 |
| 6.90 | −6.9 |

Calculate the number of milligrams of copper(II) and calcium(II) in the sample solution.

15. Many electrode systems suitable for the continuous monitoring of oxygen in natural waters have been devised. One of these consists of a silver cathode and a lead anode; the cathode is a disk and the anode is a ring. These two electrodes are separated by an insulating ring and are imbedded concentrically

in the flat end of a cast plastic rod; separate conducting wires to the two electrodes are imbedded in the rod as well. Then, the entire end of the plastic rod with its exposed electrodes is covered with a circle of filter paper saturated with $1 F$ potassium hydroxide solution; the filter paper is, in turn, covered with a polyethylene membrane (1 mil in thickness) which is held in place by a plastic collar. When this electrode system is placed into an oxygen-containing water or gas sample (and an appropriate voltage applied across the two electrodes), molecular oxygen diffuses through the polyethylene membrane, passes through the wetted filter paper, and is reduced to hydroxide ion at the silver cathode:

$$O_2 + 2\,H_2O + 4\,e \rightleftharpoons 4\,OH^-$$

When immersed into a series of solutions containing dissolved oxygen, one such electrode system produced a steady-state current of 1.05 $\mu a$ for an oxygen concentration of 1.00 milligram per liter. In order to examine the oxygen profile of a lake (having a depth of 28 feet), the electrode system was lowered to various depths, and the current reading was made. These data are recorded in the following table:

| Depth, feet | Observed current, $\mu a$ |
| --- | --- |
| Just beneath the surface | 7.98 |
| 1 | 7.76 |
| 2 | 7.56 |
| 3 | 7.40 |
| 4 | 7.03 |
| 5 | 6.40 |
| 6 | 4.41 |
| 7 | 2.94 |
| 8 | 0.53 |
| 9 | 0.01 |

Calculate the oxygen concentration in moles per liter for each depth reading and construct a plot of the results.

16. Many polarographic analyses can be performed by means of the "pilot-ion" method. A standard solution is prepared that contains known concentrations of both the substance to be determined and another species called the pilot ion, and the ratio of their polarographic wave heights is measured. Then, a known concentration of the pilot ion is added to an unknown sample solution, the ratio of the wave heights is measured, and the concentration of the substance to be determined is calculated with the aid of the data obtained with the standard solution.

When a polarogram for a $1 F$ potassium chloride solution containing 0.00532 $M$ lead(II) and 0.00422 $M$ cadmium(II) was recorded, the ratio of the diffusion current for lead(II) to that for cadmium(II) was found to be 1.22.

In order to determine the amount of a lead impurity in metallic zinc, a 9.440-gm sample of the zinc was dissolved completely in nitric acid. After the sample solution was evaporated to dryness, the residue was dissolved in water and the resulting solution was diluted to exactly 250.0 ml with a potassium chloride medium so that the final supporting electrolyte concentration was $1 F$. Next, a 25.00-ml aliquot of this sample solution and a 5.00-ml aliquot of a standard 0.00968 $M$ cadmium(II) solution in $1 F$ potassium chloride medium were pipetted into a polarographic cell. When the polarogram was recorded, the ratio of the diffusion current for lead(II) to that for cadmium(II) was 1.76. Calculate the weight per cent of lead in the zinc metal.

17. Suppose that you wish to investigate the amperometric titration of lead(II) with sulfate ion, the pertinent reaction being

$$Pb^{2+} + SO_4^{2-} \rightleftharpoons PbSO_4(s)$$

A 50.00-ml aliquot of a solution 0.00500 $F$ in lead(II) nitrate and $1 F$ in potassium nitrate was titrated with a solution containing 0.500 $F$ potassium sulfate and $1\,F$ potassium nitrate. The potential of the dropping mercury electrode was held at $-0.70$ v versus SCE, at which value the drop time and mercury flow rate were 3.96 seconds and 1.98 mg/second, respectively. If the diffusion coefficient of lead(II) in $1\,F$ potassium nitrate medium is $1.90 \times 10^{-5}$ cm$^2$/second, and if the solubility product for lead sulfate is $1.6 \times 10^{-8}$, what will be the observed current in microamperes at the equivalence point of the titration? Assume that the residual current is negligible.

18. A solution containing $2.00 \times 10^{-3}\,M$ quinone and $1.00 \times 10^{-3}\,M$ hydroquinone in a $0.1\,F$ phosphate buffer having a pH of 7.00 at 25°C shows a composite cathodic-anodic polarogram similar to that pictured in Figure 13–3B on page 435. With a particular dropping mercury electrode, the cathodic diffusion current was $+9.83\,\mu$a and the anodic diffusion current was $-4.55\,\mu$a. For the quinone-hydroquinone system, the half-wave potential is $+0.039$ volt versus SCE.

(a) If the diffusion coefficient of quinone is known to be $8.6 \times 10^{-6}$ cm$^2$/sec, what is the diffusion coefficient of hydroquinone? Assume that the capillary characteristics do not change over the potential range of interest.

(b) What should be the potential of the dropping mercury electrode when the current is zero?

(c) What should be the potential of the dropping mercury electrode at the point on the cathodic wave where the current is $+6.17\,\mu$a?

## SUGGESTIONS FOR ADDITIONAL READING

1. M. Březina and P. Zuman: *Polarography in Medicine, Biochemistry, and Pharmacy*. Wiley-Interscience, New York, 1958.
2. D. R. Crow: *Polarography of Metal Complexes*. Academic Press, New York, 1969.
3. J. Doležal and J. Zýka: Amperometric titrations. *In* F. J. Welcher, ed.: *Standard Methods of Chemical Analysis*. Sixth edition, Volume III-A, Van Nostrand, Princeton, New Jersey, 1966, pp. 377–403.
4. J. Heyrovský and J. Kůta: *Principles of Polarography*. Academic Press, New York, 1966.
5. J. Heyrovský and P. Zuman: *Practical Polarography*. Academic Press, New York, 1968.
6. I. M. Kolthoff and J. J. Lingane: *Polarography*. Second edition, Wiley-Interscience, New York, 1952.
7. J. J. Lingane: *Electroanalytical Chemistry*. Second edition, Wiley-Interscience, New York, 1958, pp. 234–295.
8. L. Meites: *Polarographic Techniques*. Second edition, Wiley-Interscience, New York, 1965.
9. L. Meites: Voltammetry at the dropping mercury electrode (polarography). *In* I. M. Kolthoff and P. J. Elving, eds.: *Treatise on Analytical Chemistry*. Part I, Volume 4, Wiley-Interscience, New York, 1959, pp. 2303–2379.
10. G. W. C. Milner: *Principles and Practice of Polarography and Other Voltammetric Processes*. Longmans, Green & Company, Ltd., London, 1957.
11. O. H. Müller: Polarography. *In* A. Weissberger and B. W. Rossiter, eds.: *Physical Methods of Chemistry*. Volume I, Part IIA, Wiley-Interscience, New York, 1971, pp. 297–421.
12. W. C. Purdy: *Electroanalytical Methods in Biochemistry*. McGraw-Hill Book Company, New York, 1965.
13. W. B. Schaap: Polarography. *In* F. J. Welcher, ed.: *Standard Methods of Chemical Analysis*. Sixth edition, Volume III-A, Van Nostrand, Princeton, New Jersey, 1966, pp. 323–376.
14. J. T. Stock: *Amperometric Titrations*. Wiley-Interscience, New York, 1965.
15. P. Zuman: *Organic Polarography*. The Macmillan Company, New York, 1964.
16. P. Zuman and C. L. Perrin: *Organic Polarography*. Wiley-Interscience, New York, 1969.

# CHEMICAL SEPARATIONS

**14**

Imagine that you have a warehouse full of apples and oranges, and that you want to know the quantities of each. Simply weighing truckloads of fruit would be futile until a separation of apples from oranges had been accomplished. At that point, uniform truckloads ("pure samples") of either 100 per cent apples or oranges could be measured. If the separation of apples from oranges was selective and careful enough to get rid of last year's potatoes and even a few tangerines lurking in the corner, the quality of your analysis of the warehouse contents would be improved.

It almost hurts to read such a homely example, but consider, on the other hand, the emotions of a chemist who has analyzed $Ba^{2+}$ gravimetrically as the sulfate without removing $Sr^{2+}$. The equivalent weight of an organic acid can be determined, and its empirical formula deduced, if a weighed sample is titrated with standard base. But what useful formula can result if the organic acid happens to be a mixture of two or three compounds?

It pays to remember to separate the apples from the oranges. Quantitative analyses, qualitative analyses, and structural analyses all must begin with a step which clearly defines the sample. Computer programmers have given us an "axiom for our times" which describes the situation with chemical systems as delicately and accurately as it does for computer systems: "Garbage in, garbage out!"

In the following sections, we shall consider chemical separations in general, stressing first that separation of compounds implies a separation of phases. Second, we shall consider mechanisms by which phases can be equilibrated and separated. As a final part of this introduction, which is designed to organize and unify all our thinking about separations, we shall construct a table of phase equilibria which amounts to a simple catalog of all possible separation techniques.

## SEPARATION OF COMPOUNDS REQUIRES SEPARATION OF PHASES

It is not possible to purify compounds by selecting individual molecules. We must work on a larger scale, and there is no way around the generalization that, if two compounds are to be separated, we must, somewhere along the line, get them into two different and separable phases. For example, if we wish to separate a carboxylic acid ($pK_a \sim 5$) from an alkane (no functional groups), we can equilibrate an ether solution of the two compounds with aqueous $0.01\ F$ sodium hydroxide in a separatory funnel (see Figure 15–10). The alkane is only slightly soluble in the aqueous phase and remains in the ether. The carboxylic acid, however, can dissociate:

$$RCO_2H + H_2O \rightleftharpoons RCO_2^- + H_3O^+; \qquad \frac{[RCO_2^-][H^+]}{[RCO_2H]} = K_a \sim 10^{-5}$$

At the pH of the aqueous phase, about 12, this dissociation is complete:

$$\frac{[RCO_2^-]}{[RCO_2H]} \sim \frac{10^{-5}}{[H^+]} = \frac{10^{-5}}{10^{-12}} = 10^7$$

The resulting carboxylate anion is far more soluble in the aqueous phase than in the ether, so that the carboxylate is partitioned quantitatively into the aqueous phase. Separation of the phases is accomplished by draining the heavier aqueous phase from the funnel. It is this phase separation which provides the desired separation of compounds—in this case, of the alkane from the acid.

Liquid-liquid extraction is by no means the only convenient method of compound and phase separation. In Chapters 7 and 8, precipitation is discussed, and examples are given of the separation of silver by precipitation as the chloride, of calcium by precipitation as the oxalate, of nickel by precipitation of the dimethylglyoximate complex, and many others. Brewers and distillers have long been aware of the possibilities of volatility and adsorptivity as tools of chemical separation. A more volatile and purified ethanol-rich phase is separated from fermented mash by distillation, and other compounds with unpleasant tastes and aromas are removed from water-ethanol solutions by adsorption on charcoal.

In precipitations, recrystallizations, and all separations based on solubility, the nature of the phase separation is clear—a solid is separated from a liquid. The separation of two liquid phases, as in the ether-aqueous base system described earlier, is also easily grasped. It is important to extend this concept to more subtly defined phases. For example, in the adsorption on charcoal mentioned above, it is best not to think of the separation of a "charcoal phase" from a liquid phase, but instead to realize that an adsorbed phase (the compounds held on the surface of the charcoal) has been separated from the liquid phase. This is not pedantic nit-picking.

If large quantities of liquid phase are passed through the charcoal bed, the adsorbed compounds will be slowly released. In a slightly different situation, the compounds in the adsorbed phase might be displaced by substances even more strongly attracted to the charcoal surface. In either case, the result is the same—the adsorbed compounds reappear in the liquid phase. This situation is most correctly viewed as an adsorbed phase-liquid phase separation.

For emphasis, we repeat: at the heart of any chemical separation are the processes of (1) phase contact and equilibration, and (2) phase separation. These steps occur in all separation techniques, and a key to understanding a given method is the identification and classification of the steps according to the nature of the phases involved and the mechanism of phase contact and separation. Similarly, if a particular method of separation is to be improved, these are the only processes worth adjusting.

*Mechanisms of phase contact.* Consideration of the ways in which phases can be brought together, equilibrated, and separated reveals four levels of complexity. These levels are summarized in Figure 14–1.

The simplest possibility is shown in Figure 14–1A, and is a single-stage operation like that discussed in the acid-alkane example used above. The phases need not be liquids only. In terms of complexity, we might deal similarly with a single-stage crystallization or gas-liquid equilibration.

A second level of complexity is reached (with no great strain . . . these distinctions are terribly simple, but well worth keeping in mind when trying to get a general view of separation processes) when fresh batches of one of the phases are repeatedly equilibrated with the other. This is shown in Figure 14–1B, and is a technique

*Figure 14–1.* A schematic representation of the techniques of phase contact. Two types of phases (**1** and **2**) are shown; sequentially used batches of each phase are assigned additional letters (**1a**, **1b**, **1c**, etc.) in the order of their introduction to the system.

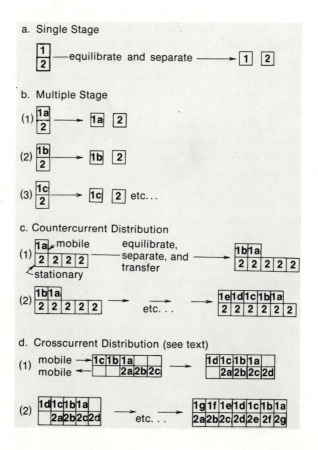

commonly employed in extractions when the aim is to increase the yield of some partially extracted product. The process can be carried out automatically in discontinuous steps, as in a Soxhlet extractor (see Figure 15–12 and its accompanying discussion), or continuously, with incremental additions and withdrawals of fresh extracting phase, as in a continuous liquid extractor (see Figure 15–11 and its accompanying discussion).

At the third level of complexity an important new feature is introduced. In countercurrent distribution (CCD), *multiple contacts* occur between pairs of phases.* In CCD, the initial equilibration occurs between only one pair of phases, but succeeding transfer operations bring about the creation of many new phase pairs, all of which are equilibrated simultaneously in later operations. As Figure 14–1C shows, it is convenient to regard one phase as stationary and the other as mobile, since it migrates through the apparatus. The mobile phase-stationary phase distinction is also characteristic of chromatographic separations (Chapter 16), which differ from this stepwise process only in that equilibrations occur between incremental regions in continuous phases instead of between individual batches of mobile and stationary phase. For example, a typical chromatographic system made up of an unpartitioned tube packed with a silica gel stationary phase and slowly washed with a hexane mobile phase can be regarded, for purposes of calculation and the understanding of separation processes, as a CCD apparatus with a large number of discrete silica gel-hexane contact units.

A fourth level of complexity is recognized but not consistently named. Here, we will use the term "crosscurrent distribution," but this is a name without any official status, and it is by no means uniformly applied by all chemists. As noted in Figure 14–1D, the distinguishing characteristic is that *both* phases are moving through the apparatus in opposite directions. This might seem to differ only trivially from CCD, in which the relative motion of the phases is the same; however, there is an important distinction, and calculations pertaining to the two types of systems differ greatly. An example of crosscurrent distribution is furnished by a distillation column in which the refluxing liquid phase streams downward through the column from the condenser to the still pot, and is continually equilibrated with upward moving vapor phase.

***Types of phases.*** It is useful to make a distinction between "bulk" and "thin-layer" phases. When a liquid and a solid precipitate are separated, or when two liquids are separated in a separatory funnel, each phase has a recognizable volume extending in three dimensions. Such extensive phases are termed "bulk phases." By contrast, when compounds are separated by adsorption on some solid material, the adsorbed "phase" can be viewed as extending in two dimensions only. The third dimension, normal to the surface of the adsorbent, extends for only a few molecular diameters. Such phases are logically termed "thin-layer phases."

Two strikingly different types of thin-layer phases can be distinguished, depending upon whether the "solute" molecules are free to diffuse within the phase. If the thin-layer phase is due simply to adsorption, then the adsorbed molecules are relatively immobile. In the other type of thin layer, the solute molecules are not held to the surface, but instead are free to diffuse within a thin layer of some liquid phase dispersed on some inert support. This distinction is further explained by Figure 14–2. Because thin layers are usually encountered in chromatography, the figure further identifies the phases as stationary and mobile.

---

* Dr. Lyman C. Craig was the major developer of liquid-liquid stepwise-operation countercurrent distribution. The whole technique bears an abbreviation honoring him, "CCD," for "Craig Countercurrent Distribution."

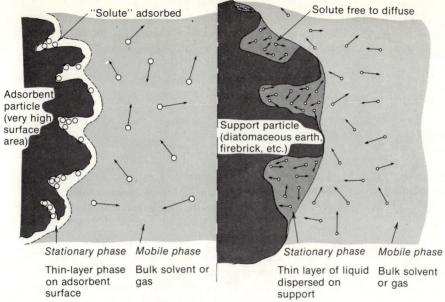

"Solute" adsorbed

Solute free to diffuse

Adsorbent particle (very high surface area)

Support particle (diatomaceous earth, firebrick, etc.)

| Stationary phase | Mobile phase | Stationary phase | Mobile phase |
|---|---|---|---|
| Thin-layer phase on adsorbent surface | Bulk solvent or gas | Thin layer of liquid dispersed on support | Bulk solvent or gas |

*Figure 14–2.* The two types of thin-layer phases.

Five different phase types can be listed as shown below:

$$\text{Bulk} \begin{cases} \text{gas} \\ \text{liquid} \\ \text{solid} \end{cases}$$

$$\text{Thin-layer} \begin{cases} \text{solute free to diffuse} \\ \text{solute adsorbed} \end{cases}$$

These phases can be arranged in a matrix as shown in Figure 14–3. This matrix, together with the discussion above on mechanisms of phase contact, provides a

| | BULK PHASES | | | THIN-LAYER PHASES | |
|---|---|---|---|---|---|
| | GAS | LIQUID | SOLID | SOLUTE FREE TO DIFFUSE | SOLUTE ADSORBED |
| GAS | gaseous diffusion | gas-liquid extraction, distillation | sublimation | gas-liquid chromatography GLC | gas-solid chromatography GSC[1] |
| LIQUID | | liquid-liquid extraction | crystallization | liquid-liquid chromatography LLC | adsorption chromatography[2] ion-exchange chromatography |

Notes:
[1] Also called "gas adsorption chromatography."
[2] Sometimes called "liquid-solid chromatography, LSC." Includes thin-layer chromatography, "TLC."

*Figure 14–3.* Phase equilibria and chemical separation techniques.

framework within which chemical separations can be considered in a quite general way. This is an advantage, because it is neither useful nor practical to discuss each method separately. In the remaining sections on chemical separations, at least one example of each mechanism of phase contact will be discussed, as well as a good selection of the possible phase pairs represented in Figure 14–3. It is important for the reader to remember this general picture in order to realize the broader significance of the specific examples chosen.

## DISTILLATION

Because of its important practical applications, distillation has been the subject of spirited study for hundreds of years. Intoxicated by their first successes, workers throughout the world have made it one of the best understood techniques of chemical separation. As a result, terminology originally developed with specific reference to distillation has been very broadly applied, and a discussion of distillation provides a good introduction to separation processes.

In addition, distillation continues to have great significance as a preparative technique. This is particularly true in analytical problems; anyone who has undertaken the analysis of trace amounts of organic materials knows the importance of carefully redistilled solvents. In spite of its importance, distillation technique is not carefully discussed in many modern chemical curricula, perhaps because analytical distillations are a thing of the past. As a result, many would-be pesticide analysts cannot distill ordinary solvents well enough to produce an acceptable blank in their analytical procedures.

Only the barest fundamentals can be covered here. The interested reader can find the theoretical treatment from which this summary was derived, and much more additional information, in reference 4 listed at the end of this chapter.

### Apparatus

The essential components of a still are depicted schematically in Figure 14–4. In addition to the still pot, heater, condenser, and product receiver, there are two components—the column and the still head—which deserve special attention.

The efficiency of the still is determined largely by the design and operation of the column and still head. Vapor leaving the still pot passes upward through these components to the condenser, which is mounted in a vertical position so that the moment vapor is condensed it drops back into the still head without further cooling. The still head has the function of dividing the condensate so that a portion ($D$) is taken off as a product and the remainder ($L$) is returned to the column as *reflux* liquid. The ratio $L/D$ is defined as the **reflux ratio**, about which we shall learn more later:

$$\text{reflux ratio} = R_D = \frac{L}{D} \qquad (14\text{--}1)$$

The column, then, has the function of providing the best possible contact between the ascending vapor and the descending reflux liquid.

Recalling that a distillation column was used as the example of crosscurrent distribution (see Figure 14–1D), the reader must realize that this column, and the repeated liquid-vapor equilibrations which it provides, is the heart of the separation procedure. If one of the phases is missing—that is, if no reflux liquid is provided—the column might as well be left out of the apparatus. This is a common fault, but many people have had the experience of performing apparently successful distillations

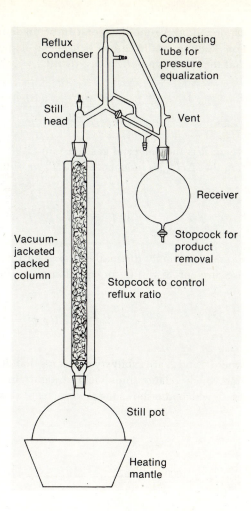

*Figure 14–4.* Schematic drawing of the essential components of a laboratory still.

Reflux condenser

Connecting tube for pressure equalization

Still head

Vent

Receiver

Vacuum-jacketed packed column

Stopcock for product removal

Stopcock to control reflux ratio

Still pot

Heating mantle

without attention to these details, a fact for which there is no shortage of explanations: (1) in most preparative applications, either the required separation is not difficult or the techniques available for product analysis are minimal (frequently both!); (2) in most stills, insulation of the column (if any) and head is so poor that there is a substantial reflux due to condensation of vapor before it reaches the condenser.

## Theory of Operation

*Vapor pressure of pure liquids.* In consideration of a technique based on the liquid-vapor phase change, the best place to begin is with the relationship between vapor pressure and temperature derived from thermodynamics, namely, the **Clausius-Clapeyron equation**,

$$(V_v - V_l) \frac{dp^\circ}{dT} = \frac{\Delta H_{vap}}{T} \tag{14-2}$$

which is valid whenever a pure liquid is in equilibrium with its vapor. In the above expression, $V_l$ and $V_v$ are the molar volumes of the liquid and vapor phases, respectively; $dp^\circ/dT$ is the slope of the saturation vapor pressure versus temperature curve at a given absolute temperature, $T$; and $\Delta H_{vap}$ is the molar heat of vaporization. A simplification can be made because $V_1$ is negligible compared to $V_v$. For example, in the case of benzene at its boiling point, $V_v/V_1 \sim 300$. If we assume that the vapor is a perfect gas, then $V_v = RT/p°$, and equation (14–2) takes the form

$$\frac{dp°}{p°dT} = \frac{d \ln p°}{dT} = \frac{\Delta H_{vap}}{RT^2} \qquad (14\text{–}3)$$

where $R$ is the universal gas constant (1.987 cal mole$^{-1}$ deg$^{-1}$). Making the assumption that $\Delta H_{vap}$ is independent of $T$, we obtain, upon integration of equation (14–3),

$$\ln p° = 2.303 \log p° = -\frac{\Delta H_{vap}}{RT} + \text{constant} \qquad (14\text{–}4)$$

Examining simply the mathematical form of equation (14–4), we can see that vapor pressure curves will fit an equation of the form

$$\log p° = C - \frac{A}{T} \qquad (14\text{–}5)$$

where $C$ and $A$ are arbitrary constants that depend on the compound under consideration. Such relationships yield straight lines on graphs having axes as in Figure 14–5, which also shows $\log p°$ versus $T^{-1}$ relationships for a few specific compounds.

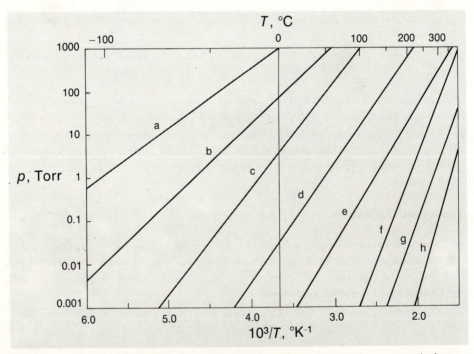

*Figure 14–5.* Saturation vapor pressure as a function of temperature for some typical compounds: a, butane, $C_4H_{10}$; b, acetone, $C_3H_6O$; c, water; d, n-dodecane, $C_{12}H_{26}$; e, mercury vapor, Hg; f, n-hexacosane, $C_{26}H_{54}$; g, adenine, $C_5H_5N_5$; and h, glycine, $H_2NCH_2CO_2H$. The last two, at least, are normally regarded as very involatile solids, glycine existing at least partially in the zwitterionic form $H_3N^+CH_2CO_2^-$.

Definite integration of equation (14–3) over an arbitrary interval provides the useful relation

$$\ln \frac{p_1^\circ}{p_2^\circ} = 2.303 \log \frac{p_1^\circ}{p_2^\circ} = \frac{\Delta H_{vap}}{2.3R} \left( \frac{T_1 - T_2}{T_1 T_2} \right) \qquad (14\text{–}6)$$

which can be used to evaluate $\Delta H_{vap}$ from values of $p^\circ$ at two different temperatures, or which can be used to estimate $p^\circ$ at any $T$, provided $\Delta H_{vap}$ and the boiling point at atmospheric pressure are known. In this connection, it is useful to remember **Trouton's rule**, which relates $\Delta H_{vap}$ and the boiling point at atmospheric pressure, $T_b$:

$$\frac{\Delta H_{vap}}{T_b} \sim 23.3 \text{ cal mole}^{-1} \text{ deg}^{-1} \qquad (14\text{–}7)$$

This is, of course, only a rough approximation, and a particularly poor one for acids, alcohols, and hydrogen-bonding liquids in general. In such cases, the Trouton constant will be empirically found to be much higher, often in excess of 26 cal mole$^{-1}$ deg$^{-1}$.

Extensive tabulations of vapor pressure versus temperature are listed among the references at the end of this chapter. A great deal of theoretical and practical information about the variation of vapor pressure with molecular structure and about related matters is summarized in the classic text by Purnell, which is also cited.

*Vapor pressure of binary mixtures.* A familiar starting point for this discussion is **Raoult's law**, which relates the partial vapor pressure $(p_A)$ of a substance in an ideal mixture to its mole fraction in the liquid phase $(X_A)$ and its saturation vapor pressure $(p_A^\circ)$ at the temperature of the mixture:

$$p_A = X_A \, p_A^\circ \qquad (14\text{–}8)$$

Consider a second substance, B, having a mole fraction $X_B$ in the liquid, partial vapor pressure $p_B$, and a saturation vapor pressure $p_B^\circ$. Then, using $Y_A$ and $Y_B$ to represent the mole fractions in the vapor phase, we can write

$$\frac{Y_A}{Y_B} = \frac{p_A}{p_B} = \frac{X_A p_A^\circ}{X_B p_B^\circ} \qquad (14\text{–}9)$$

*If* substances A and B happen to have equal saturation vapor pressures, then the mole fractions of each substance in the vapor and liquid phases will be equal. In the far more likely case that the saturation vapor pressures are unequal (which would mean that A and B have unequal boiling points), the vapor phase is enriched with the more volatile constituent. The mole fraction ratios are related by a coefficient, $\alpha$, termed the **relative volatility**, and defined by the equation

$$\frac{Y_A}{Y_B} = \alpha \frac{X_A}{X_B} \qquad (14\text{–}10)$$

where A and B are always chosen such that $\alpha > 1$. That is, A is always chosen to be the more volatile component.

Because we are dealing with a binary (two-component) mixture, equation (14–10) can be recast if we make the substitutions $X_B = 1 - X_A$ and $Y_B = 1 - Y_A$:

$$\frac{Y_A}{1 - Y_A} = \alpha \frac{X_A}{1 - X_A} \qquad (14\text{–}11)$$

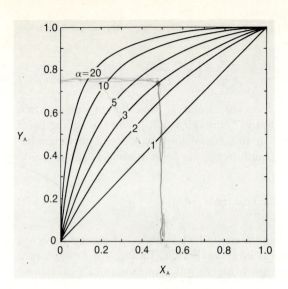

*Figure 14–6.* $Y_A$ as a function of $X_A$ and $\alpha$.

This expression is easily rearranged to give $Y_A$ as a function of $X_A$ and $\alpha$; the resulting relationship is shown graphically in Figure 14–6. It is easily seen that as $\alpha$, the relative volatility, becomes large, the degree of enrichment of the more volatile component in the vapor phase increases greatly. Another view of this same enrichment is provided by Figure 14–7, which shows the liquid- and vapor-phase composition lines for a mixture of A and B with $\alpha = 3$, $T_{bA} = 353°\text{K}$ (80°C), and $T_{bB} = 393°\text{K}$ (120°C). The reader should study the two graphs and understand their equivalence by noting, for example, that $X_A = 0.50$ corresponds to $Y_A = 0.75$ on both curves. The $Y_A$ versus $X_A$ plot is, in some ways, more general, inasmuch as it is independent of $T$ and correctly indicates the vapor-phase enrichment for any pair of liquids having $\alpha = 3.0$ (the possibilities are infinite; for example, this would apply for two liquids having $T_{bA} = 331°\text{K}$ and $T_{bB} = 369°\text{K}$ or $T_{bA} = 402°\text{K}$ and $T_{bB} = 448°\text{K}$). On the other hand, the composition versus temperature plot carries additional information in that the liquid composition line gives the boiling point of the mixture.

Comparison of equations (14–9) and (14–10) makes it clear that, *for ideal solutions,*

$$\alpha = \frac{p_A^{\circ}}{p_B^{\circ}} \tag{14–12}$$

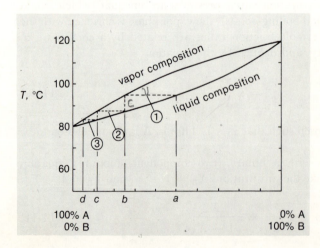

*Figure 14–7.* $X_A$ and $Y_A$ as a function of $T$ for $\alpha = 3.0$.

The saturation vapor pressures can be estimated from the Clausius-Clapeyron equation; $\Delta H_{vap}$, which is required in that calculation, can be estimated from Trouton's rule if, at least, the boiling point of the pure liquid at atmospheric pressure is known. Thus, it is possible to derive an approximate, but nonetheless useful, expression for $\alpha$ as a function of $T_{bA}$ and $T_{bB}$:[*]

$$\log \alpha = 8.9 \frac{T_{bB} - T_{bA}}{T_{bB} + T_{bA}} \tag{14-13}$$

*Column operation and theoretical plates.* The dotted lines in Figure 14–7 trace the idealized course of events in a distillation column. Imagine that the still pot contains an equimolar mixture of A and B (composition *a*, Figure 14–7). The vapor phase in equilibrium with this mixture will have composition *b*. Notice that vaporization is represented by a horizontal tie-line and condensation by a vertical tie-line on graphs of this type. Condensation of that vapor followed by revolatilization will yield a vapor of composition *c*; another stage will yield composition *d*, and so on. It is exactly this process of condensation and revolatilization which occurs during the repeated vapor-liquid equilibria within a distillation column.

The efficiency of a separation is determined by analysis of the vapor at the top of the column (that is, the composition of the liquid in the condenser) and by consideration of how many condensation and revolatilization steps would be required to reach that degree of purity. In the case shown in Figure 14–7, for example, for a still which yielded a product having composition *d* from starting material of composition *a*, it would be said that, theoretically, this degree of enrichment could be achieved in three stages of volatilization and condensation. Of course, the actual process might be considerably more complex because of nonideality in the physical processes occurring in the still, but an unambiguous measurement of efficiency is furnished by this comparison to the theoretical standard. The imaginary individual stages are called **theoretical plates**.

Not all of these theoretical plates are furnished by the fractionating column. The very first volatilization stage occurs in the still pot itself, and, thus, the vapor entering the condenser even in the absence of a column is one fractionation stage removed from the starting material. If a particular distillation apparatus performs with an efficiency equivalent to *n* theoretical plates, the column in that apparatus is said to contain $(n - 1)$ theoretical plates.

In Figure 14–8 is shown another representation of the same mixture and process depicted in Figure 14–7. On a graph of this type, volatilization steps are represented by a vertical line drawn upward from the 45° reference line to the corresponding vapor composition. Condensations are represented by a horizontal line drawn to the right. The theoretical plate numbers and composition letters in Figure 14–8 correspond exactly to those in Figure 14–7. The reader should verify this and note that, whereas enrichment of the more volatile component leads generally downward and to the left on a phase diagram, it leads upward and to the right on a plot of $X_A$ versus $Y_A$. Because only $\alpha$ need be known for its construction, the use of the latter is considerably more common in practice. Reference to Figure 14–6 emphasizes how the decreasing curvature which accompanies a decrease in $\alpha$ will drastically increase the number of steps necessary to achieve any required degree of purification.

A typical application of the stepping-off technique is represented by the heavier dashed lines in Figure 14–8. The number of theoretical plates required for the production of 98 mole per cent pure A from a starting material containing only 10 mole

---

[*] A. Rose: *Ind. Eng. Chem., Industrial Edition* 33:594, 1941.

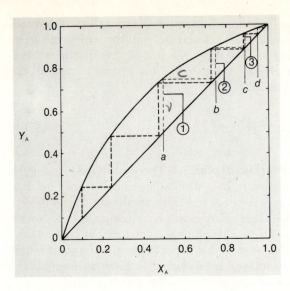

*Figure 14–8.* $Y_A$ versus $X_A$ for a mixture with $\alpha = 3.0$. The lighter dashed lines trace out the same process shown in Figure 14–7. The heavier dashed lines represent a fractionation discussed in the text.

per cent A has been graphically found to be six. Recalling that $\alpha$ can be calculated from boiling-point data, we see that this method furnishes a reasonably convenient technique for determining the number of theoretical plates required for a given separation even if only the boiling points of the substances are known.

There is a simple mathematical equivalent to this graphic determination of $n$. The **Fenske equation** relates $n$, the number of theoretical plates required in a distillation column; $\alpha$; $X_{AS}$, the mole fraction of component A in the still pot; and $Y_{AD}$, the mole fraction of A in the distillate, as follows:

$$\frac{Y_{AD}}{1 - Y_{AD}} = \alpha^{n+1} \frac{X_{AS}}{1 - X_{AS}} \tag{14–14}$$

Problem 7 at the end of this chapter will guide the reader through the derivation of the Fenske equation.

## Practical Considerations

*Oversimplifications in the preceding theoretical treatment.* The preceding sections have made great use of the term "ideal." Until now, azeotropes have not even been mentioned; however, as a reminder, the classic example of the ethanol-water system can be briefly reviewed. The phase diagram is shown in Figure 14–9, and displays a shape differing markedly from the ideal behavior shown in Figure 14–7. Starting with a liquid containing 50 mole per cent water, the mixture would be only a few theoretical plates away from a product boiling at 78.15°C and having 89.43 mole per cent (about 95 per cent by weight) ethanol, but not all the theoretical plates in the world would bring the system to 100 per cent ethanol. In fact, a distillation which started with 99 mole per cent ethanol would yield the 89 mole per cent azeotrope as a distillate—at least until all the water was used up. The further purification of ethanol is interesting because it also involves azeotropes. If benzene is added to the system, a ternary azeotrope with a boiling point of 65°C is formed containing by weight 74 per cent benzene, 18.5 per cent alcohol, and 7.5 per cent water. In this way, ethanol and water can be removed on an equimolar basis until no water

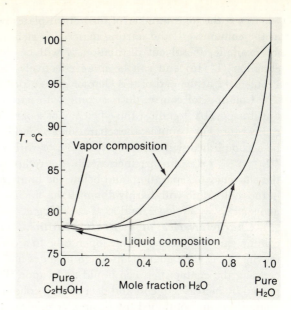

*Figure 14–9.* Phase diagram for the water-ethanol system.

remains in the still pot. Finally, distillation at 68.3°C of an azeotrope consisting by weight of 67.6 per cent benzene and 32.4 per cent ethanol eventually leaves only ethanol in the still pot. The pure ethanol can be collected by distillation at 78.5°C. The formation of azeotropes is perhaps not as frequent as suggested by the example of the ethanol-water system, but neither is it rare enough to be completely forgotten. The occurrence of azeotropes is difficult to predict, but many azeotropes are tabulated in references 2 and 6 cited at the end of this chapter.

Less drastic deviations from ideality can occur in systems which do not form azeotropes, and it is clear that, for theoretical plate-counting and for consideration of the feasibility of various distillations, it would, in general, be best to have the phase diagram and to avoid, whenever possible, reliance on a simple calculation of $\alpha$ based on boiling points. Such calculations remain useful as a rough guide and serve quite well when ideality can be reasonably expected, as, for example, in a mixture of very similar compounds.

The relationships and methods of plate-counting derived above apply, strictly speaking, only when the still is operated under total reflux ($R_D = \infty$). Any removal of product will decrease the efficiency of the column by changing the liquid composition, and, hence, the vapor composition at every stage of the fractionation. Since stills are not often operated for amusement but instead to yield some product, this sort of "ideality" can be justified only by the extreme complexity of a more exact treatment, which must, in addition, take into account the changing composition of the liquid in the still pot as the product is removed. An approximate relationship derived with these factors taken into account and based on the assumptions that the starting mixture is equimolar in A and B, that a product purity of 95 per cent for the first 40 per cent distilled is sufficient, and that $n \sim R_D$ is*

$$n = \frac{2.85}{\log \alpha} = \frac{T_{bB} + T_{bA}}{3(T_{bB} - T_{bA})} \qquad (14\text{–}15)$$

---

* A. Rose: *op. cit.*

where $n$ is the total number of theoretical plates required [therefore, $(n-1)$ plates in the column]. If the starting material is rich in the more volatile component, as, for example, in solvent distillation, $n$ will be different from what is found with equation (14–15) and can be more effectively calculated using equation (14–14). Problem 12 at the end of this chapter gives a practical example.

Columns, of course, do not come with variable numbers of theoretical plates, but it is possible to adjust the efficiency of separation to some extent by varying the reflux ratio. For example, the separation accomplished with $n = R_D = 29$, could also be accomplished with $n = 22$, $R_D = 33$, or with $n = 36$, $R_D = 24$. Above the limit $(R_D)_{max} = \frac{3}{2} n_{min}$ further increases in the reflux ratio cannot substantially increase the efficiency of separation, and below the limit $(R_D)_{min} = \frac{3}{2} n_{max}$ further decreases in the reflux ratio will sharply diminish the efficiency of separation. In the expressions above, $n_{max} = 3.6/\log \alpha$, and $n_{min} = 2.3/\log \alpha$.

*Characteristics of typical columns.* The construction of a distillation column can vary from a simple, open tube to a complex design involving high-speed moving parts. The goal of all design changes has been to improve efficiency by increasing the contact between liquid and vapor. This achievement has been accompanied by increases in cost and decreases in column throughput capacity which have guaranteed the continued use of simpler models. Four typical designs are shown in Figure 14–10. Each is insulated by a vacuum jacket in order to approach adiabatic operation as closely as possible. A column which "leaks" a great deal of heat will provoke premature condensation of vapor and have overabundant reflux at the bottom, a dribble at the top, and almost no vapor available for production of a distillate. In the open tubular column, equilibration can take place only at the interface between a freely falling liquid film and the rapidly rising vapor. Much the

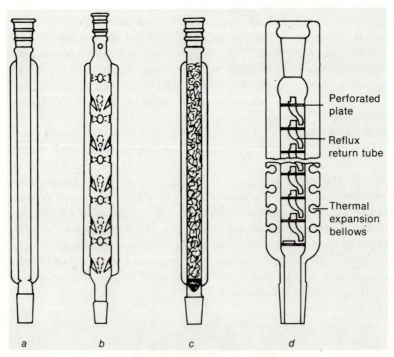

Perforated
plate

Reflux
return tube

Thermal
expansion
bellows

a    b    c    d

*Figure 14–10.* Sectional views of various types of distillation columns: (a) Open tubular, (b) Vigreux, (c) Packed, (d) Oldershaw perforated plates. Typical dimensions are given in Table 14–1.

same comment might be applied to the Vigreux column; but, in a packed column, chances for prolonged contact between liquid and vapor are clearly increased. Books have been written about various types of column packings—a typical and broadly applicable choice would be tiny glass helices which, when packed into the column, offer an enormous surface area without undue resistance to vapor or liquid flow. Columns are often constructed which actually include plates as part of the liquid-vapor contact units. Unfortunately, theoretical performance is not achieved, as will be seen in Table 14–1. In the Oldershaw column, vapor travels up through perforations in the plates, each of which is covered by a liquid layer through which the vapor must pass. Whenever the liquid layer on a plate becomes more than about one millimeter deep, it overflows by way of a downspout to the plate below. Not shown is a spinning-band column, in which a long, flat metal or Teflon band runs down the center of an open tubular column. Spinning the band at high speed (2500 rpm) throws vapor molecules into good contact with the liquid film on the column wall.

The performance of a column is summarized by noting the length (or, since the column is used vertically, *height*) of column equivalent to a theoretical plate. This figure of merit is abbreviated H.E.T.P., which signifies the height equivalent to a theoretical plate. For example, a column with an overall length of one meter which yields 20 theoretical plates would have H.E.T.P. = 5 cm. The performance of a column is not only a function of its design and of the reflux ratio at which it is operated but also of a number of other factors. In general, columns give better performance at low throughputs. In such circumstances, the residence time of vapor-phase species in the column is longer and equilibration is consequently improved. Pretreatment of the column is also important. For example, packed columns can be

Table 14–1. Distillation Column Efficiency*

| Column type and internal diameter | Throughput,† ml/min | H.E.T.P. at total reflux, cm |
|---|---|---|
| *Open tube* | | |
| 6 mm | 8.0 | 60.6 |
| | 5.0 | 30.3 |
| | 1.9 | 15.1 |
| | 0.45 | 5.05 |
| | 0.17 | 1.73 |
| *Vigreux* | | |
| 8 mm | ~6 | 10 |
| 15 mm | ~18 | 17 |
| 25 mm | ~50 | 30 |
| *Packed* | | |
| (12-mm tube with | 3.7 | 3.8 |
| 3-mm glass helices) | 11.3 | 4.3 |
| *Oldershaw* | | |
| 27 mm | 25 | 4.4 |
| (2.5-cm spacing | 42 | 4.0 |
| between plates) | 50 | 3.9 |
| | 58 | 4.1 |
| | 67 | 4.5 |
| | 80 | Flooding |
| *Spinning band* | | |
| 6 mm | 1.0 | 1.0 |

* From Chapter III, reference 4.
† Reflux-liquid flow at total reflux.

"flooded" if the boil-up rate is increased until the vapor rising in the column actually prevents reflux liquid from returning to the still pot. The slugs of liquid which accumulate in the column wet it thoroughly, thus promoting a uniform flow of reflux liquid and, consequently, increased efficiency under the more moderate conditions used during actual distillation. Table 14–1 summarizes the efficiencies actually attained by several typical column designs under various operating conditions.

## QUESTIONS AND PROBLEMS

1. The vapor pressure of $n$-dodecane is one torr at 47.8°C, 10 torr at 90.0°C, and 100 torr at 146.2°C. Estimate graphically the temperature at which the vapor pressure is (a) $10^{-3}$ torr and (b) 300 torr.
2. For the compounds listed below, (a) estimate $\Delta H_{vap}$ given their respective boiling points and (b) compute the value of the Trouton's law constant in each case using the given $\Delta H_{vap}$ values.

| Compound | $T_b$, °C | $\Delta H_{vap}$, kcal/mole |
|---|---|---|
| propane | −42.07 | 4.487 |
| $n$-pentane | 36.07 | 6.160 |
| $n$-heptane | 98.43 | 7.575 |
| benzene | 80.10 | 7.353 |
| toluene | 110.62 | 8.00 |
| ethyl benzene | 136.19 | 8.60 |
| krypton | −153.23 | 2.158 |
| xenon | −108.1 | 3.021 |
| methyl acetate | 56.3 | 7.270 |
| ethyl propanoate | 97.6 | 8.182 |
| $n$-propyl acetate | 100.4 | 8.202 |
| water | 100.0 | 9.717 |
| ethanol | 78.5 | 9.22 |
| $n$-propanol | 97.2 | 9.88 |
| $n$-pentanol | 131. | 10.6 |

3. Explain the range in Trouton's law constants observed in problem 2. What value of the Trouton's law constant would you use to estimate $\Delta H_{vap}$ of (a) neon, (b) $n$-butanol, and (c) $n$-butyl acetate?
4. Considering the data in problem 2, which organic compound could be expected to show a vapor pressure versus temperature curve like that of water?
5. Hexadecane has a boiling point of 287.5°C at atmospheric pressure. Estimate its vapor pressure at 150°C. (Note: problem 2 shows that a good estimate of the Trouton's law constant for an alkane is 20 cal mole$^{-1}$ deg$^{-1}$.)
6. In petroleum analysis, the heavy alkane fraction is commonly recovered from a light solvent by means of the simple procedure of evaporating the solvent in a gas stream until the sample reaches constant weight. This is analogous to drying a solid sample until it reaches constant weight and thus becomes free of atmospheric water. (a) Consider the case in which the solvent is $n$-hexane ($T_b = 68.95$°C) and the sample happens to be pure hexadecane ($T_b = 287.5$°C). Assume that the sample is evaporated to constant weight at 40°C. When the evaporating mixture becomes 50 mole per cent hexadecane, approximately how many hexadecane molecules will be lost for every $10^6$ hexane molecules? (Note: assume a Trouton's law constant of 20 cal mole$^{-1}$ deg$^{-1}$ for alkanes.) (b) Consider a sample of pure hexadecane at 40°C.

Assume that this material is in equilibrium with a gas-phase volume of 0.1 ml and that the gas-phase volume is swept by a ventilating nitrogen stream at the rate of 0.1 ml/sec. Estimate the rate at which the hexadecane sample is lost by evaporation into the ventilating gas stream, thus proving to yourself that, if the sample is small enough, even "involatile" compounds can be lost by evaporation.

7. By considering the individual steps in a distillation column, "derive" the Fenske equation (equation 14–14).

(a) Write an expression for the composition of the first condensate (material volatilized in the pot and condensed at the base of the column).

(b) Use the above result as a starting composition and write an expression for the composition of the condensate of vapor in equilibrium with it.

(c) Continue the sequence, remembering that the condenser condenses vapor volatilized in the top plate of the column.

8. Dibutylphthalate, a component of many commercial plastics, is an exceedingly common contaminant in solvents. The saturation vapor pressure of dibutyl-phthalate at 110.6°C is 0.13 torr, and its boiling point is 340.0°C. Calculate α, the relative volatility, of dibutylphthalate and toluene at the boiling point of toluene (110.6°C). Compare this value of α to the approximation obtained by use of equation (14–13). Does equation (14–13) over- or under-estimate the difficulty of the removal of traces of dibutylphthalate from toluene by distillation?

9. Equation (14–13) should provide a better estimate of α when the boiling points of the two substances are closer together than those in problem 8. Consider the separation of $n$-hexane ($T_b = 68.95°C$) from $n$-heptane ($T_b = 98.42°C$, $p°$ at 68.95°C = 260 torr).

(a) Calculate α directly from equation (14–12) and compare this value to the approximation obtained from equation (14–13).

(b) The value of α calculated from equation (14–12) is based on the vapor pressure of $n$-heptane at the boiling point of $n$-hexane, and is thus applicable only for the case of traces of heptane in hexane, because only in this situation will the mixture boil at the boiling temperature of hexane. An equimolar mixture of heptane and hexane boils at 79.2°C, at which tempera-ture the saturation vapor pressure of hexane is 972 torr and that of heptane is 373 torr. Calculate α at the boiling point of this mixture and compare it to the values obtained in part (a).

10. (a) By constructing and using a graph like Figure 14–6, determine the number of theoretical plates required for the preparation of hexane containing less than six weight per cent toluene from an equimolar mixture of hexane and toluene (α = 3.2).

(b) When half of the total possible hexane product has been obtained, the mole fraction of hexane in the still pot will be about 0.34. At this stage, how many theoretical plates will be required to obtain hexane of greater than 90 weight per cent purity?

(c) If the still is operated at an efficiency of four theoretical plates, what is the minimal mole fraction of hexane which can be remaining in the still pot when the distillation is stopped, if the purity specification (94 weight per cent) is to be met by the product?

11. An approximate solution for problem 10 can be obtained from equation (14–15) without the need to construct a graph. Compare the required number of theoretical plates found from equation (14–15) with the required number of theoretical plates determined graphically. Does the equation over- or under-estimate the difficulty of the separation?

12. Uncontaminated solvents are required for ultratrace analyses of pesticides in natural samples. In one case, reagent-grade hexane is found to contain the equivalent of 25 $\mu$g of DDT per gm of solvent. If the level of contamination is to be reduced to less than the equivalent of 1 ng of DDT per gm of solvent, how many theoretical plates are required in the solvent still? Assume that α = 100 for the pesticide residue-hexane system, and assume that the average molecular weight of the pesticide residues is 200. Note that equation (14–15) applies to starting material and product purities very different from those specified in this problem.

13. With reference to the steps linking equations (14–2) and (14–3), calculate $V_v/V_1$ for water and for carbon disulfide at their boiling points (100 and 46.25°C, respectively). For which of these compounds is the approximation $(V_1 \ll V_v)$ introduced in the text more justified? Why?

14. A particular still is found to yield separations equivalent to ten theoretical plates, but the experimenters report that the column has an efficiency equivalent to nine theoretical plates. Why?

15. Explain succinctly what is wrong with the still sketched below.

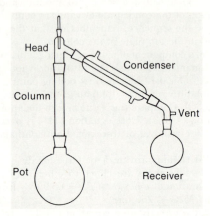

16. Sketch a design for a still head different from the one in Figure 14–4.

17. In a particular apparatus, the still head is adjusted so that all the condensate is taken off (that is, there is no reflux). The condensate is observed to accumulate at a rate of 20 ml/min. The still head is then adjusted so that 5 ml/min of product is withdrawn. What is the reflux ratio?

18. For a reflux ratio of $R_D = 20$, a particular column is found to have an efficiency equivalent to 20 theoretical plates. Could this column be used to separate an equimolar mixture of two components boiling at 100 and 106°C, respectively?

## SUGGESTIONS FOR ADDITIONAL READING

1. H. G. Cassidy: *Fundamentals of Chromatography. In* A. Weissberger, ed.: *Technique of Organic Chemistry*. Volume X, Wiley-Interscience, New York, 1957.

2. L. H. Horsley: *Azeotropic Data*. Advances in Chemistry Series, no. 6, no. 35, American Chemical Society, Washington, 1952, 1962.

3. T. E. Jordan: *Vapor Pressure of Organic Compounds*. Interscience, New York, 1954.

4. E. S. Perry and A. Weissberger, eds.: *Technique of Organic Chemistry*. Second edition, Volume IV (*Distillation*), Wiley-Interscience, New York, 1965.

5. H. Purnell: *Gas Chromatography*. John Wiley and Sons, New York, 1962, pp. 32–45.

6. J. Timmermans: *The Physico-Chemical Constants of Binary Systems in Concentrated Solutions*. 4 volumes, Wiley-Interscience, New York, 1960.

7. R. C. Weast, ed.: *Handbook of Chemistry and Physics*. Chemical Rubber Co., Cleveland. Annual editions.

# PHASE EQUILIBRIA AND EXTRACTIONS

**15**

Extraction is used for the isolation or enrichment of certain compounds or groups of compounds in chemical mixtures. For example, trace organic compounds present in a public water supply can be concentrated and isolated by means of the extraction of large volumes of water with relatively small volumes of non-polar solvents (*e.g.*, *n*-hexane). Trace metal impurities in important industrial products can be separated from overwhelming major constituents by selective extraction with specific reagents. Complex mixtures of organic compounds can be simply classified into acids, bases, and neutral species by means of two extraction steps.

After such obvious advertising of interesting practical applications, the reader can be excused for some skepticism as continued page-turning shows that much of this chapter deals with quantitative treatments of various methods of phase contact, rather than coming firmly to grips with the purification of public water supplies. Persistence in these details, however, will be rewarded with a vital understanding of the workings of partition equilibria wherever they are encountered. For example, a substantial discussion of countercurrent distribution is provided, not because it is any longer a keystone of biochemical analysis, but because it is one of the best introductions to the chromatographic techniques which are discussed in the next two chapters.

## PHASE EQUILIBRIA

### Thermodynamics

Consider some solute, Z, that is equilibrated between an aqueous phase and an organic solvent phase. The equilibrium can be simply represented as

$$Z(aq) \rightleftharpoons Z(org)$$

Using concepts developed in Chapter 3, we can express the free-energy change for the process in the following way:

$$\Delta G = \Delta G^0 + RT \ln (a_Z)_{org} - RT \ln (a_Z)_{aq} \tag{15-1}$$

$$\Delta G = \Delta G^0 + RT \ln \frac{(a_Z)_{org}}{(a_Z)_{aq}} \tag{15-2}$$

Since $\Delta G = 0$ at equilibrium, we obtain the relationships shown below:

$$\Delta G^0 = -RT \ln \frac{(a_Z)_{org}}{(a_Z)_{aq}} \tag{15-3}$$

$$\ln \frac{(a_Z)_{org}}{(a_Z)_{aq}} = -\frac{\Delta G^0}{RT} \tag{15-4}$$

$$\frac{(a_Z)_{org}}{(a_Z)_{aq}} = e^{-\Delta G^0/RT} \tag{15-5}$$

Thus, the ratio of the activities of the solute in the two phases turns out to be a constant determined by $\Delta G^0$ and $T$. In a way entirely analogous to the treatment of equilibrium constants in Chapter 3, we term this activity ratio the **thermodynamic partition coefficient**, $\mathscr{K}_p$:

$$\mathscr{K}_p = \frac{(a_Z)_{org}}{(a_Z)_{aq}} \tag{15-6}$$

### The Partition Coefficient

As in systems previously discussed, the required activity coefficients are rarely known, so that an approximation based on concentrations alone is commonly employed. This approximate relation is

$$K_p = \frac{[Z]_1}{[Z]_2} \tag{15-7}$$

where $K_p$ is called the **partition coefficient**, and the subscripts refer to the concentrations of the solute Z in phases 1 and 2. Because dilute solutions of molecular species more closely approximate ideal behavior than those containing ionic solutes, the approximation given here is better than that made earlier for equilibrium constants. It is best, however, always to remember that we are dealing with a ternary system (two solvents and a solute), the actual behavior of which can be accurately described only by a triangular phase diagram. Our present assumption of ideality will serve well in the great majority of cases; but, whenever high concentrations of solutes are encountered, or whenever the two solvent phases are not completely immiscible, significant deviations can be expected.

## The Distribution Ratio

In the context of practical chemical procedures, it is best to carry our interest yet another step beyond ideal activity ratios. For example, in the extraction of an organic acid, HA, from water by an organic solvent, there is one especially important question. How efficient is the extraction or, in other words, how much acid is left in the water phase? The partition coefficient relates only to the ratio $[HA]_{org}/[HA]_{aq}$, which does not tell the whole story in this case, because, in the aqueous phase, the acid can dissociate. The solute in the aqueous phase is thus present in two different forms, HA and $A^-$, and questions about completeness of extraction must consider the ratio $[HA]_{org}/([HA] + [A^-])_{aq}$. In general, such a ratio of the total concentration of a solute in two phases is termed a **distribution ratio**, and is denoted by $D_c$:

$$D_c = \frac{[\text{total concentration of all forms of } Z]_1}{[\text{total concentration of all forms of } Z]_2} = \frac{c_1}{c_2} \qquad (15\text{--}8)$$

Quantitative treatment of the distribution of a dissociable acid between two phases, such as the one just discussed, is simple and interesting. We can picture the situation as follows:

organic phase (1)        HA

aqueous phase (2)        HA $\rightleftharpoons A^- + H^+$

The two equilibria are governed by the expressions

$$K_p = \frac{[HA]_1}{[HA]_2} \qquad \text{and} \qquad K_a = \frac{[A^-]_2[H^+]_2}{[HA]_2} \qquad (15\text{--}9), (15\text{--}10)$$

If we assume that the charged species are essentially insoluble in the organic phase, the distribution ratio is given by

$$D_c = \frac{[HA]_1}{[HA]_2 + [A^-]_2} \qquad (15\text{--}11)$$

Combination of equations (15–9), (15–10), and (15–11) leads to the equation

$$D_c = \frac{[HA]_1}{\dfrac{[HA]_1}{K_p} + \dfrac{K_a}{K_p}\dfrac{[HA]_1}{[H^+]_2}} \qquad (15\text{--}12)$$

which can be further simplified to yield

$$D_c = \frac{K_p[H^+]_2}{[H^+]_2 + K_a} \qquad (15\text{--}13)$$

When plotted on logarithmic coordinates, equation (15–13) will have two straight line regions:

first, if $[H^+]_2 \gg K_a$, then $D_c \approx K_p$, and $\log D_c = \log K_p = $ (const.);

second, if $K_a \gg [H^+]_2$, then

$$D_c = \frac{K_p[H^+]_2}{K_a}$$

and

$$\log D_c = \log K_p + \log [H^+]_2 - \log K_a$$

$$\log D_c = \text{(const.)} - pH$$

If we take values for $K_a$ and $K_p$ that are typical for the distribution of a fatty acid between ether and water, namely, $K_a = 10^{-5}$ and $K_p = 10^3$, the graph shown in Figure 15–1 can be constructed, showing clearly the well-known dependence of fatty acid extraction on pH.

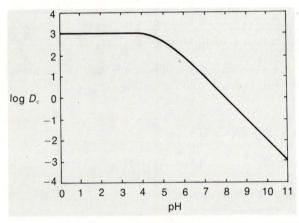

*Figure 15–1.* Distribution ratio $(D_c)$ as a function of pH for the partitioning of a carboxylic acid between ether and an aqueous buffer when $K_a = 10^{-5}$ and $K_p = 10^3$.

Another interesting situation occurs when an extracted compound polymerizes in the organic phase. Here a carboxylic acid may again serve as an example. Acetic acid is almost completely dimerized in non-polar solvents such as benzene. Choosing conditions, *viz.*, pH < 2, such that dissociation of the carboxylic acid in the aqueous phase is negligible, we can represent the overall equilibrium as shown below:

organic phase (1)            $HA \rightleftharpoons (HA)_2$

                                         $\Updownarrow$

aqueous phase (2)           $HA$
     pH < 2

The two equilibria are governed by the expressions

$$K_p = \frac{[HA]_1}{[HA]_2} \quad \text{and} \quad K_2 = \frac{[(HA)_2]_1}{[HA]_1^2} \qquad (15\text{–}9),\ (15\text{–}14)$$

The distribution ratio which we might determine, for example, by titrimetrically determining the total concentration of acetic acid in each phase, is given by

$$D_c = \frac{[HA]_1 + 2[(HA)_2]_1}{[HA]_2} \qquad (15\text{–}15)$$

Substitution of equations (15–9) and (15–14) into equation (15–15) leads to the expression

$$D_c = K_p + 2\,K_2 K_p{}^2 [HA]_2 \qquad (15\text{–}16)$$

In this case, $D_c$ depends on the total acetic acid concentration in the system.

A variation of $D_c$ causes curvature in plots of $c_1$ (the total concentration of all forms of solute in phase 1) versus $c_2$ (the total concentration of all forms of solute in phase 2). Such plots are called **partition isotherms**, even though they truly represent $D_c$, not $K_p$. Several are shown in Figure 15–2. The linear relationship depicted

*Figure 15–2.* Three types of partition isotherms.

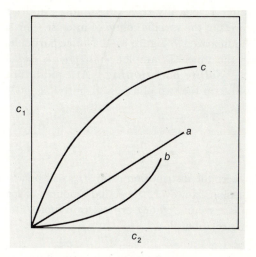

by $a$ is the behavior expected for the ideal situation when $D_c$ remains constant. Curve $b$ shows the variation in $D_c$ which would be observed for solute association (for example, dimer formation) in phase 1. Curve $c$ shows an isotherm frequently encountered when phase 1 is an adsorbed phase. The curve levels off as the concentration approaches monolayer coverage of the adsorbent.

**Tabulation of distribution ratios.** There are no tables of distribution ratios available for general reference. This frustrating situation exists because of the enormity of the task of compilation, and because distribution ratios depend so strongly on the specific details of a given experimental procedure. Distribution ratios are frequently measured and used in the development of specific techniques, but the chance that one particular combination of solute and two solvents (or solvent mixtures) would be employed by other workers for another purpose is very low. This absence of readily available information in no way diminishes the importance of the preceding discussion. An understanding of partition equilibria remains essential.

There are two ways in which distribution ratios can be sought in the literature. First, one can search through *Chemical Abstracts*, both in the subject classification of the solute compound and under the general subject-heading "partition." Second, one can refer to a ternary phase diagram, if it happens to be available for the system of interest. The estimation of a distribution coefficient valid at conditions of infinite dilution from data pertaining mainly to concentrated mixtures is somewhat complicated. For a detailed consideration of this problem, the interested reader can consult references 1, 5, and 7 listed at the end of this chapter.

## THEORY OF PHASE-CONTACT METHODS

### Single Equilibration

*The contact unit.*  For the sake of concreteness, consider the typical situation in which an aqueous solution is extracted with an organic solvent having a density less than 1.0 gm $cm^{-3}$. The organic (upper) phase is termed the "numerator phase," and all parameters associated with it are assigned the subscript 1. The aqueous (lower) phase is termed the "denominator phase" and is denoted by the subscript 2. The letter $V$ logically refers to volumes, and $c$ represents the total concentration of all forms of the specific solute of interest. The situation before and after equilibration is summarized in Figure 15–3. Notice particularly that the initial concentration of solute in the aqueous (sample) phase *prior to equilibration* is denoted as $c_0$.

*Solute partitioning.*  At equilibrium, the ratio of the total concentrations of solute in the two phases is given by

$$\frac{c_1}{c_2} = D_c \qquad (15\text{–}8)$$

where the distribution ratio, $D_c$, will be assumed to be constant throughout the discussion. However, the actual amounts of solute present in each phase depend on the phase volumes. For $c$ expressed in moles/liter and $V$ in liters, we can write the following relationships:

$$\text{moles of solute in phase 2 initially} = c_0 V_2 \qquad (15\text{–}17)$$

$$\text{moles of solute in phase 1 after equilibration} = c_1 V_1 \qquad (15\text{–}18)$$

$$\text{moles of solute in phase 2 after equilibration} = c_2 V_2 \qquad (15\text{–}19)$$

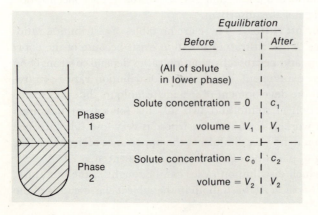

*Figure 15–3.*  Phase volumes and solute concentrations in a single equilibration stage. (See text for further discussion.)

It is very useful to define symbols for the *fractional amounts* of solute in each phase after equilibration:

$$p \equiv \text{fraction of solute in phase 1 after equilibration} = \frac{c_1 V_1}{c_0 V_2} \qquad (15\text{--}20)$$

$$q \equiv \text{fraction of solute in phase 2 after equilibration} = \frac{c_2 V_2}{c_0 V_2} \qquad (15\text{--}21)$$

Using the definitions above and setting $V_1/V_2 = V_r$ (the volume ratio), the reader should show that

$$q = \frac{1}{D_c V_r + 1} \qquad (15\text{--}22)$$

and

$$p = \frac{D_c V_r}{D_c V_r + 1} \qquad (15\text{--}23)$$

and should not proceed until these derivations have been completed.

An obvious application for calculations of this sort exists in cases where quantitative extraction is required. If $D_c$ is known or can be measured, the efficiency of a single-stage extraction is simply $p$. Note that, within limits set by the value of $D_c$ and by the size of laboratory apparatus, $p$ can be increased by increasing $V_r$. To calculate the per cent of solute extracted into phase 1, all we need to do is express $p$ in per cent:

$$\text{per cent extracted} = E = 100\,p \qquad (15\text{--}24)$$

*Example 15–1.* Calculate the concentrations and amounts in each phase after extraction of 100 ml of $10^{-2}$ $M$ acetanilide in water with 100 ml of ether if $D_c = 3.0$. What will be the effect of using 1000 ml of ether?

For the extraction with 100 ml of ether, $V_r = \dfrac{100}{100} = 1$. Using equations (15–22) and (15–23), we obtain

$$p = \frac{D_c V_r}{D_c V_r + 1} = \frac{3.0}{3.0 + 1} = \frac{3}{4}$$

$$q = \frac{1}{D_c V_r + 1} = \frac{1}{3.0 + 1} = \frac{1}{4}$$

$$\begin{aligned}
\text{amount in ether phase} \quad &= c_1 V_1 = p c_0 V_2 \\
&= (0.75)(10^{-2})(0.1) \\
&= 7.5 \times 10^{-4} \text{ mole of acetanilide}
\end{aligned}$$

$$\begin{aligned}
\text{amount in aqueous phase} &= c_2 V_2 = q c_0 V_2 \\
&= (0.25)(10^{-2})(0.1) \\
&= 2.5 \times 10^{-4} \text{ mole of acetanilide}
\end{aligned}$$

$$\begin{aligned}
\text{concentration in ether phase} &= c_1 = \frac{p c_0 V_2}{V_1} \\
&= \frac{7.5 \times 10^{-4} \text{ mole}}{0.1 \text{ liter}} = 7.5 \times 10^{-3}\ M
\end{aligned}$$

$$\begin{aligned}
\text{concentration in aqueous phase} &= c_2 = q c_0 \\
&= (0.25)(10^{-2}) = 2.5 \times 10^{-3}\ M
\end{aligned}$$

For the extraction with 1000 ml of ether, $V_r = 10$:

$$p = \frac{D_c V_r}{D_c V_r + 1} = \frac{(3.0)(10)}{(3.0)(10) + 1} = \frac{30}{31} = 0.97$$

Thus the efficiency (completeness) of the extraction is increased from 75 to 97 per cent if the volume of the ether phase is increased from 100 to 1000 ml.

### Repeated Equilibrations

*Stepwise partitioning of the solute.* The disappointingly low value for $p$ often obtained in calculations like those above leads to a consideration of repetitive extraction. Multiple extractions with fresh upper phase will remove more material. A constant fractional amount, $p$, of solute is removed from the lower phase for each equilibration. The situation through $n$ extraction steps is summarized in Figure 15–4, which must be carefully studied. The last step of Figure 15–4 gives general expressions for the fractional amounts of solute in each phase after any number, $n$, of extractions. Thus, after $n$ extractions, we have:

*solute amounts*

upper phase: $\quad pq^{n-1} \cdot$ (initial amount of solute) $= pq^{n-1}c_0 V_2 \qquad$ (15–25)

lower phase: $\quad q^n \cdot$ (initial amount of solute) $= q^n c_0 V_2 \qquad$ (15–26)

*solute concentrations*

upper phase: $\quad \dfrac{pq^{n-1}c_0 V_2}{V_1} = \dfrac{pq^{n-1}c_0}{V_r} \qquad$ (15–27)

lower phase: $\quad \dfrac{q^n c_0 V_2}{V_2} = q^n c_0 \qquad$ (15–28)

The total quantity of solute extracted is, of course, the sum of the amounts of solute in all the successive upper phases or, much more simply, the total amount of solute originally present in the lower phase minus that remaining in the lower phase after the last extraction:

$$(p + pq + pq^2 + \cdots + pq^{n-1})c_0 V_2 = (1 - q^n)c_0 V_2 \qquad (15\text{–}29)$$

*Example 15–2.* Consider again the acetanilide-ether-water system used in Example 15–1. Compare the efficiencies of the following three extraction procedures:

  I. Single extraction with 100 ml of ether
 II. Single extraction with 1000 ml of ether
III. Repetitive extraction with ten 100-ml portions of ether

We have already performed the calculations for alternatives I and II in Example 15–1, and have found efficiencies of 75 and 97 per cent, respectively. It remains to see what improvement results from dividing the 1000 ml of ether into ten equal portions as proposed in alternative III. Our interest lies mainly in the final distribution of

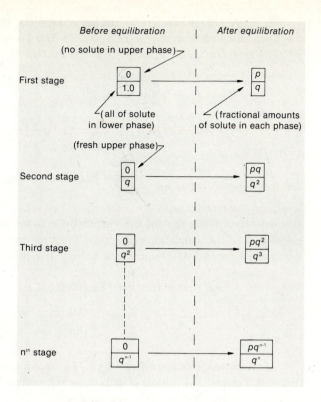

Figure 15-4. Solute partitioning in a repetitive extraction scheme.

Before equilibration | After equilibration

(no solute in upper phase)

First stage

| 0 |
| 1.0 |

| p |
| q |

(all of solute in lower phase)

(fractional amounts of solute in each phase)

(fresh upper phase)

Second stage

| 0 |
| q |

| pq |
| q^2 |

Third stage

| 0 |
| q^2 |

| pq^2 |
| q^3 |

$n^{th}$ stage

| 0 |
| q^{n-1} |

| pq^{n-1} |
| q^n |

material; but first, simply to see how the calculations are done, let us find what the concentrations and amounts of acetanilide are in each phase after the fifth extraction:

upper phase: contains fraction $pq^4$

$$\text{amount} = pq^4 c_0 V_2$$

$$= (\tfrac{3}{4})(\tfrac{1}{4})^4(10^{-2})(0.1)$$

$$= (0.75)(3.9 \times 10^{-3})(10^{-3})$$

$$= 2.9 \times 10^{-6} \text{ mole of acetanilide}$$

$$c_1 = \frac{pq^4 c_0 V_2}{V_1}$$

$$= \frac{2.9 \times 10^{-6}}{0.1} = 2.9 \times 10^{-5} \ M$$

lower phase: contains fraction $q^5$

$$\text{amount} = q^5 c_0 V_2$$

$$= (9.8 \times 10^{-4})(10^{-2})(0.1)$$

$$= 9.8 \times 10^{-7} \text{ mole of acetanilide}$$

$$c_2 = q^5 c_0$$

$$= (9.8 \times 10^{-4})(10^{-2})$$

$$= 9.8 \times 10^{-6} \ M$$

At the end of ten extractions, the key question is: What fraction of the acetanilide has been removed from the water solution? The lower phase will contain a fraction $q^{10}$. The total fraction of acetanilide extracted will be

$$\sum_{n=1}^{10} pq^{n-1} = 1 - q^{10}$$

$$= 1 - (\tfrac{1}{4})^{10} = 1 - 10^{-6}$$

$$= 0.999999$$

$$\text{or } E = 99.9999 \text{ per cent}$$

Realizing the fanciful aspects of such a number, we see that the efficiency of the third alternative surpasses that of the second by a factor of about $3 \times 10^4$. In fact, the calculation of the distribution of acetanilide after only five 100-ml extractions shows that alternative to be far superior to a single one-liter extraction:

$$E \text{ (five extractions)} = 100 \, (1 - q^5)$$

$$= 100 \, (1 - 9.8 \times 10^{-4})$$

$$\sim 100 \, (0.999) = 99.9 \text{ per cent}$$

**Maximum possible efficiency.** The maximum possible efficiency of extraction with a given total volume of extracting solvent can be found in a quite interesting way. Considering first the definition of $q$,

$$q = \frac{1}{D_c V_r + 1} = \frac{V_2}{D_c V_1 + V_2} \tag{15–22}$$

we can write an expression for the fractional amount of solute remaining unextracted after any number, $n$, of extractions with a total volume of extracting phase $\Sigma V_1$ (that is, $n$ extractions, each with solvent volume $\Sigma V_1/n$):

$$q^n = \left( \frac{V_2}{D_c \dfrac{\Sigma V_1}{n} + V_2} \right)^n \tag{15–30}$$

Taking the limit of this expression as $n \to \infty$, we obtain

$$q^\infty = e^{-D_c \Sigma V_1 / V_2} \tag{15–31}$$

The situation for $\Sigma V_1 = 5 \, V_2$, $D_c = 1$, is graphically summarized in Figure 15–5. It can be seen that the amount remaining unextracted is within a factor of five of the theoretical minimum after five steps. Twenty extractions are required to come within a factor of two of the theoretical minimum in this case.

**Separation of two partitioned solutes.** Up to now, we have been concerned with obtaining the maximum extent of extraction of just one solute. In a practical situation, it is likely that we would be equally concerned with minimizing the extent of extraction of some other solute, thus obtaining a separation of the two compounds. Clearly, the distribution ratios of the two solutes will have much to do with this, and we might naively suppose that the separability of two compounds A and B will be

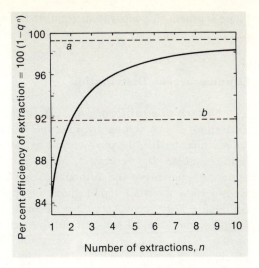

*Figure 15–5.* Per cent efficiency of extraction as a function of $n$, the number of subdivisions and separate extractions using a fixed total volume of extracting phase $\sum V_1$. The solid line represents the case where $\sum V_1 = 5V_2$ and $D_c = 1$. The dashed line *a* shows the maximum possible efficiency for this same case. The dashed line *b* shows the theoretical maximum asymptote for $D_c \sum V_1/V_2 = 2.5$. For efficiency greater than 99.9 per cent, $D_c \sum V_1/V_2 > 6.9$ is required.

determined by the ratio $D_{cA}/D_{cB}$ (here, and in the following discussion, the subscripts A and B relate to compounds A and B). The calculations below contrast two systems, both with the ratio $D_{cA}/D_{cB} = 10^3$, and show that the situation is somewhat more complicated.

$$D_{cA} = 32, D_{cB} = 0.032 \qquad D_{cA} = 10^3, D_{cB} = 1$$

For a single extraction with $V_r = 1$

$$p_A = 0.97, p_B = 0.03 \qquad p_A = 0.999, p_B = 0.50$$

For equal initial concentrations of A and B in phase 2

A obtained 97% pure         A obtained 66% pure

97% of the B removed         only 50% of the B removed

It happens that the first example above deals with the most favorable case. For $V_r = 1$, the best separation of A from B will be obtained when

$$\sqrt{D_{cA}D_{cB}} = 1 \tag{15-32}$$

Because the $D_c$ values relate only to concentrations and not to amounts, it is best to broaden our considerations by rewriting equation (15–32) in terms of $p$ and $q$, and in terms of concentrations and volumes [see equations (15–20) and (15–21)]:

$$1 = \left[\left(\frac{p_A}{q_A}\right)\left(\frac{p_B}{q_B}\right)\right]^{1/2} = \left[\left(\frac{c_{1A}V_{1A}}{c_{2A}V_{2A}}\right)\left(\frac{c_{1B}V_{1B}}{c_{2B}V_{2B}}\right)\right]^{1/2} = \sqrt{D_{cA}D_{cB}V_r^2} \tag{15-33}$$

Rearrangement of equation (15–33) shows that it is possible to optimize the conditions for a separation by adjusting $V_r$ so that

$$V_r = (D_{cA}D_{cB})^{-1/2} \tag{15-34}$$

In the second case above, application of equation (15–34) leads to an optimum $V_r$ of 0.03. Therefore, if the extraction of one liter of lower phase with 32 ml of upper phase is practical, a separation equal to the optimum result obtained in the first case can be achieved.

## Countercurrent Distribution

*Pattern of equilibrations.* In the preceding section, we have considered the separation of two solutes by selective extraction. Unhappily, even when the $D_c$ values differ by $10^3$, the purity of the extracted material does not exceed 97 per cent in the best case ($D_{cA} = D_{cB}^{-1}$). Any repeated stages of extraction can only offer the same extraction ratios and, although yield can be increased in this way, purity cannot. Consider now a second step in which the upper phase obtained in the example above, containing 97 per cent A and 3 per cent B, is equilibrated with a fresh lower phase. In this second step, 97 per cent of the A will stay in the upper phase, giving a yield of $(0.97)^2 \ 100 = 94$ per cent, whereas only 3 per cent of the B will stay in the upper phase, giving a yield of $(0.03)^2 \ 100 = 0.09$ per cent. Continuing the assumption of an equimolar initial mixture, we can easily calculate what percentage of total solute in the upper phase is A:

$$\frac{94}{94 + 0.09} \ (100) = 99.9 \text{ per cent}$$

This great increase in the purity of A in the upper phase shows that exposure to fresh lower phase is, to say the least, a very helpful step. Furthermore, if the original lower phase is again extracted with fresh upper phase, it is found to contain 99.9

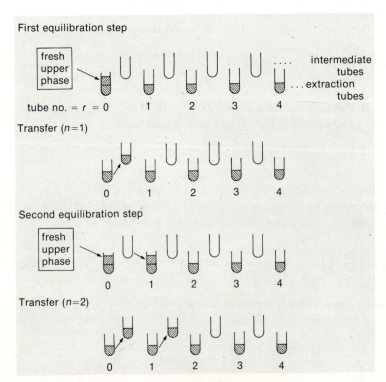

*Figure 15–6.* The pattern of phase-transfer and equilibration in counter-current distribution.

per cent pure B. Question: How far can this phenomenon be pushed? Answer: A long way; in fact, indefinitely. We have just described a countercurrent extraction process; and, although this example is but two steps long and offers high purity only because of the favorable distribution ratios chosen, it is essentially the same as chromatographic processes which involve $10^7$ and more steps and allow separation of, for example, diastereomers.

The sequence of events in an extended countercurrent distribution is shown schematically in Figure 15–6. The "extraction tubes" represent an indefinitely extending series of separatory funnels, and the "intermediate tubes" represent any convenient containers for the transfer of upper phases. At the outset, all the extraction tubes are filled with fresh lower phase except the zeroth tube, which contains the solutes to be separated. In the first step, fresh upper phase is added to the zeroth tube, and a simple extraction is performed. That upper phase is transferred to the first tube for equilibration with fresh lower phase (the essential feature of the example above), and fresh upper phase is added to the zeroth tube. In this second equilibration stage, extraction tubes 0 and 1 are shaken simultaneously. Subsequent transfer and equilibration steps have the effect of moving the upper phase through the series of extraction tubes. In general, after $n$ transfer steps the first upper phase will have reached extraction tube $n$. Because of this pattern of movement, the upper phase can be termed the "mobile phase," and the lower phase termed the "stationary phase."

**Distribution of solute.** The distribution of solute is traced quantitatively in Figure 15–7, which must be studied carefully by the reader.

In countercurrent distribution, the lateral distribution of the solute among the extraction tubes is primarily responsible for separation. The upper and lower phases in an extraction tube are combined for analysis and solute recovery. Thus, it is the total solute fraction in each tube which is of interest. Referring to Figure 15–7, and considering the distributions at $n = 1$, 2, and 3, we find the total fraction in each tube by adding the separate phase fractions:

| tube number $= r =$ | 0 | 1 | 2 | 3 |
|---|---|---|---|---|
| $n = 1$ | $q$ | $p$ | | |
| $n = 2$ | $q^2$ | $2pq$ | $p^2$ | |
| $n = 3$ | $q^3$ | $3pq^2$ | $3p^2q$ | $p^3$ |

Inspection shows that the solute fractions are the individual terms in the expansion $(q + p)^n$. Conveniently, the total solute fraction $(F_{r,n})$ in the $r$th tube after $n$ transfers can be calculated from the general form of the binomial expansion:

$$F_{r,n} = \frac{n!}{(n - r)! \, r!} \, p^r q^{(n-r)} \tag{15–35}$$

**Migration of solute.** Solutes will be carried through the apparatus by the advancing upper phase. The rate of solute migration from tube to tube will depend on the fraction of material partitioned into the mobile phase. Solutes with distribution ratios favoring the stationary phase will travel slowly. Solutes with distribution ratios favoring the mobile phase will move through the apparatus much more rapidly. Mathematically, the rate of movement must depend on $p$, and it is possible to show (see reference 6) that

$$r_{max} \approx np = \frac{nD_c V_r}{D_c V_r + 1} \tag{15–36}$$

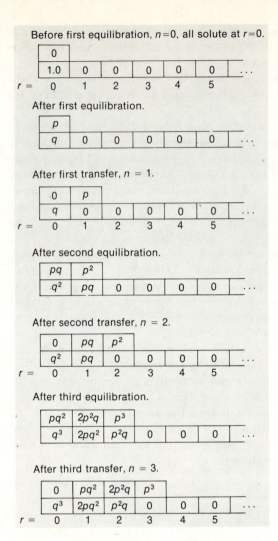

*Figure 15–7.* Fractional amounts of solutes in countercurrent distribution.

where $r_{max}$ is the number of the tube containing the maximum concentration of a solute with fraction $p$ in the mobile phase. The approximation serves very well when $n \gg r$, and is a useful guide in all cases.

Because the rates of migration depend directly on $p$, countercurrent distribution is a powerful tool for separating solutes with different $D_c$ values. Table 15–1 and Figure 15–8 demonstrate the degree of separation which can be achieved with two solutes having $D_c$ values of 3 and $\frac{1}{3}$, respectively. The second row of the table lists the terms of the binomial expansion of $(q + p)^6$ and the third and fourth rows list the calculated fractional amounts of solute. For $V_r = 1$, $p$ and $q$ have values of $\frac{3}{4}$ and $\frac{1}{4}$,

**Table 15–1.** $F_{r,n}$ Values in a Countercurrent Distribution with $n = 6$

| tube number = r = | 0 | 1 | 2 | 3 | 4 | 5 | 6 |
|---|---|---|---|---|---|---|---|
| binomial expansion term | $q^6$ | $6\,pq^5$ | $15\,p^2q^4$ | $20\,p^3q^3$ | $15\,p^4q^2$ | $6\,p^5q$ | $p^6$ |
| Fraction of substance with $D_c = 3$ | $\dfrac{1}{4096}$ | $\dfrac{18}{4096}$ | $\dfrac{135}{4096}$ | $\dfrac{540}{4096}$ | $\dfrac{1215}{4096}$ | $\dfrac{1548}{4096}$ | $\dfrac{729}{4096}$ |
| Fraction of substance with $D_c = \frac{1}{3}$ | $\dfrac{729}{4096}$ | $\dfrac{1548}{4096}$ | $\dfrac{1215}{4096}$ | $\dfrac{540}{4096}$ | $\dfrac{135}{4096}$ | $\dfrac{18}{4096}$ | $\dfrac{1}{4096}$ |

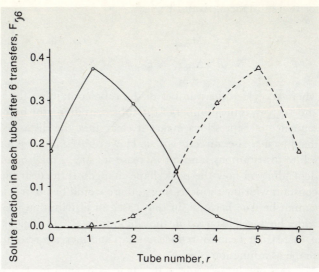

*Figure 15–8.* Fractional amounts of solute in each tube of a CCD apparatus for $n = 6$; $D_c = \frac{1}{3}$ (———○———), $D_c = 3$ (———△———).

respectively, for the solute with $D_c = 3$, whereas the $p$ and $q$ values are interchanged ($p = \frac{1}{4}$ and $q = \frac{3}{4}$) for the solute with $D_c = \frac{1}{3}$. Therefore, the distributions are mirror images of each other and all terms in the binomial expansion can be expressed as fractions with a denominator of $4^6$.

The degree of purity and the per cent recovery obtainable for each solute can be easily calculated from the data of Table 15–1. Consider that the experiment was begun with 4.096 millimoles of each substance in the zeroth tube. After six transfers, the zeroth tube still contains 729/4096 or 0.729 millimole of the slow-moving compound, but only 1/4096 or 1 micromole of the fast-moving compound. The zeroth tube thus contains 730 $\mu$moles of material, the slow-moving component being $\frac{729}{730}(100)$ or 99.86 per cent pure. The recovery of the slow-moving component is low, only $\frac{729}{4096}(100)$ or 17.81 per cent. In order to enhance the degree of recovery, the contents of tubes 0 and 1 might both be collected. In that case, $(729 + 1548)$ or 2277 $\mu$moles of the slower-moving component would be obtained along with only $(1 + 18)$ or 19 $\mu$moles of the faster-moving component. The degree of purity of the slower-moving component would be

$$\frac{2277}{19 + 2277}(100) = 99.17 \text{ per cent}$$

and the efficiency of recovery would be

$$\frac{2277}{4096}(100) = 55.59 \text{ per cent}$$

***Degree of solute separation: resolution.*** The extent to which two solutes can be separated is a measure of the resolution of a separation process. If a procedure succeeds in the separation of two closely similar substances (having nearly equal $D_c$ values in the case of countercurrent distribution), it is said to offer "high resolution." Working toward a useful and quantitative definition and treatment of resolution, we can begin by consideration of the difference between concentration maxima for two substances, A and B:

$$\Delta r_{max} = (r_{max})_A - (r_{max})_B = n(p_A - p_B) \qquad (15\text{--}37)$$

$$\Delta r_{max} = n\,\Delta p \qquad (15\text{--}38)$$

where $(r_{max})_A$ is the number of the tube containing the highest concentration of A, and $(r_{max})_B$ is the number of the tube containing the highest concentration of B. This sequence of expressions makes it clear that the distance between distribution peak maxima increases directly with $n$. However, resolution is a measure of overlap between solute distributions, and an increase in $\Delta r_{max}$ may or may not increase resolution, depending on how the peak shape changes. If the distribution peak were to maintain constant width with increasing $n$, there would be no problem at all; but we know this cannot be true because, in that case, all distribution peaks would be only one tube wide, having all started in the single tube $r = 0$. If peak width varies directly with $n$, increasing $n$ cannot reduce overlap and increase resolution. We need to study peak shape as a function of $n$.

It is helpful to begin with a new expression for $F_{r,n}$ which avoids the difficulties of calculations with factorial terms. Such calculations become particularly tedious as $n$ and $r$ become large. Fortunately, under these same circumstances the binomial distribution can be closely approximated by a normal, or Gaussian, distribution having the form

$$F_{r,n} = \frac{1}{\sqrt{2\pi}\,\sqrt{npq}}\,e^{-[(np-r)^2/2npq]} \qquad \text{(for } n > 24) \qquad (15\text{--}39)$$

Comparison with equation (2–5) given on page 15 for the normal distribution will show that the role of the standard deviation, $\sigma$, is taken by the quantity $\sqrt{npq}$ in the above expression.

Because the standard deviation is the parameter which establishes the width of any given normal distribution, we can now obtain an expression for peak width as a function of $n$. It remains only to choose a definition of the width in terms of $\sigma$. Reference to Table 2–1 (page 18 ) shows that a peak width of $4\sigma$ will include 95.46 per cent of the area. Let us arbitrarily choose this as the peak width, $w$:

$$w = 4\sigma = 4\sqrt{npq} \qquad (15\text{--}40)$$

For the example of Figure 15–8 and Table 15–1, the following results are obtained:

|  | peak separation | peak width |
|---|---|---|
|  | $\Delta r_{max} = n\,\Delta p$ | $w = 4\sqrt{npq}$ |
|  | $= n\,(0.5)$ | $= 4\sqrt{n}\,\sqrt{\frac{3}{16}}$ |
| For $n = 25$: | $\Delta r_{max} = 12.5$ | $w = 8.67$ |
| For $n = 100$: | $\Delta r_{max} = 50$ | $w = 17.3$ |

The complete distributions are graphically summarized in Figure 15–9. Note that resolution with no overlap is possible, at least in this example. In fact, the expressions above demonstrate that $\Delta r_{max}$ varies directly with $n$, whereas peak width increases with $\sqrt{n}$. Thus, increasing $n$ can decrease peak overlap because peak width increases more slowly than peak separation.

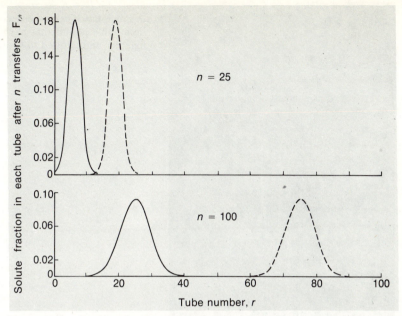

*Figure 15–9.* Fractional amount of solute in each tube of a countercurrent distribution apparatus for $n = 25$ and $n = 100$. The component with $D_c = \frac{1}{3}$ is shown by the solid line, and the component with $D_c = 3$ is shown by the dashed line. These examples represent four times and 16 times the number of steps shown for this same system in Figure 15–8.

A simple mathematical definition of **resolution**, $\mathscr{R}$, is

$$\mathscr{R} = \frac{\Delta r_{\max}}{w} = \frac{\Delta r_{\max}}{4\sigma} \tag{15-41}$$

That is, if two peaks are separated by one peak width, resolution is essentially complete and $\mathscr{R} = 1$. In the two examples discussed above, $\mathscr{R} = 12.5/8.7 = 1.44$ for $n = 25$, and $\mathscr{R} = 50/17.3 = 2.88$ for $n = 100$. The exact dependence of $\mathscr{R}$ on $n$ can be derived by substitution of equations (15–38) and (15–40) in equation (15–41):

$$\mathscr{R} = \frac{\Delta r_{\max}}{4\sigma} = \frac{n \Delta p}{4\sqrt{npq}} = \sqrt{n}\, \frac{\Delta p}{4\sqrt{pq}} \tag{15-42}$$

Resolution thus varies with $\sqrt{n}$. If $\mathscr{R}$ is to be increased by a factor $x$, $n$ must be increased by the factor $x^2$, as in the examples above, where doubling $\mathscr{R}$ required quadrupling $n$.

## PRACTICAL ASPECTS AND APPLICATIONS

### Apparatus for Extractions

A single-batch extraction can be performed readily in a separatory funnel, an example of which is shown in Figure 15–10. The two immiscible liquids are placed in

*Figure 15–10.* A separatory funnel, used in single-batch extractions, containing two immiscible liquids that have been shaken together and allowed to separate.

the separatory funnel, which is then shaken vigorously to ensure intimate contact between the phases. The liquids are allowed to separate into two layers, and the denser liquid is withdrawn through the stopcock.

***Continuous liquid-liquid extraction.*** In this technique, the liquid to be extracted has droplets of the extracting solvent passed through it continually. A relatively dense extracting solvent can simply be dropped through the liquid to be extracted, or a relatively light solvent can rise through the liquid buoyantly. Apparatus can be constructed to continue this process indefinitely, without attention, only a small volume of extracting solvent being required. The only energy input is that required to recycle and purify the solvent by a distillation process which also does the necessary work against gravity.

An apparatus adaptable to solvents of any density is shown in Figure 15–11. As shown, extracting solvents of high density can be used. The solvent is distilled from the reservoir, condensed, and dripped through the liquid to be extracted as it returns to the reservoir by way of the tube attached to the bottom of the extractor. In the case of low-density solvents, the return tube can be removed and replaced by two simple plugs, and a funnel tube with a fritted-glass-disk solvent-disperser is placed in the extractor. Condensed solvent falls into the funnel and is forced through the fritted-glass-disk by the hydrostatic head of condensate. The light solvent rises through the liquid and returns to the reservoir by overflowing in the extractor sidearm. In either case, if the extractor volume is too large, glass beads can be added to fill the space and, incidentally, to improve liquid-liquid contact by providing a tortuous path for the extracting solvent.

This technique offers the great advantage of unattended operation. However, the same distillation which ensures a constant supply of fresh solvent can cause problems. Volatile solutes may be lost, or, in a few cases, recycled with the solvent. Either way, the apparent recovery will be misleadingly low.

The question of efficiency in this extraction process is rather complex. The attainment of equilibrium requires a certain amount of mass transfer which is substantially aided by the shaking which takes place in a batch extraction. In addition, there is

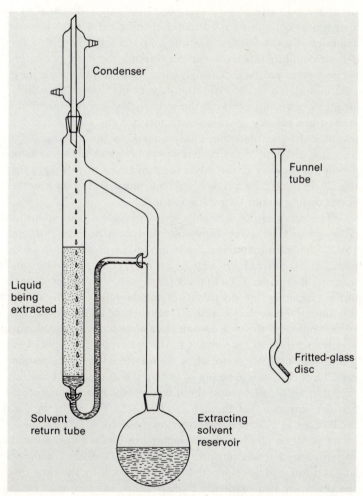

*Figure 15–11.* A continuous liquid-liquid extraction apparatus. As shown, it is arranged for use with an extracting solvent denser than the liquid to be extracted. A lighter solvent can be employed if the solvent-return tube is removed and the fritted-glass-disk funnel-tube is placed in the extractor. In that case, the extracting solvent returns to the reservoir by overflowing in the extractor sidearm.

generally an enormous interfacial area generated by the thorough mixing. The quiescent conditions prevailing in continuous extraction seem almost designed to avoid these advantages, although a rotating stirrer might easily be connected to a shaft extending down through the condenser (at least in the case of a dense extracting solvent). Systematic quantitative treatment of extraction efficiency is impossible, but a useful empirical approach is available. A known quantity of an easily determined solute can be added to the extractor and the rate of its extraction into the solvent monitored by analysis of aliquots withdrawn at various times. Determination of the solute concentration in the extracting solvent will allow calculation of the solute remaining unextracted. That amount will decrease with a characteristic half-time which can be determined from a plot of the logarithm of the fraction of solute extracted versus time. In turn, the resulting graph will allow calculation of the length of time required for any necessary degree of extraction, though, of course, due allowance must be made for any difference between the distribution ratios of the test substance and the solute to be extracted.

**The Soxhlet extractor.** The continuous extraction of solids presents a problem. An apparatus can easily be imagined in which a solvent reservoir is placed below a filter funnel containing solid material to be extracted, the solvent is boiled, and its vapor led to a condenser above the solid material by way of a bypass tube. The condensate could then percolate down through the solid, extracting it, on the way back to the solvent reservoir. In such cases, however, channels inevitably develop in the solid material, and the extraction is very inefficient. This fault can be remedied by arranging a system in which the solid is thoroughly immersed in extracting solvent, at least periodically. Soxhlet extractors are one way of accomplishing this. Another way would be to place the solid material in the apparatus shown in Figure 15–11 such that the top of the material was below the level at which extracting solvent overflowed by way of the solvent-return tube. In this way, the solid would be continually extracted by a flow of solvent, but would remain immersed, and the danger of channeling would be greatly reduced.

The operation of a Soxhlet extractor is not continuous, and, in this regard, it differs greatly from the above alternative. A Soxhlet extractor is shown in Figure 15–12. The solid material is finely ground or powdered in order to offer the greatest possible surface area for liquid contact and is then loaded into an "extraction thimble" made of filter paper. The thimble fits into the extraction chamber and allows passage of the extracting liquid while it retains the solid. Freshly distilled solvent enters and fills the extraction chamber until it reaches the highest level of the solvent-return tube. At that moment, the tube operates as a siphon, draining the extraction chamber completely as it returns the solvent to the reservoir. The operation is thus a sequence of fillings and siphonings, which is characterized as a "discontinuous" extraction process. The operation is amusing to watch, if nothing else. Chemists with small children find that they can occupy them in the laboratory simply by having an operating Soxhlet extractor for them to watch. (Playing with the light spot on a storage oscilloscope is a far better distraction, however.)

Operation of a Soxhlet extractor can become uncomfortably dramatic if mixed solvents are used without proper care. A low-boiling fraction (solvent or azeotrope) might distill into the extraction chamber, and, by the time the siphon dumps it into the reservoir, a substantially higher-boiling mixture may be present. Depending on relative volumes of high- and low-boiling solvents, the low-boiling fraction can become superheated and may be volatilized at a speed high enough to discourage even the best condenser. Another problem which arises, particularly with large extractors, is cessation of boiling when the large volume of cool solvent returns to the reservoir. Severe bumping can be encountered during the reheating, and stable boiling may

*Figure 15–12.* A Soxhlet extractor for the discontinuous extraction of solids. (Operation explained in text.)

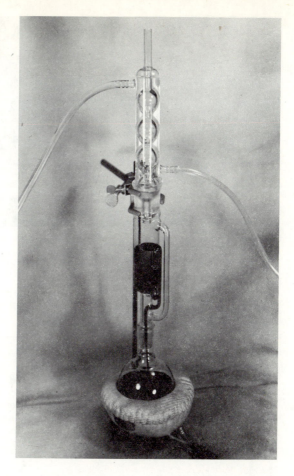

never be regained. Boiling chips are of little help because they are effective for only a few hours. A magnetically driven stirrer placed in the reservoir is the best solution. Finally, the extraction thimble itself can be a major source of contamination when minute amounts of organic material are sought.

## Apparatus for Countercurrent Distribution

The essential sequence of events in a countercurrent distribution process is shown schematically in Figure 15–6. Test tubes or separatory funnels can be employed to accomplish the task. However, because a great number of manipulations is required, some far more convenient form of apparatus is needed if the full possibilities of the technique are to be exploited by using procedures with hundreds of transfers and contact units.

The Craig extraction apparatus is perhaps the most sophisticated and efficient system available for the performance of multiple extractions. It consists of a large number of connected glass tubes—frequently as many as several hundred—mounted in a rack so that a whole series of batch extractions can be performed simultaneously. The photograph at the top of Figure 15–13 shows a group of five Craig-type tubes sealed together in a single unit. Any number of these five-tube units can be connected together to provide a complete multistage extraction apparatus. The principle of

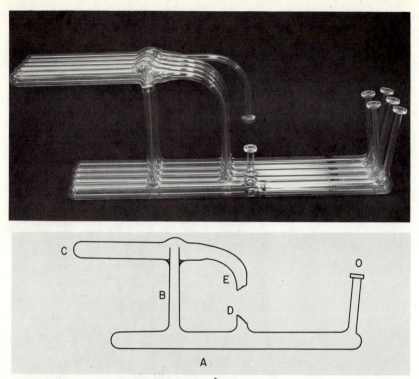

*Figure 15–13.* Craig extraction apparatus. The photograph at the top shows a unit consisting of five Craig-type tubes; any number of such units can be coupled together to give an extraction apparatus with several hundred tubes. The diagram illustrates the principle of operation of a Craig tube (see text for a complete discussion).

operation of a Craig extraction device can be most easily understood from an inspection of the drawing in Figure 15–13, which represents a single stage of the apparatus.

Each Craig tube is initially horizontal as indicated. The first of the series of Craig tubes is special. To compartment A of the first tube, one adds through opening O a portion of solvent 2 containing all the solute to be partitioned or distributed; the volume of this solution must be carefully measured so that, when the Craig tube is subsequently rotated counterclockwise about 100 degrees, none of this solution will run out through side arm B. Next, a definite volume of pure extracting solvent 1 is introduced into compartment A of the first tube such that, when the Craig tube is rotated counterclockwise, solvent 1 will not pass through side arm D (which leads in the wrong direction to an adjoining tube). To each of the other Craig tubes in the extraction apparatus, one adds only pure solvent 2 of the proper volume so that it cannot run through side arm B of that respective tube. Thus, the extraction process begins with solvent 2 containing the solute along with pure solvent 1 in the first Craig tube, whereas only pure solvent 2 is in each of the remaining Craig tubes.

The actual partitioning process is begun when the Craig apparatus is gently rocked back and forth, as one might do with a separatory funnel, keeping the Craig tubes near the horizontal position so all solvents remain in their respective A compartments. When extraction equilibrium is attained, the rack containing the Craig tubes is rotated counterclockwise about 100 degrees; this causes solvent 1, now containing some of the solute, to flow through side arm B into compartment C. Then the

Craig apparatus is returned to its original horizontal position, and solvent 1 with solute flows from compartment C of the first Craig tube, through tube E, into compartment A of the second Craig tube. A fresh portion of pure solvent 1 is added to compartment A of the first Craig tube, and the apparatus is rocked gently to begin the extraction cycle again. The mixing and transferring steps are repeated over and over, each time with a fresh batch of solvent 1 being added to the first Craig tube. Throughout the extraction procedure, each of the solvent 2 phases remains in the same Craig tube in which it was originally placed. The solvent 1 layers are passed along from one tube to the next in the series. Solute travels along the series by extraction into solvent 1.

## Extraction of Molecular Species

*Organic compounds in natural samples.* For the organic chemist interested in natural products, or for geochemists interested in organic constituents of rocks and natural water samples, extraction is an everyday procedure. As a general method for the isolation of crude organic fractions, it is the inevitable prelude to a further series of chromatographic separations. Accordingly, the practical examples taken up in Chapter 17 will extend the analyses whose beginnings are discussed here.

Many natural surface waters are now recognized as chronically contaminated by persistent pesticides or their degradation products. The levels of contamination are low, almost always below one part in $10^9$ (abbreviated ppb, for "parts per billion"). However, because of concentrative processes elsewhere in nature, even these low levels of contamination must be monitored and studied in the hope that techniques for their reduction can be found. Due to the terribly dilute solutions in which these compounds present themselves to the analyst, there is an obvious need for concentrating them prior to further analysis. Because of their volatility, chemical fragility, and chances of steam distillation at such very low concentrations, it is not practically possible to isolate these substances by removal of the water through distillation. Instead, it is found that extraction with $n$-hexane offers the best approach. Typically, a one-liter water sample will be extracted with three 25-ml portions of solvent. The extracts are "dried" with anhydrous sodium sulfate, and the 75 ml of solvent is evaporated under a stream of nitrogen gas to whatever extent is required. It is possible to reduce the volume of the $n$-hexane to only a few microliters, thus achieving enrichments approaching a million-fold. The enriched extract is further studied by high-sensitivity gas-liquid chromatography.

Solvent extraction of solid materials is the first step in the examination of organic constituents of dried plant tissues (such as leaves and shredded bark) or crushed rocks and soils. Plant tissues can sometimes be satisfactorily extracted in a separatory funnel, but the denser and finer-grained inorganic materials usually require Soxhlet extraction. With the goal of extracting as much of the organic material as possible, it is necessary to choose a solvent in which both moderately polar compounds and non-polar species such as large alkanes are quite soluble. A non-polar solvent such as hexane would not do because many polar compounds (phenolics, for example) would be inefficiently extracted. On the other hand, long-chain alkanes would be poorly extracted by methanol. Chloroform would be a good compromise, but its purification at levels required for trace analyses is relatively difficult. A good choice turns out to be a benzene-methanol mixture. The choice of solvent is not the only problem—even finely ground solids, if loosely dispersed in the extraction thimble initially, can form a dense mud, offering little chance for good phase contact. For this reason, ultrasonic extraction of solvent-dispersed inorganic material is frequently carried out simply by dipping a beaker containing the slurry into an ultrasonic cleaning bath for a few minutes. This is best done after an hour or so of stirring the solid with the solvent, and care must be taken that the input of acoustical energy does not result in an inadvertent synthesis of trace contaminants by solvent degradation; however, this is not a problem with benzene-methanol mixtures.

*Fractionation of organic mixtures.* If a sample preparation procedure has complete extraction as its goal, conditions and solvents are carefully chosen (as noted above) in order to minimize the extent to which the isolation procedure predetermines the makeup of the resulting analytical sample. Yet, it would be foolish to expect that these effects can be completely eliminated, and it often serves better to accentuate them in a way which can produce well-defined sample fractionation. Consider a mixture of organic compounds dissolved in some nonaqueous solvent. Extraction with aqueous base (see page 468) will remove molecules with dissociable acidic functional groups. Careful choice of the pH can influence the selectivity of this extraction. The opposite possibility exists for the extraction of basic species into an acidic aqueous solution. The neutral species remain in the organic solvent.

This kind of fractionation can be coupled with sample isolation. For example, in the investigation of lipid fractions in single-celled organisms, the first step is cell disruption, usually accomplished by pressure gradients or ultrasonic energy. In either case, the broken cells are left in an ethanol or water solution. The solution is made strongly alkaline (10 per cent by weight potassium hydroxide) and is allowed to stand overnight. In this step, most ester linkages (triglycerides) are hydrolyzed, and the freed acids are, of course, completely neutralized by the strong base. Extraction with a non-polar solvent such as *n*-heptane removes only the so-called "non-saponifiable" lipids (sterols) and leaves behind the great bulk of cellular material. Large masses of cell debris can be a problem during the extraction, particularly after the hydrolysis. Before the extraction, it is convenient to introduce a purely mechanical separation step, namely, removal of the debris by centrifugation. A fraction containing fatty acids can be obtained by acidification of the aqueous phase, followed by a second heptane extraction.

## Extraction of Metal Ion Complexes

Ionic species do not have appreciable solubility in non-polar liquids, but ions bound up in neutral complexes with organic ligands offer a hydrophobic exterior and possess substantial solubility in some organic solvents. The distribution ratios for such systems greatly favor extraction of the complexes into the organic phase. If an aqueous solution of two metal ions is equilibrated with an organic solvent containing a ligand which forms a complex with only one of the ionic species, the metal ions can be separated easily and with excellent resolution.

*Complex equilibria in two-phase systems.* There are a number of individual processes which must be considered.

1. Complex formation, here simplified by consideration of only the overall formation constant for the extracted species:

$$\mathrm{M}^{n+} + n\,\mathrm{L}^- \rightleftharpoons \mathrm{ML}_n; \qquad \beta_n = \frac{[\mathrm{ML}_n]}{[\mathrm{M}^{n+}][\mathrm{L}^-]^n} \qquad (15\text{--}43)$$

2. Dissociation of the ligand, which is usually a weak acid:

$$\mathrm{HL} \rightleftharpoons \mathrm{L}^- + \mathrm{H}^+; \qquad K_\mathrm{a} = \frac{[\mathrm{L}^-][\mathrm{H}^+]}{[\mathrm{HL}]} \qquad (15\text{--}44)$$

3. Partitioning of the complex between the two phases:

$$\text{ML}_n(\text{aq}) \rightleftharpoons \text{ML}_n(\text{org}); \qquad K_p = \frac{[\text{ML}_n]_{\text{org}}}{[\text{ML}_n]_{\text{aq}}} \qquad (15\text{--}45)$$

4. Partitioning of the undissociated ligand between the two phases:

$$\text{HL}(\text{aq}) \rightleftharpoons \text{HL}(\text{org}); \qquad K_p' = \frac{[\text{HL}]_{\text{org}}}{[\text{HL}]_{\text{aq}}} \qquad (15\text{--}46)$$

A better view can be obtained by the combined representation given below. Notice that all steps involving charged species occur in the aqueous phase, and that partitioning of the complex into the organic phase has the effect of driving complex formation to completion.

$$
\begin{array}{lll}
\text{organic phase} & \text{HL} & \text{ML}_n \\
\hline
& \updownarrow & \uparrow \\
\text{aqueous phase} & \text{HL} & \\
& \updownarrow & \\
\text{M}^{n+} + (n)\ \text{L}^- & \rightleftharpoons \text{ML}_n & \\
& + & \\
& \text{H}^+ &
\end{array}
$$

In consideration of the efficiency of this procedure as a technique of metal ion extraction, the quantity of interest is the distribution ratio for all forms of the metal ion in each phase. For complete accuracy, this expression must include the solubility of the "bare" metal ion in the organic phase and the presence in the aqueous phase of complexes that contain more or less than $n$ ligands per metal atom. In many circumstances these species cannot be ignored, so that the list of equilibria given above would require expansion. For the moment, however, these complications can be neglected without seriously distorting the discussion. Following equation (15–8), the expression for $D_c$ takes the form

$$D_c = \frac{(c_\text{M})_{\text{org}}}{(c_\text{M})_{\text{aq}}} = \frac{[\text{ML}_n]_{\text{org}}}{[\text{M}^{n+}]_{\text{aq}} + [\text{ML}_n]_{\text{aq}}} \qquad (15\text{--}47)$$

Evaluation of $D_c$ in terms of experimental conditions requires extensive substitution in equation (15–47). A good beginning is the derivation of an expression for $[\text{ML}_n]_{\text{org}}$. Combination of equations (15–43) and (15–45) yields

$$[\text{ML}_n]_{\text{org}} = K_p[\text{ML}_n]_{\text{aq}} = K_p\beta_n[\text{M}^{n+}]_{\text{aq}}[\text{L}^-]_{\text{aq}}^n \qquad (15\text{--}48)$$

Equation (15–44) allows evaluation of $[\text{L}^-]_{\text{aq}}$ as a function of $[\text{H}^+]$ and $[\text{HL}]$ in the aqueous phase. Equation (15–46) relates $[\text{HL}]_{\text{org}}$ and $[\text{HL}]_{\text{aq}}$ so that $[\text{L}^-]_{\text{aq}}$ can be expressed in terms of $K_a$, $K_p'$, $[\text{H}^+]_{\text{aq}}$, and $[\text{HL}]_{\text{org}}$:

$$[\text{ML}_n]_{\text{org}} = \frac{K_p\beta_n(K_a)^n}{(K_p')^n}[\text{M}^{n+}]_{\text{aq}}\frac{[\text{HL}]_{\text{org}}^n}{[\text{H}^+]_{\text{aq}}^n} \qquad (15\text{--}49)$$

Adoption of the assumption that $[M^{n+}]_{aq} \gg [ML_n]_{aq}$ allows simplification of equation (15–47) and expression of $D_c$ in terms of $[H^+]_{aq}$ and $[HL]_{org}$, two experimentally controllable variables:

$$D_c \approx \frac{[ML_n]_{org}}{[M^{n+}]_{aq}} = \frac{K_p \beta_n (K_a)^n}{(K'_p)^n} \frac{[HL]^n_{org}}{[H^+]^n_{aq}} \tag{15–50}$$

In cases where $[HL]_{org}$ is fixed and only the pH of the aqueous phase is varied, equation (15–50) takes the form

$$D_c = (\text{constant}) \beta_n [H^+]^{-n}_{aq} \tag{15–51}$$

The distribution ratio and the completeness of extraction thus depend on the overall formation constant, the number of ligand molecules bound to the metal atom, and the pH. Equation (15–51) indicates that the slope of a plot of $D_c$ versus pH depends critically on $n$. Figure 15–14 shows how the per cent extraction of a metal complex with $n = 2$ varies with pH.

Separation of two metal ions, each of which forms a complex with the ligand, should be possible in certain circumstances. A metal ion that forms a complex with a very large formation constant can be extracted at a low pH, which allows its separation from a second metal ion that forms a complex with a relatively small formation constant. For example, in Figure 15–14, extraction at pH 6.5 will remove 99.9 per cent of species A and only 0.1 per cent of species B. The formation constants for $AL_2$ and $BL_2$ must differ by a factor of $10^6$ or more to allow this degree of separation; different factors are required for other values of $n$, as shown in Problem 20 at the end of this chapter. A second extraction at pH 9.5 will allow separation of species B from any uncomplexed metal ions in the aqueous phase. Unfortunately, this potentially powerful technique for metal ion separation is hampered by the natural

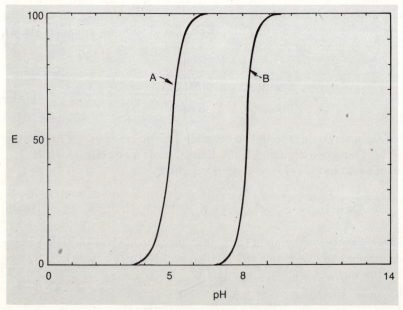

*Figure 15–14.* Curves showing the dependence of E on the pH of the aqueous phase for the extraction of two metal ion complexes, $AL_2$ and $BL_2$ ($\beta_{2A} = 10^6 \beta_{2B}$).

PHASE EQUILIBRIA AND EXTRACTIONS

tendency for related ionic species to have formation constants that differ by a factor considerably smaller than $10^6$. Fractionation can be achieved, but practical techniques often rely on the introduction of "masking agents" that tie up interfering metal ions, often by forming charged complexes which are not soluble in the organic phase.

*Extraction of bis(dimethylglyoximato)nickel(II).* Bis(dimethylglyoximato)nickel(II), previously mentioned in Chapter 8, is an excellent example of a metal ion complex that undergoes solvent extraction processes like those discussed above. Dimethylglyoxime is a weak acid

$$\begin{array}{ccc} H_3C-C\!\!=\!\!NOH & & H_3C-C\!\!=\!\!NO^- \\ | & \rightleftharpoons & | \\ H_3C-C\!\!=\!\!NOH & & H_3C-C\!\!=\!\!NOH \end{array} + H^+; \quad pK_a = 10.66$$

and the complex that it forms with nickel(II) is quite stable, having the structure and overall formation constant given below:

$$\log \beta_2 = 17.72$$

bis(dimethylglyoximato)nickel(II)

The partition coefficient for distribution of the complex between water and chloroform is $10^{2.51}$. However, we cannot proceed to an expression for $D_c$ in terms of $[H^+]_{aq}$ without knowing $K_p'$, the partition coefficient for the undissociated ligand. This sort of impasse is frequently met in attempts to determine the feasibility of a given procedure; but, in this instance, we can skirt this difficulty by assuming some "worst-case" values for $K_p'$ and examining the result. The partition coefficient for the free ligand, with its polar groups exposed, can scarcely be greater than $K_p$ for the complex itself, so let us choose $K_p' = 10^{2.51}$. If we assume that $[HL]_{org}$ can have the reasonable value of $10^{-2}$ $M$, the pH required to achieve a $D_c$ of $10^3$ can be computed by solving equation (15–50) for $[H^+]_{aq}$ and setting $D_c = 10^3$:

$$[H^+]_{aq}^n = \frac{K_p \beta_n (K_a)^n}{(K_p')^n D_c} [HL]_{org}^n$$

$$[H^+]_{aq}^2 = \frac{(10^{2.51})(10^{17.72})(10^{-10.66})^2}{(10^{2.51})^2 (10^3)} (10^{-2})^2$$

$$[H^+]_{aq} \sim 10^{-6.5} M; \quad pH \sim 6.5$$

In practice, bis(dimethylglyoximato)nickel(II) is quantitatively extracted into chloroform with $[HL]_{org} \sim 10^{-2}$ $M$ from an aqueous phase with a pH of only 4.5. Therefore, $K_p'$ must actually be approximately 3, not 300 as assumed above.

The extraction procedure is commonly employed when nickel must be determined as a minor constituent of alloys. In this application, it is frequently necessary to add other ligands to the aqueous phase in order to prevent precipitation of the oxides or hydroxides of such species as iron(III) or chromium(III). Citrate and tartrate anions form unextractable complexes with these metals, and are thus effective "masking agents." These anions also form complexes of low stability with nickel(II), and these competing reactions have the effect of raising the minimum pH for quantitative extraction of bis(dimethylglyoximato)nickel(II) from 4.5 to 6.5. The analysis can be very simply completed if the intensity of the red color which the nickel complex imparts to the chloroform layer is measured spectrophotometrically.

*Other complex extraction systems.* Particularly in radiochemical analyses, the popularity of separation procedures based on extraction has been very great. Examples far too numerous to name here are nicely covered in a well-organized book by Morrison and Freiser, which is cited as reference 4 at the end of this chapter.

## Ion-Pair Extraction

An ion pair, for example $\{R_4As^+Cl^-\}$, where R represents an organic moiety such as a phenyl group, resembles a metal complex such as those discussed above in that it is electrically neutral, and thus is soluble in organic solvents. Ion pairs and electrically neutral metal complexes differ sharply, however, in the nature of the forces which hold them together and, consequently, in most other structural details as well. For a quantitatively extractable complex, atoms from the ligand occupy the first coordination sphere of the metal ion and are bound to it strongly. The ions in an ion pair are each surrounded by solvent molecules, and the ions with their solvation spheres are held together simply by electrostatic attraction. In water, such ion pairs are unstable, and have very short lifetimes. However, if the concentration of one of the paired species is high enough, the steady-state concentration of pairs will be sufficient to allow extraction of the ion pairs into a non-polar phase. In the latter environment, the ion pair is much more stable, as will be shown in the following example.

*Extraction of $\{H^+FeCl_4^-\}$ into ether.* Iron(III) can be separated from most monovalent and divalent cations by its extraction from aqueous $6\,F$ hydrochloric acid into diethyl ether. Although the species $Fe(H_2O)_6^{3+}$ may be regarded as the parent form of iron(III) in aqueous media, the following equilibria prevail in the presence of hydrochloric acid:

$$Fe(H_2O)_6^{3+} + Cl^- \rightleftharpoons FeCl(H_2O)_5^{2+} + H_2O$$

$$FeCl(H_2O)_5^{2+} + Cl^- \rightleftharpoons FeCl_2(H_2O)_4^+ + H_2O$$

$$FeCl_2(H_2O)_4^+ + Cl^- \rightleftharpoons FeCl_3(H_2O)_3 + H_2O$$

$$FeCl_3(H_2O)_3 + Cl^- \rightleftharpoons FeCl_4(H_2O)_2^- + H_2O$$

When the concentration of hydrochloric acid is high enough, say $4\,F$ or greater, the position of equilibrium lies strongly in favor of $FeCl_4(H_2O)_2^-$, which is usually abbreviated as $FeCl_4^-$. In water, there is only a very slight tendency for $FeCl_4^-$ to combine with a proton to form the ion pair $\{FeCl_4(H_2O)_2^-H^+\}$. In other words, in an aqueous medium, whose high dielectric constant promotes the separation of charged species, the ion pair is essentially completely dissociated. However, diethyl ether is a relatively non-polar solvent with a low dielectric constant in which

{FeCl$_4$(R$_2$O)$_2$$^-$H$^+$} is stable and soluble. As indicated, the two solvent molecules in the first coordination sphere might be either water or ether (R = H or C$_2$H$_5$). In addition, the species actually present in the ether phase vary from {FeCl$_4$(R$_2$O)$_2$$^-$H$^+$}$_2$ up to {FeCl$_4$(R$_2$O)$_2$$^-$H$^+$}$_4$, depending on the total concentration of iron present. It should be noted that HCl is extracted into the ether phase and that HCl is probably involved in the mechanism of the extraction process. Nevertheless, iron(III) can be quantitatively (99 per cent) extracted into diethyl ether in a single extraction from aqueous 6 F hydrochloric acid. The success of the extraction depends critically on the concentration of hydrochloric acid, for there must be enough chloride to convert the iron(III) in the aqueous phase to FeCl$_4$$^-$ and enough acid to permit the formation of {FeCl$_4$(R$_2$O)$_2$$^-$H$^+$}$_n$ in the nonaqueous phase.

A plot of the percentage of extraction of iron(III) into diethyl ether for various concentrations of hydrochloric acid is shown in Figure 15–15. At HCl concentrations

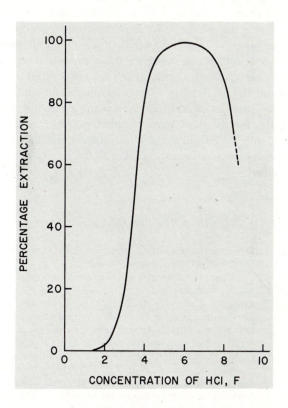

*Figure 15–15.* A plot of the percentage extraction of iron(III) from aqueous hydrochloric acid solutions of various concentrations into diethyl ether. The data pertain to equal volumes of the two phases.

greater than 6 F, the solubility of diethyl ether in the aqueous phase increases substantially, and eventually changes the polarity of that phase to the point that the distribution ratio is substantially decreased. Diisopropyl ether is considerably less soluble in the aqueous phase, and this decrease in the distribution ratio is consequently noticed only at higher acid concentrations. With the still less soluble $\beta,\beta'$-dichloroethyl ether, no decrease is observed in the distribution ratio even at a hydrochloric acid concentration of 12 F. Among the other elements which can be separated from aqueous hydrochloric acid solutions by extraction into diethyl ether are: antimony-(V), 81 % from 6 F HCl; arsenic(III), 68% from 6 F HCl; gallium(III), 97% from 6 F HCl; germanium(IV), 40–60% from 6 F HCl; gold(III), 95% from 6 F HCl; molybdenum(VI), 80–90% from 6 F HCl; platinum(II), 95% from 3 F HCl; and thallium(III), 90–95% from 6 F HCl.

## CONCLUSION

This chapter has covered phase equilibria and extractions from both the theoretical and the practical points of view. The theoretical treatment is linked closely to the same thermodynamic foundation already familiar from a study of reaction equilibria, but the expressions and calculations relevant to chemical separations *appear* more complicated because they deal with multiphase systems and must include the subscripts and terms necessary for a clear designation and accurate treatment of phase equilibria. It is, in addition, necessary for the reader to keep track of the physical movement of phases in various procedures and types of apparatus. Considerations of this sort will be new to many readers, and there is always a strong temptation to gloss over something which is new, or to trivialize it so that it fits artificially into the reader's collection of things already understood.

The practical sections of the chapter should convince the reader that the topics of phase equilibria and extractions merit careful study. Chemical separations constitute one of the most important and rapidly growing parts of modern chemical technology. In many analytical schemes, separation steps are particularly important because they can influence the composition and validity of the sample itself. More broadly, chemical separations are required in all phases of research and production, and the availability of new techniques for compound purification and identification has brought rapid progress to many areas of work. These new separation techniques are, for the most part, chromatographic, and are fully discussed in the next two chapters. The principles introduced in Chapters 16 and 17, however, cannot be assimilated and appreciated until the general material in this chapter is firmly understood.

## QUESTIONS AND PROBLEMS

1. Define and contrast: (a) thermodynamic partition coefficient, (b) partition coefficient, (c) distribution ratio.
2. For what type of species will the partition coefficient and the distribution ratio usually be equal?
3. Consider a diprotic acid ($H_2A$) with dissociation constants $K_{a1}$ and $K_{a2}$. Derive an expression for the distribution ratio, $D_c$, for partitioning of this acid between an organic solvent and an aqueous phase of controlled pH. Your result will be analogous to equation (15–13), and should give $D_c$ as a function of $K_p$ (the partition coefficient for distribution of $H_2A$ between the aqueous and organic phases), $K_{a1}$, $K_{a2}$, and $[H^+]_{aq}$.
4. Given equation (15–16) on page 489, tell what data you would collect and how you would treat these data in order to determine $K_p$ and $K_2$ for a system like that described in the text.
5. Which of the partition isotherms in Figure 15–2 represents (a) the distribution of a dissociable acid HA between an organic solvent and an aqueous buffer, (b) the distribution of a dissociable acid between an organic solvent and an aqueous phase whose pH is controlled only by the dissociation of HA itself?
6. Derive equations (15–22) and (15–23).
7. Calculate the fraction of a solute A extracted from 100 ml of an aqueous phase into 50 ml of an originally pure immiscible organic solvent, if the distribution ratio of the solute, $D_c$, is 80 and if A exists as a monomeric species in each phase.
8. Experiment shows that 90 per cent of a substituted phenol is removed from a water sample by extraction with an equal volume of benzene. What percentage of the substituted phenol will be extracted if the volume of benzene is doubled?

9. Consider some solute-water-organic solvent system with a distribution ratio of 10. The volume of water sample from which the solute is to be extracted is 25 ml. Determine the fraction of solute remaining unextracted after (a) one extraction with 250 ml of the organic solvent and (b) three extractions with 25 ml of the organic solvent; (c) express the preceding results in terms of per cent efficiency of extraction.

10. Suppose that you desire to extract a given solute from one solvent into a second, immiscible solvent, the distribution ratio for the solute being only 3.50. If the volume of the first phase, initially containing all the solute, is 10 ml, calculate the number of successive extractions with fresh 10-ml portions of the second solvent needed to extract a minimum of 99 per cent of the solute from the original solvent.

11. Long-chain alkanoic acids have a distribution ratio of about 0.1 for partitioning between hexane and an aqueous buffer with a pH of 10. The distribution ratio for partitioning of some polar lipids in the same solvent system is about 25. (a) Determine the value of the ratio (hexane volume/buffer volume) which will optimize the separation of the lipids from the acids. (b) It is very important to isolate an alkanoic acid-fraction free of the polar lipids mentioned above. Tell how you would increase the purity of the acid fraction after extracting it from the hexane.

12. Compute the minimum value of $V_r = \Sigma V_1/V_2$ for 99.9 per cent extraction of a substance with $D_c = 1$, assuming that a differential extraction process is possible (that is, an infinite number of separate extractions can be performed with a total volume of extracting solvent equal to $\Sigma V_1$).

13. If five is taken as the maximum convenient number of repeated extractions, and if $\Sigma V_1$, the total volume of extracting solvent, is equal to $V_2$, the volume of the aqueous phase, what is the minimum $D_c$ required for 99.9 per cent extraction efficiency?

14. If a Craig countercurrent extraction apparatus is employed to separate two solutes, A and B, with distribution ratios $D_{cA} = 30$ and $D_{cB} = 15$, respectively, calculate the total number of tubes needed such that the tube containing the maximum concentration of A is 10 tubes removed from the tube containing the maximum concentration of B. Calculate the number of the tube in which A is at its maximum concentration. What is the fraction of A present in that tube? Calculate the number of the tube in which B is at maximum concentration. What is the fraction of B present in that tube? Assume that $V_r = 1$.

15. Given equation (15–42), derive an approximate expression for $n$ in terms of $\Delta p$ and . What are the limitations on application of such a formula? For $D_{cA} = 0.707$ and $D_{cB} = 1.414$, what value of $n$ do you find for $\mathscr{R} = 2.0$? Assume that $V_r = 1$.

16. Consider two adjacent solute zones with equal maximum concentrations in a Craig countercurrent distribution apparatus. Define cross-contamination as the mole per cent of solute A in the solute B zone, when the zones are divided in the valley halfway between their concentration maxima. Calculate the per cent cross-contamination when $\mathscr{R} = 1.0$, 2.0, and 3.0. Make a sketch explaining your calculation and showing the degree of overlap between zones and how this overlap depends on $\mathscr{R}$.

17. In a particular continuous liquid-liquid extractor, the following concentrations of propanol were found in the aqueous phase after various total extraction times: $t = 0$ (start of experiment), $[C_3H_7OH] = 0.35\,M$; $t = 1$ hr, $[C_3H_7OH] = 0.088\,M$; $t = 4$ hr, $[C_3H_7OH] = 0.0014\,M$. How long would it take to achieve 99.9 per cent extraction efficiency?

18. Sketch a practical liquid-liquid continuous extraction apparatus which allows stirring of the extracted phase even when a solvent less dense than the sample is being used.

19. A sample of foliage is to be studied for its organic base content. Tell how you would work-up an organic base fraction from plant material.

20. Two metal ions, $A^{3+}$ and $B^{2+}$, are to be separated by being selectively extracted into an organic solvent containing a ligand that forms extractable complexes $AL_3$ and $BL_2$ with the metal ions. If the overall complex formation constant for $AL_3$ is $\beta_{3A} = 10^{10}$, and if $D_{cA}$ is $10^3$ at pH 8, what is the maximum value for $\beta_{2B}$ if quantitative separation ($p_A \geq 10^3 p_B$) is to be achieved? Assume that $V_r = 1$.

21. Estimate, from Figure 15–15, the distribution ratio for the extraction of iron-(III) into diethyl ether from $4\,F$ hydrochloric acid. How many successive extractions of the aqueous phase with equal volumes of pure diethyl ether would be required for 99 per cent extraction of iron(III) at a $4\,F$ hydrochloric acid concentration level?

22. The technique of solvent extraction is a powerful method for the evaluation of equilibrium constants. The equilibrium constant for the reaction

$$I_2 + I^- \rightleftharpoons I_3^-$$

was determined in this manner.

   (a) The partition coefficient for the distribution of molecular iodine between water and carbon tetrachloride was established by measurement of the equilibrium concentrations of $I_2$ in the two phases. Calculate the partition coefficient obtained in a typical experiment, if the titration of 100.0 ml of the aqueous phase required 13.72 ml of $0.01239\,F$ sodium thiosulfate solution and if the titration of 2.000 ml of the carbon tetrachloride layer required 23.87 ml of the same sodium thiosulfate solution.

   (b) Next, molecular iodine was partitioned between carbon tetrachloride and an aqueous $0.1000\,F$ potassium iodide solution. The titration of a 2.000-ml portion of the carbon tetrachloride phase with a $0.01239\,F$ sodium thiosulfate solution required 17.28 ml of the titrant, whereas the titration of a 5.000-ml aliquot of the aqueous phase required 25.93 ml of the sodium thiosulfate solution. Calculate the equilibrium constant for the formation of triiodide ion.

   (c) Why is molecular iodine so much more soluble in carbon tetrachloride than in water? Why is triiodide not extracted into carbon tetrachloride?

23. A study was made of the distribution of formic acid, HCOOH, between water and benzene at 25°C. Various amounts of formic acid were dissolved in water having a pH of 1.0, and the aqueous solutions were shaken in contact with benzene until equilibrium was attained. For each experiment, aliquots of the aqueous and benzene phases were taken, and the total *formal* concentration of formic acid in each phase was determined by means of a titration with a standard sodium hydroxide solution. From five different experiments, the following data were collected:

| HCOOH in aqueous phase, $F$ | 1.632 | 3.436 | 5.115 | 6.863 | 8.852 |
|---|---|---|---|---|---|
| HCOOH in benzene phase, $F$ | 0.003117 | 0.008418 | 0.01514 | 0.02402 | 0.03629 |

   (a) Assuming that monomeric formic acid is the only important species in water at pH 1, determine what is the principal form of formic acid in benzene. Draw its structure.

   (b) Evaluate the equilibrium constant for the association of monomeric formic acid in benzene

$$n\,HCOOH\ (benzene) \rightleftharpoons [HCOOH]_n\ (benzene)$$

   where $[HCOOH]_n$ is the predominant form of formic acid in benzene.

   (c) From experiments at 75°C, the equilibrium constant for the preceding association reaction was found to be 2.47. Calculate the value of $\Delta H$ for this reaction, and discuss the significance of the result.


## SUGGESTIONS FOR ADDITIONAL READING

1. E. W. Berg: *Physical and Chemical Methods of Separation.* McGraw-Hill Book Company, New York, 1963.

2. L. C. Craig and D. Craig: Laboratory extraction and countercurrent extraction. *In* A. Weissberger, ed.: *Technique of Organic Chemistry*. Second edition, Part I, Volume III (*Separation and Purification*), Wiley-Interscience, New York, 1956, Chapter II.
3. H. Irving and R. J. P. Williams: Liquid-liquid extraction. *In* I. M. Kolthoff and P. J. Elving, eds.: *Treatise on Analytical Chemistry*. Part I, Volume 3, Wiley-Interscience, New York, 1959, pp. 1309–1365.
4. G. H. Morrison and H. Freiser: *Solvent Extraction in Analytical Chemistry*. John Wiley and Sons, New York, 1957.
5. R. H. Perry, C. H. Chilton, and S. D. Kirkpatrick: *Chemical Engineers' Handbook*. Fourth edition, McGraw-Hill Book Company, New York, 1963.
6. H. Purnell: *Gas Chromatography*. John Wiley and Sons, New York, 1962.
7. R. E. Treybal: *Liquid Extraction*. Second edition, McGraw-Hill Book Company, New York, 1963.

# CHROMATOGRAPHY

# 16

   "Definitions" of *procedures* can be distracting instead of informative, and it seems most helpful instead to begin this chapter with an "example" of chromatography. That chosen here relates to a simple laboratory liquid-solid chromatographic column; but the same fundamental separation process occurs in all chromatographic systems, although the mechanisms of interactions of chemical species with the chromatographic media will differ.

   A chromatographic column is shown in Figure 16-1. In practice, it might be a glass tube about one centimeter in internal diameter. A length of it, about 15 cm, would be packed with stationary phase; some additional length at the top would be left empty to provide a reservoir for the mobile phase; and a stopcock would be attached to the bottom to regulate the flow of mobile phase. Although the stationary phase can take many different forms, let us assume for the moment that it is some adsorbent (perhaps alumina or silica gel), "activated" by being dried at 150°C and having an average particle diameter of about 100 $\mu$m, which has been packed into the column by pouring in a slurry and drawing off the excess solvent through the stopcock. The mobile phase should be some solvent which easily dissolves the sample without reacting with it, and which is not so polar that it displaces adsorbed sample molecules from the stationary phase.

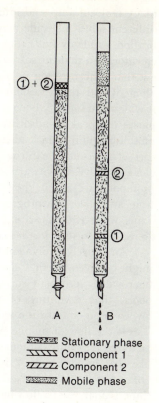

*Figure 16–1.* Diagrammatic representation of the retention and elution of a two-component mixture on a chromatographic column; *A*, just after addition of sample; *B*, during elution. Component 2 is more strongly retained than 1.

Stationary phase
Component 1
Component 2
Mobile phase

Consider the separation of a binary mixture on this column. Before the sample is added, the stopcock is opened briefly and any mobile phase above the top of the stationary-phase bed is allowed to drain through the column. Then, the sample (a mixture of components 1 and 2) is dissolved in a minimum volume of mobile phase, and applied to the top of the column. Following this, the stopcock is again opened briefly, this time to draw the portion of mobile phase containing the sample into the very topmost portion of the column, as shown in Figure 16-1A. At this moment, the sample (solute) molecules are distributed between an adsorbed phase and the solution phase in ratios determined by their individual adsorption isotherms. Let us say that component 2 is more strongly adsorbed and, therefore, that a smaller proportion of it is in the solution phase. Next, a volume of mobile phase (the **eluent**) is added to the empty space above the adsorbent bed, and the process of **elution** is begun by opening the stopcock so that this mobile phase slowly passes through the column (say, at 0.1 ml/min).

From this time onward, chaos prevails. If the reader will pardon childish simplicity and the attribution of eyes and emotions to individual molecules, which, in reality, are responding only to purely physical forces, a memorable description of these events is possible (we shall become more precise quite soon enough). Imagine a molecule of solute, sitting happily in the solution phase just after the sample has been added and the adsorbed phase-solution phase equilibrium has been established. This molecule cannot get onto the adsorbent because its surface is too crowded, and to do so would disturb the equilibrium. Suddenly, someone opens the stopcock, and the elevator in which this molecule is riding starts to go down. That is the end of equilibrium. Looking around as it descends, our molecule soon sees an uncrowded adsorbent surface and settles down on it, trying to bring the concentration of adsorbed-phase molecules up and into equilibrium with the solution-phase concentration. A different solute molecule, which had been among those adsorbed in the initial equilibration, suddenly finds that all its solution-phase colleagues have gone down on the elevator, and that the solvent around it is "empty." It jumps into solution, striving to get that concentration up into equilibrium with the crowded adsorbent. As it rides along, it finds the adsorbent surface quite crowded by our friend above and others like it which had initially been in the solution phase. Soon, however, this second molecule passes the peak of adsorbent crowding, and the forces which continually strive to establish equilibrium lead it to sit down on the adsorbent again. It is by exactly this sort of molecular leapfrog that solutes pass through the column. As the solvent elevator moves downward at a constant rate, molecules which are strongly adsorbed (component 2) are in the mobile phase for only a small fraction of any increment of time, and make slow progress relative to less strongly adsorbed species (component 1). Thus, we arrive at the situation depicted in Figure 16–1B, which shows that two solute zones have formed in the column.

An entirely equivalent description of a separation by partition chromatography might be given. In that case, we would think of sample molecules not as looking for uncrowded adsorbent surfaces but, instead, as looking for uncrowded swimming pools—that is, volumes of stationary phase in which the solute concentration is too low to be in equilibrium with the surrounding mobile phase.

If the flow of mobile phase is maintained, and if each drop of mobile phase is analyzed for solute content the instant it emerges from the column, a chromatogram like that in Figure 16–2 is obtained. In practice, such chromatograms are easily obtained with the aid of a **detector**, a special analytical instrument arranged for continuous analysis of a flowing fluid stream and having an electrical output signal which is in some known way proportional to the chemical input. Components of a chemical mixture are characterized by the degree to which they are retained on a given column, and the most common measurement of this is the **retention time**, as noted in Figure 16–2. Equally characteristic is the **retention volume**, which is a measure of the amount of mobile phase required for elution of a given component. Because even columns which are identically prepared usually differ slightly in their retention

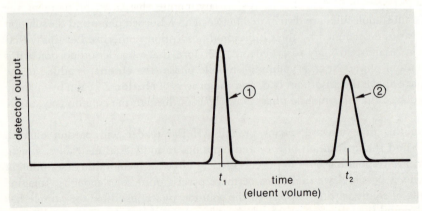

*Figure 16–2.* Graphic representation of output of the analyzing device (or detector) during a chromatographic separation; the retention time for component 1 is $t_1$, and for component 2 it is $t_2$.

characteristics, it is convenient to use **relative retention times**, or **relative retention volumes**, in which the retention of any particular component is expressed relative to that of some standard compound co-injected with the sample.

## Solute Migration

No solute can move through the column faster than the eluent which carries it, and the extent to which any compound falls short of this maximum velocity is measured by **R**, the **retention ratio**, or **retardation factor**. If the average molecule of mobile phase requires 15 minutes for passage through a 15-cm column, the eluent velocity ($v$) would be 1.0 cm/min or 0.016 cm/sec. (The instantaneous velocity of any given molecule can vary widely from this value and, because of tortuosity in the flow path, the true distance travelled during passage through the column will be somewhat greater than 15 cm. This "velocity" refers only to the average component parallel to the bulk flow.) If a particular solute zone takes an hour to pass through the column under the same conditions, its velocity would be 0.004 cm/sec or one-fourth the velocity of the mobile phase. The retardation factor is thus 0.25. The retardation factor or retention ratio offers a measure of the fraction of time which the average solute molecule spends in the mobile phase. Thus, we can write

$$\mathbf{R} = \frac{t_M}{t_M + t_S} \tag{16-1}$$

where $t_M$ is the time in the mobile phase and $t_S$ is the time in the stationary phase. In this case, since $\mathbf{R} = 0.25$, a solute molecule spends 25 per cent of its time in the mobile phase and travels 25 per cent as fast as the eluent. Notice that, if a substance is entirely unretained and spends no time in the stationary phase, it will have $\mathbf{R} = 1.0$ and will travel with the same velocity as the mobile phase. If the molecules of a substance spend, on the average, half their time in each phase, we have $t_M = t_S$ and $\mathbf{R} = 0.5$. It is interesting to recall that we have compared the mobile phase to an elevator always moving at a constant speed. Therefore, because all eluted species must travel the same column length, they must spend exactly the same length of time in the mobile phase. It is only $t_S$ which varies from component to component, $t_M$ remaining constant.

In chromatography, the **partition coefficient** ($K$) is defined by the relation

$$K = \frac{c_S}{c_M} \tag{16-2}$$

where $c_S$ and $c_M$ are the solute concentrations in the stationary and mobile phases, respectively. It is, unfortunately, practically the inverse of the **distribution ratio**, $D_c$, used in countercurrent distribution in which the solute concentration in the mobile phase appears in the numerator. In order to relate $K$ and $\mathbf{R}$, we can formulate a relationship between $K$ and the phase times ($t_S$ and $t_M$) by noting that the times depend not only on the relative solute concentrations in the two phases, but, in addition, on the volumes within which these concentrations must be dispersed. Within any region of the column, we have

$$\begin{pmatrix} \text{ratio of times spent by} \\ \text{solute in each phase} \end{pmatrix} = \begin{pmatrix} \text{ratio of quantities of} \\ \text{solute in each phase} \end{pmatrix} = \begin{matrix} \text{(partition coefficient)} \\ \times \text{(ratio of phase volumes)} \end{matrix}$$

$$\frac{t_S}{t_M} = \frac{c_S' \, V_S'}{c_M' \, V_M'} = K\left(\frac{V_S}{V_M}\right) \tag{16-3}$$

where the primed quantities pertain to the region under consideration and where $(V_S/V_M)$ is set apart in order to emphasize that we have simply assumed the *ratio* of phase volumes in that region to be characteristic of the column as a whole. Rearrangement and substitution into equation (16–1) yields

$$\mathbf{R} = \frac{1}{1 + \dfrac{t_S}{t_M}} = \frac{1}{1 + K\left(\dfrac{V_S}{V_M}\right)} = \frac{V_M}{V_M + KV_S} \tag{16–4}$$

which expresses the retention ratio in terms of the partition coefficient and column parameters.

There is a remarkable similarity between equation (16–4) and equation (15–36), which we derived earlier (page     ) for stepwise countercurrent distribution. For countercurrent distribution we have written

$$r_{max} = np = n\left(\frac{D_c V_r}{D_c V_r + 1}\right) = n\left(\frac{c_M V_M}{c_M V_M + c_S V_S}\right) = n\left(\frac{1}{1 + K\dfrac{V_S}{V_M}}\right) \tag{16–5}$$

In chromatography, $\mathbf{R}$ expresses a relative rate, not an absolute maximum and, in countercurrent distribution, the equivalent quantity would be $r_{max}/n$. Division of both sides of equation (16–5) by $n$ produces an identity with equation (16–4) and demonstrates the fundamental similarity of the two countercurrent techniques, one carried out in discrete steps and the other as a continuous process.

A relationship between $\mathbf{R}$ and the **retention volume**, $V_R$, can be derived if we note that an eluent molecule introduced at the top of the column will appear at the outlet after the passage of a volume $V_M$ of mobile phase. A solute travelling with half the eluent velocity ($\mathbf{R} = 0.5$) will take twice as long to traverse the column, so that $V_R = 2V_M$. A substance travelling at one-tenth of the eluent velocity will require ten times longer, and thus $V_R = 10V_M$. Evidently,

$$\frac{V_R}{V_M} = \frac{1}{\mathbf{R}} \tag{16–6}$$

Then, substituting into equation (16–4), we obtain

$$\mathbf{R} = \frac{V_M}{V_R} = \frac{V_M}{V_M + KV_S} \tag{16–7}$$

Therefore,

$$V_R = V_M + KV_S \tag{16–8}$$

In some forms of chromatography, it is easily possible to determine the mobile-phase volume, $V_M$. In such cases, the **adjusted retention volume**, $V_R'$, is sometimes used:

$$V_R' = V_R - V_M = KV_S \tag{16–9}$$

This parameter has the feature of being directly proportional to $K$. In either situation, these expressions are useful because they reveal the simple relationship between $V_R$ and $K$.

Fragments of chromatographic technique appear throughout the history of chemistry, but the first chromatographic experiments which included all the essential elements of the present method are due to the biochemist Mikhail Tswett, who published his first paper on the subject in 1903. Tswett is sometimes described as "Russian," other times as "Polish." He was in fact a Russian national, born in Italy, the son of a Russian father and an Italian mother, educated at the University of Geneva (Dr. Sc., 1896), and a member of the faculty of the Veterinary School at Warsaw. The substances which Tswett investigated were plant pigments, and the technique employed was adsorption chromatography with a calcium carbonate stationary phase and a petroleum ether mobile phase. The chromatographic zones were "detected" simply by their natural colors, and Tswett's name for the technique (given in 1906) was logically derived from the Greek word for color, *chromatos*. (At least one recent author has made the amusing suggestion that Tswett was also propelled toward this name by the fact that his own last name, in Russian, means "color.") Tswett published many more papers, but his later work was unfortunately cut short by the First World War.

The technique was lost until its rediscovery, again by biochemists, in 1931. This hiatus is remarkable. The scientific community of the decade 1900–1910 did not miss the developments in theoretical physics published by A. Einstein, and, though the comparison seems absurd now, the two scientists were perhaps equally poorly known at that early stage.

Liquid-liquid partition chromatography was introduced in 1942 by the biochemists A. J. P. Martin and R. L. M. Synge. These workers developed the first general theory of chromatography and suggested that the combination of a gaseous mobile phase with a liquid stationary phase would have important advantages. Another remarkable hiatus followed and, although some forms of chromatography became quite popular, no work on the gas-liquid combination was reported until Martin, this time in collaboration with A. T. James, described the technique in 1952. It was the beginning of a truly explosive period of development which continues still. This technique was of such tremendous importance that by 1956 organic chemistry laboratories around the world were employing gas-liquid chromatography. The literature now includes thousands of papers each year on the subject, and countless more in which the technique is applied. In retrospect, the crucial development was the early work of Martin and Synge on partition chromatography, and in 1954 they shared the Nobel Prize in chemistry.

## PLATE THEORY OF CHROMATOGRAPHY

In their 1942 paper introducing partition chromatography, Martin and Synge developed the first successful theoretical treatment of chromatographic zone broadening and migration. Here we can briefly review their treatment as it pertains to zone broadening, a simple discussion of migration having been given earlier. For a physical model, Martin and Synge utilized the theoretical-plate approach which had been so successful in the elucidation of distillation column processes (Chapter 14). This approach to chromatography can be strongly criticized because it requires some very unrealistic assumptions about column operation, and gives no view of the physical processes of diffusion and so on that actually occur in the column. A result of these weaknesses is that zone spreading is *experimentally* found to depend on variables (mobile-phase velocity, for example) which do not even appear in the plate theory.

For these reasons, chromatographic theory has undergone much further development. The measure of column efficiency which the simple plate theory provides, however, remains in universal use, with "theoretical plate" being a term as common in chromatographic practice as in distillation. For this reason, we briefly review the development of the plate theory here. Interested readers will find the detailed discussion from which this treatment was drawn in the text by Purnell, cited as supplemental reading at the end of this chapter.

### Relation to Countercurrent Distribution

*The physical model.* Imagine that the chromatographic column is subdivided into individual contact units corresponding to the individual extraction tubes in a countercurrent distribution apparatus. Each unit contains an amount of mobile phase, $\Delta V_M$, and of stationary phase, $\Delta V_S$. In the entire column there are $N$ of these units and, because a complete stationary phase-mobile phase equilibration occurs in each, they are called **theoretical plates**. Thus, the subdivided chromatographic column is equivalent to a countercurrent distribution apparatus with $N$ extraction stages. The mobile phase corresponds to the upper phase in countercurrent distribution, and moves through the apparatus in the same way. There is a difference in the way in which the separation process is observed. In countercurrent distribution, the contents of each tube are analyzed after a certain number of transfers $(n)$. In chromatography, the contents of the $N$th "tube" are continually analyzed by means of the detector, and $n$, the number of mobile-phase volumes $(\Delta V_M)$ passed through the column, increases as the process is carried out.

*Calculation of zone spreading.* A sample is added to the column and enters the first theoretical plate. A fractional amount $p$ remains in the mobile phase, while a fractional amount $q$ enters the stationary phase. After equilibration is complete, a small volume $(\Delta V_M)$ of mobile phase is added to the column, and the amount $p$ is transferred to the second plate. As the process continues, a pattern of solute distribution exactly like that shown in Figure 15–7 develops and, in general, the fractional amount of solute in any plate $r$ after the addition of $n$ volume increments $(\Delta V_M)$ of the mobile phase is

$$F_{r,n} = \frac{n!}{(n-r)!\,r!}\,p^r q^{(n-r)} \tag{16–10}$$

In chromatography, $n$ and $r$ both take very large values. Since the mobile phase passes through the column and a great many mobile-phase volumes are required to elute a compound, $n \gg r$; therefore,

$$q^{(n-r)} \sim q^n = (1-p)^n = e^{-np} \tag{16–11}$$

and

$$\frac{n!}{(n-r)!} \sim n^r \tag{16–12}$$

Substitution of equations (16–11) and (16–12) into equation (16–10) yields

$$F_{r,n} = \frac{(np)^r e^{-np}}{r!} \tag{16–13}$$

Equation (16–13) can be simplified by means of Stirling's approximation:

$$r! = \frac{\sqrt{2\pi r}\, r^r}{e^r} \tag{16–14}$$

A form of equation (16–10) with all factorial terms removed is obtained:

$$F_{r,n} = \frac{(np)^r e^{-np} e^r}{\sqrt{2\pi r}\, r^r} \qquad \text{(for } n \gg r) \tag{16–15}$$

Any chromatographic solute zone will be spread over a number of theoretical plates in the relative amounts specified by equation (16–15). If we denote the plate containing the largest solute fraction as $r_{\max}$, equation (16–15) can be used to calculate the solute fraction at the zone maximum or "top" of the chromatographic peak. Equation (16–5) provides the relationship $r_{\max} = np$, and can be substituted into equation (16–15) to yield

$$F_{r_{\max}, n} = \frac{1}{\sqrt{2\pi np}} = \frac{1}{\sqrt{2\pi r_{\max}}} \tag{16–16}$$

Consider the situation in which the peak is at the end of the column, that is, $r_{\max} = N$. If $m$ moles of solute were introduced to the column, the quantity of material $(Q_{N,n})$ in the $N$th theoretical plate is

$$Q_{N,n} = F_{N,n}(m) = \frac{m}{\sqrt{2\pi N}} \tag{16–17}$$

If the solute moves through a column of $N$ plates in time $t_R$, its rate of movement (plates per unit time) is $N/t_R$. We have calculated above the quantity of material in the $N$th plate when it contains the solute maximum. As that peak leaves the column, the maximum rate of solute escape, $S_{\max}$, will be (dimensions shown for clarity)

$$S_{\max} = Q_{N,n}\left(\frac{\text{moles}}{\text{plate}}\right)\frac{N}{t_R}\left(\frac{\text{plates}}{\text{unit time}}\right) = \frac{Q_{N,n}}{t_R}\, N\left(\frac{\text{moles}}{\text{unit time}}\right) = \frac{Nm}{\sqrt{2\pi N}\, t_R} \tag{16–18}$$

This expression can be solved for $N$:

$$N = \frac{2\pi (S_{\max})^2 t_R^2}{m^2} \tag{16–19}$$

The quantity $m$ (moles introduced to the column) is proportional to the peak area. Assuming that chromatographic peak height is recorded in millivolts (output from some solute detector) and peak width in seconds, and assuming that the peak is triangular (choosing some other shape only changes the factor $\frac{1}{2}$), we can write

$$m = \frac{kh t_w}{2} \tag{16–20}$$

where $k$ is a constant of proportionality having dimensions mole mv$^{-1}$ sec$^{-1}$, $t_w$ is the time width at the base of the peak (sec), and $h$ is the peak height (millivolts). (Recognize that the constant, $k$, converts the observed voltage to some material flux in mole sec$^{-1}$.) The maximum rate of solute escape, $S_{max}$, is proportional to the peak height; that is,

$$S_{max} = kh \qquad (16\text{–}21)$$

where $k$ and $h$ are just as defined above. Substitution of equations (16–20) and (16–21) into equation (16–19) yields

$$N = \frac{2\pi(kh)^2 t_R{}^2}{(\frac{1}{2}kht_w)^2} = 8\pi \left(\frac{t_R}{t_w}\right)^2 \qquad (16\text{–}22)$$

If we replace the assumption of triangular peaks with the more realistic Gaussian peak shape, we obtain

$$N = 16 \left(\frac{t_R}{t_w}\right)^2 \qquad (16\text{–}23)$$

where $t_w$ is defined as shown in Figure 16–3. Recasting this expression by substituting $t_w = 4\tau$, where $\tau$ is the standard deviation of the peak profile in time units, we obtain

$$N = \left(\frac{t_R}{\tau}\right)^2 \qquad (16\text{–}24)$$

Equations (16–23) and (16–24) furnish a way in which separation efficiency can be very easily determined from the basic experimental data. In addition, the direct

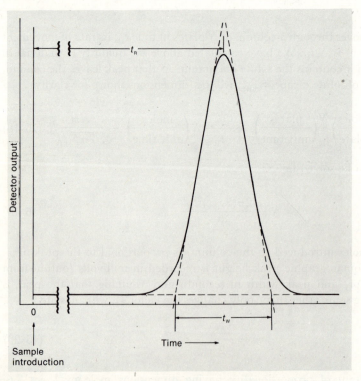

*Figure 16–3.* Definition of peak width, $t_w$, used in determining the efficiency of a chromatographic column. For chromatographic peaks which are perfectly Gaussian in shape, $t_w$ is four standard deviations.

relationship between peak width and retention time is revealed. The efficiency of any column is best judged by $H$, the height equivalent to a theoretical plate, which is given by the simple relation

$$H = \frac{L}{N} \tag{16-25}$$

where $L$ is the length of the column. Substituting for $N$, we obtain

$$H = \frac{\tau^2 L}{t_R^2} \tag{16-26}$$

## Shortcomings of the Plate Theory

*Incorrect assumptions.* First, we have assumed that $K$ is a constant independent of solute concentration in the stationary and mobile phases. This amounts to our assuming the linear partition isotherm $a$ shown in Figure 15–2; such a straight-line relationship reflects a constant $c_S/c_M$ ratio (constant $K$), which is not always the case. Particularly in adsorption chromatography, the adsorption isotherm tends to bend over at high $c_M$ values, as shown by curve $c$ in Figure 15–2. This can be understood in terms of crowding of adsorbed species on the adsorbent surface. In general, however, chromatographic separations are carried out with small amounts of material and the linear-isotherm assumption does not cause serious trouble. This is particularly true in partition chromatography, for which the plate theory was developed.

Second, it is assumed that equilibration is rapid compared to the movement of the mobile phase. In fact, it is assumed that, the moment a mobile-phase increment $(\Delta V_M)$ is added to the column and all prior mobile-stationary equilibria are upset, diffusion and redistribution of the solute in these phases take place instantaneously. However, diffusion is never instantaneous, and, particularly at high mobile-phase flow-rates, material might be swept along in the mobile phase from one plate to the next before equilibration is complete. If this occurs, it will cause a reduction in efficiency, that is, a lower value of $N$ as determined from equation (16–23).

Third, it is assumed that spreading of the chromatographic zone by longitudinal diffusion from one theoretical plate to another does not occur. In effect, it is assumed that diffusion does not occur. To say that this contradicts the second assumption is something of an understatement. Effects of longitudinal diffusion are of particular importance at low mobile-phase flow-rates, when substantial time is available for mobile-phase constituents to drift aimlessly from plate to plate, thereby broadening the chromatographic zone.

Fourth, the column is assumed to consist of a number of discrete volume elements; and fifth, it is assumed that the mobile phase is added in $\Delta V_M$ increments rather than continuously. Both these "assumptions" are, of course, completely untrue. The observation is, however, that more sophisticated treatments give very similar results.

Because the plate theory assumes a linear isotherm and an ideal diffusion situation, it is termed a "linear ideal" model for the chromatographic process. Other models which overcome various shortcomings of this treatment are termed "non-linear ideal," "linear non-ideal," and "non-linear non-ideal."

*Important variables excluded.* From the preceding discussion, it is clear that mobile-phase velocity has an important effect, but nowhere is it considered in the plate theory. Similarly, the dimensions of the phases are of great importance because they determine the distances over which diffusion must occur, and these variables are also excluded. These faults derive from the principal weakness of the treatment, namely, the failure to consider the physical processes which actually occur during zone migration.

## RATE THEORY OF CHROMATOGRAPHY

A treatment which avoids the assumption of instantaneous equilibrium and other shortcomings of the plate theory must pay close attention to the *rates* at which equilibrium can, in reality, be attained under typical chromatographic conditions. In addition, the *rates* of diffusion in the mobile and stationary phases have to be considered. This focus on kinetic aspects has earned for improved chromatographic theories the general designation, "rate theory," though it would be more precise to use the term *linear non-ideal*. The first comprehensive exposition of such a theory was provided by the Dutch chemists van Deemter, Klinkenberg, and Zuiderweg, in 1956, and the general equation for plate height as a function of mobile-phase velocity is sometimes referred to as the "van Deemter equation." There has been much additional development, in large part due to the work of the American chemist J. Calvin Giddings. The interested reader will find an extensive and highly readable discussion of the details of modern chromatographic theory is his book, cited as supplemental reading at the end of this chapter.

*Organization and central idea.* In the rate theory, each of the mechanisms which can contribute to zone broadening is considered separately. A mathematical expression for the relationship between plate height and variables important in the zone-broadening mechanism is derived and, with due attention to the details of their interaction, these separate expressions are combined to yield a general plate-height equation; that is, total $H = (H$ due to diffusion$) + (H$ due to slow equilibration$) + (H$ due to non-uniform flow patterns$)$ or, more simply,

$$H = H_d + H_e + H_f \qquad (16\text{--}27)$$

In reality, all these mechanisms act together, and the effects of any one are nearly impossible to isolate. Understanding their combination, however, is not difficult. For example, a solute zone carried along in a moving gas stream will tend to spread out, its sharp boundaries becoming blurred as time passes. A zone which was initially infinitely thin will eventually take on a Gaussian distribution, and the width of the diffusion-spread zone can be characterized in terms of the standard deviation of that distribution. Slow equilibration and flow patterns within the column will also produce spreading into a Gaussian profile. In each case, it is possible to calculate the standard deviation of the resulting zone. Using statistical principles (see Chapter 2), we have a rule for the combination of standard deviations when numbers are added or subtracted. For the sum of three terms

$$\text{sum} = X + Y + Z \qquad (16\text{--}28)$$

we can write

$$(\sigma_{\text{sum}})^2 = \sigma_X{}^2 + \sigma_Y{}^2 + \sigma_Z{}^2 \qquad (16\text{--}29)$$

The calculated standard deviation of the sum ($\sigma_{\text{sum}}$), while understandably larger than any of the individual standard deviations ($\sigma_X$, $\sigma_Y$, or $\sigma_Z$), is considerably less than ($\sigma_X + \sigma_Y + \sigma_Z$). This occurs because errors can frequently offset each other. If, for example, the individual X chosen is on the high side of its mean, Y and Z are

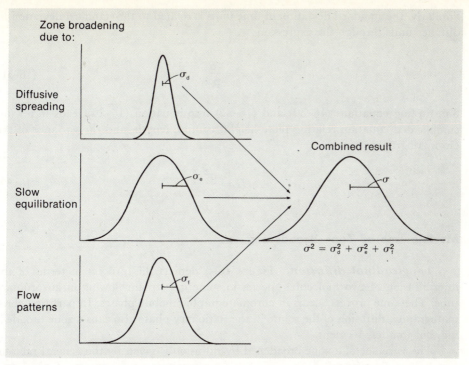

*Figure 16–4.* Combination of the different mechanisms of chromatographic zone broadening to yield the observed zone profile which results from the simultaneous operation of all these effects.

likely to be on the low side of their means, and the sum not drastically affected. In other words, the chance of observing a large positive or negative deviation simultaneously in all three variables is very low. Figure 16–4 depicts the chromatographic parallel. The reader should recognize that the variance of the final zone profile ($\sigma^2$) is the sum of the individual variances, and that the resulting combined standard deviation is $\sqrt{\sigma_d{}^2 + \sigma_e{}^2 + \sigma_f{}^2}$, where the subscripts d, e, and f represent the zone-broadening mechanisms of diffusion, slow equilibration, and flow patterns, respectively. In terms of physical processes, we can easily understand this method of combination by realizing that the individual broadening mechanisms can offset each other, at least partially. For example, a solute molecule speeded up (moved far from the center of the zone) by diffusive spreading might be delayed by slow equilibration or by non-uniform flow effects.

We can relate the standard deviation, $\sigma$, which represents the size of the zone in the column, to $H$ and $N$ by means of equation (16–26), derived earlier from plate theory:

$$H = \frac{\tau^2 L}{t_R{}^2} \qquad (16\text{–}26)$$

In this equation, the standard deviation, $\tau$, is expressed in time units, so that $\sigma$ (which has units of length) cannot be directly substituted for $\tau$. The retention time is simply the column length divided by the rate at which the zone travels through the column (recall that **R**, a ratio, is dimensionless):

$$t_R = \frac{L}{\mathbf{R}v} \qquad (16\text{–}30)$$

Similarly, the standard deviation in time units is related to the standard deviation in distance units through the expression

$$\tau = \frac{\sigma}{\mathbf{R}v} \qquad (16\text{–}31)$$

Substituting equations (16–30) and (16–31) into equation (16–26) leads to an uncomplicated equation relating plate height, column length, and the variance of the chromatographic zone:

$$H = \frac{\sigma^2}{L} \qquad (16\text{–}32)$$

### Mechanisms of Zone Broadening

*Longitudinal diffusion.* Figure 16–5 depicts this process in terms of the interdiffusion of two molecular species in some gas stream flowing in an ordinary tube. The same process occurs in chromatographic mobile phases. In partition chromatography, diffusion of the solute in the stationary phase also causes zone broadening, but to a lesser extent.

The variance of a zone broadened by diffusion is given by the general relation

$$\sigma_d{}^2 = 2Dt \qquad (16\text{–}33)$$

where $D$ is the diffusion coefficient for interdiffusion of the two molecular species (*i.e.*, mobile-phase molecules and solute molecules) and $t$ is the time over which diffusion has occurred. For diffusive spreading in the mobile phase, $(\sigma_{dM})^2 = 2D_M t_M$. It has already been noted that all eluted species spend exactly the same length of time, $t_M$, in the mobile phase and that $t_M = L/v$; thus, $(\sigma_{dM})^2 = 2D_M L/v$. Therefore, the

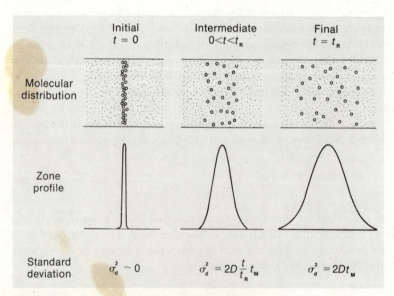

*Figure 16–5.* Spreading of the chromatographic zone due to diffusion, as seen at three different times during elution.

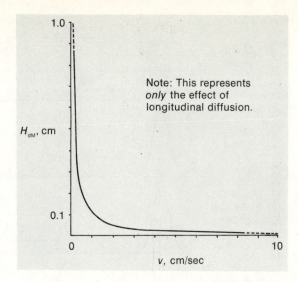

*Figure 16–6.* Plate height due to longitudinal diffusion of solute in the mobile phase, where $D_M$ is taken to be $5 \times 10^{-2}$ cm$^2$ sec$^{-1}$, a typical value for a gaseous mobile phase.

Note: This represents *only* the effect of longitudinal diffusion.

contribution to the plate height due to longitudinal diffusion in the mobile phase $(H_{dM})$ is given by

$$H_{dM} = \frac{(\sigma_{dM})^2}{L} = \frac{2D_M}{v} \qquad (16\text{--}34)$$

Equation (16–34) shows that, as $v$ decreases, $H_{dM}$ will increase. This happens because a lower velocity gives a longer time for passage through the column and, during that increased time, more zone spreading can occur.

For diffusive spreading in the stationary phase, we have $(\sigma_{dS})^2 = 2D_S t_S$, where $t_S = t_M(1 - \mathbf{R})/\mathbf{R}$, and $t_M = L/v$. These relations lead to the following expression for the plate-height contribution due to diffusive spreading in the stationary phase:

$$H_{dS} = \frac{(\sigma_{dS})^2}{L} = \frac{2D_S(1 - \mathbf{R})}{v\mathbf{R}} \qquad (16\text{--}35)$$

Notice from this equation that $H$ attributable to diffusive spreading in the stationary phase also varies as $v^{-1}$. In the common case where the mobile phase is a gas and the stationary phase a liquid, $D_M \sim 10^5 D_S$, and diffusive spreading in the stationary phase can be neglected. A graph showing how $H$ is affected by either or both of the diffusive spreading mechanisms appears in Figure 16–6.

*Slow equilibration.* In an "ideal" system, the transfer of solute molecules between the mobile and stationary phases is instantaneous. However, such ideality is not achieved for two reasons. First, there is kinetic control of the rate at which molecules can cross the interface, an effect generally termed "sorption-desorption kinetics." Second, there is kinetic control of the rate at which molecules can arrive at the interface and become available for transfer. This second effect is due to the finite rate of diffusion of solute molecules in both the stationary and mobile phases, and is called "diffusion-controlled kinetics." These combined effects force the chromatographic zone in the stationary phase to lag behind that in the mobile phase, trying, as it were, to "catch up," but being prevented from doing so by the rate at which solute molecules can be moved about. The observed chromatographic zone is a combination of the mobile-phase and stationary-phase populations and, when they are offset, the resultant zone is considerably broadened. The degree of offset depends on the velocity of the mobile phase, so that the variance of the chromatographic zone

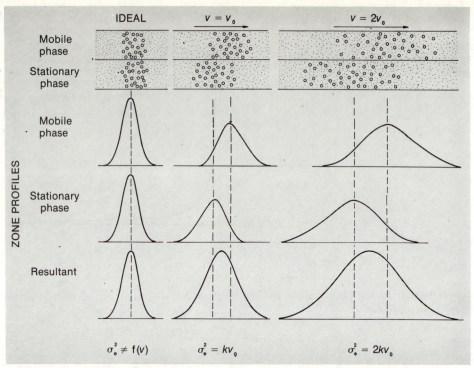

*Figure 16–7.* Spreading of the chromatographic zone due to slow equilibration of solute between the mobile and stationary phases. The first set of profiles depicts the situation assumed by the plate theory, which artificially requires that zone variance be independent of mobile-phase velocity. The second and third sets of profiles realistically depict the situations at some arbitrary mobile-phase velocity $v_0$ and at twice that value. As noted, $\sigma_e^2$ is directly proportional to mobile-phase velocity.

due to slow equilibration must vary directly with $v$. The essential points of this discussion are depicted graphically in Figure 16–7.

Quantitative expressions for $H_e$, like those derived above for $H_d$, can be developed; however, while not particularly complex, the necessary discussion is too lengthy for inclusion here. Suffice it to state that the contribution to plate height arising from slow equilibration has three terms, all of which depend directly on $v$. The first term relates to sorption-desorption kinetics and will not be further discussed. The second and third terms relate to solute diffusion in the stationary and mobile phases and logically depend on some other variables besides $v$. Considering first the effect of diffusion-controlled kinetics in the stationary phase, we can note that the degree of stationary-phase zone lag will depend not only on the mobile-phase velocity but also on the time required for an average solute molecule to diffuse to the mobile-phase interface from any point within the stationary phase. If the stationary phase is very thin, even a solute molecule at the bottom of the layer can reach the surface quickly. In addition, the diffusion coefficient of the solute in the stationary-phase liquid must be important. With these facts in mind, it can be shown that

$$H_{es} \propto \frac{d^2 v}{D_s} \qquad (16\text{–}36)$$

where $d$ is the thickness (or depth) of the stationary liquid layer. Qualitatively, equation (16–36) can be understood as follows: (a) An increase in plate height ($H$)

corresponds to a greater degree of zone broadening; (b) a doubling of stationary-layer thickness ($d$) will quadruple plate height because the average time required for solute diffusion in the stationary phase will be quadrupled; (c) an increase in mobile-phase velocity ($v$) causes an increase in zone broadening as shown in Figure 16–7 and, because equation (16–32) shows that $H \propto \sigma^2$, $H_{eS}$ is directly proportional to $v$; and (d) an increase in $D_S$ will allow faster diffusion on the stationary phase, thereby decreasing $H$. Zone broadening due to diffusion in the stationary phase cannot occur at all in adsorption chromatography, but this advantage is completely overcome by the much greater effect of sorption-desorption kinetics in that technique.

Very similar arguments apply to the development of a plate-height term for diffusion in the mobile phase (see, however, the comments below on flow patterns in the mobile phase). If the time required for a solute molecule to diffuse from the center of a flow channel to the stationary-phase interface is excessive, $\sigma_e^2$ will be increased because this slow diffusion step will prevent rapid attainment of equilibrium. The average time required for diffusion is directly proportional to the square of the distance and inversely proportional to the diffusion coefficient. The seriousness of this effect again varies directly with the velocity of the mobile phase, so we can write

$$H_{eM} \propto \frac{d_p^2 v}{D_M} \tag{16–37}$$

where $d_p$ is the average diameter of the stationary-phase particles and is, accordingly, the parameter which determines the dimensions of the flow channels. Even when the mobile phase is a gas, and therefore $D_M \gg D_S$, the fact that $d_p^2 \gg d^2$ can make this term important.

*Flow patterns.* A packed chromatographic column is a maze through which the mobile phase travels. There are countless pathways by which a molecule might travel from one end of the column to the other. These pathways are not all the same; some are a bit longer than average and some are a bit shorter. Molecules which travel a short path will emerge, all other factors being equal, a bit sooner than average. If there is a Gaussian distribution of flow-path lengths, this will lead to a Gaussian distribution of travel times. That is, a zone which started as infinitely thin at the column inlet will be dispersed to a Gaussian profile with some characteristic $\sigma_f^2$ at the column outlet. To a first approximation, the $H_f$ term is independent of mobile-phase velocity and is a function only of the structure and arrangement of the column packing.

Diffusion in the mobile phase has an interesting effect on $\sigma_f^2$. A molecule caught up in a long flow path is not constrained to stay in that pathway through the entire column length; it might be "rescued" by diffusing sideways into a more normal flow path. This diffusion-induced "path-switching" has the effect of diminishing $\sigma_f^2$. Theoretical consideration of this effect requires study of the coupling between diffusion in the mobile phase and flow pathways and is quite complex—it is called the "coupling theory of eddy diffusion." One of the results of this theory is easily grasped; because $\sigma_f^2$ depends partially on the amount of path-switching that can occur, $\sigma_f^2$ becomes dependent on the time available for this diffusion to occur. This time is simply $t_M$, which depends in turn on $v$. If $v$ is large, the time available for path-switching is short and $\sigma_f^2$ is increased. Thus, $H_f$ becomes slightly dependent on $v$.

## Dependence of Plate Height on Mobile-Phase Velocity

*Combination of H terms.* Using equation (16–32) and the model provided by equation (16–29), we can combine the individual terms in the foregoing discussion through simple addition:

$$H = \frac{\sigma^2}{L} = \frac{(\sigma_{dM})^2 + (\sigma_{dS})^2 + (\sigma_{eM})^2 + (\sigma_{eS})^2 + (\sigma_f)^2}{L} \tag{16–38}$$

Equation (16–38) can be rewritten as follows:

$$H = H_{dM} + H_{dS} + H_{eM} + H_{eS} + H_f \tag{16–39}$$

Inspection of equations (16–34), (16–35), (16–36), and (16–37) shows that the first four terms of equation (16–39) can be combined as the first two terms of equation (16–40). An expression for $H_f$ completes the equation:

$$H = \frac{B}{v} + Cv + \cfrac{1}{\cfrac{1}{A} + \cfrac{1}{C_M v}} \tag{16–40}$$

In equation (16–40), each contribution to $H$ is expressed in terms of a constant and some power of the mobile-phase velocity. Specifically, $B = 2D_M$; $C$ includes constants depending on sorption-desorption kinetics, $d^2$, $D_S$, $d_p^2$, and $D_M$; and $A$ depends on the structure and particle size of the column packing. The last term represents the plate-height contribution from flow effects and thus includes, in addition to the $A$ factor, a velocity-dependent term $C_M v$ derived from the coupling theory. Figure 16–8 shows a plot of $H$ versus $v$ and indicates how each term contributes to the observed variation of plate height as a function of mobile-phase velocity. Notice that the variation of $H$ can be fit to an equation having the form

$$H = A' + \frac{B'}{F} + C'F \tag{16–41}$$

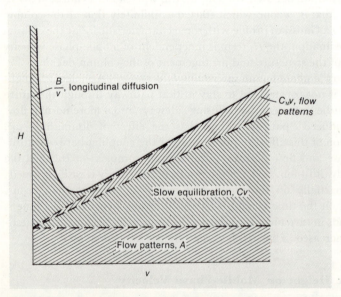

*Figure 16–8.* A graph showing the dependence of plate height on mobile-phase velocity. The contributions of various peak-broadening mechanisms are indicated.

where $F$ is the mobile-phase flow-rate (ml/unit time), and the new set of constants required by the change from $v$ to $F$ is indicated by the primed parameters $A'$, $B'$, and $C'$, which play the same role here as in equation (16–40).

**Optimum velocity.** Figure 16–8 is one of those frustrating graphs without numbers on the axes. We are forced to be vague because the velocities and plate heights depend so critically on column parameters such as $D_M$, $D_S$, $d_p$, and $d$. The particle diameter ($d_p$) and the diffusion coefficient for solute species in the mobile phase ($D_M$) form the basis for defining "reduced variables" which bypass some of the differences between various forms of chromatography. If, for velocity, we substitute the **reduced velocity**, $v = d_p v/D_M$, and for plate height we substitute the **reduced plate height**, $h = H/d_p$, a graph generally applicable to all types of chromatography can be constructed as shown in Figure 16–9. The graph indicates, for example, that a

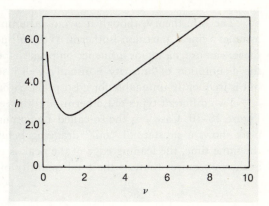

*Figure 16–9.* Graph of reduced plate height versus reduced velocity. Because reduced variables are employed, the plot is generally applicable to all forms of chromatography [Redrawn from J. C. Giddings: *Dynamics of Chromatography. Part 1, (Principles and Theory).* Dekker, New York, 1965, p. 64. By courtesy of Marcel Dekker, Inc.]

minimum plate height of about three particle diameters can be expected in well-optimized columns. Diffusion coefficients in liquids are usually in the range from $10^{-7}$ to $10^{-6}$ cm² sec⁻¹, whereas those in gases are usually in the range from $10^{-2}$ to $10^{-1}$ cm² sec⁻¹. Thus, in chromatography with a liquid mobile phase, the optimum velocity will be about

$$v_{opt} = \frac{v_{opt} D_M}{d_p} = \frac{(1)\ (10^{-6}\ \text{cm}^2\ \text{sec}^{-1})}{10^{-2}\ \text{cm}} = 10^{-4}\ \text{cm sec}^{-1}$$

whereas, in chromatography with a gaseous mobile phase, $v_{opt} \sim 5$ cm sec⁻¹.

When it is important to obtain the maximum number of theoretical plates in a given separation, the optimum flow rate can be quite accurately found. First, equation (16–41) can be satisfactorily approximated by

$$H = \frac{B'}{F} + C'F \tag{16–42}$$

Next, since we desire to minimize $H$ by choosing some optimum flow rate, equation (16–42) can be differentiated and set equal to zero

$$\frac{dH}{dF} = -\frac{B'}{F^2} + C' = 0 \tag{16–43}$$

and the resulting expression can be solved for the optimum flow rate:

$$F_{\text{opt}} = \sqrt{\frac{B'}{C'}} \qquad (16\text{–}44)$$

Given two independent sets of observed values for $H$ and $F$, we can determine $B'$ and $C'$ by substitution in equation (16–42) and the solution of simultaneous equations. Examples of such calculations will be presented in Chapter 17.

## OTHER FACTORS IN ZONE BROADENING

### Non-linear Chromatography

The rate theory, though it avoids assumptions of instantaneous diffusion, still assumes a linear partition isotherm. We shall now see that non-linearity in this isotherm can exert a major influence on peak shape. It is probably worth repeating that the assumption of linearity is usually well justified for partition chromatography, but is frequently untenable for adsorption chromatography.

Two different types of isotherms and the resulting peak shapes are illustrated in Figure 16–10. Case $a$ is the so-called Langmuir adsorption isotherm. The resulting peak shows a substantial "tail" dragging behind the maximum concentration. At the same time, the leading edge of the peak is particularly abrupt. The shape of the isotherm requires that, as the total quantity of solute increases, the fraction in the

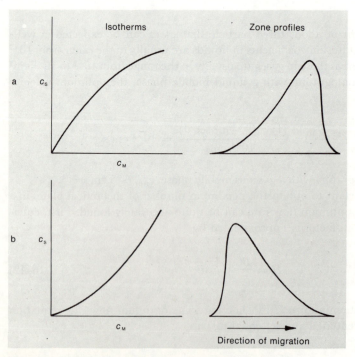

*Figure 16–10.* Non-linear partition isotherms and the resulting chromatographic zone profiles.

mobile phase increases and that, accordingly, the areas of highest concentration migrate with the greatest velocity. Thus, a Gaussian peak is transformed into the shape shown in a way that can be remembered thus: "the fast-moving high-concentration center catches up with the relatively slow-moving front and moves far ahead of the relatively slow-moving tail." A mirror-image argument describes case *b*, which is sometimes observed when a column is overloaded in gas-liquid partition chromatography. The same shape can be attributed to limited solubility of solute in the stationary phase. In either case, the solute apparently acts as its own stationary phase as total concentrations become very high.

## Tailing Due to Active Sites

Although the foregoing examples can both be regarded as anomalies associated with overloading of the stationary phase, the effect known as "tailing" is quite different. The shape of the chromatographic zone is just as shown for case *a* in Figure 16–10, but the experimental conditions preclude isotherm non-linearity as a cause. For example, in gas-liquid partition chromatography, tailing can sometimes be observed even at very low sample loads. Paradoxically, increasing the sample size *reduces* the tailing, further evidence that isotherm non-linearity is not the explanation.

Tailing of this sort is caused by a combination of two processes. The first is the "normal" partitioning or adsorption of the sample. By "normal" we mean the interaction being deliberately sought in order to bring about separation. Naturally, this interaction possesses good sorption-desorption kinetics at the chosen column velocity. The second process is the adsorption of solute molecules at a few particularly active sites which have very bad sorption-desorption kinetics. Once a molecule is adsorbed at such a site, it is released only after the peak of the chromatographic zone is well past. These sites are scarce and only a few molecules encounter them, but the cumulative effect is quite noticeable. The tailing appears to decrease at higher sample loads because the active sites become saturated, and the percentage of solute molecules affected by them becomes so small that the tail is not noticeable.

The effect of a large sample suggests a method to minimize tailing caused by active sites—the sites need only be "covered up." One technique of long standing is to wash the chromatographic medium with some highly polar compound (such as glacial acetic acid) which will be so strongly adsorbed at the active sites that it will never be released. A more modern approach which finds wide application in the treatment of solid supports for gas-liquid chromatography is chemical modification of active sites (in this case, —OH groups) through reaction with some reagent. For example, treatment with dimethyldichlorosilane replaces the highly polar hydroxyl groups with silyl ether linkages to a non-polar methyl silane surface:

$$
\begin{array}{c}
-Si-OH \quad Cl \quad CH_3 \\
O \qquad + \quad Si \\
-Si-OH \quad Cl \quad CH_3
\end{array}
\longrightarrow
\begin{array}{c}
-Si-O \quad CH_3 \\
O \qquad Si \\
-Si-O \quad CH_3
\end{array}
+ \; 2\,HCl
$$

Solid supports prepared in this way are said to be "silanized," and offer much better tailing characteristics than untreated supports. Obviously, the technique does not work when the nature of the active site is changed to some non-reactive form—a metal ion, for example. Supports rich in sites of this type must be avoided when one works with polar (readily adsorbed) samples.

## Zone Broadening Outside the Column

*Feed volume.* Our treatments of zone broadening have all assumed an infinitely thin initial zone. This is not practically attainable, and all sample inputs are characterized by some finite "feed volume." This occurs for several reasons. A primary cause is the design of the sample inlet. If the sample mixes with and is diluted by the mobile phase in the inlet volume and is then swept onto the column, the effective input volume of the sample can be very large, independent of the actual amount of sample introduced. Second, if the actual amount of sample is large, its volume can be substantial. This is particularly true in gas-liquid chromatography, where a liquid sample is vaporized in the inlet section of the chromatograph. Three $\mu l$ of methanol, when vaporized at 250°C and 2 atm pressure, will occupy 1.6 ml, a relatively enormous volume. Finally, input volume of the sample poses a problem in the analysis of gas samples by means of gas-solid chromatography. For example, the detection of trace atmospheric constituents might easily require air samples of at least 10 ml.

Quantitative treatment of these effects is simple and relates closely to our discussion of zone broadening within the column. For the moment, we shall assume that the sample can be swept from the inlet as a plug of material, with a resulting column input like that shown in Figure 16–11. (In a poorly designed inlet, the flow patterns

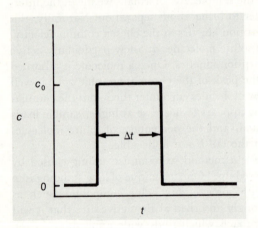

*Figure 16–11.* Concentration - versus - time profile for a well-designed sample introduction system.

of the mobile phase can lead to a much worse situation, with the inlet acting as a mixing chamber—this case will be discussed below.) The width of the input zone expressed in time ($\Delta t$) is determined by the sample volume, $V$, and the flow rate, $F$, of the mobile phase:

$$\Delta t = \frac{V}{F} \tag{16-45}$$

In order to determine the effect of this input-zone profile on overall chromatographic performance, we must calculate a standard deviation (in the time dimension) for the input zone. Theoretical considerations show that

$$\tau_i^2 = \frac{(\Delta t)^2}{12} \tag{16-46}$$

where $\tau_i$ is the required standard deviation. The extent to which this factor contributes to the total zone width can be calculated from the relation

$$\tau^2 = \tau_i{}^2 + \tau_c{}^2 \qquad (16-47)$$

where $\tau_c$ represents the net effect of all zone-broadening processes in the column and $\tau$ represents the resultant standard deviation for the zone. [Note that equation (16–47) is another application of the rule for combination of standard deviations.]

*Example 16–1.* A particular liquid chromatograph, when used with 1-ml samples, produces a final peak width of 100 sec. What peak width can be expected with a 20-ml sample input if $F = 1$ ml min$^{-1}$?

We are given that $w = 4\tau = 100$ sec; therefore, $\tau = 25$ sec. For the case with $\Delta t = V/F = 1$ ml/(1 ml min$^{-1}$) $= 60$ sec, it follows that $\tau_i{}^2 = (60)^2/12 = 300$ sec$^2$. Rearrangement of equation (16–47), and substitution of the known values of $\tau$ and $\tau_i$, allows calculation of $\tau_c{}^2$:

$$\tau_c{}^2 = \tau^2 - \tau_i{}^2 = (25)^2 - 300 = 325 \text{ sec}^2$$

In the proposed case of a 20-ml input sample, $\Delta t = V/F = 20$ ml/(1 ml min$^{-1}$) $= 1200$ sec. Then, $\tau_i{}^2 = (1200)^2/12 = 1.2 \times 10^5$ sec$^2$. The resulting final zone variance is computed from equation (16–47),

$$\tau^2 = \tau_i{}^2 + \tau_c{}^2 = (1.2 \times 10^5) + (325) = 120,325 \text{ sec}^2$$

and the peak width is given by

$$w = 4\tau = 1400 \text{ sec}$$

**Mixing chambers.** Mixing chambers are a fault of design or construction all too common in chromatographs. For example, in the simple laboratory column shown in Figure 16–12, the volume indicated is one into which a solute-containing zone may enter and mix with the entire mobile phase accessible. Thus, the zone is diluted

*Figure 16–12.* A chromatographic column showing a mixing volume which can cause zone broadening.

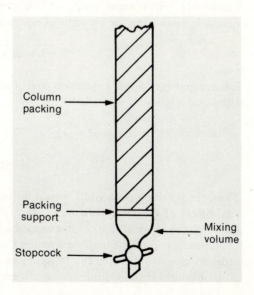

Column packing

Packing support

Stopcock

Mixing volume

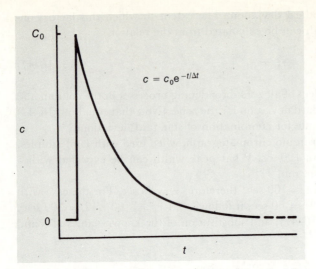

and is only slowly "rinsed out" of the volume. For such a volume, an infinitely thin input zone results in the output shown in Figure 16–13. The output concentration, $c$, follows the expression

$$c = c_0 e^{-t/\Delta t} \qquad (16\text{–}48)$$

where $c_0$ is the maximum solute concentration due to mixing of the zone in the mixing chamber and where $\Delta t = V/F$, $V$ being the volume of the mixing chamber and $F$ being the flow rate of the mobile phase. In this case, theoretical considerations show that the effective variance contribution, $\tau_m^2$, is simply $(\Delta t)^2$.

*Example 16–2.* The column shown in Figure 16–12 produces chromatographic peaks 300 sec wide when operated at a flow rate of 1 ml min$^{-1}$. The volume of the mixing chamber shown is 1 ml. If this chamber were eliminated, how wide would the peaks be?

We are given that $w = 300$ sec; therefore, $\tau = 75$ sec. We can evaluate the contribution from the mixing chamber as follows: $\tau_m = V/F = 1$ ml/(1 ml min$^{-1}$) = 60 sec. If this contribution is eliminated, only $\tau_c$, the column contribution to zone broadening, will remain. First, we can write an expression for the combination of $\tau_c$ and $\tau_m$, following the example of equation (16–47):

$$\tau^2 = \tau_c{}^2 + \tau_m{}^2$$

Rearrangement of this expression and insertion of the known values of $\tau$ and $\tau_m$ allows calculation of $\tau_c$:

$$\tau_c{}^2 = \tau^2 - \tau_m{}^2 = (75)^2 - (60)^2 = 2025 \text{ sec}^2$$

$$\tau_c = 45 \text{ sec}$$

Thus, the resulting peak width would be $4\tau_c$ or 180 sec, a substantial reduction.

**Effect of detector volume.** Many detectors used in chromatography have a finite volume in which the chromatographic zones are sensed. For example, a spectrophotometric cell might be placed at the end of a liquid chromatographic column in order to measure the intensity of some colored product derived from the chromatographic zones. If the volume of the cell greatly exceeds the mobile-phase volume in which some zone is eluted, the detector output will make the zone appear to be smeared-out and substantially broadened. In the worst case, this detector volume will act as a mixing chamber, and the effect on the total zone width can be calculated as above. In a well-designed detector, however, laminar flow of the mobile phase is preserved, and the contribution to zone broadening can be represented by a detector variance

$$\tau_d{}^2 = \frac{(\Delta t)^2}{12}$$

where $\Delta t = V/F$, with $V$ being the detector volume and $F$ being the flow rate of the mobile phase.

*Example 16–3.* A particular gas chromatographic column has $\tau_c = 5$ sec and is used at a flow rate of 1 ml min$^{-1}$. Calculate the maximum detector volume for a zone broadening of less than 1 per cent.

Our requirement is that $\tau$ be no more than 5.05 sec (1 per cent broadening). Thus, the necessary $\tau_d$ can be obtained as follows:

$$\tau^2 = \tau_c{}^2 + \tau_d{}^2$$

$$\tau_d{}^2 = \tau^2 - \tau_c{}^2 = (5.05)^2 - (5.00)^2 = 0.4016 \text{ sec}^2$$

$$\tau_d = 0.63 \text{ sec}$$

Relationships between $V$, $\Delta t$, and $\tau_d{}^2$ allow completion of the problem:

$$\tau_d{}^2 = \frac{(\Delta t)^2}{12} = \frac{V^2}{12 F^2}$$

$$\tau_d = \frac{V}{\sqrt{12}\, F}$$

$$V = \tau_d \sqrt{12}\, F = (0.63 \text{ sec})(3.46)(16.7\ \mu l\ \text{sec}^{-1}) = 36.3\ \mu l$$

This is an extremely stringent requirement, but such detectors are available.

## RESOLUTION

Resolution is a measure of the extent of overlap between two adjacent peaks. As such, it is a measure of the success of a given separation. In Chapter 15 (page 501), we defined resolution in the context of countercurrent distribution as

$$\mathscr{R} = \frac{\text{peak separation}}{\text{peak width}} = \frac{\Delta r}{4\sigma} \tag{16–50}$$

where $\Delta r$ is a measurement of the distance between the maxima of two solute zones, and $4\sigma$ is the effective zone width. There is no need to revise the definition for application to chromatographic zones. For countercurrent distribution, $\Delta r$ and $\sigma$ both are expressed in terms of numbers of extraction tubes, but $\mathscr{R}$ itself is a dimensionless ratio, and in chromatography we need only be careful that the distance of separation and the zone width are both expressed in the same units. In some forms of chromatography, distance might be convenient; in most cases of elution chromatography performed on columns, time units are most convenient.

The properties of $\mathscr{R}$ are reviewed in Figure 16–14, which shows the appearance

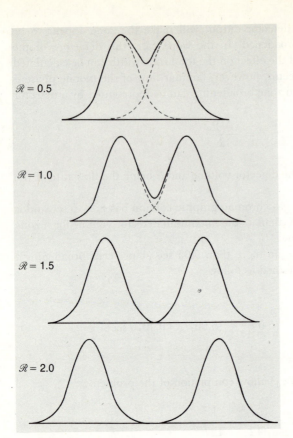

*Figure 16–14.* Pairs of chromatographic peaks showing various degrees of resolution, that is, various values of the parameter $\mathscr{R}$, defined and discussed in the text.

of chromatograms representing various degrees of resolution. The extent of cross-contamination which would result in each case (if we assume peaks of equal size) can be easily calculated from the normal curve of error. If a pair of peaks with $\mathscr{R} = 0.5$ (that is, two peaks separated by one-half peak width or $2\sigma$) is split down the middle, and each half is collected and analyzed, it will be found that each peak contains 84.13 per cent of its major component and 15.87 per cent of its minor component. This large degree of overlap is shown by the first pair of peaks in Figure 16–14. For $\mathscr{R} = 1.0$, a resolution which is considered sufficient for most practical purposes, each peak contains 97.73 per cent of the major component and 2.27 per cent of the minor component. The degree of resolution required for quantitative recovery in the usual sense, and the resolution referred to as "complete" by many chromatographers, is $\mathscr{R} = 1.5$. In this instance, the peak maxima are separated by six standard deviation units, and each peak contains 99.87 per cent of the major component and only 0.13 per cent of the minor component. The case in Figure 16–14 for which $\mathscr{R} = 2.0$ makes the interesting point that $(\mathscr{R} - 1)$ peaks can be inserted in the center of any pair resolved with resolution $\mathscr{R}$. For example, if $\mathscr{R} = 4.0$, three peaks can be added between the resolved pair without degrading resolution below $\mathscr{R} = 1.0$.

For countercurrent distribution, we found that $\mathscr{R}$ varies directly with $\sqrt{n}$, where $n$ is the number of mobile-phase increments added to the apparatus. In the case of chromatography, it is similarly useful to relate $\mathscr{R}$ to experimental variables. We can write

$$\mathscr{R} = \frac{\text{distance between zone maxima}}{4\sigma} \qquad (16\text{–}51)$$

The distance traveled by any zone is $L$ or $\mathbf{R}vt$, and so the distance between zone maxima, $\Delta L$, is $(\Delta\mathbf{R})vt$. Making the further substitution from equation (16–32) that $\sigma = \sqrt{HL}$, we obtain

$$\mathscr{R} = \frac{(\Delta\mathbf{R})vt}{4\sqrt{HL}} \qquad (16\text{–}52)$$

which can be simplified by substitution of $vt = L/\mathbf{R}$

$$\mathscr{R} = \frac{\Delta\mathbf{R}}{4\mathbf{R}}\sqrt{\frac{L}{H}} \qquad (16\text{–}53)$$

or, since $H = L/N$,

$$\mathscr{R} = \frac{\Delta\mathbf{R}}{4\mathbf{R}}\sqrt{N} \qquad (16\text{–}54)$$

Not unexpectedly, $\mathscr{R}$ is proportional to $\sqrt{N}$, indicating again that a doubling of resolution will require the number of theoretical plates to be quadrupled. In addition, the dependence of $\mathscr{R}$ on the relative retention $(\Delta\mathbf{R}/\mathbf{R})$ is not surprising, but it is important to examine this situation in greater detail. Although the derivation is lengthy and must be omitted here, the preceding equation can be rewritten as

$$\mathscr{R} = \frac{\sqrt{N}\,\Delta K}{4K}(1 - \mathbf{R}) \qquad (16\text{–}55)$$

Now it can be seen that, for a given relative difference in partition coefficients, $\Delta K/K$, $\mathscr{R}$ depends not only on $\sqrt{N}$, but also on $(1 - \mathbf{R})$. That is, the smaller $\mathbf{R}$ is, the better the resolution. Zones that travel rapidly will be poorly separated; or, to put it differently, zones that travel rapidly require a substantially greater number of theoretical plates for attainment of the same resolution. For example, inserting the values $\mathscr{R} = 1.0$, $\Delta K/K = 0.1$, and $\mathbf{R} = 0.5$ in the expression above, we find that $N = 6400$. For the same $\mathscr{R}$ and $\Delta K/K$, but with $\mathbf{R} = 10^{-2}$, we obtain $N = 1600$. The situation revealed in this discussion is summarized in Figure 16–15, which presents a plot of $N$, in this case the required number of plates for $\mathscr{R} = 1.0$, as a function of $\Delta K/K$ for various values of $\mathbf{R}$. The particularly sharp dependence of $N$ on $\mathbf{R}$ as $\mathbf{R}$ becomes large leads to a definition of the number of effective theoretical plates which is well worth keeping in mind:

$$\text{number of effective theoretical plates} = N(1 - \mathbf{R})^2$$

It might appear that this situation condemns certain separations to the realm of impossibility. This is not necessarily so. It must be remembered that, while $\mathbf{R}$ and $K$ are not independent, $\mathbf{R}$ and the ratio $\Delta K/K$ are. For example, in many cases, $\mathbf{R}$ can be substantially reduced if the temperature of the chromatographic column is decreased. In other situations, one can decrease $\mathbf{R}$ by changing the stationary phase (increasing its affinity for the solutes).

## Chromatographic Systems

In this chapter, we have examined theories which account for the spreading of chromatographic zones. These theories are fundamental to any understanding of

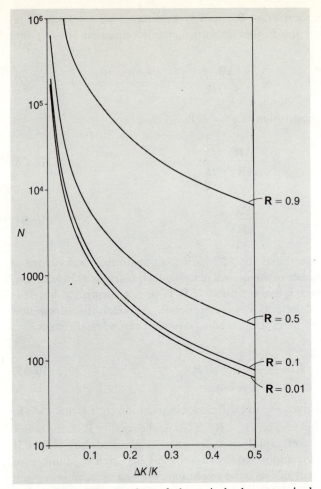

*Figure 16–15.* The number of theoretical plates required for a separation with resolution $\mathcal{R} = 1.0$ plotted as a function of $\Delta K/K$ and **R**. Notice that $N$ increases greatly as **R** increases, and that, accordingly, maximum resolution can only be obtained in conjunction with substantial retention.

chromatography. Because they are applicable to all forms of chromatography, they serve the immensely valuable function of impressing the reader with the unity of this field. In addition, they have great practical value in giving a good explanation of the importance and effect of many different experimental variables. However, it is important not to become preoccupied with theoretical aspects, or to take the view that everything of importance in chromatography goes on in the column. We have already seen that a detector and recording system are vital adjuncts in a chromatographic measurement, and a moment's thought will show that the chromatographic process is only part of an analytical *system* which couples the steps of separation and quantitative measurement. It is these integrated systems which find such extensive practical application in modern chemical analysis. Four specific examples will be discussed in Chapter 17: thin-layer chromatography, an adsorption technique; gas-liquid chromatography, an extensively developed partition technique; ion-exchange chromatography; and molecular exclusion chromatography, a partition technique.

# QUESTIONS AND PROBLEMS

[Questions marked with an asterisk (*) have very short answers and should take little time. They are marked as study guides rather than problems.]

1.* The partition coefficient, $K$, for substance A in a particular chromatographic column is greater than that for substance B. Which compound is more strongly retained in the chromatographic column?

2. The time required for passage of the mobile phase through a particular column is 25 min. What is the value of **R** for some solute which has a retention time of 261 min? How much time does the solute spend in the mobile phase and in the stationary phase?

3. The value of **R** for a particular solute on a certain chromatographic column is 0.1. The volume of mobile phase in the column, $V_M$, is 2.0 ml. What is the value of $t_S$ for the solute when the flow rate of the mobile phase is 10 ml/min?

4. For the system described in problem 3, what is the value of $K$ for the solute if $V_S$ is 0.5 ml?

5. A column 10 cm long is operated with a mobile-phase velocity of 0.01 cm/sec. Component A requires 40 min for elution. What fraction of the time required for its elution does A spend in the mobile phase? What is the value of **R** for this compound?

6. In gas chromatography, the velocity of the mobile phase can be measured directly if one injects some solute like methane which is entirely unretained by the stationary phase. On a capillary column 50 m long, the retention time of methane is 71.5 sec and the retention time of $n$-heptadecane is 12.6 min. (a) What is the velocity of the mobile phase? (b) What is the value of **R** for the $n$-heptadecane zone? (c) What is the velocity of the $n$-heptadecane zone?

7.* Write an expression for the retention time ($t_R$) of some component in terms of the times ($t_M$ and $t_S$) which that component spends in the mobile and stationary phases.

8. Under fixed conditions in a particular gas-liquid partition chromatographic column, substance A is eluted with **R** = 0.5 and $V_R = 100$ ml. The flow rate of the mobile phase must remain constant, but $V_S$ (the amount of liquid stationary phase) can be changed from its existing value of 1.5 ml. By what factor must $V_S$ be changed in order to double $V_R$? Is this factor generally applicable any time $V_R$ is to be doubled, or does it apply only in this particular case?

9.* Recast equation (16–8) in terms of flow rate of the mobile phase ($F$, ml/sec) and the retention time ($t_R$, sec) instead of the retention volume ($V_R$, ml).

10.* Following the example of equation (16–9), write an expression for the adjusted retention time ($t'_R$) in terms of $t_R$ and $t_M$.

11. Two components have adjusted retention times (see problem 10) of 15 and 20 min on a particular chromatographic column. On a different column with a larger stationary-phase volume, the first component has an adjusted retention time of 22 min. What will be the adjusted retention time of the second component on this column?

12. In a particular liquid-liquid chromatographic column, compound A has $K = 10$, and compound B has $K = 15$. The column has $V_S = 0.5$ ml and $V_M = 1.5$ ml, and is operated with a mobile-phase flow-rate of 0.5 ml/min. Calculate $V_R$, $t_R$, and **R** for each component.

13. In the column described in problem 12, the volume of the inert support is 15 ml. Calculate the mobile-phase velocity given a total column length of 30 cm.

14.* In countercurrent distribution, the upper phase (which is analogous to the mobile phase in chromatography) is present in only one tube at the start of a separation. In chromatography, the column is always *filled* with mobile phase, even before a sample is introduced. Does this have any effect on our use of countercurrent distribution as a model for chromatography? Explain your answer.

15.* Make a schematic sketch of a chromatographic column, showing the imaginary divisions between theoretical plates and labeling the incremental phase volumes $\Delta V_M$ and $\Delta V_S$. Use the sketch to show how the mobile phase moves through the column as $n$ is increased.

16. In the derivation of equation (16–22) which relates $N$, $t_R$, and $t_w$, it is assumed that the chromatographic peaks are triangular. Repeat the derivation, assuming instead that the peaks are rectangular with a width $t_w$ centered on retention time $t_R$.

17. The peak corresponding to 16 ng (nanogram, $10^{-9}$ gm) of methane on a particular gas-liquid chromatogram has an area of 100 mv sec. (a) Calculate the detector-sensitivity constant, $k$ (mole mv$^{-1}$ sec$^{-1}$), defined in equation (16–20). (b) When the same detector has an output of 3 mv, what is the methane flux in gm/sec?

18.* A chromatographic column is tested and found to produce a peak having a Gaussian shape and a width of 40 sec at a retention time of 25 min. How many theoretical plates does the column have under the conditions of the test?

19. Some chromatographic columns can be operated at efficiencies corresponding to $10^5$ theoretical plates. Calculate the peak widths obtained from such a column at retention times of 100, 1000, and $10^4$ sec. Assume a Gaussian peak-shape.

20.* If the column described in problem 18 is two meters long, what is the height equivalent to a theoretical plate in this case?

21. Explain in your own words the differences between the plate theory and the rate theory of chromatography.

22. Explain in your own words why the rate theory treats the individual zone-broadening mechanisms in terms of their standard deviations, and tell how those standard deviations combine.

23. Some dye is dumped into a well. One month later the dye first appears in another well 12 meters away. In the absence of any possible effects of ground-water flow, and if the soil is assumed to be uniform in texture, how long will it be before the dye appears in a third well 78 meters distant from the first?

24.* If the length of a column is doubled, by what factor will zone broadening due to longitudinal diffusion increase? Why?

25.* Explain briefly why: (a) $H_{dM}$ depends on $v$; (b) $H_{dS}$ depends on **R** and $v$; (c) mobile-phase velocities must be much lower in liquid-liquid chromatography than in gas-liquid chromatography; (d) $H_{eS}$ depends on $d^2$.

26. Two gas-liquid chromatographic columns differ only in that one has a greater amount of liquid stationary phase than the other. Will their optimum mobile-phase velocities differ? If so, how and why?

27.* Explain why uniformity of column packing and the use of a narrow range of particle sizes are important in obtaining highest column efficiency.

28.* Explain why $H_{dM}$ and $H_{dS}$ can be combined in a single term in equation (16–40).

29. The diffusion coefficient ($D_M$) for $n$-octane in helium at 30°C and 1 atmosphere pressure is 0.248 cm$^2$ sec$^{-1}$. In nitrogen, the same constant has the value of 0.0726 cm$^2$ sec$^{-1}$. Notice which terms in equation (16–40) are affected by this difference, and sketch two lines on a graph of $H$ versus $v$ (see Figure 16–8) showing how this difference will change the shape and location of the curve representing the dependence of column efficiency on mobile-phase velocity. Which gas will allow faster mobile-phase velocities?

30. A particular gas-liquid chromatographic column two meters in length has an efficiency of 2450 theoretical plates at a flow rate of 15 ml/min and an efficiency of 2200 theoretical plates at a flow rate of 40 ml/min. What is the optimum flow rate, and approximately what efficiency should be obtainable at that flow rate?

31. A particular gas-liquid chromatographic column two meters in length is tested at three different flow rates using helium as the mobile phase and found to have the following performance characteristics:

| methane | $n$-octadecane | |
|---|---|---|
| $t_R$ | $t_R$ | $t_w$ |
| 18.2 sec | 2020 sec | 223 sec |
| 8.0 sec | 888 sec | 99 sec |
| 5.0 sec | 558 sec | 68 sec |

(a) Determine the mobile-phase velocity for each of the runs. (b) Determine the number of theoretical plates and the value of $H$ for each of the runs. (c) By solving simultaneous equations, find the values for the constants in an equation of the form $H = A + B/v + Cv$ and graph the result. (d) Through what range of mobile-phase velocities can 90 per cent of the column efficiency be retained? (e) What is the optimum mobile-phase velocity? (f) A particular separation requires 1100 theoretical plates. What is the fastest mobile-phase velocity at which this can be achieved, and how much time will be saved by running at the maximum flow rate instead of at the optimum?

32. Following the description in the text (page 537), sketch the peak shapes (detector response versus elution time) expected from tailing due to active sites (a) when a "low" quantity of sample is used (a peak corresponding to zone profile *a* shown in Figure 16–10 is obtained), (b) when an intermediate quantity of sample is used, and (c) when a large quantity of sample is used.

33. The peak obtained from the injection of 100 ng of a polar compound on a particular column has a height of 10 cm, but shows severe tailing. Injection of 10 ng gives a "bump" about 2 mm in height. Explain this apparent "non-linear" sensitivity. Can detection limits generally be determined by the use of large samples? If not, why?

34. Transmission of sterols (*e.g.*, cholesterol) through a heated glass vacuum line connecting a gas chromatograph and a mass spectrometer is poor. The chromatograph is known to produce symmetrical peaks, but by the time they reach the mass spectrometer they have long tails. To what is this due, and how might the tailing be reduced?

35. A particular chromatograph used in the analysis of air samples produces sulfur dioxide peaks 25 sec wide with a sample input volume of 1 ml and a mobile-phase flow-rate of 40 ml/min. More sensitivity is required and could be obtained if the sample volume is increased; but the peak width must be held below 1 min. What is the maximum possible sample volume?

36. Through an error in construction, a 0.1-ml mixing volume is introduced just before the detector in an ion-exchange chromatograph. The mobile-phase flow-rate is 0.5 ml/min. The observed peak width is 140 sec. What peak width could be obtained after elimination of the mixing volume?

37. A particular chromatographic column has an efficiency corresponding to 4200 theoretical plates and has retention times for octadecane and 2-methylheptadecane of 15.05 and 14.82 min, respectively. To what degree can these compounds be resolved on this column? How many theoretical plates would be required for unit resolution at those retention times?

38. On a liquid-solid chromatographic column one meter in length, operating with an efficiency of $10^4$ theoretical plates, the retention times of α-cholestane and β-cholestane are 4025 and 4100 sec, respectively. If these two compounds are to be separated with unit resolution, how many theoretical plates will be required? How long a column of this same type will be required in order to achieve this resolution if $H = 0.1$ mm?

## SUGGESTIONS FOR ADDITIONAL READING

1. J. C. Giddings: *Dynamics of Chromatography*. Part I (*Principles and Theory*), Dekker, New York, 1965, Chapters 1, 2, and 7.

2. A. T. James and A. J. P. Martin: Gas-liquid partition chromatography: the separation and micro-estimation of volatile fatty acids from formic acid to dodecanoic acid. Biochem. J., *50*:679, 1952.

3. A. J. P. Martin and R. L. M. Synge: A new form of chromatogram employing two liquid phases. Biochem. J., *35*:1358, 1941.

4. H. Purnell: *Gas Chromatography*. John Wiley and Sons, New York, 1962, Chapter 7.

5. J. C. Sternberg: Extracolumn contributions to chromatographic band broadening. *In* J. C. Giddings and R. A. Keller, eds.: *Advances in Chromatography*. Volume 2, Dekker, New York, 1966.

# CHROMATOGRAPHIC SYSTEMS

**17**

In this chapter we will discuss four techniques that find wide application in modern chemical analysis. The first, thin-layer chromatography (TLC), is close to the ultimate in simplicity and convenience. In terms of its separation principle, thin-layer chromatography furnishes an example of adsorption chromatography. In terms of its method of operation, it is the only representative of plane (as opposed to column) chromatography that we shall discuss. The remaining techniques—gas-liquid chromatography (GLC), ion-exchange chromatography, and molecular-exclusion chromatography—offer three different separation mechanisms.

## THIN-LAYER CHROMATOGRAPHY

### Principles

*Technique of operation and separation.* Mikhail Tswett's first chromatograms have an important feature in common with this technique. Tswett eluted his columns only until the colored bands of pigment were visibly separated on the column. After that separation, he performed his "analysis" simply by noting the sizes and positions of the bands in the column. To recover the material in any part of the column, he used a wooden plunger to push the cylinder of adsorbent from the column, and after that he cut the zone of interest away from the bulk of the adsorbent and recovered the compound by means of extraction. Crude as this technique might seem, it is also simple and, in chemical analysis, there is no greater virtue save accuracy itself. The chromatographic medium itself furnishes the visual record of the separation—there are no liquid or gas fractions to collect and analyze, and no detector outputs to plot or record—and the chemist sees at a glance what is happening.

In 1944, A. J. P. Martin and co-workers developed an even simpler technique in which the column was replaced by a strip of paper. The cellulose fibers of the paper (which contain a considerable amount of water) serve as a stationary phase, and an organic solvent mobile phase travels across the paper by capillary flow (as a liquid migrates in a blotter). Colored solutes can be seen directly, and many others can be detected if the paper chromatogram is sprayed with some chemical reagent which will mark the zones by color formation. The astonishing simplicity of paper chromatography led to its wide acceptance and application. It has the great disadvantage, however, of offering no choice of stationary phase. If the compounds to be separated are not retained by hydrous cellulose fibers, some other technique must be used.

Thin-layer chromatography allows a reasonable choice of adsorbents and has a few additional advantages (such as speed and freedom from contamination) over paper chromatography. At the same time, it offers comparable simplicity. In this technique, the adsorbent is spread on a glass plate (or sheet of plastic or aluminum foil) in a layer about 0.25 mm thick. The plate is placed on one edge in a chamber to which mobile phase is added to a depth of a few millimeters. From that edge and submerged portion of the plate, the mobile phase slowly covers the rest of the plate by capillary flow. The sample is spotted on the plate a short distance above the submerged edge and is eluted by the flow of mobile phase, which carries sample components upward through the adsorbent bed. Development of the chromatogram is stopped when the leading edge of the mobile phase (known in thin-layer chromatography as the "solvent front") gets near the top of the plate. The individual chromatographic zones are located and characterized by means of their $R_f$ values:

$$R_f = \frac{\text{distance traveled by solute}}{\text{distance traveled by mobile phase}} \qquad (17\text{--}1)$$

Though closely related to $R$, the retardation factor (page 521), $R_f$ is not exactly equal. A ratio of velocities is expressed by $R$, a ratio of distances by $R_f$. In fact, the solvent front migrates a bit faster than the bulk of the mobile phase, so that $R \sim 1.15\,R_f$.

Because a verbal description of the details of a technique like thin-layer chromatography is inevitably unsatisfactory, we have been as brief as possible. As a supplement, Figures 17–1 to 17–3 review the procedure and should be carefully studied.

## The Adsorbent Bed

*Properties of adsorbents.* We can use this opportunity for a general discussion of adsorbents used in liquid chromatography, not necessarily limiting ourselves to the thin-layer technique. An invaluable reference in this more general context is the book by L. R. Snyder, cited at the end of this chapter.

*Figure 17–1.* Photograph of a thin-layer chromatogram in the process of development. The letters indicate: *A*, mobile phase; *B*, point of solute application; *C*, solute zones; and *D*, solvent front. Development is carried out in an enclosed chamber in order to ensure saturation of the atmosphere with solvent vapor. As a further precaution against unsaturation of the gas phase, the chamber should be lined with mobile-phase soaked filter paper, but this has been omitted here for the sake of visibility.

The most important characteristic of any adsorbent is its **activity**, that is, the extent to which it will retain solute species. The activity is a composite of many factors—strength and density of active sites on the adsorbent surface, surface area, and water content of the adsorbent material all play important roles. In addition, the *apparent* activity of an adsorbent in any given chromatographic application depends strongly on the mobile phase. Interactions between solutes and the mobile phase and between the mobile phase and the adsorbent can be as strong as solute-adsorbent interactions, and the degree of retention of any given solute in an adsorption chromatographic separation can accordingly be grossly affected by the nature of the mobile phase. Generalizations are elusive, and it is best to focus on some individual adsorbents. The great majority of adsorption chromatographic separations are performed with either silica or alumina. A third "adsorbent" discussed below is cellulose, included here because of its utility in thin-layer chromatography. Cellulose is not truly an adsorbent, but rather a partitioning medium.

SILICA. Silica, silica gel, and silicic acid are all terms commonly applied to the material produced by acidification of silicate solutions followed by washing and drying of the resulting gel. The product has an enormous surface area ($\sim$500 m²/gm), and

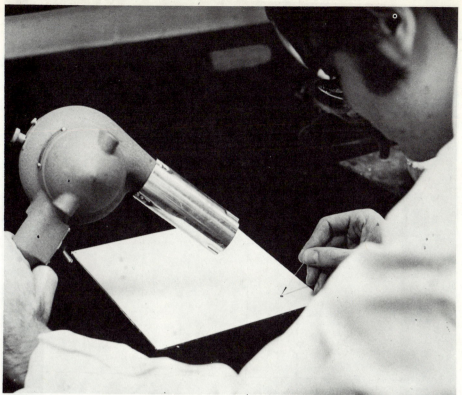

*Figure 17–2.* The process of solute application in thin-layer chromatography. The solutes are dissolved in a suitable solvent which is applied to the layer in tiny drops in order to keep the initial zone as small as possible. In order to get an appreciable quantity of material in one spot, drops are repetitively applied to the plate, the solvent being evaporated with a hot-air blower as shown.

the surface consists of Si—OH groups spaced at intervals of about 5 Å. These groups are the active sites, and the activity of a given batch of silica adsorbent depends on the number and availability of these sites. Unless surface-adsorbed water has been driven off by activation of the silica through heating at 150 to 200°C in air, many of the sites are masked. At higher temperatures, activity is lost because further dehydration occurs due to loss of water from adjacent hydroxyl groups, the product being an inactive Si—O—Si linkage. These steps can be reversed by exposure of the silica to water or water vapor. Above 400°C, surface area is permanently lost by the knitting together of surfaces, as shown in Figure 17–4, and high activity cannot be regained by any degree of rehydration. The active sites interact with polar solutes chiefly by means of hydrogen bonding, with the solute acting as an electron donor. Thus, alkenes are more strongly retained than alkanes and, because of the acidic nature of silica gel, basic substances are particularly strongly held.

ALUMINA. In sharp contrast to silica, alumina shows a steady increase in activity on heating up to 1000°C. This is good evidence that the active sites on alumina, whatever their nature, are not hydroxyl groups. In fact, alumina exhibits retention characteristics which can be interpreted in terms of three types of solute-adsorbent interactions. In the first type, adsorbates with easily polarizable electrons (Lewis bases) interact with the very strong positive fields around the $Al^{3+}$ ions in the alumina surface. Interactions of the second type result in preferential retention of acidic solutes,

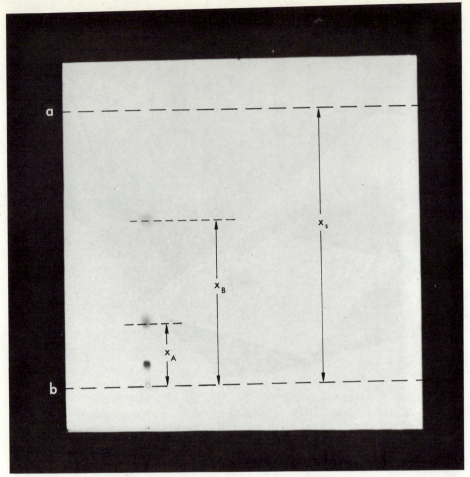

*Figure 17–3.* A developed thin-layer chromatogram. In this case, the solutes happen to be dyes which are visible without further treatment. Other methods for detection of solute zones are reviewed in the text. The indicated distances allow calculation of the $R_f$ values for the solutes. Line *a* marks the starting line, line *b* the solvent front; $R_{fA} = x_A/x_s$ and $R_{fB} = x_B/x_s$.

apparently by proton donation to basic sites ($O^{2-}$ ions) in the alumina surface. Finally, certain aromatic species are apparently retained by charge-transfer interaction, possibly by electron transfer to the positively charged $Al^{3+}$ sites mentioned earlier. It seems generally true that, when two solutes differ chiefly in their electronic structure, alumina is the preferred adsorbent. Thus, it is superior to silica in its resolution of related aromatic and olefinic hydrocarbons, for example.

Highly active alumina can be produced by overnight air drying of the material at 400°C. It is common practice to produce alumina of lower activity by redistribution of a known amount of water on the adsorbent surface. In practice, this is easily done by adding a known amount of liquid to a weighed amount of alumina and shaking the mixture to speed equilibration. Chromatographic alumina commonly has a surface area of 150 m²/gm. A monolayer of water amounts to $3.5 \times 10^{-4}$ ml/m². Thus, the addition of about 2.5 weight per cent of water to alumina will produce an adsorbent with about half a monolayer of water, and this is found to be a good general-purpose adsorbent. There is, in fact, a widely used scale of alumina activities based on this rehydration technique. Known as the **Brockmann scale** after its developer, it defines fully activated alumina as activity grade I; alumina with 3 per cent

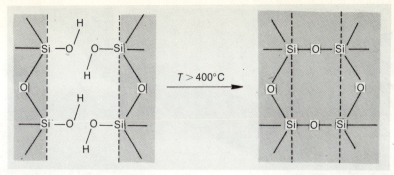

*Figure 17–4.* Mechanism for the dehydration and consequent deactivation of silica at temperatures above 400°C. (Redrawn, with permission, from L. R. Snyder: *Principles of Adsorption Chromatography.* Dekker, New York, 1968, p. 158, by courtesy of Marcel Dekker, Inc.)

added water as activity grade II; with 6 per cent water as grade III; with 10 per cent water as grade IV; and with 15 per cent water as grade V. Notice that grade V corresponds to the presence of about three monolayers of water. The utility of such an adsorbent is somewhat questionable. Certainly the largest effects on adsorbent activity are expected in the range from 0 to 6 per cent water.

Alumina, particularly the fully activated material, is far more likely than silica to catalyze chemical reactions in chromatographed solutes. Since alumina is somewhat alkaline, base-labile molecules are especially likely to be affected; double-bond migrations and even ring-expansions have been reported. This tendency to sabotage is one reason why alumina takes second place to silica in popularity as an adsorbent.

MODIFIED ADSORBENTS. A technique particularly useful in thin-layer chromatography is the addition of silver nitrate (5 to 15 per cent by weight) to silica gel layers. The resulting mixed adsorbent shows strong selectivity, preferentially retaining species which contain double bonds and are thus able to form $\pi$-complexes with the silver ions. The relative retention even depends on the *number* of double bonds in each molecule, and the adsorbent finds wide application in the characterization of lipid fractions.

CELLULOSE. The situation for cellulose is complex. It has been commonly regarded as a partitioning medium, the cellulose structure serving chiefly to hold microscopic pools of water into which solutes can be partitioned. It is inescapable, however, that the cellulose must also function to some degree as an adsorbent, and any rationalizations of observed patterns of retention must take this behavior into account. In general, cellulose is useful for the chromatographic separation of substances that are too polar to escape irreversible adsorption on more active media such as silica and alumina. The use of paper chromatography for the separation of such substances has persisted long after thin-layer chromatography has taken over most other applications. Thin-layer plates can be spread with layers of microcrystalline cellulose, and thus mimic the properties of paper, at the same time offering increased speed of separation.

*Preparation of the adsorbent bed.* Thin layers are prepared by spreading water slurries of the adsorbent on a carefully cleaned supporting material, usually glass. The home-made apparatus used for this operation in the authors' laboratories is shown in Figure 17–5. Being made of plastic, this apparatus is not corroded by silver nitrate. It has the additional advantage of being designed so that the top surface of each plate is pressed upward against the spreader guide, thus guaranteeing

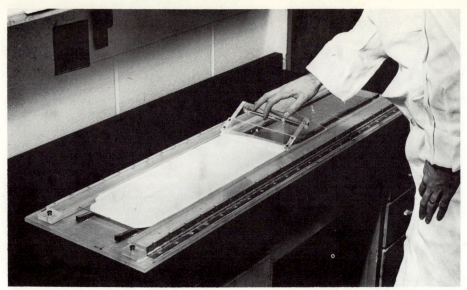

*Figure 17–5.* Preparation of TLC plates.

a constant layer thickness. This constancy is important; if it is not maintained, the rate of solvent migration will vary and cause erratic results. After the spreading operation is complete, the plates are allowed to dry in air for about 30 minutes before being activated. The time required for drying and activation depends somewhat on the thickness of the layer, which is usually 0.25 mm for analytical separations or 1.0 mm for preparative-scale separations.

The particle size of the adsorbent is very small, generally less than 15 $\mu$m. The adsorbent bed is thus about 30 particles thick. Layers of pure adsorbent accordingly have very poor mechanical characteristics and are prone to disaggregation. A binder can be incorporated to cure this problem, and the resulting plates can (in the case of a starch binder) even be written on without flaking. The most common binder is calcium sulfate, which is added to the adsorbent as the hemihydrate, $CaSO_4 \cdot \frac{1}{2}H_2O$ (plaster of Paris), in the amount of 5 to 20 per cent by weight. When mixed with water, the binder is hydrated, and rapidly sets to form the rigid dihydrate, $CaSO_4 \cdot 2H_2O$ (gypsum). Dehydration of the dihydrate begins at 100°C, a phenomenon that quite severely limits the conditions used for activation of the adsorbent. Drying at 110°C for 30 minutes is commonly employed, but above 130°C the binder is dehydrated so fast that plates are useless after only 30 minutes. Common starch (added to the extent of about 5 per cent by weight) is a superior binder in that it provides more durable layers which, in addition, allow much faster migration of the mobile phase. Starch has the great disadvantage of being an organic material which reacts with many of the more drastic zone-detection reagents.

### The Mobile Phase

***The role of the mobile phase in adsorption chromatography.*** It is dangerous to regard the mobile phase merely as the solvent which carries the sample through the chromatographic bed. At the same time the mobile phase carries out that function, it also takes a very active role in the adsorption process by competing with the sample components for space in the adsorbed layer of molecules. If the solvent is strongly adsorbed, it will be displaced by the sample only with difficulty. In this case, the solute will spend a relatively great amount of time in the mobile phase and will travel through the chromatographic bed rapidly. Following this example to its logical conclusion, we see that, for a given solute and adsorbent, the rate of solute migration depends on the strength with which the solvent is adsorbed, and *not on the solubility of the sample in the mobile phase*. This fact will often upset the ordinary view of solvent-solute pairing. For example, alkanes are more soluble in heptane than in methanol ("like dissolves like"), but methanol will move alkanes (or anything else, for that matter) through an adsorption chromatographic column much faster than heptane, simply because the methanol is so strongly adsorbed that the alkanes do not get a chance to "sit down."

Defining *solvent strength* in terms of the solvent-adsorbent interaction makes this parameter independent of the solute. As a result, for any given adsorbent, we can construct a table that lists solvents according to their strength simply by obtaining a measurement of the strength of their adsorption. Such a table is called an **elutropic series**. Snyder has done this quantitatively by defining a **solvent strength parameter**, $\varepsilon^\circ$, which is proportional to the solvent adsorption energy per unit area. Table 17–1 gives values of $\varepsilon^\circ$ for a number of common solvents.

The solvent-strength parameter can be quite accurately adjusted by the use of solvent mixtures. As an example, Figure 17–6 shows the dependence of $\varepsilon^\circ$ on the composition of benzene-methanol mixtures. The theory underlying such a relationship is complex. Although we will not go into this theory here, we do raise the

| Table 17–1. Solvent-Strength Parameters for Alumina* | |
|---|---|
| **Solvent** | $\varepsilon^\circ$ |
| Fluoroalkanes | −0.25 |
| *n*-Pentane | 0.00 |
| Isooctane | 0.01 |
| *n*-Decane | 0.04 |
| l-Pentene | 0.08 |
| Carbon tetrachloride | 0.18 |
| Xylene | 0.26 |
| Isopropyl ether | 0.28 |
| Toluene | 0.29 |
| Benzene | 0.32 |
| Ethyl ether | 0.38 |
| Chloroform | 0.40 |
| Tetrahydrofuran | 0.45 |
| Methyl ethyl ketone | 0.51 |
| Acetone | 0.56 |
| Dioxane | 0.56 |
| Ethyl acetate | 0.58 |
| Aniline | 0.62 |
| Diethylamine | 0.63 |
| Acetonitrile | 0.65 |
| Pyridine | 0.71 |
| Isopropanol, *n*-propanol | 0.82 |
| Ethanol | 0.88 |
| Methanol | 0.95 |
| Ethylene glycol | 1.11 |

*Reprinted from L. R. Snyder: *Principles of Adsorption Chromatography*. Dekker, New York, 1968, p.194. Courtesy of Marcel Dekker, Inc.

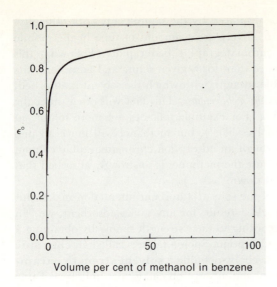

*Figure 17–6.* The solvent-strength parameter ($\epsilon^\circ$) as a function of solution composition for benzenemethanol mixtures. (Plotted from data given in L. R. Snyder: *Principles of Adsorption Chromatography.* Dekker, New York, 1968, p. 378 by courtesy of Marcel Dekker, Inc.)

topic in order to make two important points. First, practically speaking, the solvent is *the* variable in adsorption chromatography. Obviously, the sample cannot be changed, and there are only two adsorbents (silica and alumina) that deserve mention. Accordingly, the choice and regulation of solvent strength is worth a great deal of attention. Second, Figure 17–6 shows that $\varepsilon^\circ$ for benzene is sharply affected by the presence of even very small amounts of methanol, a more polar solvent. The matter of solvent purity, therefore, has tremendous importance in chromatography, and considerable care must be taken.

### Sample Application

Figure 17–2 shows how a micropipet is used to apply a spot of sample solution to a thin-layer chromatography plate. The maximum amount of material which should be applied in a single spot is about 200 $\mu$g for 0.25-mm thick layers. For preparative-scale separations, the sample is spread out in a line of closely spaced dots or applied as a streak across the bottom of the plate by some specialized device. In this way, sample loads of about 1 mg per centimeter of streak length can be handled on plates with adsorbent layers that are 1 mm thick. After elution, the zones are scraped from the plate and the separated sample components are recovered by liquid extraction of the adsorbent.

### Developing the Chromatogram

A simple method for the development of a thin-layer chromatogram is shown in Figure 17–1. In some cases, two-dimensional development offers distinct advantages. In the latter technique, which is summarized in Figure 17–7, a single spot of the sample solution is applied to one corner of a plate. The first development causes compounds to migrate upward and allows at least partial resolution of the mixture. The plate is removed from the developing tank and dried; but the chromatogram, at that stage a row of partially resolved zones up one side of the plate, is not treated for zone detection in any way. Instead, the plate is turned on its side and eluted again,

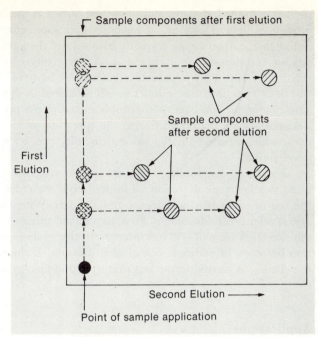

*Figure 17–7.* Schematic view of the process of two-dimensional development; see Figure 17–8 and its accompanying discussion for an example.

the resulting flow of mobile phase being at right angles to the first elution. If the same solvent were used, the result would simply be a row of spots arranged diagonally across the plate; instead, different conditions are chosen for the second elution, so that a greater number of components can be separated.

## Detection of Chromatographic Zones

Many compounds either are colored and can be directly seen on the chromatographic plate or are naturally fluorescent and can be seen as bright zones of fluorescence when an ultraviolet lamp is shined on the plate. However, the great majority of compounds are colorless and require some special detection technique. One universally applicable method is the spraying of the plate with concentrated sulfuric acid, followed by heating to 100°C. Any organic compounds can then be seen as black charred zones. (The technique is not applicable to layers with a starch binder.) Chromic or nitric acid can be added to speed the charring. A second nearly universal technique is exposure of the plate to iodine vapor. This is easily accomplished by placing a few iodine crystals along with the plate in an empty development chamber. The iodine vapor *does not* react with the organic compounds, but is preferentially condensed in the solute zones, apparently being bound to the organic molecules purely by van der Waals interactions. Thus, the zones appear as orange spots against a yellow background. The process is entirely reversible—once the plate is removed from the tank, the spots disappear in a few minutes—so that this technique is particularly useful when solute zones must be recovered for further analysis.

Organic compounds which absorb ultraviolet light at wavelengths above 230 nm can be detected through the use of fluorescent indicators. The entire plate is treated with the indicator, and the whole surface is thus fluorescent when exposed to ultraviolet light. However, in regions where some compound absorbs the incident radiation, the fluorescence is greatly reduced and the solute zones are seen as dark spots against the fluorescent background. A very convenient fluorescent indicator is a mixture of zinc silicate, which is excited by light in the wavelength region from 230 to 290 nm, and zinc sulfide, which is excited in the region from 330 to 390 nm. These inorganic compounds are simply mixed (1 per cent or less by weight) with the stationary-phase material prior to the spreading of the layers. Fluorescent organic dyes, which operate on the same principle, can be sprayed onto the plate after development.

For many compounds, there are specific reagents which can be sprayed on the plate to allow zone detection by formation of some distinctive color; an example is the use of ninhydrin to detect $\alpha$-amino acids. When applicable, these specific reagents offer great advantages. First, unresolved species which do not form colors with the detection reagent will not interfere in the analysis. Second, the process of detection becomes, in addition, one of identification. If the spot does not react properly, the chemist is alerted to the fact that something has gone awry.

## Applications

An interesting example is furnished by the chromatogram shown in Figure 17–8. In this case, the compounds separated have the structures (and, for our later convenience, the relative positions on the plate) given below:

$$C_{18}H_{35}-O-CH_2$$
$$C_{18}H_{35}-O-CH$$
$$C_{18}H_{35}-O-CH_2$$

$$C_{18}H_{35}-O-CH_2$$
$$C_{18}H_{35}-O-CH$$
$$C_{18}H_{37}-O-CH_2$$

$$C_{18}H_{35}-O-CH_2$$
$$C_{18}H_{37}-O-CH$$
$$C_{18}H_{37}-O-CH_2$$

$$C_{18}H_{37}-O-CH_2$$
$$C_{18}H_{37}-O-CH$$
$$C_{18}H_{37}-O-CH_2$$

$$C_{18}H_{35}-O-CH_2$$
$$C_{18}H_{35}-O-CH$$
$$C_{17}H_{33}-\overset{\text{O}}{\underset{\|}{C}}-O-CH_2$$

$$C_{18}H_{35}-O-CH_2$$
$$C_{18}H_{35}-O-CH$$
$$C_{17}H_{35}-\overset{\text{O}}{\underset{\|}{C}}-O-CH_2$$

$$C_{18}H_{37}-O-CH_2$$
$$C_{18}H_{37}-O-CH$$
$$C_{17}H_{33}-\overset{\text{O}}{\underset{\|}{C}}-O-CH_2$$

$$C_{18}H_{37}-O-CH_2$$
$$C_{18}H_{37}-O-CH$$
$$C_{17}H_{35}-\overset{\text{O}}{\underset{\|}{C}}-O-CH_2$$

$$C_{18}H_{35}-O-CH_2$$
$$C_{17}H_{33}-\overset{\text{O}}{\underset{\|}{C}}-O-CH$$
$$C_{17}H_{33}-\overset{\text{O}}{\underset{\|}{C}}-O-CH_2$$

$$C_{18}H_{37}-O-CH_2$$
$$C_{17}H_{33}-\overset{\text{O}}{\underset{\|}{C}}-O-CH$$
$$C_{17}H_{33}-\overset{\text{O}}{\underset{\|}{C}}-O-CH_2$$

$$C_{18}H_{35}-O-CH_2$$
$$C_{17}H_{35}-\overset{\text{O}}{\underset{\|}{C}}-O-CH$$
$$C_{17}H_{35}-\overset{\text{O}}{\underset{\|}{C}}-O-CH_2$$

$$C_{18}H_{37}-O-CH_2$$
$$C_{17}H_{35}-\overset{\text{O}}{\underset{\|}{C}}-O-CH$$
$$C_{17}H_{35}-\overset{\text{O}}{\underset{\|}{C}}-O-CH_2$$

$$C_{17}H_{33}-\overset{\text{O}}{\underset{\|}{C}}-O-CH_2$$
$$C_{17}H_{33}-\overset{\text{O}}{\underset{\|}{C}}-O-CH$$
$$C_{17}H_{33}-\overset{\text{O}}{\underset{\|}{C}}-O-CH_2$$

$$C_{17}H_{33}-\overset{\text{O}}{\underset{\|}{C}}-O-CH_2$$
$$C_{17}H_{33}-\overset{\text{O}}{\underset{\|}{C}}-O-CH$$
$$C_{17}H_{35}-\overset{\text{O}}{\underset{\|}{C}}-O-CH_2$$

$$C_{17}H_{33}-\overset{\text{O}}{\underset{\|}{C}}-O-CH_2$$
$$C_{17}H_{35}-\overset{\text{O}}{\underset{\|}{C}}-O-CH$$
$$C_{17}H_{35}-\overset{\text{O}}{\underset{\|}{C}}-O-CH_2$$

$$C_{17}H_{35}-\overset{\text{O}}{\underset{\|}{C}}-O-CH_2$$
$$C_{17}H_{35}-\overset{\text{O}}{\underset{\|}{C}}-O-CH$$
$$C_{17}H_{35}-\overset{\text{O}}{\underset{\|}{C}}-O-CH_2$$

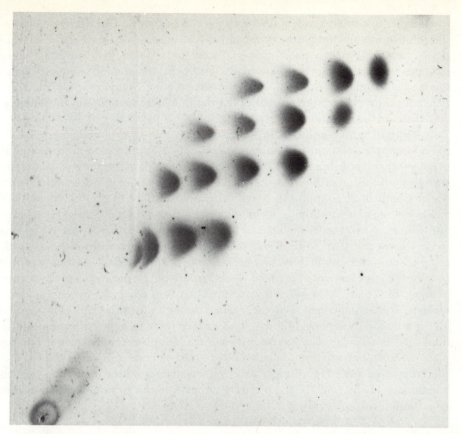

*Figure 17–8.* Two-dimensional thin-layer chromatogram of a lipid mixture. As noted, the right-hand portion of the plate has been impregnated with silver nitrate in order to facilitate separation of double-bonded species. The specific compounds and experimental conditions are discussed in the text. (Reproduced with permission from the paper by H. H. O. Schmid, W. J. Baumann, J. M. Cubero, and H. K. Mangold: Biochem. Biophys. Acta *125*:189, 1966.)

Notice that the compounds differ only in the number of ester linkages (0 to 3 from top to bottom in each vertical column) and in the number of double bonds (3 to 0 from left to right in each horizontal row). A thin-layer chromatographic separation based on the number of ester linkages, that is, on solute polarity, is achieved by chromatography on silica gel with an eluent chosen such that the least polar material has an $R_f \sim 0.8$. For the situation illustrated in Figure 17–8, the mobile phase was a 10 volume per cent solution of ethyl ether in petroleum ether (the latter is not really an ether, and is so named only because of its high volatility—it is a mixture of light hydrocarbons). A separation based on the number of double bonds is obtained by chromatography on silica gel impregnated with about 5 weight per cent silver nitrate, the same solvent as before being employed. In this example, the two separations were combined in an ingenious way. When the thin layers were spread, the spreader (see Figure 17–5) was modified by inclusion of a subdivision so that two silica gel slurries, one containing silver nitrate and the other not, could be simultaneously applied side by side. As shown in Figure 17–8, the first chromatographic separation was carried out by elution of the sample up the plain silica gel side of the plate. The second development, perpendicular to the first, quickly moved the solute zones on to the silver nitrate-containing portion of the plate. There, the separation dependent on the number of double bonds was easily achieved as shown. The relative positions of the solute zones on the plate correspond to the relative positions of the structures shown above. The least polar class ·of compounds (those with no ester linkage) migrates the farthest up the plate in the first step, and those compounds with no double bonds migrate the farthest in the second step. Spraying the plate with 70 per cent sulfuric acid saturated with potassium dichromate followed by charring at 200°C served for detection of the zones.

A second example illustrates the role which thin-layer chromatography often plays in sample isolation. In Chapter 15 (page 508), a procedure for the extraction of a non-saponifiable lipid fraction was outlined. In work in the authors' laboratories, the object of this procedure has been the eventual isolation of a sterol fraction from the non-saponifiable lipids. In a typical experiment, 27.15 gm of wet algal cells yielded 94 mg of neutral lipids, including trace amounts of sterols, organic pigments, and alkanes. Next, the 94-mg neutral lipid fraction was further separated by means of chromatography on a silica gel column. Column chromatography was used in place of thin-layer chromatography because only a crude separation was sought and because the large amount of material would require the preparation and development of many TLC plates. The column was eluted first with *n*-heptane in order to remove alkanes. The second eluent, benzene, carried through the sterols and a few pigments, giving a fraction weighing 14.6 mg. This smaller fraction, now stripped of material markedly more polar than sterols, was chromatographed on a silica gel thin-layer plate, a one-to-one mixture of diethyl ether and *n*-heptane serving as the eluent. In this latter step, any sterols present were well resolved from the interfering pigments, and the sterol band could be scraped from the plate for further examination by means of gas-liquid chromatography.

## GAS-LIQUID CHROMATOGRAPHY

### General Aspects

As its name implies, this form of chromatography makes use of a gaseous mobile phase and a liquid stationary phase. It is frequently lumped together with gas-solid chromatography, and there are many books published with the title, "Gas Chromatography," indicating a survey of both techniques.

The essential parts of a gas chromatograph are shown in Figure 17–9. The

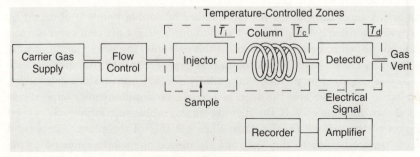

*Figure 17–9.* Block diagram of a gas chromatograph.

mobile phase is known as the "carrier gas," and it is convenient to follow the operation of the instrument by tracing the flow of carrier gas. Inertness is the main requirement for a carrier gas. Although air was used as the carrier gas in some of the very earliest experiments, it has been largely superseded by nitrogen and helium; and a typical gas chromatography laboratory is cluttered with high-pressure gas cylinders. It is necessary that the flow of the mobile phase be accurately controlled. This is accomplished with the aid of special valves which can, depending on their type, be adjusted either to provide a constant flow-rate of gas independent of downstream flow resistance or, alternatively, to provide a constant output carrier-gas pressure. The required pressure is seldom more than a few atmospheres. Flow rates depend critically on column size, but are typically about 20 to 50 ml min$^{-1}$.

On its way to the column, the carrier gas passes through the injector. This small volume is heated to cause rapid vaporization of the sample, which is injected with a syringe through a small silicone rubber disc known as the "septum." The injector is commonly about 50°C hotter than the column. The volume of sample solution injected is usually several microliters or less. The carrier gas sweeps the sample vapor directly onto the column, which is connected to the injector as closely as possible. In the column, the liquid stationary phase is dispersed as a thin layer on an inert supporting material. Columns are ordinarily between 30 cm and 5 m in length, and most have an internal diameter of about 2.5 mm. The inert supporting material has an average particle diameter of about 160 $\mu$m. Because the liquid film is very unevenly distributed on the supporting material, it is meaningless to talk about film thickness. Instead, the extent of liquid phase *loading* is specified by the weight of liquid in comparison to the support; for example, a column with 2 per cent liquid phase has 2 gm of liquid for every 100 gm of support. Liquid-phase loadings are usually 0.5 to 5 per cent in analytical columns. Column temperatures depend entirely on the volatility of the sample and can range from −196°C (liquid-nitrogen temperature) to 350°C. Both extremes are rare. A precisely thermostated oven for control of the column temperature is a vital part of the chromatograph. Unfortunately, only a few liquid phases can be taken above 250°C without themselves vaporizing and swamping the detector.

The limitation on column temperature has the effect of setting a lower limit on sample volatility. For example, the largest straight-chain alkane that can be chromatographed without difficulty is $n$-triacontane ($C_{30}H_{62}$). We can gain some idea of the generosity of this limitation by noting that the boiling point of $n$-triacontane is 450°C at atmospheric pressure and that at 304°C its vapor pressure is only 15 torr. If great attention is given to uniform heating of all parts of the chromatograph and if specially designed columns are used, it is possible to chromatograph normal alkanes as large as $C_{60}H_{122}$. This cannot be taken as an indication of the ultimate limit of gas-liquid chromatography, however, because most compounds do not possess the thermal stability of alkanes and are thermally degraded before their vapor pressures become high enough (greater than about 5 torr) to allow successful chromatography.

The detector is an instrument for the continuous measurement of the solute content of the carrier-gas stream. The detector sensitivity (that is, the minimum detectable sample flow) establishes the sensitivity of gas-liquid chromatography as an analytical technique, and it is remarkably high. Detectors in common use can measure a flow of $10^{-10}$ gm sec$^{-1}$, whereas the most sensitive types can detect a solute flow of $10^{-14}$ gm sec$^{-1}$. From the detector, the carrier gas (and sample, if it has not been consumed by the detector) is simply vented to the atmosphere. The electrical signal from the detector carries the vital information about solute zones. This information is usually displayed on a graphic recorder which automatically presents a graph of solute flow versus time.

## Solute Retention and Column Operation

*Retention volume.* In Chapter 16, we derived an expression relating retention volume $(V_R)$ and the partition coefficient $(K)$,

$$V_R = V_M + KV_S \qquad (17\text{--}2)$$

where $V_M$ and $V_S$ are the volumes of the mobile and stationary phases, respectively. In gas chromatography, a special circumstance prevails—the mobile phase is a compressible fluid with a volume dependent on pressure and temperature. In practice, the carrier-gas flow is measured at room temperature and the prevailing atmospheric pressure and must be corrected to take these factors into account. Correction for the difference in temperature inside and outside the column can be made by means of the equation

$$F_c = F_a \frac{T_c}{T_a} \qquad (17\text{--}3)$$

where $F$ is the flow rate of carrier gas (ml min$^{-1}$) and $T$ is the absolute temperature (°K), the subscripts a and c referring to ambient laboratory conditions and column conditions, respectively. Correction for the difference between atmospheric and column pressures requires first a knowledge of the average pressure within the column. The average column pressure $\bar{p}$ is not simply $(p_i + p_o)/2$, where $p_i$ and $p_o$ are the column inlet and outlet pressures, respectively. The proper expression is†

$$\bar{p} = \frac{2}{3}\left(\frac{p_i^3 - p_o^3}{p_i^2 - p_o^2}\right) \qquad (17\text{--}4)$$

or

$$\frac{p_o}{\bar{p}} = j = \frac{3}{2}\left[\frac{(p_i/p_o)^2 - 1}{(p_i/p_o)^3 - 1}\right] \qquad (17\text{--}5)$$

The **corrected retention volume**, $V_R^\circ$, is the retention volume measured at average column pressure and is given by

$$V_R^\circ = \frac{p_o}{\bar{p}} V_R = jV_R = jF_c t_R \qquad (17\text{--}6)$$

where $t_R$ is the retention time.

We have earlier noted [equation (16–9), page 522] that a direct proportionality exists between $V_R'$, the **adjusted retention volume**, and $K$. For a non-compressible mobile phase

$$V_R' = KV_S = V_R - V_M \quad \text{(non-compressible mobile phase)} \qquad (17\text{--}7)$$

In gas chromatography, the adjusted retention volume also requires correction, and the **net retention volume** $(V_N)$ is given by

$$V_N = jV_R' = KV_S \quad \text{(compressible mobile phase)} \qquad (17\text{--}8)$$

---

† A derivation of this equation is given in most specialized texts on gas chromatography. See, for example, H. Purnell: *Gas Chromatography*. John Wiley and Sons, New York, 1962, p. 67.

In practice, $V_R'$ is easily determined directly, allowing simple calculation of $V_N$. It is only necessary to inject some substance which passes through the column unretained and to watch for the detector signal. For many detectors, a few microliters of air serves this purpose; otherwise, one can use a tiny amount of methane obtained by flushing the injecting syringe in the gas stream from a laboratory gas cock. The time between injection and the appearance of a signal is $t_M$, the time required for passage of the mobile phase through the column. Because the air or methane is unretained and spends no time in the stationary phase, it follows that

$$V_M = t_M F_c \qquad (17\text{--}9)$$

and that

$$V_R' = (t_R - t_M) F_c = t_R' F_c \qquad (17\text{--}10)$$

where $t_R'$ is the **adjusted retention time**.

*Dependence of retention volume on column temperature.* The partition coefficient, $K$, is a thermodynamic quantity which depends on temperature as do all equilibrium constants:

$$\ln K = -\frac{\Delta G^0}{RT} \qquad (17\text{--}11)$$

where $\Delta G^0$ is the standard free-energy change for the gas-liquid partitioning equilibrium and $R$ is the universal gas constant. Rearranging equation (17–8) and substituting equation (17–10), we can write

$$K = \frac{V_N}{V_S} = j \frac{V_R'}{V_S} = j \frac{t_R' F_c}{V_S} \qquad (17\text{--}12)$$

If we consider the behavior of various compounds in a single column, so that $V_S$ is a constant, we can write equation (17–12) in terms of arbitrary constants and natural logarithms and obtain

$$\ln K = \ln V_N + (\text{const})_1 = \ln V_R' + (\text{const})_2 = \ln t_R' + (\text{const})_3 \qquad (17\text{--}13)$$

Combining equations (17–11) and (17–13), we see that each of the retention variables in equation (17–13) depends logarithmically on the reciprocal of the absolute temperature:

$$\ln V_N \propto \ln V_R' \propto \ln t_R' \propto \frac{1}{T} \qquad (17\text{--}14)$$

A more precise statement of this relationship takes the form

$$\log t_R' = \frac{A}{T} + B \qquad (17\text{--}15)$$

where $A$ and $B$ are arbitrary constants.

*Example 17–1.* For a particular column, the adjusted retention time ($t_R'$) of *n*-pentacosane is 20 min at $T_c = 220°C$ and 30 min at 210°C. If $t_R'$ must not exceed 1 hr, what is the lowest possible column temperature?

To evaluate the constants $A$ and $B$, we must solve two simultaneous equations obtained by substituting the known values of $t_R'$ and $T$ into equation (17–15):

$$\log 20 = \frac{A}{493} + B$$

$$\log 30 = \frac{A}{483} + B$$

$$A = 4140 \quad \text{and} \quad B = -7.1$$

Insertion of these values of $A$ and $B$ and of the maximum $t_R'$ of 60 min into equation (17–15) allows calculation of the desired result:

$$\log 60 = \frac{4140}{T} - 7.1$$

$$T = 466°\text{K} = 193°\text{C}$$

***Dependence of retention volume on vapor pressure—homologous series of compounds.*** Each compound in a *homologous series* differs from its predecessor by one —$CH_2$— group. Thus, the *n*-alkanoic acids, for example, form a homologous series, as do the *n*-alkanes. Within any such series, there is a remarkable regularity in vapor pressures such that, at any given $T$,

$$\log p_n° = k_1 + k_2 n \tag{17–16}$$

where $p_n°$ is the saturation vapor pressure of the compound with $n$ carbon atoms and where $k_1$ and $k_2$ are empirical constants. This imposes a regularity on values of $K$ for the members of any homologous series. If we assume ideal solution behavior such that, for any compound, $p_M = X_S p_n°$, where $p_M$ is the solute partial pressure in the mobile phase and $X_S$ is the mole fraction of solute in the stationary phase, then we find that K is inversely proportional to the saturation vapor pressure:

$$K = \frac{c_S}{c_M} \propto \frac{X_S}{p_M} = \frac{1}{p_n°} \tag{17–17}$$

Expressing the relationship shown in equation (17–17) in logarithmic terms and noting that the values of $k_2$ in equation (17–16) are always negative, we obtain

$$\log K \propto -\log p_n° \propto n \tag{17–18}$$

Combining equations (17–13) and (17–18), we can write

$$\log t_R' = An + B \tag{17–19}$$

Figure 17–10 demonstrates this relationship for the homologous series of methyl esters of *n*-alkanoic acids. An interesting application of this expression is given in problem 12 at the end of this chapter.

***Relative retention and retention indices.*** When compounds are to be identified by their retention characteristics, one technique is to report **relative retention**, $r$:

$$r_{a,b} = \frac{t_{Ra}'}{t_{Rb}'} = \frac{V_{Ra}'}{V_{Rb}'} = \frac{V_{Na}}{V_{Nb}} = \frac{K_a}{K_b} \tag{17–20}$$

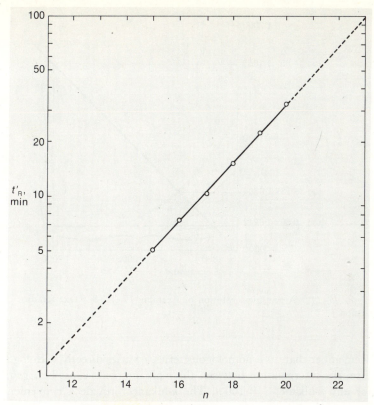

*Figure 17–10.* Relationship between $\log t'_R$ and $n$ for compounds in the homologous series $H_3C(CH_2)_nCO_2CH_3$. Although the absolute values of $t'_R$ are dependent on experimental conditions, the logarithmic interdependence shown here will hold for members of the series on any given column.

Notice that, although $r_{a,b}$ can be very simply determined from adjusted retention times, it is equal to the fundamental quantity $K_a/K_b$, and is thus independent of $V_M$, $V_S$, $p_i$, $p_o$, and $F_c$. It is, however, dependent on the nature of the liquid phase and on column temperature. The compound $b$ is usually some widely available standard. For example, retention characteristics of sterols are reported relative to cholesterol. One worker in laboratory A might read that chemists working in laboratory B had found some sterol of unknown structure having $r_{x,\text{cholesterol}} = 1.25$ for a specified column type and temperature. Using a column with the same liquid phase at the same temperature, chemists in laboratory A could check to see if, by chance, they had the same new compound. In fact, it would not be too unusual for two different compounds to both have the same $r$. Measurements with two or three different liquid phases would be required before any confidence in the identity is justified.

The **retention index** ($I$) furnishes the best method for reporting any compound's gas chromatographic characteristics. It might be noted, for example, that benzene has a retention index of $I = 650$ on a particular column. This means that, under the conditions employed, benzene behaves like an $n$-alkane with 6.50 carbon atoms. This observation would be made by injecting benzene together with $n$-hexane and $n$-heptane and noting that benzene was eluted between the two alkanes. For a chromatogram run at constant column temperature, it would not be exactly halfway.

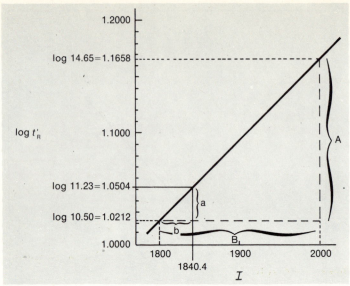

*Figure 17–11.* A graphical solution of Example 17–2. (See text for discussion.)

Remember that, in a homologous series, $n$ varies directly with log $t_R'$ and that, therefore, the interpolation between the $n$-alkane peaks must be *logarithmic*. Notice that, for any $n$-alkane, $I = 100\ n$. The $n$-alkanes provide a convenient measuring scale against which all other compounds can be compared.

*Example 17–2.* An unknown compound has an adjusted retention time $(t_R')$ of 11.23 min. The adjusted retention times of $n$-octadecane $(C_{18}H_{38})$ and $n$-eicosane $(C_{20}H_{42})$, co-injected with the unknown, are 10.50 and 14.65 min, respectively. Calculate the retention index $(I)$ of the unknown.

A graphical solution is shown in Figure 17–11. This graph is invaluable in illustrating these calculations, but ordinarily the arithmetic solution is used. According to the law of similar triangles,

$$\frac{a}{b} = \frac{A}{B}$$

Notice that $I_{\text{unknown}} = 1800 + b$, and that a, A, and B are all known:

$$a = \log 11.23 - \log 10.50 = 0.0292$$

$$A = \log 14.65 - \log 10.50 = 0.1446$$

$$B = 2000 - 1800 = 200$$

$$b = \frac{aB}{A} = \frac{0.0292}{0.1446}\,(200) = 40.4$$

$$I_{\text{unknown}} = 1800 + b = 1840.4$$

The retention-index system has many advantages. It is easy to remember the definition of the index in terms of $n$-alkane retention characteristics. The retention index is almost as simple to calculate as an $r$ value, and it conveys some immediate qualitative idea of retention characteristics. Given the $I$ values of a number of unknown compounds, homologous series are obvious. For example, the higher homolog of the compound in Example 17–2 should appear at about $I = 1940$, the lower homolog at about $I = 1740$. Finally, retention indices are less influenced by column temperature than are relative retentions. Equal column temperatures should still be used for comparisons whenever possible, but valuable information can be gained from measurements taken at temperatures as much as 30°C apart.

*Programmed column temperature.* A look at Figure 17–10 shows that, for the column temperature chosen, the smallest ester having a retention time greater than one minute is methyl undecanoate, and the largest member of the homologous series having a retention time of less than 100 min is methyl docosanoate. Even if chemical patience extended beyond 100 min, the peak width at such long retention times would be so large that sensitivity would be substantially reduced. From a practical standpoint, not more than nine members of the homologous series could be observed under the conditions used and, even then, the total time required for the separation would be an hour. An excellent alternative mode of operation is available. If the column temperature is increased linearly during a chromatographic analysis, the members of any homologous series will be eluted at approximately equally spaced intervals ($n \propto t_R'$ rather than $n \propto \log t_R'$). This technique, called **programmed-temperature gas chromatography**, offers great advantages for the analysis of mixtures which cover a wide range of compound volatilities. The comparison with isothermal column operation is further illustrated by Figure 17–12, which shows the

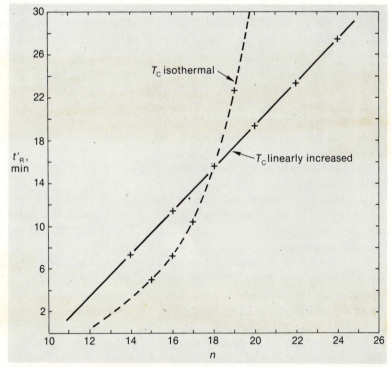

*Figure 17–12.* Adjusted retention time ($t_R'$) as a function of carbon number ($n$) for the homologous series of methyl esters shown in Figure 17–10. In the present situation, the possibility of linear temperature-programming has been introduced, and the graph demonstrates that many more members of the homologous series can be separated in a much shorter time if the column temperature is increased linearly. The dashed line shows the relationship of Figure 17–10 plotted against this linear $t_R'$ scale.

advantage of programmed-temperature analysis of the homologous series of methyl esters discussed in connection with Figure 17–10. Notice that the retention-time scale is linear, not logarithmic.

Because $t'_R$ is not *exactly* proportional to $n$, the rigorous calculation of retention indices in programmed-temperature chromatography is not possible. However, over any small range (3 carbon numbers), simple non-logarithmic interpolation allows approximate calculation of the retention index.

***Carrier-gas velocity and column efficiency.*** The rate theory of chromatography satisfactorily explains the dependence of plate height, $H$, on mobile-phase velocity, $v$, and relationships exactly like that illustrated in Figure 16–8 (page 534) can generally be observed for gas-liquid chromatography columns. At the optimum carrier-gas velocity, $H$ will typically be about 0.4 mm for analytical columns with an internal diameter of 2.5 mm. Determination of the optimum carrier-gas velocity is easy, and the novice chromatographer carries out the measurement with great zeal, ever thereafter secure in the knowledge of obtaining the highest possible performance. If that performance is higher than required, this elegance can lead to a waste of time. In any practical case, the carrier-gas flow should be just as fast as possible. Frequently, the analysis time can be greatly reduced without any serious loss in resolution.

## The Stationary Phase

***Practical requirements.*** The stationary liquid phase must have molecular characteristics such that $K$ is substantially greater than unity. That is, good retention must be obtained by having a liquid which is an effective solvent for the species to be separated. The liquid phase must have a low volatility so that at high column temperatures it does not "bleed," thereby swamping the detector and eventually ruining the column. The liquid must be dispersed on the solid support as a thin film so that diffusion times of solute molecules within the stationary phase are not excessively long. These rudimentary requirements lead to the common use of high molecular weight polymers dispersed on inorganic supporting particles as gas chromatographic stationary phases.

***Liquid phases.*** Selection of a liquid phase is not difficult, but there are a few points which deserve emphasis.

1. Among the many hundreds of liquid phases that have been used, there are not more than 20 which differ significantly in their practical characteristics. Therefore, the chemist need not attempt to choose the proper phase from 600, but only from 20 candidates.

2. Even so, it is important to avoid oversimplifications. Perhaps the worst idea is that the polarity of the liquid phase is independent of solute structure. What we casually term "polarity" has many origins—polarizable electrons, acidic or basic functional groups, dipole moments, and so on. Accordingly, there are many different possible solute-solvent interaction mechanisms which must be considered individually.

3. Choosing a liquid phase remains a largely empirical process. One of the best approaches to selection of a liquid phase is a careful study of the published experiences of other workers who have attempted separation of the same or similar compounds.

When we choose a solvent to dissolve some organic compound, the inevitable guiding rule is "like dissolves like." This can apply at a specific level—ketones dissolve ketones—or in a more general way—polar solvents dissolve polar compounds. Polarity is the most discussed characteristic of the liquid phase in chromatography, but its details seldom receive careful attention. Here, we will take an empirical approach.

Data to be used in the following discussion are shown in Table 17–2, which lists some of the most popular liquid phases. Squalane, a saturated hydrocarbon which will serve as our standard of comparison, appears at the top of the list. The retention indices of five specific compounds are listed in the remaining columns of the table. For liquid phases other than squalane, the index for each compound is calculated by adding the $\Delta I$ value tabulated to the $I$ value given for squalane. For example, the retention index of benzene on Apiezon L is $(653 + 32)$ or 685. Notice that all the $\Delta I$ values shown are positive. That is, all the liquid phases listed are more polar than

**Table 17–2. Retention Characteristics of Liquid Phases Used in Gas-Liquid Chromatography\***

| Liquid Phase\*\* | Retention Index ($I$) | | | | |
|---|---|---|---|---|---|
| | Benzene | Butanol | 2-Pentanone | Nitropropane | Pyridine |
| 1 Squalane | 653 | 590 | 627 | 652 | 699 |
| | $-----$ for the phases below, $I = I_{squalane} + \Delta I -----$ | | | | |
| | $\Delta I$ | $\Delta I$ | $\Delta I$ | $\Delta I$ | $\Delta I$ |
| 2 Apiezon L | 32 | 22 | 15 | 32 | 42 |
| 3 SE-30 | 15 | 53 | 44 | 64 | 41 |
| 4 Dioctyl Sebacate | 72 | 168 | 108 | 180 | 123 |
| 5 OV-17 | 119 | 158 | 162 | 243 | 202 |
| 6 QF-1 | 144 | 233 | 355 | 463 | 305 |
| 7 OV-225 | 228 | 369 | 338 | 492 | 386 |
| 8 Carbowax 20 M | 322 | 536 | 368 | 572 | 510 |
| 9 DEGS | 492 | 733 | 581 | 833 | 791 |

\* From a paper by W. O. McReynolds: J. Chrom. Sci., *8*:685, 1970.
\*\* Structures of liquid phases are shown in Figure 17–13.

squalane in that they retain solutes other than *n*-alkanes more strongly than does squalane. Some quite remarkable effects show up in the table. For example, butanol is eluted approximately with *n*-hexane on a squalane column ($I = 590$, Table 17–2), but interacts so strongly with diethylene glycol succinate polyester (DEGS) that it is eluted after *n*-tridecane ($I = 590 + 733 = 1323$, Table 17–2). The chemical structures of the various liquid phases are shown in Figure 17–13, which gives in addition the maximum column temperature at which each liquid phase can be used without excessive bleeding. The popularity of the silicone polymers as liquid phases is due in large part to their high-temperature stability.

Any given solute can interact with a solvent in a number of different ways, but each of the five compounds listed in Table 17–2 serves as a guide to the importance of a particular mode of polar interaction. Benzene can interact with a liquid phase through an electron-donation mechanism; it is observed that the selective retention of benzene parallels the retention of other potential electron donors such as alkyl halides. Butanol, and alcohols generally, are proton donors; participation in hydrogen-bond formation with the liquid phase is a particularly strong mechanism of interaction for these compounds, which will be preferentially retained by liquid phases with many proton-accepting sites. Selective retention of 2-pentanone is associated with strong dipole-dipole interactions with a given liquid phase, and is an

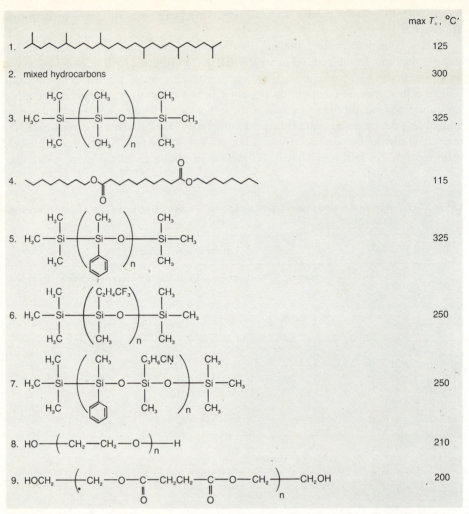

1.                                                  125

2. mixed hydrocarbons                            300

3.                                      325

4.                                    115

5.                                    325

6.                                    250

7.                                    250

8. HO—(CH₂—CH₂—O)ₙ—H             210

9.                                    200

*Figure 17–13.* Chemical structures of the liquid phases listed in Table 17–2. The values of $(T_c)_{max}$ indicate the highest temperature at which a given liquid phase can be used without vaporizing to so great an extent that the solute detector is swamped.

indication that any compounds containing carbonyl groups will be retained selectively. Retention of nitropropane can be taken as evidence of the electron-donating properties of a liquid phase; liquids which selectively retain nitropropane can be expected to selectively retain electron acceptors in general. Pyridine is able to interact with the liquid phase by sharing its non-bonding electrons; in this way, pyridine can participate in the formation of hydrogen bonds.

If the retention indices for all five compounds varied uniformly, it would be possible to arrange the liquid phases of Table 17–2 in order of their polarity. However, Figure 17–14 shows that this does not occur. In fact, the lines in this graph are not only nonparallel, they actually cross. Thus, butanol is the first compound eluted from chromatographic columns containing liquid phases 1, 2, 3, and 5, but is third in columns containing liquid phase 4, which can therefore be said to retain proton donors selectively. A look at the chemical structures of the liquids (Figure 17–13) shows the reason for this. Imagine that two compounds are unresolved on an SE-30

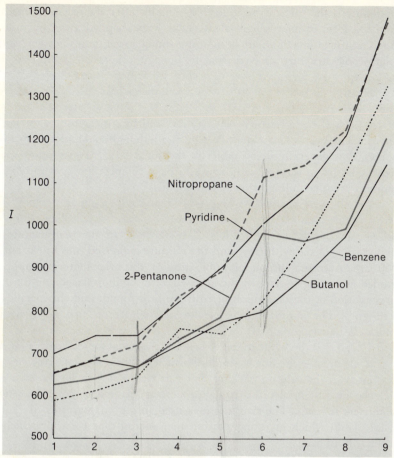

*Figure 17–14.* A graph of retention indices versus the arbitrarily assigned phase numbers shown in Table 17–2. (See text for discussion.)

column (liquid phase 3). If one of the compounds is an alcohol and the other an aromatic hydrocarbon, we can expect them to be resolved on a dioctyl sebacate column. If the unresolved pair consists of a ketone and an aromatic hydrocarbon, neither liquid phase 4 nor 5 offers the right kind of increased polarity (notice that the benzene and pentanone lines in Figure 17–14 are almost parallel for liquid phases 3, 4, and 5). It might, in fact, be better to resort to the Apiezon L column, which should retain the aromatic hydrocarbon more strongly than the ketone. A QF-1 column shows remarkable selectivity for electron acceptors and ketones, but not for hydrogen-bonding or electron-donating molecules. This can be readily explained by the nature of the trifluoromethyl groups, which are unique to the QF-1 column. Thus, the ketone-aromatic hydrocarbon pair could be nicely separated on the QF-1 column, which also offers the nearly unique possibility of separating the ketone-alcohol pair. In practice, it is specific considerations like these which most strongly influence the choice of liquid phase for a given separation—general concepts of liquid "polarity" are practically useless in that they gloss over the details of solute-solvent interactions. Retention indices like those in Table 17–2 are available for virtually all liquid phases.

**Solid supports.** Diatomite, the material which goes under the name diatomaceous earth in North America and kieselguhr in Europe—which serves as a filtration aid, inert filler in dynamite, main constituent of firebricks, and even as the abrasive ingredient of toothpaste—is simply silica in another one of its many natural forms.

Diatomite is the siliceous residue that remains after the organic constituents of diatoms, a form of microscopic algae, have been stripped away by time and decay. It has a surface area of about 20 m² gm⁻¹ (much lower than silica gel) and a beautiful microscopic structure derived from its biologic origin. Solid supports for gas chromatography are prepared from diatomite by heat treatment.

Heating diatomite to 900°C with a sodium carbonate flux produces a very fragile and glassy particulate material in which the mineral structure has been largely destroyed and the adsorptive activity reduced by a decrease in the surface area to about 0.3 m² gm⁻¹. The material is white and, though sold under many different trade names, it is generally identifiable by the suffix "W." The support can be further treated by being washed with acid and base. Silanization (page 537) is usually required to reduce tailing of polar compounds.

An alternate method of heat treatment yields a support which is superior in terms of mechanical strength and, therefore, ease of handling, but which is very inferior in its surface adsorption characteristics. In the manufacture of firebrick, diatomite is mixed with clay, and the mixture is pressed into bricks and fired above 900°C. For use as a chromatographic support, this brick material can be crushed and graded according to size. In the absence of the sodium carbonate flux, much better preservation of the diatomite microstructure is found, and the specific area is about 4 m² gm⁻¹. The material is pink in color, and commercial products generally carry the suffix "P." Deactivation of this support even by careful silanization is essentially impossible, so that the white material is much preferred for use with polar samples.

Zone broadening due to non-uniform flow patterns depends very much on the regularity of the column packing. For this reason, the support particles are usually size-classified by means of a sieving process so that in any batch the range of particle diameters is small. The size fractions used in chromatography are summarized in Table 17–3, which gives in addition the column diameters for which each size is ordinarily utilized.

*Coating of the support.* Preparation of stationary phases for gas-liquid chromatography might serve as an example of a laboratory procedure that is simple in theory, but difficult in practice. The liquid phase is taken up in some solvent, the solid support is added to this solution, and the solvent is evaporated, leaving a coating of liquid phase on the solid support. The problem is to avoid uneven distribution of the liquid and, particularly with white diatomite supports, to do this without extensive fracturing of the support by stirring. Procedures which avoid solvent evaporation are more effective. The interested reader can find detailed accounts in the list of references at the end of this chapter.

After the carrier solvent for the liquid phase has been removed, the liquid-coated support is packed into a column by pouring in the particles while the column is vibrated and a gas stream is flowing to aid transport of the particles. Then the column is

**Table 17–3. Sizes of Solid Supports Used in Gas-Liquid Chromatography**

| Size Fractions | | | Internal Diameter of Column for which Support is Typically Used, mm |
|---|---|---|---|
| U. S. Standard Screens | | Range of Particle Diameters, $\mu$m | |
| Upper mesh size | Lower mesh size | | |
| 45 | 60 | 250–350 | >10 |
| 60 | 80 | 177–250 | 5.5 |
| 80 | 100 | 149–177 | 2.5–5.5 |
| 100 | 120 | 125–149 | 2.5 |

"conditioned" by slow heating to a temperature that is 50°C above its intended maximum temperature of actual use, although the temperature maxima listed in Figure 17–13 should not be exceeded. During this first heating of the column, substantial bleeding of solvent and liquid phase can be expected, so the column must not be attached to the chromatographic detector if gross contamination of the detector is to be avoided.

*Open tubular columns.* In columns of very small diameter, the inert supporting material is omitted, and the stationary liquid phase is simply coated on the inside wall of the column tube itself. Band broadening due to flow effects in the column packing is entirely avoided in this way, and the column has, in addition, a much lower resistance to carrier-gas flow. Together, these features substantially increase the speed and efficiency with which separations can be effected. The openness of the column allows enormous lengths (often up to 100 m) and, consequently, astonishing numbers of theoretical plates—as many as $10^6$. An example of open tubular column performance is provided by the chromatogram in Figure 17–15, which shows a chromatogram of the volatile constituents of human urine. It often happens that high-resolution chromatography reveals the extreme complexity of natural mixtures such as physiological fluids or petroleums. The chromatogram in Figure 17–15 reveals the presence of over 200 different components and demonstrates dramatically the value of high-resolution open tubular columns. The very small inside diameter of such columns, typically 0.25 mm, has led many workers to call them "capillary" columns, although the aspect of column construction which deserves emphasis is not the smallness of the diameter but, rather, the openness of the bore.

## Injectors

The sample introduction system, like the column, must come into contact with the chemical sample without modifying or degrading it. (Requirements for the design and operation of detectors are not as stringent because, even if the sample is chemically changed by the detector, the separation has already taken place, and useful information on the rate of solute flow can still be obtained.) If the sample is injected as a solution and if, inside the injection port, droplets of the solution come in contact with hot metal surfaces, chemical degradation of heat-labile species is virtually assured. In order to avoid this decomposition, many injection ports are glass-lined, the ability of glass to catalyze degradation reactions being substantially less than that of stainless steel. A still-better system of "on-column injection" allows for injection of the sample directly into the column packing. The use of glass tubing for columns is very helpful and is mandatory for the successful chromatography of some particularly sensitive samples. The use of glass columns is, however, not as important as the use of glass injectors; the difference is the opportunity for liquid contact in the injector, whereas, within the column, the sample is always in the vapor state.

The silicone rubber septum through which the syringe needle is pushed must withstand temperatures up to 300°C without degradation and, at the same time, must be soft enough to be penetrated easily at much lower temperatures. These are difficult requirements, and the available compromises are all somewhat unsatisfactory. The major problem is that most septa do degrade at least slightly at temperatures above approximately 225°C. The result is "septum bleed," which is at least as troublesome as bleeding of the liquid phase. It is particularly vexing in programmed-temperature operation, where septum bleed appears not as a steady background, but instead as a sequence of highly confusing peaks which often cannot be distinguished from sample peaks. This happens because materials baked off the septum at 300°C are swept

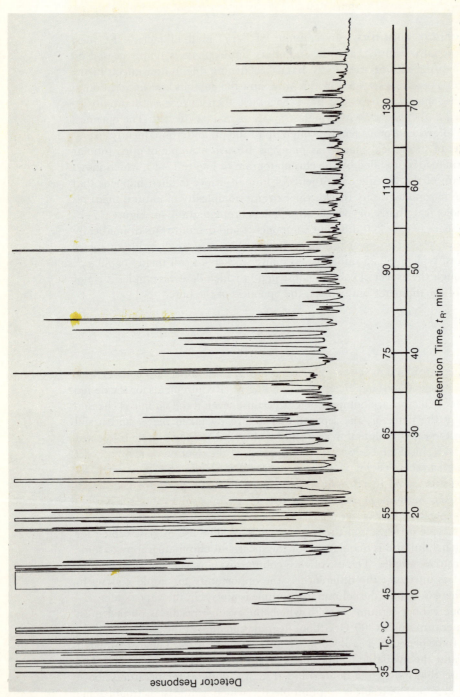

*Figure 17–15.* A high-resolution gas-liquid chromatogram of the volatile components of human urine. Obtained by means of a glass open-tubular column 0.25 mm (inside diameter) × 40 m (length) coated with SF-96, a liquid phase having the same structure as SF-30 (no. 3, Figure 17–13), but a shorter chain length (it is, thus, a liquid at room temperature). Helium carrier gas was passed through the column at 2 ml/min, and the column temperature was programmed from an initial value of 35° to 80°C at 1°C/min, and from 80° to 140°C at 2°C/min. See text for further discussion. (Chromatogram provided by M. V. Novotny and M. L. McConnell, Department of Chemistry, Indiana University.)

by the carrier gas only as far as the first cold region of the column and are "injected" back into the carrier-gas stream through the process of temperature programming.

## Detectors

Detectors are the vital measuring devices in chromatographic analytical systems which combine both separation and measurement. The remarkably high sensitivity which the popular types of detectors possess allows the successful application of gas chromatography to hundreds of interesting chemical problems not otherwise approachable. This high sensitivity also allows the use of samples so small that effective isotherm linearity is virtually guaranteed. On the other hand, detectors generally should be viewed as some of the greatest destroyers of information currently in use. The point is basically philosophical, but of tremendous practical importance. A detector is a "transducer"—the input is a "chemical signal," solute zones in a flowing gas stream; the output is an electrical signal, a single current or voltage proportional to the sample flux. While it is sometimes convenient to think of molecules as billiard balls, the individual identities of the input molecules are of great interest in the present situation. Because it measures only gross sample flux, the detector destroys information about molecular identities. The ultimate detector would give a different kind of output signal for each different form of molecular input.

*Flame-ionization detector.* A cross-sectional view of a flame-ionization detector is shown in Figure 17–16. Several design variants having different construction exist, but the principle of operation is the same in all cases. In the base of the

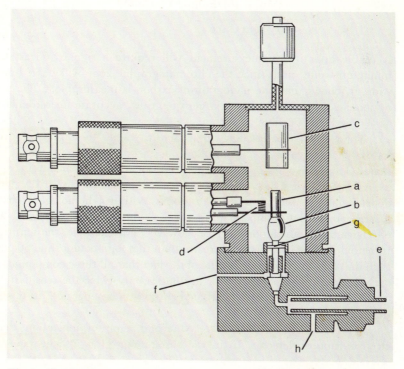

*Figure 17–16.* Cross-sectional view of a flame-ionization detector for use in gas chromatography. *a*, Anode, flame tip; *b*, glass section for electrical insulation of flame tip; *c*, cathode; *d*, ignitor coil; *e*, column; *f*, air supply; *g*, air diffuser; *h*, hydrogen inlet. (Courtesy of Varian Aerograph.)

detector, the carrier gas is mixed with hydrogen (in some instances, hydrogen is used as the carrier gas, and this step is not required). The hydrogen sustains a flame at the flame tip, the required oxidant being supplied by a steady stream of air or oxygen. The usual gas composition flowing into the detector is about 1 part carrier gas, 1 part hydrogen, and 10 parts air. A small heating coil is frequently installed as shown in order to light the flame. The detector operates by measuring the concentration of ions in the flame, the ion concentration being directly proportional to the amount of carbon in the flame.

The ion concentration is determined by measuring the conductivity of the flame between the anode and the cathode. If the electric field between these electrodes is large, 100 V/cm or more, virtually every ion formed in the flame is collected, the ion current is independent of changes in the electric field strength, and the detector is said to be operating in the saturation region. In practice, the voltage applied between the two electrodes is about 300 V. A realistic minimum detectable current with most amplifier systems is about $1 \times 10^{-13}$ amp. Knowing in addition that the efficiency of ion production and collection is about $10^{-6}$ mole of ions collected per mole of carbon introduced to the flame, we can make an interesting calculation concerning detector sensitivity. Let us convert the minimum detectable current of $1 \times 10^{-13}$ amp into the minimum detectable flow-rate of carbon:

minimum detectable flow-rate of carbon

$$= \left(\frac{1 \times 10^{-13} \text{ coulomb}}{\text{sec}}\right)\left(\frac{1 \text{ ion}}{1.6 \times 10^{-19} \text{ coulomb}}\right)\left(\frac{1 \text{ atom introduced}}{10^{-6} \text{ ion collected}}\right)$$

$$\times \left(\frac{1 \text{ mole C}}{6 \times 10^{23} \text{ atoms}}\right)\left(\frac{12 \text{ gm C}}{\text{mole C}}\right)$$

$$\sim 10^{-11} \text{ gm C/sec}$$

A usable quantitative measurement will require a signal at least ten times this minimum value; and, since the chromatographic zone will be at least ten seconds wide, it follows that one nanogram of carbon is measurable.* The range of linear response extends to very much larger quantities, accurate measurements of as much as 100 $\mu$g being possible.

The flame-ionization detector is sensitive only to compounds which produce ions in the flame; in practice, this means carbon compounds with C—C or C—H bonds. Thus, the detector is insensitive to inorganic gases, such as $O_2$ and $N_2$, as well as to CO, $CO_2$, and $CS_2$. The sensitivity of the detector toward a given compound depends on the number of carbon atoms. Within a homologous series like the $n$-alkanes, this relationship is quite precisely applicable. Structural and chemical effects can play an important role, however. For example, the sensitivity for 2,2,4-trimethylpentane is 15 per cent greater than that for $n$-octane, and 30 per cent greater than that for ethylbenzene, in spite of the fact that all three compounds contain eight carbon atoms. The sensitivity for acetone is only twice the sensitivity for methane, and it seems that, in general, carbonyl carbons "do not count" in the determination of sensitivity.

**Electron-capture detector.** A second type of ionization detector is the electron-capture detector. In it, the chromatographic effluent gas is ionized by a

---

* Sensitivities three orders of magnitude better can be achieved by increasing the efficiency of ion production and collection to about $10^{-4}$ and by measuring carbon ion currents as small as 0.1 per cent of the hydrogen flame background. The calculations given above reflect the practical capabilities of commercial instruments in wide use.

stream of particles from some radioactive source, ordinarily either $TiH_2$ containing some $_1^3H$ or nickel foil containing some $_{28}^{63}Ni$ (both isotopes are beta-emitters, though alpha-emitters can be used). The ions are collected and their concentration thereby measured by an electrode and amplifier system like that used in the flame-ionization detector. The principle of operation differs greatly, however, in that solute zones are detected by the *decrease* which they cause in the otherwise steady ion current. This decrease occurs because the extent of ionization depends sharply on the concentration of free electrons in the detector, and some chemical species are extremely efficient at capturing free electrons. The minimum sample flux detectable for substances with high electron affinities, for example, halogen-substituted compounds, is on the order of $10^{-13}$ gm/sec, and the detector is thus considerably more sensitive than the flame-ionization detector *for these species*. Electron-capture detectors are sensitive to compounds containing halogens, phosphorus, lead, or silicon, in addition to others such as polynuclear aromatics, nitro compounds, and some ketones. It happens that pesticides generally contain phosphorus or chlorine, so the detector is thus ideally suited for the measurement of low levels of these compounds. It is also possible to deliberately introduce halogens into species to which the detector would not ordinarily respond. For example, acids can be esterified with fluorinated alcohols; alcohols and amines can be treated with fluorinated acid anhydrides.

***Thermal-conductivity detector.*** Until recently, this form of detector was certainly the most widely used, not only because of its simplicity and universal applicability but also because of its relatively low cost. The cost differential between the flame-ionization and thermal-conductivity detectors is now down to about $300; and, since the former offers much higher sensitivity and far better quantitative accuracy, it is definitely to be preferred for the analysis of organic compounds.

A thermal-conductivity detector operates by measuring solute zone-induced changes in the thermal conductivity of the column effluent. Thus, the sensitivity depends only on the difference in thermal conductivity between the pure carrier gas and the carrier gas mixed with some solute. In view of the low solute concentrations encountered, this difference is inevitably small, yet the appearance of a zone can be sensed quite accurately by measuring the change in resistance of a wire heated by passage of a constant electric current and surrounded by the column effluent. An increase or decrease in the thermal conductivity of the surrounding gas will cause the wire temperature to decrease or increase, with a consequent decrease or increase of the resistance of the wire. The sensitivity is enhanced if one chooses a carrier gas with a thermal conductivity differing from that of the solute to the greatest extent possible, a consideration dictating the use of hydrogen or helium for maximum sensitivity with large organic compounds. Trace constituents of gas samples can be determined by using the major component as carrier gas. For example, the retention characteristics of nitrogen and krypton are very similar, and the determination of trace amounts of krypton in nitrogen (or air) is very difficult because the krypton is "buried" in the tail of the nitrogen peak. If such a sample is analyzed, with nitrogen being used as the carrier gas, the nitrogen in the sample is "invisible" to the thermal-conductivity detector, and the small amount of krypton can be successfully detected.

## Applications of Gas Chromatography

**"Gas chromatographic polarimetry."** During 1967, J. W. Schopf, then a graduate student at Harvard University, working on a thesis in micropaleontology, took the summer "off" in order to investigate the organic constituents of some three-billion-year-old sedimentary rocks which he was studying. He was astonished to find a mixture of amino acids resembling those present in modern organisms. Everyone wondered whether these compounds might be "chemical fossils" demonstrating that at that time, when the earth was only 1.5 billion years old, and when, it was thought by many, its crust had barely started forming, terrestrial life had already begun and evolved to the point where its biochemistry strongly resembled modern forms. An alternative hypothesis was that the amino acids were relatively recent—perhaps only a few hundred million years old—and had been introduced to the ancient rocks by percolating ground waters. Schopf had done his work in K. Kvenvolden's laboratories at the National Aeronautics and Space Administration's Ames Research Center, near San Francisco; within a year, Kvenvolden provided strong evidence that the alternative hypothesis was correct.

Amino acids have one asymmetric carbon atom (bonded to four different chemical groups) and exist in two enantiomeric forms, R and S, as shown below:

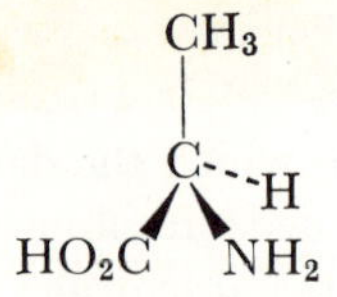

R-alanine 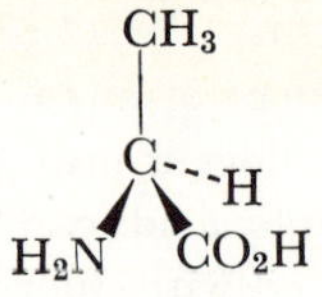          S-alanine

Nearly all naturally occurring amino acids have the S configuration, but the enantiomers can be interconverted by the process of *inversion*, which takes place naturally at a very slow rate when the acids are released by the death and decay of the organism or plant in which they were incorporated. Kvenvolden was aware of this, and had been thinking about the possibilities of measuring the age of recent sediments by determining the extent to which inversion had proceeded. The big problem was that determination of the relative concentrations of the enantiomers required measurement of the extent to which a fairly concentrated solution of a purified amino acid would rotate the plane of polarized light. Unfortunately, the amount of crude amino acid mixture, let alone a purified single compound, which could be conveniently isolated from a sediment was far below the amount required for this *polarimetric* measurement.

By luck, an alternative technique of measurement was being developed by a number of gas chromatographers around the world, one of whom was G. Pollock, working in a laboratory literally just upstairs at the Ames Research Center. Based on gas chromatography, the Pollock method requires only micrograms of sample and can be applied to complex mixtures of amino acids. The first step of the procedure is esterification of the amino acids with pure (only one enantiomer) R-2-butanol. The resulting ester has two asymmetric centers and can exist as a mixture of two diastereomers, RR and RS, where the first letter refers to the configuration of the alcohol and the second to the configuration of the amino acid. In the second step, the esters are treated with trifluoroacetic anhydride in order to reduce the polarity of the amino group by formation of an amide, thus increasing the volatility of the derivatized amino acid to a point allowing successful gas chromatography. Using an open tubular column having the characteristics given in the caption of Figure 17–17, Pollock obtained the chromatogram shown from a mixture of diastereomeric amino acid derivatives. Notice that each amino acid yields a pair of peaks. Experiments proved that each peak was one of the diastereomers, and that the retention characteristics of the diastereomers, which differ only in their configuration at one asymmetric carbon atom, were apparently great enough that the gas chromatographic column could resolve them. Thus, the relative amounts of R and S enantiomers for a number

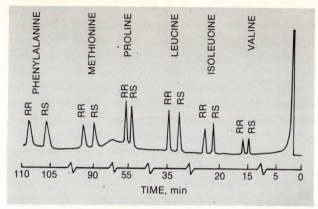

*Figure 17–17.* Chromatogram of a mixture of N-trifluoroacetyl-(R and S amino acid)-R-2-butanol esters. *Column:* 0.5 mm × 46 m open tubular, coated with Carbowax 1540 (same chemical structure as substance 8 in Figure 17–13), $N=50,000$ theoretical plates. *Carrier gas:* helium at 7.94 ml/min. *Column temperature:* isothermal at 100°C for 25 min, then linearly increased to 140°C at 1°/min, then isothermal at 140°C. (Redrawn, with permission, from a paper by G. E. Pollock and V. I. Oyama: J. Gas Chrom., 4:126, 1966.)

of different amino acids could be determined in a single chromatographic run from a comparison of the relative heights of the RR and RS peaks for each amino acid derivative. Only a few nanograms of each acid are required.

Examination of the amino acids from the three-billion-year-old rocks showed that they were almost entirely of the S configuration. It is inconceivable that in three billion years the inversion reaction would not have proceeded to completion, resulting in an equimolar mixture of the two enantiomers. Therefore, the amino acids must be recent contaminants introduced by ground water within the past few hundred thousand years. Unfortunately, they reveal nothing about the biochemistry of three billion years ago. Another striking finding based on this technique has come from Kvenvolden's laboratory recently. Analysis of the configuration of the amino acids resulting from hydrolysis of a meteorite extract shows that they are equimolar in R and S enantiomers; therefore, they are not terrestrial contaminants and are apparently not biologically synthesized, but must instead be derived from organic chemical reactions which occurred elsewhere in the solar system.

***Preparation of isotopically labeled compounds.*** In a reaction catalyzed by strong base, the hydrogen atoms bound to carbon atoms on either side of a carbonyl group will exchange with deuterium atoms, yielding a labeled product:

$$-CH_2-\underset{\underset{O}{\|}}{C}-CH_2- \xrightarrow[D_2O]{OD^-} -CD_2-\underset{\underset{O}{\|}}{C}-CD_2-$$

When a very high percentage of deuterium incorporation is desired, the procedure must be repeated several times in order to reduce the effects of dilution of the deuterium source by protons derived from the compound being labeled. It is a rather frustrating procedure because great precautions must also be taken to avoid contamination of the $D_2O$ with atmospheric $H_2O$. If a gas-liquid chromatographic support is coated with a mixture of a liquid phase and sodium hydroxide, and treated with several large injections (100 $\mu$l) of $D_2O$, all exchangeable hydrogens will be converted to deuterium. Now, if a ketone is injected, the NaOD in the column will catalyze the hydrogen-deuterium exchange reaction of the ketone. In the first increment of column, a compound may become only partially labeled with deuterium; however, as the zone travels through the column, it is continually exposed to fresh 100 per cent NaOD, so that by the time it is eluted the labeling is complete. Only a fraction of the column effluent is sent to the detector. When the detector signals the elution of a desired compound, the labeled product is trapped from the remainder of the column effluent by passage of the carrier gas through a chilled U-tube.

Of course, the preparative separation of mixtures need not be carried out in conjunction with labeling. Very large-scale chromatographs have been built for preparative applications.

***Trace atmospheric constituents.*** There are no stable fluorine compounds released to the atmosphere by natural geochemical equilibria. Halocarbons and sulfur hexafluoride are, however, quite important industrial compounds, and substantial quantities are released into the atmosphere by uncontrolled evaporative losses. For trichlorofluoromethane the estimated amount is about $2 \times 10^5$ tons per year, and for sulfur hexafluoride the amount is about 100 tons per year. J. E. Lovelock, the inventor of the electron-capture detector, has shown that the resultant concentrations in the atmosphere are on the order of $10^{-11}$ part by volume for $CCl_3F$ and $10^{-13}$ part by volume for $SF_6$. These fantastically low concentrations can be measured if one employs an electron-capture detector, and repeated analyses show that the concentration variations reveal patterns of atmospheric circulation. Air which has recently passed over heavily industrialized territory is rich in these compounds, whereas, at the other extreme, air which has just passed over the ocean is particularly clean.

## ION EXCHANGE

Certain minerals and resins in contact with an aqueous solution exhibit the property of exchanging ions with that solution. To the extent that one type of ion in solution is exchanged in preference to another type, the phenomenon of ion exchange provides the basis for a chromatographic method for the separation of ionic species.

Ion exchange was first recognized and studied over a hundred years ago, when it was found that certain clay minerals in soil can remove potassium and ammonium salts from water with the release of an equivalent amount of the corresponding calcium salt. Since that time many naturally occurring ion-exchanger substances have been identified and studied, most of them inorganic substances. Numerous practical applications have been developed; one example is the softening of water by zeolites, which are aluminum silicates with chemical formulas such as $Na_2Al_2Si_4O_{12}$.

In more recent years, primary attention has been shifted from the naturally occurring, inorganic ion exchangers to synthetic organic ion-exchange resins. Compared to the inorganic exchangers, these resins are far superior in regard to mechanical strength, chemical stability, exchange capacity, and rate of ion exchange. Furthermore, ion-exchange resins have been synthesized with a wide variety of desirable exchange characteristics and specificities. Most chemical applications of ion exchangers now involve the use of the synthetic organic resins rather than the naturally occurring types of ion-exchange substances.

Ion exchangers may be classified as being of two principal types, cationic and anionic. **Cation exchangers** possess acidic functional groups, such as the sulfonic acid group ($-SO_3H$), having replaceable hydrogen ions which are exchanged for other cations in the sample solution. A straightforward example of a cation-exchange reaction may be represented by

$$R—SO_3^-H^+ + Na^+ \rightleftharpoons R—SO_3^-Na^+ + H^+$$

in which R signifies all the ion-exchange resin molecule, except for the sulfonic acid group, and $Na^+$ represents the cation which is removed from the solution phase. As the exchange reaction proceeds in the forward direction, the solution acquires hydrogen ions in place of other cations. **Anion exchangers** are resins having basic functional groups, one common example being the $—NH_3^+OH^-$ group. Hydroxide ion, which is on the surface of the resin particle, can enter into an exchange reaction with another anion in the sample solution, as typified by the equilibrium

$$R—NH_3^+OH^- + Cl^- \rightleftharpoons R—NH_3^+Cl^- + OH^-$$

In the latter reaction, the liberation of hydroxide ions causes the pH of the solution phase to increase.

## Preparation and Types of Ion-Exchange Resins

*Cation exchangers.* One of the most widely used ion-exchange resins consists of a cross-linked sulfonated polystyrene polymer. Synthesis of this resin begins with the catalyzed polymerization of a mixture of styrene and divinylbenzene.

styrene          divinylbenzene

This reaction yields spherically shaped beads of cross-linked polystyrene. Next, the polymer beads are sulfonated to produce a substance of the following general structure:

When this resin is brought into contact with an aqueous solution, the sulfonic acid groups ionize to provide the hydrogen ions which enter into exchange reactions with other cations in the solution.

Various cation-exchange resins differ structurally from each other in their hydrocarbon skeletons or acidic functional groups. The same cross-linked polystyrene polymer can be used with a variety of functional groups in addition to the sulfonic acid group itself. Among the other functional groups employed in cation exchangers are the following:

| | |
|---|---|
| Carboxylic acid | $—COOH$ |
| Phosphonic acid | $—PO_3H^-$ or $—PO_3H_2$ |
| Phosphinic acid | $—HPO_2H^-$ or $—HPO_2H_2$ |
| Phenolic | $—OH$ |
| Arsonic acid | $—AsO_3H^-$ or $—AsO_3H_2$ |
| Selenonic acid | $—SeO_3H$ |

The strengths of the acidic groups on the organic matrix differ considerably from each other—a factor greatly influencing the selectivity of the exchange reactions.

**Anion exchangers.** One of the most useful of the anion-exchange resins incorporates the same organic matrix as does the sulfonic acid cation exchanger, but it has as its functional groups the quaternary amine group, $—CH_2N(CH_3)_3{}^+Cl^-$, here being represented with chloride as a counter ion. This functional group is a relatively strong base. Weak base anion-exchange resins frequently contain tertiary or secondary amine functions rather than quaternary ones, as in the functional groups $—CH_2NH(CH_3)_2{}^+$ and $—CH_2NH_2(CH_3)^+$, respectively.

### Characterization of Ion-Exchange Resins

An ion-exchange resin may be characterized by its capacity and by its acid or base strength. The theoretical specific capacity of an ion-exchange resin is the number of acidic or basic functional groups per gram of dry resin. On the other hand, the practical specific capacity is the number of such groups which actually enter into the exchange process per gram of dry resin under some specified conditions.

Practical specific capacity is always lower than theoretical specific capacity, and the extent to which one differs from the other is dependent upon the structure and composition of the resin and also upon the composition of the aqueous solution. Practical specific capacities of commercially available ion-exchange resins range from 1 to 10 milliequivalents per gram.

### Ion-Exchange Equilibria

Let us consider the exchange of two monovalent cations, $A^+$ and $B^+$, between a resin phase (r) and a solution phase (s):

$$A_s{}^+ + B_r{}^+ \rightleftharpoons A_r{}^+ + B_s{}^+$$

We can write the usual concentration equilibrium expression as

$$k_{A/B} = \frac{[A^+]_r[B^+]_s}{[A^+]_s[B^+]_r} \tag{17–21}$$

where $k_{A/B}$ is defined as the **selectivity coefficient**. A more fundamental approach is to express the equilibrium condition in terms of a **corrected selectivity coefficient**, $k_{A/B}^a$, which can be written in terms of concentrations and activity coefficients as

$$k_{A/B}^a = \frac{[A^+]_r[B^+]_s}{[A^+]_s[B^+]_r} \cdot \frac{f_{A_r^+}}{f_{B_r^+}} \cdot \frac{f_{B_s^+}}{f_{A_s^+}} \qquad (17\text{--}22)$$

The selectivity coefficient is constant, therefore, only if the activity-coefficient ratios in the resin and in the solution are constant. The activity-coefficient ratio in the solution phase generally is constant for dilute aqueous solutions. However, the activity-coefficient ratio in the resin phase does vary considerably, depending upon the composition of the resin. Unfortunately, there is no simple method for the measurement or calculation of the activities in the resin phase. Consequently, ion-exchange equilibria are most commonly described in terms of selectivity coefficients, even though their numerical values are not constant over wide ranges of experimental conditions.

Selectivity coefficients for the uptake of cations by a cation-exchange resin are such that the relative affinities of the resin for cations are in the following order:

$$Pu^{4+} \gg$$

$$La^{3+} > Ce^{3+} > Pr^{3+} > Eu^{3+} > Y^{3+} > Sc^{3+} > Al^{3+} \gg$$

$$Ba^{2+} > Pb^{2+} > Sr^{2+} > Ca^{2+} > Ni^{2+} > Cd^{2+} > Cu^{2+} >$$

$$Co^{2+} > Zn^{2+} > Mg^{2+} > UO_2^{2+} \gg$$

$$Tl^+ > Ag^+ > Rb^+ > K^+ > NH_4^+ > Na^+ > H^+ > Li^+$$

An exchange reaction will occur between an aqueous solution and a resin if the cation initially in the solution *precedes*, in this list, the cation initially on the resin.

In general, the resin exhibits a preference for

    (1)  the ion of higher charge,

    (2)  the ion with the smaller solvated equivalent volume, and

    (3)  the ion which has greater polarizability.

These criteria are closely analogous to the selectivity rules for the adsorption of ions from solution onto the surface of a colloidal particle, as described in Chapter 7. The first preference noted above is purely electrostatic in origin. Multiply charged ions have stronger electrostatic fields and are more strongly bound to a site of opposite charge. The second preference is related in an interesting way to resin structure. The backbone of the resin is an aromatic hydrocarbon polymer and is strongly hydrophobic. When placed in water, the resin beads swell because water enters to solvate the ionic functional groups. This swelling creates a substantial pressure in the bead, the position of equilibrium being determined by the point of balance between the forces of solvation on one hand and contraction of the stretched-out hydrophobic polymer on the other.

Slipping into an emotional model again, we might say, "The resin loves to squeeze out a big hydrated ion and substitute a small one." Interestingly, this preference for a small solvated ion diminishes sharply as the degree of cross-linking of the polymer is decreased. The cross-links can be viewed as the springs that create the pressure which "squeezes out" the larger ions.

Selectivity can be built into an ion-exchange process if one of the ions in the aqueous phase is complexed, because the concentration of the free ion in the aqueous phase is greatly reduced.

### Separations by Ion-Exchange Chromatography

The separation of two ionic species present at low concentrations, when another species is present in large excess, is of particular importance in the analytical use of ion-exchange resins. Let us consider a solution containing trace concentrations of cations B and C and a high concentration of cation A, all three ions being capable of undergoing exchange reactions with a resin phase. For simplicity, we shall assume that each of the ions possesses the same charge, $n+$, although they need not be mono-valent cations. In addition, we shall assume that the resin initially has all its exchange sites occupied by A.

When the sample solution, containing A, B, and C, is introduced onto an ion-exchange column, the exchange equilibria may be formulated as

$$B_s^{n+} + A_r^{n+} \rightleftharpoons B_r^{n+} + A_s^{n+}$$

and

$$C_s^{n+} + A_r^{n+} \rightleftharpoons C_r^{n+} + A_s^{n+}$$

where the subscripts r and s denote the resin and solution phases, respectively. For each exchange reaction, we can write a corresponding selectivity coefficient:

$$k_{B/A} = \frac{[B^{n+}]_r}{[B^{n+}]_s} \cdot \frac{[A^{n+}]_s}{[A^{n+}]_r} \qquad \text{and} \qquad k_{C/A} = \frac{[C^{n+}]_r}{[C^{n+}]_s} \cdot \frac{[A^{n+}]_s}{[A^{n+}]_r} \qquad (17\text{--}23,\ 17\text{--}24)$$

Furthermore, we can describe the partitioning of each trace cation, B and C, between the resin phase and the solution phase by means of a **concentration distribution ratio**, $D_c$. Accordingly, the pertinent concentration distribution ratios, $D_{cB}$ and $D_{cC}$, may be defined as

$$D_{cB} = \frac{[B^{n+}]_r}{[B^{n+}]_s} \qquad \text{and} \qquad D_{cC} = \frac{[C^{n+}]_r}{[C^{n+}]_s} \qquad (17\text{--}25,\ 17\text{--}26)$$

The equations for the selectivity coefficient and for the concentration distribution ratio may now be combined to yield

$$D_{cB} = k_{B/A} \frac{[A^{n+}]_r}{[A^{n+}]_s} \qquad \text{and} \qquad D_{cC} = k_{C/A} \frac{[A^{n+}]_r}{[A^{n+}]_s} \qquad (17\text{--}27,\ 17\text{--}28)$$

However, since A is present in large excess and only trace concentrations of B and C exist, the amount of A in the resin phase remains virtually constant, so the two preceding equations reduce to the much simpler expressions

$$D_{cB} = (\text{constant})_B \frac{1}{[A^{n+}]_s} \qquad \text{and} \qquad D_{cC} = (\text{constant})_C \frac{1}{[A^{n+}]_s}$$

$$(17\text{--}29,\ 17\text{--}30)$$

Equations (17–29) and (17–30) reveal that the concentration distribution ratio for each ion, B and C, is independent of the respective solution concentration of that trace cation, if we assume that activity effects may be neglected. Therefore, the uptake of each trace ion by the resin is directly proportional to its solution concentration. However, the concentration distribution ratios are *inversely* proportional to the solution concentration of A, which is to be expected since A competes with B and C for exchange sites in the resin phase. Note that the interaction between ions at trace concentration levels can be ignored; thus, the individual concentration distribution ratio of B or C in a mixture of these two cations can be considered equal to the concentration distribution ratio of B or C in a solution by itself.

In order to accomplish any separation of B from C, it is necessary that one of these cations be taken up by the resin in distinct preference to the other. In other words, the concentration distribution ratios for B and C must differ appreciably from each other. This requirement is strictly analogous to that encountered in separations by other chromatographic techniques. Let us now define the ratio of the $D_c$ values for cations B and C as the **separation factor**, $\alpha_{B/C}$:

$$\alpha_{B/C} = \frac{D_{cB}}{D_{cC}} \qquad (17\text{–}31)$$

Substituting from equations (17–27) and (17–28), we can show that

$$\alpha_{B/C} = \frac{k_{B/A}}{k_{C/A}} \qquad (17\text{–}32)$$

or

$$\alpha_{B/C} = k_{B/C} \qquad (17\text{–}33)$$

Thus, it is possible by means of ion-exchange techniques to separate one trace component B from another trace component C, even in the presence of a large excess of an adsorbable cation A and regardless of what the concentration of A is, as long as the selectivity coefficient, $k_{B/C}$, involving the two trace components is adequately large. It is likewise significant that the selectivity coefficient, $k_{B/C}$, for two trace cations of equal charge is completely independent (in the ideal situation) of the concentration of A. Consequently, if the selectivity coefficient, $k_{B/C}$, is unfavorable for the separation of B from C, any variations of the concentration of A will fail to improve the completeness of separation. Put in practical terms, if one desires to separate B from C by eluting an ion-exchange column with a solution of A, the sharpness of the separation will depend only on $k_{B/C}$ and not on the concentration of A.

The situation is entirely different if B and C have dissimilar charges, b+ and c+, because it can be demonstrated mathematically that the separation factor, $\alpha_{B/C}$, does depend on the concentration of A.

The theoretical aspects of ion-exchange chromatography differ in no way from the general treatment provided in Chapter 16. The concentration distribution ratios derived above can be used to determine the retention volume as previously noted. Because the mobile phase is a liquid, the optimum mobile-phase velocities are very low. Critical diffusion steps take place over distances of about one resin-bead diameter, and the variable $d_p$ in the reduced plate-height expression (page 535) is the diameter of the resin beads. In columns with the highest resolving power, the plate height is about five bead diameters.

## Applications

*Separation of rare-earth metal ions.* Situations frequently arise in which the separation factor is so near unity that apparently little or no separation could be accomplished by ion-exchange techniques. In such instances, it is often possible to achieve the separation through the addition of some reagent which will complex one or both trace ions in the aqueous solution. Since the equilibrium constants for the formation of complexes of similar ions, even those of identical charge, may differ significantly, one can alter the relative concentrations of the two adsorbable cations in the solution and the relative affinities of each ion for the resin phase.

A classic example of the application of this approach is the successful separation of the rare-earth (lanthanide) ions. Prior to the time that these elements were first separated by ion-exchange chromatography, the only feasible method for the isolation and purification of the rare earths involved tens or even hundreds of tedious fractional precipitations. Although this precipitation technique presumably did provide pure rare-earth compounds, careful reexamination of these preparations through the application of modern physical-analytical procedures has often revealed these supposedly pure compounds to be relatively impure. If a solution containing rare-earth ions, such as $La^{3+}$, $Ce^{3+}$, $Eu^{3+}$, $Gd^{3+}$, $Tb^{3+}$, $Er^{3+}$, $Tm^{3+}$, $Yb^{3+}$, or $Lu^{3+}$, is introduced into a cation-exchange column, all these ions will initially be adsorbed on the resin phase. The selectivity coefficients for the rare-earth cations are very similar in magnitude because each of these ions has a $3+$ charge and nearly the same ionic (solvated) radius. For reasons discussed previously, it is impossible to separate these cations by elution of the column with a solution of a simple salt such as sodium chloride or ammonium chloride. On the other hand, if citrate is added to the eluent, a remarkably sharp separation of the rare-earth ions is achieved; citrate forms a complex of sufficiently different stability with each cation so that, one at a time, the rare earths can be eluted from the column and collected separately in receiving flasks. The separation is still so difficult, however, that the complete elution of the rare-earth ions may require a period of over 100 hours.

*Separation of transuranium elements.* Figure 17–18 presents an ion-exchange elution curve for the separation of six actinide elements. In this situation, a radioactive solution containing fermium(III), einsteinium(III), californium(III), berkelium(III), curium(III), and americium(III) was placed on a cation-exchange column. The column was subsequently eluted with an aqueous solution of ammonium lactate. Notice that the progress of the separation was monitored through the measurement of the alpha-particle activity of the eluate. The experimental conditions necessary to achieve this separation were particularly critical—a column only 5 cm high and 4 mm in diameter, a $0.4\,F$ ammonium lactate eluent at pH $4.2 \pm 0.2$ and 87°C, and an eluent flow rate of approximately 18 $\mu$l per minute. However, by stringent control of experimental conditions, it was possible to collect individual drops of the eluate containing just one of these actinide elements.

*Removal of interfering ionic species.* The technique of ion exchange is a very efficient method for the removal of cations which interfere in the determination of an anion. In the gravimetric determination of sulfur as barium sulfate, many cations, such as iron(III), sodium, and ammonium, are extensively coprecipitated and cause large negative errors. However, if one passes the sulfate sample solution through an ion-exchange column containing a sulfonated resin in its hydrogen form, it is possible to replace all cations with equivalent amounts of hydrogen ion and then to form the desired barium sulfate precipitate in the absence of the interfering ions.

In much the same way, undesired anions can be separated from cations. For example, iron(III) can be separated from aluminum if one adds to the sample

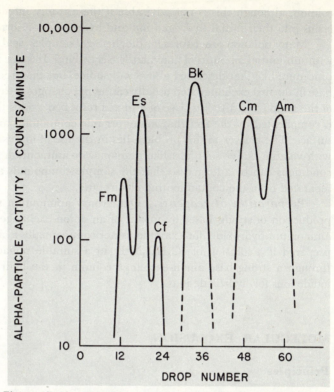

*Figure 17–18.* Ion-exchange chromatographic separation of fermium (III), einsteinium (III), californium (III), berkelium (III), curium (III), and americium (III). (Redrawn, with permission, from the paper by S. G. Thompson, B. G. Harvey, G. R. Choppin, and G. T. Seaborg: J. Amer. Chem. Soc., 76:6229, 1954. Copyright by the American Chemical Society.)

solution a large excess of thiocyanate ion, $SCN^-$, and passes the resulting solution through an anion-exchange column. In the presence of excess thiocyanate ion, iron(III) forms anionic complexes, whereas aluminum remains as a weakly complexed cation.

*Concentration of dilute species for analysis.* Still another use of ion-exchange resins is to increase the concentration of ions initially present at low concentrations. In the determination of calcium and magnesium in river or lake water, a specified volume of the water is passed through a properly prepared column at the site of the river or lake, the column is taken to the laboratory, the desired constituents are eluted with a minimal volume of a properly chosen eluent, and the analyses are then performed.

*Deionization and desalting.* A somewhat different, but very important, application of ion-exchange resins is in the preparation of deionized water. This is generally accomplished as a two-step process in which the water to be purified is passed successively through a cation exchanger in its hydrogen form and then through an anion exchanger in its hydroxide form. Foreign cations are replaced by hydrogen ions in the first step, and extraneous anions by hydroxide ions in the second step. The hydrogen ions and hydroxide ions combine, of course, to form water. This process is used commercially in many places instead of distillation to prepare pure water. It is particularly useful in the laboratory, either as a replacement for distillation or as an additional purification step after distillation to prepare water that is even more highly purified. Both cation and anion exchangers may be mixed within the same column, although this is not advisable if there is any intention of regenerating the resins into their initial hydrogen ion and hydroxide ion forms for repeated use.

Many isolation procedures for biochemical samples yield a product contaminated with substantial amounts of nonvolatile electrolytes. In some instances, a deionization procedure like that described above will suffice, but the production of strong acid or base in the first exchange step usually causes pH changes great enough to modify some of the sample constituents. Use of a mixed resin bed avoids this, but is expensive and is complicated by the fact that many organic compounds are adsorbed on the resin surface. The answer to this problem lies in the use of ion-exchange chromatography with volatile buffer solutions like ammonia or ammonium carbonate. In this way, conditions can be adjusted so that the sample compounds are chromatographically separated from the contaminating electrolytes.

*Preparation of reagents.*   Quaternary ammonium salts can be used in the production of strong bases by means of an anion-exchange procedure. Tetrabutyl-ammonium hydroxide, for example, is not commercially available, but can be easily prepared if a solution of $(C_4H_9)_4N^+I^-$ in a suitable nonaqueous solvent is passed through a strong-base anion-exchange column in the hydroxide form to exchange iodide ions for hydroxide ions.


# MOLECULAR EXCLUSION

## Principles

We have observed that ion-exchange resins prefer to retain the ion with the smallest hydrated size. This preference can be explained in terms of cross-links in the resin that act as springs to squeeze out larger particles. An extreme example of ion-selection based on size occurs in some natural (zeolite) ion-exchangers. These materials have pores of rigidly fixed size which are so small that ions must enter "bare," or stripped of their hydration spheres. Large ions such as $Cs^+$ or polyatomic species are completely excluded from the interior of the exchanger. When this exclusion occurs, the large ions pass unretarded through the ion-exchange chromatographic column and are cleanly separated from the smaller ionic species which have access to the interior of the exchanger.

An analogous separation based on size takes place in molecular-exclusion chromatography. A chromatographic column is packed with beads of material having well controlled porosity. There are three ways to account for the total volume within the chromatographic column: (1) $V_b$, the volume occupied by the inert matrix of the beads themselves; (2) $V_i$, the volume of mobile phase inside the porous beads; and (3) $V_o$, the volume of mobile phase outside the porous beads. If some molecule has a size small enough that all the mobile phase, both inside and outside of the beads, is accessible to it, then the volume of mobile phase required to effect its elution from the column will be $(V_i + V_o)$. On the other hand, molecules which are so large that they are completely excluded from the porous beads can be eluted with a mobile-phase volume of only $V_o$. Because the pore size cannot be perfectly controlled, there is an intermediate range of molecular sizes which has access to only some of the pores. For these species the retention volume will be greater than $V_o$ but less than $(V_o + V_i)$, with the *larger* molecules being eluted first.

Figure 17–19 illustrates the principle of separation of molecules by means of molecular-exclusion chromatography. A representative chromatogram is shown in Figure 17–20, in which the unusual relationship between molecular size and retention

*Figure 17–19.* An intermediate stage in the elution of a molecular-exclusion chromatographic column. The large gray circles represent the beads of porous material. The black circles represent molecules of various sizes. Because they are partially or completely excluded from the mobile phase inside the porous beads, the larger molecules make faster progress through the column.

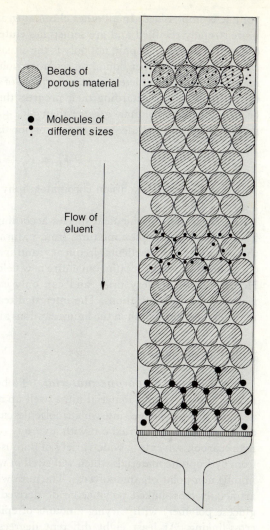

Beads of porous material

Molecules of different sizes

Flow of eluent

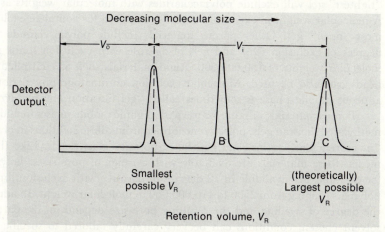

Decreasing molecular size →

$V_o$

$V_i$

Detector output

A

B

C

Smallest possible $V_R$

(theoretically) Largest possible $V_R$

Retention volume, $V_R$

*Figure 17–20.* A representative molecular-exclusion chromatogram in which peak A represents a large molecule completely excluded from the porous beads, peak C represents a small species able to diffuse freely throughout the porous material, and a peak B represents some substance of intermediate size having only partial access to the interior of the porous beads.

volume is emphasized. In all other chromatographic techniques, larger molecules are more strongly retained and are sometimes eluted only with difficulty. In molecular-exclusion chromatography, not only is the situation reversed, but there is—theoretically at least—an upper bound on retention volume. Because the matrix of the porous material is never completely inert, adsorption of solute species can occur and will have the effect of further retarding their progress through the column. When adsorption occurs, retention volumes greater than $(V_o + V_i)$ can sometimes be observed. In general, however, an adequate expression for the retention volume ($V_R$) is

$$V_R = V_o + KV_i \tag{17-34}$$

where, in molecular-exclusion chromatography, $K$ is the fraction of $V_i$ accessible to a given solute.

A separation of chemical species according to size is often very useful. In the fractionation of complex mixtures from natural sources and in polymer chemistry, this task arises with particular frequency, and specialists in these disciplines have joined chromatographers in creating an entire new field of research in the period since 1960. Here we must be very brief, and can only cover the broadest aspects of column technology and applications. The interested reader will find detailed information in the references cited both in the figure captions and at the end of this chapter.

## Practical Aspects

*Nature of the porous material.* To be useful in molecular-exclusion chromatography, a porous material must swell up and imbibe the liquid phase, creating a solvent-filled "sponge" into which molecules can diffuse. Because molecular-exclusion chromatography is carried out with a variety of liquid phases ranging from water to hydrocarbon solvents, a wide variety of porous materials must be available, ranging from hydrophilic materials which will swell in water to lipophilic materials which will imbibe non-polar organic solvents. The most widely used water-loving material is an artificially cross-linked polysaccharide derived by treatment of dextran (a natural glucose polymer) with varying amounts of epichlorohydrin to control the extent of cross-linking. At least eight different degrees of cross-linking are available; the "tightest" gel will exclude polysaccharides with molecular weights as small as 700. A molecular weight in excess of 200,000 is required for complete exclusion from the most "open" gel. Other porous materials include polyacrylamide (ten different degrees of exclusion size) and gels of agarose with exclusion limits ranging up to molecular weights of 150,000,000. Rigid materials such as controlled-porosity glass beads can also be used. Molecular-exclusion chromatography performed with an aqueous mobile phase is sometimes called "gel filtration chromatography."

Porous materials suitable for use with organic mobile phases include exhaustively methylated dextran gels, polystyrene formed from dilute solutions in order to produce "macroreticular" resins, and controlled-porosity glass beads. Like the hydrophilic media, these materials are available with a range of pore sizes. Polystyrene resins, for example, are available in 12 degrees of porosity with exclusion limits for styrene polymers ranging from 2500 to 410,000,000 in molecular weight. In addition, because the degree of swelling and the resulting pore size depend on the particular organic solvent chosen, conditions can be very carefully tailored to specific requirements. Indeed, one can even make the pore size change during the course of the chromatographic elution by employing a solvent gradient—that is, a gradual change in the composition of the mobile phase. Molecular-exclusion chromatography with an organic mobile phase is often referred to as "gel permeation chromatography."

***Columns and detectors; controlling the flow of mobile phase.*** Gel filtration chromatography with dextran gels is commonly carried out with a simple glass column 2.5 cm in diameter and 50 cm in length. In such a column, $V_o$ will be 50 to 100 ml and $(V_o + V_i)$ will be 200 to 250 ml. Sample sizes of a few milligrams are placed on the column simply by the addition of solutions to the top of the column. Detection of the solute zones as they emerge from the column can be achieved by spectrophotometric monitoring of the eluate, by measurement of the refractive index of the eluate, or by the collection of aliquots for later analysis. Mobile phase is allowed to flow by gravity through gel filtration columns at a rate of about 3.5 ml per hour for each square centimeter of column cross section. Thus, for a column with a diameter of 2.5 cm, the flow rate is around 16 ml per hour, and the time required for elution of the smallest molecules will be about 16 hours. Faster flow rates cannot be sustained because the soft gel is deformed by the shear forces of the mobile-phase stream; so the column fails, either by extrusion of the gel or by plugging of the column.

In gel permeation chromatography with organic solvents, the technique of high-speed liquid chromatography (discussed in the conclusion of this chapter) is much more widely employed. This is possible because the polystyrene resins used in work with organic solvents are far more rigid and have much better mechanical properties than the dextran beads used in gel filtration.

## Applications

***Desalting.*** In our discussion of ion exchange, we have already noted that there is frequently very extensive use of buffer solutions during the isolation of bio-chemical samples in order to control the pH in a way which either maintains enzyme activity or allows certain separations to be made. Consequently, the end result of many isolation schemes is an aqueous solution of the desired molecule *plus* electrolytes from various buffers. If a dextran gel is available which will exclude the sample molecules, "desalting" of the sample solution can be accomplished by means of gel filtration. It is only necessary that the salt-sample solution be applied to the column and that elution with pure water be performed. The relatively small ions of the buffers are not excluded from the gel and are eluted far behind the much larger sample molecules.

***Determination of molecular weight.*** Although molecular exclusion takes place, strictly speaking, as a function of molecular *size* rather than molecular weight, the use of standards having a chemical nature similar to an unknown allows molecular weights to be determined by interpolation.

For example, the relationship between $V_R$ or $K$ and the molecular weights of proteins transmitted by a certain dextran gel column is shown in Figure 17–21. The dextran gel had a nominal exclusion limit of 800,000 in molecular weight. The diameter of the gel beads was 50 to 75 $\mu$m, the dimensions of the column were 2.5 cm (diameter) $\times$ 50 cm (length), and the buffer (pH 7.5) used as a mobile phase passed through the column at a rate of 15 to 18 ml per hour. The relationship between protein molecular weight and $V_R$ is nicely demonstrated, and it is clear that the molecular weight of an unknown protein could be estimated from its $V_R$ on this column. The retention characteristics of glycoproteins differ systematically, and it can be seen that careful calibration is required if the technique is to be applied to some different class of compounds. In addition to the protein results, Figure 17–21 shows that the experimenter has determined $V_o$ by measurement of the retention volume of a soluble blue dextran (molecular weight, approximately 2,000,000) and has determined $(V_i + V_o)$

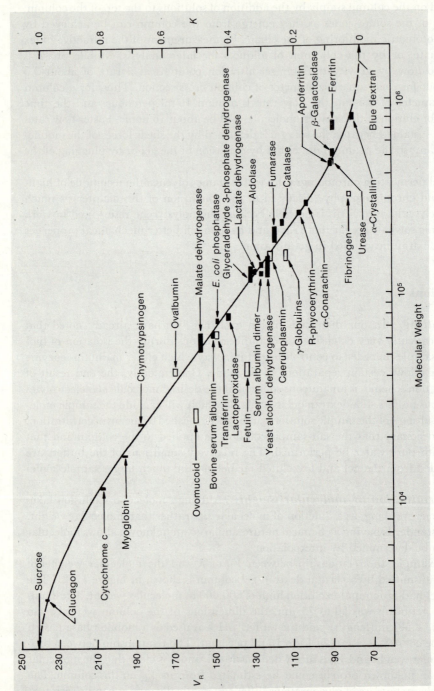

*Figure 17–21.* Plot of retention volume ($V_\mathrm{R}$) and partition coefficient ($K$) against the logarithm of molecular weight for a series of proteins (solid rectangles) and some glycoproteins (open rectangles). The lengths of the bars indicate uncertainties in the known molecular weights, and the widths indicate uncertainties of ±1 ml in $V_\mathrm{R}$. Experimental details in text. (Redrawn, with permission, from the paper by P. Andrews: Biochem. J. 96:595, 1965.)

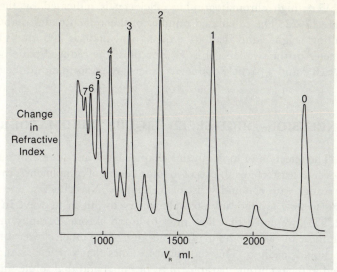

*Figure 17–22.* Gel permeation chromatogram of the epichlorohydrin-2,2-di (4'-hydroxyphenyl) propane copolymer discussed in the text. (Redrawn, with permission, from the paper by W. Heitz, B. Bömer, and H. Ullner: Makromol. Chem. *121*:102, 1969.)

by observing the retention volume of sucrose (molecular weight, 342). For "tighter" gels, $V_o$ can be determined through the use of some smaller molecule. Hemoglobin is sometimes used, and frequent practitioners of this art can be recognized by their scarred thumbs.

The use of gel permeation chromatography for determining the distribution of molecular weights for products of a polymerization reaction is demonstrated by Figure 17–22. The porous material used in this case was a soft polystyrene gel crosslinked with 2 per cent divinylbenzene. The column dimensions were 5 cm (diameter) × 200 cm (length). As can be deduced from the chromatogram, molecules completely excluded from the gel had a retention volume of 830 ml, indicating that $V_o$ was 830 ml. The observed retention volume of benzene was 3050 ml, indicating that $V_i$ was (3050 − 830) or 2220 ml. Tetrahydrofuran was used as an eluent, and a flow rate of 200 ml per hour was obtained when a pressure of about 2.7 atm was applied to the column. A differential refractometer was utilized to measure the difference between the refractive index of the pure eluent and that of the eluate. The sample was a mixture of oligomers having the general formula

$$H_2C\!-\!CHCH_2\!-\!\left[\!O\!-\!\bigcirc\!-\!\underset{CH_3}{\overset{CH_3}{C}}\!-\!\bigcirc\!-\!OCH_2CHCH_2\!-\!\right]_n\!O\!-\!\bigcirc\!-\!\underset{CH_3}{\overset{CH_3}{C}}\!-\!\bigcirc\!-\!OCH_2CH\!-\!CH_2$$

The major series of numbered peaks in Figure 17–22 corresponds to the oligomers having values of $n$ as indicated. Note again that elution occurs in order of *decreasing* molecular size. The minor series of intervening peaks indicate the presence of oligomers in which the chain is terminated by a phenol group. Such a detailed view of the distribution of polymerization products could have been obtained in no other way, so that the use of gel permeation chromatography by polymer chemists is growing rapidly.

Very high chromatographic efficiency is indicated by the chromatogram in Figure 17–22. The calculated number of theoretical plates is about 10,000, and the resulting height equivalent to a theoretical plate is 200 $\mu$m. The average particle diameter in the swollen gel is 74 $\mu$m; thus, the reduced plate height (page 535) is between 2 and 3 particle diameters, a level of performance difficult to exceed.

## CONCLUSION—HIGH-SPEED LIQUID CHROMATOGRAPHY

The selection of topics in this chapter has been quite arbitrary. We have chosen to cover at least a few subjects with some degree of completeness rather than to provide the reader with a laundry list of techniques. Nevertheless, one deliberate omission seems to flow around any obstacle we try to put in its way. In Chapter 16 (pages 532–535), we noted that flow rates in liquid chromatography must be much slower than those in gas chromatography because of the much slower diffusion which occurs in a liquid medium. This statement assumes that particle sizes of materials used in liquid and gas chromatography must be equal and that solutes must therefore diffuse as far in liquid chromatography as they do in gas chromatography. The reader can correctly judge this assumption to be arbitrary—its only support is the technical restraint that very high pressures are required to force liquid to flow through beds of very fine particles.

Very rapid developments have taken place since 1968. Pumps capable of output pressures in excess of 300 atm are frequently used to push liquid mobile phases through columns in which the average particle diameter is only 20 $\mu$m. Special chromatographic media are used in which the particles themselves have solid, inert cores. Consequently, chromatographic partitioning takes place only in the "skin" (perhaps a 1-$\mu$m thickness) of the particle, thus eliminating any possibility of stagnant pools of mobile phase being trapped deep within the pores of chromatographic particles. A universally applicable detector, offering the convenience and reliability of the flame-ionization and thermal-conductivity detectors which have helped to make gas chromatography so widely applied, will possibly be the next development.

An atmosphere of rapid progress and the challenge of many important problems likely to be soluble by advanced techniques pervades the area of chromatography. We hope that the reader has sensed this and is anxious to learn more.

## QUESTIONS AND PROBLEMS

1. Compound A migrates 7.6 cm from its point of application on a thin-layer chromatographic plate, whereas in the same time the solvent front migrates 16.2 cm beyond the point of sample application. (a) Calculate $R_f$ for compound A. (b) On an identical plate, the solvent front has moved 14.3 cm beyond the point of sample application; where should compound A be located on this plate?
2. Silica sample A has an $R_f$ for quinazoline of 0.50 with a one-to-three benzene-methanol solvent. With the same sample and solvent system, silica sample B has an $R_f$ of 0.40. Which silica is more active?
3. If you want to produce maximum activity in a particular silica sample, how do you treat it?
4. Diatomite is a form of $SiO_2$ with a surface area per unit mass of less than 50 m²/gm. Should it be more or less active than silica gel?

5. If you want to produce maximum activity in an $Al_2O_3$ adsorbent, how do you treat it?

6. Consider an alumina sample with a specific surface area of 123 m²/gm. Calculate the "number of monolayers" (realizing the fanciful aspects of such a number) corresponding to each step of the Brockmann activity scale.

7. Given the requirement that a separation is to be performed by means of thin-layer chromatography, choose the adsorbent best suited for each of the following:

   (a) the separation of a mixture of molecules according to their functional groups
   (b) the separation of methyl octadecanoate from methyl octadecenoate
   (c) the fractionation of a mixture of aromatic hydrocarbons
   (d) the separation of

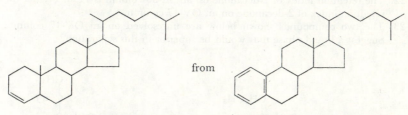

from

   (e) the separation of lysine from "neutral" amino acids

8. What solvent will have approximately the same strength ($\varepsilon°$) as 1 per cent methanol in benzene?

9. A particular compound is eluted from an alumina column too rapidly when acetone is used as the mobile phase. Will chloroform make the compound migrate faster or slower?

10. A friend complains that he is always losing the alkane fraction from his samples. He is sure that the alkanes are present when he starts his procedure; however, by the time he does thin-layer chromatography on zinc sulfide-impregnated plates, the spot for the alkanes cannot be found. He is making a big mistake—what is it?

11. A particular gas chromatographic column is operated with an inlet pressure of 45 psig (psig = gauge pressure) and an outlet pressure of 1 atm. What is the average pressure inside the column?

12. Given the following retention times for alkanes on an SE-30 column at 150°C, calculate the retention time of heptadecane under the same conditions: $CH_4$, $t_R = 25$ sec; $C_{13}H_{28}$, $t_R = 1.20$ min; $C_{15}H_{32}$, $t_R = 6.35$ min.

13. A certain gas chromatographic column is operated isothermally at 165°C with a flow rate of helium carrier gas of 24 ml/min measured at 23°C and 740 torr pressure. The column inlet pressure is 25 psig. Calculate the corrected retention volume for some compound which has a retention time of 18.72 min.

14. A given gas-liquid chromatographic column is loaded with 1.272 gm of a packing which is described as "3 per cent SE-30 on Chromosorb W." The density of SE-30 is about 0.92 gm/ml. Calculate $K$ for a compound which has a net retention volume of 273 ml on this column.

15. A particular gas chromatographic peak has adjusted retention times of 24.21 and 17.24 min at 180 and 200°C, respectively. The column in use has an efficiency corresponding to 5000 theoretical plates. What is the highest column temperature which can be used such that the peak width will still be greater than 25 sec? Assume that $t_R$ does not differ greatly from $t_R'$.

16. The methyl ester of the 22-carbon $n$-alkanoic acid has adjusted retention times of 72.0 and 58.3 min on a particular column at temperatures of 180 and 200°C, respectively. Will it be possible to elute this compound in less than 45 min, if the liquid phase in the column has a maximum operating temperature of 210°C?

17. Compound A is co-injected into a column along with $n$-tetracosane and $n$-hexacosane. The observed retention times are A, 10.20 min; $n$-$C_{24}H_{50}$, 9.81 min; and $n$-$C_{26}H_{54}$, 11.56 min. Calculate the retention index of A.

18. Under the conditions in problem 17 above, a peak appears at an observed

retention time of 12.02 min. Might it be a homolog of A? (What is its retention index?)

19. In the identification of cockroach pheromones, a peak was observed to be eluted at a column temperature of 155.2°C during a chromatographic run with temperature programming from 100 to 200°C at 2°C per minute. The elution temperatures of *n*-tetradecane and *n*-hexadecane under the same conditions were 141 and 162°C, respectively. Estimate the retention index of the peak.

20. Between which two *n*-alkanes is benzene eluted from a column containing squalane as the liquid phase? Between which two *n*-alkanes is benzene eluted from a column containing OV-17 as the liquid phase? Suggest a reason for the difference.

21. As a matter of interest, compare the retention index of nitropropane on squalane to its retention index on DEGS.

22. The retention index of 2-decanone on an SE-30 column is 1150. Predict the retention index of 2-decanone on an OV-17 column.

23. The two compounds shown below are unresolved on an OV-17 column. Suggest a liquid phase that would be superior in this separation.

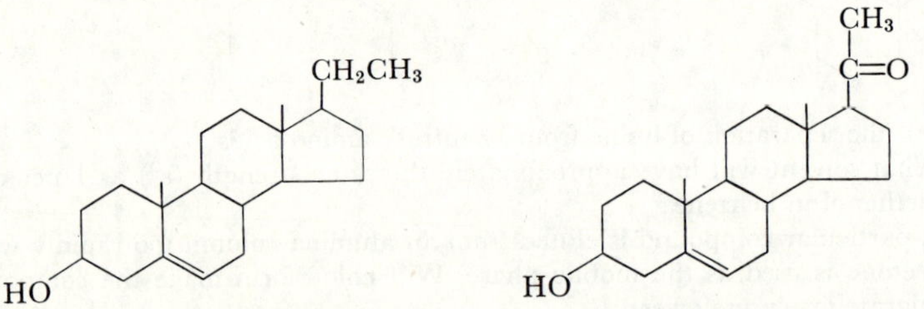

CH₂CH₃

CH₃
C=O

HO

HO

24. A colleague is running cholesterol quite satisfactorily on chromatograph X. One day, he moves his column over to chromatograph Y and finds that the cholesterol peak appears much earlier than before. After days of checking flow rates, column temperatures, and so on, he finally traps the peak as it is eluted and discovers that the "cholesterol" has the second structure shown below:

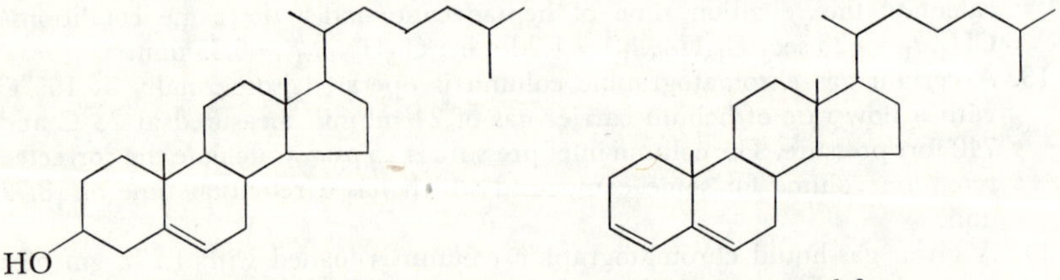

HO

cholesterol

compound from
chromatograph Y

To which part of the chromatograph should he turn his attention? What might he do to fix things up?

25. Predict the relative molar responses (sensitivities) in a flame-ionization detector for the following compounds: decane, decanone, diethyladipate, and 2,2,5-trimethylheptane.

26. It is planned to do a gas chromatographic analysis of an aqueous butanoic acid solution of a concentration which turns out to be just below the limit of detection for a flame-ionization detector. This problem can be solved if one uses a different detector and a special derivative. How?

27. A friend is analyzing trace gases using a gas chromatograph with a thermal-conductivity detector and helium carrier gas. A series of carbon dioxide standards is prepared by mixing small amounts of carbon dioxide with carefully purified helium. When these standards are injected into the column in

order to calibrate the detector, the detector-output peak profiles shown below are obtained. Explain what might be wrong.

sample input:      10 ng       6 ng       2 ng       1 ng      pure He

peak profile:
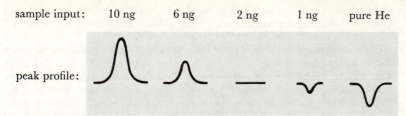

28. In our discussion of ion-exchange chromatography, we restricted ourselves to the separation of trace cations, B and C, having the same charges, and we discovered that the concentration of eluent A has no effect on the separation of B and C. In the event, however, that all three ions—A, B, and C—have different charges, $a+$, $b+$, and $c+$, respectively, the ion-exchange equilibria may be written as follows:

$$aB_s^{b+} + bA_r^{a+} \rightleftharpoons aB_r^{b+} + bA_s^{a+}$$
$$aC_s^{c+} + cA_r^{a+} \rightleftharpoons aC_r^{c+} + cA_s^{a+}$$

(a) Write expressions for the selectivity coefficients, $k_{B/A}$ and $k_{C/A}$, corresponding to each of these equilibria.

(b) Express the concentration distribution ratios, $D_{cB}$ and $D_{cC}$, for each trace cation in terms of the appropriate selectivity coefficients.

(c) Write simplified expressions for each concentration distribution ratio, assuming that the quantity of A in the resin phase remains essentially constant.

(d) Using the concentration distribution-ratio expressions derived in (c), plot the logarithm of $D_{cB}$ versus the logarithm of the concentration of A, assuming that B has a charge of $2+$ and that A has a charge of $1+$. Construct a similar plot of $D_{cC}$ versus log [A], based on the assumption that C has a charge of $3+$ and A a charge of $1+$.

(e) Superimpose the two plots obtained in (d) such that the ion-exchange adsorption isotherms for species B and C intersect. What information can be obtained from this graph in terms of the separation of B and C by means of ion-exchange chromatography? In particular, what happens to the separation factor, $\alpha_{B/C}$, as the concentration of A is varied?

29. (a) What mobile-phase pH and what type of ion exchanger would you choose in order to attempt the ion-exchange chromatographic fractionation of a mixture of carboxylic acids?

(b) Which acid would be eluted first, one with a $pK_a$ of 4 or one with a $pK_a$ of 5? Explain your answer.

(c) Which acid would be eluted first, propanoic or octadecanoic? Explain your answer. (These acids have approximately equal $pK_a$ values.)

30. Should the ion-exchange chromatographic separation of a mixture of amines be possible? If so, under what conditions? Predict the order of elution of the following compounds: ammonia, *n*-propylamine, methylethylamine, trimethylamine, tetramethylammonium chloride. Refer to Appendix 2.

31. Suggest two different approaches to the ion-exchange chromatographic separation of amino acids. Given the goal of obtaining maximum separation of amino acids which differ only in their aliphatic side chains, speculate on which approach would be preferred and tell why.

32. How much trouble would it cause the water-softening industry if people *liked* $Ca^{2+}$ and $Mg^{2+}$ in their water and wanted $Na^+$ and $K^+$ removed? Explain your answer.

33. Generally speaking, should the affinity of an ion-exchange resin for various species increase or decrease as one moves vertically downward in the periodic table of elements? Explain your answer.

34. A small amount of radioactive silver-108 ion is adsorbed in an ion-exchange column. It is necessary that the silver ion be eluted very rapidly in an absolute

minimal volume of eluent. The form of the resin after the elution is of no importance. Suggest a solution to this problem.

35. A cation-exchange column is saturated with very strongly held $Fe^{3+}$. It is desired to recover the $Fe^{3+}$ and to convert the resin to the $H^+$ form. A friend suggests washing the column with $10\,F$ sulfuric acid, but you choose $6\,F$ hydrochloric acid instead. Why?

36. The normal retention time of trace amounts of $Na^+$ on a certain ion-exchange chromatographic column is 2.4 hr. A molecule of mobile phase (which is unretained by the resin) passes through the column in 23 min. It has been determined experimentally that a mobile-phase $Na^+$ concentration of 0.080 $M$ is sufficient to keep all the exchange sites in the $Na^+$ form. While the column is being washed with a 0.080 $M$ $Na^+$ solution, a trace of radioactive sodium ion is added to the eluent at the column inlet. When does the radioactive sodium ion first appear in the eluate? Explain your answer.

37. A molecular-exclusion chromatographic column is made from an old 50-ml buret. Having no other information, what statement can you immediately make about elution volumes for this column?

38. Imagine that you can make a porous material with any range of pore sizes that you wish. In order to obtain maximum resolution of a particular pair of compounds, do you create a material with a wide range of pore sizes or a very narrow range of pore sizes? Why?

39. Gel filtration is sometimes used in large-scale pharmaceutical preparation work. When large-scale procedures are frequently repeated, it often pays to adopt some highly specialized conditions. Can you conceive of any situation, for example, where it might be advantageous to construct a column in which materials with large and small pores are mixed? Discuss the characteristics of such a column with respect to solutes of various sizes.

40. Explain in your own words why particles with solid, inert cores having thin shells of chromatographic media around them offer advantages to the chromatographer? What possible disadvantages can you imagine?

## SUGGESTIONS FOR ADDITIONAL READING

1. K. H. Altgelt and L. Segal, eds.: *Gel Permeation Chromatography*. Dekker, New York, 1971.
2. D. D. Bly: Gel permeation chromatography. Science *168*: 527, 1970.).
3. P. R. Brown: *High Pressure Liquid Chromatography*. Academic Press, New York, 1973.
4. The use of GLC in the preparation of labeled compounds has been reviewed by H. Elias: Labeling by exchange on chromatographic columns. *In* J. C. Giddings and R. A. Keller, eds.: *Advances in Chromatography*. Volume 7, Dekker, New York, 1968, pp. 243–292.
5. E. Heftmann, ed.: *Chromatography*. Second edition, Reinhold Publishing Corporation, New York, 1967.
6. F. Helfferich: *Ion Exchange*. McGraw-Hill Book Company, New York, 1962.
7. C. Horvath: Columns in gas chromatography. *In* L. S. Ettre and A. Zlatkis, eds.: *The Practice of Gas Chromatography*. Wiley-Interscience, New York, 1967.
8. J. C. Kirchner: *Thin-Layer Chromatography*. Volume XII of *Technique of Organic Chemistry*, E. S. Perry and A. Weissberger, eds. Wiley-Interscience, New York, 1967.
9. J. J. Kirkland, ed.: *Modern Practice of Liquid Chromatography*. Wiley-Interscience, New York, 1971.
10. I. M. Kolthoff and P. J. Elving, eds.: *Treatise on Analytical Chemistry*. Part I, Volumes 2 and 3, Wiley-Interscience, New York, 1959.
11. Further information about determination of amino acid configurations can be obtained from the series of papers by K. Kvenvolden, *et al.* Recent references are: Proc. Nat. Acad. Sci., *68*:486, 1971; Science *169*:1079, 1970.

12. A. B. Littlewood: *Gas Chromatography*. Second edition, Academic Press, New York, 1970.
13. H. Purnell: *Gas Chromatography*. John Wiley and Sons, New York, 1962, Chapters 3 and 10.
14. K. Randerrath: *Thin-Layer Chromatography*. Academic Press, New York, 1966.
15. H. Setemann: Principles of gel chromatography. *In* J. C. Giddings and R. A. Keller, eds.: *Advances in Chromatography*. Volume 8, Dekker, New York, 1969, pp. 3–45.
16. L. R. Snyder: *Principles of Adsorption Chromatography*. Dekker, New York, 1968.
17. E. Stahl, ed.: *Thin-Layer Chromatography, A Laboratory Handbook*. Second edition, Springer, New York, 1969.

# INTRODUCTION TO SPECTRO-CHEMICAL METHODS OF ANALYSIS

# 18

Some of the earliest means used for the characterization of objects and substances were based upon the observation of color. Even today, the description of any object, from automobiles to computers, usually includes a statement regarding color. Color arises from the absorption and emission of light by matter, each form of matter displaying its own absorption and emission properties and, therefore, its own color. From this, it is a simple extension to realize that chemical species can be characterized by means of their absorption or emission behavior. In fact, this behavior is so important that an entire subdiscipline of chemistry called **spectrochemical analysis** has been developed around it. In this chapter, we will examine the fundamental principles of spectrochemistry; in the following three chapters, we will consider a number of specific techniques which employ these principles for purposes of qualitative and quantitative chemical analysis.

# WHAT IS SPECTROCHEMICAL ANALYSIS?

The term "spectrochemical" derives from two other words: spectrum and chemical. In spectrochemical analysis, we employ the spectrum of electromagnetic radiation to determine chemical species and to characterize their interactions with electromagnetic radiation. As shown in Figure 18–1, a spectrum is a plot of some measurable property of the radiation, $f(\nu)$, as a function of the frequency of the radiation, $\nu$. From a spectrum, two important pieces of information can be obtained. First, from the shape of the spectrum, a chemical species can often be identified qualitatively. Second, from the magnitude of $f(\nu)$ at chosen frequencies, the amount of a chemical species present can be determined quantitatively.

These statements become more meaningful when it is recalled that, for a photon (quantum) of electromagnetic radiation, frequency is related to energy through the Planck equation,

$$E = h\nu \tag{18–1}$$

where $E$ is the energy of the photon, $\nu$ is its frequency, and $h$ is Planck's constant ($6.624 \times 10^{-27}$ erg sec). Therefore, a photon of electromagnetic radiation has a definite energy and can cause transitions between the quantized energy states in atoms, molecules, and other chemical species. To cause such a transition, the energy of the photon must be equal to the difference between the energy states involved in the transition. Thus, we can examine chemical species by using electromagnetic radiation as a probe, the frequency of the radiation being related to the energy change associated with the observed transition.

Because energy states differ among chemical species, it is to be expected that the energy changes involved in the transitions will also differ. This implies that a spectrum will be a highly individual property of each substance, and that observation of the spectrum can be employed to advantage for identification of the substance. In effect, the spectrum is a map of the transitions which occur between energy states of the

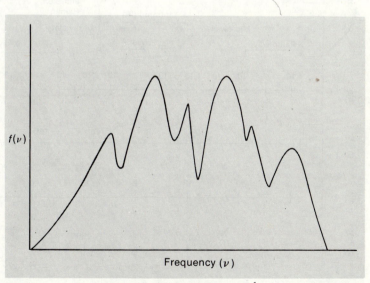

*Figure 18–1.* Generalized spectrum consisting of a plot of some function of frequency, $f(\nu)$, versus $\nu$, the frequency of electromagnetic radiation. In spectrochemical analysis, $f(\nu)$ is usually related to the power of the radiation at each frequency.

chemical species. Furthermore, the number of times each transition occurs during a fixed interval is related to the total number of chemical species which can undergo that transition. Therefore, if the measured parameter, $f(\nu)$, in a spectrum can be related to the total number of transitions, the spectrum can be used to determine the concentration of species present.

There is no restriction on the frequency of electromagnetic radiation employed in spectrochemistry. In fact, spectrochemical analysis can utilize frequencies ranging from those of audio waves (10 to 10,000 Hz)* to those of gamma rays ($10^{22}$ Hz). Yet, over this enormous range of frequencies, the principles of spectrochemical analysis remain unchanged, the only difference being the magnitude of the energy changes which are probed. These energy changes are often associated with particular types of transitions—rotational, vibrational, electronic, nuclear, and so on. Figure 18–2 shows the range of frequencies commonly used in spectrochemistry, along with the kind of transition probed in each frequency region.

Although we can associate definite types of transitions with each frequency region delineated in Figure 18–2, another reason for establishing these frequency regions is that the instrumentation required to make meaningful measurements in each region is different. For example, the optical instrumentation useful for the measurement of electronic transitions in the ultraviolet and visible spectral regions cannot be used at all for observation of nuclear transitions in the $\gamma$-ray region or of molecular rotations in the microwave region. For this reason, spectrochemical analysis has grown exceedingly diverse and now includes a number of individual branches,

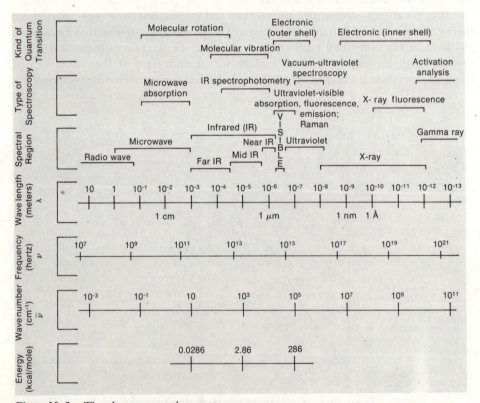

*Figure 18–2.* The electromagnetic spectrum.

---

* 1 Hz (hertz) = 1 cycle per second.

each of which deals with a specific spectral region. Because the experimental measurements and tools are different for each spectral region, it is customary to consider each region individually, so that the study of spectrochemistry often becomes an examination of seemingly unrelated techniques. In our discussion, however, we will retain the thought that all spectrochemical techniques have many common underlying principles that can be applied equally to all areas.

## A SPECTROCHEMICAL VIEW OF ELECTROMAGNETIC RADIATION

Because the interaction of radiation with chemical species is the foundation upon which spectrochemical analysis is based, several important characteristics of electromagnetic radiation must be understood before we proceed further. To assist in this discussion, it will be helpful to visualize an electromagnetic wave.

Figure 18–3 portrays a sinusoidally oscillating electromagnetic wave, traveling through space in an arbitrary direction $x$. As its name implies, the wave consists of

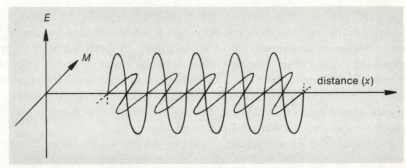

*Figure 18–3.* Portrayal of an electromagnetic wave, showing the electric ($E$) and magnetic ($M$) vectors; the wave is represented at some instant in time and is traveling through space in direction $x$.

oscillating electric ($E$) and magnetic ($M$) fields, which are orthogonal to each other and which travel at a constant velocity ($c$), equal to approximately $3 \times 10^{10}$ cm per sec in vacuum. Let us assume that the wave has a constant frequency of oscillation ($\nu$) and that it can be observed at some instant in time. Because the wave travels at a constant velocity ($c$), the spacing between its maxima—that is, its **wavelength**— will be a constant which is characteristic of the wave. This wavelength will simply be the distance the wave travels during one period ($1/\nu$) of its oscillation. Therefore, the wavelength ($\lambda$) can be calculated from the following well-known formula:

$$\text{distance} = \text{velocity} \times \text{time}$$

$$\lambda = c \times \frac{1}{\nu} = \frac{c}{\nu} \tag{18–2}$$

Because of this relationship to frequency ($\nu$), wavelength ($\lambda$) is often used as the horizontal axis for a spectrum such as that illustrated in Figure 18–1. Wavelength is expressed in terms of meters (m), centimeters (cm), micrometers ($\mu$m), nanometers (nm), or angstrom units (Å), depending on the spectral region. The relationship among these units is given in Table 18–1. Note that wavelength is inversely proportional to frequency and, therefore, to energy as well. Because it is sometimes desirable to refer to the energy of a spectrochemical transition, a quantity proportional to energy is often used instead of wavelength or frequency. This quantity,

**Table 18–1**

| | |
|---|---|
| 1 cm | $= 10^{-2}$ m |
| 1 $\mu$m | $= 10^{-6}$ m |
| 1 nm | $= 10^{-9}$ m |
| 1 Å | $= 10^{-10}$ m |

called a **wavenumber**, has units of cm$^{-1}$ (reciprocal centimeters), is given the symbol $\tilde{\nu}$, and is defined by the expression

$$\tilde{\nu} = \frac{1}{\lambda} = \frac{\nu}{c} \qquad (18\text{–}3)$$

where $\lambda$ is the wavelength in cm. Although $\tilde{\nu}$ is *proportional* to frequency, it is in fact *not* a frequency and should never be so called. Properly formulated, frequency is expressed in units of sec$^{-1}$ or Hz (hertz), whereas $\tilde{\nu}$ is expressed in units of reciprocal length. These three quantities—wavelength, frequency, and wavenumber—can be used interchangeably when spectra are displayed. Generally, the mode of display is chosen according to convenience, although conversions between units can be easily made by applying the formulas presented above.

Several other characteristics of electromagnetic waves are important to spectrochemical analysis. One such characteristic, monochromaticity, refers to the spectral purity of the wave. For an idealized wave such as that depicted in Figure 18–3, only a single frequency exists. Such a wave is said to be **monochromatic**, which literally means "single colored." Actually, few truly monochromatic waves are ever employed in spectrochemistry. More often, the radiation used contains a range of frequencies spread over a certain spectral interval. To describe the breadth of this frequency range, the term **bandwidth** or **spectral bandwidth** is used. Although bandwidth properly refers only to a frequency range, it is used commonly to denote a wavelength interval. As we shall find later, the bandwidth of electromagnetic radiation is often of considerable importance in spectrochemical measurements and can affect both the qualitative and quantitative validity of an analysis.

Another important property of an electromagnetic wave is its degree of **polarization**. Nonpolarized electromagnetic waves will exhibit a random direction of polarization of their electric and magnetic fields about the axis of propagation of the wave. With reference to Figure 18–3, this means that the electric and magnetic fields, which always remain orthogonal to each other, have a variable and unpredictable orientation within a plane perpendicular to the direction of travel of the wave. If, however, all the electric (or magnetic) field oscillations are in a single plane, the wave is said to be **plane polarized**, as illustrated in Figure 18–3. In addition, if this plane appears to rotate at a constant rate around the axis of propagation as the wave travels, the wave is said to be **circularly polarized**. Although we will not explore these concepts further, they form the basis for several useful spectrochemical techniques—polarimetry, optical rotatory dispersion (ORD), and circular dichroism (CD). These methods depend on the ability of certain *optically active* chemical species to alter the direction of polarization of an electromagnetic wave and are used in the analysis and characterization of this special class of substances.

In all spectrochemical measurements, it is important to determine the amplitude and frequency of the electromagnetic radiation. Unfortunately, the accurate measurement of both amplitude and frequency is possible only for radiation at microwave frequencies or lower, because of the limited frequency-response of available detectors. In spectral regions of higher frequency, the variable that can be measured is the **radiant power** ($P$), which is proportional to the *square* of the wave amplitude. Radiant power has spectrochemical importance because it is the amount of energy transmitted in the form of electromagnetic radiation per unit time. If $E$ is the energy of a photon, the radiant power can be expressed by the relation

$$P = E\Phi = h\nu\Phi \qquad (18\text{–}4)$$

where $\Phi$ is the photon flux—that is, the number of photons (quanta) per unit time.

The radiant power of a beam of electromagnetic radiation is often referred to as its *intensity*. Actually, **intensity** is properly defined as radiant power from a point source per unit solid angle, typically with units of watts per steradian. Although one can correctly speak of the intensity of a source of radiation, it is incorrect to describe radiation striking a sample in terms of its intensity, especially if the radiation is **collimated**—that is, if it appears to originate from an infinitely distant source. Unfortunately, the term *intensity* is widely used in a qualitative sense or even as an equivalent to *radiant power*. To avoid these ambiguities, we will use the term *intensity* only in its correct sense.

## TYPES OF INTERACTIONS OF RADIATION WITH MATTER

Let us consider how radiation can interact with chemical species to provide information about the species. We will specifically exclude several types of interactions from our discussion. These interactions, termed reflection, refraction, and diffraction, are used extensively in spectrochemical instrumentation and are better treated in a text on that topic. Here, we will concern ourselves with the ways in which radiation interacts more directly with the chemical sample itself. Four distinct examples of this sort of interaction can be cited—absorption, luminescence, emission, and scattering.

### Absorption

If a beam of electromagnetic radiation is sent into a chemical sample, it is possible for the sample to absorb a portion of the radiation. This phenomenon is depicted in Figure 18–4A, which shows a beam of radiation having a radiant power $P_0$ being directed into a sample. Each specific frequency, $\nu_1$, $\nu_2$, and so on, which comprises the beam of radiation will, of course, have its own energy, $h\nu_1$, $h\nu_2$, and so on. If the chemical sample contains a species whose energy states differ by any of these exact energies, the sample will absorb radiation at those frequencies. This behavior is illustrated in Figure 18–4B, in which a chemical species having energy levels **G** and **E** is portrayed. If the species (atom, molecule, or ion) exists in the lower (ground) energy state **G** before its encounter with the beam of radiation, it can be **excited** to the upper state **E**. In this excitation, the species must absorb a quantity of energy $h\nu_1$, which is exactly equal to the difference in energy between states **G** and **E**.

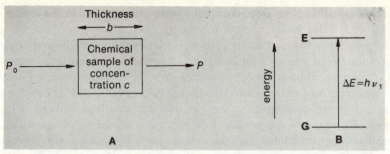

*Figure 18–4.* *A,* Diagram showing the relationship between the incident ($P_0$) and transmitted ($P$) radiant power for an absorbing substance at a concentration $c$ in a spectrophotometer cell with a sample path length $b$.

*B,* Diagram illustrating the change in energy ($\Delta E$) of a species during absorption of a photon of energy $h\nu_1$. In the act of absorption, the species is promoted from the ground state (**G**) to an excited state (**E**).

Energy required to excite the species to the upper energy state **E** is drawn from the beam of radiation, so that the total radiant power of the beam is diminished at the frequency of absorption, $\nu_1$. Therefore, after encountering a number of absorbing species in the sample, the beam will exit from the sample with a reduced radiant power $P$. It should be recognized that *only* those frequencies capable of being absorbed by the sample will be attenuated in a purely absorbing sample; all other frequencies will pass through the sample with no power loss. This result suggests the possibility that components in the sample can be identified from the absorption spectrum, that is, from the frequency components of the radiation which are absorbed. Understandably, the diminution in the radiant power of the beam at each frequency should be related to the number of absorbing chemical species present in the sample. This information, which provides the basis for quantitative analysis, will be treated in detail later in this chapter.

## Luminescence

When a quantity of radiant energy (a photon) is absorbed from a beam of radiation by a chemical species, the species is promoted to an *excited state,* **E**, as shown in Figure 18–4B. However, the excited-state species has a limited lifetime, and would prefer to rid itself of this extra energy and return to the ground state **G**. To attain this state of tranquillity, an excited chemical species must dispose of a quantity of energy equal to the difference in energy between the excited state and the ground state. This energy can be released in several ways: it can be transferred to other species, it can be converted into other forms of energy (such as thermal or electrical energy), or it can be emitted in the form of electromagnetic radiation. When the energy gained by a chemical species during absorption is emitted in the form of radiation, the process is called **fluorescence** or, more generally, **luminescence**. This process is portrayed in Figure 18–5.

In Figure 18–5A, a beam of electromagnetic radiation of power $P_0$ is sent into a chemical sample, just as in the previously considered case of absorption. If any components of the sample have suitable energy levels, a portion of the incoming radiation will be absorbed so that the transmitted beam will have a slightly reduced power $P$. Thus, the difference in power ($P_0 - P$) between the incident ($P_0$) and transmitted ($P$)

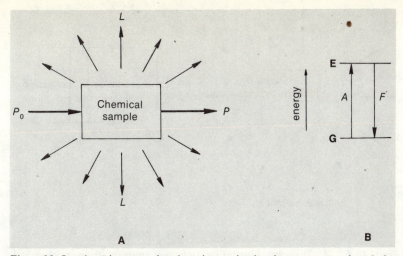

*Figure 18–5.* *A*, Diagram showing isotropic luminescence produced by absorption of radiation from incident beam of power $P_0$. Radiant power ($L$) of luminescence is some fraction of the radiant power absorbed ($P_0 - P$).

*B*, Changes in energy of a chemical species during absorption ($A$) and resonance fluorescence ($F$). Resonance fluorescence is a special luminescence process.

beams is used to excite chemical species present in the sample. This process is portrayed as the absorption step labeled $A$ in Figure 18–5B. The excited species now present will spontaneously undergo deactivation, one possible way being by emission of radiation. If the energy is emitted immediately, the radiated photon will be equal in energy and frequency to the radiation which was initially absorbed. This so-called **resonance fluorescence** is one type of luminescence and is labeled $F$ in Figure 18–5B. A number of other possible luminescence processes exist which involve quite complex pathways for energy loss. However, most of these are peculiar to particular spectral regions and will be discussed later.

Although the incoming radiation ($P_0$) and the transmitted radiation ($P$) are directional, the luminescent radiation ($L$) has an equal probability of traveling in any direction, as indicated in Figure 18–5A, and is thus said to be **isotropic**.

Like absorption, the phenomenon of luminescence can be used for both qualitative and quantitative analysis. Clearly, the radiant power of luminescence will depend both on the concentration of the luminescing chemical species and on the frequency of the incident radiation.

## Emission

Species in a chemical sample can, of course, be excited by means other than the absorption of radiant energy. Thermal, chemical, electrical, and other forms of energy can all be used to excite atoms, molecules, and ions to higher energy states. If this excitation results in the liberation of electromagnetic radiation from the sample, the process is termed **emission**. This process is portrayed in Figure 18–6A, in which energy in a nonradiant form is introduced into a sample. Provided that the energy is of sufficient magnitude, a number of species in the sample can be excited to a higher energy state **E**, as shown in Figure 18–6B. Following excitation, the species are in a situation similar to that which exists during the processes of absorption and

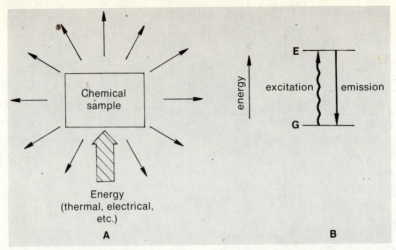

*Figure 18–6.* *A*, Diagram illustrating isotropic emission from a chemical species excited through the input of thermal, electrical, or other energy.

*B*, Energy changes that occur during excitation of and emission by a chemical species; the wavy line used for excitation denotes a radiationless process.

luminescence, where their excited-state energy can be lost by either radiational or nonradiational means. If, as in the case of luminescence, the energy is released through radiation, it will be liberated in all directions. This emission process is very important in spectrochemical analysis, not only for the examination of chemical samples but also in the generation of radiation used in the analysis. Because other forms of energy can be conveniently converted to radiant energy by this means, the process of emission is essential to the operation of sources of radiation, as will be discussed later. As one would surmise, luminescence is a special kind of emission process.

### Scattering

Unlike the processes of absorption, emission, and luminescence, the scattering of electromagnetic radiation need not involve a transition between quantized energy states of an atom or molecule. Rather, the process of scattering, as its name implies, outwardly appears to cause a randomization in the direction of a beam of radiation. Actually, the events involved in scattering are considerably more complex. Because scattering phenomena are becoming more and more important to spectrochemical analysis, let us review here, albeit in a simplified manner, the basic nature of the scattering process.

If a beam of electromagnetic radiation impinges upon a particle which is small with respect to the wavelength of the radiation, the particle will experience an intense disturbance caused by the oscillating electric and magnetic fields of the passing radiation. In effect, while the radiation is passing, the particle finds itself in a strong field, whose polarity alternates at the frequency of the radiation. If the particle is polarizable—that is, if charges within it can separate under the influence of the field— the induced charges in the particle will also oscillate as the field changes polarity. This effect is illustrated schematically in Figure 18–7, in which a small particle is shown to have an oscillating dipole induced in it by a passing electromagnetic wave.

The oscillating dipole induced in the small particle now produces a field of its

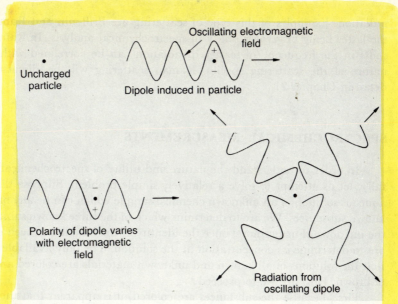

*Figure 18–7.* Classical portrayal of Rayleigh scattering. Oscillating dipole induced in the particle acts as a secondary source to produce scattered radiation of the same wavelength (frequency) as that incident on the particle.

own, which oscillates at the same frequency as the incoming radiation. This field produced by the oscillating dipole acts as a source of radiation itself, so that radiation of the same frequency and wavelength as the incident radiation is sent in all directions from the particle. This type of scattering, called **Rayleigh scattering**, is characteristic of scattering from small particles such as atoms and molecules.

The physical nature of Rayleigh scattering is well understood. Although most of its characteristics need not concern us here, one of its more important properties is a dependence on the frequency of the incoming radiation. The intensity of Rayleigh-scattered radiation increases in proportion to the *fourth power* of the frequency of the incident radiation. This relationship is responsible for many natural phenomena, such as the blue sky and red sunset. Thus, in experimental systems where scattering is deliberately sought, radiation of high frequency or short wavelength is desirable. In situations where scattering is to be minimized, it is advantageous to employ radiation with a lower frequency or longer wavelength.

In addition to Rayleigh scattering, several other forms of radiation scattering deserve brief mention here. Scattering which occurs from particles that are large compared to the wavelength of incoming radiation is called **Mie scattering**.* Mie-scattered radiation, like that produced in Rayleigh scattering, contains the same frequencies as the incident radiation. However, the angular distribution of the radiation scattered during a Mie process is not uniform. In fact, the angular distribution of radiation scattered from large particles can be used as a means of determining particle size. Mie scattering is important in the analytical techniques of turbidimetry and nephelometry, where the concentration of components in a suspension is determined by measurement of the radiation scattered from the suspended particles.

Other forms of scattering which are of spectrochemical importance involve a shift in the frequency of the scattered radiation with respect to that of the incident

* Mie is pronounced as if it were spelled mee.

radiation. Examples of this type of scattering are Brillouin and Raman scattering, the latter being of greater utility in spectrochemical analysis. In Raman scattering, shifts in the frequency of scattered radiation can be correlated with the chemical nature of the scattering species. Raman scattering will be discussed in greater detail in Chapter 21.

## SPECTROCHEMICAL MEASUREMENTS

In order to understand the nature and utility of spectrochemical analysis more fully, let us attempt to solve a relatively simple problem. Suppose that we have an aqueous solution of an unknown chemical sample which can be any of three possible known substances. We are to determine which of the three known substances matches the unknown. Furthermore, once the identity of the sample has been established, we are to determine its concentration in the solution. The only available information is that the solutions of all known and unknown materials are colored a different shade of brown. How are we to proceed?

First, because the substances are colored, it is apparent that they must absorb visible electromagnetic radiation (that is, light). It is likely that we could differentiate among the substances by means of their absorption spectra—in other words, by means of their differing absorption at various wavelengths or frequencies of light. Choosing this method as a reasonable initial approach, we must assemble some equipment, however rudimentary, to help us perform the necessary measurements. One such experimental setup is shown schematically in Figure 18–8.

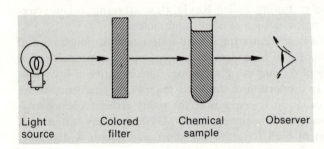

Light source   Colored filter   Chemical sample   Observer

*Figure 18–8.* Schematic diagram of a simple instrumental setup for qualitative and quantitative determinations by means of visual observation of absorption.

Our choice of using visible light in this measurement determines the type of measuring apparatus we need. In this case, a tungsten-filament light bulb can be used to illuminate the sample. However, because the light bulb emits a broad range (continuum) of frequencies of visible radiation, it is necessary to isolate certain portions of its emission spectrum in order to examine the absorption of a sample at selected frequencies. For this purpose, a simple colored glass filter might suffice. This glass filter operates by absorbing most frequencies except those within a narrow band, as shown in Figure 18–9. Those frequencies not absorbed are transmitted through the filter and can be used to illuminate the sample. Although certainly not truly "single-colored," the narrow band of frequencies passing through the filter is often said to be "monochromatic" because of its narrowed spectral bandwidth. To permit a study of the absorption of light by the sample at different frequencies, the filter can simply be exchanged for another which transmits radiation in the desired frequency region.

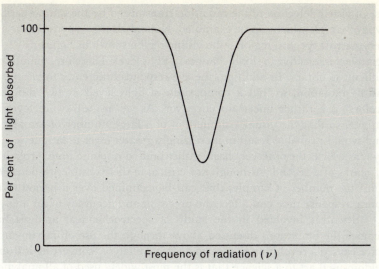

*Figure 18–9.* Idealized absorption spectrum of a colored glass filter.

To operate our simple instrument, we will sequentially insert the known and unknown solutions into the sample chamber, illuminate them with light passing through various filters, and visually observe the light which is transmitted by the solutions. Although the absolute power of the transmitted light will probably be different for each sample, the relative powers of transmitted light at different frequencies must be the same for the unknown sample and its known counterpart. Thus, the unknown can be identified with the known which appears similar to the unknown in color when viewed with light transmitted by each filter. This is true because the two species have identical chemical compositions and, consequently, identical spectral properties.

To determine the concentration of our now-identified substance in the sample solution, we need solutions containing various known concentrations of the desired species. Our procedure will be to insert the various standard solutions into the instrument and visually match the amount of light transmitted through the unknown sample to that of one of the standards. For this work, we should select a filter that passes frequencies absorbed most strongly by the sample. This selection can be made if we place various filters into the instrument and alternately insert and remove the sample. That filter which causes the greatest apparent difference in the observed radiation with and without the sample will be the correct filter to use.

This very simple example shows the utility of spectrochemical analysis, albeit at a rather elementary level. It indicates the sort of components needed to perform an analysis and illustrates three additional points:

1. Components used in any spectrochemical instrument are defined by the spectral (frequency) region being employed.
2. Spectrochemical analysis often, though not always, requires empirical standardization. That is, standard or known samples are often needed for comparison with an unknown sample or for calibration of the measuring system.
3. The quality of spectrochemical measurements relies greatly on the quality and sophistication of the measuring apparatus employed.

This last point is important. Apparatus used in the present example is of the most elementary type and, consequently, has several limitations. For example, the use of visual detection of the radiation transmitted by the sample seriously limits the type of sample which can be examined. Not only is the response of the eye limited to a very narrow spectral region—the visible region shown in Figure 18–2—but the eye is relatively insensitive to small changes in light level. Therefore, quantitative results are difficult to obtain. In addition, the eye responds differently to various frequencies of visible radiation, so that a comparison of light levels at two different frequencies (colors) is a rather uncertain endeavor. As we are all aware, eyes are subject to fatigue, so that the visual examination of a large number of samples is undesirable and would probably result in increasingly greater error as fatigue set in. Finally, the eye (as well as the observer) has a rather limited response time, probably on the order of tenths of a second. Although not critical in this example, response time will often limit the number of samples that can be examined over a period of time. Furthermore, response time can affect the precision and accuracy of an analysis.

Scientists involved in the study of spectrochemical analysis have sought to remove the limitations discussed above through the use of instruments. Instruments have caused a revolution in spectrochemical analysis, converting it from a mistrusted, seldom-used art to a science that is the most widely used of all analytical techniques. In its most general form, an instrument for use in spectrochemical analysis can be visualized as shown in Figure 18–10.

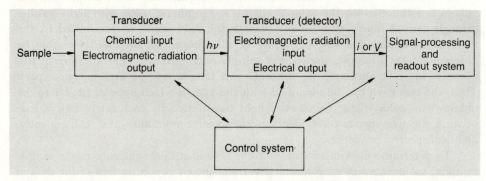

*Figure 18–10.* Generalized spectrochemical instrument; arrows indicate the flow of information through the instrument.

## A Generalized Spectrochemical Instrument

In the generalized instrument of Figure 18–10, an input device is used to convert chemical information in a sample into information in the form of electromagnetic radiation. Therefore, this device is a **transducer** which *encodes* the chemical information into another form, namely, electromagnetic information. In the simple instrument presented in Figure 18–8, this transducer consists of a light bulb and a sample chamber. In this transducer, radiation from the light bulb passes through the sample, where its frequency (spectral) characteristics are altered in accordance with the chemical composition of the sample. Thus, the transmitted radiation contains information about the chemical sample but encoded in a spectral form.

This concept of a transducer is highly general and can be applied to other kinds of chemical and physical instrumentation as well. A loudspeaker is an electrical input-acoustical output transducer, whereas a glass electrode and flame-ionization detector are both chemical input-electrical output transducers. These examples illustrate that a transducer changes the *form* of the information but, ideally, not its content.

**INTRODUCTION TO SPECTROCHEMICAL METHODS OF ANALYSIS**

To decode or extract the chemical information encoded in the radiation, a second transducer, commonly called the detector, is required. In the spectrochemical instrument, the function of a detector is to provide a measurable electrical signal which is proportional to some property, usually radiant power, of the radiation incident upon the detector. In effect, the detector converts the information present in the electromagnetic radiation to another form, generally electrical, which is more amenable to signal-processing techniques.

The electrical signal from the detector, which contains the encoded information about the sample, must next be rendered interpretable to a human observer. For this operation, a readout system is employed; the final output of the readout device can take many forms, from a simple meter deflection to columns of numbers or graphical displays from a computer. However, its function remains the same—to convert the information present in the incoming electrical signal to a form which is meaningful.

A final but indispensable component of our generalized spectrochemical instrument is the control system. This device is responsible for coordinating all the events and operations that take place in the instrument, from selection and introduction of the sample to interpretation of the readout signal. In most conventional instruments, this control function is mainly performed by a human operator. To a varying extent, the operator will be assisted by partially automated control systems, although he will still be primarily responsible for monitoring and decision-making. In future spectrochemical instruments, it can be expected that the control function will be automated to an ever greater extent, with on-line digital computer systems furnishing the necessary manipulatory and decision-making power.

***An absorption spectrophotometer.*** As an example of a specific type of instrument that can be employed to solve the problem posed earlier, but which is less limited than the crude apparatus of Figure 18–8, let us consider the device shown in Figure 18–11. Components within the dashed box work together as a chemical input–electromagnetic radiation output transducer, as called for in Figure 18–10. For the

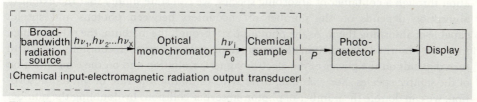

*Figure 18–11.* Block diagram of a versatile spectrochemical instrument for the measurement of absorption in the ultraviolet-visible spectral region.

instrument in Figure 18–11, a broad-band source of radiation replaces the ordinary light bulb of the earlier system. This broad-band source, which can be any of several available types (such as a tungsten or deuterium lamp), emits continuous radiation over a wide spectral range. To select the frequencies of radiation to be sent through the sample, a device called an optical monochromator is employed. This monochromator performs the same function as the glass filter used before, except that it permits one to choose a narrower range of frequencies and to vary the frequencies which are passed. From the monochromator, selected radiation is sent to the sample, where the radiant power ($P_0$) of the radiation is altered according to the chemical nature of the sample. That portion of the radiation transmitted by the sample ($P$) is converted by the photodetector into a proportional electrical signal, generally in the form of a voltage or current. Finally, the electrical signal is displayed on a suitable meter or recorder, possibly calibrated in terms of the concentration of the sample. In the present situation, it is assumed that all control operations are carried out manually.

Because the instrumentation requirements for various spectral regions can be very different, there is often little similarity in the design and appearance of instruments used for different methods of spectrochemical analysis, even though all instruments fit the generalized block diagram of Figure 18–10. Instruments employed for one spectral region or another vary widely in size, cost, complexity, and even principle of operation.

## FUNDAMENTAL LAWS OF SPECTROCHEMISTRY

We have briefly discussed the nature of four processes commonly employed in spectrochemical analysis: absorption, luminescence, emission, and scattering. In this section, we will derive and discuss some of the quantitative relationships governing these processes. These relationships, which will be used extensively in subsequent chapters, are fundamental to all spectrochemical techniques and supply the foundation upon which much of quantitative spectrochemistry is based.

### Quantitative Laws of Absorption

Quantitative methods of analysis based upon the absorption of radiant energy by matter require the measurement of radiant power and a quantitative understanding of the laws which govern the extent of absorption. The power of an electromagnetic wave is quantized, and the wave can be viewed as a beam of photons, with the number of photons per unit time being proportional to radiant power.*

Adopting this view and considering the transmission of electromagnetic radiation through a container of absorbing species, we can state simply that the extent of absorption depends on the number of encounters between photons and species capable of absorbing them. As the photons pass through the medium, the rate at which they are absorbed depends on the number of photon-absorber collisions, which in turn depends on the power of the electromagnetic radiation and on the concentration of the sample species. For example, if we wish to double the rate of photon absorption, we can double the number of photon-absorber collisions by doubling the number of photons or by doubling the number of absorbers. In either case, collisions will occur twice as often, and the rate of absorption will go up by a factor of two. If the population densities of both photons and absorbers were doubled, they would bump into each other four times as often, and the rate of absorption would increase by a factor of four.

The relationships between radiant power, concentration, and rate of absorption are embodied in two laws: Lambert's law, which expresses the dependence of the rate of absorption on the power of the photon beam; and Beer's law, which relates the absorption rate to the concentration of absorbing species in the sample.

*Combined Lambert-Beer law.* As a beam of photons passes through a system of absorbing species, the rate of photon absorption is directly proportional to

---

* To speak of an electromagnetic wave as a beam of photons is to oversimplify the situation somewhat. For example, an understanding of scattering phenomena or of deviations from Beer's law caused by changes in refractive index (discussed later) requires that the electromagnetic radiation be treated as a wave.

the power of the photon beam and to the concentration (or partial pressure) of the absorbers.

This restatement of the relationship between the number of collisions and the extent of absorption can be expressed mathematically by means of the differential equation

$$-\frac{dP}{dx} = kcP \qquad (18\text{--}5)$$

where $P$ represents the radiant power at any point $x$ in the absorbing medium, $c$ is the concentration of species capable of absorbing a photon, and $k$ is a constant characteristic of the nature of the absorbing species and of the energy of the photons. Rearrangement and separation of variables in equation (18–5) yield

$$-\frac{dP}{P} = -d\ln P = kc\,dx \qquad (18\text{--}6)$$

To extend this microscopic picture to a calculation of the absorption of radiation in a medium of some finite thickness $b$, we can stipulate that $P_0$ is the power of the incident radiation at $x = 0$ and that $P$ represents the power of the transmitted radiation emerging from the absorbing medium at $x = b$. We can then integrate equation (18–6) along the entire radiation path length:

$$-\int_{P_0}^{P} d\ln P = kc\int_{0}^{b} dx \qquad (18\text{--}7)$$

Carrying out the indicated integration, we obtain

$$\ln P_0 - \ln P = kcb \qquad (18\text{--}8)$$

or

$$\ln \frac{P_0}{P} = kcb \qquad (18\text{--}9)$$

Use of base-ten logarithms instead of natural logarithms requires only that the value of $k$ be changed. Accordingly,

$$\log \frac{P_0}{P} = k'cb \qquad (18\text{--}10)$$

where $k = 2.303k'$.

Several features of equation (18–10) have experimental significance. First, it can be seen that concentration can be determined from a measurement of relative beam power with and without the absorber in the beam. Absolute measurements of beam power are not necessary, a fact which considerably simplifies absorption spectrophotometric procedures. Second, it is clear that the path length of the radiation through the sample medium must be accurately known. This condition is usually met if one places the sample in a **spectrophotometer cell** which is as transparent as possible to the radiation being employed. The composition, size, and form of the cell depend upon the nature and concentration of the sample and on the spectral region being utilized for the measurements. For example, in atomic absorption spectrometry the cell is often a flame, whereas in infrared absorption spectrophotometry the cell will frequently be a thin space between two salt plates.

In practice, equation (18–10) is usually written in the form

$$\log \frac{P_0}{P} = abc \qquad (18\text{–}11)$$

which is the most familiar expression of the combined Lambert-Beer law, and which is often referred to simply as Beer's law. Because $P_0$ and $P$, the powers of the incident and transmitted radiation, appear as a ratio, any units of radiant power can be used; even completely empirical units such as a meter reading are fully satisfactory. If the parameter, $b$, commonly known as the **sample path length**, is expressed in centimeters and the concentration factor, $c$, in grams of absorbing substance per liter of solution, the constant $a$ (which is equivalent to $k'$) is designated as the **absorptivity\*** and has units of liter $gm^{-1}$ $cm^{-1}$.

Frequently, it is desirable to specify $c$ in terms of molar concentration, with $b$ remaining in units of centimeters. In the latter situation, the Lambert-Beer law is rewritten as

$$\log \frac{P_0}{P} = \varepsilon bc \qquad (18\text{–}12)$$

where $\varepsilon$ (in units of liter $mole^{-1}$ $cm^{-1}$) is called the **molar absorptivity**.

The quantity $\log (P_0/P)$ is defined as **absorbance\*** and is given the symbol $A$. Because $A$ is directly proportional to the concentration of the absorbing species, some instruments for absorption measurements are calibrated to read directly in absorbance units. The ratio $(P/P_0)$ is called the **transmittance**, $T$, whereas $100(T)$ is the **per cent transmittance**. It is not uncommon to find commercial instruments calibrated directly in units of transmittance or per cent transmittance.

*Deviations from the Lambert-Beer law.* The Lambert-Beer law is, of course, applicable only when its component relations are valid. However, because there are no known exceptions to Lambert's law, all apparent deviations from the combined law are due to the concentration factor $c$. The applicability of the Lambert-Beer law can be tested for any particular system if one measures the absorbance for each of a series of samples of known concentration of the absorbing species. A plot of the experimental data in terms of absorbance ($A$) versus concentration ($c$) will yield a straight line passing through the origin if the Lambert-Beer law is obeyed. More often than not, a plot of the data over a wide range of concentration of an absorbing substance will give a graph such as that shown in Figure 18–12, indicating that the Lambert-Beer law is applicable only up to concentration $c_1$. Nevertheless, if a suitable calibration curve is prepared from a series of samples containing known concentrations of the absorbing species, it is still possible to determine the concentration of the absorbing substance in an unknown.

Deviations from Beer's law are of three types—real, chemical, and instrumental. Real deviations arise from changes in the refractive index of a medium which occur because of variations in the concentrations of its components. A rigorous derivation of Beer's law assumes a constant refractive index for the absorbing medium; any error in this assumption produces a consequent uncertainty in the experimental results. Generally, errors caused by changes in refractive index are minimal, so that *real*

---

\* Absorptivity ($a$) and absorbance ($A$) are terms recommended originally by the Joint Committee on Nomenclature in Applied Spectroscopy, established by the Society for Applied Spectroscopy and the American Society for Testing Materials (see H. K. Hughes: Anal. Chem., *24*:1349, 1952). In the older literature, absorptivity was frequently referred to as absorbancy index, specific extinction, or extinction coefficient; and absorbance was often called optical density, absorbancy, or extinction.

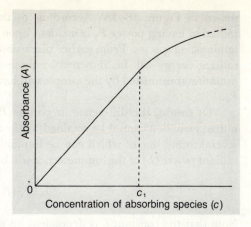

*Figure 18–12.* Relationship between absorbance ($A$) and concentration ($c$) of an absorbing substance. Such a graph is often called a *Beer's law plot* or, more generally, a *working curve*.

deviations from Beer's law are neglected in most absorption spectrochemical analyses.

Chemical deviations from Beer's law are caused by shifts in the position of a chemical or physical equilibrium involving the absorbing species. Consider, for example, the following reaction between an absorbing species A and another species B to form species C and D:

$$A + B \rightleftharpoons C + D$$

Obviously, any change affecting the position of this equilibrium will alter the concentration of A. In fact, a certain change in the original concentration of A itself *might* not result in a proportional change in the final concentration of A available to absorb radiation, because the position of equilibrium depends on other species besides A alone. To avoid this problem, conditions should be established so that any equilibrium involving A is relatively unaffected by changes in the concentration of A. For example, the concentration of either C or D could be made sufficiently great to keep the position of equilibrium shifted well toward the formation of A. Specific examples of this behavior will be encountered in later chapters.

Use of non-monochromatic radiation is the most common instrumental cause of deviations from Beer's law. Because Beer's law is rigorously applicable only for absorption of radiation of a single frequency, some error will nearly always exist when real instrument components are used. In most spectral regions, it is difficult or impossible to obtain truly monochromatic radiation. Specific examples of this sort of error will be discussed when we consider absorption of radiation in particular spectral regions.

Some experimental procedures used in absorption spectrophotometry involve calculations based upon the Lambert-Beer law, and others do not. As already implied, it is frequently possible to perform highly accurate quantitative determinations even when the chemical system departs from this fundamental law of absorption. For a successful analysis, the chief requirements are that the radiation-absorbing properties of a chemical system be measurable and reproducible.

## Quantitative Law of Luminescence

Like absorption, the phenomenon of luminescence is useful for quantitative spectrochemical analysis. To develop the quantitative relationships pertaining to luminescence of a sample, let us consider an experimental arrangement such as that shown in Figure 18–5A. According to this diagram, a beam of monochromatic radiation having power $P_0$ is incident upon the chemical sample, which contains the luminescing species. From earlier discussion, we know that a portion of the incident radiant energy will be absorbed by those species capable of doing so, and that the radiation transmitted by the sample will have a power $P$, less than the incident power $P_0$.

Of course, the difference in power $(P_0 - P)$ between the incident and transmitted radiation cannot be retained by the absorbing species but will be dissipated in several forms, one of which can be luminescence. If constant conditions prevail, the radiant power $(L)$ of the luminescence will be some fraction $(k)$ of the absorbed power:

$$L = k(P_0 - P) \qquad (18\text{–}13)$$

Note that the constant $k$ is dependent on the nature of the luminescing species and its environment and is related to the **quantum efficiency** for luminescence, a term to be discussed in detail later. Although the latter equation is not very useful in its present form for practical quantitative measurements, we can modify it by using the Lambert-Beer law.

First, let us write the Lambert-Beer law in exponential form:

$$P = P_0(10^{-abc}) \qquad (18\text{–}14)$$

Substituting this expression into equation (18–13), we obtain

$$L = k[P_0 - P_0(10^{-abc})] = kP_0(1 - 10^{-abc}) \qquad (18\text{–}15)$$

From this modified relationship, it is apparent that the radiant power of the luminescence is proportional to the power of the radiation incident on the sample, regardless of the concentration of the chemical species. This is a very important practical point in quantitative measurements based on luminescence. Because a luminescence signal will increase in proportion to the power of the incident radiation, one can measure the luminescence of very small quantities of chemical species merely by increasing the source intensity. For this reason, the primary source used for luminescence measurements is usually made as intense as possible without the radiation becoming powerful enough to cause decomposition or alteration of the sample.

Equation (18–15) can be further modified to increase its utility for practical measurements. To do this, let us first consider only the exponential quantity $10^{-abc}$. This term can be expanded in a Taylor series to yield

$$10^{-abc} = e^{-2.3(abc)} = 1 - 2.3(abc) + \frac{[2.3(abc)]^2}{2!} - \frac{[2.3(abc)]^3}{3!} + \cdots \qquad (18\text{–}16)$$

If we substitute this expression into equation (18–15), the result is

$$L = kP_0(1 - 10^{-abc}) = kP_0\left\{1 - 1 + 2.3(abc) - \frac{[2.3(abc)]^2}{2!} + \frac{[2.3(abc)]^3}{3!} - \cdots\right\}$$

$$\qquad (18\text{–}17)$$

or

$$L = kP_0\left\{2.3(abc) - \frac{[2.3(abc)]^2}{2!} + \frac{[2.3(abc)]^3}{3!} - \cdots\right\} \qquad (18\text{–}18)$$

This seemingly complex relationship becomes considerably more meaningful when it is realized that, in many luminescence measurements, the quantity $abc$ is quite small. In fact, if $abc$ is less than 0.01, the term $(abc)^2$ is less than 1 per cent of $abc$. Under these circumstances, all the bracketed terms except the first in equation (18–18) become negligible. Therefore, if $abc < 0.01$,

$$L = 2.3\, kP_0 abc \qquad\qquad (18\text{–}19)$$

or

$$L = k' P_0 abc \qquad\qquad (18\text{–}20)$$

In the analysis of most chemical species, the radiant power of luminescence is proportional not only to the radiant power incident on the sample $(P_0)$, but to the absorptivity of the species $(a)$, to the path length of absorbed radiation through the sample $(b)$, and to the concentration of the species $(c)$. Because this latter relationship is applicable only for small values of $abc$, and because $a$ and $b$ are fixed for a specific analysis, the relationship between the radiant power of luminescence and the concentration of the chemical species will be expected to be that shown in Figure 18–13.

In Figure 18–13, the radiant power of luminescence is seen to be essentially proportional to the concentration of the luminescing species up to a value $c_1$, above

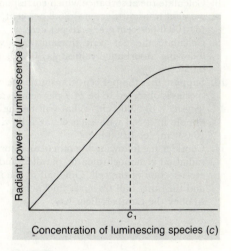

*Figure 18–13.* Relationship between radiant power of luminescence $(L)$ and the concentration $(c)$ of the luminescing species. Note that $c_1$ is the concentration beyond which non-linearity becomes significant.

which point the approximation that $abc < 0.01$ is no longer valid. Beyond concentration $c_1$, the curve bends toward the concentration axis, as predicted by equation (18–18). In practice, a **working curve** such as that shown in Figure 18–13 is also affected by concentration quenching (to be discussed in a later section) and by any factors which alter the validity of the Lambert-Beer law.

Notice from an inspection of equation (18–20) that several parameters can be varied to increase the yield of luminescence obtained from a chemical sample. Although the absorptivity $(a)$ for a given species is fixed, one can obtain a greater luminescence $(L)$ by increasing the path length $(b)$ or by increasing the source power $(P_0)$. As mentioned previously, it is most common to increase source power. Reasons for this choice will become obvious when we later consider the subject of concentration quenching.

## QUESTIONS AND PROBLEMS

1. Define or explain each of the following terms: spectrum, radiant power, intensity, wavelength, wavenumber, monochromatic, spectral bandwidth, polarization, photon, absorption, luminescence, emission, scattering, ground state, excited state, excitation, isotropic, transducer, source, readout system, absorptivity, molar absorptivity, absorbance, transmittance, per cent transmittance, Beer's law.

2. Calculate the frequency, $\nu$, in reciprocal seconds (hertz) corresponding to each of the following wavelengths of electromagnetic radiation: (a) 222 nm, (b) 530 nm, (c) 17 Å, (d) 0.030 cm, (e) $1.30 \times 10^{-7}$ cm, (f) 6.1 $\mu$m.

3. Calculate the wavenumber, $\bar{\nu}$, in $cm^{-1}$ for each of the wavelengths listed in problem 2. To what spectral region does each of these wavenumber values correspond?

4. Calculate the wavelength, in centimeters and in nanometers, corresponding to each of the following frequencies of electromagnetic radiation: (a) $1.97 \times 10^9$ hertz, (b) $4.86 \times 10^{15}$ hertz, (c) $7.32 \times 10^{19}$ hertz.

5. Calculate the wavenumber, in $cm^{-1}$, for each of the frequencies in problem 4 and give the spectral region in which each is found.

6. Calculate the energy in ergs of a photon of each of the following wavelengths: (a) 803 nm, (b) 3.68 $\mu$m, (c) 9.95 Å, (d) 11.5 cm.

7. Calculate the energy in ergs of a photon of wavelength 2615 Å. What is the total energy of a mole of such photons? If 1 mole of photons of wavelength 100 nm has a total energy of 286 kcal, what is the energy in kilocalories of 1 mole of photons of wavelength 2615 Å? Calculate the number of ergs in 1 kcal.

8. Calculate the absorbance which corresponds to each of the following values of per cent transmittance: (a) 36.8 per cent, (b) 22.0 per cent, (c) 82.3 per cent, (d) 100.0 per cent, (e) 4.20 per cent.

9. Calculate the per cent transmittance which corresponds to each of the following absorbance values: (a) 0.800, (b) 0.215, (c) 0.585, (d) 1.823, (e) 0.057.

10. Suppose that the per cent transmittance of a sample containing an absorbing species is observed to be 24.7 per cent in a spectrophotometer cell with a path length, $b$, of 5.000 cm. What will be the per cent transmittance of the same sample in a cell with each of the following path lengths: (a) 1.000 cm, (b) 10.00 cm, (c) 1.000 mm?

11. Consider the following hypothetical absorption cell. Assume that the incident radiation is monochromatic, parallel, and of uniform radiant power over the cross-section of the cell. One half of the incident beam traverses thickness $b_1$ and the other half traverses thickness $b_2$. The absorbing substance is assumed to obey the Lambert-Beer law.

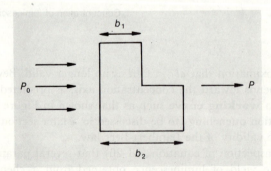

(a) Show that the following general relation is valid for the observed absorbance as a function of $b_1$ and $b_2$, the absorptivity $a$ of the absorbing substance, and the concentration $c$ of the absorbing substance:

$$A = \log 2 - \log [10^{-ab_1c} + 10^{-ab_2c}]$$

(b) What will be the apparent slope of a plot of absorbance ($A$) versus concentration ($c$) for a situation in which $c$ approaches 0, and one in which $c$ approaches $\infty$?

12. The *polarizability* of a species is a measure of the ease of displacement of positive and negative charges with respect to each other when that species is placed in an electric field. What effect do you suppose a change in polarizability would have on the power of radiation scattered from, say, a molecule? What would happen if the polarizability varied in a periodic manner?

13. Name two detectors of electromagnetic radiation familiar to you. Name two readout devices found in the home. For example, a television set is a sophisticated combination of a detector and readout device for radio-frequency waves, with the antenna serving as a detector and the set itself performing both signal-processing and readout (picture tube) functions. In view of this, identify two transducers found in a television system.

14. In what spectral region would the radiation from a "black light" be found?

## SUGGESTIONS FOR ADDITIONAL READING

1. E. J. Bair: *Introduction to Chemical Instrumentation.* McGraw-Hill Book Company, New York, 1962.
2. G. L. Clark, ed.: *The Encyclopedia of Spectroscopy.* Reinhold Publishing Corporation, New York, 1960.
3. G. W. Ewing: *Instrumental Methods of Chemical Analysis.* Third edition, McGraw-Hill Book Company, New York, 1969.
4. I. M. Kolthoff and P. J. Elving, eds.: *Treatise on Analytical Chemistry.* Part I, Volumes 5 and 6, Wiley-Interscience, New York, 1959.
5. G. H. Morrison, ed.: *Trace Analysis: Physical Methods.* Wiley-Interscience, New York, 1965.
6. R. A. Sawyer: *Experimental Spectroscopy.* Dover, New York, 1963.
7. D. A. Skoog and D. M. West: *Principles of Instrumental Analysis.* Holt, Rinehart and Winston, New York, 1971.
8. H. A. Strobel: *Chemical Instrumentation.* Addison-Wesley Publishing Company, Reading, Massachusetts, 1973.
9. F. J. Welcher, ed.: *Standard Methods of Chemical Analysis.* Sixth edition, Volume III-A, Van Nostrand, Princeton, New Jersey, 1966, pp. 3–282.
10. H. H. Willard, L. L. Merritt, and J. A. Dean: *Instrumental Methods of Analysis.* Fourth edition, Van Nostrand, Princeton, New Jersey, 1965.
11. J. D. Winefordner, ed.: *Spectrochemical Methods of Analysis.* Wiley-Interscience, New York, 1971.

# SPECTROSCOPY IN THE ULTRA-VIOLET AND VISIBLE REGIONS— INSTRUMENTATION AND MOLECULAR ANALYSIS

# 19

In this chapter we will examine methods of spectrochemical analysis which rely upon the absorption, emission, or luminescence of visible and ultraviolet radiation. Although it might seem awkward to discuss together two apparently different spectral regions, the distinction between ultraviolet and visible radiation, unlike that between most other regions of the spectrum, is really quite artificial, being based primarily on the spectral response of the human eye.

Interestingly, the human eye responds only to a narrow band of wavelengths of electromagnetic radiation. This band, called the visible region of the spectrum (Figure 18–2, page 602), extends from approximately 400 to 700 nm. Not by coincidence, this range of wavelengths is the region of greatest intensity of solar radiation reaching the earth. Therefore, the traditional distinction between the ultraviolet and visible spectral regions is based on a physiological rather than a chemical or physical response. In spectrochemical analysis, however, we are concerned with chemical and physical phenomena, so that the physiological distinction is unimportant. From a spectrochemical standpoint, the nature of the transitions arising from the absorption, emission, or luminescence of ultraviolet and visible radiation is the same. In addition, the instrumentation used in these spectral regions is similar, so that it is simpler to treat the regions together.

When ultraviolet or visible radiation interacts with atoms, molecules, or ions, transitions can occur between energy levels associated with valence or outer-shell electrons. Because these are the electrons involved in chemical bonding, spectroscopic observations made with ultraviolet or visible radiation can often be correlated with structural characteristics of a molecule or with association among atoms. Clearly, ultraviolet or visible spectra will differ greatly depending upon whether atoms or molecules are being observed and upon whether the chemical sample exists in the gaseous, liquid, or solid state.

Because of these variations, we will discuss atomic and molecular spectroscopy separately in this book. In the present chapter, after a consideration of the basic instrumentation required for ultraviolet-visible spectrometry, the qualitative and quantitative aspects of molecular spectroscopy based on absorption and luminescence of ultraviolet-visible radiation are examined. At the end of this chapter, we will consider spectrophotometric titrations and reaction-rate methods. Although these latter techniques need not be restricted to the ultraviolet-visible region, they most frequently employ ultraviolet-visible radiation and so will be treated here.

In the next chapter, we will investigate methods of atomic or elemental analysis which utilize ultraviolet-visible radiation, including atomic emission, absorption, and fluorescence flame spectrometry. In addition, there is a section on the uses of electrical discharges such as arcs and sparks as emission sources for elemental analysis.

## INSTRUMENTATION FOR ULTRAVIOLET-VISIBLE SPECTROMETRY

Referring back to Figure 18–10 on page 612, let us consider what instrumentation is required for spectrochemistry in the ultraviolet-visible region. To begin, the chemical input–electromagnetic radiation output transducer, which converts chemical information about the sample into the form of electromagnetic radiation, will vary in kind and complexity, depending on whether we desire to measure emission, absorption, or luminescence and on whether we wish to perform an atomic or molecular analysis. Accordingly, we will defer discussion of the configuration as well as certain components of this transducer to later sections which deal with specific techniques. However, one component of this transducer that is common to all ultraviolet-visible spectrochemical techniques is a **frequency selector**. A frequency selector separates or disperses electromagnetic radiation into relatively narrow bands of wavelengths or frequencies which can then be examined individually or simultaneously to determine the encoded information about the sample.

To determine the radiant power of the beam at each isolated band of frequencies, an electromagnetic radiation input–electrical output transducer (**detector**) is employed, as discussed in Chapter 18, page 613. Detectors, like frequency selectors, for the ultraviolet-visible region are of several types, any of which can be utilized for a specific kind of analysis. Let us examine these frequency selectors and detectors in some detail.

### Frequency Selectors

Three broad categories of frequency selectors are commonly used in the ultraviolet-visible spectral region—monochromators, polychromators, and filters. Each has advantages and disadvantages which must be carefully weighed for any particular application. Important differences among these frequency selectors are cost, ability to separate adjacent spectral intervals (**resolving power**), and amount of transmitted radiation of the chosen frequency (**optical speed**).

Resolving power is defined as the ratio of the wavelength ($\lambda$) being examined to the smallest difference between adjacent wavelengths ($\Delta\lambda$) which can be resolved. Therefore, the higher the resolving power ($\lambda/\Delta\lambda$) of a frequency selector, the greater will be its ability to separate two nearby spectral intervals. Optical speed has no similar quantitative definition, but is employed in a qualitative sense. To express optical speed, various parameters such as the transmission factor (discussed below) and the f/ number (used with cameras) are utilized.

*Filters.* Generally, filters are the least expensive frequency selectors. Two types of filters exist—absorption filters and interference filters. Absorption filters function by absorbing some part of the spectrum of incident radiation, so that the transmitted radiation is deficient in that portion of the spectrum. Often, individual absorption filters are placed in series so that only narrow spectral bands are transmitted. This is illustrated in Figure 19–1, which shows the fraction of radiation

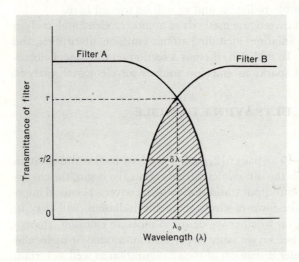

*Figure 19–1.* Transmittance curve for a combination filter (shaded region). Filters A and B, whose transmittance curves are shown, combine to produce a filter with a peaked transmittance curve; $\tau$ is the transmission factor for the combination filter, $\lambda_0$ is the nominal (central) wavelength of the combination filter, and $\delta\lambda$ is the spectral bandwidth of the combination filter.

transmitted by a pair of absorption filters as a function of the wavelength of radiation. Combining the two filters results in one having a peaked transmittance curve. In Figure 19–1, the **spectral bandwidth** ($\delta\lambda$) of the filter is defined as the range of transmitted wavelengths measured at half the peak height. For absorption filters, as well as for all other frequency selectors, the spectral bandwidth is a measure of resolving power. Absorption filters typically have spectral bandwidths ranging from 30 to 50 nm, so their resolving power is not great. The wavelength at the peak of the transmittance curve for the combination filter ($\lambda_0$) is termed the **central wavelength** or **nominal wavelength** of the filter. Another parameter depicted in Figure 19–1 is the **transmission factor** ($\tau$), which indicates the optical speed of the wavelength selector and which is defined as the ratio of the output radiant power ($P$) to the input power ($P_0$) at the central wavelength:

$$\tau = \frac{P}{P_0} \tag{19–1}$$

This transmission factor is identical in definition to transmittance, as discussed earlier. Transmission factors for absorption filters are usually small, on the order of 0.05 to 0.2, so that absorption filters have low optical speeds.

Interference filters, as the name implies, operate on the principle of interference of waves of electromagnetic radiation. Although somewhat more expensive than absorption filters, interference filters often have considerably narrower spectral bandwidths and greater transmission factors. It is not uncommon to find interference filters with spectral bandwidths between 5 and 20 nm, and transmission factors greater than 0.6.

Whenever absorption or interference filters are used, it is necessary to employ a different filter for each spectral interval to be examined. Therefore, filters are most often used in applications in which only a single spectral interval is needed, such as in quantitative spectrometry.

*Monochromators.* When a rapid and convenient way is needed to vary or scan the spectral region under observation, a monochromator is chosen instead of a filter. A monochromator (literally a device which produces a single color), like a filter, isolates only one spectral interval at a time. However, unlike a filter, a monochromator permits the spectral interval to be set anywhere in the ultraviolet-visible region. With a monochromator, the spectral bandwidth can be as small as 0.01 nm; the transmission factor for a monochromator is a function of spectral bandwidth but is generally less than that of a filter system.

Obviously, the major advantage of a monochromator is that a very narrow band of wavelengths can be selected and isolated for examination. Let us see how this is accomplished. Basically, a monochromator consists of a dispersing device, focusing optics, and a pair of slits. Figure 19–2 reveals the arrangement of these components for a typical prism monochromator. Radiation arriving at the entrance slit is sent to the prism through a lens. Within the prism, the radiation is separated by refraction into its components, ultraviolet radiation being refracted (bent) most and visible red light being refracted least. Each refracted component of the radiation is then focused at the **focal plane**, where the dispersed spectrum appears. A movable exit slit placed at the focal plane can then be positioned to allow selection of any wavelength or frequency in the dispersed spectrum. In practice, the dispersed spectrum is rarely scanned by moving the exit slit. Instead, the prism is rotated so that the desired portion of the dispersed spectrum falls on the exit slit. With suitable calibrated mechanical linkages, the position of the prism can be related to the central wavelength impinging upon the exit slit; the numerical value of this wavelength can then be displayed for observation.

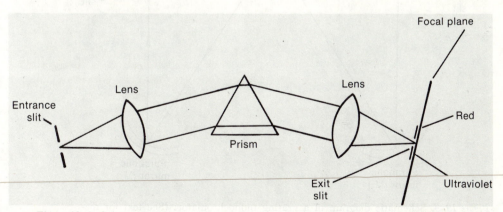

*Figure 19–2.* Schematic diagram of a simple prism monochromator. (See text for discussion.)

Prism monochromators of the kind shown in Figure 19–2 produce a spectral dispersion which is nonlinear with respect to location on the focal plane. Therefore, considerable care is necessary to ensure that the desired central wavelength actually falls on the exit slit. In addition, high-quality prisms must be free of bubbles and defects and must be as transparent as possible. To overcome these problems, a diffraction grating is often substituted for the prism as the dispersing element in a monochromator. With a grating monochromator, radiation is dispersed linearly in wavelength. Moreover, the efficiency of a grating monochromator can be higher and its cost lower than for a corresponding prism instrument.

A grating disperses electromagnetic radiation by utilizing the phenomenon of **diffraction**. Although a thorough discussion of diffraction is beyond the scope of this book, suffice it to say here that diffraction can be produced by grooved reflecting (mirror-like) surfaces rather than by transparent optical materials. Therefore, compared with a prism, a grating is not as strongly affected by imperfections in the bulk material from which it is made. In addition, diffraction gratings can be inexpensively fabricated by replication from carefully made master gratings. This reduces the cost of replicate gratings, making it practical to utilize large gratings to increase the optical speed of the monochromator. A typical grating monochromator is depicted in Figure 19–3.

The grating monochromator shown in Figure 19–3 is only one of a number of types in common use today. In this monochromator, radiation from the entrance slit is rendered parallel (**collimated**) by a parabolic mirror ($M_2$) after being reflected from a flat mirror ($M_1$). The parallel radiation is then dispersed into its component wavelengths by the diffraction grating. Next, the parallel, dispersed rays are focused onto the plane of the exit slit by a second parabolic mirror ($M_3$). Thus, one can select the wavelengths which pass through the exit slit merely by moving the entrance or exit slit or by rotating the grating. Generally, the latter method is chosen because it is simpler and more convenient, as is true for a prism monochromator. Notice that a

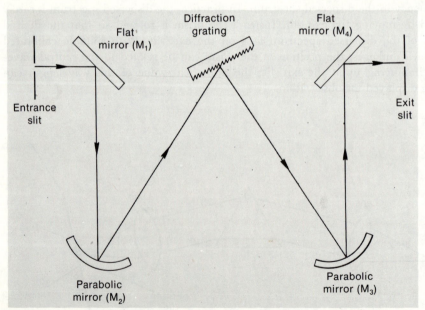

*Figure 19–3.* Schematic diagram of a grating monochromator. (See text for discussion.)

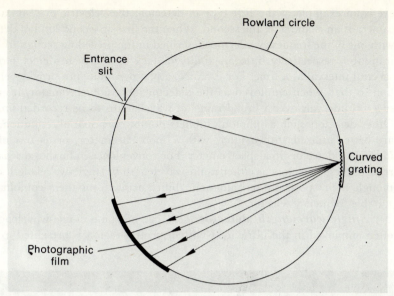

*Figure 19–4.* Diagram of a curved-grating (Rowland circle) spectrograph. Note that the curved grating, photographic film, and entrance slit all lie on the same circle. The radius of curvature of the grating is equal to the diameter of the circle.

grating monochromator frequently uses only reflecting surfaces in its optical system, whereas a prism monochromator (Figure 19–2) utilizes transmitting optics. Although this distinction between prism and grating systems is not always found, it is common. When a choice is possible, reflecting optics are often preferred because of their freedom from undesirable optical effects such as chromatic aberration.

*Polychromators.* Whereas filters and monochromators serve to isolate specific, narrow spectral regions for measurement, a polychromator allows the simultaneous observation of many or all spectral elements in a beam of radiation. Physically, a polychromator (literally a device producing many colors) resembles a monochromator in that it has an entrance slit, focusing optics, and a dispersing device. In fact, one type of polychromator, called a **spectrometer**, is little more than a monochromator with several exit slits. In a spectrometer, a radiation detector is placed at every exit slit so that each isolated spectral interval can be individually and simultaneously observed.

Another often-encountered form of polychromator is a **spectrograph**. A spectrograph has no exit slit but, rather, employs a continuous radiation detector such as a photographic plate or film placed along the focal plane. An example of a spectrograph can be seen in Figure 19–4. In contrast to a spectrometer, the spectrograph of Figure 19–4 produces a continuous record of *all* components of a spectrum, so that a great deal of information is gathered over a given time interval. In effect, a spectrograph resembles a spectrometer having an infinite number of exit slits and detectors. However, a spectrograph requires a specific type of detector such as a photographic emulsion which is capable of recording a continuous spectrum.

## Detectors for Ultraviolet-Visible Radiation

Radiation detectors used in the ultraviolet-visible spectral region fall into two categories, those which provide spatial resolution and those which do not. The human eye is an example of the first type of detector; the "electric eye" on an automatic camera is an example of the second. When the first or second kind of detector is used with one of the frequency selectors described in the preceding section, this distinction translates, respectively, into an ability or lack of ability to detect more than one spectral interval at a time. For instance, a photographic film provides spatial resolution so that, when employed as the detector for a polychromator, it enables many spectral intervals over a broad range of wavelengths to be recorded simultaneously. Other detectors such as photomultipliers do not offer spatial resolution, so that they are better utilized with a filter, with a monochromator, or as one of the several individual detectors in a spectrometer. For convenience in further discussion, we will refer to these detectors as either multi-wavelength or single-wavelength devices, even though their spectral-discrimination ability arises from their combination with a suitable frequency selector.

*Single-wavelength detectors.* A list of some single-wavelength detectors often employed in the ultraviolet-visible spectral region is presented in Table 19–1.

These detectors differ in their sensitivity, linearity, spectral response, response time, and price. Ideally, each detector should be extremely sensitive and should respond linearly to the radiant power incident upon it. Furthermore, the detector should respond equally to all spectral frequencies and should respond rapidly to a change in the level of incident radiation. The degree to which each detector approaches these ideals is summarized in Table 19–1.

Notice that the human eye has been included among the detectors listed in Table 19–1. Although the human eye is inherently a multi-wavelength detector, it has often been used in relatively simple single-wavelength measurements (recall the example in Chapter 18, page 610). In fact, the earliest observations of many important spectroscopic phenomena were made with the human eye as a detector. An example is the discovery by Fraunhofer of the dark lines in the solar spectrum, an observation which led, after considerable development, to the modern analytical technique of atomic absorption spectrometry. However, the human eye is very inferior to many of the other single-wavelength detectors in terms of sensitivity, spectral response, and linearity, so that it is rarely employed in present-day spectrometry. Other detectors, primarily the photomultiplier tube and the vacuum photodiode, are excellent in all categories of behavior and are consequently found in most modern spectrochemical instruments.

A phototube and a photomultiplier tube (the latter sometimes called a multiplier phototube) both operate on the principle of the **photoelectric effect**. This effect, first explained by Einstein, involves the absorption of a photon by a substance and the subsequent emission of an electron from the material. Some elements, notably the alkali metals, release electrons quite readily and exhibit strong photoelectric properties. In the vacuum phototube, radiation enters through a transparent window and strikes such a photosensitive surface (the photocathode). The resulting ejected electrons (photoelectrons) travel through vacuum to a positively charged anode, where they are collected. Measurement of the current that flows from the anode indicates the number of photoelectrons ejected and, indirectly, the power of the incident radiation.

A photomultiplier tube operates similarly except that, for each photoelectron produced, many electrons appear at the anode. This multiplication is accomplished with a series of electrodes, termed dynodes, each of which releases electrons very readily. A photoelectron ejected from the photocathode travels to a dynode held at a more positive voltage (approximately 100 volts). When the photoelectron (which is greatly accelerated by the positive voltage) strikes the surface of the dynode, several (between 1 and 6) secondary electrons are kicked out of the dynode. These electrons then travel to second, third, fourth dynodes, and so on, each of which is held at a successively higher positive voltage and each of which releases several secondary electrons for each incident electron. In this way, a single photoelectron released by the photocathode can produce a burst of $10^5$ to $10^8$ electrons at the anode. This multiplicative feature makes the photomultiplier so much more sensitive than the phototube that it is possible to detect the arrival of a single photon at the photocathode. Of course, the photomultiplier is more costly than the phototube. As with all the detectors listed in Table 19–1, the choice depends upon the requirements of a specific application, being based on a compromise between price and performance.

*Multi-wavelength detectors.* In the ultraviolet-visible region, the photographic emulsion is the most common multi-wavelength detector. Since the emulsion is usually supported on glass or plastic, the actual detector takes the form of a photographic plate or film, respectively. Although other multi-wavelength detectors are now appearing, photographic detection is still dominant. An example of a spectrum (that of sodium) recorded on photographic film is displayed in Figure 19–5.

As mentioned earlier, the use of a photographic emulsion offers the advantages of being able both to detect a *range* of wavelengths simultaneously and to resolve the spectral intervals within the range extremely well. Furthermore, a photographic

*Figure 19–5.* A photographically recorded emission spectrum of sodium. Notice the relative simplicity of this "line" spectrum, which shows three doublets; the numbers are the wavelengths of the lines in nanometers (nm).

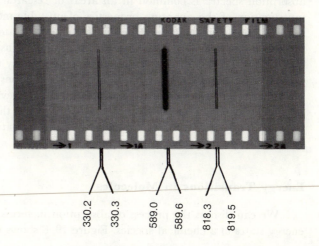

330.2  330.3    589.0  589.6   818.3  819.5

emulsion directly *integrates* the radiation incident upon it, so that extremely weak radiation levels can be detected if sufficiently long exposure times are employed. Moreover, a spectrum recorded on an emulsion can be stored for extended periods of time if suitable precautions are taken. These advantages over single-wavelength detectors would seem to make the emulsion an attractive choice for the observation and recording of any spectrum.

Unfortunately, photographic detection has several undesirable features. First, because the exposing, developing, fixing, and drying of a photographic emulsion is a lengthy process, the time needed to obtain a spectrum is quite long, usually from several minutes to a few hours. Compare this delay with the relatively rapid response times of the single-wavelength detectors in Table 19–1. Second, even after the photographic processing is completed, spectral information present in the emulsion must be converted to a proportional electrical signal, often by means of a tedious operation requiring use of a **microphotometer**. Third, photographic detection is notoriously nonlinear with respect to integrated radiant power. Therefore, it is necessary to calibrate each individual emulsion for its response to radiation if quantitative results are to be achieved. Fourth, photographic emulsions usually have a limited storage time, both before and after exposure. This characteristic necessitates either occasional recalibration of the emulsion or immediate examination of the developed emulsion with a microphotometer.

Because of these disadvantages and inconveniences, photographic emulsions are not frequently used in modern spectrochemical instruments except in those applications where direct integration and recording of continuous spectra are essential. Examples of such situations will be described in the next chapter.

In this section we have briefly considered specific examples of two building blocks common to all instruments for ultraviolet-visible spectrometry—the frequency selector and the radiation detector. As we investigate various analytical methods which employ ultraviolet-visible radiation, it will be of interest to observe the ways in which these components are used. It should become obvious that the spectrochemical techniques discussed in this and other chapters are remarkably similar, differing primarily in their methodology and application.

## MOLECULAR ANALYSIS BY ULTRAVIOLET-VISIBLE ABSORPTION SPECTROPHOTOMETRY

Characterization of chemical species by means of their ultraviolet or visible absorption spectra is common in all areas of research and development, from the investigation of the quantum properties of unstable molecules in the upper atmosphere to the determination of the number of cobalt atoms present in a molecule of vitamin $B_{12}$. Ultraviolet-visible spectrophotometry is useful for both qualitative and quantitative analysis, although we will focus primarily upon the latter.

For the majority of quantitative determinations, ultraviolet-visible spectrophotometry is performed with liquid samples. However, the technique is equally applicable to gaseous or solid samples. As we shall see, the state of the sample plays an important role in governing the nature of the transitions observed in ultraviolet-visible absorption spectra.

### Energy Transitions in Molecules

We can discuss the process of absorption in terms of a transition between two energy states of a chemical species. Figure 19–6 shows three electronic energy levels

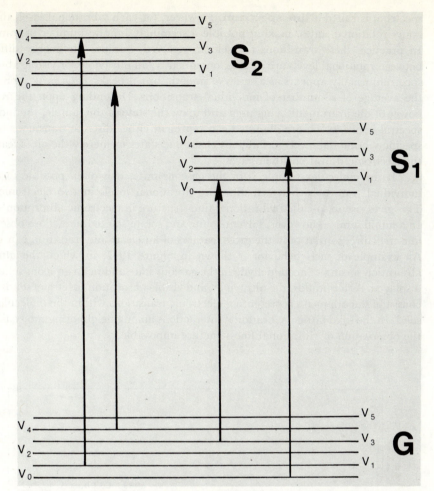

*Figure 19–6.* Electronic energy-level diagram for a hypothetical molecule, showing the energy changes (arrows) that occur upon absorption of electromagnetic radiation; **G** represents the ground electronic state, $S_1$ and $S_2$ are excited electronic states, and $v_0$, $v_1$, $v_2$, and so on denote the various vibrational levels for each electronic state.

of a molecule: the ground state (**G**) and two excited singlet states ($S_1$ and $S_2$). In turn, each electronic state has associated with it a number of vibrational levels labelled $v_0$, $v_1$, $v_2$, and so on. Because the energy difference between electronic levels is considerably greater than the thermal energy ($kT$) at room temperature, the Boltzmann distribution law indicates that most molecules will reside in the ground electronic state. However, the difference in energy between vibrational levels is much smaller. Therefore, at or near room temperature, most molecules are found in the ground electronic state but might be in excited vibrational states.

Absorption of ultraviolet or visible radiation usually involves raising a molecule from one of several vibrational levels in the ground electronic state to a vibrational level of an excited electronic state. Several transitions of this type are indicated in Figure 19–6. Each of these transitions corresponds to a definite energy change and, therefore, can be caused only by absorption of a photon having exactly that energy. Thus, we would expect the ultraviolet or visible spectrum of a molecule to exhibit a set of sharp absorption lines, one for each of the discrete transitions. This sort of spectrum is called a **line spectrum**. However, for each vibrational level, there are many rotational states, making possible a seemingly infinite number of transitions. In practice, these transitions are seldom resolved, because the energy differences between rotational levels are small. Consequently, an ultraviolet or visible absorption spectrum usually appears as a series of smooth, broad peaks, each peak representing the average of a number of individual transitions. Depending upon the resolving power of the instrument being used and upon the state of the sample, the individual spectral bands which are observed can represent either discrete transitions between specific vibrational levels in different electronic states or merely the gross features of the electronic transitions themselves.

If the molecular sample is in the gaseous state, it is often possible to observe individual transitions between different vibrational levels in two electronic states. This gives rise to so-called vibrational fine-structure in electronic absorption spectra. In a liquid sample, however, solvent-solute and solute-solute interactions obscure this fine-structure, so that only the gross features of an electronic transition can be seen. An example of such behavior is shown in Figure 19–7, in which the ultraviolet absorption spectra of acetaldehyde in the gaseous state and in an aqueous solution are displayed. As for liquids, the ultraviolet and visible absorption spectra of solids usually consist of smooth peaks representing electronic transitions within the molecules of the solid. In the solid lattice, vibrational interactions among neighboring molecules make the observation of vibrational fine-structure impossible.

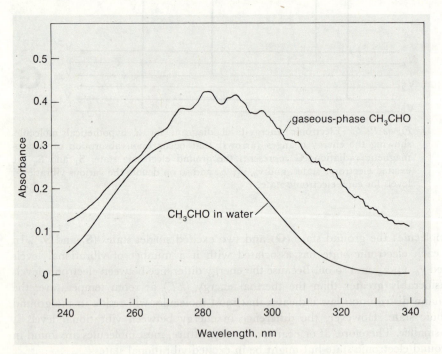

*Figure 19–7.* Ultraviolet absorption spectra for acetaldehyde ($CH_3CHO$) in the gaseous state and in an aqueous solution. Vibrational fine-structure, in the form of a number of readily resolved, small humps on top of the electronic absorption band, is clearly visible in the gas-phase spectrum. In an aqueous solution, solute-solvent and solute-solute interactions obscure this fine-structure, so that only a smooth absorption band is seen. The spectrum of acetaldehyde in the gaseous state has been displaced upward 0.1 absorbance unit to keep the two curves from overlapping. Notice that the wavelength of the absorption maximum for acetaldehyde is influenced by its environment.

Remember that, in order for a molecule to absorb ultraviolet or visible radiation, the energy differences between various electronic levels of the species must be identical to the energies present in frequency components of the incident radiation. These energy differences, which are proportional to the frequency of the absorbed radiation, can be expressed by means of the relationship

$$\Delta E = \frac{286,000}{\lambda} \tag{19-2}$$

where $\Delta E$ is the energy difference between states in kilocalories per mole and $\lambda$ is the wavelength in angstroms of the radiation responsible for the transition between those states. For example, radiation having a wavelength of 2860 Å will cause a transition between states which differ in energy by 100 kilocalories per mole. This energy difference is comparable to the strengths of some common chemical bonds; the energy of a C—H bond is approximately 98 kcal/mole and the energy of a N—H bond is approximately 92 kcal/mole. Thus, radiation causing transitions in the ultraviolet-visible spectral region is capable of dissociating certain molecules. The process whereby bonds are cleaved through absorption of radiation is termed **photolysis**. Obviously, photolysis must be avoided when ultraviolet-visible spectrophotometry is being used for quantitative determinations. Also, the likelihood of photolysis renders thermal or electrical excitation of molecules rather inefficient, so that emission of ultraviolet or visible radiation by molecules is seldom observed, except in a luminescence process (to be discussed later in this chapter).

## Advantages of Spectrophotometry

Ultraviolet and visible molecular spectrophotometry offers several significant advantages as an analytical technique. First, its sensitivity is excellent. It is frequently possible to quantitatively determine less than $10^{-7}$ to $10^{-6}$ $M$ concentrations of a molecular species. This virtue is important in two areas of application. In one area, namely **trace analysis**, ultraviolet-visible spectrophotometry permits the quantitative measurement of minor sample constituents which are present at levels near 1 part per million by weight. In another area, termed **microanalysis**, major components can be quantitatively determined in a sample of extremely small size. Compared with traditional titrimetric and gravimetric methods, spectrophotometry is rapid and convenient. With most modern spectrophotometric instrumentation, it is possible to analyze five to ten samples per minute. Moreover, it is relatively easy to automate spectrophotometric procedures, from the first step of sample introduction to the final calculation of the concentrations of several constituents in a sample. Finally, spectrophotometric methods yield, in addition to analytical data, fundamental information about the structure of molecules and the nature of chemical bonding. In fact, such studies have been responsible for much of our present knowledge of the quantum properties of matter.

## Spectrophotometric Measurements

Let us now examine the basic instrumentation and methodology of ultraviolet-visible spectrophotometry. Because we are dealing with absorption, quantitative work is based upon Beer's law. Also, instrumentation requirements are defined by the considerations presented in Chapter 18 (see Figures 18–10 and 18–11). A typical ultraviolet-visible spectrophotometer is shown in Figure 19–8.

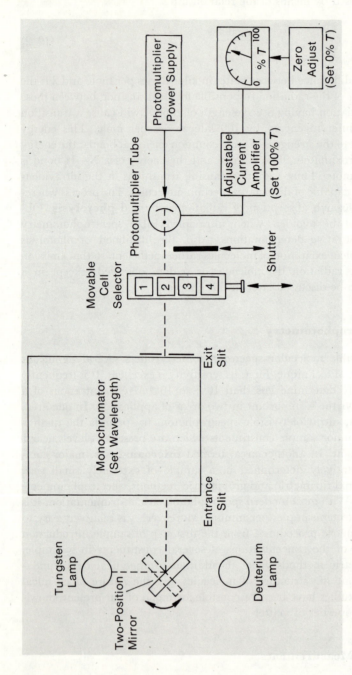

*Figure 19–8.* Schematic diagram of a generalized single-beam ultraviolet-visible spectrophotometer having two sources and a movable cell selector. Selection of the desired source is accomplished with the aid of a two-position mirror. To set the readout meter to 0 per cent transmittance, the shutter is pushed in front of the photomultiplier tube (to block the light path) and the zero adjust control is manipulated. To set the readout meter to 100 per cent transmittance, the shutter is pulled open, a spectrophotometer cell containing a blank solution is positioned in the light path, and the adjustable current amplifier is manipulated. To obtain the per cent transmittance for an unknown solution, it is only necessary to position the cell containing it in the light path, to open the shutter, and to read the meter.

Most probably, the source of radiation in the spectrophotometer will be a tungsten, deuterium (hydrogen), or quartz-halogen lamp. Because each of these sources generates its maximum radiation intensity in a different region of the ultraviolet-visible spectrum, it is not uncommon to find two lamps incorporated into the more sophisticated spectrophotometers, each lamp being used in its optimum spectral region. Each of these sources emits a broad spectrum of radiation, so that it is necessary to isolate a portion of their output spectra for illumination of the chemical sample. For this purpose, a frequency selector such as a monochromator or filter is employed. Instruments which utilize a monochromator as a spectral dispersing device are termed **spectrophotometers**. In contrast, **filter photometers** employ either absorption or interference filters for wavelength selection. A filter photometer which operates only in the visible region is often called a **colorimeter**. In our discussion we will use the term *spectrophotometer* in a general way to designate all these devices, although in a specific case the correct distinction should be made.

After isolation of appropriate wavelengths, the selected radiation is used to illuminate the sample cell. Radiation transmitted through the sample cell then falls upon a suitable detector, such as a photomultiplier or photodiode, which in turn converts the radiation to a proportional electrical signal for readout. For protection from excessive and extraneous radiation, the detector is often equipped with a shutter. Whenever a sample is being introduced into or removed from the spectrophotometer, the shutter should be closed to avoid damage to the photodetector.

Let us now consider the measurements necessary to perform a quantitative determination with the instrument of Figure 19–8. From Beer's law, we know that the concentration of the desired chemical species (**analyte**) is proportional to its absorbance at a specific frequency or wavelength; in turn, the absorbance is the logarithm of the ratio of incident radiant power to transmitted radiant power:

$$A = abc = \log \frac{P_0}{P} \tag{19–3}$$

To evaluate concentration, we must determine the radiant power incident on the sample $(P_0)$ and the radiant power transmitted by the sample $(P)$. In the simplest case, we can make these measurements by first removing the sample and its container from the sample compartment of the spectrophotometer. When this is done, the detector will measure all the radiation incident upon the sample cell. For convenience of interpretation, the readout device can be set to read 100 per cent transmittance when all the incident radiation $(P_0)$ falls on the detector. Similarly, the readout device can be adjusted to read 0 per cent transmittance when no light falls upon the detector—that is, when the shutter is closed. After these two preliminary settings are made, the readout system will automatically display the desired value of transmittance $(P/P_0)$ when the sample is reinserted into the spectrophotometer. The absorbance $(A)$ can then be obtained if we compute the negative logarithm of the transmittance. However, many modern spectrophotometers provide readout directly in terms of both transmittance and absorbance.

In practice, the simple procedure outlined above can produce considerable error. To understand the difficulty, consider a typical sample cell for ultraviolet-visible spectrophotometry as shown in Figure 19–9. A properly designed cell will have accurately fixed dimensions, will be made of a suitable transparent material (such as quartz or Vycor glass), and will have parallel, optically clear windows on opposite sides. Ideally, such a cell should have no effect upon the measurements taken.

However, as portrayed in Figure 19–10, there can be an *apparent* absorption of

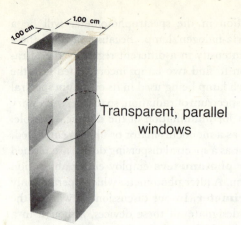

Transparent, parallel
windows

radiation even when the spectrophotometer cell contains none of the analyte species. This apparent absorption is caused by reflection of incident radiation from the cell walls or by scattering of incident radiation from dust, fingerprints, or other foreign material present on the cell walls or suspended in the solvent; absorption of the incident radiation by another component in the sample or by the solvent itself produces the same effect. These radiation losses decrease the effective radiant power incident upon the analyte from the original value $(P_0)$ to a lower value $(P_0)_{effective}$. For correct absorbance measurements, it is essential to employ $(P_0)_{effective}$ for the 100 per cent transmittance (zero absorbance) point. This can be accomplished if a sample blank is utilized.

A sample blank is merely a solution identical to the sample in all respects

$$(P_0)_{effective} = P_0 - \left[ (P_0)_{scatter} + (P_0)_{absorption} + (P_0)_{reflection} \right]$$

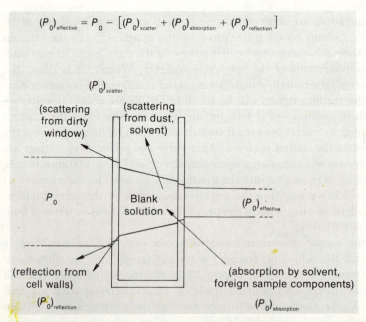

Figure 19–10. Radiation losses in a spectrophotometer cell in which no analyte is present. The width of the radiation beam is a schematic indication of the radiant power. (See text for discussion.)

except that it contains none of the analyte species. When placed in a spectrophotometer cell, the blank should scatter, absorb, refract, and reflect radiation in the same way as the sample solution. Thus, the radiant power transmitted by the blank should be equal to that effectively incident upon the analyte, $(P_0)_{effective}$. Of course, this procedure assumes a perfect match of the cells in which the sample and blank solutions are held, a criterion that can never be exactly met but can be approached through the purchase of so-called matched cells. The difficulty of having exactly matched cells can be the factor which limits the accuracy of high-quality spectrophotometers.

Therefore, correction for undesirable radiation losses can be accomplished if the 100 per cent transmittance point is set with a blank solution in the cell compartment. To simplify this procedure, many commercial spectrophotometers have specially designed cell compartments in which several cells can be held and rapidly inserted into the radiation path. With such an arrangement, a cell containing the blank can be placed into position to set $(P_0)_{effective}$ (the 100 per cent transmittance level). Next, one or more cells containing solutions to be analyzed can be sequentially inserted to obtain respective values for $P$ (or per cent transmittance). With care, this method is capable of yielding spectrophotometric measurements precise to $\pm 1$ per cent.

## Spectrophotometric Errors

Several sources of error exist in ultraviolet-visible spectrophotometry. Among the sources of error are those arising from non-obeyance of Beer's law (see Chapter 18, page 616). Others involve the characteristics of ultraviolet-visible spectrophotometric instrumentation. Instrumental errors include variations in source power, detector response, electrical noise, and cell positioning; and there is always the human error associated with the reading of absorbance or transmittance scales. Although these errors can often be minimized or eliminated through the use of calibration procedures, the cumulative error from all sources is usually between 0.2 and 1 per cent. Let us first consider the scale-reading error.

Many existing ultraviolet-visible spectrophotometers present their readout in terms of transmittance on a meter, chart recorder, or similar system. These readout devices all generate a common source of error, that of reading the transmittance exactly. For a properly designed meter, the reading error will be constant and will probably equal the width of the needle on the meter face. For a recorder, the constant readout error will probably be the width of the pen trace. If we assume that the transmittance reading obtained from each of these devices has a constant uncertainty, how much error will this uncertainty cause in the recorded concentration value of the species we are seeking to determine?

To calculate the error in concentration, we must examine the effect that an uncertainty in transmittance has on the value of the concentration. To do this, let us write Beer's law in the form

$$T = \frac{P}{P_0} = 10^{-abc} \tag{19-4}$$

which shows that the relationship between transmittance and concentration is an exponential one. From the plot of this relationship in Figure 19-11, we can determine the effect that a constant error in transmittance will have on the calculated value of $c$.

Consider first a spectrophotometric measurement made on a solution with a low concentration ($c_1$) of the desired species. According to Figure 19-11, this solution will have a high transmittance value ($T_1$). Assuming that the uncertainty of reading a

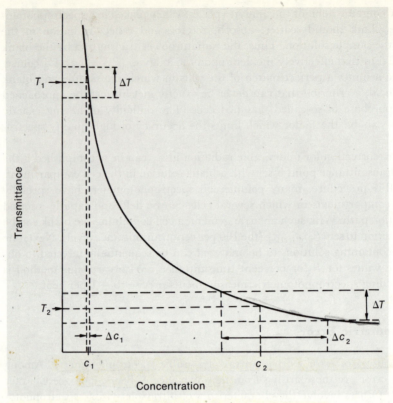

*Figure 19–11.* Plot of exponential form of Beer's law showing the relative error in concentration ($\Delta c/c$) produced by a constant error in transmittance ($\Delta T$) for different transmittance values. (See text for discussion.)

meter, recorder, or other device corresponds to a constant absolute error in $T$, we can determine from Figure 19–11 the resulting error in $c$. From the extrapolated lines in Figure 19–11, it is seen that the absolute error in concentration ($\Delta c_1$) is small when a solution has a high transmittance. However, because the concentration ($c_1$) itself is low, the relative error in concentration ($\Delta c_1/c_1$) is quite large.

Let us compare the preceding situation with that encountered when a high concentration ($c_2$) of the analyte is present. This solution will have a relatively low transmittance value ($T_2$). If a constant error in $T$ is again assumed, there is a large absolute error ($\Delta c_2$) in the concentration; and, even though the concentration itself is large, the relative error ($\Delta c_2/c_2$) is also large.

These results suggest that, somewhere between the extremes of samples having high and low transmittances, we might expect a transmittance value for which the relative error in concentration ($\Delta c/c$) is minimal. In fact, the relative error is minimal at 36.8 per cent transmittance. We can reach this conclusion in a quantitative way by recalling the definition of absorbance:

$$A = \log\left(\frac{1}{T}\right) = \left(\frac{1}{2.303}\right)\ln\left(\frac{1}{T}\right) = 0.434\ln\left(\frac{1}{T}\right) \tag{19–5}$$

If this equation is differentiated with respect to $T$, the result is

$$\frac{dA}{dT} = -\frac{0.434}{T} \tag{19–6}$$

or

$$dA = -\frac{0.434}{T}\,dT \tag{19-7}$$

In order to evaluate the relative error in absorbance and, therefore, in concentration, it is necessary to divide each side of equation (19–7) by the absorbance ($A$). Thus,

$$\frac{dA}{A} = -\left(\frac{0.434}{TA}\right)dT = \left(\frac{0.434}{T\log T}\right)dT \tag{19-8}$$

To find the relative error in concentration, let us substitute for $A$ from Beer's law:

$$\frac{d(abc)}{abc} = \frac{ab\,dc}{abc} = \frac{dc}{c} = \left(\frac{0.434}{T\log T}\right)dT \tag{19-9}$$

Equation (19–9) shows that the *relative* error in concentration ($dc/c$) is governed directly by the *absolute* uncertainty in the transmittance ($dT$) as well as by the reciprocal of a product of terms ($T\log T$) involving the transmittance itself. A plot of this equation showing the relative error in concentration ($dc/c$) as a function of per cent transmittance for a constant reading error ($dT$) of 1 per cent is shown in Figure 19–12.

From Figure 19–12, it is apparent that the relative error in concentration is reasonably low for transmittances between 20 and 70 per cent, but that at very low or high transmittance values the error increases rapidly. These observations verify our earlier qualitative conclusions. We can find the minimum relative error in concentration by differentiating equation (19–9) and setting the expression equal to zero. As stated earlier, this minimum error occurs at a transmittance of 36.8 per cent or an absorbance of 0.434. Therefore, for spectrophotometers having a meter or recorder

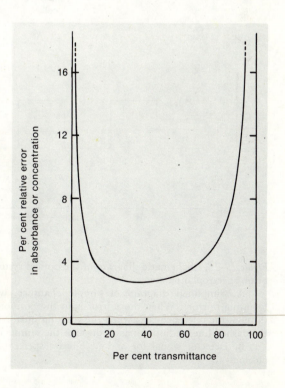

*Figure 19–12.* Per cent relative error in absorbance or concentration, as a function of per cent transmittance, due to a 1 per cent absolute error in the observed per cent transmittance. (See text for discussion.)

readout device, measurements of greatest accuracy can be made if readings are limited to the range from 20 to 70 per cent transmittance (or from an absorbance of 0.7 to 0.2), with values closest to 36.8 per cent transmittance or 0.434 absorbance being optimal. These restrictions can usually be met through appropriate selection of the path length of the sample cell ($b$) or dilution of the sample solution taken for analysis.

*It must be emphasized that the foregoing considerations apply only to measurements made with spectrophotometers having a constant reading error in transmittance.* Although many present-day instruments fall into this category, the trend is toward spectrophotometers which incorporate scale-expansion capability or digital readout, in which readout error is not the accuracy-limiting factor. In fact, spectrophotometric measurements limited by readout error are the *least* accurate and can always be improved if one simply substitutes a more sensitive readout device. In this case, electrical noise can become the limiting source of error. For those instruments in which random noise is

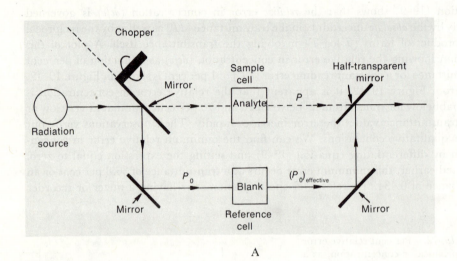

A

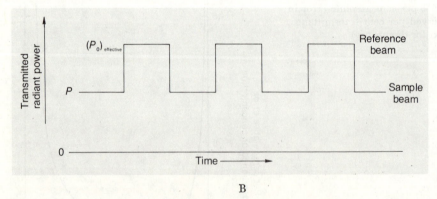

B

*Figure 19–13.* Schematic diagram of the optical system and operation of a double-beam spectrophotometer.

A. Simplified diagram of sample chamber, with beam-switching chopper (rotating mirror). When the mirror is in the position shown, radiation is directed through the reference cell; when the rotating mirror reaches the position shown by the dashed line, radiation passes through the sample cell.

B. Detector signal for a double-beam spectrophotometer. (See text for discussion.)

the accuracy-limiting factor, the optimum results are obtained near 11 per cent transmittance (or an absorbance of 0.96).[a-c]

Other errors in ultraviolet-visible spectrophotometric methods arise from drift in the output power of the source or variation in the wavelength response of the detector. These errors can be minimized if a spectrophotometer having double-beam optics is employed. The optical system and operation of a typical double-beam spectrophotometer are illustrated in Figure 19–13. In the optical system of Figure 19–13A, the incident radiation $(P_0)$ is alternately sent to two cells, one containing the sample solution and the other a reference solution. This reference solution serves as a blank and should be identical to the sample solution except that it contains none of the analyte species. Alternating radiation transmitted by the sample and reference cells then falls upon the detector, producing an electrical signal such as that shown in Figure 19–13B. Because the species being determined absorbs a larger fraction of the incident radiation than does the blank, the radiant power transmitted by the sample cell $(P)$ is smaller than that transmitted by the reference cell $(P_0)_{effective}$. Therefore, the detector output will vary from a level representing $(P_0)_{effective}$ to a lower level representing $P$. The periodic variation from sample to blank will produce an output from the detector having the appearance of a square wave and of amplitude proportional to $[(P_0)_{effective} - P]$. In effect, a double-beam spectrophotometer rapidly and automatically interchanges the sample and blank solutions in the spectrophotometer to help compensate for variations in source power or detector response.

As an example of the effectiveness of a double-beam spectrophotometer in correcting for instrument errors, consider a change in source power $(P_0)$. If $P_0$ should increase, the radiation transmitted by the sample, and therefore the signal from the detector, would appear to increase. However, this change will be largely compensated by a proportional increase in the radiant power transmitted by the reference cell, so that the difference $[(P_0)_{effective} - P]$ is less susceptible to a variation in the source power. Analogous considerations apply to any variation of detector response. For this reason, double-beam spectrophotometers are used almost always when it is important to scan a large region of the spectrum.*

## Deviations from Beer's Law

Deviations from the fundamental law of absorbance are especially prominent in ultraviolet-visible spectrophotometry and, therefore, deserve consideration here. Of the instrumental deviations, the most common is that arising from the use of non-monochromatic radiation. The derivation of the Lambert-Beer law assumes that

---

[a] J. D. Ingle, Jr., and S. R. Crouch: Anal. Chem., *44*:1375, 1973.
[b] H. K. Hughes: Appl. Opt., *2*:937, 1963.
[c] J. D. Ingle, Jr.: Anal. Chem., *45*:861, 1973.

* In a double-beam spectrophotometer, it is the quantity $[(P_0)_{effective} - P]$ rather than $P$ which is measured. As is true for a single-beam spectrophotometer, the instrument is set to read 100 per cent transmittance by placement of the same blank solution in both the sample and reference beams. The readout system will then directly register $[(P_0)_{effective} - P]/(P_0)_{effective}$. This quantity is related to transmittance $(T)$ in the following way:

$$\frac{(P_0)_{effective} - P}{(P_0)_{effective}} = 1 - \frac{P}{(P_0)_{effective}} = 1 - T$$

Therefore, the magnitude of the output from a double-beam spectrophotometer increases with a decrease in transmittance. In most commercial instruments, this inversion is corrected automatically by a simple reversal of the readout scale. In these systems, a conversion to logarithmic units allows direct readout of absorbance on a special meter or recorder.

monochromatic radiation is employed. If this requirement is not satisfied, two situations can arise. In the first and most common case, the molar absorptivity of the analyte changes over the range of wavelengths of radiation incident upon the sample. Of course, the radiant powers at each of these wavelengths will be linearly additive. However, to satisfy Beer's law, the *logarithms* of the radiant powers must add, so that a deviation from Beer's law results. This problem is difficult to avoid in spectrophotometric measurements, and necessitates a compromise between the use of a wide spectral bandpass to obtain high levels of radiant power and a narrow bandpass to have adherence to Beer's law.

In the second situation involving non-monochromatic radiation, the range of wavelengths incident upon the sample is so much wider than the absorption band of the analyte that some wavelengths pass nearly unattenuated through the sample. This can be considered as the extreme of the preceding case. Here, the molar absorptivity of the analyte changes drastically (and, in fact, approaches zero) over part of the band of radiation employed. Such a situation is depicted in Figure 19–14.

In Figure 19–14, a beam of radiation consisting of several wavelengths (represented by $\lambda_1$ and $\lambda_2$) is incident upon a sample. However, the sample is capable of absorbing radiation only at wavelength $\lambda_1$; all other wavelengths pass unattenuated through the sample. If the incident radiation at wavelength $\lambda_1$ has a power $(P_0)_{\lambda_1}$ and that at any other wavelength ($\lambda_2$) is of power $(P_0)_{\lambda_2}$, the observed transmitted power $(P_{obs})$ will contain the correct transmitted power $(P)_{\lambda_1}$ at wavelength $\lambda_1$ as well as the unattenuated component $(P_0)_{\lambda_2}$. Therefore, regardless of the concentration of the analyte, $(P_0)_{\lambda_2}$ will generate a constant level of transmitted background radiation, thereby making the sample solution appear to transmit more radiation than would be predicted by Beer's law. Thus, the absorbance of the sample solution would increase less rapidly than expected, and a plot of absorbance versus concentration would look like that shown in Figure 19–15. Curvature seen in Figure 19–15 is negligible except at high values of concentration or absorbance. This is because the background radiation at wavelength $\lambda_2$ is sufficiently small that its contribution to the transmitted power $(P_{obs})$ is not significant until the absorbance of the sample solution is large, that is, until $(P)_{\lambda_1}$ is small. Such behavior is termed a *negative deviation* from Beer's law because the plot of absorbance versus concentration bends toward the concentration axis. Note that, although a curve such as Figure 19–15 does not obey Beer's law, it can still be used for analytical purposes because it accurately represents the change in observed absorbance with sample concentration.

As suggested in Chapter 18 (page 617), chemical deviations from Beer's law are generally encountered when the absorbing substance undergoes a change in its degree of dissociation, hydration, or complexation upon dilution or concentration. For

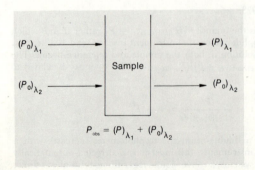

*Figure 19–14.* Illustration showing how the use of non-monochromatic radiation can lead to a negative deviation from Beer's law; $\lambda_1$ refers to the wavelength of absorbed radiation, $\lambda_2$ refers to the wavelength of non-absorbed radiation. (See text for discussion.)

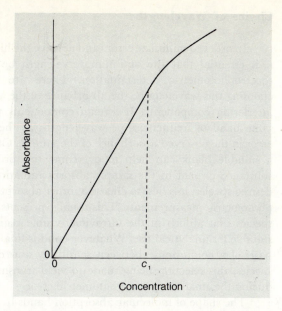

*Figure 19–15.* Plot showing negative deviation from Beer's law; $c_1$ is the maximum concentration for which Beer's law is valid. (See text for discussion.)

example, the extent of dissociation of a weak acid changes with concentration, so that Beer's law will not be applicable to a solution of a weak acid in which the radiation is absorbed by the anion of the weak acid. Dilution or concentration of a potassium chromate solution causes a shift in the position of equilibrium between chromate ion and dichromate ion:

$$2 \, CrO_4^{2-} + 2 \, H^+ \rightleftharpoons Cr_2O_7^{2-} + H_2O$$

Therefore, chromate solutions fail to follow Beer's law unless the pH is sufficiently high to keep essentially all the anions in the chromate form. If this is not done, the absorbance of chromate at 372 nm will show a negative Beer's law deviation. By contrast, the dichromate absorbance at 348 nm will deviate in a positive direction, as illustrated in Figure 19–16.

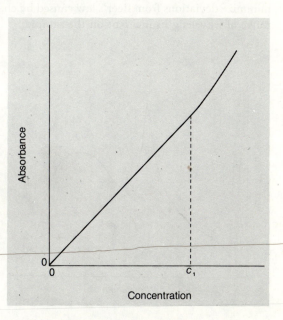

*Figure 19–16.* Plot showing positive deviation from Beer's law; $c_1$ is the maximum concentration for which Beer's law is valid. (See text for discussion.)

## Choice of Wavelength

In order to minimize errors and achieve the highest possible analytical sensitivity, it is essential that the optimum wavelength be selected for an ultraviolet-visible spectrophotometric determination. There are two important considerations in choosing this wavelength: the absorbance of the analyte itself and the absorbance of interfering components. If foreign components present in the sample absorb the same band of frequencies or wavelengths as the desired species, there will be an error in the observed absorbance of the sample. Of course, the preparation and use of a suitable blank can help to overcome this problem. However, if an interfering substance present in the sample absorbs radiation at the same wavelengths as the desired species, it is best to choose another absorption band of the analyte for spectrophotometric measurements. This usually presents no problem, since most molecular species that absorb in the ultraviolet-visible region have several absorption bands suitable for analytical use. Whatever band is finally chosen should be selected on the basis of a compromise between sensitivity and concentration range of the desired species, this selection being made to yield transmittance readings in the range producing the smallest spectrophotometric error.

The shape of molecular absorption bands in the ultraviolet-visible region varies greatly and must be considered in the final selection of the particular wavelength used for a spectrophotometric analysis. Suppose, for example, that the desired species has the hypothetical absorption spectrum shown in Figure 19–17. This absorption spectrum consists of an intense, sharp peak and a less intense, broad band. Obviously, selection of the wavelength corresponding to the maximum of the sharp peak will provide greater sensitivity. However, a small variation in the position of the frequency selector will cause a shift in the wavelengths of radiation incident upon the sample cell and, consequently, a large change in the absorbance reading. On the other hand, if the central portion of the broad band is employed for analysis, any shift in the wavelengths of incident radiation will have a much smaller effect on the absorbance. Because a slight wavelength shift is not uncommon in absorption spectrophotometers, it is better to choose a wavelength for analysis which lies in the central portion of a broad band rather than one either at the top of a sharp peak or on the sloping side of a peak or band. In addition, the use of a broad absorption band will minimize deviations from Beer's law caused by changes in molar absorptivity over the range of wavelengths incident on the sample.

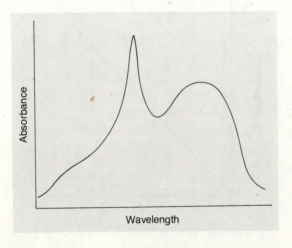

*Figure 19–17.* Hypothetical absorption spectrum of an analyte. To minimize the spectrophotometric error, absorbance measurements should be made at wavelengths corresponding to the flat portion of the broad band. (See text for discussion.)

## Analysis of Multicomponent Samples

It is often possible to determine in the same sample the concentrations of several species having overlapping absorption spectra. The procedure relies upon the fact that, if two or more species contribute to the absorbance at a fixed wavelength, the total absorbance equals the sum of the individual absorbances. Thus, if two components $x$ and $y$ contribute to the absorbance at wavelength $\lambda_1$, we have

$$(A_{\text{total}})_1 = (A_x)_1 + (A_y)_1 \qquad (19\text{--}10)$$

Applying the Lambert-Beer law to each of the two components and choosing a spectrophotometer cell with a sample path length ($b$) of 1.00 cm, we can rewrite the preceding relation as

$$(A_{\text{total}})_1 = a_{x_1}c_x + a_{y_1}c_y \quad \text{(all at } \lambda_1) \qquad (19\text{--}11)$$

A similar equation can be formulated for the total absorbance at another wavelength $\lambda_2$:

$$(A_{\text{total}})_2 = a_{x_2}c_x + a_{y_2}c_y \quad \text{(all at } \lambda_2) \qquad (19\text{--}12)$$

Numerical values of the absorptivities $a_{x_1}$ and $a_{x_2}$ can be determined at $\lambda_1$ and $\lambda_2$, respectively, by measurements upon a standard solution of $x$—that is, upon a solution in which $c_x$ is known and $c_y$ is zero. Similarly, the values of $a_{y_1}$ and $a_{y_2}$ at the same two wavelengths can be determined with a standard solution of $y$. Note that, as long as the same sample cell is used throughout the procedure, no error arises from the assumption that the path length in the cell is equal to 1.00 cm.

When we measure the absorbance of an unknown mixture of two species at two different wavelengths, we have two equations containing only $c_x$ and $c_y$ as unknowns. These equations can be solved simultaneously for the composition of the unknown sample. Preferably, one of the wavelengths should be chosen such that $x$ absorbs much more than $y$ and the other wavelength such that $y$ absorbs to a much greater extent than $x$.

In principle, this procedure can be extended to samples containing several ($n$) species; it is only necessary to make $n$ measurements of the absorbance of the sample solution at $n$ different wavelengths. These data provide $n$ equations containing $n$ unknowns, each unknown being the concentration of one species in the mixture. However, this procedure is not very satisfactory for samples containing more than three components because accuracy declines drastically.

## Selected Determinations

Because there are many examples of useful analyses based upon the absorption of ultraviolet or visible radiation, those procedures discussed here have been chosen arbitrarily and are meant only to be illustrative. We shall consider the determination of metal ions, organic compounds, and species of current ecological interest. All the methods are of practical value in modern analytical laboratories, and each procedure possesses one or more features that are applicable to other determinations as well.

In addition, ultraviolet-visible spectrophotometry is valuable for the characterization of chemical equilibria in solution, because measurements of absorbance can be made without altering or disturbing a system under examination. This approach is often utilized to determine the formulas of complex ions in solution and to evaluate the formation constants for these species. Although we will not develop here the details of the theory and experimental procedures used in such studies, several problems at the end of this chapter are concerned with this subject.

*Metal ion analysis—determination of iron.* Many analytical methods have been devised for the determination of iron by means of ultraviolet-visible spectrophotometry. Some procedures involve the direct determination of iron(III), whereas others take advantage of highly colored complexes formed between iron(II) and various organic ligands.

Among the most important reagents for the spectrophotometric determination of iron are 1,10-phenanthroline (orthophenanthroline) and its substituted analogs, such as 4,7-diphenyl-1,10-phenanthroline (bathophenanthroline), plus several other related ligands:

1,10-phenanthroline          4,7-diphenyl-1,10-phenanthroline

These organic ligands react with iron(II) to yield intensely colored three-to-one complexes (see page 289) and, in addition, they form relatively pale-colored complexes with iron(III). In practice, a sample solution containing iron, which is almost invariably present as iron(III), is treated with an appropriate reducing agent to convert iron(III) to iron(II) and the organic ligand is added to form the desired complex. Hydroxylamine and hydroquinone are two common reductants used in these procedures. Bathophenanthroline forms a red complex with iron(II), its absorption maximum occurring at approximately 535 nm, and the familiar iron(II)-orthophenanthroline complex has an absorption maximum at 510 nm.

Compared with orthophenanthroline, the use of bathophenanthroline for the determination of iron offers two important advantages. First, the molar absorptivity of the iron(II)-bathophenanthroline complex, 22,350 liter mole$^{-1}$ cm$^{-1}$, is approximately twice the value for the corresponding tris(1,10-phenanthroline)iron(II) species. Second, the former complex can be extracted into water-immiscible solvents such as isoamyl alcohol and *n*-hexyl alcohol. This approach permits the iron(II)-bathophenanthroline complex to be concentrated into a small volume of the nonaqueous phase. Since the organic phase can be subjected directly to a spectrophotometric measurement, very high sensitivity can be achieved. Typically, 0.1 $\mu$g of iron per milliliter of solution can be detected by using orthophenanthroline, and as little as 0.005 $\mu$g per milliliter by employing bathophenanthroline. These methods are so sensitive that precautions must be taken to ensure that no iron is present as an impurity in the reagents used for the determination.

*Spectrophotometric determination of organic and biological compounds.* Perhaps the most common application of ultraviolet or visible absorption spectrophotometry is the determination of organic or biological compounds. The high absorptivities and characteristic absorption spectra of many of these species often make spectrophotometry the technique of choice for both qualitative and quantitative purposes. Furthermore, the presence of specific functional groups can sometimes be ascertained from the absorption spectrum of a compound. A partial list of the absorption wavelengths of a number of functional groups is given in Table 19–2.

As is generally true in the ultraviolet-visible region, the absorption bands listed in Table 19–2 are fairly broad. For example, an aldehyde group having a maximum

**Table 19-2. Wavelengths of Maximum Absorbance for Certain Organic Functional Groups**

| Absorbing Group | | $\lambda_{max}$, nm |
|---|---|---|
| Olefin: | $\diagup C = C \diagdown$ | 190 |
| Conjugated olefin: | $+ C = C +_2$ | 210–230 |
| | $+ C = C +_3$ | 260 |
| | $+ C = C +_4$ | 300 |
| | $+ C = C +_5$ | 330 |
| Acetylene: | $- C \equiv C -$ | 175–180 |
| Benzene: | | 184, 202, 255 |
| Naphthalene: | | 220, 275, 312 |
| Ether: | $-O-$ | 185 |
| Thioether: | $-S-$ | 194, 215 |
| Amine: | $-NH_2$ | 195 |
| Nitro: | $-NO_2$ | 210 |
| Azo: | $-N = N-$ | 285–400 |
| Ketone: | $\diagup C = O$ | 195, 270–285 |
| Thioketone: | $\diagup C = S$ | 205 |
| Aldehyde: | $-CHO$ | 210 |
| Carboxyl: | $-COOH$ | 200–210 |
| Bromide: | $-Br$ | 208 |
| Iodide: | $-I$ | 260 |

absorptivity at 210 nm will exhibit appreciable absorption at wavelengths as long as 230 nm and as short as 200 nm. In addition, the position of a certain band will change depending on the structure of the molecule in which the group is found. Thus, the benzene absorption at 255 nm is shifted to 262 nm in toluene and to 265 nm in chlorobenzene.*

Quantitative determinations by measurement of ultraviolet or visible absorption are not limited to unknown constituents which themselves undergo suitable electronic transitions. For example, no absorptions are found for alcohol (—OH) groups throughout the entire wavelength range from 200 to 1000 nm. However, many alcohols react with phenyl isocyanate to form phenyl alkyl carbamates,

$$\text{phenyl isocyanate} \quad \bigcirc -N = C = O + C_2H_5OH \longrightarrow \bigcirc -\underset{H}{N} - C \diagup^{O}_{OC_2H_5}$$

phenyl isocyanate                    ethyl-N-phenylcarbamate

which can be determined quantitatively by measurement of the absorbance at 280 nm.

A common method for the determination of amino acids involves their reaction with ninhydrin to produce an intensely blue-colored compound whose absorbance can be measured at 575 nm. This method is employed in several commercial amino-acid

---

* For a more detailed treatment of the relationship between organic structure and ultraviolet-visible absorption spectra, consult H. H. Jaffe and M. Orchin: *Theory and Applications of Ultraviolet Spectroscopy.* John Wiley and Sons, New York, 1962; and A. I. Scott: *Interpretation of the Ultraviolet Spectra of Natural Products.* Pergamon Press, Oxford, 1964.

analyzers, in which proteins under investigation are hydrolyzed into their component amino acids which are, in turn, separated and measured spectrophotometrically.

**Automated clinical spectrophotometry.** Spectrophotometry is the mainstay of many of the modern automated clinical analyzers. Using rapid, sequential sample insertion or specially designed flow cells, one of these instruments can spectrophotometrically analyze samples at rates exceeding 60 per hour. In these instruments all operations, including sample dilution, reagent addition, and digestion, are performed automatically before the spectrophotometric determination is made. Often a number of separate determinations can be performed on each sample, at a level of precision and accuracy equal to or exceeding that attainable by a trained technician. An example of a simple, single-channel clinical analyzer can be seen in Figure 19–18.

**Water-pollution analysis—measurement of phosphate.** One of the most sensitive techniques available for the determination of phosphate in water is the so-called *molybdenum blue* method. In this procedure, phosphate reacts with ammonium molybdate to form ammonium phosphomolybdate, $(NH_4)_3P(Mo_3O_{10})_4$. In turn, the latter compound can be reduced with a wide variety of reagents, including hydroquinone, tin(II), and iron(II), to yield an intensely colored substance called molybdenum blue, the concentration of which can be measured spectrophotometrically to provide a determination of phosphate. Molybdenum blue is a complex polymer which consists of a mixture of molybdenum(VI) and molybdenum(V), but it is not a stoichiometrically well-defined compound. It is interesting to note that compounds containing the same element in two different oxidation states display characteristically intense colors and correspondingly high absorptivities. Both of the two main reactions in this procedure, namely, the formation of ammonium phosphomolybdate and the subsequent reduction to molybdenum blue, can yield incorrect products. Consequently, these reactions must be carried out under carefully prescribed conditions with respect to the acidity of the solution, the amount of ammonium molybdate reagent used, and the temperature and time of the reaction.

**Air-pollution analysis—measurement of sulfur dioxide.** Sulfur dioxide is probably the most important of all air pollutants because of its ubiquitous nature and its adverse effect on the upper respiratory tract. One of the standard methods for the detection and measurement of sulfur dioxide is the spectrophotometric procedure devised by West and Gaecke.[a] In this procedure, one collects the sulfur dioxide by bubbling the air sample through a $0.1 \, F$ sodium tetrachloromercurate(II) solution to form the stable, nonvolatile product disulfitomercurate(II):

$$HgCl_4{}^{2-} + 2\,SO_2 + 2\,H_2O \rightarrow Hg(SO_3)_2{}^{2-} + 4\,Cl^- + 4\,H^+$$

The stabilized product is then treated with the acid-bleached dye *p*-rosaniline [(4-amino-3-methylphenyl)bis(4-aminophenyl)methanol] in the presence of formaldehyde to yield a bright red-violet substance whose absorbance is measured at 569 nm. This procedure is suitable for the accurate determination of sulfur dioxide in air at concentrations as low as 0.005 ppm by volume; the only known interference is nitrogen dioxide ($NO_2$), which, if present at concentrations above 2 ppm, must be removed prior to the analysis.

# MOLECULAR ANALYSIS BY LUMINESCENCE SPECTROMETRY

Although luminescence spectrometry is not used as much as absorption spectrophotometry, the former provides the basis for some of the most sensitive

---

[a] P. W. West and G. C. Gaecke: Anal. Chem., *28*:1816, 1956.

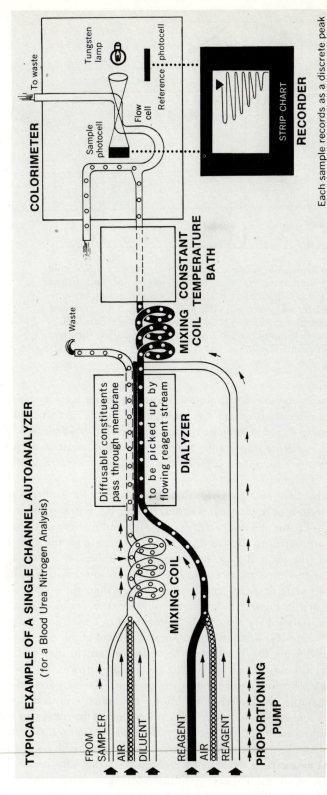

*Figure 19–18.* Schematic diagram of a single-channel Technicon Auto-Analyzer, an automated clinical instrument employing spectrophotometry. (Courtesy of Technicon Corporation, Tarrytown, New York.)

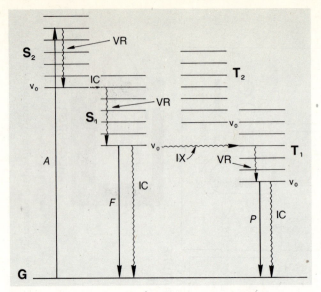

*Figure 19–19.* Electronic energy-level diagram for a molecule with ground state (**G**) and excited singlet (**S**) and triplet (**T**) states. Radiationless transitions between states are represented by wavy arrows; $A$ is absorption, $F$ is fluorescence, $P$ is phosphorescence, VR is vibrational relaxation, IX is intersystem crossing, and IC is internal conversion. (See text for discussion.)

molecular analytical techniques. **Luminescence** is the emission of radiation from a species after that species has absorbed radiation. As we shall see, absorption is a necessary, but not sufficient, prerequisite for luminescence. To understand luminescence and its importance to spectrochemical analysis, let us consider the molecular events leading to a luminescent transition by referring to the energy-level diagram of Figure 19–19.

This diagram resembles that shown earlier for absorption, except that we will consider not only the singlet states, $S_1$ and $S_2$, but also the associated triplet states. $T_1$ and $T_2$. An electronic triplet state, it should be recalled, has two electrons whose spins are unpaired. By contrast, in a singlet state all the electrons in a molecule are paired; the ground state of a molecule is therefore almost always a singlet state. Because two unpaired electrons have a slightly lower energy than two paired electrons, a triplet state is somewhat lower in energy than the corresponding singlet state, as illustrated in Figure 19–19. To simplify matters, let us examine only one absorption process (labeled $A$), which will be assumed to be a transition from the ground electronic state (**G**) to an excited vibrational level of an excited singlet state, say $S_2$.

Following this transition, the molecule can rid itself of the absorbed energy by several alternative pathways. The particular pathway used will be governed by the kinetics of various competing processes, some of which are indicated in Figure 19–19. For example, the molecule residing in an excited vibrational level of electronic state $S_2$ can lose energy by radiating a photon equal in energy to the difference between its existing state and the ground state. However, in solution this radiative loss has a much smaller rate constant ($k \simeq 10^8$ $sec^{-1}$) than the competing process of **vibrational relaxation** (labeled VR in Figure 19–19). Vibrational relaxation involves the transfer of vibrational energy to neighboring molecules and, in solution, occurs very rapidly ($k \simeq 10^{13}$ $sec^{-1}$). By comparison, an excited molecule in the gas phase will suffer far fewer collisions, so that vibrational relaxation is much less efficient; in the gas phase, it is common to see the emission of a photon equal in energy to that absorbed. This process is termed **resonance fluorescence** and will be important in our discussion of atomic fluorescence in the next chapter.

In solution, an excited molecule will rapidly relax to the lowest vibrational level of the electronic state in which it resides, in the present case $S_2$. At this point, the kinetically favored event is **internal conversion** (labeled IC in Figure 19–19), which shifts the molecule from the lowest vibrational level of $S_2$ to an excited vibrational level of the lower singlet state $S_1$. This occurrence is made possible by the relatively small energy difference between the excited singlet states of most molecules and the high degree of coupling that exists between their vibrational levels. Following internal conversion, the molecule is rapidly deactivated by vibrational relaxation to the lowest vibrational level of state $S_1$. Because vibrational relaxation and internal conversion have such high rates, *most excited molecules in solution will decay to the lowest vibrational level of the lowest excited singlet state before any analytically useful radiation is emitted.*

Once a molecule reaches the lowest vibrational level of the lowest excited singlet state, it can return to the ground state in several ways, among which is the emission of radiation. Internal conversion can still occur, of course. However, it is less probable since for most molecules the energy separation between the first excited singlet state and the ground state is greater than that between excited states. The radiative loss of energy from a singlet to the ground state is termed **fluorescence** and is labeled $F$ in Figure 19–19. Because several other processes compete with fluorescence, it is necessary to employ a figure of merit, termed the **fluorescence quantum efficiency**, to indicate the fraction of excited molecules that fluoresce. Depending on the specific molecule under investigation and its environment, quantum efficiencies can range from near unity to approximately zero, and are governed by the rates of the various processes competing for further energy loss. These processes have rate constants on the order of $10^7$ to $10^9$ sec$^{-1}$, so that the lifetime of an excited singlet state is usually between $10^{-9}$ and $10^{-7}$ second.

Fluorescence is obviously an analytically useful process and its basic characteristics can be deduced from the preceding discussion. Because fluorescence almost always occurs after some loss of vibrational or electronic energy, the wavelength of the fluoresced radiation will be longer (that is, its frequency will be lower) than that of the absorbed radiation. For this reason, the fluorescence spectrum of a molecule is shifted to longer wavelengths from the absorption spectrum. The lifetime of the fluorescence can also be of importance and will be equal to that of the excited singlet state (between $10^{-9}$ and $10^{-7}$ second).

As an alternative to internal conversion and fluorescence, a molecule in state $S_1$ can undergo **intersystem crossing** (IX in Figure 19–19), which involves an electron spin-flip within the molecule to place it in an excited vibrational level of triplet state $T_1$. Following this, vibrational relaxation will be very rapid, dropping the molecule to the lowest vibrational level of $T_1$. Because intersystem crossing is a spin-forbidden process (that is, because it is relatively improbable from quantum mechanical considerations), it can compete with fluorescence and internal conversion only for certain molecules. In these cases, the rate of intersystem crossing approximates that of the competing processes ($k$ between $10^6$ and $10^9$ sec$^{-1}$).

Any transition from triplet state $T_1$ to the ground state $G$ is also a spin-forbidden process. For this reason, the lifetime of the triplet state is relatively long, ranging from $10^{-6}$ to 10 seconds, depending on the specific molecule. The particular lifetime will be governed by the dominant mode of energy loss; if radiative, this process is termed **phosphorescence** ($P$ in Figure 19–19) but, if nonradiative, it again involves internal conversion and vibrational relaxation. Phosphorescence is therefore a luminescence process in which a molecule undergoes a transition from a triplet state to the ground state. Because of its spin-forbidden character, phosphorescence has a much longer lifetime than fluorescence; the lifetime is equal to the triplet-state lifetime of between $10^{-6}$ and 10 seconds.

As mentioned earlier, triplet state $\mathbf{T_1}$ is lower in energy than singlet state $\mathbf{S_1}$ and is therefore closer to the ground state. For this reason, internal conversion to the ground state is much more efficient than phosphorescence for most molecules, and at room temperature is the dominant pathway for the loss of triplet-state energy. To utilize phosphorescence for analytical work, therefore, it is customary to cool the sample, often to liquid nitrogen temperatures (77°K), in an effort to increase the **phosphorescence quantum efficiency**. The latter is defined as the fraction of excited molecules which phosphoresce.

From the preceding discussion, it should be clear that the two luminescence phenomena—fluorescence and phosphorescence—are competitive and, furthermore, must compete with other modes of energy loss by an excited molecule. Whichever process dominates is dependent upon both the type of molecule being examined and its environment. In some cases, the environment can be controlled to increase the yield of the desired event; the yield of either event can be determined from its quantum efficiency. We will consider means of optimizing quantum efficiencies in a later section.

### Luminescence Instrumentation

A block diagram for a generalized luminescence spectrometer is shown in Figure 19–20. Because luminescence is an isotropic property—that is, because there is emission in all directions—it is possible to detect the emitted radiation at any desired direction from the sample. In order to minimize the likelihood of interference from radiation used to excite the sample molecules, the luminescence is often observed at right angles to the exciting radiation. Although this 90° orientation is not found in all luminescence spectrometers, it is by far the most common configuration. Since the

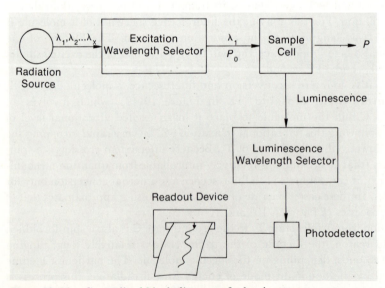

*Figure 19–20.* Generalized block diagram of a luminescence spectrometer. (See text for discussion.)

luminescence power is proportional to the incident source power (see Chapter 18, page 618), the exciting radiation is usually provided by an intense source such as a xenon arc lamp. A wavelength selector such as a filter or monochromator is used to obtain a narrow band of wavelengths from the source suitable for excitation of the desired molecular species in the sample. When excited, these molecules can either fluoresce or phosphoresce, and the spectrum of the luminescence radiation is determined with the aid of another wavelength selector and an appropriate photodetector and readout system.

There are two classes of luminescence spectrometers: those which employ filters and those which employ monochromators for spectral isolation. Filter instruments are called **fluorimeters** or **phosphorimeters**, depending on which luminescence process is being studied. Similarly, the monochromator instruments are termed **spectrofluorimeters** and **spectrophosphorimeters**, respectively. Obviously, instruments which utilize monochromators are more versatile and provide greater selectivity; because a narrower wavelength band is isolated for both excitation and observation of the luminescence spectrum, there is less chance of interference by other sample components. Also, with the monochromator systems it is possible to obtain either a complete luminescence or a complete excitation spectrum. One obtains an **excitation spectrum** by scanning the desired wavelength region with the excitation monochromator while the emission monochromator is held at a fixed wavelength. Thus, an excitation spectrum is similar, but not identical, to a spectrophotometric absorption spectrum. An excitation spectrum indicates what wavelengths can be used to *pump* the specific transition being observed with the emission monochromator.

Compared with spectrofluorimeters and spectrophosphorimeters, the filter devices are simpler to operate and less expensive. In addition, because an absorption or interference filter provides a greater optical speed than a monochromator, a filter fluorimeter or phosphorimeter provides higher sensitivity for quantitative determinations. Both types of luminescence spectrometers are widely used. Of course, filter instruments are not useful for obtaining structural information about a luminescent species, because only one spectral interval can be examined.

## Advantages of Fluorimetry and Phosphorimetry

**Fluorimetry** and **phosphorimetry** are terms used to denote luminescence spectrometry based on fluorescence or phosphorescence, respectively. Both techniques have several advantages over absorption spectrophotometric methods. First, fluorimetry and phosphorimetry offer greater selectivity and freedom from spectral interferences. This is because there are far more chemical species which absorb ultraviolet or visible radiation than those which fluoresce or phosphoresce. Moreover, in fluorimetry and phosphorimetry, one can vary not only the absorption wavelength but also the emission wavelength, so that it is possible to further reduce spectral interferences by judicious choice of *both* excitation (absorption) and luminescence wavelengths.

Second, fluorimetry and phosphorimetry are generally more sensitive than absorption methods. This is because in the luminescence techniques a direct measurement of the power of the emitted radiation can be made. By contrast, it is necessary in absorption methods to determine the *difference* between two large radiation levels, the incident power $P_0$ and the transmitted power $P$. Because it is always easier to measure a small signal against no background than it is to measure the difference between two large signals, fluorimetry and phosphorimetry provide greater sensitivity than absorption spectrophotometry. This added sensitivity gives the luminescence methods still another advantage compared with absorption spectrophotometry. Whereas Beer's law plots are often linear over a 10-to-100-fold range of concentration, it is not uncommon to find the relationship between luminescence power and concentration being linear over three or four orders of magnitude in concentration. Although this extended linear range is not necessary for quantitative analysis, it is often quite useful practically and requires fewer points on a calibration curve.

One additional advantage is often enjoyed in phosphorimetry—that of **time-resolution**. Because phosphorescence has a relatively long lifetime, it is possible to discriminate among different molecular species on the basis of their luminescence-time behavior. This provides an added dimension useful in both qualitative identification and quantitative determination of particular species by minimizing interferences from fluorescence and phosphorescence of other sample constituents as well as fluorescence from the analyte itself. In addition, scattering from the sample is eliminated through the use of time-resolution. To obtain time-resolution in phosphorimetry, two common mechanical configurations are employed. In one arrangement, a slotted can rotates around the sample cell at varying speeds. Thus, the sample is illuminated by the primary source with the can in one position; the phosphorescent radiation can reach the detector only after the can has rotated through an angle of 90°. Changing the rotation rate of the can then varies the time delay between absorption by the sample and measurement of its phosphorescence. The second common configuration employs two rotating slotted disks, located on opposite sides of the sample cell. If the slots are displaced from each other, one can alter the time delay between excitation of the sample and measurement of its luminescence merely by adjusting the rotation rate of the disks.

Despite the obvious advantages of time-resolution, phosphorimetry is not as widely utilized as fluorimetry. This fact is due to the added complexity of phosphorimetric instrumentation, the smaller number of species that phosphoresce, and the inconvenience of having to cool the samples to obtain adequate phosphorescence quantum efficiencies. Until recently, it was necessary in phosphorimetry to employ special solvent mixtures which would form clear, rigid glasses upon cooling.[a] One such solvent frequently utilized is EPA, a 5:5:2 mixture of diethyl *e*ther, iso*p*entane, and ethyl *a*lcohol. Recently, the use of frozen "snows" of aqueous solutions has been investigated with some success.[b]

### Errors in Fluorimetry and Phosphorimetry

As discussed earlier, the power of fluorescent or phosphorescent radiation emitted by a sample is a direct function of the quantum efficiency. Thus, the quantum efficiency for the desired luminescence process must be constant and reproducible if a successful fluorimetric or phosphorimetric method of analysis is to be developed. When the quantum efficiency is reduced appreciably, the luminescence is said to be **quenched**. Unfortunately, many extraneous substances can affect the quantum efficiency and can quench luminescence. In particular, heavy atoms or paramagnetic species strongly affect the rate of intersystem crossing which, in turn, alters the quantum efficiency for fluorescence or phosphorescence, thereby causing an error in the analysis. In phosphorimetry, of course, it is highly desirable to increase intersystem crossing, whereas in fluorimetry it is not. Therefore, to avoid quenching in most fluorimetric procedures, heavy atoms or paramagnetic species must be excluded from the sample solution. Oxygen, being paramagnetic, is a particularly serious offender and is usually removed by the bubbling of nitrogen through solutions to be analyzed.

[a] J. D. Winefordner and P. A. St. John: Anal. Chem., *35*:2212, 1963.
[b] R. J. Lukasiewicz, J. J. Mousa, and J. D. Winefordner: Anal. Chem., *44*:1339, 1972.

Unlike absorption methods, fluorimetry and phosphorimetry involve the direct measurement of a radiation signal. Although it is possible to compensate for drifts in such parameters as source power and cell positioning in absorption spectrophotometry, such compensation is not as convenient for the luminescence techniques. For example, any drift or variation with wavelength of the primary source power reflects itself in a corresponding drift in the luminescence power. In most high-quality luminescence spectrometers, this problem is overcome either through adequate stabilization of the source power or through the monitoring of source power and the application of appropriate corrections.

Any factors which affect absorption will also influence fluorescence and phosphorescence since absorption is necessary before either of these processes can occur. Therefore, all sources of error previously discussed for absorption spectrophotometry apply equally to luminescence methods of analysis. In addition, a phenomenon called the **inner-filter effect** which is peculiar to luminescence processes must be considered. The inner-filter effect can be understood with the aid of Figure 19–21.

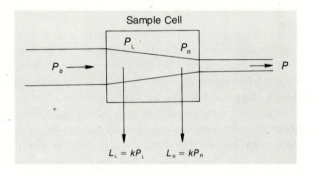

*Figure 19–21.* Pictorial representation of the *inner-filter effect* caused by attenuation of the radiation beam passing through the sample; $P_L$ and $P_R$ are the powers of the transmitted radiation at the left and right sides of the cell, respectively, and $L_L$ and $L_R$ are the powers of luminescence from the left and right sides of the cell, respectively. The power of the radiation incident upon the sample cell is $P_0$, and the power of the radiation transmitted through the cell is $P$. (See text for discussion.)

In Figure 19–21, assume that the common 90° viewing configuration is used and that the exciting radiation enters the cell from the left. Because the sample solution will absorb the exciting radiation as it passes through the cell, the power of the radiation will be less near the right side of the cell ($P_R$) than at the left side ($P_L$). Since the luminescence power is directly proportional to the power incident on the sample, the luminescence power will be less on the right side of the cell ($L_R$) than on the left ($L_L$). Furthermore, the luminescence from the right side of the cell will not be directly proportional to concentration, because the radiation exciting the molecules near that side of the cell does not remain constant but rather varies with concentration. The inner-filter effect generally limits the maximum concentration of the sample species which can be determined by means of fluorimetry or phosphorimetry. In addition, it can cause the shape of a plot of luminescence power versus concentration to be extremely nonlinear or even ambiguous, as shown in Figure 19–22. The nonlinearity depicted in Figure 19–22 is, of course, aggravated at high concentrations by the breakdown of the assumption that $abc < 0.01$ (see Chapter 18, page 619).

## Luminescence Determinations

Most luminescence determinations are performed on samples of clinical, biological, or forensic interest, although inorganic luminescence analysis is not uncommon.

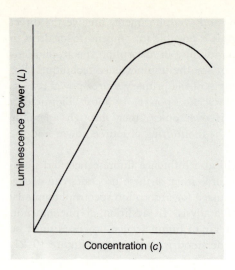

*Figure 19–22.* Dependence of luminescence power ($L$) on the concentration ($c$) of the luminescing species. The anomalous curvature of the plot is caused by the inner-filter effect. (See text for discussion.)

Many drugs possess rather high quantum efficiencies for luminescence, so that their determination by means of fluorimetry or phosphorimetry is both sensitive and practicable. Quinine, for example, can be detected at levels below one part per billion by weight and is often used as a calibration standard for fluorescence analysis. Many inorganic ions can be determined by means of fluorimetry or phosphorimetry if complexed with a luminescing organic ligand. Such ligands can be quite selective in their affinities for certain metals or nonmetals, so that luminescence methods for the determination of inorganic substances can be both specific and extremely sensitive. Examples of both organic and inorganic luminescence analyses are presented below.

**Fluorimetric drug analysis.**[a,b] Quinine is just one of a large family of drugs which can be sensitively determined by means of fluorimetry or phosphorimetry. Like quinine, some of these physiologically active agents fluoresce directly; others can form luminescent complexes with other organic substances. As an example, consider the fluorimetric determination of lysergic acid diethylamide (LSD), whose structure is

The increasingly common unauthorized use of LSD as a hallucinogenic agent has made it mandatory for law enforcement agencies to devise methods for the detection and determination of LSD in the presence of other drugs. LSD is one of the most active of all hallucinogens; as little as 50 $\mu$g taken orally is sufficient to produce hallucinatory effects. Thus, the method of analysis for LSD must be extremely sensitive. Because LSD is known to be highly fluorescent, a luminescence method of analysis meets this criterion. In the established fluorimetric procedure, a sample of blood plasma or urine (approximately 5 ml) is made slightly alkaline and extracted with a 98:2 volume-to-volume mixture of $n$-heptane and isoamyl alcohol. After

---

[a] J. Axelrod, R. O. Brady, B. Witkop, and E. V. Evarts: Ann. N.Y. Acad. Sci., 66: 435, 1957.
[b] G. K. Aghajanian and H. L. Bing: Clin. Pharmacol. Therap., 5: 611, 1964.

extraction of LSD into the organic phase, the LSD is extracted back into an aqueous solution of 0.004 $F$ hydrochloric acid. The organic solvent is then separated, and the fluorescence of the acid extract is measured directly, an excitation wavelength of 335 nm and a fluorescence wavelength of 435 nm being used.

This procedure is extremely selective, with little interference from other hallucinogens or from metabolites of LSD. With care, this technique can be used to measure plasma concentrations of LSD as low as one nanogram per milliliter, more than sufficiently sensitive to detect a *single* oral dose of LSD.

*Fluorimetric air-pollution analysis.*[a,b]   Some of the most insidious of the man-made components of air pollution fall into the category of carcinogens (cancer-causing agents). Many of these carcinogens are polynuclear aromatic hydrocarbons formed during combustion and are often present at appreciable concentration levels in automobile exhaust and cigarette smoke. One such carcinogen is 3,4-benzopyrene. Because benzopyrene is extremely carcinogenic, it is important to have available a highly sensitive method for measuring its concentration in urban air. The extremely high fluorescence intensity of benzopyrene makes fluorimetry an ideal technique for this application. The excitation and fluorescence spectra of 3,4-benzopyrene are shown in Figure 19–23. Even in very complex mixtures of carcinogens and other

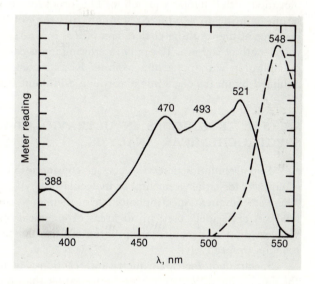

*Figure 19–23.* Excitation and luminescence spectra for benzopyrene in sulfuric acid. The solid line is the *excitation spectrum*, a plot of meter reading (which is proportional to fluorescence power) measured at 545 nm as the exciting wavelength is varied. The dashed line is the *fluorescence spectrum*, a plot of meter reading as a function of the wavelength of luminescence while the exciting wavelength is fixed at 520 nm. (Reproduced, with permission, from the paper by E. Sawicki, T. R. Hauser, and T. W. Stanley: Intern. J. Air Pollution. 2: 253, 1960.)

materials, the fluorescence maximum between 545 and 548 nm is very characteristic of benzopyrene and can be used for its identification and determination. For the spectra in Figure 19–23, the fluorescence of benzopyrene was measured in a sulfuric acid medium; other solvents are less satisfactory. For routine determinations of benzopyrene, the air sample to be examined is bubbled through a sulfuric acid solution. The resulting solution is then analyzed fluorimetrically, an excitation wavelength of 520 nm and a fluorescence wavelength of 545 nm being used. In addition, performing the analysis at a reduced temperature ($-190°C$) has been found to increase both the sensitivity and specificity. With the aid of low-temperature fluorescence measurements, it has been found that as much as 130 $\mu g$ of benzopyrene is present

[a] E. Sawicki, T. R. Hauser, and T. W. Stanley: Intern. J. Air Pollution *2*: 253, 1960.
[b] S. Udenfriend: *Fluorescence Assay in Biology and Medicine.* Volume 2, Academic Press, New York, 1969, p. 568.

in the exhaust fumes released from an automobile over a period of 90 minutes. Additionally, the smoke of one cigarette contains approximately 10 nanograms of benzopyrene, although only a small portion of this amount is inhaled and absorbed. Benzopyrene has even been found in one alcoholic beverage (Kirsch), but only to the extent of about 0.4 ng/ml.

*Inorganic fluorimetric analysis.* Although most inorganic ions do not fluoresce directly, many of these species do form chelate complexes with organic molecules, some of which are highly fluorescent. Elements which have been analyzed in this way are Al, Au, B, Be, Ca, Cd, Cu, Eu, Ga, Gd, Ge, Hf, Hg, Mg, Nb, Pd, Rh, Ru, S, Sb, Se, Si, Sm, Sn, Ta, Tb, Th, Te, W, Zn, and Zr. Although a large number of organic molecules form fluorescent chelate complexes with these elements, three of the more common reagents are 8-hydroxyquinoline, 2,2′-dihydroxyazobenzene, and dibenzoylmethane. Unfortunately, some of these chelating agents are rather nonspecific and form complexes with many inorganic ions, so that interferences are common. Often, a prior separation procedure such as ion-exchange chromatography must be employed. However, the sensitivity of a fluorimetric determination often adequately compensates for the lack of specificity.

An example of an inorganic fluorimetric determination is the routine clinical method for the measurement of magnesium ion in serum and urine after complexation with 8-hydroxyquinoline. This procedure entails addition of the sample of serum or urine to a buffered solution (pH 6.5) of 8-hydroxyquinoline. The fluorescence of the resulting chelate complex is measured at 510 nm with an excitation wavelength of 380 nm. The only interfering ion is calcium which can be removed by precipitation with potassium oxalate. Results compare very favorably with those obtained by means of atomic absorption and flame emission spectrometry.

## SPECIAL TECHNIQUES IN ULTRAVIOLET-VISIBLE SPECTROCHEMICAL ANALYSIS

Two techniques deserving special consideration but which are not employed exclusively for either elemental or molecular analysis will be discussed in this section. These techniques, spectrophotometric titrimetry and reaction-rate methods of analysis, are widely used in all areas of chemical analysis and utilize ultraviolet-visible radiation to monitor the progress of a titration and the initial rate of a chemical reaction, respectively.

Spectrophotometric titrimetry and reaction-rate methods of analysis both involve the monitoring of *changes* in absorbance rather than the measuring of absolute values of absorbance or transmittance. This feature minimizes the seriousness of any scattering, reflection, or refraction of radiation from the cells and sample, and simplifies the instrumentation required for absorbance measurements. In addition, spectrophotometric titrations and reaction-rate methods of analysis make use of chemical reactions to introduce more specificity into the absorption measurements; in general, interfering substances are those which interfere with the chemical reaction rather than those which interfere by absorbing radiation at the same wavelengths as the species being determined.

### SPECTROPHOTOMETRIC TITRIMETRY

Because all titrations involve the disappearance and formation of chemical species, the progress of a titration can often be monitored through measurement of the absorbance of the solution being titrated. This frequently permits more sensitive end-point detection as well as the automation of titrations. Spectrophotometric titrimetry can be performed whenever (a) the substance being titrated, (b) the substance formed in the titration, or (c) the titrant itself exhibits distinctive absorption characteristics. In these cases, a spectrophotometric titration curve (which is a plot of absorbance as a function of added titrant) consists essentially of two straight lines, their point of intersection being the equivalence point. In other situations in which none of the reactants or products absorbs radiation in the ultraviolet-visible region, it is sometimes feasible to use an acid-base, metallochromic, redox, or fluorescent indicator which does exhibit suitable absorption or luminescence properties. In these instances, the shape of the titration curve will depend on the nature of the indicator and on its interaction with reactant or product species.

When no indicator is employed, spectrophotometric titration curves having several different shapes are possible. Consider the general reaction

$$A + T \rightarrow C$$

in which A is the species being titrated, T is the titrant, and C is the product of the reaction. If just one of these three species—A, T, or C—absorbs radiation at the wavelength which is examined spectrophotometrically, the resulting titration curve will appear as shown in Figures 19–24A, 19–24B, and 19–24C, respectively. However,

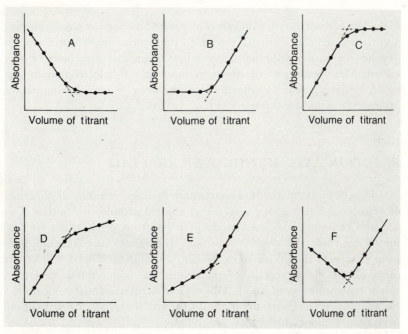

*Figure 19–24.* Spectrophotometric titration curves for the hypothetical reaction

$$A + T \rightarrow C$$

in which A is the species being titrated, T is the titrant, and C is the product of the reaction.

Curve A: only the species being titrated (A) absorbs. Curve B: only the titrant (T) absorbs. Curve C: only the product of the titration (C) absorbs. Curve D: A does not absorb, and the molar absorptivity of C is greater than that of T. Curve E: A does not absorb, and the molar absorptivity of T is greater than that of C. Curve F: C does not absorb, and the molar absorptivity of T is greater than that of A.

if more than one species absorbs at a given wavelength, the shape of the titration curve will depend upon the relative molar absorptivities of the absorbing components of the system, as illustrated in Figures 19–24D, 19–24E, and 19–24F.

Notice that, near the equivalence point in each figure, the experimental data do not fall perfectly on the two straight-line portions of the titration curve. This behavior is typical of spectrophotometric titration curves and is due to the fact that no reaction is ever 100 per cent complete. The extent to which the experimental data deviate from the straight lines will increase as the equilibrium constant for the titration reaction decreases. In a manual spectrophotometric titration, one always takes experimental data on each side of the equivalence point, well away from the curved region, and then extrapolates the straight-line portions to their point of intersection to obtain the equivalence point. For greatest accuracy, it is necessary to correct the absorbance readings for the effect of dilution which occurs during the titration; otherwise, the titration curve will show no straight-line portions, and extrapolation to the equivalence point will be uncertain. Of course, linearity in a titration curve presumes that Beer's law is obeyed.

An example of a chemical system that follows the behavior depicted in Figure 19–24B is the titration of arsenic(III) in a neutral medium with a standard triiodide solution:

$$HAsO_2 + I_3^- + 2\,H_2O \rightleftharpoons H_2AsO_4^- + 3\,I^- + 3\,H^+$$

If the absorbance due to triiodide is monitored during the course of this titration, the absorbance remains virtually zero until the equivalence point is reached. However, beyond the equivalence point, the absorbance increases linearly as excess $I_3^-$ is added. Many examples of titrations amenable to spectrophotometric monitoring could be cited; some applications in the area of complexometric titrimetry are presented in Chapter 6 (page 189 ).


## REACTION-RATE METHODS OF ANALYSIS

Reaction-rate methods of analysis are finding increasing application in a number of areas, especially in the analysis of clinical samples by means of enzymatic reactions. These methods are based upon the principle that, if a species to be determined can react with another substance, the initial rate of reaction is approximately proportional to the initial concentration of the desired species. Therefore, measurement of the initial reaction rate permits an unambiguous determination of the initial reactant concentration. This type of analysis can be performed very rapidly, because it is unnecessary to wait until the reactants have come to equilibrium. This is especially important for slow reactions.


### Basis of the Method

For many reactions, the initial reaction rate can be shown to be directly proportional to the initial reactant concentration. As an example, let us consider the generalized enzymatic reaction

$$S + E \underset{k_{-1}}{\overset{k_1}{\rightleftharpoons}} [SE] \underset{k_{-2}}{\overset{k_2}{\rightleftharpoons}} P + E$$

where E is the enzyme, S is the substrate (reactant), and P is the product of the enzymatic reaction; for purposes of this discussion, the nature of the intermediate complex, SE, need not be specified.

Let us write the rate equation for the disappearance of the substrate:

$$-\frac{d[S]}{dt} = k_1[S][E] - k_{-1}[SE] \tag{19-13}$$

Now, the sum of the concentrations of the free enzyme, E, and the substrate-enzyme complex, SE, is equal to the initial concentration of the enzyme, $E_0$:

$$[E_0] = [E] + [SE] \tag{19-14}$$

Substitution of equation (19–14) into equation (19–13) yields

$$-\frac{d[S]}{dt} = k_1[S][E_0] - [SE]\{k_1[S] + k_{-1}\} \tag{19-15}$$

If only the *initial* stage of the overall forward reaction is considered, we can neglect the reverse reaction between P and E to re-form the substrate-enzyme complex. For this situation, we can write the following steady-state expression for the substrate-enzyme complex:

$$\frac{d[SE]}{dt} = k_1[E][S] - k_{-1}[SE] - k_2[SE] = 0 \tag{19-16}$$

Next, if equation (19–16) is solved for [SE], and if a substitution for [E] is made from equation (19–14), we have

$$[SE] = \frac{[S][E_0]}{[S] + K_M} \tag{19-17}$$

where $K_M$, the so-called **Michaelis constant**, is $(k_{-1} + k_2)/k_1$. If equation (19–17) is substituted into equation (19–15) and the resulting expression rearranged, we obtain

$$-\frac{d[S]}{dt} = \frac{k_2[S][E_0]}{[S] + K_M} \tag{19-18}$$

*Determination of enzyme.* Two approaches can be employed to determine the initial concentration of enzyme, $[E_0]$. One method is to *saturate* the enzyme with substrate; that is, the initial concentration of the substrate, [S], is made very large compared with $K_M$. In this case, equation (19–18) takes the simple form

$$-\frac{d[S]}{dt} = k_2[E_0] \tag{19-19}$$

Therefore, the rate of the reaction remains *constant* for a considerable time until [S] falls to the level of $K_M$. To perform the actual determination, two reaction mixtures must be prepared; the first contains a known concentration of enzyme, the second contains the unknown concentration of enzyme, and each mixture has the *same* (large) substrate concentration. By comparing the reaction rates for the two systems, one can compute the unknown enzyme concentration.

However, it is not necessary that [S] be greater than $K_M$. If measurements are made of just the *initial* reaction rate, the concentration of substrate will not have time to change significantly. Therefore, all that is really required for the determination is for the two reaction mixtures to have the *same* initial substrate concentration (whatever its value), because the *initial* rate of reaction will be directly proportional to the enzyme concentration.

**Determination of substrate.** If, at the start of the reaction, [S] is chosen to be much less than $K_M$, equation (19–18) becomes

$$-\frac{d[S]}{dt} = \frac{k_2[S][E_0]}{K_M} = K[S][E_0] \tag{19–20}$$

where $K = k_2/K_M$. Equation (19–20) indicates that the initial rate of reaction is proportional to both the initial enzyme and substrate concentrations. However, if the enzyme concentration is the same for all samples and standards to be analyzed, the value for the enzyme concentration can be incorporated into the constant $K$, so that equation (19–20) can be rewritten as

$$-\frac{d[S]}{dt} = K'[S] \tag{19–21}$$

Thus, to determine the concentration of substrate in an unknown, two reaction mixtures are prepared, one containing the unknown and the other containing a known amount of substrate. Then, after addition of the same amount of enzyme, the initial reaction rate of each mixture is measured. The rate of reaction of the known mixture provides a value of the constant $K'$, which can be used to find the concentration of the unknown sample from its initial reaction rate. This procedure succeeds because, when only the initial reaction rates are measured, the concentrations of the two reactants remain virtually constant.

An example of a reaction-rate method is the determination of glucose by measurement of the rate of selective oxidation of glucose in the presence of the flavoprotein enzyme, glucose oxidase:

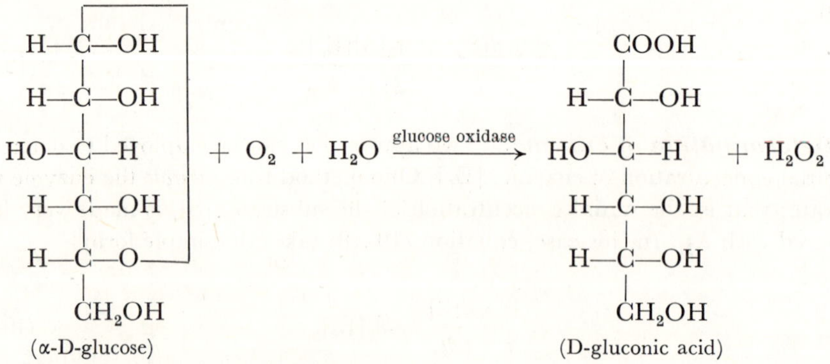

To monitor the progress of this oxidation, the following reaction can be utilized:

$$H_2O_2 + 2\,H^+ + 3\,I^- \rightarrow I_3^- + 2\,H_2O$$

Hydrogen peroxide formed in the enzymatic oxidation of glucose reacts with iodide in the presence of a molybdenum(VI) catalyst to yield triiodide which absorbs

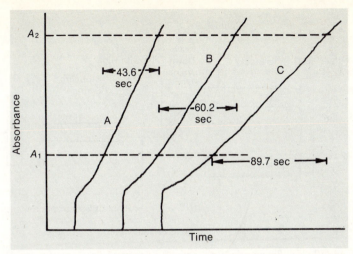

*Figure 19–25.* Spectrophotometric reaction-rate curves for the determination of glucose. Different amounts of glucose were added to 10 ml of a solution containing fixed concentrations of enzyme (glucose oxidase), iodide ion, and molybdenum(VI) catalyst. Curve A: 200 μg of glucose. Curve B: 150 μg of glucose. Curve C: 100 μg of glucose. (Redrawn, with permission, from the paper by H. V. Malmstadt and G. D. Hicks, *Anal. Chem.*, *32*: 395, 1960.)

strongly at 360 nm. Spectrophotometric monitoring at 360 nm provides reaction-rate curves such as those shown in Figure 19–25. The initial reaction rate is determined by observation of the time required for the triiodide concentration to change from that corresponding to absorbance $A_1$ to that corresponding to absorbance $A_2$. Because the time required is very short compared to the time needed for the reaction to reach completion, the observed change is linear with respect to time and does not display the curvature expected for a pseudo first-order process. Therefore, the initial rate is given by

$$-\frac{d[S]}{dt} = \frac{\Delta c}{\Delta t} = K'[S] \tag{19–22}$$

where $\Delta t$ is the time interval required for a fixed concentration change ($\Delta c$) to occur. The value of $\Delta c$ is preset by the selection of the two absorbances $A_1$ and $A_2$. The initial substrate concentration is thus inversely proportional to $\Delta t$. Both automated and manual methods have been developed for measuring the initial rate of the reaction.

## Measurement of Initial Reaction Rates

For relatively slow reactions, such as the oxidation of glucose described above, manual mixing and stirring of the reactants are usually sufficient to initiate the reaction of interest. All the reagents except the glucose can be added to a suitable sample cell placed in the light path of a spectrophotometer. Then an aliquot of the unknown glucose solution can be pipetted into the cell, the solution stirred, and the initial reaction rate measured.

For faster reactions, special mixing techniques are often necessary to enable the reactants to be combined in a sufficiently short time. The most widely used technique for these fast reactions is the so-called **stopped-flow procedure** which is performed with apparatus similar to that shown in Figure 19–26. In a stopped-flow system, two

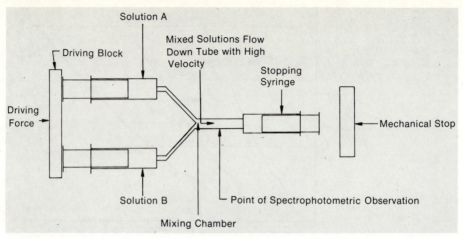

*Figure 19–26.* Schematic diagram of a stopped-flow system for the spectrophotometric monitoring of the rates of rapid reactions. (Redrawn, with permission, from the paper by R. M. Reich: Anal. Chem., *43*(12): 85A, 1971 .)

syringes (each containing a solution of one of the reactants) are driven at the same rate so that the solutions flow together through a mixing chamber. The flow velocity through the chamber is sufficiently high to mix the reagents very rapidly. The thoroughly mixed solution fills the stopping syringe, pushing its plunger outward. When the plunger of the stopping syringe strikes a mechanical barrier, the mixed solution suddenly halts and can be monitored at a suitable point of observation. In a properly designed stopped-flow spectrophotometric measuring system, the rates of reactions having half-lives as short as a few milliseconds can be measured.

For either slow or fast reactions, it is most convenient for interpretation of results to measure initial reaction rates in order to determine the concentration of the desired reactant. This minimizes the effects of temperature changes, side reactions, and other uncontrolled variables. When spectrophotometric monitoring of the reaction is employed and changes in the absorbance of the product are followed, it is possible to determine the initial reaction rate from the initial slope of a plot of absorbance versus time. For slow reactions, this plot can be simply displayed on a strip-chart recorder. For fast reactions, a high-speed recording device such as an oscilloscope must be used.

For convenience, ease of automation, and reliability, many reaction-rate procedures employ sophisticated systems which provide a direct readout of the initial reaction rate or, if calibrated properly, of the reactant concentration itself.*

## QUESTIONS AND PROBLEMS

1. Define or explain each of the following terms: resolving power, optical speed, transmission factor of a filter, microphotometer, photographic emulsion, photolysis, colorimeter, analyte, spectrophotometer, fluorimeter, phosphorimeter, spectrofluorimeter, sample blank, collimate, luminescence, resonance

---

*For further detail on automated reaction-rate measurements, refer to H. V. Malmstadt, E. A. Cordos, and C. J. Delaney: Anal. Chem., *44*(12):26A, 79A, 1972.

fluorescence, internal conversion, intersystem crossing, vibrational relaxation, triplet state, and inner-filter effect.

2. What is the difference between a *spectrograph* and a *spectrometer*? What is the difference between a *spectrometer* and a *monochromator*? Give at least two advantages of each of these systems.

3. What is the distinction between *trace analysis* and *microanalysis*?

4. Some simple instruments for absorption spectrophotometry employ test tubes as sample cells. Comment on this practice. Will a double-beam optical system compensate for errors introduced by use of test-tube cells? Why or why not?

5. Suppose that you have been presented with a sample of anthracene dissolved in carbon disulfide. The strongest absorption band for anthracene lies at 252 nm; but, unfortunately, carbon disulfide absorbs very strongly in the ultraviolet region below 380 nm. How would you proceed to determine the concentration of anthracene? Why will a double-beam spectrophotometer not provide perfect compensation for solvent absorption? (Hint: consider the path length, $b$, through the sample and reference cells.) For other absorption bands of anthracene, consult L. Meites, ed.: *Handbook of Analytical Chemistry*. McGraw-Hill Book Company, New York, 1963, p. **6**-90.

6. Define the fluorescence quantum efficiency ($\varphi_F$) by means of an expression that incorporates rate constants for the processes depicted in Figure 19-19. Do the same for the phosphorescence quantum efficiency ($\varphi_P$). For assistance, consult the references listed at the end of the chapter.

7. Explain how the time-resolution capability of a rotating-can phosphorimeter can eliminate interference from scattering and fluorescence by the sample.

8. For some molecules, the absorption and fluorescence spectra are nearly mirror images of each other. Using the energy-level diagram of Figure 19-19, explain why this is so.

9. Explain why the luminescence spectrum of a molecule will be similar but not identical to the absorption spectrum of the molecule.

10. If an ultraviolet-visible spectrophotometer has an inherent absolute error of 0.50 per cent in the observed per cent transmittance, calculate the relative error in concentration ($dc/c$) for solutions having the following transmittances: (a) 0.095, (b) 0.803, (c) 0.631, (d) 0.492. Perform similar calculations for solutions having the following absorbances: (e) 0.195, (f) 0.796, (g) 0.482, (h) 1.449.

11. For a spectrophotometer which reads directly in absorbance and which has, therefore, a constant error in absorbance, what is the absorbance value for which $dc/c$ is minimal? What experimental problems do you foresee in obtaining measurements at this absorbance level?

12. List the differences between fluorescence and phosphorescence.

13. Spectrophotometry is a valuable tool for the evaluation of equilibrium constants. For example, the equilibrium constant for the reaction

$$AuBr_4^- + 2\,Au + 2\,Br^- \rightleftharpoons 3\,AuBr_2^-$$

can be established by preparation of a mixture of $AuBr_4^-$ and $AuBr_2^-$ in contact with a piece of pure gold metal, followed by spectrophotometric measurement of the concentration of $AuBr_4^-$ from its absorption maximum at 382 nm. In one experiment, a solution *initially* containing a *total* of $6.41 \times 10^{-4}$ milliequivalent per milliliter of dissolved gold (present as both $AuBr_4^-$ and $AuBr_2^-$) in 0.400 $F$ hydrobromic acid was allowed to equilibrate in the presence of pure gold metal. The absorbance of the resulting solution was found to be 0.445 in a 1-cm cell at 382 nm. In separate experiments, the absorbance of an $8.54 \times 10^{-5}$ $M$ $AuBr_4^-$ solution in 0.400 $F$ hydrobromic acid was determined to be 0.410 at 382 nm, and $AuBr_2^-$ was observed to exhibit no absorption at this wavelength.

(a) Calculate the molar absorptivity of $AuBr_4^-$ at 382 nm.

(b) Calculate the equilibrium concentrations of $AuBr_4^-$ and $AuBr_2^-$.

(c) Evaluate the equilibrium constant for the reaction, neglecting activity effects.

(d) Predict how ionic strength will affect the magnitude of the equilibrium constant based on concentrations alone.

14. Compute the per cent relative uncertainty in the calculated concentration of a substance due to a spectrophotometric reading error, if the transmittance of the sample solution is 0.237 and if the absolute uncertainty in the transmittance reading is 0.3 per cent. If the preceding transmittance reading was obtained in a 1.00-cm spectrophotometer cell, calculate what the sample path length should have been to ensure the smallest possible uncertainty in the calculated concentration.

15. The titanium(IV)-peroxide complex has an absorption maximum at 415 nm, whereas the analogous vanadium(V)-peroxide complex exhibits its absorption maximum near 455 nm. When a 50-ml aliquot of $1.06 \times 10^{-3} M$ titanium(IV) was treated with excess hydrogen peroxide and the final volume adjusted to exactly 100 ml, the absorbance of the resulting solution (containing $1 F$ sulfuric acid) was 0.435 at 415 nm and 0.246 at 455 nm when measured in a 1-cm cell. When a 25-ml aliquot of $6.28 \times 10^{-3} M$ vanadium(V) was similarly treated and diluted to 100 ml, the absorbance readings were 0.251 at 415 nm and 0.377 at 455 nm in a 1-cm cell. A 20-ml aliquot of an unknown mixture of titanium(IV) and vanadium(V) was treated as the previous standard solutions were, including the dilution to 100 ml, and the final absorbance readings were 0.645 at 415 nm and 0.555 at 455 nm. What were the titanium(IV) and vanadium(V) concentrations in the original aliquot of the unknown solution?

16. Thioacetamide ($CH_3CSNH_2$) undergoes an acid-catalyzed hydrolysis to yield hydrogen sulfide gas which can be employed for the homogeneous precipitation of metal sulfides (see page 129 ). The rate of hydrolysis in a solution containing a small concentration of thioacetamide and a relatively high concentration of acid can be expressed by the pseudo first-order equation

$$ -\frac{d[CH_3CSNH_2]}{dt} = k'[CH_3CSNH_2] $$

The rate constant $k'$ can be evaluated if the logarithm of the absorbance of thioacetamide solutions at 262 nm is plotted as a function of time. In one experiment, a solution initially containing $1.00 \times 10^{-4} F$ thioacetamide and $0.25 F$ hydrochloric acid was heated at $90°C$, and the absorbance was measured as a function of time, the following data being obtained:

| Time in Minutes | Absorbance |
| --- | --- |
| 0 | 1.10 |
| 3 | 0.99 |
| 6 | 0.87 |
| 9 | 0.77 |
| 12 | 0.69 |
| 15 | 0.60 |

Evaluate the rate constant $k'$, and calculate the *initial* rate of hydrolysis of thioacetamide in moles/liter-minute.

17. Construct the idealized spectrophotometric titration curve which would result from each of the following hypothetical situations. Ignore dilution effects, but be sure to show quantitatively how the absorbance varies with the volume of added titrant.

   (a) The titration of $0.0001\ M$ B with $0.01\ M$ A, according to the reaction

$$ A + B \rightleftharpoons C + D $$

   Both the titrant (A) and the product D absorb at the chosen wavelength, but the molar absorptivity of A is exactly twice that of D.

   (b) The titration of $0.0001\ M$ B with $0.01\ M$ A, according to the reaction

$$ A + B \rightleftharpoons C + D $$

The substance being titrated (B) and the product C both absorb at the selected wavelength, but the molar absorptivity for B is twice that of C.

(c) The titration of 0.0001 $M$ B with 0.01 $M$ A, according to the reaction

$$A + B \rightleftharpoons C + D$$

The substance being titrated (B) and the titrant (A) both absorb at the selected wavelength, but the molar absorptivity of A is twice that of B.

18. A 5.00-ml aliquot of a standard iron(III) solution, containing 47.0 mg of iron per liter, was treated with hydroquinone and orthophenanthroline to form the iron(II)-orthophenanthroline complex, and finally diluted to exactly 100 ml. The absorbance of the resulting solution was measured in a 1-cm spectrophotometer cell and found to be 0.467 at 510 nm. Calculate the per cent transmittance of the solution and calculate the molar absorptivity of the iron(II)-orthophenanthroline complex.

19. A mixture of dichromate and permanganate ions in 1 $F$ sulfuric acid was analyzed spectrophotometrically at 440 and 545 nm as a means for the simultaneous determination of these two species, and the observed values of the absorbances were 0.385 and 0.653, respectively, at each wavelength for a 1-cm cell. Independently, the absorbance in a 1-cm cell of an $8.33 \times 10^{-4} M$ solution of dichromate in 1 $F$ sulfuric acid was found to be 0.308 at 440 nm and only 0.009 at 545 nm. Similarly, a $3.77 \times 10^{-4} M$ solution of permanganate, placed in a 1-cm cell, exhibited an absorbance of 0.035 at 440 nm and 0.886 at 545 nm. Calculate the molar absorptivity of dichromate at 440 nm, the molar absorptivity of permanganate at 545 nm, and the concentrations of dichromate and permanganate in the unknown mixture.

20. Dissociation constants for acid-base indicators can be evaluated by means of spectrophotometry. The acid dissociation constant for methyl red was determined as follows. A known quantity of the indicator was added to each of a series of buffer solutions of various pH values, and the absorbance of each solution was measured at 531 nm, at which wavelength only the red acid form of the indicator absorbs radiation. The experimental data were as follows:

| pH of Buffer | Absorbance |
|---|---|
| 2.30 | 1.375 |
| 3.00 | 1.364 |
| 4.00 | 1.274 |
| 4.40 | 1.148 |
| 5.00 | 0.766 |
| 5.70 | 0.279 |
| 6.30 | 0.081 |
| 7.00 | 0.017 |
| 8.00 | 0.002 |

Calculate the acid dissociation constant for methyl red indicator.

21. The *method of continuous variations* or *Job's method* is a procedure commonly used to determine the number of ligands coordinated to a metal ion. In this technique, the sum of the moles of metal ion and ligand is kept *constant*, as is the solution volume, but the individual quantities of metal and ligand are varied. A plot is constructed of the absorbance due to the metal-ligand complex as a function of the mole fraction of the metal ion. Such a plot consists of two straight-line portions, intersecting at a point which corresponds to the mole fraction of metal ion in the unknown complex. Suppose that the method of continuous variations was employed to establish the identity of the complex formed between iron(II) and 2,2'-bipyridine

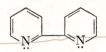

2,2'-bipyridine

and that the following data were obtained:

| Mole Fraction of Iron(II) | Absorbance at 522 nm |
|---|---|
| 0.08 | 0.231 |
| 0.12 | 0.346 |
| 0.17 | 0.491 |
| 0.22 | 0.632 |
| 0.28 | 0.691 |
| 0.36 | 0.615 |
| 0.45 | 0.531 |
| 0.56 | 0.422 |
| 0.65 | 0.334 |
| 0.75 | 0.241 |
| 0.83 | 0.162 |
| 0.91 | 0.087 |

(a) Construct a plot of absorbance versus the mole fraction of iron(II).

(b) Determine the formula of the complex formed between iron(II) and 2,2'-bipyridine by extrapolating the two straight-line portions to their point of intersection and interpreting the result.

(c) If the sum of the concentrations of iron(II) and 2,2'-bipyridine (in all their forms in solution) in this experiment remained constant at $2.74 \times 10^{-4}$ $M$ and all absorbances were measured in a 1-cm cell, calculate the molar absorptivity of the complex.

22. A $0.150$ $F$ solution of sodium picrate in a $1$ $F$ sodium hydroxide medium was observed to have an absorbance of $0.419$, due only to the absorption by picrate anion. In the same spectrophotometer cell and at the same wavelength as in the previous measurement, a $0.300$ $F$ solution of picric acid was found to have an absorbance of $0.581$. Calculate the dissociation constant for picric acid.

23. Suppose that you desire to determine spectrophotometrically the acid dissociation constant for an acid-base indicator. A series of measurements is performed in which the *total* concentration of indicator is $0.000500$ $F$. In addition, all the spectrophotometric measurements are obtained with a cell of 1-cm path length and at the same wavelength. Other components, in addition to the indicator, are introduced as listed in the following table of data, but none of these exhibits any measurable absorption. Calculate the dissociation constant for the indicator.

| Solution Number | Other Component | Absorbance |
|---|---|---|
| 1 | $0.100$ $F$ HCl | 0.085 |
| 2 | pH 5.00 buffer | 0.351 |
| 3 | $0.100$ $F$ NaOH | 0.788 |

24. A solution known to contain both ferrocyanide and ferricyanide ions was examined spectrophotometrically at a wavelength of 420 nm, where only ferricyanide absorbs. A portion of the solution was placed into a 1-cm cell and found to have a transmittance of $0.118$. The molar absorptivity of ferricyanide at 420 nm is 505 liter $mole^{-1}$ $cm^{-1}$. A platinum indicator electrode was inserted into the solution of ferricyanide and ferrocyanide, and its potential was observed to be $+0.337$ v versus the normal hydrogen electrode. Calculate the concentrations of ferricyanide and ferrocyanide in the original solution.

25. A two-color acid-base indicator has an acid form which absorbs visible radiation at 410 nm with a molar absorptivity of 347 liter $mole^{-1}$ $cm^{-1}$. The base form of the indicator has an absorption band with a maximum at 640 nm and a molar absorptivity of 100 liter $mole^{-1}$ $cm^{-1}$. In addition, the acid form does not absorb significantly at 640 nm, and the base form exhibits no measurable absorption at 410 nm. A small quantity of the indicator was added to an aqueous solution, and absorbance values were observed to be

0.118 at 410 nm and 0.267 at 640 nm for a 1-cm spectrophotometer cell. Assuming that the indicator has a p$K_a$ value of 3.90, calculate the pH of the aqueous solution.

26. In the so-called *mole-ratio method* for determining the identities of colored metal ion complexes, a series of solutions is prepared in which the concentration of the metal ion is kept constant while the concentration of the ligand is varied. Then a plot is constructed of the absorbance at a suitable wavelength versus the *ratio* of total moles of ligand to total moles of metal cation. From the point of intersection of the straight-line portions of the resulting curve, the formula of a metal-ion complex can be found. In a study of the chloro complexes of cobalt(II) formed in acetone, D. A. Fine (J. Amer. Chem. Soc., *84*:1139, 1962) prepared solutions containing a constant cobalt(II) concentration of $3.75 \times 10^{-4}\, M$ and various concentrations of lithium chloride. When absorbance readings were taken in a 5-cm cell at 640 nm, the following data were obtained:

| Chloride Concentration, $M \times 10^4$ | Absorbance at 640 nm |
|---|---|
| 1.99 | 0.118 |
| 3.83 | 0.225 |
| 4.95 | 0.292 |
| 6.08 | 0.352 |
| 6.83 | 0.405 |
| 7.50 | 0.423 |
| 8.30 | 0.422 |
| 9.08 | 0.403 |
| 9.84 | 0.386 |
| 10.50 | 0.369 |
| 11.44 | 0.361 |
| 12.45 | 0.372 |
| 13.36 | 0.382 |
| 14.29 | 0.393 |

(a) Construct a plot of absorbance versus the mole ratio of chloride to cobalt(II). What can you deduce regarding the identities of any chloro complexes of cobalt(II)?

(b) Extrapolate the straight-line portions of the curve to their points of intersection. Assuming that each point of intersection corresponds to the absorbance of a pure complex ion, calculate the molar absorptivity of each complex ion identified above. What do you think about the validity of this approach?

(c) Using experimental data from the plot of absorbance versus mole ratio, along with the results obtained in parts (a) and (b), what can you determine about the value(s) for the formation constant(s) for any of the chloro complexes of cobalt(II)?

27. In the fluorimetric determination of the excretion of therapeutic doses of penicillin, a sample of urine is first extracted with chloroform. A 10.00-ml portion of the penicillin-containing chloroform extract is then added to 5.00 ml of a benzene solution of 10 mg of 2-methoxy-6-chloro-9-($\beta$-aminoethyl)-aminoacridine,

$$\text{HNCH}_2\text{CH}_2\text{NH}_2$$

along with 2.00 ml of acetone and 5.00 ml of a 1 volume per cent solution of glacial acetic acid in benzene. During a waiting period of 1 hour, the aminoacridine forms a condensation product with penicillin. Next, the condensation product is isolated by means of a series of extractions. Finally, an acidified aqueous solution of the condensation product is placed in a cell, the solution is irradiated with ultraviolet light at 365 nm, and the yellow luminescence at 540 nm is measured with a detector, the signal appearing in the form of a current reading on a galvanometer. In the analysis of 10.00 ml of urine according to this procedure, the galvanometer readout device registered 28.78 microamperes. A sample blank (containing no penicillin) gave a reading of 9.13 microamperes, whereas two different standard solutions containing 0.625 and 1.500 $\mu$g of penicillin per 10.00 ml gave galvanometer readings of 18.89 and 32.54 microamperes, respectively, when treated according to the above procedure. Calculate the penicillin content of the urine sample in micrograms per milliliter.

28. In the spectrophotometric measurement of reaction rates, the *variable-time method* is often used. This method, depicted in Figure 19–25, involves the determination of the time that elapses between the crossing of two absorbance thresholds. Because the fixed difference between the thresholds is the vertical axis on the reaction-rate curve, the elapsed time is reciprocally related to the rate and, indirectly, to the initial reactant concentration. For the glucose determination illustrated in Figure 19–25, what would the quantity of glucose be for a sample which produced an elapsed time of 42.0 seconds?

29. Some reaction-rate curves are not as smooth as those shown in Figure 19–25, owing to disturbances in the chemical system or the measuring apparatus. Such a curve might have the following appearance:

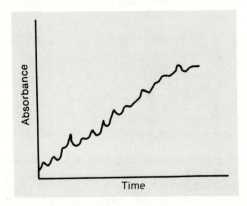

Why would the variable-time method described in problem 28 probably yield an erroneous value for the concentration of glucose if a reaction-rate curve such as the above were utilized?

30. Another approach taken toward the automatic measurement of reaction rates is the *derivative method*, in which the electronically calculated time-derivative of the absorbance is continuously plotted. Ideally, this derivative should be directly proportional to the slope of the reaction-rate curve and also to the initial reactant concentration. List some advantages and disadvantages of the derivative method, while considering the reaction-rate curve shown in problem 29.

## SUGGESTIONS FOR ADDITIONAL READING

*General References*

1. R. E. Dodd: *Chemical Spectroscopy.* Elsevier, Amsterdam, 1962.
2. I. M. Kolthoff and P. J. Elving, eds.: *Treatise on Analytical Chemistry.* Part I, Volume 5, Wiley-Interscience, New York, 1959, pp. 2707–3078.
3. D. A. Skoog and D. M. West: *Principles of Instrumental Analysis.* Holt, Rinehart and Winston, New York, 1971, Chapters 2, 3, 4, and 8.

4. F. J. Welcher, ed.: *Standard Methods of Chemical Analysis*. Sixth edition, Volume III-A, Van Nostrand, Princeton, New Jersey, 1966, pp. 3–37, 78–104.
5. H. H. Willard, L. L. Merritt, and J. A. Dean: *Instrumental Methods of Analysis*. Fourth edition, Van Nostrand, Princeton, New Jersey, 1965, Chapters 3 and 4.
6. J. D. Winefordner, ed.: *Spectrochemical Methods of Analysis*. Wiley-Interscience, New York, 1971, Chapters 5–8.

*Ultraviolet-Visible Spectrophotometry*

1. R. P. Bauman: *Absorption Spectroscopy*. John Wiley and Sons, New York, 1962.
2. C. R. Hare: Visible and ultraviolet spectroscopy. *In* T. H. Gouw, ed.: *Guide to Modern Methods of Instrumental Analysis*. Wiley-Interscience, New York, 1972, Chapter 5.
3. E. B. Sandell: *Colorimetric Determination of Traces of Metals*. Third edition, Wiley-Interscience, New York, 1959.

*Fluorescence and Phosphorescence*

1. G. G. Guilbault, ed.: *Fluorescence; Theory, Instrumentation, and Practice*. Dekker, New York, 1967.
2. D. M. Hercules, ed.: *Fluorescence and Phosphorescence Analysis*. Wiley-Interscience, New York, 1966.
3. S. Udenfriend: *Fluorescence Assay in Biology and Medicine*. Academic Press, New York, Volume 1, 1962; Volume 2, 1969.
4. J. D. Winefordner, S. G. Schulman, and T. C. O'Haver: *Luminescence Spectrometry in Analytical Chemistry*. Wiley-Interscience, New York, 1972.

# SPECTROSCOPY IN THE ULTRA-VIOLET AND VISIBLE REGIONS— ATOMIC ELEMENTAL ANALYSIS

# 20

In this chapter we will consider techniques for elemental analysis which utilize ultraviolet and visible radiation. Among these are flame spectrometric methods based on atomic emission, absorption, and fluorescence; DC arc spectroscopy; and high-voltage spark spectroscopy. Because flame spectrometric methods are most widely used, we will deal with them in greatest detail.

## ELEMENTAL ANALYSIS

In order to perform an elemental analysis by means of a spectrochemical technique, one must be able to observe spectral information that is unambiguously characteristic of each element being determined. Such information would obviously not be produced by translational, rotational, or vibrational motions of molecules, because these motions involve several or all of the atoms in the molecule. Similarly, the electronic spectra of most molecules do not directly reflect elemental composition, because the electronic energy levels between which transitions occur are combinations of the atomic energy levels of several elements.

There appear to be two ways to perform a successful elemental analysis through spectrochemical means: (1) by observing transitions between atomic levels which are not involved in bonding and which are characteristic of a specific element or (2) by separating the atoms comprising a molecule, isolating them in the gas phase, and observing the electronic transitions characteristic of the free atoms.

In the first category, energy states not affected by bonding include those of inner-shell electrons and of the atomic nucleus. Transitions between the innermost electronic levels involve energy differences corresponding to x-ray frequencies and are highly characteristic of the atom undergoing the transition. A technique which utilizes these transitions for elemental analysis is **x-ray fluorescence spectrometry**.* The only other commonly observed electronic transitions which are free from bonding effects are those involving the f electrons of the rare-earth elements. These electrons do not participate in chemical bonding and are capable of undergoing low-energy transitions corresponding to radiation in the ultraviolet-visible region. These electronic transitions are relatively unaffected by vibrational or rotational motions and, therefore, produce narrow-line spectra which are easily distinguishable from most molecular spectra. This characteristic was important in establishing one of the early methods for the identification of the rare earths, which are difficult to distinguish from each other by chemical means.

Nuclear transitions are also free from the effects of chemical bonding and can be utilized for elemental analysis. Naturally radioactive elements undergo distinctive nuclear transitions resulting in the emission of alpha particles, beta particles, and gamma rays. The energy spectra of these radiations can be used to identify and determine the elements. This procedure provides the basis for radiotracer experiments in biology, biochemistry, and medicine. For nonradioactive elements, a technique called **neutron activation analysis**† can be employed to yield similar results. In neutron activation analysis, the sample to be analyzed is bombarded with slow (thermal) neutrons, some of which are captured by the atomic nuclei to produce radioactive isotopes. Because these artificially activated nuclei emit characteristic radiation just as the naturally radioactive elements do, they can then be determined. Neutron activation analysis is an extremely sensitive technique and can detect as little as $10^{-15}$ gm of an element. However, its use requires the availability of a source of slow neutrons, so that it is somewhat restricted in its application.

In the second category of spectrochemical techniques for elemental analysis, atoms are separated from each other and are caused to produce characteristic absorption, emission, or fluorescence spectra independently. This separation of atoms is surprisingly easy and can be achieved in a variety of ways. Among the spectrochemical techniques employing this principle are mass spectroscopy and those methods to be discussed in this chapter.

## Atomic Spectrochemical Analysis

Among the techniques which employ ultraviolet-visible radiation for elemental analysis, the methods used to produce free atoms vary considerably. However, in all cases, the atom-producing medium must be energetic—so energetic, in fact, that the liberated atoms are often appreciably excited. Therefore, the device producing the

---

* X-ray fluorescence spectrometry is an important tool for the routine elemental analysis of many industrial samples. For a detailed treatment, refer to the book by L. S. Birks: *X-Ray Spectrochemical Analysis*. Second edition, John Wiley and Sons, New York, 1969.

† For a detailed treatment of neutron activation analysis, consult the monograph by P. Kruger: *Principles of Activation Analysis*. Wiley-Interscience, New York, 1971.

atoms serves not only as an **atom reservoir** but often as an **excitation source** as well. Generally, electrical atomization devices such as the arc or spark are more energetic than a thermal atomization system such as a flame, so that the number of excited atoms is larger with the former. For this reason, the arc and spark are nearly always used as excitation sources from which atomic emission is observed. In contrast, flames can be employed as atom reservoirs for emission, absorption, and fluorescence spectrometry.

Regardless of whether atomic emission, absorption, or fluorescence is used for an analysis, the observed spectral features are similar. Because the atoms are essentially isolated from each other, their spectra consist of narrow lines, generally less then 0.1 Å in width. Furthermore, the spectra are relatively simple, especially for the lighter elements. Consider, for example, the emission spectrum for mercury displayed in Figure 20–1. This spectrum of mercury atoms excited in an electrical discharge shows a number of spectral lines, well separated from each other and located at specific wavelengths, the most prominent of which have been marked. These emission lines are highly characteristic of mercury and can be used for its qualitative detection in a sample. Even if other elements are present in the sample, the identification of mercury is still possible, because each element produces a characteristic spectrum of narrow lines which does not obscure the line spectra of other elements.

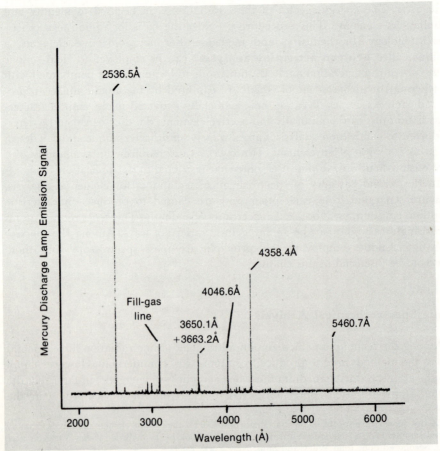

*Figure 20–1.* Emission spectrum of mercury atoms excited in a low-power electrical discharge. Prominent, characteristic spectral lines are indicated. A fill-gas emission line is produced by excitation of an inert gas present in the mercury discharge lamp.

Atomic spectrochemical analysis is therefore a powerful tool for qualitative elemental analysis. Furthermore, the power of each spectral line can be related to the concentration of a specific component of the sample. Generally, quantitative spectrochemical analysis is performed empirically, with calibrated standards and fixed experimental conditions.

## FLAME SPECTROMETRIC ANALYSIS

Let us begin our discussion of atomic spectroscopy with the flame spectrometric methods, at present the most widely used of all techniques for elemental analysis. In these methods, a chemical flame is employed as an atom reservoir, and the emission, absorption, or fluorescence properties of the liberated atoms are examined spectrometrically. Because the procedures used to measure emission, absorption, and fluorescence differ somewhat, the field of flame spectrometry is commonly divided into three areas—flame emission spectrometry, atomic absorption spectrometry, and atomic fluorescence flame spectrometry.

We have seen in Chapter 18 (pages 605 to 608) that emission, absorption, and luminescence are distinctly different. Yet, in flame emission, atomic absorption, and atomic fluorescence flame spectrometry, the form of the chemical sample is the same—namely, *free atoms in a flame*. In order for these techniques to be employed successfully, the generation of free atoms in a flame must be efficient, reproducible, and predictable. However, the process of converting a chemical sample into free atoms in a flame is exceedingly complex and depends upon a number of factors that must be carefully controlled.

### Formation of Atoms in a Flame

As shown in Figure 20–2, a typical system for the production of free atoms in flame spectrometry consists of a **spray chamber** and a **burner**. A sample solution is

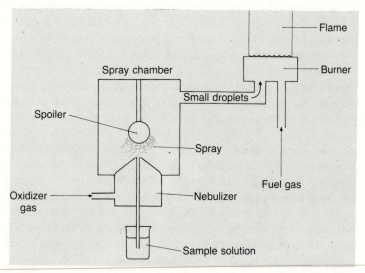

*Figure 20–2.* Premix burner-nebulizer system used in flame spectrometry.

aspirated into the spray chamber with the aid of a **nebulizer** which is controlled by the flow of an oxidizer gas such as air. The cloud of droplets so produced strikes an obstruction in the spray chamber termed a **spoiler**, which traps the larger droplets or breaks them into smaller droplets. Small droplets are carried by the oxidizer gas from the spray chamber into the burner, where the droplets and oxidizer gas are mixed with a fuel gas. Next, the droplets pass through holes in the top of the burner along with the oxidizer-fuel gas mixture which, when ignited, forms the flame.

In the flame, a complex sequence of events must occur to convert the species dissolved in the sample droplets into free atoms. First, the droplets must evaporate (desolvate) to form small particles of the dried solute. Then the solute particles must vaporize to yield free atoms. In the process of vaporization, the solid particles of solute can be transformed into not only free atoms but molecules and molecular fragments as well; however, the latter species cannot contribute to atomic emission, absorption, or fluorescence. Furthermore, some of the free atoms are ionized in the flame and cannot undergo the desired electronic transitions. Clearly, if analytical results of high sensitivity and reproducibility are to be obtained, all factors affecting atom production should be optimized and held as constant as possible. What are some of the experimental variables influencing the liberation of atoms in a flame?

*Nebulization.* The efficiency of nebulization (spraying) controls the fraction of the sample solution converted into tiny droplets which can ultimately reach the flame. In turn, nebulization is governed by such parameters as the viscosity and surface tension of the sample solution, the flow rate of the nebulizer gas, and the design of the nebulizer itself. For example, the use of organic solvents which lower the viscosity of the sample solution frequently enhances the nebulization process. On the other hand, sample solutions containing high solute concentrations have relatively high viscosities which cause a decrease in the amount of sample nebulized and a consequent reduction in the number of free atoms in the flame.

*Desolvation.* In the flame, desolvation of the nebulized droplets critically influences the number of free atoms. Desolvation is affected by a number of experimental parameters, including the flame temperature, the nature of the solvent, and the residence time of the droplets in the flame. Thus, it is desirable to employ a combination of oxidizer and fuel gases that burns at a high temperature, and to use a slow rate of gas flow so that the droplets spend as much time in the flame and undergo as much desolvation as possible. The use of organic solvents often enhances desolvation as well, since many of these solvents evaporate more rapidly than water and, additionally, tend to cool the flame less as they evaporate.

*Vaporization.* Vaporization of the solid particles of solute which remain after desolvation is the final step leading to the formation of free atoms. Vaporization is controlled by many of the same variables as desolvation. An increase in flame temperature or the residence time of particles in the flame will increase the fraction of particles that completely vaporize. However, vaporization is strongly affected by the nature of the vaporizing particle itself. For example, a particle of aluminum oxide vaporizes much more slowly than a particle of sodium chloride of similar size. Therefore, if we compare the behavior of a cloud of sodium chloride particles and a cloud of aluminum oxide particles of the same size, the fraction of sodium chloride particles which vaporize after a specified time will be greater than the fraction of aluminum oxide particles at any given flame temperature. This effect becomes very important in the flame spectrometric analysis of complex solutions. Certain concomitants in a solution can significantly alter the process of vaporization and atom production of an element under observation. Such interferences are not uncommon and will be discussed later.

*Ionization.* Any ionization which occurs in the flame reduces the total number of atoms available for observation by means of flame spectrometry. Ionization is usually important only in high-temperature flames (above 2000°K) and for elements having relatively low ionization energies. Alkali and alkaline earth metals are particularly susceptible to ionization, as we shall discuss in detail later.

From the foregoing, we see that atom formation in a flame is a complex process and that the use of a flame for analytical purposes requires control of a number of variables, many of which are interdependent. In the selection of optimum conditions for a flame spectrometric analysis, the choice of the flame and burner are particularly critical.

## Flames, Burners, and Nebulizers

There are several desirable characteristics of an analytical flame which derive from the considerations of the preceding section. Of course, the flame should be stable, safe, and inexpensive to maintain; it should also have a relatively high temperature and a slow rise-velocity, both of which enhance the efficiency of desolvation and vaporization and lead to a larger emission, absorption, or fluorescence signal. In addition, a flame should provide a reducing atmosphere. In the flame, many metals tend to form stable oxides. Because these oxides are refractory and not readily dissociated at ordinary flame temperatures, it is necessary to reduce the oxides to promote the formation of free atoms. This reduction can be enhanced in almost any flame if the flow rate of fuel gas is greater than that needed for stoichiometric combustion. Under these conditions, the flame is said to be **fuel rich**. Fuel-rich flames produced by hydrocarbon fuels such as acetylene provide excellent reducing atmospheres because of the presence of many carbon-containing radical species.

Two basic types of burners have been developed for use in flame spectrometry in an effort to meet the above criteria. These two types—termed total consumption and premix—differ both in construction and in the mixtures of fuel and oxidizer gases which can be used.

*Total-consumption burner.* A schematic diagram of a total-consumption burner is shown in Figure 20-3. This burner derives its name from the fact that all the nebulized solution reaches the flame. In operation, a flow of oxidizer gas leaves the burner through an orifice concentrically surrounding the nebulizer tube. This gas flow creates a slight vacuum at the burner tip so that sample solution is drawn up the nebulizer tube into the high-velocity flow of oxidizer gas; solution leaving the nebulizer tube is thereby shattered into tiny droplets. Fuel gas exits the total-consumption burner through a second circular orifice concentrically surrounding both the nebulizer tube and the oxidizer port. Turbulence in the flowing gases produced by this arrangement intimately mixes the fuel and oxidizer gases for complete combustion.

A total-consumption burner has several distinct advantages. First, it is both safe and inexpensive to operate. Because the fuel and oxidizer gases are not mixed until they leave the burner, no combustible mixture is formed except within the flame itself. This makes it possible to employ fuel and oxidizer mixtures, such as hydrogen-oxygen and acetylene-oxygen, which would otherwise be extremely dangerous to handle. These mixtures provide flames with extremely high temperatures (greater than 3000°K) and with excellent oxide-reducing properties, especially the fuel-rich acetylene flame. Because all the nebulized sample solution reaches the flame in the total-consumption burner, the efficiency of this burner should be extremely high. However, many of the larger droplets produced during nebulization leave the

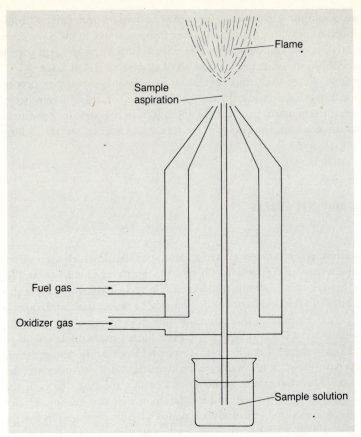

*Figure 20–3.* Cutaway diagram of a total-consumption nebulizer-burner.

flame without being completely desolvated, and many large desolvated salt particles never fully vaporize, so that the number of atoms released in the flame is not as great as would be expected.

Another disadvantage of the total-consumption burner is that the turbulence of the flame makes the resulting signal, be it emission, absorption, or fluorescence, unsteady. Furthermore, the high degree of turbulence permits droplets, desolvated particles, and atoms a relatively short residence time in the flame. All these factors serve to decrease both the number of free atoms formed and the magnitude of the signal which can be derived from the atoms which are freed.

***Premix burner.*** The second type of burner often used in flame spectrometry is the premix burner, an example of which was shown in Figure 20–2. Because the fuel and oxidizer gases are premixed before leaving the burner, a major advantage of the premix burner is the great stability of the flame it produces. For some premix burners, a Meker-type burner head is used to produce a cylindrical flame. In other burners, especially those employed in atomic absorption spectrometry, the exit orifice of the burner is a slot, so that a narrow flame with a long path length is obtained.

The stability of the flame from a premix burner is further enhanced through the use of a spray chamber, as described earlier. Although this arrangement reduces the flow of sample solution to the flame, a much greater fraction of droplets reaching the flame are desolvated and finally converted into free atoms.

The recent introduction of oxidizer and fuel gas mixtures such as nitrous oxide-acetylene has made the premix burner competitive with or superior to the total-consumption burner. A nitrous oxide-acetylene flame approaches 3000°K in temperature and has excellent reducing characteristics. In addition, the nitrous oxide-acetylene flame burns slowly enough to allow droplets, particles, and free atoms a long residence time in the flame. Also commonly used in the premix burner is a mixture of air and acetylene. Although lower in temperature than the nitrous oxide-acetylene flame, the air-acetylene flame generates less background radiation to interfere with an atomic emission, absorption, or fluorescence signal.

Among the disadvantages of the premix burner are that it is sometimes unsafe, often expensive, and rather inefficient in its use of sample solution. In particular, the fuel and oxidizer gases must necessarily be combined inside the burner to form a potentially explosive mixture. If this mixture is inadvertently ignited, the resulting detonation can do considerable damage. To avoid this problem, commercially available premix burners are very carefully designed.

## Interferences in Flame Spectrometry

Interferences common to all three flame spectrometric techniques can be classified as vaporization interferences, spectral interferences, or ionization interferences. A vaporization interference occurs when some component of the sample influences the rate of vaporization of particles containing the desired species. The origin of a vaporization interference can be a chemical reaction which alters the vaporization behavior of solid particles, or it can be a physical process in which the vaporization of a host or matrix controls the release of atoms being observed. Spectral interferences originate because the line or band spectrum of an undesired species overlaps that of the element being determined. Ionization interferences arise from changes in the extent of ionization of atoms caused by the presence of other elements in the sample. Let us now examine some commonly encountered interferences.

*Phosphate interference in the determination of calcium.* A well documented case of vaporization interference is the effect of phosphate on the flame spectrometric determination of calcium. It has been found that a calcium solution containing phosphate will generate a smaller flame spectrometric signal than a solution of identical calcium concentration but containing no phosphate. This behavior can be seen in Figure 20–4, in which signals arising from phosphate-containing calcium solutions have been plotted against the phosphate concentration. Phosphate decreases the calcium signal linearly; however, beyond a well-defined point no further depression of the signal is observed. The point at which phosphate ceases to depress the calcium signal occurs at a definite calcium-to-phosphate ratio. This suggests the formation of a stoichiometric compound between calcium and phosphate which vaporizes more slowly than calcium alone. Further evidence for this is the observation that the extent to which phosphate depresses the calcium signal is greatest at points low in the flame. If the calcium signal is measured high in the flame, the calcium-containing solute particles have a longer time to vaporize so that a greater fraction of the phosphate-bound calcium atoms are released.

The interference of phosphate in the determination of calcium can be minimized by means other than viewing the calcium signal high in the flame. A spray chamber-nebulization system, such as that often used with the premix burner, can help to minimize this problem. Smaller droplets leaving the spray chamber-nebulizer produce, upon desolvation, tinier salt or solute particles which require less time to vaporize, so that the phosphate interference is reduced. Similarly, a very hot flame such as that produced by a nitrous oxide-acetylene mixture will minimize the phosphate interference by increasing the rate of particle vaporization.

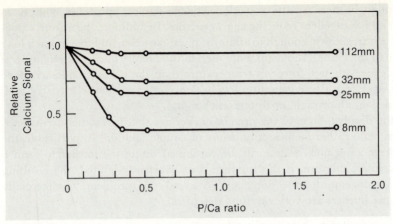

*Figure 20–4.* Illustration of the phosphate interference in the flame spectrometric determination of calcium. Various curves represent signals measured at different indicated heights in the flame. The calcium concentration was constant in all determinations. (Reproduced, with permission, from the paper by C. T. J. Alkemade and M. H. Voorhuis: Z. anal. Chem., *163*: 91, 1958.)

Finally, the vaporization interference can be minimized or eliminated through the use of substances called "releasing agents" which enhance the release of calcium atoms from the slowly vaporizing phosphate-containing particle. For example, large amounts of lanthanum ion, added to a solution containing calcium and phosphate, preferentially combine with phosphate, leaving the calcium atoms free to vaporize. Another releasing agent is EDTA which, when added to the sample solution, complexes calcium to prevent the formation of the calcium-phosphate species. When the solution is sprayed into a flame, the calcium-EDTA complex is readily destroyed, thereby releasing the calcium atoms. A third type of releasing agent merely provides a matrix or substrate which decomposes or vaporizes rapidly. For example, a particle composed primarily of glucose tends to break up readily in the flame because of its organic nature. If a large amount of glucose is added to a solution containing phosphate and an alkaline earth metal, the desolvated particle will consist primarily of glucose with calcium and phosphate spread throughout the particle. When the particle breaks up, the calcium-phosphate species, if such exists, will then be extremely small and readily vaporized.

*Vaporization interference caused by aluminum oxide.* Aluminum oxide ($Al_2O_3$) is highly refractory—that is, it vaporizes extremely slowly at ordinary flame temperatures. Therefore, when aluminum ion is present in appreciable concentration in a sample solution, particles of aluminum oxide are likely to be formed which do not vaporize completely before leaving the flame. Atoms trapped in these particles will not be available for analysis. This same interference is caused by other elements that form refractory oxides, such as titanium, zirconium, vanadium, and tantalum. To minimize these physical interferences, the methods discussed above for chemical interferences can be employed; one of the most effective procedures is the addition of a complexing agent such as EDTA to the solution.

*Spectral interferences.* For reasons that will become evident later, spectral interferences are somewhat more common in flame emission spectrometry than in the techniques of atomic absorption and atomic fluorescence flame spectrometry. One example of a spectral interference is encountered in the determination of barium in the presence of large amounts of calcium by means of flame emission spectrometry.

Unfortunately, the most sensitive barium emission line (5535.6 Å) lies within a broad band of wavelengths emitted by a CaOH molecular fragment. Therefore, radiation from the flame is attributable to both the barium line and the CaOH band and cannot be unambiguously related to the barium concentration in the sample. To circumvent this problem, a high-resolution spectral dispersing device is employed to scan across the barium line of interest. In this case, the spectrum would be scanned past the barium line at 5535.6 Å, which would appear as a measurable peak on top of the broad spectral envelope of the CaOH band.

Although *interelement* spectral interferences of the kind discussed above are of greatest importance in flame spectrometry, other spectral interferences can arise from the flame itself. In particular, the high background radiation produced by most acetylene-fueled flames is especially troublesome, although the wavelength regions where the background is highest are well known and can be avoided whenever possible.

**Ionization interferences.** Unlike a vaporization interference, an ionization interference results in an enhancement rather than a depression of the observed signal. This effect is most often seen during the determination of an easily ionizable element such as rubidium, because of a shift in the equilibrium

$$Rb \rightleftharpoons Rb^+ + e$$

In Figure 20–5, the signal from a fixed concentration of rubidium is plotted versus the concentration of potassium added to the sample. Potassium, being an alkali metal, also ionizes readily

$$K \rightleftharpoons K^+ + e$$

thereby increasing the electron concentration in the flame. In turn, this increased electron concentration shifts the rubidium ionization equilibrium back toward the formation of rubidium atoms. As greater amounts of potassium are added to the rubidium-containing solution, the rubidium signal continues to increase until essentially all the rubidium is in the free atomic state. Beyond a certain point, the addition of more potassium produces no further increase in the rubidium signal.

It is clear from Figure 20–5 that a reliable rubidium analysis could not be performed on a sample containing an unknown amount of potassium. To avoid this problem, samples can be doped with an excess of another alkali metal, such as lithium, whose concentration is not sought. This added element ionizes in the flame just as rubidium and potassium to greatly increase the electron concentration and to

*Figure 20–5.* An ionization interference due to the presence of potassium in the flame spectrometric determination of a constant concentration of rubidium. (Reproduced from the paper by S. Fukushima: Mikrochim. Acta, 1960, p. 332.)

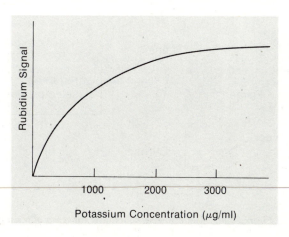

shift the ionization equilibria of all other elements toward the free-atom state. This approach has the doubly beneficial effect of increasing the number of free atoms and removing the interference.

### Sample Preparation

Because flame spectrometric methods usually rely on a burner-nebulizer system for the production of free atoms, samples to be analyzed must be in the form of a solution. Although this requirement presents no problem for many samples, others are dissolved only with great difficulty. Digestion, fusion, or combustion might be required before a suitable sample solution can be obtained. For example, clinical samples are usually digested first with hot nitric acid, perchloric acid, or a mixture of the two acids to destroy the organic matter; then, after addition of reagents to prevent interferences, the solution can be diluted to the desired concentration. Other samples, particularly silicates, are extremely hard to dissolve and must be treated with hydrofluoric acid or be fused with an appropriate flux to render them soluble. A large number of detailed procedures have been developed for the treatment of hard-to-dissolve samples. These procedures can be found in references cited at the end of this chapter.

## FLAME SPECTROMETRIC TECHNIQUES

Although the characteristics of atom formation in a flame and the existence of interferences are common to all three flame spectrometric methods, each of the methods has its own particular capabilities and limitations. In the following discussion, we will consider each of the methods independently, drawing comparisons where appropriate.

### Flame Emission Spectrometry

Flame emission spectrometry is instrumentally the simplest of the flame spectrometric techniques. Recalling the generalized spectrochemical instrument of Chapter 18 (page 612), we see that the chemical input-electromagnetic radiation output transducer in flame emission spectrometry is the flame itself. This fact becomes clear when the generalized instrument of Figure 18–10 is compared with the schematic diagram of a typical flame emission spectrometer shown in Figure 20–6. In the latter

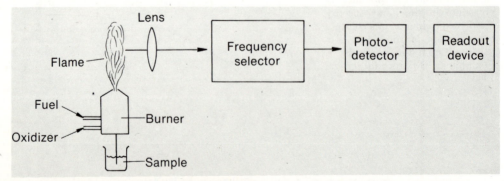

*Figure 20–6.* Generalized, schematic diagram of a simple flame emission spectrometer.

instrument, the atoms liberated in the flame are excited to emit their characteristic radiation. This radiation, which can be used for both quantitative and qualitative analysis, is focused by a simple lens onto a frequency selector such as a filter or monochromator. The frequency-selected radiation is then detected and converted to an electrical signal by a suitable photodetector such as a photomultiplier tube. The resulting electrical signal, which is proportional to the power of the emitted radiation, is displayed on a readout device.

*Excitation and emission.* To obtain the greatest possible sensitivity in flame emission spectrometry, the temperature of the flame should be extremely high. This can be understood in terms of the Boltzmann distribution equation

$$N_m^* = N_m \frac{g_u}{B(T)} e^{-E_u/kT} \tag{20-1}$$

where $N_m^*$ is the number of excited atoms of species $m$ in the flame, $N_m$ is the number of free $m$ atoms in the flame, $g_u$ is the statistical weight (degeneracy) of the excited atomic state, $B(T)$ is the partition function of the atom over all states, $E_u$ is the energy of the excited state, $k$ is the Boltzmann constant, and $T$ is the absolute temperature.

From the Boltzmann expression, it can be seen that a flame of higher temperature will produce a greater number of atoms in the excited state. The power of the emitted radiation is proportional to the number of excited atoms according to the relation

$$P_T = h\nu_0 N_m^* A_T \tag{20-2}$$

where $P_T$ is the total power (in watts) radiated by the atoms in the flame, $h$ is Planck's constant ($6.624 \times 10^{-34}$ joule·sec), $\nu_0$ is the frequency (in sec$^{-1}$) of the peak of the spectral line being observed, and $A_T$ is the Einstein coefficient (number of transitions each excited atom undergoes per second).† In effect, this equation states that the total power radiated by the atoms in the flame ($P_T$) is equal to the energy of each transition ($h\nu_0$) multiplied by the number of transitions that occur per second ($N_m^* A_T$). From this expression, it appears that quantitative analysis by means of flame emission spectrometry should be relatively straightforward and that the detected radiation could be unambiguously related to the concentration of a specific element. Unfortunately, this is not true. Although it is possible to relate the amount of detected radiation to the concentration of atoms in the flame, it is far more difficult to relate either of these quantities to the concentration of an element in a sample solution. Uncertainties existing in the atom-production system make this *absolute* approach to flame emission spectrometry impractical.

*Flame emission working curves.* Instead, quantitative analysis by means of flame emission spectrometry is usually performed with the aid of a series of working or calibration curves, one for each element to be determined. An example of such a working curve for potassium is shown in Figure 20–7. As predicted by the preceding equation and by the considerations of Chapter 18, the relationship between the total power of emitted radiation (or the readout value) and solution concentration is linear up to relatively high concentrations—in this case approximately 85 parts per million by weight of potassium ion. However, above this concentration, the curve bends toward the concentration axis because of a phenomenon known as **self-absorption**. Self-absorption is the absorption of emitted radiation by cooler atoms

---

† The Einstein coefficient for spontaneous emission is a characteristic constant of each specific transition and reflects the lifetime of the energy state from which the transition originates. For most atoms, $A_T$ is on the order of $10^8$ transitions per excited atom per second.

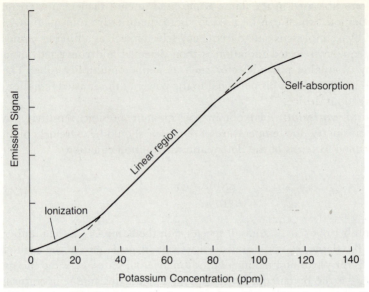

*Figure 20–7.* Working curve for determination of potassium by means of flame emission spectrometry.

near the edge of the flame; in other words, atoms near the center of the flame, being hotter, emit radiation which can be absorbed by atoms of the same element residing at the edge of the flame. This effect is most significant when the atom concentration of the flame is high. In fact, it can be shown theoretically that, whereas the relationship between emitted power and concentration should be linear at low concentrations, the power of the emitted radiation will increase only in proportion to the square root of the concentration at higher concentration values. Just as in the case of deviations from Beer's law, the onset of self-absorption at high concentrations does not preclude quantitative analysis by means of flame emission spectrometry. It is only necessary to construct a working curve such as that shown in Figure 20–7.

At low concentrations of potassium ion, the curve in Figure 20–7 again bends, but away from the concentration axis, in this case because of ionization. For a low total concentration of potassium in the flame, the number of electrons added to the flame by ionization of potassium will be small. This low electron concentration causes the ionization equilibrium

$$K \rightleftharpoons K^+ + e$$

to be shifted toward the right. Therefore, at low potassium concentrations, a smaller fraction of the total number of potassium species residing in the flame exists as free atoms. This causes the atomic emission to be lower than that expected, producing a bend in the working curve at low potassium ion concentrations and reducing the sensitivity for the determination of potassium. A similar situation prevails for any other alkali or alkaline earth metal that one wishes to determine. Like ionization interferences, this problem can be overcome by adding an excess of another easily ionizable element whose determination is not desired.

***Flame emission instruments.*** Some of the high-temperature flames often used in flame emission spectrometry are listed in Table 20–1, along with their maximum temperatures. With these extremely hot flames, excellent sensitivities can be obtained for the determination of a great number of elements. Minimum detectable

**Table 20–1.** **Common Flame-Gas Mixtures Employed in Flame Spectrometry***

| Fuel Gas | Oxidizer Gas | Measured Temperature (°C)† |
|----------|--------------|----------------------------|
| Acetylene | Air | 2125–2400 |
| Acetylene | Nitrous oxide | 2600–2800 |
| Acetylene | Oxygen | 3060–3135 |
| Hydrogen | Oxygen | 2550–2700 |
| Hydrogen | Air | 2000–2050 |

* From the compilation by R. N. Kniseley. *In* J. A. Dean and T. C. Rains, eds.: *Flame Emission and Atomic Absorption Spectrometry.* Vol. 1, Dekker, New York, 1969, p. 191. (Courtesy of Marcel Dekker, Inc.)

† Exact temperature depends on the flow rates of the fuel and oxidizer gases and on the design of the burner.

concentrations of these elements, together with the wavelength of the most sensitive emission line for each, are included in Figure 20–15 (to be discussed in detail later). Most elements can be readily detected at concentration levels less than one part per million (ppm). Furthermore, using flame emission spectrometry, it is possible with care to perform quantitative analyses at levels of precision between 1 and 5 per cent. In the simplest instruments (called **flame photometers**) for flame emission spectrometry, which are designed for routine, quantitative determinations involving no spectral interferences, the frequency selector is merely a filter. The lower cost, convenience, and large radiation throughput of a filter all combine to provide high sensitivity and operational simplicity. It is necessary only to insert the sample solution into the atomizing system and to observe the emission signal on a meter or strip-chart recorder; calibration is possible so that the readout can be directly in concentration units.

To perform qualitative analyses by means of flame emission spectrometry, a monochromator must be used as the frequency selector. In operation, the monochromator is set to scan across the wavelength region of interest, and the emission lines characteristic of each element appear as peaks against the flame background. By observing the wavelengths at which emission peaks do occur, one can identify the elements present from compiled listings or tables such as Figure 20–15. A flame emission spectrum for a sample containing sodium, magnesium, and calcium is shown in Figure 20–8. Although the concentration of each element in solution was identical, the lines identified with each element differ in intensity and wavelength. The simplicity of such a line spectrum usually makes qualitative analysis by means of flame emission spectrometry relatively straightforward, although complications can arise from the flame background.

Figure 20–9 shows typical background emission spectra of three common flame-gas mixtures used in flame emission spectrometry: oxygen-hydrogen, oxygen-acetylene, and nitrous oxide-acetylene. These spectra differ in complexity, in intensity, and in the wavelengths of maximum emission, and are strongly affected by the fuel-oxidizer gas-flow ratio. For elements present in a sample at low concentration, qualitative or quantitative analysis can be seriously hindered by the background spectrum of the flame. For this reason, it is good practice to record a background emission spectrum for a blank solution and to compare this background spectrum with that of the actual sample solution. If care is taken, this procedure enables very accurate and sensitive qualitative and quantitative flame emission analyses to be performed.

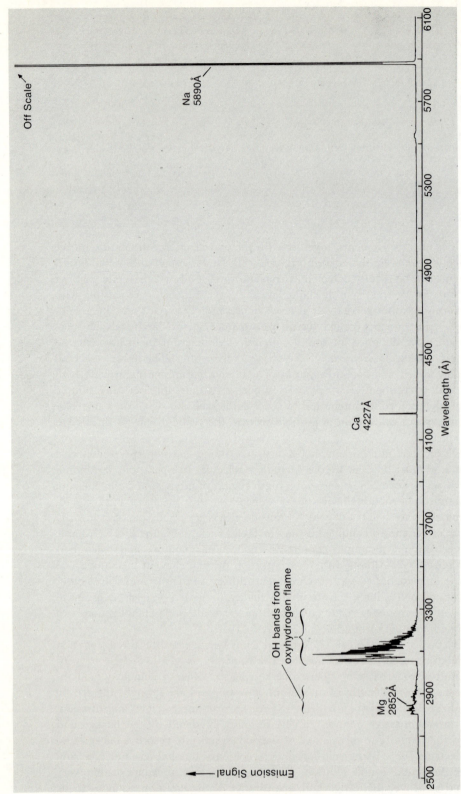

*Figure 20-8.* Flame emission spectrum of a solution which contains 10 ppm each of the elements sodium, magnesium, and calcium. Prominent, analytic-ally useful spectral lines are indicated. Notice that the emission line at 2852 Å for magnesium is nearly buried by the —OH emission bands from the oxygen-hydrogen (oxyhydrogen) flame.

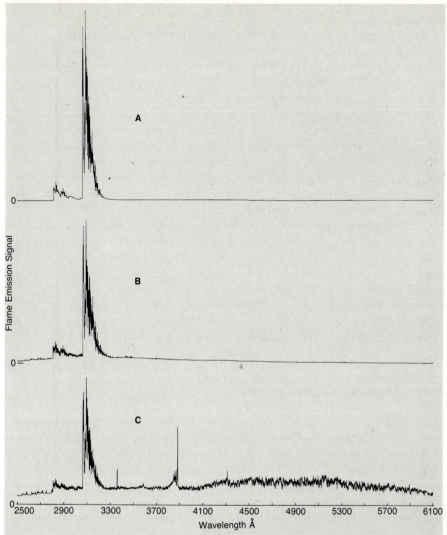

*Figure 20–9.* Background emission spectra for the oxygen-hydrogen (A), oxygen-acetylene (B), and nitrous oxide-acetylene (C) flames. The spectra have all been recorded at the same sensitivity; however, the actual band intensities are strongly dependent on the flow rates of the fuel and oxidizer gases.

Another common complication which can arise in flame emission spectrometry is a spectral interference between elements in a sample, as discussed earlier (page 680). A number of elements possess emission lines lying only small fractions of an angstrom from each other, so that more than one emission line might be measured simultaneously. To minimize spectral interferences, nonroutine flame emission spectrometry should usually be performed with a high-quality monochromator, preferably one capable of resolving wavelength intervals as close together as 0.1 Å. This resolution should be compared to that required for ultraviolet-visible molecular spectrophotometry, in which bandwidths of 100 Å are often adequate. Of course, in some applications such as the determination of sodium and potassium in clinical samples discussed below, few other elements exist in appreciable concentration in the sample. This permits the use of a low-background flame and a low-resolution spectral dispersion system; even an interference filter provides adequate spectral isolation in some cases, as mentioned earlier.

*Applications of flame emission spectrometry.* Flame emission spectrometry is widely used to determine the concentrations of sodium, potassium, calcium, and magnesium in clinical samples. The convenience, accuracy, sensitivity, and speed of this technique allow it to lend itself well to routine analyses. To perform the analysis, the sample is first dissolved with nitric or perchloric acid if considerable protein is present (as in blood serum). A releasing agent (lanthanum) and an ionization suppressant (lithium) are then added, and the solution is accurately diluted to an appropriate volume with high-purity deionized water. Many body fluids contain a significant quantity of phosphate, so that the use of a releasing agent is essential. Finally, the prepared sample solution is analyzed with a flame emission spectrometer, such as a flame photometer having separate channels (detectors) or replaceable filters for each element to be determined.

Flame emission spectrometry is also valuable for the determination of metal ions in waste water and for the measurement of water hardness. In these applications, a qualitative analysis is often desired, in which case a preliminary spectrum is obtained with a monochromator-equipped flame emission spectrometer. Elements present are ascertained from the resulting spectrum, and suitable emission lines are selected for quantitative analysis. After the preparation of standards and the construction of working curves, the unknown sample is introduced into the flame emission spectrometer and a quantitative analysis is performed.

## Atomic Absorption Spectrometry

Fraunhofer, in his famous discovery of the dark lines in the solar spectrum, made the first observation of atomic absorption. However, until 1955 atomic absorption had never been used for analytical purposes. In that year, Dr. Alan Walsh, of C.S.I.R.O.,* Australia, applied the principle of atomic absorption to the analysis of metals. The work started by Dr. Walsh has now grown into the world's most widely used technique for elemental analysis. Since 1955, the growth of atomic absorption spectrometry has been truly phenomenal; one worker in the field recently noted that, if the present expansion of the technique continued unabated, by the year 2000 the entire surface of the earth would be covered with atomic absorption spectrometers.

In principle, atomic absorption spectrometry is similar to molecular absorption spectrophotometry. In atomic absorption spectrometry, the absorption spectra of isolated atoms are used to obtain information on the kind and number of atoms present in a chemical sample. This information, in turn, indicates the elemental composition of the sample. The instrumentation used in atomic absorption spectrometry provides a clear indication of the characteristics of this powerful technique.

*Instrumentation for atomic absorption spectrometry.* As shown in Figure 20–10, the instrumentation for atomic absorption spectrometry resembles that used in ultraviolet-visible spectrophotometry. In an atomic absorption spectrometer, the flame is illuminated with a primary source of radiation, a portion of which is absorbed by ground-state atoms. Radiation transmitted by the flame passes through a monochromator and on to a photodetector-readout system which displays the final signal in terms of either transmittance or absorbance. Although similar in appearance to an ultraviolet-visible spectrophotometer, an atomic absorption spectrometer differs with respect to both the sample cell and the source of primary radiation.

---

\* Commonwealth Scientific and Industrial Research Organization.

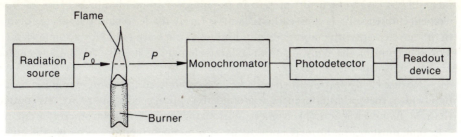

Figure 20–10. Simplified schematic diagram of an atomic absorption spectrometer.

SAMPLE CELL. In atomic absorption spectrometry, the sample cell is simply the flame itself. Because atomic absorption spectrometry relies upon Beer's law, the sensitivity of the technique depends on the path length of primary radiation through the flame. For this reason, **slot burners** which provide a long path length have been developed for atomic absorption spectrometry. Slot burners are of the premix type, but differ from conventional premix burners in that the exit orifice is a narrow slot between 5 and 10 centimeters in length rather than an array of tiny holes. The premixed fuel and oxidizer gases leave the bowl of the burner through this slot, thereby defining the geometrical shape of the resulting flame. As shown in Figure 20–11, the source of primary radiation is directed down the long axis of the flame.

PRIMARY SOURCE. Unlike the broad absorption bands found in molecular spectrophotometry, the absorption spectra for atomic species consist of extremely narrow lines, usually on the order of 0.01 Å wide. For Beer's law to be valid, the bandwidth of the radiation to be absorbed by the atoms must be narrower than the absorption line for the absorbing species. This means that either the line width of the primary radiation source or the bandpass of the frequency selector (monochromator) must be less than 0.01 Å. However, because all but the most expensive monochromators have bandpasses of 0.1 Å or more, it is the primary source which must provide radiation of a sufficiently narrow bandwidth. The source of narrow-band radiation most commonly used in atomic absorption spectrometry is the **hollow-cathode lamp**. In the hollow-cathode lamp, a low-power electrical discharge is sustained between an inert electrode (anode) and a second electrode (cathode) made from the element to be determined. Atoms from the cathode are thereby excited, to produce a very pure line spectrum of the desired element in addition to the line spectrum of the

Figure 20–11. Illustration of a typical slot burner; $P_0$ is the radiant power incident upon the flame, $P$ is the radiant power transmitted by the flame, and $b$ is the radiation path length through the long axis of the flame.

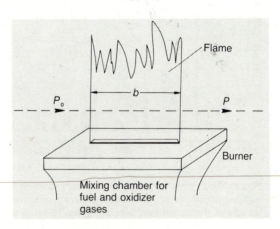

inert fill-gas (argon or neon) present in the hollow-cathode lamp. Because of inter-element influences, a hollow-cathode lamp is rarely made to emit the spectra of more than four elements; the most stable and intense lamps are made for a single element. The spectrum of the emission from a three-element (Ca, Mg, Al) hollow-cathode lamp is shown in Figure 20–12.

Because each spectral line from a hollow-cathode lamp has an extremely narrow bandwidth, spectral interferences in atomic absorption spectrometry are less common than in flame emission spectrometry. In flame emission spectrometry, frequency selection must be performed entirely by the monochromator; because the monochromator passes a relatively broad range of wavelengths, it is not unlikely that the monochromator will pass, in addition to the line of the desired element, a line of an interfering element or an emission line or band from the flame itself. However, in atomic absorption spectrometry, the line spectrum of the hollow-cathode lamp governs the effective bandwidth so that fewer spectral interferences result.

MONOCHROMATOR.    In general, the wavelength selector used in atomic absorption spectrometry can be simpler than that used in flame emission spectrometry. Because the range of wavelengths detected in atomic absorption spectrometry is determined primarily by the hollow-cathode source and not by the monochromator, the monochromator serves primarily to minimize the detected background radiation from the flame and to remove extraneous lines emitted by the hollow-cathode fill-gas. Usually, a monochromator with a bandpass of 0.5 Å is sufficient to prevent significant non-linearity of Beer's law plots caused by background radiation from the flame.

Another problem in atomic absorption spectrometry is that excited-state atoms in the flame undergo emission. Unless eliminated, this emission signal will cause a concentration-dependent background and a consequent curvature of Beer's law plots. This difficulty is usually solved through the use of a **chopper**, of which there are two

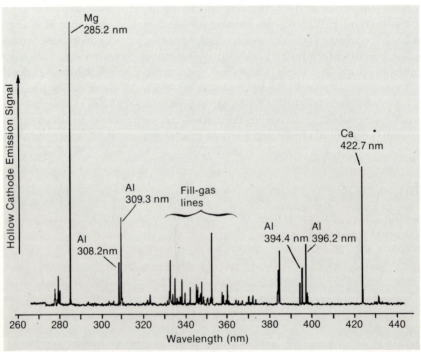

*Figure 20–12.*    Spectrum of the emitted radiation from a three-element (Ca, Mg, Al) hollow-cathode lamp. Note the presence of spectral lines from the fill-gas. Wavelengths of the prominent lines are given in nanometers (nm).

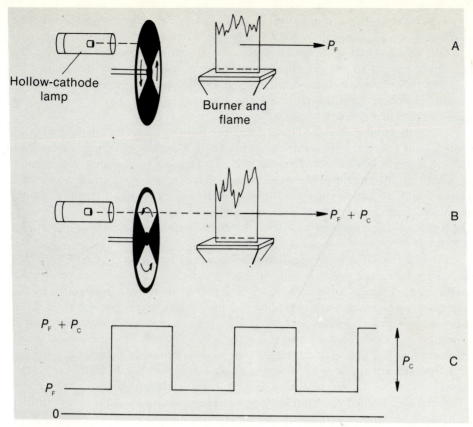

*Figure 20–13.* Elimination of emission background in atomic absorption spectrometry through the use of a rotating chopper; $P_F$ is the radiant power emitted by the flame and $P_C$ is the desired radiant power from the hollow-cathode lamp transmitted by the flame. (See text for discussion.)

types: in electronic chopping, the radiation from the hollow-cathode source is alternately turned on and off, whereas in mechanical chopping a rotating shutter is used to achieve the same effect.

The operation of a rotating-shutter chopper is illustrated in Figure 20–13. In Figure 20–13A, the rotating shutter is shown positioned to block the radiation from the hollow-cathode lamp. Under these conditions, the only radiation reaching the photodetector is that emitted by the flame ($P_F$). When the shutter is opened, as shown in Figure 20–13B, the radiation reaching the detector is the *sum* of the emission from the flame ($P_F$) and the radiation from the hollow-cathode lamp transmitted by the flame ($P_C$). As the rotating shutter alternately passes and blocks the radiation from the hollow-cathode source, the photodetector signal will appear as shown in Figure 20–13C. The alternating, square-wave portion of this signal will have an amplitude proportional to the radiation from the hollow-cathode lamp transmitted by the flame.

## Comparison between Atomic Absorption and Flame Emission Spectrometry

Although flame emission and atomic absorption spectrometry both employ a flame as an atom reservoir, the two methods differ in several important respects.

***Sensitivity.*** For a number of years, a controversy existed over the question of whether atomic absorption spectrometry or flame emission spectrometry was superior for elemental analysis. It is now recognized that for the most part the techniques are not competitive but complementary. Elements best determined by means of flame emission spectrometry are generally not those best determined by atomic absorption spectrometry—and the converse is true. The reason for this difference is rather simple.

Elements having low excitation energies—that is, elements which are easily excited—will emit very efficiently when placed in a high-temperature flame. This emission signal, when measured against the relatively low background of the flame, provides a sensitive method for the detection of that element. By contrast, an element having a high excitation energy will not be efficiently excited in a chemical flame; instead, most of the atoms of this element will reside in the ground electronic state. These ground-state atoms are amenable to determination by means of an absorption technique.

From the Boltzmann distribution relationship given earlier, it can be seen that ease of excitation is correlated with the energy difference between an excited state and the ground state of an atom. Consequently, the general rule can be stated that in most cases *flame emission spectrometry is more sensitive for the determination of elements having resonance spectral lines\* between 400 and 800 nm, whereas atomic absorption spectrometry is more sensitive for elements whose resonance lines lie between 200 and 300 nm.* Elements whose resonance lines are between 300 and 400 nm (and certain other elements as well) can be determined equally well by means of either technique. Figure 20–15, to be discussed later, shows the elements best determined by each procedure.

It has been argued that atomic absorption spectrometry should be more sensitive than flame emission spectrometry for the determination of all elements. This is because most atoms of any element will reside in the ground electronic state even at flame temperatures, so that a greater number of atoms is capable of absorption than of emission. In fact, for many elements the sensitivity of atomic absorption spectrometry is inferior to that of flame emission spectrometry. This is largely due to the necessity in atomic absorption spectrometry, as in all absorption methods, to measure a small difference between two large signals, a feat which is always more difficult than the measurement of a small signal by itself.

***Interferences.*** As mentioned earlier, vaporization interferences should be equally troublesome in both atomic absorption and flame emission spectrometry. Although exaggerated claims have frequently been made that interferences are less in atomic absorption spectrometry, the observed differences can be traced to the customary use of total-consumption burners for emission spectrometry and premix burners for absorption spectrometry. Today, high-quality flame emission spectrometers employ premix burners.

***Qualitative and quantitative analysis.*** When qualitative analysis is desired, flame emission spectrometry is clearly superior to atomic absorption spectrometry. Qualitative analysis with flame emission spectrometry merely requires that the entire emission spectrum of the flame be scanned, whereas atomic absorption spectrometry necessitates the use of a different hollow-cathode lamp for each element to be detected. Of course, qualitative analysis by means of atomic absorption spectrometry is possible if a continuous primary source is employed, but then a monochromator having an extremely narrow bandpass must be used if adequate sensitivity is to be obtained.

---

\* A resonance spectral line is one which arises from a transition between the ground state and an excited state.

***Effect of changes in flame temperature.*** In a typical flame, a large fraction of the atoms of most elements will be in the ground electronic state, although the exact fraction will depend on the flame temperature. If the flame temperature should vary, the resulting change in the number of excited atoms will have a relatively greater effect on the excited-state population, because of its initially smaller number, than on the ground-state population. Therefore, it would seem that flame emission spectrometry would be more strongly affected by changes in flame temperature than would atomic absorption spectrometry. However, the dominant effect of such a change is an alteration in the extent of atom formation, so that the two techniques are affected nearly equally. As discussed earlier, atom formation in the flame depends upon the desolvation of droplets, the vaporization of the resulting solute particles, and the establishment of favorable conditions for equilibria involving the atoms. The overall temperature-dependence of all these processes can, in many cases, outweigh that of the excited-state and ground-state populations.

***Instrumental requirements.*** Instrument systems for atomic absorption spectrometry are generally more expensive than those for flame emission spectrometry. A typical atomic absorption spectrometer costs between $3000 and $15,000, whereas simple special-purpose (clinical) instruments for flame emission spectrometry are often priced at less than $2000. Of course, the more sophisticated atomic absorption systems employ choppers, complex signal-processing equipment, instrumental correction for curvature of Beer's law plots, and direct, digital readout of concentration. With a high-quality atomic absorption spectrometer, precision levels of 0.1 per cent can often be obtained in quantitative analysis. Figure 20–14 shows a schematic diagram of a modern atomic absorption spectrometer.

From the foregoing discussion, it should now be clear that atomic absorption and flame emission spectrometry are indeed complementary rather than competitive. In recognition of this, most modern atomic absorption spectrometers incorporate a provision for analysis by means of flame emission spectrometry. Well established procedures are available for the analysis of most elements in the periodic table by either one technique or the other. The periodic table shown in Figure 20–15 indicates the

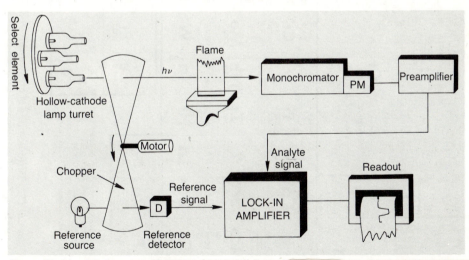

*Figure 20–14.* Modern atomic absorption spectrometer, with a multiple hollow-cathode lamp turret which can be rotated to select the desired lamp and a sophisticated signal-processing system; PM is a photomultiplier tube.

**Legend (Sample box):**

Atomic Absorption Wavelength — 357.9 —nm
Atomic Absorption Detection Limit — 0.002 —μg/ml
**Cr**
Atomic Emission Wavelength — 425.4 —nm
Atomic Emission Detection Limit — 0.004 —μg/ml
Sample

- Elements best analyzed by atomic absorption shown in outline
- Elements best analyzed by flame emission shaded gray
- Elements suitable for analysis by either method shown in black
- Elements unsuitable for analysis by either absorption or emission in shaded boxes
- Criteria for preference based on significant difference in detection limit

Each element box lists: Atomic Absorption Wavelength / Detection Limit, Symbol, Atomic Emission Wavelength / Detection Limit.

**H**

| Element | AA λ | AA DL | AE λ | AE DL |
|---|---|---|---|---|
| Li | 670.8 | 0.001 | 670.8 | 0.00002 |
| Be | 234.9 | 0.002 | 234.9 | 0.1 |
| Na | 589.0 | 0.0008 | 589.0 | 0.00005 |
| Mg | 285.2 | 0.0003 | 285.2 | 0.005 |
| K | 766.5 | 0.003 | 766.5 | 0.0005 |
| Ca | 422.7 | 0.0005 | 422.7 | 0.0001 |
| Sc | 391.2 | 0.1 | 402.4 | 0.08 |
| Ti | 364.3 | 0.1 | 399.8 | 0.2 |
| V | 318.4 | 0.02 | 437.9 | 0.01 |
| Cr | 357.9 | 0.002 | 425.4 | 0.004 |
| Mn | 279.5 | 0.001 | 403.1 | 0.005 |
| Fe | 248.3 | 0.004 | 372.0 | 0.02 |
| Co | 240.7 | 0.002 | 345.4 | 0.03 |
| Ni | 232.0 | 0.005 | 352.5 | 0.02 |
| Cu | 324.7 | 0.004 | 324.7 | 0.01 |
| Zn | 213.9 | 0.001 | 481.1 | 8.0 |
| Ga | 287.4 | 0.05 | 417.2 | 0.01 |
| Ge | 265.2 | 0.2 | 265.2 | 0.4 |
| As | 193.7 | 0.1 | 193.7 | 12.0 |
| Se | 196.0 | 0.1 | NA | NA |
| B | 249.7 | 2.5 | 518.0* | 0.05 |
| C | NA | NA | | |
| N | NA | NA | | |
| O | NA | NA | | |
| Al | 396.2 | 0.04 | 396.2 | 0.005 |
| Si | 251.6 | 0.1 | 251.6 | 3.0 |
| P | NA | NA | 526.0* | 0.05 |
| S | NA | NA | 392.0 | 2.0 |
| Rb | 780.0 | 0.005 | 780.0 | 0.001 |
| Sr | 460.7 | 0.004 | 460.7 | 0.0002 |
| Y | 410.2 | 0.4 | 597.2* | 0.03 |
| Zr | 360.1 | 4.0 | 360.1 | 5.0 |
| Nb | 405.9 | 1.0 | 405.9 | 1.2 |
| Mo | 313.3 | 0.03 | 390.3 | 0.1 |
| Tc | | | | |
| Ru | 349.9 | 0.06 | 372.8 | 0.3 |
| Rh | 343.5 | 0.02 | 343.5 | 0.03 |
| Pd | 247.6 | 0.01 | 363.5 | 0.05 |
| Ag | 328.1 | 0.0005 | 328.1 | 0.008 |
| Cd | 228.8 | 0.0006 | 326.1 | 0.8 |
| In | 303.9 | 0.03 | 451.1 | 0.003 |
| Sn | 224.6 | 0.03 | 284.0 | 0.1 |
| Sb | 217.6 | 0.03 | 252.8 | 0.6 |
| Te | 214.3 | 0.05 | 238.3 | 200.0 |
| Cs | 852.1 | 0.10 | 455.5 | 0.05 |
| Ba | 553.5 | 0.02 | 553.5 | 0.002 |
| La | 550.1 | 5.0 | 441.8* | 0.01 |
| Hf | 307.3 | 14.0 | 531.2 | 19.0 |
| Ta | 271.5 | 3.0 | 474.0 | 4.0 |
| W | 400.9 | 3.0 | 400.9 | 0.6 |
| Re | 346.0 | 0.6 | 346.0 | 0.1 |
| Os | 305.9 | 0.3 | 442.0 | 1.7 |
| Ir | 284.9 | 1.0 | 550.0* | 0.4 |
| Pt | 265.9 | 0.05 | 265.9 | 3.8 |
| Au | 242.8 | 0.02 | 267.6 | 0.5 |
| Hg | 253.7 | 0.2 | 253.7 | 10.0 |
| Tl | 276.8 | 0.02 | 535.1 | 0.02 |
| Pb | 217.0 | 0.01 | 405.8 | 0.1 |
| Bi | 223.1 | 0.05 | 223.1 | 2.0 |

Other groups: He, Ne, F, Cl, Br, Kr, Xe, I, Po, At, Rn, Fr, Ra, Ac

**Lanthanides:**

| Element | AA λ | AA DL | AE λ | AE DL |
|---|---|---|---|---|
| Ce | NA | NA | 494.0 | 98.0 |
| Pr | 495.1 | 4.0 | 495.1 | 0.07 |
| Nd | 463.4 | 0.6 | 492.5 | 0.7 |
| Pm | | | | |
| Sm | 429.7 | 2.0 | 476.0 | 0.2 |
| Eu | 459.4 | 0.1 | 466.2 | 0.001 |
| Gd | 368.4 | 4.0 | 622.0 | 0.07 |
| Tb | 432.7 | 2.0 | 534.0* | 0.03 |
| Dy | 421.2 | 0.2 | 404.6 | 0.05 |
| Ho | 410.4 | 0.1 | 410.4 | 0.1 |
| Er | 400.8 | 0.1 | 400.8 | 0.07 |
| Tm | 371.8 | 0.4 | 371.8 | 0.08 |
| Yb | 398.8 | 0.02 | 398.8* | 0.006 |
| Lu | 331.2 | 3.0 | 451.9 | 1.3 |

**Actinides:**

| Element | AA λ | AA DL | AE λ | AE DL |
|---|---|---|---|---|
| Th | NA | NA | 492.0 | 11.0 |
| Pa | | | | |
| U | 358.5 | 23.0 | 544.8 | 5.5 |
| Np, Pu, Am, Cm, Bk, Cf, Es, Fm, Md, No, Lw | | | | |

The elements Na, K, and Rb were determined in an air-acetylene flame. All others in a nitrous oxide-acetylene flame. Detection limits in many cases can be improved by the addition of an easily ionized substance such as sodium or potassium.

*Band Emission

*Figure 20–15.* Periodic table of elements detectable by means of flame emission and atomic absorption spectrometry. Courtesy of Instrumentation Laboratory, Inc., Lexington, Massachusetts.

elements determinable by means of flame emission or atomic absorption spectrometry, which of the two methods is better, the optimum wavelengths for use with each technique, and the minimum concentrations detectable by each method. It is possible to detect most elements at the parts-per-million to parts-per-billion concentration level with either flame emission or atomic absorption spectrometry. This high degree of sensitivity, the excellent precision attainable, and the convenience of using solution samples have made the combination of these two techniques the most popular current method for elemental analysis.

## Atomic Fluorescence Flame Spectrometry

Atomic fluorescence flame spectrometry is the newest flame spectrometric method of analysis. Although the observation of the fluorescence of metal atoms was first made in the 1890's by R. W. Wood, it was not until 1964 that Dr. J. D. Winefordner and co-workers introduced atomic fluorescence as an analytical technique. Since that time, atomic fluorescence flame spectrometry has been shown to be competitive with atomic absorption and flame emission techniques in terms of sensitivity, precision, and operational convenience. At present, atomic fluorescence flame spectrometry is still in the developmental stage. No successful commercial instruments designed expressly for this method have yet been introduced, although the number of applications using specially constructed instruments continues to increase. Unlike flame emission but like atomic absorption spectrometry, atomic fluorescence flame spectrometry exhibits its greatest sensitivity for elements having high excitation energies.

*Instrumentation for atomic fluorescence flame spectrometry.* As shown in Figure 20–16, instrumentation used in atomic fluorescence flame spectrometry is similar in configuration to that used for molecular fluorescence spectrometry. However, because atoms in a flame can absorb radiation of only specific and characteristic wavelengths, it is usually unnecessary to employ an excitation monochromator. Instead, the source radiation is focused directly upon the flame to excite the atomic fluorescence. The fluorescent radiation is then dispersed with the aid of a frequency selector and is detected by a suitable photodetector. Finally, the resulting signal is displayed on an appropriate readout device.

In atomic fluorescence flame spectrometry, as in molecular fluorescence methods, the fluorescence power is directly proportional to the intensity of the primary source. It is therefore desirable in atomic fluorescence flame spectrometry to select a primary source which provides intense radiation at those wavelengths absorbed by atoms in

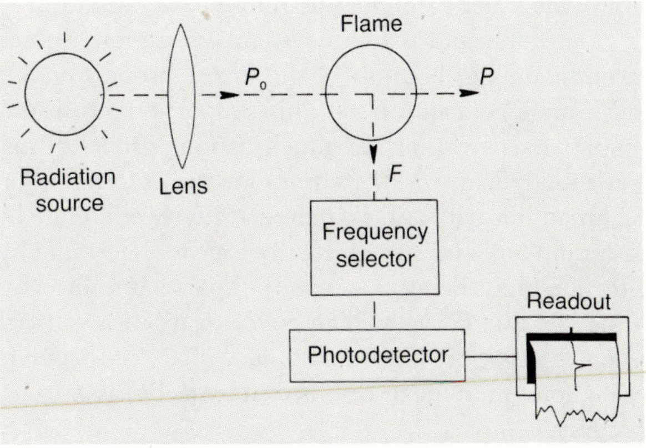

*Figure 20–16.* Schematic diagram of an atomic fluorescence flame spectrometer; $P_0$ is the incident power, $P$ is the transmitted power, and $F$ is the fluorescence.

the flame. Furthermore, to enable qualitative analysis to be performed, the primary source should emit radiation over a broad spectral range to excite fluorescence from the maximum number of elements. Unfortunately, although most continuous primary sources, such as the xenon arc lamp, are extremely powerful (hundreds of watts or more), their power is spread over a large spectral range so that little power is available in the very narrow band of wavelengths absorbed by atoms. Thus, xenon arc lamps have been used with limited success as primary sources in atomic fluorescence flame spectrometry; better results have been obtained with a source called an electrodeless discharge lamp.

An **electrodeless discharge lamp** consists of a glass envelope containing a small amount of an element or the iodide salt of that element in a low-pressure inert-gas atmosphere. When the lamp is thermostated at several hundred degrees centigrade and placed in a radio-frequency or microwave field, the vaporized atoms within the envelope are efficiently excited and emit their characteristic spectra. The line width of the resulting radiation is sometimes broader than that emitted by a hollow-cathode lamp, but the radiant power is far greater, generally on the order of 5 to 10 watts. To increase the efficiency with which the electrodeless discharge lamp can excite fluorescence, a lens is often used to focus the radiation upon the flame.

Because the radiant power of atomic fluorescence is directly proportional to the quantum efficiency for fluorescence, the composition of the flame is far more critical than that for either atomic absorption or flame emission spectrometry. Acetylene-fueled flames are efficient for the atomization of samples, but do not provide a high quantum efficiency for fluorescence. This is because the radical and molecular species present in the flame serve as efficient quenchers for excited atoms, thereby diminishing the amount of fluorescence. By contrast, hydrogen-fueled flames exhibit higher quantum efficiencies for fluorescence, but produce more vaporization interferences. For example, a hydrogen-argon-air flame has been found to provide extremely high quantum efficiencies for fluorescence and has been responsible for some of the lowest detection limits yet reported. This conflict between solute-vaporization efficiency and quantum efficiency has been one of the most important factors limiting the application of atomic fluorescence flame spectrometry.

Scattering of source radiation by the flame can be a serious cause of interference in atomic fluorescence flame spectrometry. Rayleigh and Mie scattering (discussed in Chapter 18, page 609) occur at all wavelengths of radiation from the source and are therefore indistinguishable from resonance atomic fluorescence. To compensate for scattering to the greatest possible extent, the following procedure is usually employed. A blank solution, identical to the sample but containing none of the analyte, is nebulized into the flame. The observed radiation is assumed to be due entirely to extraneous scattering and is subtracted from a similarly detected signal for the sample to provide a net value for the fluorescent radiation.

Detection and readout systems for atomic fluorescence flame spectrometry can be quite simple, because of the direct proportionality between fluorescence power and sample concentration. This proportionality usually holds over a wide range of sample concentrations, as much as four orders of magnitude. Figure 20–17, which shows analytical working curves for the determination of several elements, indicates the broad linearity and extreme sensitivity of atomic fluorescence flame spectrometry. Cadmium and zinc are especially easy to determine by means of this technique, their detection limits being as low as $10^{-5}$ part per million by weight. However, the bending of the working curve at high concentrations is so marked that the slope of the curve becomes negative. This situation, which can occur for any element, might lead to ambiguous analytical results but can be avoided through use of dilute sample solutions.

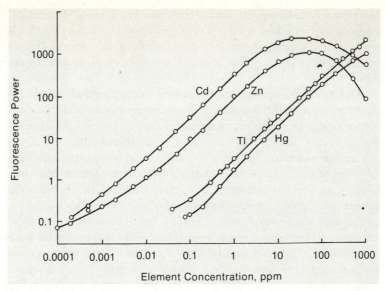

*Figure 20–17.* Working curves for the determination of cadmium, zinc, thallium, and mercury by means of atomic fluorescence flame spectrometry. (Reproduced, with permission, from the paper by J. M. Mansfield, J. D. Winefordner, and C. Veillon: Anal. Chem., *37*: 1051, 1965.)

***Advantages of atomic fluorescence flame spectrometry.*** As already suggested, atomic fluorescence flame spectrometry is an extremely sensitive analytical technique. Some representative detection limits, listed in Table 20–2, show this technique to be for many elements more sensitive than either atomic absorption or flame emission spectrometry (refer to Figure 20–15, page 694). This high sensitivity results largely from a combination of the most desirable characteristics of atomic

**Table 20–2.** Limits of Detection in Atomic Fluorescence Flame Spectrometry*

| Element | Wavelength (nm) | Detection Limit, ppm | Element | Wavelength (nm) | Detection Limit, ppm |
|---------|-----------------|----------------------|---------|-----------------|----------------------|
| Ag | 328.1 | 0.001 | Mg | 285.2 | 0.001 |
| Al | 396.2 | 0.1 | Mn | 279.5 | 0.006 |
| As | 193.7 | 0.1 | Mo | 313.3 | 0.5 |
| Au | 267.6 | 0.005 | Na | 589.6 | 0.008 |
| Be | 234.9 | 0.01 | Ni | 232.0 | 0.003 |
| Bi | 306.8 | 0.005 | Pb | 405.8 | 0.01 |
| Ca | 422.7 | 0.02 | Pd | 340.5 | 0.04 |
| Cd | 228.8 | 0.000001 | Rh | 369.2 | 0.16 |
| Co | 240.7 | 0.005 | Sb | 231.1 | 0.05 |
| Cr | 357.9 | 0.05 | Se | 196.0 | 0.04 |
| Cu | 324.7 | 0.001 | Si | 204.0 | 0.1 |
| Fe | 248.3 | 0.008 | Sn | 303.4 | 0.05 |
| Ga | 417.2 | 0.007 | Sr | 460.7 | 0.03 |
| Ge | 265.1 | 0.1 | Te | 214.3 | 0.005 |
| Hg | 253.7 | 0.0002 | Tl | 377.6 | 0.006 |
| In | 451.1 | 0.1 | V | 318.4 | 0.07 |
|  |  |  | Zn | 213.8 | 0.00002 |

* From the compilations in J. D. Winefordner and R. C. Elser: Anal. Chem., *43*(4):24A, 1971, and J. D. Winefordner: Accounts Chem. Res., *4*:259, 1971.

absorption and flame emission spectrometry. Atomic fluorescence, like atomic emission, is detected by observation of the desired radiation above a low background. However, like atomic absorption, atomic fluorescence does not rely upon the energy in the flame for excitation of atoms, but rather employs a more powerful auxiliary source.

One of the most important potential applications of atomic fluorescence flame spectrometry is in the area of multi-element analysis. As in flame emission spectrometry, it should be feasible to use atomic fluorescence to determine several elements simultaneously merely by employing a polychromator and a suitable detection system. The detection system can consist of a photographic film or plate or can be composed of a number of individual photodetectors set at appropriate wavelengths of fluorescence for the elements of interest. Either a single broad-band source or a number of separate sources for individual elements can be used to excite atomic fluorescence from each element. Several instrument systems of this kind have been proposed, and at least one has been considered for commercial development.

***Atomic fluorescence flame spectrometry with tunable lasers.*** It has been suggested that the ideal primary source for atomic fluorescence flame spectrometry would be a tunable laser. The high power, narrow bandwidth, and directionality of the laser would seem to be perfect for exciting the fluorescence from metal atoms in a flame. Also, the tunability of the laser would allow the sequential excitation and fluorescence of a number of elements, so that multi-element analysis would be simplified. With this capability, atomic fluorescence flame spectrometry would clearly be the technique of choice for most elemental analyses. However, a sufficiently inexpensive, practical, tunable laser system suitable for use in atomic fluorescence work has yet to be developed. Research in this area is currently underway, however, and atomic fluorescence spectrometers with tunable laser sources can be expected.

## FLAMELESS ATOMIZATION

In the foregoing discussion, only flames were considered as atomization devices. However, flames are imperfect atom reservoirs for several reasons. They are somewhat dangerous to operate, and necessitate the storage and handling of potentially hazardous fuel and oxidizer gases, a requirement which makes them undesirable in industrial, teaching, and clinical laboratories. Flames are also relatively expensive to employ because large volumes of fuel and oxidizer gases are consumed. In addition, spectroscopic techniques based on the use of flames require relatively large amounts of sample solution, are often troubled by interferences, and are generally restricted to samples of relatively low concentration. Certainly, it would be desirable if a cheaper, safer, and more efficient atomizer could be found. A number of atomizers having these characteristics have been developed and are being used more and more widely.

The new flameless atomizers take several forms. In general, they consist of a rod, a loop, a boat, or a trough, made of conductive carbon or metal upon which the sample is placed. The conductive sample support is electrically heated by passage of a large current that vaporizes and partially atomizes the sample. These devices are often extremely efficient atomizers; they utilize extremely small samples; and their use reduces the amount of sample preparation necessary for an analysis. Although flameless atomization techniques can be used for atomic fluorescence, until now most applications have been to atomic absorption because of the ready availability of atomic absorption spectrometers.

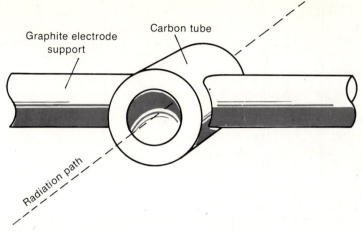

Graphite electrode
support

Carbon tube

Radiation path

*Figure 20–18.* Carbon-tube type of flameless atom reservoir. Courtesy of Varian Techtron, Walnut Creek, California.

A typical flameless atomization system, illustrated in Figure 20–18, consists of a small carbon tube supported between two graphite electrodes. In operation, 1 to 50 $\mu$l of a liquid sample is placed within the tube and the entire assembly is heated by passage of a low current through it. At this current, the temperature of the tube is just sufficient to remove the solvent from the sample. Then, a higher current is passed through the assembly to ash the sample if it contains a volatile matrix. Finally, the sample itself is vaporized by passage of a current of several hundred amperes through the tube. At this current, the tube can reach temperatures greater than 3000°K, sufficient for the vaporization and atomization of most elements.

In commercial flameless atomization systems, all the operations described above are performed automatically according to a programmed sequence. If the sample is observed by means of atomic absorption spectrometry, the detected signal appears as shown in Figure 20–19. Note the "false absorption" which appears during the sample ashing and drying periods. This "false absorption" is caused by scattering of radiation from the hollow-cathode source by the smoke and ash produced during ashing. If the ashing and drying periods are not separated in time from atomization of the sample, an erroneous absorption reading will result. This signal overlap is one of the present problems plaguing flameless atomizers. To minimize this difficulty, special instrumental arrangements for "background subtraction" must be employed. A discussion of this instrumentation is, unfortunately, beyond the scope of this text.

As atom reservoirs, flameless atomization systems have several significant advantages over chemical flames. Perhaps the greatest virtue is that of absolute sensitivity. Although the techniques of flame spectrometry are capable of providing high sensitivity for very low concentrations, they require at least two milliliters of sample solution for a reliable result; at a detection limit of one part per billion, this volume of solution corresponds to $2 \times 10^{-9}$ gram of a particular element. By contrast, flameless atomization systems are often able to detect as little as $10^{-14}$ gram of an element. Absolute detection limits by weight obtained through the use of the carbon-tube atomizer in atomic absorption spectometry are shown in Table 20–3; these can be compared with the detection limits for the flame methods given in Figure 20–15 and Table 20–2.

At present, the most serious problem encountered in flameless atomization involves interelement interferences. Because the regions of atom observation in the

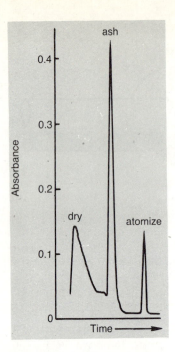

*Figure 20–19.* Atomic absorption signal produced by atomization of vanadium in fuel oil (diluted 1:10 with xylene) using the carbon-tube atomizer shown in Figure 20–18. Total sample volume, 5 $\mu$l; concentration of vanadium in sample, 1.4 ppm. (See text for discussion.)

flameless systems are not in direct contact with the heated atomizer surface, they do not reach the high temperatures found in most flames, so that atom recombination often occurs and interferences are common. The design of the carbon-tube atomizer shown in Figure 20–18 reduces interferences somewhat by partially confining the vaporized sample within the heated orifice of the tube. Atomization is thereby enhanced and atom recombination reduced. Furthermore, the atomized sample can

**Table 20–3. Absolute and Concentration Detection Limits Attainable with a Flameless Atomizer***

| Element | Absolute Grams | Concentration, ng/ml in 5 $\mu$l Sample | Element | Absolute Grams | Concentration, ng/ml in 5 $\mu$l Sample |
|---------|----------------|-----------------------------------------|---------|----------------|-----------------------------------------|
| Ag | $2 \times 10^{-13}$ | 0.04 | Li | $5 \times 10^{-12}$ | 1.0 |
| Al | $3 \times 10^{-11}$ | 6.0 | Mg | $6 \times 10^{-14}$ | 0.012 |
| As | $1 \times 10^{-10}$ | 20 | Mn | $5 \times 10^{-13}$ | 0.1 |
| Au | $1 \times 10^{-11}$ | 2.0 | Mo | $4 \times 10^{-11}$ | 8.0 |
| Be | $9 \times 10^{-13}$ | 0.18 | Na | $1 \times 10^{-13}$ | 0.02 |
| Bi | $7 \times 10^{-12}$ | 1.4 | Ni | $1 \times 10^{-11}$ | 2.0 |
| Ca | $3 \times 10^{-13}$ | 0.06 | Pb | $5 \times 10^{-12}$ | 1.0 |
| Cd | $1 \times 10^{-13}$ | 0.02 | Pd | $2 \times 10^{-10}$ | 40 |
| Co | $6 \times 10^{-12}$ | 1.2 | Pt | $2 \times 10^{-10}$ | 40 |
| Cr | $5 \times 10^{-12}$ | 1.0 | Rb | $6 \times 10^{-12}$ | 1.2 |
| Cs | $2 \times 10^{-11}$ | 4.0 | Sb | $3 \times 10^{-11}$ | 6.0 |
| Cu | $7 \times 10^{-12}$ | 1.4 | Se | $1 \times 10^{-10}$ | 20 |
| Eu | $1 \times 10^{-10}$ | 20 | Sn | $6 \times 10^{-11}$ | 12 |
| Fe | $3 \times 10^{-12}$ | 0.6 | Sr | $5 \times 10^{-12}$ | 1.0 |
| Ga | $2 \times 10^{-11}$ | 0.4 | Tl | $3 \times 10^{-12}$ | 0.6 |
| Hg | $1 \times 10^{-10}$ | 20 | V | $1 \times 10^{-10}$ | 20 |
| K | $9 \times 10^{-13}$ | 0.18 | Zn | $8 \times 10^{-14}$ | 0.016 |

* Courtesy of Varian Techtron, Walnut Creek, California.

provide a larger signal by residing within the region of observation for a longer time. A small hydrogen flame surrounding the atomizer has been found to further minimize vaporization interferences by providing a reducing environment. However, even with these precautions, some elements are not amenable to flameless atomization techniques. For example, boron cannot be determined with the carbon-tube atomizer because boron carbide, an extremely refractory compound, is formed.

Other problems encountered with present flameless atomizers are sample carryover and limited useful lifetime. Carbon-tube atomizers are often porous and tend to absorb a portion of each sample, carrying the portion over to the next determination. This difficulty can be largely circumvented by the use of pyrolytic carbon, which is less permeable than other forms of carbon, and by the placement of a drop of an inert, pure organic liquid such as xylene inside the tube before each determination. Apparently, xylene forms a coating on the carbon surface and prevents penetration of the sample solution.

Although longer-lived carbon-tube atomizers are now being introduced, most are usable for only a hundred determinations or less. This limited lifetime necessitates the frequent replacement of the tube and, unless the replacement tube is identical to the original, a readjustment of the atomizing-current program. This readjustment, when required, is often critical because of its influence on interelement interferences. A recent innovation to prolong the carbon-tube lifetime involves passing methane gas through the heated tube. This procedure results in the pyrolysis of the methane and the deposition of a fresh carbon surface on the tube.

## ELECTRICAL-DISCHARGE SPECTROMETRY

Topics to be discussed in this section usually come under the heading "optical emission spectroscopy." Although the techniques do utilize emitted radiation in the optical (ultraviolet-visible) region of the spectrum, the term is hardly descriptive. At present, a large number of different electrical discharges are employed to excite atomic emission. Among these are the DC arc, AC arc, high-voltage spark, radio-frequency and microwave plasmas, plasma jet, and plasma torch. Because the most frequently used are the DC arc, the high-voltage spark, and the radio-frequency plasma, we will confine our discussion to these three sources. Each of these discharges excites a sample in a different way and by a slightly different mechanism. In addition, the instrumentation differs considerably so that it will be most convenient to examine each discharge separately.

### DC Arc

A DC arc source consists of a high-current (5 to 30 amperes), low-voltage (10 to 25 volts) electrical discharge supported between two electrodes. Because of the high temperature of the discharge (2000 to 4000°K), a sample placed within the discharge is partially vaporized into free atoms which are excited and emit characteristic spectra.

Excitation of a sample by a DC arc discharge is partly thermal and partly electrical in origin. Unfortunately, it is difficult to control the variables—primarily arc current and arc resistance—which govern the temperature of the discharge and, thus, the excitation of the sample. For example, refractory samples vaporize slowly in the arc, thereby changing the character and resistance of the discharge. In addition, the high temperature of the discharge gradually erodes the electrodes between which the discharge occurs, also causing the arc resistance to change. Consequently, there is a lack of reproducibility from sample to sample, an instability of the discharge during the analysis of a single sample, and a strong dependence on the sample matrix. These characteristics limit the attractiveness of the DC arc as an excitation source for quantitative analysis. However, the extreme sensitivity and the ease of handling certain samples make the DC arc an attractive source for some applications.

*DC arc spectra.* Radiation emitted by a DC arc discharge has several distinguishable components, including atomic spectra, molecular spectra, and background emission from the electrode material itself. Because the DC arc operates at an extremely high temperature, the electrodes are heated to incandescence and emit intense and spectrally continuous background radiation, resembling that from a blackbody.

Molecules and radicals are quite efficiently excited in a DC arc discharge and produce their distinctive band spectra, often with superimposed vibrational fine-structure. These band spectra often cover broad spectral regions, and can obscure much of the atomic radiation emitted by the discharge. One notable example is the presence of the so-called cyanogen bands, emitted by CN radicals formed by reaction of carbon electrodes with atmospheric nitrogen.

Atomic line spectra from a DC arc discharge are considerably different in appearance from those obtained in flame emission spectrometry, primarily because of the difference in temperature between the two sources. In a flame, atoms are usually raised only to lower excited electronic levels so that the emission spectra are quite simple. By contrast, the thermal and electrical energy available in the DC arc is sufficient to cause population of extremely high energy levels; thus, a considerably greater number of transitions is possible and a more complex emission spectrum is obtained, especially for the heavier elements. For example, the emission spectrum of iron contains more than 4000 lines. In the case of uranium, the number of emission lines is so great that the spectrum appears to be nearly continuous and obscures all but the most intense emission lines of other elements present.

To further complicate the situation, many emission lines of ions are observed from the DC arc source. The high temperature of the DC arc is capable of ionizing a number of elements, notably the alkali metals and alkaline earths, having relatively low ionization energies. For these elements, the observed emission spectra exhibit not only atomic emission lines but ion emission lines as well.

Because of the high excitation efficiency and complex emission spectra obtained with the DC arc source, two facts are evident. First, the DC arc is an excellent source for qualitative analysis. Second, to allow qualitative analysis to be performed efficiently and reliably, a high-quality spectral dispersion device must be employed to separate the desired atomic emission lines from the large number of other spectral features present.

*Dispersion and detection of spectra.* To permit observation of a large number of emission lines from atoms and to allow correction for any spectral background, a high-dispersion polychromator is usually used in DC arc spectrometry. Compared with monochromators commonly employed in flame spectrometry, the DC arc polychromator is physically quite large in order to provide greater dispersion and easier separation of desired lines from background and band radiation.

Photographic and photoelectric detection systems are both employed in DC arc spectrometry, the photographic system being most common for qualitative analysis and the photoelectric detector system for quantitative applications. As discussed in Chapter 19 (page 629), photographic detection has an advantage in that it permits examination of a broad spectral range at one time. This makes it possible to detect a large number of elements simultaneously and, furthermore, to obtain readings of the background emission adjacent to the spectral lines being observed. Also, a photographic emulsion automatically integrates the emission from the arc, thereby averaging out any arc instability. An example of a photographically recorded spectrum for sodium was presented in Figure 19–5 (page 629).

When photoelectric detection is employed, it is usually necessary to electronically integrate the emission signal for a period of time to minimize the effect of arc instability. In photoelectric detection systems, a number of photomultipliers are placed along the focal curve or plane of a polychromator. For quantitative work, a background reading must be obtained at a wavelength immediately adjacent to each spectral line being observed; thus, two photodetectors are needed for each spectral line of interest. This, of course, greatly increases the cost of a photoelectric detection system. Such a **direct-reading spectrometer** can cost up to $100,000. Its use is only justified for applications in which certain elements must be determined frequently and rapidly. One such application is in the metals industry, where analyses of steel and aluminum alloys, for example, must be performed routinely and rapidly to maintain proper process control.

*Qualitative analysis using a DC arc.* DC arc spectrometry is one of the most sensitive of all techniques for elemental analysis. Concentrations between parts per million and parts per billion can be detected *simultaneously* for as many as 70 elements. To perform a qualitative analysis using DC arc excitation and photographic readout, the wavelength scale on the photographic emulsion must be calibrated. To do this, one can photograph in two separate experiments the emission spectrum of a known element (such as iron) on the top half of the emulsion, and the emission spectrum of the sample on the bottom half of the emulsion. As mentioned earlier, iron has a very rich arc spectrum, consisting of a large number of lines whose wavelengths are accurately known. Therefore, the iron spectrum serves as a calibration scale from which the wavelengths and identities of the spectral lines of a chemical sample can be determined.

A sometimes inconvenient feature of the DC arc discharge which can be turned to advantage in qualitative analysis is fractional distillation of the sample. Because of the extremely high temperatures found in the DC arc discharge, solid samples are often melted and fractionally vaporized. More volatile species such as alkali metals, alkaline earths, and other light elements vaporize first, with heavier or more refractory elements vaporizing later. By successive exposure of several spectra above and below each other on the same photographic emulsion, the emission spectra of these lighter elements can be detected before the complex spectra of heavier elements can interfere. An extension of this method is found in the "carrier distillation" technique, in which a low-boiling element such as gallium is deliberately added to the sample. When it vaporizes, the low-boiling element carries with it a number of other elements, thereby separating them from more refractory species.

Because of the high temperature gradient found in a DC arc discharge, the incidence of self-absorption is high. Atoms in the core of the discharge, being at a higher temperature than those toward the outside of the discharge, emit radiation which is absorbed by atoms at the outer edge of the discharge. Furthermore, spectral line broadening can occur at the high temperatures and electric fields found in the central portion of the DC arc, so that radiation emitted by atoms within this region has a broader spectral bandwidth than the absorption line width of the atoms surrounding the arc. Under these conditions, the central portion of the band of radiation emitted by the atoms within the arc is absorbed by those atoms surrounding the arc. This extreme example of self-absorption, termed **self-reversal**, can complicate qualitative analysis if its existence is not recognized. A self-reversed line, such as that shown in Figure 20–20, can appear to be two separate lines on either side of the true

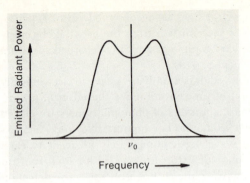

*Figure 20–20.* Profile of a self-reversed spectral line; the center of the line is at frequency $\nu_0$. (See text for discussion.)

emission line. (Compare this self-reversed line with the sodium doublet line at 589 nm in Figure 19–5.)

**Quantitative analysis by DC arc spectrometry.** Quantitative analysis with a DC arc source rarely achieves precision levels of better than 5 to 10 per cent because of arc instability and matrix interferences. Because of arc instability, there is nonuniform atomization of the sample, so that it is difficult to obtain a precise quantitative estimate of its composition. The intensity of atomic emission is dependent on the sample matrix, because the sample matrix strongly affects the resistance and mode of excitation of the DC arc. Fortunately, matrix effects can be minimized if one employs a **matrix buffer**, usually a low-boiling substance such as lithium chloride. When the buffer is added in large quantity to a sample, the vaporization and excitation behavior is dependent primarily on the nature of the buffer and not on the sample matrix.

Another way to minimize the effects of arc instability and the sample matrix is through use of an **internal standard**, an element whose vaporization and excitation characteristics very closely match those of the element to be determined. The internal standard is added in constant concentration to all samples and standards. Any instabilities or variations in the arc will affect the intensities of emission lines for the sample and internal standard equally. Analytical working curves can then be plotted as the *ratio* of the line intensities of the internal standard and sample versus the concentration ratio of the two. Use of an internal standard provides improvements in precision up to an order of magnitude.

**Sample preparation.** Solids, liquids, and gases can all be analyzed with the aid of the DC arc source. Special electrodes, standards, and procedures have been developed for many different types of samples. We will discuss only a few representative examples.

CONDUCTIVE METAL SAMPLES. These are the most convenient samples to analyze by means of DC arc spectrometry. Conductive metals are usually cast or machined into appropriate electrodes which can be used directly in the DC arc discharge. At times, both electrodes are formed from the conductive sample; at other times, the sample is employed as a cathode and another material, often spectroscopically pure graphite, as the anode (counter electrode).

Two electrode configurations can be used for conductive metal samples. In the arrangement termed *point-to-point* excitation and illustrated in Figure 20–21A, the sample electrode is cast or rolled into the form of a rod, the tip of which is sharpened to increase the electrical field. In the *point-to-plane* configuration shown in Figure 20–21B, the sample electrode is cast into the form of a disk, and the arc is struck between the planar disk and a pointed counter electrode. In all procedures, extreme care must be exercised to prevent contamination of samples, especially because of the high sensitivity of DC arc spectrometry.

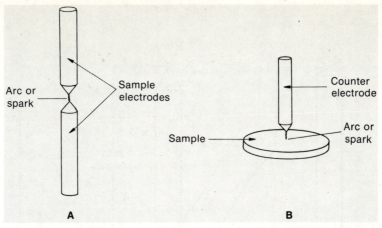

*Figure 20–21.* Point-to-point and point-to-plane electrode configurations used to excite a conductive metal sample in an electrical discharge. (See text for discussion.)

POWDERS AND NONCONDUCTIVE SOLIDS. Although many samples are non-conductive, they can be pulverized and mixed with a conductive material such as high-purity graphite powder. The mixture of sample and conductive material must be as homogeneous as possible and must be packed carefully into a specially formed sample electrode. Several sample electrodes used for powders or nonconductive samples are shown in Figure 20–22. These electrodes are usually made of high-purity graphite; in addition, some of the electrodes are shaped to reduce conduction of heat from the sample cup so that vaporization will be improved. Typical counter electrodes are also depicted in Figure 20–22. Counter electrodes are often pointed, in an attempt to confine the arc discharge to a small area of the sample electrode.

LIQUID SAMPLES. It is neither convenient nor safe to introduce liquid samples directly into a DC arc discharge. Therefore, liquid samples must be handled with the aid of special techniques or electrodes. Three common procedures involve the use of the flat-topped electrode, the dipping-wheel electrode, and the porous-cup electrode.

In the first procedure, a cylindrical, flat-topped electrode formed from low-porosity graphite is employed. The electrode is electrically preheated to a temperature just sufficient to volatilize the solvent from the liquid sample, which is introduced from a dropper onto the electrode surface. As each drop evaporates, a new portion of sample solution is added to the electrode surface until an appreciable crust of dried sample material is built up. An arc is struck between two such electrodes prepared for each sample to provide extremely high sensitivities for the elements present in the original solution. This procedure is quite time-consuming and tedious, but provides extremely high sensitivity and respectable levels of precision.

The dipping-wheel electrode is most often used for viscous liquid samples, such as lubricating oils. As shown in Figure 20–23, the dipping-wheel electrode system consists of a slowly rotating disk and a counter electrode. The disk rotates through a pool of the liquid to be analyzed, bringing it into position directly beneath the counter electrode. An arc struck between the disk and the counter electrode then excites the elements in solution which are carried up by the disk from the bath. With viscous liquid samples, such as oils, the dipping-wheel electrode offers both excellent sensitivity and precision for trace metal analysis.

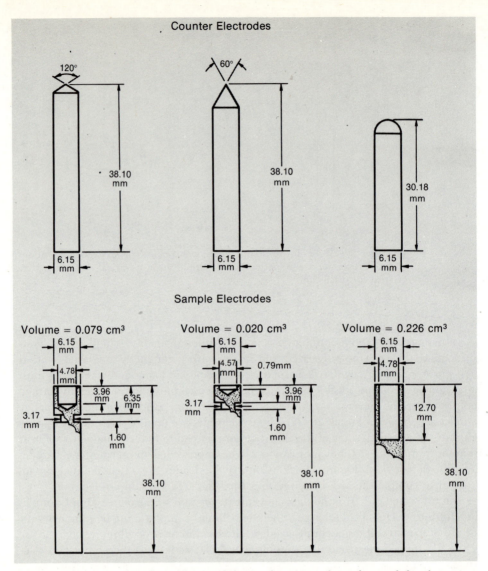

Figure 20–22. Examples of sample-containing and counter electrodes used for the arc or spark discharge analysis of powdered samples. Dimensions are in millimeters (mm).

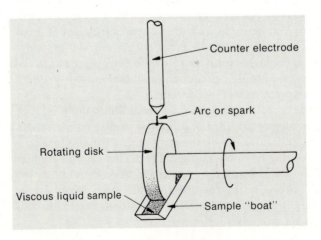

Figure 20–23. Dipping-wheel electrode arrangement for the analysis of viscous liquids in an electrical discharge.

The porous-cup electrode consists of a tubular electrode having a porous graphite bottom. A liquid sample placed in the porous cup slowly seeps through the bottom of the electrode, and an arc is struck between it and the counter electrode beneath to excite the elements present in the sample. Although this procedure is less sensitive and precise than either of the two preceding methods, the minimal handling of the sample makes this an attractive technique for solution analysis.

ANALYSIS OF GASES.   Analysis of gases in a DC arc is a relatively simple process and merely involves directing the gases into an arc struck between two inert electrodes. To increase sensitivity and reduce the occurrence of interferences from dust particles and other foreign material, the arc is often surrounded by a chamber. One such chamber is the Stallwood jet, into which the gases to be analyzed are introduced tangentially. The gases swirl about the arc, so that elements present in the gases enter the arc periodically and are efficiently excited. Electrodes are introduced into the Stallwood jet through appropriately placed holes in the top and bottom of the chamber. A Stallwood jet enclosure can also be used for the analysis of solid samples when it is desired to employ a controlled gas environment around the arc to eliminate cyanogen bands and to minimize self-absorption.

## High-Voltage Spark Spectrometry

The emission characteristics of a high-voltage spark differ markedly from those of the DC arc. Unlike the continuous DC arc, the spark appears intermittently and utilizes extremely high voltages and currents. An example of the wave form of a typical single spark discharge is shown in Figure 20–24. A high-voltage spark is an oscillatory discharge, often having positive and negative peak currents exceeding 1000 amperes. At the onset of the discharge, the sample electrode is usually negative, so that the sample will be efficiently vaporized by positive-ion bombardment. During subsequent oscillations of the spark, the electrodes change polarity so that the sample is recycled between the sample and counter electrodes to provide efficient excitation. This enhances the quantitative analytical utility of the spark.

The extremely high currents found in a single spark discharge cause localized but extremely intense heating of the sample, with temperatures in excess of 10,000°K sometimes being attained. Such heating populates the very high energy electronic levels of atoms in the discharge. Thus, the emission spectra are even more intense

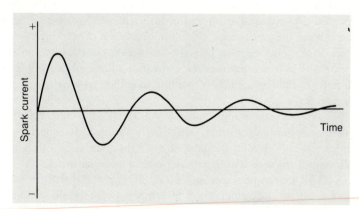

*Figure 20–24.*   Time behavior of the current in a single oscillatory high-voltage spark discharge.

and complex than those produced by the DC arc. This increases the difficulty of detecting individual emission lines and renders qualitative analysis more tedious.

Although the duration of a single spark discharge is quite short (usually about 100 microseconds), the period between discharges is relatively long (tens of milliseconds or longer). This delay gives the electrodes a reasonably long time to cool off between spark discharges, resulting in less fractional distillation and sample consumption, but also in lower sensitivity, than when the DC arc is used. Because a relatively long time exists between individual spark discharges, the atoms vaporized from the sample have sufficient time to diffuse or blow away before the appearance of the next discharge. Because there is little accumulation of atomic vapor around the electrode, self-absorption and self-reversal occur less frequently.

The intermittent nature of the high-voltage spark makes it a superior source for quantitative analysis. Because the spark starts anew for each discharge, it is not confined to specific "hot spots" on the electrode, but instead tends to randomly and reliably sample the entire electrode surface. This improved sampling provides better statistics for quantitative analysis. With care, relative precision levels between 1 and 5 per cent can be routinely obtained with the high-voltage spark.

*Instrumentation for high-voltage spark spectrometry.* Dispersion and detection systems employed in high-voltage spark spectrometry are, for the most part, identical to those in DC arc spectrometry. Also, the procedures and precautions used in both techniques are very similar. However, because quantitative analysis is more frequently performed with the high-voltage spark, a direct-reading spectrometer is often used. With such a system, 20 or more elements can be routinely determined at precision levels below 5 per cent.

Sample handling and electrode preparation are similar in both high-voltage spark and DC arc spectrometry; all techniques discussed earlier can be used for high-voltage spark spectrometry. However, because of the intermittent nature of the high-voltage spark, it is possible to analyze a solution by spraying it directly into the discharge. The high energy of the discharge is capable of evaporating solvent from the sample droplets, vaporizing the dry sample, and exciting the resulting atoms quite efficiently.

## Radio-Frequency Plasma

The radio-frequency plasma is a flame-like electrical discharge which shows increasing promise as an excitation source for qualitative and quantitative elemental analysis. A schematic diagram of a radio-frequency plasma appears in Figure 20–25. A high current, oscillating at a radio frequency between 10 and 50 MHz, generates an intense magnetic field which interacts with the charged species present within the coil. When no sample is present, an inert supporting gas such as argon sustains the plasma. Electrons produced by ionization of the supporting gas move with the radio-frequency field and, in their oscillation, collide with neutral supporting gas atoms and ionize them.

A cooled quartz tube, placed within the radio-frequency coil, is used to contain the plasma. Argon gas, flowing tangentially through the quartz tube, forces the ionized gases partially beyond the coil, so that the plasma is visible above the quartz tube. Elemental emission is ordinarily viewed in this upper region.

This source possesses an extremely high temperature (above 5000°K), so that vaporization and atomization of samples occur readily and the intensity of atomic emission is high. Furthermore, the source is far more stable than either the high-voltage spark or the DC arc, and can provide analytical results precise to $\pm$ 1 per

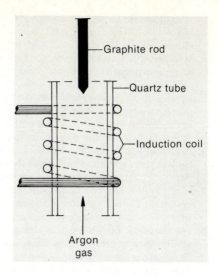

*Figure 20–25.* Schematic diagram of an induction-coupled radio-frequency plasma source. The graphite rod is to illustrate how the plasma can be started by induction heating. After initiation and stabilization of the plasma, the rod would be removed and atomic emissions viewed from that location. (Reproduced, with permission, from V. A. Fassel, "Electrical Flame Spectroscopy," Colloq. Spectrosc. Int., Plenary Lect. Rep. 16th, Adam Hilger, London, 1972, p. 63.)

cent. Liquid samples can be sprayed directly into the plasma, so that sample preparation is minimized. Furthermore, the time required for an analysis is greatly reduced so that it becomes feasible to employ simple, inexpensive, single-channel monochromator systems.

The radio-frequency plasma is simple, safe, and inexpensive to operate and maintain. Its high sensitivity and multi-element capability make it a strong contender to replace some of the less attractive, currently used emission sources.

## QUESTIONS AND PROBLEMS

1. The observed widths of atomic spectral lines, when measured with a high-dispersion spectral isolation device, are greater than would be predicted from quantum mechanical considerations. This broadening arises from a number of sources, the most prominent in flames, arcs, or sparks being *Doppler broadening* and *Stark broadening*. Doppler broadening occurs because the atoms being observed are in constant motion. Those traveling toward the point of observation appear to radiate at a higher frequency, whereas those traveling away emit radiation of a lower frequency. A large number of emitting atoms all moving randomly therefore generate a broader spectral profile having a Gaussian shape. Stark broadening is caused by the interaction of the radiating atoms with an electric field. The electric field produces a splitting of the energy levels of each atom according to the quantized orientation of an atom within the field. An inhomogeneous field, such as found in a flame or electrical discharge, will thus cause an apparent broadening of the spectral line. On the basis of the preceding discussion, which spectral lines do you feel would be broader, those generated within an arc, spark, or flame? What effect will this broadening have on the shape of an analytical working curve? Which of the three flame spectrometric techniques will be most affected by broadening? Which of the three techniques will be least affected? Explain.

2. Explain why the simultaneous analysis of several elements by means of atomic absorption spectrometry would be more difficult than by either atomic fluorescence or flame emission spectrometry.

3. In Figure 20–8 (page 686), the emission line for magnesium is seen to be nearly obscured by the —OH band background emission of the flame.

Will this make the determination of magnesium in the oxygen-hydrogen flame impossible? Why or why not? Which of the three flame spectrometric techniques will be affected most and which least by this spectral interference?

4. What effect will a change in the flow rate of fuel gas produce in a practical flame spectrometric analysis? What effect will a change in the flow rate of oxidizer gas have?

5. In atomic absorption spectrometry, a slot burner is usually employed to provide a long path length through the absorbing atoms. What would be the optimum flame shape in flame emission spectrometry? What would be the optimum flame shape in atomic fluorescence flame spectrometry? Explain.

6. Recalling the discussion of molecular fluorescence in Chapter 19 (page 655), explain the extreme curvature of the working curves for atomic fluorescence flame spectrometry at high analyte concentrations.

7. If the hollow-cathode lamp primarily determines the spectral bandwidth of detected radiation in atomic absorption spectrometry, as stated on page 690, why is a monochromator needed at all?

8. For the alkali metals such as sodium, potassium, and lithium, many of the atomic species present in a flame are ionized. In these cases, why is ionic emission not stronger than atomic emission?

9. Most metals are in an ionic form in a solution, whereas in a flame the emission from neutral atoms is observed. To go from ions to atoms, these species must acquire one or more electrons. From where do these electrons come?

10. A novel flameless atom reservoir is now being used for the atomic absorption determination of mercury contamination in water supplies. The atomizer consists of a reduction cell and an enclosed sample cell. In operation, a 10.00-ml sample of water is placed in the reduction cell and is diluted to 100 ml; then 25 ml of concentrated sulfuric acid is added, and 10 ml of a 10 per cent tin(II) sulfate solution in 0.25 F sulfuric acid is introduced as a reducing agent. Mercury is reduced to the elemental (atomic) state and is carried to an absorption cell by a stream of air bubbling through the solution in the reduction cell. Finally, using a mercury hollow-cathode lamp as a source, the absorbance due to the mercury atoms is measured at a wavelength of 2537 Å; the absorbance reaches a maximum level in approximately 3 minutes.

In such a determination, the following absorbance values were obtained for a series of standard mercury-containing solutions:

| Total Amount of Mercury in Standard ($\mu g$) | Absorbance |
| --- | --- |
| 0.00 | 0.002 |
| 0.30 | 0.090 |
| 0.60 | 0.175 |
| 1.00 | 0.268 |
| 2.00 | 0.440 |

Two water samples, when treated according to the above procedure, gave absorbance values of 0.040 and 0.305, respectively. What was the total amount of mercury in each of the samples? What was the concentration (in $\mu g$/ml) of mercury in each of the original water samples?

11. In Figure 20–15 (page 694), the halogens and rare gases are conspicuously absent from the list of elements which can be determined by means of flame emission or atomic absorption spectrometry. Explain why these elements are not amenable to flame spectrometric analysis. Could these elements be determined by means of DC arc or high-voltage spark spectrometry? Why or why not?

12. Predict how chemical and physical interferences would affect atomic fluorescence flame spectrometry, compared with flame emission and atomic absorption spectrometry.

13. Would either the DC arc or the high-voltage spark be suitable as flameless atomization systems for use in atomic absorption spectrometry? Why or why not?

14. Which would be superior—the DC arc or the high-voltage spark—as an electrical discharge source for the spectrometric analysis of a disk of a very inhomogeneous alloy? Explain and justify your answer.

15. In the flame emission spectrometric analysis of clinical samples, lithium is often employed both as an ionization suppressant and as an internal standard for the determination of sodium and potassium. Because the lithium emission is affected by variations in flame composition and temperature in a manner similar to that of sodium and potassium, the effects of these variations can be minimized if the observed sodium or potassium emission is compared with that produced by a constant amount of lithium added to each solution. To construct a working curve for sodium, the ratio of the emission signal for sodium to that for lithium is plotted versus the concentration of sodium; the same approach can be followed for potassium. Construct working curves for the following data:

| Standard Solution | Relative Signals (arbitrary units) | | |
| --- | --- | --- | --- |
| | Na | K | Li |
| (a) 0.1 ppm Na, 0.1 ppm K, 1000 ppm Li | 0.11 | 0.15 | 86 |
| (b) 0.5 ppm Na, 0.5 ppm K, 1000 ppm Li | 0.52 | 0.68 | 80 |
| (c) 1.0 ppm Na, 1.0 ppm K, 1000 ppm Li | 1.2 | 1.5 | 91 |
| (d) 5.0 ppm Na, 5.0 ppm K, 1000 ppm Li | 5.9 | 7.7 | 91 |
| (e) 10.0 ppm Na, 10.0 ppm K, 1000 ppm Li | 10.5 | 14 | 81 |

What are the sodium and potassium concentrations in a sample which produces the following emission signals: Na, 1.4; K, 0.73; Li, 95? Now, plot *normal* working curves (that is, graphs of just the emission signal for sodium or potassium versus the concentration of the respective element); determine the sodium and potassium concentrations in the unknown sample from these working curves. How much error would have been incurred if the internal standard technique had *not* been employed? Was the internal-standard approach essential in this determination? Why or why not?

16. The ionization of an alkali-metal atom in a flame

$$M \rightleftharpoons M^+ + e$$

can be treated as a simple dissociation equilibrium.
(a) If $p$ is the partial pressure (in atm) of metal atoms present before ionization, if $x$ is the fraction of metal atoms which ionize, and if $K$ is the equilibrium constant for the ionization, formulate the equilibrium expression for the ionization of an alkali metal.
(b) The effect of flame temperature ($T$) on the value of $K$ is given by the Saha equation

$$\log K = -\frac{5041E}{T} + \frac{5 \log T}{2} - 6.49 + \log \frac{g_{M^+} g_e}{g_M}$$

where $E$ is the ionization energy of the atom in electron volts, and $g_{M^+}$, $g_e$, and $g_M$ are statistical weights for the pertinent species. For alkali metals, the final term in the Saha equation turns out to be zero. If the ionization energy of lithium is 5.390 electron volts, calculate the value of $K$ at $T = 2000°K$ and $3500°K$.
(c) If it is assumed that $p$ is $10^{-6}$ atm, calculate the fraction of lithium atoms ionized at $2000°K$ and at $3500°K$. Which of these two temperatures would be better for the flame emission spectrometric determination of lithium? Why?
(d) Would the ionization of flame gases affect the results of the preceding calculations? Explain.

17. To determine the lead content of urine by means of atomic absorption spectrometry, the method of standard addition can be utilized. Exactly 50.00 ml of urine was pipetted into each of two 100-ml separatory funnels. To one funnel was added 300 $\mu$l of a standard solution containing 50.0 mg of lead per liter. Then the pH of each mixture was adjusted to 2.8 by dropwise addition of hydrochloric acid. Next, 500 $\mu$l of a freshly prepared 4 per cent solution of ammonium pyrrolidine dithiocarbamate in methyl-$n$-amyl ketone was added to each funnel, and the aqueous and nonaqueous phases were thoroughly shaken to extract the lead. Finally, the lead-containing organic phase was examined by means of atomic absorption spectrometry, a lead hollow-cathode lamp with an emission line at 283.3 nm being used. If the absorbance of the extract from the unknown urine sample was 0.325, and if the absorbance of the extract from the unknown spiked with a known amount of lead was 0.670, what was the lead concentration (in mg/liter) of the original urine sample?

18. Atomic absorption spectrometry can be employed to determine traces of wear metals in used lubricating oils. To perform an analysis of a sample of used lubricating oil, 5.000 grams of the thoroughly mixed oil is weighed into a 25.00-ml volumetric flask, the oil is dissolved in 2-methyl-4-pentanone, and additional 2-methyl-4-pentanone is added up to the calibration line on the flask. Then the resulting solution is aspirated into an air-acetylene flame; for the determination of copper and lead, separate hollow-cathode lamps with emission lines at 324.7 and 283.3 nm, respectively, are used. A series of standard solutions, containing known amounts of copper and lead in the correct mixture of *unused* lubricating oil and 2-methyl-4-pentanone, can be used to establish calibration curves. Utilizing the following information, determine the *weight per cent* of copper and lead in a 5.000-gm sample of used lubricating oil:

| | Absorbance | |
|---|---|---|
| Solution | at 283.3 nm (due to Pb) | at 324.7 nm (due to Cu) |
| (a) standard: 19.5 $\mu$g Pb/ml, 5.25 $\mu$g Cu/ml | 0.356 | 0.514 |
| (b) standard: 4.00 $\mu$g Pb/ml, 4.00 $\mu$g Cu/ml | 0.073 | 0.392 |
| (c) standard: 12.1 $\mu$g Pb/ml, 6.27 $\mu$g Cu/ml | 0.220 | 0.612 |
| (d) standard: 8.50 $\mu$g Pb/ml, 1.05 $\mu$g Cu/ml | 0.155 | 0.101 |
| (e) standard: 15.2 $\mu$g Pb/ml, 2.40 $\mu$g Cu/ml | 0.277 | 0.232 |
| (f) unknown | 0.247 | 0.371 |

## SUGGESTIONS FOR ADDITIONAL READING

*Flame Methods*

1. F. Burriel-Martí and J. Ramírez-Muñoz: *Flame Photometry: A Manual of Methods and Applications*. Elsevier, New York, 1957.
2. J. A. Dean: *Flame Photometry*. McGraw-Hill Book Company, New York, 1960.
3. J. A. Dean and T. C. Rains, eds.: *Flame Emission and Atomic Absorption Spectrometry*. Dekker, New York, Volume I (Theory), 1969; Volume II (Components and Techniques), 1971.

4. R. Herrmann and C. T. J. Alkemade: *Chemical Analysis by Flame Photometry.* Translated by P. T. Gilbert, Jr., Wiley-Interscience, New York, 1963.
5. R. Mavrodineanu, ed.: *Analytical Flame Spectroscopy.* Springer-Verlag, New York, 1970.
6. R. Mavrodineanu and H. Boiteux: *Flame Spectroscopy.* John Wiley and Sons, New York, 1965.
7. W. Slavin: *Atomic Absorption Spectroscopy.* Wiley-Interscience, New York, 1968.
8. J. D. Winefordner, ed.: *Spectrochemical Methods of Analysis.* Part I (Flame Spectrometric Methods), Wiley-Interscience, New York, 1971.

*Electrical Discharge Methods*

1. L. H. Ahrens and S. R. Taylor: *Spectrochemical Analysis.* Second edition, Addison-Wesley Publishing Company, Reading, Massachusetts, 1961.
2. W. R. Brode: *Chemical Spectroscopy.* Second edition, John Wiley and Sons, New York, 1943.
3. P. W. J. M. Boumans: *Theory of Spectrochemical Excitation.* Plenum Publishing Company, New York, 1966.
4. I. M. Kolthoff and P. J. Elving, eds.: *Treatise on Analytical Chemistry.* Part I, Volume 6, Wiley-Interscience, New York, 1959.
5. R. A. Sawyer: *Experimental Spectroscopy.* Third edition, Dover, New York, 1963.

# INFRARED AND RAMAN VIBRATIONAL SPECTROMETRY

In this chapter, we shall consider in detail two spectrochemical methods—Raman spectrometry and infrared spectrophotometry—which can be used to probe the nature of molecular vibrations. Although these techniques differ in a number of respects, they can be employed in a complementary way to obtain information of both analytical and structural importance. Because both methods are based on the observation of transitions between vibrational energy levels of molecules, let us first examine the nature of molecular vibrations.

## MOLECULAR VIBRATIONS

Because chemical bonds between atoms in a molecule are not completely rigid, the individual atoms are in constant motion, giving rise to vibrations of the molecule as a whole. If a molecule could be observed by means of extremely high-speed motion-picture photography, these vibrations would appear to occur in a completely random, disorganized way. However, all the vibrational motion of a molecule can be resolved into so-called normal modes, each having its own spatial characteristics and vibrational frequency.

## Normal Modes

The motion of any body can be resolved into directional components along three mutually perpendicular axes in space. Therefore, for a molecule consisting of $N$ atoms, all of which are in constant motion, there are $3N$ possible ways—that is, **degrees of freedom**—in which the atoms can move with respect to each other. However, three of these degrees of freedom involve *translational* movement of the molecule as a whole along each one of the mutually perpendicular spatial axes. In addition, for a non-linear molecule, there are three more degrees of freedom corresponding to *rotation* of the molecule about each of the three axes. This leaves $3N - 6$ degrees of freedom for *vibrational* motion.

These $3N - 6$ remaining degrees of freedom represent the number of possible **normal modes** of vibration of the molecule. That is, the molecule will have $3N - 6$ identifiable (independent) vibrations. A linear molecule, of course, can only rotate in two dimensions, so that it will have $3N - 5$ normal modes. As an illustration, consider the vibrational motion of a simple triatomic molecule such as water. Water, being nonlinear, has $(3N - 6)$ or 3 normal modes of vibration, as shown in Figure 21–1.

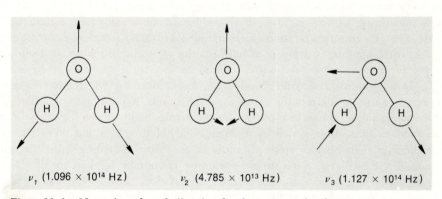

$\nu_1$ (1.096 × 10$^{14}$ Hz)    $\nu_2$ (4.785 × 10$^{13}$ Hz)    $\nu_3$ (1.127 × 10$^{14}$ Hz)

*Figure 21–1.* Normal modes of vibration for the water molecule.

Notice that all three atoms move to prevent a shift in the center of gravity of the molecule which, if allowed to occur, would involve translation rather than vibration. To identify normal modes, descriptive names are often used. In Figure 21–1, the vibrations corresponding to frequencies $\nu_1$, $\nu_2$, and $\nu_3$ are termed symmetric stretching, scissoring, and asymmetric stretching, respectively. For larger molecules, additional terms such as bending, twisting, wagging, rocking, and deformation are needed. Further descriptions and pictorial representations of normal modes can be found in many of the references listed at the end of the chapter.

Each normal mode has a specific vibrational frequency; those for water are included in Figure 21–1. To examine this point in more detail, let us consider a simple mechanical model. Imagine that the bond between a pair of atoms has the properties of a spring. According to Hooke's law, if one atom is displaced a distance $x$ from its equilibrium position with respect to the other atom, the force, $F$, which operates to restore the atoms to their original positions is

$$F = -kx$$

where $k$ is termed the **force constant**. If the atoms are pulled away from or pushed toward each other, the spring will oppose their displacement and cause them to oscillate harmonically about their equilibrium position at a **fundamental vibration frequency** $v$ (in Hz) given by

$$v = \frac{1}{2\pi}\sqrt{k\left(\frac{m_1 + m_2}{m_1 m_2}\right)} = \frac{1}{2\pi}\sqrt{\frac{k}{\mu}} \qquad (21\text{-}1)$$

where $m_1$ and $m_2$ are the masses of the atoms (in gm) and $k$ is again the force constant (in dynes/cm) of the bond (spring); the ratio $m_1 m_2/(m_1 + m_2)$ is called the **reduced mass** ($\mu$) of the diatomic oscillator.

This simple model has been found to be surprisingly accurate in describing the vibrations of many diatomic molecules. Gaseous hydrogen chloride (HCl), in going from the ground vibrational level to the first excited vibrational level, absorbs infrared radiation at 2886 cm$^{-1}$, which corresponds to a frequency ($v$) of $8.658 \times 10^{13}$ Hz. If this frequency—which can be shown to be identical to the vibrational frequency of the hydrogen-chlorine bond in the above model—is substituted into equation (21-1) along with values for the masses of the atoms, the force constant ($k$) is calculated to be $4.8 \times 10^5$ dynes/cm. What becomes significant about this result is that force constants for *single* bonds all range from about $2 \times 10^5$ to $8 \times 10^5$ dynes/cm, the average value being $5 \times 10^5$ dynes/cm. In conjunction with equation (21-1), this average value can be used to *predict* the vibrational frequencies for absorption peaks corresponding to transitions between the ground and first excited vibrational levels for diatomic molecules with single bonds.

Double bonds and triple bonds have force constants which are, respectively, about two and three times the values for single bonds. According to equation (21-1), therefore, atoms connected by double or triple bonds have higher vibrational frequencies than those joined by a single bond. Hydrogen atoms vibrate at higher frequencies than heavier atoms, because of the effect of hydrogen on the reduced mass of the oscillator. Such simple reasoning cannot be applied to all molecules, particularly large and complicated ones, but it does provide a valuable, intuitive foundation for the study and analytical use of molecular vibrations.

Although each normal mode of vibration of a molecule will have an assignable frequency, several modes can have identical frequencies and will not be distinguishable. In addition, each mode will have its own spectroscopic characteristics. To understand this, let us examine the quantized nature of transitions between vibrational levels.

## Vibrational Transitions

If, for simplicity, we consider a diatomic molecule, only certain quantized vibrational energy states can exist, as given by the relation

$$E = \left(v + \frac{1}{2}\right)hv \qquad (21\text{-}2)$$

where $E$ is the energy of the state having a vibrational quantum number v, $h$ is Planck's constant, and $v$ is the frequency of vibration from equation (21-1). For a given vibration, $v$ is constant, so that the value of $E$ is determined by v, which can only have integral values such as 0, 1, 2, 3, and so on.

If the molecule undergoes a transition from one energy state with quantum number $v_1$ to another with quantum number $v_2$, the change in energy will be

$$\Delta E = E_2 - E_1 = \left(v_2 + \frac{1}{2}\right)h\nu - \left(v_1 + \frac{1}{2}\right)h\nu = (\Delta v)h\nu \qquad (21\text{-}3)$$

Therefore, a transition between vibrational energy levels involves quantized changes equal to integral ($\Delta v$) multiples of $h\nu$. Note that $\Delta v$ can be either positive or negative, corresponding to a gain or loss in vibrational energy, respectively. Although $\Delta v$ can strictly have any integral value, quantum mechanical considerations indicate that most transitions in infrared and Raman spectrometry must involve changes ($\Delta v$) of $\pm 1$, except in special cases to be discussed later. When $\Delta v$ is $+1$, $\Delta E = h\nu$, so that the gain in vibrational energy is related by Planck's constant to the fundamental vibration frequency $\nu$. It is this frequency that is most important in spectrochemical analysis.

To probe vibrational transitions with electromagnetic radiation, the radiation must interact with some electrical property of the molecule which is altered during a vibration. This can happen in two ways—through a change in dipole moment or a change in polarizability. These phenomena provide the foundations for infrared and Raman spectrometry, respectively.

## INFRARED SPECTROPHOTOMETRY

### Interaction of Infrared Radiation with Molecules

In the preceding section, it was indicated that the normal modes of vibration of typical diatomic molecules have frequencies on the order of $10^{14}$ Hz. For example, the scissoring ($\nu_2$) vibration of the water molecule has a frequency of $4.785 \times 10^{13}$ Hz. Using the relationship introduced in Chapter 18 (page 603),

$$c = \lambda \nu \qquad (21\text{-}4)$$

we can calculate the wavelength of radiation corresponding to this frequency:

$$\lambda = \frac{c}{\nu} = \frac{2.998 \times 10^8 \text{ m sec}^{-1}}{4.785 \times 10^{13} \text{ sec}^{-1}} = 6.27 \ \mu\text{m} \qquad (21\text{-}5)$$

This wavelength lies in the infrared region of the spectrum. If infrared radiation with a wavelength of 6.27 $\mu$m can interact (couple) with this particular scissoring motion, it should be possible to cause or observe a transition between vibrational levels of the water molecule at this wavelength.

In order for infrared radiation to interact with a vibrating molecule, the molecule must undergo a change in dipole moment during the vibration. If such a change occurs, the oscillating electromagnetic field of the infrared radiation can couple with the oscillating field of the dipole to increase or decrease the amplitude of the vibration. If the amplitude is increased, the molecule will have absorbed energy from the infrared radiation. If the amplitude is decreased, a quantity (photon) of infrared radiation will be emitted. Because most molecules reside in the lowest (ground) vibrational state at room temperature, the absorption of infrared radiation is usually of more utility to spectrochemical analysis than is infrared emission.

Suppose, for example, that a simple diatomic dipolar molecule having a vibrational frequency $\nu_0$ is passed by a beam of radiation of the very same frequency. As the radiation passes the oscillating dipole, the dipole will experience electrostatic forces from the electromagnetic field of this radiation which will compress or stretch the dipole, depending upon the relative directions in which the field and dipole oscillate. Because the field oscillates at the natural resonant frequency of the dipole, interactions between the field and the dipole will increase the vibrational amplitude of the molecule. Energy needed to produce this increase in amplitude is drawn from the passing beam, resulting in absorption. Those normal modes of vibration which produce a change in dipole moment and can couple with infrared radiation are said to be **infrared active**.

Some molecular vibrations are incapable of interacting with infrared radiation, since they produce no change in dipole moment. In fact, homonuclear diatomic molecules such as $O_2$, $N_2$, and $H_2$ cannot absorb infrared radiation at all, because they have a dipole moment of zero, which is not altered by vibration.

## Infrared Spectral Region

Because vibrating molecules are capable of absorbing infrared radiation at certain resonant frequencies, the infrared absorption spectrum of a molecule presents an accurate picture of its infrared-active normal modes of vibration. Furthermore, these normal modes are largely unaffected by neighboring species, so that the infrared spectrum is highly characteristic of a specific molecule. This behavior should be contrasted with ultraviolet or visible molecular spectra, in which the observed electronic transitions are greatly influenced by intermolecular complexation, solvation, and other effects. Therefore, infrared absorption spectrophotometry is a powerful tool for the qualitative identification of molecules.

According to Figure 18–2 (page 602), the infrared region extends from approximately 0.8 to 200 $\mu$m (wavelength) or from 12,000 to 5000 cm$^{-1}$ (wavenumbers). However, this range can be divided into several smaller intervals, based on the types of vibrations which occur and the nature of the instrumentation required to observe the vibrations.

*Near-infrared region.* The near-infrared region, running from approximately 0.7 to 2.5 $\mu$m, is immediately adjacent to the visible region. Absorption peaks observed in the near-infrared region include stretching vibrations between hydrogen and other atoms as well as overtone bands and combination bands.

**Overtone bands** and **combination bands** are forbidden in the model of the classical harmonic oscillator presented earlier. These bands arise from anharmonicity in molecular vibrations and are usually weaker than the fundamental vibrational bands at longer wavelengths. Overtone bands are absorptions in which the vibrational quantum number (v) changes by more than $\pm 1$. For example, the fundamental absorption band corresponding to the C—H stretching vibration in the —CH$_3$ group occurs at 2960 cm$^{-1}$ (3.38 $\mu$m). Therefore, the first overtone band ($\Delta v = \pm 2$) of this vibration would be found at 5920 cm$^{-1}$ (1.69 $\mu$m) in the near-infrared region.

Combination bands originate in the quantum-mechanical mixing of two vibrations, and appear at the sum and difference frequencies of the two combined vibrations. As an example, a weak absorption band is observed in the near-infrared spectrum of acetylene at 4092 cm$^{-1}$ (2.44 $\mu$m), which is not due to a fundamental vibration. Careful examination has shown this band to be a combination of the symmetrical stretching vibration at 3372 cm$^{-1}$ (2.97 $\mu$m) and the bending vibration at 730 cm$^{-1}$ (13.7 $\mu$m). As in this situation, combination bands corresponding to the sum of two fundamental frequencies are observed more often than those due to the difference between two frequencies. Notice that, for overtone and combination bands, it is the frequencies or wavenumbers that add or subtract, not the wavelengths. Because of the relatively low intensity of these bands, the near-infrared spectral region is not of great analytical utility.

*Fundamental region.* Between 2.5 and 50 $\mu$m, fundamental vibrations are most often observed. Absorption spectra in this region are highly characteristic of individual species, so that both qualitative and quantitative analyses can be performed. In particular, the interval between 2.5 and 15 $\mu$m is extensively used for the analysis and characterization of organic compounds, because of the large number of absorption peaks of organic functional groups which appear there. For this reason, much of our later discussion will pertain to this region.

From 15 to 50 $\mu$m, the absorption bands which are observed arise chiefly from bending vibrations involving heavy atoms or cyclic compounds. Although sometimes useful for the investigation of inorganic substances, this spectral range is not often used for analysis and will not be further considered in any detail.

A good example of the analytical usefulness of infrared spectrophotometry in the wavelength interval from 2.5 to 15 $\mu$m is provided by Figure 21–2, which compares the infrared spectra of nitrobenzene and toluene. Despite the structural similarity of these compounds, their spectra are distinctly different. On the spectra of Figure 21–2, there are two horizontal axes, one calibrated in $\mu$m (wavelength) and the other in cm$^{-1}$ (wavenumbers). Both units are in common use, the choice being partly based on instrumental convenience. For spectroscopists interested in the study of fundamental molecular vibrations, the wavenumber is preferred, because of its direct proportionality to frequency and energy. Vertical axes in Figure 21–2 are labeled in per cent transmittance, an almost universal practice. However, some spectra are calibrated in terms of absorbance for convenience in quantitative work.

*Far-infrared region.* From 50 to 500 $\mu$m, called the far-infrared spectral region, low-frequency vibrations and rotations are observed. However, the types of vibrations and rotations which produce absorption peaks here are usually not of great analytical utility. Most analytically useful rotations occur at microwave frequencies, so that the extra expense of a far-infrared spectrometer can usually not be justified. A relatively new kind of instrument, called a Fourier transform infrared spectrometer, now makes the determination of far-infrared spectra much easier; this device can be expected to increase the attractiveness of this spectral region for analytical purposes.

## Infrared Instrumentation

As mentioned earlier, most molecules reside in the ground vibrational state at room temperature, so that absorption of infrared radiation is generally more sensitive and useful than emission. In addition, except for a few special cases, infrared fluorescence is not a very efficient process. Therefore, one would expect the instrumentation used for infrared measurements to employ absorption, and to be similar in form to that used in ultraviolet-visible spectrophotometry. However, because the transmission characteristics of most materials for infrared radiation differ from those for ultraviolet-visible radiation, the instrument components are different in the two situations. A block diagram of a typical infrared absorption spectrophotometer is shown in Figure 21–3, and should be compared to the generalized spectrochemical instrument of Figure 18–10 and the absorption spectrophotometer of Figure 18–11. The only obvious difference between these absorption instruments is in the location

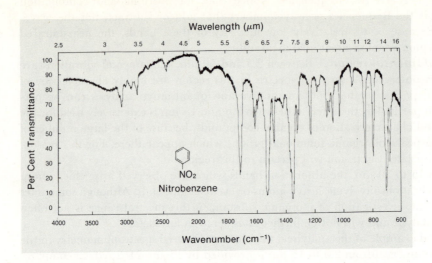

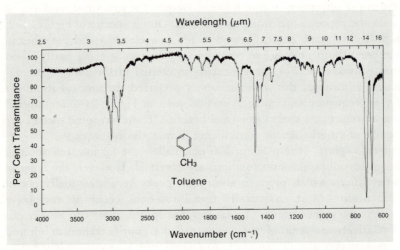

*Figure 21–2.* Infrared spectra of nitrobenzene and toluene, two compounds of similar structure whose spectra differ considerably.

of the chemical sample. For infrared spectrophotometry, the chemical sample is located before rather than after the monochromator. This arrangement minimizes the effect of stray and background radiation, a problem which can be quite serious in infrared spectrophotometry, and also protects the detector from any radiation not selected by the monochromator. The ubiquitous character of thermally produced infrared radiation renders this protection imperative.

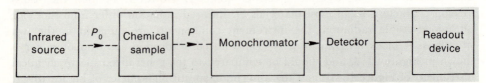

*Figure 21–3.* Simplified schematic diagram of an infrared spectrophotometer.

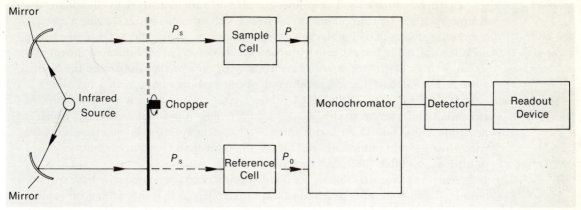

*Figure 21–4.* Schematic diagram of a double-beam infrared spectrophotometer.

Infrared spectrophotometers almost universally employ double-beam optical systems to minimize the effects of absorption by atmospheric carbon dioxide and water and to reduce the importance of radiation scattering. Scattering of radiation in the infrared region is especially troublesome, because infrared radiation is of approximately the same wavelength as the dimensions of dust particles. Figure 21–4 illustrates the design of a double-beam infrared spectrophotometer. Although a number of different double-beam instruments are in common use, the principles of operation are the same as those discussed in detail in Chapter 19 (page 641). In a double-beam infrared spectrophotometer, the reference cell usually contains merely the sample solvent, although any other sample component whose absorption spectrum is not sought can be included.

In addition to minimizing the effects of atmospheric absorption, scattering, and reflection losses, double-beam infrared spectrophotometers can be used to offset solvent absorption. Most organic and inorganic solvents exhibit in the infrared region of interest some characteristic absorption bands which might obscure the absorption spectrum of the desired species. If solvent is present in both the reference and sample cells of a double-beam spectrophotometer, solvent absorption bands will not appear in the recorded spectrum. Of course, this approach assumes that the reference and sample cells are identical and that the path lengths through the cells are the same. However, this procedure has limitations. In régions of high solvent absorption, the radiation transmitted by both the sample and reference cells will be relatively low, so that little radiation will reach the detector. Because many detectors of infrared radiation are inherently insensitive, this loss of radiation lowers the instrument response and tends to make the readout system appear sluggish. Therefore, solvents having the lowest possible absorption in the spectral region of interest should always be employed in infrared spectrophotometry. It must never be thought that a double-beam spectrophotometer can compensate for all solvent effects.

## Components of Infrared Instruments

Many of the instrument components shown in Figures 21–3 and 21–4 are peculiar to the infrared spectral region. Although we will not deal with any of these components in great detail, it is important to have some knowledge of infrared instrumentation in order to appreciate the capabilities and limitations of the technique.

*Infrared sources.* The primary source of radiation used to illuminate the sample in infrared spectrophotometry is usually a continuous source, having a spectral output resembling that of a blackbody. Simple sources, such as heated tungsten wires, are often utilized in less expensive instruments. However, in sophisticated instrumentation, sources of greater intensity are employed, such as the Globar and the Nernst glower. The Globar is a rod of sintered silicon carbide about 5 cm long, which is electrically heated to 1300–1700°C. The Nernst glower consists of a mixture of zirconium and yttrium oxides, shaped into a hollow rod about 3 cm long, which is heated to 1500–2000°C. Although these latter sources are more intense than a heated wire, their spectral outputs are still far less in power than the sources commonly used in ultraviolet-visible spectrophotometry.

Sources such as the Globar, the Nernst glower, and the heated wire have spectral outputs which peak at wavelengths of approximately 2 to 3 $\mu$m, falling off rapidly toward longer and shorter wavelengths. This output, displayed in Figure 21–5, is

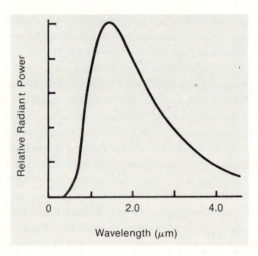

*Figure 21–5.* Spectrum of the radiant power emitted by a typical infrared source, the Nernst glower, at a temperature of 2000°K.

further evidence of the need for a double-beam optical system in infrared spectrophotometry. Such a system, besides correcting for the possible errors noted earlier, helps eliminate the effects of changes in the output power of the source as a function of wavelength. Because the power of the source is low at certain wavelengths, high-quality spectrophotometers frequently incorporate control systems which automatically cause the fraction of radiation reaching the detector to be varied according to wavelength.

Notice in Figure 21–5 that the Globar, the Nernst glower, and the hot wire would not be efficient sources for either the near-infrared or far-infrared regions. Spectrophotometers designed for these spectral regions must be equipped with special sources.

*Dispersing devices for infrared radiation.* Spectral dispersing devices employed in infrared spectrophotometry are very similar in design to those used in the ultraviolet-visible region. However, because glass does not transmit infrared radiation efficiently, lenses and prisms, if they are utilized in the spectral dispersing device, must be fabricated from infrared-transmitting materials such as rock salt or cesium bromide. A list of infrared-transmitting materials and their regions of greatest transmission is presented in Table 21–1. To obviate the need to use the costly and

**Table 21-1.   Some Infrared-Transmitting Materials**

| Material | Wavelength Range of Usable Transmittance in the Infrared Region |
|---|---|
| Fused silica[b] | to   4.0 $\mu$m |
| Sodium chloride | to 17.0 $\mu$m |
| Potassium bromide | to 26.0 $\mu$m |
| Cesium iodide | to 40.0 $\mu$m |
| Calcium fluoride[b] | to 10.0 $\mu$m |
| Barium fluoride[b] | to 13.5 $\mu$m |
| Irtran 2[a] (ZnS)[b] | to 13.0 $\mu$m |
| Silver chloride[b] | to 25.0 $\mu$m |

[a] Trademark of Eastman Kodak Company.
[b] Relatively insensitive to water and water vapor.

fragile materials listed in Table 21–1, modern monochromators designed for infrared spectrometry employ reflecting rather than transmitting (refracting) optics. Gratings and curved mirrors are therefore much more common in modern instruments than are prisms and lenses. In some applications of infrared absorption, the spectral dispersing device is eliminated entirely.

*Infrared detectors.*   Like sources, infrared detectors are useful only over certain wavelength ranges. For the fundamental infrared region (2.5 to 50 $\mu$m) most often employed in analysis, common detectors are thermocouples or those of the solid-state or pneumatic variety. Solid-state detectors and thermocouples detect infrared radiation by sensing the heat produced when the radiation is absorbed by a blackened surface. Pneumatic detectors operate by sensing the increase in pressure which occurs when a gas is warmed by the incoming radiation. All these detectors have a relatively low sensitivity. This, coupled with the low intensity of the usual infrared sources, would seem to make infrared spectrophotometers relatively insensitive devices. However, this problem is circumvented by allowing the instrument a relatively long time (5 to 15 minutes) to examine each sample. This tradeoff between sensitivity and response time is common in chemical instrumentation and can often be used to advantage whenever it is necessary to obtain increased sensitivity or speed of analysis. Fortunately, the recent introduction of new, pyroelectric detectors, available on a few instruments, now makes it possible to obtain infrared spectra at high sensitivity in a relatively short time.

## Non-Dispersive Infrared Photometers

Although the analog of the simple ultraviolet-visible filter photometer does not exist for infrared work, non-dispersive systems are often used for the infrared analysis of specific substances. These simple, relatively inexpensive instruments have found increasing use in the analysis of process streams and atmospheric contaminants. An example of such a system is shown in Figure 21–6. In this device, infrared radiation from a suitable source is directed by means of two mirrors, $M_1$ and $M_2$, to two identical sample cells. The sample of interest is placed or passed into one cell, while the other cell contains a reference material. In the measurement of atmospheric contaminants, the reference cell would probably contain a sample of pure air. Radiation transmitted by the sample and reference cells falls upon two matched detectors, $D_1$ and $D_2$. Signals from these two detectors are sent to a suitable readout device, such as a meter or recorder, which displays their difference. This difference

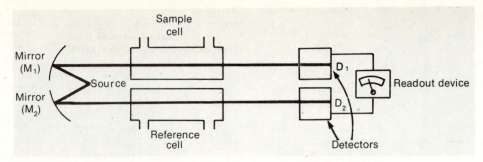

*Figure 21–6.* Schematic diagram of a non-dispersive infrared photometer.

signal represents the radiation transmitted by the sample, corrected for background absorption.

### Handling of Samples for Infrared Spectrophotometry

Infrared spectrophotometry can be used to examine samples in the liquid, solid, or gaseous state. Of course, the methods of sample handling and the nature of the spectrophotometer cells will vary depending upon the state of the sample. Because liquid samples are most common, let us first consider the procedures for handling them.

*Liquid samples.* Because of the large number of absorption bands commonly found in infrared spectra, multicomponent analysis is inconvenient. Instead, individual compounds are isolated whenever possible and analyzed by themselves. Also, because there are no known solvents that are entirely free of absorption bands in the infrared region, liquid samples are preferably examined for qualitative analysis in their "neat" (pure) form. Because of the high absorptivities of most samples in the infrared region, the analysis of pure samples necessitates the use of absorption cells having extremely short path lengths, usually between 0.01 and 1 mm.

A cell for liquids generally consists of two infrared-transmitting plates, held at constant spacing, between which the sample is introduced. The plates are often made of crystalline sodium chloride, or one of the other infrared-transmitting materials listed in Table 21–1, and they must be ground or polished for maximum transmission of radiation. One form of infrared cell for liquid samples is shown in Figure 21–7. To introduce samples into this cell, a syringe is used. To clean the cell, the sample is removed, the cell is purged with a suitable solvent, and the cell is dried by blowing dry nitrogen through it. Because of the water-soluble nature of most cell windows, air should never be drawn through a cell to dry it, for this procedure tends to cool the cell windows and promote the condensation of atmospheric water vapor on them.

Because of their water-soluble character, most infrared cells must be stored and handled under anhydrous conditions. Fortunately, special (although rather expensive) materials, such as silver chloride and infrared-transmitting plastics (Irtran), which are impervious to water can be used for the windows of cells for aqueous samples. When possible, aqueous samples are avoided in infrared spectrophotometry because of the strong absorption of water molecules in this region.

Special variable-path-length and demountable cells are available for infrared spectrophotometry. Variable-path-length cells are especially useful when a large number of samples having widely different absorptivities are to be determined quantitatively. Variable-path cells are more costly than conventional cells, however,

*Figure 21–7.* A typical infrared cell for holding liquid samples. Courtesy of Barnes Engineering Company, Stamford, Connecticut.

and are therefore less frequently used. By contrast, demountable cells are relatively inexpensive and are excellent for the examination of extremely viscous samples which are difficult to introduce into regular fixed cells; a drop of the viscous liquid can be placed onto one of two transmitting plates, and the plates can be pressed together and mounted in a special holder. Of course, the path length of such a demountable cell is not known and must be determined through calibration if quantitative work is to be performed.

To determine the path lengths of infrared cells, two techniques are available. In the first procedure, a pure liquid having a known absorptivity is placed in the cell; from the measured absorbance, the path length ($b$) is calculated from Beer's law. The second technique involves a measurement of interference fringes produced by multiple reflections of infrared radiation from the windows of an empty cell. This technique requires that the cell windows (plates) be flat and parallel. If these criteria are met, the empty cell is introduced into an infrared spectrophotometer and its "spectrum" is recorded. This "spectrum" will appear as a series of waves (interference fringes) whose spacing is dependent upon the separation of the plates in the cell. Then the number of interference fringes ($n$) between wavenumbers $\tilde{\nu}_1$ and $\tilde{\nu}_2$ is counted, and the path length is calculated from the relation

$$b = \frac{n}{2}\left(\frac{1}{\tilde{\nu}_1 - \tilde{\nu}_2}\right) \qquad (21\text{–}6)$$

As an example of the latter procedure, consider Figure 21–8 which shows the pattern of fringes produced by an empty cell. Counting the interference fringes, we see that there are 10 peaks or valleys between 1918 and 1150 cm$^{-1}$. From the preceding equation, we have

$$b = \frac{10}{2}\left(\frac{1}{1918 - 1150}\right) = 0.0065 \text{ cm} \qquad (21\text{–}7)$$

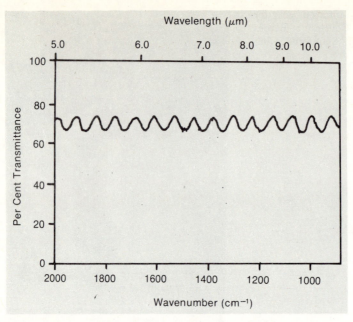

*Figure 21–8.* Interference fringe pattern produced by the reflection of radiation from the flat, parallel windows of an empty infrared cell with a path length ($b$) of 0.0065 cm.

A less convenient but more precise method for determining the path length $b$ from the pattern of interference fringes is to plot the number of fringes ($n$) counted from some arbitrary starting point versus the wavenumber ($\tilde{\nu}$). The slope ($\Delta n / \Delta \tilde{\nu}$) of the resulting line will then be equal to $2b$, as can be seen from the preceding equations.

*Gaseous samples.* Gases to be analyzed by means of infrared spectrophotometry are placed into sample cells similar in design to those used for liquids but having much longer path lengths (commonly 5 or 10 cm). A picture of an infrared cell for

*Figure 21–9.* A typical cell having a path length of 5 cm for the examination of gaseous samples by means of infrared spectrophotometry. Courtesy of Barnes Engineering Company, Stamford, Connecticut.

gases appears in Figure 21–9. As with liquid sample cells, gas cells must have windows transparent to infrared radiation. However, gas cells must also be sealed to prevent any leakage. In order to investigate the infrared absorption of minor components in a gas, cells having extremely long effective path lengths (between 1 and 100 meters) are occasionally used. To obtain such a long effective path length, the ends of these cells are replaced by polished, mirrored surfaces which reflect the infrared radiation back and forth through the sample.

*Solid samples.* The easiest way to handle a solid in infrared spectrophotometry is to dissolve it in a suitable solvent and to employ a sample cell used for liquids. A number of solvents can be used for the dissolution of solid samples, although all exhibit some absorption in the infrared region. Three of the most common solvents are carbon tetrachloride, chloroform, and carbon disulfide, the infrared absorption spectra of which are displayed in Figure 21–10. Carbon tetrachloride and carbon disulfide complement each other as solvents for solid samples. Regions in which carbon tetrachloride absorbs strongly are not the same as those in which carbon disulfide exhibits its maximum absorption. Therefore, if a sample is dissolved in each solvent and if infrared spectra are recorded, the absorption bands of the sample can be observed without strong solvent interferences by reference to one or the other of the spectra. Lists of other useful solvents and the infrared regions in which they absorb appreciably can be found in the references at the end of this chapter.

Two procedures exist for the handling of solids that are not soluble in any common solvents. Both techniques involve grinding the solid sample into particles smaller than the wavelength of infrared radiation in an effort to reduce scattering. The pulverized sample is then incorporated into a suitable matrix, transparent to infrared radiation.

In the **pressed-pellet** technique, a pulverized solid sample is intimately mixed with a quantity of highly purified and desiccated salt powder. Potassium bromide is the salt most commonly used, although cesium bromide and potassium iodide are occasionally employed. The sample-salt mixture is placed in a specially designed die and subjected to extremely high pressures (perhaps 80,000 psig) under vacuum. This produces a highly transparent disk which can be inserted directly into the sample compartment of an infrared spectrophotometer. The pressed-pellet technique can produce surprisingly good spectra for qualitative purposes, but quantitative analysis is hindered by the difficulty of producing pellets of constant and known thickness and by the scattering of infrared radiation due to particles of excessive size.

In the preparation of a **mull**, the pulverized solid sample is mixed and further ground with a viscous, infrared-transparent liquid until a thin paste is formed. This paste is spread between two sodium chloride plates which comprise the sample cell. Viscous liquids most often used in the preparation of a mull are Nujol and Fluorolube. These oils are complementary in their infrared transmission characteristics; Nujol is a hydrocarbon mineral oil having characteristic C—H vibrational bands, whereas Fluorolube is a saturated fluorocarbon that exhibits no C—H absorption. This technique ordinarily yields poor quantitative results, because of the difficulties in preparing a mull of known concentration and in spreading a mull of known thickness between the two plates.

## Special Sample-Handling Techniques

Self-supporting polymer films are relatively easy to analyze by means of infrared spectrophotometry, because they can be placed directly in the sample beam of a

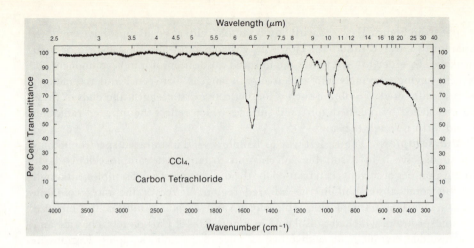

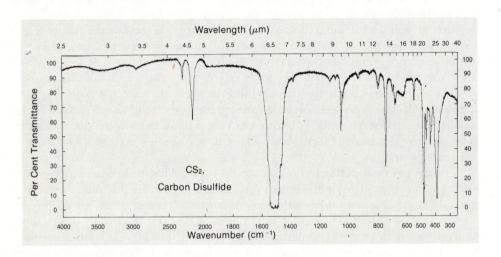

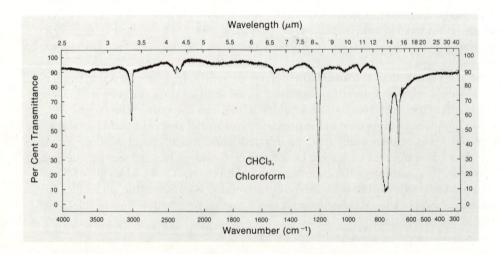

*Figure 21–10.* Infrared transmission spectra for carbon tetrachloride, carbon disulfide, and chloroform, three common solvents used in infrared spectrophotometry. For all spectra, the upper horizontal scale represents wavelength ($\mu$m), whereas the lower horizontal scale is in wavenumbers.

spectrophotometer. On the other hand, for polymer films attached to non-transparent substrates, and for certain other samples, a technique termed **attenuated total reflectance spectrometry** has been developed.

Attenuated total reflectance is based upon the total reflection of a beam of radiation from the boundary or interface between two media of different refractive indexes. If a beam of radiation is traveling through a medium of relatively high refractive index and encounters a medium of lower refractive index, the beam can be reflected without loss from the interface between the two media. However, in this reflection, the beam travels a very short distance into the medium of lower refractive index. If a chemical sample is placed at this interface, the infrared radiation will penetrate the sample slightly and experience some absorption. Therefore, the reflected beam will be altered in its spectral characteristics in a manner similar to that observed in conventional absorption measurements. The spectrum of reflected radiation can then be analyzed to provide information similar to that obtained from a conventional transmission spectrum. One type of system for attenuated total reflectance spectrometry is portrayed in Figure 21–11.

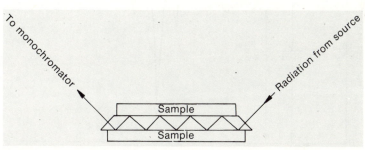

*Figure 21–11.* Attenuated total reflectance cell used to produce multiple encounters between the sample and the beam of radiation.

A number of commercial systems are available which permit cells for the attenuated total reflectance technique to be placed in the sample compartment of an infrared spectrophotometer. Some of these systems involve only a single reflection of the infrared radiation, whereas others involve multiple reflections and, consequently, offer greater sensitivity. Because of these different experimental arrangements, attenuated total reflectance (ATR) spectrometry masquerades under several other names, including multiple internal reflection (MIR), frustrated internal reflectance (FIR), and frustrated multiple internal reflection (FMIR) spectrometry.

Attenuated total reflectance spectrometry has been especially useful for the analysis of polymer films on cans, paint coatings, and other non-transparent samples. Because the depth of penetration of the infrared radiation into the sample is small, this technique is especially sensitive to the surface characteristics of a material; for this reason, attenuated total reflectance spectrometry can be used in the study of adsorbed films.

## Infrared Quantitative Analysis

As in the case of ultraviolet-visible spectrophotometry, quantitative analysis in the infrared region is based upon Beer's law. However, there are some problems in infrared spectrophotometry that are worthy of mention.

In the analysis of mixtures of chemical compounds, it is quite common to encounter situations in which there are overlapping absorption bands. Therefore, the technique for the analysis of multicomponent systems in ultraviolet-visible spectrophotometry (page 645) must be utilized. If there are $n$ components in a sample, $n$ measurements of absorbance at $n$ different wavelengths must be made. Solving the $n$ resulting simultaneous equations then makes it possible to compute the concentration of each component. Of course, the absorptivities of all species must be known at each wavelength selected for the measurements.

In infrared spectrophotometry, even the determination of a value for absorbance can pose some difficulty. Let us examine the infrared spectrum shown in Figure 21–12.

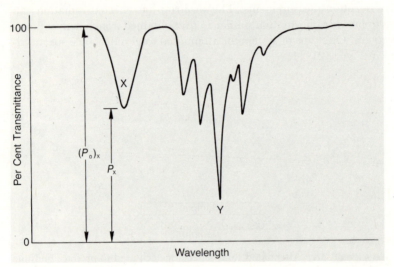

*Figure 21–12.* Hypothetical infrared spectrum exhibiting two bands (X and Y) whose effective transmittance values are to be determined; $(P_0)_X$ and $P_X$ are the effective values for the incident and transmitted radiant power, respectively, for band X.

Because the base line for band X is at precisely 100 per cent transmittance, it is a simple matter to evaluate the absorbance, $A_X$, for band X from the relation

$$A_X = \log \frac{(P_0)_X}{P_X} \qquad (21\text{–}8)$$

where $(P_0)_X$ is the per cent transmittance at the base line and $P_X$ is the per cent transmittance at the minimum of the band. However, the base line for band Y is not at 100 per cent transmittance; in addition, this band is superimposed on one or more other bands, so that the base line is uncertain. Two procedures have been devised to determine the absorbance value for a band such as Y.

Figure 21–13 illustrates the so-called **base-line method** for the determination of $P_0$ and $P$ for a band such as Y. In a practical situation, a series of samples of known concentration could be prepared, the infrared spectrum recorded for each sample, and the values of $P_0$ and $P$ measured for each spectrum by means of the base-line method. Then the absorbance ($A$) for each sample could be computed from equation (21–8); finally, a working curve, consisting of a plot of absorbance versus concentration, could be constructed for use in the analysis of an unknown.

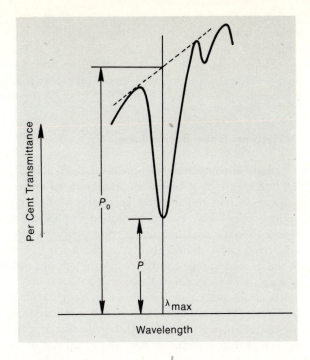

*Figure 21–13.* Illustration of the *base-line method* for the determination of the effective transmittance of a band such as band Y in Figure 21–12; $\lambda_{max}$ is the wavelength at the peak of the absorption band. (See text for discussion.)

A somewhat different way to evaluate the absorbance for a band such as Y in Figure 21–12 is the **empirical-ratio method**. This approach is useful when the absorption band of interest lies on the side of another band known to be caused either by a major constituent of the sample or by an internal standard. Figure 21–14 shows how $P_0$ and $P$ are measured. In the example shown, $P_0$ was taken as the per cent transmittance at the maximum of the main absorption band; this sharply defined

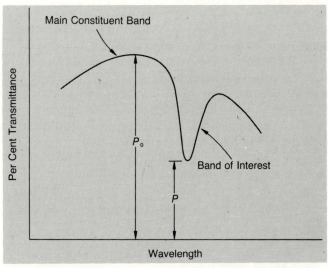

*Figure 21–14.* Hypothetical infrared absorption spectrum, showing the *empirical-ratio method* for the determination of the effective transmittance of a desired band which lies on the side of a band due to a major sample constituent. (See text for discussion.)

point helps to offset any drift in the wavelength setting of the spectrophotometer. As before, a calibration curve can be prepared from data obtained for a set of standard samples.

Using these techniques and exercising due care, one can perform quantitative analyses by means of infrared spectrophotometry at precision levels between 1 and 5 per cent.

## Deviations from Beer's Law

Both instrumental and chemical deviations from Beer's law are common in infrared spectrophotometry. Because infrared sources are not especially intense and detectors are not very sensitive, spectral resolution is frequently sacrificed for increased optical speed. This loss of resolution implies the use of broad-band rather than monochromatic radiation and a consequent deviation from Beer's law. Accordingly, a working curve should always be constructed over the entire concentration range of the desired sample constituent. Although the resulting curve might be nonlinear, quantitative results can still be obtained.

Certain molecular vibrations are strongly affected by the presence of neighboring molecules. This problem is aggravated in infrared spectrophotometry by the common use of concentrated solutions. If the vibrational transition being examined is affected by concentration changes, Beer's law will not be obeyed. As an example, consider the stretching vibration of the —OH group. In a dilute solution, this vibration will be relatively independent of neighboring molecules, provided that a non-polar solvent is employed. However, at higher concentrations, hydrogen bonding between —OH groups will become significant and will lower the —OH stretching frequency. If the —OH stretching band were being used for analysis, this frequency shift would change the apparent absorbance at the observed wavelength by an amount not predicted by Beer's law. The high incidence of these intermolecular effects again makes it mandatory to employ working curves in infrared quantitative analysis.

## RAMAN SPECTROSCOPY

Raman spectroscopy began with the discovery by Sir C. V. Raman in 1928 that light scattered from chemical species is shifted in wavelength (frequency). Since that time, Raman spectroscopy has become a powerful technique for the chemical and physical characterization of substances. However, until the advent of the laser, Raman spectroscopy received little attention as a tool for analysis. Today, Raman techniques provide a viable alternative to infrared spectrophotometry for use in structure determinations, multi-component qualitative analysis, and quantitative measurement of trace constituents.

Although Raman scattering is applicable to the investigation of transitions between rotational, electronic, or vibrational energy levels, we will be concerned primarily with changes in vibrational energy. Vibrational energy changes observed with Raman spectroscopy are similar to those discussed at the beginning of this chapter.

### The Raman Effect

To understand the Raman effect, we must recall the discussion of scattering presented in Chapter 18 (page 608). Rayleigh scattering, as applied to molecules, involves the partial polarization of electrons in a molecule in resonance with a passing electromagnetic wave. The electric field produced by the polarized molecule oscillates at the same frequency as the passing wave, so that the molecule acts as a source of radiation itself, sending out radiation of that frequency in all directions. Of course, this argument assumes that the polarizability* of the molecule remains constant. If the polarizability of the molecule should change, the intensity of the scattered radiation would vary in a manner which corresponds to this change in polarizability.

Let us suppose that a molecule undergoes a change in polarizability during one of its normal modes of vibration. Such a normal mode provides the basis for the Raman effect and is said to be **Raman active**. As the polarizability of the molecule increases and decreases during this vibration, the amplitude of the scattered radiation will change, as shown in Figure 21–15, at a frequency equal to that of the molecular vibration.

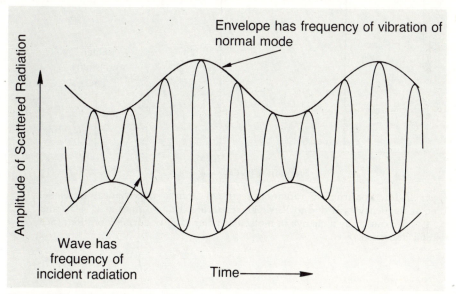

*Figure 21–15.* Waveform of radiation scattered from a molecule while the molecule undergoes a periodic change in polarizability during, for example, a vibration. (See text for discussion.)

Even if the incident radiation is of one frequency (monochromatic), the Raman-scattered radiation will not be. As discussed in Chapter 18 (page 603), truly monochromatic radiation can be represented by a sinusoidally varying electric (or magnetic) field having a constant peak amplitude. However, the amplitude-modulated wave of Figure 21–15 can be shown to consist of three frequency components—one line at the frequency of the incident, monochromatic radiation, and two components shifted to either side of this line by an amount equal to the frequency of the molecular vibration. This is represented schematically in the spectrum of Figure 21–16. Notice that the frequency ($\nu_v$) of the Raman-active molecular vibration is the difference between the observed frequency of either Raman-shifted line and the frequency of the incident radiation. Information similar to that sought in infrared spectrophotometry is thereby obtained.

---

* Polarizability of a molecule is related to the ease with which charges in the molecule can be separated under the influence of an external electric field.

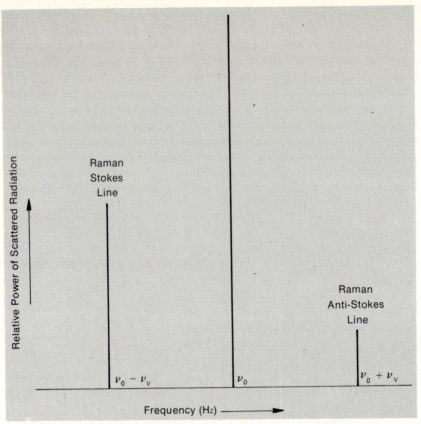

*Figure 21–16.* Spectrum of radiation scattered from a hypothetical molecule. Line at frequency $\nu_0$ is due to Rayleigh scattering. Raman lines at $\nu_0 + \nu_v$ and $\nu_0 - \nu_v$ are caused by a change in molecular polarizability during vibration. (See text for discussion.)

In the preceding classical description of the Raman effect, it would appear that the two Raman-shifted lines would be of equal intensity. Employing a more quantum mechanical approach, we can see that this is not true. Consider a molecule initially in its ground vibrational state, as portrayed in the energy-level diagram of Figure 21–17A. When a beam of radiation of frequency $\nu_0$ impinges upon the molecule, the molecule will be polarized, going into oscillation as described earlier. When oscillating, the molecule has an energy greater than that of the ground vibrational state ($v_0$). The extra energy gained by the molecule in its oscillation is equal to $E_0$, the energy of the incident radiation. It is important to recognize that, although the molecule has been energized by the incident radiation, it is not excited in the conventional sense. That is, the molecule has not been excited to any particular quantized level and remains energized only as long as the passing wave can interact with it. As the wave passes, the polarized molecule ceases to oscillate and returns to an ordinary quantized level. If this quantized level is one other than that originally occupied by the molecule ($v_0$), its energy loss ($E_s$) will be less than the energy ($E_0$) originally gained. As indicated in Figure 21–17A, ($E_0 - E_s$) is equal to the difference in energy ($E_v$) between the ground vibrational level and an excited vibrational state. The energy difference $E_v$ is extracted from the incident radiation, so that the scattered radiation will contain a frequency $\nu_s$ equal to the difference between the frequency

$(\nu_0)$ of the incident radiation and that $(\nu_v)$ of the Raman-active molecular vibration. This frequency $(\nu_s)$ corresponds to the low-frequency line shown in Figure 21–16.

Raman scattering from molecules already in an excited vibrational state is illustrated in Figure 21–17B. In this case, the vibrationally excited molecule is set into oscillation by the incident radiation, so that the molecule is temporarily increased in energy by an amount $E_0$. As the radiation passes, the molecule immediately loses this excess energy, but need not return to the same vibrationally excited state as before. If the molecule returns to the ground vibrational level, the lost energy will be added to that of the scattered radiation, giving it a total energy $(E_s')$ equal to $(E_0 + E_v)$. Therefore, the frequency $(\nu_s')$ of this scattered radiation will equal the sum of the frequencies of the incident radiation $(\nu_0)$ and the molecular vibration $(\nu_v)$. The frequency $\nu_s'$ corresponds to the high-frequency line displayed in Figure 21–16.

From the above considerations, a Raman spectrum should appear as shown in Figure 21–16. Of course, the line at $\nu_0$ corresponds to Rayleigh-scattered radiation, whereas the lines on either side of $\nu_0$ arise from a vibrational Raman effect. Because many more molecules are in the ground vibrational state than in an excited state at room temperature, the process represented in Figure 21–17A is much more probable than that in Figure 21–17B. Therefore, the low-frequency $(\nu_0 - \nu_v)$ Raman line will be considerably more intense than the high-frequency $(\nu_0 + \nu_v)$ line; the more-intense, low-frequency Raman line is called the **Stokes line**, and the less-intense, high-frequency line is named the **anti-Stokes line**.

Although vibrational transitions were considered in this discussion, the Raman effect is equally applicable to any other transition that involves a change in polarizability. Rotational and electronic transitions have been studied by means of Raman spectroscopy. In addition, lattice vibrations in solids and spin-coupled vibrations in magnetic materials have been found to cause Raman-like shifts in scattered monochromatic radiation. However, vibrational Raman scattering is by far the most frequently employed for analytical purposes.

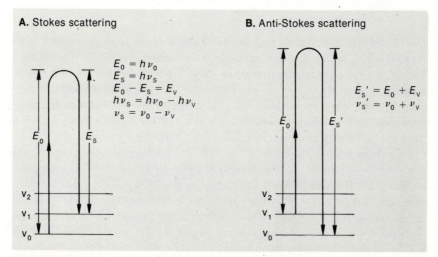

**A. Stokes scattering**

$$E_0 = h\nu_0$$
$$E_s = h\nu_s$$
$$E_0 - E_s = E_v$$
$$h\nu_s = h\nu_0 - h\nu_v$$
$$\nu_s = \nu_0 - \nu_v$$

**B. Anti-Stokes scattering**

$$E_s' = E_0 + E_v$$
$$\nu_s' = \nu_0 + \nu_v$$

*Figure 21–17.* Pictorial diagram of the energy-level changes in a molecule during Raman scattering; $E_0$ is the energy of incident radiation, $E_s$ and $E_g'$ are the respective energies of the Stokes and anti-Stokes scattered radiation, and $v_0$, $v_1$, and $v_2$ are vibrational energy levels. A. Stokes scattering. B. Anti-Stokes scattering. The curved lines denote the temporary gain and loss of energy of the molecule as it is polarized and set into oscillation by the passing electromagnetic wave. (See text for discussion.)

As in Rayleigh scattering and luminescence, the power of Raman-scattered radiation is directly proportional to the power of the incident, polarizing source radiation. Also, like infrared absorption, the transitions which are usually observed by means of Raman spectroscopy involve changes in vibrational levels corresponding to $\Delta v = \pm 1$.

## Raman Spectra

To produce a Raman spectrum, monochromatic radiation is sent into a chemical sample and the spectrum of the scattered radiation is determined. If properly performed, this procedure will yield a spectrum such as that for carbon tetrachloride shown in Figure 21–18. Several features of this spectrum are important. First,

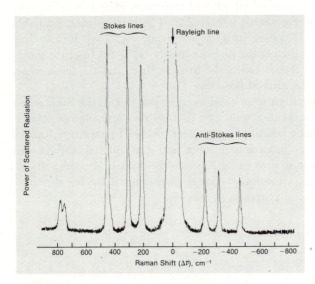

*Figure 21–18.* Raman spectrum of pure liquid carbon tetrachloride, obtained with a helium-neon laser source. The volume of sample was only 3 $\mu$l. The Raman shift ($\Delta \tilde{v}$) is the difference in wavenumbers (cm$^{-1}$) between the Rayleigh line and a Raman-shifted line. (Redrawn, with permission, from the article by B. J. Bulkin: J. Chem. Educ., *46*: A781, 1969.)

notice the relatively small magnitude of the Raman lines compared to that of the Rayleigh line which, in this example, exceeded the range of the readout system. Except in very unusual situations, this disparity will always exist and necessitates the measurement of the relatively weak Raman lines in close proximity to the Rayleigh line. Second, observe the difference in power between the low-frequency Stokes lines and the feeble anti-Stokes lines. Because of their greater magnitude, the Stokes lines are ordinarily employed for analytical purposes.

In Figure 21–18 the horizontal axis of the Raman spectrum is calibrated in wavenumbers (cm$^{-1}$). A wavelength scale is never used in Raman spectroscopy, because the **Raman shift** ($\Delta \tilde{v}$)—that is, the difference in frequency (or wavenumbers) between the Rayleigh line and a Raman-shifted line—can be directly correlated with a molecular vibration frequency or with the position of an infrared absorption band. Notice that the magnitude of the Raman lines in Figure 21–18 is expressed as the power of the scattered radiation, rather than by a ratio such as transmittance. This mode of presentation is adequate for both qualitative and quantitative analysis because, as we shall see, the power of the Raman-scattered radiation is directly proportional to the concentration of scattering species. However, the measurement of these extremely small Raman lines near the enormous Rayleigh peak requires rather specialized instrumentation.

## Instrumentation for Raman Spectroscopy

A Raman spectrometer consists essentially of a source of monochromatic radiation, a sample cell, a frequency selector, a detector, and a suitable readout system, as diagrammed in Figure 21–19. Raman-scattered radiation is ordinarily observed at right angles to the incident source radiation in an effort to minimize the effect of the latter on the detected spectrum. Both the quality and design of each component of a Raman spectrometer are quite critical, so that the system is rather expensive. Because of the importance of each of these components, we will discuss them separately.

*Figure 21–19.* Block diagram of a Raman spectrometer.

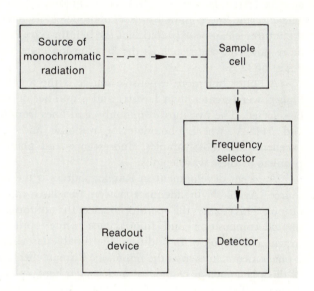

*Raman sources.* Raman-scattered radiation is dwarfed by Rayleigh scattering. Therefore, the primary source of radiation in Raman spectroscopy must be as monochromatic as possible to permit the spectral separation of Raman lines from the Rayleigh line. In addition, the source should be extremely intense in order to produce detectable Raman lines. For structural studies, it is also desirable that the incident radiation be polarized, since an examination of the depolarization of Raman-scattered radiation provides information about molecular symmetry.

The wavelength of the monochromatic source must be chosen with some care. Although the Raman effect is independent of the frequency of incident radiation, there are several practical matters to consider. Raman scattering, like Rayleigh scattering, increases as the fourth power of the frequency of incident radiation. Therefore, it would seem desirable to choose the shortest possible wavelength from the monochromatic source. Unfortunately, at wavelengths below approximately 400 nm, photodecomposition of some samples can occur. In addition, some samples fluoresce when illuminated with short-wavelength radiation. Thus, a compromise must be made between a source wavelength which produces the greatest degree of Raman scattering and one which minimizes photodecomposition and fluorescence. This optimum wavelength will vary from sample to sample. Colored samples present a further problem by absorbing either the incident or Raman-scattered radiation. To avoid this difficulty, the source wavelength should be chosen at a point well removed from the absorption bands of any sample constituents.

Until recently, a high-intensity mercury discharge lamp was the source customarily employed in Raman spectroscopy. Because of the limitations of this source, however, Raman spectroscopy was not widely employed and required extremely elaborate and costly instrumentation. Today, the laser has all but replaced every other Raman source. The monochromaticity, high power, and polarization of laser radiation has caused a rebirth of Raman spectroscopy as an analytical tool.

Two varieties of lasers are in common use as Raman sources. The helium-neon laser, having a radiation output at 6328 Å, is a relatively inexpensive source, although it has limited power (approximately 80 milliwatts). Its monochromatic output is in the red portion of the visible spectrum, so that little fluorescence or photodecomposition need be feared. However, at this relatively long wavelength, Raman scattering is less efficient, and the commonly used photodetectors have rather poor sensitivities. In the future, it can be expected that the helium-neon laser will be largely replaced by more powerful lasers such as the krypton ion laser, whose most useful spectral line lies at 5682 Å.

A better but more expensive source is the argon ion laser. It has a relatively high power (greater than 1 watt) and a number of output spectral lines which can be selected. The two most commonly used lines from the argon ion laser lie at 4880 and 5145 Å, where photodetector response and scattering efficiency are both excellent. For most samples, fluorescence and photodecomposition will not be a problem at these wavelengths.

A recent development in Raman sources is the introduction of a tunable laser source. Available in adequate power, this laser has a selectable, monochromatic output. This enables the operator to select the optimum wavelength needed to prevent photodecomposition and fluorescence while optimizing Raman scattering and instrument response. In addition, the tunable laser makes possible the utilization of a phenomenon known as the resonance Raman effect. Although a detailed discussion of this phenomenon cannot be undertaken here, suffice it to say that the resonance Raman effect is capable of producing up to one-hundred-fold improvements in sensitivity. Instruments designed to take advantage of the resonance Raman effect are just now being introduced commercially.

***Raman sample cells.*** Before the introduction of laser sources to Raman spectroscopy, illumination of the sample by the primary source and collection of the scattered radiation was a difficult problem. However, the availability of laser sources has made possible the development of many new types of sample cells.

Two cell configurations often employed for macro samples are shown in Figure 21–20. In the system of Figure 21–20A, the laser radiation is directed through the bottom of a transparent cell and reflected from a mirror to traverse the cell twice. Scattered radiation is collected perpendicular to this beam and is focused with a lens onto the frequency selector. This arrangement is simple, yet provides adequate sensitivity.

To achieve higher sensitivity, the sample cell shown in Figure 21–20B is sometimes used. Two flat, parallel sides of the cell are coated with a reflecting material so that the laser radiation is reflected back and forth a number of times within the cell. Scattered radiation is again focused upon the frequency selector with the aid of a suitable lens. Although providing greater sensitivity, this arrangement requires an expensive cell and necessitates accurate alignment of the cell each time it is placed into the Raman spectrometer.

Two techniques which have proved successful for the study of ultra-micro samples are diagrammed in Figure 21–21. In the axial excitation-transverse viewing

*Figure 21–20.* Cell configurations used for the examination of macro samples by means of Raman spectrometry. (See text for discussion.)

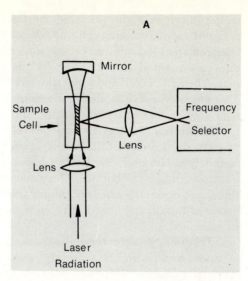

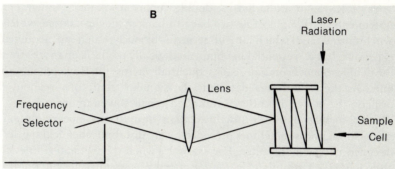

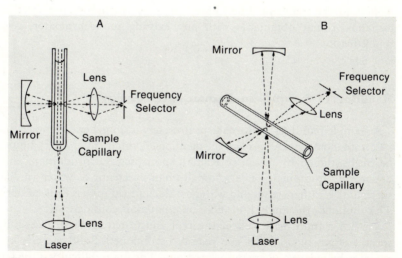

*Figure 21–21.* Sample cell configurations useful for the Raman spectrometric study of ultramicro samples. A. Axial excitation-transverse viewing mode; minimum sample size is 0.04 $\mu$l. B. Transverse excitation-transverse viewing mode; minimum sample size is 0.008 $\mu$l. [Reproduced, with permission, from the article by S. K. Freeman and D. O. Landon in Spex Speaker, Volume XIII, No. 4, December, 1968.]

mode, shown in Figure 21–21A, the laser beam enters the bottom of a specially designed capillary-tubing cell, the bottom of which serves as a lens to collimate* the laser radiation through the cell. The high power-density of the laser radiation produces a great deal of scattering which is efficiently collected by means of a mirror placed behind the cell and a lens situated between the cell and the frequency selector. This microcell provides excellent sensitivity, is simple to use, and can be employed with liquid samples as small as 0.04 $\mu l$.

In the transverse excitation-transverse viewing technique portrayed in Figure 21–21B, the scattered radiation is again collected at right angles to the incident laser beam. However, the laser radiation enters the cell from the side rather than from one end or the bottom. This makes alignment of the cell easier, and permits the use of open-end capillary tubing which can be filled with a liquid sample by means of capillary action. With this arrangement, samples as small as 0.008 $\mu l$ can be examined.

*Frequency selector.* An extremely high-quality frequency selector is necessary to separate the relatively weak Raman-scattered lines from the powerful Rayleigh radiation. To obtain adequate resolution, and to reduce interference from "stray" Rayleigh radiation, a double monochromator is often employed. This system can be thought of as two back-to-back monochromators, which are carefully synchronized and which act in series. Obviously, such an arrangement is quite expensive, but it is essential to obtain analytically useful Raman spectra. In operation, the double monochromator scans sequentially across the Rayleigh line and the Raman lines. As the radiation is detected, the signal is sent to a readout device which displays the power of scattered radiation versus wavenumber.

To further diminish stray radiation and to enhance resolution, the double monochromator is sometimes preceded by an absorbing chemical species or a specially designed narrow-band filter. For example, when the 5145.36 Å line of an argon ion laser is used to stimulate Raman scattering, a small cell containing iodine vapor can be employed to selectively absorb this wavelength and to remove a large fraction of the Rayleigh-scattered radiation. With this aid, sensitivity is enhanced to such an extent that it becomes possible to observe rotational Raman transitions as close as 9 cm$^{-1}$ to the Rayleigh line. This degree of resolution and freedom from stray radiation is not obtainable with any existing monochromator, regardless of its sophistication.

## Handling of Samples in Raman Spectroscopy

Sample handling in Raman spectroscopy is, in general, simpler than in infrared spectrophotometry. As previously mentioned, small (microliter) liquid samples can be easily studied by means of Raman spectroscopy. In addition, aqueous samples are readily examined, because the Raman spectrum of water (unlike its infrared spectrum) is rather weak, and water-soluble optical materials need not be employed. This simplifies the analysis of inorganic materials considerably.

With a multi-path cell of the type illustrated in Figure 21–20B, the Raman spectra of gaseous samples can readily be obtained. Solid samples are also easy to examine by means of Raman spectroscopy. The main difficulty is the large amount of scattering produced by the surfaces of solid particles. This scattering increases the intensity of the Rayleigh peak and renders the examination of Raman peaks more

---

* Collimation means that the laser radiation is rendered parallel so that it travels with nearly equal power throughout the entire length of the sample cell.

difficult. However, the combination of a monochromatic laser source and a double monochromator minimizes these difficulties. Although a number of specially designed holders for solid samples have been developed, the simplest technique merely involves packing the powdered sample into a capillary tube (similar to a melting-point capillary) and exciting the sample in one of the configurations shown in Figure 21–21.

Special sample-handling procedures have been developed for a number of applications. With suitable sample holders or non-fluorescing glass vessels, coatings on substrates, fibers, films, and coarsely ground samples can be examined directly. This freedom from extensive sample preparation gives Raman spectroscopy a distinct advantage over many other techniques such as infrared spectrophotometry. Because the Raman effect is inherently a scattering phenomenon, opaque materials can be characterized if the incident radiation is directed obliquely at the sample; the Raman-scattered radiation is then measured at right angles.

## Quantitative Analysis by Raman Spectroscopy

Before the introduction of laser sources, Raman spectroscopy was not often used for quantitative purposes. However, modern instrumentation has greatly simplified the analysis of multi-component mixtures and the determination of micro quantities of substances.

As indicated earlier, the power ($P_r$) of Raman-scattered radiation is directly proportional to the molar concentration ($c$) of the scattering constituent in a sample:

$$P_r = Jc \tag{21–9}$$

In this equation, $J$ is the specific or molar "intensity" of the Raman line under observation. For accurate results, the measured parameter, $P_r$, should be the integrated power of the Raman line rather than just its magnitude. Integration serves to compensate for the broadening of Raman lines which sometimes occurs with increases in the concentration of the scattering species.

To obtain the highest precision and accuracy, an internal standard is ordinarily employed in quantitative Raman work. For nonaqueous samples, carbon tetrachloride serves this purpose well and aids in compensating for instrumental drift. If the sample is illuminated with monochromatic (but non-absorbable) radiation, $P_r$ can be taken as the ratio of the integrated power of the Raman line for the sample constituent to that of the internal standard.

Raman lines are ordinarily so narrow that overlapping of lines is minimal and base-line correction is unnecessary. However, if a base-line correction is desired, the procedures discussed for infrared spectrophotometry can be employed.

## Applications of Raman Spectroscopy

Raman spectroscopy can be applied to a number of problems not amenable to solution by means of infrared spectrophotometry. Notable examples are in the areas of inorganic analysis and the determination of air pollutants.

*Inorganic analysis.* The ability to handle aqueous solutions in Raman spectroscopy allows the quantitative determination of several complex anions, some of which are difficult to measure by other means. For example, a procedure has been developed for the determination of small quantities of nitrate in the presence of large amounts of nitrite.

The symmetric stretching vibration of the nitrate ion is strongly Raman active and exhibits a Stokes line at 1055 cm⁻¹. Because nitrite is the major constituent of the sample, it is assumed to be present at a constant concentration and serves as an internal standard. If the ratio of the integrated power of the nitrate band at 1055 cm⁻¹ to the integrated power of the nitrite band at 810 cm⁻¹ is measured, a working curve can be constructed by plotting this ratio versus the concentration of nitrate. With this approach, it is possible to determine the nitrate impurity in nitrite samples at levels below 0.2 per cent by weight.

**Remote analysis of air pollutants.** A unique application of Raman spectroscopy to the determination of air pollutants involves the collection and characterization of radiation scattered by distant atmospheric contaminants. The necessary instrumentation is shown schematically in Figure 21–22. To permit the

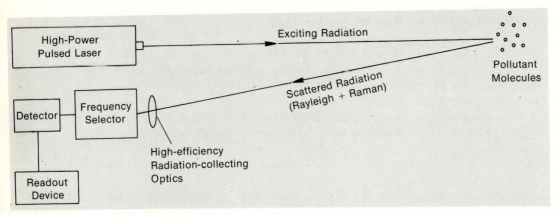

*Figure 21–22.* Schematic diagram of a remote Raman spectrometer used for the detection and mapping of air pollutants. (See text for discussion.)

observation of Raman scattering from remote molecules, an extremely powerful, pulsed laser is used as an excitation source. Radiation scattered from the pollutant molecules then returns to a Raman spectrometer located near the laser. In the spectrometer, the scattered radiation is selected and measured according to frequency, and the Raman spectrum is displayed on a suitable readout device. If a high-efficiency radiation-collecting system is employed on the front end of the frequency selector, pollutants at extremely low concentrations can be detected. Obviously, the minimum detectable concentration depends upon the distance between the remote Raman instrument and the detected pollutants.

By utilizing sophisticated electronics and a time-resolved detector system, this instrument is capable of mapping pollutants as a function of distance from it. This capability is based upon the fixed velocity of electromagnetic radiation. Radiation produced by the laser has a very short pulse width, on the order of tenths of a microsecond. Because this radiation is directed away from the Raman spectrometer, the scattered radiation reaching the spectrometer at any specified time after the laser pulse must have traveled a known round-trip distance. By restricting the observation of the scattered radiation to a very short time interval, the instrument can detect and measure only that radiation scattered by molecules at a known distance from the spectrometer. The resulting Raman spectrum can be used for both qualitative identification and quantitative determination of the pollutants present in that remote region. To map pollutants at different distances from the spectrometer, it is only necessary to vary the delay between production of the laser pulse and the time at which the Raman spectrum is determined. This is most easily accomplished by electronically turning on (gating) the photodetector.

Although the remote Raman spectrometer is a complex instrument, it makes possible the three-dimensional mapping of pollutants from a single, stationary sampling site. The advantages of this method for continuous monitoring of the quality of air are obvious.

## CORRELATION OF MOLECULAR STRUCTURE AND SPECTRA

In the foregoing discussion, we have considered Raman and infrared spectrometry only as they apply to qualitative and quantitative analysis. However, these techniques are also powerful tools for the elucidation of molecular structure and bonding. Because the features of infrared and Raman spectra often can be correlated with the vibration of specific groups, the presence or absence of Raman and infrared bands can be used to identify the substituent groups of a molecule. Taken as a whole, a Raman or infrared spectrum provides a useful molecular "fingerprint" which is unique for a particular species.

Although infrared and Raman bands arise from molecular vibrations, each normal mode of vibration does not necessarily lead to the appearance of a band or line. Vibrations which do not cause a change in dipole moment will not produce infrared bands; those that cause no change in molecular polarizability will not produce Raman lines.

When a molecular vibration is both infrared and Raman active, considerable correlation will exist between the Raman and infrared spectra. However, the magnitudes of the observed bands in the two spectra will often be different. In fact, most vibrations giving rise to intense infrared absorption bands do not produce strong Raman lines. Conversely, strong Raman bands usually arise from molecular vibrations that generate weak infrared signals. For example, the O—H stretching vibration which appears so strongly in an infrared spectrum produces a rather weak Raman peak. In contrast, the vibrations of the S—S bond are observable in a Raman spectrum but do not appear at all in an infrared spectrum. The fact that highly symmetric vibrations, such as those of C=C and C≡C bonds, show up strongly in a Raman spectrum makes Raman spectroscopy particularly suited to the study of skeletal vibrations of organic molecules. Because the vibrations of polar groups are infrared active, infrared spectrophotometry is more useful for the identi-fication of *substituents* on organic molecules.

### Group Frequencies

For the large number of vibrations that are both infrared and Raman active, it is often possible to associate observed bands with the vibrations of specific groups of atoms. Because the vibrations of these groups are largely independent of the nature of the rest of the molecule, the bands tend to appear consistently in the same region of the spectrum. Therefore, by examining an infrared or Raman spectrum, one might be able to determine the presence of specific functional groups on an organic molecule and even to assign tentative identity to the molecule itself. To aid in this assignment, tables and charts of the vibrational frequencies of functional groups have been compiled. Of these charts, one of the most useful is that reproduced in Figure 21–23

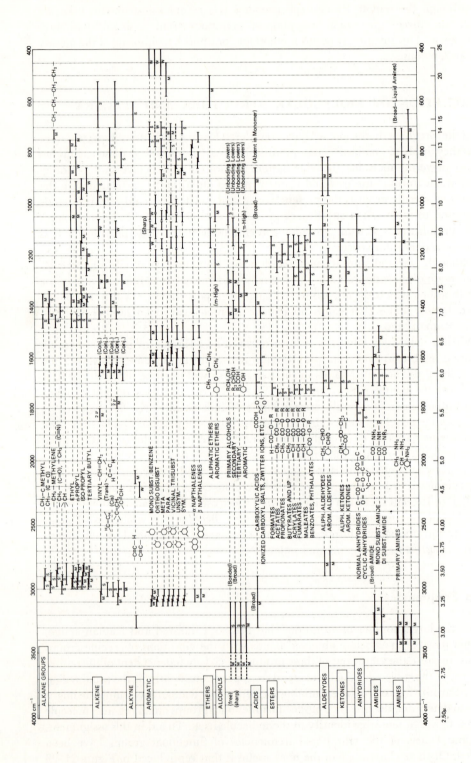

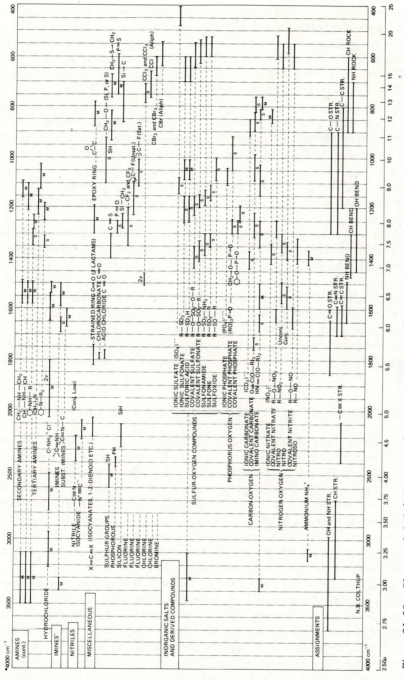

*Figure 21–23.* Characteristic infrared group frequencies. (Reproduced from N. B. Colthup: J. Opt. Soc. Am.: *40*:397, 1950.)

and attributable to Dr. N. B. Colthup. In the Colthup chart, vibrational frequencies of various groups are listed according to their spectral position in wavenumbers (cm$^{-1}$). Each group frequency, it is seen, covers a relatively broad spectral range. This is caused by the finite dependence of the group frequency upon the structure of the rest of the molecule.

Although compilations such as the Colthup chart can be useful for a determination of the presence of functional groups and the general structure of an unknown species, such tables seldom allow an unambiguous identification of the species. The greatest asset in determining the identities of molecular species from their infrared and Raman spectra is vast experience. The subtle relationships between group frequencies and vibrations of a molecule lead to spectral features which can only be understood and utilized after considerable practice.

Without this practical experience, the best correlation between an infrared or Raman spectrum and the structure of a molecule can be obtained by spectral matching, in which the spectrum of the unknown compound is compared to spectra of known materials. Because of the uniqueness of each vibrational "fingerprint," this matching usually provides an unambiguous identification of the unknown. Unfortunately, this procedure is rather time-consuming, even when the number of possibilities for the unknown species is reduced by the identification of certain group frequencies.

To increase the speed of spectral matching, a number of new techniques have been developed. Probably the simplest of these is the manual sorting of punched keysort cards. However, large libraries of standard spectra are now encoded on magnetic tape for use with digital computers. Using these libraries and sophisticated file-searching techniques, a computer can rapidly identify unknown samples or at least reduce the possible choices to a small number which can readily be examined by hand. Coupled with other structure-elucidation techniques, such as nuclear magnetic resonance spectroscopy, these procedures are the mainstay of many industrial laboratories whose function it is to identify unknown organic and inorganic materials.

## COMPARISON OF INFRARED AND RAMAN SPECTROMETRY

Although infrared and Raman spectrometry provide similar information, considerable difference exists in the type of information obtained and in the difficulty of obtaining it. Because of the large library of standard infrared spectra which has been compiled, infrared spectrophotometry is still more heavily employed in elucidating the structures of organic compounds. Of course, this attractiveness is enhanced by the strong infrared absorption of organic functional groups. In contrast, Raman spectroscopy is often the technique of choice for the examination of inorganic substances whose infrared spectra are difficult to interpret and whose spectra are most convenient to obtain in aqueous solution.

Until recently, a major deterrent to the use of Raman spectroscopy was its cost. Raman spectrometers ordinarily cost several times as much as infrared instruments of comparable quality. However, the recent introduction of inexpensive, table-top Raman instruments should reverse this trend. The relative simplicity of Raman spectra and the sensitivity of Raman spectroscopy as a method of quantitative analysis ensure the future of this technique as an alternative to infrared spectrophotometry. In novel applications such as remote Raman spectroscopy discussed earlier, Raman techniques offer exceptional promise. Also, in applications involving the observation of short-lived or transient species, Raman spectroscopy is bound to gain in utility.

## QUESTIONS AND PROBLEMS

1. What will be the wavelength (in $\mu$m) and wavenumber (in cm$^{-1}$) of the infrared radiation which can be absorbed by a molecule with the following normal-mode vibrational frequencies: (a) $7.236 \times 10^{13}$ Hz, (b) $2.459 \times 10^{14}$ Hz, (c) $4.785 \times 10^{13}$ Hz, (d) $1.096 \times 10^{14}$ Hz, (e) $1.127 \times 10^{14}$ Hz, (f) $3.279 \times 10^{13}$ Hz, (g) $9.431 \times 10^{13}$ Hz, and (h) $6.003 \times 10^{13}$ Hz.

2. Using the Hooke's law model for the vibration between two atoms, calculate the resonant vibrational frequency (in Hz) of each of the following diatomic groups from the force constants $(k)$ given:
   (a) a $C{=}O$ bond in formaldehyde ($k = 12.3 \times 10^5$ dynes/cm)
   (b) a $C{-}C$ bond in benzene ($k = 7.6 \times 10^5$ dynes/cm)
   (c) a $C{\equiv}N$ bond in $CH_3CN$ ($k = 17.5 \times 10^5$ dynes/cm)
   (d) a $C{-}C$ bond in $C_2H_6$ ($k = 4.5 \times 10^5$ dynes/cm)
   (e) a $C{-}H$ bond in $C_2H_6$ ($k = 5.1 \times 10^5$ dynes/cm)

3. What information can you infer from the relative magnitudes of the force constants and vibrational frequencies of the diatomic groups in problem 2?

4. Which of the normal-mode vibrations listed in problem 2 would you expect to be Raman active? Which would be infrared active? For those which are Raman active, calculate the Raman shift for each corresponding Raman line. For those which are infrared active, calculate the expected wavelength (in $\mu$m) where the corresponding infrared absorption band would appear.

5. Give the wavenumber (in cm$^{-1}$) and wavelength (in $\mu$m) of the first and second harmonic vibrations (overtones) for each of the vibrations of problem 2. Are all of these harmonic vibrations at frequencies which correspond to the infrared spectral region? If not, which lie outside the infrared region? In what region do they lie?

6. A transition between two vibrational energy levels of a molecule produces an infrared absorption peak centered at 6.43 $\mu$m.
   (a) Calculate the wavenumber of this band (in cm$^{-1}$) and the frequency of the vibration (in Hz).
   (b) Calculate the energy separation ($\Delta E$) between the two vibrational levels.
   (c) From the Boltzmann relation given in Chapter 20 (page 683), calculate the ratio of the number of molecules in the lower of these two vibrational levels to the number in the upper level at 300°K (room temperature). Assume that the statistical weights $(g_u)$ of the vibrational levels are equal.
   (d) From your answer to part (c), what can you say about the relative magnitudes of the Raman Stokes and anti-Stokes lines for this transition?

7. The interference-fringe patterns shown on page 748 were obtained for several empty cells. Calculate the path length $(b)$ of each cell.

8. Because Beer's law involves a ratio,

$$\log \frac{P_0}{P} = abc$$

it would seem at first glance that the precision of the measured absorbance in infrared spectrophotometry would be independent of the source power. Why is this not true?

9. Comment on the effect of material suspended in a sample that is to be quantitatively analyzed by means of both infrared spectrophotometry and Raman spectroscopy. Is it possible to compensate for the error caused by scattering, reflection, and so on? If so, how? If not, why not?

10. Explain how you would set up a non-dispersive infrared photometer for the analysis of nitrogen oxides in the atmosphere. What would you place in the reference compartment? Would you need a filter of any sort? If so, describe it. How would you eliminate the interference of another atmospheric contaminant whose determination is not sought but which has infrared absorption bands overlapping those of the nitrogen oxides? (For assistance, consult the references at the end of the chapter.)

11. Comment on the applicability of the non-dispersive infrared photometer to the determination of ozone in the atmosphere. Why is the technique unlikely to be successful?

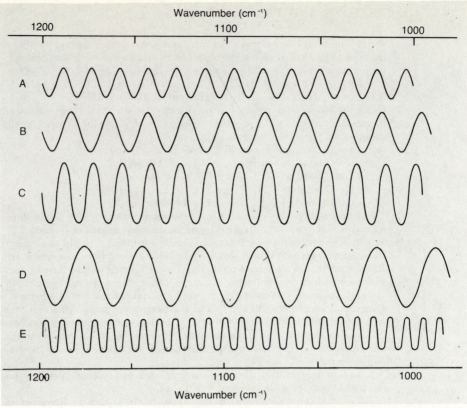

Wavenumber (cm$^{-1}$)

12. In the attenuated total reflectance spectrum of polyethylene shown below, calculate the effective transmittance and absorbance values for bands A, B, C, D, and E by using the base-line method. Which band would you choose for a quantitative determination? Explain your choice.

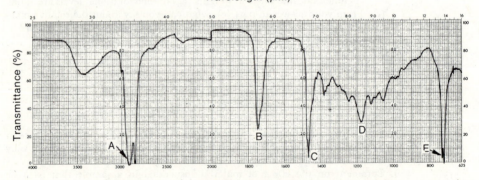

# SUGGESTIONS FOR ADDITIONAL READING

*Books*

1. N. L. Alpert, W. E. Keiser, and H. A. Szymanski: *Theory and Practice of Infrared Spectroscopy*. Plenum Publishing Company, New York, 1970.
2. N. B. Colthup: Infrared and Raman spectroscopy. *In* T. H. Gouw, ed.: *Guide to Modern Methods of Instrumental Analysis*. Wiley-Interscience, New York, 1972, pp. 195:229.
3. C. E. Meloan: *Elementary Infrared Spectroscopy*. The Macmillan Company, New York, 1963.
4. W. J. Potts: *Chemical Applications of Infrared Spectroscopy*. Volume 1, John Wiley and Sons, New York, 1963.
5. C. N. R. Rao: *Chemical Applications of Infrared Spectroscopy*. Academic Press, New York, 1963.
6. J. E. Stewart: *Infrared Spectroscopy ; Experimental Methods and Techniques*. Dekker, New York, 1970.

*Papers*

1. B. J. Bulkin: Raman spectroscopy. J. Chem. Educ., *46*:A781, 1969; *46*: A859, 1969.
2. J. R. Fontana, G. Hassin, and B. E. Kincaid: Remote sensing of molecular pollutants in the atmosphere by stimulated Raman emission. Nature (London), *234*:292, 1971.
3. H. J. Hendra and C. J. Vear: Laser Raman spectroscopy: A review. Analyst, *95*:321, 1970.
4. R. E. Hester: Anal. Chem., *44*:490R, 1972. A biannual review.
5. D. E. Irish and H. Chen: The application of Raman spectroscopy in chemical analysis. Appl. Spectrosc., *25*:1, 1971.
6. R. S. McDonald: Anal. Chem., *44*:241R, 1972. A biannual review.
7. H. J. Sloane: The technique of Raman spectroscopy: A state-of-the-art comparison to infrared. Appl. Spectrosc., *25*:430, 1971.
8. The Spex Speaker, available from Spex Industries, Inc., Box 798, Metuchen, New Jersey, especially the following:
   (a) Volume XI, No. 1, April, 1966: Laser Raman Spectroscopy.
   (b) A. Weber, Volume XI, No. 4, December, 1966: Raman Spectroscopy Revisited.
   (c) S. K. Freeman and D. O. Landon, Volume XIII, No. 4, December, 1968: Microsampling Techniques in Laser Raman Spectroscopy.
   (d) Volume XV, No. 3, September, 1970: Laser Raman Notes.

# APPENDIX 1

## SOLUBILITY PRODUCT CONSTANTS*

**(All values are valid at or near room temperature)**

| Substance | Formula | $K_{sp}$ |
|---|---|---|
| Aluminum hydroxide | $Al(OH)_3$ | $2 \times 10^{-32}$ |
| Barium arsenate | $Ba_3(AsO_4)_2$ | $7.7 \times 10^{-51}$ |
| Barium carbonate | $BaCO_3$ | $8.1 \times 10^{-9}$ |
| Barium chromate | $BaCrO_4$ | $2.4 \times 10^{-10}$ |
| Barium fluoride | $BaF_2$ | $1.7 \times 10^{-6}$ |
| Barium iodate | $Ba(IO_3)_2 \cdot 2H_2O$ | $1.5 \times 10^{-9}$ |
| Barium oxalate | $BaC_2O_4 \cdot H_2O$ | $2.3 \times 10^{-8}$ |
| Barium sulfate | $BaSO_4$ | $1.08 \times 10^{-10}$ |
| Beryllium hydroxide | $Be(OH)_2$ | $7 \times 10^{-22}$ |
| [a]Bismuth hydroxide | $BiOOH$ | $4 \times 10^{-10}$ |
| Bismuth iodide | $BiI_3$ | $8.1 \times 10^{-19}$ |
| Bismuth phosphate | $BiPO_4$ | $1.3 \times 10^{-23}$ |
| Bismuth sulfide | $Bi_2S_3$ | $1 \times 10^{-97}$ |
| Cadmium arsenate | $Cd_3(AsO_4)_2$ | $2.2 \times 10^{-33}$ |
| Cadmium hydroxide | $Cd(OH)_2$ | $5.9 \times 10^{-15}$ |
| Cadmium oxalate | $CdC_2O_4 \cdot 3H_2O$ | $1.5 \times 10^{-8}$ |
| Cadmium sulfide | $CdS$ | $7.8 \times 10^{-27}$ |
| Calcium arsenate | $Ca_3(AsO_4)_2$ | $6.8 \times 10^{-19}$ |
| Calcium carbonate | $CaCO_3$ | $8.7 \times 10^{-9}$ |
| Calcium fluoride | $CaF_2$ | $4.0 \times 10^{-11}$ |
| Calcium hydroxide | $Ca(OH)_2$ | $5.5 \times 10^{-6}$ |
| Calcium iodate | $Ca(IO_3)_2 \cdot 6H_2O$ | $6.4 \times 10^{-7}$ |
| Calcium oxalate | $CaC_2O_4 \cdot H_2O$ | $2.6 \times 10^{-9}$ |
| Calcium phosphate | $Ca_3(PO_4)_2$ | $2.0 \times 10^{-29}$ |
| Calcium sulfate | $CaSO_4$ | $1.9 \times 10^{-4}$ |
| Cerium(III) hydroxide | $Ce(OH)_3$ | $2 \times 10^{-20}$ |
| Cerium(III) iodate | $Ce(IO_3)_3$ | $3.2 \times 10^{-10}$ |
| Cerium(III) oxalate | $Ce_2(C_2O_4)_3 \cdot 9H_2O$ | $3 \times 10^{-29}$ |
| Chromium(II) hydroxide | $Cr(OH)_2$ | $1.0 \times 10^{-17}$ |
| Chromium(III) hydroxide | $Cr(OH)_3$ | $6 \times 10^{-31}$ |
| Cobalt(II) hydroxide | $Co(OH)_2$ | $2 \times 10^{-16}$ |
| Cobalt(III) hydroxide | $Co(OH)_3$ | $1 \times 10^{-43}$ |
| Copper(II) arsenate | $Cu_3(AsO_4)_2$ | $7.6 \times 10^{-36}$ |
| Copper(I) bromide | $CuBr$ | $5.2 \times 10^{-9}$ |
| Copper(I) chloride | $CuCl$ | $1.2 \times 10^{-6}$ |
| Copper(II) iodate | $Cu(IO_3)_2$ | $7.4 \times 10^{-8}$ |
| Copper(I) iodide | $CuI$ | $5.1 \times 10^{-12}$ |
| Copper(I) sulfide | $Cu_2S$ | $2 \times 10^{-47}$ |
| Copper(II) sulfide | $CuS$ | $9 \times 10^{-36}$ |

* See footnotes at end of table.

| Substance | Formula | $K_{sp}$ |
|---|---|---|
| Copper(I) thiocyanate | $CuSCN$ | $4.8 \times 10^{-15}$ |
| Gold(III) hydroxide | $Au(OH)_3$ | $5.5 \times 10^{-46}$ |
| Iron(III) arsenate | $FeAsO_4$ | $5.7 \times 10^{-21}$ |
| Iron(II) carbonate | $FeCO_3$ | $3.5 \times 10^{-11}$ |
| Iron(II) hydroxide | $Fe(OH)_2$ | $8 \times 10^{-16}$ |
| Iron(III) hydroxide | $Fe(OH)_3$ | $4 \times 10^{-38}$ |
| Lead arsenate | $Pb_3(AsO_4)_2$ | $4.1 \times 10^{-36}$ |
| Lead bromide | $PbBr_2$ | $3.9 \times 10^{-5}$ |
| Lead carbonate | $PbCO_3$ | $3.3 \times 10^{-14}$ |
| Lead chloride | $PbCl_2$ | $1.6 \times 10^{-5}$ |
| Lead chromate | $PbCrO_4$ | $1.8 \times 10^{-14}$ |
| Lead fluoride | $PbF_2$ | $3.7 \times 10^{-8}$ |
| Lead iodate | $Pb(IO_3)_2$ | $2.6 \times 10^{-13}$ |
| Lead iodide | $PbI_2$ | $7.1 \times 10^{-9}$ |
| Lead oxalate | $PbC_2O_4$ | $4.8 \times 10^{-10}$ |
| [b]Lead oxide | $PbO$ | $1.2 \times 10^{-15}$ |
| Lead sulfate | $PbSO_4$ | $1.6 \times 10^{-8}$ |
| Lead sulfide | $PbS$ | $8 \times 10^{-28}$ |
| Magnesium ammonium phosphate | $MgNH_4PO_4$ | $2.5 \times 10^{-13}$ |
| Magnesium arsenate | $Mg_3(AsO_4)_2$ | $2.1 \times 10^{-20}$ |
| Magnesium carbonate | $MgCO_3 \cdot 3H_2O$ | $1 \times 10^{-5}$ |
| Magnesium fluoride | $MgF_2$ | $6.5 \times 10^{-9}$ |
| Magnesium hydroxide | $Mg(OH)_2$ | $1.2 \times 10^{-11}$ |
| Magnesium oxalate | $MgC_2O_4 \cdot 2H_2O$ | $1 \times 10^{-8}$ |
| Manganese(II) hydroxide | $Mn(OH)_2$ | $1.9 \times 10^{-13}$ |
| [c]Mercury(I) bromide | $Hg_2Br_2$ | $5.8 \times 10^{-23}$ |
| [c]Mercury(I) chloride | $Hg_2Cl_2$ | $1.3 \times 10^{-18}$ |
| [c]Mercury(I) iodide | $Hg_2I_2$ | $4.5 \times 10^{-29}$ |
| [d]Mercury(II) oxide | $HgO$ | $3.0 \times 10^{-26}$ |
| [c]Mercury(I) sulfate | $Hg_2SO_4$ | $7.4 \times 10^{-7}$ |
| Mercury(II) sulfide | $HgS$ | $4 \times 10^{-53}$ |
| [c]Mercury(I) thiocyanate | $Hg_2(SCN)_2$ | $3.0 \times 10^{-20}$ |
| Nickel arsenate | $Ni_3(AsO_4)_2$ | $3.1 \times 10^{-26}$ |
| Nickel carbonate | $NiCO_3$ | $6.6 \times 10^{-9}$ |
| Nickel hydroxide | $Ni(OH)_2$ | $6.5 \times 10^{-18}$ |
| Nickel sulfide | $NiS$ | $3 \times 10^{-19}$ |
| Palladium(II) hydroxide | $Pd(OH)_2$ | $1 \times 10^{-31}$ |
| Platinum(II) hydroxide | $Pt(OH)_2$ | $1 \times 10^{-35}$ |
| Radium sulfate | $RaSO_4$ | $4.3 \times 10^{-11}$ |
| Silver arsenate | $Ag_3AsO_4$ | $1 \times 10^{-22}$ |
| Silver bromate | $AgBrO_3$ | $5.77 \times 10^{-5}$ |
| Silver bromide | $AgBr$ | $5.25 \times 10^{-13}$ |
| Silver carbonate | $Ag_2CO_3$ | $8.1 \times 10^{-12}$ |
| Silver chloride | $AgCl$ | $1.78 \times 10^{-10}$ |
| Silver chromate | $Ag_2CrO_4$ | $2.45 \times 10^{-12}$ |
| Silver cyanide | $Ag[Ag(CN)_2]$ | $5.0 \times 10^{-12}$ |
| Silver iodate | $AgIO_3$ | $3.02 \times 10^{-8}$ |
| Silver iodide | $AgI$ | $8.31 \times 10^{-17}$ |
| Silver oxalate | $Ag_2C_2O_4$ | $3.5 \times 10^{-11}$ |
| [e]Silver oxide | $Ag_2O$ | $2.6 \times 10^{-8}$ |

| Substance | Formula | $K_{sp}$ |
|---|---|---|
| Silver phosphate | $Ag_3PO_4$ | $1.3 \times 10^{-20}$ |
| Silver sulfate | $Ag_2SO_4$ | $1.6 \times 10^{-5}$ |
| Silver sulfide | $Ag_2S$ | $2 \times 10^{-49}$ |
| Silver thiocyanate | $AgSCN$ | $1.00 \times 10^{-12}$ |
| Strontium carbonate | $SrCO_3$ | $1.1 \times 10^{-10}$ |
| Strontium chromate | $SrCrO_4$ | $3.6 \times 10^{-5}$ |
| Strontium fluoride | $SrF_2$ | $2.8 \times 10^{-9}$ |
| Strontium iodate | $Sr(IO_3)_2$ | $3.3 \times 10^{-7}$ |
| Strontium oxalate | $SrC_2O_4 \cdot H_2O$ | $1.6 \times 10^{-7}$ |
| Strontium sulfate | $SrSO_4$ | $3.8 \times 10^{-7}$ |
| Thallium(I) bromate | $TlBrO_3$ | $8.5 \times 10^{-5}$ |
| Thallium(I) bromide | $TlBr$ | $3.4 \times 10^{-6}$ |
| Thallium(I) chloride | $TlCl$ | $1.7 \times 10^{-4}$ |
| Thallium(I) chromate | $Tl_2CrO_4$ | $9.8 \times 10^{-13}$ |
| Thallium(I) iodate | $TlIO_3$ | $3.1 \times 10^{-6}$ |
| Thallium(I) iodide | $TlI$ | $6.5 \times 10^{-8}$ |
| Thallium(I) sulfide | $Tl_2S$ | $5 \times 10^{-21}$ |
| [f]Tin(II) oxide | $SnO$ | $1.4 \times 10^{-28}$ |
| Tin(II) sulfide | $SnS$ | $1 \times 10^{-25}$ |
| Titanium(III) hydroxide | $Ti(OH)_3$ | $1 \times 10^{-40}$ |
| [g]Titanium(IV) hydroxide | $TiO(OH)_2$ | $1 \times 10^{-29}$ |
| Zinc arsenate | $Zn_3(AsO_4)_2$ | $1.3 \times 10^{-28}$ |
| Zinc carbonate | $ZnCO_3$ | $1.4 \times 10^{-11}$ |
| Zinc ferrocyanide | $Zn_2Fe(CN)_6$ | $4.1 \times 10^{-16}$ |
| Zinc hydroxide | $Zn(OH)_2$ | $1.2 \times 10^{-17}$ |
| Zinc oxalate | $ZnC_2O_4 \cdot 2H_2O$ | $2.8 \times 10^{-8}$ |
| Zinc phosphate | $Zn_3(PO_4)_2$ | $9.1 \times 10^{-33}$ |
| Zinc sulfide | $ZnS$ | $1 \times 10^{-21}$ |

Although water appears in the formulas of a number of substances, it is not included in the solubility-product expression.

[a] $BiOOH \rightleftharpoons BiO^+ + OH^-$; $K_{sp} = [BiO^+][OH^-]$

[b] $PbO + H_2O \rightleftharpoons Pb^{2+} + 2 OH^-$; $K_{sp} = [Pb^{2+}][OH^-]^2$

[c] All mercury(I) compounds contain the dimeric species $Hg_2^{2+}$. Therefore, the solubility reaction and solubility-product expression are represented in general by:

$$(Hg_2)_mX_n \rightleftharpoons m\,Hg_2^{2+} + n\,X^{-2m/n}; \qquad K_{sp} = [Hg_2^{2+}]^m[X^{-2m/n}]^n$$

[d] $HgO + H_2O \rightleftharpoons Hg^{2+} + 2 OH^-$; $K_{sp} = [Hg^{2+}][OH^-]^2$

[e] $\frac{1}{2} Ag_2O + \frac{1}{2} H_2O \rightleftharpoons Ag^+ + OH^-$; $K_{sp} = [Ag^+][OH^-]$

[f] $SnO + H_2O \rightleftharpoons Sn^{2+} + 2 OH^-$; $K_{sp} = [Sn^{2+}][OH^-]^2$

[g] $TiO(OH)_2 \rightleftharpoons TiO^{2+} + 2 OH^-$; $K_{sp} = [TiO^{2+}][OH^-]^2$

More comprehensive tables of solubility-product data are given in the following references:

1. L. G. Sillén and A. E. Martell: *Stability Constants of Metal-Ion Complexes*. Chem. Soc. (London), Spec. Publ. No. 17 (1964); No. 25 (1971).
2. L. Meites, ed.: *Handbook of Analytical Chemistry*. McGraw-Hill Book Company, New York, 1963.

# APPENDIX 2

EQUILIBRIUM CONSTANTS FOR PROTON-TRANSFER REACTIONS OF ACIDS AND THEIR CONJUGATE BASES IN WATER*

| System | | Proton-Transfer Equilibrium | $pK_a$ | $pK_b$ |
| Acid | Base | (Acid $\rightleftharpoons$ Conjugate Base + H$^+$) | for acid | for conjugate base |
| --- | --- | --- | --- | --- |
| Acetic | | $CH_3COOH \rightleftharpoons CH_3COO^- + H^+$ | 4.76 | 9.24 |
| | $\alpha$-Alanine | $\underset{+NH_3}{CH_3CHCOOH} \rightleftharpoons \underset{+NH_3}{CH_3CHCOO^-} + H^+$ | 2.34 | 11.66 |
| $\alpha$-Alanine | | $\underset{+NH_3}{CH_3CHCOO^-} \rightleftharpoons \underset{NH_2}{CH_3CHCOO^-} + H^+$ | 9.87 | 4.13 |
| | Ammonia | $NH_4^+ \rightleftharpoons NH_3 + H^+$ | 9.26 | 4.74 |
| | Aniline | | 4.60 | 9.40 |
| Arsenic | | $H_3AsO_4 \rightleftharpoons H_2AsO_4^- + H^+$ <br> $H_2AsO_4^- \rightleftharpoons HAsO_4^{2-} + H^+$ <br> $HAsO_4^{2-} \rightleftharpoons AsO_4^{3-} + H^+$ | 2.19 <br> 6.94 <br> 11.50 | 11.81 <br> 7.06 <br> 2.50 |
| [a]Arsenious | | $H_3AsO_3 \rightleftharpoons H_2AsO_3^- + H^+$ | 9.29 | 4.71 |

* See footnotes at end of table.

| Acid | Reaction | $pK_a$ | $pK_b$ |
|---|---|---|---|
| **Benzoic** | $C_6H_5COOH \rightleftharpoons C_6H_5COO^- + H^+$ | 4.20 | 9.80 |
| [b]**Boric** | $H_3BO_3 \rightleftharpoons H_2BO_3^- + H^+$ | 9.24 | 4.76 |
| **Bromoacetic** | $BrCH_2COOH \rightleftharpoons BrCH_2COO^- + H^+$ | 2.90 | 11.10 |
| *n*-**Butanoic** | $CH_3CH_2CH_2COOH \rightleftharpoons CH_3CH_2CH_2COO^- + H^+$ | 4.82 | 9.18 |
| **Carbonic** | $H_2CO_3 \rightleftharpoons HCO_3^- + H^+$ | 6.35 | 7.65 |
|  | $HCO_3^- \rightleftharpoons CO_3^{2-} + H^+$ | 10.33 | 3.67 |
| **Chloroacetic** | $ClCH_2COOH \rightleftharpoons ClCH_2COO^- + H^+$ | 2.86 | 11.14 |
| **Chromic** | $H_2CrO_4 \rightleftharpoons HCrO_4^- + H^+$ | −0.98 | 14.98 |
|  | $HCrO_4^- \rightleftharpoons CrO_4^{2-} + H^+$ | 6.50 | 7.50 |
| **Citric** | $HOOCCH_2\overset{\text{OH}}{\underset{\text{COOH}}{C}}CH_2COOH \rightleftharpoons HOOCCH_2\overset{\text{OH}}{\underset{\text{COO}^-}{C}}CH_2COOH + H^+$ | 3.13 | 10.87 |
|  | $HOOCCH_2\overset{\text{OH}}{\underset{\text{COO}^-}{C}}CH_2COOH \rightleftharpoons HOOCCH_2\overset{\text{OH}}{\underset{\text{COO}^-}{C}}CH_2COO^- + H^+$ | 4.77 | 9.23 |
|  | $HOOCCH_2\overset{\text{OH}}{\underset{\text{COO}^-}{C}}CH_2COO^- \rightleftharpoons {}^-OOCCH_2\overset{\text{OH}}{\underset{\text{COO}^-}{C}}CH_2COO^- + H^+$ | 6.40 | 7.60 |

| System | | Proton-Transfer Equilibrium (Acid ⇌ Conjugate Base + H⁺) | $pK_a$ for acid | $pK_b$ for conjugate base |
|---|---|---|---|---|
| Acid | Base | | | |
| | Diethylamine | $(CH_3CH_2)_2NH_2^+ \rightleftharpoons (CH_3CH_2)_2NH + H^+$ | 10.93 | 3.07 |
| | Dimethylamine | $(CH_3)_2NH_2^+ \rightleftharpoons (CH_3)_2NH + H^+$ | 10.77 | 3.23 |
| | Ethanolamine | $HOCH_2CH_2NH_3^+ \rightleftharpoons HOCH_2CH_2NH_2 + H^+$ | 9.50 | 4.50 |
| | Ethylamine | $CH_3CH_2NH_3^+ \rightleftharpoons CH_3CH_2NH_2 + H^+$ | 10.67 | 3.33 |
| | Ethylenediamine | $^+NH_3CH_2CH_2NH_3^+ \rightleftharpoons NH_2CH_2CH_2NH_3^+ + H^+$ | 7.18 | 6.82 |
| | | $NH_2CH_2CH_2NH_3^+ \rightleftharpoons NH_2CH_2CH_2NH_2 + H^+$ | 9.96 | 4.04 |
| Ethylenediamine-tetraacetic ($H_4Y$) (see page 170 for more information) | | $H_4Y \rightleftharpoons H_3Y^- + H^+$ | 1.99 | 12.01 |
| | | $H_3Y^- \rightleftharpoons H_2Y^{2-} + H^+$ | 2.67 | 11.33 |
| | | $H_2Y^{2-} \rightleftharpoons HY^{3-} + H^+$ | 6.16 | 7.84 |
| | | $HY^{3-} \rightleftharpoons Y^{4-} + H^+$ | 10.26 | 3.74 |
| | | (in 0.1 F KCl, 20°C) | | |
| Formic | | $HCOOH \rightleftharpoons HCOO^- + H^+$ | 3.75 | 10.25 |
| | Glycine | $^+NH_3CH_2COOH \rightleftharpoons {}^+NH_3CH_2COO^- + H^+$ | 2.35 | 11.65 |
| Glycine | | $^+NH_3CH_2COO^- \rightleftharpoons NH_2CH_2COO^- + H^+$ | 9.60 | 4.40 |
| | Hydrazine | $H_2NNH_3^+ \rightleftharpoons H_2NNH_2 + H^+$ | 7.99 | 6.01 |
| Hydrazoic | | $HN_3 \rightleftharpoons N_3^- + H^+$ | 4.72 | 9.28 |
| Hydrocyanic | | $HCN \rightleftharpoons CN^- + H^+$ | 9.21 | 4.79 |
| Hydrofluoric | | $HF \rightleftharpoons F^- + H^+$ | 3.17 | 10.83 |
| Hydrogen sulfide | | $H_2S \rightleftharpoons HS^- + H^+$ | 6.99 | 7.01 |
| | | $HS^- \rightleftharpoons S^{2-} + H^+$ | 14.96 | −0.96 |

| Name | Equilibrium | | |
|------|-------------|------|------|
| **Hydroxylamine** | $HONH_3^+ \rightleftharpoons HONH_2 + H^+$ | 5.98 | 8.02 |
| **Hypochlorous** | $HClO \rightleftharpoons ClO^- + H^+$ | 7.53 | 6.47 |
| **Iodic** | $HIO_3 \rightleftharpoons IO_3^- + H^+$ | 0.79 | 13.21 |
| **Lactic** | $CH_3CHCOOH \rightleftharpoons CH_3CHCOO^- + H^+$ <br> $\quad\quad\;$—OH $\quad\quad\quad$ —OH | 3.86 | 10.14 |
| **Malonic** | $HOOCCH_2COOH \rightleftharpoons HOOCCH_2COO^- + H^+$ <br> $HOOCCH_2COO^- \rightleftharpoons\ ^-OOCCH_2COO^- + H^+$ | 2.86 <br> 5.70 | 11.14 <br> 8.30 |
| **Methylamine** | $CH_3NH_3^+ \rightleftharpoons CH_3NH_2 + H^+$ | 10.72 <br> (in 0.5 $F$ $CH_3NH_3^+NO_3^-$) | 3.28 |
| **Methylethylamine** | $(CH_3)(CH_3CH_2)NH_2^+ \rightleftharpoons (CH_3)(CH_3CH_2)NH + H^+$ | 10.85 | 3.15 |
| **dNitrilotriacetic** <br> **(H₃Y)** | $H_3Y \rightleftharpoons H_2Y^- + H^+$ <br> $H_2Y^- \rightleftharpoons HY^{2-} + H^+$ <br> $HY^{2-} \rightleftharpoons Y^{3-} + H^+$ | 1.89 <br> 2.49 <br> 9.73 | 12.11 <br> 11.51 <br> 4.27 |
| **Nitrous** | $HNO_2 \rightleftharpoons NO_2^- + H^+$ | 2.80 <br> (in 1 $F$ $NaClO_4$) | 11.20 |
| **Oxalic** | $H_2C_2O_4 \rightleftharpoons HC_2O_4^- + H^+$ <br> $HC_2O_4^- \rightleftharpoons C_2O_4^{2-} + H^+$ | 1.19 <br> 4.21 | 12.81 <br> 9.79 |
| **Phenol** | OH— $\bigcirc$ $\rightleftharpoons$ $^-O$— $\bigcirc$ $+ H^+$ | 9.98 | 4.02 |

| System | | Proton-Transfer Equilibrium | $pK_a$ | $pK_b$ |
|---|---|---|---|---|
| Acid | Base | (Acid $\rightleftharpoons$ Conjugate Base + H$^+$) | for acid | for conjugate base |
| **Phosphoric** | | $H_3PO_4 \rightleftharpoons H_2PO_4^- + H^+$ | 2.13 | 11.87 |
| | | $H_2PO_4^- \rightleftharpoons HPO_4^{2-} + H^+$ | 7.21 | 6.79 |
| | | $HPO_4^{2-} \rightleftharpoons PO_4^{3-} + H^+$ | 12.32 | 1.68 |
| **Phosphorous** | | $H_3PO_3 \rightleftharpoons H_2PO_3^- + H^+$ | 1.29 (18°C) | 12.71 (18°C) |
| | | $H_2PO_3^- \rightleftharpoons HPO_3^{2-} + H^+$ | 6.70 (18°C) | 7.30 (18°C) |
| _o_-**Phthalic** | | | 2.95 | 11.05 |
| | | | 5.41 | 8.59 |
| **Picric** | | | 0.29 | 13.71 |
| | **Piperidine** | | 11.20 (in 0.5 $F$ KNO$_3$) | 2.80 |
| **Propanoic** | | $CH_3CH_2COOH \rightleftharpoons CH_3CH_2COO^- + H^+$ | 4.87 | 9.13 |
| | _n_-**Propylamine** | $CH_3CH_2CH_2NH_3^+ \rightleftharpoons CH_3CH_2CH_2NH_2 + H^+$ | 10.74 (20°C) | 3.26 (20°C) |

| Name | Equilibrium | | |
|---|---|---|---|
| Pyridine | (pyridinium, N⁺–H) ⇌ (pyridine, N) + H⁺ | 5.23 | 8.77 |
| Salicylic | (benzene ring with COOH and OH) ⇌ (benzene ring with COO⁻ and OH) + H⁺ | 2.97 | 11.03 |
| Succinic | $HOOCCH_2CH_2COOH \rightleftharpoons HOOCCH_2CH_2COO^- + H^+$ | 4.21 | 9.79 |
| | $HOOCCH_2CH_2COO^- \rightleftharpoons {}^-OOCCH_2CH_2COO^- + H^+$ | 5.64 | 8.36 |
| Sulfamic | $NH_2SO_3H \rightleftharpoons NH_2SO_3^- + H^+$ | 0.99 | 13.01 |
| Sulfuric | $HSO_4^- \rightleftharpoons SO_4^{2-} + H^+$ | 1.92 | 12.08 |
| Sulfurous | $H_2SO_3 \rightleftharpoons HSO_3^- + H^+$ | 1.76 | 12.24 |
| | $HSO_3^- \rightleftharpoons SO_3^{2-} + H^+$ | 7.21 | 6.79 |
| Tartaric | HOOCCH–CHCOOH (with OH OH) ⇌ HOOCCH–CHCOO⁻ (with OH OH) + H⁺ | 3.04 | 10.96 |
| | HOOCCH–CHCOO⁻ (with OH OH) ⇌ ⁻OOCCH–CHCOO⁻ (with OH OH) + H⁺ | 4.37 | 9.63 |
| Triethylamine | $(CH_3CH_2)_3NH^+ \rightleftharpoons (CH_3CH_2)_3N + H^+$ | 10.77 [in 0.4 F (C₂H₅)₃NH⁺NO₃⁻] | 3.23 [in 0.4 F (C₂H₅)₃NH⁺NO₃⁻] |
| Trimethylamine | $(CH_3)_3NH^+ \rightleftharpoons (CH_3)_3N + H^+$ | 9.80 | 4.20 |
| Tris(hydroxymethyl)-aminomethane | $(HOCH_2)_3CNH_3^+ \rightleftharpoons (HOCH_2)_3CNH_2 + H^+$ | 8.08 | 5.92 |

Unless stated otherwise, all data pertain to 25°C and zero ionic strength. However, for numerical problems in this book, the constants may be applied to other conditions without correction for differences in temperature or ionic strength.

Each listed $pK_a$ value refers to a proton-transfer equilibrium in which an acid reacts with water (acting as a Brønsted-Lowry base). For example, $pK_a$ for acetic acid is 4.76; this value pertains to the equilibrium

$$CH_3COOH + H_2O \rightleftharpoons CH_3COO^- + H_3O^+$$

Correct proton-transfer equilibria for other acids can be similarly written.

Each listed $pK_b$ value refers to a proton-transfer equilibrium in which a base reacts with water (acting as a Brønsted-Lowry acid). For example, $pK_b$ for acetate is 9.24; this value pertains to the equilibrium

$$CH_3COO^- + H_2O \rightleftharpoons CH_3COOH + OH^-$$

Correct proton-transfer equilibria for other bases can be formulated in the same way.

For acids such as arsenic acid ($H_3AsO_4$) that undergo stepwise proton-transfer reactions, the use of additional subscripts to distinguish among the several $pK_a$ values is necessary; for arsenic acid, we can write $pK_{a1} = 2.19$, $pK_{a2} = 6.94$, and $pK_{a3} = 11.50$. Similar subscripts can be used to designate $pK_b$ values for bases that can accept more than one proton, *e.g.*, carbonate ion.

Notice that the *sum* of $pK_a$ for an acid and $pK_b$ for its conjugate base must be 14.00 for water solutions; see page 88 .

[a] Arsenious acid behaves essentially as a monoprotic acid in water. To indicate this behavior, the abbreviated proton-transfer equilibrium is often written

$$HAsO_2 \rightleftharpoons AsO_2^- + H^+$$

[b] Boric acid behaves as a monoprotic acid in water. To convey this behavior, the abbreviated proton-transfer reaction is frequently written

$$HBO_2 \rightleftharpoons BO_2^- + H^+$$

[c] Ethylenediaminetetraacetic acid is a tetraprotic acid with the structural formula

but, for purposes of brevity, it is usually represented as $H_4Y$.

[d] Nitrilotriacetic acid is a triprotic acid with the structural formula

but, for the sake of brevity, it is often represented as $H_3Y$.

Comprehensive data concerning acid-base equilibria are given in the following references:

1. L. Meites, ed.: *Handbook of Analytical Chemistry*. McGraw-Hill Book Company, New York, 1963.
2. L. G. Sillén and A. E. Martell: *Stability Constants of Metal-Ion Complexes*. Chem. Soc. (London), Spec. Publ. No. 17 (1964); No. 25 (1971).

# APPENDIX 3

## LOGARITHMS OF STEPWISE AND OVERALL FORMATION CONSTANTS FOR METAL ION COMPLEXES*

| Ligand | Metal Ion | $\log K_1$ | $\log K_2$ | $\log K_3$ | $\log K_4$ | $\log K_5$ | $\log K_6$ | Conditions |
|---|---|---|---|---|---|---|---|---|
| Acetate, $CH_3COO^-$ | $Ag^+$ | 0.73 | −0.09 | | | | | $3\,F\,NaClO_4$ |
| | $Cd^{2+}$ | 1.30 | 0.98 | 0.14 | −0.42 | | | $3\,F\,NaClO_4$ |
| | $Cu^{2+}$ | 1.79 | 1.15 | | | | | |
| | $Hg^{2+}$ | | 8.43 $(\beta_2)$ | | | | | $2\,F\,NaClO_4$ |
| | $Pb^{2+}$ | 2.19 | 0.72 | 0.61 | | | | |
| Acetylacetonate,  | $Al^{3+}$ | 8.6 | 7.9 | 5.8 | | | | 30°C |
| | $Cd^{2+}$ | 3.83 | 2.76 | | | | | 30°C |
| | $Co^{2+}$ | 5.40 | 4.11 | | | | | 30°C |
| | $Cu^{2+}$ | 8.22 | 6.73 | | | | | 30°C |
| | $Fe^{2+}$ | 5.07 | 3.60 | | | | | 30°C |
| | $Fe^{3+}$ | 11.4 | 10.7 | 4.6 | | | | |
| | $Mg^{2+}$ | 3.63 | 2.54 | | | | | 30°C |
| | $Mn^{2+}$ | 4.18 | 3.07 | | | | | 30°C |
| | $Ni^{2+}$ | 6.06 | 4.71 | 2.32 | | | | 20°C |
| | $UO_2^{2+}$ | 7.74 | 6.43 | | | | | 30°C |
| | $Zn^{2+}$ | 4.98 | 3.83 | | | | | 30°C |
| Ammonia, $NH_3$ | $Ag^+$ | 3.37 | 3.84 | | | | | |
| | $Cd^{2+}$ | 2.65 | 2.10 | 1.44 | 0.93 | −0.32 | −1.66 | $2\,F\,NH_4NO_3$; 30°C |
| | $Co^{2+}$ | 1.99 | 1.51 | 0.93 | 0.64 | 0.06 | −0.74 | 30°C |
| | $Co^{3+}$ | 7.3 | 6.7 | 6.1 | 5.6 | 5.05 | 4.41 | $2\,F\,NH_4NO_3$; 30°C |

* See footnotes at end of table.

| Ligand | Metal Ion | $\log K_1$ | $\log K_2$ | $\log K_3$ | $\log K_4$ | $\log K_5$ | $\log K_6$ | Conditions |
|---|---|---|---|---|---|---|---|---|
| Ammonia, $NH_3$ (continued) | $Cu^+$ | 5.93 | 4.93 | | | | | $2\,F\,NH_4NO_3$; 18°C |
| | $Cu^{2+}$ | 4.31 | 3.67 | 3.04 | 2.30 | −0.46 | | $2\,F\,NH_4NO_3$; 18°C |
| | $Hg^{2+}$ | 8.8 | 8.7 | 1.00 | 0.78 | | | $2\,F\,NH_4NO_3$; 22°C |
| | $Ni^{2+}$ | 2.36 | 1.90 | 1.55 | 1.23 | 0.85 | 0.42 | $1\,F\,NH_4NO_3$ |
| | $Zn^{2+}$ | 2.18 | 2.25 | 2.31 | 1.96 | | | 30°C |
| Bromide, $Br^-$ | $Bi^{3+}$ | 4.30 | 1.25 | 0.32 | 0.10 | | | $1\,F\,HNO_3$ |
| | $Cd^{2+}$ | 2.23 | 0.77 | −0.17 | | | | |
| | $Hg^{2+}$ | 8.94 | 7.94 | 2.27 | 1.75 | | | $0.5\,F\,NaClO_4$ |
| | $Pb^{2+}$ | 1.65 | 0.75 | 0.88 | 0.22 | | | $3\,F\,NaClO_4$ |
| | $Zn^{2+}$ | 0.22 | −0.32 | −0.64 | −0.26 | | | $0.7\,F\,HClO_4$; 20°C |
| Chloride, $Cl^-$ | $Bi^{3+}$ | 2.43 | 2.00 | 1.35 | 0.43 | 0.48 | | |
| | $Cd^{2+}$ | 2.00 | 0.70 | −0.59 | | | | |
| | $Cu^+$ | | 4.94 ($\beta_2$) | | | | | |
| | $Fe^{3+}$ | 1.48 | 0.65 | −1.0 | 1.05 | | | |
| | $Hg^{2+}$ | 6.74 | 6.48 | 0.95 | 1.05 | | | $0.5\,F\,NaClO_4$ |
| | $Pb^{2+}$ | 1.10 | 1.16 | −0.40 | −1.05 | | | |
| Cyanide, $CN^-$ | $Ag^+$ | 5.48 | 19.85 ($\beta_2$) | | | | | $3\,F\,NaClO_4$ |
| | $Cd^{2+}$ | | 5.14 | 4.56 | 3.58 | | 19.09 ($\beta_6$) | $5\,F\,CaCl_2$ |
| | $Co^{2+}$ | | | | | | | |
| | $Cu^+$ | | 24.0 ($\beta_2$) | 4.59 | 1.70 | | | $0.1\,F\,NaNO_3$; 20°C |
| | $Hg^{2+}$ | 18.00 | 16.70 | 3.83 | 2.98 | | | |
| | $Ni^{2+}$ | | | | 30.3 ($\beta_4$) | | | |
| | $Zn^{2+}$ | | | | 16.72 ($\beta_4$) | | | |
| Ethylenediamine, $H_2NCH_2CH_2NH_2$ | $Ag^+$ | 4.70 | 3.00 | | | | | $0.1\,F\,NaNO_3$; 20°C |
| | $Cd^{2+}$ | 5.63 | 4.59 | 2.07 | | | | $1\,F\,KNO_3$ |
| | $Co^{2+}$ | 5.93 | 4.73 | 3.30 | | | | $1\,F\,KCl$ |
| | $Co^{3+}$ | 18.7 | 16.2 | 13.81 | | | | $1\,F\,NaNO_3$; 30°C |
| | $Cu^+$ | | 10.8 ($\beta_2$) | | | | | |
| | $Cu^{2+}$ | 10.75 | 9.28 | | | | | $1\,F\,KNO_3$ |
| | $Fe^{2+}$ | 4.28 | 3.25 | 1.99 | | | | $1\,F\,KCl$; 30°C |

_(continued from preceding page)_

| Ion | log $K_1$ | log $K_2$ | log $K_3$ | log $K_4$ | log $K_5$ | log $K_6$ | Ionic medium |
|---|---|---|---|---|---|---|---|
| $Hg^{2+}$ | 14.3 | 9.0 | | | | | 0.1 *F* $KNO_3$ |
| $Ni^{2+}$ | 7.72 | 6.36 | 4.33 | | | | 1 *F* KCl |
| $Zn^{2+}$ | 5.77 | 5.06 | 3.28 | | | | 20°C |

**Ethylenediaminetetraacetate (EDTA),** $(^-OOCCH_2)_2NCH_2CH_2N(CH_2COO^-)_2$ [see Table 6–1, page 173]

**Fluoride, F⁻**

| Ion | log $K_1$ | log $K_2$ | log $K_3$ | log $K_4$ | log $K_5$ | log $K_6$ | Ionic medium |
|---|---|---|---|---|---|---|---|
| $Al^{3+}$ | 6.13 | 5.02 | 3.85 | 2.74 | 1.63 | 0.47 | 0.53 *F* $KNO_3$ |
| $Ce^{3+}$ | 3.99 | | | | | | |
| $Fe^{3+}$ | 5.17 | 3.92 | 2.91 | | | | 0.5 *F* $NaClO_4$ |

**[a]Hydroxide, OH⁻**

| Ion | log $K_1$ | log $K_2$ | log $K_3$ | log $K_4$ | Ionic medium |
|---|---|---|---|---|---|
| $Al^{3+}$ | 8.98 | | | 32.43 ($\beta_4$) | 3 *F* $NaClO_4$ |
| $Bi^{3+}$ | 12.42 | | | | 1 *F* $KNO_3$ |
| $Cd^{2+}$ | 6.38 | | | | |
| $Co^{2+}$ | 2.80 | | | | |

**[a]Hydroxide, OH⁻** _(continued)_

| Ion | log $K_1$ | log $K_2$ | log $K_3$ | log $K_4$ | Ionic medium |
|---|---|---|---|---|---|
| $Cu^{2+}$ | 6.66 | | | | |
| $Fe^{2+}$ | 4.5 | | | | |
| $Fe^{3+}$ | 10.95 | | | | 1 *F* $NaClO_4$ |
| $Hg^{2+}$ | 10.77 | 10.74 | | | 3 *F* $NaClO_4$ |
| $Ni^{2+}$ | 3.08 | | | | |
| $Pb^{2+}$ | 6.9 | | 13.95 ($\beta_3$) | | |
| $Zn^{2+}$ | 4.34 | | 14.23 ($\beta_3$) | 1.26 | 3 *F* $NaClO_4$ |

**8-Hydroxyquinolate (oxinate),**

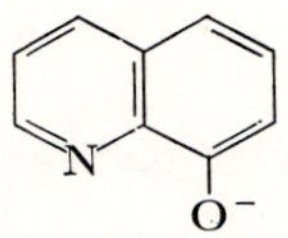

| Ion | log $K_1$ | log $K_2$ | log $K_3$ | Ionic medium |
|---|---|---|---|---|
| $Cd^{2+}$ | 9.43 | 7.68 | | 50% dioxane |
| $Ce^{3+}$ | 9.15 | 7.98 | | 50% dioxane |
| $Co^{2+}$ | 10.55 | 9.11 | | 50% dioxane |
| $Cu^{2+}$ | 13.49 | 12.73 | | 50% dioxane |
| $Fe^{2+}$ | 9.83 | 9.01 | | 70% dioxane |
| $Fe^{3+}$ | | | 38.00 ($\beta_3$) | 50% dioxane |
| $Mg^{2+}$ | 6.38 | 5.43 | | 50% dioxane |
| $Mn^{2+}$ | 8.28 | 7.17 | | 50% dioxane |
| $Ni^{2+}$ | 11.44 | 9.94 | | 50% dioxane |
| $Pb^{2+}$ | 10.61 | 8.09 | | 50% dioxane |
| $UO_2^{2+}$ | 11.25 | 9.64 | | 50% dioxane |
| $Zn^{2+}$ | 9.96 | 8.90 | | 50% dioxane |

| Ligand | Metal Ion | $\log K_1$ | $\log K_2$ | $\log K_3$ | $\log K_4$ | $\log K_5$ | $\log K_6$ | Conditions |
|---|---|---|---|---|---|---|---|---|
| Iodide, $I^-$ | $Bi^{3+}$ | | 1.33 | 1.06 | 0.92 | | 19.4 $(\beta_6)$ | 2 $F$ NaClO$_4$; 20°C |
| | $Cd^{2+}$ | 2.10 | | | | | | |
| | $Cu^+$ | | 8.85 $(\beta_2)$ | | | | | 0.5 $F$ NaClO$_4$ |
| | $Hg^{2+}$ | 12.87 | 10.95 | 3.67 | 2.37 | | | 1 $F$ NaClO$_4$ |
| | $Pb^{2+}$ | 1.26 | 1.54 | 0.62 | 0.50 | | | |
| Nitrilotriacetate (NTA), $N(CH_2COO)_3^{3-}$ | $Ba^{2+}$ | 6.41 | | | | | | 20°C |
| | $Ca^{2+}$ | 8.17 | 3.43 | | | | | 20°C |
| | $Cd^{2+}$ | 9.54 | 5.7 | | | | | 0.1 $F$ KCl; 20°C |
| | $Co^{2+}$ | 10.6 | 3.9 | | | | | 0.1 $F$ KCl; 20°C |
| | $Cu^{2+}$ | 12.68 | | | | | | 0.1 $F$ KCl; 20°C |
| | $Fe^{2+}$ | 8.83 | | | | | | 0.1 $F$ KCl; 20°C |
| | $Fe^{3+}$ | 15.87 | 8.45 | | | | | 0.1 $F$ KCl; 20°C |
| | $Mg^{2+}$ | 7.00 | | | | | | 20°C |
| | $Mn^{2+}$ | 7.44 | 3.7 | | | | | 0.1 $F$ KCl; 20°C |
| | $Ni^{2+}$ | 11.26 | 4.7 | | | | | 0.1 $F$ KCl; 20°C |
| | $Pb^{2+}$ | 11.39 | | | | | | 0.1 $F$ KNO$_3$; 20°C |
| | $Sr^{2+}$ | 6.73 | | | | | | 20°C |
| | $Zn^{2+}$ | 10.67 | 3.0 | | | | | 0.1 $F$ KCl; 20°C |
| Oxalate, $C_2O_4^{2-}$ | $Al^{3+}$ | | 13 $(\beta_2)$ | 3.3 | | | | 18°C |
| | $Cd^{2+}$ | 4.00 | 1.77 | | | | | |
| | $Co^{2+}$ | 4.79 | 1.91 | | | | | |
| | $Cu^{2+}$ | 6.19 | 4.04 | 0.70 | | | | |
| | $Fe^{2+}$ | | 4.52 $(\beta_2)$ | | | | | |
| | $Fe^{3+}$ | 9.4 | 6.80 | 4 | | | | 0.5 $F$ NaClO$_4$ |
| | $Mg^{2+}$ | 3.82 | 4.38 $(\beta_2)$ | | | | | |
| | $Mn^{2+}$ | | 1.43 | | | | | |
| | $Ni^{2+}$ | 5.16 | 1.35 | | | | | |
| | $Pb^{2+}$ | | 6.54 $(\beta_2)$ | | | | | |
| | $Zn^{2+}$ | 5.00 | 2.36 | | | | | |

| Ligand / Ion | | | | Conditions |
|---|---|---|---|---|
| **Pyridine, $C_5H_5N$** | | | | |
| $Ag^+$ | 2.00 | 2.11 | 2.50 ($\beta_4$) | 0.1 $F$ $KNO_3$ |
| $Cd^{2+}$ | 1.14 | 2.14 ($\beta_2$) | | 0.5 $F$ $HNO_3$ |
| $Co^{2+}$ | | 0.4 | | 0.5 $F$ $KNO_3$ |
| $Cu^{2+}$ | 2.52 | 1.86 / 1.31 | 0.85 | 0.5 $F$ $HNO_3$ |
| $Hg^{2+}$ | 5.1 | 4.9 | | 0.5 $F$ $HNO_3$ |
| $Ni^{2+}$ | 1.78 | 1.05 | | 0.1 $F$ $KCl$ |
| $Zn^{2+}$ | 1.41 | −0.30 / 0.50 | 0.32 | |
| | | | | |
| **Thiocyanate, $SCN^-$** | | | | |
| $Ag^+$ | 8.82 | 8.39 ($\beta_2$) / 1.23 | 0.28 | 5 $F$ $NaNO_3$ |
| $Cu^+$ | | 11.00 ($\beta_2$) / −0.10 | −0.42 | 0.5 $F$ $NaClO_4$ |
| $Fe^{3+}$ | 2.14 | 1.31 | | |
| $Hg^{2+}$ | | 17.26 ($\beta_2$) / 2.71 | 1.72 | |
| | | | | |
| **Thiosulfate, $S_2O_3^{2-}$** | | | | |
| $Ag^+$ | 8.82 | 4.64 / 0.69 | | 20°C |
| $Cd^{2+}$ | 3.92 | 2.52 / 1.44 | | 0.8 $F$ $Na_2SO_4$ |
| $Cu^+$ | 10.35 | 1.92 | | 3 $F$ $NaClO_4$ |
| $Hg^{2+}$ | | 29.27 ($\beta_2$) / 2.40 | 1.35 | |
| $Pb^{2+}$ | 2.56 | 2.32 / 1.46 | −0.09 | |

Unless indicated otherwise, the data pertain strictly to 25°C and zero ionic strength; however, for all numerical problems in this book, these constants can be used without corrections for differences in temperature and ionic strength. Note that in several instances the logarithms of $\beta_n$ values are listed instead of the logarithms of $K_n$ values; for example, an entry such as 8.43 ($\beta_2$) denotes that $\log \beta_2 = 8.43$. Discussions of the formulation and significance of stepwise ($K_n$) and overall ($\beta_n$) formation constants are presented in Chapter 6, pages 163–165.

[a] Reactions involving metal cations and hydroxide ions do not fit into the usual classification of complex-formation equilibria but should be thought of as examples of simple Brønsted-Lowry proton-transfer reactions.

Each metal ion species in the above tabulation is represented by the general symbol $M^{n+}$, where $n$ is the charge of the metal cation. However, it should be recalled that a metal ion is hydrated (solvated). For example, although the iron(III) ion is abbreviated as $Fe^{3+}$, the actual species may be more reasonably written as $Fe(H_2O)_6^{3+}$. A usual set of complexation equilibria involves the stepwise substitution of a (hypothetical monodentate) ligand, L, for one of the water molecules, as in the following reactions:

$$Fe(H_2O)_6{}^{3+} + L \rightleftharpoons Fe(H_2O)_5L^{3+} + H_2O$$

$$Fe(H_2O)_5L^{3+} + L \rightleftharpoons Fe(H_2O)_4L_2{}^{3+} + H_2O \cdots$$

$$\cdots Fe(H_2O)L_5{}^{3+} + L \rightleftharpoons FeL_6{}^{3+} + H_2O$$

In the case of reactions between a metal ion such as $Fe(H_2O)_6{}^{3+}$ and $OH^-$, however, one should visualize a process in which the hydroxide ion (Brønsted-Lowry base) accepts a proton from one of the water molecules (Brønsted-Lowry acid) coordinated to iron(III), $e.g.$,

$$Fe(H_2O)_6{}^{3+} + OH^- \rightleftharpoons Fe(H_2O)_5OH^{2+} + H_2O$$

If the water molecules bound to iron(III) are omitted from the preceding equilibrium, we have

$$Fe^{3+} + OH^- \rightleftharpoons FeOH^{2+}$$

Clearly, no actual exchange or substitution of $OH^-$ for $H_2O$ is involved in such an equilibrium.

It may be noted that the so-called acid dissociation constants for several metal cations can be calculated from the data in this table. Thus, the first step in the acid dissociation of $Fe(H_2O)_6{}^{3+}$ can be formulated as follows:

$$Fe(H_2O)_6{}^{3+} + H_2O \rightleftharpoons Fe(H_2O)_5OH^{2+} + H_3O^+$$

In abbreviated form, this reaction becomes

$$Fe^{3+} + H_2O \rightleftharpoons FeOH^{2+} + H^+$$

Now, the equilibrium constant for this acid dissociation of iron(III) may be calculated as follows:

$$Fe(H_2O)_6{}^{3+} + OH^- \rightleftharpoons Fe(H_2O)_5OH^{2+} + H_2O; \quad K_1 = 9.1 \times 10^{10}$$

$$2\,H_2O \rightleftharpoons H_3O^+ + OH^-; \quad K_w = 1.0 \times 10^{-14}$$

$$\overline{Fe(H_2O)_6{}^{3+} + H_2O \rightleftharpoons Fe(H_2O)_5OH^{2+} + H_3O^+; \quad K_{a1} = K_1 K_w = 9.1 \times 10^{-4}}$$

or

$$Fe^{3+} + OH^- \rightleftharpoons FeOH^{2+}; \quad K_1 = 9.1 \times 10^{10}$$

$$H_2O \rightleftharpoons H^+ + OH^-; \quad K_w = 1.0 \times 10^{-14}$$

$$\overline{Fe^{3+} + H_2O \rightleftharpoons FeOH^{2+} + H^+; \quad K_{a1} = K_1 K_w = 9.1 \times 10^{-4}}$$

Notice that the acid strength of $Fe(H_2O)_6{}^{3+}$ or $Fe^{3+}$ in water is greater than that of acetic acid, for example, and almost as great as the second dissociation constant for sulfuric acid ($K_{a2} = 1.2 \times 10^{-2}$).

Good sources of information about formation constants for metal ion complexes are:

1. L. G. Sillén and A. E. Martell: *Stability Constants of Metal-Ion Complexes.* Chem. Soc. (London), Spec. Publ. No. 17 (1964); No. 25 (1971).
2. K. B. Yatsimirskii and V. P. Vasil'ev: *Instability Constants of Complex Compounds* (translated from Russian). Consultants Bureau, New York, 1960.
3. L. Meites, ed.: *Handbook of Analytical Chemistry.* McGraw-Hill Book Company, New York, 1963.

# APPENDIX 4

## STANDARD AND FORMAL POTENTIALS FOR HALF-REACTIONS*

(All values pertain to 25°C and are quoted in volts with respect to the normal hydrogen electrode, taken to have a standard potential of zero.)

| Half-Reaction | $E^0$ |
|---|---|
| *Aluminum* | |
| $Al^{3+} + 3\,e \rightleftharpoons Al$ | $-1.66$ |
| $Al(OH)_4^- + 3\,e \rightleftharpoons Al + 4\,OH^-$ | $-2.35$ |
| *Antimony* | |
| $Sb_2O_5 + 6\,H^+ + 4\,e \rightleftharpoons 2\,SbO^+ + 3\,H_2O$ | $+0.581$ |
| $Sb + 3\,H^+ + 3\,e \rightleftharpoons SbH_3$ | $-0.51$ |
| *Arsenic* | |
| $H_3AsO_4 + 2\,H^+ + 2\,e \rightleftharpoons HAsO_2 + 2\,H_2O$ | $+0.559$ |
| $HAsO_2 + 3\,H^+ + 3\,e \rightleftharpoons As + 2\,H_2O$ | $+0.248$ |
| $As + 3\,H^+ + 3\,e \rightleftharpoons AsH_3$ | $-0.60$ |
| *Barium* | |
| $Ba^{2+} + 2\,e \rightleftharpoons Ba$ | $-2.90$ |
| *Beryllium* | |
| $Be^{2+} + 2\,e \rightleftharpoons Be$ | $-1.85$ |
| *Bismuth* | |
| $BiCl_4^- + 3\,e \rightleftharpoons Bi + 4\,Cl^-$ | $+0.16$ |
| $BiO^+ + 2\,H^+ + 3\,e \rightleftharpoons Bi + H_2O$ | $+0.32$ |
| *Boron* | |
| $H_2BO_3^- + 5\,H_2O + 8\,e \rightleftharpoons BH_4^- + 8\,OH^-$ | $-1.24$ |
| $H_2BO_3^- + H_2O + 3\,e \rightleftharpoons B + 4\,OH^-$ | $-1.79$ |
| *Bromine* | |
| $2\,BrO_3^- + 12\,H^+ + 10\,e \rightleftharpoons Br_2 + 6\,H_2O$ | $+1.52$ |
| $Br_2(aq) + 2\,e \rightleftharpoons 2\,Br^-$ | $+1.087$[a] |
| $Br_2(l) + 2\,e \rightleftharpoons 2\,Br^-$ | $+1.065$[a] |
| $Br_3^- + 2\,e \rightleftharpoons 3\,Br^-$ | $+1.05$ |
| *Cadmium* | |
| $Cd^{2+} + 2\,e \rightleftharpoons Cd$ | $-0.403$ |
| $Cd(CN)_4^{2-} + 2\,e \rightleftharpoons Cd + 4\,CN^-$ | $-1.09$ |
| $Cd(NH_3)_4^{2+} + 2\,e \rightleftharpoons Cd + 4\,NH_3$ | $-0.61$ |
| *Calcium* | |
| $Ca^{2+} + 2\,e \rightleftharpoons Ca$ | $-2.87$ |
| *Carbon* | |
| $2\,CO_2 + 2\,H^+ + 2\,e \rightleftharpoons H_2C_2O_4$ | $-0.49$ |
| *Cerium* | |
| $Ce^{4+} + e \rightleftharpoons Ce^{3+} \quad (1\,F\ HClO_4)$ | *+1.70* |
| $Ce^{4+} + e \rightleftharpoons Ce^{3+} \quad (1\,F\ HNO_3)$ | *+1.61* |
| $Ce^{4+} + e \rightleftharpoons Ce^{3+} \quad (1\,F\ H_2SO_4)$ | *+1.44* |
| *Cesium* | |
| $Cs^+ + e \rightleftharpoons Cs$ | $-2.92$ |
| *Chlorine* | |
| $Cl_2 + 2\,e \rightleftharpoons 2\,Cl^-$ | $+1.3595$ |
| $2\,ClO_3^- + 12\,H^+ + 10\,e \rightleftharpoons Cl_2 + 6\,H_2O$ | $+1.47$ |
| $ClO_3^- + 2\,H^+ + e \rightleftharpoons ClO_2 + H_2O$ | $+1.15$ |

* See footnotes at end of table.

| Half-Reaction | $E^0$ |
|---|---|
| $HClO + H^+ + 2\,e \rightleftharpoons Cl^- + H_2O$ | $+1.49$ |
| $2\,HClO + 2\,H^+ + 2\,e \rightleftharpoons Cl_2 + 2\,H_2O$ | $+1.63$ |
| *Chromium* | |
| $Cr_2O_7^{2-} + 14\,H^+ + 6\,e \rightleftharpoons 2\,Cr^{3+} + 7\,H_2O$ | $+1.33$ |
| $Cr^{3+} + e \rightleftharpoons Cr^{2+}$ | $-0.41$ |
| $Cr^{2+} + 2\,e \rightleftharpoons Cr$ | $-0.91$ |
| $CrO_4^{2-} + 4\,H_2O + 3\,e \rightleftharpoons Cr(OH)_3 + 5\,OH^-$ | $-0.13$ |
| *Cobalt* | |
| $Co^{3+} + e \rightleftharpoons Co^{2+}$ | $+1.842$ |
| $Co(NH_3)_6^{3+} + e \rightleftharpoons Co(NH_3)_6^{2+}$ | $+0.1$ |
| $Co(OH)_3 + e \rightleftharpoons Co(OH)_2 + OH^-$ | $+0.17$ |
| $Co^{2+} + 2\,e \rightleftharpoons Co$ | $-0.277$ |
| $Co(CN)_6^{3-} + e \rightleftharpoons Co(CN)_6^{4-}$ | $-0.84$ |
| *Copper* | |
| $Cu^{2+} + 2\,e \rightleftharpoons Cu$ | $+0.337$ |
| $Cu^{2+} + e \rightleftharpoons Cu^+$ | $+0.153$ |
| $Cu^{2+} + I^- + e \rightleftharpoons CuI$ | $+0.86$ |
| $Cu^{2+} + 2\,CN^- + e \rightleftharpoons Cu(CN)_2^-$ | $+1.12$ |
| $Cu(CN)_2^- + e \rightleftharpoons Cu + 2\,CN^-$ | $-0.43$ |
| $Cu(NH_3)_4^{2+} + e \rightleftharpoons Cu(NH_3)_2^+ + 2\,NH_3$ | $-0.01$ |
| $Cu^{2+} + 2\,Cl^- + e \rightleftharpoons CuCl_2^-$ | $+0.463$ |
| $CuCl_2^- + e \rightleftharpoons Cu + 2\,Cl^-$ | $+0.177$ |
| *Fluorine* | |
| $F_2 + 2\,e \rightleftharpoons 2\,F^-$ | $+2.87$ |
| *Gold* | |
| $Au^{3+} + 2\,e \rightleftharpoons Au^+$ | $+1.41$ |
| $Au^{3+} + 3\,e \rightleftharpoons Au$ | $+1.50$ |
| $Au(CN)_2^- + e \rightleftharpoons Au + 2\,CN^-$ | $-0.60$ |
| $AuCl_2^- + e \rightleftharpoons Au + 2\,Cl^-$ | $+1.15$ |
| $AuCl_4^- + 2\,e \rightleftharpoons AuCl_2^- + 2\,Cl^-$ | $+0.926$ |
| $AuBr_2^- + e \rightleftharpoons Au + 2\,Br^-$ | $+0.959$ |
| $AuBr_4^- + 2\,e \rightleftharpoons AuBr_2^- + 2\,Br^-$ | $+0.802$ |
| *Hydrogen* | |
| $2\,H^+ + 2\,e \rightleftharpoons H_2$ | $0.0000$ |
| $2\,H_2O + 2\,e \rightleftharpoons H_2 + 2\,OH^-$ | $-0.828$ |
| *Iodine* | |
| $I_2(aq) + 2\,e \rightleftharpoons 2\,I^-$ | $+0.6197$[b] |
| $I_3^- + 2\,e \rightleftharpoons 3\,I^-$ | $+0.5355$ |
| $I_2(s) + 2\,e \rightleftharpoons 2\,I^-$ | $+0.5345$[b] |
| $2\,IO_3^- + 12\,H^+ + 10\,e \rightleftharpoons I_2 + 6\,H_2O$ | $+1.20$ |
| $2\,ICl_2^- + 2\,e \rightleftharpoons I_2 + 4\,Cl^-$ | $+1.06$ |
| *Iron* | |
| $Fe^{3+} + e \rightleftharpoons Fe^{2+}$ | $+0.771$ |
| $Fe^{3+} + e \rightleftharpoons Fe^{2+} \quad (1\,F\,HCl)$ | $+0.70$ |
| $Fe^{3+} + e \rightleftharpoons Fe^{2+} \quad (1\,F\,H_2SO_4)$ | $+0.68$ |
| $Fe^{3+} + e \rightleftharpoons Fe^{2+} \quad (0.5\,F\,H_3PO_4 - 1\,F\,H_2SO_4)$ | $+0.61$ |
| $Fe(CN)_6^{3-} + e \rightleftharpoons Fe(CN)_6^{4-}$ | $+0.36$ |
| $Fe(CN)_6^{3-} + e \rightleftharpoons Fe(CN)_6^{4-} \quad (1\,F\,HCl\ or\ HClO_4)$ | $+0.71$ |
| $Fe^{2+} + 2\,e \rightleftharpoons Fe$ | $-0.440$ |
| *Lead* | |
| $Pb^{2+} + 2\,e \rightleftharpoons Pb$ | $-0.126$ |
| $PbSO_4 + 2\,e \rightleftharpoons Pb + SO_4^{2-}$ | $-0.3563$ |
| $PbO_2 + SO_4^{2-} + 4\,H^+ + 2\,e \rightleftharpoons PbSO_4 + 2\,H_2O$ | $+1.685$ |
| $PbO_2 + 4\,H^+ + 2\,e \rightleftharpoons Pb^{2+} + 2\,H_2O$ | $+1.455$ |
| *Lithium* | |
| $Li^+ + e \rightleftharpoons Li$ | $-3.045$ |
| *Magnesium* | |
| $Mg^{2+} + 2\,e \rightleftharpoons Mg$ | $-2.37$ |
| $Mg(OH)_2 + 2\,e \rightleftharpoons Mg + 2\,OH^-$ | $-2.69$ |

| **Half-Reaction** | $E^0$ |
|---|---|
| *Manganese* | |
| $Mn^{2+} + 2\ e \rightleftharpoons Mn$ | $-1.18$ |
| $MnO_4^- + 4\ H^+ + 3\ e \rightleftharpoons MnO_2 + 2\ H_2O$ | $+1.695$ |
| $MnO_4^- + 8\ H^+ + 5\ e \rightleftharpoons Mn^{2+} + 4\ H_2O$ | $+1.51$ |
| $MnO_2 + 4\ H^+ + 2\ e \rightleftharpoons Mn^{2+} + 2\ H_2O$ | $+1.23$ |
| $MnO_4^- + e \rightleftharpoons MnO_4^{2-}$ | $+0.564$ |
| $Mn^{3+} + e \rightleftharpoons Mn^{2+}$    $(8\ F\ H_2SO_4)$ | *$+1.51$* |
| *Mercury* | |
| $2\ Hg^{2+} + 2\ e \rightleftharpoons Hg_2^{2+}$ | $+0.920$ |
| $Hg^{2+} + 2\ e \rightleftharpoons Hg$ | $+0.854$ |
| $Hg_2^{2+} + 2\ e \rightleftharpoons 2\ Hg$ | $+0.789$ |
| $Hg_2SO_4 + 2\ e \rightleftharpoons 2\ Hg + SO_4^{2-}$ | $+0.6151$ |
| $HgCl_4^{2-} + 2\ e \rightleftharpoons Hg + 4\ Cl^-$ | $+0.48$ |
| $Hg_2Cl_2 + 2\ e \rightleftharpoons 2\ Hg + 2\ Cl^-$    $(0.1\ F\ KCl)$ | *$+0.334$* |
| $Hg_2Cl_2 + 2\ e \rightleftharpoons 2\ Hg + 2\ Cl^-$    $(1\ F\ KCl)$ | *$+0.280$* |
| $Hg_2Cl_2 + 2\ K^+ + 2\ e \rightleftharpoons 2\ Hg + 2\ KCl(s)$ | *$+0.2415$* |
| (saturated calomel electrode) | |
| *Molybdenum* | |
| $Mo^{6+} + e \rightleftharpoons Mo^{5+}$    $(2\ F\ HCl)$ | *$+0.53$* |
| $Mo^{4+} + e \rightleftharpoons Mo^{3+}$    $(4\ F\ H_2SO_4)$ | *$+0.1$* |
| *Neptunium* | |
| $Np^{4+} + e \rightleftharpoons Np^{3+}$ | $+0.147$ |
| $NpO_2^+ + 4\ H^+ + e \rightleftharpoons Np^{4+} + 2\ H_2O$ | $+0.75$ |
| $NpO_2^{2+} + e \rightleftharpoons NpO_2^+$ | $+1.15$ |
| *Nickel* | |
| $Ni^{2+} + 2\ e \rightleftharpoons Ni$ | $-0.24$ |
| $NiO_2 + 4\ H^+ + 2\ e \rightleftharpoons Ni^{2+} + 2\ H_2O$ | $+1.68$ |
| *Nitrogen* | |
| $NO_2 + H^+ + e \rightleftharpoons HNO_2$ | $+1.07$ |
| $NO_2 + 2\ H^+ + 2\ e \rightleftharpoons NO + H_2O$ | $+1.03$ |
| $HNO_2 + H^+ + e \rightleftharpoons NO + H_2O$ | $+1.00$ |
| $NO_3^- + 4\ H^+ + 3\ e \rightleftharpoons NO + 2\ H_2O$ | $+0.96$ |
| $NO_3^- + 3\ H^+ + 2\ e \rightleftharpoons HNO_2 + H_2O$ | $+0.94$ |
| $NO_3^- + 2\ H^+ + e \rightleftharpoons NO_2 + H_2O$ | $+0.80$ |
| $N_2 + 5\ H^+ + 4\ e \rightleftharpoons N_2H_5^+$ | $-0.23$ |
| *Osmium* | |
| $OsO_4 + 8\ H^+ + 8\ e \rightleftharpoons Os + 4\ H_2O$ | $+0.85$ |
| $OsCl_6^{2-} + e \rightleftharpoons OsCl_6^{3-}$ | $+0.85$ |
| $OsCl_6^{3-} + e \rightleftharpoons Os^{2+} + 6\ Cl^-$ | $+0.4$ |
| $Os^{2+} + 2\ e \rightleftharpoons Os$ | $+0.85$ |
| *Oxygen* | |
| $O_3 + 2\ H^+ + 2\ e \rightleftharpoons O_2 + H_2O$ | $+2.07$ |
| $H_2O_2 + 2\ H^+ + 2\ e \rightleftharpoons 2\ H_2O$ | $+1.77$ |
| $O_2 + 4\ H^+ + 4\ e \rightleftharpoons 2\ H_2O$ | $+1.229$ |
| $H_2O_2 + 2\ e \rightleftharpoons 2\ OH^-$ | $+0.88$ |
| $O_2 + 2\ H^+ + 2\ e \rightleftharpoons H_2O_2$ | $+0.682$ |
| *Palladium* | |
| $Pd^{2+} + 2\ e \rightleftharpoons Pd$ | $+0.987$ |
| $PdCl_6^{2-} + 2\ e \rightleftharpoons PdCl_4^{2-} + 2\ Cl^-$ | $+1.288$ |
| $PdCl_4^{2-} + 2\ e \rightleftharpoons Pd + 4\ Cl^-$ | $+0.623$ |
| *Phosphorus* | |
| $H_3PO_4 + 2\ H^+ + 2\ e \rightleftharpoons H_3PO_3 + H_2O$ | $-0.276$ |
| $H_3PO_3 + 2\ H^+ + 2\ e \rightleftharpoons H_3PO_2 + H_2O$ | $-0.50$ |
| *Platinum* | |
| $PtCl_6^{2-} + 2\ e \rightleftharpoons PtCl_4^{2-} + 2\ Cl^-$ | $+0.68$ |
| $PtBr_6^{2-} + 2\ e \rightleftharpoons PtBr_4^{2-} + 2\ Br^-$ | $+0.59$ |
| $Pt(OH)_2 + 2\ H^+ + 2\ e \rightleftharpoons Pt + 2\ H_2O$ | $+0.98$ |
| *Plutonium* | |
| $PuO_2^+ + 4\ H^+ + e \rightleftharpoons Pu^{4+} + 2\ H_2O$ | $+1.15$ |

| Half-Reaction | $E^0$ |
|---|---|
| $PuO_2^{2+} + 4\,H^+ + 2\,e \rightleftharpoons Pu^{4+} + 2\,H_2O$ | $+1.067$ |
| $Pu^{4+} + e \rightleftharpoons Pu^{3+}$ | $+0.97$ |
| $PuO_2^{2+} + e \rightleftharpoons PuO_2^+$ | $+0.93$ |
| *Potassium* | |
| $K^+ + e \rightleftharpoons K$ | $-2.925$ |
| *Radium* | |
| $Ra^{2+} + 2\,e \rightleftharpoons Ra$ | $-2.92$ |
| *Rubidium* | |
| $Rb^+ + e \rightleftharpoons Rb$ | $-2.925$ |
| *Selenium* | |
| $SeO_4^{2-} + 4\,H^+ + 2\,e \rightleftharpoons H_2SeO_3 + H_2O$ | $+1.15$ |
| $H_2SeO_3 + 4\,H^+ + 4\,e \rightleftharpoons Se + 3\,H_2O$ | $+0.740$ |
| $Se + 2\,H^+ + 2\,e \rightleftharpoons H_2Se$ | $-0.40$ |
| *Silver* | |
| $Ag^+ + e \rightleftharpoons Ag$ | $+0.7995$ |
| $Ag^{2+} + e \rightleftharpoons Ag^+ \quad (4\,F\,HNO_3)$ | $+1.927$ |
| $AgCl + e \rightleftharpoons Ag + Cl^-$ | $+0.2222$ |
| $AgBr + e \rightleftharpoons Ag + Br^-$ | $+0.073$ |
| $AgI + e \rightleftharpoons Ag + I^-$ | $-0.151$ |
| $Ag_2O + H_2O + 2\,e \rightleftharpoons 2\,Ag + 2\,OH^-$ | $+0.342$ |
| $Ag_2S + 2\,e \rightleftharpoons 2\,Ag + S^{2-}$ | $-0.71$ |
| *Sodium* | |
| $Na^+ + e \rightleftharpoons Na$ | $-2.714$ |
| *Strontium* | |
| $Sr^{2+} + 2\,e \rightleftharpoons Sr$ | $-2.89$ |
| *Sulfur* | |
| $S + 2\,H^+ + 2\,e \rightleftharpoons H_2S$ | $+0.141$ |
| $S_4O_6^{2-} + 2\,e \rightleftharpoons 2\,S_2O_3^{2-}$ | $+0.08$ |
| $SO_4^{2-} + 4\,H^+ + 2\,e \rightleftharpoons H_2SO_3 + H_2O$ | $+0.17$ |
| $S_2O_8^{2-} + 2\,e \rightleftharpoons 2\,SO_4^{2-}$ | $+2.01$ |
| $SO_4^{2-} + H_2O + 2\,e \rightleftharpoons SO_3^{2-} + 2\,OH^-$ | $-0.93$ |
| $2\,H_2SO_3 + 2\,H^+ + 4\,e \rightleftharpoons S_2O_3^{2-} + 3\,H_2O$ | $+0.40$ |
| $2\,SO_3^{2-} + 3\,H_2O + 4\,e \rightleftharpoons S_2O_3^{2-} + 6\,OH^-$ | $-0.58$ |
| $SO_3^{2-} + 3\,H_2O + 4\,e \rightleftharpoons S + 6\,OH^-$ | $-0.66$ |
| *Thallium* | |
| $Tl^{3+} + 2\,e \rightleftharpoons Tl^+$ | $+1.25$ |
| $Tl^+ + e \rightleftharpoons Tl$ | $-0.3363$ |
| *Tin* | |
| $Sn^{2+} + 2\,e \rightleftharpoons Sn$ | $-0.136$ |
| $Sn^{4+} + 2\,e \rightleftharpoons Sn^{2+}$ | $+0.154$ |
| $SnCl_6^{2-} + 2\,e \rightleftharpoons SnCl_4^{2-} + 2\,Cl^- \quad (1\,F\,HCl)$ | $+0.14$ |
| $Sn(OH)_6^{2-} + 2\,e \rightleftharpoons HSnO_2^- + H_2O + 3\,OH^-$ | $-0.93$ |
| $HSnO_2^- + H_2O + 2\,e \rightleftharpoons Sn + 3\,OH^-$ | $-0.91$ |
| *Titanium* | |
| $Ti^{2+} + 2\,e \rightleftharpoons Ti$ | $-1.63$ |
| $Ti^{3+} + e \rightleftharpoons Ti^{2+}$ | $-0.37$ |
| $TiO^{2+} + 2\,H^+ + e \rightleftharpoons Ti^{3+} + H_2O$ | $+0.10$ |
| $Ti^{4+} + e \rightleftharpoons Ti^{3+} \quad (5\,F\,H_3PO_4)$ | $-0.15$ |
| *Tungsten* | |
| $W^{6+} + e \rightleftharpoons W^{5+} \quad (12\,F\,HCl)$ | $+0.26$ |
| $W^{5+} + e \rightleftharpoons W^{4+} \quad (12\,F\,HCl)$ | $-0.3$ |
| $W(CN)_8^{3-} + e \rightleftharpoons W(CN)_8^{4-}$ | $+0.48$ |
| $2\,WO_3(s) + 2\,H^+ + 2\,e \rightleftharpoons W_2O_5(s) + H_2O$ | $-0.03$ |
| $W_2O_5(s) + 2\,H^+ + 2\,e \rightleftharpoons 2\,WO_2(s) + H_2O$ | $-0.043$ |
| *Uranium* | |
| $U^{4+} + e \rightleftharpoons U^{3+}$ | $-0.61$ |
| $UO_2^{2+} + e \rightleftharpoons UO_2^+$ | $+0.05$ |
| $UO_2^{2+} + 4\,H^+ + 2\,e \rightleftharpoons U^{4+} + 2\,H_2O$ | $+0.334$ |
| $UO_2^+ + 4\,H^+ + e \rightleftharpoons U^{4+} + 2\,H_2O$ | $+0.62$ |

| Half-Reaction | $E^0$ |
|---|---|

*Vanadium*

$VO_2^+ + 2\ H^+ + e \rightleftharpoons VO^{2+} + H_2O$      $+1.000$

$VO^{2+} + 2\ H^+ + e \rightleftharpoons V^{3+} + H_2O$      $+0.361$

$V^{3+} + e \rightleftharpoons V^{2+}$      $-0.255$

$V^{2+} + 2\ e \rightleftharpoons V$      $-1.18$

*Zinc*

$Zn^{2+} + 2\ e \rightleftharpoons Zn$      $-0.763$

$Zn(NH_3)_4^{2+} + 2\ e \rightleftharpoons Zn + 4\ NH_3$      $-1.04$

$Zn(CN)_4^{2-} + 2\ e \rightleftharpoons Zn + 4\ CN^-$      $-1.26$

$Zn(OH)_4^{2-} + 2\ e \rightleftharpoons Zn + 4\ OH^-$      $-1.22$

The *standard potential* for a redox couple is defined on page 267 as the potential (sign and magnitude) of an electrode consisting of that redox couple under standard-state conditions measured in a galvanic cell against the normal hydrogen electrode at 25°C.

*Formal potentials*, properly designated by the symbol $E^{0\prime}$ and defined on page 271 are italicized in the above table. The solution condition to which each formal potential pertains is written in parentheses following the half-reaction.

[a] The half-reaction and standard potential

$$Br_2(aq) + 2\ e \rightleftharpoons 2\ Br^-; \qquad E^0 = +1.087\ v$$

pertain to the system in which the activity of dissolved molecular bromine, $Br_2$, as well as the activity of the bromide ion, is unity in water. Actually, this is an impossible situation because the solubility of $Br_2$ in water is only about 0.21 $M$ at 25°C.

On the other hand, the half-reaction and standard potential

$$Br_2(l) + 2\ e \rightleftharpoons 2\ Br^-; \qquad E^0 = +1.065\ v$$

apply to an electrode system in which excess *liquid* bromine is in equilibrium with an aqueous solution containing bromide ion at unit activity. It follows that an aqueous solution in equilibrium with liquid bromine will be saturated with respect to molecular bromine at a concentration (activity) of 0.21 $M$.

Thus, these two standard potentials are different because the former refers to the (hypothetical) situation in which the concentration (activity) of $Br_2(aq)$ is taken to be unity, whereas the latter refers to the physically real situation for which the concentration (activity) of $Br_2(aq)$ is only 0.21 $M$.

[b] The reason for the difference between the two entries

$$I_2(aq) + 2\ e \rightleftharpoons 2\ I^-; \qquad E^0 = +0.6197\ v$$

and

$$I_2(s) + 2\ e \rightleftharpoons 2\ I^-; \qquad E^0 = +0.5345\ v$$

is essentially the same as that stated in the preceding footnote. The first half-reaction requires (hypothetically) an aqueous molecular iodine concentration or activity of unity, whereas the second half-reaction specifies that excess *solid* iodine be in equilibrium with an aqueous, iodide solution of unit activity. Since the solubility of molecular iodine in water at 25°C is approximately 0.00133 $M$, it is impossible to ever have an aqueous solution containing 1 $M$ molecular iodine. Thus, the standard potential for the second half-reaction will be considerably less oxidizing or less positive than the value for the first half-reaction.

Among the best sources of standard and formal potentials are the following references:

1. W. M. Latimer: *Oxidation Potentials.* Second edition, Prentice-Hall, Englewood Cliffs, New Jersey, 1952.
2. J. J. Lingane: *Electroanalytical Chemistry.* Second edition, Wiley-Interscience, New York, 1958, pp. 639–651.
3. L. Meites, ed.: *Handbook of Analytical Chemistry.* McGraw-Hill Book Company, New York, 1963.

# APPENDIX 5

## TABLE OF LOGARITHMS

|     | 0 | 1 | 2 | 3 | 4 | 5 | 6 | 7 | 8 | 9 |
|-----|---|---|---|---|---|---|---|---|---|---|
| 1.0 | 0.0000 | 0.0043 | 0.0086 | 0.0128 | 0.0170 | 0.0212 | 0.0253 | 0.0294 | 0.0334 | 0.0374 |
| 1.1 | 0.0414 | 0.0453 | 0.0492 | 0.0531 | 0.0569 | 0.0607 | 0.0645 | 0.0682 | 0.0719 | 0.0755 |
| 1.2 | 0.0792 | 0.0828 | 0.0864 | 0.0899 | 0.0934 | 0.0969 | 0.1004 | 0.1038 | 0.1072 | 0.1106 |
| 1.3 | 0.1139 | 0.1173 | 0.1206 | 0.1239 | 0.1271 | 0.1303 | 0.1335 | 0.1367 | 0.1399 | 0.1430 |
| 1.4 | 0.1461 | 0.1492 | 0.1523 | 0.1553 | 0.1584 | 0.1614 | 0.1644 | 0.1673 | 0.1703 | 0.1732 |
| 1.5 | 0.1761 | 0.1790 | 0.1818 | 0.1847 | 0.1875 | 0.1903 | 0.1931 | 0.1959 | 0.1987 | 0.2014 |
| 1.6 | 0.2041 | 0.2068 | 0.2095 | 0.2122 | 0.2148 | 0.2175 | 0.2201 | 0.2227 | 0.2253 | 0.2279 |
| 1.7 | 0.2304 | 0.2330 | 0.2355 | 0.2380 | 0.2405 | 0.2430 | 0.2455 | 0.2480 | 0.2504 | 0.2529 |
| 1.8 | 0.2553 | 0.2577 | 0.2601 | 0.2625 | 0.2648 | 0.2672 | 0.2695 | 0.2718 | 0.2742 | 0.2765 |
| 1.9 | 0.2788 | 0.2810 | 0.2833 | 0.2856 | 0.2878 | 0.2900 | 0.2923 | 0.2945 | 0.2967 | 0.2989 |
| 2.0 | 0.3010 | 0.3032 | 0.3054 | 0.3075 | 0.3096 | 0.3118 | 0.3139 | 0.3160 | 0.3181 | 0.3201 |
| 2.1 | 0.3222 | 0.3243 | 0.3263 | 0.3284 | 0.3304 | 0.3324 | 0.3345 | 0.3365 | 0.3385 | 0.3404 |
| 2.2 | 0.3424 | 0.3444 | 0.3464 | 0.3483 | 0.3502 | 0.3522 | 0.3541 | 0.3560 | 0.3579 | 0.3598 |
| 2.3 | 0.3617 | 0.3636 | 0.3655 | 0.3674 | 0.3692 | 0.3711 | 0.3729 | 0.3747 | 0.3766 | 0.3784 |
| 2.4 | 0.3802 | 0.3820 | 0.3838 | 0.3856 | 0.3874 | 0.3892 | 0.3909 | 0.3927 | 0.3945 | 0.3962 |
| 2.5 | 0.3979 | 0.3997 | 0.4014 | 0.4031 | 0.4048 | 0.4065 | 0.4082 | 0.4099 | 0.4116 | 0.4133 |
| 2.6 | 0.4150 | 0.4166 | 0.4183 | 0.4200 | 0.4216 | 0.4232 | 0.4249 | 0.4265 | 0.4281 | 0.4298 |
| 2.7 | 0.4314 | 0.4330 | 0.4346 | 0.4362 | 0.4378 | 0.4393 | 0.4409 | 0.4425 | 0.4440 | 0.4456 |
| 2.8 | 0.4472 | 0.4487 | 0.4502 | 0.4518 | 0.4533 | 0.4548 | 0.4564 | 0.4579 | 0.4594 | 0.4609 |
| 2.9 | 0.4624 | 0.4639 | 0.4654 | 0.4669 | 0.4683 | 0.4698 | 0.4713 | 0.4728 | 0.4742 | 0.4757 |
| 3.0 | 0.4771 | 0.4786 | 0.4800 | 0.4814 | 0.4829 | 0.4843 | 0.4857 | 0.4871 | 0.4886 | 0.4900 |
| 3.1 | 0.4914 | 0.4928 | 0.4942 | 0.4955 | 0.4969 | 0.4983 | 0.4997 | 0.5011 | 0.5024 | 0.5038 |
| 3.2 | 0.5051 | 0.5065 | 0.5079 | 0.5092 | 0.5105 | 0.5119 | 0.5132 | 0.5145 | 0.5159 | 0.5172 |
| 3.3 | 0.5185 | 0.5198 | 0.5211 | 0.5224 | 0.5237 | 0.5250 | 0.5263 | 0.5276 | 0.5289 | 0.5302 |
| 3.4 | 0.5315 | 0.5328 | 0.5340 | 0.5353 | 0.5366 | 0.5378 | 0.5391 | 0.5403 | 0.5416 | 0.5428 |
| 3.5 | 0.5441 | 0.5453 | 0.5465 | 0.5478 | 0.5490 | 0.5502 | 0.5514 | 0.5527 | 0.5539 | 0.5551 |
| 3.6 | 0.5563 | 0.5575 | 0.5587 | 0.5599 | 0.5611 | 0.5623 | 0.5635 | 0.5647 | 0.5658 | 0.5670 |
| 3.7 | 0.5682 | 0.5694 | 0.5705 | 0.5717 | 0.5729 | 0.5740 | 0.5752 | 0.5763 | 0.5775 | 0.5786 |
| 3.8 | 0.5798 | 0.5809 | 0.5821 | 0.5832 | 0.5843 | 0.5855 | 0.5866 | 0.5877 | 0.5888 | 0.5899 |
| 3.9 | 0.5911 | 0.5922 | 0.5933 | 0.5944 | 0.5955 | 0.5966 | 0.5977 | 0.5988 | 0.5999 | 0.6010 |
| 4.0 | 0.6021 | 0.6031 | 0.6042 | 0.6053 | 0.6064 | 0.6075 | 0.6085 | 0.6096 | 0.6107 | 0.6117 |
| 4.1 | 0.6128 | 0.6138 | 0.6149 | 0.6160 | 0.6170 | 0.6180 | 0.6191 | 0.6201 | 0.6212 | 0.6222 |
| 4.2 | 0.6232 | 0.6243 | 0.6253 | 0.6263 | 0.6274 | 0.6284 | 0.6294 | 0.6304 | 0.6314 | 0.6325 |
| 4.3 | 0.6335 | 0.6345 | 0.6355 | 0.6365 | 0.6375 | 0.6385 | 0.6395 | 0.6405 | 0.6415 | 0.6425 |
| 4.4 | 0.6435 | 0.6444 | 0.6454 | 0.6464 | 0.6474 | 0.6484 | 0.6493 | 0.6503 | 0.6513 | 0.6522 |
| 4.5 | 0.6532 | 0.6542 | 0.6551 | 0.6561 | 0.6571 | 0.6580 | 0.6590 | 0.6599 | 0.6609 | 0.6618 |
| 4.6 | 0.6628 | 0.6637 | 0.6646 | 0.6656 | 0.6665 | 0.6675 | 0.6684 | 0.6693 | 0.6702 | 0.6712 |
| 4.7 | 0.6721 | 0.6730 | 0.6739 | 0.6749 | 0.6758 | 0.6767 | 0.6776 | 0.6785 | 0.6794 | 0.6803 |
| 4.8 | 0.6812 | 0.6821 | 0.6830 | 0.6839 | 0.6848 | 0.6857 | 0.6866 | 0.6875 | 0.6884 | 0.6893 |
| 4.9 | 0.6902 | 0.6911 | 0.6920 | 0.6928 | 0.6937 | 0.6946 | 0.6955 | 0.6964 | 0.6972 | 0.6981 |
| 5.0 | 0.6990 | 0.6998 | 0.7007 | 0.7016 | 0.7024 | 0.7033 | 0.7042 | 0.7050 | 0.7059 | 0.7067 |
| 5.1 | 0.7076 | 0.7084 | 0.7093 | 0.7101 | 0.7110 | 0.7118 | 0.7126 | 0.7135 | 0.7143 | 0.7152 |
| 5.2 | 0.7160 | 0.7168 | 0.7177 | 0.7185 | 0.7193 | 0.7202 | 0.7210 | 0.7218 | 0.7226 | 0.7235 |
| 5.3 | 0.7243 | 0.7251 | 0.7259 | 0.7267 | 0.7275 | 0.7284 | 0.7292 | 0.7300 | 0.7308 | 0.7316 |
| 5.4 | 0.7324 | 0.7332 | 0.7340 | 0.7348 | 0.7356 | 0.7364 | 0.7372 | 0.7380 | 0.7388 | 0.7396 |
| 5.5 | 0.7404 | 0.7412 | 0.7419 | 0.7427 | 0.7435 | 0.7443 | 0.7451 | 0.7459 | 0.7466 | 0.7474 |

|      | 0 | 1 | 2 | 3 | 4 | 5 | 6 | 7 | 8 | 9 |
|------|--------|--------|--------|--------|--------|--------|--------|--------|--------|--------|
| 5.6 | 0.7482 | 0.7490 | 0.7497 | 0.7505 | 0.7513 | 0.7520 | 0.7528 | 0.7536 | 0.7543 | 0.7551 |
| 5.7 | 0.7559 | 0.7566 | 0.7574 | 0.7582 | 0.7589 | 0.7597 | 0.7604 | 0.7612 | 0.7619 | 0.7627 |
| 5.8 | 0.7634 | 0.7642 | 0.7649 | 0.7657 | 0.7664 | 0.7672 | 0.7679 | 0.7686 | 0.7694 | 0.7701 |
| 5.9 | 0.7709 | 0.7716 | 0.7723 | 0.7731 | 0.7738 | 0.7745 | 0.7752 | 0.7760 | 0.7767 | 0.7774 |
| 6.0 | 0.7782 | 0.7789 | 0.7796 | 0.7803 | 0.7810 | 0.7818 | 0.7825 | 0.7832 | 0.7839 | 0.7846 |
| 6.1 | 0.7853 | 0.7860 | 0.7868 | 0.7875 | 0.7882 | 0.7889 | 0.7896 | 0.7903 | 0.7910 | 0.7917 |
| 6.2 | 0.7924 | 0.7931 | 0.7938 | 0.7945 | 0.7952 | 0.7959 | 0.7966 | 0.7973 | 0.7980 | 0.7987 |
| 6.3 | 0.7993 | 0.8000 | 0.8007 | 0.8014 | 0.8021 | 0.8028 | 0.8035 | 0.8041 | 0.8048 | 0.8055 |
| 6.4 | 0.8062 | 0.8069 | 0.8075 | 0.8082 | 0.8089 | 0.8096 | 0.8102 | 0.8109 | 0.8116 | 0.8122 |
| 6.5 | 0.8129 | 0.8136 | 0.8142 | 0.8149 | 0.8156 | 0.8162 | 0.8169 | 0.8176 | 0.8182 | 0.8189 |
| 6.6 | 0.8195 | 0.8202 | 0.8209 | 0.8215 | 0.8222 | 0.8228 | 0.8235 | 0.8241 | 0.8248 | 0.8254 |
| 6.7 | 0.8261 | 0.8267 | 0.8274 | 0.8280 | 0.8287 | 0.8293 | 0.8299 | 0.8306 | 0.8312 | 0.8319 |
| 6.8 | 0.8325 | 0.8331 | 0.8338 | 0.8344 | 0.8351 | 0.8357 | 0.8363 | 0.8370 | 0.8376 | 0.8382 |
| 6.9 | 0.8388 | 0.8395 | 0.8401 | 0.8407 | 0.8414 | 0.8420 | 0.8426 | 0.8432 | 0.8439 | 0.8445 |
| 7.0 | 0.8451 | 0.8457 | 0.8463 | 0.8470 | 0.8476 | 0.8482 | 0.8488 | 0.8494 | 0.8500 | 0.8506 |
| 7.1 | 0.8513 | 0.8519 | 0.8525 | 0.8531 | 0.8537 | 0.8543 | 0.8549 | 0.8555 | 0.8561 | 0.8567 |
| 7.2 | 0.8573 | 0.8579 | 0.8585 | 0.8591 | 0.8597 | 0.8603 | 0.8609 | 0.8615 | 0.8621 | 0.8627 |
| 7.3 | 0.8633 | 0.8639 | 0.8645 | 0.8651 | 0.8657 | 0.8663 | 0.8669 | 0.8675 | 0.8681 | 0.8686 |
| 7.4 | 0.8692 | 0.8698 | 0.8704 | 0.8710 | 0.8716 | 0.8722 | 0.8727 | 0.8733 | 0.8739 | 0.8745 |
| 7.5 | 0.8751 | 0.8756 | 0.8762 | 0.8768 | 0.8774 | 0.8779 | 0.8785 | 0.8791 | 0.8797 | 0.8802 |
| 7.6 | 0.8808 | 0.8814 | 0.8820 | 0.8825 | 0.8831 | 0.8837 | 0.8842 | 0.8848 | 0.8854 | 0.8859 |
| 7.7 | 0.8865 | 0.8871 | 0.8876 | 0.8882 | 0.8887 | 0.8893 | 0.8899 | 0.8904 | 0.8910 | 0.8915 |
| 7.8 | 0.8921 | 0.8927 | 0.8932 | 0.8938 | 0.8943 | 0.8949 | 0.8954 | 0.8960 | 0.8965 | 0.8971 |
| 7.9 | 0.8976 | 0.8982 | 0.8987 | 0.8993 | 0.8998 | 0.9004 | 0.9009 | 0.9015 | 0.9020 | 0.9025 |
| 8.0 | 0.9031 | 0.9036 | 0.9042 | 0.9047 | 0.9053 | 0.9058 | 0.9063 | 0.9069 | 0.9074 | 0.9079 |
| 8.1 | 0.9085 | 0.9090 | 0.9096 | 0.9101 | 0.9106 | 0.9112 | 0.9117 | 0.9122 | 0.9128 | 0.9133 |
| 8.2 | 0.9138 | 0.9143 | 0.9149 | 0.9154 | 0.9159 | 0.9165 | 0.9170 | 0.9175 | 0.9180 | 0.9186 |
| 8.3 | 0.9191 | 0.9196 | 0.9201 | 0.9206 | 0.9212 | 0.9217 | 0.9222 | 0.9227 | 0.9232 | 0.9238 |
| 8.4 | 0.9243 | 0.9248 | 0.9253 | 0.9258 | 0.9263 | 0.9269 | 0.9274 | 0.9279 | 0.9284 | 0.9289 |
| 8.5 | 0.9294 | 0.9299 | 0.9304 | 0.9309 | 0.9315 | 0.9320 | 0.9325 | 0.9330 | 0.9335 | 0.9340 |
| 8.6 | 0.9345 | 0.9350 | 0.9355 | 0.9360 | 0.9365 | 0.9370 | 0.9375 | 0.9380 | 0.9385 | 0.9390 |
| 8.7 | 0.9395 | 0.9400 | 0.9405 | 0.9410 | 0.9415 | 0.9420 | 0.9425 | 0.9430 | 0.9435 | 0.9440 |
| 8.8 | 0.9445 | 0.9450 | 0.9455 | 0.9460 | 0.9465 | 0.9469 | 0.9474 | 0.9479 | 0.9484 | 0.9489 |
| 8.9 | 0.9494 | 0.9499 | 0.9504 | 0.9509 | 0.9513 | 0.9518 | 0.9523 | 0.9528 | 0.9533 | 0.9538 |
| 9.0 | 0.9542 | 0.9547 | 0.9552 | 0.9557 | 0.9562 | 0.9566 | 0.9571 | 0.9576 | 0.9581 | 0.9586 |
| 9.1 | 0.9590 | 0.9595 | 0.9600 | 0.9605 | 0.9609 | 0.9614 | 0.9619 | 0.9624 | 0.9628 | 0.9633 |
| 9.2 | 0.9638 | 0.9643 | 0.9647 | 0.9652 | 0.9657 | 0.9661 | 0.9666 | 0.9671 | 0.9675 | 0.9680 |
| 9.3 | 0.9685 | 0.9689 | 0.9694 | 0.9699 | 0.9703 | 0.9708 | 0.9713 | 0.9717 | 0.9722 | 0.9727 |
| 9.4 | 0.9731 | 0.9736 | 0.9741 | 0.9745 | 0.9750 | 0.9754 | 0.9759 | 0.9763 | 0.9768 | 0.9773 |
| 9.5 | 0.9777 | 0.9782 | 0.9786 | 0.9791 | 0.9795 | 0.9800 | 0.9805 | 0.9809 | 0.9814 | 0.9818 |
| 9.6 | 0.9823 | 0.9827 | 0.9832 | 0.9836 | 0.9841 | 0.9845 | 0.9850 | 0.9854 | 0.9859 | 0.9863 |
| 9.7 | 0.9868 | 0.9872 | 0.9877 | 0.9881 | 0.9886 | 0.9890 | 0.9894 | 0.9899 | 0.9903 | 0.9908 |
| 9.8 | 0.9912 | 0.9917 | 0.9921 | 0.9926 | 0.9930 | 0.9934 | 0.9939 | 0.9943 | 0.9948 | 0.9952 |
| 9.9 | 0.9956 | 0.9961 | 0.9965 | 0.9969 | 0.9974 | 0.9978 | 0.9983 | 0.9987 | 0.9991 | 0.9996 |

# APPENDIX 6

## FORMULA WEIGHTS OF COMMON COMPOUNDS

| | | | |
|---|---|---|---|
| $AgBr$ | 187.78 | $FeS_2$ | 119.98 |
| $AgBrO_3$ | 235.78 | $FeSO_4 \cdot 7\ H_2O$ | 278.05 |
| $AgCl$ | 143.32 | $Fe_2(SO_4)_3$ | 399.87 |
| $Ag_2CrO_4$ | 331.73 | $Fe(NH_4)_2(SO_4)_2 \cdot 6\ H_2O$ | 392.14 |
| $AgI$ | 234.77 | $HBr$ | 80.92 |
| $AgNO_3$ | 169.87 | $HCHO_2$ (formic acid) | 46.03 |
| $Ag_3PO_4$ | 418.58 | $HC_2H_3O_2$ (acetic acid) | 60.05 |
| $AgSCN$ | 165.95 | $HC_7H_5O_2$ (benzoic acid) | 122.13 |
| $Ag_2SO_4$ | 311.80 | $HCl$ | 36.46 |
| $AlBr_3$ | 266.71 | $HClO_4$ | 100.46 |
| $Al_2O_3$ | 101.96 | $H_2C_2O_4 \cdot 2\ H_2O$ (oxalic acid) | 126.06 |
| $Al(OH)_3$ | 78.00 | $HI$ | 127.91 |
| $Al_2(SO_4)_3$ | 342.15 | $HNO_3$ | 63.01 |
| $As_2O_3$ | 197.84 | $H_2O$ | 18.02 |
| $As_2O_5$ | 229.84 | $H_2O_2$ | 34.01 |
| $As_2S_3$ | 246.04 | $H_3PO_4$ | 98.00 |
| $BaCl_2$ | 208.25 | $H_2S$ | 34.08 |
| $BaCl_2 \cdot 2\ H_2O$ | 244.28 | $H_2SO_3$ | 82.08 |
| $BaCO_3$ | 197.35 | $H_2SO_4$ | 98.08 |
| $BaC_2O_4$ | 225.36 | $Hg_2Br_2$ | 561.00 |
| $BaF_2$ | 175.34 | $HgCl_2$ | 271.50 |
| $BaI_2$ | 391.15 | $Hg_2Cl_2$ | 472.09 |
| $Ba(IO_3)_2$ | 487.15 | $Hg_2I_2$ | 654.99 |
| $BaO$ | 153.34 | $HgO$ | 216.59 |
| $Ba(OH)_2$ | 171.36 | $KBr$ | 119.01 |
| $BaSO_4$ | 233.40 | $KBrO_3$ | 167.01 |
| $Bi_2O_3$ | 465.96 | $KCl$ | 74.56 |
| $Bi_2S_3$ | 514.15 | $KClO_3$ | 122.55 |
| $CaCl_2 \cdot 2\ H_2O$ | 147.02 | $KClO_4$ | 138.55 |
| $CaCO_3$ | 100.09 | $KCN$ | 65.12 |
| $CaF_2$ | 78.08 | $KCNS$ | 97.18 |
| $Ca(NO_3)_2$ | 164.09 | $K_2CO_3$ | 138.21 |
| $CaO$ | 56.08 | $K_2CrO_4$ | 194.20 |
| $Ca(OH)_2$ | 74.10 | $K_2Cr_2O_7$ | 294.19 |
| $Ca_3(PO_4)_2$ | 310.18 | $K_3Fe(CN)_6$ | 329.26 |
| $CaSO_4$ | 136.14 | $K_4Fe(CN)_6 \cdot 3\ H_2O$ | 422.41 |
| $Ce(HSO_4)_4$ | 528.41 | $KHC_4H_4O_6$ (tartrate) | 188.18 |
| $CeO_2$ | 172.12 | $KHC_8H_4O_4$ (phthalate) | 204.23 |
| $Ce(NH_4)_4(SO_4)_4 \cdot 2\ H_2O$ | 632.57 | $KHCO_3$ | 100.12 |
| $CO_2$ | 44.01 | $KHSO_4$ | 136.17 |
| $Cr_2O_3$ | 151.99 | $KI$ | 166.01 |
| $CuO$ | 79.54 | $KIO_3$ | 214.00 |
| $Cu_2O$ | 143.08 | $KMnO_4$ | 158.04 |
| $CuS$ | 95.60 | $KNO_2$ | 85.11 |
| $Cu_2S$ | 159.14 | $KNO_3$ | 101.11 |
| $CuSO_4 \cdot 5\ H_2O$ | 249.68 | $K_2O$ | 94.20 |
| $FeCl_3$ | 162.21 | $KOH$ | 56.11 |
| $Fe(NO_3)_3 \cdot 6\ H_2O$ | 349.95 | $K_3PO_4$ | 212.28 |
| $FeO$ | 71.85 | $K_2PtCl_6$ | 486.01 |
| $Fe_2O_3$ | 159.69 | $K_2SO_4$ | 174.27 |
| $Fe_3O_4$ | 231.54 | $LiCl$ | 42.39 |
| $Fe(OH)_3$ | 106.87 | $Li_2CO_3$ | 73.89 |

| | | | |
|---|---|---|---|
| $Li_2O$ | 29.88 | $Na_3PO_4$ | 163.94 |
| $LiOH$ | 23.95 | $Na_3PO_4 \cdot 12\ H_2O$ | 380.12 |
| $MgCO_3$ | 84.32 | $Na_2S$ | 78.04 |
| $MgNH_4AsO_4$ | 181.27 | $Na_2SO_3$ | 126.04 |
| $MgNH_4PO_4$ | 137.32 | $Na_2SO_4 \cdot 10\ H_2O$ | 322.19 |
| $MgO$ | 40.31 | $Na_2S_2O_3$ | 158.11 |
| $Mg(OH)_2$ | 58.33 | $Na_2S_2O_3 \cdot 5\ H_2O$ | 248.18 |
| $Mg_2P_2O_7$ | 222.57 | $P_2O_5$ | 141.94 |
| $MgSO_4$ | 120.37 | $PbCl_2$ | 278.10 |
| $MnO_2$ | 86.94 | $PbCrO_4$ | 323.18 |
| $Mn_2O_3$ | 157.87 | $PbI_2$ | 461.00 |
| $NH_3$ | 17.03 | $Pb(IO_3)_2$ | 557.00 |
| $NH_4Cl$ | 53.49 | $Pb(NO_3)_2$ | 331.20 |
| $NH_4NO_3$ | 80.04 | $PbO$ | 223.19 |
| $NH_4OH$ | 35.05 | $PbO_2$ | 239.19 |
| $(NH_4)_2SO_4$ | 132.14 | $PbSO_4$ | 303.25 |
| $NO$ | 30.01 | $SO_2$ | 64.06 |
| $NO_2$ | 46.01 | $SO_3$ | 80.06 |
| $N_2O_3$ | 76.01 | $Sb_2O_3$ | 291.50 |
| $Na_2B_4O_7 \cdot 10\ H_2O$ | 381.37 | $Sb_2O_5$ | 323.50 |
| $NaBr$ | 102.90 | $Sb_2S_3$ | 339.69 |
| $NaBrO_3$ | 150.90 | $SiCl_4$ | 169.90 |
| $NaCHO_2$ (formate) | 68.01 | $SiF_4$ | 104.08 |
| $NaC_2H_3O_2$ (acetate) | 82.03 | $SiO_2$ | 60.08 |
| $NaCl$ | 58.44 | $SnCl_2$ | 189.60 |
| $NaCN$ | 49.01 | $SnCl_4$ | 260.50 |
| $Na_2CO_3$ | 105.99 | $SnO_2$ | 150.69 |
| $Na_2C_2O_4$ | 134.00 | $SrCO_3$ | 147.63 |
| $NaHCO_3$ | 84.00 | $SrO$ | 103.62 |
| $Na_2HPO_4$ | 141.96 | $SrSO_4$ | 183.68 |
| $NaHS$ | 56.06 | $TiO_2$ | 79.90 |
| $NaH_2PO_4$ | 119.97 | $UO_3$ | 286.03 |
| $NaI$ | 149.89 | $U_3O_8$ | 842.09 |
| $NaNO_2$ | 69.00 | $WO_3$ | 231.85 |
| $NaNO_3$ | 84.99 | $ZnO$ | 81.37 |
| $Na_2O$ | 61.98 | $Zn_2P_2O_7$ | 304.68 |
| $Na_2O_2$ | 77.98 | $ZnSO_4$ | 161.43 |
| $NaOH$ | 40.00 | $ZnSO_4 \cdot 7\ H_2O$ | 287.54 |

# APPENDIX 7

## ANSWERS TO NUMERICAL PROBLEMS

(Except in those problems in the text having stated numerical values for the pertinent equilibrium constants, the equilibrium-constant data compiled in the various preceding appendixes have been used to determine the answers to problems given below.)

### CHAPTER 2

2. (a) 9.6 parts per thousand
   (b) 0.083 part per thousand
   (c) 0.99 part per million
   (d) 1.76 parts per million
   (e) 2.19 parts per thousand
   (f) 0.17 part per thousand
5. (a) 60.10 per cent
   (b) 0.26 per cent
   (c) 0.43 per cent
6. $-0.56$ per cent; $-0.92$ per cent
8. (a) 2.91 ppm
   (b) 1.68 ppm
9. (a) 21.19 per cent
   (b) $2.9 \times 10^{-5}$ per cent
   (c) 55.86 per cent
   (d) 54.00 per cent
10. 0.0548
11. 1.83 hr
12. 8 mg
13. 0.642
14. $5.82 \times 10^{-3}$
15. (a) 0.5 mg
    (b) 100
    (c) 49.8 per cent
16. (a) 7671 hr; 7536 hr
    (b) Yes, only 0.13 per cent chance
    (c) 2.27 per cent
17. $\pm 4.1$ ppm
18. (a) $\pm 0.58$ per cent
    (b) $\pm 0.73$ per cent
    (c) $\pm 0.82$ per cent
19. No
20. (a) 12.0104
    (b) 0.0012
    (c) 0.00038
    (d) 0.0012

21. Yes
22. Inconclusive
23. $c = 3.922A - 0.0973$
24. (a) output $= 0.7201 + 9.916c$
    (b) 0.467 ppm
    (c) $0.024_6$; $\pm 0.068_4$
25. Yes
    (a) 120
    (b) $17.18_5$
    (c) $\pm 38.87$


# CHAPTER 3

1. (a) $0.02570\,F$
   (b) $0.1192\,F$
   (c) $0.007597\,F$
   (d) $0.3122\,F$
   (e) $0.08541\,F$
2. (a) 0.2574 gm
   (b) 17.72 gm
   (c) 0.5531 gm
   (d) 12.97 gm
   (e) 0.6394 gm
3. $0.0053\,M$
8. 4.25
9. $0.0247\,M$
10. $1.72 \times 10^{-4}\,M$
11. 7.65 gm
12. $3.16 \times 10^{-7}\,M$
14. 0.388 gm
15. (a) $+78.12$ kcal
    (b) $+31.1$ kcal
    (c) $-16.73$ kcal
    (d) $-24.14$ kcal
    (e) $-17.05$ kcal
    (f) $+59.9$ kcal
    (g) $+12.1$ kcal
16. (a) $K = 1.05 \times 10^{489}$
    (b) $K = 6.62 \times 10^{5}$
    (c) $K = 1.15 \times 10^{-39}$
    (d) $K = 4.46 \times 10^{18}$
    (e) $K = 9.12 \times 10^{-17}$
    (f) $K = 1.82 \times 10^{230}$
    (g) $K = 4.46 \times 10^{22}$
19. 22.1 kcal
21. $[PCl_3] = [Cl_2] = 1.89\,M$;
    $[PCl_5] = 0.11\,M$
22. (b) $K_p = 1.01 \times 10^{-6}$;
    $K = 2.86 \times 10^{-8}$
23. (a) 0.018
    (b) 0.033

      (c)  0.26

      (d)  1.50

24.  (a)  0.825

      (b)  0.747

      (c)  0.462

      (d)  0.587

      (e)  0.813

25.  (a)  0.161

      (b)  0.426

      (c)  0.474

26.  (b)  $S$ (in $MgCl_2$) $= 2.40 \times 10^{-5}\,F$;
          $S$ (in $H_2O$) $= 1.02 \times 10^{-5}\,F$

28.  (a)  0.630

      (b)  0.708

      (c)  0.603

      (d)  0.884

      (e)  0.424

29.  (b)  $K_{ap} = 1.618 \times 10^{-10}$

30.  $f_{Na^+} = 0.710$; $f_{HPO_4{}^{2-}} = 0.219$

## CHAPTER 4

1.  (a)  $4.26 \times 10^{-7}\,M$

     (b)  $1.48 \times 10^{-5}\,M$

     (c)  $6.03 \times 10^{-12}\,M$

     (d)  $2.88 \times 10^{-9}\,M$

     (e)  $7.41 \times 10^{-4}\,M$

2.  (a)  $1.91 \times 10^{-11}\,M$

     (b)  $7.25 \times 10^{-5}\,M$

     (c)  $2.95 \times 10^{-12}\,M$

     (d)  $1.55 \times 10^{-2}\,M$

     (e)  $5.13 \times 10^{-8}\,M$

3.  (a)  1.86

     (b)  4.68

     (c)  10.13

     (d)  $-0.38$

     (e)  6.38

5.  (a)  0.40

     (b)  12.67

     (c)  2.47

     (d)  4.67

     (e)  1.19

     (f)  9.24

     (g)  9.44

     (h)  6.11

     (i)  7.17

     (j)  1.40

     (k)  2.34

     (l)  10.57

6.  $[H^+] = 1.62 \times 10^{-7}\,M$; pH $= 6.79$

7.  1.00 to 27.6

8. 1.00 to 7.25

11. $[CO_3{}^{2-}] = 0.11\ M$;
    $[HCO_3{}^-] = 0.47\ M$

12. 1.00 to 1.70; pH $= 5.17$

13. pH $= 1.24$;
    $[HPO_4{}^{2-}] = 6.2 \times 10^{-8}\ M$

14. $[H^+] = 4.5 \times 10^{-12}\ M$;
    $[NH_4{}^+] = 2.2 \times 10^{-3}\ M$;
    $[NH_3] = 0.265\ F$

15. $[NH_4{}^+] = 0.283\ M$;
    $[NH_3] = 7.78 \times 10^{-7}\ M$;
    pH $= 3.70$

16. $[H^+] = 1.89 \times 10^{-3}\ M$;
    $[C_2H_5COO^-] = 1.89 \times 10^{-3}\ M$;
    $[C_2H_5COOH] = 0.265\ M$

17. $[C_2H_5COO^-] = 0.340\ M$;
    $[C_2H_5COOH] = 1.26 \times 10^{-6}\ M$;
    pH $= 10.30$

18. 5.06

19. $[H^+] = 2.74 \times 10^{-2}\ M$;
    $[NO_2{}^-] = 2.74 \times 10^{-2}\ M$;
    $[HNO_2] = 0.473\ M$

20. $[F^-] = 2.10 \times 10^{-2}\ M$;
    $[HF] = 0.649\ M$;
    pH $= 1.68$

21. $[NH_3] = 0.0260\ F$;
    $[NH_4{}^+] = 6.9 \times 10^{-4}\ M$;
    $[H^+] = 1.45 \times 10^{-11}\ M$;
    $[OH^-] = 6.9 \times 10^{-4}\ M$

22. (b) $[HC_2O_4{}^-] = 0.00834\ M$;
        $[H_2C_2O_4] = 0.000639\ M$;
        $[C_2O_4{}^{2-}] = 0.00967\ M$
    (c) $[H_2C_2O_4] = 3.58 \times 10^{-6}\ M$;
        $[HC_2O_4{}^-] = 2.31 \times 10^{-3}\ M$;
        $[C_2O_4{}^{2-}] = 1.42 \times 10^{-3}\ M$

23. (b) $[H_3Cit] = 0.0181\ M$;
        $[H_2Cit^-] = 0.281\ M$;
        $[HCit^{2-}] = 0.100\ M$;
        $[Cit^{3-}] = 8.32 \times 10^{-4}\ M$
    (c) pH $= 5.04$;
        $[H_3Cit] = 0.00165\ M$;
        $[H_2Cit^-] = 0.134\ M$;
        $[HCit^{2-}] = 0.249\ M$;
        $[Cit^{3-}] = 0.0173\ M$

24. pH $= 6.11$;
    $[H_2Al^+] = 1.72 \times 10^{-6}\ M$;
    $[HAl] = 0.0100\ M$;
    $[Al^-] = 1.72 \times 10^{-6}\ M$

26. 180.0 gm

27. 0.4021 $F$

29. Points on titration curve correspond to pH 2.54, 4.35, 6.15, 8.14, 11.12, and
    12.22; titration error $= -0.23$ per cent; titration error $= +0.12$ per cent

30. Points on titration curve correspond to pH 11.33, 9.86, 9.44, 9.08, 8.66, 5.08, 1.64, and 1.24; titration error $= +0.63$ per cent; titration error $= -0.055$ per cent
32. $1.84 \times 10^{-5} F$
33. (a) $-0.99$ per cent
    (b) 8.43
    (c) 0.0836
34. (a) 336
    (b) $1.25 \times 10^{-5}$
    (c) 8.76
38. Points on titration curve correspond to pH 4.00, 4.11, 4.32, 5.74, 6.96, 7.00, 9.00, and 9.25
39. 6.005 gm
40. $0.3449 F$
41. $0.14 F$ $H_2SO_4$; $0.16 F$ $H_3PO_4$
42. $1.158 F$ acid; $1.242 F$ base
43. 118.7
44. 4.804 per cent
45. $0.1835 F$; $0.06805 F$
46. 35.52 per cent
47. 0.2932 per cent
48. 28.00 ml
49. $[HPO_4{}^{2-}] = 0.120\ M$;
    $[H_2PO_4{}^-] = 0.0138\ M$
50. $[CO_3{}^{2-}] = 0.048\ M$;
    $[HCO_3{}^-] = 0.024\ M$

## CHAPTER 5

12. 19.99 per cent ethylamine;
    24.34 per cent diethylamine;
    55.67 per cent triethylamine
13. 43.30 per cent 2-naphthoic acid;
    38.68 per cent 1-hydroxy-2-naphthoic acid
14. 5.173 per cent sulfanilamide;
    5.085 per cent sulfathiazole
15. (a) 10.7, 31.7, and 0.52 pH units
    (b) 8.45 pH units
16. (b) $3.45 \times 10^{-3}\ M$; $2.51\ F$

## CHAPTER 6

5. $Zn(NH_3)^{2+}$, 0.24; $Zn(NH_3)_2{}^{2+}$, 0.18; $Zn(NH_3)_3{}^{2+}$, 0.15; $Zn(NH_3)_4{}^{2+}$, 0.05;
   $[Zn(NH_3)^{2+}] = 3.87 \times 10^{-4}\ M$; $[Zn(NH_3)_2{}^{2+}] = 2.90 \times 10^{-4}\ M$;
   $[Zn(NH_3)_3{}^{2+}] = 2.42 \times 10^{-4}\ M$; $[Zn(NH_3)_4{}^{2+}] = 8.10 \times 10^{-5}\ M$
8. (b) $Ni(NH_3)_6{}^{2+}$, $Ni(NH_3)_5{}^{2+}$, $Ni(NH_3)_4{}^{2+}$, $Ni(NH_3)_3{}^{2+}$, $Ni(NH_3)_2{}^{2+}$, $Ni(NH_3)^{2+}$, $Ni^{2+}$
   (c) $[Ni^{2+}] = 4.25 \times 10^{-11}\ M$; $[Ni(NH_3)^{2+}] = 7.33 \times 10^{-9}\ M$; $[Ni(NH_3)_2{}^{2+}] = 4.35 \times 10^{-7}\ M$; $[Ni(NH_3)_3{}^{2+}] = 1.16 \times 10^{-5}\ M$; $[Ni(NH_3)_4{}^{2+}] = 1.48 \times 10^{-4}\ M$; $[Ni(NH_3)_5{}^{2+}] = 7.85 \times 10^{-4}\ M$; $[Ni(NH_3)_6{}^{2+}] = 1.56 \times 10^{-3}\ M$
9. $[Cu(NH_3)^{2+}] = 7.65 \times 10^{-3}\ M$; $[Cu(NH_3)_2{}^{2+}] = 0.0895\ M$; $[Cu(NH_3)_3{}^{2+}] = 0.246\ M$; $[Cu(NH_3)_4{}^{2+}] = 0.123\ M$

10. $0.589\,F$

11. $[Zn(NH_3)_2^{2+}] = 7.85 \times 10^{-7}\,M$; $[Zn(NH_3)_3^{2+}] = 2.56 \times 10^{-5}\,M$

12. $[Ag^+] = 1.19 \times 10^{-9}\,M$;　$[Ag(NH_3)^+] = 1.10 \times 10^{-6}\,M$;　$[Ag(NH_3)_2^+] = 0.00300\,M$.

13. $[Ag^+] = 3.56 \times 10^{-19}\,M$;　$[CN^-] = 0.0428\,M$;　$[Ag(CN)_2^-] = 0.0461\,M$

14. $[Ag^+] = 0.0166\,M$; $[CN^-] = 1.88 \times 10^{-10}\,M$; $[Ag(CN)_2^-] = 0.0417\,M$

15. $[H_4Y] = 4.49 \times 10^{-20}\,M$; $[H_3Y^-] = 1.42 \times 10^{-12}\,M$; $[H_2Y^{2-}] = 9.74 \times 10^{-6}\,M$; $[HY^{3-}] = 0.0213\,M$; $[Y^{4-}] = 3.69 \times 10^{-3}\,M$

16. $3.92 \times 10^{-8}\,M$

17. $7.55 \times 10^{-19}\,M$

18. $pH = 9.56$; $[H_4Y] = 1.74 \times 10^{-20}\,M$; $[H_3Y^-] = 6.31 \times 10^{-13}\,M$; $[H_2Y^{2-}] = 4.96 \times 10^{-6}\,M$; $[HY^{3-}] = 0.0125\,M$; $[Y^{4-}] = 2.50 \times 10^{-3}\,M$

19. $[NH_3] = 5.09 \times 10^{-4}\,M$; $[NH_4^+] = 1.85 \times 10^{-4}\,M$

20. (c) $[H_3Y] = 2.07 \times 10^{-6}\,M$; $[H_2Y^-] = 3.43 \times 10^{-4}\,M$; $[HY^{2-}] = 1.47 \times 10^{-2}\,M$; $[Y^{3-}] = 3.60 \times 10^{-8}\,M$

21. $5.62 \times 10^{10}$

22. $1.49 \times 10^{-12}$

25. $0.008832\,F$

26. $0.0844\,M$

27. $+0.45$ per cent

28. $28.9$ mg

29. $195.7$ mg

30. $111.1$ ppm $CaCO_3$; $35.76$ ppm $Ca^{2+}$; $5.30$ ppm $Mg^{2+}$

31. $[Fe^{3+}] = 0.02378\,M$; $[Al^{3+}] = 0.02788\,M$

32. $31.45$ per cent

33. $78.40$ per cent

34. $7.886$ per cent


## CHAPTER 7

5.　(a) $6$ cm$^2$
　　(b) $12$ cm$^2$
　　(c) $24$ cm$^2$

12.　(a) $1.78 \times 10^{-4}$
　　(b) $2.52 \times 10^{-13}$
　　(c) $1.42 \times 10^{-16}$
　　(d) $1.24 \times 10^{-11}$

13.　(a) $8.8 \times 10^{-14}$
　　(b) $7.5 \times 10^{-3}$
　　(c) $1.3 \times 10^{-5}$
　　(d) $2.6 \times 10^{-3}$
　　(e) $6.2 \times 10^{-4}$

15. $4.4 \times 10^{-12}\,M$

16. $3.5 \times 10^{-5}\,M$

18. $3.2 \times 10^{-4}$ mg

19. $[Ag^+] = 1.39 \times 10^{-13}\,M$

20. $11.5\,F$

21. $1.08 \times 10^{-16}$

22. $2.07 \times 10^{-5}\,F$

23. (a) $1.4 \times 10^{-4}\,F$

25. $2.53 \times 10^{-5} F$
26. $7.29 \times 10^{-4}$
27. $0.096 F$
28. $0.017 F$
29. $2.05 \times 10^{-4} F$
30. $S = 1.82 \times 10^{-4} F$; $[SO_4^{2-}] = 6.05 \times 10^{-7} M$; $[HSO_4^-] = 1.82 \times 10^{-4} M$
31. $[Ag^+] = 1.65 \times 10^{-4} M$; $[Pb^{2+}] = 7.94 \times 10^{-6} M$; $[IO_3^-] = 1.81 \times 10^{-4} M$; 2.26 mg
32. $5.6 \times 10^{-4} F$
33. $[Pb^{2+}] = 1.8 \times 10^{-4} M$; $[Sr^{2+}] = 4.3 \times 10^{-3} M$; $[HSO_4^-] = 4.4 \times 10^{-3} M$; $[SO_4^{2-}] = 8.8 \times 10^{-5} M$
34. $1.0 \times 10^{-8} M$
35. $1.1 \times 10^{-13}$
36. (a) $4.9 \times 10^{-6} M$
    (b) $5.2 \times 10^{-6} F$
    (c) $Ni(OH)_2(aq)$
37. $5.3 \times 10^{-3} M$; $6.5 \times 10^{-3} M$; $1.73 \times 10^{-4} M$
38. $1.4 \times 10^{-9}$
39. (a) $[Sr^{2+}] = 1.61 \times 10^{-3} M$; $[Ca^{2+}] = 2.30 \times 10^{-5} M$
40. $2.09 \times 10^{-3} F$
41. (a) AgCl
    (b) $[Cl^-] = 5.05 \times 10^{-6} M$
    (c) 50 per cent
42. (a) $Cr(OH)_3$
    (b) 7.30; 4.48
    (c) 5.48 to 7.30
43. $[Ba^{2+}] = 7.2 \times 10^{-4} M$; $[Pb^{2+}] = 1.24 \times 10^{-7} M$; $[IO_3^-] = 1.45 \times 10^{-3} M$
44. (a) $5.46 \times 10^{-5}$
    (b) $7.39 \times 10^{-3} M$
45. $1.31 \times 10^{-6} M$
46. (a) $3.2 \times 10^{-10}$
    (b) $2.6 \times 10^{-6} M$
47. $0.325 F$
48. $2.15 \times 10^{-2} M$

## CHAPTER 8

5.  11.98 per cent
6.  16.18 per cent
7.  31.65 per cent
8.  46.80 per cent
9.  25.74 per cent
10. $0.3270$ gm $K_2CO_3$; $0.6730$ gm $KHCO_3$
11. 31.20 per cent
12. 1.38 per cent $Na_2O$; 2.67 per cent $K_2O$
13. 77.8 per cent Ag; $0.0197$ gm $PbCl_2$
16. 13.61 ml
17. $0.08592 F$
18. 91.00 per cent
19. 26.96 weight per cent
20. 22.70 per cent

21. 0.09824 $F$ AgNO$_3$; 0.1330 $F$ KSCN
22. 66.02 per cent NaOH; 30.22 per cent NaCl; 3.76 per cent H$_2$O
23. 8.1 × 10$^{-5}$ $M$
25. 2.241 × 10$^{-5}$ $M$
29. [Cl$^-$] = 1.00 × 10$^{-5}$ $M$; pCl = 5.00; −47.9 per cent; +5.9 per cent
30. (a) 4.3 × 10$^{-3}$ $M$; (b) 0.12 $M$
31. (a) [SCN$^-$] = 6.5 × 10$^{-7}$ $M$; [IO$_3$$^-$] = 0.020 $M$
    (b) −64.8 per cent; (c) 0.0864 $F$

## CHAPTER 9

7. (a) +0.925 v
   (b) −0.950 v
   (c) +0.495 v
   (d) +0.325 v
   (e) −1.342 v
8. +0.336 v, $K$ = 2.0 × 10$^{12}$, copper electrode is anode;
   −0.64 v, $K$ = 2 × 10$^{-13}$, platinum electrode is cathode;
   +0.414 v, $K$ = 1.59 × 10$^{13}$, left-hand platinum electrode is anode;
   −1.361 v, $K$ = 8.0 × 10$^{-47}$, left-hand platinum electrode is cathode;
   +0.487 v, $K$ = 6.3 × 10$^{13}$, zinc electrode is anode;
   −0.109 v, $K$ = 1.4 × 10$^{-9}$, platinum electrode is cathode
9. (a) 0.328
   (b) 3.23 × 10$^4$
   (c) 4.95 × 10$^{18}$
   (d) 0.122
   (e) 6.75 × 10$^{71}$
10. 1.9 × 10$^{-8}$
11. 1.29 × 10$^{-3}$ $M$
12. 708
13. 1.0 × 10$^{-12}$
14. 1.48 × 10$^{-19}$
15. 3.6 × 10$^{-7}$
16. +1.68 v
17. +1.51 v
18. +1.50 v
19. (b) $E^0$ = −0.73 v; $K$ = 1.8 × 10$^{-25}$
    (c) −0.73 v
    (d) copper electrode
    (e) 4.6 × 10$^{-16}$ $M$
20. (a) $E^0$ = +0.010 v; $K$ = 4.8
    (b) [PtCl$_4$$^{2-}$] = 3.0 × 10$^{-6}$ $M$; [PtCl$_6$$^{2-}$] = 4.3 × 10$^{-11}$ $M$
    (c) −0.71 v
    (d) +0.33 v
23. (a) 6.3 × 10$^{16}$
    (b) 1.4 × 10$^9$
    (c) 8.0 × 10$^{61}$
    (d) 3.6 × 10$^{-11}$
24. 3.08 × 10$^{-12}$
25. [Co$^{2+}$] = 2.5 × 10$^{-3}$ $M$; [Tl$^+$] = 0.500 $M$
26. 1.20 × 10$^{-7}$ $M$

27. (a) $[Ce^{3+}] = [Fe^{3+}] = 0.00833\ M$; $[Fe^{2+}] = 0.0042\ M$; $[Ce^{4+}] = 2.1 \times 10^{-15}\ M$

(b) $[Ce^{3+}] = [Fe^{3+}] = 0.0081\ M$; $[Fe^{2+}] = 2.2 \times 10^{-15}\ M$; $[Ce^{4+}] = 0.0038\ M$

28. 0.271 v

29. $2.6 \times 10^{-3}\ M$

30. (c) $+1.257$ v

(d) $+1.183$ v

31. (a) 1.691 v

(b) lead electrode

32. $[V^{3+}] = 4.54 \times 10^{-5}\ M$; $[V^{2+}] = 0.0750\ M$; $[Cd^{2+}] = 0.0375\ M$

33. $[SnCl_4{}^{2-}] = 2.2 \times 10^{-5}\ M$; $[SnCl_6{}^{2-}] = 2.15 \times 10^{-3}\ M$; $[Fe^{2+}] = 4.30 \times 10^{-3}\ M$; $[Fe^{3+}] = 1.41 \times 10^{-11}\ M$

34. $0.0191\ M$

35. $0.490\ M$

36. 7.07

39. (a) $+0.46$ v

(b) $1.30 \times 10^{-14}\ M$

(c) 40.00 ml

(d) $+1.16$ v

(e) $-1.1$ per cent

40. $[V^{3+}] = 0.300\ M$; $[VO^{2+}] = 0.400\ M$

41. (b) $1.76 \times 10^{-7}\ M$

(c) 50.0057 ml

(d) $+0.011$ per cent

42. $K_{HgY} = 2.5 \times 10^{19}$; $K_{PbY} = 1.4 \times 10^{13}$

44. (a) $1.74 \times 10^{15}$

(b) $+0.550$ v

(c) $3.43 \times 10^{-10}\ M$

(d) 5.00 ml

(e) $+0.722$ v


## CHAPTER 10

2. $0.01492\ F$

3. $-0.136$ v

4. $K = 23.5$; $[OH^-] = 8.4\ M$

5. $0.02186\ F$

6. $0.05676\ F$

8. 85.24 per cent

9. (a) 17.86 per cent

(b) 8.359 per cent FeO; 16.25 per cent $Fe_2O_3$

10. $[HCOOH] = 0.03617\ F$; $[CH_3COOH] = 0.1036\ F$

11. 37.93 mg

12. 59.73 per cent

13. 3.332 gm

14. 0.5842 gm

15. 8.780 gm

16. $2.81 \times 10^{-6}$

19. $[Fe^{3+}] = 0.0385\ M$; $[VO_2{}^+] = 0.0142\ M$

20. $0.02642\ F$

21. $0.08130\,F$

22. $[C_4H_6O_6] = 0.00940\,F$; $[HCOOH] = 0.139\,F$

23. 118.3 ml

24. 0.7607 weight per cent

25. (a) $+0.99$ v
    (b) $+0.49$ v
    (c) $-0.42$ per cent
    (d) $+0.96$ v; $+0.30$ v; $-0.00025$ per cent

26. $[H_5IO_6] = 1.06 \times 10^{-9}\,M$; $[H_4IO_6^-] = 2.44 \times 10^{-4}\,M$; $[H_3IO_6^{2-}] = 1.07 \times 10^{-5}\,M$; $[IO_4^-] = 0.00976\,M$

27. 26.68 ml

28. $0.002810\,M$

29. $S = 0.180\,M$; $K = 18.0$

30. 0.9575

31. (a) $+1.174$ v
    (b) $+1.24$ v
    (c) $+1.45$ v

32. $K = 4.90 \times 10^{36}$; $[Br_2] = 0.00435\,M$

33. $0.1383\,F$

34. $0.1175\,F$

35. 0.05668 per cent

36. 0.6672 gm

37. 12.76 per cent

38. 2

40. 14.59 mg glycerol; 32.72 mg ethylene glycol

41. 39.26 per cent

42. 22.44 mg; 4.51 fluid ounces

43. $0.03177\,F$

44. 3.64 per cent

45. 48.41 per cent diethylsulfide; 50.59 per cent di-$n$-butylsulfide

46. 23.80 mg

47. $0.1158\,F$

48. 69.32 per cent

49. 97.47 per cent pure

50. 5.334 per cent


**CHAPTER 11**

1. (a) $\pm 7.5$ per cent
   (b) $\pm 0.129$ mv

2. (d) $0.07304\,F$
   (e) $K_a = 4.32 \times 10^{-6}$; pH $= 9.01$

5. (a) $9.064 \times 10^{-6}\,M$
   (b) $+1.5$ per cent

6. 46.67 per cent; 46.87 per cent; $-0.43$ per cent

8. (a) 2.815 millimoles

10. $0.000337\,M$

11. $1.0 \times 10^{13}$

12. 2.25

13. $7.11 \times 10^{-5}\,M$

14. $3.759 \times 10^{-5}\,M$

15. (a) theoretical volume = 31.09 ml; experimental volume = 30.93 ml
   (b) −0.1939
   (c) $6.79 \times 10^{-6}\ M$
   (d) $4.183 \times 10^{-3}\ M$
   (e) $1.31 \times 10^{-18}$
16. 4.05
17. $1.69 \times 10^{-9}$
18. (b) 3.66
19. $K_{sp} = 8.26 \times 10^{-17}$ for AgI; $K_{sp} = 6.83 \times 10^{-9}$ for $PbI_2$
20. (a) $7.75 \times 10^{-4}\ M$
   (b) $7.17 \times 10^{-4}\ M$ to $8.40 \times 10^{-4}\ M$

## CHAPTER 12

1. (a) +0.350 v
   (b) +0.120 to +0.299 v; +0.475 v
2. −0.053 v; 0.0986; 154 mg
3. $9.55 \times 10^{-8}\ M$
4. 0.148 v
5. (a) −0.365 v; 288
   (b) 354 mv
6. 0.1360
7. 13.01 mg
8. 17.1 electrons; reduce two picric acid molecules to form 2,2′,4,4′-tetra amino-3,3′-dihydroxyhydrazobenzene
9. 0.413 per cent
10. 22.64 mg acetamide; 72.88 mg N,N-dimethylacetamide
11. 1.094 gm
12. $[Ni^{2+}] = 0.06232\ M$; $[Co^{2+}] = 0.3022\ M$
13. 57.97 mg
14. 4.15 ml per gallon
15. $[Br^-] = 0.06042\ M$; $[Cl^-] = 0.1159\ M$
16. 89.46 per cent
17. (b) 53.04 mg
18. 117.6 mg
19. 22.94 mg $NH_4Cl$; 5.394 per cent
20. 55.47 per cent

## CHAPTER 13

1. −2
2. $OsO_4 + 8\,H^+ + 5\,e \rightleftharpoons Os^{3+} + 4\,H_2O$
3. $1.09 \times 10^{-5}\ cm^2/sec$
5. $1.77 \times 10^{-4}\ M$
6. 0.42 per cent
7. $1.26 \times 10^{10}$
8. $1.15 \times 10^{-3}\ M$
9. $Cd(NH_3)_4^{2+}$; $\beta_4 = 9.01 \times 10^6$
10. two ligands per copper; $\beta_2 = 8.51 \times 10^{20}$
11. $4.30 \times 10^{-3}\ M$

12. $2.147 \times 10^{-3}$ per cent
13. 28.65 mg
14. 13.83 mg of $Cu^{2+}$; 11.74 mg of $Ca^{2+}$
15. $2.37 \times 10^{-4}$; $2.31 \times 10^{-4}$; $2.25 \times 10^{-4}$; $2.20 \times 10^{-4}$; $2.09 \times 10^{-4}$;
    $1.90 \times 10^{-4}$; $1.31 \times 10^{-4}$; $8.74 \times 10^{-5}$ $1.58 \times 10^{-5}$; $2.98 \times 10^{-7}$ $M$
16. 1.93 per cent
17. 1.34 microamperes
18. (a) $7.35 \times 10^{-6}$ cm$^2$/sec
    (b) $+0.049$ v $vs.$ SCE
    (c) $+0.025$ v $vs.$ SCE

## CHAPTER 14

1. (a) $-35.1°C$
   (b) $175.3°C$
2. (a) 5.38; 7.20; 8.65; 8.23; 8.95; 9.54; 2.794; 3.847; 7.676; 8.639; 8.704; 8.695;
       8.195; 8.629; and 9.417 kcal/mole, respectively
   (b) 19.4; 19.9; 20.4; 20.8; 20.8; 21.0; 17.99; 18.30; 22.06; 22.07; 21.96; 26.04;
       26.21; 26.68; and 26.23 kcal mole$^{-1}$ deg$^{-1}$, respectively
3. (a) $\sim$18 kcal mole$^{-1}$ deg$^{-1}$
   (b) $\sim$26.3 kcal mole$^{-1}$ deg$^{-1}$
   (c) $\sim$22.0 kcal mole$^{-1}$ deg$^{-1}$
4. $n$-propanol
5. 28.9 torr, from Trouton's law and equation (14–6); actual $p°$ at 150°C is 10.0
   torr
6. (a) 880
   (b) $3.5 \times 10^{-7}$ gm/sec
8. From equation (14–12), $\alpha = 5846$; from equation (14–13), $\alpha = 112$; the latter
   overestimates the difficulty because it assumes an equimolar mixture rather than
   a trace contamination
9. (a) From equation (14–12), $\alpha = 2.92$; from equation (14–13), $\alpha = 2.33$
   (b) From equation (14–12), $\alpha = 2.60$
10. (a) 3 plates are more than sufficient
    (b) 3
    (c) $\sim$0.15
11. $n = 5.64$, an $apparent$ overestimate; however, the graphic estimates are based on
    total reflux, and the value obtained from equation (14–15) is therefore a better
    practical guide
12. 2.2 plates; from equation (14–14), $n = 1.19$
13. $V_v/V_l$ for $H_2O = 1700$, for $CS_2 = 515$; assumption better for $H_2O$
15. no provision for reflux
17. 3
18. no; from equation (14–15), $n_{min} = 32.4$

## CHAPTER 15

5. (a) $a$
   (b) $b$
7. 0.976
8. 94.7 per cent

9. (a) $9.90 \times 10^{-3}$
   (b) $7.51 \times 10^{-4}$
   (c) 99.010 per cent; 99.925 per cent
10. 3.06 extractions required; therefore, 4 extractions must be performed
11. (a) 0.632
12. 6.91
13. 14.905
14. $n = 331$; $(r_{max})_A = 320$; for compound A, $F_{320, 331} = 0.1235$; $(r_{max})_B = 310$; for compound B, $F_{310, 331} = 0.09036$
15. limitations are that $0.2 \leqslant p \leqslant 0.8$, $pq \sim 0.2$, and $n \sim 3.2(\mathscr{R}/\Delta p)^2$; for separation of A and B, $n \sim 435$
16. 2.27 per cent; $3.2 \times 10^{-3}$ per cent; $9.9 \times 10^{-8}$ per cent; see Table 2-1
17. 5 hr
20. $10^{12}$
21. $D_c \sim 4$; $n = 2.86$, so 3 extractions would be required
22. (a) 86.99
    (b) 750
23. (a) a dimer
    (b) 76.1
    (c) $\Delta H = -14.12$ kcal; the energy of a hydrogen bond in the formic acid dimer is 7.06 kcal/mole

# CHAPTER 16

2. $\mathbf{R} = 0.0958$; $t_M = 25$ min; $t_S = 236$ min
3. 108 sec
4. 36
5. 0.4167; 0.4167
6. (a) 69.93 cm/sec
   (b) 0.0945
   (c) 6.47 cm/sec
8. a factor of 3; not generally applicable
11. 29.33 min
12. for component A: 6.5 ml, 13 min, 0.231; for component B: 9 ml, 18 min, 0.167
13. 0.167 cm/sec
17. (a) $10^{-11}$ mole $mv^{-1}$ $sec^{-1}$
    (b) $4.8 \times 10^{-8}$ gm/sec
18. 22,500
19. 1.265 sec; 12.65 sec; 126.5 sec
20. 0.089 mm
23. 1267.5 days
24. $\sqrt{2}$
30. 21.8 ml/min; 2625 theoretical plates
31. (a) 11.0, 25.0, 40.0 cm/sec
    (b) 1310, 1286, 1068 theoretical plates; 1.526, 1.555, 1.872 mm
    (c) $A = 0.56$, $B = 7.2$, $C = 0.0283$ for $v$ in cm/sec and $H$ in mm
    (d) $8.8 \leqslant v \leqslant 28.8$ cm/sec
    (e) 15.95 cm/sec
    (f) 37.7 cm/sec, a time-saving of a factor of 2.36
35. 31.51 ml

36. 131.5 sec
37. $\mathcal{R} = 0.25$; 67,500 plates required for $\mathcal{R} = 1.0$
38. 47,000 plates; $L = 4.70$ m

## CHAPTER 17

1. (a) 0.47
   (b) 6.47 cm from origin
2. B
6. I, "no" $H_2O$; II, 0.70 monolayer; III, 1.39 monolayers; IV, 2.32 monolayers; V, 3.48 monolayers
11. 2.84 atm
12. 45.28 min
13. 335.7 ml
14. 6779
15. 259°C
16. no; at 210°C, $t_R' = 52.81$ min ($>45$ min)
17. 2448
18. yes; its retention index is 2648 (two additional $-CH_2-$ units)
19. 1535
22. 1268
23. QF-1
24. get a glass-lined injector
25. 10, 9, 8, $\sim$10.5
36. 2.4 hr after its introduction
37. The largest elution volume should be less than 50 ml

## CHAPTER 18

2. (a) $1.35 \times 10^{15}$ Hz
   (b) $5.66 \times 10^{14}$ Hz
   (c) $1.76 \times 10^{17}$ Hz
   (d) $1.00 \times 10^{12}$ Hz
   (e) $2.31 \times 10^{17}$ Hz
   (f) $4.92 \times 10^{13}$ Hz
3. (a) $4.50 \times 10^4$ cm$^{-1}$; ultraviolet
   (b) $1.89 \times 10^4$ cm$^{-1}$; visible
   (c) $5.86 \times 10^6$ cm$^{-1}$; x-ray
   (d) 33.3 cm$^{-1}$; far infrared
   (e) $7.70 \times 10^6$ cm$^{-1}$; x-ray
   (f) $1.64 \times 10^3$ cm$^{-1}$; middle infrared
4. (a) 15.2 cm; $1.52 \times 10^8$ nm
   (b) $6.17 \times 10^{-6}$ cm; 61.7 nm
   (c) $4.10 \times 10^{-10}$ cm; $4.10 \times 10^{-3}$ nm
5. (a) 0.0656 cm$^{-1}$; microwave
   (b) $1.62 \times 10^5$ cm$^{-1}$; ultraviolet
   (c) $2.44 \times 10^9$ cm$^{-1}$; x-ray

6. (a) $2.48 \times 10^{-12}$ erg
   (b) $5.40 \times 10^{-13}$ erg
   (c) $2.00 \times 10^{-9}$ erg
   (d) $1.73 \times 10^{-17}$ erg
7. $7.602 \times 10^{-12}$ erg; $4.579 \times 10^{12}$ ergs; 109.4 kcal; $4.186 \times 10^{10}$ ergs/kcal
8. (a) 0.434
   (b) 0.658
   (c) 0.0846
   (d) 0
   (e) 1.377
9. (a) 15.8
   (b) 61.0
   (c) 26.0
   (d) 1.50
   (e) 87.7
10. (a) 75.6
    (b) 6.10
    (c) 97.2
11. (b) $\frac{1}{4}a(b_1 + b_2)$; $ab_1$


**CHAPTER 19**

10. (a) 0.0223
    (b) 0.0284
    (c) 0.0172
    (d) 0.0143
    (e) 0.0174
    (f) 0.0170
    (g) 0.0136
    (h) 0.0421
13. (a) 4800 liter mole$^{-1}$ cm$^{-1}$
    (b) $9.27 \times 10^{-5}\ M$; $3.63 \times 10^{-4}\ M$
    (c) $3.23 \times 10^{-6}$
14. 0.88 per cent; 0.694 cm
15. $[\text{Ti(IV)}] = 2.74 \times 10^{-3}\ M$; $[\text{V(V)}] = 6.38 \times 10^{-3}\ M$
16. $k' = 0.040$ min$^{-1}$; initial rate $= 4.0 \times 10^{-6}$ mole liter$^{-1}$ min$^{-1}$
18. 34.1 per cent transmittance; $1.11 \times 10^4$ liter mole$^{-1}$ cm$^{-1}$
19. For $Cr_2O_7^{2-}$, $\varepsilon = 370$ liter mole$^{-1}$ cm$^{-1}$; for $MnO_4^-$, $\varepsilon = 2350$ liter mole$^{-1}$ cm$^{-1}$; $[Cr_2O_7^{2-}] = 9.73 \times 10^{-4}\ M$; $[MnO_4^-] = 2.73 \times 10^{-4}\ M$
20. $K_a = 7.98 \times 10^{-6}$
21. (b) $Fe(C_{10}H_8N_2)_3^{2+}$
    (c) $1.05 \times 10^4$ liter mole$^{-1}$ cm$^{-1}$
22. $K_a = 0.470$
23. $K_a = 6.08 \times 10^{-6}$
24. $[Fe(CN)_6^{3-}] = 1.84 \times 10^{-3}\ M$; $[Fe(CN)_6^{4-}] = 4.51 \times 10^{-3}\ M$
25. 4.80
26. (a) $CoCl_2$ and $CoCl_3^-$ are present
    (b) $\varepsilon = 235$ liter mole$^{-1}$ cm$^{-1}$ for $CoCl_2$; $\varepsilon = 191$ liter mole$^{-1}$ cm$^{-1}$ for $CoCl_3^-$
    (c) $\beta_2 = 2.67 \times 10^{10}$
27. 0.126 $\mu$g/ml
28. 212 $\mu$g

## CHAPTER 20

10. 0.12 and 1.16 $\mu$g, respectively; 0.012 and 0.116 $\mu$g/ml, respectively
15. Na$^+$, 1.13 ppm; K$^+$, 0.44 ppm; $\sim$5 per cent error; no, because of the relatively low precision of the readings, which were only to two significant figures
16. (b) $K = 1.48 \times 10^{-12}$ at 2000°K and $4.07 \times 10^{-6}$ at 3500°K
    (c) 0.00121 at 2000°K; 0.83 at 3500°K
17. 0.283 mg/liter
18. 0.0076 per cent copper; 0.0271 per cent lead


## CHAPTER 21

1.  (a) 4.143 $\mu$m; 2413 cm$^{-1}$
    (b) 1.219 $\mu$m; 8203 cm$^{-1}$
    (c) 6.265 $\mu$m; 1595 cm$^{-1}$
    (d) 2.735 $\mu$m; 3656 cm$^{-1}$
    (e) 2.660 $\mu$m; 3756 cm$^{-1}$
    (f) 9.143 $\mu$m; 1094 cm$^{-1}$
    (g) 3.179 $\mu$m; 3146 cm$^{-1}$
    (h) 4.994 $\mu$m; 2002 cm$^{-1}$
2.  (a) $5.230 \times 10^{13}$ Hz
    (b) $4.394 \times 10^{13}$ Hz
    (c) $6.426 \times 10^{13}$ Hz
    (d) $3.381 \times 10^{13}$ Hz
    (e) $9.147 \times 10^{13}$ Hz
4.  All normal-mode vibrations are Raman active; infrared active normal-mode vibrations are (a), (c), and (e); the Raman shifts are 1744, 1465, 2143, 1128, and 3051 cm$^{-1}$, respectively; the infrared bands appear at 5.732, 4.665, and 3.278 $\mu$m, respectively
5.  First overtones: (a) 3488 cm$^{-1}$ and 2.867 $\mu$m, (b) 2930 cm$^{-1}$ and 3.413 $\mu$m, (c) 4286 cm$^{-1}$ and 2.333 $\mu$m, (d) 2256 cm$^{-1}$ and 4.433 $\mu$m, (e) 6102 cm$^{-1}$ and 1.639 $\mu$m; second overtones: (a) 5232 cm$^{-1}$ and 1.911 $\mu$m, (b) 4395 cm$^{-1}$ and 2.275 $\mu$m, (c) 6429 cm$^{-1}$ and 1.555 $\mu$m, (d) 3384 cm$^{-1}$ and 2.955 $\mu$m, (e) 9153 cm$^{-1}$ and 1.093 $\mu$m
6.  (a) 1555 cm$^{-1}$; $4.66 \times 10^{13}$ Hz
    (b) $3.09 \times 10^{-13}$ erg
    (c) 1744
7.  Cell A, 0.033 cm; cell B, 0.024 cm; cell C, 0.032 cm; cell D, 0.016 cm; cell E, 0.056 cm
12. Band A, cannot be measured because it exceeds the range of the instrument; band B, $T = 0.27$, $A = 0.57$; band C, $T = 0.051$, $A = 1.29$; band D, $T = 0.60$, $A = 0.22$; band E, $T = 0.039$, $A = 1.40$; band B is best to use because it falls within the transmittance range of least error, it is easiest to measure, and it has the least ambiguous base line

# INDEX

Note: In this index, page numbers in *italic* refer to illustrations;
page numbers followed by (t) refer to tables.